Stored Product Protection
Volume 1

Stored Product Protection

**Proceedings of the 6th International Working Conference
on Stored-product Protection**

17–23 April 1994
Canberra, Australia

Edited by E. Highley, E.J. Wright, H.J. Banks and B.R. Champ

Volume 1

CAB INTERNATIONAL

CAB INTERNATIONAL
Wallingford
Oxon OX10 8DE
UK

Tel: Wallingford (01491) 832111
Telex: 847964 (COMAGG G)
E-mail: cabi@cabi.org
Fax: (01491) 833508

A catalogue record for this book is available from the British Library.

ISBN   0 85198 932 2

Pre-press production by Arawang Information Bureau Pty Ltd, Canberra, Australia
Printed and bound in the UK at the University Press, Cambridge

Preface

The International Working Conference on Stored-product Protection (IWCSPP), held every four years, is the premier world forum for presentation of research results bearing on the safe storage of durable foodstuffs, of which cereal grains, pulses, and oilseeds make up the largest component. The 6th IWCSPP was held in Canberra, Australia from 17–23 April 1994, under the auspices of CSIRO (Australia's national R&D agency), the Australian Centre for International Agricultural Research (ACIAR), and the Australian grain handling and storage industry. These two volumes place on permanent record the Proceedings of the Conference.

There were over 400 participants in the Canberra conference. Apart from Australia, there were substantial contingents from Brazil, Canada, China, France, Germany, India, Indonesia, Israel, Pakistan, the Philippines, the U.K., and the USA. A further 20 countries were represented. A list of participants is given at the end of Volume 2.

The overall objective of the 6th IWCSPP was to bring participants abreast of advances in research and development on the preservation of stored durable commodities since the 5th IWCSPP, held in Bordeaux, France in 1990.

Much had happened in the four years: methyl bromide had come under a cloud; pressure from consumers and environmentalists to reduce chemical additives to foodstuffs had become more intense; that mycotoxin contamination of food and feedstuffs was a serious threat to human and animal health had gained general acceptance. Recognising these and related issues, the theme *Stored-product Protection — a Time of Challenge* was adopted for the conference. This theme was sustained by participants.

Some 250 papers were presented during the conference. These proceedings contain, in addition to most of those papers, reports of a series of workshops on specialist topics — trapping, application of inert dusts, extension and small-scale and farm storage, standards, expert systems, and appropriate storage — that were also part of the scientific program. The workshops were well supported, as was a full-day field excursion to storage facilities in part of the southern New South Wales wheatbelt close to Canberra. Further insights into the grain handling and storage systems in the east and west of Australia were gained by those delegates who participated in the pre- and postconference tours organised.

A trade exhibition held in association with the conference attracted over 30 local and transnational companies and agencies.

The Organising Committee wishes to thank all those who contributed to the success of the conference, not least the authors and presenters of what was an extremely stimulating assemblage of papers. It sincerely appreciates the financial and other assistance provided by various public and private organisations and agencies, the Permanent Committee for IWCSPP, Capital Conferences Pty Ltd and the National Convention Centre, Canberra.

My personal thanks go to all members of the Organising Committee for their efforts in making our conference an outstanding success. This success will be greatly augmented by the publication of these proceedings which, with the support of CAB International, will disseminate conference presentations and discussions to the widest international audience.

I must express my appreciation particularly to the CSIRO team led by Jane Wright, Julie Carter, and Jan van Graver for the superb arrangements they have put in place, to Gail Hawke, Nicole Rooney, and Kerry Meagher of Capital Conferences for their professional administration of the conference, and to Ed and Kerry Highley for the marvellous documentation they have provided us with, which will include timely production of the proceedings. A special thank you must go also to the participants in the industry exhibition and of course to all the speakers and other attendees who made the conference so successful. Finally, it is my pleasure to record that the 7th IWCSPP will be held in Beijing, China in 1998. Professor Jin Zuxun, President of the Nanjing University of Food Economics, and a member of the Permanent Committee, will convene a local committee to organise the conference.

B.R. Champ
Chairman
6th IWCSPP Organising Committee

6th International Working Conference on Stored-product Protection

Organising Committee

Executive Committee

Dr Bruce Champ	Coordinator, ACIAR Postharvest Research Program (Chairman)
Dr Jonathan Banks	Section Leader, CSIRO Stored Grain Research Laboratory
Mrs Julie Carter	Liaison Officer, CSIRO Stored Grain Research Laboratory
Mr Mick Catley	Senior Assistant Director, Plant Quarantine and Inspection Branch, Department of Primary Industries and Energy
Mr Jan van S. Graver	Experimental Scientist, CSIRO Stored Grain Research Laboratory
Mr Ed Highley	Communications Consultant, ACIAR Postharvest Research Program
Dr Jane Wright	Senior Research Scientist, CSIRO Stored Grain Research Laboratory

Other Members

Dr Mervyn Bengston	Chief Entomologist, Queensland Department of Primary Industries
Mr Ian Broadfoot	General Manager (Business Development), GRAINCO(Queensland)
Mr Melville Connell	General Manager, Grain Elevators Board of Victoria
Mr Tony Connon	Australian Wheat Board (Honorary Auditor)
Dr Ron Wills	Director, Australian Wheat Board Academy of Grain Technology

Session and Workshop Conveners

Sessions

Storage Engineering	Mr Jack Ford – Dr Dirk Maier
Sampling and Trapping	Mr Paul Cogan – Dr Jane Wright
Fumigation and Controlled Atmospheres	Mr Peter Annis – Dr Nick Price – Mr Colin Waterford
Grain Protectants	Dr Mervyn Bengston – Dr Larry Zettler
Grain Quality	Dr Ben Juliano – Dr Colin Wrigley
Insect Biology	Dr Jonathan Donahaye – Dr Barry Longstaff
Storage Fungi and Mycotoxins	Dr Ailsa Hocking – Dr David Miller
Integrated Commodity Management	Dr Chris Haines – Dr Geoff Norton
Quarantine and Regulatory Issues	Mr Mick Catley
Biological Control	Mr David Rees
Physical Control	Mr James Darby
Inert Dusts	Dr Jim Desmarchelier

Workshops

Trapping	Mr Paul Cogan – Dr Jane Wright
Application of Inert Dusts	Mr Alan McLaughlin
Extension and Small-scale and Farm Storage	Ms Margaret Annis
Standards	Dr Francis Fleurat-Lessard
Expert Systems	Dr Paul Flinn – Dr Barry Longstaff – Dr Lincoln Smith
Appropriate Storage	Mr Mervyn Adams

The International Working Conferences on Stored-product Protection

Permanent Committee

J.L. Zettler, USA — President
V.F. Wright de Malo, USA — Secretary/Treasurer
B. Amoaka-Atta, Liberia
H.J. Banks, Australia
J.H. Boczek, Poland
B.R. Champ, Australia
R. Davis, USA — Honorary President
F. Fleurat-Lessard, France
Jin Zuxun, China

Z. Kozakiewicz-Lawrence, U.K.
S. Partoatmodjo, Indonesia
N. Paster, Israel
M.R. Sartori, Brazil
S.C. Saxena, India
T.A. Taylor, Nigeria
N.D.G. White, Canada
T. Yoshida, Japan

Origins of the IWCSPP*

The First International Working Conference on Stored-product Entomology was held in Savannah, Georgia, USA from 7–11 October 1974. Stored-product entomology was the subject of the first three conferences, after which it was decided to broaden their scope, so as to attract, as well as entomologists, engineers, mycologists and others working in postharvest research.

It would true to say that the original idea for what became the IWCSPP was born of a feeling among U.S. stored product entomologists that they were adrift in a backwater of the noble profession of U.S. entomology. In the 1960s and 1970s, those of us at the Stored-Product Insects Research and Development Laboratory seemed always to be hosting entomologists from all parts of the world, who were travelling with remarkable frequency and regularity to various laboratories to learn what were the new problems and what progress was being made in their solution. There was no such round of visits to other laboratories for us. Also, the U.S. Department of Agriculture (USDA), a major employer of U.S. stored product entomologists, professed no great interest in encouraging wider publication of research results or in providing funds for attendance at scientific meetings. Food manufacturers, another large employer of stored-product entomologists, were not eager to have their problems and solutions published or discussed at scientific societies. Thus, U.S. stored-product entomologists played a minor role in professional entomology meetings. Even at the annual general meeting of the Entomological Society of America one rarely heard more than 5 or 6 papers from the stored-product entomology fraternity. Branch and State meetings suffered similarly.

During a discussion at the Miami meeting of the Entomological Society of America in 1970, while Dr John Brower, Dr Phillip Harein, Dr S.R. Loschiavo, Mr E.W. Tilton, and myself were lamenting our lot within the profession even more vocally than usual, it was suggested that a specialist stored-product entomology meeting be held. U.S. stored product entomologists were polled on this during 1971 and we found that there was great interest in such a meeting. In 1972, when I attended the International Congress of Entomology in Canberra, Australia — where Mr Bill Bailey and

his associates of the Stored Grain Research Laboratory, CSIRO Division of Entomology, put on a fine program on stored-product entomology — I was able to survey a large group of the world's entomologists on their desire for a stored-product conference. There was good support.

At the USDA Stored-Product Insects Research and Development Laboratory, we decided to investigate the possibility of holding such a conference in Savannah. A committee composed of Drs J.H. Brower, Edward G. Jay, Patrick P.T. Lum, Michael A. Mullen and myself undertook to investigate the possibility. We soon encountered problems. We learned that our control over the conference would be greatly restricted if it were held at the laboratory. We would, for example, have to submit all invitees for government approval, as well as the program agenda and the program dates. Our committee was broadened when it was joined by three members of the Department of Entomology, University of Georgia: Drs Horace O. Lund, Head of Department, Preston E. Hunter and U. Eugene Brady. We next consulted an attorney and incorporated our committee as a non-profit educational corporation with a three-year lifetime in the State of Georgia. This allowed us to establish a bank account and to be free of taxes, and facilitated soliciting of funds. The USDA agreed to provide services-in-kind, such as postage, stationery, photocopying and typing assistance.

Through a series of committee meetings, a program agenda was established with five areas of general interest:

1 Tropical Stored-Product Entomology
   Symposium convener W.H. Jepson, England
   Panel discussion leader Fred Ashman, Malawi

2 Biology, Ecology and Integrated Control
   Symposium convener J.H. Boczek, Poland
   Panel discussion leader S.C. Saxena, India

3 Pesticides, Toxicity and Insect Resistance
   Symposium convener E.J. Bond, Canada
   Panel discussion leader C.E. Dyte, England

4 Radiation and other Physical Means of Insect Control
   Symposium convener Moshe Calderon, Israel
   Panel discussion leader F.L. Watters, Canada

5 Pesticide Residues, Tolerances and Registration
   Symposium convener L.S. Henderson, USA
   Panel discussion leader E.E. Turtle, FAO, Italy

Each topic area was divided into a symposium composed of 4 or 5 invited speakers, and a panel for discussion of these topics that was moderated by a scientist knowledgeable on

---

* A contribution by Dr Robert Davis, Vice President of Pest Management Consultants and Associates, Inc., Savannah, Georgia 31410, USA, who was the Foundation Chairman of the Permanent Committee for the International Working Conferences on Stored-product Protection.

the symposium subject. Questions and some debate from the floor were permitted. Contributed papers were neither encouraged nor discouraged. The intention was that the conference participants would utilise their time in listening and then discussing amongst themselves in a structured manner, problems and solutions related to the important topics brought to the conference by the speakers.

The committee sought funds from agribusinesses in the USA and elsewhere. Many agribusinesses contributed directly to scientists they had worked with or were working with in various countries. They advised us of this fact so that we could plan our expenditure of funds accordingly. Many others provided direct financial support to the Organising Committee. All funds collected by the committee went to support the program and were used to bring speakers to the conference. The costs of the conference per se were paid out of the participants' registration fees. These fees also covered the cost of publishing and mailing the conference proceedings at the conclusion of the meeting.

The program for the First International Working Conference on Stored-product Entomology consisted of 50 symposium and panel presentations and 27 submitted papers. Attendance exceeded all expectations, with 214 agricultural scientists, marketing specialists and administrators from 27 nations. Special invitation addresses were presented by Mr T.W. Edminster, Administrator, Agricultural Research Service, USDA, and Dr Curtis W. Sabrosky, President, XV International Congress of Entomology.

Towards the end of the conference, at an informal meeting of participants, a Permanent Committee for International Working Conferences on Stored-product Entomology was elected. The committee was charged with maintaining contact with participants and with establishing a mechanism for holding similar meetings in the future. The membership of the inaugural committee was:

J.H. Boczek, Poland
E.J. Bond, Canada
M. Calderon, Israel
D.P. Childs, USA, Secretary
M. Connell, Australia
R. Davis, USA, Chairman
H. Dell'Orto Trivelli, Chile
C.E. Dyte, England
A.H.M. Kamel, Egypt
S.C. Saxena, India
T. Ajibola Taylor, Nigeria

S. Utida, Japan
P. Wheatley, England
K. Whitney, USA

Conference participants also passed a resolution on 'Preservation of Staple Food Stocks Following Production', as follows:

This Conference, being cognisant of the urgent need for action to provide and preserve a food supply to all mankind, confirms the great importance of losses during storage and transportation — representing wastage of human endeavor, natural resources and energy — which could be greatly reduced by implementation of methods currently known in our applied science.

It was the consensus that each conference participant was free to take whatever action he deemed appropriate regarding the utilisation and distribution of the resolution. The resolution was sent to a number of Heads of States and to the United Nations Secretary General.

The sentiments in the resolution remain as valid today as they were then, and the International Working Conferences on Stored-product Protection have become an important means of discussing and disseminating information on the latest methods for preserving staple food and feedstuffs.

Robert Davis

Proceedings of Previous Conferences

Copies of the proceedings of previous IWCSPP are available from:

Dr Valerie F. Wright
2950 West 12th Avenue
Manhattan, Kansas 66502
USA
Fax: 1 913 532 6232

The costs are, in US$:

1st IWCSPE	7.50
2nd IWCSPE	10.00
3rd IWCSPP	30.00
4th IWCSPP	80.00
5th IWCSPP	100.00

Contents

Volume 1

Preface — v

The International Working Conferences on Stored-product Protection — vii

6th IWCSPP Conference Summary – Bruce Champ — xvii

Opening Address – John Kerin — xix

FUMIGATION AND CONTROLLED ATMOSPHERES — 1

Keynote Address
Fumigation — an endangered technology? – H.J. Banks — 2

Carbon dioxide — more rapidly impairing the glycolytic energy production than nitrogen? – C. S. Adler — 7
A comparison of the efficacy of CO_2-rich and N_2-rich atmospheres against the granary weevil *Sitophilus granarius* (L.) (Coleoptera: Curculionidae) – C. S. Adler — 11
Numerical modelling of the movement of carbon dioxide through stored-wheat bulks – S.K. Alagusundaram, D.S. Jayas, W.E. Muir, N.D.G. White, and R.N. Sinha — 16
Modified atmosphere storage of bagged maize outdoor using flexible liners: a preliminary report – D.G. Alvindia, F.M. Caliboso, G.C. Sabio and A.R. Regpala — 22
Sealed storage technology on Australian farms – A.S. Andrews, P.C. Annis and C.R Newman — 27
Time to population recovery as a means for specifying low oxygen dosages – P.C. Annis — 37
A same-day test for detecting resistance to phosphine – C.H. Bell, N. Savvidou, K.A. Mills, S. Bradberry and M.L. Barlow — 41
A preliminary evaluation of carbon dioxide under high pressure for rapid fumigation – F.M. Caliboso, H. Nakakita and K. Kawashima — 45
The feasibility of increasing the penetration of phosphine in concrete silos by means of carbon dioxide – Y. Carmi, Y. Golani, H. Frandji and E. Shaaya — 48
Mortality of snails, *Cernuella virgata* and *Cochlicella acuta*, exposed to fumigants, controlled atmospheres or heat disinfestation – J. Cassells, H.J. Banks and P.C. Annis — 50
Application of pressure-swing absorption (PSA) and liquid nitrogen as methods for providing controlled atmospheres in grain terminals – J. Cassells, H.J. Banks and R. Allanson — 56
Fumigation of a 7000 t bulk of wheat with phosphine using the Phyto-Explo® system to assist gas circulation – B. Chakrabarti, C. R. Watson, C. H. Bell, T.J. Wontner-Smith and J. Rogerson — 64
Technical study of controlled atmosphere with carbon dioxide in brick silo for safe storage of wheat – Cheng Fu-chang, Mei Bao-Liang, Yu Jian, Dou He-tong and Tang Shun-gong — 68
Improved procedures for fumigation of oaten hay in shipping containers – C.P.F. De Lima, R.N. Emery and P. Jackson — 71
Carbonyl sulphide as a fumigant for control of insects and mites – J.M. Desmarchelier — 78
Improved procedures for measurement of fumigants – J.M. Desmarchelier — 83
The influence of temperature on the sensitivity of two nitidulid beetles to low oxygen concentrations – J. E. Donahaye, S. Navarro and M. Rindner — 88
Methyl isothiocyanate used as a grain fumigant – V. Ducom — 91
A Western Australian farm survey for phosphine-resistant grain beetles – R.N. Emery — 98
Effects of low oxygen phosphine fumigations on adult *Rhyzopertha dominica* – F.M. Johnston and C.P. Whittle — 104
Response of *Liposcelis bostrychophila* and *L. entomophila* (Psocoptera) to carbon dioxide – E.C.W. Leong and S.H. Ho — 108
Inheritance of phosphine resistance in *Sitophilus oryzae* (L.) (Coleoptera, Curculionidae) – Li Yan-sheng and Li Wen-zhi — 113
Study of circumfluent fumigation with phosphine for killing stored-grain insects in silos – Lu Jian-hua, Zhao Zeng-hua, Liu Qing, Hu Shu-tian and Qi Jin-shen — 116
Comparative toxicity of carbon dioxide to two *Callosobruchus* species – G. Mbata, C. Reichmuth and T. Ofuya — 120
A new method of using low levels of phosphine in combination with heat and carbon dioxide – D.K. Mueller — 123

A new method to control stored-product insects using carbon dioxide with high pressure followed by sudden pressure loss – H. Nakakita and K. Kawashima — 126

The future of hermetic storage of dry grains in tropical and subtropical climates – S. Navarro, J.E. Donahaye and S. Fishman — 130

Western Australian Fumigation Practice Survey (1992) – C.R Newman — 139

Biogeneration of carbon dioxide for use in modified atmosphere storage of sorghum grains – K.L. Patkar, C. M. Usha, H. S. Shetty, N. Paster and J. Lacey — 144

The current status of phosphine fumigations in India – S. Rajendran and K.S. Narasimhan — 148

A new phosphine releasing product – C. Reichmuth — 153

Uptake of phosphine by stored-product pest insects during fumigation – C. Reichmuth — 157

Carbon dioxide under high pressure of 15 bar and 20 bar to control the eggs of the Indianmeal moth *Plodia interpunctella* (Hübner) (Lepidoptera: Pyralidae) as the most tolerant stage at 25°C – C. Reichmuth and R. Wohlgemuth — 163

Studies on the effect of carbon dioxide in insect treatment with phosphine – Y.L. Ren, I.G. O'Brien and C.P. Whittle — 173

The impact of temperature, moisture content, grain quality and their interactions on changes in storage vessel atmospheres – R. Reuss, K. Damcevski and P.C. Annis — 178

Low-cost detector for the continuous monitoring of phosphine fumigation – R. Ryan, S.Waddell, P.W. Alexander, K. Bowles, L. Cherkson, J. Morgan and D. B. Hibbert — 183

Controlled release of phosphine — an update – D. Schonstein, W. Shore, R. Ryan and S. Waddell — 188

A survey of phosphine and methyl bromide resistance in Malaysia – Z. Sulaiman, M. Rahim, M.E. Faridah and M. Rasal — 192

Effectiveness of carbon dioxide under reduced pressure against some insects infesting dried fruit – L. Süss and D.P. Locatelli — 194

Evolution of phosphine from aluminium phosphide formulations at various temperatures and humidities – Tan Xianchang — 201

Carbon dioxide fumigation of processed dried vine fruit (sultanas) in sealed stacks – C. Tarr, S.J. Hilton, J. van S. Graver and P.R. Clingeleffer — 204

The fumigation of bag-stacks with phosphine under gas-proof sheets using techniques to avoid the development of insect resistance – R.W.D. Taylor and A.H. Harris — 210

Effects of different speed of build up and decrease of pressure with carbon dioxide on the adults of the tobacco beetle *Lasioderma serricorne* (Fabricius) (Coleoptera: Anobiidae) – C. Ulrichs — 214

Response of the pea weevil *Bruchus pisorum* (L.) to phosphine – C.J. Waterford and R.G. Winks — 217

New aluminium phosphide formulations for controlled generation of phosphine – C.J. Waterford, C.P. Whittle and R.G. Winks — 226

Correlation between phosphine resistance and narcotic response in *Tribolium castaneum* (Herbst) – C.J. Waterford and R.G. Winks — 231

Fumigation of dried vine fruit for export – P. Williams — 236

Phosphine fumigation of stored field peas for insect control – P. Williams and C. P. Whittle — 240

Measurement of resistance to grain fumigants with particular reference to phosphine – R.G. Winks and E.A. Hyne — 244

Control of the common clothes moth *Tineola bisselliella* (Hummel) (Lepidoptera: Tineidae) and other museum pests with nitrogen – A. Wudtke and C. Reichmuth — 251

Fumigation and Controlled Atmospheres — Session Summary — 255

STORAGE ENGINEERING — 257

Keynote Address
Developments in silo design for the safe and efficient storage and handling of grain – A. W. Roberts — 259

Temperature studies on steel silos in North Africa – El H. Bartali — 281

Development of a programmable aeration controller – P. A. Gibbs — 286

Observations on large-scale outdoor maize storage in jute and woven polypropylene sacks in Zimbabwe – L. Kennedy and A.D. Devereau — 290

Quality enhancement of stored grain by improved design and management of aeration – J-C. Lasseran, G. Niquet and F. Fleurat-Lessard — 296

Chilled aeration and storage of U.S. crops — a review – D.E. Maier — 300

Pest management of stored maize using chilled aeration — a mid-west United States perspective – L. J. Mason, D.E. Maier, W.H. Adams and J. L. Obermeyer — 312

Mobile drive-over hoppers and stackers for filling and emptying grain bunkers – F.M. Miller — 318

Using controlled aeration for insect and mould management in the south-western United States – R.T. Noyes, G.W. Cuperus and P.Kenkel — 323

Closed loop fumigation systems in the south-western United States – R.T. Noyes and P. Kenkel — 335

Engineering input in the design of on-farm storage in India – B.D Shukla and K.K Singh — 342

Design chart for in-store maize drying under tropical climates – S. Soponronnarit, P. Wongvirojtana
and A. Nathakaranakule   346
Advances in research on in-store drying – G.S. Srzednicki and R.H. Driscoll   351
Modelling heat and mass transfer phenomena in bulk stored grains – G.R. Thorpe   359
Grain aeration system controlled by computer – Wu Zidan and Li FuJan   368

Storage Engineering — Session Summary   **371**

SAMPLING AND TRAPPING   373

Keynote Address

The use of sex pheromones to control *Ephestia kuehniella* Zeller (Mediterranean flour moth) in flour mills
by mass trapping and attracticide (lure and kill) methods – P. Trematerra   375

Grain storage in a small-farm ecosystem: Angoumois grain moth movement and management – R.J. Barney
and P.A. Weston   383
The use of multiple trapping methods to assess population size: an evaluation – J.H. Brower, L. Smith and
E.P. Wileyto   385
The use of a managed bulk of grain for the evaluation of PC, pitfall beaker, insect probe and WBII probe
traps for trapping *Sitophilus granarius*, *Oryzaephilus surinamensis* and *Cryptolestes ferrugineus* –
P.M. Cogan and M.E. Wakefield   390
New trends in stored-grain infestation detection inside storage bins for permanent infestation risk
monitoring – F. Fleurat-Lessard, A-J. Andrieu and D.R. Wilkin   397
Acoustical monitoring of stored-grain insects: an automated system – D.W. Hagstrum, P.W. Flinn and
D.Shuman   403
Responses of *Tribolium castaneum* to different pheromone lures and traps in the laboratory – A. Hussain,
T.W. Phillips and M.T. AliNiazee   406
Response of *Prostephanus truncatus* and *Teretriosoma nigrescens* to pheromone-baited flight traps –
G.E. Key, B.J. Tigar, E. Flores-Sanchez and M. Vazquez-Arista   410
Development of immunoassays for quantitative detection of insects in stored products – G. B. Kitto,
F.A. Quinn and W.E. Burkholder   415
Development of pheromone-baited insect traps – M.A. Mullen   421
Effect of single and multiple species release on the capture of *Plodia interpunctella* and *Cadra cautella* in
pheromone-baited traps – M.A. Mullen   425
Monitoring field populations of *Lyctocoris campestris*, a predator of stored-grain insects: assessment
of different trap designs – M.N. Parajulee and T.W. Phillips   429
Comparison between two methods of insect sampling in stored wheat – P.R.V.S. Pereira, F.A. Lazzari,
S.M.N. Lazzari and A.A. Almeida   435
Using pheromones for location and suppression of phycitid moths and cigarette beetles in Hawaii—
a five-year summary – L.H. Pierce   439
Improved early detection of internal infestation by flotation using product-adapted salt solutions –
K. Richter and P. Tchalale   444
The use of various insect traps for studying psocid populations – R. Roesli and R. Jones   448
Trapping stored-product insects using an unbaited multifunnel trap – P. Trematerra, G. Rotundo and
A. De Cristofaro   451
The potential of insect self-marking for the interpretation of trap catch – E.P. Wileyto   455
The statistical interpretation of insect self-marking and trapping – E.P. Wileyto   459
The detection of insects in grain during transit—an assessment of the problem and the development
of a practical solution – D.R. Wilkin, D. Catchpole and S. Catchpole   463
Trapping *Trogoderma variabile* (Coleoptera: Dermestidae): a comparison of traps and techniques for
adult and larval monitoring – E.J. Wright and R.L. Delves   470

Sampling and Trapping — Session Summary   **475**

INSECT BIOLOGY   477

Keynote Address

Pheromones of stored-product insects: current status and future perspectives – T.W. Phillips   479

Studies on the relative susceptibility of varieties and germplasm lines of sesame to infestation by *Tribolium
castaneum* (Herbst) (Coleoptera: Tenebrionidae) – R.K. Murali Baskaran, M.S. Venugopal and
C.V. Sivakumar   487
A comparison of the demography of four major stored grain coleopteran pest species and its implications
for pest management – S. J. Beckett, B. C. Longstaff and D. E. Evans   491
A new host of the groundnut seed beetle, *Caryedon serratus* (Ol.), in Israel – M. Calderon   498

Dynamics and expansion of populations of stored product beetles – Z. Ciesielska   500

Bioassays with bruchid beetles: problems and (some) solutions – P.F. Credland   509

The distribution and PCR-based fingerprints of *Rhyzopertha dominica* (F.) in Canada – P.G. Fields
    and T.W. Phillips   517

Some stored-product insects of increasing importance in China – Guan Lianghua, Chen Lanfen,
    Xie Gengfa and Yang Shaojun   523

Factors affecting survival and development of *Sitophilus oryzae* (L.) in rice grain pericarp layers –
    Y. Haryadi and F. Fleurat-Lessard   525

Molecular and morphological markers for diagnosis of *Sitophilus oryzae* and *S. zeamais* (Coleoptera:
    Curculionidae) – P. Hidayat, R.H. ffrench-Constant and T.W. Phillips   528

Pheromone biology and factors affecting its production in *Tribolium castaneum* – A. Hussain, T.W. Phillips,
    T. J. Mayhew and M.T. AliNiazee   533

Stored agricultural product protection in Croatia – I. Kalinovic and M. Ivezic   537

Pheromone biology of the lesser grain borer, *Rhyzopertha dominica* (Coleoptera: Bostrichidae) –
    T.J. Mayhew and T.W. Phillips   541

The measurement of resistance to *Acanthoscelides obtectus* (Say) (Coleoptera: Bruchidae) in seeds
    of *Phaseolus vulgaris* L. – C.J. Moss and P.F Credland   545

Function and composition of cuticular hydrocarbons of stored-product insects – J. Nawrot, E. Malinski
    and J. Szafranek   553

Factors affecting oviposition and orientation by female *Plodia interpunctella* – T.W. Phillips and
    M.R. Strand   561

Studies of responses of stored-product pests, *Prostephanus truncatus* (Horn) and *Sitophilus zeamais* Motsch.,
    to food volatiles – V. Pike, J.L. Smith, R.D. White and D.R. Hall   566

Influence of synthetic Sitophilate, the aggregation pheromone of *Sitophilus granarius* (L.) (Coleoptera:
    Curculionidae), on dispersion and aggregation behaviour of the granary weevil – R. Plarre   570

Distribution and status of Psocoptera infesting stored products in Australia – D. Rees   583

Hormonal control of reproduction in the female pyralid moth *Plodia interpunctella* (Hübner) (Lepidoptera:
    Phycitidae) – E. Shaaya, D. Silhacek, P. Shirk, H. Rees, G. Zimowska and S. Plotkin   588

Role of ultrasound production and chemical signals in the courtship behaviour of *Ephestia cautella* (Walker),
    Ephestia kuehniella Zeller and *Plodia interpunctella* (Hübner) (Lepidoptera: Pyralidae) – P. Trematerra
    and G. Pavan   591

Effect of maize variety and storage form on oviposition and development of the maize weevil, *Sitophilus
    zeamais* Motschulsky (Coleoptera: Curculionidae) – K.A. Vowotor, N.A. Bosque-Pérez and J.N. Ayertey   595

Life history data for *Sitotroga cerealella* (Olivier) (Lepidoptera: Gelechiidae) in farm-stored corn and the
    importance of sub-optimal environmental conditions in insect population modelling for bulk
    commodities – D.K. Weaver and J.E. Throne   599

Influence of planting date, harvest date, and maize (corn) hybrid on preharvest infestation of maize by
    Sitotroga cerealella – P.A. Weston   605

Variable longevity in the rusty grain beetle, *Cryptolestes ferrugineus* – N.D.G. White and R.J. Bell   608

Effect of food volume and photoperiod on initiation of diapause in the warehouse beetle, *Trogoderma
    variabile* Ballion (Coleoptera: Dermestidae) – E.J. Wright and A.P. Cartledge   613

Insect Biology — Session Summary   **617**

Author Index   **619**

Volume 2

INERT DUSTS   621

Silica aerogels as alternative protectants of maize against *Prostephanus truncatus* (Horn) (Coleoptera:
    Bostrichidae) infestations – A. Barbosa, P. Golob and N. Jenkins   623

Structural treatment with amorphous silica slurry: an integral component of GRAINCO's IPM strategy –
    B.W. Bridgeman   628

Efficacy of an amorphous silica dust against bean bruchids – D.P. Giga and P. Chinwada   631

Effect of zeolite on the development of *Sitophilus zeamais* (Motsch.) – Y. Haryadi, R. Syarief, M. Hubeis
    and I. Herawati   633

Effects of Dryacide® on the physical properties of grains, pulses and oilseeds – K. Jackson and D. Webley   635

Laboratory trials on desiccant dust insecticides – A. McLaughlin   638

Combination of cooling with a surface application of Dryacide® to control insects – P.J. Nickson,
    J.M. Desmarchelier and P. Gibbs   646

Effectiveness of Insecto®, a new diatomaceous earth formulation, in suppressing several stored-grain
    insect species – Bh. Subramanyam, C. L. Swanson, N. Madamanchi and S. Norwood   650

Inert Dusts — Session and Field Trip Workshop Summary   **660**

GRAIN QUALITY

661

Keynote Address
Concerns for quality maintenance during storage of cereals and cereal products – B.O. Juliano

663

Keynote Address
Maintenance of grain quality during storage — prediction of the conditions and period of 'safe' storage –
C.W. Wrigley, P.W. Gras and M.L. Bason

666

Valuing Australian wheat quality characteristics in selected Asian markets – F.Z. Ahmadi-Esfahani and
R.G. Stanmore

671

Modelling the effects of temperature, water activity and storage atmosphere on the viability of stored maize
and paddy – M. L. Bason, P.W. Gras and H.J. Banks

677

A mathematical model for stockpile management – E. Boyapati and A. Oates

684

Infestations by *Sitophilus granarius* (L.) and *Rhyzopertha dominica* (F.) on durum wheat, and their influence
on the rheological characteristics of the semolina – G. Domenichini, M. Pagani and D. Fogliazza

689

Effect of modified atmosphere storage on wheat seed germination vigour and on physiological criteria of
the ageing process – F. Fleurat-Lessard, D. Just, P. Barrieu, J.-M. Le Torc'h, P. Raymond and P. Saglio

695

Comparison of methods for moisture content determination on soybeans – F.A. Lazzari

701

Modification of the nutritional quality of nitrogen content of Leguminosae seed damaged by *Acanthoscelides
obtectus* (Say) (Coleoptera: Bruchidae) – C. Regnault–Roger, C. Watier and A. Hamraoui

704

Effect of rice storage conditions on the quality of milled rice – D. M. Trigo-Stockli and J. R. Pedersen

706

Functional properties of stored grains after microwave treatment – A.M. Zain and L.H. Ooi

712

Grain Quality — Session Summary

715

GRAIN PROTECTANTS

717

Keynote Address
Grain protectant chemicals: present status and future trends – F.H. Arthur

719

Keynote Address
Grain protectants: trends and developments – J.M. Desmarchelier

722

Trials of grain protectants on stored maize under Philippine conditions – M.A. Acda, P.B. Sayaboc,
A.G. Gibe and C.B. Gragasin

729

Use of methoprene without adulticide as a grain protectant – R. Allanson and B. Wallbank

734

Using a PCR diagnostic for detection of insecticide resistance in *Tribolium castaneum* populations –
D. Andreev, T. Phillips, R. Beeman and R. ffrench-Constant

737

Effectiveness of pyrethroids as protectants of raw agricultural commodities stored in southeast Georgia, USA –
F.H. Arthur

741

Repellent and phagodeterrent activity of *Sphaeranthus indicus* extract against *Callosobruchus chinensis* –
J. K. Baby

746

Analysis of bioassay data using the Wadley's Problem technique in probit analysis — a neglected option –
M. Bengston and A.C. Strange

749

Recent developments in grain protectants for use in Australia – M. Bengston and A.C. Strange

751

Resistance considerations for choosing protectants – P.J. Collins

755

Efficacy of several mixtures of grain protectants on paddy and maize – G.J. Daglish

762

Insect growth regulators for the control of stored-grain insect pests – M.J. Dales, S. Harding, N. Freeman
and H. Gaffney

765

Development of a closed system for application of grain protectants – M.A. Ebert, J.L. McLeod and
B.A. Smith

770

Effect of the chitin-synthesis-inhibitor, chlorfluazuron, on immature development of *Rhyzopertha
dominica* (F.) (Coleoptera: Bostrichidae) – J.A. Elek

773

Prevention of beetle infestation of dried fish – P. Golob, A. Gueye-Ndiaye and S. Johnson

777

Residues of grain protectants on paddy – Ma. Gragasin, B. Cristina, M.A. Acda, A.G. Gibe and
P.D. Sayaboc

782

Are residual insecticide applications to store surfaces worth using? – I. Gudrups, A. Harris and M. Dales

785

Potential of common herbs as grain protectants: repellent effect of herb extracts on the granary weevil,
Sitophilus granarius (L.) – S. Ignatowicz and B. Wesolowska

790

Field evaluation of a test kit for monitoring insecticide resistance in stored-grain pests – A. Jermannaud

795

The fate of residues of deltamethrin in treated wheat during its transformation into food products –
A. Jermannaud and J. M. Pochon

798

Introduction of the neem tree in Mexico, in vitro propagation and validation of its properties against
stored-product insects – J. Leos-Martínez, R.P. Salazar-Saenz and O.G. Alvarado-Gómez

804

Chemical control testing on foodstuff mites – G. C. Lozzia, I. E. Rigamonti and F. Ottoboni   809

The influence of temperature and modified atmosphere on effectiveness of *Lavandula angustifolia* Mill. oil
for controlling *Tyrophagus putrescentiae* – Lungshi Li, Xiaowei Zhang and Yiquan Guo   817

Toxicity of *Annona squamosa* Linn. seed oil extract on *Tribolium castaneum* (Herbst) (Coleoptera:
Tenebrionidae) – M.A. Malek and R.M. Wilkins   819

A new bioassay detecting for IGR activity with larvae of *Tribolium freemani* Hinton (Coleoptera:
Tenebrionidae) – H. Nakakita, P. Sittisuang and T. Suzuki   824

Persistence of grain protectants in maize – S. H. Ong, M. Rahim and Z. Sulaiman   828

Cyfluthrin plus piperonyl butoxide — a promising new stored product protectant – R. Pospischil and
G. Smith   830

Organophosphorous and synergised synthetic pyrethroid insecticides as grain protectants for stored maize –
M. Rahim, Z. Sulaiman and S.H. Ong   833

Antifeedant effect of Mediterranean plant essential oils upon *Acanthoscelides obtectus* (Say) (Coleoptera),
bruchid of kidney beans, *Phaseolus vulgaris* L. – C. Regnault-Roger and A. Hamraoui   837

Dynamics of insect populations in stored shelled corn (maize) treated with pirimiphos-methyl and
thiabendazole – J.D. Sedlacek, P.A. Weston, B.D. Price and P.L. Rattlingourd   841

Rapid testing for insecticide residues in stored products using immuno- and enzyme assays – J.H. Skerritt,
A.S. Hill, S.L. Edward, H.L. Beasley, N. Lee, D.P. McAdam and A.J. Rigg   843

Efficacy of pithraj (*Aphanamixis polystachya*) seed extracts against stored-product pests – F.A. Talukder
and P.E. Howse   848

Effectiveness of residual insecticides against warehouse beetle, *Trogoderma variabile* Ballion –
B.E. Wallbank   853

Grain protectants and pesticide residues – D.J. Webley   857

An assessment of Damfin to control an established infestation of saw-toothed grain beetle in malting barley –
D.R. Wilkin, T.J. Binns and T. Hoppe   863

Correlation of probit parameters of malathion-resistant *Tribolium castaneum* (Herbst) (Coleoptera:
Tenebrionidae) determined by topical application and residual methods – J.L. Zettler and F.H. Arthur   872

Grain Protectants — Session Summary   **876**

INTEGRATED COMMODITY MANAGEMENT   877

Keynote Address

Decision support systems for integrated management of stored commodities – D.R. Wilkin and
J.D. Mumford   879

Food aid: a substitute for domestic production and commercial imports? – F.Z. Ahmadi-Esfahani and
C.G. Locke   884

Adding value to Australian wheat: present problems and future prospects – F.Z. Ahmadi-Esfahani and
P. H. Jensen   890

Some effects of grain cleaning on mites, insects and fungi – D.M. Armitage   896

Loss assessment and loss prevention in wheat and storage — technology development and transfer
in Pakistan – U.K. Baloch, M. Irshad and M. Ahmed   902

The effect of maize cob selection and the impact of field infestation on stored maize losses by the larger
grain borer (*Prostephanus truncatus* (Horn) Coleoptera: Bostrichidae) and associated storage pests –
C. Borgemeister, C. Adda, B. Djomamou, P. Degbey, A. Agbaka, F. Djossou, W.G. Meikle and
R.H. Markham   906

Integrated pest management in the GRAINCO, Queensland, Australia, storage system – B.W. Bridgeman
and P.J. Collins   910

Insect control in farm-stored grains—the 'Grainsafe' extension project 995 – K.S. Bullen, P. Collins
and A.S. Andrews   915

Sustainable postharvest systems in developing countries — framework for intervention – C.P.F. De Lima   918

Field validation of a decision support system for farm-stored grain – P.W. Flinn and D.W. Hagstrum   921

Dividing the harvest: an approach to integrated pest management in family stores in Africa – C. Henckes   925

Recent advances in the biology and control of *Prostephanus truncatus* (Coleoptera: Bostrichidae) –
R.J. Hodges   929

U.S. stored-wheat pest management practices: producers, elevator operators, and mills – P. Kenkel,
R.T. Noyes, G.W. Cuperus, J. Criswell, S. Fargo and K. Anderson   935

Decision support systems for pest management in grain stores – B.C. Longstaff   940

Technologies for storage and preservation of coffee beans in India – K.S. Narasimhan, S. Rajendran,
M. Jayaram and N. Muralidharan   946

An analysis of the importance of liposcelids in tropical large-scale storage – V. Pike   950

Insect losses on sorghum stored in selected Malian villages, with particular emphasis on varietal
differences in grain resistance – A. Ratnadass, S. Berté, D. Diarra and B. Cissé   953

Storage systems for maize (*Zea mays* L.) in Nigeria from five agro-ecological zones – J. Udoh, T. Ikotun, and K. Cardwell — 960

Integrated Commodity Management — Session Summary — **966**

STORAGE FUNGI AND MYCOTOXINS — **969**

Keynote Address
Fungi and mycotoxins in grain: implications for stored product research – J. D. Miller — 971

Effect of extracts from nine plant species found in Africa on the mycelial growth of *Aspergillus flavus* Link – K.F. Cardwell and L. Dongo — 978
The effect of *Sitophilus zeamais* on fungal infection, aflatoxin production, moisture content and damage to kernels of stored maize – O.S. Dharmaputra, H. Halid, Sunjaya and Koo Soek Khim — 981
Aspergillus flavus and *Penicillium islandicum* on milled rice collected from different parts of the postharvest handling chain – O. S. Dharmaputra and I. Retnowati — 985
Application of mathematical modelling techniques for predicting mould growth – A. M. Gibson, M.J. Eyles, A.D. Hocking and D.J. Best — 988
Effect of preincubation of fungal conidia in modified atmosphere on subsequent germination and growth on a solid medium – I. Haasum and P. V. Nielsen — 992
Characterisation of aflatoxins B_1, B_2, G_1, and G_2 in groundnuts and groundnut products – Y. Haryadi and E. Setiastuty — 996
Taxonomy: the key to mycotoxin identification in food and feedstuffs – Z. Kozakiewicz — 999
Respiration and losses in stored wheat under different environmental conditions – J. Lacey, A. Hamer and N. Magan — 1007
Occurrence of *Fusarium* toxins in stored maize in southern Brazil – F. A. Lazzari — 1014
Estimating the social costs of the impacts of fungi and aflatoxins in maize and peanuts – A.S.G. Lubulwa and J.S. Davis — 1017
Environmental factors and tenuazonic acid production by *Alternaria* spp. isolated from sorghum – N. Magan and E. Baxter — 1043
Production of polyclonal antibodies against polypeptides from an aflatoxin strain of the fungus *Aspergillus flavus*, a pathogen of stored grain – N. Paster, M. Menasherov, R. Salomon and E. Kuttin — 1047
Levels of aflatoxins in grains from Santa Catarina State, Southern Brazil – V.M. Scussel and W.R. Baratto — 1051
Effect of physical treatments on moulding and aflatoxin production in maize – H.S. Shetty, P. Vijaya, C.M. Usha, K.L. Patkar and J. Lacey — 1054
The impact of insect pests on aflatoxin contamination of stored wheat and maize – A. K. Sinha — 1059
Preharvest contamination of maize by *Aspergillus flavus* – P. Siriacha, P. Tonboonek, A. Wongurai, and S. Kositcharoenkul — 1064
Traditional storage of pandanus nuts in the Papua New Guinea highlands – J. van S. Greve, A.D. Hocking and A.K Sharp — 1068
Preharvest origins of toxigenic fungi in stored grain – D. T. Wicklow — 1075

Storage Fungi and Mycotoxins — Session Summary — **1082**

BIOLOGICAL CONTROL — **1085**

Keynote Address
Can biological control resolve the larger grain borer crisis? – R.H. Markham, C. Borgemeister and W.G. Meikle — 1087

The dispersion pattern of *Teretriosoma nigrescens* Lewis (Coleoptera: Histeridae) after its release and monitoring of the occurrence of its host *Prostephanus truncatus* (Horn) (Coleoptera: Bostrichidae) in the natural environment in Togo – J. Boeye, A. Biliwa, H.U. Fischer, J. Helbig and J. Richter — 1098
Suppression of insects in stored wheat by augmentation with parasitoid wasps – P.W. Flinn, D.W. Hagstrum and W.H. McGaughey — 1103
Biological control in the context of an integrated management strategy for the larger grain borer, *Prostephanus truncatus* (Horn) (Coleoptera: Bostrichidae) and associated storage pests – R.H. Markham, F. Djossou, J.M. Hirabayashi, P. Novillo, V.F. Wright, R.M. Rios, F.J. Trujillo, W.G. Meikle and C. Borgemeister — 1106
Bacillus thuringiensis variety *tenebrionis* (DSM-2803) in the control of coleopteran pests of stored wheat – S. G. Mummigatti, A.N. Raghunathan and N.G.K. Karanth — 1112
Ability of the predator *Teretriosoma nigrescens* Lewis (Coleoptera: Histeridae) to control larger grain borer (*Prostephanus truncatus*) (Horn) (Coleoptera: Bostrichidae) under rural storage conditions in the southern region of Togo – P. Mutlu — 1116
Life history, predatory biology, and population ecology of *Lyctocoris campestris* (F.) (Heteroptera: Anthocoridae) – M.N. Parajulee and T.W. Phillips — 1122

Research on multiplication of *Beauveria bassiana* fungus and preliminary utilisation of Bb bioproduct for
pest management in stored products in Vietnam – Pham Thi Thuy, Le Doan Dien and Nguyen Giang Van   1132
Host specificity of *Teretriosoma nigrescens* Lewis (Coleoptera: Histeridae) – M. Pöschko   1134
Studies on biological control of *Ephestia kuehniella* (Zeller) (Lepidoptera: Pyralidae) with *Trichogramma
evanescens* Westwood (Hymenoptera: Trichogrammatidae) — host-finding ability in wheat under
laboratory conditions – M. Schöller, C. Reichmuth and S.A. Hassan   1142
Computer simulation model for biological control of maize weevil by the parasitoid *Anisopteromalus
calandrae* – L. Smith   1147
The functional response of *Uscana lariophaga* Steffan (Hymenoptera: Trichogrammatidae) under different egg
distributions of its host *Callosobruchus maculatus* (Fab.) (Coleoptera: Bruchidae) – F.A.N. van Alebeek   1152
The role of semiochemicals in host location by *Uscana lariophaga*, egg parasitoid of *Callosobruchus
maculatus* – A. van Huis, C. Schütte, M.H. Cools, Ph. Fanget, H. van der Hoek and S.P. Piquet   1158

Biological Control — Session Summary   **1165**

QUARANTINE AND REGULATORY ISSUES   1167

Decision making in regulatory entomology: the case of *Trogoderma variabile* in Western Australia –
M.J. Butcher   1169
Insects found in stored products entering the port of Ravenna, Italy during 1976–91 – A. Contessi   1173
An integrated approach to stored-grain protection in Western Australia – K.R Dean   1179
The changing role of AQIS in the regulation of grain exports from Australia – D. Heinrich and J. Dean   1183
Factors influencing current U.K. strategies to meet quarantine requirements for export grain – M.P. Kelly
and D.R. Wilkin   1186
GRAINCO (Queensland, Australia) attains 'certification assurance' accreditation – P. Wilson and
B. Bridgeman   1192

Quarantine and Regulatory Issues — Session Summary   **1195**

PHYSICAL CONTROL   1197

Commodity disinfestation treatments with heat – N.W. Heather   1199
Radiation disinfestation of used packagings: irradiation trials with electron beams – S. Ignatowicz and
I.H.M. Zaedee   1201
Detection of irradiated insect pests in stored products: locomotor activity of irradiated adult beetles –
S. Ignatowicz, B. Wesolowska and I.H. Zaedee   1209
The effect of grain movement on *Liposcelis decolor* (Pearman), *Liposcelis bostrychophila* Badonnel
(Psocoptera: Liposcelidae) and *Cryptolestes ferrugineus* (Stephens) (Coleoptera: Cucujidae) infesting
bulk-stored barley – D. Rees, T. van Gerwen and T. Hillier   1214

Physical Control — Session Summary   **1220**

WORKSHOP REPORTS   1223

Appropriate Storage   1225
Expert Systems   1227
On-farm and Small-scale Storage and Extension   1228
Standards   1230
(Reports of other workshops are included in the appropriate Session Summaries)

LATE PAPERS   1231

Field evaluation of a cylinder trap design for monitoring *Ephestia cautella* – T.G. Bowditch, J.L. Madden
and B.F. Brassington   1233
Effect of storage and thermal treatment on quality of rain-damaged wheat – P.W. Gras, M.L. Bason
and J.D. Tomlinson   1235
Effectiveness of SIROFLO® in horizontal storages – R.G. Winks and G.F. Russell   1238
Effectiveness of SIROFLO® in vertical storages – R.G. Winks and G.F. Russell   1245
A brief history of the entomological problems of wheat storage in Australia – J. van S. Graver and
R.G. Winks   1250

List of Participants   **1259**

Trade Exhibitors   **1272**

Author Index   **1273**

6th IWCSPP Conference Summary

Bruce Champ
Chairman of the Organising Committee

On behalf of the organising committee I shall attempt to summarise the activities of the past six days. It has been a great pleasure to welcome you all to Canberra and particularly to have the conference opened by Mr John Kerin, formerly Minister for Primary Industries and Energy and for Trade and Overseas Development, and acknowledged as a great friend of agriculture in Australia. Our sincere thanks must be conveyed to CSIRO and ACIAR, the organisations which have contributed so much to making the conference a reality. Notwithstanding, the conference has truly been a team effort by the many faces and groups which make up the Australian grain industry, and it is extremely gratifying to see participation by such a cross-section of the industry in both the conference and the trade exhibition. Equally, the attendance of our overseas colleagues has ensured that the widest range of technologies is available for discussion, and thus accessible to protect food and feedstuffs in storage.

The working conferences have a simple purpose — 'to provide the opportunity for international scientists to discuss current basic and practical stored-product research and to identify research needs for the future'. The organising committee feels this has been achieved. Thus, as the conference draws to a close, we come to the final session titled 'Looking to the future'. All the papers have been presented and the discussion sessions and workshops are complete. The various chairmen have presented their summaries distilling from the various deliberations the messages for the future. Their reports will be included in the proceedings of the conference.

The organising committee implemented a policy of requiring all contributed papers to be presented in poster format even if presented orally. This appears to have been very useful and well received by participants. Similarly, the co-location of the industry exhibition with the poster displays has been very successful not only in its own right but as a venue for morning and afternoon breaks and other informal discussions.

A number of major issues have been identified. Some of these should be revisited briefly and, of course, the specific recommendations that address the title of this session 'Looking to the future' should be brought to your attention.

Before doing this, it would be appropriate to comment generally on the conference. It has been extremely gratifying for the organising committee to see so many participants. There is a wide range of disciplines represented, covering research and development in both the commercial and public sectors and, of course, from all around the world. The conference was arranged at a difficult time on the world scene, with a recession both in economic terms and particularly in resources available to agriculture, and more so to the postharvest sector. Furthermore, Australia is at the end of the world, so to speak, increasing travel costs considerably. Against this background we have had a record attendance. The message from this is clear — the global network on stored product protection is alive and thriving.

I am sure that through the conference sessions, the industry exhibition, and the social events, many new friendships and business contacts have been forged and old friendships consolidated, and that in 'looking to the future' our activities will be that much more effective and profitable.

On a different note, I should like to refer to an increasingly obvious deficiency in current research strategy that has emerged in discussions during the conference, and which I believe is impeding our potential to look to the future. This concerns the importance of ensuring that our research base and the pool of information generated is sustained and not eroded. With resources becoming more limited and less funding available for basic research, particularly through core funding of organisations, there is a real danger that building the base of tomorrow's technology is at risk. It is not sufficient to depend on limited-term funding from industry for solving immediate problems — there must also be long-term, stable support through the public sector for research and development. The economic rationalism that has done so much damage to agriculture must be abandoned and the entrepreneurial management approaches of the last decade recognised for the failures they have been. All must strive to ensure that a balance between basic and strategic research is maintained in our programs and that we convince those who control the finances of the necessity for this.

Still in this context, it was encouraging to see the interest in the conference in applied science, but disturbing to see the disappointingly small attendances at biology sessions and the parallel dearth of contributed papers in this area. Let us hope that the biology base on which many of our activities depend is not being neglected. Certainly, there is a perception that pure biology studies do not have a high profile and therefore do not attract the funding that appears to be a requirement in operating our research and development programs today. The situation must be rectified.

With reference to more specific issues, my summary is conditioned by a problem common to all of us — it was not possible to attend every session.

The **fumigation and controlled atmospheres** sessions appeared to attract the strongest support. The concerns of the discussion groups were reflected in the following *recommendations*:

- That government and other regulating authorities, in cooperation with manufacturers and distributors, provide accurate information for the effective training of fumigators.

- That adoption of effective fumigation practices be fostered by

  - appropriate extension programs arranged by government

  - effective training of fumigators and regulation of their activities

  - identification of incentives that ensure fumigations are carried out in the best possible manner

  - effective preparation of enclosures based on pressure testing to accepted standards

  - monitoring of gas concentrations during fumigation to ensure minimum effective dosages are exceeded.

These recommendations reinforce earlier initiatives in the region by the ASEAN Food Handling Bureau in organising a series of 'Suggested Recommendations for Fumigation of Grains in the ASEAN Region'. The working party convened for developing these ASEAN recommendations will provide a platform for promoting the new recommendations at least in Southeast Asia.

The complementary sessions on **grain protectants** were also well supported. While changes in market attitudes to pesticide residues are being reflected in a reduction in use of protectants, they have also promoted development of new age materials. Currently in the pipeline are one organophosphorous compound, three pyrethroids and a chitin inhibitor. Additionally, concern was expressed that when the registration of older materials lapsed, the lack of publicity on challenges to their re-registration could result in the loss of these tried and proven materials simply by default. It was *recommended*:

- that specific attention be focused on this problem to ensure continued availability of pest management chemicals.

Inert dusts are certainly a growth area. The materials have been with us for a long time but are now only realising their full potential. There was considerable interest in their applications as demonstrated during the field trip on the second day of the conference.

The sessions on **storage fungi and mycotoxins** including the mini symposium on 'Changing perspectives of the origins of mycotoxin contamination' highlighted an area that has emerged as a major constraint to safe storage and provision of food to consumers. Previously, most attention was focused on aflatoxins after harvest but this has now changed to examining preharvest aspects of the problem and involving a wider range of fungi, particularly the *Fusarium* species. As with other areas, the need to involve other disciplines has been identified as mandatory if the problems are to be managed effectively. In this case we are dealing with materials that may be toxins and may adversely affect feeding in animals as well as being carcinogens or immuno-suppressants. Hence, efforts to manage the problem must be interfaced with human and animal health and nutrition studies. Moreover, increasing awareness and acceptance of

the mycotoxin problem in these areas could release resources of an order not available in agriculture to address the problem. There are specialist groups which will ensure that this will happen.

The sessions on the vitally important topics of **engineering** and **grain quality** were well supported but, considering the pivotal role that both play in stored product protection, still greater attention to them is commended, as engineering provides inherent permanency in solutions to problems and quality provides the measure of success in achieving this.

Integrated commodity management and **physical** and **biological control** are commanding increasing attention and this trend must continue.

With the workshops, that on **trapping** maintained the high level of interest it always attracts. Progress in interpretation of trap catches was highlighted, as was the use of pheromones in control programs. It was evident that the other workshops were equally successful. The **extension** workshop was very well supported and analysis of the exercises undertaken is referred to elsewhere in this proceedings. The **standards** workshop identified where draft standards were needed and the key people to draw these up for pesticide resistance, airtight storage, aeration, and population dynamics. With **expert systems**, a pressing need was identified for networking in this topic so that information and ideas could be exchanged — some form of electronic bulletin board. There was very considerable interest in the **appropriate storage** workshop which was concerned with tying together our activities in systems for storage and commodity management.

Thus, there has been a significant body of information generated at the conference and considered in plenary and workshop discussions. We shall publish the proceedings of the conference in association with the Commonwealth Agricultural Bureaux International, at which time the full impact of the Conference will be realised. There will be effective follow-ups as the participants, particularly those in the workshops, will expedite arrangements through existing channels to implement the findings of their discussions.

This conference is now finished and on behalf of the organising committee, I pass the meeting to the President of the Permanent Committee — Dr Larry Zettler.

* * * * *

6th IWCSPP Collaborating Sponsors

Australian International Development Assistance Bureau
Australian Wheat Board
The Commonwealth Industrial Gases Limited
Co-operative Bulk Handling (Western Australia)
Dryacide Australia Pty Ltd
Elsevier Science Ltd
GRAINCO (Queensland)
Grain Elevators Board (Victoria)
Grains Research and Development Corporation
International Conference Support Scheme (DITARD/IEAust)
NSW Grain Corporation Ltd
South Australian Co-operative Bulk Handling

Opening Address

John Kerin*

This conference is concerned primarily with grains, the staple food of people and most of their domestic animals. For reasons that I will mention shortly, the conference theme is 'A Time of Challenge', but of course the provision of food in the right quantities, at the right place, and at the right time is always a challenge somewhere on earth, and storage specialists have a paramount role to play in meeting this challenge.

Feeding the world's population is perhaps humanity's greatest ongoing problem, and one that has been tackled vigorously, and I think with some success, since the spectre of the 'population time bomb' first emerged about 30 years ago.

The response to exploding world population led to the 'Green Revolution', during which enormous gains in crop productivity were made through the activities of organisations such as the International Rice Research Institute. Countries such as the Philippines and Indonesia which had food deficits and needed to import rice, the staple foodstuff of most of their populations, now have food surpluses. Rice production in Vietnam is accelerating and it seems likely that that country will, by the turn of the century, be second only to Thailand among the world's rice exporters.

But the gains have not all been due to production factors, as is well known to this audience. Indeed, plant breeders and other production orientated people seem to have consistently forgotten that when the crop is harvested it generally has to be stored for shorter or longer periods. Early in the green revolution it became painfully obvious that while many of the new crop varieties developed grew rapidly and had high yields, they also had very poor storage characteristics and were particularly attractive to the insects, moulds, and rodents that make a living infesting foodstuffs. So began the development of a range of modern postharvest technologies, many of which will be discussed during this conference.

These technologies have played a key role in ensuring food security in many of the countries in this region. I think here immediately of Australia's nearest neighbour, Indonesia, whose National Logistics Agency (Bulog) maintains buffer stocks of rice using controlled atmosphere storage techniques developed and refined in association with CSIRO and the Australian Centre for International Agricultural Research (ACIAR), two of the Australian agencies much involved in organising this meeting.

The maintenance of food supplies is a general, ongoing challenge, but the organisers of this conference had in mind some specific current issues when they gave the 6th International Working Conference on Stored-product Protection the epithet 'A Time of Challenge'.

These days — and another indicator of just how much things have improved — markets and consumers are demanding not just food, but what they construe as good food. This generally means food free of contaminants such as insects, moulds, and preservatives or other introduced chemicals. I believe — and I am sure most of you do too — that they are entitled to expect this, but of course it makes the storage specialist's life even more difficult.

Health and environmental agencies around the world are homing in on chemical pesticides and fumigants used to protect stored grain and are imposing severe limitations on their use. I think it is safe to say that in doing this they have the general support of consumers.

Perhaps the writing was on the wall for chemical protectants in any case, seeing as stored-product insect pests have so strong a propensity to develop resistance to them.

Methyl bromide, a particularly useful grain fumigant, especially for quarantine applications, has been identified as an atmospheric ozone depleter and international agencies are rapidly phasing it out. Finding suitable replacement technologies is an immediate challenge on which many stored-product specialists are currently focusing their attention. The methyl bromide problem, and possible solutions, are the subject of many papers to be presented during the coming week.

Despite these challenges, I do not think that we need to be pessimistic about the future of stored-product protection. Stored-product specialists have shown themselves to be innovative and committed to the task. The fact that you people have come here — many of you from the other side of the world — indicates a high level of interest in tackling the problems involved.

---

* The Hon. John Kerin is a former Minister for Primary Industries and Energy, and for Trade and Overseas Development, in the Australian Government.

Here in Australia, we are proud, and justifiably I believe, in our contributions to postharvest technology. During this conference, we want to share them with our overseas colleagues, in a personal way, but of course we will be picking their brains for some of their secrets too.

This conference is about sharing current technology for the common good and providing a forum for discussion on what are the problems, which have the highest priority for attention, how we should address them with current technology, and where new knowledge is needed to provide satisfactory solution of the problems.

There is a dangerous notion being promulgated in some quarters that we know it all, and that research — the search for knowledge — can be put on the back burner. I believe that nothing could be further from the truth. The work of the Stored Grain Research Laboratory — Australia's national grains postharvest facility — is testimony to this. It could never have achieved what it has without the basic research that is part of its program. Through its ongoing support, the Australian grain industry recognises this.

The participants in this conference come from many countries. In some developing countries, the main issue is still to halt the depredations of stored-products insect pests. In developed countries, the focus is now more on issues of quality and the tailoring of products to specific markets. The perspectives may be different, but the objective is the same — the preservation of food and feedstuffs. Current projections on world food supply and demand indicate that this objective is imperative.

I commend the Australian Centre for International Agricultural Research (ACIAR), CSIRO, and the Australian grain handling industry for their efforts in organising what seems certain to be a successful meeting on a topic of perennial, international importance. I welcome all delegates, from all parts of the world, to this conference.

Fumigation and Controlled Atmospheres

Fumigation — an endangered technology?

H.J. Banks*

Abstract

Fumigation is becoming an endangered technology. Continued efforts will be required to prevent it becoming extinct. There are many pressures on use of particular chemicals, ranging from development of resistance from the target pests to increasing demands for registration and reregistration to maintain use. Bond (1984) was able to list a wide range of materials as fumigants for stored products, yet ten years later there are only two left in widespread use — phosphine and methyl bromide. Even these remaining materials are under pressure. Methyl bromide has been listed as an ozone-depleting material and will be subject to controls from 1 January 1995 in developed countries. Environmental restrictions on release of phosphine make its use difficult or economically impossible in some parts of the world. There have been some recent developments in our knowledge of phosphine which may assist in keeping this fumigant available. These include studies on its atmospheric fate and detection of its natural occurrence in humans.

The technique of fumigation is sufficiently important to stored product protection for much effort and expense to be justified in supporting existing fumigants and their registration. Also, reregistration and modernisation of those few other fumigants that remain available, though little used, is needed to maintain the choice of materials for fumigation. There may yet be new fumigants to be discovered or recognised.

Introduction

Fumigation, as a technology, is widely used for disinfestation of durable foodstuffs. There is no doubt, from the point of view of stored product protection, that the technology is versatile and valuable. Usually, fumigation gives a quick, low cost and effective solution to problems of insect attack or pest presence in commodities such as stored grain, oilseeds, pulses, coffee and cocoa beans, and dry products derived therefrom.

However, there is a diverse range of pressures, which have led to increasing restriction of the practice of fumigation in many parts of the world. These include problems with environmental contamination, health concerns both for the public and workforce, particularly with respect to potential or suspected carcinogenicity of some fumigants, effects on treated products, including production of residues, and general market/consumer aversion to use of chemicals. Many of the forces leading to increased restriction or regulation of fumigation are not purely technical in nature, but they are potent and real nonetheless. .

---

* Stored Grain Research Laboratory, CSIRO Division of Entomology, GPO Box 1700, Canberra ACT 2601, Australia.

The science and practice of fumigation is now left with two principal fumigants, phosphine and methyl bromide. In two decades the choice of fumigants has apparently been reduced from a wide range of candidates. We must now ask whether this trend will continue and whether now fumigation can be considered an endangered technology. One where only a few extra restrictions and some unexpected discoveries or changes in attitude could lead to extinction of fumigation as presently carried out in some or all regions of the world.

In this paper, I highlight the current importance of fumigation for protection of stored products worldwide. This is followed by consideration of the shrinking list of available fumigants, with a summary of why they have been discontinued, then a discussion of the threats to our remaining accepted materials. Some recent developments are also noted which suggest that all is not lost and there may be hope for continued use of fumigation. Perhaps even new materials may be added to the list.

World Use of Fumigants on Durables

Globally, only two fumigants for durable commodities remain in widespread use. These are phosphine and methyl bromide.

Many countries use fumigation with either, or both, of these materials as their principal method of disinfestation of cereal grain stocks for domestic use, or import and export. Exact figures for the global use of these fumigants are not available, but it is possible to approximate the tonnage of grain and similar commodities treated from estimates of methyl bromide and aluminium phosphide, the precursor of phosphine, for agricultural and post harvest use.

Table 1 gives an estimate for recent usage of methyl bromide and phosphine on durable commodities (foodstuffs). These figures must be taken as broad approximations only. Nevertheless, they indicate the magnitude of use and current reliance, globally, on fumigation of durables. They are remarkably consistent with those given by Muller (1992).

Dichlorvos is also still available for treatment of raw commodities, but its use is more restricted. Dichlorvos can be regarded as a fumigant and its vapour is extremely toxic to insect pests of stored grain. However, it is usually applied as an aerosol or as a grain protectant and is not considered in detail here.

Many countries have become reliant on fumigation as their primary means of controlling or eliminating infestation in cereal grain and similar foodstuffs. Two examples will suffice to illustrate this: Japan, which requires fumigation with methyl bromide to treat imported grains found to be infested, to meet phytosanitary requirements, and Australia, which uses phosphine increasingly as a residue-free system to protect grain in store against infestation. Table 2 and Figure 1 show tonnages and proportions treated in Japan and Australia respectively.

Table 1. Estimates of annual global fumigant usage on durable foodstuffs (1992 data)

Fumigant	Estimated usage (t)	% grain treated
Phosphine	1900[a]	47[b]
Methyl bromide	6700[c]	15[d]

[a] Based on worldwide metal phosphide production of 4060 tonnes per year (Muller 1992) and assuming this to be equivalent to 2380 phosphine per year, with 20% discount for uses other than fumigation of durables, particularly rodent control. Use of cylinder-supplied phosphine ignored.

[b] Based on a world grain harvest of 1350Mt and assuming a treatment rate of 2 g/t phosphine with 50% of grain treated retreated once.

[c] Based on a total commodity usage of 8400 t (Watson et al. 1992) and assuming half of this was for non-foodstuffs, notably timber and perishables, and that CIS, PRC and India used 2500 t in total on durable foodstuffs not included in data of Watson et al. (1992), and that usage on durable foodstuffs, apart from grains, could be neglected.

[d] Based on a world grain harvest of 1350 Mt and a treatment rate of 25 g/t and no retreatments, and usage on durable foodstuffs other than grains ignored.

Table 2. Methyl bromide treatments of grains imported into Japan (1992)

Grain	Tonnage (Mt)	Treated (%)
Wheat	6.1	15
Barley	1.5	13
Maize	16.5	67
All grains	23.6	51

Data source: Japan Plant Quarantine Data Base (1992)

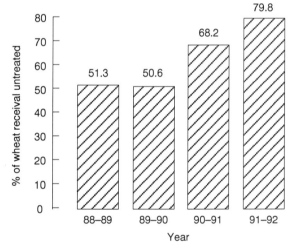

Fig. 1. Percentage of wheat received untreated into the central handling system in Australia. Most of this is subsequently fumigated with phosphine. Data source: Australian Wheat Board.

Overall, protection of durable foodstuffs in much of the world is carried out with the aid of fumigation. The level of fumigant use is such that the technique must be recognised as a major tool. Therefore, loss, or even severe restriction of its use, would have very serious repercussions. It is thus important to consider whether the two remaining materials in widespread use, and the others still available in some countries, are under threat and to assess the likelihood that their use will be curtailed, or even banned.

Status of Fumigants (Past and Present)

Many materials have, in the past, been applied as commodity fumigants for control of insect pests. In 1984 Bond was

able to list more than a dozen materials, many of which were in widespread use for this purpose. Since then the use of many of these has been discontinued, at least in some regions of the world.

The reasons why particular materials are no longer available vary widely, but several have been lost through lack of interest internationally, resulting from want of support for continued national or international registration. Despite lack of international registration or banning of use on commodities in some countries, the same materials are still accepted in others (e.g. carbon tetrachloride in parts of francophone Africa, ethylene dibromide in India) where local considerations may outweigh the risks identified in other countries.

It is instructive to consider the threats to the fumigants listed by Bond (1984) so as to be warned of what might happen and, where possible, to take action to preserve the technique of fumigation. I am not suggesting here that those involved in stored product protection, either as researchers or practitioners, should seek to defend a fumigant in the face of good evidence against it, but only that data should be provided to ensure that materials are not lost unnecessarily.

'Bond's list', with slight modification, is given in Table 3, together with the major threats to continued use of these fumigants. It will be noted, in many cases, that these threats have become reality in that they have resulted in banning or deregistration of particular materials at least in some countries. Paradoxically, one of the main reasons why some materials have fallen into disuse has been the success and effectiveness of phosphine and methyl bromide. Both these materials, in specific circumstances, have properties which are close to ideal for a fumigant. These include low cost, high penetrant ability, ease of airing, lack of effect on end use qualities of most commodities, and acceptability of residues resulting from their use. If phosphine and methyl bromide had never been available, it is likely that materials with recognised major problems in current use, e.g. carbon disulphide and hydrogen cyanide, would still be available, albeit without the wide application of phosphine and methyl bromide. Amongst other drawbacks, carbon disulphide is notoriously flammable, while hydrogen cyanide has poor penetrant ability, significant storability problems and a severely adverse public image.

Table 3. 'Bond's list'. Fumigants listed in Bond (1984), with current status and threats, real or alleged, to continued use.

Fumigant	Threat[a]	Status[b]
Acrylonitrile	Suspect carcinogen, residues	
Carbon disulphide	Lack of interest	*
Carbon tetrachloride	Ozone depletor, residues	
Chloropicrin	Almost forgotten	*
Dichlorvos	Residues, alleged carcinogen	*
Ethylene dibromide	Environmental contamination, fertility effects, alleged carcinogen	
Ethylene dichloride	Not very effective, alleged carcinogen	
Ethylene oxide	Suspect carcinogen, residues	
Ethyl formate	Almost forgotten	*
Hydrogen cyanide	Lapsed Codex Alimentarius registration	*
Methallyl chloride	No food registration	
Methyl bromide	Ozone depletor, alleged carcinogen	*
Methyl formate	Almost forgotten	*
Phosphine	(see text)	*
Sulphuryl fluoride	No food registration	
Trichloroethylene	Not very effective, residues	

[a] 'Suspect carcinogen' refers to A2 status in ACGIH (1993). 'Alleged carcinogen' refers to statement 'Substance identified by other sources as a suspected or confirmed human carcinogen' in ACGIH (1993) or more recent studies.

[b] Materials marked '*' are either still in common use or may no longer be registered but do not at present appear to have data on their effects technically sufficient to deny registration.

Threats to Methyl Bromide and Phosphine

Methyl bromide

There are a number of threats to the continued use of methyl bromide as a commodity fumigant. These include:

- status as an ozone-depleting substance;
- increasing regulation of chemicals generally;
- alleged carcinogenicity;
- concern over residues in commodities;
- toxicity to humans.

Undoubtedly, the main current threat arises from its listing as a recognised ozone-depleting substance at the Copenhagen meeting (1992) of the Parties to the Montreal Protocol on Substances that Deplete the Ozone Layer ('Montreal Protocol'). Article 2H of the Protocol states:

> Each Party shall ensure that for the twelve-month period commencing on 1 January 1995, and in each twelve-month period thereafter, its calculated level of consumption of the controlled substance in Annex E does not exceed, annually, its calculated level of consumption in 1991. Each Party producing the substance shall, for the same periods, ensure that its calculated level of production of the substance does not exceed, annually, its calculated level of production in 1991. However, in order to satisfy the basic domestic needs of the Parties operating under paragraph 1 of Article 5, its calculated level of production may exceed that limit by up to ten per cent of its calculated level of production in 1991. The calculated levels of consumption and production under this Article shall not include the amounts used by the Party for quarantine and pre-shipment applications.

While use for 'quarantine and pre-shipment' fumigation is currently exempted from restriction, there is no guarantee that such exemptions will continue, particularly in the light of strengthening concern over the state of the ozone layer and hardening evidence with regard to the role of bromine in ozone depletion. At least some commodity fumigations are likely to fall within the categories of quarantine or pre-shipment treatments. The restrictions on methyl bromide are due to be reconsidered in late 1995 by the Parties to the Montreal Protocol, and it is not certain that this exemption will continue.

Another major and continuing concern is over the possible carcinogenicity of methyl bromide. Methyl bromide is categorised by the American Council of Governmental Industrial Hygienists (ACGIH) as a 'substance identified by other sources as a suspected or confirmed human carcinogen'. Titles and contents of scientific publications quite frequently refer to the substance as mutagenic or carcinogenic (e.g. Danse et al. 1984; Djalali-Behzad et al. 1981; Singh et al. 1982).

The other threats to methyl bromide are generally concerned with public reaction to use of chemicals. Methyl bromide is highly toxic to humans, a fact reflected in the low current maximum permissible levels for workspace environments (e.g. ACGIH 1993). Any accident with methyl bromide could result in pressure to lower these levels still further. Even the current levels are such that extensive airing periods, and safety zones, may be required, adding to the inconvenience and cost of treatments. There is also often a direct linkage between permissible workspace levels and those tolerated in the environment around treatments. Lowering of these already very low levels could make monitoring of those levels technically very difficult as well as imposing further costs on the system. The overall cost and difficulty, both logistic and technical, resulting from regulations relating to workspace and environmental permissible levels, have already resulted in trends away from methyl bromide use. Thus, the adoption of nitrogen-based CA treatments at the grain export terminal in Newcastle, Australia (Cassells et al., these proceedings), was driven in part by the cost and inconvenience of carrying out fumigation with methyl bromide, while restrictions on fumigation with methyl bromide in Germany have become sufficiently onerous for alternatives, including use of organophosphorous pesticides, to become a preferable pest control option in many circumstances.

While it is clear that there are some severe threats to the long term use of methyl bromide on durable commodities, there have also been some developments which may help prolong its use, although possibly at considerable cost. These involve development of sealing systems for a variety of storage types (Ripp et al. 1984), including bag stacks (Annis 1990), to contain the gas, and progress towards practical scrubbing systems for removal of methyl bromide from air streams (Anon. 1994). Both address the problem of potential adverse impact of methyl bromide on the environment and workspaces.

Overall, the long term prognosis for the continued use of methyl bromide as a fumigant on stored products generally appears poor, and it would be prudent to consider alternatives as a matter of urgency.

Phosphine

While there is quite widespread concern over the future of methyl bromide, this is not so with phosphine. However, even phosphine cannot be regarded as immune from attack. We must consider both the threats to phosphine as well as any developments that may help to permit its continued use.

Threats include:

- toxicity and effects on humans;
- environmental and workspace restrictions;
- resistance by pests;
- accidents.

The recent publication of data which apparently showed genotoxic effects of phosphine on humans, in this case fumigators, at low levels of exposure (Garry et al. 1989; Alranya et al. 1988) has highlighted the vulnerability of phosphine to unexpected new data. Considerable damage has been done to the reputation of phosphine and it can be expected that the papers which refer to the alleged genotoxic effects will continue to be widely cited, even though the actual effects appear unproven.

Phosphine is known to be very toxic to mammals and, in consequence, permissible workspace atmospheric concentrations are set at very low levels, in most countries at 0.3 ppm v/v (e.g. ACGIH 1993), but some even lower at 0.1 ppm (e.g. Germany). The required monitoring of these concentrations already imposes substantial costs on the conduct of a fumigation. Any extension to phosphine of the general historical trend towards continued lowering of maximum permissible levels will substantially increase these costs. A further consequence of low permissible levels of phosphine in the workplace is a still lower maximum permissible level in the environment around fumigation. Different regions have different formulae to determine the latter value, but, where it is set, it is typically at least 10x less than the workspace value. The need to meet such requirements has already made phosphine fumigations in some parts of the world (e.g. North Carolina, Germany) extremely difficult to carry out.

The development of resistance to phosphine by pests, particularly in the Indian subcontinent, has made effective phosphine fumigation much more difficult to achieve. The problems of resistance to phosphine are discussed elsewhere in this Conference (Winks and Hyne, these proceedings). The level of resistance which has developed so far has been insufficient to prevent the effective use of phosphine but does require considerable improvements in application technology for successful treatments, including both better containment of the gas and prolonged exposure times. The development of high levels of resistance to phosphine, and particularly development of the ability to withstand increased exposure periods, would certainly jeopardise the use of phosphine as currently carried out in at least most tropical countries. The apparent ease with which resistance to phosphine develops suggests that such a resistance pattern is not impossible. This contrasts with the situation with methyl bromide, where only slight increases in tolerance have been observed to date (Champ 1986).

A further, and possibly more serious, threat is that of the effect that an accident in use would have on phosphine. Fumigation, including that with phosphine, is a potentially hazardous process requiring careful, skilled application to ensure a safe, effective outcome. Any accident arising from poor practice, including fires arising from incorrect application, or storage and handling of formulations, is likely to result in calls for further regulation and restriction of use of phosphine.

There have been several developments in our knowledge of phosphine which are likely to lend positive support to the continued use of phosphine as a fumigant. These involve its atmospheric fate, natural occurrence and improved application technology.

With the continued use of methyl bromide under threat from its influence on the ozone layer, there has been an immediate need to define more closely the fate of phosphine emitted from fumigations. While it can be expected that phosphine would eventually be oxidised to innocuous phosphorus oxyacids, the rate that this occurs is important as it determines the quantity of the gas that can enter the upper atmosphere. Recent studies (Frank and Rippen 1987; Dévai et al. 1988) have shown phosphine at ground level is attacked by hydroxyl radicals and that the atmospheric lifetime is very short.

The identification of phosphine at low levels as a normal constituent of human faeces, and thus of the lower gut, shows that humans are likely to be constantly exposed to low levels of phosphine. This data may help to decide safe levels of exposure, but it should also be noted that Gassmann and Glindemann (1993) suggest, disturbingly, and without experimental support, that phosphine in the gut may be a cause of colon cancer.

Finally, continued use of phosphine has been strongly supported by recent technical developments in containment and application technology. Recent advances in sealing of grain storages and bag stacks (van S. Graver and Annis 1994) have allowed much reduced dosages of phosphine and substantially reduced the rate of uncontrolled leakage. In situations where sealing is not feasible or economic, the SIROFLO® technique (Winks 1993) offers an alternative which creates much lower concentrations in stores than 'traditional' application methods, with consequent decrease in risk of exposure to high concentrations in the vicinity of a treatment. SIROFLO® concentrations typically are 50 ppm or less in the storage, while normal concentrations in stores treated by addition of aluminium phosphide tablets frequently exceed 500 ppm at some stage. SIROFLO® exposures also typically use less phosphine in total than conventional treatments in poorly sealed systems.

Alternative Fumigants

'Old' fumigants

Bond's list (Table 3) contains several materials which have fallen into disuse because of inconvenient properties and lack of commercial support rather than because they have toxicological or other features which are sufficient to actively prevent their use. These are carbon disulphide, chloropicrin, hydrogen cyanide and ethyl and methyl formate. The existence of these materials, and their historical use as fumigants, gives some hope that a replacement could be found for methyl bromide or phosphine, should the use of either be restricted or banned. Both carbon disulphide and hydrogen cyanide have lost their former status under the Codex Alimentarius as approved materials with agreed international tolerances. In the case of carbon disulphide this was a 'guideline' tolerance, but with HCN the position lost was a well established Maximum Residue Limit. Reinstatement of these materials will require provision of a full toxicological data package and extensive field trials. Since both are commonly available chemicals without patent protection, it appears unlikely that commercial companies, normally the providers of supporting data for registration, will do this. Competitors can easily benefit from such registration without financial penalty. It is a very real question as to what organisations could, or should provide the required information, given that this would probably require investment of tens of millions of dollars.

Public organisations may be the appropriate bodies to seek to obtain or support registration of 'old' fumigants, to maintain their use for public benefit. This could be the impetus for effective international collaboration.

New fumigants

It has been a well known assumption that no new fumigants will be developed for stored product protection. There have been recent developments that have challenged this belief. It remains true that no new fumigants have been registered for some decades. However, there is active work on use of ozone (Yoshida 1975), carbonyl sulphide (Desmarchelier, these Proceedings) and methyl isothiocyanate (Ducom, these proceedings) and there is no reason to believe that further candidates will not be identified. It may also be that well known fumigants in other fields, such as methallyl chloride or sulphuryl fluoride, may yet find application in disinfestation of stored foodstuffs.

As with reregistration of old fumigants, the cost and effort required to gather the necessary data are major disincentives to registration of new materials as fumigants for use on foodstuffs.

Conclusion

In summary, fumigation *is* an endangered technology. There is no doubt that it is a most valuable tool for the protection of stored products. Some would argue that it is *the* most valuable, and critical to the continued protection of food stocks globally. Nevertheless there are many pressures from a wide variety of sources which are actively threatening particular materials. Neither of the two most widely used and accepted fumigants, phosphine and methyl bromide, are secure in the long term. Methyl bromide use will be banned by 2001, at least in the USA (EPA 1993), unless current legislation is challenged and changed. Phosphine use appears to be assured at present, though the historical trend of continued loss of fumigants for stored product protection should warn us that we cannot assume it will be always so. We should not be complacent.

It falls to those involved with the science of stored product protection to defend the practice of fumigation. We need to provide the data necessary to ensure that the continued use of fumigation is well based, and to counter and anticipate criticism. When faced with evidence of problems we need to appreciate them and, if possible, find ways to overcome them.

Without continuing and consolidated effort internationally, fumigation will not only be endangered, it will become extinct.

References

ACGIH. 1993. 1993–1994 Threshold limit values for chemical substances and physical agents and biological exposure indices. The American Conference of Governmental Industrial Hygienists, Cincinnatti, USA, 124 p.

Alvanya, M.C.R., Rush, G.A., Steward, P. and Blair, A. 1988. Proportionate mortality study of workers in the grain industry. Journal of the National Cancer Institute, 78, 247.

Annis, P.C. 1990. Sealed storage of bag stacks: status of the technology. In: Champ, B.R., Highley, E. and Banks, H.J., ed., Fumigation and controlled atmosphere storage of grain: proceedings of an international conference, Singapore, 14–18 February 1989. ACIAR Proceedings No. 25, 203–210.

Anon. 1994. Report of the Methyl Bromide Technical Options Committee. In: Report of the Technology and Economics Assessment Panel, Chapter 6, Nairobi, United Nations Environment Programme, 29 p.

Bond, E.J. 1984. Manual for fumigation for insect control. Rome, FAO Plant Production and Protection Paper No 54, 432 p.

Champ, B.R. 1986. Occurrence of resistance to pesticides in grain storage pests. In: Champ, B.R. and Highley, E., ed., Pesticides and humid tropical grain storage systems: proceedings of an international seminar, Manila, Philippines, 27–30 May 1985. ACIAR Proceedings No. 14, 229–255.

Danse, L.H.J.C., van Velsen, F.L. and van der Heijden, C.A. 1984. methyl bromide: carcinogenic effects in the rat prestomach. Toxicity and Applied Pharmacology, 72, 262–271.

Dévai, I., Felföldy, L., Wittner, I. and Plósz, S. 1988. Detection of phosphine: new aspects of the phosphorous cycle in the hydrosphere. Nature, 333, 343–345.

Djalali-Behzad, G., Hussain, S., Osterman-Golkar, S. and Segerbäck, D. 1981. Estimation of genetic risks of alkylating agents VI. Exposure of mice and bacteria to methyl bromide. Mutation Research, 84, 1–9.

EPA (Environmental Protection Agency) 1993. Regulatory action under the Clean Air Act on methyl bromide. United States Environmental Protection Agency, Office of Air and Radiation Stratospheric Protection Division, Washington DC, USA, Update, Winter 1993, 1p.

Frank, R. von and Rippen, G. 1987. Verhalten von Phospine in der Atmosphäre. [Fate of phosphine in the atmosphere]. Lebensmitteltechnik, Juli/August 1987, 409–411.

Garry, V.F., Griffith, J., Danzl, T.J., Nelson, R.L., Whorton, E.B., Krueger, L.A. and Cervenka, J. 1989. Human genotoxicity: pesticide applicators and phosphine. Science, 246, 251–255.

Gassmann, G. and Glindemann,D. 1993. Phosphane (PH3) in the biosphere. Angewandte Chemie, International Edition England, 32, 761–763.

Muller, D.K. 1992. Malathion update. Fumigants and Pheromones, 29, 7.

Muller, D.K. 1993. An alternative to methyl bromide in store products. Fumigants and Pheromones, 31, 5.

Ripp, B. E., Banks, H.J., Bond, E.J., Calverley, D.J., Jay, E.G. and Navarro, S., ed. 1984. Controlled atmospheres and fumigation in grain storages. Proceedings of an international symposium on practical aspects of controlled atmosphere and fumigation in grain storages, Perth, Australia, 11–22 April 1983, xiv + 798.

Singh, H.B., Salas, L.J. and Stiles, R.E. 1982. Distribution of selected gaseous organic mutagens and suspect carcinogens in ambient air. Environmental Science Technology, 16, 872–880.

van S. Graver, J. and Annis, P. 1994. Suggested recommendations for the fumigation of grain in the ASEAN Region. Part 3. Phosphine fumigation of bag-stacks sealed in plastic enclosures: an operations manual. Kuala Lumpur, Malaysia, ASEAN Food Handling Bureau/Canberra, Australia, ACIAR, 79 p.

Watson, R.T., Albritton, D.L., Andersen, S.O. and Lee-Bapty, S. 1992. Methyl bromide its atmospheric science, technology, and economics. Nairobi, United Nations Environment Programme, 41 p.

Winks, R.G. 1993. The development of SIROFLO in Australia. In: Navarro, S. and Donahaye, E., ed., Proceedings of the international conference on controlled atmosphere and fumigation in grain storages, Winnipeg, Canada, June 1992, 399–410.

Yoshida, T. 1975. Lethal effect of ozone gas on the adults of Sitophilus oryzae (Coleoptera: Curculionidae) and Oryzaephilus surinamensis (Coleoptera: Cucujidae). Science Report of the Faculty of Agriculture, Okayama University, 45, 10–15.

Carbon dioxide — more rapidly impairing the glycolytic energy production than nitrogen?

C. S. Adler*

Abstract

To compare the effects of different controlled atmospheres on anaerobic energy production of insects, wheat grains infested with pupae of the granary weevil *Sitophilus granarius* (L.) were exposed to pure nitrogen or pure carbon dioxide for times between 2 hours and 21 days at $20 \pm 1°C$, 76% r.h. Under anoxic conditions, 10–12 pupae (approx. 50 mg) were removed from each grain sample, weighed and their lactate content was determined using a standardised enzyme test kit.

It could be demonstrated that granary weevil pupae produce lactate under anoxic conditions. The contents of lactate increased most strongly within the first 24 hours of anoxia. Later the lactate production decreased to almost zero, suggesting an inhibition of this metabolic pathway. The lactate levels in pupae fumigated with CO_2 were only about one third of those found in N_2-fumigated pupae. This may be attributed to an acidification caused by carbonic acid and lactic acid which in N_2-fumigated pupae is attained by the accumulation of greater amounts of lactic acid alone. A higher concentration of hydrogen ions could inhibit the production of phosphoenol-pyruvate out of glycerol-3-phosphate by causing a shortage of free NAD^+ which is needed in this glycolytic reaction. Compared with nitrogen, the inhibition of glycolysis at lower lactate levels, corresponding to a lower energy yield, could be a reason for the more rapid lethal action of CO_2-rich atmospheres.

Introduction

Energy production from nutrients is always connected to a partial or complete oxidation of the substrate with electrons or hydrogen ions being transferred to an acceptor, or down a chain of acceptors. In this process, the primary or intermediate acceptors are reoxidised while the terminal acceptor accumulates in the reduced state. Oxygen is an ideal terminal acceptor because water can be utilised, accumulated and excreted without damaging the organism (Urich 1990). Eucaryotic cells first utilised oxygen for respiration about 1.4 billion years ago. This oxygen had previously been produced by the photosynthesis of prokaryotes, themselves using other terminal acceptors in an originally anoxic atmosphere (Wegener 1988).

Even animals that live in habitats extremely poor in oxygen, such as parasites of the intestinal tract, depend at least once in their ontogenesis on the availability of oxygen (Wegener 1988). As Wegener states, there are several good reasons for the uniformity in which oxygen is used as a terminal acceptor:

1. Different nutritive compounds, such as lipids, proteins and carbohydrates can be utilised, whereas without oxygen only glycolysis and, in a few cases, the breakdown of certain amino acids can be utilised for the production of the energy-conserving adenosine triphosphate (ATP).
2. Only in the presence of oxygen can substrates be metabolised with a maximum energy yield. In contrast, the energy yield of anoxic metabolism is rather low.
3. The main foodstuffs, carbohydrates and lipids, can be broken down completely to the harmless and readily excretable end-products carbon dioxide and water. Anaerobic metabolism, on the other hand, usually gives rise to acidic compounds and protons that may interfere with cellular functions and that are excreted or accumulated with much more difficulty.

If an organism is frequently confronted with a lack of oxygen in its habitat it can react in one or several of the following ways.

Migration

Migration to sources of oxygen can be observed in many aquatic organisms that actively move into water layers with a high oxygen content. Extreme examples are some fish species (e.g. of the genus *Gambusia*) that can survive in eutrophic, oxygen-free water by swallowing air.

Energy conservation

Carrying out microcalorimetric studies with the desert locust *Locusta migratoria* L. and the tobacco hornworm *Manduca sexta* (Joh.), Moratzky et al. (1992) could prove that these insects reduce their heat production to less than 5% of the normal values after 4 or 5.5 hours under anoxia, respectively. This enormous conservation of energy can therefore be seen as a major factor rendering insects much more tolerant to the lack of oxygen than vertebrates, for example (for mechanisms of energy conservation see Gäde 1985).

Anoxic energy production

Glycolysis does not require oxygen. Therefore, this reaction can be utilised under anoxic conditions. To balance the redox state, the resulting pyruvate can be reduced to lactate by a reaction with NADH, in this way replacing oxygen as a terminal acceptor. This is known to happen in the muscles of terrestrial animals during periods of high activity and insufficient oxygen supply (described for the saltatoric muscle in the hind leg of *L. migratoria* in Gäde 1985). The production of lactate will supply an organism with 3 mol ATP per glycosylic unit while a total of 36 mol ATP is produced if pyruvate can be transformed into malate and broken down in the Krebs cycle under consumption of oxygen.

Many aquatic animals are able to produce excretable end-products of anaerobic metabolism like propionate (via succinate) while at the same time achieving a higher energy

* Federal Biological Research Centre for Agriculture and Forestry, Institute for Stored Product Protection, Königin-Luise Straße 19, D-14195 Berlin, Germany.

yield (7 mol ATP) (Fig. 1). The production of propionate, however, has not yet been described for insects.

The mud-inhabiting larvae of *Chironomus riparius* (formerly *C. thummi thummi*) are known to excrete ethanol (Zebe 1977). But generally, the production of lactate seems to be the main metabolic pathway of anoxic energy production in insects.

Adult cerambycid beetles *Rhagium inquisitor* L. surviving arctic winters enclosed in ice were found to have high lactate levels after being kept in pure nitrogen gas at 5°C for a month (Zachariassen and Pasche 1976). Also, soil-inhabiting larvae of the fly *Callitroga macellaria* F. were found to accumulate lactate under experimental anoxia (Gäde 1985).

Research findings on the respiratory physiology of stored-product insects have so far been published mainly by scientists from the Volcani Center, Bet Dagan, Israel. Navarro and Friedlander (1975) exposed pupae of *Ephestia cautella* for 24 hours to 10 % O_2 and different concentration of CO_2 (20–89%). Proportional to the increase of CO_2 in the experimental atmosphere, they could find a rising level of lactate in the haemolymph. In atmospheres of nitrogen and small amounts of oxygen, the lactate levels rose sharply if the oxygen content was reduced below 3%. In another survey, the same authors found a significant reduction in glucose levels under hypoxic conditions, whereas under hypercarbia these levels remained constant (Friedlander and Navarro 1979). The amounts of citrate were found to be reduced under both hypoxic and hypercarbic atmospheres which could be a consequence of disrupting the Krebs cycle.

Till now, no studies have been published on the physiological effects of controlled atmospheres on the granary weevil, *Sitophilus granarius*, though this insect is known to be quite tolerant to hypoxia and hypercarbia. The pupa of this species was chosen for the following experiments because pupae are the most tolerant developmental stage and because they combine both high body volume and a low degree of sclerotisation.

The levels of lactate in pupae fumigated with pure nitrogen or pure carbon dioxide were determined by high performance liquid chromatography (HPLC) and by a standardised enzymatic test. The HPLC study was carried out only with pupae fumigated for 3 weeks to detect the presence of lactate and to receive a first estimate of the maximum amount of lactate to be expected after a treatment with controlled atmospheres.

The results of this study aided calculation of the appropriate dilution factors for the enzymatic analysis. The enzymatic test, more accurate in quatitative terms, was carried out with pupae being exposed to the respective atmospheres for various periods. This was done in order to determine the influence of exposure time and atmosphere on the lactate production.

Materials and Methods

Wheat grains infested with 28–31-day-old developmental stages (mainly pupae) from a culture kept at $25 \pm 1°C$, 75 ± 5 % r.h., were placed in wire mesh cages. Three cages were placed into a 500 mL Dressel flask that was then purged with pure nitrogen or pure CO_2 (flow rate 1000 mL/minute). Before use, the gases had been humidified to 76% r.h. by passing them through a saturated sodium chloride solution (Winston and Bates 1960). When the Drechsel flask contained an atmosphere with 0.1% O_2 or less, the taps of its gas inlet and gas outlet tubes were closed. The flasks were then stored at 201°C for various periods. The following exposure times were chosen:

N_2-fumigation: 2 hours, 4 hours, 1 day, 4 days, 7 days, 14 days and 21 days;

CO_2-fumigation: 1 day, 4 days, 7 days, 14 days and 21 days.

At the end of the exposure period, the flasks were checked for their oxygen content. In a glove box continuously flushed with pure N_2, the bottles were opened and the pupae were removed from the grain. About 50 mg of intact whitish pupae were weighed within the glove box using a CAHN model 4400 electrobalance and were then blended after addition of 200 µL of 3 N perchloric acid. This helped to stop all enzymatic processes immediately. Untreated control samples were weighed and blended in ambient air following the same procedure.

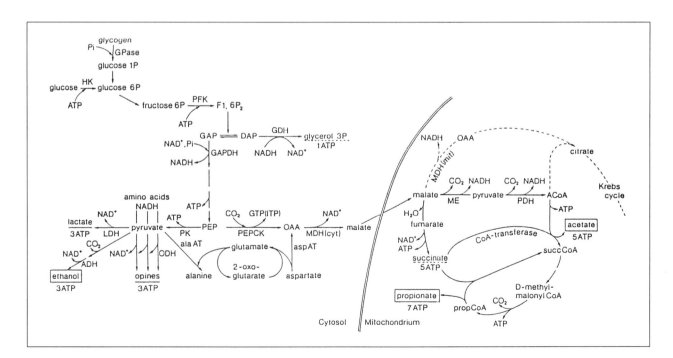

Fig. 1.   Scheme of anaerobic energy metabolism (from Wegener 1988)

The homogenate, together with 200 µL of twice distilled water (to rinse the blender), was transferred into an Eppendorf vessel and neutralised by 200 µL 2.55-m potassium hydroxide solution. All probes were then centrifuged (2 times 5 minutes at 12 000 rpm and 4°C) to remove all particles (for details see Adler 1993). Samples were taken from the liquid fraction for both the HPLC-determination and the standardised enzymatic l-lactate (+) test (test kit from Boehringer Mannheim). In this test lactate is determined through the addition of NAD resulting in:

L-lactate (+) + NAD pyruvate + NADH.

This reaction takes place in the presence of lactate anhydrogenase. The resulting NADH causes a change in absorption of light (at = 340 nm) and thus allows an indirect determination of lactate contents by photometry (Bergmeyer and Graßl 1983).

Results

It was demonstrated that granary weevil pupae produce lactate under anoxic conditions. In the HPLC determination, a distinctive peak was recorded after the appropriate retention time (rt = 12.93). Lactate levels for pupae exposed for 21 days to pure nitrogen or pure carbon dioxide were 195 ± 5 mg/100 mL and 65 ± 5 mg/100 mL, respectively.

Further peaks were recorded after retention times close to those of -ketoglutarate (rt = 8.27 instead of 8.18), succinate (rt = 11.84 instead of 12.26) and fumarate (rt 15.28 instead of 15.24).

Lactate contents as determined by the enzymatic analysis are presented in Figure 2.

Discussion

The two corresponding values from HPLC determination are considerably lower than those attained in the enzymatic survey which may have been caused by filtering the samples with a molecular sieve prior to injection into the HPLC column. The accuracy of the enzymatic determination was secured by a lactate sample with known concentration (real concentration 19.9 mg/100 mL, determined concentration: 20.2 mg/100 mL).

The lactate levels measured by Navarro and Friedlander (1975) in the haemolymph of *Ephestia cautella* pupae after 24 hours exposure to N_2 with less than 3% O_2 (287 mL/100 mL) are a little higher than those found in the present study (200

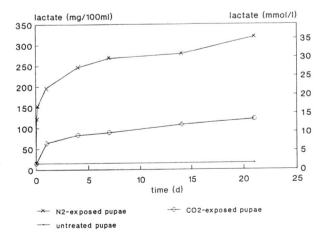

Fig. 2. Lactate contents in *S. granarius* pupae exposed to pure N_2 or pure CO_2 gas. Data from enzymatic analysis (100 mL liquid fraction corresponds to about 100 g fresh weight).

mg/100 mL after 24 hours exposure to pure N_2). It has to be considered, however, that the amounts of lactate may be higher in haemolymph than in the whole body tissue. Moreover, the residual oxygen in their controlled atmosphere (<3 %) could be responsible for this difference as well as different rates of anoxic energy production in the pupae of two different insect orders.

Another interesting point in the findings of Navarro and Friedlander is that lactate levels shot up only when the O_2 contents were reduced below 3%. If this value is exceeded, lethal exposure periods are extremely prolonged in several stored-product species (Lindgren and Vincent 1970). Therefore, this seems to correspond to the 'critical partial pressure' of oxygen where the energy metabolism is converted from aerobic to anaerobic (discussed in Grieshaber et al. 1988).

In both the enzymatic and the HPLC type of lactate determination, lactate levels in pupae exposed to CO_2 were much lower than those of pupae exposed to N_2 for the same time span. As the results of the enzymatic test show, after 24 hours exposure, the level of accumulated lactate in CO_2-treated individuals amounted to only one third of the value detected in N_2-exposed pupae. This relation remained constant through all exposure periods tested (see Fig. 2). Another remarkable fact is that, in both controlled atmospheres, the greatest increase in lactate took place within the first 24 hours whereas the production was reduced later, resulting in a more or less hyperbolic function of lactate generation (Fig. 2). These findings immediately give rise to two questions:

1. Why is so much less lactate produced in an atmosphere of pure carbon dioxide?

2. Why does the production of lactate decrease in both atmospheres after the first 24 hours?

Both questions may be answered by the following hypothesis: CO_2, readily dissolving in all body liquids forms carbonic acid and causes an acidification on the cellular level. Together with lactate, itself being an acidic compound, the excessive amount of hydrogen ions present could directly or indirectly inhibit glycolysis. In vertebrates, a pathologically high accumulation of lactate in body tissues causes a severe disturbance in the redox balance known as lactic acidosis.

In his extensive studies on the induced tolerance of *Tribolium castaneum* Herbst to anoxia and hypercarbia, Donahaye (1991) found that the population selected for tolerance against hypercarbia had a significantly higher body mass than unselected individuals of the same laboratory strain. This result is supported by a comparative fumigation of 10 different strains of *S. granarius*, where Adler (1993) found that the average adult body weight of strains more tolerant to a treatment with 95% CO_2 (rest air) was significantly higher than that of susceptible strains. A higher body mass correlates with more body liquids which could allow a longer exposure to CO_2 before the same level of acidification is reached.

Studies on the selection of *Sitophilus* weevils for tolerance against atmospheres high in CO_2-contents were carried out by Bond and Buckland (1979) and Navarro et al. (1985). Donahaye (1991) selected *T. castaneum* for tolerance not only against hypercarbia but in a second study also against hypoxic atmospheres. In all these experiments the selection was successful under laboratory conditions with extreme moisture contents (≥95% r.h.) and all authors mention the susceptibility of fumigated individuals to desiccation. These findings stress the importance of body water and fit into the picture of acidification being caused by hypercarbia or anoxia, respectively.

In pupae exposed to nitrogen, glycolysis take place until a much higher level of lactate alone has caused a similar acidification as carbonic and lactic acid cause in CO_2-treated pupae.

A step in glycolysis that could be directly affected by acidification is the reaction from glycerolaldehyde 3-phosphate to phosphoenol pyruvate where hydrogen ions are released (Fig. 1). Nicotin adenosyl diphosphate (NAD^+) needed as an acceptor in this reaction would not be available in an acidic environment.

Donahaye (1991) attributed the mortality of treated individuals to desiccation and the exhaustion of triglycerides as energy reserves. This exhaustion theory would, however, not explain the more rapid lethal action of hypercarbia.

It remains debatable what the ultimate cause of insect mortality is, but compared with hypoxia, the rapid inhibition of the glycolytic energy production by hypercarbia could be a reason for the faster lethal action of atmospheres high in CO_2.

Acknowledgments

The author is indebted to the working group of Professor Dr Irene Zerbst and Dr Rolf Nitcai, Department of Biology, Free University, Berlin, for their assistance in the HPLC study.

References

Adler, C. 1993. Zur Wirkung modifizierter Atmosphären auf Vorratsschädlinge in Getreide am Beispiel des Kornkäfers *Sitophilus granarius* (L.) (Col., Curculionidae), Dissertation, Aachen, Verlag Shaker, 146 p.

Bergmeyer, J. and Graßl, M., ed., 1983. Methods of enzymatic analysis. Verlag Chemie GmbH, Chapter 3.12, 582–588.

Bond, E.J. and Buckland, C.T. 1979. Development of resistance of carbon dioxide in the granary weevil.. Journal of Economic Entomology, 72, 770–771.

Donahaye, E. 1992. The potential for stored-product insects to develop resistance to modified atmospheres. In: Fleurat-Lessard, F. and Ducom, P., ed., Proceedings of the Fifth International Working Conference on Stored-product Protection, Bordeaux, September 1990, 989–997.

Friedlander, A. and Navarro, S. 1979. The effect of controlled atmospheres on carbohydrate metabolism in the tissue of *Ephestia cautella* (Walker) pupae. Insect Biochemistry, 9, 79–83.

Gäde, M. 1985. Anaerobic energy metabolism. In: Hoffmann, K.,ed., Environmental physiology and biochemistry of insects. Springer Verlag Berlin, Heidelberg, New York, Tokyo, 119–136.

Grieshaber, M.K., Kreutzer, U. and Pörtner, H.O. 1988. Critical pO_2 of euroxic animals. In: Acker, H., ed., Oxygen sensing in tissues. Springer Verlag Berlin, Heidelberg, New York, 37–48.

Lindgren, D.L. and Vincent, L.E. 1970. Effect of atmospheric gases alone or in combination on the mortality of granary and rice weevils. Journal of Economic Entomology, 6, 1926–1929.

Moratzky, T., Burkhardt, G., Weyel, W. and Wegener, G. 1992. Mikrokalorimetrische Untersuchungen an adulten Insekten unter normoxischen und anoxischen Bedingungen. In: Pfannenstiel H.J., ed., 1992 Verhandlungen der Deutschen Zoologischen Gesellschaft, G. Fischer Verlag Stuttgart, Jena, New York, 150.

Navarro, S., Dias, R. and Donahaye, E. 1985. Induced tolerance of *Sitophilus oryzae* adults to carbon dioxide. Journal of Stored Products Research, 21, 207–213.

Navarro, S. and Friedlander, A. 1975. The effect of carbon dioxide anaesthesia on the lactate and pyruvate levels in the hemolymph of *Ephestia cautella* (Wlk.) pupae. Comparative Biochemistry and Physiology, 50b, 187–189.

Urich, K. 1990. Vergleichende Biochemie der Tiere. G. Fischer Verlag Stuttgart, New York, 710 p, chapter 18, Oxidativer Stoffwechsel, 614–622.

Wegener, G. 1988. Oxygen availability, energy metabolism, and metabolic rate in invertebrates and vertebrates. In: Acker, H., ed., Oxygen sensing in tissues. Springer Verlag Berlin, Heidelberg, New York, 13–35.

Winston, P.W. and Bates, D.H. 1960. Saturated solutions for the control of humidity in biological research. Ecology, 41, 232–237.

Zachariassen, K.E. and Pasche, A. 1976. Effect of anaerobiosis on the adult cerambycid beetle, *Rhagium inquisitor* L. Journal of Insect Physiology, 22, 1365–1368.

Zebe, E 1977. Anaerober Stoffwechsel bei wirbellosen Tieren. Rheinisch Westfälische Akademie der Wissenschaften, Vorträge N 269, Westdeutscher Verlag, 51–70.

A comparison of the efficacy of CO_2-rich and N_2-rich atmospheres against the granary weevil *Sitophilus granarius* (L.) (Coleoptera: Curculionidae)

C. S. Adler*

Abstract

For temperatures below 15°C and above 22°C almost no literature data could be found on the efficacy of N_2 or CO_2 fumigations against the granary weevil. Laboratory experiments were therefore carried out at 5, 10, 25 and 30°C, 75% r.h., with the following gas mixtures (% by vol.): 98% N_2, 2% O_2; 32% N_2, 8% O_2, 60% CO_2 and 8% N_2, 2% O_2, 90% CO_2.

The results of these experiments, and literature data, were graphed to show the exposure times needed to achieve 95–100% mortality of different life stages. Compared was the influence of temperature on the efficacy of atmospheres consisting of 97–100% N_2 (rest O_2) or 60–100% CO_2 (rest N_2, O_2), respectively.

The resulting functions of mortality over temperature, seem to indicate that atmospheres with high contents of CO_2, while causing more rapid death below 20°C, do not produce shorter LT values than hypoxic atmospheres at 25°C or above. At high grain temperatures, factors other than minimum fumigation time may therefore be more important determinants of the choice of an atmosphere to be used in a treatment.

Introduction

Studies on the efficacy of controlled atmospheres (CA) on stored-product insects have been carried out intensively since around 1970. Due to the many different factors affecting insect mortality and the various ways to present mortality values (see Table 1) it is very difficult, however, to draw general information from this vast mass of accumulated data. Literature findings were reviewed by Bailey and Banks (1980) and Banks et al. (1990). A first graphical approach to give an overview on dosage schedules to control the main stored-product insect species was the work published by Annis (1987). Banks and Annis (1990) studied the comparative advantages of high CO_2 and low O_2 types of controlled atmospheres. The present paper aims to give a more detailed picture on the effects of temperature on nitrogen or carbon dioxide fumigations. Because stored-product insects vary in their susceptibility to CA treatments, this study concentrates on only one species: the granary weevil *Sitophilus granarius* (L.) (Coleoptera: Curculionidae).

The granary weevil is a very common grain pest in moderate climates and proved to be among the species least sensitive to CA fumigations. This fact has made *S. granarius* the main test species when studies on the efficacy of controlled atmospheres have to be carried out for registration purposes in Germany.

In general, it may be suggested that lethal exposure times calculated from experiments with the granary weevil should be safe for fumigations aimed to control other, less tolerant stored-product pest species. However, this cannot be guaranteed, since too often insect biology has held surprises for generalising entomologists. As Banks and Annis (1990) pointed out, *Trogoderma granarium* Everts may be another species quite tolerant to controlled atmospheres.

Material and Methods

Literature review

Literature data (see Tables 2 and 3 for sources) fulfilling the following requirements were tabulated:
1 species: *Sitophilus granarius*,
2 fumigation atmosphere: 97–100 % N_2 or 60–100% CO_2/(balance: O_2, CO_2 or N_2),
3. achieved mortality: 95–100 %.

From these data two graphs were drawn showing the efficacy of atmospheres containing high contents of either N_2 or CO_2 on different life stages of the granary weevil at various temperatures.

Laboratory experiments

Because almost no data could be found at temperatures below 15°C or above 22°C, fumigation experiments were carried out in the laboratory at 5 ± 1°C, 10 ± 1°C, 25 ± 1°C and 30 ± 1°C, 75 ± 5 % r.h.

The following gas mixtures were produced manometrically from pure components (Adler and Reichmuth 1988) (in % by volume):
98% N_2, 2% O_2;
32% N_2, 8% O_2, 60% CO_2;
8% N_2, 2% O_2, 90% CO_2.

Insect cultures

To study the effects of the gas mixtures on developmental stages of defined age, insects were cultured as follows: At 25°C, 75% r.h., adult granary weevils were placed weekly on uninfested wheat for an oviposition period of 3 days, leaving age groups of developmental stages with a deviation of ±1.5 days. At 25°C, this culture technique produces 5 preadult stages. Details of this procedure are given in Adler (1991). From each preadult group, 70 kernels of infested wheat were placed into a wire mesh cage. Fifty adult weevils, 1–3 weeks old, and some 5 g of wheat, were filled into a slightly larger cage completing the set of developmental stages.

Before fumigation experiments at 5 and 10°C, the test insects were adapted to cold by a storage at 15°C for 24 hours before exposing them to the desired temperature. The lethal effects of cold alone were studied by keeping untreated samples at the experimental temperature and at 25°C.

* Federal Biological Research Centre for Agriculture and Forestry, Institute for Stored Product Protection, Königin-Luise Straße 19, D–14195 Berlin, Germany.

Fumigation

A set of developmental stages and adult granary weevils was placed into a 2L Dressel flask and fumigated with one of the three gas mixtures mentioned above. Before use, each gas mixture was heated or cooled to the desired temperature and purged through a saturated sodium chloride solution to adjust to a close-to-optimal moisture content of 75–76% r.h. (see Winston and Bates 1960). Each Dressel flask was purged with the respective gas mixture until the atmosphere in the vessel was completely replaced by the desired atmosphere. The flask was then closed and stored in a climate chamber for exposure periods between 1 and 12 weeks in experiments at 5 or 10°C and between 3 and 13 days at 25 or 30°C. After this time, grain and insects were transferred into petri dishes. These were stored at 25°C, 75% r.h., and emergence and mortality were recorded for up to 12 weeks.

Table 1. Factors impairing the comparability of data on the efficacy of CAs

Insect biology	Variations in the susceptibility of different species, variations in relative susceptibility on the subspecies level, medium age of stages, age deviations due to variations in developmental speed, general health conditions, behavioural aspects
Experimental set up	Gas composition, moisture conditions, gas flow rates, fumigation technique (e.g. continuous, pulsed, one purge), time span for temperature adaptation, maintenance of atmospherical composition during insect exposure, effects of culture medium or microorganisms on insects or atmosphere, number of individuals and effects of respiratory processes on the CA and its moisture content, reliability of data (number of individuals and number of replicates)
Presentation of data	Exposure time leading to complete kill or 100 % reduction in emergence, LT_{99}, LT_{95}, LT_{50} or other LT-values

Table 2. Literature data on lethal exposure times of *S. granarius* caused by 97-100% N_2[a]

Author/year	Temp. °C	r.h. %	N_2 %	O_2 %	CO_2 %	Mort. %	Eggs	L1/L2	L3/L4	L4/P	adults
									Lethal exposure time (days)		
Reichmuth 1987/1990	15	75	99.5	0.5		95					6.3
			99	1			30	40	40	45	10.8
			98	2			30	40	60	52	13.1
			97	3			30	31	65	53	6.9
Adler & Reichmuth 1988	15	75	98	2		95	(>42)	35	(>42)	(>42)	21
Lindgren & Vincent 1970	16	65	100			95					12.5
Krishnamurthy et al. 1986	20	70	99.5	0.5		100					10
Reichmuth 1987/1990	20	75	99.5	0.5		95					4.4
			99	1			20	13	30	30	13.0
			98	2			12	13	31	25	7.4
			97	3			17	12	35	30	5.3
Adler 1991/1993	20	75	99	1		99					
strain F-L							14.9	7.6	19.1	20.4	9.0
strain USA-L							14.7	5.1	18.6	22.2	6.6
strain CN-L1							11.7	8.5	22.5	28.8	6.3
strain CN-L2							12.3	8.5	21.5	24.1	7.6
strain AUS-L							17.2	6.0	17.2	21.8	4.2
strain AUS-f							14.5	7.8	22.5	28.4	4.3
strain GB-L1							13.0	6.6	14.7	20.5	4.3
strain GB-L2							14.8	10.6	18.6	19.5	8.0
strain D-L							15.6	9.9	23.8	22.5	7.3
strain D-f							13.0	8.2	28.4	26.6	5.5
Adler & Reichmuth 1988	20	75	98	2		95	28	21	(>42)	(>42)	(<14)
Shejbal et al. 1973	22	70	99.9	0.1		100					2
			99.7	0.3							3
			99.5	0.5							4
			99.2	0.8							9
			99.0	1.0							10
Lindgren & Vincent 1970	27	65	100			95	8.5	(4.5)[b]	—	8	3.5
Bailey & Banks 1980	30	70	98.7	1.3		100	(<14)	(<14)	(<14)	(<14)	(<14)

[a] Literature data mentioned only if 95–100% mortality achieved.
[b] Larval age not discriminated.

Results

Literature data are given in Tables 2 and 3. The times needed to achieve complete kill of a certain developmental stage with a given gas mixture in the laboratory experiments are presented in Table 4. These data were added into the compilation of literature values which are displayed graphically in Figures 1 and 2.

Because developmental speed varies strongly in the granary weevil and because larval age is seldom accurately defined in the literature, only young, more sensitive 1st and 2nd instar larvae and old, more tolerant 3rd and 4th instar larvae were discriminated. For better readability, single values were omitted. Mean values and standard deviations of all data at a certain temperature are listed in Table 5.

Discussion

Due to great differences in developmental speed in *Sitophilus granarius*, 4th instar larvae and even some hatching weevils may have influenced the results obtained by fumigating pupal stages in the experimental results and cited literature data. Nevertheless, this stage proved to be most tolerant to both nitrogen and carbon dioxide treatments. For two reasons this fact is not surprising:

- pupae are known to have a very low oxygen uptake (Bailey 1969); and

- the larval stages have accumulated reserves of carbohydrates that the pupal stage can use as a substrate for anoxic energy production through glycolysis (Adler 1993).

Only in the experiments at 5°C did eggs survive fumigations with 60 % CO_2 and 90 % CO_2 longer than did pupae. Because no values could be derived from literature for this temperature, further studies are needed to verify these findings.

As presented in Table 4, 60% CO_2 was not less effective in the laboratory experiments than was 90% CO_2. This is supported by Fleurat-Lessard and Le Torc'h (1991) who found that atmospheres of only 50% CO_2 and 4% or 20% oxygen (rest N_2) produced mortality rates in *S. granarius* larvae similar to those produced by an atmosphere of 100% CO_2.

Figure 1 shows that in N_2-fumigations the longest LT-values were found around 10°C. Below this temperature the cold may have had an additional effect causing mortality of pupal and larval stages in shorter periods. However, this effect was not detected in CO_2-fumigated developmental stages where lethal times increased with decreasing temperature (Fig. 2). On the contrary, old larvae and eggs in particular seem to be protected from the toxic effects of CO_2 at 5°C better than at 10°C.

Comparing untreated samples exposed to 5°C with those kept at 25°C, it was noticed that the cold alone caused complete kill of eggs and young larvae within 4 to 5 weeks and high mortality rates among late larvae and pupae (while the mortality of adults remained low). This leads to the conclusion that low temperatures not only slow down the effect of controlled atmospheres but also that the effects of cold and CA treatment are antagonistic at temperatures around 5°C.

Table 3. Literature data on the lethal exposure times of *S. granarius* caused by 60-100% CO_2[a].

Author/ year	Temp. °C	r.h. %	N_2 %	O_2 %	CO_2 %	Mort. %	Eggs	Lethal exposure time (days)			
								L1/L2	L3/L4	L4/P	adults
Reichmuth 1987/1990	15	75	16	4	80	95	45	15	39	40	6.2
			12	3	85		21	11	24	34	3.9
			8	2	90		13	12	15	27	6.7
			4	1	95		30	9	22	35	8.0
			2	0.5	97.5		-	-	-	-	5.2
Lindgren & Vincent 1970	16	65			100	95					11.5
Reichmuth 1987/1990	20	75	16	4	80	95	(23?)	(18?)	(40?)	(38?)	2.2
			12	3	85		10	2	16	26	2.3
			8	2	90		10	5	7	19	2.9
			4	1	95		16	2	20	24	4.7
			2	0.5	97.5		-	-	-	-	3.2
Adler 1991/1993	20	75	4	1	95	99					
strain F-L							14.3	2.6	15.5	15.5	4.4
strain USA-L							13.6	3.0	14.0	20.2	4.4
strain CN-L1							12.1	3.8	17.9	22.5	5.5
strain CN-L2							11.9	4.4	15.6	19.5	5.8
strain AUS-L							14.7	4.7	13.0	19.3	4.0
strain AUS-f							15.4	3.2	15.0	23.2	3.8
strain GB-L1							15.0	2.7	9.2	11.3	3.4
strain GB-L2							14.1	4.0	15.0	18.3	8.1
strain D-L							15.7	2.5	19.9	18.5	4.5
strain D-f							13.2	3.3	18.6	19.7	5.5
Fluerat-Lessard & Le Torch 1991	25	75			100	95			9		
Lindgren & Vincent 1970	27	65			100	95	8.5	(4.3)[b]		8	3.5

[a]literature data mentioned only if 95–100% mortality achieved

[b]Larval age not discriminated

Generally, at lower temperatures CO_2-atmospheres are significantly faster than other CAs in lethal action. Clearly, this is due to the toxic properties of CO_2 being less affected by low temperatures. CO_2 has a high solubility in water and the solubility of gases increases with decreasing temperature. Therefore, at temperatures decreasing from 15°C to 5°C the higher passive uptake of CO_2 may to some extent compensate for the temperature-triggered slow down in insect metabolism. In eggs this effect may not be detectable because this stage has a smaller ratio of body and respiratory surface to body volume than the later stages.

In contrast to the findings on *S. oryzae* adults (Banks and Annis 1990), the type of controlled atmosphere seems to cause only minor differences in the lethal exposure period of adult *S. granarius*.

At 25°C and above, the toxic effects of CO_2 seem to cause no faster mortality of granary weevil stages than the reduction of O_2 by a treatment with N_2.

As Banks and Annis (1990) demonstrated, many factors influence the question whether atmospheres of high CO_2 or low O_2 are advantageous under certain practical conditions. For example, one of the great advantages of CO_2 is that it may be diluted by air down to 40% while still being efficient against insects. If, however, N_2- and CO_2-treatments of *S. granarius* are compared solely from the point of view of achieving disinfestation in the shortest possible time, the

Table 4. Lethal exposure times, results of the experimental study

	Temp. °C	r.h. %	N_2 %	O_2 %	CO_2 %	Mort. %	Lethal exposure time (days)				
							eggs	L1/L2	L3/L4	L4/P	adults
High N_2- contents	5	75	98	2	-	100	42	28	63	70	21
	10						42	56	70	77	14
	25						5	5	13	(>13)	7
	30						3	3	10	10	7
High CO_2- contents	5	75	32	8	60	100	56	21	42	49	18
			8	2	90		56	21	38.5	49	18
	10		32	8	60		31.5	(<21)	24.5	45.5	14
			8	2	90		35	(<21)	28	45.5	14
	25		32	8	60		5	5	10	13	3
			8	2	90		7	7	5	13	3
	30		32	8	60		3	3	10	7	3
			8	2	90		3	3	7	7	3

Table 5. Lethal exposure times of *S. granarius* caused by 97–100 % N_2-(rest O_2) and 60–100 % CO_2-(rest N_2/O_2) atmospheres, literature and experimental data[a]

	Temp. °C	Exposure period (days) causing 95–100 % mortality									
		eggs		L1/L2		L3/L4		L4/P		adults	
		av.	s.d.	av.	s.d.	av.	s.d.	av.	s.d.	av.	s.d.
High N_2- contents	5	42		49	±7	63	±7	70		21	
	10	42		56	±7	70	±14	77	±7	14	
	15	30		36.5	±3.8	55	±10.8	50	±3.6	9.3	±2.8
	16									12.5	
	20	15.6	±4.1	9.8	±3.9	23.3	±5.9	24.6	±3.7	7.1	±2.4
	22									5.6	
	25	5		5		13		(>13)		7	
	27	8.5		4.5[b]				8		3.5	
	30	3		3		10		10		7	
High CO_2-contents	5	56		21		40.3	±5.8	49		18	
	10	33	±3.0	(<21)		26.3	±3	45.5	±3.5	14	
	15	27.3	±20.4	11.8	±2.5	25	±8.8	34	±4.6	6	±1.4
	16									11.5	
	20	14.2	±3.1	3.3	±1	15.1	±3.7	19.8	±3.6	4.7	±1.4
	25	6	±1	6	±1	8.3	±1.3	13		4.8	±2.5
	27	8.5		4.3[a]				8		3.5	
	30	3		3		8.5	±1.5	7		3	

[a]average values (av.) and standard deviations (s.d.), s.d.-values only in case of contradictory data; - see figures 1 and 2.
[b]Larval age not discriminated

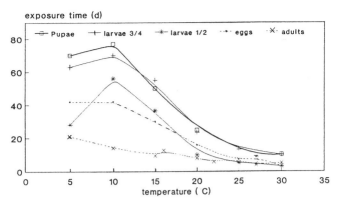

Fig. 1. Lethal exposure times of *S. granarius* produced by 97–100% N_2 (rest O_2)

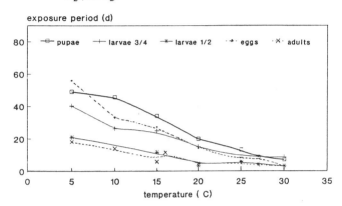

Fig. 2. Lethal exposure times of *S. granarius* produced by 60–100% CO_2 (rest N_2/O_2)

present study seems to support the assertion that CO_2-fumigations will give a more rapid disinfestation at temperatures below 20°C but not at 25°C or above. Therefore, at high grain temperatures, factors other than minimum fumigation time may determine the choice of an atmosphere to be used in a treatment.

Acknowledgments

The author wishes to thank Dr Christoph Reichmuth for his advice and help and Mrs Sylvia Krause for her technical assistance.

References

Adler, C. 1991. Efficacy of controlled atmospheres on ten strains of the granary weevil *Sitophilus granarius* (L.) from different places of origin. In: Fleurat-Lessard, F. and Ducom, P., ed., Proceedings of the Fifth International Working Conference on Stored-product Protection, Bordeaux, September 1990, 2, 727–736.

Adler, C. 1993. Zur Wirkung modifizierter Atmosphären auf Vorratsschädlinge in Getreide am Beispiel des Kornkäfers *Sitophilus granarius* (L.) (Col., Curculionidae), Dissertation, Aachen, Verlag Shaker, 146 p.

Adler, C. and Reichmuth, Ch. 1988. Der Kornkäfer *Sitophilus granarius* (L.), Coleoptera, Curculionidae, seine Biologie und seine Bekämpfung in Getreide, insbesondere mit modifizierten Atmosphären. — Mitteilungen der Biologischen Bundesanstalt für Land- und Forstwirtschaft Berlin (Dahlem) 239, 100 p.

Annis, P.S. 1987. Towards rational controlled atmosphere dosage schedules; — a review of the current knowledge. In: Donahaye, E. and Navarro, S, ed., Proceedings of the Fourth International Working Conference on Stored-product Protection, September 1986, Tel Aviv, Israel, Maor-Wallach Press, 128–148.

Bailey, S.W. 1969. The effects of physical stress in the grain weevil *Sitophilus granarius*. Journal of Stored Products Research, 5, 311–324.

Bailey, S.W. and Banks, H.J. 1980. A review of recent studies of the effects of controlled atmospheres on stored product pests. In: Shejbal, J. ed., Controlled atmosphere storage of grains. Amsterdam, Elsevier, 101–108.

Banks, J. and Annis, P.C. 1990. Comparative advantages of high CO_2 and low O_2 types of controlled atmospheres for grain storage. - In: Calderon, M. und Barkai-Golan, R., eds., Food preservation by modified atmospheres. Boca Raton, CRC Press. Florida, USA 1990, 93–122.

Banks, H.J., Annis, P.C. and Rigby, G.R. 1990. Controlled atmosphere storage of grain: the known and the future. In: Fleurat-Lessard, F. and Ducom, P. eds., Proceedings of the Fifth International Working Conference on Stored-product Protection, Bordeaux, September 1990, 2, 695–706.

Fleurat-Lessard, F. and Le Torc'h, J. M. 1991. Influence de la teneur en oxygéne sur la sensibilité de certains stades juvéniles de *Sitophilus oryzae* et *Sitophilus granarius* au dioxide de carbone. Entomologia Experimentalis et Applicata, 58, 37–47

Krishnamurthy, T.S, Spratt, E.C. and Bell, C.H. 1986. The toxicity of carbon dioxide to adult beetles in low oxygen atmospheres. Journal of Stored Products Research, 22, 145–151.

Lindgren, D.L. and Vincent, L.E. 1970. Effect of atmospheric gases alone or in combination on the mortality of granary and rice weevils. Journal of Economic Entomology, 6, 1926–1929.

Reichmuth, Ch. 1987. Low oxygen content to control stored product insects. In: Donahaye, E. and Navarro, S, ed., Proceedings of the Fourth International Working Conference on Stored-product Protection, September 1986, Tel Aviv, Israel, Maor-Wallach Press, 194–207.

Reichmuth, Ch. 1990. Toxic gas treatment responses of insect pests of stored products and impact on the environment. In: Champ, B.R., Highley, E. and Banks, H.J., eds., Fumigation and controlled atmosphere storage of grain. ACIAR Proceedings 25, 56–69.

Shejbal, J., Tonolo, A. and Careri, G. 1973. Conservation of cereals under nitrogen. SNAM Progretti S.p.A., Rome, Italy, 1133–1144.

Winston, P.W. and Bates, D.H. 1960. Saturated solutions for the control of humidity in biological research. Ecology, 41, 232–237.

Numerical modelling of the movement of carbon dioxide through stored-wheat bulks

S.K. Alagusundaram*, D.S. Jayas*, W.E. Muir†, N.D.G. White†, and R.N. Sinha†

Abstract

A step-by-step approach was taken in an attempt to model the three-dimensional movement of introduced CO_2 gas through stored-wheat bulks. Experimental data on the distribution of CO_2 in wheat bulks contained in three 1.42 m diameter bins were collected to validate the model. These bins were equipped with three different partially perforated floors (0.3 m diameter floor opening near the centre, 1.14×0.36 m rectangular floor opening, and 0.3 m diameter floor opening near the wall). In the experiments the grain surfaces were left open or covered with polyvinylidene chloride sheets. Dry ice was used to create high CO_2 concentrations in the wheat bulk. A model that assumes pure diffusion as the transport mechanism did not predict the CO_2 distribution well.

Later the model was modified to predict CO_2 distribution using experimentally determined flow coefficients (apparent flow coefficient) during the dry ice sublimation period and then a diffusion coefficient afterwards. Although model predictions improved in this approach, there were still high errors during the initial time period. The model predictions were further improved when the sorption of CO_2 by the wheat was included. The sorption rates for the predicted CO_2 concentrations were calculated by extrapolating from the available sorption data which are for a 100% initial CO_2 concentration. The importance of including the sorption for accurate model predictions is demonstrated by assuming various rates of sorption from 0 to 100% of the measured sorption rate. The need for sorption data at lower initial concentrations, empirical data on the gas loss from grain storage structures, and the inclusion of convection terms in the model is highlighted.

Introduction

Computer simulation models of physical parameters are integral parts of research and development in many fields of science. These models, based on physical principles and validated against measured experimental data, become powerful design and management tools. They eliminate the need for time- and labour-intensive experimental studies. Furthermore, experimental data are specific to the system on which they were observed and cannot be generalised. For example, the temperatures measured in grain bulk-stored near Winnipeg (Canada) cannot be extrapolated to predict the temperatures of grain bulk-stored near Darwin (Australia) as the weather conditions in these two places are entirely different from each other. Winnipeg experiences temperate weather

conditions while Darwin experiences tropical conditions. On the other hand, mathematical models with fewer assumptions may be more general in nature. In a stored-grain ecosystem, for example, the damaging insects, mites, and microorganisms grow and multiply under optimum conditions of grain temperature, moisture content, and intergranular gas composition. A farmer, manager of a commercial storage facility, or a grain storage scientist who must determine the management practices required to safeguard the grain before any damage occurs should know the changes in these abiotic factors in relation to the local conditions. Numerous mathematical models have been developed to simulate the grain temperatures and moisture contents (Jayas 1994) but only limited research work has been done to model the changes in the intergranular gas composition (Singh et al. 1983; Jayas et al. 1988; Nguyen 1986). The intergranular gas composition, initially at atmospheric level, is altered due to the respiration of the insect pests, microflora and the grain, or by artificially introducing different compositions of atmospheric gases as in controlled atmosphere (CA) storage.

The success of CA storage relies on the uniformity of distribution of the introduced gases in the grain bulk and the maintenance of these gases for the minimum required exposure period. A knowledge of the distribution and maintenance of the introduced gases is important for the design and management of CA storage systems. The application and management of CA storage systems are often based on experience and rules of thumb. A comprehensive mathematical model that will predict with accuracy the distribution and maintenance of the introduced gases would be helpful in efficiently designing and operating CA storage systems. The present work was undertaken to develop a finite element model for predicting the three-dimensional distribution of carbon dioxide (CO_2) in stored-wheat bulks. In this paper we will explain the step by step procedure we followed in developing and validating the model and the assumptions we made at each step.

Experimental Data for Model Validation

Carbon dioxide concentrations were measured at various locations (Fig. 1) in wheat bulks contained in three experimental bins, 1.42 m in diameter and 1.47 m tall. The CO_2 data were collected from 5 levels in each bin (beginning at the floor, and then spaced 0.33 m apart in the vertical direction). At each level there were 11, 13, and 12 sampling points for Bins 1, 2, and 3, respectively (Fig. 1). Metal boxes of size $0.5 \times 0.5 \times 0.37$ m for Bins 1 and 3 and $1.22 \times 0.46 \times 0.36$ m for Bin 2 were mounted centrally under the floor openings. Known quantities of dry ice were placed in these boxes. The CO_2 concentrations were measured using a gas chromatograph at 1, 3, 6, 9, 12, and 21 hours after the introduction of dry ice. The grain surfaces were left open or covered with a polyvinylidene chloride sheet. Three replicates were conducted for each experimental combination.

The grain was aerated using a 1.5 kW centrifugal fan for about 1 hour immediately after each replicate and for about 15

* Department of Agricultural Engineering, 438 Engineering Building, University of Manitoba, Winnipeg, Canada R3T 5V6.
† Agriculture and Agri-Food Canada Research Station, 195 Dafoe Road, Winnipeg, Canada R3T 5V6.

minutes just before the next replicate to bring the intergranular CO_2 concentrations to atmospheric level. The grain was left undisturbed for about 24 hours between replicates. Grain samples were collected after each experiment for determining the moisture content. The moisture contents of the wheat samples were determined by drying triplicate samples of about 15 g each in an air convection oven at 130°C for 19 hours (ASAE 1992). The moisture content of the wheat did not vary appreciably during the course of the experiments. The average moisture content of the wheat used in the experiments was 12.6 ± 0.4% (wet basis). In addition to the CO_2 samples, the grain temperatures were also monitored at all sampling times. The grain temperatures were measured at five locations in Bin 1 and at 15 locations in each of Bins 2 and 3 (Fig. 1). The grain temperatures were fairly constant in any given experiment. The maximum observed deviation from the mean grain temperature in any experiment was ±2.7°C.

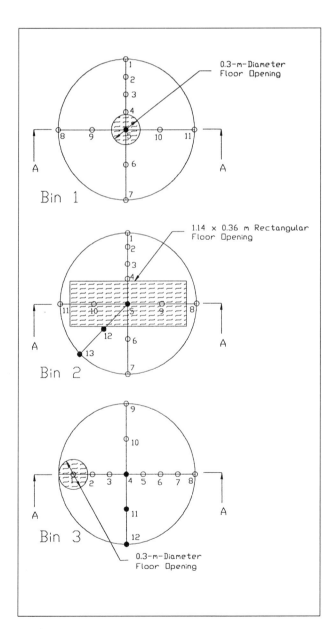

Fig. 1. Schematic diagram of the temperature and gas sampling locations in 1.42 m diameter × 1.47 m tall bins.(● = temperature and gas samples; O = only gas samples)

Model Development and Validation

Step 1: The diffusion model

In a model of the gas transport through a porous medium, the first and the simplest assumption is that the mechanism of transport is by pure diffusion. In addition to this basic assumption we also assumed that the sorption of CO_2 by the wheat was negligible and that there was no gas loss from the grain bin. The partial differential equation governing the diffusion of a gas through porous material under these assumptions is given as:

$$\nabla(D\nabla C) = \frac{\partial C}{\partial \tau} \tag{1}$$

subject to the boundary conditions:

$$C = C_{s1} \qquad on\ S1 \tag{2}$$

$$\frac{\partial C}{\partial \eta} = 0 \qquad on\ S2 \tag{3}$$

and the initial condition:

$$C(x,y,z, \quad \tau = 0) = C_i \ on\ \Omega \tag{4}$$

For the movement of CO_2 through a stored-grain bulk, the various notations used in the above equations are:

C = concentration of CO_2 at time $\tau > 0$ (g/m^3),
C_i = initial CO_2 concentration in the grain bulk (g/m^3),
C_{s1} = specified CO_2 concentration on the boundary S1 (g/m^3),
D = Diffusion coefficient of CO_2 through wheat bulk (m^2/s),
Ω = problem domain (the grain bulk),
η = outward pointing normal,
τ = time (seconds), and
$S1$ and $S2$ = boundary segments.

Boundary segment $S1$ may be made of more than one segment. For example, $S1$ may represent the surface of the grain where concentration may be specified as constant at the atmospheric level in a ventilated head space. Another portion of $S1$ may represent the portion of a grain boundary where CO_2 is injected. Similarly $S2$ may comprise the bin wall, bin floor other than the floor opening, and a grain surface covered with a sheet impermeable to CO_2. $S1$ and $S2$ together make the total boundary of the domain Ω.

The system of equations 1–4 was solved using the finite element method (Segerlind 1984; Rao 1982), resulting in a set of algebraic equations:

$$\left[\frac{[K1]}{\Delta\tau} + \theta[K2]\right]\{C\}_{n+1} = \left[\frac{[K1]}{\Delta\tau} - (1-\theta)[K2]\right]\{C\}_n \tag{5}$$

where:

$[K1]$ = matrix of constants,
$[K2]$ = matrix containing diffusion coefficients,
$\{C\}_n$ = vector of nodal CO_2 concentrations at time τ,
$\{C\}_{n+1}$ = vector of nodal CO_2 concentrations at time $\tau+\Delta\tau$,
θ = a constant.

By choosing the value of θ between 0 and 1, the following schemes can be obtained (Wood and Lewis 1975) (0—forward difference scheme; 0.5—Crank-Nicholson scheme; 0.667—Galerkin's scheme; and 1.0—backward difference scheme). Equation 5 can be solved to obtain the CO_2 concentrations at time $\tau + \Delta\tau$ by using the CO_2 concentrations at time τ.

Simulation of CO_2 distribution

A FORTRAN program was written to solve the system of equations (equation 5). For simulating the distribution of CO_2

in the grain bulk, half of the grain bulk (cut along section A-A of Fig. 1) was discretised into linear elements with 445 nodes for Bin 1, 430 nodes for Bin 2, and 390 nodes for Bin 3. The measured CO_2 concentrations near the floor opening (sampling point 5 of Bin 1, the average of sampling locations 5, 9, and 10 of Bin 2, and sampling location 1 of Bin 3) were specified for the nodes lying inside or on the boundary of the floor opening. To include this boundary condition in the program, the measured CO_2 concentrations at these locations were fitted using procedure NLIN of SAS (SAS 1982) to an equation of the form:

$$\%CO_2 = a\,e^{-b.t} \qquad (6)$$

where:

$\%CO_2$ = measured CO_2 concentration at the inlet boundary,

a and b = empirical constants, and

t = time (hours).

For tests with uncovered top grain surfaces, the measured CO_2 concentrations near the surface of the grain were specified at the nodes lying on this boundary, and when the grain surface was covered with a plastic sheet, this boundary was assumed impermeable to flow of CO_2 ($\partial C/\partial \eta = 0$). The bin wall and the bin floor, excluding the floor opening, were assumed impermeable to CO_2. A diffusion coefficient of 4.15 mm^2/second for red spring wheat at 12% moisture content (Singh et al. 1985) was used in the simulations. It was assumed that the diffusion coefficient was independent of the direction of diffusion (Singh et al. 1984), and of concentration (Cunningham and Williams 1980). The predicted CO_2 concentrations were much lower than the measured concentrations at every sampling point and at all times. The mean relative percent error of prediction (e) was calculated as:

$$e = \frac{1}{N} \sum \frac{|M_i - P_i|}{M_i} \times 100 \qquad (7)$$

where:

Mi = measured CO_2 concentration at sampling point i,

Pi = predicted CO_2 concentration at sampling point i, and

N = number of data points.

The relative error for replicate 1 of Bin 1 with an uncovered top grain surface ranged from 71% at 3 hours to 31% at 21 hours. Similar high relative errors were observed for other test combinations.

The governing partial differential equation and the associated boundary conditions (equations 1–3), on which the model was based, assume that the mechanism of transport is by diffusion only. In the experiments, 180 to 740 g of dry ice were placed in the metal boxes under the floor openings. When the dry ice sublimated into CO_2 gas, a pressure must have developed inside the box which would have caused a bulk movement of CO_2 through the grain bulk. This may have caused the large errors in prediction. To account for the bulk movement of CO_2 a model of forced mass transport must be included in the model.

Step 2: Model of bulk movement of CO_2

It was hypothesised that replacing the diffusion coefficient with an experimentally determined flow coefficient of CO_2 through wheat bulks (apparent flow coefficient, $Dapp$) during the sublimation period of the dry ice would improve the accuracy of the model. To achieve this, laboratory experiments were conducted to determine the D_{app} values (Alagusunda-

ram 1993). The natural logarithm of D_{app} values and time were linearly related as:

$$In\left(D_{app}\right) = A + B\,In(t) \qquad (8)$$

The constants A and B were separate linear functions of grain temperature in the range of −10 to 30°C and expected pressure drops across the grain column due to the sublimation of dry ice were in the range of 77 to 310 kPa. No definite pattern of variation of the constants A and B with an increase in the grain moisture content (in the range 11–18.5% w.b.) was observed. The D_{app} values in the horizontal direction were greater than in the vertical direction and were related as follows:

$$\frac{D_{app}H}{D_{app}V} = 4.011 - 0.212\,In(t) \qquad (9)$$

Using these empirical equations, D_{app} was determined for the experimental conditions. Estimated D_{app} was used in place of D for the first 3 hours of simulation. It was assumed that at the end of this arbitrarily chosen time period the pressure created by the expansion of the CO_2 was dissipated and that the movement of CO_2 through the wheat bulk was purely due to the concentration gradient.

The relative error (Table 1) was calculated using equation 7. At sampling times 6 hours and afterwards, the predicted CO_2 concentrations were close to the measured values in all three bins with 180 or 370 g of dry ice. In most of the experiments, and particularly in Bin 1 with 540 g of dry ice and an open grain surface, the errors were lower in the first four levels (0, 0.33, 0.66, and 0.99 m above the floor) than in all five levels (shown as 4 L and 5 L, respectively, in Table 1). This might be because, in the model predictions, the grain surface was assumed perfectly levelled. But at sampling level 5, which was only 0.05 m below the grain surface, small undulations in the grain surface would cause a relatively large difference in the measured CO_2 concentrations. For example, a difference of 0.02 percentage points in a CO_2 concentration of 0.05% causes an error of 40% while at a CO_2 concentration of 10% the error is only 0.2%.

During the initial periods after the introduction of the dry ice (sampling times 1 hour and 3 hours), the relative errors were very high (Table 1). At these sampling times the predicted CO_2 concentrations at heights of 0.33 m and more above the floor were higher than the measured concentrations. For example, in replicate 1 of Bin 1 with 180 g of dry ice and an open grain surface the error at a height of 0.33 m above the floor was 53% and that near the floor was only 17%. Similar high errors at heights of 0.33 m and above were observed in the other two bins. The reason for this cannot be explained.

The errors were high at all sampling times in Bin 2 with 740 g of dry ice and a covered grain surface (Table 1). At all sampling times the predicted CO_2 concentrations were much higher than the observed values. That observed values were low was probably the result of sorption of CO_2 by the wheat. A mass balance, assuming no loss from the bin, indicated that in Bin 2 with 740 g of dry ice and covered grain surface, about 330 g of CO_2 had been absorbed by the wheat. So, based on the hypothesis that inclusion of CO_2 sorption by the wheat might improve the predictions of the model, we decided to include it.

Table 1. Mean relative percent errors between the measured and predicted CO_2 concentrations in wheat bulk. Predictions were based on mass flow during the first 3 hours and molecular diffusion during the remaining time period.

Bin	Top grain surface	Mass of dry ice (g)	Replicate	1 hour	3 hours	6 hours 5L[a]	6 hours 4L[ba]	9 hours 5L	9 hours 4L	12 hours 5L	12 hours 4L	21 hours 5L	21 hours 4L	24 hours 5L	24 hours 4L
Bin 1	Open	180	1	35	20		6	9	8	7	6	10	4		
			2	35	16		9	9	10	9	9	8	4		
			3	35	13		7	9	7	10	8	17	7		
			4[d]	34	13		7	10	7	10	8	18	8		
	Covered	180	1	54	25		14			16	11	11	10		
			2	29	33		14			13	9	11	9		
			3	35	31		12			15	10	9	8		
			4[d]	39	30		12			14	10	10	9		
		540	1	103	64	34	17	44	18	53	17	53	23		
			2	104	60	44	17	40	10	41	8	58	16		
			3	110	48	26	14	21	9	22	13	21	13		
			4[d]	106	56	33	15	31	10	31	10	37	17		
Bin 2	Open	370	1	21	29		10			7	11	21	23		
			2	24	22		9			8	9	22	27		
			3	20	10		19			10	15	11	10		
			4[d]	19	13		11			5	9	18	17		
	Covered	370	1	28	15		22			13	18	6	8		
			2	31	16		22			15	20	7	9		
			3	23	15		10			7	9	10	9		
			4	22	12		11			8	11	7	7		
		740	1	97	113	122	70			96	69	102	80	101	80
			2	118	165	157	93			116	88	124	101	129	104
			3	113	101	102	56			85	60	93	74	96	75
			4[d]	105	121	124	71			96	70	103	82	105	83
Bin 3	Open	180	1	127	48		27	22	21	18	18	11	11		
			2	103	50		23	16	18	14	15	9	10		
			3	159	55		30	26	22	22	18	18	13		
			4[d]	151	61		32	19	21	16	17	14	14		
	Covered	180	1	93	56		35	26	27	17	18	12	11		
			2	189	49	26	17	17		14	13	8	7		
			3	144	65	36	22	22		14	14	10	10		
			4[d]	127	55	30	20	21		14	15	9	9		

$$\text{Mean relative percent error (\%)} = \frac{1}{N}\sum \frac{|\text{Measured} - \text{Predicted}|}{\text{Measured}} \times 10$$

[a] the mean relative percent errors were calculated for all 5 levels; [b] the mean relative percent errors were calculated for bottom 4 levels; [c] Bins are illustrated in Fig. 1; [d] The measured data for the three replicates were averaged and compared with the simulation results.
N number of data points (25 in Bin 1, 50 in Bin 2, and 45 in Bin 3)

Step 3: Including sorption of CO_2 by wheat in the model

The governing partial differential equation that includes the sorption or desorption (sink or source term) of CO_2 by the wheat can be written as:

$$\nabla(D\nabla C) + q = \frac{\partial C}{\partial \tau} \qquad (10)$$

where q is the sorption or desorption of CO_2 by the wheat ($g/m3/s$). The solution of this equation using the Galerkin weighted residual method will result in the following set of algebraic equations (Segerlind 1984; Rao 1982):

$$\left[\frac{[K1]}{\Delta\tau} + \theta[K2]\right]\{C\}_{n+1} - \left[\frac{[K1]}{\Delta\tau} - (1-\theta)[K2]\right]\{C\}_n =$$

$$\left[\theta\{F\}_{n+1} + (1-\theta)\{F\}_n\right] \qquad (11)$$

where F is a vector that contains the sorption term.

Estimation of sorption of CO_2 by the wheat

Cofie-Agblor et al. (1992) measured the sorption of CO_2 by wheat at various moisture contents (12, 14, 16, and 18%) and at various temperatures (0, 10, 20, and 30°C). They measured the sorption of CO_2 by wheat using an initial intergranular CO_2 concentration of 100%. Other than the work by Cofie-Agblor et al. (1992) we could find no other detailed study on the sorption of CO_2 by wheat. So we decided to extrapolate linearly their data to estimate the sorption at lower concentrations. Based on a study on the characteristics of CO_2 sorption by several grains, Yamamoto and Mitsuda (1980) concluded that the sorption of CO_2 by grains is completely reversible. The sorption and desorption curves were symmetric to the time axis, indicating that the sorption and desorption are two opposite phenomena. If the CO_2 concentration at any time is lowered from the original CO_2 concentration it is essential to account for the desorption of CO_2 by the grain. The value of q

for each element in the domain was estimated using the following equation:

$$q = (-1)^i \cdot \left(\frac{C_n + C_{n-1}}{2} \right) \frac{SCO_2 \rho_w}{\rho_{co_2} \Phi \, 86400} \tag{12}$$

where:

q = sorption or desorption of CO_2 by wheat (g/m^3/s),

SCO_2 = rate of sorption of CO_2 by wheat, linearly extrapolated from data of Cofie-Agblor et al. (1992) (g/ kg/day),

ρ_w = bulk density of wheat (kg/m^3),

ρCO_2 = density of CO_2 gas (g/m^3),

C_n = average predicted CO_2 concentration of an element at the present time step (g/ m^3)

C_{n-1} = average predicted CO_2 concentration of an element at the previous time step(g/ m^3)

Φ = porosity of wheat,

$i = 1$ when $C_{n-1} > C_n$, and

$i = 2$ when $C_{n-1} < C_n$

Simulation results

In Bins 1 and 3 with 180 g of dry ice and with open or covered grain surfaces, the inclusion of sorption of CO_2 by the wheat slightly increased the accuracy of prediction in the first few hours after the introduction of dry ice (sampling times 1 and 3 hours), and reduced the accuracy of prediction later. At 21 hours after the introduction of dry ice, for example, in Bin 1 with covered grain surface the error increased from 9% when sorption = 0 to 11% when sorption > 0. It is possible that at low CO_2 concentrations (180 g dry ice will create a CO_2 concentration of approximately 10% in the intergranular space of Bins 1 and 3), the rate of sorption may be lower than the values obtained by linearly extrapolating the CO_2 sorption rate at 100% concentration.

In Bin 1 with 540 g of dry ice, and in Bin 2 with 370 or 740 g of dry ice, the accuracy of prediction was increased at all sampling times, indicating that including the sorption of CO_2 by wheat is essential for accurate model predictions. The errors with 370 g of dry ice in Bin 2 were around 10% at 21 hours after the introduction of dry ice. But with 540 g of dry ice in Bin 1 and 740 g dry ice in Bin 2, the errors were high even after the inclusion of sorption and desorption in the model. As mentioned earlier, linear extrapolation of sorption data at 100% initial CO_2 concentration might not be the

correct way to estimate sorption at lower concentrations. Further experimental data on the sorption of CO_2 by wheat at lower initial concentrations are needed.

To demonstrate the effect of including the sorption of CO_2 by wheat on the accuracy of model prediction, we simulated the CO_2 concentrations using various sorption rates (0 to 100% of sorption measured by Cofie-Agblor et al. (1992) in steps of 10%). The simulated CO_2 concentrations were compared with the measured concentrations averaged from three replicates of the 740 g dry ice experiment in Bin 2 with a covered grain surface. Table 2 shows the errors at various sampling times with various sorption values. The accuracy of prediction at 24 hours was the best (an error of 7%) with 60% of actual sorption. At 70 and 100% of actual sorption rate the predicted CO_2 concentrations were close to the measured values at 21 and 12 hours, respectively.

Of the total amount of CO_2 sorbed by grains in 24 hours, 50 to 60% is sorbed in the first 4 to 6 hours (Yamamoto and Mitsuda 1980). Cofie-Agblor et al. (1992) observed that nearly 80% of the total amount of CO_2 sorbed by wheat occurred in the first 12 hours. From Table 2 it can be seen that the error was the minimum at 12 hours with 100% of the measured sorption rate. It is possible that at lower concentrations the sorption of CO_2 by wheat may be same as for 100% CO_2. Thus, using a high sorption rate in the initial few hours and a low sorption rate afterwards might give accurate model predictions. Further experimental evidence is required before such an approach is taken in the model.

Step 4 : The future model

Predicting CO_2 concentrations using an apparent flow coefficient in the beginning and a diffusion coefficient afterwards improved the accuracy of prediction but there were still high errors of prediction in the initial time periods. Theoretically, a differential equation of the form:

$$\nabla(D\nabla C - uC) + q = \frac{\partial C}{\partial \tau} \tag{13}$$

which includes both diffusion and convection terms can be solved to accurately predict the CO_2 distribution in the grain bulk. In this equation u stands for the average interstitial velocity of flow in x, y, and z coordinate directions. The values of u can be computed using the pressure at the inlet boundary

Table 2. Mean relative percent errors between the predicted CO_2 concentrations and the measured concentrations (average of 3 replicates) with 740 g of dry ice in Bin 2 with covered grain surface, and various sorption rates.

Sorption rate (% of actual q[a])	Mean relative percent error									
	1 hour	3 hours	6 hours		12 hours		21 hours		24 hours	
			5L[b]	4 L[c]	5 L	4 L	5 L	4 L	5 L	4 L
10	105	119	117	67	87	64	89	70	88	70
20	105	116	111	64	78	57	74	59	72	0
30	104	113	104	60	69	51	60	47	55	43
40	103	110	97	56	59	44	45	35	39	30
50	103	108	90	52	50	37	30	24	23	17
60	102	105	84	49	41	31	16	12	7	4
70	102	102	77	45	31	24	4	3	11	11
80	101	99	71	41	23	18	12	14	27	23
90	101	97	64	38	13	11	28	23	43	36
100	100	94	57	34	7	6	43	34	59	49

[a]the measured sorption data of Cofie-Agblor et al. (1992) were used to estimate the actual q
[b]the mean relative percent errors were calculated for all 5 levels
[c]the mean relative percent errors were calculated for bottom 4 levels

(the pressure created by the expansion of the CO_2) or using the inlet velocity (when liquid CO_2 is vaporised and pumped into the grain bulk). In both cases care must be taken to account for the anisotropicity of the grain to the gas flow in horizontal and vertical directions. Resistance of grains and oilseeds to the bulk flow of gas is lower for horizontal flow than for vertical flow (Kumar and Muir 1986; Jayas et al. 1987; Alagusundaram et al. 1992). In a grain bulk with uniform temperatures equation 13 along with the calculated velocity components can be used during the CO_2 application period and equation 1 should be used afterwards. CO_2 distribution and maintenance in a farm bin is further influenced by natural convection currents caused by temperature differences in the grain bulk and gas loss from the store due to weather changes and leaks in the bin wall. These factors should be taken into account for accurate model predictions. In our current research we are attempting to solve equation 13 and to validate it against the CO_2 data measured in wheat bulks under isothermal conditions. Finally, the model will be further modified to include the CO_2 transport by natural convection currents and gas loss from the granary. The resulting model will be validated against CO_2 distribution data measured in farm bins.

Summary

A pure diffusion model was not sufficient to accurately model the distribution of CO_2 in a grain bulk, especially during the application period. Although, predicting the CO_2 concentrations using the experimentally determined apparent flow coefficients improved the accuracy of model predictions compared with the assumption of pure diffusion, another way to approach the problem is to solve the convection-diffusion equation. Including the sorption phenomenon in the model is important for accurate model predictions. We now have only limited data on the sorption of CO_2 by food grains which are for an initial CO_2 concentration of 100%. In a CA treatment we seldom observe 100% CO_2 concentration in the grain bulk. The sorption of CO_2 by food grains at various initial concentrations is essential for the model. Models of natural convection and gas losses from the grain storage structure are essential for predicting CO_2 distribution in farm granaries.

Acknowledgments

We thank Messrs Jack G. Putnam, and Danny D. Mann, and Dale P. Muir for their technical assistance. This project was funded by the Natural Sciences and Engineering Research Council of Canada.

References

Alagusundaram, K. 1993. Movement of CO_2 gas, introduced as solid formulation, through stored wheat bulks. Unpublished Ph.D. dissertation. University of Manitoba, Winnipeg, Manitoba, Canada. 120 p.

Alagusundaram, K., Jayas, D.S., Chotard, F. and White, N.D.G. 1992. Airflow pressure drop relationships of some specialty seeds. Sciences des Aliments, 12, 101–116.

ASAE. 1992. Standards 1992. ASAE Standard S 352.2. Moisture measurement — unground grain and oilseeds. American Society of Agricultural Engineers, St. Joseph, MI.

Cofie-Agblor, R., Muir, W.E., Cenkowski, S. and Jayas, D.S. 1992. Carbon dioxide gas sorption by stored wheat. In Proceedings of the Controlled Atmosphere and Fumigation Conference, Winnipeg, MB. In press.

Cunningham, R.E., and Williams, R.J.J. 1980. Diffusion in gases and porous media. Plenum Press, New York, N.Y., 275 p.

Jayas, D.S. 1994. Mathematical modelling of heat, moisture, and gas transfer in stored-grain ecosystems. In: Jayas, D.S., White, N.D.G., and Muir, W.E., ed., Stored-Grain Ecosystems, Marcel Dekker, Inc., New York, NY, in press.

Jayas, D.S., Sokhansanj, S., Moysey, E.B. and Barber, E.M. 1987. The effect of airflow direction on the resistance of canola (rapeseed) to airflow. Canadian Agricultural Engineering, 29, 189–192.

Jayas, D.S., Muir, W.E. and White, N.D.G. 1988. Modelling the diffusion of carbon dioxide in stored grain. ASAE Paper No. 88-6013. American Society of Agricultural Engineering, St. Joseph, MI., 12 p.

Kumar, A. and Muir, W.E. 1986. Airflow resistance of wheat and barley affected by airflow direction, filling method and dockage. Transactions of the ASAE, 29, 1423-1426.

Nguyen, T.V. 1986. Movement of Fumigants in bulk grain. In: Champ, B.R., Highley E., ed., Pesticides and humid tropical grain storage systems. Proceedings of an international seminar, Manila, Philippines, 27–30 May 1985. ACIAR Proceedings No. 14, 195–201.

Rao, S.S. 1982. The finite element method in engineering. Pergamon Press, New York, N.Y., 625 p.

SAS. 1982. SAS Users Guide: Statistics. Statistical Analysis Systems Inc., Cary, NC., 584p.

Segerlind, L.J. 1984. Applied finite element analysis. 2nd edition. John Wiley and Sons, New York, N.Y., 427 p.

Singh (Jayas), D., Muir, W.E. and Sinha, R.N. 1983. Finite element modelling of carbon dioxide diffusion in stored wheat. Canadian Agricultural Engineering, 25, 149–152.

Singh (Jayas), D., Muir, W.E. and Sinha, R.N. 1984. Apparent coefficient of diffusion of carbon dioxide through samples of cereals and rapeseed. Journal of Stored Products Research, 20, 169–175.

Singh (Jayas), D., Muir, W.E., and Sinha, R.N. 1985. Transient method to determine the diffusion coefficient of gases. Canadian Agricultural Engineering, 27, 69–72.

Wood, E.L., and Lewis, R.W. 1975. A comparison of time marching schemes for the transient heat conduction equation. International Journal of Numerical Methods in Engineering, 9, 679–689.

Yamamoto, A., and Mitsuda, H. 1980. Characteristics of carbon dioxide gas adsorption by grain and its components. In: Shejbal, J., ed., Controlled atmosphere storage of grains. Elsevier Scientific Publishing Company, Amsterdam, 247–258.

Modified atmosphere storage of bagged maize outdoors using flexible liners: a preliminary report

D.G. Alvindia, F.M. Caliboso, G.C. Sabio and A.R. Regpala*

Abstract

Field trials were carried out in the Philippines for outdoor storage of maize sealed in flexible liners. Two storage atmospheres were evaluated: carbon dioxide (CO_2)-enriched and hermetic storage.

Tests were carried out to determine insect infestation, moisture content, microbial infection, weight loss, grain quality and seed germination.

Insect infestation was completely prevented in the CO_2-enriched cubes while few live insects were noted in the hermetic cubes. Insect-damaged kernels and weight loss were minimised. Grain moisture content remained stable after three months although mould growth was noted at the top surface of the stacks. Grain quality was preserved and seed germination was not affected.

Introduction

In the Philippines, few farmers and cooperatives store their grains in well-designed, purpose-built structures. Those without storage facilities usually stack bagged grain in barns, under the eaves of their residences and in the open. Depending on climatic conditions, such stacks are covered with tarpaulins, or plastic sheets. Typically, these storage sites lead to losses caused by insect infestation, moulding, rodents and birds. Another option for the farmers is to sell their produce immediately after harvest. However, the price of grains is depressed during harvest and safe temporary storage is necessary until the market price is more favourable.

Advances in storage technology has led to the development of flexible PVC liners to envelope stacks of bagged grains for outdoor storage (Donahaye and Navarro 1989). The development of 'storage cubes' was designed for the hermetic storage of grains in situations where permanent structures are not available (Donahaye et al. 1991). It is also designed for the application of modified atmospheres to prevent reinfestation of the commodity during prolonged storage.

Field evaluation of storage 'cubes' under carbon dioxide (CO_2)-enriched and hermetic conditions have been conducted to determine its applicability under Philippine conditions.

Materials and Methods

Site and duration of storage

Storage trials were conducted at Farmers' Cooperative Incorporated (FCI), Kisolon Sumilao, Bukidnon from 28 October 1991 to 31 January 1992 and 2 April 1992 to 8 July 1992. Details of the trials are given in Table 1.

Grain supplies

Newly-harvested locally-grown yellow maize bagged in 50 kg polypropylene sacks was provided by the Farmers Cooperative Incorporated. Maize was dried down below 14% m.c. using a mechanical drier.

Storage structures

The storage cubes were manufactured by Haogenplast Ltd, Israel. These consisted of a lower floor–wall section and an upper roof–wall section. The lower floor–wall section was made of white flexible 0.83-mm-thick polyvinyl chloride (PVC) sheeting while the top cover was made of white nylon reinforced chlorinated polyethylene (CPE) plastic sheet which provided protection from degradation by UV radiation (Donahaye, pers. comm.). The storage cubes are equipped with ports to allow application of modified atmospheres or conventional fumigants by gravity displacement (Donahaye and Navarro 1989). The lower floor–wall and the upper roof–wall are sealed together by a gas-proof zipper. The zippers are covered by a protective over-flap. Dimensions of the 20-t 'cubes' are 4450 × 3360 × 2000 mm (length × width × height) with a maximum storage volume of 29.9 m^3, and a weight of 76 kg when empty. Provision was made, by means of tension straps and buckles attached to the liner, to keep the walls of the cubes under tension even when they are not filled to capacity. This ensures that there are no folds of material at point of contact around the floor level, thus affording a large measure of protection from rodent penetrations (Navarro and Donahaye 1986).

The control maize was stacked on the wooden pallets and covered with tarpaulin plastic sheets.

Loading

Stacks were set up on a selected area which was cleared of sharp objects. Stacks were built directly on the lower floor--wall section which were spread out straight on the ground. Loading was manually done with 35 bags (5 × 7) per layer and nine layers to a height of 1.9 m. Stacks were constructed in a pyramid shape to allow rain water to run-off immediately on the sides of the cubes.

Grain sampling

Samples were collected during loading and immediately after the trial. Three composite samples of 1 kg were collected from all bags in the stacks using sampling spears. An additional 500 g was collected from the 15 individually marked bags.

Instrumentation

T-type thermocouple cables and 3-mm-diameter plastic tubings were installed at different locations inside the cubes to monitor temperatures and gas composition during storage (Fig. 1).

---

* Food Protection Department, National Post Harvest Institute for Research and Extension, CLSU Compound 3120, Nueva Ecija, Philippines.

Table 1. Details of storage trials.

Trial no.	Stack code	Treatment	Volume (t)	Duration (days)
I	C1	CO$_2$-treated	18.45	93
	C2	Hermetic	15.02	93
	C3	Control	4.75	93
II	C4	Hermetic	17.00	97
	C5	CO$_2$-treated	16.62	97
	C6	Control	4.58	97

Insulation

The top surface of each stack was insulated by adding one layer of dry bagged maize cobs. This was designed to prevent or reduce heat-flow and temperature fluctuations in grains.

Gassing

Food grade CO$_2$ at the rate of 1.5 kg/t was used to establish the modified atmosphere. The gas applicator was made of a flexible metal tube attached to the gas tank valve that was fixed onto the inlet valve of the cube.

Parameters

Moisture content determinations were carried out by drying grain samples for one hour in the oven (Anon. 1982). Live insects were sieved from the composite and representative samples. These were sorted according to group and species.

The extent of fungal infection was determined by plating 30 seeds randomly taken from composite samples into each media of Aspergillus Flavus Parasiticus Agar (Pitt et al. 1983); Diglycerol Glucose Agar (Hocking and Pitt 1980), Dichloran Rose Bengal Chloramphenicol Agar (King et al. 1979) and Dichloran Chloramphenicol Agar (Nash and Snyder 1962).

Temperature was recorded daily between 0800–0900 hours; 1200–1300 hours and 1600–1700 hours using the Anritsu type-T model HL600. CO$_2$ levels were monitored right after gassing and after 1,3, 5, 9, 11 and 15 days. Thereafter, weekly monitoring was done. The CO$_2$ level in the hermetic cubes was monitored weekly.

Quality evaluation was carried out to determine the effect of modified and hermetic atmosphere storage on maize. The calculation of quality parameters was determined by hand counting the number of insect damaged, discoloured, mouldy and germinated kernels in each 500 g grain sample. Viability was estimated using the rag-doll method. The actual weight loss was calculated from the difference between the weight of bagged maize at the start and at the end of the storage trial.

Data were statistically analysed using the Multi-Factor Analysis of Variance (AVMF).

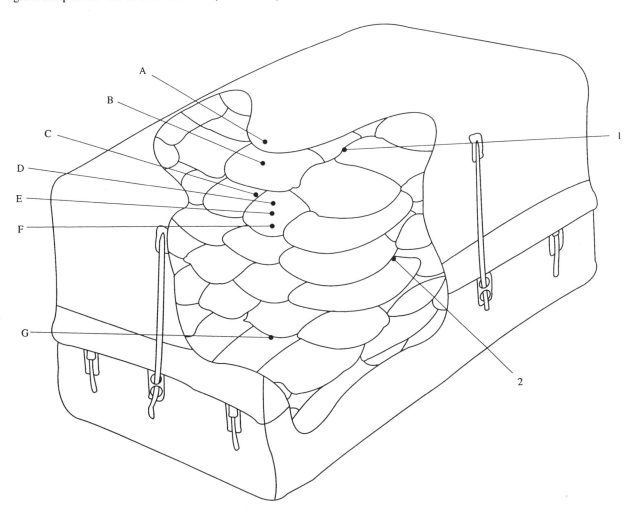

Fig. 1. Diagram of experimental stack, showing the different temperature and gas-sampling points. Tempereature sensors are located at: (A) above liner; (B) below liner; (C) below insulation; (D) 5 cm inside uppermost bag; (E) 10 cm in; (F) 20 cm in; (G) core of the stack. Gas concentration tubings are located at: (1) below liner; (2) middle core.

Results

Percentage moisture content

The moisture content of maize in both treatments and controls did not significantly change over the treatment period of 95–97 days (Table 2).

Table 2. Moisture content of maize at start and end of treatments with modified atmospheres in sealed flexible storage cubes.

Stack code	Treatment	Initial	Final	Difference
C1	CO_2-treated	13.16	12.16	ns
C2	Hermetic	12.56	12.56	ns
C4	CO_2-treated	11.59	11.93	ns
C5	Hermetic	11.44	11.72	ns
C3	Control	13.26	12.23	ns
C6	Control	11.98	11.82	ns

ns $P > 0.050$

Insect infestation

Initial and final counts of live insects did not significantly change in CO_2-enriched and hermetic stacks. However, the number of live insects increased significantly over time in the control stacks (Table 3).

The number of live infestations recorded in the hermetic stacks were: *Sitophilus zeamais* (1), *Tribolium castaneum* (1) in the first trial and *Rhyzopertha dominica* (1) in the second trial. The live infestations recorded in the control stacks include all the major pests of maize in storage *(Sitophilus zeamais, Rhyzopertha dominica, Tribolium castaneum, Carpophilus hemipterus* and *Latheticus oryzea)*.

Microbial infection

The various species of fungi isolated from maize and their extent of infection before and after the trial is summarised in Table 4. Stacks of newly-harvested maize were initially infected with *Aspergillus flavus*, other *Aspergillus* species *(fumigatus* and *nidulans)*, *Eurotium chevalieri*, *Penicillium citrinum* and *Fusarium moniliforme.*

Results indicate that infection with *Aspergillus flavus* in the CO_2-treated and hermetic stacks of newly harvested maize fell by 100% and 70%, respectively after 3 months of storage. Levels of infection by other *Aspergillus* species in maize increased 14 times in the CO2-treated stacks and 4 times in the hermetic stacks. Likewise, infection by *Fusarium moniliforme* increased to 7.5 and 30 times in the CO_2-enriched and hermetic stacks, respectively. Level of infection by various fungi in the control stacks generally increased after storage.

Storage and ambient temperatures

Temperatures recorded inside the CO_2-treated and hermetic stacks (Figs 2 and 3) were generally lower when compared to the ambient. Point a, which measured the temperature above liner showed greater fluctuations due to differences in time of day and weather conditions. In spite of the temperature fluctuations outside the storage cubes, temperature inside the cubes remained stable whereas in the control stacks, wide temperature fluctuations were recorded (Fig. 4).

Carbon dioxide levels

The CO_2 level of more than 35% was maintained for 11 weeks in the modified atmosphere cube (C1) while highest CO_2 level recorded in the hermetic maize stacks (C2) was 7% (Fig. 5).

Grain quality

The quality of maize under CO_2-enriched atmospheres did not significantly change after 3 months of storage. A significant increase in discoloured and insect-damaged kernels were noted on the hermetic stack C4. But other quality parameters where not significantly changed. Discoloured and insect-damaged grains in the control stacks significantly increased over the storage period.

Maize germination

Germination of maize did not significantly change during storage irrespective of storage technique employed.

Weight loss

The extent of percentage weight loss in maize stored under CO_2-enriched atmosphere (0.252–0.265%) and hermetic condition (0.229–0.379%) was lower than conventionally stored maize (5.073–5.611%) (Table 5).

Table 3. Average insect density in 1 kg samples of maize before and after storage.

Treatment	Trial 1		Trial II	
	Initial	Final	Initial	Final
CO_2-treated	0 a	0 a	0 a	0 a
Hermetic	0 a	0.67 a	0 a	0.33 a
Control	1.71 a	68.66 b	0.30 a	17.33 b

Means followed by the same letter are not significantly different at 5% level of significance.

Table 4. Percentage fungal infection in maize stacks at various storage atmospheres during Trial 1 and II.

Fungal species	CO_2-treated		Hermetic		Control	
	Initial	Final	Initial	Final	Initial	Final
Aspergillus flavus	23	0	23	7	11	7
Other *Aspergillus* spp.	0	14	0	4	4	7
E. chevalieri	40	1	2	11	2	39
P. citrinum	2	0	0	0	2	6
F. moniliforme	2	15	1	30	2	28

Table 5. Change in actual weight in maize stored under various conditions

Stack code	Treatment	Initial weight	Final weight	Variance (kg)	% weight loss
C1	CO_2-treated	18450	18401	49	0.265
C2	Hermetic	15020	14963	57	0.379
C4	Hermetic	17000	16961	39	0.229
C5	CO_2-treated	16620	16578	42	0.252
C3	Control	4750	4509	241	5.073
C6	Control	4580	4323	257	5.611

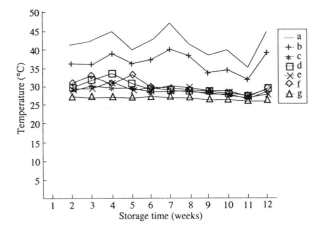

Fig. 2. Temperature recorded from seven different points in the CO_2-treated maize (Stack 5); a, above the liner; b, above insulation; c, below insulation; d, e, f, 5, 10, 20 cm in from the top surface of the same bag; g, core of the stack.

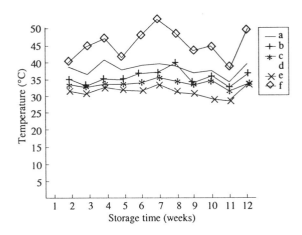

Fig. 4. Temperature recorded from six different points in the control maize (Stack 6); a, above the liner; b, top surface of the centre bag at the uppermost layer of the stack; c, d, e, 5, 10, 20 cm in from the top surface of the same bag; f, core of the stack.

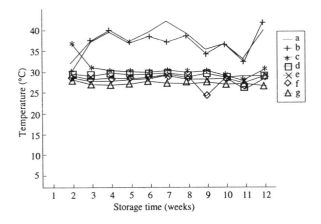

Fig. 3. Temperature recorded from seven different points in the hermetically stored maize (Stack 4); a above the liner; b, above insulation; c, below insulation; d, e, f, 5, 10, 20 cm in from the top surface of the same bag; g, core of the stack.

Discussion

The results provided information on the use of PVC flexible liners for modified atmosphere storage of maize in the Philippines. They also demonstrated the insect control capacity of the technology as part of a quality preservation system.

The standard assessment that there should be no live insects after storage was attained in the CO_2-enriched stacks. The complete disinfestation of stacks in the CO_2-enriched cubes demonstrated the enclosures gastightness, to hold the required standard and gas concentrations of 35% for at least 10 days (Annis et al. 1984). In the hermetic cubes, light infestation was evident (0.33–0.67 live insects/kg) after storage. Despite the

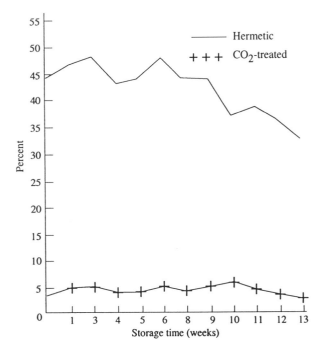

Fig. 5. Carbon dioxide concentrations recorded in CO_2-treated and hermetic maize stacks.

presence of live insects, no significant increase of insect damaged kernels was noted.

Mould growth was visible in the CO_2-enriched and hermetic storage cubes. Mould growth in the CO_2-enriched cubes was limited to the insulation while mould growth on the hermetic cubes was noted on the insulation and on the bags immediately underneath. Moulding was attributed to the inadequate drying of insulation.

There was no significant increase in the moisture content of maize stored under modified and hermetic atmospheres. Weight loss in the modified atmosphere cubes was calculated at 0.25 and 0.30% in the hermetic stacks. These values should be compared with loss evaluation of 5.34% in the conventional storage.

The quality of maize was not significantly affected by modified atmosphere storage. It appears that storage of maize under CO_2-enriched and hermetic atmospheres is better than the conventional practice. However, further work is required under more tightly controlled conditions to quantify the effect of modified atmosphere on maize quality.

The effect of modified atmospheres on maize germination suggest that it can be used for seed storage purposes. The result corroborates the findings of Bason et al. (1987) and Gras and Bason (1989) that carbon dioxide levels of 7.5 to 60% had no significant effect on the mean life-span retention period of maize.

Conclusion

Storage of maize for three months in cubes under CO_2-enriched atmospheres prevented insect infestation. Meanwhile, few live insects were recorded in the hermetic cubes. Both storage systems appeared to have advantage in minimise insect-damaged kernels and weight loss.

Modified atmosphere storage of maize under plastic enclosures provides significant advantages of maintaining quality and germinability of maize. The effect of modified atmospheres under plastic enclosures outdoors on the grain microflora would require further research before any conclusion can be drawn.

Acknowledgments

These field trials were carried out with the support of Agricultural Research Organization, Israel. The Farmers' Cooperative Incorporated, Kisolon, Bukidnon provided the test commodity and other logistic support.

Thanks are also due to the National Food Authority of the Philippines officials and staff of Cagayan de Oro City and Malaybalay, Bukidnon for their support and cooperation during the implementation of the field trials.

References

Annis, P.C., Banks, H.J. and Sukardi 1984. Insect control in stacks of bagged rice using carbon dioxide treatment and an experimental PVC-membrane enclosure. CSIRO Australia Division of Entomology Technical Paper No. 22.

Anon. 1982. Approved methods of the American Association of Cereal Chemists, revised ed. St. Paul, Minnesota, American Association of Cereal Chemists.

Donahaye, E. and Navarro, S. 1989. Flexible PVC liners for hermetic or modified atmosphere storage of stacked commodities. In: Champ, B.R., Highley, E. and Banks, H.J., ed., Fumigation and controlled atmosphere storage of grain: proceedings of an international conference, Singapore, 14–18 February 1989. ACIAR Proceedings No. 25, 271.

Donahaye, E., Navarro, S., Ziv, A., Blauschild, Y. and Wirrasinghe, D. 1991. Storage of paddy in hermetically sealed plastic liners in Sri Lanka. Tropical Science, 31, 109121.

Gras, P.W. and Bason, M.L. 1989. Biochemical effects of storage atmospheres on grain and grain quality. In: Champ, B.R., Highley, E. and Banks, H.J., ed., Fumigation and controlled atmosphere storage of grain: proceedings of an international conference, Singapore, 14–18 February 1989. ACIAR Proceedings No. 25, 83–91.

Hocking, A.D. and Pitt, J.I. 1980. Dichloran–glycerol medium for enumeration of xerophilic fungi from low moisture foods. Applied and Environmental Biology, 39, 488–492.

Navarro, S. and Donahaye, E. 1986. Plastic structures from temporary storage of grain. In: Semple, R.L. and Frio, A.S., ed., Research and development systems and linkages for a viable grain post-harvest industry in the tropics. Proceedings of the 8th ASEAN Technical Seminar on Grain Post-harvest Technology, Manila, Philippines, 189–194.

King, A.D., Hocking, A.D. and Pitt, J.I. 1979. Dichloran–rose bengal medium for enumeration and isolation of moulds from foods. Applied and Environmental Biology, 37, 959–964.

Nash, S.M. and Snyder, W.C. 1962. Quantitative estimations by plate counts of propagules of the bean rot, *Fusarium* in field soils. Phytopathology, 52, 567–572.

Pitt, J.I., Hocking, A.D. and Glenn, D.R. 1983. An improved medium for the detection of *Aspergillus flavus* and *Aspegillus parasiticus*. Journal of Applied Bacteriology, 54, 109–114.

Sealed storage technology on Australian farms

A.S. Andrews,* P.C. Annis† and C.R Newman§

Abstract

Australian farmers adopted commercially-available sealed storage technology from the early 1980s onwards. Widespread on-farm use was, however, restricted mainly to Western Australia (WA) until the early 1990s. Government policy, technology transfer methods, and grain quality standards demanded by buyers are key factors determining adoption rates of sealed storage in different regions.

WA silo manufacturers supply over 90% of their silos as sealed units, compared with less than 10% in the eastern states of Australia (January 1994 figures). Sealed storage use on farms is, however, increasing rapidly in the eastern grain belt of Australia. Most manufacturers construct sealed silos from sheet steel, in both on-ground (flat-bottomed) and elevated (cone-based) configurations. Fully-welded designs use light steel plate.

This paper summarises the evolution of sealing methods used on agricultural silos up to 400 m^3 in volume, and outlines compromises between cost, performance and ease of operation. The authors stress the importance of effective technology transfer for successful design and operation of sealed storage. Inadequate maintenance is identified as a major constraint to effective adoption of sealed storage technology on farms.

Background

Cereal grain production is a major export earning industry in Australia. Wheat dominates all grain types in both the domestic and export markets. The average gross value of wheat exports during the five year period spanning 1988–89 to 1992–93 was approximately $A2010m (Australian Bureau of Statistics 1994, pers. comm. 28 February). Australia generally exports around 70–80% of its annual wheat crop.

Australian farmers now hold an increasing amount of grain on-farm for extended periods before sale or consumption by livestock. Deregulation of the domestic grain market and increasing production of 'non-traditional' grain crops as alternatives to cereal grains such as wheat, barley and oats are key factors driving the move to greater on-farm storage. The authors estimate that over the next 5 years at least one-third of non-export stocks of grain will remain on-farm after each harvest for periods greater than 3 months. In Western Australia, on-farm storage levels are expected to increase at a slower rate.

Grain insects are a major threat to the Australian grain industry. The climate in most Australian grain-growing areas favours insect development throughout much of the year. This contrasts markedly with many other grain-producing areas of the world. Until the late 1980s contact insecticides dominated grain insect control strategies, both on-farm and in most bulk handling systems. Insect resistance development, tighter market standards on chemical residues, and environmental concerns have placed this reliance on chemical controls under increasing pressure.

Market Standards

Australia depends heavily on its international markets for its grain sales. Although a low volume supplier by international standards, Australia has a reputation for high quality grains that are free of insect and chemical contamination. Chemical residue standards imposed by buyers are progressively tightening, with many customers specifying a nil tolerance for previously acceptable insecticides. Grain treated with phosphine fumigant is accepted by all non-specialist markets, though residue and contamination limits still apply.

Insect Control Options

Options available to control grain insects and meet the market's tightening standards on chemical residues include:

- restricting by legislation the range of chemicals and fumigants available commercially;
- developing 'new-generation' chemical insecticides and fumigants;
- adopting more effective systems for applying existing chemical insecticides;
- using efficient fumigation techniques more widely;
- developing physical or 'non-chemical' methods of controlling insects; and
- applying controlled atmosphere (CA) storage techniques e.g. CO_2 or N.

A combination of these options will emerge in the medium term, with an increasing emphasis on physical controls, CAs and fumigants. Widespread adoption of integrated pest management (IPM) strategies, while desirable, is unlikely at the farm level in the near future.

In the immediate future many farmers will make the transition to fumigation or CAs in fully-sealed silos as a key strategy in supplying their buyers with grain that is free of insects and unacceptable chemical residues.

Australia's major state-based bulk handling organisations now use fumigants such as phosphine gas (PH_3) in combination with other physical controls as their principal insect control tools. Many Australian grain growers use PH_3 to disinfest their grain, particularly just before delivery to buyers. But with the exception of Western Australia (WA), most growers do not use fumigants routinely on newly stored grain. Reasons include:

- contact insecticides are readily available to grain growers in all states except WA and control a wide range of insect pests for extended periods;
- phosphine provides no residual protection on grain stored in unsealed structures;

---

* Department of Primary Industries, P.O. Box 102, Toowoomba, Queensland 4350, Australia.
† Stored Grain Research Laboratory, CSIRO Division of Entomology, GPO Box 1700, Canberra, ACT 2601 Australia.
§ Agriculture Protection Board, Bougainvillea Avenue, Forrestfield, Western Australia 6058, Australia.

- uncontrolled fumigation (i.e. fumigation in unsealed structures) usually kills adult and larval stages of insects only, so insects reinfest the grain rapidly.

Fumigation on Farms

Australian farmers have traditionally fumigated grain in unsealed metal silos, bins or trucks. In most areas of Australia this is still a regular practice. Bullen et al. (1991) reported the findings from a survey of grain growers' fumigation practices in southeastern Queensland. Of 28 growers who fumigated with phosphine, none used sealed storages, and all application rates claimed by growers were below the recommended label rate of 5 g/m³ for unsealed structures[1]. The fumigation periods reported by most growers were less than the 7–10 days (depending upon temperature) recommended on the fumigant label.

Unpublished data from a survey of grain growers in southeastern Queensland who fumigated grain in small (< 150 m³) unsealed silos showed that the fumigation practices used by these growers would not have achieved the concentration–time (CT) relationships needed to kill all insect stages (G. White, pers. comm., August 1992).

Some growers hang bags of fumigant tablets over the inlet of aeration fans and blow gas generated by the tablets into storages via the aeration ducting. A monitored field trial of this method showed that concentrations varied widely throughout the 120 m³ unsealed trial silo and that the highest concentration recorded was 4 ppm (P. Collins, pers. comm., January 1994). This result confirmed the non-effectiveness of the method for controlling insects regardless of the time period over which it is applied.

Newman (these proceedings) reported the fumigation practices of growers in the central wheat belt of Western Australia. Of the respondents who stored grain in unsealed structures, 85% used phosphine fumigants to control insects and applied an average of 5.8 g/t/year spread between one and three applications.

Each of the surveys and trials reported above revealed that poor practices dominated on-farm fumigation methods. But even poor fumigation practice often achieves significant adult mortality because of the lower phosphine tolerance of adult and larval stages in comparison with eggs and pupae. In fact, over 90% of growers surveyed by Bullen et al. (1991) classed their fumigations as 'successful', meaning that they killed sufficient adult insects to pass quality inspections during delivery to buyers or central storages. These fumigations are control failures from an entomological perspective because the concentrations and exposure times are insufficient to control all stages and species. Banks (1985) noted that such failures may contribute to insects developing resistance to the fumigant being used.

Annis (1992) described a computer model which simulated the performance of PH_3 fumigations in 'small' silos (<300 t wheat). The model has been further refined by Annis to more closely predict the field performance of storages. Figure 1 shows concentration versus time plots for sealed and unsealed silos of 100 m³ volume. The output confirms the poor field performance of unsealed structures when used for fumigation.

Sealed silos on farms

Most silos on Australian farms are constructed from bolted or riveted steel sheeting (typically 1 to 2 mm thick). Some manu-

facturers use fully-welded steel sheet construction. One company uses a novel manufacturing technique to 'spiral-form' the silo barrel from a continuous coil of steel sheet.

Until recently, most silo manufacturers in eastern Australia made no allowance for sealing their storages either during construction or later. Traditional designs are usually difficult and costly to seal to the standard required for effective fumigation.

This paper concentrates on sealed silo technology for farm silos up to 400 m³ in volume. Sealed storage allows operators to:
- rapidly and completely disinfest stored grain using fumigants (for example, phosphine gas) and controlled atmospheres (for example, high carbon dioxide);
- reduce fumigant dose rates, thus reducing costs;
- avoid unacceptable residues from contact insecticides; and
- minimise likelihood of insect reinfestation.

Silo operators who store 'non-traditional' crops are usually quick to recognise the benefits of integrating sealed silos into their stored-grain management systems. For example, commodities such as mung bean have strict market limits on quality and attract insects that are relatively tolerant of poor fumigation or CA practices.

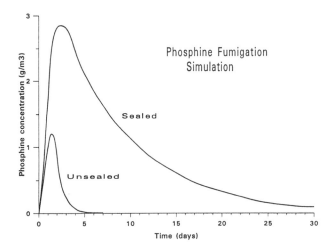

Fig. 1. Concentration versus time plots for PH_3 fumigations in 100 m³ sealed and unsealed silos. Commodity type: wheat @ 860 kg/m³; temperature: 25°C; moisture content - 12% wet basis; sorption loss - 7.4% per day; relative humidity - 56.7%; total silo volume - 100 m³; capacity used - 100%; total gas volume within storage - 43.4 m³; mixing time - 4 days; leakage loss (sealed) - 5.0% of total silo volume per day; leakage loss (unsealed) - 150%; dosage rate - 1.50 g PH_3/m³ of total silo volume i.e. 150 g PH_3 total.

Historical Development

Assessments of the suitability of materials and techniques for sealing small (<400 m³) silos have been reported from at least the mid-1970s. For example, Banks and Annis (1977), Banks and Annis (1980), and Williams and Murphy (1981) discussed materials and methods associated with sealing small structures.

Banks and Annis (1977) indicated that sealing of silos under 300 t capacity was unlikely to be viable using materials and methods available at that time. Since this early paper, however, advances in sealing technology, combined with tighter market demands and ongoing insect resistance development, have swung the economics in favour of sealed storage.

Western Australian silo manufacturers took the lead in developing sealed silos for use on farms from around 1980. The limited range of insecticidal grain protectants available to farmers in WA forced them to consider fumigation in sealed storages as a key tool in their insect control strategies.

---

[1] Amendments to the label recommendations since this survey remove the provision for fumigating in unsealed structures.

Two State government departments, the WA Department of Agriculture and the Agriculture Protection Board of WA (APB), stimulated this activity through their widespread promotion of sealed silo technology. The APB coordinates regulatory and advisory activities related to agricultural pests, including stored grain insects. Cooperative Bulk Handling of WA (CBH) undertook a major sealing program at their receival points during the same period, sending a clear message to growers that fumigants were the preferred option for future insect control.

Other grain-producing areas of Australia devoted few resources to on-farm stored-grain insect management through the use of sealed storage. This was due largely to the wide range of chemical-based control options available to growers in those States.

Sealed silos are now the standard line offered by most WA manufacturers. Newman (1989) reported that 95% of new silos sold in WA were sealed, and a telephone survey revealed that 65% of growers had at least one sealed silo on their property.

In other parts of Australia, expertise in sealed silo technology is concentrated amongst relatively few manufacturers.

Barriers to Adoption

Fumigation or CA in sealed storages has important advantages over other methods of controlling stored-grain insects. Despite this, a range of technical and non-technical barriers currently limit their wider adoption in most States of Australia, as summarised in Table 1.

Banks (1991) provided an overview of insect control options using fumigants or CAs on farms. He summarised that widespread successful use on farms was restricted by a lack of suitably sealed structures on farms, appropriate literature and training for operators, and market incentives.

Retro-Sealing Versus Sealing During Construction

Field experience by the authors over more than 10 years in WA and to a lesser extent in the eastern states of Australia shows that retro-sealed units are more likely to fail than factory-sealed units. Figure 2 shows areas which must receive specific attention (both on new and retro-sealed silos) to achieve reliable long-term performance. Common failure points on retro-sealed units are roof panel joints, roof/wall junctions, filling hatches and grain outlets.

Silo sealing contractors operate on farms in some areas of Australia. Experienced operators seal to a higher standard of reliability than could be achieved by most farmers. Many WA farmers undertook the task of sealing their own silos in the early 1980s but few achieved the standard of gastightness needed for efficient fumigation. The cost of contract services is usually offset by the greater long-term reliability of the finished storage.

During the early years of silo sealing activity in Western Australia, most contractors entered the industry with an unrealistic understanding of the expertise and financial outlay on equipment needed to seal silos effectively. As a result, contracts were often quoted too cheaply to effectively seal the units. Thin layers of sealant paints and the use of silicones or tapes to temporarily seal silo openings consistently resulted in failure within 12 months.

Some silos designs are relatively difficult to seal, putting high demands on labour and materials. This prices them out of the market or seriously reduces profitability of manufacturers who do not modify their designs appropriately. Experience over the last 15 years shows that custom-designed sealed silos are more cost-effective to manufacture than adaptations of existing designs. Purpose-built units also deliver better long-term performance in the field.

Manufacturers in WA progressively refined their designs through the 1980s to produce units they believe to be an acceptable compromise between function and cost. Small

Table 1. Summary of technical and non-technical barriers to adoption of sealed silos on farms.

Potential stimulus to demand for sealed silos	Potential barriers to stimulus
Widespread commercial availability of sealed silos for farm use	Resistance by manufacturers to changes in manufacturing methods and designs
	Variable demand by consumer
	Poor understanding of sealing technology by manufacturers
Restricted availability of grain protectants for on-farm use	Wide range of contact insecticides registered for on-farm use in most Australian states
Grain injection at point-of-sale if unacceptable chemical residues are present	Acceptance standards and sampling methods used by domestic buyers and handlers vary widely and are relatively imprecise
Premium prices paid for residue-free grain	Existing market for premium-priced residue-free commodities is limited in size
Details of reduced insect control costs achieved by using sealed storage transferred to operators	Growers unwilling to invest capital to seal existing storages or purchase new units
	Relative cheapness of fumigants allows growers to perform several inefficient, high dosage fumigations before delivery of grain
Extension programs to deliver accurate information on operation and maintenance aspects of sealed silos	Lack of widely available extension information targeted at growers, traders and manufacturers of grain storage and management equipment
	Limited funds to promote grain management practices (including sealed silo technology) within the rural sector
Industry-wide adoption of Integrated Pest Management (IPM) strategies for stored-grain insect control	Deregulated grain industry in Australia encourages on-farm holding of grain on farms under variable quality storage conditions

capacity silos for on-farm use do not vary greatly in their designs between manufacturers in WA, due in part to the highly competitive market.

Sealing Methods and Materials

Wall, roof and floor panels

Silicone-based sealants are used widely to seal sheet-metal panels during silo manufacture. Sealant is usually applied to sheets before riveting or bolting them together. One design spiral-forms the main barrel of the silo from a continuous coil of sheet metal — no sealant is needed to establish a gastight seal along the lapped seam thus created.

Flexible membrane paints were widely used by growers and contractors to seal both new and existing silos during the early years of on-farm sealed silo usage. Some manufacturers also used flexible membranes when constructing farm silos but few continued with them due to problems such as poor adhesion and practical difficulties in applying them internally. They continue to be used on retro-sealing jobs — usually applied on the outside of structures.

Sealants

Most sealants used for sealing work are either silicone-based 'mastics' or paintable 'flexible acrylic membrane' paints. Flexible membranes are used extensively to seal large central storage structures and existing on-farm steel silos. CBH in WA has sealed more than 70% of their sealable receival points using flexible acrylic membranes as a key material.

Acrylic sealants must be applied to manufacturers' specifications, with particular attention to thickness. The specified depth is often difficult to achieve with a brush, necessitating two or more applications. Airless spray guns apply acrylic membranes very successfully and most acrylic membranes will bridge gaps up to 5 mm wide.

On larger expansion joints or areas which are difficult to reach, a fabric backing is necessary. Early experience in WA showed that some of the fabrics and tapes used as substrates created relatively inflexible joints leading to subsequent failure.

Sealant adhesion is a problem on new steel or on the dusty internal surfaces of existing grain stores. Failure to adequately prepare surface according to the sealant manufacturers' recommendations often leads to poor performance. Field experience shows that external application of the product on weathered steel needs less preparation than on internal surfaces of silos.

Hatches and outlets

Loading hatches, access points and grain outlets must usually be either radically redesigned or retro-fitted with purpose-designed sealable units to ensure easy and reliable long-term operation on farms.

Early experiments which used silicone, tape or acrylic membranes to seal existing hatches and outlets caused problems for operators when loading or out-loading. Purpose-built hatches and outlets for on farm silos incorporate rubber strips for instant sealing when closed. These are far more con-

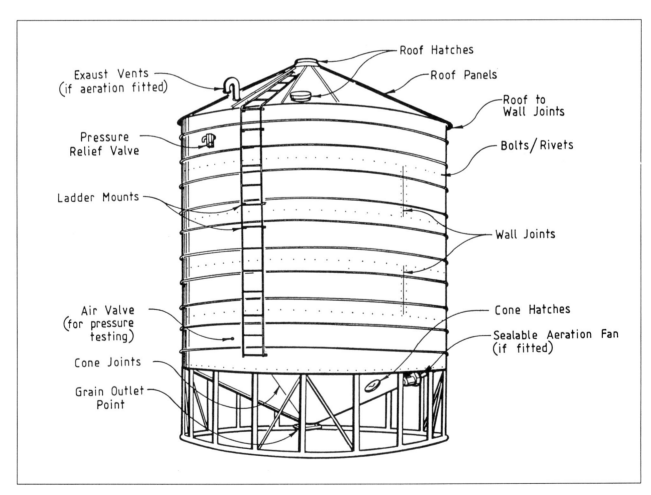

Fig. 2.   Major areas needing attention during retro-sealing of existing farm silos or design of purpose-built units.

venient and reliable than sealing with tape or silicone each time the access point is closed.

Figure 3 gives examples of successful sealable grain outlets used on farm silos. Slide gates fitted with quick-release cams or screws are popular in eastern states. Their accessibility, clearance and high discharge rates are a major advantage over other designs when used with large conveyors.

Many sealable hatches have been developed and trialed during the evolution of sealed storage on farms. Most manufacturers now use variations of the cost-effective design shown in Figures 4 and 5.

Pressure Relief

Sealed storages experience wide swings in internal pressure due to variations in ambient conditions. Changes in incident solar radiation levels, temperature and barometric pressure can produce internal pressure deficits or excesses which may structurally damage a storage. Figure 6 is typical of relief valves fitted to all sealed farm storages to protect against pressure damage.

The most likely source of pressure-related damage in sealed storage is caused by implosion due to a pressure deficit within the storage. Additional protection against implosion is available by using a vacuum relief valve, as shown in Figure 7.

Hybrid Designs

Unpredictable weather at harvest in some areas of Australia, notably Queensland, often leads to grain being harvested at higher moisture contents than is safe for storage. High ambient temperatures and intense solar radiation encourage insect activity, moisture migration and quality loss in stored grain.

Aeration is a grain management technique which passes controlled amounts of ambient air through stored grain. It stabilises temperature and moisture levels in the grain, prevents localised temperature increases ('hot spots') and discourages moisture migration. Aeration also lowers the temperature of the grain bulk during long term storage, reducing insect activity.

Silo designs which allow both aeration and fumigation are an advantage in areas where climatic conditions predispose harvested grain to variations in moisture content and heating. Figure 8 shows typical features on an aeratable/sealable silo for farm use. Designs similar to this are now made by several Australian manufacturers.

Aeratable/sealable silo configurations need a higher level of management than a standard sealed silo. The 'decision tree' of Figure 9 assists grain growers and storage operators to determine sequences of fumigation and aeration practice which give best protection to their stored grain.

Pneumatic conveying has traditionally been considered incompatible with the structural requirements of sealed storage. Recent developments in this area have produced further hybrid designs incorporating the advantages of both technologies, as shown in Figure 10.

Until recently, the operating flexibility sought by experienced managers of farm-stored grain often conflicted with the perceived constraints imposed by sealed storages. Hybrid silo designs now allow multiple functions to be built in at the design stage. Figure 11 gives a commercial example of this approach.

Sealing Standards

Pressure testing

Silo manufacturers, users and advisers in Australia use half-life pressure decay tests to determine whether storages are sealed sufficiently for effective fumigation or CA use. The valve illustrated in Figure 6 is suitable for gastightness testing in addition to its primary role of pressure relief.

The standard adopted for small (<300 m^3) farm silos in WA from the early 1980s specified 3 minutes as the minimum half-life decay from an excess internal pressure of 25 mm water gauge (that is, from 25 mm to 12.5 mm). Most WA manufacturers still use this as a de facto industry standard.

Studies by the CSIRO Stored Grain Research Laboratory (SGRL) show that a 3 minute decay is acceptable in storages if they are tested when full of grain but not when they are empty (P.C. Annis, pers. comm. August 1993). A buffering effect created by the greater air volume in empty storages results in

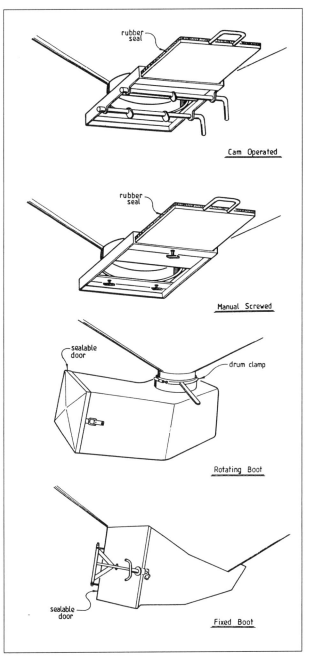

Fig. 3. Typical grain outlet configurations used on farm silos.

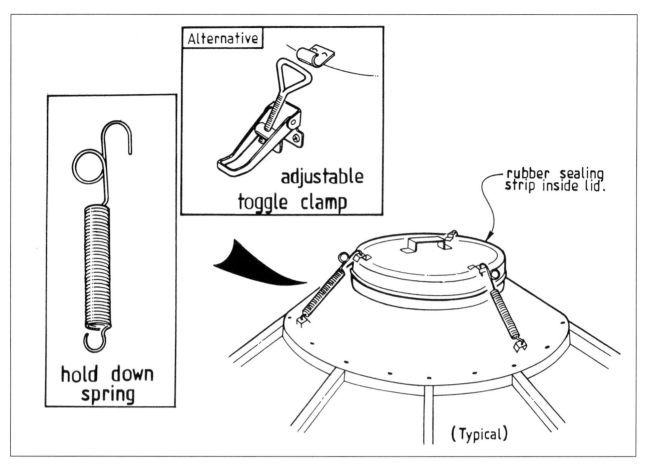

Fig. 4. Australian manufacturers of sealed farm silos use spun metal lids and metal bands widely as loading or inspection hatches (lid/band units manufactured by Andersons Metal Spinners, Kelmscott, WA).

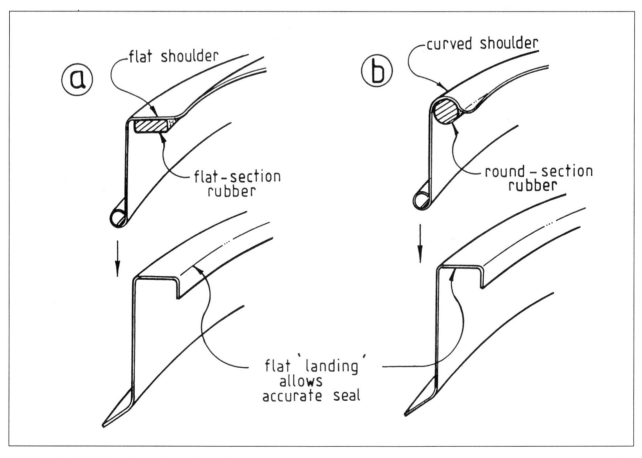

Fig. 5. Detail of alternative sealing strip arrangements in the spun lid shown in Figure 4. Option (b) gives a more reliable seal but requires extra care during installation of the round-section rubber.

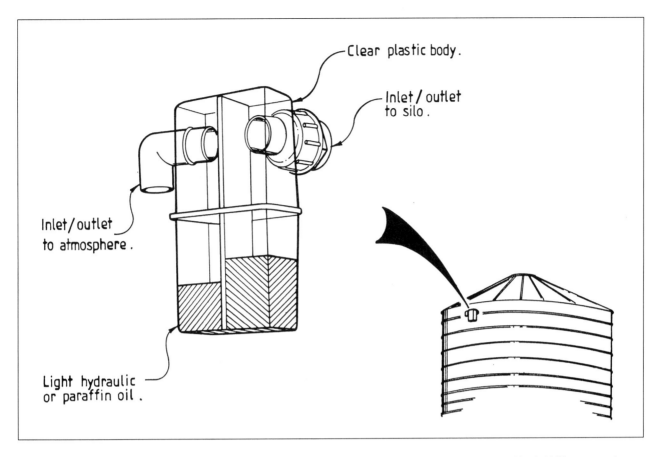

Fig. 6. Typical pressure/vacuum relief valve used on farm silos. The unit is also suited to gastightness testing using half-life pressure decays (Manufacturer: Acrifab, WA).

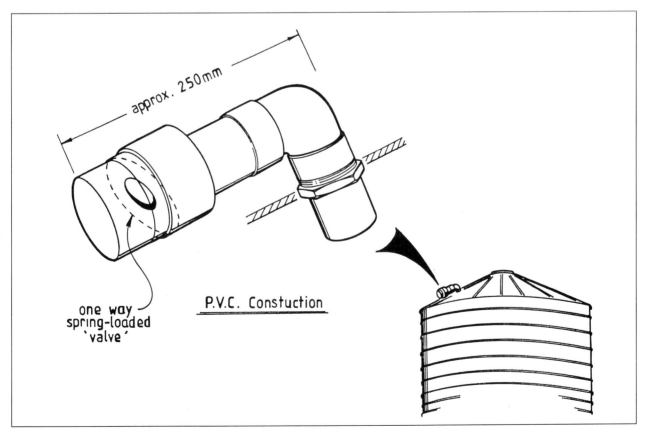

Fig. 7. Vacuum relief valve suited to implosion protection on small farm silos. The one-way valve does not provide protection against over-pressure events.

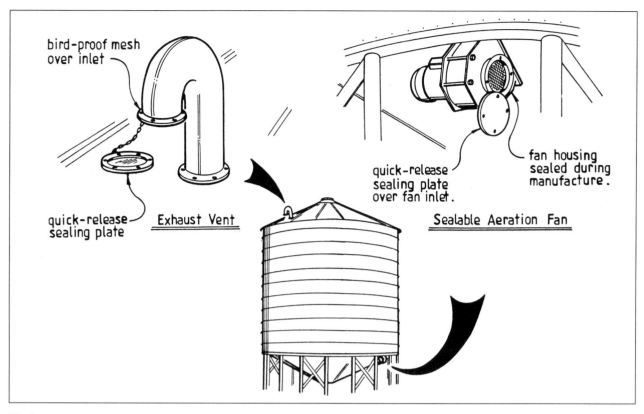

Fig. 8. Features of a typical hybrid sealable/aeratable silo. The storage is sealed to allow disinfestation with fumigants, then unsealed for long-term aeration.

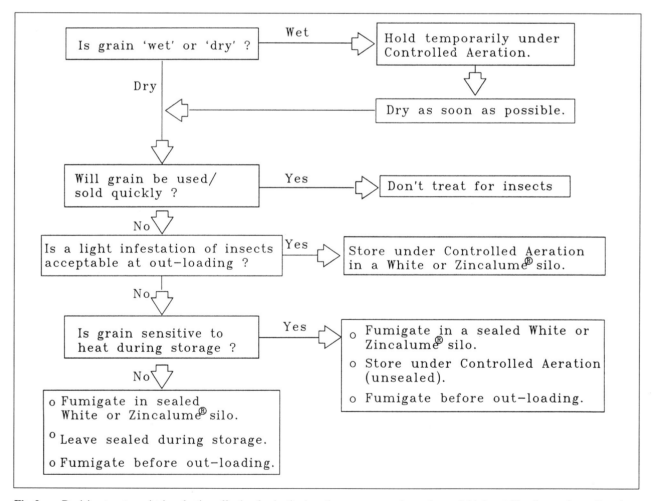

Fig. 9. Decision tree to assist in selecting effective fumigation/aeration sequences when using sealable/aeratable silos on farms (based on information from H.J. Banks, pers. comm., July 1993).

longer decay times than observed in the same silo when full of grain.

SGRL, in collaboration with a network of Grain Storage Associations representing Australian silo manufacturers, propose industry-wide adoption of a national testing standard for small farm silos. The draft proposal calls for half-life pressure testing to the following specification:

- maximum of 3 minutes for storages filled to capacity with grain;
- maximum of 5 minutes for empty or partially filled storages

Manufacturers groups are considering a proposal to adopt higher standards if testing is done in the more stable conditions associated with undercover silo assembly areas in factories.

Sealed silo quality standards

Advisory groups across Australia recommend buyers specify in their purchase agreement that the silo must meet the standard pressure test after installation or on completion of sealing. Most manufacturers of factory-assembled silos in WA

Fig. 10. Sealable grain cyclones fitted to a pneumatically-filled silos. Cyclones shown in both sealed (foreground) and unsealed (background) modes.

are, however, unwilling to guarantee compliance with a standard pressure test at delivery on-site. They argue that a guarantee cannot be given because of the unstable weather conditions that exist in the field.

The rough road conditions over which some silos are delivered also predisposes them to loss of seal during transit. This reinforces the argument that on-site testing of silos is the only guarantee of performance.

The Grains Research and Development Corporation (GRDC) in Australia, an industry representative body funding RD+E activities, has suggested that a manual of instruction and a standard guarantee be written for use by all farm silo manufacturers.

Operation, Maintenance and Servicing

Soon after sealed silos were introduced in WA it became obvious from inspections that maintenance was being neglected. A postal survey in 1985 found 44% of silos were not achieving gastightness standards (Newman 1987). This survey involved a telephone call to the grower followed by a visual inspection.

A physical inspection and pressure test of silos on farm were carried out in 1989 (Newman 1989). This survey found 56% of silos were failing to meet gastightness standards. The main reasons for failure were leaking seals at the inlet and outlet ports or low oil in pressure relief valve. A further survey found 75% of silos tested were leaking excessively (Newman 1994).

Valve oil

Incorrect oils were used in some valves leading to premature failure e.g. vegetable oil was recommended during early 1980s, but solidified when phosphine bubbled through it. The current recommendations are paraffin oil in valves with silicone in joints, and light hydraulic oil in plastic valves (Newman 1987).

Fig. 11. An on-farm bank of 140 m³ hybrid silos incorporating several grain management and handling features: — sealable for fumigation or CA; white Colorbond® construction for temperature stability; pneumatic filling; provision for aeration equipment.

Technology Transfer

From the inception of sealed silos in WA, a concerted effort was made to transfer information on the use of sealed silos to the farming community (Chantler 1983), largely through the activities of the APB and the WA Department of Agriculture.

Information is delivered to silo operators through pamphlets, field days, static displays and direct advice from APB district officers based throughout the state. No full-time extension officers work specifically on sealed storage technology transfer in WA. Other States devoted minimal extension resources to on-farm sealed storage programs during the 1980s.

Recent initiatives funded by the GRDC, have improved the flow of information since the early 1990s. For example, the GRAINSAFE program run by the Department of Primary Industries (Queensland) and described by Bullen et. al. (these proceedings) transfers technology on stored-grain insect control to farmers and grain traders.

GRAINSAFE has impacted significantly on the adoption rate of sealed storage on farms in eastern Australia. Sealed silos now represent a major proportion of sales amongst those eastern Australian manufacturers with expertise in sealing techniques. Up to 90% of some manufacturer's sales are sealed units (GRDC 1994).

Despite this recent upsurge in sales of sealed silos, industry-wide understanding of the principles of sealed storage operation and maintenance is still at a low level, even in WA. An urgent need to translate fundamental and applied research findings into user-friendly literature for manufacturers and growers still exists.

Stored Grain Quality

Moisture

A common argument used against sealed storage is the potential for grain damage due to moisture trapped within the sealed structure. The authors agree that damage to moist grain placed in a fully-sealed storage is likely to occur faster than in a similar unsealed unit. But the argument points more to the need for a professional approach to grain management than a case against sealed storage. Placing over-moist grain in any storage structure, sealed or unsealed, predisposes it to a high risk of rapid quality loss.

Grain must be stored at a moisture level appropriate to its storage temperature. In WA, where the average bulk temperature of stored grain is likely to be relatively high, growers are advised to only store grain at moisture contents below 12%.

Even low levels of uncontrolled insect infestation can lead to moisture problems in silos. This is usually linked to failure to fumigate effectively at the time of loading.

Germination

Individual farmers throughout Australia occasionally suggest that phosphine fumigation in both sealed and unsealed silos affects grain germination. The authors are unaware of scientific studies to support this anecdotal theory.

Conclusion

Effective sealing and fumigation or CA treatment of grain in farm silos has the following advantages:
* less likelihood of control failures,
* lower dose rates of fumigants or CA gases,
* longer intervals between treatments,

* no unacceptable chemical residues on grain.

A major barrier to wider adoption of sealed silos on farms is lack of market incentive to change existing fumigation practices. Factors that would catalyse a more rapid transition to fumigation and CA use in sealed silos on farms include:
* restricted range of grain protectants available to growers;
* more precise sampling procedures at receival points;
* tighter market standards on chemical residues;
* greater consumer demand for 'chemical free' foods.

Each of these has already occurred within Australia, but not to the extent needed for growers to collectively change their insect control practices. With the exception of Western Australia, a wide range of chemical options is still available to grain handlers to solve their immediate insect control needs.

A broader, longer-term view is needed if insect resistance development and loss of export markets is not to occur. Change will occur more rapidly when clear economic benefits are demonstrated and understood by growers, particularly if these benefits accrue in a short time frame. Many Australian growers have adopted sealed silos in their insect management programs, but most continue to use contact insecticides and fumigate in unsealed structures. So long as the minimum acceptance criteria set by buyers are able to be met using these methods, there will be little incentive to change.

A significant gap exists between state-of-the-art practices in sealed silo technology and their adoption by end-users. Accurate information must be transferred not only to growers but to silo manufacturers, suppliers of sealing consumables, chemical resellers, advisers and grain buyers.

Initiatives by the industry funding body GRDC are helping to address these needs. The GRAINSAFE stored grain management extension project and establishment of Grain Storage Associations throughout Australia have already made significant advances.

References

Annis, P.C. and Banks, H.J. 1992. A predictive model for phosphine concentrations in grain storage structures. Proceedings of International Conference on Controlled Atmosphere and Fumigation in Grain Storages, Winnipeg, Canada, June 1992. Department of Stored Products, Agricultural Research Organisation Bet Dagan Israel.

Banks, H.J. and Annis, P.C. 1977. Suggested procedures for controlled atmosphere storage of dry grains. Division of Entomology Technical Paper No. 13, CSIRO Australia. 23 p.

Banks, H.J. and Annis, P.C. 1980. Conversion of existing grain storage structures for modified atmosphere use. In: Controlled Atmosphere Storage of Grains, Proceedings of International Symposium, Rome, 12 to 15 May 1980, Elsevier, 461–473.

Banks, H.J. 1985. Application of fumigants for disinfestation of grain and related products. In: Champ, B.R., and Highley, E., ed., Pesticides and Humid Tropical Grain Storage Systems. ACIAR Proceedings No. 14, 291–298.

Banks, H.J. 1991. Technology for using grain fumigants and controlled atmospheres on farms. Proc. Farm Grain Storage Workshop, GRDC, Gatton College, Qld. July 1991, 55–61.

Bullen, K., Andrews, A. and White, G. 1991. Grain storage and protection practices on Queensland farms. Proc. Farm Grain Storage Workshop, GRDC, Gatton College, Qld., July 1991, 89–97.

Chantler, D. 1983. The Adoption of silo sealing by Western Australian farmers. In: Ripp, B.E. et al., ed., Proceedings of International Symposium on Controlled Atmosphere and Fumigation in Grain Storages, Perth 1983. Elsevier, Amsterdam, 683–705.

GRDC 1994. Sealed silos take off. Ground Cover, Issue 5 1994. Grains Research and Development Corporation, Canberra, Australia, 1.

Newman, C.R., ed., 1987. Sealed grain stores in Western Australia — a review. Report of Workshop Agriculture Protection Board, Forrestfield (WA September 11 1987), 32 p.

Newman, C.R. 1989. Sealed grain storage in Western Australia — a progress report. In: Champ, B.R., Highley, E., and Banks, H.J., ed., 1990. Fumigation and controlled atmosphere storage of grain. Singapore, 14–18 February 1989, ACIAR Proceedings No. 25, 272. (Abstract only).

Williams, P. and Murphy, G. 1981. Sealing farm silos for fumigation of insect pests. Proceedings First Australian Stored Grain Pest Control Conference, Melbourne, May 1981, 3–4 to 3–8.

Time to population recovery as a means for specifying low oxygen dosages

P.C. Annis*

Abstract

Dosage regimes for gaseous toxicants are usually given in terms of concentration and an exposure time. These regimes are conveniently based on the effective dose known to give some defined level of mortality (ED_x). The ED_x can be closely defined and confidence intervals calculated when experimental conditions are well controlled. While the ED_x is useful scientifically it is not so useful for grain managers. Their task is to integrate treatments with a range of operational considerations. This paper describes the integration of a model of mortality of *Sitophilus oryzae* under low oxygen atmospheres (<1% O_2) with demographic models of this species. The purpose of the combined model is to allow grain managers to optimise exposure periods in terms of available time and grain temperature. This allows them to determine the minimum duration insect-free storage expressed as time to population recovery.

Introduction

One of the aims of a grain manager is to have grain of known provenance and quality available for dispatch when required. In many cases this means having grain free of detectable insects at the time it leaves the storage facility. It is a difficult task to predict when, and if, insect numbers will become detectable by outturn inspection procedures at some time in the future. The task is made harder because of gaps in our knowledge of insect population dynamics at very low densities. Furthermore, the detection of no insects at inspection is at best semi-quantitative. In stored-product insects, densities of 1500/t are about the lowest useable with reasonable population models (Longstaff 1981), whereas the average insect density entering the Australian bulk handling system has been estimated to be about 8 insects/t (White 1985).

The mortality response of insects to toxicants is conventionally described by probit analysis, on the assumption that there is a linear relationship between the probit of mortality and some transformation (normally logarithmic or linear) of an expression of dosage applied (time, concentration or the product of concentration and time). Therefore, if the original population number is known, it should be possible to combine models of population dynamics with the results of probit analysis to predict population numbers at any time after treatment. Although this is theoretically possible and its use has been reported previously (Banks 1986), there are several simplifying assumptions required that may limit its practical use.

This paper discusses the process used to determine times of low oxygen exposure required that should ensure that no insects will be detected when the grain is presented for inspection some time later. The effects of simplifying assumptions are considered, along with gaps in our overall knowledge, and the potential effects these may have on the outcome are discussed.

In Australia there are about 10 species of insects most likely to be found during outturn inspection of grain. There are very few comprehensive dosage–response data for the effects of low oxygen atmospheres for these insects. There are, however, enough data to suggest that pupae of *Sitophilus* spp. and *Rhyzopertha dominica* are the most difficult to kill under low oxygen conditions (Annis and Dowsett 1993). There are sufficient data on these species to make crude estimates of probit response lines at temperatures in the range 21–27°C. The demography of *Sitophilus oryzae* is well documented (Longstaff and Evans 1983) and it is possible to model the time of occurrence and duration of the immature stage (Annis and Banks 1993).

There are no single or adequate sets of data that relate all the processes needed to make the necessary models which will give the interactions between initial population, projected population size, percentage kill, temperature and relative humidity. There are, however, several individual studies that may be linked and approximations that can justifiably be made. The aim of the study reported here was to make conservative estimates of the maximum probable number of live insects that may occur at a specified time after treatment. This means that all assumptions must be conservative in terms of the estimated period of insect-free storage.

The major assumptions are as follows:

- Any infestation will include *S. oryzae*, a species known to be highly tolerant of low oxygen atmospheres.

- Any treatment giving substantial mortality in *S. oryzae* pupae will give complete mortality in all other stages and species.

- At the time of treatment, even if no insects were detected, the actual number would be just below the limit of detection.

- The same low level infestation (i.e. just below the limit of detection) will be acceptable at a future outturn.

- It is known that the number of eggs produced per insect is reduced at low population densities (Longstaff 1981), and that there is likely to be a time lag until widely dispersed pairs of insects find each other. As this effect is not quantified, it is not considered in the population models used, and reproduction rates at low population densities are always assumed to be at optimum levels.

- Temperature/time/mortality data are far from comprehensive, therefore a common slope is assumed for the time/mortality, and a second common slope for the temperature/mortality, curves.

---

* Stored Grain Research Laboratory, CSIRO Division of Entomology, GPO Box 1700, Canberra ACT 2601, Australia.

Data Sources

Effects of temperature on the duration of immature stages

Eastham and Segrove (1947) present a comprehensive set of data on the duration of immature stages of *S. granarius* over the temperature range 15–30°C and 40–70% r.h. As no similar data exists for *S. oryzae*, the data for *S. granarius* was used. The duration of the immature stages in this data set could be well represented by the equation

$$P_i = 10^{(0.6835 + 27.441/T - 0.0002r - 0.072330r/T)} \quad ... (1)$$

where P_i is the duration of the immature stages (days), T is temperature in °C and r is the relative humidity. The immature period was taken as the basis for a physiological time period, and used to scale a beta function that described the survivorship curve of Birch (1953). Time is corrected to physiological time units on the basis of P_i compared to 28 days (a nominal standard immature period), by a scaling factor t_f where $t_f = 28/P_i$ and therefore the corrected physiological time $t_c = t \, t_f$

$$f = 1 - BETADIST(t_c, 3.155, 3.802, A, B) \quad ... (2)$$

where f is the fraction of the whole population surviving at t_c (in weeks), BETADIST is a Microsoft Excel spreadsheet function that returns the cumulative beta probability density, t_c is the value at which to evaluate the function over the interval $A < t_c < B$, A and B lower and upper bounds to the interval of t_c.

Fecundity based on temperature and relative humidity

The composite data in Longstaff (1981) which cover the number of eggs laid over the reproductive life of *S. oryzae* are well fitted by the polynomial in T and r given below. These data cover the range 45–70% r.h. and 15–35°C. The original data are corrected to a uniform population density of 1 female/500 grains.

$$L = 10.85 + 155.1r - 2.235r^2 + 0.01103r^3 - 495.7T + 23.91T^2 - 0.3727T^3 + 0.000109rT^3 \quad ... (3)$$

Where L is the lifetime egg production per female at population density of 1 female/500 grains, T is the temperature, r is the relative humidity.

Density dependent fecundity

The equation relating fecundity to population density is from Longstaff 1981 and applies to *S. oryzae*.

$$F = 6.821(\log_e N)^{2.322} N^{-0.3992} \quad ... (4)$$

This function has been shown to be applicable from approximately $10–10^7$ weevils/million grains. The assumption used in the model is that, below 10 weevils/million grains, the fecundity remains at 18.9 eggs/female/week; the value calculated at a density of 10 weevils/million grains. This assumption is conservative and presumes that no matter what the initial distance between insects mating will occur immediately (no searching time involved). This equation was derived from data collected at 27°C and 65% r.h. (14% m.c. grain).

Time mortality data

There are three sets of data available with useable information on pupal survival of *S. oryzae* under low oxygen conditions (Lindgren and Vincent 1970: Storey 1975a: Annis and Dowsett 1993). Most other published data lack adequate definition of developmental stages and/or results to allow their incorporation into any model. The reported LT_{50} and LT_{95} from the useful papers are shown in Figure 1, which indicates that the gradient between log(time) and probit mortality does not vary widely no matter what the developmental stage or temperature. It was possible to model the LT values of pupae as a surface with a probit mortality axis, a linear temperature axis, a logarithmic time axis and an interaction term between log time and temperature. The fit was not perfect but was well

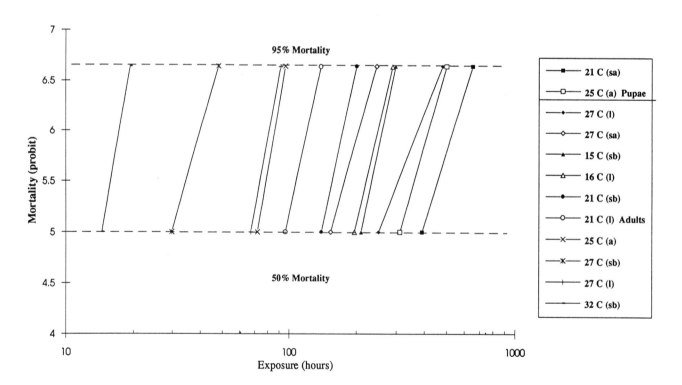

Fig. 1. LT_{50} and LT_{95} for *Sitophilus oryzae* adults and pupae at various temperatures reported by several authors: a. Annis (1987); l. Lindgren and Vincent (1970); sa Storey (1975a); sb Storey (1975b).

within the needs of the model (Fig. 2). The resulting equation was:

$$M = 100\{NORMSDIST[(-15.57 + 0.2201*T + 1.871(LOG_{10}(t)) + 0.10807TLOG_{10}(t)] - 5\} \quad ...(5)$$

Where M is the calculated percentage mortality in the pupal stage, T is the temperature in °C, t is the exposure period in hours and NORMSDIST is the Excel function that returns the standard normal distribution function.

The component parts of the model were combined to calculate the number of live individuals in 1-day age cohorts. This was updated in 7 day time steps for natural mortality, egg production and mortality, if a treatment has taken place. An Excel spreadsheet was used as a convenient framework for calculation but many other computer applications and language packages would be equally applicable.

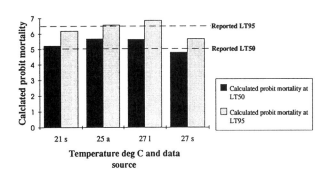

Fig. 2. LT$_{50}$ and LT$_{95}$ of *Sitophilus oryzae* adults and pupae modelled at various temperatures using equation (5) and compared with the reported values a. Annis (1987); l. Lindgren and Vincent (1970); s Storey (1975a).

Results and Discussion

Combination of the empirical models allows estimates of the minimum time until a population returns to its original size (see Fig. 3). This gives the grain manager an indication of the minimum safe storage period. It can be seen that this period remains more or less constant for any given exposure time over the range 22–32°C and tends to give longer protection as temperatures fall below 22°C. Because this model is purely empirical it cannot be used outside the range of data used to create its components which was 17.5–32.5°C.

A major limitation of this analysis is that no allowance is made for a lag in reproduction or reduction in fecundity due to very low population densities. These phenomena have been reported for *Sitophilus granarius* by Maclagan (1932) who stated that at population densities below 1 insect/200 grains the number of progeny per weevil began to fall. These findings were confirmed (and some additional data added) by Longstaff (1981) who showed that this phenomenon probably continues down to densities of about 1 insect /20000 grains (about 1 insect/750 g of wheat). There is no information for less than this density although it is likely that it will continue until some very low population densities where reproduction will no longer occur. There is, unfortunately, no information on which to model what happens at the low densities expected after an effective treatment. This means that in real treatments, it could be expected that the protection will be somewhat longer than that estimated.

Conclusions

It is possible to produce a mathematical model of population growth following treatment with low oxygen atmospheres at different temperatures. This model relies on combining a

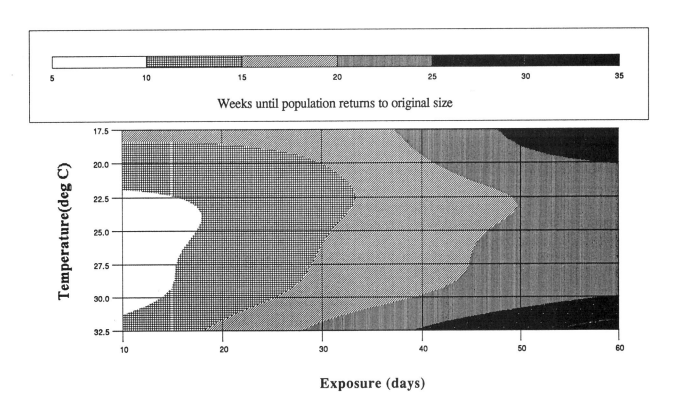

Fig. 3. Estimated time for a population of *Sitophilus oryzae* to return to the pre-treatment numbers after exposure to < 1% O$_2$ for various times over a range of grain temperatures.

series of empirical equations that describes mortality and the components that determine subsequent population growth. There are obvious limitations to this approach, but it provides more information than was previously available for determining exposure regimes for different temperatures. There are still important gaps in the knowledge required to complete the modelling process. Until these are filled there is little value in improving the model or setting up experiments to validate its output.

The current model, while having many limitations, will be useful because it gives grain managers an idea of the minimum period of effective protection given by an exposure to a low oxygen (<1%) treatment. Predictions by the model will be conservative in terms of the period of protection and this period is likely increase as more knowledge becomes available.

References

Annis, P.C. 1987. Towards rational controlled atmosphere dosage schedules: a review of current knowledge. In: Donahaye, E. and Navarro, S., ed., Proceedings 4th International Working Conference on Stored-product Protection, Tel Aviv, Israel, September 1986, 128–148.

Annis, P.C. and Banks, H.J. 1993. A predictive model for phosphine concentration in grain storage structures. In: Navarro, S. and Donahaye, E., ed., Proceedings International Conference on Controlled Atmosphere and Fumigation in Grain Storages, Winnipeg, Canada, June 1992. Jerusalem, Caspit Press, 299–312.

Annis, P.C. and Dowsett, H.A. 1993. Low oxygen disinfestation of grain: exposure periods needed for high mortality. In: Navarro, S., and Donahaye, E., ed., Proceedings International Conference on Controlled Atmosphere and Fumigation in Grain Storages, Winnipeg, Canada, June 1992. Jerusalem, Caspit Press, 71–83.

Banks, H.J. 1986. Impact, physical removal and exclusion for insect control in stored products. In: Donahaye, E. and Navarro, S., ed., Proceedings 4th International Working Conference on Stored-Products Protection, Tel Aviv, Israel, September 1986, 165–184.

Banks, H.J. 1986. Impact, physical removal and exclusion for insect control in stored products. In: Donahaye, E. and Navarro, S., ed., Proceedings 4th International Working Conference on Stored-Products Protection, Tel Aviv, Israel, September 1986, 165–184.

Birch, L.C. 1953. Experimental background to the study of the distribution and abundance of insects I. The influence of temperature moisture and food on the innate capacity for increases of three grain beetles. Ecology, 34, 698–711.

Eastham, L.E.S. and Segrove, F. 1947. The influence of temperature and humidity on instar length in *Calandra granaria*. Linnean Journal Experimental Biology, 12, 35–42.

Lindgren, D.L., and Vincent, L.E. 1970. Effects of atmospheric gases alone or in combination on the mortality of granary and rice weevils. Journal of Economic Entomology, 63, 1926–1929.

Longstaff, B.C. 1981. Density-dependent fecundity in *Sitophilus oryzae* (L.) (Coleoptera: Curculionidae). Journal of Stored Products Research, 17, 73–76.

Longstaff, B.C. and Evans, D.E. 1983. The demography of the rice weevil *Sitophilus oryzae* (L.) (Coleoptera: Curculionidae). Submodels of age-specific survivorship and fecundity. Bulletin of Entomological Research, 73, 333–344.

Maclagan, D.S. 1932. The effects of population density upon the rate of reproduction with special reference to insects. Proceedings of the Royal Society B, 111, 437–454.

Storey, C.L. 1975a. Mortality of *Sitophilus oryzae* (L.) and *S. granarius* (L.) in atmospheres produced by an exothermic inert atmosphere generator. Journal of Stored Products Research, 11, 217–221.

Storey, C.L. 1975b. Mortality of adult stored-product insects in an atmospheres produced by an exothermic inert atmosphere generator. Economic Entomology, 68, 316–318.

White, G.C. 1985. Population dynamics of *Tribolium castaneum* (Herbst) with implications for control strategies. PhD thesis, Department of Entomology, University of Queensland.

A same-day test for detecting resistance to phosphine

C.H. Bell, N. Savvidou, K.A. Mills, S. Bradberry and M.L. Barlow*

Abstract

A new test to identify resistance has been developed for three species of stored-product beetles, based on their mobility on paper cones during exposure to phosphine. Adults of the reference strains of *Cryptolestes ferrugineus* and *Tribolium castaneum* were all knocked down within 3 hours and those for *Sitophilus oryzae* within 4.5 hours in 0.31–0.42 mg/L phosphine. Homozygous resistant adults of all the species remained active for more than 20 hours with no overlap between susceptible and homozygous resistant populations.

When an F_1 was obtained by pooling the progeny of single pair crosses of homozygous resistant and susceptible individuals, a regression line intermediate between susceptible and resistant populations was obtained for each species, lying close to the susceptible population. There was considerable overlap between susceptible and heterozygous populations but it was still possible to set discriminatory exposure times, based on total knockdown of susceptibles, as follows: *C. ferrugineus*, 160 minutes at 0.36 ± 0.05 mg/L; *S. oryzae*, 272 minutes at 0.39 ± 0.03 mg/L; *T. castaneum,* 162 minutes at 0.38 ± 0.04 mg/L.

For one of the species, *T. castaneum*, 6 strains recently collected from provender mills in the U.K., which were identified as resistant by the standard FAO mortality test, were subsequently shown to be resistant by the new test method, with similar proportions of insects responding in both tests.

Introduction

Since the FAO survey of pesticide resistance in stored-product pests in the 1970s (Champ and Dyte 1976), the potential for problems arising from resistance to phosphine has been well known. Initially the levels of phosphine resistance detected were quite low, both in terms of incidence among different populations and in terms of the extent to which tolerance levels were increased among individual beetles. During the 1980s, however, very much higher tolerances were encountered among resistant populations, firstly in Bangladesh and India (Tyler et al. 1983; Taylor 1986), and later in parts of Africa, South America and Indonesia (Taylor 1989). Subsequent tests revealed that such populations also displayed resistance in the naturally tolerant immature stages and that there were serious implications for control (Price and Mills 1988; Mills et al. 1990). Furthermore, it has been established that high levels of resistance can result from selection with phosphine at any stage of the life cycle (Rajendran 1992).

In spite of growing resistance problems, phosphine is still effective in achieving control, provided that adequate gas concentrations can be maintained so that the exposure period can be extended. For this to be possible the storage structure needs to be sealed to pressure test standards (Banks and Annis 1980),

or there needs to be some means of maintaining a constant gas concentration throughout a storage structure for the necessary exposure period (Winks 1990). The latter is possible with the continuous-flow systems developed in Australia and the U.K., which are based on a cylinder-based supply of phosphine in carbon dioxide.

With continuous-flow systems it is possible to extend an exposure period to any chosen duration, although this may increase costs sharply. It would be of considerable value if the resistance status of the pest population were known at the start of a treatment so that the need for any special measures could be ascertained. At present the discriminatory dose test for common beetle species relies on a 14-day mortality assessment following a 20-hour exposure to phosphine. A 15-day delay between discovery of an infestation and arranging a treatment is often unacceptable, and information on resistance is unlikely to be obtained in time to influence the control strategy. The possibility that resistance may be identified by the different activity levels of individuals under gas, a likely side effect of reduced uptake or active exclusion (Price 1984), has already been partly explored (Reichmuth 1992). In the current study a rapid diagnostic test has been developed for some common beetle pests by which the presence of resistance in populations can be identified within a few hours.

Methods

The insects used were from mixed-age populations. The reference insects were taken from laboratory stock cultures and the homozygous resistant insects from a population previously selected for phosphine resistance by the standard FAO mortality test. Heterozygotes were produced from single-pair crosses between virgin adults of the reference strain and the homozygous strain. Crosses were set up over several consecutive weeks and the progeny from all the crosses then combined. Every 3–4 weeks, depending on the life-cycle of the species, the heterozygotes were transferred to fresh food to prevent any of their progeny from emerging and contaminating the population with susceptible and homozygous resistant insects.

The insects were tested on paper cones (6.5 cm diameter, 3 cm high) in glass crystallising dishes (7 cm diameter, 4 cm deep). Molten paraffin wax was used to seal the base of the cones to the dishes to prevent the insects from getting under the edge of the paper. For glass-climbing insects, such as *Sitophilus oryzae*, the walls of the dishes were coated with a thin coating of fluon (an aqueous suspension of polytetrafluoroethylene) to prevent them from escaping. Tests were conducted in 6-litre glass desiccators, each fitted with a flat glass top, to enable clear observation of the insects, and a central dosing port fitted with a rubber septum.

For the initial tests to establish the discriminating times, 3 replicates of 20 insects were observed for each strain. The dishes were placed on a wire-mesh shelf inside the desiccator, approximately 6 cm from the top, and left to acclimatise in the test conditions of 25°C, 60% r.h., for at least 30 minutes before sealing with the glass top. The desiccators were then dosed with phosphine, using a gastight syringe through the central dosing port, and stirred with a magnetic stirrer for approximately 5 minutes. Timing was started from the introduction of fumigant. The insects were observed at intervals of 2–5

* Central Science Laboratory, Ministry of Agriculture, Fisheries and Food, London Road, Slough, Berks., SL3 7HJ, U.K.

minutes until all the reference insects were knocked down. Subsequently, a number of field strains of *Tribolium castaneum,* obtained from provender mills in the U.K., was tested for resistance using the FAO resistance test method. Six strains which were identified as resistant were then included in parallel tests. One test was observed every 2–5 minutes as before (exploratory test), and the other, which used 3 replicates of 50 insects, was left for a fixed period, based on the approximate discriminating knockdown times for each species, and then assessed for knockdown (discriminating test). In both cases each field strain was tested alongside the reference strain.

At the end of each test the concentration in the desiccators was analysed by gas chromatography, using a flame photometric detector, standardised with a cylinder formulation of phosphine in nitrogen.

Results

The probit regression lines for all the species showed a clear distinction between the reference and the homozygous resistant insects, with at least 12 hours between the last reference insect and the first resistant insect being knocked down. The regression lines for the heterozygotes all lay between those of the reference strain and those of the resistant strain, albeit very near to the reference lines (Figs. 1–3). Due to the closeness of the reference and heterozygote lines there was considerable overlap in their response times, although for *T. castaneum* and *Cryptolestes ferrugineus* there was always at least 1 hour between the last reference insect and the last heterozygote insect being knocked down. The results for *S. oryzae* (Fig. 3) showed less consistency among replicate tests and there was a greater degree of overlap between heterozygote and reference strain results than for the other two species.

The six field strains of *T. castaneum* identified as resistant by the FAO test were also shown to be resistant by the knockdown test and all but one regression line lay to the right of the mean reference line (Fig. 4). Generally, the level of activity in the knockdown tests was higher than the level of survival in the FAO tests. One strain (line 5 in Fig. 4), collected from Northfields, gave inconsistent results in both tests. Initially, one survivor was recorded in the FAO test and the strain was classified as resistant. However, a re-test showed no

resistance. In the knockdown test no insects remained active in the exploratory test but one insect remained active at the end of the discriminating test (Table 1).

Table 1. A comparison between the results from the FAO test and those from the immobilisation test for *Tribolium castaneum.*

Strain	Insects surviving/standing (%)			
	FAO method		Knockdown method	
	1st test	Re-test	Exploratory	Discriminating
Calne	9.5	12.8	70.5	21.3
Chard	34.0	16.6	46.7	46.6
Melksham	9.5	18.2	63.2	17.3
Northfields	0.5*	susceptible	susceptible	0.7*
Wells	36.5	20.6	55.0	36.6
Wrexham	25.5	32.2	49.0	53.3

*Both these figures represent one survivor in each case.

Discussion

Previous attempts to identify resistance to phosphine, based on the ability of insects to remain active under gas, have been hampered by the difficulty of determining whether or not insects have become immobilised, and the complication of narcosis at high gas concentrations. Reichmuth (1992), for example, testing at concentration levels well above the 0.4 mg/L tested here, assessed resistance on the basis of insects not becoming 'motionless' or 'narcotised' within about half an hour, and presents results for *Rhyzopertha dominica* (F.). To date, narcotic concentration thresholds have not yet been identified for this species but it is known that narcotic thresholds for susceptible and resistant strains of *T. castaneum* differ widely, that for the susceptible strain lying near 0.5 mg/L, with much higher levels for resistant strains (Winks 1985).

The uncertainty associated with the narcotic responses of insects and the concentration levels at which they occur indicate that a resistance test based on such criteria is hard to interpret. For each species, consideration would have to be given to variation of narcotic response among strains and

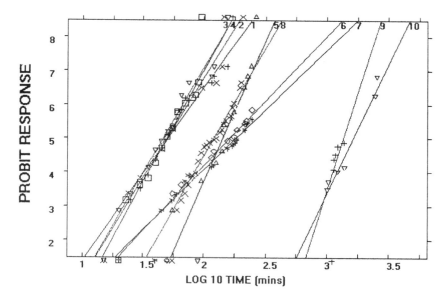

Fig. 1. Knockdown responses for *Tribolium castaneum* reference strain, homozygous resistant strain and heterozygotes which resulted from the crossing of these two, at concentrations of 0.4 + 0.03 mg/L. Lines 1–4 = reference strain, lines 5–8 = heterozygotes, lines 9–10 = homozygous resistant strain.

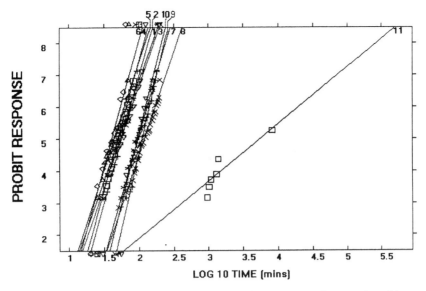

Fig. 2.   Knockdown response for *Cryptolestes ferrugineus* reference strain, homozygous resistant strain and heterozygotes which resulted from the crossing of these two, at concentrations of 0.36 $^+$ 0.05 mg/l. Lines 1–6 = reference strain, lines 7–10 = heterozygotes, line 11 = homozygous resistant strain.

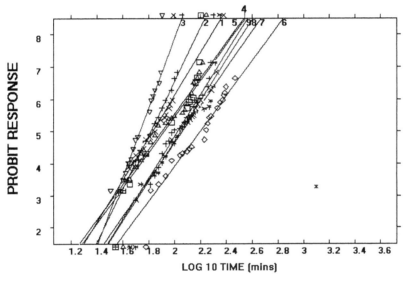

Fig. 3.   Knockdown responses for *Sitophilus oryzae* reference strain, homozygous resistant strain and heterozygotes which resulted from the crossing of these two, at concentrations of 0.39 $^+$ 0.03 mg/l. Lines 1–5 = reference strain, lines 6–9 = heterozygotes, * = first homozygous resistant individual knocked down.

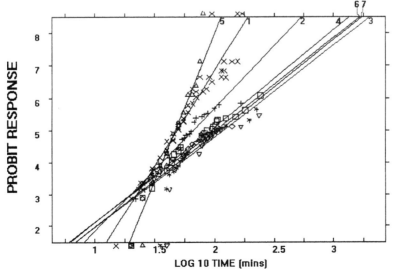

Fig. 4.   Knockdown responses for six field strains of *T. castaneum* compared with the mean response line for the reference strain at concentrations of 0.38 $^+$ 0.04 mg/l. Line 1 = mean response line for the reference strain, lines 2–7 = field strains identified as resistant by the FAO mortality test.

among genotypes within a strain. For these reasons the current study is based on a knockdown response at concentrations below that which initiate narcosis in *T. castaneum*. Knockdown is more clearly defined from the normal movements of an insect, and the use of paper cones renders the assessment of knockdown much easier. Knocked-down individuals are defined as those unable to retain their grip on the sloping side of the cone and which fall to the bottom and are unable to stand. The number of test insects was initially only 20 per replicate for the experimental observations which provided the regressions of knockdown against time. This was done to aid accurate observation in obtaining the provisional discriminatory knockdown times, though no limit need be placed on the number of individuals in a discriminatory dose test . A larger sample size would enable the detection of low incidences of resistant individuals.

For *T. castaneum*, good agreement was obtained between the results of the current test method and those of the FAO discriminatory dose test. The correlation extended to the point that a strain showing a marginal result in the FAO test also gave a marginal result in the new knockdown test. Thus, these preliminary results indicate that the new test method is likely to be as efficient as the FAO test in identifying resistance . It has the considerable advantage of being able to give a result within a working day so that the findings can be used to formulate a fumigation strategy. It has also made it possible to differentiate heterozygous insects from susceptible ones, provided that a sufficient number of heterozygotes is present in the population sample at the time of testing. This gives the added advantage of being able to detect the presence of resistance in a population at a relatively early stage in the selection process, before homozygous genotypes become established in response to inadequate phosphine fumigations. In order to make full use of the advantages presented by this test more information is required on the mortality responses of all life stages of heterozygous resistant insects.

At present only three species have been tested using the new method and further tests are required on these with a larger number of strains, both resistant and susceptible according to the FAO method, to establish a more reliable discriminating time between resistance and susceptibility. It is clearly desirable also that the test method is investigated against other species for which an FAO-based discriminatory dose test is available.

Acknowledgment

This work was funded by MAFF Pesticide Safety Division and its support is gratefully acknowledged.

References

Banks, H. J. and Annis, P. C. 1980. Conversion of existing grain storage structures for modified atmosphere use. In: Shejbal, J., ed., Controlled atmosphere storage of grains. Developments in Agricultural Engineering 1. Amsterdam, Elsevier, 461–473.

Champ, B. R. and Dyte, C. E. 1976. Report of the FAO global survey of pesticide susceptibility of stored grain pests. Rome, FAO Plant Production and Protection Series No. 5.

Mills, K. A., Clifton, A. L., Chakrabarti, B. and Savvidou, N. 1990. The impact of resistance on the control of insects in stored grain by phosphine fumigation. Brighton Crop Protection Conference (Pests and Diseases), 1990, 1181–1187.

Price, N. R. 1984. Active exclusion of phosphine as a mechanism of resistance in *Rhyzopertha Dominica* (F.) (Coleoptera: Bostrychidae). Journal of Stored Products Research, 20, 163–168.

Price, L. A. and Mills, K. A. 1988. The toxicity of phosphine to the immature stages of resistant and susceptible strains of some common stored product beetles, and implications for their control. Journal of Stored Products Research, 24, 51–59.

Rajendran, S. 1992. Selection for resistance to phosphine or methyl bromide in *Tribolium castaneum* (Coleoptera: Tenebrionidae). Bulletin of Entomological Research, 82, 119–124.

Reichmuth, C. 1992. Schnelltest zur resistenzbestimmung gegenuber phosphorwasserstoff bei vorratsschadlichen insekten. Mitteilungen der Deutschen Gesellschaft fur Allgemeine Angewandte Entomologie, 8, 245–247.

Taylor, R. W. D. 1986. Response to phosphine of field strains of some insect pests of stored products. Proceedings of the GASGA Seminar on Fumigation Technology in Developing Countries. TDRI, Slough, 18–21 March 1986, 132–140.

Taylor, R. W. D. 1989. Phosphine — a major grain fumigant at risk. International Pest Control, 31, 10–14.

Tyler, P. S., Taylor, R. W. and Rees, D. P. 1983. Insect resistance to phosphine fumigation in food warehouses in Bangladesh. International Pest Control, 25, 10–21.

Winks, R. G. 1985. The toxicity of phosphine to adults of *Tribolium castaneum* (Herbst.): Phosphine-induced narcosis. Journal of Stored Products Research, 21, 25–29.

Winks, R. G. 1990. Recent developments in fumigation technology, with emphasis on phosphine. In: Champ, B. R., Highley, E., and Banks, H. J., ed., Fumigation and controlled atmosphere storage of grain: proceedings of an international conference, Singapore, 14–18 February 1989. ACIAR Proceedings No. 25, 144–151.

A preliminary evaluation of carbon dioxide under high pressure for rapid fumigation

F.M. Caliboso,[*] H. Nakakita[†] and K. Kawashima[§]

Abstract

Fumigation is a widely used practice for commodity disinfestation but there are few available fumigants. In addition, there are environmental concerns in regard to the use of methyl bromide, and a low level of resistance to phosphine exists in stored-product insects.

Numerous studies have been carried out on the use of carbon dioxide for grain fumigation but most require lengthy exposure intervals, typically 10 days or longer. The current work involved the use of carbon dioxide at high pressures and concentrations for short time intervals.

Exposure of relevant development stages of major pests of stored grain to high concentrations of carbon dioxide (around 98%) for short intervals (5–20 minutes) produced complete mortality. In general, beetle species were more difficult to control than moths, and eggs were the most tolerant stage. At pressures up to 30 kg/cm^2 and 5–20 minutes exposure there was no significant effect on the viability of rice. The technique appears to have considerable potential for quarantine disinfestation.

Introduction

Fumigation is a widely used practice for commodity disinfestation but there are few available fumigants. In addition, there are environmental concerns in regard to the use of methyl bromide and a low level of resistance to phosphine currently exists.

Numerous studies have been carried out on the use of carbon dioxide for grain fumigation but most strategies require lengthy exposure intervals, typically 10 days or longer.

Exposure to CO_2 requires over 10 days to kill insects at 20–25°C (Annis 1987; Jay et al. 1990). Although exposure periods can be reduced as temperatures are raised (Bailey and Banks 1980; Jay 1986), the duration is still impracticable for quarantine treatments. Likewise, alternative methods such as holding insects under high vacuum alone requires over a week for acceptable kill even at warm temperatures (Calderon and Navarro 1968; Cline and Highland 1987; Locatelli and

Traversa 1989). A combination of CO_2 under low pressure and raised temperatures can achieve complete kill at best within 12 hours at 40°C, but longer (18–54 hours) at lower temperatures (Locatelli and Daolio 1993).

The new process of using carbon dioxide under high pressure offers an exciting and novel alternative to conventional CO_2 application. It was first described by Stahl et al. (1985). According to Gerard et al. (1988) and Pohlen et al. (1989), the quality of the treated commodities is not adversely affected. The required amounts of CO_2 are minute compared with the natural carbon dioxide emanating from the surface of the earth (Reichmuth 1990). The high pressure CO_2 treatment combines most of the advantages of CA technology and at the same time addresses its most serious drawback that is it requires extremely short durations for lethal exposure, within the range of a few minutes. This renders the new method highly promising for quarantine treatment and rapid disinfestation of valuable products.

Survival of eggs of *Plodia interpunctella* was completely prevented when these were exposed to high pressure CO_2 at 30 bar for 10 minutes (Gerard et al. 1988). Likewise, adult and immature *Lasioderma serricorne* were controlled after 15 minutes at 30 bar and 90 minutes at 20 bar.

The current work involved the application of carbon dioxide at high pressures for short time intervals.

Materials and Methods

Relevant developmental stages of four species of storage insects, i.e., *Sitophilus oryzae*, *Rhyzopertha dominica*, *Tribolium castaneum* and *Corcyra cephalonica* were challenged with carbon dioxide released at high pressures for short intervals.

Eggs, larvae and pupae of the test insects were exposed to carbon dioxide under pressures of 5, 10, 20 and 30 kg/cm^2 for 5 or 10 minutes in the pressurised chamber illustrated on Figure 1. The cylinder measured 17.3 cm high with external diameter of 3.8 cm to yield a volume of 30 mL. Pure (>99%) food-grade carbon dioxide was used. Pressure was adjusted by carefully moving the first regulator; the second regulator is further used for finer adjustment to attain the desired level of pressure. Although the actual CO_2 level inside the chamber was not measured and monitored, CO_2 concentration was calculated to be more than 90%.

Each larval or pupal sample contained 10 individuals However for the egg stage, each replicate had 20 individuals. Each treatment combination was replicated three times for the larval and pupal stages while the trials on eggs were replicated six times.

The insects were brought back to their respective culture rooms immediately after treatment. All experiments were conducted at 20–25°C and 70–75% r.h.

The cultures were assessed for adult emergence. This was continued for 30 days in the case of eggs to ensure that there had no extreme delay in eclosion had occurred. On the other

---

* Food Protection Department, National Postharvest Institute for Research and Extension, Muñoz, Nueva Ecija, Philippines.

† Stored Product Entomology Laboratory, National Food Research Institute, Tsukuba-shi, Ibaraki-ken 305, Japan.

§ Crop Production and Postharvest Technology Division, Japan International Research Center for Agricultural Sciences, Ministry of Agriculture, Forestry and Fisheries, 1-2 Ohwashi, Tsukuba-shi, Ibaraki-ken 305, Japan.

hand, mortality of the larvae and pupae was assayed 48 hours after exposure. Moribund insects were considered dead. The development of the remaining live pupae to adult was also observed.

Results and discussion

Among the various species tested, the rice moth (*C. cephalonica*) was the most susceptible. It was highly sensitive to the treatment with the pupae succumbing to the lethal atmosphere even at the lowest dosage of 5 kg/cm^2 pressure and exposure period of 5 minutes (Table 1). All the coleopteran species tested exhibited similar levels of tolerance, with complete control achieved at 30 kg/cm^2 within 5 minutes (Table 2).

In terms of developmental stage, the egg was the most tolerant, followed by the larva. A pressure of 30 kg/cm^2 was required to prevent egg survival. The pupal stage was the most susceptible, in most cases a complete kill resulting from exposure to 20 kg/cm^2 pressure for 5 minutes.

Increasing pressure had a more pronounced effect on insect mortality than did duration of exposure. As evidence of this, doubling the pressure from, say, 10 to 20 kg/cm^2 resulted in 50% or more reduction in survival; the same degree of increase in exposure period failed to produce a corresponding quantum response in insect survival or mortality.

Gerard et al. (1988) reported 100% mortality in eggs of *Plodia interpunctella* after exposure to 20 bar (4.04 kg/cm^2) for 30 minutes or 30 bar (6.06 kg/cm^2) for 10 minutes at room temperature. In the same study complete kills of beetle species such as *Stegobium paniceum* and *Tribolium confusum* required 120 minutes exposure at 20 bar. Increasing this to 30 bar shortened the exposure time necessary for complete kill to 15 minutes in *S. paniceum* and 40 minutes in *T. confusum*. A further increase in pressure to 40 bar (8.08 kg/cm^2) correspondingly cut the lethal time to 10 miutes in both species. Although the pressures utilised in this work were much lower than the ones reported in the current work, high levels of

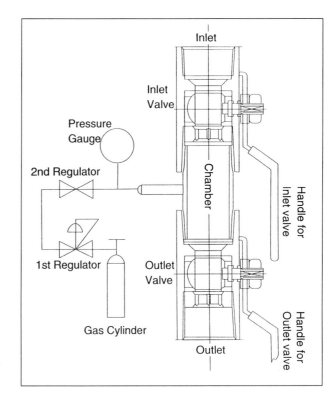

Fig. 1. Schematic diagram of pressurized cylinder and experimental set-up for CO_2 treatment under high pressure.

mortality were likewise obtained. This could be attributed to the low temperature —10°C— utilised by Gerard et al.

Experiments were also carried out to determine the effect of the new technique on the viability of rice seeds. Mean lifespan of the seeds was retained at carbon dioxide pressures of 10 and 20 kg/cm^2 for up to 60 minutes exposure while it declined slightly at 30 kg/cm^2 after a 10 minutes duration although this was not significant.

Table 1. Mean percent[a] survival of immature stage of *Corcyra cephalonica* exposed to high pressure carbon dioxide.

	Egg[b]		Larva[b]		Pupa[b]	
	Exposure time		Exposure time		Exposure time	
Pressure (kg/cm^2)	5 min	10 min	5 min	10 min	5 min	10 min
5	53.33	48.33	96.67	96.67	0.00	0.00
10	40.00	33.33	63.00	71.67	0.00	0.00
20	0.00	0.00	0.00	0.00	0.00	0.00
30	0.00	0.00	0.00	0.00	0.00	0.00

[a]Mean of 3 replicates.
[b]Means of control are 61.67%, 100% and 93.33% for egg, larva and pupa, respectively.

Table 2. Mean percent[a] survival of immature stage of *Tribolium castaneum* exposed to high pressure carbon dioxide.

	Egg[b]		Larva[b]		Pupa[b]	
	Exposure time		Exposure time		Exposure time	
Pressure (kg/cm^2)	5 min	10 min	5 min	10 min	5 min	10 min
5	60.00	63.33	76.67	66.67	16.67	16.67
10	75.00	63.33	16.67	13.33	3.33	6.67
20	60.00	23.33	0.00	0.00	0.00	0.00
30	0.00	0.00	0.00	0.00	0.00	0.00

[a]Mean of 3 replicates.
[b]Means of control are 68.33%, 100% and 100% for egg, larva and pupa, respectively.

Conclusion/Recommendation

Survival of any stage of all species herein investigated can be completely suppressed by CO_2 treatment released at a high pressure of 30 kg/cm^2 for 5 minutes.

The mechanism of action of CO_2 released at high pressure on target insects is not yet fully understood. However, it is well known that the solubility of gases increases at higher pressures. Thus, the toxic action of carbon dioxide could have been enhanced by its greater solubility in the insect hemolymph leading to a better penetration of, and possibly greater interaction with, the sites of action. Stahl et al. (1985) explained that the treatment acts by increasing the respiration and solution in intestinal liquids. In addition, it destroys the cell membranes during rapid decompression.

The results obtained further support the strong potential of this novel procedure as a feasible alternative for quick disinfestation of agricultural commodities, in particular for quarantine treatment of exported and imported grain commodities. In addition, this technique poses minimal risk to workers, leaves no harmful residue and is environmentally safe.

References

Annis, P.C. 1987. Towards rational controlled atmosphere dosage schedules: a review of current knowledge. Proceedings 4th International Working Conference on Stored Product Protection, Tel Aviv, Israel, 1986, 128–148.

Bailey, S.W. and Banks, H.J. 1980. A review of recent studies of the effects of controlled atmospheres on stored product pests. Proceedings International Symposium on Controlled Atmosphere Storage of Grains, Rome, Italy 1980. Amsterdam, Elsevier, 101–118.

Calderon, M. and Navarro, S. 1968. Sensibility of three stored product insect species exposed to different low pressure. Nature, 218, 190.

Cline, D.L. and Highland, H.A. 1987. Survival of four species of stored product insects confined with food in vacuumised and unvacuumised film pouches. Journal of Economic Entomology, 80, 73–76.

Gerard, D., Kraus, J. and Quirin, K.W. et al. 1988. (Residue free insect control using natural carbon dioxide under high pressure). Gordian, 88, 90–94.

Jay, E.G. 1986. Factors affecting the use of carbon dioxide for treating raw and processed agricultural products. GASGA Seminar on Fumigation Technology in Developing Countries. London, Tropical Development and Research Institute, 173–189.

Jay, E.G., Banks, H.J. and Keever, D.W. 1990. Recent developments in controlled atmosphere technology. Proceedings of International Conference on Fumigation and Controlled Atmosphere Storage of Grain, Singapore, 1989, ACIAR Proceedings No. 25 134–143.

Locatelli, D.P. and Daolio, E. 1993. Effectiveness of carbon dioxide under reduced pressure against some insects infesting packaged rice. Journal of Stored Products Research, 29, 81–87.

Locatelli, D.P. and Traversa, S. 1989. Confezionamento sottovuoto: valutazione dell'efficacia su infestanti le derrate. Technica Molitoria, 40, 585–589.

Pohlen, W., Rau, G. und Finkenzeller, E. 1989. Erste praktische Erfahrungen mit einem Verfahren zur Druckentwesung mit Kohlendioxid. Pharmazeutische Industrie, 8, 917–918.

Reichmuth, C. 1990. New Techniques in fumigation research today. Proceedings 5th International Working Conference on Stored-product Protection, Bordeaux, France 1990, 2, 709–725.

Stahl, E., Rau, G. and Adophi, H. 1985. Entwesung von Drogen durch Kohlendioxid-Druck-Behandlung (PEX-Verfahren). Pharmazeutische Industrie, 47, 528–530.

The feasibility of increasing the penetration of phosphine in concrete silos by means of carbon dioxide

Y. Carmi, Y. Golani, H. Frandji and E. Shaaya*

Abstract

Treatments with CO_2 (alone or combined with various gases) are considered to be less effective in concrete than metal silos due to reactivity between CO_2 and concrete. However, field studies conducted on wheat in concrete silos up to 44 m in depth showed that fumigation of the grain was effective. Considerable concentrations of phosphine were detected at the bottom of the bins during the 10-day treatment period, and total mortality of the insects, which heavily infested the wheat, was achieved.

Introduction

The use of fumigant mixtures containing phosphine (PH_3) and carbon dioxide (CO_2) has been proved to be effective in treating large masses of grain in deep silo bins. Good penetration of PH_3 to depths of 10, 22 and 37 m has been achieved (Carmi et al. 1990, Carmi and Shaaya, 1992). These experiments were conducted in metal structures with a satisfactory level of gastightness. However, in Israel, large amounts of grain are stored in concrete structures, which are considered to be less suitable for treatments involving CO_2, owing to its absorption by and chemical reactivity with concrete (Snelson 1987; Newman 1989). Application of the advantages of PH_3 + CO_2 mixtures in treatments of concrete structures is desirable. Reichmuth (1993) achieved considerable penetration of phosphine to 18m depth of grain in a concrete silo using a PH_3 + CO_2 mixture, in conditions of less than optimal gastightness.

Experimental

A PH_3 + CO_2 mixture was tested for fumigation of infested wheat in a 44-m deep concrete bin. A dosage of 2 g/m^3 of magnesium phosphide was applied to the top of the bin, followed by the application of 200 g/m^3 CO_2 as dry ice. The fumigation lasted for 10 days, during which gas samples were taken from depths of 0.5 and 2 m below the surface, and from

the bottom of the bin. Phosphine concentrations were determined by the Bedfont EC 80 phosphine monitor, and CO_2 levels were determined by Drager Detector tubes.

Results and Discussion

Considerable amounts of phosphine had already penetrated to the bottom after 24 hours; a peak level of 490 ppm was reached after 48 hours and the final level was 180 ppm after 10 days (Table 1). Though the concentrations of PH_3 in the upper region were higher, the amount that reached the bottom was in the lethal range and was enough to achieve 100% kill of the insects in the grain. Yongsheng (1992) found that 130 ppm of PH_3 mixed with 8–10% CO_2 was effective in control of insects and mites. Hashem and Reichmuth (1989) found that 0.53 mg/L (360 ppm) of PH_3 killed eggs of *R. dominica* in 72 hours. Mueller (1993) found that the combination of 65–100 ppm PH_3 + 4–6% CO_2 at a temperature of 32–37°C was effective in controlling insects in flour mills. In our trial, large amounts of CO_2 penetrated to the bottom, thus making an effective combination with PH_3 concentrations which were already in the lethal range. This trial showed that, in spite of the disadvantages of concrete structures for CO_2 treatments, good results can nevertheless be achieved by using the PH_3 + CO_2 mixture. In future studies optimisation of the treatment should be sought, through better sealing of the concrete surfaces to reduce its effect on the gas.

Conclusion

The method described here has the following advantages
1. An effective strategy for fumigation of deep grain bins.
2. A low dose of phosphine.
3. A single, short and simple application.
4. Inexpensive, with no sophisticated equipment required.
5. Suitable for a wide range of bin structures if they have a fair level of gastightness.

References

Carmi, Y., Golani, Y. and Frandji, H. 1990. Fumigation of a silo bin with a mixture of magnesium phosphide and carbon dioxide by surface application. Proceedings 5th International Working Conference on Stored Product Protection, Bordeaux, France, Vol 2, 767–774.

Carmi, Y. and Shaaya, E. 1992. Penetration of phosphine in deep grain bin by aid of carbon dioxide. Proceedings XIX International Congress of Entomology, Beijing, June 1992.

Hashem, M.Y. and Reichmuth, Ch. 1989. The efficiency of phosphine against eggs of lesser grain borer. Nachrichtenbl. Deut, Pflanzenschutzd. (Braunschweig), 41, 1989.

* The Volcani Center, ARO, Institute for Technology and Storage of Agricultural Products, Department of Stored Products, POB 6, Bet Dagan 50250, Israel.

Table 1. Concentrations of phosphine (ppm) and carbon dioxide (%) in a 44 m deep concrete wheat bin after fumigaion with $PH_3 + CO_2$ mixture: $2 \text{ g/m}^3 + 200 \text{ g/m}^3$.

Depth (m)	Days							
	1		2		3		4	
	PH_3	CO_2	PH_3	CO_2	PH_3	CO_2	PH_3	CO_2
0.5	1750	7.0	1280	4.0	1115	2.0	980	1.5
2.0	730	18.0	690	6.0	1050	2.0	825	1.0
44 (bottom)	270	9.0	490	10.0	360	4.0	180	2.0

Mueller, D.K. 1993. A new method of using phosphine in combination with heat and carbon dioxide. Practical Use of Fumigants and Pheromones. International Technical Conference and Workshop, Lubeck, Germany, 1–3 December 1993.

Newman, C.J.E. 1989. Specification and design of enclosures for gas treatment. In: ACIAR Proceedings No. 25, 108–130.

Reichmuth. Ch. 1993. Controlled atmospheres in stored Product protection. Practical Use of Fumigants and Pheromones. International Technical Conference and Workshop, Lubeck, Germany, 1–3 December 1993.

Snelson, J.T. 1987. Grain protectants. ACIAR Monograph No. 3. Published by the Australian Centre for International Agricultural Research. 448 p.

Yongsheng, Yan, 1992. A study on mixed fumigation of aluminium phosphide and carbon dioxide for controlling stored grain pests. Proceedings XIX International Congress of Entomology, Beijing, June 1992.

Mortality of snails, *Cernuella virgata* and *Cochlicella acuta*, exposed to fumigants, controlled atmospheres or heat disinfestation.

J. Cassells, H.J. Banks and P.C. Annis*.

Abstract

Introduced snails from the Mediterranean region have become a major problem for the grain industry in southern Australia in recent years. Apart from the losses and inconvenience caused by their feeding and effects on harvesting machinery, some snails may become mixed with harvested grain or legumes. This causes downgrading of the commodity or the requirement for further handling to remove the snails. The snails also present a quarantine problem with some export countries. A range of fumigants, controlled atmospheres and heat disinfestation were assessed as methods for killing *Cernuella virgata*, the Mediterranean white snail, and *Cochlicella acuta*, the small conical snail, in stored grain. The snails were exposed while inactive or aestivating, the form in which they are found contaminating harvested grain.

Except for low oxygen atmospheres, snails require a higher dosage rate of all fumigants tested than those needed to eliminate all insect stages in grain. A 100% mortality was obtained by exposures of 5.4 g/m^3 phosphine for 10 days, 30 g/m^3 methyl bromide for 24 hours, greater than 150 g/m^3 carbon disulphide for 24 hours and 10 g/m^3 hydrogen cyanide for 24 hours. Controlled atmospheres of 80% carbon dioxide for 10 days of low oxygen atmospheres of 1% at ≥25°C for 10 days were required for 100% mortality. Heat disinfestation at peak temperatures of 67°C killed all *Cernuella virgata* while only 57°C was required to kill *Cochlicella acuta*. Disinfestation using low oxygen atmospheres or heat appears the most practical for use with a wide range of stored grains. Phosphine is notably ineffective, requiring excessive dosages for extended periods. The snails became more sensitive to fumigants and other treatments with increasing time in aestivation.

Introduction

Introduced snails from the Mediterranean area have become a major agricultural problem in southern and Western Australia (Baker 1989).The Mediterranean white snail, *Cernuella virgata* (da Costa), and the small conical snail, *Cochlicella acuta* (Müller), were first recorded in South Australia (SA) in 1920 at Millicent and 1953 at Minlaton, respectively (Baker 1986, and references therein). In recent seasons these snails have extended their range from their former infestation focus on the Eyre and Yorke Peninsulas of SA into Victoria and New South Wales. In the 1992–93 season, with a dry mid winter and an above average rainfall in spring and early summer, snail numbers have increased to plague proportions.

White snails cause extensive damage to crops through their feeding. The behaviour of the snails increases the chance that they are harvested with the grain. To avoid dehydration and

temperature extremes near the ground in summer the snails climb up the stalks of the crop. They attach themselves to the plant and go into a period of dormant activity called aestivation (Pomeroy 1968). They may then be harvested with the crop causing it to be downgraded or rejected. The snails also make harvesting difficult as they clog machinery. In heavy infestations a third or more of harvested grain or legume may be composed of snails, requiring extensive grain cleaning to give an acceptable grade of commodity. These species of snails are subject to quarantine in many countries and so present a problem in certifying the grain free of pests at the export terminal. Previous work by Richardson and Roth (1965) had shown that aestivating *Cochlicella barbara* (Linnaeus) were very resistant to methyl bromide, hydrogen cyanide, phosphine and other fumigants. It has therefore become critically important to determine what treatments will kill the snails and to provide a range of alternatives to cover a variety of commodities, including those sensitive to particular treatments, such as malting barley.

The response of *C. virgata* and *C. acuta* to fumigants, controlled atmospheres (CA) and heat disinfestation was assessed. Phosphine (PH_3), methyl bromide (MeBr) carbon disulphide (CS_2) and hydrogen cyanide (HCN) were used as fumigants. Carbon dioxide (CO_2) and low oxygen (O_2) controlled atmospheres were tested.

Materials and methods

Snails

In January of 1985 and 1986 approximately 10000 *C. virgata* were collected from Bulgowan, 12 km west of Weetulta, SA. They were collected from fenceposts, stems and stalks of plants and placed for transportation in dry pesticide-free barley. All snails were aestivating at the time of collection.

The *C. virgata* obtained in February 1993 were shipped to Canberra mixed with barley in sacks. The snails were from the Wallaroo and Kadina area, SA and approximately 30000 snails were received. A large proportion of the snails broke aestivation in transit, though became quiescent again in Canberra. *C. acuta* was also available in large numbers in 1993. Approximately 25000 snails from Port Giles, SA were received in February 1993.

On arrival in Canberra, the snails were sorted into classes based on their approximate size and stored in either pesticide-free barley (9% moisture content) in 1985–86, or wheat (10% moisture content) in 1993, at 25°C and ambient relative humidity. The snails were assessed as being alive or dead by appearance and by apparent density, determined by rolling the snail around in the palm of the hand. Dead snails are light and roll differently from live snails. The reliability of this method for determining live snails was approximately 97% for *C. virgata* and 85% for *C. acuta*, as determined by dissection of a sample.

* Stored Grain Research Laboratory, CSIRO Division of Entomology, GPO Box 1700, Canberra, ACT 2601, Australia.

For fumigation tests, a sample size of 27 was used consisting of a ratio for *C. virgata* of 3 small:12 small to medium:9 medium to large:3 large. This ratio reflects the size distribution of the snails obtained in the initial 1985 sample. For *C. acuta* the size mixture was 18 small : 8 medium : 1 large. Snail samples were weighed and placed in 70×50 mm crystallising dishes containing 100 g of 60% r.h. barley or wheat. Each fumigation was conducted at 25°C in a 2.5 L desiccator with a stirrer in the base and a Quickfit cone screwthread adaptor top with a rubber septum. Two randomly selected samples were placed in each desiccator (both of *C. virgata* in 1985–86 experiments and one each of *C. virgata* and *C. acuta* in 1993). Controls were conducted for every fumigation time period. The concentration of the fumigant in each desiccator was measured within an hour of dosing and again just before opening the desiccator at the end of the exposure period.

After all treatments the snails were left overnight to air and removed from the grain the next day. The number of living snails remaining after treatment was determined by wetting the snails lightly with water. This stimulated most of the snails to emerge from their shells within an hour. Emergence of the snails was observed over a 4-hour period and after 24 hours. The snails were stored on petri dishes, with meshed ventilation holes in their lids, at 25°C and ambient relative humidity, typically around 30%. After 14 days the mortality was reassessed using the wetting technique. Those which did not emerge were dissected. Shells with dry and decomposed contents were scored as dead. A very few snails (<15%) were apparently alive, but did not emerge. These were scored as alive for subsequent analysis.

Fumigants

PH_3 was obtained from a concentrated source generated from an aluminium phosphide pellet. The concentration of the source was determined with a Tracor MT–150 gas chromatograph (1.8 m, 80/100 mesh Porapak Q column) using a Gow Mac gas density detector . The dosage rates were 0.8, 1.5, 2.5 and 5.0 g/m^3 PH_3 for each exposure time of 4, 6, 8 or 10 days. The PH_3 concentration in the desiccators was determined using a Tracor MT-220 gas chromatograph (0.82 m 120/150 mesh Porapak Q column) with a flame photometric detector.

The MeBr dosage rates were 6, 12, 24 and 32 g/m^3 for 24 hours and 3, 6, 12 and 16 g/m^3 for 48 hours. The MeBr dose was obtained from the headspace above a liquid source in a sealed vial with a Mininert valve top. The MeBr concentration in the desiccators was measured using a Shimadzu 6AM series gas chromatograph (2 m, 20% OV101 80/100 mesh GasChromQ column) with a flame ionisation detector.

The dosage rates for CS_2 were 25, 50, 100 and 150 g/m^3 for 24 hours. CS_2 was injected as a liquid into the desiccator after a volume of air, equivalent to the volume of the resulting gas, had been removed. The CS_2 concentration in the desiccators was determined using the flame ionisation detector.

The dosage rates for HCN were 5, 10, 20 and 40 g/m^3 for 24 hours and 2.5, 5, 10 and 20 g/m^3 for 48 hours. The HCN was created in the desiccators by the reaction of sodium cyanide with 4M sulphuric acid. A volume of air, equivalent to the volume of the resulting gas, was removed before the sulphuric acid was injected into the sodium cyanide in the desiccator. The HCN concentration in the desiccators was determined, in the 1993 experiments, using a Varian Model 3300 gas chromatograph (15 m, Megabore DB-WAX capillary column) with a thermal conductivity detector.

Controlled atmospheres

Dosage rates for low O_2 treatments were 0.5 and 1.0% O_2 with the balance nitrogen for a fumigation period of 5 and 10 days. These concentrations were obtained by mixing air and nitrogen flows using massflow controllers, Brooks 5850 TR Series with Model 5876 and 5872.6 control units. The gas was passed through a glycerol–water mixture to humidify the air to 60% r.h. and through the desiccators at a flow rate of 100 mL/minute. The experiments were conducted at temperatures of 15, 20, 25 and 30°C.

Dosage rates for CO_2 were 30, 40, 60 and 80% for each fumigation time of 2, 5, 7 and 10 days. The desiccators were dosed by injecting 700 mL CO_2 or larger volumes into the desiccator with the screw cap off to allow displaced air to escape. Experiments were also conducted at temperatures of 15, 20, 25 and 30°C. Low O_2 and CO_2 concentrations were measured using a Fisher Gas Partitioner Model 1200 gas chromatograph (1.8 m, 80/100 mesh Porapak Q, and a 3.3 m, 60/80 mesh Molecular Sieve 13X columns) with a thermal conductivity detector.

Heat disinfestation

For heat disinfestation trials in 1986 and 1993 a sample size of 25 was used for *C. virgata*, with the size ratio of 1 small : 7 small to medium : 10 medium to large : 7 large. In 1993, a sample size of 25 was also used for *C. acuta* with a size ratio of 18 small : 6 medium : 1 large. The snails were counted out, weighed and then placed in 100 g of 60% r.h. barley or wheat overnight. The snails were added to 900 g of barley or wheat before treatment to make a 1000 g sample. The 1000 g samples were heated in a laboratory-scale fluidised bed to 50, 55, 60, 65 or 70°C and then cooled quickly to 35°C (Dermott and Evans 1978). Controls for the experiment consisted of fluidised and non-fluidised samples. Each treatment was conducted in triplicate.

Results

Probit analysis was conducted on the PH_3 and CO_2 results giving LD_{99} for mean concentration × time (CT) dosage as shown in Tables 1 and 2. The results were corrected for control mortality before probit analysis using Abbot's correction (Finney 1971). The relationship between LD_{50} and the length of time in storage of the snails is given Figures 1 and 2. The effect of temperature on the response of the snails to CO_2 is given in Figure 3. This experiment was conducted 120 days after the collection of the snails.

The response of the snails to MeBr is shown in Figure 4. Results for *C. acuta* were suitable for probit analysis, giving an LD_{99} mean CT dosage rate of 16.8 g day/m^3 with 5% upper and lower fiducial limits of 26.6 g day/m^3 and 13.7 g day/m^3 respectively.

The dosage rates of CS_2 below 100 g day/m^3 were unsuccessful in killing all the snails, as shown in Figure 5. The 1985 *C. virgata* results were suitable for probit analysis giving a LD_{99} mean CT dosage rate of 110.5 g day/m^3 with 5% upper fiducial limits of 143.7 g day/m^3 and lower limits of 95.7 g day/m^3.

The effect of HCN on the snails is shown in Figure 6. There was some survival of *C. virgata*, recently collected from the field, even at the high initial dose of 20 g/m^3 HCN (mean concentration of 8 g/m^3) for a fumigation time of 2 days. Sorption of HCN in these experiments was high, with less than 50% of the initial dose remaining in the headspace at the end of the treatment.

Mortality of 100% was obtained at low O_2 concentrations of 0.5 and 1.0% for 5 and 10 days at temperatures 25°C. At 20°C, 4% of the *C. acuta* survived at 1% O_2 while at 15°C survival at 1% O_2 occurred for both snails. *C. acuta* also survived at the

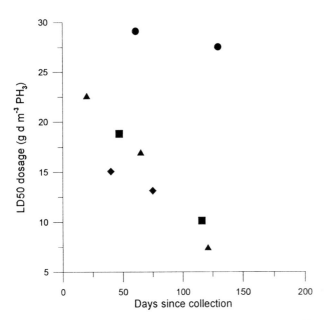

Fig. 1. The relationship between phosphine LD_{50} dosages at 25°C and the length of time in storage of *Cernuella virgata* and *Cochlicella acuta*. *C. virgata* ◆1985, ▲1986, ■1993, *C. acuta* ●1993.

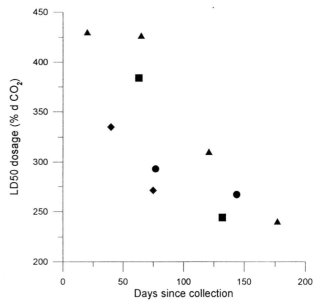

Fig. 2. The relationship between carbon dioxide LD_{50} dosages at 25°C and the length of time in storage of *Cernuella virgata* and *Cochlicella acuta*. *C. virgata* ◆1985, ▲1986, ■1993, *C. acuta* ●1993.

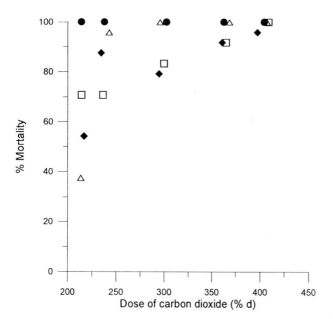

Fig. 3. The effect of temperature on response of *Cernuella virgata* to carbon dioxide, □15°C, ◆20°C, △25°C, ●30°C.

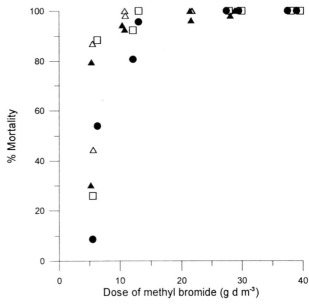

Fig. 4. Response of *Cernuella virgata* and *Cochlicella acuta* to methyl bromide at 25°C. *C. virgata* ▲34 days stored 1986, △76 days stored 1986, □18 days stored 1993, *C. acuta* ● 32 days stored 1993.

0.5% O_2 level. This work was conducted after the snails had been in storage for more than 90 days.

Probit analysis of the heat disinfestation results gave LD_{99} temperatures as shown in Table 3. No additional mortality was incurred by the fluidising action of the method.

Discussion

C. virgata were the most difficult to kill with PH_3 in the 1986 experiment conducted 20 days after collection of the snails. On this occasion a mean concentration of 4.4 g/m³ PH_3 for 10 days would have been required to kill all *C. virgata*. This value is higher than the dosage rate of 0.26 g/m³ PH_3 for 10 days

required to kill all stages of insect grain pests (C. J. Waterford, personal communication 1993). *C. acuta* exhibited an even greater tolerance to PH_3 than *C. virgata*. A higher dosage rate of 5.4 g/m³ PH_3 for 10 days would be required for 100% kill of these snails. Maintenance of such a high PH_3 concentration in a store for 10 days is likely to be impractical on technical, cost and environmental grounds, ruling out phosphine as a possible treatment.

C. acuta and *C. virgata* exhibited similar responses to CO_2. A mean concentration of 80% CO_2 for 10 days or a more easily obtainable level of 60% CO_2 for 15 days would be required to kill all snails, whereas a dosage rate of 70% initially with 35% CO_2 for 15 days will kill all insect stages (Annis 1987). The CO_2 levels would require continuous

Table 1. Exposure of *Cernuella virgata* and *Cochlicella acuta* to phosphine, showing dosages required for 99% mortality and its change with length of time since collection of the snails.

Snail	Year	Days since collection	LD$_{99}$		
			Dosage CT (g.day/m^3)	upper 5% fiducial limit	lower 5% fiducial limit
C. virgata	1985	40	24.7	27.9	22.6
C. virgata	1985	75	22.8	26.1	20.6
C. virgata	1986	20	43.4	55.8	36.7
C. virgata	1986	65	31.6	40.4	26.9
C. virgata	1986	121	15.7	19.8	13.7
C. virgata	1993	47	42.0	51.4	36.4
C. virgata	1993	116	30.1	41.3	24.7
C. acuta	1993	61	53.4	67.2	45.7
C. acuta	1993	130	50.8	63.7	43.8

Table 2. Exposure of *Cernuella virgata* and *Cochlicella acuta* to carbon dioxide, showing dosages required for 99% mortality and its change with length of time since collection of the snails.

Snail	Year	Days since collection	LD$_{99}$		
			Dosage CT (% d)	upper 5% fiducial limit	lower 5% fiducial limit
C. virgata	1985	40	735.0	1198.9	576.0
C. virgata	1985	75	632.9	988.5	501.6
C. virgata	1986	20	800.2	907.6	725.2
C. virgata	1986	65	786.0	927.2	698.2
C. virgata	1986	121	647.5	794.8	565.8
C. virgata	1986	177	731.8	1411.5	553.5
C. virgata	1993	63	796.0	917.5	715.1
C. virgata	1993	132	610.1	745.6	531.6
C. acuta	1993	77	574.2	667.0	515.0
C. acuta	1993	144	556.5	702.0	480.2

Table 3. Peak heat disinfestation temperatures required for 99% mortality of *Cernuella virgata* and *Cochlicella acuta*.

Snail	Year	Days since collection	LD$_{99}$		
			Peak grain temperature (°C)	upper 5% fiducial limit	lower 5% fiducial limit
C. virgata	1986	55	66.7	67.9	65.7
C. virgata	1993	74	62.7	64.6	61.9
C. acuta	1993	89	57.0	60.0	55.4

Table 4. Summary of treatment options for obtaining 100% kill of *Cernuella virgata* and *Cochlicella acuta* in a stored product environment.

Treatments	Dosage
Rapid treatments	
Methyl bromide	CT = 720 g.hour/m^3 or 30 g/m^3 for 1 day (average concentration)
Hydrogen cyanide	CT = 240 g.hour/m^3 or 10 g/m^3 for 1 day (average concentration)
Heat disinfestation	67°C (peak temperature)
Slower treatments	
Carbon dioxide	60% for 15 days (constant)
Low oxygen	1% at 25°C for 10 days
Impractical treatments	
Phosphine	CT = 1296 g.hour/ m^3 or 5.4 g/m^3 for 10 days (average concentration)
Carbon disulphide	CT = 3600 g.hour/m^3 or 150 g/m^3 for 1 day (average concentration)

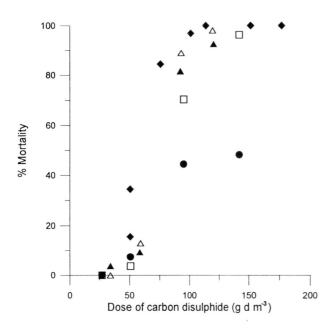

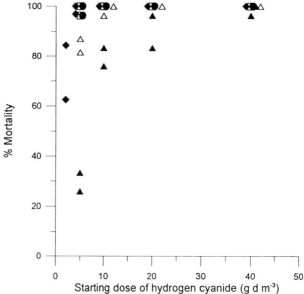

Fig. 5. Response of *Cernuella virgata* and *Cochlicella acuta* to carbon disulphide at 25°C. *C. virgata* ◆ 95 days stored 1985, ▲ 34 days stored 1986, △ 76 days stored 1986, ☐ 18 days stored 1993, *C. acuta* ● 32 days stored 1993.

Fig. 6. Response of *Cernuella virgata* and *Cochlicella acuta* to hydrogen cyanide at 25°C. *C. virgata* ◆ 102 days stored 1985, ▲ 34 days stored 1986, △ 76 days stored 1986, ☐ 95 days stored 1993, *C. acuta* ●109 days stored 1993.

topping up to compensate for leakage at present Australian standards of sealing of CA stores but are technically feasible.

The increase in sensitivity of the snails to PH_3 and CO_2 with the length of time that they are held in storage was found in each year's collection of snails. The LD_{50} dosage rates (Figs 1 and 2) gave the best representation of the mortality trends in the different experiments. This response presents an option for fumigating the grain later in the storage period when the snails would be more sensitive to the treatment. The scattering of the LD_{50} values with different years reflects the condition of the snails before collection.

The MeBr, CS_2 and HCN experimental results with a smaller number of dosage rates were not suitable for probit analysis. A reasonable estimate of the dosage rates required to kill all snails can be determined from the graphs. MeBr requires a mean dosage rate of 30 g/m^3 for 1 day (*CT* product of >720 g.hours/m^3). This level is about 4 times higher than required to eliminate common insect grain pests and is likely to give unacceptable germination loss from malting barley (Strong and Lindgren 1959). The highest dosage rate of CS_2 used obtained a 100% kill for only the longest stored *C. virgata*. *C. acuta* were also very tolerant to CS_2 suggesting that concentrations well in excess of a mean 150 g/m^3 for 1 day would necessary to obtain 100% kill. HCN would require, as shown in Figure 6, an initial dose of at least 40 g/m^3 for fumigation period of 1 day (mean dosage rate of 10 g.days/m^3 after accounting for sorption losses on the grain). This dosage rate would be difficult to obtain in the field due to the high rate of sorption of HCN (Lubatti and Harrison 1944). The snails exhibited increased sensitivity to these fumigants with time in storage, a response also observed by Richardson and Roth (1965), with *C. barbara*. The dosage rates required are also higher than those used for insect pests of grain. The snails' greater tolerance to fumigants may be due to the adaptive mechanisms that they use to prevent dehydration. These mechanisms, the epiphragm that they secrete across the mouth of the shell (Barnhart 1983) and the reduction of respiration rate and metabolism with aestivation (Godan 1983), may aid in reducing fumigant uptake.

Heat disinfestation will successfully kill all *C. virgata* at a temperature of 67°C. This temperature is slightly higher than

the 65°C required to kill insects (Evans and Dermott 1981). The smaller species *C. acuta* was killed at a lower temperature.

At lower temperatures, the snails, like insects (Banks and Annis 1990), are less susceptible to CO_2 and low O_2. This is a factor that would have to be taken into account when determining the duration of fumigation. Oxygen concentrations of 1% will kill all snails at temperatures 25°C. Given that the time required to kill all insect stages is 4 weeks at a grain temperature of 23°C and an O_2 of 1.2% (Banks and Annis 1990) this procedure would give 100% snail mortality.

Conclusion

The results of this study suggest a range of treatment options for *C. virgata* and *C. acuta*. Selection from within these options will be based on the availability of fumigants, industrial safety, level of sealing, cost of treatment, time available and commodity being fumigated. The treatments are summarised in Table 4.

Acknowledgments

This work was funded under the Stored Grain Research Laboratory Agreement with the Australian Wheat Board and the Bulk Handling Authorities, with additional funding from the Australian Barley Board. The authors would like to thank Geoff Baker, Barry Kelly and Don Hawk for their advice and assistance with snail collection.

References

Annis, P.C. 1987. Towards rational controlled atmosphere dosage schedules: a review of current knowledge. In: Donahaye, E. and Navarro, S., ed., Proceedings of the Forth International Working Conference on Stored-product Protection, Tel Aviv, Israel, September 1986, 128–148.

Baker, G.H. 1986. The biology and control of white snails (Mollusca: Helicidae), introduced pests in Australia. CSIRO Division of Entomology, Technical Paper No. 25.

Baker, G.H. 1989. Damage, population dynamics, movement and control of pest helicid snails in southern Australia. In: Henderson, I., ed., Slugs and snails in world agriculture. Monograph No. 41:

Proceedings British Crop Protection Council Symposium Guilford, UK, 10–12 April 1989, 175–185.

Banks, H.J. and Annis, P.C. 1990. Comparative advantages of high CO_2 and low O_2 types of controlled atmospheres for grain storage. In: Calderon, M. and Barkai-Golan, R., ed., Food preservation by modified atmospheres, 93–122.

Barnhart, M.C. 1983. Gas permeability of the epiphragm of a terrestrial snail, *Otala lactea*. Physiological Zoology, 56, 436–444.

Dermott, T. and Evans, D.E. 1978. An evaluation of fluidised-bed heating as a means of disinfesting wheat. Journal of Stored Products Research, 14, 1–12.

Evans, D.E. and Dermott, T. 1981. Dosage-mortality relationships for *Rhyzopertha dominica* (F.) (Coleoptera: Bostrychidae) exposed to heat in a fluidised bed. Journal of Stored Products Research, 17, 53–64.

Finney, D.J. 1971. Probit analysis. Cambridge, University Press, 333 p.

Godan, D. 1983. Pest slugs and snails: biology and control. Berlin, Springer, 445 p.

Lubatti, O.F. and Harrison, A. 1944. Determination of fumigants. XVII. Comparison of the sorption of hydrogen cyanide, ethylene oxide, trichloroacetonitrile, and methyl bromide by wheat. Journal of the Society of Chemical Industry, 63, 353–359.

Pomeroy, D.E. 1968. Dormancy in the land snail, *Helicella virgata* (Pulmonata : Helicidae). Australian Journal of Zoology, 16, 857–869.

Richardson, H.H. and Roth, H. 1965. Methyl bromide, sulfuryl floride, and other fumigants against quarantinable *Cochlicella* and *Theba* snails. Journal of Economic Entomology, 58, 690–693.

Strong, R.G. and Lindgren, D.L. 1959. Effect of methyl bromide and hydrocyanic acid fumigation on the germination of barley. Journal of Economic Entomology, 52, 319–322.

Application of pressure-swing absorption (PSA) and liquid nitrogen as methods for providing controlled atmospheres in grain terminals

J. Cassells*, H.J. Banks* and R. Allanson†

Abstract

The use of controlled atmospheres as a method for providing pest-free grain is becoming more popular with increasing market demand for grain free of residues. Two systems for providing low oxygen atmospheres were trialed at GrainCorp export terminals in New South Wales, Australia. The pressure-swing absorption system (PSA), at the Port Kembla terminal, and a liquid nitrogen-based supply at the Newcastle terminal. The test bins at both terminals were sealed to a level that gave greater than 5 minutes for a pressure half-life (full bin). They were originally equipped for fumigation with methyl bromide under recirculation.

The PSA unit, with a gas output of 1.47 m^3/minute at 0.7% oxygen, purged two 13 660 m^3 bins in series down to an oxygen concentration of <1%. The first bin in the series was purged down to 1.5% oxygen in 7 days. The concentration in both bins was maintained for 19 days below 1.5% and typically below 1.0% by gas provided on a pressure demand system. Due to the good sealing of the bins, the maintenance flow rate corresponded to the calculated maintenance rate required to compensate for gas loss due to diurnal temperature changes and resulting pressure cycling.

Vaporised liquid nitrogen was used to purge three 1954 m^3 bins to an oxygen concentration of <1% throughout. The efficiency of purge was increased by the use of a higher flow rate. A flow rate of 6.02 m^3/minute purged a bin down to <1% oxygen in 3.5 hours. The concentrations in the bins were maintained below 2% with a flow rate of about 50 L/minute. Individual bins required slightly different maintenance rates and entry position of the maintenance gas at either the top or base of the bin to hold the concentration in them below 1.0%. The maintenance rates were higher than those required to compensate for gas loss due to diurnal temperature changes alone, suggesting other leakage factors such as wind were involved in gas loss.

Both the liquid nitrogen and PSA systems were successfully demonstrated as practical systems for CA generation in the grain terminals, though the PSA system was of insufficient output to provide the required atmosphere rapidly. On the basis of a 4-week exposure at <1% oxygen, the resulting cost, including hire charges of the two systems, was $A0.34/t for the PSA system and $A0.65/t for the liquid nitrogen system. Although the material cost of phosphine fumigations is lower ($A0.20/t) the operational advantages of nitrogen in the terminals make it the preferred option. Liquid nitrogen is now being used routinely at the Newcastle terminal to provide controlled atmosphere disinfestation for 29 000 t of storage capacity.

Introduction

Nitrogen, carbon dioxide or a mixture thereof can be added to grain storages in a controlled fashion to displace or dilute the storage atmosphere to give a gaseous composition which is insecticidal. Further gas may be added to maintain this composition. The technique is known as controlled atmosphere (CA) storage.

It is well established (Banks and Annis 1990) that atmospheres containing less than 2% oxygen are capable of disinfesting stored grain. The technology of application of such low oxygen (O_2) CAs, using liquid nitrogen as a gas source, was developed in Australia and Italy in the 1970s. Commercial field trials were successfully carried out (Banks 1979; Tranchino et al. 1980) but these did not lead to the adoption of the technique. The remarkable success of phosphine fumigation (recirculation, surface application and SIROFLO®) displaced the need for nitrogen atmospheres in Australia, with phosphine fumigations generally both cheaper and easier to apply than nitrogen treatments. The sealed technology and standards for grain stores developed for nitrogen-based CA technology proved useful in improving phosphine technology.

With the increasing difficulties and regulations faced by fumigants, for example the anticipated phasing out of methyl bromide due to it being an ozone-depleting material, it is appropriate that CA systems, including nitrogen-based atmospheres, be re-evaluated. Nitrogen-based CA has several operational advantages over fumigants that may make its use preferable in certain circumstances.

These include:
- no use of acutely toxic materials
- little or no airing required before grain can be moved
- reduced hazard area in comparison with fumigants
- compatible with 'organic' and truly residue-free grain
- no sorption problems
- no environmental restrictions or obligatory monitoring
- capable of being produced from air on-site.

The trials detailed here were carried out collaboratively between GrainCorp Operations Limited and the Stored Grain Research Laboratory at two grain export terminals in New South Wales, Australia. These trials aimed to demonstrate the application of nitrogen atmospheres, to gather local costing and performance data, and specifically:
- to confirm the quantity of nitrogen required to create and maintain storage atmospheres at <1% O_2 throughout the planned 28-day exposure; and
- to determine what modifications would be required to rig a silo block for routine commercial use with nitrogen-based CA.

Two methods of nitrogen generation were assessed: the pressure-swing absorption system (PSA) at the Port Kembla

* Stored Grain Research Laboratory, CSIRO, Division of Entomology, GPO Box 1700, Canberra ACT 2601, Australia.
† GrainCorp Operations Limited, P.O. Box A268, Sydney South, NSW 2000, Australia.

terminal and the liquid nitrogen-based supply at the Newcastle terminal. The bins treated at both these terminals were sealed to specification for methyl bromide fumigation.

The general design of the process, flow rates and subsequent calculations adopted for the trials follow those of Banks and Annis (1977).

Materials and Methods

Pressure-swing absorption system trial — Port Kembla

The trial was conducted on two 10000 t capacity white-painted steel bins, Bins A1 and A2, at the Port Kembla terminal. The bins contained Australian Standard White wheat from the 1990–91 harvest season. The capacity of the bins, contents and pressure decay times are given in Table 1. These bins were equipped for methyl bromide fumigation by recirculation, fitted with a pressure relief valve and required no additional sealing.

Table 1. Dimensions, tonnages and pressure tests for PSA syste trial, Bins A1 and A2, Port Kembla terminal.

Parameters	Bin A1	Bin A2
Bin volume (m^3)	13660	13660
Bin height (m)	35.9	35.9
Rated wheat capacity (t)	11576	11576
Wheat tonnage (t)	9392	9958
Actual wheat volume (m^3)	11083	11750
Headspace volume (m^3)	2577	1910
Total free air in bin (est. m^3)	6789	6375
Full bin pressure decay time, 2–1.5 kPa (minutes)	80	30
Grain moisture content (%, w.b.)	9	9

The PSA unit, hired from The Commonwealth Industrial Gases Pty Ltd, consisted of an air compressor and dryer, an air receiver tank, two absorber beds, and a nitrogen receiver tank. In the absorber beds, under pressure, nitrogen is separated from the oxygen by a carbon molecular sieve. The rated output from this unit was 1.33 m^3/minute at 0.8% O_2 concentration, but slightly increased performance was obtained in practice (1.47 m^3/minute at 0.7% O_2).

The two bins were purged in series. The PSA outlet was connected to the annular diffuser, originally installed to distribute methyl bromide, in the base of Bin A2 using 2" spiral wound PVC hose. The outlet at the top of Bin A2 was connected to the top of Bin A1, with Bin A1 venting from the diffuser at the base of the bin. The flow from the outlet of the PSA unit was measured using Rotameters and a Datametrics Model 800-LM hot-wire flow meter. The calculated purge time for Bin A2, following Banks (1984), was 9.8 days, assuming free mixing in the headspace and plug flow in the bulk, and taking into account the 86% filling ratio of the bin. Bin A1 was expected to require a longer purge time as some dispersion of the purging front would probably occur during passage through Bin A2.

After the initial purge period, the outlet from Bin A1 was closed and the PSA system was fitted with a regulating system to add gas on demand to maintain a fixed minimum pressure in the bins. The outlet regulator of the unit was set at 1.9 kPa, allowing gas into the bins when the gas pressure dropped below this value. The pressure levels in the bins were measured by a micromanometer (P.P.F.A.–EC060) connected to an outlet in the base of Bin A1. The oxygen concentrations in the bins were monitored for a further 8 days after shutdown of the

PSA unit to determine the natural leakage rate of air into the bins.

Liquid nitrogen based trial — Newcastle

The trials were conducted in Bins W1, X1 and X3, situated on the south-eastern corner of the Newcastle terminal. The bins contained wheat from the 1991–92 harvest season. Capacities, tonnages and pressure tests for each bin are given in Table 2. The bins were constructed of reinforced concrete. They were part of a 4×8 nested cell block, with each trial bin comprising part of the exterior wall. These bins were designed to be of gastight construction, but they had not been tested or used as sealed bins since commissioning in 1972.

Table 2. Dimensions, tonnages and pressure tests for liquic nitrogen based trial, Bins W1, X1 and X3, Newcastle terminal.

Parameters	Bins W1	Bin X1	Bin X3
Bin volume (m^3)	1954	1954	1954
Bin height (+ conical section, m)	27.7	27.7	27.7
Rated wheat capacity (t)	1655	1655	1655
Wheat tonnage (t)	1506	1506	1526
Actual wheat volume (m^3)	1780	1780	1800
Headspace volume (m^3)	175	175	152
Total free air in bin (est. m^3)	850	850	836
Full bin pressure decay time, 1000–500 Pa (minutes)	20	5	11
Grain moisture content (%, w.b.)	10.0	9.2	10.0 (est)

Bins W1 and X1, equipped for recirculation fumigation with methyl bromide, each had an exterior return duct and diffusers for the gas on the inside of the bin cone. These required no modification to accept the nitrogen gas, except for fitting an inlet to the duct work between the closed valve leading to the recirculation fan and the diffusers. Bin X3 had no internal duct work. Nitrogen was introduced directly into the grain via a 2" diameter inlet fitted above the slide on the outloading valve. A small shedder plate was welded to the valve body on the inside to prevent blockage of the inlet by grain. The wall-to-roof and the wall-to-floor joins in Bins W1 and X1 were sealed with acrylic sealer (Duraflex), but Bin X3 was untreated. All bins met or exceeded the specified 5 minute (full) pressure half-life test.

Each bin was purged by introducing nitrogen gas into the base of the bin. Liquid nitrogen was vaporised in an electrically driven forced draft heat exchanger (max. capacity 16 m^3/minute gas) before being fed to the bins in 2" PVC pipe-work. Nitrogen flows were regulated by a gate valve downstream of the vaporiser.

Bins W1 and X1 were purged in parallel at a relatively slow purge rate (~3 m^3/minute each). The input flows were monitored by a Rotameter (W1) or Datametrics 800-LM hot-wire flow meter (X1). Bin X3 was purged at about 6 m^3/minute, with an Annubar pitot tube to measure flows. About 1.3 m^3 of N_2/t of grain was expected to be required to purge the 90% full bins (following Banks and Annis 1977). The amount of liquid nitrogen required to purge a 91% full, 1954 m^3 bin, at 15°C, was calculated to be 2.3 t.

During the purge, the valve at the top of the bin was left fully open until the headspace oxygen concentration had fallen below 15%. The valve was then partially closed so as to create a small back pressure of about 50 Pa. This prevented influx of air under the windy conditions experienced during the purge and consequently decreased purge times. The purge was discontinued and the bin sealed when the headspace concentration fell below 2%. At this point the atmosphere in the grain bulk was < 1% O_2 throughout.

Gas for concentration maintenance was taken directly from the boil off of the storage vessel. This vessel was equipped with a pressure-raising circuit to maintain a set pressure above the stored liquid. Initially, the maintenance gas for Bins X1 and W1 was piped via 15 mm diameter garden hose into the inloading valve at the top of each bin. Maintenance flow was measured by a variable cross-section flow meter (Gapmeter). During the later part of the maintenance on these bins the flow was switched to the base of the bins. Bin X3 maintenance was applied via the base of the bin. The oxygen concentrations in Bins W1 and X1 were monitored for a further 12 days after shutdown of maintenance gas to determine the natural leakage rate of air into the bins. Bin W1 was left sealed, but Bin X1 had the 2" diameter gate valve at the top bin opened.

Gas and temperature monitoring systems

The gas in the bins was sampled through 3 mm O.D. nylon tubing. The sampling points in the bins for the two trials are shown in Table 3. Each line was sampled sequentially through a series of solenoid valves. The lines were purged using a diaphragm pump for a 12-minute period to ensure complete flushing. The oxygen concentration was measured by a Neotronics Otox 90 meter and at the end of the sampling period the results were recorded on a computer and a chart recorder. As a calibration check, lines were run from a nitrogen cylinder (instrument grade) and from ambient air. A flow meter was placed in line to check for any blockages.

Temperatures were measured using type T copper–constantan thermocouples. A thermocouple was laid at the same point as each gas line. The ambient temperature was measured at a point under the eves of the monitoring shed. The temperatures were recorded on a Data Electronics Datataker DT100 data logger.

Table 3. Gas and temperature sampling point positions in the bins at the Port Kembla and Newcastle terminals.

Port Kembla terminal	Newcastle terminal
Headspace, 2/3 down from bin top	Headspace, 2/3 down from bin top
3 m into grain	$1/3$ down from bin top
Bottom of bin	$1/3$ up from bin bottom
Bin A1 only, at bottom ring	Bottom of bin

Results

Pressure-swing absorption system trial

The PSA system purged the bins at a flow rate of 1.47 m^3/minute and an oxygen concentration of 0.7%. The first bin in the series, Bin A2, was purged down to 1.5% oxygen in 7 days. The purging efficiency, calculated on basis of plug flow in the grain bulk and free mixing in the headspace (E_3, of Banks 1979), of Bin A2 to a 1.5% O_2 concentration was 64%. Bin A1 had been purged at all sample points to below 1.0%, with a single point in the bin base, furthest from the gas inlet, at <3% after 36 days of purging. The maintenance phase was then started and continued for 19 days. Over the maintenance phase the O_2 concentration of the PSA output gas decreased to 0.5% with a flow rate averaging 352 m^3/day. The changing oxygen concentration in the bins is shown in Figure 1. The times taken for each position to reach an oxygen concentration

of 1.5% and the final equilibrium O_2 concentrations at the end of the maintenance phase are given in Table 4.

Table 4. Time taken for the PSA system to reduce each sampling position in Bins A1 and A2 to an oxygen concentration of 1.5% and equilibrium oxygen concentrations obtained at the end of the trial.

Sampling points	Time to 1.5% [O_2] (days)	Equilibrium [O_2] (%)
A2 – bottom	9.8	0.5
A2 – 3 m into grain	1.9	0.5
A2 – headspace	7.1	0.5
A1 – headspace	10.4	0.5
A1 – 3 m into grain	11.6	0.5
A1 – bottom	37.5	1.17

Figure 2 depicts the flow of gas into the bins during the maintenance phase and the corresponding changes of pressure within Bin A1. On most days the flow was cut off by the pressure-sensing system at about 9.00 a.m. and recommenced between 3.00 and 4.00 p.m. Over the maintenance phase the running hours were approximately 7.8 hours/day, as determined from monitors on the compressor and measurements of the compressor motor temperatures.

The average temperatures in the bins over the trial period at the various sampling points are shown in Table 5. Temperature swings in the headspace, the calculated ventilation rates and actual compensatory maintenance rates required to cover gas leakage associated with these temperature swings (Banks and Annis 1984) are given in Table 6.

The weighted average of the various sampling points in the bin was taken to give a bin average O_2 decay rate. The decay rates, calculated as described by Banks (1984), and the time it would take each position to increase from 1% to 5% oxygen concentration, are given in Table 7.

Liquid nitrogen trial

The flow rates and the resulting purge data and efficiency are given in Table 8. At the end of purge the concentration of oxygen in the headspace of Bins W1 and X3 was 1.1% and of Bin X1, 2.7%. The change in oxygen concentrations in Bin X3 during the purge is shown in Figure 3. The other bins gave similar purging profiles. The resulting O_2 concentrations during the maintenance phase and the changes in flow rates and entry position for the maintenance gas into the bins are shown in Figure 4.

The average temperatures over the trial period are given in Table 9. The headspace temperature swings and the calculated gas loss and maintenance rate to compensate for this in Bin W1 are given in Table 6.

The bin average oxygen decay rates and the time for the O_2 concentration to increase from 1 to 5% are given in Table 7.

Table 5. Average temperatures in the Port Kembla terminal Bins A1 and A2 over trial period.

Sampling points	Temperatures (°C)	
	Bin A1	Bin A2
Headspace	15.6	15.5
3 m into the grain	23.3	20.0
Bottom	15.9	16.2
Ring	16.5	–

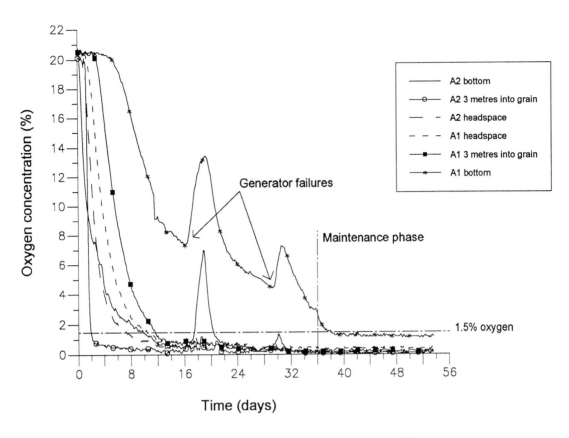

Fig. 1.   Oxygen concentrations in Bins A1 and A2, Port Kembla terminal, during purge and maintenance.

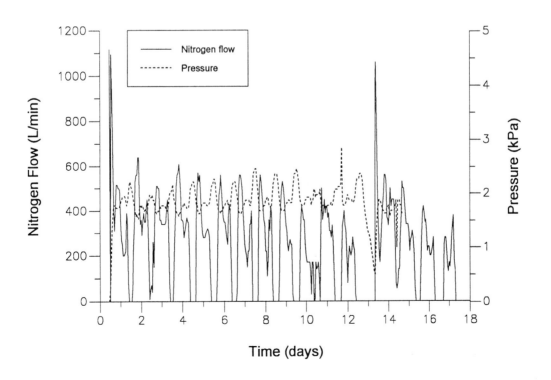

Fig. 2.   Pressure in Bin A1 and flow rate from PSA system during the maintenance phase, Port Kembla terminal.

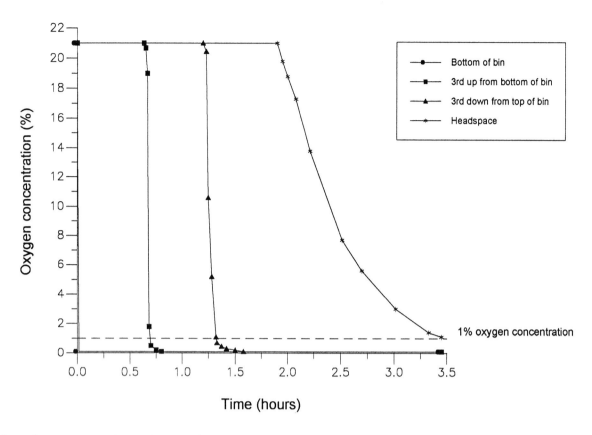

Fig. 3. Oxygen concentrations during purge of Bin X3, Newcastle terminal.

Table 6. Temperature factors involved in gas loss, gas loss rates expected, and observed and calculated maintenance rates.

Factors	Port Kembla Bin A1	Newcastle Bin X1
Average temperature decrease in headspace (°C)	11.3	8.3
Maximum temperature decrease in headspace (°C)	21.5	12.7
Equivalent ventilation rate (%/day)	1.5	0.59
Maximum ventilation rate (%/day)	2.9	0.90
Gas loss from temperature swings (av. m^3/day)	102	5.0
Gas loss from temperature swings (max. m^3/day)	195	7.7
Calculated maintenance rate (m^3/day)[a]	306–585	15–23
Observed maintenance rate (m^3/day)	352	65

[a]Maintenance rated expected, calculated as 3× gas loss rate from temperature swings. This is equivalent approximately to that expected assuming free mixing in the headspace of air leaking in and no effects in the grain bulk.

Table 7. The bin average oxygen decay rates in the Port Kembla terminal Bins A1 and A2 and Newcastle terminal Bins W1 and X1, and the time taken for the oxygen concentration in the bins to increase from 1 to 5%.

Sampling position	O_2 decay rate/day	Days to 5% [O_2]
A1	0.006	37.2
A2	0.020	11.1
W1 (sealed)	0.021	10.7
X1 (valve at top open)	0.027	8.4

Table 9. Average temperatures in Newcastle terminal Bins W1, X1 and X3 over trial period.

Sampling points	Temperatures (°C)		
	Bin W1	Bin X1	Bin X3
Headspace	16.3	15.9	18.9
1/3 down from bin top	–	18.0	19.5
Bottom	15.2	15.3	17.4
Ambient	14.0	14.0	17.6

Table 8. Purge data for the liquid N2 trial, Newcastle terminal Bins W1, X1 and X3

Bin No.	Purge rate (m^3/minute)	Purge time (h:min)	Liquid nitrogen consumed (t)	Purging efficiency (E$_3$, %)
W1	2.71	10:28	2.02	70
X1	2.98	10:28	2.22	55
X3	6.02	3:27	1.48	91

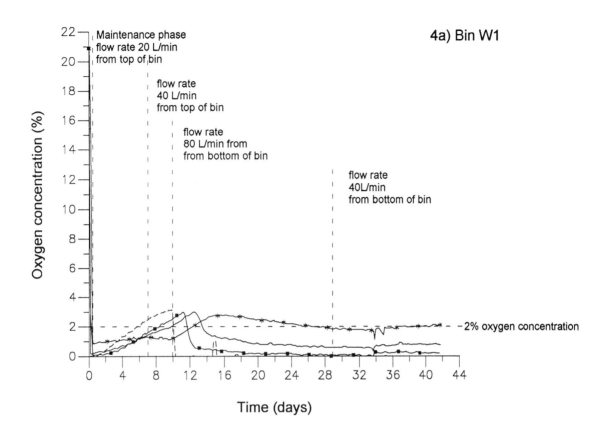

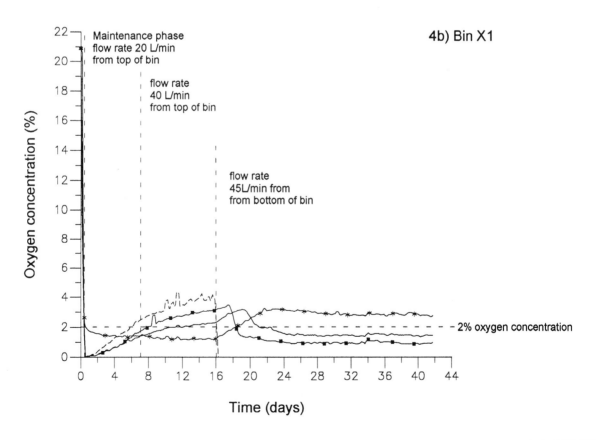

Fig. 4. Oxygen concentrations and position of maintenance gas line during maintenance phase in bins at the Newcastle terminal: (a) Bin W1;
(b) Bin X1; and (c) Bin X3. — — — Bottom of bin, —■— 1/3 up from bottom of bin, ———— 1/3 down from top of bin, —*—
headspace.

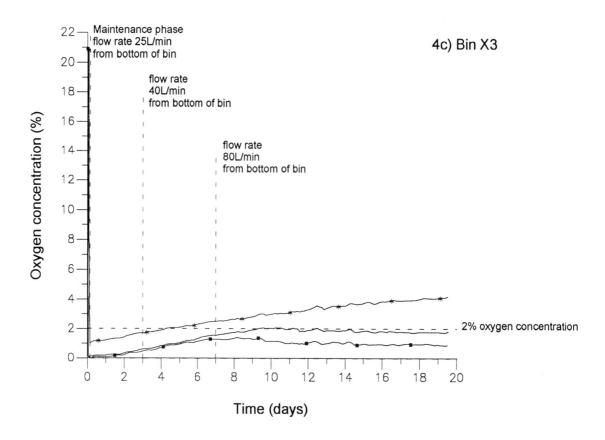

Fig. 4. – Cont'd

Discussion

The PSA unit purged the bulk of Bin A2 (the first bin to be purged) down to 0.7% O_2, close to the time expected (observed 11.5 days, calculated 9.8 days). However, in areas below the entry point of the gas, where mixing was slower or leakage occurred, the reduction of the O_2 concentration took longer. The low purging efficiency, 64%, indicates a substantial degree of mixing occurred in the grain bulk, presumably resulting from the low purge rate. Generator failures during the trial allowed the influx of air through the inlets and outlets raising the oxygen concentration and delaying the lowering of oxygen to specified levels in the bottom of Bin A1. Levels below 1.5% were eventually reached at all points in both bins (Fig. 1).

Maintenance using a pressure-demand system kept the O_2 concentration below 1% in all areas that had reached this level. The setting of the regulating system at 1.9 kPa was the lowest that could obtained with the equipment available. A lower pressure setting (e.g. 100 Pa) may have been sufficient to maintain the low O_2 levels. As the bins were well sealed the maintenance rate was within the calculated rate ($3\times$ loss from temperature swings (calc.), i.e. assuming free mixing in headspace and no effect on the bulk concentration) to compensate for gas loss due to temperature swings in the headspace (Table 6).

The PSA unit output was insufficient to purge two large bins (10000 t capacity each) rapidly and was not sized optimally for the application. It is not known what effect a slow reduction of oxygen has on insect tolerance to low oxygen atmospheres. For simple logistic reasons more rapid purging

would have been preferred. Nevertheless, the trial showed, for the first time, that PSA systems could be used for generating N_2 CA atmospheres in grain storages if required.

The purge of the Newcastle bins using liquid nitrogen gave the expected pattern with dispersed plug flow in the bulk and some mixing in the headspace. The higher flow rate used in Bin X3 resulted in reduced mixing in the headspace and consequent increase in the efficiency of purging (Table 8). Increased efficiency with increased purging rate has been noted previously (Tranchino et al. 1980; Banks et al. 1980).

The initial input of the maintenance gas through the top of the bin was found in the case of Bins W1 and X1 to be unsatisfactory, with better results obtained from adding the N_2 at the base of the bin. A flow of >40 L/minute was required to keep the O_2 concentrations constant. Unlike Bins W1 and X1, even a higher maintenance rate of 80 L/minute from the base of Bin X3 still resulted in a rising headspace O_2 concentration. The mixing of the air leaking into the headspace with the low O_2 atmosphere resulted in a higher maintenance rate being required to flush out that extra volume of gas. Maintenance of this bin from the top may be more appropriate with the air leaking in being directly displaced by the maintenance gas. Leakage in Bins W1 and X1 may have been greater from the base of the bin than the headspace and also reduced by the sealing carried out on the wall to roof and wall to floor joins.

Although the concrete bins passed the pressure test of a minimum of 5 minutes they were less gastight than the steel bins (see Tables 1 and 2). Ventilation rates due to temperature swings in the headspace would be an important component in determining maintenance rates. The actual maintenance rates required were greater than the calculated rate, solely due to temperature swings in the headspace. Other factors such as wind are expected to contributed to gas leakage and additional nitrogen input for maintenance (Banks and Annis 1984).

As oxygen levels of <18% become dangerous to humans, for safety reasons the bins were vented on completion of the trial. Although some (minimal) precautions are necessary to ensure adequate O_2 around CA treatment, the 'risk area' is usually small. The low oxygen decay rates, especially in the steel bins, would have allowed the grain to be protected for some time after shut down of maintenance. The venting of the grain would then occur naturally at outloading.

The setting up of the bins for CA required only a small capital outlay. Apart from the need initially for a well-sealed storage, the basic requirements were ducting, flow meters, a pressure sensor and an oxygen sensor to detect when the outlet from the bin has reached the required O_2 concentration to determine the end of purge. Up to a level where all leakage resulted from temperature and barometric cycling, expenditure in increasing the gastightness of the bin would be returned in reducing the maintenance rates.

The resulting costs of the two systems, taking into account hire charges and extrapolating running costs for a treatment time of 4 weeks [recommended dosage time at 23°C and 1% O_2 (Banks and Annis 1990)], were \$0.34/t for the PSA system and \$0.65/t for the liquid nitrogen system. In comparison with phosphine, at \$0.20/t for materials, the use of N_2 CAs is more expensive. However, this cost comparison does not take into account the cost of monitoring and logistic restrictions associated with phosphine fumigation. With these factors taken into account, use of nitrogen compares favourably with phosphine at the export terminals treated in these trials.

Conclusion

These trials were conducted at terminals where there is a particular requirement for a reliable supply of insect-free grain, to avoid the large costs associated with disruption of supply and demurrage incurred when insects are detected in a terminal on outloading. Both processes of atmosphere supply, liquid nitrogen and PSA systems, were demonstrated as feasible, though the more rapid pulldown achieved by the liquid-based system was clearly more convenient in the situations the two systems were used. Nitrogen CA treatments have been considered unsuitable in export terminals, but this is not so. Nitrogen treatments inherently require longer than phosphine or methyl bromide fumigations to achieve 100% kill of all stages of insect pests of stored products and do not allow quick throughput of grain through the terminal if applied only after infestation is detected. However, with correct management it is possible to apply such CA treatments even where timing can be critical. Treatments are applied immediately a particular parcel of grain is received, before loading to ship, thus maximising chances of a successful outturn in insect-free condition.

The demand for residue-free grain, increasing regulations on the use of fumigants and availability of a method for fumigating commodities that are not recommended for treatments with phosphine or methyl bromide have all been factors in the consideration of N_2 treatments in situations were timing can be critical. The location of the terminals close to liquid nitrogen supplies, avoiding large transportation costs, and the availability of on-site nitrogen generators have made the use of CA more attractive. Although CAs are more expensive than phosphine fumigation in material costs, the changing market and environmental climate have resulted in low oxygen CA becoming a more acceptable alternative.

On the basis of this work, the Newcastle terminal management set aside 29 000 t of storage which is regularly fumigated using controlled atmospheres. Over 60 000 t of grain have been treated and exported from the terminal to date without stoppage for presence of insects on inspection of grain as it is loaded to ship.

Acknowledgments

The authors wish to thank the GrainCorp Operations Limited staff at the Port Kembla and Newcastle terminals for arrangement of facilities and assistance during these trials.

References

Banks, H.J. 1979. Recent advances in the use of modified atmospheres for stored product pest control. In: Proceedings of the Second International Working Conference on Stored-product Entomology, Ibadan, Nigeria, September 1978, 198–217.

Banks, H.J. 1984. Modified atmospheres—generation using externally supplied gases. In: Champ, B.R., and Highley, E., ed., Proceedings of the Australian Development Assistance Course on the Preservation of Stored Cereals, 2, 544–557.

Banks, H.J. and Annis, P.C. 1977. Suggested procedures for controlled atmosphere storage of dry grain. CSIRO Australian Division of Entomology Technical Paper No. 13, 23 p.

Banks, H.J. and Annis, P.C. 1984. Importance of processes of natural ventilation to fumigation and controlled atmosphere storage. In: Ripp, B.E. et al., ed., Controlled atmosphere and fumigation in grain storages. Amsterdam, Elsevier, 299–323.

Banks. H.J. and Annis, P.C. 1990. Comparative advantages of high CO_2 and low O_2 types controlled atmospheres for grain storage. In: Calderon, M. and Barkai-Golan, R., ed., Food preservation by modified atmospheres. Roca Baton, Florida, CRC Press, 93–122.

Banks, H.J., Annis, P.C., Henning, R.C. and Wilson, A.D. 1980. Experimental and commercial modified atmosphere treatments of stored grain in Australia. In: Shejbal, J., ed., Controlled atmosphere storage of grain. Amsterdam, Elsevier, 207–224.

Tranchino, L., Agostinelli, P., Costantini, A. and Shejbal, J. 1980. The first Italian large scale facilities for the storage of cereal grains in nitrogen. In: Shejbal, J., ed., Controlled atmosphere storage of grain. Amsterdam, Elsevier, 445–459.

Fumigation of a 7000 t bulk of wheat with phosphine using the Phyto-Explo[1] system to assist gas circulation

B. Chakrabarti[*], C. R. Watson[†], C. H. Bell[*], T. J. Wontner-Smith[*] and J. Rogerson[†]

Abstract

A 7000 t bulk of wheat in an improvised store was treated with phosphine at the rate of 1.9 g/t following the sinking of three pairs of temporary shafts into the grain using the Phyto-Explo pneumatic hammer device. Each pair of shafts was linked via plastic (land drainage) piping to a fan which was operated for the first four days after dosing to push air through the shafts and assist gas distribution.

After 12 days a further 0.5 g phosphine/t was added adjacent to the foremost fan and to the ventilation ducts opening at the rear of the store, bringing the total dosage to 2.4 g/t, still substantially less than the normal recommended dosage of 3–5 g/t for floor-stored grain.

Apart from one corner in the direct path of the wind, where concentrations were held down, and one sample point giving high readings near a dosing position, concentration–time products after 15 days did not vary by more than a factor of 3, ranging from 88 to 249 g.hours/L. Such dosages would achieve a very high level of control of storage insects over this timescale, even at low temperature.

Introduction

Cereal grain grown in U.K. is stored on farms, in bins from 100–300 t capacity and floor-stores of 200–800 t. Deep silo bins of 250–2000 t and floor-stores from 500–60000 t capacity are used by grain merchants and for the storage of grain by the Intervention Board. Purpose-built floor-stores are a comparatively recent development. None of these structures is designed for gas retention. Often a building constructed for a quite different purpose is used for grain storage. Aircraft hangars or old munitions factories are examples of buildings used for bulk grain storage. They have many disadvantages. Internal stanchions are often present at regular intervals throughout the building and create problems in covering the grain surface for fumigation.

The treatment of grain in ships' holds in transit between one country and another is also widely practised and offers further problems in achieving adequate distribution of gas throughout grain depths of up to 30 m (Redlinger et al. 1979, 1982). For bulk commodities such as grain, oilseeds etc. phosphine is now the most commonly used fumigant worldwide. If the structure is reasonably gastight, a successful fumigation, though dependent on various factors, i.e. temperature, sorption, gas distribution, weather conditions etc., should destroy all the target pests including pre-adult stages.

Due to the diverse nature of storage structures, each storage facility demands a different approach to dosing and there may be further problems in achieving and maintaining an adequate concentration of phosphine over the period necessary for the control of insects. Using standard sealing methods it is difficult to maintain phosphine concentrations above the minimum level for effective action for the 16 days required at temperatures below 15°C to control all stages of *Sitophilus granarius* (L.), the most phosphine-tolerant species of the common grain pests (Anon 1984; Hole et al. 1976). Also, in a large bulk of floor-stored grain, phosphine concentrations tend to build up slowly at deeper levels. It is very common for a small upward draught in a bulk of grain to be apparent during treatments. Although it has been claimed by the manufacturers of aluminium phosphide preparations that phosphine travels 3 m/day up to 20 m deep in a grain silo, in practice this hardly ever seems to be achieved. To enable a successful treatment to be carried out, infested grain would have to be turned from one bin to another while phosphide preparations are being added, and to retain phosphine for the requisite period, every effort is necessary to seal the bin top, bottom and all the cracks and crevices in the sides. This practice for bin fumigation is, however, no longer encouraged due mainly to the high concentrations of phosphine that can rapidly develop in parts of the workplace during application, and the powdery residues that remain in the grain.

A novel method of dosing has been developed by 'Desinsectisation Moderne' for use in deeper grain bulks (tall silos, ship's holds etc.). The process is called the Phyto-Explo System and is a patented method for the introduction, distribution and maintenance of phosphine or any other fumigant in bulk cereals (Vacquer and Vacquer 1991). This technique has been investigated in recent trials in the U.K.

Experimental

In the Phyto-Explo process an expandable, corrugated shaft of about 63 mm diameter, wholly or partially perforated, is slid over a metal pipe and fitted to a metal probe which is introduced into the grain using a pneumatic hammer. When it reaches the desired depth, the probe is withdrawn leaving the shaft expanded in position (Fig. 1). Aluminium phosphide tablets or pellets are put in dust-retaining nylon socks and introduced into the shaft. Phosphine generated in the shaft spreads in the grain mass through the perforations in the shaft and, if the seal is adequate, will spread evenly throughout the bulk. For a large bulk of grain in a floor-store, a number of shafts in a matrix may assist distribution (Igrox 1993). A further refinement has been the use of fans to draw phosphine

---

[1] Phyto-Explo® is the registered trade name and patented system of Desinsectisation Moderne (D.M.) France, 34 Rue du Contrat Social, Rouen 76000, France.

[*] Central Science Laboratory, Ministry of Agriculture, Fisheries and Food, London Road, Slough, Berks SL3 7HJ, U.K.

[†] Igrox Ltd, White Hall, Worlingworth, Woodbridge, Suffolk IP13 7HW, U.K.

generated on the surface into the grain mass, with the objective of achieving a more rapid and even distribution of gas.

In a recent trial, a 7000 t bulk of wheat in a floor-store was treated with phosphine using the Phyto-Explo system. The grain bulk was contained by corrugated iron walls on three sides, sloping down to the floor at the front. There did not appear to be any sealant at the joints, and the walls were far from gastight. The outer wall of the store was constructed of corrugated aluminium sheeting and there was a gap of 0.5 m from the grain retaining walls. The surface of the bulk was very uneven with several peaks and troughs and the depths of grain ranged from 3–8 m.

Altogether 6 shafts were introduced into the grain from the surface, each with only the bottom half metre perforated, 4 in the front half and the other 2 in the back half of the grain. Each shaft was positioned halfway between the wall and the centre of the store (Fig. 2). The shafts were linked in pairs to a fan with the air-inlet attached to a perforated suction pipe laid along the ridge of the grain under the sheeting.

The grain mass was dosed along the centre of the ridge with 'Detia' bag-chains at the rate of about 1.9 g of phosphine/t of grain, and was covered with 150 micron polythene sheeting with the fans underneath. The edges of the sheeting were buried under the grain all the way round and joints were rolled and then stapled together.

After dosing, the use of the fans to inject phosphine was continued for 4 days and the gas concentrations were monitored on-line from 23 positions using a Hewlett Packard 5880 gas chromatograph fitted with a flame-photometric detector and housed in a purpose-built mobile laboratory. The sheeting near the ridge of the bulk was slit open after 12 days and an additional dose of about 0.5 g/t was introduced to the area adjacent to the four shafts near the slope, and to the ventilation ducts at the rear of the store to extend the treatment period. Then the bulk was resealed, the fans connected to the four shafts were switched on and the test was continued up to 15 days.

The results obtained were compared with those obtained in an earlier program investigating application methods for fumigating grain with phosphine (Bell et al. 1991).

Results

A typical data set for concentration–time (*CT*) products of phosphine obtained from fumigating a small bulk of grain are presented in Table 1. In spite of dosing both at the grain surface and below the bulk in the ventilation ducts, the gas concentrations and *CT*s obtained varied widely.

A very different result was obtained in the treatment of the larger bulk using the Phyto-Explo system where some gas circulation was provided by fans and the pattern of sunken shafts. In spite of the surface dosing method, concentrations of phosphine were generally lower near the surface. However, the distribution of phosphine throughout the bulk occurred rapidly, concentrations ranging from 6×10^{-4} g/m^3 to 3.8 g/m^3 within 10 hours and reaching at all points a threshold level of 0.05 mg/L within 15 hours. By 55 hours the phosphine level ranged between 0.5 g/m^3 and 2.5 g/m^3. After 8 days, phosphine levels at positions near the windward edge of the store started dropping below the threshold level. In normal circumstances, redosing would occur after 7 days. In this case, to create a worst case scenario and to see how long phosphine would remain within the bulk without further dosing, fumigation was continued and redosing was carried out only on the 12th day when the levels of gas at most positions had dropped below or near the threshold (Table 2).

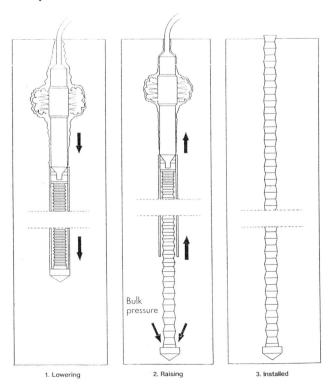

Fig. 1. Principle of fumigation shaft installation.

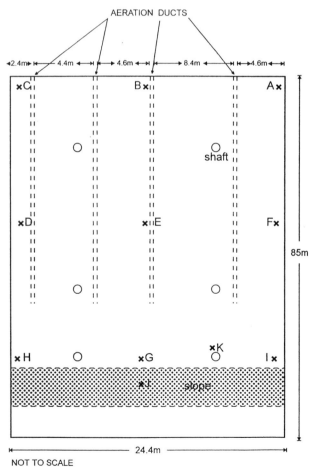

Fig. 2. Gas sampling and shaft positions with aeration ducts, marked with dotted lines.

Table 1. A typical trial on 900 t of barley with both surface and in-duct dosing at the rate of 5g/t using retrievable bag formulation (after Bell et al. 1991).

Gas sampling positions in a rectangular bulk	Depth in grain (m)	Grain temperature (°C)	Corrected % mortality of *S. granarius*	CT product (gh/m^3)
Top of slope, at front of bulk	0.0 (surface)	6.1	99	154
	1.0	—	100	200
	2.0 (bottom)	7.0	100	88
Centre of bulk	0.0 (surface)	7.7	100	367
	2.0	12.7	100	416
	4.0 (bottom)	13.0	100	477
Back of bulk in corner	0.0 (surface)	5.6	11	Trace only
	1.0	—	18	Trace only
	2.5 (bottom)	8.1	0	Trace only
On diagonal between centre and front corner	0.0 (surface)	5.4	100	317
	2.0	8.2	100	318
	4.0 (bottom)	—	100	314
Back of bulk near rear door	0.0 (surface)	—	76	82
	3.5 (bottom)	—	0	Trace only
Centre at side of bulk	0.0 (surface)	—	98	21
	3.0 (bottom)	—	98	21
Centre part way down front slope	0.0 (surface)	—	—	312
	1.0	7.9	100	—
	1.5 (bottom)	—	—	71

Table 2. Phosphine concentrations at different positions on the 12th day before redosing and on the 15th day before the termination of the test using the Phyto-Explo system.

Position	Depth from surface (m)	Phosphine concentration (g/m^3)	
		12th day	15th day
A	3.25	0.148	1.080
	1.0	0.197	0.309
B	4.5	0.032	0.890
	1.0	0.003	0.328
C	3.5	0.002	0.176
	1.0	0.002	0.335
D	4.0	0.001	0.064
	1.0	0.002	0.070
E	6.0	0.038	0.025
	3.0	0.036	0.006
F	3.5	0.200	0.045
	1.0	0.182	0.035
G	6.0	0.019	0.116
	3.0	0.022	0.028
H	4.0	0	0.033
	1.0	0	0.177
I	4.0	0.083	0.011
	1.0	0.063	0.213
J	4.0	0.015	0.209
	0.5	0.017	0.004
K (highest peak)	0.5	0.245	0.007
In duct.	—	0.236	1.149

Table 3. Concentration-time products of phosphine obtained after 7, 12 and 15 days in the test using the Phyto-Explo system, dosing at the rate of 2.4 g/t of grain.

Position	Depth from surface, m	CT product (gh/m³)			Remarks
		7 days	12 days	15 days	
A	3.25	133	186	249	
	1.0	66	104	123	
B	4.5	152	186	211	
	1.0	136	157	166	
C	3.5	103	104	230	
	1.0	88	90	108	
D	4.0	105	107	116	
	1.0	83	86	88	
E	6.0	134	166	167	
	3.0	117	138	139	
F	3.5	116	157	161	
	1.0	109	149	152	
G	6.0	79	101	106	Line blocked
	3.0	159	193	205	
H	4.0	59	59	68	
	1.0	43	47	53	
I	4.0	135	171	184	
	1.0	126	141	153	
J	4.0	102	156	168	
	0.5	11	25	26	Line damaged, later restored
K (highest peak)	0.5	350	420	423	
In duct		160	198	232	

The highest *CT* product was recorded near the highest peak of the bulk at position K (Table 3), probably because of its proximity to a dosing point. Positions C, D and H were all on the windward side of the store and recorded lower concentrations and *CT* products. The lowest *CT* product was recorded near the surface at position J but it was suspected that the sampling line to this point was damaged.

Discussion

From the data it can be seen that good gas distribution can be achieved by the Phyto-Explo System. To retain phosphine at the desired level for the required period to kill tolerant strains without redosing, a more gastight store than the one investigated here would be necessary. However, the system does allow low doses of phosphine to be used effectively in less than gastight situations. The six shafts passing gas drawn from the grain surface to the bottom of the bulk enabled gas to distribute rapidly throughout the bulk. Fewer shafts, say three or four, may well have been sufficient, as the gas distribution in the rear of the store where only two shafts were present was just as good as at the front.

The *CT* products achieved after 15 days were sufficient to achieve control of tolerant grain pests at all positions except position H. This was the corner position most affected by the wind prevailing throughout much of the trial. However, as the grain temperature ranged up to 20°C at several points, the expected level of survival would have been minimal.

The Phyto-Explo System could also prove useful for localised hot-spot treatments but no work has been done in this area. For whole bulks, if the required dosage is applied in two halves at an interval of 7 days, the phosphine concentration may be maintained for the necessary period. Uneven distribution of phosphine and insufficient exposure period are the most common causes for ineffective treatments which may lead to the development of resistance. Both factors can be tackled using the Phyto-Explo System.

References

Anon. 1984. Standard 18. Phosphine fumigation of stored products. EPPO Bulletin, 14, 598–599.

Bell, C. H., Chakrabarti, B. and Mills, K. A. 1991. New application methods for the use of phosphine to disinfest bulk grain. HGCA Research Report No. 41, Home-Grown Cereals Authority, London.

Hole, B. D., Bell, C. H., Mills, K. A. and Goodship, G. 1976. The toxicity of phosphine to all developmental stages of thirteen species of stored product beetles. Journal of Stored Products Research, 12, 235–244.

Igrox Ltd. 1993. Stored grain fumigation with phosphine and the Phyto-Explo System. Worlingworth, Suffolk, 20 p.

Redlinger, L. M., Leesch, J. G., Davis, R., Zettler, J. L., Gillenwater, H. B. and Zehner, J. M. 1982. In-transit shipboard fumigation of wheat on a tanker. Journal of Economic Entomology, 75, 1147–1152.

Redlinger, L. M., Zettler, J. L., Leesch, J. G., Gillenwater, H. B., Davis, R. and Zehner, J. M. 1979. In-transit shipboard fumigation of wheat. Journal of Economic Entomology, 72, 642–647.

Vacquer, B. and Vacquer, F. 1991. Phyto-Explo: granular bulk penetrating tool. In: Fleurat-Lessard, F. and Ducom, P., ed., Proceedings of the Fifth International Working Conference on Stored-product Protection, Bordeaux, France, September 1990, III, 1831.

Technical study of controlled atmosphere with carbon dioxide in brick silo for safe storage of wheat

Cheng Fu-chang, Mei Bao-Liang, Yu Jian, Dou He-tong and Tang Shun-gong*

Abstract

Carbon dioxide was piped into a sealed brick silo until well distributed in the grain mass. The silo atmosphere was recirculated until there was an even distribution of 70% carbon dioxide in the grain mass. Fourteen days after ventilation was stopped, over 35% CO_2 still remained in the grain mass. Using this method, wheat free of insect and moulds can be safely stored.

Introduction

In 1971 Pujidao Grain Storage, Tianjin carried out the first CO_2 treatment of bulk paddy rice in China. The method involved treatment of the bulk under tarpaulins. It has been in use since then on bulk grain. The technical study described here aimed to provide information on treatment with CO_2 of grain held in brick silo bins. This had not been investigated previously in China. The study was carried out using two bins known as the 'test' and 'control' bins. The test bin was extensively modified and the control bin was sealed but not modified. Information from Australia on similar treatments was not directly applicable as it referred to much larger-scale applications.

Work started in 1990 to make the test silo bin sealed and to equip it with a multifunctional gas ducting system. In both the trial and control bins, CO_2 was applied through the duct at the base of the bin, forcing the air out of the bin, until the concentration was over 70%. The study, completed in 1992, met the assessment criteria of a panel of scientists and fulfilled the Chinese requirements for grain stores.

Materials and Methods

Structure of the silo bins

Figure 1 gives a general view of the silo bins at Junliangchen Mill. The brick silo bins are 14 m high, 6.63 m inside diameter and 7.45 m outside diameter. They are equipped with steel bottom cones with a 0.3 m diameter grain outlet. The bin roof is of concrete with a rectangular manhole, 0.56 × 0.57 m, and an aeration vent of 0.46 m diameter. The ductwork fitted was multifunctional, allowing recirculation of fumigant, aeration and fumigant application.

In the sealing trial the inner wall of the test bin was rendered with two coats of a mixture of sand, hydrated lime and cement. Three further coats were applied using a mixture of cement with a little sand. The control bin was coated, inside and out,

Fig. 1. Brick silos in Junliangchen Mill.

with two layers of the latter mixture. The outer wall of the test bin was also rendered with two coats of this mixture. The holes in the roof were all fitted with easily removable steel plates (0.005 m thickness) bedded onto 0.03 m thick silicone rubber and secured with nuts.

Testing for gastightness

Table 1 gives the results of the gastightness tests after sealing.

Grain

The test bin contained 220 t hard red wheat, 11.0% m.c., inloaded on 4 August 1992. The control bin contained 220 t soft red wheat, 11.0% m.c., inloaded on 5 August 1992.

Test equipment used

Centrifugal blower (CQ18–I), multifunctional duct system (see Figure 2), CO_2 gas detector (CYES–IIO2), Air sampler (CD–I).

* Tianjin–Junliangchen Mill, Tianjin 300301, China.

Table 1. Test results, pressure tests to determine gastightness.

	Initial pressure (mm H_2O)	Pressure limits (mm H_2O)	Decay time (seconds)
Test bin	302	200–100	103
	302	100–50	98
	302	50–25	99
Control bin	288	200–100	67
	288	100–50	56
	288	50–25	58

Fig. 2. A part of the multifunctional duct system.

CO_2 addition

After aeration using the duct system, CO_2 was introduced into the base of each bin, displacing the bin atmosphere gradually through the aeration vent in the roof. It was planned to seal the gas inlet and outlet when the concentration in the bin reached 70% CO_2.

In the test silo bin 422.5 kg was introduced on 8 August 1992 at a rate of 0.73 kg/min. When the 7 sampling points detected an average CO_2 level of >80% the inlet and outlet were sealed. After 120 hours the CO_2 level had dropped below 35% at some points. Supplementary CO_2 was added into the headspace on 15 August (109.5 kg). On 21 August (218 kg) was added with recirculation to keep the CO_2 concentration above 35% for 14 days.

In the control silo bin, 586.5 kg of CO_2 was added on 7 August into the duct at the base of the bin at a rate of 0.79 kg/min, with the bin sealed when the average of the 7 sampling points exceeding 80%. After 38 hours the CO_2 concentration had dropped below 35% at (250 kg) was added on 9 August. Because of the rate of loss of CO_2, no further gas was added.

Results and Analysis

The results of CO_2 analysis in summary (Table 2) showed that the Australian standard for CO_2 Controlled Atmosphere (CA) was met in the test bin (i.e. 70% or greater initially, with no less than 35% 10–14 days later).

Results from the control bin in Table 2 showed that this bin could not reach the Australian standard.

Details of CO_2 levels at the 7 sampling points in the test bin are given in Table 3. In this bin the highest concentration was 90.7% and the lowest 82.3% after purging, a difference of 8.4%. The concentrations in the control bin (Table 4) were less even with 93% and 81% as highest and lowest concentration after purging, a difference of 12%.

Table 3, test bin results, showed that when the average at the 7 sampling points was over 35% the maximum was 67.2%, the minimum was 24.6% and the difference 42.6%. The corresponding values for the control bin (Table 4) were 85.6%, 20.6% and 65.0% respectively. This shows that there was a more even distribution of CO_2 in the test bin than in the control bin.

The quantity of CO_2 lost from the control bin was high at high CO_2 concentration (43.5%) and lower at low concentration (31%). Corresponding figures for the test bin were 30.4% and 9.1% respectively. This shows that it is important not to add CO_2 to give too high a concentration initially.

Recirculation was very important as indicated by data in Table 3 for before and after recirculation.

CO_2-based CA can safely protect grain without grain or environmental contamination.

With CO_2-based CA applied to 30 silo bins, avoiding downgrading of the grain, it is calculated that ¥144 000 per year can be saved.

The reason why the trial silo bin retained CO_2 better than the control one, as shown by results given above, is because it was better sealed. The difference, caused by the high density of CO_2 relative to air, was more apparent at low CO_2 concentrations than at high concentrations.

Conclusions

Brick silos with a pressure half life of 103 seconds and equipped with sealing and ductwork as described can be used to store grain with CO_2 CA technology.

Table 2. Summary of CO_2 treatment and results.

Bin	Sealing treatment		First CO_2 addition		Second CO_2 addition		Third CO_2 addition	
	Render on inner and outer walls—number of coats	Extra render on inner wall—number of coats	Quantity (kg)	Average CO_2 concentration (%)	Quantity (kg)	Average CO_2 concentration (%)	Quantity (kg)	Average CO_2 concentration (%)
Test bin	2	3	422.5 (8 Aug)	87.0	109.5 (15 Aug)	42.1	218	42.4
				69.9 after 26 hours		34.3 after 72 hours	(21 Aug)	36.5 after 43 hours
				47.5 after 96 hours				
Control bin	2	0	586 (7 Aug)	87.1	250 (9 Aug)	68.7	–	–
				62.9 after 24 hours	(9 Aug)	35.2 after 70 hours		
				49.1 after 48 hours				

Table 3. CO_2 concentrations (%) in test and control bins.

Time	Sampling point							Average
	1	2	3	4	5	6	7	
Test bin								
1500, 8 Aug. 1992	83.1	89.8	90.1	90.7	83.4	82.4	89.8	87.02
1700, 9 Aug. 1992	48.3	67.2	91.8	90.4	46.0	65.3	80.0	69.85
1500, 10 Aug. 1992	33.8	53.3	88.1	90.1	33.1	53.2	72.0	60.51
1500, 11 Aug. 1992	27.5	43.3	81.0	86.1	27.2	43.3	67.2	53.65
1500, 12 Aug. 1992	21.6	36.7	74.7	84.2	19.7	36.2	59.5	47.51
1500, 13 Aug. 1992	18.5	31.9	69.3	78.7	18.2	32.0	54.1	43.24
1500, 15 Aug. 1992[a]	13.3	23.7	69.4	82.3	33.0	23.1	31.9	42.10
1500, 16 Aug. 1992	28.5	30.2	62.4	71.9	26.9	28.7	39.4	41.14
1500, 17 Aug. 1992	24.7	29.0	56.3	67.2	24.6	26.9	39.3	38.28
1500, 18 Aug. 1992	21.7	24.4	51.2	63.5	19.0	24.4	36.7	34.30
1500, 19 Aug. 1992	9.5	24.2	49.4	61.5	15.1	23.4	32.9	32.78
1500, 20 Aug. 1992	17.0	21.8	45.5	52.9	17.2	22.0	33.9	30.02
0800, 21 Aug. 1992	15.7	21.0	45.2	52.5	15.8	20.9	32.2	29.04
1000, 21 Aug. 1992	20.2	19.5	20.7	50.9	19.0	19.0	18.0	23.90
1500, 21 Aug. 1992[b]	19.9	20.1	93.3	93.7	17.6	18.0	93.0	50.80
1900, 21 Aug. 1992	38.5	38.6	35.2	73.7	41.5	35.8	33.8	42.44
1000, 22 Aug. 1992	35.6	38.5	38.5	59.2	36.5	33.1	36.8	39.74
1000, 23 Aug. 1992	32.5	33.9	40.8	48.3	31.4	33.0	35.8	36.52
1000, 24 Aug. 1992	28.0	31.5	39.7	44.4	28.6	30.1	34.4	33.81
1000, 25 Aug. 1992	24.5	28.9	38.7	42.3	24.1	27.9	32.9	31.32

[a]109.5 kg added into headspace

[b]Reading after 218 kg addition and 2 hours recirculation

Time								
Control bin								
1500, 7 Aug. 1992	81.0	88.0	92.0	93.0	86.0	83.5	86.0	87.07
1500, 8 Aug. 1992	47.8	61.5	74.3	90.3	60.2	48.1	58.1	62.90
1500, 9 Aug. 1992	38.0	35.4	67.4	91.3	45.5	29.7	36.5	49.11
1700, 9 Aug. 1992[a]	42.5	5.0	92.3	90.0	70.8	44.6	67.3	68.71
1500, 10 Aug. 1992	34.1	45.9	77.1	2.1	46.1	34.0	44.3	55.37
1500, 11 Aug. 1992	25.6	33.8	60.2	91.3	33.2	25.4	33.0	43.21
1500, 12 Aug. 1992	21.4	23.9	49.0	85.6	23.6	20.6	22.5	35.22
1500, 13 Aug. 1992	18.5	17.9	39.4	78.5	17.6	18.2	18.4	29.78
1500, 14 Aug. 1992	16.1	15.9	33.0	71.6	15.7	16.3	15.9	26.30
1500, 15 Aug. 1992	13.2	13.4	29.3	67.7	12.5	13.6	13.3	23.28
1500, 16 Aug. 1992	4.3	13.0	12.4	62.1	12.5	12.3	12.3	18.41
1500, 17 Aug. 1992	3.0	10.5	22.6	56.8	10.4	11.0	11.0	17.90
1500, 18 Aug. 1992	6.4	8.7	19.2	49.4	10.0	10.4	10.2	16.32
1500, 20 Aug. 1992	2.6	8.5	14.6	38.2	8.8	9.2	9.2	13.01
1500, 22 Aug. 1992	1.3	8.0	15.4	35.5	7.4	7.6	7.5	11.81
1000, 23 Aug. 1992	1.1	7.3	14.1	33.1	6.7	7.0	7.0	10.90

[a]250 kg CO_2 added into headspace

The study showed that the test bin achieved >70% CO_2 and still had >35% CO_2 after 14 days. The study cost ¥22010 for materials and ¥2200 for labour.

References

Grain and oil storage edition group. 1987. China National Financial and Economics, Grain and Oil Storage, 226–253.

Rui Sun, Lianghua Guan, Chengguang Zhang, Chunhe Liu and Jianhua Lui. 1993. Ducting systems for silo fumigations in Tianjin municipality, China. In: Navarro, S. and Donahaye E., ed., Proceedings of an International Conference on Controlled Atmospheres and Fumigation in Grain Storages. Winnipeg, Canada, June 1992. Caspit Press Ltd., Jerusalem, 343–351.

Zhou Jing-Xing and Jiang Yong-Jia. 1985. Grain storage technology. Peoples Republic of China, Zheng Zhou Grains College Library, 85–87.

Improved procedures for fumigation of oaten hay in shipping containers

C.P.F. De Lima, R.N. Emery and P. Jackson*

Abstract

Oaten hay is compressed and placed in ISO shipping containers for export as animal feed. The standard procedure for fumigation using schedules for grain does not provide complete control, and therefore investigations were carried out on fumigation procedures specific to oaten hay.

The trials reported here were carried out under summer and winter conditions in 12.2 m (40 ft) ISO shipping containers of 67 m^3 volume. Over 100 trials were conducted with and without tarpaulins depending on the level of gastightness of the containers. The fumigants tested were methyl bromide and phosphine.

New fumigation protocols were developed and tested. The results show that effective control is achieved by the uniform dispersal of fumigant throughout the container. There have been no rejections of hay fumigated under the new protocols despite a 10-fold increase in exports since 1987.

Introduction

Export of oaten hay (*Avena sativa*) to Japan is a highly promising industry for Western Australia. It has shown a ten-fold increase from an experimental tonnage of 477 t in 1987 to over 48000 t in 1993. The market potential for Australian hay is assessed at 250000 t/annum.

Oaten hay is shipped to Japan for animal feed. The hay is compressed ('double dumped') to improve the weight/volume ratio for shipping in 6.1 m (20 ft) or 12.2 m (40 ft) containers. The most common reason for the rejection of Australian hay has been the presence of live insects, mainly pests of stored grain.

The hay is routinely fumigated before export using phosphine or methyl bromide. These fumigations were not always successful and caused significant monetary losses due to high treatment costs in Japan. The cost of a fumigation in Australia is approximately $70 per container, compared with a rejected container in Japan at $1300/container.

Industry therefore requested investigation of the reasons for failure and development of more effective treatments.

The work program was implemented to meet industry requirements, and the outputs reflect the demands made by industry and other relevant bodies at the time of the research work.

---

* Department of Agriculture, 3 Baron-Hay Court, South Perth, Western Australia.

Existing protocols

Methyl bromide

The application of methyl bromide involves introducing the gas (after a measured volume of the liquid has passed through a heat exchanger) into the container through a slightly opened door. Containers are fumigated in this manner under tarpaulins either singly or in batches of five. The dose rate is 32 g/m^3 for an exposure period of 24 hours. This dose rate should give a *CT* product of 200 g.hours/m^3

Phosphine

Phosphine fumigation requires the placement of 10 aluminium phosphide sachets giving a dose rate of 1.68 g/m^3 (1206 ppm) in the front, near the door, before the container is sealed. Fumigations are done under tarpaulins on concrete pads with an exposure period of 7–10 days.

Materials and Methods

In this paper, only trials conducted in 12.2 m (40 ft), 67 m^3 containers in 'double dumped' bales are reported. These bales are 500 mm long × 400 mm wide × 400 mm high and weigh 42–45 kg depending on moisture differences and compression by the baler. Each container carries approximately 588 bales giving a load of 24.7–26.5 t/container. The gross product volume is 47 m^3 under the most tightly packed conditions, achieving a load factor of 70%.

A purpose-built PVC application pipe was used for administration of both methyl bromide and phosphine sachets. The pipe is the same length as the container and has 4 mm diameter holes drilled at 100 cm intervals starting 200 cm from the door end of the container. The pipe was introduced along the top of the bales through the centre of the container and left in place for the duration of the fumigation.

The trials were conducted in two series. In the first series, containers packed with hay ready for export were fitted with 3 nylon gas sampling lines along the front, middle and rear end of each container. The required treatments were applied, and observations made as frequently as the working environment allowed. The container was released for shipment on the immediate request of the exporter.

In the second series, containers were specially rented for experimental use. Fifty gas sampling lines were introduced into each container to measure gas concentrations uniformly throughout the container; 15 positions in the airspace between bales and 35 positions in the centre of bales. Each container was 'partitioned' into 5 sections along the length and 3 along the width giving a total of 15 sections, which when further partitioned into top, middle and bottom gave 45 units of sampling, each of approximately 1.48 m^3. The 30 sampling units along the length on each side of the container had gas probes placed in the centre of the bales, while the central units had 5 probes placed in bales occupying the 'centre' position along the entire length of the container. Thus, in each of the 5 cross-sectional partitions of the container, there were 10 sampling lines, 7 in bales (top, middle and bottom) and 3 in air. Temperatures were measured by placing thermistor probes in bales uniformly at 7 locations throughout the container. The probes were connected to a digital data logger placed on the top surface of the container. One external probe measured the ambient temperature outside the container. Temperature records were taken continuously at hourly intervals and the data obtained at the end of each trial were down-loaded onto the hard disk of an IBM computer. All gas sampling lines were drawn to the top surface of the container directly above their position in order to minimise the dead volume in the line.

Gas concentrations were measured at regular intervals by using direct reading instruments (Riken™ interferometer for methyl bromide; and an EC80 Phosphine Monitorô for phosphine), gas detector tubes (Auer™) and gas chromatographs (Varian™ 3400 Series with a flame ionisation detector for methyl bromide; and a thermionic specific detector for phosphine). Cumulative concentration by time products were calculated using the method described by Bond (1989).

The gastightness of test containers was measured by introducing compressed air from a G-size cylinder into the sealed container through an air hose fitted with a shut-off valve. A modified 'finger device' (Sharp and Cousins 1982) of hard plastic tubing was fitted under the door jamb of the sealed container to admit compressed air and to sense the pressure developed. The pressure was measured by a Dwyer Magnahelicô Differential Pressure Gauge series 2000, 250 PaC and required that the needle on the gauge register a deflection of 250 Pa before the gas is turned off. As the pressure drops, the time taken (in seconds) for the pressure to drop from 200 to 100 Pa is measured on a stop watch and is used as the pressure decay value. If the container cannot achieve a 250 Pa deflection, it is deemed to have failed the pressure test and is classed as '0' requiring that fumigations be done under tarpaulins.

The standard treatments of phosphine and methyl bromide were first tested for effectiveness. When these were found to be deficient, improvements were tested through a series of experiments and introduced step by step until a cost-effective method was developed.

Results and Discussion

Problems with existing protocols

Methyl Bromide

The existing protocols are unsatisfactory, Figure 1 shows that placing the application pipe only at the front of a container prevents uniform distribution of the gas. Clearly the concentration at the back of the container will never reach the required *CT* product of 200 g.hour/m^3 and insect control failure will result.

Phosphine

Figure 2 shows that placing the phosphine tablets or sachets only at the front of the container is unsatisfactory as this results in uneven distribution of the gas. Furthermore, the poor state of gastightness of the container makes it impossible for gas concentrations to be held above the required 100 parts per million for 7–10 days.

Improvements to existing protocols

Methyl bromide

A new application technique was developed which involved a purpose-designed 40 mm open ended PVC application pipe described above. Initial application rates were 32 g/m^3 in accordance with the existing protocol. Figure 3 shows more uniform distribution of the gas through the front, middle and back of the container than that shown in Figure 1. It appears that it will take several days before the required *CT* product of 200 g.hours/m^3 is reached.

Another problem identified during the project was that poor gastightness of containers also resulted in fumigation failure. Figure 4 shows that although more even distribution was achieved through the use of the application pipe, the required exposure dose of 200 g.hour/m^3 could not be reached in a leaky untarped container even with an application rate of 100 g/m^3.

Figure 5 shows the improvement resulting from the use of a tarpaulin to cover a leaky container. The 200 g.hours/m^3 dose was achieved after only 6.5 hours.

Some containers are suitable for fumigation without a tarpaulin, provided a pressure test is carried out. Containers which are able to hold the pressure test for as little as 5 seconds can be fumigated without a tarpaulin. Figure 6 shows a

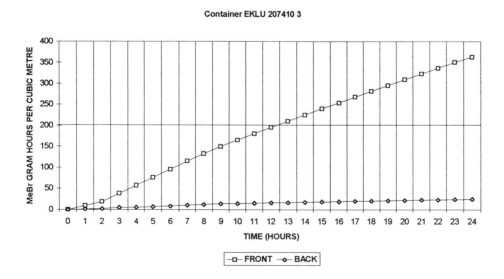

Fig. 1.　*CT* product for MeBr fumigation using existing protocol (0 second pressure test, tarpaulin used, 32 g/m^3 dose).

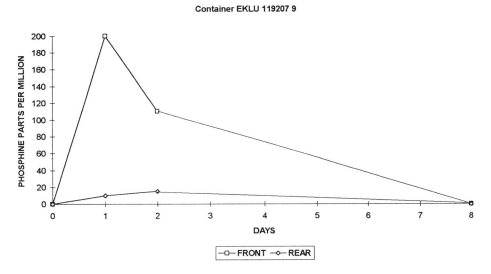

Fig. 2. *CT* product for phosphine fumigation using existing protocol (0 second pressure test, tarpaulin used, 1.68 g/m^3 dose).

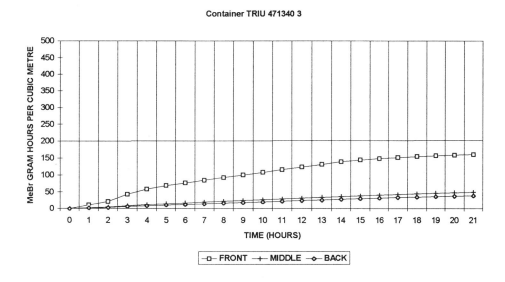

Fig. 3. MeBr fumigation using application pipe (6 second pressure test, tarpaulin not used, 32g/m^3 dose)

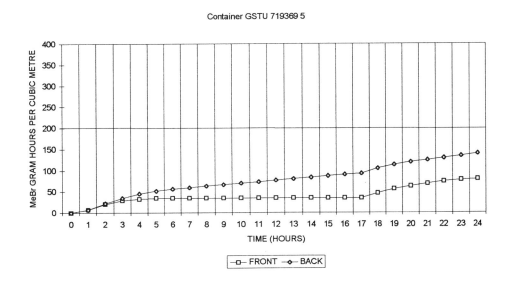

Fig. 4. MeBr fumigation using application pipe in poorly sealed container (0 second pressure test, tarpaulin not used, 100 g/m^3 dose)

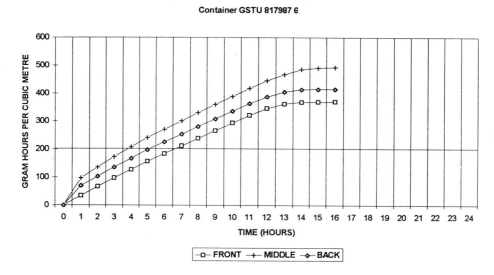

Fig. 5. MeBr fumigation using application pipe in poorly sealed container using tarpaulin (0 second pressure test, 100 g/m^3 dose)

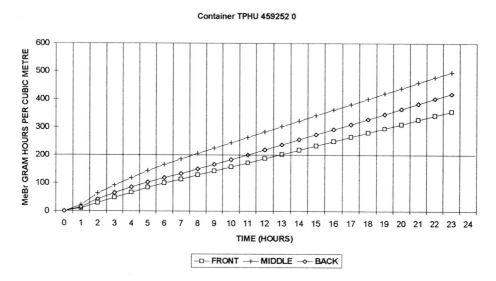

Fig. 6. MeBr fumigation using application pipe in slightly sealed container (3 second pressure test, tarpaulin not used, 48 g/m^3 dose)

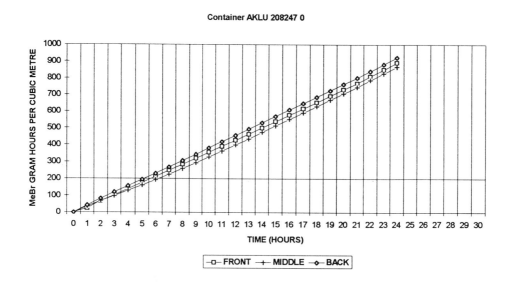

Fig. 7. MeBr fumigation using application pipe in well sealed container (9 second pressure test, tarpaulin not used, 48 g/m^3).

container which held the pressure test for 3 seconds under experimental conditions; insect control was achieved after 12.5 hours. Over 100 containers of varying levels of gastightness have now been tested to confirm these findings.

The existing protocol recommended a dosage of 32 g/m^3, but more than half of the gas is adsorbed by the hay resulting in an inadequate fumigation. A series of trials was carried out to determine a dosage which would overcome the effect of adsorption and maintain sufficient concentration of gas in the airspace to achieve disinfestation. Figure 7 shows the improvement in airspace concentrations resulting from a dosage of 48 g/m^3.

The project had now established that sufficient concentrations of gas could be achieved in the airspace between hay bales throughout the container. The question remained whether the gas would penetrate as readily to the centre of the bale at every position in the container. To answer this 35 probes were placed in the centre of bales which were uniformly distributed throughout the container. A further 15 probes were placed in the airspace between bales.

Figure 8 demonstrates that gas penetrated to the centre of the bales as rapidly as it did through the airspace. Equilibration was achieved within 3 hours at all points within the container and the recommended *CT* product was achieved after 9 hours.

Phosphine

Improvements to the phosphine fumigation protocol required better distribution of the aluminium phosphide sachets. The PVC application pipe described above was used to place four sachets at the front, three in the middle and three at the back of the container. The container shown in Figure 9 held the required pressure for 10 seconds. Clearly, there was uniform distribution of the gas and the concentration was held above 100 parts per million after 7 days. This fumigation successfully controlled all developmental stages of all known stored-product pests.

Having established that phosphine concentration could be held above 100 parts per million for 7 days and that even distribution of the gas could be achieved, the question remained whether phosphine readily penetrated the double dumped hay bales. The 50-point experiment shown in Figure 8 was repeated using phosphine. Figure 10 shows that phosphine readily penetrated the hay bales and that equilibration was achieved within 1 day.

Conclusions

Pressure testing of containers to ensure adequate gastightness is essential for fumigations without tarpaulins.

Effective fumigation cannot be achieved unless special procedures are followed to ensure even dispersal of the fumigant throughout the container.

Both phosphine and methyl bromide readily penetrate into the centre of double-dumped hay bales.

The results of the project have already been put to practical use. Immediate benefits have been a substantial reduction in the number of rejections in 1990–91, and no rejections between 1991–94 even though exports have been steadily increasing each year.

Recommendations

1. Pressure decay test

The containers should be measured for gastightness using the following procedure:

1.1. Use a Dwyer Magnahelic Æ Differential Pressure Gauge series 2000, 250 PaC with piping specified and approved by the Department of Agriculture.

1.2. Introduce the industrial grade compressed air (available in G-size cylinders) into the container until the needle on the gauge registers a deflection of 250 Pa. Then turn off the gas.

1.3. If 250 Pa cannot be achieved the container is deemed to have failed the pressure test and must be fumigated under gastight tarpaulins.

1.4. Gas leakage from the container is indicated by pressure decay as shown on the gauge. Start the stop-watch when the needle drops to the 200 Pa mark.

1.5. Stop the watch when the needle drops to the 100 Pascal mark.

1.6. The time taken is the pressure test value in seconds for the container.

1.7. Pressure testing may be done on full or empty containers. For full containers subtract two seconds from the registered time to give the true value of the container.

1.8. Only containers achieving a pressure test decay time of greater than 5 seconds are deemed to have passed the pressure test and are suitable for fumigation without tarpaulins.

2. Fumigation in containers that meet the pressure test

2.1. Methyl bromide

Tarpaulins are not required for containers that exceed a pressure decay test value of 5 seconds.

Methyl bromide should be applied in the standard way through a heat exchange unit by a licensed operator. The pipe delivering the fumigant should be connected to a 40 mm open ended PVC pipe of the same length of the container. The pipe should have 4 mm diameter holes drilled at 100 cm intervals starting 200 cm from the door end of the container. The pipe should be introduced along the top of the bales through the centre of the container and should be left in place for the duration of the fumigation. This will ensure even distribution of the gas.

Methyl bromide should be applied at the rate of 48 g/m^3 for the following conditions:

Exposure period	Ambient temperature
12 hours	>25°C
24 hours	10-25°C

2.2. Phosphine

Containers that hold pressure for at least 10 seconds are considered to have passed the pressure test and may be fumigated without the use of tarpaulins.

Phosphine may be applied as 10 sachets (34 g each) of aluminium phosphide and inserted into the container using the PVC application pipe described for methyl bromide fumigations. A weighted string can be dropped down the pipe and the sachets attached to the end. This will allow removal of the pipe immediately and the sachets at the conclusion of the fumigation. The sachets should be distributed as four sachets to the rear of container, three sachets to the middle of container and three sachets at the door end of container. At the end of the fumigation period the sachets must be retrieved and disposed of in accordance with the label. The phosphine should be

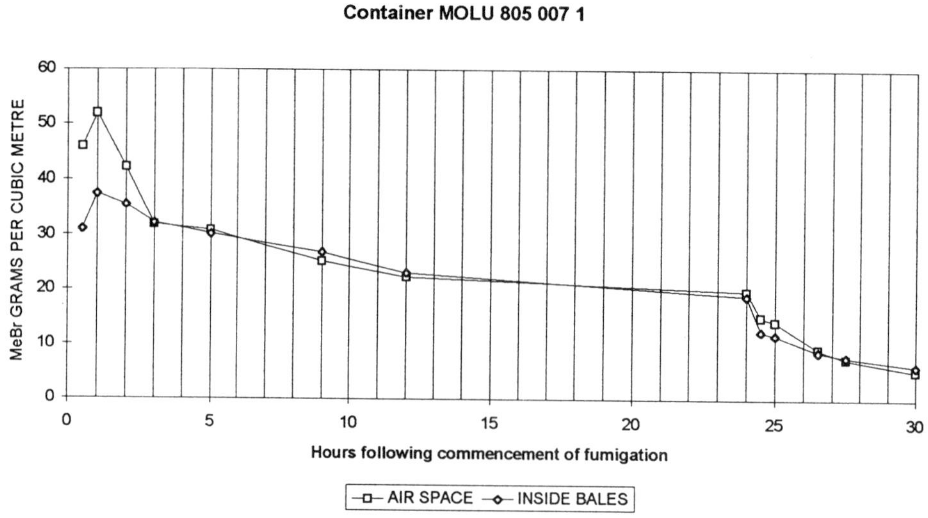

Container MOLU 805 007 1

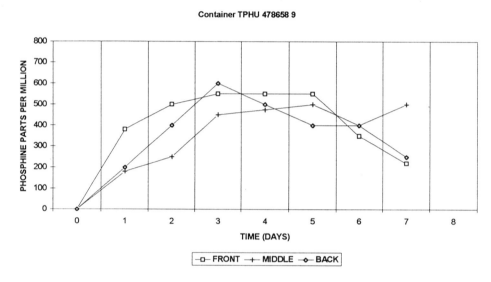

Fig. 8.   Comparison of MeBr concentrations in the airspace and within hay bales (8 second pressure test, tarpaulin not used, 64 g/m^3 dose)

Container TPHU 478658 9

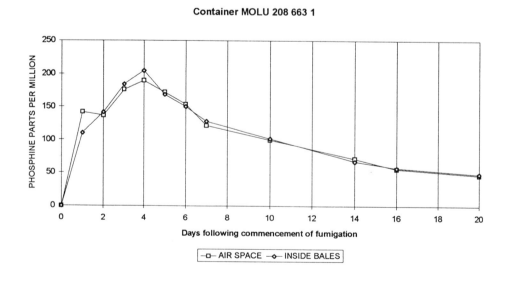

Fig. 9.   Phosphine concentration over time in container fumigated with improved protocol (10 second pressure test, tarpaulin not used, 1.68 g/m^3 dose)

Container MOLU 208 663 1

Fig. 10.   Comparison of phosphine concentrations in the airspace and within hay bales (16 second pressure test, tarpaulin not used, 1.68 g/m^3

retained for the following exposure periods (Winks et al. 1980)

Exposure Period	Ambient Temperature
7 days	>25°C
10 days	15-25°C

Do not attempt phosphine fumigations if temperatures are below 15°C.

3.Fumigation with tarpaulins in containers unable to meet the pressure test

3.1.Methyl bromide

Containers that do not hold pressure for 5 seconds are considered to have failed the pressure test. The following fumigation specifications then apply.

The containers must be placed on an impervious floor (e.g. cement/concrete) with a minimum of 50 cm edge all around the container. The tarpaulins should be free from holes and made from heavy gauge plastic (greater than 1500 micron) or bitumen-canvas. Tarpaulins should be secured to the sides of the container with ropes. This is to prevent any bellows action in the presence of strong winds which has the effect of pumping the gas out of the container. The tarpaulin should be weighted down with 15 cm diameter sand snakes placed in contact with the container body. The overall cover on the container should be tight to prevent any flexing of the tarpaulins.

Methyl bromide should be applied as described in Section 2.1.

3.2.Phosphine

Containers which cannot hold pressure for at least 10 second are consider to have failed the pressure test and must be fumigated under tarpaulins. Apply the phosphine as described in Section 2.2 and the tarpaulins as described in Section 3.1.

Acknowledgments

Financial support by the Rural Industries Research and Development Corporation is gratefully acknowledged.

References

Bond, E.J. 1989. Manual of fumigation for insect control. Food and Agriculture Organisation Plant Protection Paper. 54

Sharp, A.K. and Cousins, E.R. 1982. 'Contestor', an automatic pressure-decay timer. Commonwealth Scientific and Industrial Research Organisation, Food Research Quarterly. 42, 14–17.

Winks R.G. et al. 1980. Dosage recommendations for the fumigation of grain with phosphine. Science Communication Unit, Standing Committee on Agriculture Technical Report Series, 8.

Carbonyl sulphide as a fumigant for control of insects and mites

J.M. Desmarchelier*

Abstract

The reasons why new fumigants are required are outlined. Principally, these reasons are actual or threatened withdrawal of use because of problems relating to mammalian toxicity or environmental protection. The properties of carbonyl sulphide (COS) as a fumigant are discussed.

Bioassay data are given for 11 species of insects, namely *Tribolium castaneum* (Herbst), *T. confusum* du Val, *Rhyzopertha dominica* (F.), *Oryzaephilus surinamensis* (L.), *Ephestia cautella* (Walker), *Bactocera tyroni* (Froggatt), *Lepidoglyphus destructor* (Schrank), *Coptotermes acinaformis* (Froggart), *Crytotermes domesticus* (Haviland) and *Liposcelis bostrychophila* Badonnel.

All external stages were controlled by a 24-hour exposure to 25 mg/L, or by lower doses. The most susceptible stages were adult psocids and adult *R. dominica*; the most tolerant stages were eggs of *T. castaneum*, *O. surinamensis* and *E. cautella*. An exposure of 24 hours to 25 mg/L also controlled all internal stages of *R. dominica*. The hardest species to control was *S. oryzae*. Dosages required to control internal stages were examined for exposure periods of from 6 hours to 7 days, where single dosages were applied to infested wheat, and the grain held for different periods.

Chemical data on the sorption of carbonyl sulphide are compared with data from phosphine and methyl bromide. Levels of carbonyl sulphide in the intergranular air space decay more quickly than do levels of phosphine, but more slowly than do levels of methyl bromide. Carbonyl sulphide was blown through and out of a column of grain as easily as was phosphine, and was blown out of grain more easily than was methyl bromide. Carbonyl sulphide had no adverse effect on the germination of wheat.

Procedures for analysis by gas-chromatography are discussed. Carbonyl sulphide can be separated from other gases on packed columns or on wide-bore columns, but it lacks the specific sensitivity to at least one detector of fumigants such as phosphine or methyl bromide.

Data are presented on the chemistry of carbonyl sulphide and its environmental fate (each of which has been reviewed), and on its mammalian toxicity.

Preamble

Fumigants are widely used for insect disinfestation of commodities, buildings and soils. No compound is ideal for use as a fumigant; for example, all are highly toxic to mammals and many are flammable. The use of many fumigants, including acrylonitrile, ethylene dibromide and ethylene oxide, has been stopped or restricted recently because of problems with human toxicity. Methyl bromide is under threat of restricted use, and possible withdrawal, because it apparently depletes the ozone layer of the earth's atmosphere. There is, therefore, an urgent need for new fumigants. This paper discusses the properties of carbonyl sulphide with respect to its use as a fumigant. This usage has been patented by CSIRO (International Patent Application PCT/AU93/00018). One reason for patent protection is to increase the probability of the expenditure necessary for registration, as discussed in my keynote address on grain protectants at this conference.

This paper discusses the properties of carbonyl sulphide (COS) with respect to its use as a fumigant, under the following 8 headings:

1. the efficacy of carbonyl sulphide against insects and mites;
2. the chemistry and industrial uses of carbonyl sulphide;
3. the environmental fate of carbonyl sulphide;
4. mammalian toxicity of carbonyl sulphide and worker safety;
5. analysis and sorption of carbonyl sulphide;
6. movement of carbonyl sulphide through wheat;
7. the effect of carbonyl sulphide on seed germination; and
8. overview.

The Efficacy of Carbonyl Sulphide against Insects and Mites

Insects tested were *Tribolium castaneum* (Herbst), *T. confusum* du Val, *Sitophilus oryzae* (L.), *Rhyzopertha dominica* (F.), *Oryzaephilus surinamensis* (L.), *Ephestia cautella* (Walker), *Bactocera tyroni* (Froggatt), *Lepidoglyphus destructor* (Schrank), *Coptotermes acinaciformis* (Froggatt), *Cryptotermes domesticus* (Haviland) and *Liposcelis bostrychophila* Badonnel.

Bioassays for external stages were conducted in sealed glass jars, fitted with a Mininert valve. The humidity in the jars was 55–60%, except for assays on psocids and mites, where 5 g of 14% moisture wheat was added to jars, to raise the humidity to approximately 70%, and for assays with fruit fly, where 1 mL of water was added to each jar. Assays were performed in triplicate, using 30–50 insects and results discarded when control mortality exceeded 5%. After exposure periods of 6 or 24 hours, insects were placed on wheat for assessment of mortality 7 days after the end of dosing. Bioassays for internal stages of *R. dominica* and *S. oryzae* used cohorts aged 0–1 weeks, 2–3 weeks and 4–5 weeks. This was achieved by leaving 50 adults to oviposit for 1 week on 300 g of 11.8% moisture wheat at 25°C, removing adults and allowing to stand for the appropriate period (e.g. 2 weeks for assays on internal stages aged 2–3 weeks). Fumigant was added to 300 g of wheat in a 700 mL sealed glass jar. Assays on immature stages were replicated 6–9 times. The progeny were allowed to develop for 9 weeks after oviposition at 30°C, and assessed relative to control.

Results for external stages are summarised in Table 1 which records assay conditions, LC_{95} values and 'minimum effective doses', i.e. the lowest tested dose that gave 100% mortality from at least 100 insects. All tested external stages were completely controlled by a 24 hour exposure to a dosage of 25 mg/L. The most susceptible species tested were adult

* Stored Grain Research Laboratory, CSIRO Division of Entomology, GPO Box 1700, Canberra, ACT 2601, Australia.

psocids and adult *R. dominica*; the most tolerant stages were eggs of *T. castaneum, O. surinamensis* and *E. cautella*.

Results for internal stages of *S. oryzae* and *R. dominica* (including eggs) are summarised in Table 2. An exposure of 24 hours to dosages of 16–25 mg/L controlled all stages of *R. dominica* (100% mortality in 5/6 replicates). Eggs and adults of this species are also susceptible to carbonyl sulphide (Table 1). *S.oryzae* was harder to control than *R.dominica* (Table 2), and a 24-hour exposure to 24 mg/L a dosage that completely controlled *R.dominica*, controlled only an average of 80.7% of *S. oryzae*.

Both the dosage and period of exposure were varied in experiments on immature *S. oryzae*, and results are recorded in Table 2. Thus, a 6-hour exposure to a dosage of 200 mg/L gave 98.1% control of *S. oryzae*, whereas 100% control was obtained by a 24-hour exposure to a dosage of 60 mg/L, a 72-hour exposure to a dosage of 30 mg/L and an 168-hour exposure to a dosage of 20 mg/L. In the last example, measured concentrations of carbonyl sulphide fell from 17 mg/L 3 hours after dosing, to 7 mg/L, 168 hours after dosing.

In summary, a single dosage of 25 mg/L and an exposure period of 24 hours controlled the external stages of all tested insects and all stages, including internal stages, of *R. dominica*. However, control of internal stages of *S. oryzae* required either a higher dosage or a longer exposure period.

The Chemistry and Industrial Uses of Carbonyl Sulphide

Pure carbonyl sulphide is a colourless, odourless gas with a boiling point of –50.2°C. Its chemistry has been reviewed (Ferm 1957). Its vapour density at 1 atmosphere and 25°C is 2.485. Carbonyl sulphide, according to Ferm, is 'stable, but can undergo decomposition, hydrolysis, oxidation and reduction' and, as shown in Section 3, these reactions enable carbonyl sulphide to play a major, indeed almost certainly essential, role in the natural sulphur cycle in plants and soils.

Carbonyl sulphide has a water solubility of 1.4 g/L at 25°C (Kluczewski et al. 1985). At the same temperature, water solubilities of methyl bromide and carbon tetrachloride are, respectively,13.4 g/kg and 0.28 g/kg (Worthing 1979). The solubility of phosphine is given by Fluck (1973) as 22.8 mL per 100 ml [0.35 g/L] at 17°C, based on the data of Weston (1954), but the original article covers a much wider range of temperatures.

Carbonyl sulphide has been extensively studied for a number of reasons, including its natural role in the sulphur cycle (cf Section 3), its presence in pyrolysis products, its use as a chemical feedstock (see e.g. Leiber and Berk 1985), its presence as radioactive carbonyl sulphide in releases from nuclear reactors cooled with carbon dioxide (Kluczewski et al. 1985), and its formation, in mammalian systems, in the metabolism of some sulphur-containing chemicals.

The Environmental Fate of Carbonyl Sulphide

An important requirement of any fumigant is that it does not harm the environment. The environmental fate of carbonyl sulphide has been reviewed (Payton et al. 1978; Kluczewski et al.1985; Mihalopoulos et al. 1989). It occurs uniformly in the troposphere at a concentration of 1.3 g/m^{3}. It has been suggested as the major natural sulphur species in the atmosphere. Emissions from natural and anthropogenic sources are estimated as being between 1 and 10 million tonnes per year (Mihalopoulos et al. 1989).

Ferm (1957) outlines preparation of carbonyl sulphide from S (sulphur), CS_2 (carbon bisulphide) and KCNS (potassium thiocyanate), and reactions to form SO_2 (sulphur dioxide), SO_4 (sulphate), H_2S (hydrogen sulphide), CS_2 (carbon bisulphide) and S (sulphur).

Carbonyl sulphide forms part of the natural sulphur flux, e.g. in soils (Staubes et al. 1989) and marshes (Steudler and Peterson 1985). It is formed in the anaerobic degradation of manure and compost, and is suggested to be an intermediate in the bacterial sulphur cycle (Elliott and Travis 1973).

As an example of the natural flux of carbonyl sulphide, Steudler and Peterson (1985) found an average yearly emission from a New England, USA salt marsh of 34.3 g of sulphur/m^2/hour (i.e. 64.3 g of carbonyl sulphide/m^2/hour). A mean value of carbonyl sulphide flux from soils has been estimated as 540 ngm^2/hour (Mihalopoulos et al. 1989).

Taylor et al. (1983) studied decomposition velocities on plants, which took up carbonyl sulphide more slowly than they did sulphur dioxide, but more quickly than carbon disulphide. Carbonyl sulphide is less phytotoxic to bean plants (*Phaseolus vulgaris*) than sulphur dioxide or hydrogen sulphide (Taylor and Selvidge 1984). The estimated concentration by time product at which a decline in photosynthesis occurs is 4.8 mg/hour/m^3, at concentrations in the range 0.36–4.9 mg/m.

Mihalopoulos et al. (1989) showed that carbonyl sulphide levels in the atmosphere over land, as distinct from oceanic sites, increase with height above ground and postulate that plants act as a sink for carbonyl sulphide. Mikalopoulos et al. (1989) also summarise data on atmosphere levels, which do not appear to be rising, stratospheric photo-oxidation (0.1–0.16 Mt/year) and reaction with hydroxy radicals (slow; half life of about 17 years for typical concentrations of hydroxy radicals). Hoffman (1990) discusses the postulate that background sulphuric acid levels in the stratosphere are caused by 'a sulphur-bearing compound that is chemically inert and water-insoluble, such as carbonyl sulphide'.

It is clear that carbonyl sulphide is an intermediate — probably essential — in the atmospheric sulphur cycle. This in itself does not exclude the possibility of localised environmental problems from large-scale usage. Nitrate ion, for example, is an intermediate in the nitrogen cycle, but is also a major pollutant of aquifers. Nonetheless, fumigation with carbonyl sulphide is highly unlikely to cause non-localised environmental damage of the type caused by methyl bromide.

Toxicity of Carbonyl Sulphide to Mammals and Worker Safety Aspects

A good fumigant should be safe to consumers and workers, and parameters such as flammability and acute and chronic toxicity are important.

Sax and Lewis (1989) summarise the properties of carbonyl sulphide and other chemicals. From these authors, carbonyl sulphide has a flammability range of 12–28.5%, V/V, in air. In comparison, the authors list the flammability limits of methyl bromide as 13.5–14.5%, those for phosphine as above 1%, those for carbon bisulphide as 1.3–50%, those for ethylene dichloride as 6.2–15.9% and those for ethyl formate as 2.7-13.5%. Thus, carbonyl sulphide is flammable, but insecticidal concentrations are well below the 'lower explosion limit'.

Sax and Lewis (1989) state that carbonyl sulphide is a 'poison by intraperitoneal route. Mildly toxic by inhalation, Narcotic in high concentration. An irritant'. A fatal accident with carbonyl sulphide, caused by high concentrations of carbonyl sulphide in an enclosed space, is discussed by Thiess et al. (1968), who also give acute toxicity data on cats, rabbits

Table 1. Toxicity of carbonyl sulphide to insects and mites

Species	Stage	Exposure (hours)	Temperature (°C)	LC_{95}	Minimum effective[a] dose (mg/hour/L)
Rhyzopertha dominica	adult	6	25	38	68.7
Tribolium castaneum	adult	6	25	82	101
	adult	24	25		297
Sitophilus oryzae	adult	6	25	99	100
	adult	24	25		264
Oryzaephilus surinamensis	adult	6	30	198	240
	adult	24	30		240
Tribolium confusum	adult	6	25	111	146
Lepidoglyphus destructor	adult	6	27		120
	adult	24			240
Liposcelis bostrychophila	adult	6	25		22.5
Coptotermes acinaformis	adult	24	30		288
Coptotermes domesticus	adult	6			360
Tribolium castaneum	pupae	6	30	290	360
	pupae	24	30	490	600
Ephestia cautella	pupae	24	27		240
Bactocera tyroni	pupae	6	27		360
	pupae	24	27	440	600
Rhyzopertha dominica[b]					
Sitophilus oryzae[b]					
Coptotermes acinaciformis	nymphs	24	27		600
Tribolium castaneum	larvae	6	25	270	300
	larvae	24	30	-	480
Ephestia cautella	larvae	6	30		240
	larvae	24	39	410	480
Oryzaephilus surinamensis	larvae	6	30	210	300
	larvae	24	30		360
Bactocera tyroni	larvae	6	27		180
	larvae	24	27		360
Rhyzopertha dominica	eggs 0–1 day	24	30	145	192
	eggs 0–1 day	6	30	102	144
	eggs 2–3 days	24	30		144
	eggs 4–5 days	24	30		120
Tribolium castaneum	eggs 0–1 day	24	30	520	600
		6	30	430	480
		48	30		360
Oryzaephilus surinamensis	eggs 0-1 day	24	30	495	600
		6	30	390	420
Bactocera tyroni	eggs -8h	24	30	460	600
Ephestia cautella	eggs 0-1d	24	30	450	600
		6	30		720

[a] The minimum tested dose that caused 100% mortality in assays against at least 100 insects.

[b] cf. experiments on internal stages, Table 2.

and guinea pigs. Klemenc (1943) reports data on mice. In these studies there was a cut-off point at which no effect was observed. This was a 6-hour exposure to 0.03%, V/V, in cats, rabbits and guinea pigs (Thiess et al. 1968) and indefinite exposure to 0.09%, V/V in rats (Klemenc 1943).

In a 7-week exposure of 50 ppm carbonyl sulphide to rabbits, Hugod (1981) and Hugod and Astrup (1981) concluded that carbonyl sulphide had no significant effect on mycocardial ultrastructure. Chengelis and Neal (1979, 1980) showed that carbonyl sulphide is degraded by the enzyme carbonic anhydrase to hydrogen sulphide. This supports the previous classification of Thiess et al. (1968) that carbonyl sulphide, along with hydrogen sulphide and hydrogen cyanide, was an 'either-or' poison, meaning that recovery to exposure was complete, except where fatalities have occurred.

As carbonyl sulphide is a major contaminant of hydrocarbons, etc., much work has been done on its removal from air streams. The usual method involves an amine dissolved or suspended in a liquid scrubbing medium such as water or oil (Ferm 1957). Other scrubbers include ion-exchange resins and absorbent clays containing monoethanolamine (Ferm 1957). Thus, carbonyl sulphide scrubbing is already a commercial process which could be adapted to grain storage, if required.

Analysis and Sorption of Carbonyl Sulphide

Procedures for analysis of standard concentrations are outlined by Ferm (1957). Carbonyl sulphide can be chromatagraphically resolved on packed columns (e.g. Tangerman 1986) or on capillary columns (e.g. Desmarchelier, these proceedings). It is relatively insensitive to the electron-capture and flame-ionisation detectors, but shows some selectivity to the flame photometric detector (sulphur mode) and the photoionisation detector. In each case, however, more carbonyl sulphide is required than carbon bisulphide to achieve the same machine response.

In addition to direct chromatographic analysis of carbonyl sulphide, Leiber and Berk (1985) report an in situ derivitisation procedure whereby carbonyl sulphide is trapped in polyamide impregnated with diamino propane. This method

was used to detect air concentrations in the range 1–20 ppm carbonyl sulphide.

In fumigation, the amount of gas in the intergranular air space declines with time, even in sealed systems. The rate of disappearance of the fumigant is important for estimating the mean and range of concentrations over an exposure period.

Figure 1 compares the sorption of carbonyl sulphide, methyl bromide and phosphine on wheat. Carbonyl sulphide is much less strongly sorbed than methyl bromide, and slightly more sorbed than phosphine. Because the decay of carbonyl sulphide is relatively slow, concentrations that are able to affect insects are still present several days after initial dosage. This explains why the minimum effective dosage can be reduced by extending the exposure period (Section 1).

Movement of Carbonyl Sulphide through Wheat

An ideal fumigant should be able to be easily passed through bulk grain, in order to obtain good distribution and in order to be able to blow away all fumigant after fumigation.

In a comparison between methyl bromide, phosphine and carbonyl sulphide, each fumigant (20 mL) was separately blown through a 1.1 m column of wheat, of total volume 7.9 L, at an airflow rate of 200 mL/minute. Each gas was introduced via a 200 mL flask at the bottom of the column. Concentrations in the eluate, at the top of the column, are recorded in Figure 2, for all concentrations greater than 1 ppm, V/V. The 'chromatography' of phosphine and carbonyl sulphide are essentially identical with respect to retention time, peak width and degree of tailing. However, methyl bromide has a broader peak and greater tailing, consistent with stronger sorption.

Note that carbonyl sulphide was rapidly blown out of a grain column, to levels below 1 ppm in the intergranular air.

The Effect of Carbonyl Sulphide on Seed Germination

It is important that fumigants do not harm grain quality, and viability of seeds is an important quality test. Australian Standard White wheat was conditioned to 12 and 16% moisture, as determined by the ISO air oven method. Grains

Table 2. % control of all immature stages of *R. dominica* and *S. oryzae* from exposure to carbonyl sulphide at 25 °C

Species	Dosage (mg/L)	Exposure period (hour)	% mortality
R. dominica	8	24	93.1
	16	24	99.0
	25	24	99.96
		48	100
	45	24	100
S. oryzae	24	24	84.4
	200	6	98.1
	60	6	81.2
		24	99.8
		48	100
	40	48	100
	30	48	99.2
		72	100
	20	72	97.3
	20	168	100
	10	168	75.0

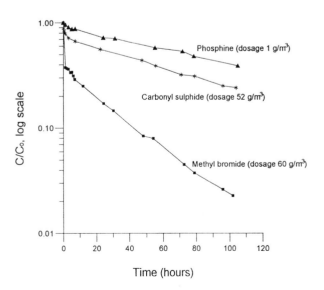

Fig. 1. Comparative sorption data for phosphine, carbonyl sulphide and methyl bromide, taken from measurement of loss of fumigant concentrations in sealed containers 95% full of wheat of 12% moisture content, wet basis, at 25°C.

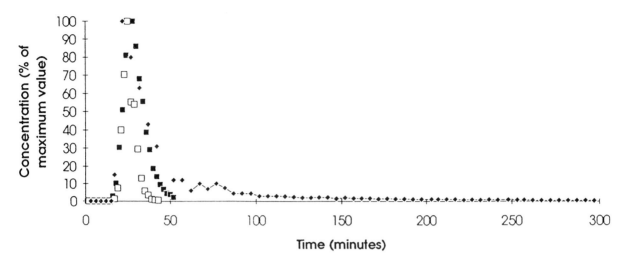

Fig. 2   Concentration of fumigant, (% of maximum concentration) eluting from a 1.1M column of wheat, at an air flow of 200mL/minute, plotted against time (in minutes). □: carbonyl sulphide; ■, phosphine; ◆, methyl bromide.

were dosed for 24 hours with concentrations of carbonyl sulphide of 0.5% V/V, 1.0% V/V and 5.0% V/V. Corresponding nominal concentration × time products were 300 mg/hour/L, 600 mg/hour/L and 3000 mg/hour/L.

No adverse effect on seed germination was detected, either in tests immediately after dosing or 3 months later. Germination was slightly higher in treated wheat than in control wheat on 11/12 occasions.

Overview

No fumigant is ideal, and most suffer at least some problems in areas of worker safety, flammability, residues and the environment. Carbonyl sulphide is no exception to this rule, but it has qualities that make it comparable with fumigants such as methyl bromide and phosphine. For example, its flammability limits are greater than those of methyl bromide but less than those of phosphine. It is environmentally superior to methyl bromide and, on the basis of greater knowledge, to phosphine. Phosphine is the best fumigant for long exposures, methyl bromide is the best for short exposures, but carbonyl sulphide is versatile being able to be used for short exposures, or for longer exposures at reduced dosages.

Acknowledgements

The Australian Wheat Board and Bulk Handling Authorities are thanked for financial assistance. Much of the work reported here was done by S.E. Allen, J. Cassells, Ren Yong Lin and Le Trang Vu, whose technical assistance is gratefully noted. H.J. Banks was responsible for initial selection of COS for testing.

References

Chengelis, C.P. and Neal, R.A. 1979. Hepatic carbonyl sulfide metabolism. Biochemical Biophysical Research Communication, 90, 993–999.

Chengelis, C.P. and Neal, R.A. 1980. Studies on carbonyl sulfide toxicity — metabolism by carbonic anhydrase. *Toxicology and* Applied Pharmacology, 55, 198–202.

Elliot, L.F. and Travis, T.A. 1973. Detection of carbonyl sulphide and other gases emenating from beef cattle manure. Soil Science Society of America, Proceedings, 37, 700–702.

Ferm, R.J. 1957. The chemistry of carbonyl sulphide. *Chemical Reviews*, 57, 621–640.

Fluck,E. 1973. The chemistry of phosphine. Fortschritte der chemischen Forschung, 35, 1–64.

Hoffman, D.J. 1990. Increase in the stratospheric background sulfuric acid aerosol mass in the last 10 years. Science, 248, 996–1000.

Hugod, C. 1981. Mycocardial morphology in rabbits exposed to various gas-phase constituents of tobacco smoke. Atheroschlerosis, 40, 181–190.

Hugod, C. and Astrup, P. 1981. Studies of coronary and aortic intimal morphology in rabbits exposed to gas phase constituents of tobacco smoke (hydrogen cyanide, nitric oxide and carbonyl sulphide). Smoking and Arterial Disease, 89–94.

Klemenc, A. 1943. Die Giftwirkung von Kohlenoxysulfid. Berichte der deutschen. chemischen. Gesellschaft, 76, 299–303.

Kluczewski, S.M., Brown, K.A. and Bell, J.N.B. 1985. Deposition of carbonyl sulphide to soils. Atmospheric Environment, 19, 1295–1299.

Leiber, M.A. and Berk, H.C. 1985. Determination of carbonyl sulphide in air by derivitization with 1,3-diaminopropane and capillary gas chromatographic analysis. Analytical Chemistry, 57, 2792–2796.

Mihalopoulos, N., Bonsang, B., Nguyen, B.C., Kanakidou, M. and Belviso, S. 1989. Field observations of carbonyl sulfide deficit near the ground: possible implication of vegetation. Atmospheric Environment, 23, 2159–2166.

Payton, T., Steele, R. and Mabey, W. 1978. Carbon disulphide, Carbonyl Sulfide. Literature Review and Environmental Assessment. EPA 600/9–78–009, Environmental Protection Agency, Office of Health and Ecological Effects, Washington DC.

Sax, N.I. and Lewis, R.J. 1989. Dangerous properties of industrial materials, 7th ed. New York. Van Nostrand Reinhold.

Staubes, R., Georgii, H.W. and Ockelmann, G. 1989. Flux of COS, DMS and CS2 from various soils in Germany. Tellus, 41B, 305–313.

Steudler, P.A. and Peterson, B.J. 1985. Annual cycle of gaseous sulphur emissions from a New England *Spartina alternifolia* marsh. Atmospheric Environment, 19, 1411–1416.

Taylor, G.E. Jr. and Selvidge, W.J. 1984. Phytotoxicity in bush bean of five sulphur-containing gases released from advanced fossil energy technologies. Journal of Environmental Quality, 13, 224–230.

Tangerman, A. (1986). Determination of volatile sulphur compounds in air at the parts per trillion level by Tenax trapping and gas chromatography. Journal of Chromatography, 366, 205–216.

Taylor, G.E., McLaughlin, S.B., Shriner, D.S. and Selvidge, W.J. 1983. The flux of sulphur-containing gases to vegetation. Atmospheric Environment, 17, 789–96.

Thiess, A.M., Hey, W., Hofmann, H.T. and Oettel, H. 1968. Zur Toxicitaet des Kohlenoxysulfids. Archiv für Toxikologie, 23, 253–263.

Weston, 1954. The solubility of phosphine in aqueous solutions. Journal of the American Chemical Society, 1027–1028.

Worthing, C.R. 1979. The pesticide manual, 6th ed. British Crop Protection Council.

Improved procedures for measurement of fumigants

J.M. Desmarchelier[*]

Abstract

The requirement for analysis of volatile chemicals has greatly increased in recent years and, as a result, analytical procedures have improved. Research, in particular by Regulatory Agencies in the United States, such as EPA or NIOSH, has resulted in sensitive detection procedures, suitable for regulatory procedures. The recent principal advances in analysis of 'volatiles' can be readily transferred to analysis of fumigants. These include: use of megabore or capillary columns, in place of packed columns; standard procedures for desorption of fumigants from commodities; use of absorbent tubes ('traps') coupled with either solvent or thermal desorption; and use of headspace chromatography. These techniques are illustrated for grain fumigants, using both literature data and unpublished data from the Stored Grain Research Laboratory. The relevance of improved techniques for measurement of fumigants is discussed. Finally, some critical comments are made on analysis of fumigants in grain.

Introduction

In recent years, great advances have been made in the analysis of 'volatile chemicals', which include solvents, air pollutants, naturally-occurring gases and fumigants. These advances are summarised yearly in the journal, *Analytical Chemistry*, and also regularly in publications from regulatory agencies in the United States (ACGIH 1983; NIOSH 1985; USEPA 1989). These advances enable use of techniques other than those traditionally used for fumigant analysis in stored products (Heuser 1973; Berck 1975), though some recent advances are outlined by Scudamore (1988). In this paper, the relevance of these advances to analysis and, indeed, use of fumigants is discussed. Examples of fumigant analysis are given from the literature and from unpublished work.

Advances in Analysis of Volatiles in Gas Chromatography

Advances in analysis of volatiles by gas chromatography can be considered under six headings, namely:
1. calibration of standards;
2. detection, including use of specific detectors;
3. separation, including use of capillary and megabore columns;
4. headspace analysis, based on concentration equilibrium between a gaseous and non-gaseous phase;
5. pre-concentration, including trapping techniques and use of membranes; and
6. laboratory accreditation and quality assurance programs.

Calibration of standards

Inaccurate standards are a major source of error in analytical chemistry and the problem is especially severe in gas analyses because of problems in accurately diluting gases. The major recent advance is the provision of calibrated gas standards, at low concentrations, by organisations such as the Environmental Protection Agency. Provision of standards for calibration are an essential part of any quality assurance program (Ratliff 1993). 'Purge and trap' techniques (cf. 5 below) have also been used to obtain a range of concentrations from one primary standard (Rhoderick and Miller 1990). In this technique, different concentrations are obtained by purging gases at measured flow rates over different timed intervals. This enables calibration of machine response to different concentrations of volatile chemicals, including the fumigants methyl bromide and ethylene dibromide.

Nonetheless, standardised official procedures for calibration of gas standards are not yet readily available in one convenient book, though the review by Scudamore (1988) is an excellent step in this direction. Infra-red spectroscopy would seem to be a convenient method of checking concentrations of gas standards, in the same way as ultra-violet spectroscopy is used for liquids and solids, but there is no officially-validated procedure for fumigants.

Detection, including specific detectors

Specific detectors have been used for many years in the analysis of pesticides and fumigants.

Table 1 lists the sensitivity of some detectors, with an indication of selectivity. Thus, the flame photometric detector, in the phosphorus mode, is capable of detecting 0.001 ng of phosphorus per second. This means that, if a phosphine peak can be passed through a detector in one second, one can detect 0.001 ng of phosphorus or 0.0011 ng of phosphine. The units of sensitivity (ngG/sec) indicate that sensitivity is related both to detector response and to resolution of the chromatography peak. However, because sensitivity in practice is limited by background 'noise' or interferences, it is often defined in terms of signal-to-noise ratio. For example, 'limits of detection' are typically given as a signal-to-noise ratio of two. In this instance, use of capillary columns (cf. 3 below) greatly reduces noise by better separation of interferences from the peak of interest.

The mass spectrometer is increasingly being used in analysis of volatiles. One obtains the whole spectrum, for purposes of identification, as well as the intensity of 'specific ions'. The machine sensitivity of the mass spectrometer varies with type of chemical, ionisation process and background interference. As an example, HCN and nitrogen have the same m/z values. Nonetheless, specific ion monitoring is very useful in reducing background noise, and thus increasing the signal-to-noise ratio. Currently, sample volume is limited with the mass spectrometer as the detector, principally because of the requirement for capillary columns. However, mass spectrometry is becoming routine in analysis of volatiles principally because it can function as a universal detector. Its

---

[*] Stored Grain Research Laboratory, CSIRO Division of Entomology, GPO Box 1700, Canberra ACT 2601, Australia.

Table 1.   Sensitivity of some GC detectors.

Detector	Sensitivity ng/sec	Specificity
Flame Photometric (P)	0.001	P only
Falme Photometric (S)	0.1	S only[a]
Thermionic	0.0001	N&P
Flame Ionisation	0.002	none
Electron Capture	0.0001	varied[b]
Photo-ionisation	0.01	varied
Conductivity (Hall)	0.005	N,S, halides
Specific ions Mass Spec	0.01	'all' [c]
Thermal Conductivity	0.3 ng/mL	'all' [d]

[a]response non-linear but area proportional to concentration squared
[b]linearity limited; easily contaminated
[c]not generally suitable for megabore columns
[d]note the different unit of concentration

use is often coupled with that of a specific detector, when extra sensitivity is required. It is frequently the case that mass spectrometry is used for confirmation of identity, and specific detectors for quantification, although the mass spectrometer can be used for quantitative analysis.

The electron capture detector is usually regarded as 'specific' for compounds containing halogens, or conjugated systems. However, it also responds to oxygen, and not to argon or nitrogen. This is useful because it is difficult to separate argon from oxygen, and argon interference is a problem in analysis of low oxygen atmospheres. Interestingly, there is little detailed information in stored-product literature on techniques for measuring low oxygen concentrations. The response of the electron capture detector to oxygen is shown in Figure 1, together with the response to several fumigants. Thus, the electron capture detector is useful for analysis of low oxygen concentrations, but it becomes saturated at higher oxygen concentrations, where less selective detectors, such as the thermal conductivity detector, can be used at concentration levels where the 'error' from argon is less important.

Specific detectors are, of course, not new but there are several detectors that are 'new' in the sense of being readily available for the first time. These include specific ion monitoring and the photo-ionisation detector. It is not the detectors alone, however, that have increased the limits of detection of volatiles but their integration with other techniques discussed below.

Separation, including use of capillary and megabore columns

Most sensitive methods of determining volatile chemicals now use capillary columns (internal diameter 0.32 mm) or 'megabore' columns (internal diameter 0.53 mm). Capillaries give better resolution but megabores have two advantages for gas analysis. First, they can be readily attached to chromatographs designed for packed columns. Second, because the flow of carrier gas is quite high, typically 10 mL/min, injection volume can be large (e.g. up to 1 mL). It should, however, be noted that some detectors require large quantities of carrier gas, so that replacement of packed columns with megabores requires a make-up gas. This appears to be especially a problem with flame-photometric detectors.

The laboratory of which I am a member routinely uses a GSQ megabore column (J & W 115–3432) for analysis of oxygen, phosphine, carbonyl sulphide, methyl chloride, carbon bisulphide, methyl bromide, and methyl isothiocyanate. The column also separates hydrogen cyanide, though

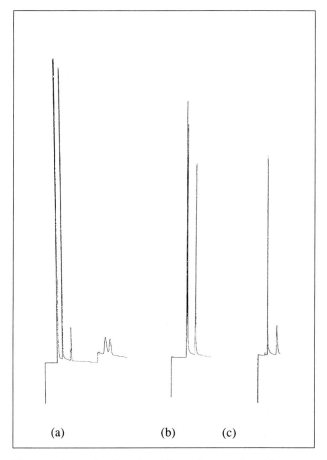

Fig. 1.   Separation, in order of increasing retention: (a) oxygen, methyl bromide, carbon bisulphide, methyl isothiocyanate and chlorpicrin on GSQ; electron capture detector. (b) phosphine and hydrogen cyanide on DBwax; thermionic detector. (c) methyl-isothiocyanate and phenylethyl-isothiocyanate on DBwax; thermionic detector..

greater sensitivity is obtained on the polar DBwax (J&W 125–7012) or on a column designed for acids, FFAP (J&W 125–3212). A typical chromatograph is shown in Figure 1a, which records the response of an electron-capture detector to oxygen, methyl bromide and carbon bisulphide. The small peaks are methyl isothiocyanate, which was the warning agent in the methyl bromide, and chlorpicrin.

In Figure 1b the separation of phosphine from hydrogen cyanide using the megabore DBwax is shown. In this case, a thermionic specific detector was used (specific for nitrogen and phosphorus). The peak shape for hydrogen cyanide (the second peak) shows less tailing than that for phosphine (the first peak) because the column, DBwax, was selected for optimal detection of hydrogen cyanide. Conversely, the peak shape of phosphine on GSQ is better than that of hydrogen cyanide. Nonetheless, the peak shape of each fumigant in Fig. 1b is acceptable, whereas it would be difficult to obtain good peak shapes for phosphine and hydrogen cyanide on the same packed column. Figure 1c shows separation of methyl isothiocyanate and phenylethyl isothiocyanate on the same column. These chemicals were obtained from unfumigated canola, using a trapping technique (cf. 5).

In Figure 1 the retention times, in minutes, are recorded. It can therefore be seen that peak half-widths of most of the chemicals are 1–2 sec (except for methyl isothiocyanate and chlorpicrin in Figure 1a, where, however, the peak width can be narrowed by higher column temperatures). Narrow peak widths are important for two reasons. First, as sensitivity is a function of mass/peak width (Table 1), reducing peak width increases sensitivity. Second, narrow peak width reduces the chance of interference.

The GSQ megabore column enables complete separation of the following fumigants: phosphine, hydrogen cyanide, carbonyl sulphide, carbon bisulphide, ethylene dichloride and chlorpicrin, as well as oxygen, possible alteration products such as methyl chloride and hydrogen sulphide, and naturally-occurring fumigants such as hydrogen cyanide, methyl isothiocyanate and other isothiocyanates.

Headspace analysis

'Headspace analysis' involves partitioning of a volatile chemical between a gaseous phase (air or nitrogen) and another phase (typically liquid), thus enabling determination of the volatile chemical in the non-gaseous phase from its concentration in the vapour phase. The method, which has been reviewed (Nunez et al. 1984), has become a standard method for analysis of volatile chemicals in water. Elimination of the use of solvents is one of the great advantages of this method.

Our laboratory has developed a variation of headspace analysis for determination of cyanide residues in grain (Vu and Desmarchelier, unpublished data). In this procedure, grain is placed with water in a sealed container and, after extraction is complete, a standard quality of dilute hydrochloric acid is injected and the cyanide in grain is determined from the concentration of hydrogen cyanide in the headspace. Analytical conditions include a DBwax megabore column and a specific detector (a thermionic detector, specific for nitrogen and phosphorus). This method is much quicker, and more sensitive, than the standard procedure (American Association of Cereal Chemists, 1983) which involves distillation of hydrogen cyanide from a solution of grain in boiling water, and determination of cyanide by argentimetric titration.

A method has been developed for headspace analysis of methyl bromide (Greve and Hogendoorn 1979) in which the fumigant is first extracted into acetone, and methyl bromide is determined from the headspace concentration in a sealed container. This method can detect residues down to 10 ppb.

An old method used by Turtle for analysis of hydrogen cyanide, and subsequently modified (American Association of Cereal Chemists 1983), involved refluxing grain in water and trapping the volatile fumigant. In the case of hydrogen cyanide, the trap was aqueous alkali. This method of refluxing grain residues in water has been adapted to headspace analysis (Nunez et al. 1984) by Heikes (1987) and Heikes and Hopper (1986). In this procedure, grain or grain products are refluxed in water under a nitrogen stream, the volatiles are absorbed on a trap (cf 5 below), and taken up in a solvent (solvent desorp-

tion), an aliquot of which is injected into the chromatograph. The method detects several halogenated fumigants at detection limits in the range 0.1–6 ng/g (ppb) and carbon bisulphide at a detection limit of 12 ng/g. These detection limits are considerably below former 'guideline' maximum residue limits which were set at 'detection limits' of 0.1 ppm to 10 ppm. Nonetheless, trapping followed by thermal desorption would result in even greater sensitivity than the solvent desorption used by Heikes. The method, however, is unsuitable for methyl bromide because recoveries of freshly added fumigant are low.

The Association of Official Analytical Chemists has, for many years, called for a multi-residue determination of fumigants. There are a number of multi-residue methods that have been tested. These include purge and trap procedures (e.g. Heikes 1987) and solvent extraction (e.g. Daft 1987). In 1988 (Schmidt et al. 1988) the Association of Official Analytical Chemists recommended further work on the method of Daft (1987). The method involves extraction into acetone/water, back-extraction into iso-octane and determination by gas-chromatography. This procedure enables detection of several fumigants down to parts per billion levels, but is unsatisfactory for methyl bromide and has not been tested on phosphine. However, a protocol for more detailed evaluation was considered acceptable by the Association but, as of 1992 'no further work has been accomplished toward initiating a collaborative study, although the Associate Referee plans to do so as time permits' (Sawyer 1992). The method of Daft (1987) uses essentially old technology, which enables it to be conveniently used by pesticide laboratories, but it suffers the problems of all methods using solvent extraction, including disposal problems and possibilities of contamination and interference.

Pre-concentration

Volatile chemicals can be concentrated on 'traps'. These are either chemical traps (e.g. charcoal, Tenax) or cold traps. Volatiles can be removed from chemical traps with a small quantity of solvents (solvent desorption) or with heat (thermal desorption). Recent advances involve direct coupling of a trap to a chromatograph, so that all the trapped material is eluted as one peak. [The intermediate step of 'thermal focusing' is also used whereby material is transferred from a larger to a smaller trap, and then onto the gas chromatograph, e.g. Burger et al. (1991) and Pankow et al. (1988). 'Thermal focusing' may employ several traps outside the gas chromatograph, or traps may be part of a multi-column system inside the chromatograph.]

Tangerman (1986) used such techniques to determine concentrations of carbonyl sulphide and carbon bisulphide in laboratory air. Gases were trapped on a tube containing Tenax at low temperatures and the tube was transferred to the injection port of the chromatograph. The detection limit of carbonyl sulphide was 1.2 ng/m^3 (3 parts per trillion). This, incidentally, was less than one thousandth of the concentration in the laboratory air. Kallio and Shibamoto (1988) determined methyl bromide and chloropicrin at similar sensitivities, after trapping the gases on a cold capillary column and then rapidly heating the column, which was directly attached to the gas chromatograph.

Membranes have been used to concentrate volatile gases (Blanchard and Hardy 1985). The method avoids the problems associated with traps and proved useful for such fumigants as acrylonitrile, carbon tetrachloride and various halogenated alkanes.

Quality assurance programs

Laboratory accreditation and quality assurance programs have long been a part of analysis of protectants (e.g. Ratliff 1993, NATA 1992). Written procedures, interlaboratory trials, etc., are a regular part of the work of pesticide laboratories. The advantages of participation in interlaboratory trials is obvious from a wide literature and also obvious to anyone who has participated in such trials. How this might be done is discussed later in this paper.

Overview

Detectors are capable of determining chemicals at less than 0.001 ng/sec, columns can give resolution of a few seconds, and trapping techniques enable all the volatiles in large quantities of air to be focused into one concentrated injection. All these techniques use standard, purchasable, items. It appears that detection limits of 1 ng/g (1 ppb) on food and 1 ng/m^3 in air (1 ppt) are readily achievable, and even lower limits of detection seem possible.

As part of a general environmental program, regulatory authorities in the United States have validated procedures that detect parts per billion levels in foodstuffs, and parts per trillion levels in air. These are limits of detection much more sensitive than those published from stored-product laboratories. They are also below the 'guideline' levels (i.e. detection limits) of 100–1000 ppb that form part of food regulations. These detection limits have implications for the regulation of fumigants in foodstuffs. It is possible that the current 'residue-free' status of fumigants will not remain intact (presuming, for the sake of the argument, that it currently exists).

For several decades, workers in stored-product laboratories were the leading experts in analysis of fumigants. Without wishing to offend any individual, this is no longer the case as 'fumigant analysis' may now be considered to be a sub-set of the general topic of 'analysis of volatiles', where regulatory agencies in the USA are at the forefront in the development, or at least in the publication, of sensitive methods of analysis.

General Critique of Fumigant Analysis

Although much work has been done on fumigant analysis, there is still no general regulatory method validated by bodies such as the (International) Association for Official Analytical Chemistry, and particularly none that uses modern techniques. From the point of view of stored products, it is unfortunate that multi-residue procedures such as those of Heikes (1987) or Daft (1987) give poor recoveries of methyl bromide and do not include data on phosphine. Perhaps it is the responsibility of workers in stored-products to ensure that commonly-used fumigants are included in multi-residue methods of analysis.

Analysis of fumigant residues, like analysis of most insecticide residues, faces the fundamental problem of 'aged' or 'weathered' residues. That is, if one analyses residues of a chemical some time after application, one does not know the 'true' answer. In addition, recovery of freshly added or 'spiked' residues is insufficient evidence that the method works on 'aged' residues. In homogeneous media, such as air or water, the problem of aging is not important, as the mixture remains homogeneous. The problem is important in heterogeneous media, such as soils or grains, where chemicals migrate from sites with low binding energy to sites from which it is difficult to dislodge the chemical. General techniques to solve this problem have long been in use (e.g. Sandall 1959) and long applied to grain protectants (e.g. Desmarchelier et al. 1977). Such techniques require the use of several regimes which vary extraction conditions, and the 'true' answer is that

obtained from several different procedures, provided no other procedure gives higher results. Gunther (1962) has claimed that results from procedures that rely on recovery of 'spiked' samples for validation are 'illusory'. On this criterion (which I personally fully support) most residue determination of fumigants, and perhaps all residue determination of fumigants, are 'illusory', i.e. the values obtained by the procedures used merely describe the results obtained by the procedures, and not necessarily the 'true' value.

If one accepts the criteria of Sandall, Gunther and others, validation of methods of analysis of aged residues of fumigants requires obtaining the same answer by several different techniques. This is a methodology we employ at SGRL, but it is very time-consuming. It is also very rare that different methods give the same answer, to within the degree of precision claimed by each method.

Validation of analytical procedures by adhering to protocols such as those of Sandall (1959) is important to remove systematic errors of analysis. However, it is still possible for analytical errors to occur with valid analytical procedures. To detect, and ultimately to reduce, such 'random' errors, interlaboratory trials are standard practice in laboratories which determine pesticides. The advantages of such trials are well published (e.g. IUPAC 1978). I know of no such interlaboratory trials for grain fumigants, and certainly none in stored-product laboratories. It would certainly be useful to conduct such trials, which would either detect significant variations between laboratories or, in the event that we all obtained the same answer, give credence to results from different laboratories. A collaborative study on fumigant residues would be difficult, but not impossible. For a start, gaseous samples can be reliably maintained, as they are in samples certified by such agencies as the U.S. Environmental Protection Agency. Second, such gaseous samples could be used for 'spiked', i.e. freshly-added, samples in fumigant analysis. Third, it would be possible to transport 'aged' residues in sealed containers, ideally at low temperature, for analysis on a pre-determined date.

In summary, I believe that we have marvellous technology available for fumigant analysis, but that we have failed to take the necessary steps to validate our results. With currently-available technology, it is feasible, and almost routine, to be able to detect 1 ppb on food and 1 ppt in air. However, if we accept the arguments of Sandall (1959) and Gunther (1962), and validation of their arguments for protectants on grain (Desmarchelier et al. 1977), we must doubt whether our procedures have been shown to give the 'true' answer and, in the absence of inter-laboratory check programs, whether existing procedures are accurately used.

Acknowledgments

The Partners to the Stored Grain Research Laboratory are thanked for financial assistance and Chris Whittle and Peter Annis for helpful comments.

References

ACGIH (American Conference of Government Industrial Hygienists) 1983. Air Sampling Instruments for Evaluation of Atmospheric Contaminants, 6th Edition (Lioy, P.J. and Lioy, M.J.Y. editors), 515 p, Cincinatti, Ohio, USA.

Berck, B. 1975. Analysis of fumigant and fumigant residues. Journal of Chromatography, 13, 256–267.

Blanchard, R.D. and Hardy, J.K. 1985. Use of a permeation sampler in the collection of 23 volatile organic priority pollutants. Analytical Chemistry, 57, 2349–2351.

Burger, B.V., Le Roux, M., Munro, Z.M. and Wilken, M.E. 1991. Production and use of capillary traps for headspace gas chromatography of airborne volatile organic compounds. Journal of Chromatography, 552, 137–151.

Daft, J. 1987. Determining multifumigants in whole grains and legumes, milled and low-fat grain products, spices, citrus fruits and beverages. Journal of the Association of Official Analytical Chemists, 70, 734–9.

Desmarchelier, J., Bengston, M., Connell, M., Minett, W., Moore, B., Phillips, M., Snelson, J., Sticka, R., and Tucker, K. 1977. A collaborative study of residues on wheat of chlorpyrifos-methyl, fenitrothion, malathion, methacrifos and pirimiphos-methyl. I. Method development. Pesticide Science, 8, 473–483.

Gunther, F.A. 1962. Instrumentation in pesticide residue determinations. Advances in Pest Control Research, 5, 191–319.

Greve, P.A. and Hogendoorn, E.A. 1979. Meded. Fac. Landbouwivet. Rijksuniv. Gent. 44, 877–884; Scudamore (1988) p244–246.

Heikes, D.L. 1987. Purge and trap method for determination of volatile halocarbons and carbon disulfide in table-ready foods. Journal of the Association of Official Analytical Chemists, 70, 215–225.

Heikes, D.L. and Hopper, M.L. 1986. Purge and trap method for determination of fumigants in whole grains, milled grain products, and intermediate grain-based foods.Journal of the Association of Official Analytical Chemists, 69, 990–998.

Heuser, S.G.B. 1973. Determination of residues arising from fumigant practice. Pesticide Science, 4, 409–416.

IUPAC (International Union of Pure and Applied Chemistry) 1978. Advances in Pesticide Science (Geissbuehler, H., Brook, G.T. and Kearney, P.C. ed.) Symposium Vic. The reliability of pesticide residues, pp615–668.

Kallio, H. and Shibamoto, T. 1988. Direct capillary trapping and gas chromatographic analysis of bromomethane and other highly volatile air pollutants. Journal of Chromatography, 454, 392–397.

NATA (National Association of Testing Authorities) (1991). General Requirements for Registration. National Association of Testing Authorities, Australia.

NIOSH (National Institute for Occupational Safety and Health) 1985. NIOSH Manual of Analytical Methods, 3rd Edition, US Government Printing Office, Washington, D.C.

Nunez, A.J., Gonzalez, L.F. and Janak, J. 1984. Pre-concentration of headspace volatiles for trace organic analysis by gas chromatography. Journal of Chromatography, 300, 127–161.

Pankow, J.F., Ligochi, M.P., Rosen, M.E., Isabelle, L.M. and Hart, K.M. (1988). Adsorption/thermal desorption with small cartridges for the determination of trace aqueous semivolatile organic compounds. Analytical Chemistry, 60, 40–47.

Ratliff, T.A. (1993). The Laboratory Quality Assurance System. Van Nostrand Reinhold, New York, USA, 250 p.

Rhoderick, G.C. and Miller, W.R. (1990). Multipoint calibration of a gas chromatography using cryogenic preconcentration of a single gas standard containing volatile organic compounds. Analytical Chemistry, 62, 810–815.

Sandall, E.B. (1959) in 'Treatise on Analytical Chemistry' (Kolthoff, I.M., Elving, P.J. ed.), Part 1, Vol. 1, Interscience, New York, U.S.A.

Sawyer, L.D. (1992) Multiresidue methods. Journal of the Association of Official Analytical Chemists, 75, 119–122.

Schmitt, R., Steller, W.A., Boyer, K.W., Fong, W.G., Myrdal, G.R., Elkins, E.R., McCully, K.A. and Albert, R.H. (1988). Committee on residues: recommendations for official methods. Journal of the Association of Official Analytical Chemists, 71, 128–132.

Scudamore, K.A. (1988). Fumigant analysis. In: Analytical Methods for Pesticides and Plant Growth Regulators, Vol. XVI, Academic Press, N.Y., 207–261.

Tangerman, A. (1986). Determination of volatile sulphur compounds in air at the parts per trillion level by Tenax trapping and gas chromatography. Journal of Chromatography, 366, 205–216.

USEPA (United States Environmental Protection Agency) 1989. Assessing Multiple Pollutant Multiple Source Cancer Risks from Urban Air Toxins. Summary of Approaches, and Insights from Completed and On-going Urban Air Toxins Assessment. EPA-450/2-89-010, 258p.

The influence of temperature on the sensitivity of two nitidulid beetles to low oxygen concentrations

J. E. Donahaye, S. Navarro and M. Rindner*

Abstract

In laboratory experiments, larval, pupal, and adult stages of the nitidulid beetles *Carpophilus hemipterus* (L.) and *Urophorus humeralis* (F.) were exposed to simulated burner-gas concentrations at three temperatures of 26°, 30°, and 35°C. The gas concentrations were: $1\%O_2$, $85\%N_2$, $14\%CO_2$; $2\%O_2$, $84.7\%N_2$, $13.3\%CO_2$; $3\%O_2$, $85\%N_2$, $12\%CO_2$—all at 75% relative humidity. For all insects submitted to the modified atmosphere (MA) containing $3\%O_2$ at 26°C, exposure time to produce 95% mortality was 196 hours. To obtain the same mortality level with the MA containing $1\%O_2$ at 35°C 60 hours was required. Comparison of exposure times required to produce 50% kill (LT_{50}) showed that the effect of temperature on treatment efficacy was most pronounced at the 1% O_2 level where for the three stages of both species tested, values of LT_{50} at 26°C were about half those at 35°C. However, at $3\%O_2$ and 35°C, LT_{50} levels were only marginally reduced.

Introduction

The problems of field infestations of nitidulid beetles in dates at harvest time in Israel have been addressed in previous investigations (Donahaye and Navarro 1989; Donahaye et al. 1991 a,b). For commercial scale fumigations, the addition of 20% carbon dioxide (CO_2) in air to methyl bromide (MB) has enabled recommended MB dosage rates to be reduced to 12 g/m^3 and reduction in exposure time from 16 to 6 hours (Navarro et al. 1989). However, decisions have already been made by some regulatory organisations to phase out the use of MB as a fumigant due to its involvement in depletion of the stratospheric ozone (USEPA 1993).

A further incentive for the introduction of non-toxic chemical control procedures into the date industry has been for the treatment of organically grown dates using MAs. At present we are experimenting with treatments in a fumigation chamber of 36.1 m^3 capacity attached to a controlled atmosphere generator using a catalytic converter running on butane or propane. This produces in-chamber concentrations of 1–3% oxygen (O_2) and 12–14% CO_2 within 6 hours. It is highly effective in causing insect emigration, but as a control treatment it is slow, and if quick turnaround is required, short exposure times may not be sufficient to produce complete mortality of residual infestations.

Recent studies (Soderstrom et al. 1992) have shown that the sensitivity of *Tribolium castaneum* (Herbst) to hypoxia and hypercarbia is strongly influenced by temperature. This laboratory study was undertaken to investigate whether treatments of the burner-gas concentrations could be applied as an alter-native to the present MB/CO_2 fumigations, by optimising temperatures to obtain complete kill within an acceptably short exposure period.

Materials and methods

Modified atmospheres

Three combinations of atmospheric gases were chosen to cover the range of atmospheres frequently obtained in the burner-gas treatment chamber, namely: $1\%O_2$, $85\%N_2$, $14\%CO_2$; $2\%O_2$, $84.7\%N_2$, $13.3\%CO_2$; $3\%O_2$, $85\%N_2$, $12\%CO_2$. These compositions were obtained from supply cylinders of O_2, N_2 and CO_2 using a gas-mixing apparatus described by Donahaye (1992). This consisted of component gases supplied in tubing at rates regulated by a series of valves and gas-flow meters, that enabled the components to be mixed in the desired combinations. After the gas supplies converged, gas in the common supply-line was led to temperature controlled incubators, adjusted to the desired temperature, and passed through a wash-bottle containing sulphuric acid to obtain a constant relative humidity (r.h.) of 75%. Finally the gas mixtures were delivered via a distribution chamber to a series of 100 mL Erlenmeyer flasks that served as exposure chambers arranged in-parallel.

Temperatures

The three exposure temperatures of 26° 30° and 35°C were chosen as being within the proven feasible range for burner-gas treatments in the chamber. The ambient temperatures during the harvest season were in the lower range, and when the cooling system was not operated, the heated gases from the exothermic converter contributed to heating of the chamber at a rate of about 1°C per hour. Any additional heat required to raise the temperature to 35°C could be supplied from an electric heater installed on the chamber wall.

Insects

The beetles *Carpophilus hemipterus* (L.) and *Urophorus humeralis* (F.) served as test insects. Larval, pupal and adult stages were obtained from cultures reared at 26°C and 70% r.h. on a synthetic food medium (Donahaye and Navarro, 1989). Larvae were taken from culture jars 7 days after egg hatch. Pupae were exposed to the treatments 1–2 days after pupation, and were obtained by daily removal of pupae from culture jars. Newly emerged adults were collected daily and held on culture medium for 7 days before exposure to treatments.

Experimental procedure

Six groups of 30–50 insects were placed in exposure flasks together with approximately 2 g synthetic food medium, and linked to the gas mixture apparatus. An additional flask served as control. Periodic removal of flasks was based on preliminary trials to cover the time ranges over which insect mortality

* Department of Stored Products, Agricultural Research Organization, P.O. Box 6, Bet Dagan, 50250 Israel.

was found to occur for each species, stage, gas-mixture, and temperature. Due to heterogeneity of response, each set of exposures was repeated five times. At the end of each exposure time a flask was removed from the apparatus and held in a constant temperature room at $30 \pm 1°C$ and $60 \pm 5\%$ r.h. Mortality was determined after 10 days, with larvae that failed to pupate and pupae that failed to reach adult emergence being considered as dead. Experimental results were subjected to probit analysis using a program written by Daum (1979).

Results

The exposure times required to produce 95% mortality (LT_{95}) for *C. hemipterus* and *U. humeralis* are given in Figures 1 and 2. Values missing in the larval stages of both species are

because mortality was prolonged, and heterogeneity of response was so great that significant regression could not be obtained.

From both figures it can be seen that the effect of temperature was most pronounced at the 1% O_2 and 2% O_2 levels, where in most cases, at 35°C, less than half the time was required to produce 95% mortality than at 26°C. At the 3% O_2 level, adult mortalities of both species were hardly affected by temperature with more than 12 days exposure required to produce LT_{95}. For *C. hemipterus*, mortalities below 48 hours at the LT_{95} level were recorded only for adults (18 hours) and pupae (46 hours) whereas for *U. humeralis* all LT_{95} mortalities were more prolonged than 48 hours.

For a comparison of sensitivity between species to the different treatments, results are given in Table 1 at the LT_{50} level.

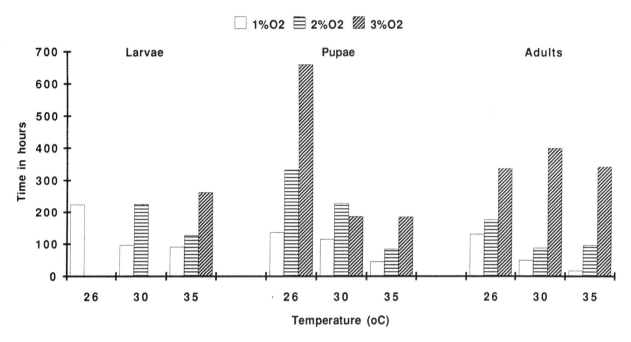

Fig. 1.   Mortalities at the LT_{95} level for three stages of *Carpophilus hemipterus* at three temperatures.

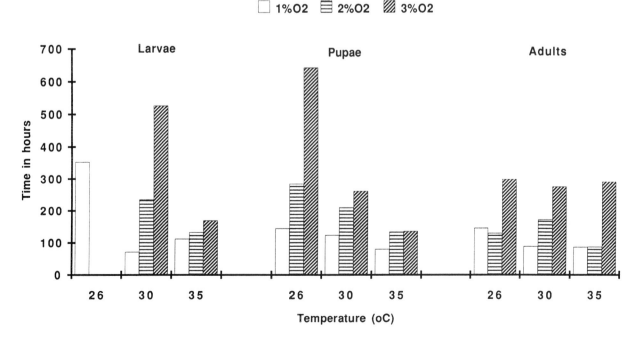

Fig. 2.   Mortalities at the LT_{95} level for three stages of *Urophorus humeralis* at three temperatures.

Table 1. Comparison of sensitivities at the LT_{50} level, of larvae, pupae and adults of *Carpophilus hemipterus* and *Urophorus humeralis* to three burner-gas concentrations at three temperatures. (NS = non-significant regression, numbers in brackets represent 95% confidence limits.)

		Temperature					
		26°C		30°C		35°C	
Gas mixture	Stage	*C. hemipterus*	*U. humeralis*	*C. hemipterus*	*U. humeralis*	*C. hemipterus*	*U. humeralis*
$1\%O_2$	Larva	112(97–136)	112(73–913)	60(51–69)	36(23–46)	60(54–67)	57(51–63)
$85\%N_2$	Pupa	78(62–109)	102(83–116)	24(17–30)	50(44–58)	26(23–30)	45(40–49)
$14\%CO_2$	Adult	69(61–77)	66(58–72)	29(25–32)	48(42–53)	11(10–12)	38(32–44)
$2\%O_2$	Larva	NS	NS	105(88–138)	89(72–123)	69(61–81)	82(70–91)
$84.7\%N_2$	Pupa	137(61–313)	109(95–126)	43(12–59)	71(64–82)	36(31–40)	60(51–68)
$13.3\%CO_2$	Adult	81(21–141)	83(77–88)	50(5–91)	82(76–87)	36(31–41)	66(58–72)
$3\%O_2$	Larva	NS	NS	317(275–405)	218(186–289)	104(82–121)	144(139–149)
$85\%N_2$	Pupa	183(134–286)	149(95–54847	108(101–114)	139(121–156)	108(99–121)	94(90–99)
$12\%CO_2$	Adult	151(134–185)	196(165–232)	138(119–180)	173(153–219)	176(152–233)	157(142–178)

Discussion

Soderstrom et al. (1992) examined the influence of temperature over the range 38–42°C on the influence of hypoxia and hypercarbia on *T. castaneum* adults for 6-hour exposures. Although the different experimental conditions make comparison difficult, their results clearly indicate that raised temperatures could be used to reduce treatment duration. Possibly the fact that the nitidulid beetles in this experiment are also field pests that develop normally at high temperatures had an attenuating influence on the effect of temperature on insect mortality under exposure to the burner-gas modified atmospheres. These results would not enable CA treatments using burner-gas to replace conventional fumigations when quick turn-around of the fruit in the treatment chamber is required.

Acknowledgment

The authors thank Mr A. Azrieli for technical assistance provided during the course of this study.

References

Daum, R.J. 1979. A revision of two computer programs for probit analysis. Bulletin Entomological Society of America, 16, 10–15.

Donahaye, E. and Navarro, S. 1989. Sensitivity of two dried fruit pests to methyl bromide alone, and in combination with carbon dioxide or under reduced pressure. Tropical Science, 29, 9–14.

Donahaye, E., Navarro, S. and Miriam Rindner. 1991a. The influence of low temperatures on two species of *Carpophilus* (Coleoptera; Nitidulidae) Zeitschrift fuer angewandte Entomologie, 111, 297–302.

Donahaye, E., Navarro, S., Miriam Rindner and Dias, R. 1991b. The influence of different treatments causing emigration of nitidulid beetles. Phytoparasitica, 19, 273–282.

Navarro, S., Donahaye, E., Dias, R., and Jay, E. 1989. Integration of modified atmospheres for disinfestation of dried fruits. Final Report of BARD project No. I–1095–86. 86 p.

Soderstrom, E.L., Brandl D.G. and Mackey, B. 1992. High temperature combined with carbon dioxide enriched or reduced oxygen atmospheres for control of *Tribolium castaneum* (Herbst) (Coleoptera: Tenebrionidae.) Journal Stored Products Research, 28, 235–238.

USEPA, 1993. Preliminary use and substitute analysis of MBr in agricultural and other uses. Draft report prepared for the Office of Air and Radiation, USEPA. May 1992, 48 p.

Methyl isothiocyanate used as a grain fumigant

V. Ducom*

Abstract

Farmers of villages in Niger, traditionally incorporate dried leaves of *Boscia senegalensis* to their silos of 'niébé' (cowpeas) to protect them against insect pests. After separation analysis and antennography the natural product responsible for this insecticide property was identified as methyl isothiocyanate (MITC). MITC is presented as crystals which sublime at ambient temperature. Its molecular weight is 73 g/mol and so lies between CH_3Br and PH_3. The biological efficacy assays of MITC on *Sitophilus granarius* (all stages) were done at 20°C (60% r.h.) in 4 hours exposure time and different concentrations. All gas concentrations were measured by gas chromatography using a TSD detector. Statistical analysis was done with Statistical Analysis System.

Studies on the biological efficacy of MITC on all stages showed that 99% of mortality is obtained between 3 and 10 g.hours/m^3 and that the pupa is the most tolerant stage. Other biological experiments showed that we obtained the same efficacy at 10°C. Sorption experiments were also done with various filling ratios (F.R.), different concentrations and four hours exposure time. Even at a very low filling ratio, the sorption is high (e.g. 26% F.R., [C] = 2 g/m^3 gives 77.8% sorption) and with filling ratios above 50%, the sorption is better than 90%. Mass spectrometric analysis has shown that MITC sorbed on grain doesn't degrade to other compounds. MITC is an interesting molecule. A very high rate of sorption allows the application of MITC either by admixture with the grain during loading or by a recirculation system. That sorption is persistent due to the very low desorption rate if the grain remains in the bin. This also allows the use of MITC in leaky silos. The results indicate that dosages between 20 and 40 g/m^3 during 24 hours following application methods at a temperature higher than 10°C, should be effective.

Introduction

Farmers of villages in Niger traditionally incorporate dried leaves of *Boscia senegalensis* to their silos of 'niébé' (cowpea, a native legume) to protect them against insect pests. After separation, analysis and antennography the natural product responsible for this insecticide property was identified (Oger et al. 1989) as methyl isothiocyanate (MITC).

Owing to its broad spectrum of activity, i.e. against soil fungi, insects, weed seeds and nematodes (Pieroh et al. 1959), MITC was commercialised as a soil fumigant by Schering in the late fifties. Today, it is no longer applied alone in its active form, but as compounds such as the soil fumigants Dazomet or Metham sodium which liberate MITC in the soil.

The fumigant properties of isothiocyanates against insects have been known for many years. After the First World War, a lot of assays were done in the laboratory to determine the biological efficacy of 2-propenyl-isothiocyanate on *Sitophilus oryzae, Tribolium confusum,* and *Plodia interpunctella* tested at a minimum concentration (Niefer et al. 1925). Both 2-propenyl and ethyl isothiocyanate killed all *Sitophilus oryzae* exposed for 24 hours at 20 g/m^3, and exposure of wheat to even higher concentrations did not prevent germination (Roarl and Cotton 1929).

Considering the insecticide properties of isothiocyanates at low concentrations and the actual application of MITC through *Boscia senegalensis* in Africa to protect stored crops, it was decided to investigate more closely this molecule in the laboratory, in order to estimate its possible use as a grain fumigant. Two series of assays were done. First, the conditions of application of MITC were studied to develop some standardised methods for measuring the gas concentrations inside the test chambers and for quantifying the sorption on grain, and thus to target the optimal range of conditions for MITC use. Then, the biological efficacy of MITC was tested at 20°C on all life stages of one major pest of stored wheat: *Sitophilus granarius*.

Materials and Methods

Materials

Methyl isothiocyanate

At room temperature (20°C), MITC (CH$_3$-N=C=S) is found as colourless crystals with a pungent horse radish-like odour, a molecular weight of 73 g/mol and a density of 3060 g/m^3. It melts at 35°C (solubility in water of 7.6 g/L at 20°C) and boils at 117–119°C (vapour pressure of 2.7 kPa and saturated concentration of 75.6 g/m^3 at 20°C).

MITC was obtained in a crystal block, which could be divided into separate parts to get the required quantities of compound.

Gas analysers

Gas chromatograph Varian 3300
Detector: Thermionic Specific detector
Column: chromosorb 101, 80/100 mesh, 2 m × 1/8 inch i.d.
GC conditions:
Gas flow rates: Air: 175 mL/minute; H$_2$: 4 mL/minute;
N$_2$: 30 mL/minute
Temperatures: Column : 200°C; Injector: 200°C;
Detector: 230°C
Attenuation 2, sensitivity 10^{-10}
Retention time: 3 minutes
Integration card: STAR CS + STAR chromatograph Workstation System (V.3)
Mass spectrometer — NERMAG R10-10C
Electronic impact (E.I.) coupled with GC Delsi Di 700
Column DB5 60 m × 0.33 mm i.d.
Conditions: E.I.: 70; evmass range: 33 to 350 a.m.u.
Temperatures: column: 200°C; Inj: 200°C;
Interface: 200°C; Source: 150°C

---

* Ministère de l'agriculture, Laboratoire National d'études des techniques de fumigation et de protection des Denrées Stockées, Chemin d'Artigues – 33150 Cenon – France.

Insects

A laboratory population of *Sitophilus granarius* was used, bred in the Laboratoire des Denrées Stockées (Cenon, France). The tested insects were grown in the dark, at a temperature of 28°C and in a relative humidity of 75%. Under these conditions, the life cycle was completed in 35 days.

At day 0, 400 adults of *Sitophilus granarius* were put inside 1 kg untreated wheat for reproduction and egg-laying. Seven days later, the grain was sieved in order to remove the adults. At day 35, the first emergencies of adults were observed. At day 52, the population was ready for the experimentation and all life stages of the insect were present.

Methods

Sublimation rate

The sublimation rate of MITC at 20°C was determined in order to estimate the loss occurring between the weighing of the crystals and their introduction in the fumigation chambers, and thus a better estimate of the real dose injected.

The evolution of MITC with time was monitored for three different weights of crystal (100, 200 and 1000 mg) with three replicates/quantity. The crystals were weighed on an electronic balance (sensitivity in the order of 0.1 mg) and the decrease in weight was continuously monitored for 250 minutes using a RS 232 linkage to a computer.

Qualitative and quantitative analysis of MITC

Since this compound is known for its extreme sorption on many materials including steel and glass, MITC has to be measured in a solvent.

A linear relationship was found with methanol (CH_3OH) as solvent for the range of MITC concentrations 0.005 to 8 g/L, when an attenuation of 2 and a sensitivity of 10^{-10} were applied. Therefore these conditions were chosen as standard. A 1 g/L solution of MITC in methanol was used as the reference for the calculation of gaseous concentrations in the samples of atmosphere taken from the fumigation chambers. Before each series of measurements, the chromatograph was calibrated using five aliquots of 5 μL of methanol, each containing 5 μg of MITC.

Gas sampling from the test chambers was done directly by removing 0.5 mL with a syringe for chambers whose volumes ranged below 2 L. For chambers of 125 and 300 L, samples were taken in 500 mL blisters, which have been previously subjected to vacuum. 5 μL were then removed with a syringe and very quickly (≤ 3 seconds) transferred onto the chromatograph, to avoid any sorption on the glass. Data were either expressed in terms of concentrations (g/m^3) or converted in terms of CT product (CTP) ($g.hours/m^3$).

Sorption: relationship between concentration-filling ratio-CTP-fumigation technique

Assays with homogenisation. The assays were done inside 2 L glass jars containing grain at various filling ratios (e.g. 0.05; 0.013 ; 0.26 ; 0.52 ; 0.75 and 0.95) under the following conditions: temperature of 20°C (±1°C); relative humidity (r.h.) of 70% (±5%) ; moisture content (m.c.) of the grain of 13.5% (±0.5%) and specific weight (s.w.) of 775 g/L (±10 g/L).

Six different concentrations of gas were tested: 1, 2, 4, 8 and 16 g/m^3 for 4 hours and 10 g/m^3 for 24 hours exposure time, with four replicates/treatment. After filling up the jars, the crystals were deposited on the surface and immediately and thoroughly mixed with the wheat.

For each treatment, concentrations were measured on the surface. For experiments lasting 4 hours, eight measures were spread out on the exposure time; for the exposure time of 24

hours, the measures were taken each hour up to 5 hours then at 8, 19 and 24 hours fumigation.

Comparison between fumigation without homogenisation and fumigation in layers. The assays were done inside food containers (125 L bins) 75% filled with wheat and MITC was applied at a rate of 50 g/m^3, either for 4 hours or for 24 hours exposure time and assays were repeated twice. The following external conditions were recorded: temperature 21°C, 60% r.h., 14% m.c. and 765 g/L s.w.

Two techniques of application were applied and compared: application of the crystals on the surface of the grain after filling up; and application during the filling up, i.e. 1 g of MITC on each layer of 10 kg wheat.

Measurements were taken from three locations (top, middle and bottom) at the same times as for the assays with homogenisation.

Influence of the initial dosage (fumigation in layers). The same procedure as described in the second technique of application (above) was followed with the same two exposure times (4 and 24 hours) but using three different initial MITC concentrations (6.5, 20 and 50 g/m^3).

Influence of a recirculation system. This experiment was done at 20°C inside a small silo of 300 L equipped with a ventilator (nominal flow rate of 150 m^3/hour; water column of 60 mm) and 95% filled with grain.

After introduction of a MITC dose of 10 g/m^3, the air–gas mixture was recirculated for 1 hour and the fumigation lasted either 4 or 24 hours. Measurements were taken at four points (top; upper middle; lower middle; bottom) at the same times as for assays with homogenisation (above).

Desorption from wheat after fumigation

Some trials were carried out to determine the best method to observe the desorption rate of MITC from the grain after the end of the fumigation exposure. Comparison was made between methanol and hexane as solvent with 50 g of wheat exposed during 4 hours to a MITC concentration of 1,2,4,8 and 16 g/m^3. The grain was spread onto the laboratory bench, then immersed in 50 mL of solvent for 0, 4, 24 and 48 hours. Methanol was found to be the best compound.

The stability of the molecule after desorption was then checked by mass-spectrometry under the conditions in the section above on gas analysers.

Biological efficacy of MITC on Sitophilus granarius *at 20°C*

Samples of 100 g wheat containing on average 200 insects were used as replicates. They were prepared 24 hours before the beginning of the experiment. The procedure described in the section on assays with homogenisation was observed, using an exposure time of 4 hours, four replicates of the treatment, controls and MITC concentrations of 1, 2, 4, 8 and 16 g/m^3.

At the end of the exposure time, the glass jars were opened up to allow desorption of fumigant from the grain and each sample was subsequently sieved to remove the adults. All adults present in a sample were transferred into a single Petri dish containing 5 g of untreated grain and kept in the dark breeding room (first series of lots). The remaining grain, which contained all hidden life stages (egg, larvae and pupae) was collected into a separate container and also kept in the breeding room (second series of lots).

After 7 days, the percentage of adult mortality was calculated for each lot of the first series. At 7, 14, 21 and 28 days, each lot of the second series was sieved again; the adults were removed and counted; the grain was collected and put back again in the breeding room. A last adult counting was done at 35 days. Each count of adults corresponded to a different initial life stage subpopulation submitted to MITC. Thus adults counted at 7 days originated from pupae; at 14 days from old larvae; at 21 and 28 days from young larvae and at 35 days from eggs.

These numbers were compared to those found in the control lots and percentages of mortality could therefore be estimated for each life-stage subpopulation. The probit procedure available from the statistics package of Statistical Analysis System (SAS Institute Inc., Cary, NC) was applied to determine the CTP which killed 50, 99 and 99.997% of each subpopulation.

Results and Discussion

Sublimation rate

A quantity of 1 g crystals needed more time to achieve complete sublimation than a quantity of 100 mg but the sublimation rate was greater. A delay of 5 seconds was necessary between the weighing and the introduction of the crystals inside the fumigation chambers. Regarding the sublimation rate, the loss of MITC could equal 0.2% of the initial quantity for 1 g and 0.15% for 100 mg. These losses were then neglected in the subsequent studies.

Sorption

Relationship between initial dosage – filling ration – CTP – 4 hour fumigation

For the experiments lasting 4 hours, an inverted exponential relationship was found between the CTP and the filling ratio at constant dosage values (with values of the coefficient of determination, $r^2 > 0.9$) (Table 1; Fig. 1).

An exponential relationship was found between the sorption percentage of MITC on the grain and the filling ratio (Table 2). Even at a very low filling ratio, the sorption is high and close to 100% with filling ratios greater than 50%.

At constant filling ratio values (i.e. same surfaces of contact), a positive linear relationship was found between the CTP and the introduced dose ($r^2 > 0.99$) and the slope coeffi-

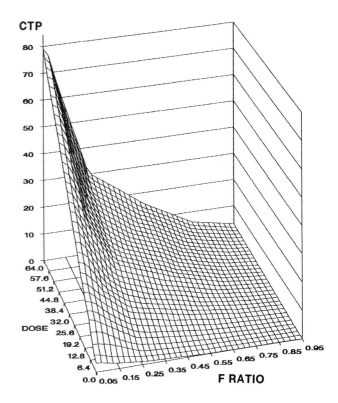

Fig. 1. CTP of methyl isothiocyanate resulting from various combinations of dosage and filling ratio. CTP = f([C], F Ratio). 4 hours fumigation; temperature = 20°C; CT g.hours/m^3; dose: dosage g/m^3; F ratio:filling ratio.

cients of these straight lines were negatively correlated to the filling ratio values (Table 3).

For the experiments lasting 24 hours, an exponential relationship was also observed between the sorption percentage and the filling ratio. The sorption percentages were of the same

Table 1. CTP (expressed in g.hours/m^3) obtained in a 4-hour fumigation with homogenisation, for various filling ratios and various concentrations of methyl isothiocyanate.

Filling ratio	Initial dosages in g/m^3						
	1	2	4	8	16	32	64
0	3.16	6.32	12.64	25.28	50.56	101.1	202.2
0.065	3.1	4.6	7	13	21	40.71	79.11
0.13	2	3	6	10	16	31.38	61.14
0.26	1	1.4	2	4	8	15.4	30.44
0.52	0.2	0.4	0.7	1.5	4	7.79	15.79
0.75	0.05	0.11	0.19	0.46	1.60	3.05	6.25
0.95	0.02	0.04	0.07	0.17	0.77	1.5	3.1

Table 2. Percentage of sorption obtained in a 4-hour fumigation (expressed in percentage of losses). Fumigation with homogenisation, various filling ratios and various concentrations.

Filling ratio	Initial dosages in g/m^3						
	1	2	4	8	16	32	64
0.065	1.9	27.2	44.6	48.6	58.5	59.7	60.9
0.13	36.7	52.5	52.5	60.4	68.4	69.0	69.8
0.26	68.4	77.8	84.2	84.2	84.2	84.8	84.9
0.52	93.7	93.7	94.5	94.1	92.1	92.3	92.2
0.75	98.4	98.2	98.5	98.2	96.8	97.0	96.9

order as those recorded during the fumigations of 4 hours for a dose of 8 g/m^3 (Table 4).

In these laboratory trials, with a 4-hour fumigation, it seems impossible, due to sorption, to obtain the CTP necessary to kill insects when the filling ratio is above 50%.

Comparison between fumigation without homogenisation and fumigation in layers

Regarding the first technique of fumigation without homogenisation, the maximum concentrations obtained at all measuring points were reached very quickly (15–20 minutes). A considerable heterogeneity was observed in the bin characterised by very low CTPs in the middle and the bottom (Table 5; Fig. 2). The maximum concentration obtained on the bottom was only 1/100th of the initial dose.

In contrast, with the second technique of fumigation in layers, maximum concentrations were obtained much more slowly, with a peak occurring only after 5 hours. Since CTPs obtained were lower on the top compared to the first technique, they were much more even. The differences between CTPs observed in the lower layer and those in the top layer could be the result of the difference of filling ratio between the top and the bottom, the empty volume being on the top.

The CTPs obtained with the first method are never enough in the grain mass to kill insects even at 24 hours. With the second method, the CTPs obtained at 24 hours are high enough, and nearly enough at 4 hours exposure time.

The quick natural diffusion of MITC associated with its rapid sorption on grain shows that an admixture layer by layer improves the even distribution of the fumigant. This reflects the traditional practice in Africa.

Influence of the initial dosages in layer fumigation

Even with a fumigation in layers, the CTPs obtained on the bottom are lower than those obtained on the top. According to sorption assays and mathematical calculations, for a filling ratio of 0.75, we would expect CTPs of 0.5, 1.85 and 4.85 g.hours/m^3, respectively, for initial dosages of 6.5, 20 and 50

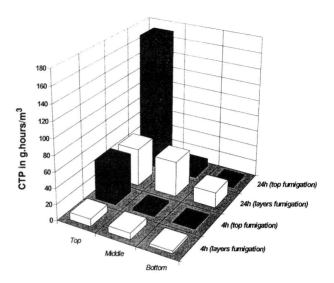

Fig. 2.   Influence of fumigation in layers on CTP.

g/m^3, 4 hours exposure time). These calculated CTPs are lower than those obtained on the top, and equal to or greater than those obtained on the middle and bottom (Table 6). The difference observed between calculated and actual values could be the result of a better availability of free gas in the top layers.

Recirculation system in MITC fumigation

The influence of a 1 hour recirculation system on the MITC sorption is important and allows the distribution of equal concentration at all levels. A CTP of 5 g.hours/m^3 (Table 7; Fig. 3) is not enough to kill insects at all stages. Nevertheless, if the fumigation was done at 10 g/m^3 in layers, the CTP obtained should have been 0.9 g.hours/m^3 (CTP = 102 exp(–0.04(F Ratio)), $r^2 = 0.98$)

Desorption from wheat after fumigation

Only preliminary experiments have been done. The rate of desorption varied considerably according to the exhaust gas method used.

- With ventilation, desorption is very quick, two-thirds MITC concentration on grain is lost in 4 hours and only traces remain after 24 hours.
- In contrast, when the grain is left in the bins, after 24 hours only 50% is lost and the actual concentrations on grain remain at that level for some days.

MITC sorbed on grain does not degrade to other compounds as mass spectrometer analysis showed.

Table 3.   Relationship between CTP and introduced dose. y = CTP (g.hours/m^3); x = initial dosage (g/m^3), 4 hours fumigation with homogenisation.

Filling ratio	Relationship between CTP and [C]	r^2
0.065	$y = 1.2x + 2.31$	0.99
0.13	$y = 0.93x + 1.62$	0.98
0.26	$y = 0.47x + 0.36$	0.99
0.52	$y = 0.25x - 0.21$	0.98
0.75	$y = 0.1x - 0.15$	0.99
0.95	$y = 0.05x - 0.1$	0.99

Table 4.   Relationship between sorption percentage and filling ratio for a 24-hour fumigation, dosage 10 g/m^3.

Filling ratio	Control	0.065	0.13	0.26	0.52	0.75	0.95
CTP	190	91.5	56.3	33.6	14.6	2.4	0.9
% of sorption		51.9	70	82	92	98.7	99.5

Table 5.   Influence of fumigation in layers on CTP (4 and 24 hours fumigation, filling ratio 0.75, dosage of 50 g/m^3)

	CTP obtained in 4 hours			CTP obtained in 24 hours		
	Top	Middle	Bottom	Top	Middle	Bottom
Top fumigation	50.1	3.2	0.7	171	10.5	2.3
Fumigation in layers	9.1	8	3.1	47.8	44.6	19

Table 6. Influence of the dosage on CTP in fumigations by layer (CTPs expressed in g.hours/m^3).

	Bin A 6.5 g/m^3			Bin B 20 g/m^3			Bin C 50 g/m^3		
	Top	Middle	Bottom	Top	Middle	Bottom	Top	Middle	Bottom
CTP 4 hours	3.6	1	0.6	9.6	1.2	0.6	9.1	8	3.1
CTP 24 hours	8	2.6	3.1	25.5	7.5	6.2	47.8	44.6	19

Table 7. Recirculation influence on methyl isothiocyanate fumigation (CTP in g.hours/m^3)

	Top	Middle top	Middle bottom	Bottom
CTP 24 hours	5.7	4.7	4.1	5.7

Biological efficacy of MITC on *Sitophilus granarius* at 21°C

In general, the slope of the curve is very high (Fig. 4). For eggs, the mortality lies between 10 and 90% with a CTP of 1.2–4; for young larvae, the mortality lies between 10 and 90% with a CTP reaching 2–4.9; for old larvae, the mortality lies between 10 and 90% with a CTP reaching 1.8–4.5; for pupae, the mortality lies between 10 and 90% with a CTP reaching 0.8–4.7.

The biological efficacy of MITC on all stages shows that 99% of mortality is obtained between 3–10 g.hours/m^3 and that the pupal stage is the most tolerant (Table 8; Fig. 4).

Other biological experiments show that the same efficacy is obtained at 10°C (Table 9).

Conclusion

MITC is an interesting molecule from several points of view. Its diffusion is very rapid but the sorption on grain takes place immediately after its introduction. As a result, if the fumigant is applied as usual on the surface of the grain after loading, the resulting concentrations are very low. With an initial dosage of 50 g/m^3, a 24 hour exposure time and a filling ratio of 0.75, the resulting CTP obtained at the top of the bin is 171 g.hours/m^3 and only 2.3 g.hours/m^3 at the bottom, which is not enough to control all stages *of S. granarius* since a CTP of 10 g.hours/m^3 is needed.

For fumigations in layers, like applications of contact insecticides, better homogeneity in terms of CTP is obtained. Nevertheless, fumigations of 24 hours with an initial dosage of 20 g/m^3 (filling ratio 0.75) give a CTP superior to those desired in the top of the bins (25.5 g.hours/m^3) while in the bottom the CTP are too low to kill insects (6.2 g.hours/m^3). As a result, it would be necessary to apply more MITC in the bottom layers and gradually less upwards.

Table 8. Biological efficacy of methyl isothiocyanate against *S. granarius* at 21°C (CTP necessary to obtain a given mortality)

	Eggs	Young larvae	Old larvae	Pupae
CT 50	2.21	3.16	2.91	1.98
CT 99	6.64	7.14	6.44	9.8
CT Probit 9	14.5	13	11.5	16

Table 9. Biological efficacy of methyl isothiocyanate against *S. granarius* at 10°C (CTP necessary to obtain a given mortality)

	Eggs	Larvae	Pupae
CT 50	0.5	1.46	0.52
CT 99	8	5.45	6.3

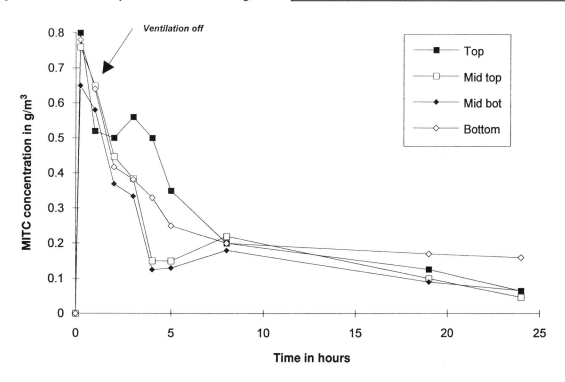

Fig. 3. Influence of recirculation system on methyl isothiocyanate concentration.

(a)

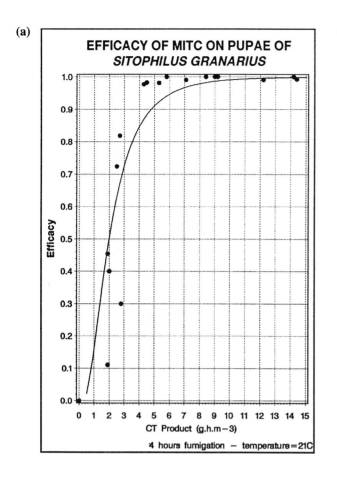

(b)

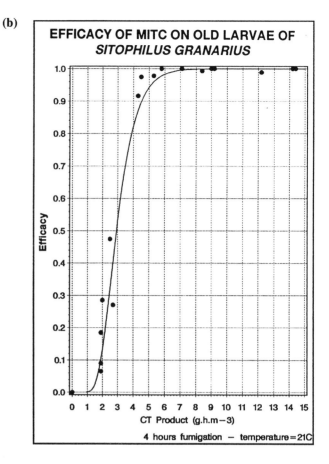

(c)

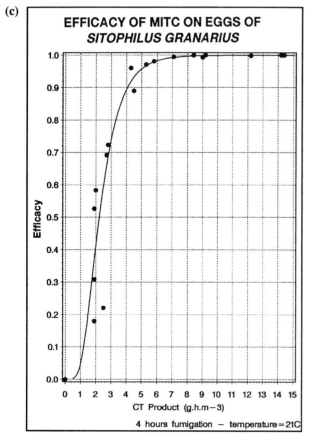

(d)

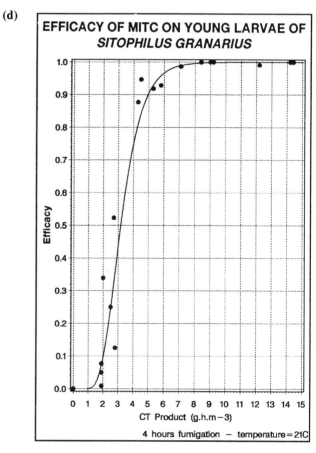

Fig. 4. Efficacy of methyl isothiocyanate on *S. granarius* at 21°C: (a) pupae; (b) larvae; (c) eggs; (d) young larvae.

The best way to obtain quickly a good homogeneity is the use of a recirculation system. In that case, fumigation at 20 g/m^3 allows control all stages of *S. granarius*.

However, its extreme sorption power gives a curative action and a protectant effect. This would allow its use in silos which are not gastight.

Knowing its anti-fungal properties, it would be interesting to carry out studies on its efficacy on the pathogen flora of grains. If it could be proved efficient, then it would be a powerful pesticide for grain, both insecticide and fungicide.

Acknowledgments

I would like to express my gratitude to Mr Patrick Ducom and Mr Daniel Richard-Molard for their advice on several aspects. I wish to thank Calliope company for its financial support.

References

Neifert, I.E., Cook, F.C., Roark, R.C., Tonkin, W.H., Back E.A. and Cotton, R.T. 1925. Fumigation against grain weevils: various volatile organic compounds. USDA Bulletin, 1313, 40 p.

Oger, J., Nammour, D. and Huignard, J. 1989. Mise en évidence de l'effet insecticide de composés souffrés sur *Bruchidius atrolineatus* (PIC) (Coleoptera: Bruchidae). Insect Science and Applications, 10, 49–54.

Pieroh, E.A. et al, 1959. Anz. Schaedlingsskd.: 32–183.

Roark, R.C. and Cotton, R.T. 1929. Tests of various aliphatic compounds as fumigants. USDA Technical Bulletin, 162, 522 p.

A Western Australian farm survey for phosphine-resistant grain beetles

R.N. Emery[*]

Abstract

A survey of 4547 Western Australian farms for phosphine resistance in rust-red flour beetle, *Tribolium castaneum* (Herbst), confused flour beetle, *Tribolium confusum*, Jacq. du Val, rice weevil, *Sitophilus oryzae* (Linnaeus), granary weevil, *S. granarius* (Linnaeus), lesser grain borer, *Rhyzopertha dominica* (Fabricius) and sawtoothed grain beetle, *Oryzaephilus surinamensis* (Linnaeus) collected separately from handling equipment, sealed and unsealed grain storages during 1991–92. A discriminating dose test detected resistance in 17.1% of *T. castaneum*, 9.4% of *S. oryzae*, 37.5% of *S. granarius*, 15.2% of *R. dominica* and 15.6% of *O. surinamensis* samples. No resistance was found in *T. confusum*. Frequency of phosphine resistance between grain handling equipment, sealed and unsealed storages was similar, although sealed storages were 2.8 times as effective as unsealed storages in minimising the incidence of grain insect infestations. Infrequent treatments with phosphine in grain-handling equipment may have resulted in more rapid development of resistance than in grain storages.

Introduction

Despite having storage conditions that are more suitable to rapid grain insect development than many of its overseas competitors, Australia has, through judicious use of protectants and management of insecticide resistance, established a reputation as an exporter of insect-free grain. The challenge now is to meet increasingly frequent market specifications which require 'residue-free' grain. In order for Western Australia to take advantage of these burgeoning markets for 'residue-free' grain, it must guarantee that its product does not contain contact insecticide residues applied at either the central handling system or on-farm.

Since 1990 all Western Australian grain exports were shipped without the use of contact protectants during storage in the central handling system. This has been brought about by strategic planning and co-operation of industry and Government over the last decade (Dean, these proceedings).

A major strategy to achieve this is through the use of controlled atmosphere (CA) storage. This significant advance in grain storage management was inaugurated in 1980 when Co-operative Bulk Handling Western Australia (CBH) decided that CA storage was the most efficient method of bulk grain storage. To this end CBH embarked on a program of sealing storage facilities which has resulted in the upgrading of over 65% (5 Mt) of its permanent storage capacity. Currently these storages are fumigated with phosphine, although other CA alternatives could be readily implemented.

Sealed storage with phosphine fumigation has also proved to be the most effective and economic method for the on-farm storage of grain. The Agriculture Protection Board of Western Australia (APB) has had an extension campaign in effect since the early 1980s, which has actively promoted this form of grain storage on-farm. It is no longer possible to purchase an unsealed farm silo in Western Australia, as all manufacturers now produce only sealed units.

Western Australia is becoming more dependent on phosphine for grain insect control both on-farm and in the central handling system. Phosphine (Worthing 1991) has many benefits over other fumigants, including ease of use, penetration and cost. There are few alternatives to phosphine which have the same advantages, particularly with the inevitable withdrawal of methyl bromide owing to its ozone-depleting properties. In order that phosphine remain available as a fumigant to the industry it is essential that any risks that could lead to its removal be identified and eliminated.

One threat to the industry is from the delivery to the central handling system of phosphine-resistant strains of grain insects which may develop in farm storages through the use of phosphine in unsealed and poorly maintained sealed storages. It has been shown that storages in this condition are unable to meet the exposure periods required for effective control. There is much evidence to suggest that the development of phosphine resistance is associated with poor fumigation techniques (Banks 1986; Friendship et al. 1986; Mills 1986; Price 1986; Taylor 1986; Tyler et al. 1983; Webley 1984).

Alternatively, some Australian researchers working with grain insect resistance have expressed concern that inefficient use of sealed silos rather than unsealed silos poses the greater threat through the development of higher levels of resistance (G.G. White, personal communication, 1990).

Entomologists involved in grain storage have been concerned for some time over the potential for grain insects to develop phosphine resistance. The U.K. Tropical Development and Research Institute (TDRI 1985) recognised the value of the FAO worldwide survey of insecticide resistance (Champ and Dyte 1976) and noted that this survey indicated the potential for a phosphine resistance problem. It considered that assessment of the status of phosphine resistance in grain insects be classed as most urgent.

In Western Australia, APB inspectors have routinely collected grain insect samples for phosphine resistance testing since 1986. However, these results do not give an accurate indication of the extent of phosphine resistance in grain insects because farms with a history of resistance are specifically chosen by APB inspectors. The objective of the APB campaign is to detect and eradicate resistant strains, not to survey the extent of resistance.

In order to accurately determine the extent and level of phosphine resistance in grain insects on farms in Western Australia, a random farm survey was implemented during 1991/92.

Western Australia is able to conduct an extensive resistance survey because of the declaration of grain insects under the Agriculture Protection Board Act and the co-operation between this group, the Western Australian Department of Agriculture and other industry bodies.

* Department of Agriculture, 3 Baron-Hay Court, South Perth, Western Australia.

This farm survey for phosphine resistance in grain insects has determined the extent of phosphine resistance throughout the grain-producing areas of Western Australia and will facilitate the establishment of a resistance-management plan which can be used to limit the development and spread of phosphine-resistant strains of grain insects. This plan will provide a model which could be used by similar industries in other parts of Australia and in other countries. As Brattsten et al. (1986) aptly stated, 'A critical prerequisite to resistance management is anticipation of resistance before control actually fails'.

Materials and Methods

Insect collection

Preliminary monitoring by the APB indicated that approximately 30% of Western Australian farms were infested with grain insects (APB Annual Reports 1981–90) and that approximately 14% of all farms were infested with phosphine-resistant insects (Table 1).These preliminary data were used to calculate that at least 5000 farms should be surveyed to achieve return of sufficient (>200) farms with resistant grain insects.

The APB database has over 12000 registered grain-producing properties throughout Western Australia. To simplify selection, every second grain-producing property in APB zones 4,5,7 and 8 (Fig. 1) was randomly selected by computer. These zones cover the major grain-producing areas of the State. Farms smaller than 500 ha were excluded as these were considered to be too small to be relevant. The total number of farms included in the survey was 5612.

Species collected for resistance testing were *Tribolium castaneum, T. confusum, Sitophilus oryzae, S. granarius, Rhyzopertha dominica* and *Oryzaephilus surinamensis*.

APB officers were given 12 months (July 1991–June 1992) to carry out the sampling of grain insects from farms. They were instructed to collect separate samples of grain insects from sealed storages, unsealed storages and grain handling equipment (headers, augers etc.) from each property. The insects were removed from the grain residue and placed in a vial with a 1:1 mixture of chemical-free rolled oats and flour. The infestation level in each of the storage types was recorded, the codes used were:

H—high infestation, large numbers of insects easily found and superficially obvious

M— moderate infestation, many insects present but not superficially obvious

L— low infestation, only a few insects present on close inspection

N— no insects found on close inspection.

Farms which did not have sealed or unsealed storages, or grain handling equipment, were recorded as A (absent) for that equipment type. The entry E (error) was used where access to the property could not be obtained. These farms were deleted from the survey, leaving a total of 4547 farms inspected.

The insect samples were then sent by post to the Department of Agriculture laboratories at South Perth for phosphine-resistance testing.

Resistance testing procedures

Grain insects were sorted by species and tested for phosphine resistance directly from the field if at least 50 live insects were present, or cultured at 26°C, 60% relative humidity (r.h.) until populations were of sufficient numbers to test. Standard procedures recommended by the Food and Agriculture Organisation (FAO) for phosphine resistance

testing were used (Anon 1975).As there were over 1000 tests to be conducted, a discriminating dose procedure was used. Phosphine was generated from pellets containing aluminium phosphide, collected over acidified water. Fumigation was carried out in 6.1 L desiccators. Insects were confined during fumigation in 50 mL polystyrene vials fitted with muslin. Phosphine was injected with a gastight syringe through a rubber septum fitted to a socket in the desiccator lid. Phosphine was mixed with air in the desiccator by a magnetic stirrer. Susceptible control insects were included in every test to ensure that there were no protocol failures (e.g. blocked syringe, broken seals).

Test insects were exposed to phosphine in the desiccators for 20 hours at 25°C, 60% r.h. then transferred to an incubator and a mixture of 1:1 flour and broken wheat was added. Mortality assessments were made 14 days after treatment. Insects were classified as dead if incapable of co-ordinated movement. The discriminating dosages shown in Table 2 are taken from the FAO method except for the *T. castaneum dosage* which was increased by 0.008 mg/L in accordance with earlier trials which had shown this dose to be more effective (Moulden 1987).

The FAO method states that the probability of a single insect in a batch of 100 being unaffected due to chance is less than 0.1. For the purpose of this paper strains were classified as resistant if more than 1 insect in 10 tested survived the discriminating dose.

Results

The resistance data in this paper are presented as a proportion of resistant strains or a proportion of farms with resistant grain insects. Many of the farm totals will not equal the sum of the sub-totals (made up of species and storage types) because some farms may have several resistant species or resistant insects in both sealed and unsealed storages. These farms will be counted only once when totals are calculated.

Phosphine resistance monitoring carried out by the APB from 1985 to 1992 is summarised in Table 1. This routine monitoring, which is biased because the APB selects properties with a history of grain insect infestation or phosphine resistance, shows that in 1991–92, 12% of farms inspected were infested with resistant insects. Over the same period this survey found 7% (Table 5).

Infestation levels in sealed storages, unsealed storages and grain handling equipment are outlined in Table 3. Infestations are lowest in grain handling equipment. More importantly, infestations of grain insects in unsealed storages are almost three times those in sealed storages. An overall infestation level of 38.31% (6.91% heavily infested) is similar to that found during routine monitoring by the APB over the same period (Agriculture Protection Board, 1992).

Table 4 summarises phosphine resistance based on strains. Of the 2238 strains tested 16% were phosphine resistant.

The frequency of resistance expressed as a percentage of farms inspected is shown in Table 5. The total percentage of farms infested with resistant insects is 7%, this is considerably lower than the figure for resistant strains shown in Table 4 (16%) because only 38% of farms are actually infested with grain insects (Table 3).

The relationship between storage type and frequency of resistance is summarised in Table 6. Of the 1861 strains collected from unsealed storages, 16.7% were resistant compared with 15.8% from sealed storages.

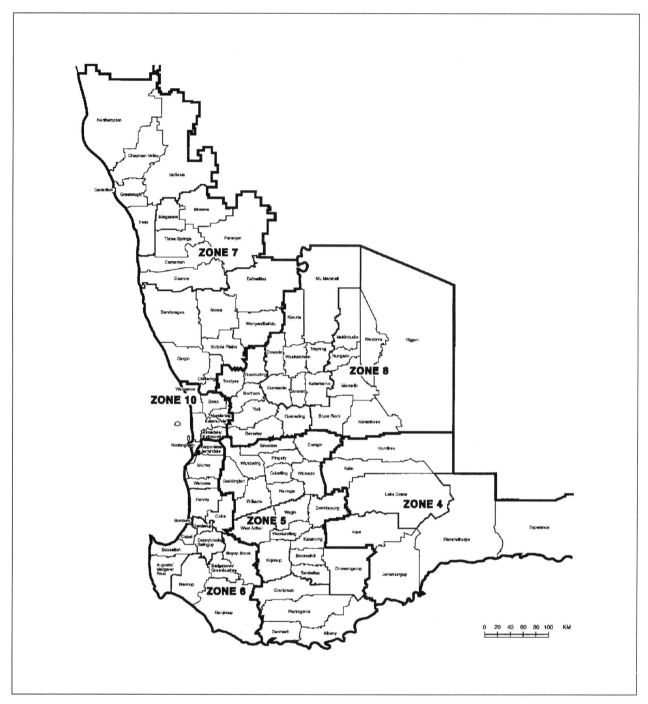

Fig. 1.   Agriculture Protection Board zones and districts.

Discussion

Infestations

The use of sealed storages on-farm has been extensively promoted in Western Australia as the most effective method of grain insect control (Newman 1987).This survey for phosphine resistance in grain insects provided a unique opportunity to compare infestation levels on sealed and unsealed farm storages (Table 3).This survey indicates that infestations in unsealed storages are 2.8 times more common than in sealed storages (39% v. 14%).Clearly, sealed storages should be the preferred option for grain insect control. Infestation levels of 3% in grain handling equipment are insignificant and recommended control methods, which include treatment with

organophosphate insecticides and amorphous silica dust, should be maintained.

The percentage of heavily infested properties shown in Table 3 is considerably higher than previous routine monitoring by the APB (7% Table 3 v. 3% APB Annual Reports 1981–1991), this has been attributed to reduced grain insect inspections and control programs by the APB during the previous year.

Population resistance

Surveys conducted by Attia and Greening (1981) in New South Wales between 1968 and 1980 detected phosphine resistance in 3% of *R. dominica*, 12%of *T. castaneum* and 29% of *T. confusum* samples. Herron (1990) found phosphine

Table 1. Monitoring of phosphine resistance on Western Australian farms 1985 to 1992.

	1985–86	1986–87	1987–88	1988–89	1989–90	1990–91	1991–92
# Farms resistant	34	72	33	77	27	7	26
# Farms susceptible	187	180	164	399	154	155	188
% Farms resistant	15	29	17	16	15	4[a]	12

[a]an experimental error resulted in test exposures of 24 hours rather than 20 hours.

Table 2. Discriminating dosages used in phosphine resistance testing.

Species	Dose (mg/L)	Exposure period (hours)
Oryzaephilus surinamensis	0.050	20
Rhyzopertha dominica	0.030	20
Sitophilus granarius	0.070	20
Sitophilus oryzae	0.040	20
Tribolium castaneum	0.048	20
Tribolium confusum	0.050	20

Table 3. Grain insect infestation levels on Western Australian farms.

	Heavily infested (%)	Moderately infested (%)	Lightly infested (%)	Insect free (%)	Farms infested (%)	No. of farms inspected
Handling	0.39	0.52	1.47	97.62	3.42	4414
Sealed	0.74	0.70	2.68	95.87	13.75	4435
Unsealed	6.44	7.63	14.91	71.03	38.69	4380
All storages	7	8	17	65	38	4547

Table 4. Summary of phosphine resistance in grain insect species.

	Species						
	Oryzaephilus surinamensis	*Rhyzopertha dominica*	*Sitophilus granarius*	*Sitophilus oryzae*	*Tribolium castaneum*	*Tribolium confusum*	Total
No. of trains resistant	102	53	18	8	188	0	369
No. of strains susceptible	553	296	30	77	910	3	1869
% strains resistant	16	15	38	9	17	0.00	16

Table 5. Percentage of farms with resistant grain insects collected from farm storages.

	Species							
Storage type	*Oryzaephilus surinamensis*	*Rhyzopertha dominica*	*Sitophilus granarius*	*Sitophilus oryzae*	*Tribolium castaneum*	*Tribolium Confusum*	% farms resistant	No. of farms inspected
Handling	0.11	0.05	0.05	0.05	0.20	0.00	0.43	4414
Sealed	0.38	0.11	0.02	0.05	0.32	0.00	0.83	4435
Unsealed	1.83	1.05	0.34	0.09	3.77	0.00	6.51	4380
All storages	2.13	1.14	0.40	0.18	4.05	0.00	7.15	4547

Table 6 Summary of phosphine resistance in grain insects collected from farm storages.

	Storage type			
	Handling	Sealed	Unsealed	Total
No. of strains resistant	20	39	310	369
No. of strains susceptible	111	207	1551	1869
% strains resistant	15	16	17	16

resistance during a 1985–86 survey in 15% of *R. dominica*, 8% of *T. castaneum and* 5% of *T. confusum*, 18% of *S. oryzae*, 14% of *S. granarius* and 2% of *O. surinamensis* samples. Herron (1990) commented that, while phosphine resistance levels had not changed markedly between these two surveys, the increase in the incidence of resistance in *R. dominica* over this period would be of concern in Western Australia.

Data collected during this Western Australian survey are summarised in Table 4. Although frequency of resistance in *R. dominica* is similar to that found by Herron (1990), the higher resistance in *T. castaneum* and *O. surinamensis* is of concern as both these pests are of high pest status in Western Australia. Resistance in *T. confusum* and *S. granarius* is of lesser consequence in Western Australia where these pests are relatively rare in grain stores intended for delivery to the central handling system. The high figure of 37% resistant *S. granarius* was attributed to the discriminating dose being too low for this species. This resulted in tolerant rather than resistant individuals surviving the test. Therefore, the discriminating dose for *S. granarius* needs to be reviewed in future trials.

Resistance on farms

The incidence of resistance presented as a fraction of strains tested may initially appear somewhat daunting (16% of all strains tested). However, Table 5 shows that, on the basis of number of farms inspected, only 7% of properties were infested with resistant insects. Continued vigilance by Western Australian farmers, combined with APB inspections and extension campaigns, are required to maintain infestations of grain insects at the current low level and to continue to minimise the chances of resistant insects contaminating the central handling system.

Effect of storage equipment on resistance

Resistant strains were found in 15% of samples collected from grain handling equipment (Table 6).This figure is similar to sealed and unsealed storages and may warrant further investigation because phosphine is rarely used to disinfest grain-handling equipment. This could indicate that either resistance has developed rapidly following a small number of treatments or that cross-infestation by resistant insects from grain storages is a problem.

Concerns that the continued inefficient use of phosphine in unsealed storages will inevitably increase the frequency of phosphine resistance have lead the Western Australian Grain Protection Committee (GPC) to consider restricting the use of phosphine to sealed storages only in this State. This would leave Western Australian farmers without sealed storages with few alternatives other than purchasing sealed storages or converting old storages because the application of contact insecticides to farm stored grain in Western Australia is highly restricted. However, Table 6 shows that phosphine resistance occurred in 17% of strains collected from unsealed storages compared with 16% from sealed storages. It would seem unlikely that 1% higher incidence of resistance in unsealed storages is sufficient to warrant action by the GPC.

Further work is required to determine if the level of resistance (i.e. resistance factor) rather than frequency of resistance, is higher in strains collected from sealed storages. Strains which have survived the discriminating dose tests during this survey have been kept in culture and a series of graded concentration tests to calculate resistance factors is planned should funds become available in the future.

Summary

The data collected during this survey could be presented in a variety of formats with alternative criteria for resistance. Interested workers are encouraged to contact the author should they require additional information.

While sealed storages had little effect on the frequency of phosphine resistance in grain insects, this type of storage is almost three times as effective in minimising the incidence of infestations.

Resistance to phosphine in grain handling equipment should be investigated further to determine if it is developing more rapidly in this situation than in grain storages.

This survey has provided base-line resistance data. Funds will be sought for another survey in the mid-1990s which will indicate any change in the status of phosphine resistance.

Acknowledgments

The skilled technical assistance of Ms H. Collie, Ms D. Hutchinson, Mr G. Mcdonald and Mr A. Szito is greatly appreciated as are Dr I. Dadour's comments on this manuscript. Financial support by the Grains Research and Development Corporation (formerly Wheat Industry Research Committee of Western Australia) is gratefully acknowledged.

References

Anon. 1975. Recommended methods for the detection and measurement of resistance of agricultural pests to pesticides 16. Tentative method for adults of some major species of stored product insects with methyl bromide and phosphine. FAO Plant Protection Bulletin, 23, 12–25.

APB (Agriculture Protection Board). 1981–1992. Annual reports. Perth, Western Australia, APB.

Attia, F.I. and Greening, H.G. 1981. Survey of resistance to phosphine in coleopterous pests of grain and stored products. General and Applied Entomology, 13, 93–97.

Banks, H.J. 1986. Fumigation of flat bulks with phosphine. Tropical Development and Research Institute. Group for Assistance on Systems relating to Grain After Harvest Seminar on Fumigation Technology in Developing Countries, 119–131.

Brattsten, L.B. Holyoke, C.W., Leeper, J.R. and Raffa, K.F. 1986. Insecticide resistance: a challenge to pest management and basic research. Science, 231, 1255–1260.

Champ, B.R. and Dyte C.E. 1976. Report of the FAO global survey of pesticide susceptibility of stored grain pests. FAO Plant Protection Service. 5.

Friendship, C.A.R., Haliday, D. and Harris, A.H. 1986. Factors causing development of resistance to phosphine by insect pests of stored produce. Tropical Development and Research Institute. Group for Assistance on Systems relating to Grain After Harvest Seminar on Fumigation Technology in Developing Countries, 141–149.

Herron, G.A. 1990. Resistance to grain protectants and phosphine in coleopterous pests of grain stored on farms in New South Wales. Journal of the Australian Entomological Society, 29, 183–189.

Mills, K.A. 1986. Phosphine dosages for the control of resistant strains of insects. Tropical Development and Research Institute. Group for Assistance on Systems relating to Grain After Harvest Seminar on Fumigation Technology in Developing Countries, 119–131.

Moulden, J.H. 1987. Grain insect infestation and insecticide resistance in Western Australia — an analysis 1973 to 1984. Western Australian Department of Agriculture Research Report 1/87.

Newman, C.R., ed. 1987. Sealed grain stores in Western Australia — a review. Agriculture Protection Board Report.

Price, N.R. 1986. The biochemical action of phosphine in insects and mechanisms of resistance. Tropical Development and Research Institute. Group for Assistance on Systems relating to Grain After

Harvest Seminar on Fumigation Technology in Developing Countries, 88–89.

Taylor, R.W.D. 1986. Response to phosphine of field strains of some insect pests of stored products. Proceedings of the GASGA Seminar on Fumigation Technology in Developing Countries, TDRI, Slough, U.K., 18–21 March 1986, 132–140.

Tropical Development and Research Institute. 1985. The development of phosphine resistance by insect pests of stored grain. Tropical Development and Research Institute Newsletter, 5.

Tyler, H.J., Taylor, R.W. and Rees, D.P. 1983. Insect resistance to phosphine fumigation in food warehouses in Bangladesh. International Pest Control, 25, 10–13.

Webley, D.J. 1984. Developments in the use of phosphine in the Australian wheat industry. Journal of Stored Products Research, 20,88–89.

Worthing, C.R., ed. 1991. The pesticide manual—a world compendium. 9th edition. The British Crop Protection Council, 1141 p.

Effects of low oxygen phosphine fumigations on adult *Rhyzopertha dominica*

F.M. Johnston and C.P. Whittle*

Abstract

The effects of low oxygen atmospheres on the mortality of adult *Rhyzopertha dominica* in flow-through and single-dose phosphine fumigations are described. Flow-through fumigations were conducted using phosphine at 0.1 mg/L applied continuously to adult phosphine-susceptible and resistant *R. dominica* over a period of 30 hours in the presence of either 1, 2, 8 or 21% oxygen atmospheres. Fumigations were conducted using a phosphine concentration of either 0.003 mg/L or 0.007 mg/L to adult phosphine-susceptible *R. dominica* with 1, 8, 12, and 21 or 1, 6, 8, and 21% oxygen atmospheres for a period of 20 hours. The results of this study showed no synergistic effect of low oxygen on the toxicity of phosphine for either the flow-through or the single-dose fumigations.

Introduction

Fumigation with phosphine is an important method for controlling stored-product pests and in recent years there has been an interest in developing techniques which might lead to improved efficacy at lower dosages. Experiments (Bond 1989) have shown that modifications to the normal atmosphere can change the susceptibility of insects to the fumigant. In particular the effect of low oxygen on the efficacy of phosphine has been investigated by several researchers, though not all results have agreed. Studies by Liang Quan (1981, 1982) on the synergy of phosphine with oxygen and carbon dioxide against common stored-product beetles in the laboratory showed that a decrease in oxygen below 12% and/or an increase in carbon dioxide above 4% enhanced the effectiveness of phosphine. On the other hand, other researchers (Bond 1963; Bond et al. 1976; Price and Walter 1981; Kashi 1981a, b, 1982) have found that the lower the oxygen concentration during fumigation the lower the insect mortality. In fumigations containing no oxygen no deaths occurred even when the dosage was increased to more than a 1000 times that used in air. Post-fumigation exposure of phosphine-treated insects to either oxygen or nitrogen atmospheres showed that insects recovering in nitrogen had a higher survival rate than those recovering in air (Bond 1963).

Combined low oxygen and low phosphine atmospheres are currently being used in China. One of the techniques which uses this combination is called the 'double low' method of insect control (Tao and Wang 1993; Xu and Wang 1993). Freshly harvested grain is covered with PVC sheeting which is then hermetically sealed. The oxygen concentration in the grain store falls due to natural respiration. The grain is then fumigated with phosphine (0.05–0.5 g/m^3) from aluminium phosphide (AlP, 1–5 g/m^3). With use of the 'double low' fumi-

gation method stored grain losses have been reduced to 0.05% as compared with 3.5% for conventional fumigations (Tao and Wang 1993). Techniques which have the potential to make phosphine more effective at lower application rates are of interest particularly with respect to cost, and environmental and safety considerations. For these reasons it was considered important to verify the role of low oxygen atmospheres in the use of phosphine. The work presented here reports on the effects of low oxygen on the mortality of adult *Rhyzopertha dominica* using phosphine flow-through and single-dose fumigations. The initial experiments were conducted using a flow-through fumigation method to ensure constant oxygen, carbon dioxide and phosphine levels. Subsequent experiments were carried out using conditions which duplicated as closely as possible those used by Liang Quan (1981).

Materials and Methods

Flow-through fumigations

Three stains of *R. dominica*, differing in phosphine susceptibility were used for the flow-through fumigation trials [for culturing details see Winks 1975)]. The strains were cRD2, a phosphine-susceptible strain, cRD316, a mid-range susceptible strain and cRD 235p10, a selected phosphine-resistant strain. The adult *R. dominica* were subjected to a constant level of phosphine, 0.1 mg/L and oxygen levels of 1, 2, 8 and 21%. Insects were exposed to a particular phosphine/oxygen regime for periods between 15 minutes and 30 hours. The insects were then assessed for end-point mortality during a 28 day recovery period. The concentrations of oxygen and phosphine were maintained during the fumigation period using mass flow controllers. Before the addition of phosphine the gas mixture was conditioned to 60% relative humidity (r.h.) by passage through an appropriate aqueous glycerol solution. The fumigations were conducted at 25°C in modified glass desiccators (Fig. 1). During the fumigation the insects were placed inside mesh cages within the desiccators. For each fumigation two cages with 50 insects for each strain and exposure time were randomly assigned to the treatment and control desiccators. The concentrations of oxygen and phosphine were monitored throughout the fumigation period using gas chromatography (see Table 4). Two replicates were conducted at each oxygen concentration.

Following fumigation the insects were placed in recovery jars, containing whole wheat, at 25°C and 60% r.h. in air. Mortality was assessed after 24 hours recovery time, and then at 7, 14, 21, and 28 days. Probit analysis was used to assess the end-point mortality against time for each of the oxygen treatment results.

Single-dose fumigations

The strain of insect used in the single-dose fumigations was the phosphine-susceptible strain cRD2. The insects were exposed to a single dose concentration of 0.003 mg/L or 0.007 mg/L phosphine for 20 hours. During the fumigation the insects were placed in desiccators which were flushed with an

---

* Stored Grain Research Laboratory, CSIRO Division of Entomology, GPO Box 1700 Canberra, ACT 2601, Australia.

appropriate oxygen concentration to establish the required atmosphere. The desiccators were then sealed and dosed with phosphine to give an atmosphere with either 0.003 mg/L or 0.007 mg/L. The 0.003 mg/L fumigations were conducted at 1,8,12, and 21% oxygen and the 0.007 mg/L fumigations at 1, 6, 8, and 21% oxygen. The oxygen atmosphere was conditioned to 60% r.h. using glycerol solutions before entering the fumigation desiccators. The fumigations were conducted at 25°C. The conditions in to these experiments duplicated (as closely as possible) those used by Liang Quan (1981,1982, personal communication). The concentrations of the oxygen and phosphine within the single-dose fumigation were monitored at intervals throughout the 20-hour fumigation using gas chromatography (see Table 4). Insects were monitored until end-point mortality was reached. Data were assessed for significant differences between the treatments using Student's t test.

Results

The results for the probit analysis on the mortality of adult *R. dominica* to flow-through low oxygen/phosphine fumigations are tabulated in Table 1. Data are shown for the time to kill 50 and 99% of the insects (shown as lethal times, LT_{50}, LT_{99}). Student's t tests performed on the LT_{50}, LT_{99} and slopes of the probit lines for each insect strain indicate no significant difference between the toxicity of phosphine to insects exposed to low or normal atmospheric concentrations of oxygen.

The results for the single-dose phosphine fumigations conducted at 0.003 mg/L and 0.007 mg/L are given in Tables 2 and 3, respectively. Student's t tests showed that there were no significant differences between the mortalities in the low oxygen/phosphine treatment and those for insects exposed to normal atmospheric oxygen and phosphine. The results indicate that phosphine at 0.007 mg/L causes, on average, a 60% mortality in adult *R. dominica* but that this is due solely to phosphine rather than a synergistic effect of low oxygen on the toxicity of phosphine. The results in Table 2 indicate that phosphine at 0.003 mg/L does not cause mortality in adult *R. dominica* after a 20-hour exposure at any of the oxygen concentrations used.

Discussion

The relationship between oxygen and phosphine has been investigated by several workers (Bond 1963; Bond et al. 1976; Liang Quan 1981, 1982). In addition, many studies have been conducted on the action of oxygen on stored-product pests and other insects (e.g. Annis 1990; Bell 1984; Storey 1980). It was the results from these studies that prompted the present investigation. The study conducted by Liang Quan (1981) investigated the effect of low phosphine and low oxygen concentrations on several stored-grain pests. The results indicated an increase in phosphine toxicity when oxygen levels were held below 8%. The ability to extend the efficacy of phosphine would have obvious environmental, occupational health and safety and commercial benefits, but the present work has been unable to confirm a synergistic effect for low oxygen on the toxicity of phosphine against adult *R. dominica*.

Initially, fumigations were carried out using a continuous flow or flow-through process. The flow-through system has the advantage that gas concentrations and humidity can be held constant during the course of the fumigation. In contrast the single-dose procedure may incur problems associated with changes in experimental conditions due to insect respiration, changes in relative humidity and the decline in phosphine concentration during the fumigation. The results from the flow-

Table 1. Probit analysis[a] of flow-through fumigation with 0.1 mg/L phosphine and oxygen treatments, 1, 2, 8 and 21%.

strain	% oxygen	slope	slope SE[b]	LT_{50}[c] (days)	LT_{99}[c] (days)
cRD2	1	1.46	0.38	1.06	2.66
cRD2	2	1.66	0.67	1.08	2.49
cRD2	8	6.62	3.87	0.33	0.68
cRD2	21	3.51	0.88	0.67	1.32
cRD316	1	0.91	0.18	2.30	4.85
cRD316	2	1.39	0.64	2.14	3.81
cRD316	8	1.51	0.52	2.09	3.63
cRD316	21	1.21	0.47	2.69	4.61
cRD235P10	1	1.29	0.33	3.64	5.43
cRD235P10	2	0.85	0.20	3.21	5.93
cRD235P10	8	0.91	0.09	3.27	5.83
cRD235P10	21	1.21	0.47	2.69	4.61

[a] Raw data corrected with Abbot's correction factor
[b] SE, standard error
[c] LT, lethal time

Table 2. End-point insect mortality (%) from single-dose fumigations with 0.003 mg/L phosphine and oxygen atmospheres of 1, 8, 12 and 21% for a 20-hour fumigation period.

Oxygen percent	without phosphine (n=6)		with phosphine (n=6)	
	mean mortality[a]	SD[b]	mean mortality[a]	SD[b]
1	2.06	0.66	0.23	0.46
8	3.09	1.90	0.65	0.71
12	2.07	0.51	0.49	0.68
21	1.88	0.77	0.46	0.74

[a] Abbot's corrected figures.
[b] SD, standard deviation

Table 3. End-point insect mortality (%) from single-dose fumigations with 0.007 mg/L phosphine and oxygen atmospheres of 1, 6, 8, and 21% for a 20-hour fumigation period.

Oxygen percent	without phosphine (n=6)		with phosphine (n=6)	
	mean mortality[b]	SD[a]	mean mortality[b]	SD[a]
1	2.81	0.90	46.9	9.75
6	1.89	0.69	51.7	9.61
8	3.66	0.96	52.3	2.48
21	3.11	1.36	47.1	7.61

[a] SD, standard deviation
[b] Abbot's corrected figures

through fumigation showed no synergistic effect of oxygen concentration on the efficacy of phosphine in any of the strains of *R. dominica* used. Subsequently, the work conducted by Liang Quan (1981, 1982) was replicated as closely as possible to test the increase in mortality in low oxygen atmospheres when using single dose fumigation (Liang Quan 1981). End-point mortalities assessed over 28 days showed no evidence of a synergistic effect for low oxygen atmospheres. Not all experimental details were duplicated exactly, and a compari-

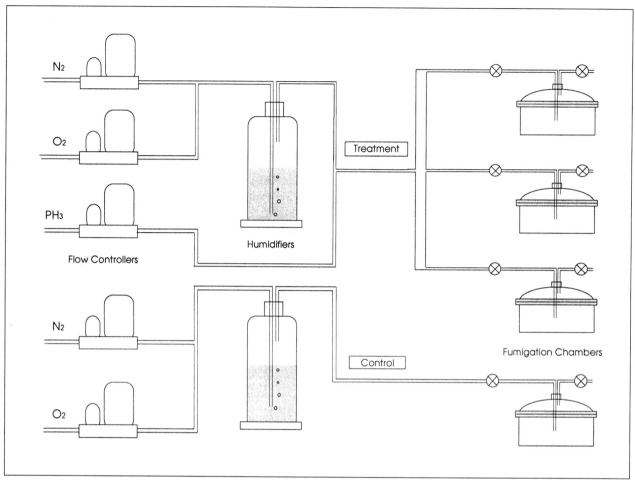

Fig. 1.   Apparatus used to fumigate insects with phosphine in low oxygen atmospheres.

son of the work by Liang Quan (1981) and that conducted here (Table 4) shows some possible sources of experimental variation.

Although different results were found in the present study compared with that of Liang Quan the reasons have not been ascertained. Many environmental factors may influence the toxic action and effectiveness of fumigants. Among these, temperature, ambient pressure, humidity, and oxygen tension are the most important. Physiological factors such as digestion of food, age, sex, activity and excitability of the insects all effect the metabolism of the insect and thus susceptibility to a given fumigant (Bond and Monro 1967). Although the use of different strains could have a bearing on the observed response of the insect to the given fumigation conditions the present work was unable to demonstrate any differences between three insect strains with respect to the effects of low oxygen. On the other hand, differences in experimental procedure may offer a simpler explanation. In the study of Liang Quan (1981) a recovery period of 21 days was used, while in the experiments conducted here 28 days was used. In one experiment carried out during our investigation, the mortalities measured at 14 days indicated a synergistic effect for low oxygen, but this effect was no longer apparent when end-point mortalities were assessed. It is possible that different oxygen concentrations may affect the speed of action of phosphine without altering the end-point mortality. That is, insects in one oxygen concentration may die faster than those fumigated in a different oxygen atmosphere. Thus, mortalities assessed before the end-point may show an apparent synergistic effect for low oxygen on phosphine toxicity not observed when the final or end-point mortalities are assessed. Further work would be required to investigate this possibility.

The successful use of low oxygen and low phosphine atmospheres for insect control in China (the 'double low' fumigation technique) may arise from the use of gastight sealing practices to enclose grain maintaining low oxygen and phosphine over a long period, thereby extending the time the insects are exposed to phosphine. Moreover, the conditions produced using the 'double low' fumigation may involve not only the decrease in oxygen but also the increase in carbon dioxide levels. Carbon dioxide above 35% is toxic to insects and between 15 and 30% synergises the toxicity of phosphine (Y.L. Ren et al., these proceedings). Both low oxygen and high carbon dioxide concentrations have been used to control insect populations in stored grain. The results of Navarro et al. (1981) indicate that insects may seek to avoid low oxygen, high carbon dioxide atmospheres. Thus the protection of grain afforded by the 'double low' technique may be due to a combination of a number of these factors.

Conclusions

The single-dose fumigations conducted in the experiments reported here closely followed the work carried out by Liang Quan (1981). While experimental conditions such as temperature, humidity and basic equipment design followed those described by Liang Quan, the procedure varied with respect to fumigation technique, fumigation equipment, insect strain and post-exposure treatment of the test insects.

The results from both flow-through and single-dose phosphine fumigation trials showed no synergistic effect of low oxygen on phosphine toxicity against adult *R. dominica*.

Table 4. Experimental conditions compared with those used by Liang Quan (1981).

Experimental condition	Present work	Liang Quan
Insect species	*Rhyzopertha dominica*	*R. dominica*
Insect strain	cRD2	unknown, phosphine-susceptible
Insect age	adults- 4–8 weeks mixed	peak of emergence
Insect rearing technique	8 weeks at 60% r.h., 25°C	unknown
Insect recovery conditions	recovery bottles, 25°C, 60% r.h., food supplied (whole wheat)	rearing bottles, 28°C, 60–80% r.h
Criterion of insect death	only slight appendage motion	loss of normal climbing ability
Time to death period	28 days	21 days
Phosphine source	gas cylinder of known concentration	evolution from AlP using sulphuric acid with dilution
Phosphine application	gas-tight syringe	gas-tight syringe
Phosphine concentration	0.1 mg/L (flow-through); 0.003 mg/L (single dose); 0.007 mg/L (single dose)	0.007 mg/L
Fumigation equipment	mass flow controllers, sealed glass desiccators,	mass flow controllers, testing bottles
Fumigation conditions	flow through, single dose	single dose
Fumigation temperature	25–26°C	20–35°C
Fumigation relative humidity	60%	50–70%; 80–90%
Fumigation exposure time	30 hours (flow- through); 20 hours (single dose)	20 hours
Oxygen concentration check	Fisher GC with TCD and Porapak Q column	Orsat gas analyser
Phosphine concentration check	Tracor GC equipped with FPD and Porapak Q column	not stated
number of insects exposed	100/test	50/cage
oxygen range tested	1–21%	1–21%
Experimental result	no synergistic effect of low oxygen on phosphine at concentration tested	synergistic effect of oxygen below 8% on phosphine at 0.007 mg/L

Acknowledgments

The authors wish to thank Mr P.C. Annis, Mr C.J. Waterford and Dr H.J. Banks for helpful advice and discussions, and Mrs J. Cassells and Ms E. Hyne for technical assistance.

References

Annis, P.C. 1990. Requirements for fumigation and controlled atmospheres as options for pest and quality control in stored grain. In: Champ, B.R., Highley E. and Banks, H.J. ed., Fumigation and controlled atmosphere storage in grain: proceedings of an international conference, Singapore, 14–18 February 1989, ACIAR Proceedings, No. 25, 70–82.

Bell, C.H. 1984. Effects of oxygen on the toxicity of carbon dioxide to storage insects. In: Ripp, B.E., Banks H.J., Bond, E.J., Calverley, D. J., Jay, E.G. and Navarro, S., ed., Practical aspects of controlled atmosphere and fumigation in grain storages. Proceedings of an international symposium, Perth, Western Australia, 1983, 67–74.

Bond, E.J. 1963. The action of fumigants on insects — the effects of oxygen on the toxicity of fumigants to insects. Canadian Journal of Biochemical Physiology, 41,993–1114.

Bond, E.J. 1989. Current scope and usage of fumigation and controlled atmospheres for pest control in stored products. In: Champ, B.R., Highley E. and Banks, H.J. ed., Fumigation and controlled atmosphere storage in grain: proceedings of an international conference, Singapore, 14–18 February 1989. ACIAR Proceedings, No. 25, 20–29.

Bond, E.J. and Monro, H.A.U. 1967. The role of oxygen in the toxicity of fumigants to insects. Journal of Stored Products Research, 3, 295–310.

Bond, E.J., Monro, H.A.U. and Buckland, C.T. 1976. The influence of oxygen on the toxicity of fumigants to *Sitophilus granarius*. Journal of Stored Products Research, 3, 289–294.

Kashi, K.P. 1981a. Responses of five species of stored product insects to phosphine in oxygen deficient atmospheres. Pesticide Science, 12, 111–115.

Kashi, K.P. 1981b. Toxicity of phosphine to five species of stored product insects in atmospheres of air and nitrogen. Pesticide Science, 12, 116–122.

Kashi, K.P. 1982. Dose mortality responses of five species of stored products insects to phosphine. International Pest Control, 24 (2), 46–48.

Liang Quan. 1981. Controlled atmosphere and the control of stored grain insects. Grain Storage, 1, 20–27.

Liang Quan. 1982. Studies on the synergy of phosphine activity in control of stored grain insects by controlled atmospheres. Grain Storage, 1, 1–11.

Navarro, S., Amos, T.G. and Williams, P. 1981. The effect of oxygen and carbon dioxide gradients on the vertical dispersion of grain insects in wheat. Journal of Stored Products Research, 17, 101–107.

Price, N.R. and Walker, C.M. 1987. A comparison of some effects of phosphine, hydrogen cyanide and anoxia in the lesser grain borer, *Rhyzopertha dominica* (F.) (Coleoptera: Bostrychidae). Comparative Biochemistry and Physiology, 86C, 33–36.

Storey, C.L. 1980. Mortality of various stored product insects in low oxygen atmospheres produced by an exothermic inert atmosphere generator. Journal of Economic Entomology, 68, 316–318.

Tao, M-C. and Wang, Y-N. 1993. 'Double low': A synthesis of controlled atmosphere and fumigation techniques for stored grain conservation. In: Navarro, S. and Donahaye, E. ed. Proceedings of an International Conference on Controlled Atmosphere and Fumigation in Grain Storages, Winnipeg, Canada, 1992, 511–514.

Winks, R. G. 1975. Laboratory culturing of some storage pests. International training course on the Preservation of Stored Cereals. Australian Development Assistance Bureau 1, 206–219.

Xu, H. and Wang, N. 1993. Present and prospective state of the 'triple low' grain storage technique. In: Navarro, S. and Donahaye, E. ed., Proceedings of an International Conference on Controlled Atmosphere and Fumigation in Grain Storages, Winnipeg, Canada, 1992, 29–35.

Response of *Liposcelis bostrychophila* and *L. entomophila* (Psocoptera) to carbon dioxide

E.C.W. Leong and S.H. Ho[*]

Abstract

Mixed-age samples of *L. bostrychophila* and *L. entomophila* were exposed to various concentrations of CO_2 (10–90%). *L. bostrychophila* was found to be the more tolerant of the two species. A linear correlation is indicated between the LT_{100}s and CO_2 concentrations for both species. Eggs are the most tolerant life stage for both species.

Mortality of 1–3 week old female liposcelids was determined at fixed CO_2 concentrations. *L. bostrychophila* was the more tolerant species when either 45 or 60% CO_2 was used. Increases in exposure periods resulted in corresponding increases in mortality in both species. Increasing the CO_2 concentration from 45 to 60% did not produce a significant change in response of *L. entomophila*. However, for *L. bostrychophila*, at 60% CO_2, an unexpected increase in exposure time was required to achieve the same level of kill. Currently recommended dosages of CO_2 are adequate for controlling both these species in well-sealed enclosures.

Introduction

Studies on the effects of fumigants (Pinniger 1985; Kalinovic 1984; S.H. Ho and R.G. Winks, unpublished data) and controlled atmospheres (Pinniger 1985; Leong 1986; Bell et al. 1990) on *Liposcelis* spp. have been limited. Leong (1986) investigated the effects of carbon dioxide (CO_2) on the mortality of *L. entomophila* at 28°C and 77% r.h. He noted that 30% CO_2 effected a 100% kill of *L. entomophila* female adults when exposed for 24 hours or more. However, at lower concentrations, mortality was significantly higher in samples exposed for longer periods. Bell et al. (1990) in his studies on the effects of CO_2 on *L. bostrychophila* noted that, at 40% CO_2, the time required to bring about control was greater than 8 and 12 days at 10 and 15°C, respectively. When a higher concentration of CO_2 (80–100%) was applied, the required time for control was 10 and greater than 7 days for the respective temperatures. These findings demonstrate the potentiating effect of increasing temperatures on CO_2 toxicity in *L. bostrychophila*. In the review paper by Annis (1987), the absence of information on the response of stored-product liposcelids to controlled atmospheres was obvious. Since then, apart from the two studies cited above, no other research has apparently been conducted. Therefore, the work reported here was undertaken to determine the response of *L. bostrychophila* and *L. entomophila* to CO_2.

Materials and Methods

The CO_2 chambers were constructed from glass desiccators (180 mm diameter) (Fig. 1). All connections were grease-sealed to prevent leakage. Experiments were carried out at 30 ± 1°C and 75 ± 3% r.h. maintained by a 50% v/v glycerol solution (Johnson 1939; Braun and Braun 1958) in the lower compartment of the chamber.

Ninety–nine percent pure CO_2 from a cylinder (Singapore Oxygen Air Liquide Pte. Ltd (SOXAL)) was slowly released into the exposure chamber until the desired percentage of CO_2 in air was registered on a Riken Interferometer Model 18. The monitoring of the gas in the chamber was achieved with a closed-system design, so as not to dilute the gas during the process (Fig. 2). The interferometer was regularly calibrated against CO_2 standards (19.8%, 29.8%, 53.9%, 74.4% and 99.9%) obtained from SOXAL. Deviations were negligible over a period of 2 months. As gas from the cylinder tended to be cold and dry, the CO_2 was first passed through warm (30°C) 50% (v/v) glycerol solution before introduction into the chambers. This humidified the gas to 75% r.h. The oxygen concentration was estimated using Jay's (1984) data. A linear regression analysis of these data showed a good predicability of the oxygen concentration based on the CO_2 concentration in air ($r^2 = 0.994$).

The concentration of CO_2 was recorded both at the beginning and end of the experiment. For experiments studying the time-to-100% kill of the liposcelids, the concentrations were checked daily and replenished when necessary. Chambers with daily deviations in concentrations of greater than 3% when checked were not considered in the final analysis.

Time-to-100% kill

Mixed-age samples of *L. bostrychophila* and *L. entomophila* (> 150 individuals per cage) were exposed to various concentrations of CO_2 (10–90%). Fifty adults were isolated in each cage and set aside for 4 weeks to obtain the required mixed-age samples. By the end of this time, adults, nymphs and eggs of various ages were present. Five cages were placed in each exposure chamber.

The LT_{100} was determined by the method of inverse sampling (Finney 1971). At each fixed concentration, various exposure periods (at 24-hour intervals) were randomly selected. Upon termination of the treatment, cages were placed in incubators at 30 ± 1°C, 75 ± 3% r.h. and inspected weekly. Initial observations indicated that eclosion occurred within 2 weeks after treatment. A 2-week holding period was thus selected to include any delays in eclosion of the surviving eggs. Samples with surviving nymphs or adults were immediately discarded, while samples without surviving nymphs or adults were set aside for the respective holding periods before egg mortality was ascertained. The experiments were repeated with the exposure periods increased or decreased (by 24 hours) depending on the outcome of the preceding results.

---

[*] Zoology Department, National University of Singapore, Lower Kent Ridge Road, Singapore 0511.

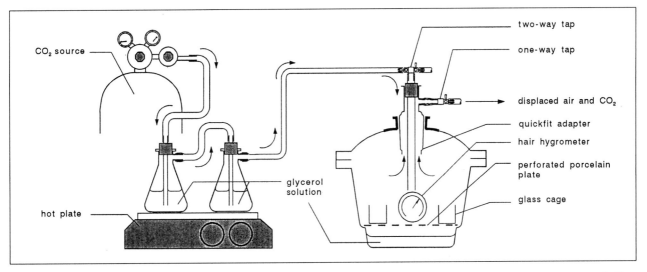

Fig. 1. Schematic diagram of the experimental set up. Pure CO_2 discharged from the cylinder is first conditioned before being channelled into the exposure chamber. The mixture of air and CO_2 displaced is then passed through soda lime before being released into the surroundings.

When the apparent LT_{100} for a particular concentration was obtained, the dosage was replicated 3 more times to confirm the time-to-100% kill. If survivors were noted in any one replicate, the exposure period was again increased by another 24 hours for the next treatment. Similarly, three replicates for confirmation were conducted. The results were analysed by linear correlation using the Maximum Likelihood Program (MLP).

End-point mortality determination

The mortalities of the two species exposed to 60% CO_2 were monitored until end-point mortality was attained or when control mortalities exceeded 20%. This was replicated at least five times. The corrected mortalities were plotted against time and the end-point mortality determined graphically.

Response of female lipscelids to fixed concentrations of CO_2

L. bostrychophila and *L. entomophila* were cultured as described by Leong and Ho (1990) and Leong (1993). Females 1–3 weeks old were used in all experiments. Batches

of at least 40 liposcelids were isolated into each cage and counted. Three cages were placed in each exposure chamber and this was replicated up to 12 times per dosage. The variation in replicates used was dependent on the availability of test specimens.

Mortality response of 1–3-week-old female liposcelids was determined at two fixed concentrations (45 and 60%) of CO_2 with increasing exposure periods from 2 to 6 hours at half-hour intervals. Controls consisting of a similar number of cages in the exposure chambers not dosed with CO_2 were set aside for 4 hours. Preliminary trials showed that mortalities in the controls did not vary amongst the various exposure periods. Replicates with high control mortalities and for which changes in CO_2 concentrations before and after the experiment exceeded 3% were disregarded in the final analysis.

Results and Discussion

Time-to-100% kill

It is obvious from the results (Fig. 3) that *L. bostrychophila* is the more tolerant of the two species to CO_2 treatment. A

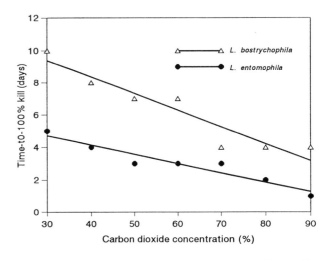

Fig. 2. Monitoring of CO_2 in the exposure chamber. The 'closed-system' adopted prevents dilution of the CO_2 concentration within the chamber.

Fig. 3. Time-to-100% kill of mixed-age individuals of *Liposcelis* spp to various concentrations of CO_2 at 31 ± 1°C and 75 ± 3% r.h.

linear correlation is suggested between the LT_{100}s and CO_2 concentrations for *L. entomophila* and *L. bostrychophila*. Although no significant difference (P > 0.05) was noted between the slopes of the two plots, significant displacements (P < 0.01) were noted (Table 1). Generally, increasing concentrations decreased the time-to-100% kill of both species. The most tolerant life stage for both species is the eggs. Neither adults nor nymphs survived the treatments. This is not surprising as the 'inactive' stages (eggs or pupae if present) have been found to be the most tolerant life stage in many insects treated with CO_2, with the exception of a few larval stages of some stored product beetles (AliNiazee 1971; Bailey and Banks 1980; Jay 1984; Ho et al. 1987; Annis 1987). Literature on the lethal effects of CO_2 on insect eggs is limited. It is possible that the toxicity of high CO_2 concentrations to the eggs of insects may be due to interference with the normal metabolic growth process of the eggs or the anaesthetic action of the gas on the embryo's nervous system. Bell (1984) reported increased sensitivity to CO_2 in early eggs of pyralid moths; however, in the presence of oxygen, sensitivity decreased as embryogenesis proceeded. Furthermore, he reported that, in the presence of oxygen, development is delayed with no evidence of cessation, suggesting that death results from progressive CO_2 poisoning or the accumulation of toxic products otherwise removed via oxidative metabolism. Interestingly, during anoxia the development of eggs virtually ceases (Price and Bell 1981), and thus survival depends on the capacity of the embryo to accumulate glycolytic products and reduce its needs for active metabolism.

The observed differences in LT_{100}s of the two liposcelid species can at best be attributed to interspecific differences in tolerance to CO_2. The shorter incubation period of *L. entomophila* eggs (3–4 days shorter than *L. bostrychophila* eggs), and the different modes of reproduction (*L. bostrychophila* being parthenogenetic), may be responsible for the observed differences. From the available information (Press and Flaherty 1973; Bell 1984), it is clear that the relation of egg age and CO_2 toxicity is a complex one, with no one simple trend that can best describe this relationship.

End-Point mortality determination

Five and 8-day post-treatment holding periods were found to best reflect the end-point mortalities of *L. entomophila* and *L. bostrychophila*, respectively (Fig. 4). To avoid high control mortalities, post-treatment holding periods longer than those suggested above are not recommended, especially for *L. entomophila*. Although mortalities recorded after a 24 hour post-treatment period for both species provide good estimates of end-point mortalities, the monitoring of mortalities over several more days is recommended (limited by high natural mortalities) to accommodate variations expected in end-point mortalities for different dosage regimes. Winks (1984, 1986) noted the importance of response time (defined as the time that elapses between the administration of a dosage of a drug or poison and the expression of the response to that dose) when assessing mortalities for different dosages of phosphine. Hence, variations amongst the replicates recorded over the duration monitored could be attributed to population variations in not only tolerance to CO_2 but also response time to the treatment.

Response of females liposcelids to fixed concentrations of CO_2

L. bostrychophila is the more tolerant (P < 0.05) species when either 45 and 60% CO_2 was used (Fig. 5). Increases in exposure periods resulted in corresponding increases in the mortality in both species. Increasing the concentration of CO_2 from 45 to 60% did not produce a corresponding increase in mortality for *L. entomophila*. For *L. bostrychophila*, at 60% CO_2 an unexpected increase in exposure time was required to achieve the same level of kill as 45% CO_2. More data are needed from further work on the response of these liposcelids to a range of CO_2 concentrations at fixed exposure periods.

In comparison with other insects reviewed by Annis (1987) in terms of LT_{100}s for various concentrations of CO_2, it is apparent that female *L. entomophila* and *L. bostrychophila* rank amongst the more susceptible. The effect of CO_2 on the

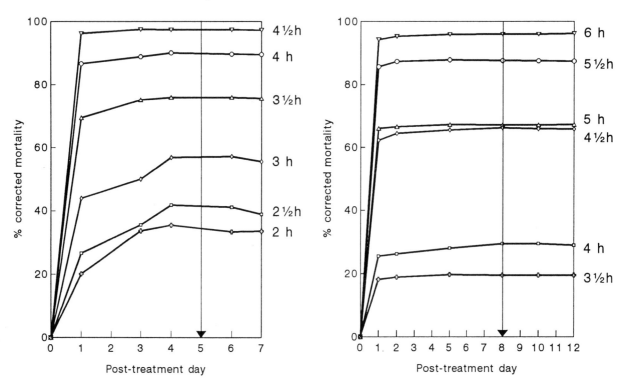

Fig. 4. End-point mortality determination in *L. entomophila* (left) and *L. bostrychophila* (right) at 60% CO_2, 31 ± 1°C and 75 ± 3% r.h. The arrows indicate the time which reflect end-point mortality.

Table 1. Linear regression of LT_{100}s against CO_2 concentration for mixed-age samples of *L. entomophila* and *L. bostrychophila* (mean ± s.e.).

	Species		d.f.	F-ratio[a]
	L. entomophila	*L. bostrychophila*		
Intercept	7.00 ± 0.61	12.50 ± 0.98	5	73.89**
Slope	-0.064 ± 0.010	-0.104 ± 0.016	5	4.618 ns
R-value	0.938	0.934	—	—

[a] Derived from Analysis of Parallelism table from the MLP output where the F-ratios for the intercept and slope were calculated from: $MS_{Displacement}/MS_{Within\ group}$ and $MS_{Parallelism}/MS_{Within\ goup}$, respectively.

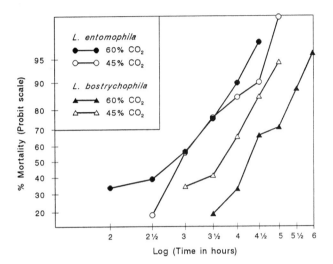

Fig. 5. Mean plots of the mortality of *L. entomophila* and *L. bostrychophila* (1–3 week old females) treated with 45% and 60% CO_2 for various exposure periods, showing the relatively linear association between the two parameters.

stimulation of prolonged spiracular opening (Hoyle 1960; Miller 1974), and thus water loss, has been suggested to contribute to the overall toxicity of CO_2 in the liposcelids (Leong 1986). Moreover, the CO_2-permeability of the integument of these soft-skinned insects (Wigglesworth 1972) coupled with their heavy dependence on the active physiological absorption of water from the atmosphere (Devine 1978) would also play a significant role in the susceptibility of liposcelids to hypercarbic atmospheres. However, the eggs of these liposcelids are noted to be amongst the most tolerant when compared with other stored-product insects' eggs (Annis 1987).

Conclusion

In conclusion, this work provides the baseline data for the response of *L. bostrychophila* and *L. entomophila* to CO_2. Although the eggs are more tolerant than many of the species recorded by Annis (1987), the latter's suggested CO_2 dosage regimes are adequate for controlling the liposcelids in well-sealed enclosures.

References

AliNiazee, M.T. 1971. The effect of carbon dioxide gas alone or in combinations on the mortality of *Tribolium castaneum* (Herbst) and *T. confusum* du Val (Coleoptera, Tenebrionidae). Journal of Stored Products Research, 7, 243–252.

Annis, P.C. 1987. Towards rational controlled atmosphere dosage schedules: a review of current knowledge. In: Donahaye, E., and Navarro, S., ed., Proceedings of the Fourth International Working

Conference on Stored-product Protection, Tel Aviv, Israel, September 1986, 128–148.

Bailey, S.W. and Banks, H.J. 1980. A review of recent studies of the effect of controlled atmospheres on stored product pests. In: Shejbal, J., ed., Controlled Atmosphere Storage of Grains: proceedings of an international symposium, Rome, 12–15 May 1980. Developments in Agricultural Engineering, 1, 101–118.

Bell, C.H. 1984. Effects of oxygen on the toxicity of carbon dioxide to storage insects. In: Ripp, B.E., Banks, H.J., Bond, E.J., Calverley, D.J., Jay, E.G. and Navarro, S., ed., Controlled Atmosphere and Fumigation in Grain Storages: proceedings of an international symposium, Perth, April 1983. Developments in Agricultural Engineering, 5, 67–74.

Bell, C. H., Spratt E. C. and Llewellin B. E. 1990. Current strategies for the use of controlled atmospheres for the disinfestation of grain under U.K. conditions. In: Champ, B.R., Highley, E., and Banks, H.J. ed., Fumigation and Controlled Atmosphere Storage of Grain: proceedings of an international conference, Singapore, 14–18 February 1989. ACIAR Proceedings No. 25, 251–253.

Braun, J.V. and Braun, J.D. 1958. A simplified method of preparing solutions of glycerol and water for humidity control. Corrosion, 14(3), 17–18.

Devine, T. L. 1978. The turnover of the gut contents (traced with inulin-carboxyl-^{14}C), tritiated water and ^{22}Na in three stored product insects. Journal of Stored Products Research, 14, 189–211.

Finney, D.J. 1971. Probit analysis, 3rd ed. Cambridge University Press, 333p.

Ho, S.H., Choo, K.W. and Lee, J.Y.Y. 1987. The effects of carbon dioxide on the mortality of four life stages of *Tribolium castaneum* (Herbst) and on some physicochemical characteristics of two varieties of rice. In: de Mesa, B.M., ed., Grain Protection in Postharvest Systems: proceedings of the 9th ASEAN technical seminar on grain postharvest technology, Singapore, 26–29 August 1986, ACPHP, 185–195.

Hoyle, G. 1960. The action of carbon dioxide gas on an insect spiracular muscle. Journal of Insect Physiology, 4, 63–79.

Jay, E.G. 1984. Imperfections in our current knowledge of insect biology as related to their response to controlled atmospheres. In: Ripp, B.E., Banks, H.J., Bond, E.J., Calverley, D.J., Jay, E.G. and Navarro, S., ed., Controlled Atmosphere and Fumigation in Grain Storages: proceedings of an international symposium, Perth, April 1983. Developments in Agricultural Engineering, 5, 493–508.

Johnson, C.G. 1939. The maintenance of high atmospheric humidities for entomological work with glycerol–water mixtures. Annals of Applied Biology, 37, 295–299.

Kalinovic, I. 1984. Efikasnost fosforovodika u suzbijanju pra_nih u_i (Insecta: Psocoptera: Liposcelidae). [Efficacy of phosphine on the booklice (Insecta: Liposcelidae)]. Znanost i Praksa Poljoprivredi Prehrambenoj Tehnologiji (Yugloslavia), 14/3–4, 239–247. Osijek.

Leong, E.C.W. 1986. Some studies on *Liposcelis entomophilus* (Enderlein)(Psocoptera: Liposcelidae) with emphasis on culturing techniques and the effects of carbon dioxide and relative humidity. BSc (Hons) thesis, Department of Zoology, National University of Singapore, 61p.

Leong, E.C.W. 1993. Comparative toxicology of *Liposcelis entomophila* and *L. bostrychophila* in relation to their management. PhD thesis, Department of Zoology, National University of Singapore, 243p.

Leong, E.C.W. and Ho, S.H. 1990. Techniques in the culturing and handling of *Liposcelis entomophilus* (Enderlein) (Psocoptera: Liposcelidae). Journal of Stored Products Research, 26, 67–70.

Miller, P.L. 1974. Respiration — aerial gas transport. In: Rockstein, M., ed., Physiology of Insecta, 2nd ed., 6, 345–402. Academic Press.

Pinniger, D.B. 1985. Recent research on psocids. In: Dodd, G.D., ed., Aspects of Pest Control in the Food Industry: proceedings of the symposium of the Society of Food Hygiene Technology Nottinghamshire, 27 September 1984. SOFHT, 37–42.

Press, J.W. and Flaherty, B.R. 1973. Hatchability of *Plodia interpunctella* eggs exposed to a carbon dioxide atmosphere: relationship of egg age to exposure time. Journal of the Georgia Entomological Society, 8(3), 210–213.

Price, N.R. and Bell, C.H. 1981. Structure and development of embryos of *Ephestia cautella* (Walker) during anoxia and phosphine treatment. International Journal of Invertebrate Reproduction, 3, 17–25.

Wigglesworth, V.B. 1972. Respiration — the elimination of carbon dioxide. In: Principles of Insect Physiology, 7th ed. English Language Book Society/ Chapman and Hall, 375.

Winks, R.G. 1984. The toxicity of phosphine to adult *Tribolium castaneum* (Herbst): time as a dosage factor. Journal of Stored Products Research, 20(1), 45–46.

Winks, R.G. 1986. The biological efficacy of fumigants: time/dose response phenomena. In: Champ, B.R., and Highley, E., ed., Pesticides and Humid Tropical Grain Storage Systems: proceedings of an international seminar, Manila, 27–30 May 1985. ACIAR Proceedings No. 14, 211–221.

Inheritance of phosphine resistance in *Sitophilus oryzae* (L.) (Coleoptera, Curculionidae)

Li Yan-sheng and Li Wen-zhi*

Abstract

Genetic crosses were carried out on a susceptible strain and a resistant strain of *Sitophilus oryzae* (L.). Time–mortality characteristics of susceptible and resistant parents and their F_1 hybrids, and response to time of the F_1-backcross progeny and the reciprocal F_1 progeny were assessed in order to determine inheritance character. Results indicated that the response of the F_1 hybrid to phosphine was closer to its susceptible strain than to its resistant parent; there is no evidence of sex linkage between the RS and SR hybrids. χ^2 analysis of the response observed of the F_1-backcross progeny and the F_2 progeny rejected the null hypothesis of monogenic inheritance for resistance. Inheritance of phosphine resistance in *S. oryzae* is more complex and resistance appears to be controlled by more than one autosomal factor, but the major gene involved is incompletely recessive. The degree of dominance (D) to phosphine resistance for the species is –0.348.

Introduction

In recent years phosphine has become a popular fumigant for the disinfestation of stored grain and other commodities. In China this fumigant has been used from the early 1960s and almost total reliance has been progressively placed on phosphine because of its ease of use, wide availability and low residues. Currently, more than 80% of stored grain is dependent on phosphine fumigation for disinfestation. The continuous and widespread use of an insecticide can result in the rapid development of resistance. High resistance to phosphine has occurred where repeated fumigations are undertaken in poorly sealed storage (Taylor 1989).

Resistance to phosphine was first reported in a strain of *Sitophilus granarius* selected in the laboratory (Monro et al. 1972). Subsequently, the FAO global survey report (Champ and Dyte 1976) detailed the presence of low levels of phosphine resistance in several species of stored grain pests. Eight of 135 samples of *Sitophilus oryzae* tested from six countries were found to be resistant. The maximum resistance factor for *S. oryzae* was 2.5 times. More recently, a survey undertaken by Taylor (1989) revealed widespread resistance to phosphine in several major stored grain pests; 5 of the 8 samples of *S. oryzae* tested from South Asia and Brazil were resistant. It seems that phosphine resistance can be selected in most strains of all species of stored-grain pests (Taylor 1989; Winks 1986). Nevertheless, despite the importance and widespread occurrence of this phenomenon there is little information on the genetics of phosphine resistance in stored-

grain pests (Ansell et al. 1990). This information is fundamental to an understanding of the evolution of resistance to phosphine and is a necessary basis for the development of procedures to manage this resistance.

In China, high-level resistance to phosphine was first detected in 1976 in a field strain of *S. oryzae* (Liang Quan 1976) from Mei county of Guangdong province, an area where phosphine had long been generated from calcium phosphide formulations. *S. oryzae* is regarded as the most important insect pest of stored grains in the southern provinces of China. In this report we present an analysis of the inheritance of phosphine resistance in adults of *S. oryzae*.

Materials and Methods

Insect strains

The discriminating concentration of 0.04 mg phosphine/L for 20 hours as recommended by the FAO (Anon. 1975) was used to define resistance. The susceptible strain used in this study was collected originally in 1983 from Tibet and the resistant strain was supplied by Guangdong Institute of Cereal Science Research in 1976 from Mei county. The susceptible strain (T–16) was originally derived from 110 single pair crosses of an insect sample from Tibet. Newly emerged virgin adults were used to set up each single pair. These insects were allowed to oviposit on wheat for 2 weeks. The parents were then tested at 0.01 mg/L (20 hours) phosphine. Offspring of insects which failed to survive were pooled to form the susceptible strain. Before this investigation, the resistant strain (G–12) was selected for four generations with 1.6 mg/L and 3.6 mg/L (20 hours) and then 116 single pairs were tested at 3.6 mg/L. The progeny of pairs which survived this concentration were pooled to form the resistant strain.

Generating phosphine

Phosphine gas was generated by the action of acid on zinc phosphide dust as described in the FAO method (Anon. 1975). Phosphine concentration was determined using colorimetrically.

Phosphine susceptibility tests

In all cases, the responses to phosphine of parental strains and the F_1 and F_2 progenies were measured as mortality occurring at a range of exposure times at a constant concentration of 0.04 mg/L (Anon. 1975). This is the lowest concentration at a 20 hours exposure period that would give >99.99% mortality of homozygous susceptible individuals. Three batches of 40 adult insects were exposed to the fumigant at each of a range of time intervals. Insects were confined within 16 mL plastic bottles sealed with mesh. Fumigations were carried out in 1 L jars into which phosphine was injected through a rubber septum. After each exposure time the insects

* Chengdu Institute of Grain Storage Research, Ministry of Internal Trade, 95 Haupaifang Street, Chengdu 610031, China.

were held in culture medium for 2 weeks at which time mortality was assessed.

Cross procedure

To determine the mode of inheritance reciprocal mass crosses were made between susceptible and resistant strains to obtain F_1, F_2 and backcross progeny (F_1–BC) following the plan outlined by Collins (1986). Virginity was assured by isolating each sex one day after emergence of adults (Halstead 1963) and the appropriate cross was made.

Reciprocal F_1 crosses were made to test for dominance and whether resistance was autosomal or sex linked. The single gene hypothesis was tested by exposing the F_2 and the F_1–BC progeny to a full range of time–mortality plots at the fixed concentration of 0.04 mg/L. If a single gene is responsible for resistance then plateaus will occur in the F_2 regression line about at 25 and 75% mortality and in the F_1–BC progeny regression line about at 50% mortality.

Statistical analyses

The results of phosphine susceptibility tests of the parental strains and their reciprocal F_1 and F_2 progenies were fitted to log time–probit mortality curves by method of Finney (1952). Goodness-of-fit was tested using χ^2(P=0.05) and LT_{50} values were estimated. The resistance factors were obtained as the simple ratio of the LT_{50} of the resistant strain against the susceptible strain. Degree of dominance ($\underline{D}$) is based on the following calculation (Stone 1968).

$$D = \frac{2LT_{50}(RS) - LT_{50}(RR) - LT_{50}(SS)}{LT_{50}(RR) - LT_{50}(SS)}$$

Relative potency analysis (Finney 1952) was used to test for differences between the reciprocal crosses of the F_1 (i.e., between progeny of R × S and S × R) and of their F_2 progeny (i.e., between progeny of RS × RS and SR × SR). For assays of the F_2 and F_1–BC progenies, differences between observed and expected responses to phosphine were tested using the χ^2 analysis described by Finney (1952). To test the hypothesis that a single gene controlled resistance the expected proportion responding at each time plot was calculated as following (Georghiou 1969, Collins 1986):

(a) for F_2 progeny: $X_y = W_{(SS)}0.25 + W_{(SR)} 0.5 + W_{(RR)} 0.25$ where X = the expected response at a given time y; and W = the observed response of SS, SR and RR genotypes at time y, obtained directly from respective regression lines.

(b) for F_1–BC to RR parent: $X_y = W_{(SR)}0.5 + W_{(RR)}0.5$

Results

Results of reciprocal crosses of the resistant (G–12) and susceptible (T–16) strains indicated the response of the F_1 hybrids to phosphine was closer to its susceptible parent than to the resistant strain at the LT_{50} (Table 1, Fig. 1).

Relative potency analysis showed that there was no evidence of sex linkage between the progeny of R × S and S × R (relative potency [95% limits] = 1.03 [0.97 – 1.09]) and maternal effects were absent. Resistance, therefore, was autosomally inherited and degree of dominance (D) was –0.348 for *S. oryzae* to phosphine resistance.

The observed response of the F_1 – BC progeny showed a lack of a plateau at 50% mortality (Fig. 1) and χ^2 analysis (χ^2 = 80.48; df = 7; p<0.005) rejected the null hypothesis of single gene control of resistance. Inheritance of phosphine resistance in *S. oryzae* appears to be controlled by more than one autosomal factor but the major gene involved is incompletely recessive. The shape of the F_2 regression line (Fig. 2)

indicated that there was no plateau at 25 and 75% mortality and X^2 analysis showed that the null hypothesis of monogenic inheritance was rejected (X^2 = 290.78; df = 7; p 0.005). Reciprocal F_2 progenies were not significantly different in their response to phosphine (relative potency [95% limits] = 1.02 [0.96–1.08]).

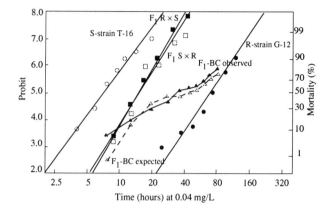

Fig.1. Comparison of the observed response of *S. oryzae* to phosphine and the expected response of the F_1–BC progeny for resistance controlled by a single gene and calculated from the response of SS, F_1 and RR phenotypes. O = SS (T–16), □ = F_1 (S × R), ■ = F_1 (R × S), ▲ = F_1–BC to RR progeny observed, △ = expected response of F–BC progeny, ● = RR (G-12).

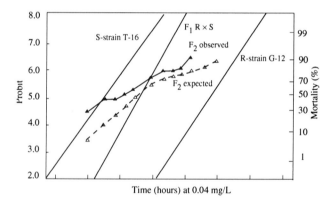

Fig. 2. Comparison of the observed response of *S. oryzae* to phosphine and the expected response of the F_2 progeny for resistance controlled by a single gene and calculated from the response of SS, F_1 and RR phenotypes. ▲ = F_2 progeny observed, △ = expected response of F_2 progeny.

Acknowledgment

We are grateful to Drs M. Bengston, P.J. Collins and G.J. Daglish for supporting this paper and for helpful comment on the manuscript.

References

Anon. 1975. Recommended methods for detection and measurement of resistance of agricultural pests to pesticide. Tentative method of adults of some major pest species of stored cereals, with methyl bromide and phosphine. FAO Method No. 16. FAO Plant Protection Bulletin, 23, 12–26.

Ansell, M.R., Dyte, C.E. and Smith, R.H. 1990. Inheritance of phosphine resistance in *Tribolium castaneum* and *Rhyzopertha dominica*. In: Fleurat-Lessard, F. and Ducom, P., ed., Proceedings

Table 1. Response to phosphine of T–16, G–12 and their reciprocal F_1 progenies of *Sitophilus oryzae*.

Strains	n	LT_{50} (95% FL) Time (hours) at 0.04 mg/L	Slope (SE)	Resistance factor
T–16 (S)	840	7.6 (7.0–8.1)	5.31 (0.17)	–
G–12 (R)	840	69.1 (65.4–73.1)	5.99 (0.69)	9
F_1 (SXR)	840	15.8 (14.8–16.7)	6.67 (0.50)	2
F_1 (RXS)	840	15.6 (14.7–16.5)	6.96 (0.79)	2

of Fifth International Working Conference on Stored-product Protection, Bordeaux, France, September 1990, 961–970.

Champ, B.R. and Dyte, C.E. 1976. Report of the FAO global survey of pesticide susceptibility of stored grain pests. Rome, Food and Agriculture Organization.

Collins, P.F. 1986. Genetic analysis of fenitrothion resistance in the sawtoothed grain beetle, *Oryzaephilus surinamensis* (Coleoptera: Cucujidae). Journal of Economic Entomology, 79, 1196–1199.

Finney, D.J. 1952. Probit analysis 2nd ed. London, Cambridge University Press.

Georghiou, G.P. 1969. Genetics of resistance to insecticides in houseflies and mosquitoes. Experimental Parasitology, 26, 224–255.

Halstead, D.G.H. 1963. External sex differences in stored-products of Coleoptera. Bulletin of Entomological Research, 54, 119–134.

Liang Quan 1976. Resistance to phosphine in four major species of stored grain pests in Guangdong, China. Sichuan Liangyou Keji. 1976, 4, 1–11 (in Chinese).

Monro, H.A.U., Upitis, E. and Bond, E.J. 1972. Resistance of a laboratory strain of *Sitophilus granarius* (L.) (Coleoptera: Curculionidae) to phosphine. Journal of Stored Products Research, 8, 199–207.

Stone, B.F. 1968. A formula for determining degree of dominance in cases of monofactorial inheritance of resistance to chemicals. Bulletin of the World Health Organization, 38, 325–326.

Taylor, R.W.D. 1989. Phosphine—a major grain fumigant at risk. International Pest Control, 31, 10–14.

Winks, R.G. 1986. The effect of phosphine on resistant insects. In: Proceedings of GASGA Seminar on Fumigation Technology in Developing Countries. London, Tropical Development and Research Institute.

Study of circumfluent fumigation with phosphine for killing stored-grain insects in silos

Lu Jian-hua*, Zhao Zeng-hua†, Liu Qing*, Hu Shu-tian* and Qi Jin-shen*

Abstract

A study of circumfluent fumigation with phosphine for killing insects was conducted in silos 14.63 m tall by 6.63 m inside diameter. Before fumigation, aluminium phosphine tablets were put on the surface of bulk red wheat in the sealed silo. A multifunctional duct system was fitted to circulate phosphine gas. The fumigation resulted in 100% mortality of *Sitophilus zeamais*, *Oryzaephilus surinamensis*, and *Tribolium castaneum*. The method was easy to operate, and the small axial fan used was sufficiently slow, gave a recirculation rate low enough to be safe and had low power consumption.

Introduction

Recirculation fumigation with phosphine has not previously been tested in large silo bins in China. In 1990 a multifunctional duct system for aeration and fumigation was tested for killing stored-grain insects using methyl bromide with recirculation. Based on this successful experience, a trial on recirculation fumigation with phosphine using the duct system was carried in 1991 and 1992. This study is complete and has passed assessment by a scientific panel.

Materials and Methods

Silo bins and grain used for the trial

The trial silo was part of Tanggu Storage. The bins were constructed of brick, rendered inside and out with cement. Bins 5, 6 and 7 were used for the trial. Between the render and the brick was a layer made of sand, hydrated lime and cement. The bins had steel concrete bases and flat concrete tops. Each bin was 14.27 m high, 7.45 m outside diameter and 6.63 m inside diameter. Each bin contained 240 t red wheat with 11 m height of the grain mass and a bin volume of 380 m³.

Grain temperatures and moisture contents, and ambient temperatures are given in Table 1.

Gastightness tests

Pressures were observed at three points in the empty bin during pressure testing. The half-life for 2000–1000 Pa was 45 seconds for each point.

* Tanggu Grain Storage, Tianjin, 300451, China.
† Tianjin Grain Bureau, Tianjin 300014, China.

Test and recirculation equipment

The duct work attached to the bins is shown in Figure 1. The trial used the following equipment: phosphine detector tubes; temperature/humidity sensor; electronic air-velocity meter; in-bin automatic temperature recording system; small axial fan. The grain parameters are given in Table 1.

Test insects

Adults of *Sitophilus zeamais*, *Oryzaephilus surinamensis* and *Tribolium castaneum* (20 of each) were sealed into phosphine-permeable envelopes. Twelve of these envelopes were distributed evenly in the upper, middle and lower parts of the wheat in the test bins.

Gas testing and phosphine dosages

Phosphine dosages (as aluminium phosphide) used in three trial bins are given in Table 1.

Phosphine concentrations were determined at 12 points in each bin, with sample points distributed in the upper, middle and lower parts of the bin. Gas samples were taken through polyethylene tubing.

Conduct of the fumigation

The wheat was loaded into the bins after placement of the insect-containing envelopes and the gas sample lines. Aluminium phosphide (AlP) tablets were then placed on the surface of the grain bulk in dishes. The phosphine concentrations were checked during the recirculation of the internal gases using a small axial fan and before and after recirculation. The fan gave one air change per hour and was run for 2 hours at specified

Table 1. Experimental parameters

Bin No.	5	6	7
Quantity of wheat (t)	240	240	240
Moisture content (%)	12	12	12
Average grain temperature (°C)	22	17	21
Date of AlP application	2/8	25/6	20/7
Temperature in headspace (°C)	30	24	27
Relative humidity in headspace (%)	80	72	73
Ambient temperature (°C)	34	27	30
AlP tablets applied (g)	940	1500	2250
Time of exposure (hours)	240	168	120
Time of first recirculation (hours)	22	8	10
Time of second recirculation (hours)	–	22	22
Time of third recirculation (hours)	–	46	46
Recirculation period (L)	2	2	2

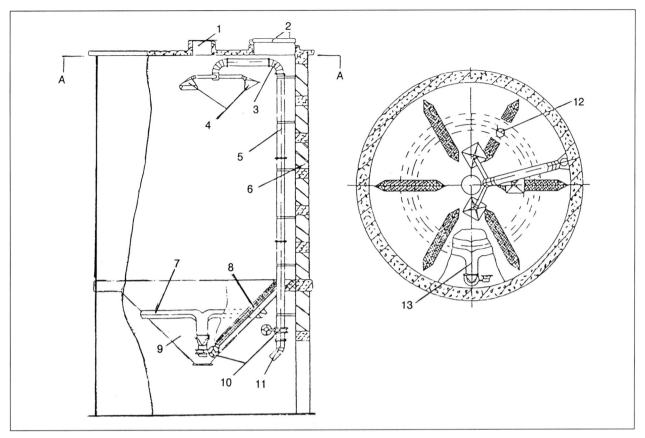

Fig. 1. Trial silo equipped with multifunctional duct system. 1. Grain inlet; 2. Sealable manhole; 3. Tube at 90° angle; 4. Nozzles; 5. Vertical pipe; 6. Silo; 7. Circular pipe surrounding the outer cone; 8. Radiating pipes inside the cone; 9. Steel base of the cone; 10. Wing nut; 11. Movable tube at 135° angle; 12. Air inlet (6 in total); and 13. Standing and returning pipe.

times after putting the aluminium phosphide tablets in the bin (Table 1). Recirculation was stopped as soon as the gas had been distributed evenly.

After the fumigation, the bins were aerated and then unloaded. The envelopes were recovered and the mortality of test insects assessed. Mortality was reassessed after 21 days incubation.

Results and Discussion

The three trial bins were all treated with different aluminium phosphide dosages and were slightly different in temperature and humidity (Table 1). Phosphine levels recorded in the trial are given in Tables 2–6.

Bin 5 was dosed at 2.5 g AlP/m^3. After 22 hours the phosphine levels varied from 0.6–2.5 g/m^3. After 2 hours of

recirculation the gas concentration was even throughout the bin at 1.2 g/m^3. The concentrations at 48 hours were even and similar to those at 24 hours, so no further mixing by recirculation was needed.

Bins 6 and 7 were dosed at 4 and 6 g AlP/m^3, respectively. Both required three periods of recirculation (2 hours each) in order to keep gas concentrations even. The first recirculation was carried out when the highest concentration reached 1.8–2 g/m^3. This occurred 8–12 hours after addition of the tablets. Further recirculation was carried out at 22 and 46 hours.

Bins 5, 6 and 7 were exposed to phosphine for 240, 168 and 120 hours, respectively. There was 100% mortality of test insects both immediately after fumigation and recovery of the envelopes, and after 21 days of incubation.

Table 2. Phosphine concentrations (g/m^3) before and after first recirculation periods

	Bin No.	Time after application of tablets (L)	Headspace	Depth of sampling point										
				Central sampling line						Lateral sampling line				
				1	3	5	7	9	10.5	1	3	5	7	9
Before recirculation	5	22	2.5	2.0	0.6	0	0	0	0	2.0	0.5	0	0	0
	6	10	1.8	0.7	0.2	0	0	0	0	0.8	0.2	0	0	0
	7	8	2.0	0.6	0.14	0	0	0	0	0.5	0.1	0	0	0
After recirculation	5	24	1.2	1.0	1.0	1.0	1.0	1.0	1.1	1.2	0.9	1.0	1.0	0.8
	6	16	0.7	0.6	0.7	0.5	0.6	0.6	0.3	0.6	0.4	0.5	0.5	0.4
	7	12	1.0	0.8	0.8	0.8	0.9	0.9	1.0	1.1	0.9	0.9	0.9	0.9

Table 3. Phosphine concentrations (g/m^3) after recirculation and 24 hours later

| Bin No. | Headspace | Depth of sampling point | | | | | | | | | | |
| | | Central sampling line | | | | | | Lateral sampling line | | | | |
		1	3	5	7	9	10.5	1	3	5	7	9
After recirculation	5	1.2	1.0	1.0	1.0	1.0	1.0	1.1	1.1	0.9	1.0	1.1	1.0
	6	1.6	1.5	1.5	1.5	1.6	1.6	1.7	1.6	1.4	1.4	1.5	1.5
	7	2.6	2.5	2.6	2.4	2.1	2.1	2.3	2.5	2.2	1.8	2.0	2.0
24 hours later	5	1.2	1.1	1.0	0.9	1.0	0.9	0.9	1.2	1.0	1.0	1.0	0.9
	6	1.5	1.4	1.5	1.5	1.5	1.6	1.6	1.5	1.5	1.3	1.4	1.5
	7	2.5	2.2	2.0	2.0	2.0	2.0	1.8	2.5	2.0	2.0	2.0	2.0

Table 4. Change in gas concentrations (g/m^3) in Bin 5, dosed at 2.5g AlP/m^3

| | | Time (hours) | | | | | | | | | |
		24	48	72	96	120	144	168	192	216	240
Headspace		1.2	1.2	1.2	1.1	0.9	0.8	0.6	0.5	0.5	0.4
Central sampling line	1	1.0	1.1	1.0	0.9	0.9	0.7	0.5	0.4	0.3	0.3
	3	1.0	1.0	1.0	0.9	0.9	0.7	0.6	0.4	0.4	0.3
	5	1.0	0.9	1.0	0.9	0.8	0.7	0.5	0.4	0.3	0.3
	7	1.0	1.0	1.0	1.0	0.9	0.7	0.7	0.6	0.3	0.2
	9	1.2	0.9	1.0	0.9	0.8	0.6	0.5	0.5	0.4	0.3
	10.5	1.1	0.9	1.0	0.9	0.8	0.7	0.6	0.5	0.4	0.3
Lateral sampling line depth below surface (m)	1	1.2	1.2	1.2	1.0	0.9	0.7	0.6	0.5	0.4	0.4
	3	0.9	1.0	1.0	0.9	0.9	0.7	0.6	0.4	0.4	0.3
	5	1.0	1.0	1.0	1.0	0.9	0.6	0.5	0.5	0.3	0.3
	7	1.1	1.0	0.9	0.8	0.8	0.6	0.5	0.4	0.4	0.3
	9	1.0	0.9	0.9	0.8	0.8	0.7	0.5	0.4	0.3	0.3
Average		1.1	1.0	1.0	0.9	0.9	0.7	0.6	0.4	0.4	0.3

Table 5. Change in gas concentrations (g/m^3) in Bin 6, dosed at 4 g AlP/m^3

| | | Time (hours) | | | | | | | | |
		12	24	36	48	72	96	120	144	168
Headspace		0.6	0.8	1.1	1.6	1.5	1.4	1.2	0.9	0.5
Central sampling line	1	0.5	0.8	1.1	1.5	1.4	1.2	1.1	0.9	0.5
	3	0.5	0.9	1.2	1.5	1.5	1.4	1.3	1.0	0.4
	5	0.4	0.8	1.2	1.5	1.5	1.3	1.2	0.8	0.4
	7	0.4	0.8	1.1	1.6	1.5	1.3	1.1	0.8	0.5
	9	0.4	0.8	1.1	1.6	1.6	1.4	1.2	0.9	0.5
	10.5	0.5	0.9	1.3	1.7	1.6	1.4	1.3	1.0	0.6
Lateral sampling line depth below surface (m)	1	0.5	0.9	1.1	1.6	1.5	1.2	1.2	0.8	0.4
	3	0.4	0.8	1.2	1.4	1.5	1.4	1.3	0.8	0.5
	5	0.4	0.8	1.3	1.4	1.3	1.3	1.2	0.7	0.5
	7	0.5	0.9	1.3	1.5	1.4	1.4	1.2	0.8	0.4
	9	0.4	0.8	1.1	1.5	1.5	1.3	1.2	0.9	0.5
Average		0.46	0.83	1.18	1.53	1.48	1.33	1.21	0.86	0.54

Table 6. Change in gas concentrations (g/m^3) in Bin 7, dosed at 6 g AlP/m^3

		Time (hours)						
		12	24	36	48	72	96	120
Headspace		1.0	2.0	2.5	2.6	2.5	2.2	2.0
Central sampling line	1	0.8	2.0	3.0	2.5	2.2	2.0	1.8
	3	0.8	2.0	3.0	2.6	2.0	2.0	1.9
	5	0.8	2.0	2.1	2.4	2.0	2.0	1.8
	7	0.9	2.0	2.0	2.1	2.0	1.9	1.7
	9	0.9	2.5	2.0	2.1	2.0	1.7	1.7
	10.5	1.0	2.5	1.8	2.3	1.8	2.0	1.8
Lateral sampling line depth below surface (m)	1	1.1	2.0	3.0	2.5	2.5	2.0	1.8
	3	0.9	2.0	2.0	2.2	2.0	1.9	1.6
	5	0.9	2.0	2.0	1.8	2.0	2.0	1.8
	7	0.9	2.0	1.8	2.0	2.0	1.8	1.8
	9	0.9	2.0	1.8	2.0	2.0	1.8	1.7
Average		0.91	2.08	2.25	2.26	2.08	1.94	1.78

Conclusion

This study has demonstrated highly efficient, economic and safe treatment with phosphine in large silo bins. It is concluded that:
- with correct design and multifunctional duct system it is possible to distribute gas evenly so as to achieve 100% mortality;
- the small, sparkless axial fan was sufficiently slow and gave sufficiently low circulation rate for the fumigation to be safe;
- the buildup of phosphine from the aluminium phosphide tablets, applied to the grain surface, could be controlled by the recirculation so as not to exceed a predetermined safe limit. With dosages of 4–6 g AlP/m^3 two to three recirculation periods of 2 hours duration were sufficient to prevent excessive concentrations;
- the time of exposure can be regulated according to dosage, with 5 days sufficient at normal rates of application and longer for the low dosage treatments; and
- the fumigation is safe, easy to carry out and will save power.

Comparative toxicity of carbon dioxide to two *Callosobruchus* species

G. Mbata*, C. Reichmuth[†] and T. Ofuya[§]

Abstract

Mortalities of eggs, larvae, pupae and adults of two bruchids, *Callosobruchus maculatus* (F.) and *Callosobruchus subinnotatus* (Pic), exposed to an inert atmosphere of 100% carbon dioxide were observed at 32°C, 70% r.h. 100% mortality of the eggs and adults of both species. All young larvae of both species died within an exposure period of 48 hours. The older larvae required an exposure period of 72 hours for 100% mortality to be achieved. The pupal stage was the most tolerant stage to carbon dioxide and exposure periods of 5 and 6 days, respectively, were required for 100% mortality of *C. maculatus* and *C. subinnotatus*, respectively.

Introduction

In west African countries, and indeed most other subtropical and tropical countries, grain legumes which are the most important sources of proteins are attacked by two bruchids, *Callosobruchus maculatus* (Fab.) and *Callosobruchus subinnotatus* (Pic). *C. maculatus* is a major pest of stored cowpea, *Vigna unguiculata* L. in these regions while *C. subinnotatus* is a major pest of bambarra groundnuts, *Vigna subterranea* (L.) Verde (Jackai and Daoust 1986; Mbata 1991). The damage is done by the larvae feeding inside the seeds. Severe infestation can lead to losses in weight of up to 30% and render the commodity unfit for human consumption.

Chemical control of bruchids by dusting with insecticides such as pirimiphos-methyl, fenithrothion and synthetic pyrethroids or by fumigation with phosphine and methyl bromide is effective (Cardona and Karel 1990). However, problems associated with the use of chemical insecticides, such as development of resistance by insects, environmental pollution, contamination of grains, etc. are inducing scientists to look for other ways of controlling insect pest populations. Controlled atmosphere storage offers a safe, residue-free method for protecting grains from insect pests (Bailey and Banks 1980). Ofuya and Reichmuth (1992) and Mbata and Reichmuth (1994) have investigated the toxicity of inert atmospheres to *C. maculatus* and *C. subinnotatus*. The present report compared the toxicity of an atmosphere of 100% CO_2 to the eggs, larvae, pupae and adults of the two bruchids.

Materials and Methods

Eggs, larvae, pupae and adults of both *C. maculatus* and *C. subinnotatus* used in this study were obtained from cultures of the insects maintained at 30°C and 70% r.h. *C. maculatus* was reared on blackeye cowpea while bambarra groundnut was used in rearing *C. subinnotatus*.

The developmental stages tested were as follows: eggs (ca. 2 days old), young larvae (5–9 days old), old larvae (14–17 days old), pupae (1–3 days old) and adults (ca. 24 hours old). The experiments were set up separately for both *C. maculatus* and *C. subinnotatus*.

To obtain eggs of the Bruchids on seeds, 10 pairs of the adults were placed in 500 mL jars containing about 300 cowpea seeds for *C. maculatus*, or bambarra seeds *C. subinnotatus*. These were allowed 24 hours to lay eggs on the seeds. Seeds with egg load of 3 or 4 eggs per seed were selected. Fifty eggs on seeds or fifty larvae or pupae in seeds or 20 adults were placed in separate wire cages of dimensions 5 cm long and 1.5 cm diam. The wire cages were closed with rubber stoppers and exposed to 100% CO_2. The CO_2 was delivered from a pressurised cylinder and the insects were exposed to the gas in gastight connected dressel flasks. The exposure period was from 1–4 days at 32°C. The treatments for each of the stages of the insects were replicated three times and there were three trials for each treatment. Replicated controls were also set up for each of the stages of the insects.

For the adults, mortality was checked 24 hours after the exposure period, sufficient time for insects anaesthetised by the intoxicating effect of the CO_2 to recover. For the eggs, larvae and pupae, the treated seeds were placed in separate Petri dishes and sufficient time allowed for adult emergence at 30°C. Percentage mortality in each stage was calculated and corrected using Abbott's (1925) mortality formula. The mortality values for the different species were compared using two-way analysis of variance.

Results and discussion

There were significant differences in the exposure periods required to obtain 100% mortality among the developmental stages of the bruchids exposed to 100% CO_2 (Table 1). The eggs and the adults were most susceptible: 100% mortalities were recorded in these two stages within 24 hour exposure periods (Figs 1 and 2). These results support earlier observations by Ofuya and Reichmuth (1992) and Mbata and Reichmuth (1994).

The susceptibilities of the larval stages of the bruchids are shown in Figure 3. When the first larval stage was exposed to the inert atmosphere, 100% mortality was recorded after 2 days of exposure. The last larval instar required a longer exposure period of 4 days for 100% mortality to be obtained. The pupae were the most tolerant stages and required exposure periods up to 5 and 6 days for 100% mortality to be

* School of Biological Sciences, Abia State University, P.M.B. 2000, UTURU. Abia state, Nigeria.

† Federal Biological Research Centre for Agriculture and Forestry - Institute for Stored Products Protection, Königin-Luise-Straße 19, D-14195, Berlin. Germany.

§ Department of Biology, Royal Holloway, University of London, Egham, Surrey TW20 OEX, England.

Table 1. Summary of ANOVA for the mortality of different stages of *C. maculatus* and *C. subinnotatus* exposed to 100% CO_2 at 32°C

Source of variation	df	Mean square at different exposure periods (hours)				
		24	48	72	96	144
Insect stage (A)	4	10509.2[**]	10052.4[**]	9788.4[**]	5944.1[**]	1325.6[**]
Insect type (B)	1	27.1	19.3	15.3	18.0	15.7
Interaction (A × B)	4	22.5	16.2	12.7	10.6	9.7
Error	20	18.5	17.0	19.3	15.7	16.3
Total	29					

**Significant at $P < 0.01$.

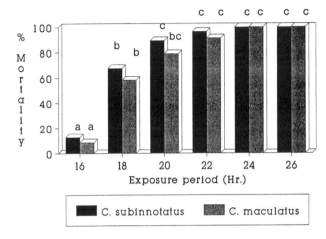

Fig. 1. Mortality of eggs of the two bruchids exposed to 100% carbon dioxide (values bearing different letters are significantly different, $P < 0.05$).

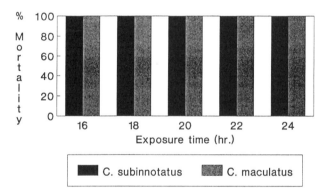

Fig. 2. Mortality of adults of the two bruchids exposed to 100% carbon dioxide (values bearing different letters are significantly different, $P > 0.05$).

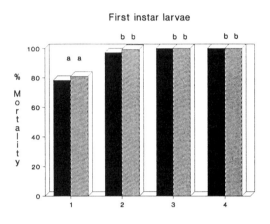

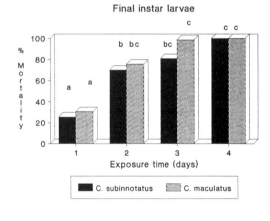

Fig. 3. Mortality of larvae of the two bruchids exposed to 100% carbon dioxide (values bearing different letters are significantly different, $P > 0.05$).

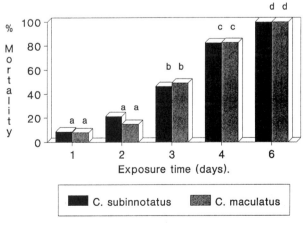

Fig. 4. Mortality of pupae of the two bruchids exposed to 100% carbon dioxide (values bearing different letters are significantly different, $P > 0.05$).

recorded in *C. maculatus* and *C. subinnotatus*, respectively (Fig 4).

Oosthuizen and Schmidt (1942) found that the eggs and the adults of *Callosobruchus chinensis* (L), a species related to those used in the present study, were more susceptible than older larvae and pupae when exposed to an atmosphere of CO_2. The high tolerance of the pupae to inert atmosphere might be attributable to the reduced metabolism in the pupae compared with the other stages.

There was no significant difference between the exposure periods required to obtain 100% mortality of similar stages of the two bruchids. Bambarra seeds infested by the two bruchids can therefore be disinfested by exposing them to 100% CO_2 for 6 days. This is in line with the recommended exposure period of 15 days for disinfesting grain with high CO_2 at commodity temperature of 25–29°C (Banks et al. 1990).

Acknowledgments

T.I. Ofuya and G.N. Mbata gratefully thank the Alexander von Humboldt Foundation, Bonn, Germany, for the award of research fellowships which enabled them to participate in the investigation reported here.

References

Abbot, W.S. 1925 A method for computing the effectiveness of an insecticide. Journal of Economic Entomology, 8, 265–267.

Bailey, S.W. and Banks, H. J. 1980 A review of recent studies of the effects of controlled atmospheres on stored products pests. In: Shejbal, J., ed., Controlled atmosphere storage of grains. Proceedings of international symposium, Rome, 1980. Amsterdam, Elsevier, 100–118.

Banks, H.J., Annis, P.C. and Rigby, G.R. 1991. Controlled atmosphere storage of grain: the known and the future. In: Fleurat-Lessard, F. and Ducom, P., ed., Proceedings of the 5th International Working Conference on Stored Product Protection, Bordeaux, France 1990, 695–706.

Cardona, C. and Karel, A.K. 1990 Key insects and other invertebrate pests of beans. In: Singh, S.R., ed., Insect pests of food legumes. Chichester, John Wiley and Sons, 157–191.

Jakai, L.E.N. and Daoust, R.A. 1986. Insect pests of cowpea. Annual Review of Entomology, 31, 95–119.

Mbata, G.N. 1991. Seasonal incidence and abundance of insect pests of stored bambarra groundnuts. In: Wolf, J., ed., The influence of climatic factors on the agricultural productivity of tropical countries. Stockholm, CTA and FIS, 452–459.

Mbata, G.N. and Reichmuth, Ch. 1994. The toxicity of inert atmospheres to the different stages of the bambarra bruchid *Callosobruchus subinnotatus* (Pic.). Mededelingen van de Faculteit van de Landbouwwetenschappen Universitet Gent., in press.

Ofuya, T.I. and Reichmuth, Ch. 1992. Mortality of the cowpea bruchid, *Callosobruchus maculatus* (Fabricius) in a highly elevated carbon dioxide atmosphere. Proceedings of the First European Conference on Grain Legumes, Angers, France, 1992, 365–366.

Oosthuzien, M.J. and Schmidth, U.W. 1942. The toxocity of carbon dioxide to cowpea weevil. Journal of Entmolological Society of South Africa, 5, 55–110.

A new method of using low levels of phosphine in combination with heat and carbon dioxide

D.K. Mueller*

Abstract

A combination of low levels of phosphine (65–100 ppm), heat (32–37°C), and carbon dioxide (4–6%) were used in three mills in the United States. Fumigations were carried out for 24 hours. Multiple species of stored-product insects in various life stages were used as bioassays. A corrosion study was conducted with copper and electronic equipment. A penetration study was conducted with 2 and 3 m deep tubes filled with wheat flour. All insect bioassays were retained for 30 days. A 100% insect mortality within bioassays was achieved in 24 hours or less. This method of fumigation holds promise as a replacement for methyl bromide fumigations in flour mills and similar structures.

Introduction

The purpose of the study was to evaluate the use of combination treatments to effectively control given populations of stored-product insects. Insects are stressed by the increased levels of carbon dioxide and heat. This allows lower levels of phosphine to be more effective in shorter periods of time.

History

Fumigations with inert gases and methyl bromide have been performed since 1929. The famous stored-product entomologist R.T. Cotton is noted for the work and patented a combination method. More recently, the Australians have performed numerous experiments with carbon dioxide and phosphine on commodities in sealed structures. The Israelis patented a method in 1979 for using four parts carbon dioxide with one part methyl bromide in grain storages. Carmi and Leesch have shown that carbon dioxide is effective in moving an atmosphere of phosphine deeper into the grain mass.

Mueller states in 'The Mallis Handbook of Pest Control' (7th ed.) that there are several ways to produce insect stress including: 1) decreasing oxygen concentration, 2) increasing carbon dioxide levels, or 3) increasing temperature. It has been published that insect respiration can be increased by 50% by increasing carbon dioxide levels to 3%. Insect respiration increased 300% when carbon dioxide levels were raised to 5%.

Materials and Methods

Fumigants

Fumigants for this study were produced from Degesch FUMI-CEL™ and Degesch FUMI-STRIP™, a magnesium phosphide formulation in a solid plate form. Each plate generates 33 g of hydrogen phosphide (phosphine) gas. These formulations have advantages over standard aluminium phosphide formulations in that gas is generated more rapidly. A minimum concentration of 100 ppm can be achieved in 7 hours at 32°C and 50% r.h. This compares to 11 hours with aluminium phosphide at the same tested dosage rates and conditions.

Dosage

A dosage rate of 6.6 g/1000 cu. ft. was initially used to treat each location. One gram of phosphine produces 25 ppm in 1000 cu. ft. The theoretical maximum concentration for the dosage rate was 165 ppm. Conventional fumigations with phosphine would have concentrations between 850 and 1500 ppm. Each of the three locations showed peak phosphine concentrations of more than 50% of the theoretical maximum concentration. These concentrations were representative of properly sealed buildings.

The first location (Mill #I) was a two storey, 60000 cu. ft. feed mill at Purdue University in West Lafayette, IN. The second location (Mill #2) was a six storey, 181000 cu. ft. flour mill in Honolulu, HI. The third location (Mill #3) was a six storey, 300000 cu. ft. flour mill in Frankenmuth, MI. All fumigations were carried out during the summer of 1993.

Carbon Dioxide

During the first combination fumigation at Purdue University (Mill #I), a total of 37 fifty pound steel cylinders of carbon dioxide were used during the fumigation. Use of cylinders was cumbersome and more expensive than larger vessels.

Carbon dioxide must be vaporised from a liquid to a gaseous state. Gas temperatures should range from 70–90°F or 20–30°C when entering the building. Special hoses and regulators are required when working with 40°F liquid carbon dioxide. Advanced knowledge of how to control and release this inert gas is necessary. Even though carbon dioxide is inert, it can be very deadly. Oxygen levels are decreased when carbon dioxide enters a building. Carbon dioxide should be treated as a hazardous fumigant similar to methyl bromide and phosphine.

Heat

Frequently mills and similar facilities have heating systems that maintain a content temperature in the building. The Purdue University location utilised electric heaters to maintain a 100°F (38°C) temperature (± 2°C). The Hawaiian Flour Mill (Mill #2) had no heating system. No heaters were available on the island of Oahu.

The ambient temperature of the flour mill was 30–31°C, temperatures necessary for the combination techniques. A steam boiler was used to heat radiators in the building and an additional steam coil-type 125000 BTU heater was utilised to enhance heating capacity. Outdoor temperatures during the Michigan fumigation (Mill #3) reached 4°C or 40°F. The other two fumigations (Mill #1 and Mill #2) were performed on warm summer days.

* Fumigation Service & Supply, Inc. Indianapolis, Indiana USA 46280-1451. Patent pending.

Results

Test Insects

Mill #1

Test insects were placed in four locations at the Purdue Feed Mill. Four species of insects were utilised. These included: Angoumois grain moth *(Sitotroga cerealella)*, red flour beetles *(Tribolium castaneum)*, warehouse beetles *(Trogoderma variable)*, and rice weevils *(Sitophilus oryzae)*. Eggs, larvae, pupae, and adults were placed in 250 mL plastic containers.

Two groups of 36 containers were placed on the first floor and two groups of 36 containers were placed on the second floor.

Test insects were retrieved from the fumigated mill beginning 20 hours after the start of fumigation. Test insects continued to be retrieved every 4 hours until 48 hours after the start of fumigation. Adult and larval stages were evaluated for all sampling intervals. All insect stages were dead starting 20 hours into the fumigation. All of the containers were taken to the lab at Purdue University and incubated for 30 days in a growth chamber. No insect activity was observed after 30 days. Control insects remained alive throughout the evaluation period.

Mill #2

Two species of stored-product insects were placed as bioassays in the flour mill fumigation in Hawaii; red flour beetles *(Tribolium castaneum)* eggs, larvae, pupae, and adults (Indiana strain and Hawaii strain), and rice weevil *(Sitophilus oryzae)* adults. A total of 150 insect cages with 10 or more insects per cage were placed on the 6 floors. Insect cages were placed approximately one meter above the floor. Insect cages were retrieved from the building beginning 13 hours after the start of fumigation. All insects within bioassays and retrieved at the 13 hour interval were dead. All insects retrieved from the flour mill at 24 hours after the start of fumigation were also dead. These insects were incubated in Hawaii and Indianapolis for 30 days with no activity observed. All indoor and outdoor insect control groups were alive 48 hours after the fumigation start and 87% were alive after 30 days.

Mill #3

During the Michigan flour mill fumigation, 3 species of stored-product insects were used in bioassays: red flour beetle eggs, larvae, pupae, and adults, rice weevil adults, and Indianmeal moth eggs *(Plodia interpunctella)*.

Insect cages were placed on each floor and controls were maintained as described previously. Over 150 cages were placed in the mill with 10 or more insects per cage.

Insect canes were retrieved from the building twelve hours after the start of the fumigation. All adult and larval specimens were dead twelve hours from the beginning of the fumigation. All insects retrieved from 12–24 hours after the start of fumigation were subsequently dead. All specimens were held at room temperature. No insect activity was observed for any stage of the three stored-product insects tested. Controls from outdoor cages and indoor cages remained alive.

Penetration Study

Phosphine and carbon dioxide are excellent penetrating gases. Phosphine is a more effective penetrator than methyl bromide. Twelve 6 inch diameter × 6 ft long PVC pipes were capped and permanently sealed on one end. Insect cages were placed in the bottom of these long tubes. Each tube was subsequently filled to a 3-foot level with wheat flour and insect cages were placed at that level. Flour was again added to the remaining 3 feet and tubes were filled completely. A 1.4 ml polyethylene bag was secured with tape over the open end of each tube. Two tubes were placed on each of 6 floors of the (Mill #2) Hawaiian Flour Mill (HFM).

Five flour-filled tubes were retrieved from the flour mill 24 hours after the start of the fumigation. Carbon dioxide and phosphine levels were measured with a Draeger tube and levels were found to be equal to the ambient concentration of the fumigated mill. Phosphine levels under polyethylene bags were 50 ppm, the carbon dioxide levels were 3%. Test insects at 3 feet and 6 feet were evaluated. All adults and larvae were dead. Eggs and pupae were incubated for 30 days.

The remaining 6 tubes were retrieved 48 hours after the start of fumigation. To affect a kill on test insects, gases had to penetrate the 1.4 mL polyethylene bag, permeate 3–6 feet of flour and inside polyethylene bioassay tubes and kill egg larvae, pupae, and adult insect specimens. Insects in control groups remained alive. After 30 days, insects in polyethylene vials positioned at the bottom of 6 foot flour tubes showed some survival in the first instar larvae stage. Some eggs had survived this penetration study.

The study was taken one step further in Michigan (Mill #3). Two 5-foot and one 10-foot tube similar to those described above were placed in the fumigated mill. All stages of insects were positioned at 10-foot, 5-foot, and I-foot levels within tubes. A 1.4 mL polyethylene bag was placed over one 5- foot tube. No bags were placed over the remaining tubes.

Observations 12 hours after the start of fumigation indicated that insects at the bottom of the 5-foot tube without a plastic bag were dead. After 24 hours the test insects at the bottom of the 5-foot tube with bag and 10-foot tube were dead. The immature stages were incubated for 30 days. No survivors appeared after 30 days. This fumigation was about 4°C or 10°F warmer than the Hawaiian study. The gas concentrations were nearly the same.

Corrosion Study

One negative characteristic of phosphine is that under certain conditions it can cause corrosion of precious metals and coppers. Many items should be removed from buildings to be fumigated when possible. Often it is impossible to remove items that could be affected by high levels of phosphine and high humidity.

Phosphine produced from magnesium phosphide or aluminium phosphide can generate phosphoric acid. When phosphoric acid and sufficient moisture are combined on precious metals, a mild acid is formed and corrosion occurs. With severe corrosion, instruments will often fail. Corrosion management is necessary when phosphine is to be used in fumigating buildings that contain printed circuits, and sensitive equipment.

Heat and carbon dioxide will lower humidity within a facility. Samples of wheat from the Mill #3) dropped from 13–11% moisture during the fumigation process, while moisture in the flour mill dropped 12% (56% vs. 44%).

By managing large peaks in phosphine gas levels, levels of H_3PO_4 can also be managed. Phosphine levels should remain in a range from 50–100 ppm with the need for magnesium phosphide fumigant to be physically removed when levels reach 150 ppm. Magnesium phosphide can be easily added or removed from buildings to achieve these target levels. Proper safety precautions are necessary.

A new method of evaluating corrosion was developed for this test. New copper pennies where placed throughout the structures to determine severity of corrosion with combination fumigation techniques. Each penny was labelled with a number and weighed prior to fumigation. Pennies were located randomly throughout each structure and placed vertically on-edge. Some pennies were suspended from ceilings with fishing line at various heights. Solid blocks of steel and copper where placed within the Purdue Feed Mill during fumigation.

Results of the corrosion study revealed small weight gains of each of the pennies. Average weight gains of 0.0009 grams for the 10 pennies in the Michigan study and average weight gains of 0.0040 grams were observed for 31 pennies in the Hawaiian study.

Electron microscope scans revealed no traces of phosphorus or other elements (except inherent impurities) on the two steel samples placed on the first and second floors of the feed mill. No traces were found in the copper sample (apart from impurities) placed on the second floor of the mill. The copper sample placed on the first floor, did show a treatment effect of about 8% phosphorus contamination (average of two readings) due to phosphine gas treatment.

Although this information is preliminary, future comparisons should allow for ratings of corrosion potential and a determination of the method's viability.

A ten-penny corrosion test is planned for each future fumigation. Average percent weight gains should be a factor that will help with corrosion management assessment programs.

Cost

Real costs of a fumigation are related to shutdown time. Many mills and processed food operations cannot afford shutdowns longer than 24 hours.

The costs of carbon dioxide are offset by reduced costs for magnesium phosphide. Additional costs for equipment rental (vaporiser and vessel) also exist. Permanent installation of such equipment would be relatively simple. Additional costs for heat would be determined by outdoor temperatures and how well a building is sealed and/or insulated. Additional steam coils or electrical heaters could be installed if necessary.

It is difficult to determine the specifics of a building without studying it extensively during the first few combination fumigations. This will require many around-the-clock gas readings throughout the building. After experience is gained in stabilising and maintaining levels of phosphine within the proper range, combination fumigations can be very successful and cost-effective.

Conclusion

The future of the fumigation business is unclear. Dozens of fumigants have been removed from the market during the past decade. Methyl bromide has been identified as a serious ozone depletion with an Ozone Depletion Potential (ODP) of 0.7. The Clean Air Act in the United States declares that any product with an ODP of 0.2 or greater will be eliminated in due course. The Montreal Protocol's 120 plus signature countries are discussing similar retribution. There may be a day in the near future when phosphine could be the only conventional fumigant available.

With this uncertainty, alternatives are needed to eliminate insects from structures. Methyl bromide provides several advantages in that it acts quickly (24-hours or less), is inexpensive, and causes minimum damage to contents of structures, although it is a less aggressive penetrator than phosphine, it provides for effective kill of pests.

A Combination Method of fumigation using maintained lower levels of phosphine, moderate heat, and higher carbon dioxide levels for a 24-hour periods in sealed structures has potential for replacing many methyl bromide applications within mills and similar structures.

Acknowledgment

I thank John Mueller (Fumigation Service & Supply, Inc., Indianapolis, IN) for his fumigation expertise, and Larry Pierce (Food Protection Services, Mililani, Hawaii) for his assistance, and to Dirk Maier, Ph.D. (Purdue University, Agricultural Engineering Department) for his assistance on the corrosion study, and Prof. Hruska (Purdue University, Metallurgical Engineering Department for his analysis of the copper and steel specimens, and to Linda Mason, Ph.D. (Purdue Entomology Department) for her assistance with the stored-product insect bioassays and their evaluation, and Degesch America, Inc. for contributing the magnesium phosphide, and Fred Whitford, Ph.D. (Purdue Training and Certification Division) for his assistance and photography, and Angie Richards (insects Limited, Inc., Indianapolis, IN) for the graphs and data collecting, and HFM Foods, Inc. (Hawaiian Flour Mills) for performing this test on their facility, especially Don Sorum, their Cereal Chemist, and Dick Kraft of Star of the West Milling for his strong support and belief in Al's sons, and to Keith Weber for getting the building sealed and the work done right and to Mike Culy, Ph.D. and Curt Hale for their detailed review.

A new method to control stored-product insects using carbon dioxide with high pressure followed by sudden pressure loss

H. Nakakita* and K. Kawashima[†]

Abstract

A newly developed method for the control of stored-product insects consisting of a rapid addition of high pressure gas into a chamber, followed by the sudden release of the gas was evaluated using carbon dioxide (CO_2) or helium (He). The effect of various pressure levels and exposure periods was assessed against *Sitophilus zeamais* Motshchulsky, *Rhyzopertha dominica* F., *Tribolium castaneum* (Herbst), and *Lasioderma serricorne* (F.). With CO_2 at 20 kg/cm^2, an exposure time of 5 minutes was sufficient to kill all adults of the four species. However, eggs of *S. zeamais* required treatment at 30 kg/cm^2 of CO_2 for 5 minutes for complete kill. The integument of insects exposed to this treatment was severely damaged because of the expansion of internally dissolved CO_2 in the body when gas pressure was rapidly equilibrated with atmospheric pressure. In contrast, the application of He had no effect on adults of *S. zeamais* even at 70 kg/cm^2. The solubility of gases in the hemolymph appears to be a significant factor in the effectivenessof this method.

Introduction

Many people concerned with control of stored-product insects once believed that methyl bromide and phosphine are ideal chemicals since both can effectively control insects by penetrating into inter- and intragranular spaces within the grain mass without leaving toxic residues. However, a recent movement on methyl bromide has greatly changed the situation since it is considered as one of major chemicals that destroy the earth's ozone layer (Singh et al. 1988). Thus, the limitation of its production level has been internationally determined in the Fourth Meeting of Parties to the Montreal Protocol at Copenhagen in 1992. In addition, although more work is needed, there is a report that phosphine may also have carcinogenic effects on humans (Garry et al. 1989). Thus, safer and more effective means of combating stored-product insects are needed.

As an alternative to conventional fumigants, modified atmospheres using carbon dioxide(CO_2) have recently been widely investigated and developed in many countries such as Australia, the USA and Israel (Shejbal 1980; Ripp et al. 1984; Champ et al. 1990; Calderon and Barkai-Golan 1990). In Japan, the Food Agency has banned the use of methyl bromide or phosphine on domestic rice since 1991 and recommended CO_2 as an alternative. Nevertheless, low temperature warehouses (15°C) have been widely used for rice storage with more than 3 Mt capacity throughout the country.

Carbon dioxide, being a natural component of the atmosphere, is a safe chemical and has been permitted for use as an additive to many types of drinks and foods. A limitation of CO_2-enriched atmospheres for insect control, however, is a longer exposure period compared with those for conventional fumigants (Annis 1989). This limits the use of CO_2 for quarantine and food processing purposes in which quick treatment is necessary. However, several researchers have suggested recently that the limitation of CO_2 can be solved by high-pressure treatment with the gas (Stahl et al. 1985; Reichmuth 1990; Le Torc'h and Fleurat-Lessard 1992). In this study, we investigated conditions for insect control using pressurised gas followed by rapid pressure loss.

Materials and Methods

Insects

The species tested were the maize weevil, *Sitophilus zeamais* Motschulsky, red flour beetle, *Tribolium castaneum* (Herbst), lesser grain borer, *Rhyzopertha dominica* (F.), and cigarette beetle, *Lasioderma serricorne* (F.). All test insects were taken from cultures maintained at 25°C and 75% r.h. in a dark room. Both species of *S. zeamais* and *R. dominica* were reared on whole brown rice; wheat feed was used for *T. castaneum* and *L. serricorne*.

Equipment

The apparatus was composed of a 30 mL cylindrical chamber of 5 mm thickness steel placed between inlet and outlet valves (Fig. 1) (Nitto Koatsu K.K.). The valves can be opened or closed using handles. Test insects, either in test tubes or in grains, were introduced by opening the inlet valve while the outlet valve remain closed. Either CO_2 or helium (He) was then discharged from a pressurised tank into the chamber after closing all valves. After the exposure period, The gas was rapidly discharged to the atmosphere by opening the outlet valve.

Mortality Test

For the test tube test, 25 adults of each of two species were placed in perforated polyethylene tubes, 13 mm in diameter and 28 mm high. For the bulk grain trial, 200 adults of *S. zeamais* were released on 200 g of brown rice and allowed to oviposit for 3 days. From this, 10 g was drawn randomly and mixed with 10 g of fresh rice to make up 20 g test samples of brown rice. Twenty kernels were infested with larvae and another 20 kernels that contained pupae were introduced to the brown rice test medium. These kernels were selected from cultures previously identified to contain larval and pupal stages of the maize weevil. Manifestations of damage such as the presence of frass were used as a basis for judgment. Ages of immature stages used were 25 days for pupae, 15 days for larvae, and 3 days for eggs. Finally 2-week-old unsexed adults were also added to the brown rice . The brown rice containing

---

* Stored Product Entomology Laboratory, National Food Research Institute, 2–1–2 Kannondai, Tsukuba, Ibaraki 305, Japan.

[†] Japan International Research Center for Agriculture and Science, 1–2 Oowashi, Tsukuba, Ibaraki 305, Japan.

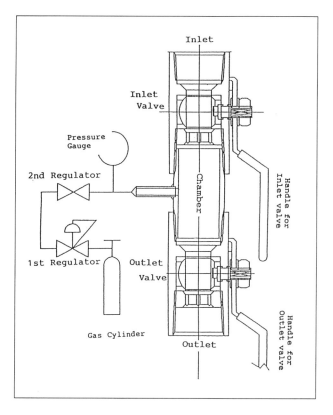

Fig. 1. Schematic diagram of equipment for high pressure gas increase followed by sudden release.

all stages of *S. zeamais* was then placed in the chamber for treatment. After exposure, the insects were transferred to the rearing room for observation. All stages of *S. zeamais* treated in the brown rice were assessed 1 month later to count the number of insect that had emerged. To compare the efficacy of CO_2 with that of another gas, helium was also applied to adults of *S. zeamais*, employing the same method.

Results

Table 1 shows the mortality of adults of four stored product beetles, *S. zeamais*, *R. dominica*, *T. castaneum* and *L. serricorne* resulting from the method after either 5 or 10 minutes exposure to highly purified CO_2 at pressures of 5, 10, 15 and 20 kg/cm^2. Increasing pressure reduced the exposure times needed to obtain higher mortalities although there were differences in susceptibility among species. *R. dominica* was the most susceptible species tested, requiring 5 kg/cm^2 for 10 minutes for a complete kill. *T. castaneum* was the most tolerant species since complete mortality was not obtained until a pressure of 15 kg/cm^2 for 10 minutes was used. All adults of the four species were killed completely in 5 minutes at 20 kg/cm^2.

Figure 2 shows the effect of the treatment on the adults of species. The integument is severely damaged as a result of the bursting of the insect's internal organs.

The eggs and larvae of *S. zeamais* exposed to elevated CO_2 for 5 min were more tolerant than pupae and adults, as shown in Table 2. Eggs were most resistant to the method, with only 21.9% killed at 20 kg/cm^2. A pressure as high as 30 kg/cm^2 was needed to kill all stages of *S. zeamais*. In contrast to CO_2,

Fig. 2. Physical damage in adult insects killed by high pressure technique: (a) *S. zeamais,* (b) *R. dominica,* (c) *T. castaneum* and (d) *L. serricorne.*

helium produced no significant lethal effect, even at pressures as high as $70 \, kg/cm^2$ (Fig. 3).

Discussion

The rapid increase in CO_2 pressure followed by a quick release of the gas after a 5–10 minutes exposure period appears to be an effective, fast and safe method for the control of stored-product insects.

Comparing it with previously studied control methods, this technique would be almost as rapid as thermal disinfestation procedures such as hot-air fluidised beds (Evans et al. 1983;

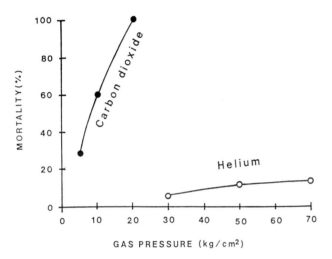

GAS PRESSURE (kg/cm^2)

Fig. 3. Comparison of mortality curves of *S. zeamais* adults exposed to high pressure CO_2 or helium for 5 minutes at 25°C. All mortality data were assessed 48 hours after treatment.

Fleurat-Lessard 1987) and microwave radiation (Kirkpatrick and Robert 1971; Nakakita et al. 1989). Increasing the pressure could correspondingly shorten the exposure time. Thus, there is a need to establish specific *PT* (pressure × time) products, similar to *CT* (concentration × time) products used for fumigants.

Although the application of the method using CO_2 gave rapid control of insects, there were some differences in susceptibilities among species, and among stages of *S. zeamais*. Of the adults of all four species tested, *T. castaneum* was the most tolerant, followed by *S. zeamais*, *R. dominica*, and *L. serricorne*. However, the differences in susceptibility among species are unlike those obtained from conventional CO_2 treatments at atmospheric pressure which involve wide ranges of susceptibility among species (Annis 1989). Application of the method to immature stages of *S. zeamais* showed that the egg was the most resistant life stage, requiring almost three times as much pressure of CO_2 as pupae or adults to achieve complete mortality. In conventional CO_2 treatments, the pupal stage is commonly the most resistant (Navarro 1987). Thus, there must be differences in responses of insects to high pressure and conventional CO_2 methods. These may be due to the different mechanisms involved in the two methods. There would be two actions occurring to achieve insect mortality with the high-pressure method: the physiological action of CO_2 at high concentrations in target sites; and the physical action of high pressure gas in tissues of the insects, causing expansion as the gas is suddenly released to the atmosphere.

Helium had almost no effect on insect mortality even at the extremely high pressure of $70 \, kg/cm^2$. The great difference in mortality between CO_2 and He treatments may be explained by the different solubilities of the gases in the hemolymph of insects, which is mainly water. Carbon dioxide has about an 80-fold greater solubility than helium at $25 \, kg/cm^2$ and 25°C: 16.5 mL and 0.21 mL for CO_2 and He, respectively, in 1 g

Table 1. Effect of CO_2 treatment using high pressure followed by sudden discharge on adults of four stored-product insects.

Species	Pressure (kg/cm^2)							
	5		10		15		20	
	Exposure time (minutes) % mortalities ± SE[a]							
	5	10	5	10	5	10	5	10
S. zeamais	28.0 ± 8.4	36.0 ± 5.6	60.0 ± 10.0	96.0 ± 1.9	100	100	100	100
R. dominica	78.6 ± 7.1	100	100	100	100	100	100	100
T. castaneum	32.0 ± 1.9	45.3 ± 2.9	69.3 ± 1.1	88.0 ± 1.9	85.3 ± 3.9	100	100	100
L. serricorne	81.3 ± 2.1	93.3 ± 2.1	100	100	100	100	100	100

[a]Mean of three replicates.
*Percentage mortality corrected from control mortality by Abbott's (1925) formula.

Table 2. Effects of CO_2 treatment using high pressure followed by sudden discharge on all life stages of *Sitophilus zeamais* (Motsch.).

Pressure (kg/cm^2)	% Mortality ± SE[a]			
	Eggs	Larvae	Pupae	Adults
0	0 (32.0 ± 2.6)[b]	0 (19.3 ± 0.5)[b]	0 (19.6 ± 0.3)[b]	0
10	28.1 ± 10.6	89.6 ± 2.4	100	100
15	32.3 ± 6.1	98.5 ± 1.4	100	100
20	21.9 ± 3.8	100	100	100
30	100	100	100	100

[a]Mean of three replicates.
[b]Value in parenthesis is number of emerged adults.
*Mortality assessed after one month for immature stages; values were calculated by dividing the total number of adults that emerged after the treatment by the total number of adults in the control.

water at normal state (Japan Chemical Society 1966). Helium does not penetrate the bodies of insects; thus no expansion occurs after sudden release of the gas from high pressure. From the viewpoint of water solubility of gases, we may find other gases that are as effective as CO_2.

Applying a gas such as CO_2 using the method described would provide a safe means of insect control leaving no harmful residues on commodities. Thus, the method may apply in places or situations where rapid treatment for insect control is needed such as in quarantine, food processing factories, etc. Further work is, however, necessary to provide information on the relationship between time and pressure, efficacy of this technique on immature stages of different species and quality effects on various commodities. In addition, pilot-scale trials should be carried out determine the commercial applicability and practicability of the technique.

Acknowledgment

We are grateful to Nitto Koatsu Co. Ltd for their kind advice on designing the equipment.

References

Abbott, W.C. 1925. A method for computing the effectiveness of an insecticide. Journal of Economic Entomology, 18, 265–267.

Annis, P.C. 1990. Requirements for fumigation and controlled atmospheres as options for pest and quality control in stored grain. In: Champ, B.R., Highley, E. and Banks, H.J., ed., Fumigation and controlled atmosphere storage of grain: proceedings of an international conference, Singapore, 14–18 February 1989. ACIAR Proceedings No. 25, 20–28.

Calderon, M. and Barkai-Golan, R., ed., 1990. Food preservation by modified atmospheres. Boca Raton, Florida, USA, CRC Press, 402 p.

Champ, B.R., Highley, E. and Banks, H.J., ed. 1990. Fumigation and controlled atmosphere storages of grain: proceedings of an international conference, Singapore, 14–18 February 1989. ACIAR Proceedings No. 25, 301 p.

Evans, D.E., Thorpe, G.R. and Sutherland, J.W. 1983. Large scale evaluation of fluid-bed heating as a means of disinfesting grain. In: Proceedings 3rd International Working Conference on Stored-product Entomology, Manhattan, Kansas, USA, 523–530.

Fleurat-Lessard, F. 1987. Control of storage insects by physical means and modified environmental conditions. In: Lawson, T.J., ed., Stored product pest control. Thornton Heath, U.K., BCPC Publications, Monograph No 37, 209–218.

Garry, V.F., Griffith, J., Danzl, T.J., Nelson, R.L., Whorton, E.B., Krugel, L.A. and Carvenka, J. 1989. Human genotoxicity: pesticide applicators and phosphine. Science, 246, 251–255.

Japan Chemical Society, ed. 1966. Chemical handbook. Tokyo, Maruzen, 623 p.

Kirkpatrick, L.R. and Robert, J.R. 1971. Insect control on wheat by use of microwave energy. Journal of Economic Entomology, 49, 33–34.

Le Torc'h, J.M. and Fleurat-Lessard, F. 1992. Effect des fortes pressions sur l'efficacite insecticide des atmospheres modifiees par CO_2 contre *Sitophilus granarius* (L.) et *S. oryzae*. In: Fleurat-Lessard, F. and Ducom, P., ed., Proceedings 5th International Working Conference on Stored-product Protection, Bordeaux, France, September 9–14, II, 847–855.

Nakakita, H., Imura, O., Nabetani, H., Watanabe, A., Watanabe, S. and Chikubu, S. 1989. Effect of microwave on susceptibilities of insects and qualities of rice. Nippon Shokuhin Kogyo Gakkaishi, 36, 267–273. (In Japanese with English summary)

Navarro, S. 1987. Application of modified atmospheres for controlling stored grain insects. In: Lawson, T.J., ed., Stored product pest control. Thornton Heath, U.K., BCPC Publications, Monograph No. 37, 229–236.

Reichmuth, C. 1990. New techniques in fumigation research today. In: Fleurat-Lessard, F. and Ducom, P., ed., Proceedings 5th International Working Conference on Stored-product Protection, Bordeaux, France, September 9–14.

Ripp, B.E., Banks, H.J., Bond, E.J., Calverley, D.J., Jay, E.G. and Navarro, S. ed. 1984. Controlled atmosphere and fumigation in grain storages. Elsvier Amsterdam. 798p.

Shejbal, J., ed. 1980. Controlled atmosphere storage of grains. Amsterdam, Elsevier, 608 p.

Sabine, P. and Reichmuth, C. 1990. Response of the granary weevil *Sitophilus granarius* (L.) (Col.: Curculionidae) to controlled atmospheres under high pressure. In: Fleurat-Lessard, F. and Ducom, P., ed., Proceedings 5th International Working Conference on Stored-product Protection, II, 911–918.

Stahl, E., Rau, G. und Adolphi, H. 1985. Entwesung von drogen durch hoklendioxide-druchbehandlung. Die Pharmazeutishe Industrie, 47, 528–530.

Singh, O.N., Borchers, R., Fabin, P., Lal. and Subbraya, B.H. 1988. Measurements of atmospheric BrOx radicals in the tropical and mid-latitude atmosphere. Nature, 334, 593–595.

The future of hermetic storage of dry grains in tropical and subtropical climates

S. Navarro*, J.E. Donahaye[†] and S. Fishman[†]

Abstract

Oxygen (O_2) depletion and carbon dioxide (CO_2) enrichment of the intergranular atmosphere form the basis for suppressing and controlling insect infestations during hermetic storage of dry grain. Traditional methods and recent improvements are reviewed, and modern structures designed for hermetic storage at the commercial and farmer levels are described. Improvements needed to render the hermetic concept more widely acceptable are enumerated, and the development of hermetic storage within flexible plastic liners is evaluated on the basis of more than a decade of experience in hot climates. A preliminary model is employed to simulate the interdependent changes in gas concentrations, insect populations and amounts of grain consumed. A theoretical ingress rate of 0.05% O_2/day was found sufficient to arrest development of residual insect infestations. Potential niches for hermetic storage applications in developing and technologically advanced countries are identified. In tropical climates aeration for cooling of grain is not feasible, reinfestation is frequent and the available contact insecticides degrade rapidly because of high temperatures. The advantages of long-term hermetic storage in technologically advanced countries, and as a medium-term, user-friendly technology in developing countries, are stressed. In sharp contrast to the use of chemicals, hermetic storage is environmentally sound and poses no risk to storage operators, consumers or non-target organisms.

Historical

Underground pits

The considerable literature on sealed storage of grain termed 'air-tight storage' or 'hermetic storage' is well summarised by Hyde et al. (1973) and De Lima (1990). From this a clear picture emerges of underground storage in pits, from prehistoric times until the present day as a traditional method that is frequently sufficiently airtight to enable insects and other aerobic organisms in the grain mass to reduce oxygen (O_2) concentrations below those permitting insect development. These pits were excavated into the soil or rock, and are sometimes lined with supporting walls of brick or cement. However, the ideal situation of O_2 depletion and carbon dioxide (CO_2) accumulation as demonstrated in laboratory experiments of Oxley and Wickenden (1963) is rarely achieved. This is generally because of gas-exchange through the pit walls and roof, and the sorption of CO_2 by the grain itself and sometimes by the pit walls (Hyde and Daubney 1960).

* Department of Stored Products, Agricultural Research Organization, P.O. Box 6, Bet Dagan, 50250 Israel.
[†] Department of Statistics, Agricultural Research Organization, P.O. Box 6, Bet Dagan, 50250 Israel.

Semi-underground structures

Large-scale construction for prolonged storage of grain surpluses in Argentina during the second world war consisted of below- and above-ground concrete lined trenches covered with flexible roofs (Anon. 1949). Later attempts at achieving hermetic storage were the Cyprus bins constructed in the 1950s (Hyde et al. 1973). These consisted of concrete lined conical pits surmounted by domed concrete-shell roofs. They were successfully used under hermetic conditions for a number of years. Improved versions of these structures were later constructed in Kenya for hermetic storage of the national grain reserve (De Lima 1990).

Above-ground small-scale structures

Another traditional method used by subsistence and small-scale farmers has been the storing of grain in sealed gourds, though these provide an incomplete hermetic seal unless treated with a sealing material (McFarlane 1970). A related method adopted in a number of tropical countries has been the adaptation of empty oil-drums and other metal drums for storage (Pattinson 1970; Sakho 1971). To prevent development of heavy infestations before control is achieved, these metal drums should be completely filled. This is because when only partially filled, the headspace volume may remain large (in relation to the grain mass) and developing populations may cause perceptible damage before O_2 concentrations are sufficiently reduced to arrest development.

Factors Affecting Insect Mortality in Hermetic Storage

The important role of low O_2 concentration rather than high CO_2 in causing mortality of stored-product insects in hermetic storage was demonstrated by Bailey (1965). Only later was the importance of the synergistic effect of concomitant O_2 depletion and CO_2 accumulation for insect control clearly demonstrated (Calderon and Navarro 1979; 1980). These synergistic and combined effects are essential for successful insect control, as shown by studies of the effects of incomplete airtightness upon insect populations (Oxley and Wickenden 1963; Burrell 1968). Furthermore, the lower the grain moisture content (m.c.) and corresponding intergranular humidity, the higher the mortality, due to the desiccation effect on insects caused by low O_2 (Navarro 1978), or elevated CO_2 concentrations (Navarro and Calderon 1973). The influence of temperature on insect respiration implies that, in warm climates, O_2 intake by insects is very intensive. Conversely, in temperate climates, insect metabolism is much slower, depletion of O_2 may be lower than its ingress, and insect control may not be achieved. This led Burrell (1980) to postulate that, for light infestations of cool grain, residual populations would provide an inoculum for reinfestation after the grain is removed from hermetic storage.

Modern-day Hermetic Storage

Above-ground rigid structures

Documented data on successful application of hermetic storage to above-ground constructions are largely lacking. Many existing silos and warehouses have been modified to provide a high degree of hermetic seal especially in Australia (Delmenico 1993). However, the objectives have been to convert these storages for modified atmosphere (MA) treatments or improved fumigations, and not for hermetic storage as such.

In contrast, the sealing of both bagged stacks and bulk grain in warehouses in China using plastic liners is part of a grain preservation regime termed 'Triple-Low'. This is an integrated approach to insect control consisting of obtaining reduced O_2 concentrations by metabolic activity within the grain bulk, in combination with phosphine and low temperature treatments (Wang et al. 1993; Xu and Wang 1993). This procedure is claimed to provide effective protection.

Above-ground flexible structures

In the early 1970s, above-ground structures were designed in England for emergency storage using flexible plastic liners supported by a weldmesh frame. These liners were made of butyl rubber, sometimes laminated with white EPDM, and consisted either of a wall-floor section, plus a roof section attached after loading, or both sections welded into a single unit. These silos were recommended for both conventional storage and hermetic storage of dry grain (Kenneford and O'Dowd 1981). However, under tropical and subtropical climates the liners were found to deteriorate, and gas permeability increased to a level where the liners could no longer be used for hermetic storage (Navarro and Donahaye 1976; O'Dowd and Kenneford 1982).

In Israel, the manufacture of PVC liners that conform to prerequisite specifications of durability to climate, gas-permeabilities, and physical properties, enabled the development of three storage systems based on the hermetic principle. These are:

- Bunker storage for conservation of large bulks of 10000–15000 t capacity (Navarro et al. 1984; Navarro et al. 1993).
- Flexible silos supported by a weldmesh frame of 50–1000 t capacity for storage of grain in bulk or in bags (Calderon et al. 1989; Navarro et al. 1990).
- Liners for enclosing stacks of 10–50t capacity termed storage cubes, and designed for storage at the farmer-cooperative and small trader level (Donahaye et al. 1991).

The problem of applying present-day technology to provide hermetic storage for subsistence farmers lies in the need to provide an easily sealable low-cost container of 50–100 kg capacity. The high surface area to volume ratio necessitates a liner with a very low permeability to gases. The most recent attempt to address this problem has been through the 'Joseph bag', which is made of a plastic-metal foil laminate, sealable by means of a hot-iron (Murray 1990).

Underground flexible structures

The main approach to achieving lower levels of O_2 and higher accumulations of CO_2 has been by lining pits with plastic liners in order to improve the hermetic seal (Donahaye et al. 1967; Dunkel et al. 1987). With a similar approach, small-scale underground storages have been developed for farmer storage of maize and dry beans in Brazil (Sartori and Costa 1975; Sartori 1987).

Experience Gained Using Flexible Liners

Our accumulated experience of hermetic storage using several types of flexible liners for above-ground storage, in-the-open, under tropical and subtropical conditions (Calderon et al. 1989; Donahaye et al. 1991; Navarro et al. 1968; Navarro and Donahaye 1993; Navarro et al. 1984, 1990, 1993), is summarised in the following sections.

Structural durability

The use of PVC-based sheeting without mesh reinforcement produces a material of suitable strength and elasticity for storing grain. This material was formulated to have a high resistance to solar UV irradiation. Rodent penetration has been recorded on only exceptional occasions involving minor damage. Our hypothesis that rodents find it difficult to gain a tooth-hold on the smooth surface has been corroborated by laboratory studies using wild-caught roof rats and house mice (Navarro, Moran, Dias and Donahaye, unpublished data).

Liners have been used continuously for over 10 years, and though they have lost some plasticity, permeability to gases decreases as the plasticisers evaporate. This characteristic renders the liners more effective with time in retaining gas concentrations, e.g., for 0.83 mm PVC, the initial permeability (expressed throughout as a measure given at a gradient from 21% O_2) decreased from 87 to 50 mL O_2/m^2/day after 4 years of exposure under Mediterranean climate (unpublished data).

Insect control

At a liner thickness of 0.83 mm and a gas permeability level of 87 mL O_2/m^2/day, there is a possibility of insect survival close to the grain-liner interface. This is especially so at the top layer of the structures where moisture content tends to be higher than the remaining parts of the bulk. However, after the minimum O_2 concentration is reached, survival is usually well below 1 insect/kg, and would require multiple sampling to detect a single insect (Navarro et al. 1984, 1993). This residual infestation is more of a problem on return to aerobic conditions and the commodity should be consumed without additional prolonged storage. This residual infestation is less serious than the danger of reinfestation by insects from the surroundings under storage by conventional methods. For grain destined for export and where freedom from insects is mandatory, a final treatment using phosphine may be undertaken if necessary. In future, this treatment may be superfluous if higher degrees of gas retention achieve complete elimination of residual infestation.

Moisture migration

Diurnal temperature fluctuations, accentuated by solar radiation on liners, followed by rapid cooling at night, cause successive moistening and drying cycles at the upper grain surface. This may result in gradual moisture accumulation, particularly during the transient seasons between summer and winter when temperature fluctuations are greatest, so that initially dry grain may rise to above critical levels enabling limited microfloral spoilage to occur. This is particularly accentuated along the peaks of bunkers where warm air rising on convection currents tends to concentrate the moisture condensation in confined areas. For bunkers of 12000–15000 t capacity constructed in recent years, the condensation phenomenon has been alleviated and almost eliminated by levelling the peaked apex (with a ridge of less than 2 m) to a slightly convex, wide apex of bunker cross-section (with a ridge of more than 6 m) that is just sufficient to permit rainwater run-off. This configuration appears to enable the dispersal of moisture migration over a much larger area. Differences in the intensity of moisture increase between bunker peaks with narrow ridges and peaked apices, and apices with a broad ridge are demonstrated in Figure 1. Although comparative results for concurrent storages are not available, results shown in this Figure 1 form a summary of observations made over 4 intermittent years at the same storage site in Israel.

For dry grain kept in 'storage cubes' in subtropical climates, moisture migration is not a pronounced phenomenon. However, for maize or paddy stored in the tropics, moisture migration is accentuated because the initial grain moisture is closer to its critical level. For this purpose, the solution to moisture migration has been under examination by placing an insulating layer between the liner and the upper layer of bagged grain. Preferably this consists of a layer of bagged agricultural wastes such as rice hulls, or straw, or if these are not available a 'felt-fibre' layer with insulating properties appears promising.

Development of a Predictive Model

In view of the complexity of the grain bulk ecosystem prevailing under hermetic conditions, we propose to use a simulation model to rapidly analyse numerous situations, and describe the critical limits of the different factors. The model is in a preliminary phase of development to simulate the interdependent changes in gas concentrations, dynamics of insects population and amounts of grain consumed. The model has been set to run numerical experiments to investigate the influence of degree of gastightness of the structure expressed as the rate of O_2 ingress through the storage membrane, size of the grain mass, volume of the treated structure, initial number of insects/kg of grain, respiration rate of the mixed insect population, and birth and death rates of the species on changes in O_2 concentrations

in the storage, changes in insect population, and amounts of grain consumed.

The preliminary version of the model is the first approximation of the system built to study the influence of the physical characteristics of the storage structure as well as initial infestation. The main state variables that define the dynamics of the system are O_2 concentration, and number of insects and loss in grain weight. The model was written using the modelling package STELLA (Pytte and Doyle 1984) available for Macintosh® personal computers. Values of structural membrane permeabilities were based on laboratory measurements. Birth and death rates were estimated from field observations obtained from storages under aerobic and hermetic conditions. At present the main assumptions of this preliminary version of the model used in the exercise are as follows:

- Oxygen is distributed uniformly throughout the grain mass and no gas stratification occurs.
- CO_2 effect on insects and CO_2 sorption by grain are ignored.
- Temperature of the grain mass is uniform, and therefore moisture migration due to temperature gradients is ignored.
- Influence of wind on the structure is negligible.
- Influences of changes in temperature and barometric pressure are ignored.
- Insect distribution is homogenous in the grain mass.
- The storage structure is cube shaped.
- No head-space volume exists, the volume of the structure is occupied by the grain mass with an interstitial air space of 45% and a bulk density of 750 kg/m^3.

Calculated changes in oxygen concentration in hermetic storages

Influence of different initial insect populations

For this exercise a fixed O_2 ingress rate equivalent to about 0.24%/day was chosen for a structure with a volume of 10 m^3. For these values, changes in oxygen concentrations in

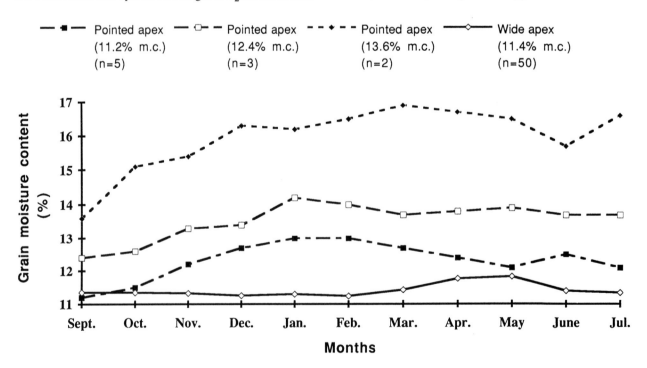

Fig. 1. Seasonal changes in moisture content of wheat as recorded at the top layers (0 –20 cm) of bunkers with pointed apices and narrow ridges (observations made over one season), and bunkers with wide apices and broad ridges (observations made over three seasons) (initial moisture contents [m.c.] given in brackets, n = number of samples taken from a specific location).

response to different initial insect populations are illustrated in Figure 2. Accordingly, a cyclic change in concentrations is obtained as a result of O_2 ingress and the ability of insects to survive at low O_2 levels. These theoretical cyclic changes in O_2 concentrations were also observed in different laboratory and field studies (Oxley and Wickenden 1963; Hyde et al. 1973; Navarro et al. 1990).

Under the conditions governing the numerical experiment, the model calculates that there is a residual insect population even after an extended storage period of one year. This is shown by the continuing fluctuations in O_2 levels before a steady-state is reached (Fig. 2). This result is corroborated by field observations that a residual population may remain when the grain is re-exposed to normal atmospheric air, though under the hermetic conditions and restricted O_2 supply their reproductive capacity is limited (Burrell 1980; Navarro et al. 1993). There are only limited data in the literature regarding respiration of mixed population of insects (Birch 1947; Calderwood 1961; Carlson 1966 1968; Chaudhry and Kapoor 1967; Keister and Buck 1974; Park 1936). Furthermore, the respiration of these species under low O_2 tension is not well documented (Donahaye 1992; Navarro 1974). To determine the changes in O_2 levels under different hermetic conditions, more information is needed on the contribution of different species to O_2 consumption at low O_2 tensions.

Influence of membrane permeability levels

To clarify the importance of membrane permeability in a specific situation, the model was run with an initial infestation of 2 insects/kg at different levels of O_2 ingress rates for a 10 m^3 cube containing grain as above. Results obtained with O_2 ingress rates of 0.05, 0.12, and 0.24% O_2/day, at an initial infestation level of 2 insects/kg, are shown in Figure 3. The calculated line for the 0.05%O_2/day ingress rate differs significantly from the lines with higher ingress rates. At the 0.05%O_2/day ingress rate, after a minimum O_2 level is reached the increase in O_2 concentration follows the O_2 ingress rate of a structure without insects. This exemplifies the importance in reducing the O_2 ingress rate levels below which the residual insect population can be eliminated from the grain. These levels of gastightness are easily obtained in the laboratory, but difficult to achieve at commercial levels, especially with existing large rigid structures.

Calculated grain losses in hermetic storages

Influence of different initial insect populations

Aerobic metabolism of insects associated with respiration involves the utilisation of carbohydrates which constitute the largest component of cereal grains. In the model, the dry-matter loss was calculated on the basis of O_2 required for oxidation of carbohydrates utilised in the process of insect metabolism. The model was run with the same parameters as used in Figure 2, and its results are shown in Figure 4. Calculated weight losses obtained with a gas ingress rate of 0.24% O_2/day at the three infestation levels indicate that losses over a 1-year period range between 0.050 and 0.058% of initial weight. Observed weight loss (count and weight method) due to insect activity in a 15 500 t capacity bulk storage held under sealed conditions over a 15-month period was 0.15% (Navarro et al. 1984). The differences between field results and the model estimate may partly derive from difficulties in obtaining accurate evaluation of the field trial. The experimental field values and the calculated values from the model indicate that although insect activity causes weight losses, these losses are within the lower range of commercially

acceptable biological losses. Furthermore, these levels in some cases may fall within the accuracy range of commercial scales.

Influence of membrane permeability levels

The model was run with an initial fixed infestation of 2 insects/kg at O_2 ingress rates of 0.05, 0.12, and 0.24% O_2/day for a 10 m^3 cube containing grain. Results in Figure 5 show that an O_2 ingress rate of 0.05%/day is sufficient to arrest the theoretical weight loss at a level of 0.018% over a 1-year storage period, whereas for higher O_2 ingress rates, the weight loss continues to rise in proportion to the O_2 ingress rate. At an ingress rate of 0.05% O_2/day insect development is arrested and therefore the possibility of a residual surviving insect population is eliminated. This low O_2 ingress level, is difficult to obtain in rigid structures, but is achievable in practice using flexible liners. It could serve as a guideline for the sealing specifications of structures appropriate to the hermetic storage method. With a permeability level of 81 mL O_2/m^2/day using a flexible liner 0.83 mm thick, a structure with a capacity of more than 10 m^3 would meet this requirement. As far as rigid structures are concerned the main drawbacks in obtaining this level of gastightness lie in their constructional limitations. When a silo is not used at its full capacity, there remains a headspace volume which renders the structure more sensitive to the influence of ambient temperature and barometric fluctuations. Even when the structure is extremely gastight, if a breather-bag is not used, gas exchange through the pressure relief valve to compensate for pressure changes due to temperature and barometric fluctuations cannot be avoided.

The size factor in hermetic storages

Experience shows that hermetic storage works best for large structures. This is obvious from the low surface area/volume ratio in large bulks compared with small bulks. Although the factor of O_2 ingress rate can be reduced by suitable modern technologies, in practice it is a goal difficult to achieve. Therefore, depending on the commercially available membrane permeabilities, engineers should aim at designing hermetic structures of sufficiently large dimensions. This is in sharp contrast to the objective of using hermetic storage at farm level in developing countries where low-permeability liners must be preferred.

To emphasise the importance of the size of the structure in hermetic storage, the model was run assuming a permeability level of 200 mL O_2/m^2/day for structures of different dimensions ranging from 1 to 1000 m^3 (Fig. 6). The model demonstrates that a tenfold increase in the volume of the bulk causes approximately a twofold decrease in the initial O_2 ingress rate.

The model is still being developed and we intend to publish a full description elsewhere.

The Need for Future Research and Development

Although the principle of hermetic storage is simple, there are still a number of aspects that require clarification or improvement in order to render it more acceptable as a storage alternative. Of these, we believe the following to be the most important.

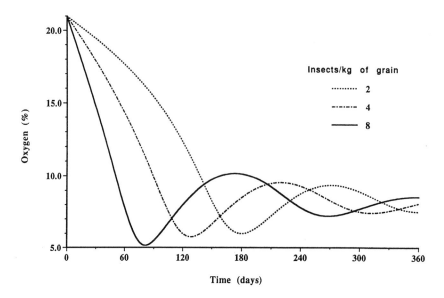

Fig. 2. Calculated oxygen concentrations in a 10 m³ grain mass containing different infestation levels of insects having an oxygen intake of 157 µL/insect/day using a sealed liner with an oxygen ingress rate of 0.24%/day.

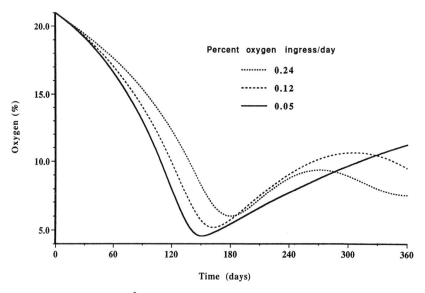

Fig. 3. Calculated oxygen concentrations in a 10 m³ grain mass containing a fixed level of initial infestation of 2 insects/kg, having an oxygen intake of 157 µL/insect/day using a sealed liner at different oxygen ingress rates.

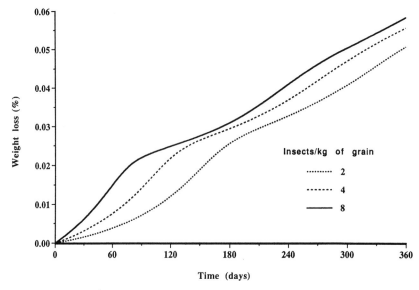

Fig. 4. Calculated weight-loss from a 10 m³ grain mass containing different infestation levels of insects having an oxygen intake of 157 µL/insect/day using a sealed liner with an oxygen ingress rate of 0.24%/day.

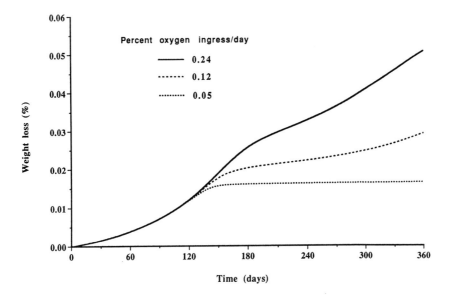

Fig. 5.   Calculated weight-loss from a 10 m³ grain mass containing a fixed level of initial infestation of 2 insects/kg, having an oxygen intake of 157 µL/insect/day using a sealed liner at different oxygen ingress rates.

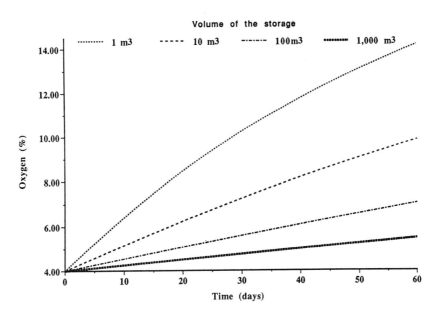

Fig. 6.   Calculated rates of oxygen ingress into grain bulks (without infestation) of different volumes confined within a sealed liner with an oxygen permeability of 200 mL/m²/day.

Influence on grain quality

The review of Banks (1981) reveals that most studies on grain quality have been carried out using modified atmospheres (MA). He concludes that for low moisture content grains, low O_2 and high CO_2 concentrations do not have a detrimental effect on germination, milling and baking properties of wheat, or organoleptic properties of rice, though for intermediate and high moisture grains, quality is affected. Our studies on long-term hermetic storage of wheat (Navarro et al. 1984, 1993) also clearly indicate that both germination and baking properties of wheat are well preserved, and milling properties of dry paddy are not affected (Donahaye et al. 1991). For maize seed, Moreno et al. (1988) found that at 14–14.5% m.c., hermetic storage did not have a detrimental effect on germination. Nevertheless, confusion has arisen in the past because of moisture migration in hermetic storage structures and the consequential deterioration of moistened

grains. This has had a negative effect on consumer acceptance that needs to be redressed both by experimentation on moisture migration and grain quality parameters in hermetic storage, and by explanation of the findings at all professional levels.

Degree of sealing

Sealing methodologies have been well developed and published (Alexander 1984; Lloyd 1984; Sutherland and Thomas 1984; Woodcock 1984; Newman 1990). However, in practice, sealing of rigid structures has been limited mainly to Australia, and elsewhere many new silos are being built to low standards of sealing that do not permit application of hermetic storage or MA treatments. The future approach to silo construction should take a more professional attitude to silo sealing. It should be noted that a high level of sealing is essential in the modern approach to fumigation for reduction in emissions (especially when using methyl bromide because of its association with ozone depletion), and also to eliminate the possibility of development of phosphine-resistant insect populations.

Headspace

A major problem in rigid structures is the volume of headspace above the grain bulk. This is not only a limiting factor in determining the rate at which O_2 concentration decreases, but also accentuates pressure differences between the interior and exterior of the silo as a result of daily temperature and barometric fluctuations of the ambient. Pressure-relief valves provide a satisfactory solution for MA storage. However, the use of breather-bags to unify internal and external pressures appears to be more desirable for hermetic storage since it reduces the gas exchange. To the best of our knowledge breather-bags have not been employed in silo bins, except for 'Harvestores' used for silage. Experimental data on the use of such breather-bags under tropical and subtropical climates would be very useful when considering the application of hermetic storage to reduce the intensity of air ingress into the storage system.

Moisture condensation

A phenomenon that discourages the use of hermetic storage in hot climates is moisture migration and condensation, and this is especially accentuated in metal silos. In conventional storages, engineers rely on designing well-ventilated headspaces to reduce the intensity of this phenomenon, and even incorporate aeration systems with exaggeratedly high airflow rates regardless of their efficacy in the tropics. In subtropical climates with a cool season, aeration systems were shown to effectively overcome this problem by equalising grain temperatures (Navarro and Calderon 1982). For this reason, when applying MA storage in metal structures the integration of aeration was proposed by Navarro and Calderon (1980). However, the efficacy of this approach was never adequately documented.

For metal silos in hot climates, moisture condensation is intensified when insect infestation causes grain heating. The most disturbing effect of moisture migration, especially in rigid constructions, is the difficulty in removing the damaged layer, usually at the top of the bulk.

So far, two approaches are known to reduce the intensity of this phenomenon: equalising grain temperatures, and insulation of the roof. Equalising grain temperatures by aeration is limited to climates with a cool season. Comparative data on the efficacy of aeration and the effect of insulation in preventing moisture migration in metal silos in the tropics are lacking.

For small-scale applications using flexible liners, the influence of insulation materials in reducing the intensity of moisture migration in subtropical (Israel) and tropical (Philippines) climates has been investigated by Donahaye and Navarro (unpublished data). A model describing moisture migration by natural convection has been proposed (Nguyen 1986). However, more experimental data are required to support the development of predictive models for hermetic storage.

Flexible liners

One approach to hermetic storage has been through the use of flexible liners. All liners have the advantage that they follow the contours of the grain-bulk surface and, with no headspace, gas exchange is restricted to the intergranular air volume. However, liner materials need improvement. Although liners

with zero permeability to gases are available, other factors such as physical characteristics, durability, resistance to penetration by pests (rodents and insects), amenability to jointing and welding, and cost of manufacture are all critical to the choice of liner. Integration of all these characteristics into a single liner is still an objective of the future.

Monitoring

Standard inspection procedures based on temperature and grain sampling may be employed. However, since the objective is to retain the hermetic seal, attention should be paid not to leave the inspection port open for more than the minimum time required. It is most desirable to support sampling evaluation of storage condition by measurement of O_2 and CO_2 concentrations. For large-scale operations modern and reliable electronic monitoring equipment is available. However, these instruments are expensive and clearly are not suitable to the small-scale farmer in developing countries. Yet the use of Orsat-like manometric gas analysis apparatus relies too heavily upon equipment maintenance. Affordable and reliable instrumentation for hermetic storage monitoring has still to be developed.

Hermetic Storage as a Future Alternative?

The need for alternative methods of prevention and control of insect infestations in stored products has become acute over the last few years. This is because conventional measures using insecticides are being questioned by environmental agencies and pressure-groups, and the choice of available permissible materials is decreasing. Of the two remaining fumigants in general use, a decision has been made by the U.S. Environmental Protection Agency to phase-out methyl bromide by the year 2001 due to its destructive effect on ozone in the stratosphere (USEPA 1993). This is coupled with mounting evidence of development of insect resistance to phosphine (Zettler 1993) indicating that even phosphine may not be economically effective in years to come. Modern, safer and acceptable technologies such as aeration, refrigerated aeration and modified atmospheres are still expensive and require adequate infrastructure. In sharp contrast to the use of chemicals, hermetic storage is an environmentally friendly technology, involving no hazard to the storage operators, consumers or non-target organisms, and as such, its application should enjoy a high level of consumer acceptance.

In developing countries

Hermetic storage may provide an answer to the need for a less costly method of storage for food security of the rural populations. This could be achieved by supplying a storage solution at the farmer level, and thereby affording the farmer protection from seasonal fluctuations in grain prices. The basic advantage of hermetic storage in developing countries is its simplicity, obviating the need for insecticidal admixture procedures or fumigations, both of which require high levels of expertise not usually possessed by the small-scale farmer. Furthermore, it is generally the only MA option since MA-generators or gas cylinders are neither affordable nor obtainable.

In technologically advanced countries

In spite of the trend towards improved sealing of existing silos in some countries (Newman 1990; Delmenico 1993), the objective has been either to obtain increased fumigation efficiency, or to convert structures for storage under modified atmospheres. The relatively slow rates of O_2 depletion in the hermetic storage process, especially when the initial population is low, renders it inappropriate to apply this method to short-term storage systems. Neither is it practical for application in rigid horizontal storage structures where the headspace is always relatively large. Clearly, it is best suited to long-term storage projects such as national grain reserves, buffer stocks and storage of grain surpluses.

The conversion of rigid structures to sealed storages should be considered for long-term large-scale storage projects. The evidence that the method is effective at the high temperatures prevailing in tropical climates is best documented by De Lima (1990). In tropical climates aeration for cooling of grain is not feasible, reinfestation is frequent and the available contact insecticides degrade rapidly because of high temperatures. Under these conditions hermetic storage may provide an advantageous solution.

Acknowledgments

The authors thank Mrs Miriam Rindner and Mr A. Azrieli for their assistance during the preparation of this manuscript. Part of this study was undertaken with the support of the U.S. Agency for International Development, under the Cooperative Development Research (CDR) project C7–053.

References

Alexander, I. 1984. Polyurethane foam for sealing grain storages. In: Ripp, B.E. et al., ed., Controlled atmosphere and fumigation in grain storages. Amsterdam, Elsevier, 145–158.

Anon 1949. Conservación de granos y almacenamiento en silos subterráneos. Ministerio. de Agricultura y Ganaderia, Buenos Aires, Argentina. 222 p.

Bailey, S.W. 1965. Air-tight storage of grain; its effect on insect pests — IV *Rhyzopertha dominica* (F.) and some other Coleoptera that infest stored grain. Journal of Stored Products Research, 1, 25–33.

Banks, H.J. 1981. Effects of controlled atmosphere storage on grain quality: a review. Food Technology in Australia, 33, 335–340.

Birch, L.C. 1947. The oxygen consumption of the small strain of *Calandra oryzae* and *Rhyzopertha dominica* as affected by temperature and humidity. Ecology, 28, 17–25.

Burrell, N.J. 1968. Miscellaneous experiments on grain storage under plastic sheeting. V. The control of insects in infested wheat in a glass-fibre and plastic bin. Agricultural Research Council, Pest Infestation Laboratory Report, 8p.

Burrell, N.J. 1980. Effect of airtight storage on insect pests of stored products. In: Shejbal, J. ed., Controlled atmosphere storage of grains. Amsterdam, Elsevier, 55–62.

Calderon, M. and Navarro, S. 1979. Increased toxicity of low oxygen atmospheres supplemented with carbon dioxide on *Tribolium castaneum* adults. Entomologia Experimentalis et Applicata, 25, 39–44.

Calderon, M. and Navarro, S. 1980. Synergistic effect of CO_2 and O_2 mixtures on two stored grain insect pests. In: Shejbal, J., ed., Controlled atmosphere storage of grains. Amsterdam, Elsevier, 79–84.

Calderon, M., Donahaye, E., Navarro, S. and Davis, R. 1989. Wheat storage in a semi-desert region. Tropical Science, 29, 91–110.

Calderwood, W.A. 1961. The metabolic rate of the flour beetle *Tribolium confusum*. Transactions of Kansas Academy of Science, 64, 150–152.

Carlson, S.D. 1966. Respiration measurement of *Tribolium confusum* by gas chromatography. Journal of Economic Entomology, 59, 335–338.

Carlson, S.D. 1968. Respiration of the confused flour beetle in five atmospheres of varying CO_2:O_2 ratios. Journal of Economic Entomology, 61, 94–96.

Chaudhry, H.S., and Kapoor, R.P.D. 1967. Studies on the respiratory metabolism of the red flour beetle. Journal of Economic Entomology, 60, 1334–1336.

De Lima, C.P.F. 1990. Airtight storage: principle and practice. In: Calderon, M. and Barkai-Golan, R., ed., Food preservation by modified atmospheres. Boca Raton, CRC Press Inc., Florida. 9–19.

Delmenico, R.J. 1993. Controlled atmosphere and fumigation in Western Australia — a decade of progress. In: Navarro, S. and Donahaye, E. ed., Proceedings International Conference on Controlled Atmosphere and Fumigation in Grain Storages, Winnipeg, Canada, June 1992, Jerusalem, Caspit Press Ltd., 3–12.

Donahaye, E. 1992. Laboratory selection of resistance by the red flour beetle, *Tribolium castaneum* (Herbst), to an atmosphere of low oxygen concentration. Phytoparasitica, 18, 189–202.

Donahaye, E., Navarro, S. and Calderon, M. 1967. Storage of barley in an underground pit sealed with a PVC liner. Journal of Stored Products Research, 2, 359–364.

Donahaye, E., Navarro, S., Ziv, A., Blauschild, Y. and Weerasinghe, D. 1991. Storage of paddy in hermetically sealed plastic liners in Sri Lanka. Tropical Science, 31, 109–121.

Dunkel, F., Sterling, R. and Meixel, G. 1987. Underground bulk storage of shelled corn in Minnesota. Tunnelling and Underground Space Technology, 2(4), 367–371.

Hyde, M. B., Baker, A.A., Ross, A.C. and Lopez, C.O. 1973. Airtight grain storage. FAO Agricultural Services Bulletin 17, 71 p.

Hyde, M. B. and Daubney, C.G. 1960. A study of grain storage fossae in Malta. Tropical Science, 2, 115–129.

Keister, M. and Buck, J. 1974. Respiration: some exogenous and endogenous effects on rate of respiration. In: Rockstein, M., ed., The physiology of Insecta, 2nd ed.. New York, Academic Press, Vol. VI, 469–509.

Kenneford, S. and O'Dowd, T. 1981. Guidelines for the use of flexible silos for grain storage in tropical countries. Tropical Stored Products Information, 42, 11–20.

Lloyd, D.J. 1984. The properties of various sealing membranes and coatings used for controlled atmosphere grain stores. In: Ripp, B.E., et al., ed., Controlled atmosphere and fumigation in grain storages. Amsterdam, Elsevier, 211–227.

McFarlane, J.A. 1970. Insect control by airtight storage in small containers. Tropical Stored Products Information 19, 10–14.

Moreno, E., Benavides, C. and Ramirez, J. 1988. The influence of hermetic storage on the behaviour of maize seed germination. Seed Science and Technology, 16, 427–434.

Murray, A. 1990. 'Wise Joseph sacks': a hermetic storage system for small-scale use. In: Champ, B.R., Highley, E. and Banks, H.J. ed., Fumigation and controlled atmosphere storage of grain: proceedings of an international conference, Singapore, 14–18 February 1989, ACIAR Proceedings No. 25, 286–287.

Navarro, S. 1974. Studies on the effect of alterations in pressure and composition of atmospheric gases on the tropical warehouse moth, *Ephestia cautella* (Wlk.), as a model for stored-products insects. Ph.D. Thesis submitted to the Senate of Hebrew University, Jerusalem, 118 p. (In Hebrew with English summary).

Navarro, S. 1978. The effect of low oxygen tensions on three stored-product insect pests. Phytoparasitica, 6, 51–58.

Navarro, S. and Calderon, M. 1973. Carbon dioxide and relative humidity: interrelated factors affecting the loss of water and mortality of *Ephestia cautella* (Wlk.) (Lepidoptera, Phycitidae). Israel Journal of Entomology, 8, 143–152.

Navarro, S. and Calderon, M. 1980. Integrated approach to the use of controlled atmospheres for insect control in grain storage. Proceedings of Symposium on Controlled Atmosphere Storage, Rome, 73–78.

Navarro, S. and Calderon, M. 1982. Aeration of grain in subtropical climates. Rome, FAO Agricultural Services Bulletin, No. 52, 119 p.

Navarro, S., Calderon, M. and Donahaye, E. 1968. Hermetic storage of wheat in a butyl rubber container. Israel Ministry of Agriculture Report of the Stored Products Research Laboratory, 102–110. (In Hebrew with English summary).

Navarro, S. and Donahaye, E. 1976. Conservation of wheat grain in butyl rubber/EPDM containers during three storage seasons. Tropical Stored Products Information, 32, 13–23.

Navarro, S. and Donahaye, E. 1993. Preservation of grain by airtight storage. 5th International Congress on Mechanisation and Energy in Agriculture, 11–14 October 1993, Kusadasi, Turkey, 425–434.

Navarro, S., Donahaye, E., Kashanchi, Y., Pisarev, V. and Bulbul, O. 1984. Airtight storage of wheat in a P.V.C. covered bunker. In: Ripp,

B.E. et al., ed., Controlled atmosphere and fumigation in grain storages. Amsterdam, Elsevier, 601–614.

Navarro, S., Donahaye, E., Rindner M. and Azrieli, A. 1990. Airtight storage of grain in plastic structures. Hassadeh Quarterly, 1(2), 85–88.

Navarro, S., Varnava, A. and Donahaye, E. 1993. Preservation of grain in hermetically sealed plastic liners with particular reference to storage of barley in Cyprus. In: Navarro, S. and Donahaye, E. ed., Proceedings International Conference on Controlled Atmosphere and Fumigation in Grain Storages, Winnipeg, Canada, June 1992 , Jerusalem, Caspit Press Ltd, 223–234.

Newman, C.J.E. 1990. Specification and design of enclosures for gas treatment. In: Champ, B.R., Highley, E. and Banks, H.J. ed. Fumigation and controlled atmosphere storage of grain: proceedings of an international conference, Singapore, February 1989, ACIAR Proceedings No. 25, 108–130.

Nguyen, T.V. 1986. Modelling temperature and moisture changes resulting from natural convection in grain stores. In: Champ, B.R. and Highley, E. ed., Preserving grain quality by aeration and in-store drying: proceedings of an international seminar, Kuala Lumpur, Malaysia, 9–11 October 1985. ACIAR Proceedings No. 15, 81–88.

O'Dowd, E.T. and Kenneford, S.M. 1982. Field performance of flexible silos in the tropics. TDRI Report No. G179.

Oxley, T.A. and Wickenden, G. 1963. The effect of restricted air supply on some insects which infest grain. Annals of Applied Biology, 51, 313–324.

Park, T. 1936. Studies on population physiology. V. The oxygen consumption of the flour beetle *Tribolium confusum* Duval. Journal of Cellular and Comparative Physiology, 7, 313–323.

Pattinson, I. 1970. Grain storage at village level. FFHC Action Program Report TAN/11., FFHC/FAO, Rome.

Pytte, A. and Doyle, J. 1984. Stella stack; model building and simulation. High Performance Systems Inc., NH.

Sakho, C.Y. 1971. Rapport sur l'opération 'Canagrenier' de conservation des grains par fûts hermetiques. Institut de technologie alimentaire, Dakar, Senegal.

Sartori, M. R. 1987. Underground storage of corn and dry beans in Brazil. Tunnelling and Underground Space Technology, 2, 373–380.

Sartori, M. R. and Costa, S.I. 1975. Armazenamento de milho a granel em silo subterraneo revestido com polietileno. Boletim do ITAL, 42, 55–69.

Sutherland, E.R. and Thomas, G.W. 1984. Silo sealing with Envelon. In: Ripp, B.E. et al., ed., Controlled atmosphere and fumigation in grain storages. Amsterdam, Elsevier, 181–209.

USEPA 1993. Preliminary use and substitute analysis of MBr in agricultural and other uses. Draft report, prepared for the Office of Air and Radiation, USEPA, May 1992, 48 p.

Wang Nanyan, Zhang Quo-Qiang, Zhang Yan-yan and Xu Hunai. 1993. The combined action of low-temperature, low-oxygen and low-phosphine concentrations in the 'Triple-Low' grain storage technique. In: Navarro, S. and Donahaye, E. ed., Proceedings International Conference on Controlled Atmosphere and Fumigation in Grain Storages, Winnipeg, Canada, June 1992. Jerusalem, Caspit Press Ltd, 271–280.

Woodcock, W. 1984. The practical side of silo sealing. In: Ripp, B.E. et al. ed., Controlled atmosphere and fumigation in grain storages. Amsterdam, Elsevier, 133–144.

Xu Huinai and Wang Nanyan 1993. Present and prospective state of the 'Triple-Low' grain storage technique. In: Navarro, S. and Donahaye, E. ed., Proceedings International Conference on Controlled Atmosphere and Fumigation in Grain Storages, Winnipeg, Canada, June 1992. Jerusalem, Caspit Press Ltd, 29–37.

Zettler, L.H. 1993. Phosphine resistance in stored-product insects. In: Navarro, S. and Donahaye, E. ed., Proceedings International Conference on Controlled Atmosphere and Fumigation in Grain Storages, Winnipeg, Canada, June 1992. Jerusalem, Caspit Press Ltd, 449–460.

Western Australian Fumigation Practice Survey (1992)

C.R Newman*

Abstract

Sealed storage for grains has been in use on farms in Western Australia for ten years and represented 44% of on-farm silo capacity in 1987. Phosphine is a commonly used stored-grain fumigant and phosphine resistance has been recorded in grain insects in the past decade. Phosphine fumigation procedures in unsealed structures created a selection pressure on the target organisms.

Earlier surveys showed that grain storages purchased as sealed units were not being maintained. Up to 56% of silos failed the standard test for gastightness.

A study of randomly selected farms from 43 shires of the Western Australian wheat belt was conducted to investigate the relationship between poorly maintained sealed grain storage and the incidence of phosphine resistance on those properties.

The survey found that sealed silo maintenance had deteriorated, with a 75% failure of silos tested. The results of the survey revealed no relationship between phosphine resistance and poor fumigation technique.

Introduction

In conjunction with a phosphine resistance survey conducted by the Department of Agriculture (Emery, these proceedings) a subset was selected to examine on-farm fumigation practice. The fumigation practice survey was conducted by the Agriculture Protection Board (APB) which has responsibility for on-farm grain insect control advice. The sample area covered 43 shires in the central wheat belt. (see Fig. 1) From the incoming samples and records returned by district officers conducting the resistance survey a selection of properties was made in two broad groups:

• sealed silos
• unsealed silos.

These two groups were divided into three subsections, comprising properties with:

(1) no insects,
(2) susceptible insects,
(3) resistant insects.

The fumigation practice survey was conducted from March to November 1992. Property owners were contacted by telephone in the evening between 7 and 9pm. The late telephone technique was used to achieve higher positive contact with the growers. No announcement of the survey was given to those telephoned. This was to ensure the responses were as close as possible to the actual practices employed rather than answers researched with the aid of literature or product labels.

All farmers contacted agreed to be interviewed. Those that did not store grain were not asked any further questions. The remainder were asked a series of questions related to their fumigation practice. The results of the interviews are recorded below. The sealed silo group was also asked if we could visit their property to test the silos. The owner was invited to be present when the test was being conducted.

In summary the questions asked were
• When do you fumigate?
• How do you fumigate and how often?
• What is the capacity of your grain store?
• How many tablets do you apply?
• Where are the tablets placed?

Sealed Silo Group

Properties were chosen by hand from lists as supplied by the Department of Agriculture resistance testing laboratory. The list was collated from samples returned to the laboratory from the farms in the phosphine resistance survey. Selection was made from the returns received between August 1991 and May 1992.

(1) No insects — first entry every fifth page of returned labels,
(2) Susceptible insects — every entry,
(3) Resistant insects — every entry.

No insects

44 farmers contacted and interviewed.

27 (61%) used phosphine and their silos were tested.

17 (38%) were not using phosphine in sealed silos for the following reasons. These silos were not tested.

8 farmers use phosphine but not in sealed silos where seed wheat or lupins is stored. (These growers had heard the rumour that phosphine reduced germination)

3 farmers stated there was no need to use phosphine in sealed silos.

4 farmers did not use phosphine on the property.

2 farmers used Dryacide® or lime in their sealed silo.

Susceptible insects

44 farmers contacted.

32 (72%) used phosphine and their silos were tested.

12 (27%) were not using phosphine in sealed silos for the following reasons. These silos were not tested.

6 farmers did not fumigate because the grain is insect free and a sealed silo will protect it.

4 farmers stored only seed grain and lupins in sealed silos and did not feel it necessary to fumigate.

1 farmer did not have a sealed silo (returned sample wrongly labelled).

1 farmer did not want his silos tested.

---

* Agriculture Protection Board, Bougainvillea Avenue, Forrestfield, Western Australia 6058.

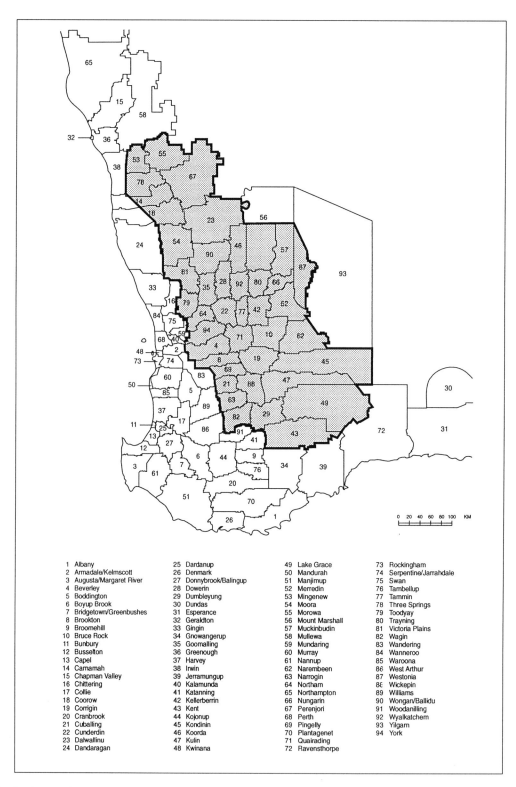

1	Albany	25	Dardanup	49	Lake Grace	73	Rockingham
2	Armadale/Kelmscott	26	Denmark	50	Mandurah	74	Serpentine/Jarrahdale
3	Augusta/Margaret River	27	Donnybrook/Balingup	51	Manjimup	75	Swan
4	Beverley	28	Dowerin	52	Merredin	76	Tambellup
5	Boddington	29	Dumbleyung	53	Mingenew	77	Tammin
6	Boyup Brook	30	Dundas	54	Moora	78	Three Springs
7	Bridgetown/Greenbushes	31	Esperance	55	Morowa	79	Toodyay
8	Brookton	32	Geraldton	56	Mount Marshall	80	Trayning
9	Broomehill	33	Gingin	57	Muckinbudin	81	Victoria Plains
10	Bruce Rock	34	Gnowangerup	58	Mullewa	82	Wagin
11	Bunbury	35	Goomalling	59	Mundaring	83	Wandering
12	Busselton	36	Greenough	60	Murray	84	Wanneroo
13	Capel	37	Harvey	61	Nannup	85	Waroona
14	Carnamah	38	Irwin	62	Narembeen	86	West Arthur
15	Chapman Valley	39	Jerramungup	63	Narrogin	87	Westonia
16	Chittering	40	Kalamunda	64	Northam	88	Wickepin
17	Collie	41	Katanning	65	Northampton	89	Williams
18	Coorow	42	Kellerberrin	66	Nungarin	90	Wongan/Ballidu
19	Corrigin	43	Kent	67	Perenjori	91	Woodanilling
20	Cranbrook	44	Kojonup	68	Perth	92	Wyalkatchem
21	Cuballing	45	Kondinin	69	Pingelly	93	Yilgarn
22	Cunderdin	46	Koorda	70	Plantagenet	94	York
23	Dalwallinu	47	Kulin	71	Quairading		
24	Dandaragan	48	Kwinana	72	Ravensthorpe		

Fig. 1. Fumigation practice survey area.

Resistant insects

26 farmers contacted. (every entry within the survey area)

20 (77%) used phosphine and their silos were tested.

6 (23%) were not using phosphine in sealed silos for the following reasons. These silos were not tested.

2 farmers ceased using phosphine 4–6 years previously.

2 farmers store seed grain and lupins in sealed silos and do not fumigate, but they do use phosphine in unsealed silos.

1 farmer fumigates in unsealed silos only because the grain in his sealed silos is moved through very quickly.

1 large property could not establish how much phosphine was used because the worker concerned had left.

From the comments provided by the respondents it is clear there remains a number of misconceptions regarding the use of phosphine and stored grain. Some growers believed there was a relationship between the use of phosphine and poor germination. Some still believe no further is treatment necessary once the grain has been placed in a sealed structure. The replies that fumigation was carried out only in unsealed structures give cause for concern. Farmers were asked when they carried out a fumigation on their stored grain.

	No. insects	Susceptible insects	Resistant insects
At loading of the silo	41%	22%	38%
At loading and when insects are found	22%	31%	14%
At loading and as a routine	11%	3%	14%
At loading and (x) months after loading		3%	
When insects are found	15%	19%	24%
When insects are found and as a routine			5%
Routine basis during the year		3%	5%
(x) months after filling	11% (8–11)	19% (1–7)	

From the table above it appears that most farmers apply the initial insect treatment when the silo is first filled. This is recommended practice published by the APB and other relevant authorities. A large number apply phosphine treatments only when insects are found and on an ad-hoc basis throughout the storage period.

The following table indicates how the fumigation is carried out. The farmers were asked how they applied the tablets to the silo.

	No. insects	Susceptible insects	Resistant insects
Auger %	30	28	38
Auger and top of grain %	22	3	19
On top of grain %	18	21	9.5
Inserted with a probe %	11	17	14.5
Auger and probe %	11	7	9.5
Placed in boot %	4		
On top and boot %		24	
Bag chute and boot %			9.5
Placed in containers %	4		

From the respondents replies 56% use the auger as a method of application for all or part of the phosphine dose. The APB recommends only one method of application of the tablets to a sealed silo. This is on top of the grain after filling and before the lid is closed and the silo sealed. This method prevents gas loss either when the tablets are placed in the auger or if the farmer takes more than one day to load the silo. When the tablets are placed on top of the grain, the air currents created by temperature gradients carry the liberated gas around the silo. Placing the tablets in the bag chute or boot removes the gas from the main air currents in the silo which may limit circulation of the liberated gas.

From the table it is noted some farmers place the tablets in containers in the silo. This is a highly dangerous practice which can lead to the generation of an explosive atmosphere inside the container.

Probing the tablets into the grain is difficult and unnecessary in a sealed silo. However it is noted this is the practice of 23% of respondents when applying later doses of phosphine.

Farmers were questioned on the amount of phosphine tablets applied to a silo and the frequency of application.

	No. insects	Susceptible insects	Resistant insects
Average dose (g/t)	1.47	1.5	1.4
Range g/t	0.09–3.1	0.16–6	0.08–6
Frequency of application (average/year)	1.5	1.4	1.8
Range	1–3	1–3	1–3
Average annual application (g/t)	2.2	2.1	2.5

Regardless of grouping, the application rates of phosphine are remarkably similar. There is no statistical difference in the annual application rate. The most disconcerting statistic is the range of dosages of phosphine. Applications rates of 0.08 g/t up to 6 g/t indicate a failure to read the label or adhere to published and label recommendations of 2 g/t of silo capacity .

From this information it could be suggested that the dose rate has not had any impact on the presence or absence of a phosphine resistance.

Silo Testing

An important aspect of the survey was to determine the gas-tightness of the respondents' silos. This statistic could then be compared to the level of resistance in insects.

All respondents were asked for permission to visit their property to test the silo from which the initial sample was obtained. Some 99% of property owners agreed to the farm visit and test. The property was visited in company with the officer who collected the sample and the farmer where possible. This turned the silo test into a demonstration and problem solving exercise in many cases.

Test results (%)	No. insects	Susceptible insects	Resistant insects
Pass	33	25	24
Fail	66	75	76

There appears to be no significant difference in silo failure between the different groups. This suggests there is no relationship between phosphine resistance and the gastightness of the silo.

On average failure could be attributed to:
- leaking seals (57%)
- leaking seals and low oil level in the pressure relief valve (13%)
- low oil levels (15%)
- damage (13%)
- damage and seals (4%)

Causes of failure were similar to those found by Newman (1989).

Unsealed Silo Group

Properties were chosen by hand from lists as supplied by the Department of Agriculture resistance testing laboratory. Selection of properties was made from the returns made to the laboratory for their phosphine resistance survey from August 1991 to May 1992.

(1) No insects — first entry every fifth page of returns.
(2) Susceptible insects — every fifth entry
(3) Resistant insects — every entry

No insects

41 farmers interviewed by telephone.

18 (46%) did not fumigate.

Susceptible insects

32 farmers interviewed by telephone.

12 (37.5%) did not fumigate.

Resistant insects

37 farmers interviewed by telephone.

2 (5.4 %) did not fumigate.

Some farmers did not want to be interviewed but this did not adversely affect the result. One owner could not be contacted despite considerable effort.

Farmers were asked when they carried out a fumigation on their stored grain.

	No. insects	Susceptible insects	Resistant insects
At loading of the silo	4%	30%	17%
At loading and when insects are found	18%	10%	26%
When insects are found	61%	35%	43%
When insects are found and as a routine			6%
Routine basis during the year	13%	10%	8%
(x) months after filling	4% (11)	15% (7–11)	

Results suggest farmers tend to use phosphine in reaction to insect infestation instead of as a prevention measure.

Farmers were asked how they applied tablets to the silo.

	No. insects	Susceptible insects	Resistant insects
Probe	70%	50%	46%
Auger	9%	30%	20%
Probe and auger	9%	10%	9%
Top of grain	3%		9%
Silo boot			3%
Door		5%	
Probe door and top	3%		3%
Fixed tubes			6%
Boot and top		5%	
Probe and boot			3%
Probe and top	3%		3%

A wide variety of techniques was chosen to apply the tablets. The most popular is the probe method, followed by application via the auger as the grain is being loaded. These techniques are acceptable as the most efficient method of tablet insertion in an unconfined airspace. This will assist gas development in the grain bulk and give a degree of insect control. A proportion of the surveyed groups persists in applying tablets to the periphery (top, boot, doors, etc.) of the grain bulk which is of limited benefit and may help select for resistance.

Farmers were asked how they applied phosphine and the frequency of application. (Replies received in terms of pack sizes. This has been converted to g/t)

	No. insects	Susceptible insects	Resistant insects
Average dose (g/t)	3.6	3.1	2.8
Range (g/t)	0.5–10	0.4–10	0.07–10
Frequency of application (avg /year)	1.7	1.6	2.3
Range	1–4	1–4	1–8
Average annual application. (g/t)	6.1	4.96	6.4

Considerable variation in rates of phosphine applied were noted. From 0.07–10 g/t were quoted by respondents. When asked more specifically how they calculated how many tablets to add to the silo, the answer was invariably 'by the amount of grain in the silo'. This conflicts with APB recommendations to dose by the capacity of the silo. It will be noted that the average annual application rate is considerably higher than that for the sealed silo group. The farmers in the unsealed silo group applied 3.5 g more phosphine per annum than those with sealed storage.

Summary

Sealed silo maintenance has decreased since an earlier survey: 27% passed the standard APB test compared with 44% found in the previous survey. (Newman 1989)

Poor rubber seals were again the major reason for failure, followed by low oil levels. Both of these problems can be remedied quickly at low cost to the farmer.

It appears the application rate of phosphine and silo failure rate has not influenced the number of phosphine-resistant properties in the sealed silo group. The unsealed silo group demonstrated more frequent use of phosphine than the sealed silo group.

A comparison of the treatment times of the two groups demonstrates a difference in the reasons for fumigation. In the sealed silo group, an average of 43% apply phosphine when insects are found, or in a combination with loading or routine measures. Some 67% apply the fumigant as a preventative treatment. In the unsealed silo group 66% apply phosphine in response to insect infestations. This action has allowed a large population to evolve, increasing the number that will be selected for resistance. Thirty-seven farmers were found to have resistant insects in the unsealed group compared with 26 farmers in the sealed group.

The majority of farmers in both surveyed groups applied phosphine in a suitable manner. However, a small proportion applied the tablets in a manner that could help select for phosphine resistance.

Despite extensive publicity since the original survey was released it appears few farmers have recognised the importance of maintaining a silo in a sealed condition. Application techniques and dose rate recommendations are two areas that require increased extension input. Fumigation in unsealed storage is impossible to achieve. The recommendation to fumigate in this manner should be removed from publications. More effective insect control can be achieved by placing phosphine tablets under a tarpaulin covering grain in a truck or stacked on a plastic sheet. A recommendation to fumigate only when grain is being removed from the property needs to be reinforced on a local level.

This survey has not demonstrated any relationship between phosphine resistance and faulty fumigation techniques. The presence of insects may then be due to the importation of insects from other sources.

Further research is suggested to study more closely:

• exact degree of gastightness of silos which failed the standard test

• hygiene of the property

• movement of resistant insects from unsealed to sealed storages.

• movement of grain onto the property.

Reference

Newman, C.R. 1989. Sealed grain storage in Western Australia: a progress report. In : Champ, B.R., Highley, E. and Banks, H.J., ed., Fumigation and controlled atmosphere storage of grain: proceedings of an international conference, Singapore 14–18 February 1989. ACIAR Proceedings, No. 25, 272.

Biogeneration of carbon dioxide for use in modified atmosphere storage of sorghum grains

K.L. Patkar*, C. M. Usha*, H. S. Shetty*, N. Paster[†] and J. Lacey[§]

Abstract

One of the safest ways of storing cereal grains is in modified atmospheres. However, if modified atmospheres are to be used in small storage bins on farms in developing countries with minimal cost, a means is required to produce carbon dioxide and to avoid the high cost of bottled gas. To achieve this, biogenerators have been developed in which CO_2 is produced through the fermentation of waste vegetable products. This paper describes the use of saw dust, wheat bran and coffee husk as possible substrates. Saw dust allowed the production of no more than 2% CO_2 in the atmosphere while those containing wheat bran accumulated up to 19% CO_2 and those containing coffee husk accumulated up to 26% CO_2 with O_2 decreased to only 0.6%. Most CO_2 was produced by materials containing 40% water. Biogenerators containing coffee husk were connected to plastic bins containing sorghum grain with insect cages placed near the top, middle and bottom of the grain. A maximum of 21% CO_2 was recorded in the bin on the ninth day of storage, with O_2 decreased to 2%. When the biogenerators were disconnected after 14 days storage, all insects had died in the test bins but none in the control bins. Biogenerators for CO_2 production could easily be adapted for use with grain stored either in grain bins or in underground pits in rural communities in developing countries.

Introduction

Loss of food grains in developing countries after harvest is a major problem because it leads to food shortages and malnutrition. Insects are perhaps the most important agents of spoilage in stored food grains, closely followed by fungi. Both cause losses of dry matter and quality. Chemical preservation of grain is still widely practised but their toxicity and implication in ozone depletion has led to increased emphasis on non-chemical methods of preservation. Of the different methods of grain preservation in use, modified atmosphere (MA) storage has been shown to be promising in creating lethal conditions both for insects and fungi in stored grain. This method is already used to control insect infestation in large scale grain storage in Australia (Annis 1987; Ripp et al. 1990) but conditions required for the inhibition of fungi are more extreme.

In developing countries, the creation of MAs using commercial gases can be expensive, but a cheaper alternative can be to produce them through the fermentation of plant waste materials using only the natural microflora (Paster et al. 1990). CO_2 production by peanut shells and wheat bran was compared during fermentation at different water contents in specially designed structures. Wheat bran at 40% water content produced 25% CO_2 in the atmosphere after 48 hours incubation and maintained a concentration of about 20% up to the 12th day of incubation. Peanut shells had produced slightly less CO_2 than wheat bran after 48 hours but the concentration then decreased gradually to about 12%. CO_2 production by orange peel with 80% water content, fermenting in biogenerators connected to bins containing maize grain, was similar to that by wheat bran with 35% water content after 2–7 days incubation but was significantly greater after 10 days. The CO_2 concentration in the grain was increased from less than 0.1% to about 18%, but that of O_2 was decreased from 21% to only about 10%. In experiments in Costa Rica (Paster et al. 1991), concentrations of 18.8% CO_2 and 4.7% O_2 were attained in maize grain from biogenerators containing wheat bran with 35% water content. These concentrations were sufficient to kill all insects within 9 days. Paster et al. (1991) therefore concluded that biogenerators utilising waste plant material, connected to storage bins, could be used to restrict insect damage in the small grain bins of subsistence farmers.

The present investigation aimed to test the ability of some waste plant materials commonly available in India, to produce MAs in grain bins sufficient to control storage insects.

Materials and Methods

Grain bins

Plastic bins, each holding about 25 kg of sorghum grain, were fitted with plastic tubes (0.5 cm diameter) for gas sampling at three different heights, near the top, middle and bottom of the grain bulk. Three insect cages, each containing 25 adult *Tribolium castaneum* (Herbst) 7–10 days old, were introduced at different depths in the grain bins to determine their mortality in treated and untreated bins during storage. The water content of the grain samples was determined both before and after the experiment as described by ISTA (1985).

Biogenerator construction

Biogenerators for the production of high CO_2 and low O_2 atmospheres were constructed following the design of Paster et al. (1990). A plastic container about 24 cm high and 16 cm internal diameter, with a volume of about 5 L, was fitted with a false floor of wire mesh, 4 cm above the bottom, to allow excess water to drain through to a drainage tube fixed at the bottom of the container. Another tube, also connected to the bottom of the biogenerator, transferred the gas mixture produced to a grain bin. A hole in the top, 0.5 cm diameter, allowed air access to the biogenerator to prevent anoxia.

* Department of Applied Botany, University of Mysore, Mysore 570006, India.
† Department of Stored Products, ARO- The Volcani Center, 50250 Bet Dagan, Israel.
§ Institute of Arable Crops Research, Rothamsted Experimental Station, Harpenden, Herts AL5 2JQ, U.K.

Substrates for biogeneration of carbon dioxide

Biogenerators were filled with sawdust (SD), wheat bran (WB), coffee husk (CH), or a 1:3 mixture of WB and SD as substrates for the production of CO_2. Lots of 800 g of each substrate, wetted to 30, 35 or 40% water content, were allowed to equilibrate at 4°C for 8 days with frequent mixing. After preliminary tests, biogenerators filled with CH or WB:SD (1:3), each containing 40% water, were selected for grain storage tests. Biogenerators containing these substrates were connected using plastic tubing (0.5 cm inner diameter) to grain bins containing about 25 kg sorghum. Each biogenerator was connected to a separate grain bin. A grain bin connected to an empty biogenerator served as a control.

Measurement of CO_2 and O_2 concentrations

Air samples were withdrawn from biogenerators and grain bins through the air sampling tubes. Instruments for measuring CO_2 (Riken, Japan) and O_2 (Gowmac, Ireland) were connected in series and air was drawn through both samplers using the peristaltic pump of the CO_2 analyser. Two replicate bins were prepared with each fermentation substrate and the experiment was repeated twice.

Results

Concentrations of CO_2 and O_2 produced in biogenerators containing four different substrates are given in Figures 1 and 2. After 2 days incubation, the CO_2 concentration in biogenerators containing wheat bran (WB) with 30% water content was significantly less than that found with 35 or 40% water content. By contrast, O_2 concentrations after 2 and 4 days incubation, with 35 and 40% water contents, were significantly smaller than those with 30% water content. Subsequently, there were no differences in CO_2 and O_2 concentrations with water content between biogenerators

containing wheat bran. The largest concentrations of CO_2 in a WB biogenerator were found after 2 days incubation with water contents of 35 and 40% and after 4 days with 35% water content.

Biogenerators containing sawdust (SD) produced maxima of 3.0, 5.0 and 5.5% CO_2 with water contents of 30, 35 and 40%, respectively, after 8 days incubation. There were no significant differences in CO_2 production with water content. The smallest O_2 concentration similarly occurred after 8 days and, again, there were no significant differences between different water contents.

Biogenerators containing coffee husk (CH) with 30, 35 and 40% water contents produced, respectively, 23, 25 and 26 % CO_2 after 2 days incubation. CO_2 production did not differ significantly with water content during incubation. After 12 days incubation, the CO_2 concentration in CH biogenerators at the three water contents ranged between 20 and 22%. O_2 concentrations decreased as CO_2 increased and, as with CO_2, there were no significant differences in O_2 concentration with water content. O_2 concentrations were never more than 2.5% during the 12 days incubation.

Largest CO_2 concentrations were found with biogenerators containing WB+SD (1:3) which reached 21% after 4 days incubation. At the same time, O_2 concentration with 40% water content declined to 1.5% after 2 days incubation. The CO_2 concentration produced by WB+SD with 40% water content did not differ significantly from that obtained with WB+SD with 30% water content after 2 and 4 days incubation, while O_2 concentrations with WB+SD at 35 and 40% water contents were significantly smaller than those with WB+SD at 30% water content.

Concentrations of CO_2 and O_2 in biogenerators containing CH and WB+SD (1:3) and in grain bins containing sorghum connected to these biogenerators are shown in Figure 3. CO_2 production in CH biogenerators were significantly greater than in those containing WB+SD except after 6 and 10 days incubation. The O_2 concentrations differed significantly

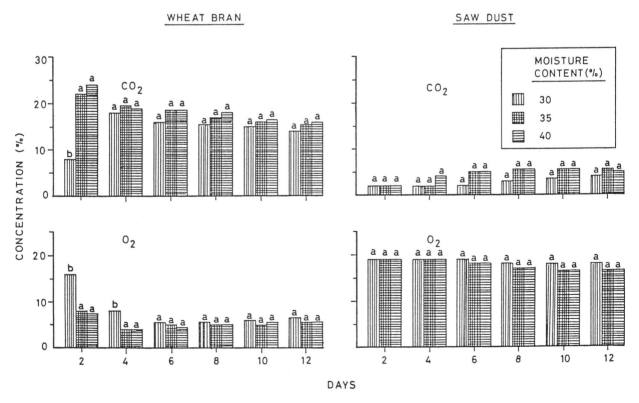

Fig. 1. Carbon dioxide and oxygen concentrations over wheat bran and sawdust at different water contents in biogenerators (within the same group, values with common letter do not differ significantly at P≤0.05).

between the two substrates after 2, 10 and 12 days incubation. Concentrations of CO_2 and O_2 in the grain were smaller than those in the biogenerators to which they were connected. Concentrations of CO_2 and O_2 in grains bins connected to CH biogenerators were significantly greater than those in grain bins connected to WB +SD biogenerators, both at the beginning and end of experiments. Control bins contained <0.1% CO_2 and 20.8% O_2.

Transfer of MAs from the biogenerators caused no change in grain water contents and there were no off-odours. All the

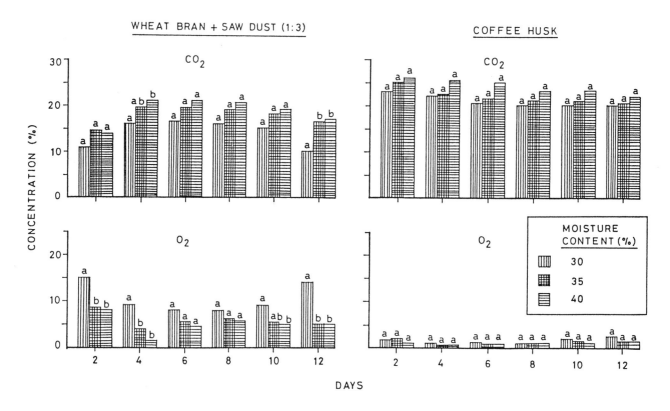

Fig. 2.   Carbon dioxide and oxygen concentrations over wheat bran: sawdust (1:3) mixtures and over coffee husk at different water contents in biogenerators (within the same group, values with common letter do not differ significantly at P≤0.05).

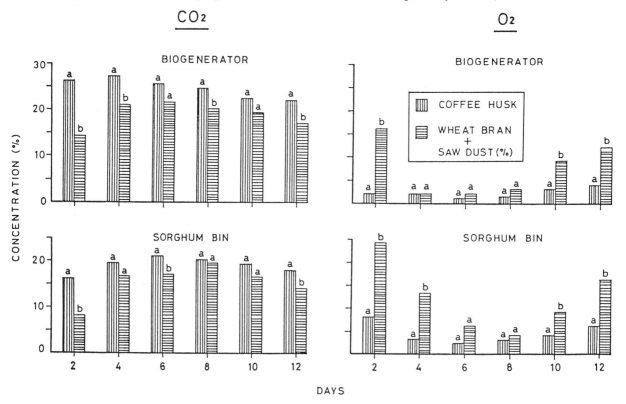

Fig. 3.   Carbon dioxide and oxygen concentrations in biogenerators over wheat bran: sawdust (1:3) mixtures and in connected sorghum grain bins (within the same group, values with the common letter do not differ significantly at P≤0.05).

test insects (*Tribolium castaneum*) introduced into the grain bins were killed at all three positions in grain bins connected to biogenerators but mortality in the control bins was always <2%.

Discussion

This study has confirmed that MAs can conveniently be produced by fermenting waste plant materials and that the MA produced in the biogenerators could successfully be transferred to grain bins containing sorghum. Of the three substrates used in the present study to generate MAs, CH, WB and a mixture of WB+SD produced an MA sufficient to completely kill *Tribolium castaneum* adults. Calderon and Navarro (1979) calculated that the exposure time and gas concentrations necessary to kill 95% of *T. castaneum* adults were 5 days at 22 % CO_2 and 5% O_2. The CO_2 and O_2 concentrations obtained in the present study are close to these and, with the longer exposure period (12 days), caused 100% mortality. The results also agree well with those of Paster et al. (1990–1991) and clearly indicate the potential of this system for producing MAs for small-scale storage on farms in developing countries.

Further work is needed to evaluate the effectiveness of biogenerators for controlling moulds and other insect species. However, fungi are likely to be inhibited only when there is more than 50% CO_2 in the atmosphere together with only 0.2% O_2 (Lacey 1994). Conversely, artificial inoculation of the substrate in biogenerators with fast growing microbial species could perhaps give larger concentrations of CO_2 and more rapid elimination of O_2 than the natural inoculum used in these tests. Least CO_2 produced in the sawdust charged biogenerator and this was insufficient to be lethal to any storage insects. Significantly greater concentrations of CO_2 were produced when wheat bran and sawdust were mixed in the ratio of 1:3 than by the individual components. This could

have resulted from more vigorous microbial growth in the wheat bran and the production of enzymes which better degraded both substrates. With suitable development, this model plant-waste-material biogenerator and storage bin could be effectively used with farm-level storage structures in rural, semi-arid regions of India, even where sorghum grains are stored in the underground pits. Biogenerated MAs could easily be transferred into the pits and could significantly decrease storage losses.

References

Annis, P.C. 1987. Towards rational controlled atmosphere dosages schedules: a review of current knowledge. In Donahaye, E. and Navarro, S., ed., Proceedings of the 4th International Working Conference on Stored–product Protection, Tel-Aviv, Israel, 128 –149.

Calderon, M. and Navarro, S. 1979. Increased toxicity of low oxygen atmospheres supplemented with carbon dioxide on *Tribolium castaneum* adults. Entomologin Experimentalis of Applicata, 25, 39–44.

Lacey, J. 1994. Effects of intergranular gas composition and fumigants on mould growth and mycotoxin production. In: Navarro, S. and Donahaye, E., ed., Proceedings of the International Conference on Controlled Atmosphere and Fumigation in Grain Storages, Winnipeg, 10–13 June 1992. Jerusalem, Caspit Press Ltd.

Paster, N., Calderon, M., Menasherov, M., and Mora, M. 1990. Biogeneration of modified atmospheres in small storage containers using plant waste. Crop Protection, 9, 235–238.

Paster, N., Calderon, M., Menasherov, M., Barak, V. and Mora, M. 1991. Application of biogenerated modified atmospheres for insect control in small grain bins. Tropical Science, 32, 355–358.

Ripp, B.S., de Largie, T.A. and Barry, C. B. 1990. Advances in the practical application of controlled atmospheres for the preservation of grain in Australia. In: Calderon, M. and Barkai–Golan, R., ed., Food preservation by modified atmospheres. Boca Raton, Florida, CRC Press, 151–187.

The current status of phosphine fumigations in India

S. Rajendran and K.S. Narasimhan*

Abstract

India uses about 80% of its total production of 2500 tonnes of aluminium phosphide formulations for the protection of stored products. Grain in bag-stacks are fumigated under cover with aluminium phosphide tablets, up to seven times depending on the storage period. Whole warehouses are fumigated only occasionally. Cured tobacco, in bales, is invariably treated with phosphine to control *Lasioderma serricorne*. Phosphine is also used to treat coffee seeds, especially monsooned arabicacherry coffee which is susceptible to *Araecerus fasciculatus*. Currently, dosage rates for various commodities vary from 3–6 tablets/t with a 5 day exposure period.

Field experiments revealed better gas retention by 0.15 mm PVC covers than the currently used 0.25 mm LDPE covers. For fumigation under prevailing storage conditions, 34 g aluminium phosphide pouches were found to be superior to tablet formulations. Field trials on fumigation of coffee seeds with aluminium phosphide tablets and tobacco bales with a magnesium phosphide bag formulation are reported. A survey on phosphine resistance indicated that *Tribolium castaneum* was commonly resistant, whereas *Sitophilus oryzae* was found to be occasionally so. Cases of control failures due to occurrence of resistant strains are documented. The steps taken to delay and/or to overcome the resistance problem including revised dosage schedules, according to ambient temperatures, and improvements in sealing methods are discussed.

Introduction

In India various stored products such as food grains, milled products, pulses, oilseeds, tobacco and animal products are disinfested by fumigating with phosphine evolved from aluminium phosphide tablet preparations. India is one of the leading manufacturers of metal phosphide formulations. The fumigant was introduced in this country during the 1960s. The first field trial on grains was reported by Lallan Rai et al. (1964). Phosphine, which is cheaper and easy to handle and use, replaced liquid fumigants such as ethylene dichloride — carbon tetrachloride mixture and mixtures containing ethylene dibromide. Of late there are reports criticising unsatisfactory fumigation practices involving phosphine in countries on the Indian subcontinent. They point out the spread of phosphine-resistant strains from such places (Taylor and Halliday 1986). Therefore it was intended to review the

state of affairs of phosphine applications with regard to the protection of stored products in India. This paper outlines current fumigation practices involving phosphine, discusses resistance status and reports on recent field trials to improve application techniques.

Grain Fumigation

Wheat and rice, the two major staple food commodities, are stored and handled in jute bags (95 kg net weight). The bags containing the grains undergo at least 12 handlings from the start of procurement to reaching retail stores. Government agencies and the co-operatives keep the grain-stocks in their warehouses and the surplus stocks are stored in the open in CAP (cover and plinth) storages. Outdoor storage involves wheat and paddy only. Facilities are also available to store 4 lakh tonnes of paddy or wheat in metal and concrete silos (Shivanna 1990). The bag-stack grain storage system facilitates ventilation throughout the storage period and aids rapid distribution as well as dissipation of fumigants but at the same time it favours cross-infestation by crawling and flying insect pests. In many of the warehouses there is rapid turnover of stocks and the system of first-in first-out is not followed. The major pests encountered in grain stacks and storage premises are *Sitophilus oryzae* (L.), *Rhyzopertha dominica* (F.), *Tribolium castaneum* (Herbst), *Oryzaephilus surinamensis* (L.,) *Trogoderma granarium* Everts and *Ephestia cautella* (Walker).

Besides routine prophylactic treatment of bag-stacks and storage premises with malathion (0.15 g/m^2) and dichlorvos (0.2 g/m^2), the grain stocks are fumigated with phosphine under fumigation covers or sheets. Black low-density polyethylene (LDPE) covers of 0.25 mm thickness and, in a few places, rubberised fabric with aluminium finish on one side are used for fumigation of bag- stacks. The cover is weighted down to the floor with sandsnakes, mud, loose sand or merely gum tape or newspaper strips. The stacks are fumigated on more than one occasion at a dosage of 3 aluminium phosphide tablets per tonne with 5 days exposure period. The longer the stack remains in the storage depot the more it is fumigated. As many as seven fumigations with phosphine are permitted for stored grains.

The exposure period is rarely extended to 10 days or more which is necessary when the temperature is 20°C or less and when *Sitophilus* spp. or phosphine-resistant strains are present. Palliative treatments with higher application rates i.e., > 10 tablets/t and/or inadequate exposure periods i.e., less than 5 days are not ruled out. Most of the warehouses with gabled asbestos roofs are not suitable for whole-godown treatment. Nevertheless, occasionally fumigation of entire warehouse (shed-fumigation) with shell-type roof is carried out using aluminium phosphide tablets.

Reports supported by gas concentration data on whole godown treatments are, however, lacking. Detection and gas monitoring devices for phosphine are hard to come by and hence they are rarely used to assess the safety of the working environment when fumigation operations are under way. Phosphine concentrations as high as 2 ppm have been

---

* Infestation Control and Protectants Department, Central Food Technological Research Institute, Mysore 570 013, India.

estimated in the workplace during grain fumigation. In the absence of suitable gas mask canisters, transient symptoms of occupational exposure in the workers involved in fumigation work have been noticed (Misra et al. 1988).

Quality checks are periodically made on aluminium phosphide tablet formulations supplied by the manufacturers as per the standards laid by the Bureau of Indian Standards (BIS 1980). Accordingly the tablet formulation should weigh 3 g, must contain not less than 56% aluminium phosphide by mass and the tablet should not decompose in 30 minutes to liberate phosphine in 100% humidified chamber. The aluminium phosphide content is determined by the reaction of phosphine carried by a stream of nitrogen with potassium permanganate. As the rate of nitrogen carrier gas flow has not been specified in the method of BIS (1980) chances of variable results exist, as pointed out by Rajen (1990). The optimum flow rate of nitrogen to carry phosphine to potassium permanganate is 15–25 mL/minute. Tablets manufactured in India were analysed at this institute for the user organisations in the last 10 years. Aluminium phosphide content ranged from 56 to 70%. Fumigation sheets/covers should also conform to the criteria stipulated in Indian standard (BIS 1991). There are various tests exemplified by fumigation retention test, tests for workmanship, strength of joints, tensile strength, ease of repair, blocking test, flex test, accelerated ageing test, etc.

During fumigation of bag-stacks with aluminium phosphide tablets, 50% of the tablets are distributed on the top and 50% all around the stack. It is known that the decomposition of the tablets is influenced by ambient humidity and temperature. Even in the presence of favourable conditions like 25°C and 75% r.h., it has been reported that undecomposed aluminium phosphide up to 5% persists in the spent dust (Friendship 1989). The latter is a source of contamination of grain. A 34 g aluminium phosphide bag formulation which releases 11 g phosphine, manufactured in India, has been recently cleared for use in the country. The authors conducted field trials in a storage depot at New Delhi and studied the comparative rate of evolution of phosphine from tablet and bag formulations. Bag-stacks of 11.7% moisture content of wheat, each of 135 t, were fumigated under 0.25 mm thick LDPE cover at 1.5 g phosphine/m^3 for 11 days at 17±2°C and 62±10% r.h.. Nylon tubes, 3 mm diameter, were positioned at the top, middle and bottom levels of the stacks to monitor the gas concentrations at regular intervals using a Miran 104 Gas Analyser. The rate of liberation of phosphine from the bag formulation was moderate and the trend is more acceptable from the point of achieving 100% insect kill (Fig. 1). The bag formulation is more convenient to use and dispose of than the tablet formulation and is preferable for use under the storage conditions in this country.

In an another experiment the performance of 0.15 mm plain polyvinylchloride (PVC) fumigation cover was compared with that of other covers currently used by government agencies i.e. 0.25 mm black LDPE and rubberised fabric with aluminium finish. All 3 covers tested were new and had not been used previously. The test fumigations were conducted on 135 t wheat stacks with a phosphine dosage of 1.5 g/m^3 at 17 ± 2°C and 62 ± 10% r.h. The PVC cover showed higher retention of the gas than LDPE or rubberised fabric covers (Table 1). Laboratory tests on permeability of phosphine through various films/fabrics revealed that PVC sheets were superior to LDPE sheets and rubberised fabrics (Kashi et al. 1977). In contrast, Banks (1984) while discussing the lack of data on permeability of fumigation sheets, quotes higher loss rates of phosphine through PVC than LDPE. Nevertheless PVC sheets have better functional qualities such as flexibility and durability. Plain and reinforced PVC sheets of various thicknesses are manufactured in India.

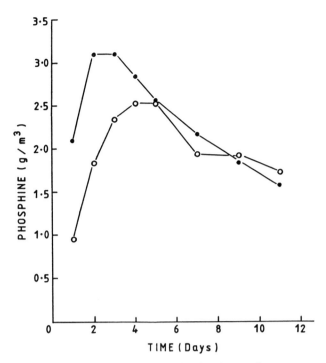

Fig. 1. Rate of liberation of phosphine from bag-stacks.

Detection and monitoring devices for phosphine are not commercially made in India. Phosphine indicator paper which can detect the presence of phosphine in the workplace at the hygienic level of 0.3 ppm was developed by Muthu et al. (1973) and improved by Kashi and Muthu (1975). Rajendran and Narasimhan (unpublished data) further improved the indicator paper for phosphine with rapid signalling action and longer shelf-life. The paper can be used for detection of leakage and to ascertain the safety of the working environment.

Fumigation at Rural Level

About 70–75% of grains produced in India are stored in the rural areas by the farmers (Sadana et al. 1988). Grains are stored indoors and outdoors in metal bins and in different types of traditional storage structures. Previously, only liquid fumigants, particularly ethylene dibromide, were used for grain protection. Recent surveys indicate that aluminium phosphide tablets are commonly used by the farmers although their traditional storage structures are not sufficiently gastight to retain phosphine a high vapour pressure fumigant, and the structures are often inside the living room (Thakre et al. 1988; Sadana et al. 1988). Lately, unit packages or capsules containing 3 aluminium phosphide tablets suitable for fumigation of one ton of grains and other food commodities have been introduced into the market in India (Banerjee and Deshmukh 1992).

Fumigation of tobacco

India is the fourth largest producer of tobacco in the world; annual production of flue-cured tobacco is about 120000 t. Tobacco plays an important place in the Indian economy (Rao 1992). Cured tobacco in bales is attacked by the cigarette beetle *Lasioderma serricorne* (F.) and the infestation problem is greater in the hot and humid coastal region in Andhra Pradesh which is the main production centre for tobacco. Tobacco bales, 100–180 kg net weight, are at present fumigated under cover with aluminium phosphide tablets at the recommended dosage of 1 g phosphine/ m^3 for 4 or 5 days

Table 1. Gas retention and other parameters of fumigation covers tested on 135 t wheat stacks dosed at 1.5 g PH₃/m³ with 7 days exposure at 17±2°C and 62±10% r.h.

Particulars	PVC	LDPE	Rubberised
Particulars PVC LDPE Rubberised Price (size 9.7 m x 6.4 m x 5.2 m) in Rupees	6000	4000	10000
Weight (g/m2)	210	220	600
Thickness (μm)	150	250	–
Phosphine concentration on the final day (g/m³)	3.0	2.2	1.3
Cumultative CT achieved	470	380	275
Handling characteristics	Very satisfactory	Satisfactory	Poor

(Philip Morris 1991). In addition to usual fumigation covers, use of an overlapping 0.05 mm polythene sheet has been recommended for improved gas retention during phosphine fumigations (ITC 1985).

In order to achieve effective gas concentrations quickly at all locales and also to avoid possible contamination by spent dust from aluminium phosphide tablets, fumigation tests were conducted with 34 g magnesium phosphide bag formulation. In a trial at Ghaziabad in Uttar Pradesh, 22.6 t of tobacco (11.6% moisture) in bales was successfully fumigated under a reinforced PVC cover of 0.15 mm thickness with magnesium phosphide bags at 16±3°C and 80±10% r.h. The dosage worked out to 2 g phosphine/m³ with 10 days exposure period. The concentration profile as determined by the infrared gas analyser was satisfactorily high throughout the period (Table 2). Analysis of spent bag revealed a very low level of 0.5% magnesium phosphide, as against the initial level of 66%.

Table 2. Gas concentrations recorded during fumigation of tobacco bales with 34 g magnesium phosphide bags at 2 g phosphine/m³ dosage at 16±3°C and 80±10% r.h.

Experiment day	Average PH₃ concentration (g/m³)	Remarks
0.5	0.96	Cumulative Ct = 410 g.hours/m³
1.0	1.11	
1.5	1.64	100% kill of all stages of *L. serricorne* observed
2.0	1.75	
3.0	1.99	
4.0	2.13	
6.0	2.06	
8.0	1.98	
10.0	1.75	

Fumigation of green coffee

Coffee seeds especially monsooned ones stored in the hot and humid coastal regions are prone to infestation by the coffee bean weevil *Araecerus fasciculatus* (De Geer). Generally, coffee stocks are fumigated with methyl bromide. For the first time, a field fumigation trial with phosphine was carried out on a 20 t bag-stack of arabica-cherry-coffee in a storage depot at Mysore. Reinforced PVC cover of 0.15 mm thickness was used and the application rate was 1.5 g phosphine/m³ (3.4 tablets/t) with 7 days exposure period. Average gas concentration at the time of termination was 0.8 g/m³ and the estimated CT was 288 g hours/m³. At the above effective dosage the quality of coffee was not impaired.

Resistance and Control Failures

Insects collected from grain storage depots located in different parts of the country were screened for resistance to phosphine and methyl bromide following the FAO method (1975). *T. castaneum* from all the eight places were found to be resistant to phosphine; the level of kill recorded at the discriminating dose was 0–20% only. In the case of *S. oryzae,* insects from 3 of 5 locations showed resistance and the kill ranged from 0–10% for the resistant strains. All the insects were susceptible to methyl bromide. Furthermore, *R. dominica* from wheat samples from a silo at Bombay was found to be resistant to phosphine by a factor of 380 at LD50. Wheat in the silo had been fumigated with phosphine on three occasions during a 1 year storage period. A similar case of field occurrence of high level phosphine resistance in the same species was reported earlier (Rajendran, 1989).

Three distinct cases of control failures in field fumigations owing to the occurrence of phosphine-resistant strains have been recorded (Table 3). In a warehouse at New Delhi a bag-stack of wheat was fumigated under 0.25 mm LDPE cover with aluminium phosphide tablets at 1.5 g phosphine/m³ for 11 days at 15–19°C and 50–72% r.h. Despite the high concentration of 1.6 g/m³ (analysed by MIRAN 104 Gas Analyser) on the final day, active adults of *T. castaneum* and *S. oryzae* were noticed. Inspection of post-fumigation samples revealed that other life stages were also alive. Fumigation with phosphine alone at 3 tablets/t and 5 days of exposure had been the practice in the warehouse for many years. In an another instance, survival of adults of *Liposcelis* spp. was observed after fumigation of milled rice under cover at 1.5 g phosphine/m³ for 7 days.

In the last case, survival of all stages of *L. serricorne* was observed after fumigation of the tobacco bales at 1.25 g phosphine/m³ (5 tablets/t) for 7 days at 27–33°C and 65–90% r.h. The recommended dosage for the control of cigarette beetle in tobacco bales is 1 g/m³ (EPPO 1982; Philip Morris 1991). At 25°C, phosphine concentration exceeding 0.3 g/m³ throughout the exposure period of 96 hours has been claimed to be sufficient for all the stages of *L. serricorne* (Hole et al. 1976). In the present case, gas concentration on the 4th day was 0.4 g/m³ which declined to 0.2 g/m³ on the 7th day. Repeated applications under less-retentive fumigation covers had probably resulted in high-level resistance to phosphine. Further trials in the warehouse revealed that a dosage of 2 g phosphine/m³ for not less than 10 days was effective against the resistant strain.

Conclusions

The climatic conditions in India allow rapid multiplication of insects in short durations and the current grain storage system permits quick reinfestation of fumigated commodities. Hence,

Table 3 Instances of control failures during wheat (at New Delhi), rice (Mysore) and tobacco bale (Ongole) fumigations with phosphine

	Commodity		
	Wheat (Case1)	Rice (Case 2)	Tobacco (Case 3)
Moisture content (%)	11.7	13.3	12.1
Quantity (t)	135	180	18.5
Volume (m³)	190	243	71
Fumigation cover	LDPE, 0.25 mm	LDPE, 0.25 mm	HDPE woven laminated with polyethylene
Phosphine dosage (g/m³)	1.5	1.5	1.25
Exposure period (days)	11	7	7
Temperature range (°C)	15–19	25–28	27–33
Relative humidity (%)	50–72	70–80	65–90
Final concentration (g/m³)	1.6	1.5	0.2
Cumulative CT (g.hour/m³)	550	260	85
Survivors	*S. oryzae, T. castaneum* (all stages)	*Liposcelis* sp.	*L. serricorne* (all stages)

fumigants and insecticides are repeatedly applied. Intentional and/or ignorant palliative treatments which occur both in commercial treatments and at central/state storages need to be discouraged. It is recommended to give phosphine dosages preferably on volume basis at 1.5 g/m³ for grains (3–4 g/m³ for paddy) and coffee seeds and 2 g/m³ for tobacco bales with a minimum of 7 days exposure period at and above 25°C and 10 days and above at lower temperatures. More than 15 days exposure period will be necessary when *Sitophilus* spp. and *T. granarium* or phosphine-resistant strains are present. Low phosphine dosages of less than 1 g/m³ with still longer exposures are suggested for the buffer stocks of bagged grains and grains in silos. Such long-term phosphine treatment techniques have been attempted and proved effective in Thailand and China (Sukprakarn et al. 1986; Liang Quan 1990). It is unlikely that any major changes in the grain storage system will take place in the near future in India. Therefore, phosphine application techniques have to be improved to avoid repetitive treatments. Sealing materials and methods for cover fumigations need to be evaluated. For instance, at the Indian Grain Storage Institute, Hapur (Sone Lal, personal communication 1993) experiments are in progress on securing the fumigation sheet with rubber piping against a plastic trough embedded all around the bag-stack, as practised in China for improved gastightness during sheeted fumigations (Liang Quan 1990). Technical personnel involved in the preservation of stored products need to be informed/ instructed about the slow action of phosphine on insects, the importance of extended exposure periods and the resistance problem.

References

Banerjee, K. and Deshmukh, P.M. 1992. The fumigation by using aluminium phosphide (phosphine)—a perspective. Pest Management, August 1992.

Banks, H.J. 1984. Fumigation and the properties of gases. In: Champ. B.R. and Highley, E., ed., Proceedings of the Australian Development Assistance Course on the Preservation of Stored Cereals, Vol.2, CSIRO Division of Entomology, Australia, 739–753.

BIS (Bureau of Indian Standards) 1980. Specification for aluminium phosphide formulations. IS: 6438–1980. Bureau of Indian Standards, New Delhi, 10 p.

BIS (Bureau of Indian Standards) 1991. Thermoplastics fumigation covers. IS 13217: 1991. Bureau of Indian Standards, New Delhi, 8 p.

EPPO (European Plant Protection Organization) 1982. EPPO recommendations on fumigation standards. 2nd Edition, EPPO Publication Series, C 65, 15 p.

FAO 1975. Recommended methods for the detection and measurement of resistance of agricultural pests to pesticides. Tentative method for adults of some major beetle pests of stored cereals with methyl bromide and phosphine. FAO method No. 16. FAO Plant Protection Bulletin, 23, 12–25.

Friendship, R. 1989. Fumigation with phosphine under gas-proof sheets. ODNRI Bulletin, No.26, 22 p.

Hole, B.D., Bell, C.H., Mills, K.A. and Goodship, G. 1976. The toxicity of phosphine to all developmental stages of thirteen species of stored product beetles. Journal of Stored Products Research, 12, 235–244.

ITC (India Tobacco Company) 1985. Storage pest control recommendations. India Tobacco Company Limited - ILTD Division, 12 p.

Kashi, K.P. and Muthu, M. 1975. A mixed indicator strip for phosphine detection. Pesticide Science, 6, 511–514.

Kashi, K.P., Muthu, M. and Majumder, S.K. 1977. Rapid evaluation of phosphine permeability through various flexible films and coated fabrics. Pesticide Science, 8, 492–496.

Lallan Rai, Sarid, J.N. and Ramasivan, T. 1964. Fumigation of foodgrains in India with hydrogen phosphide. Series III. Sacked wheat fumigated under rubberised gas-proof-covers. Bulletin of Grain Technology, 2, 75–88.

Liang Quan, 1990. The current status of fumigation and controlled atmosphere storage technologies in China. In: Champ, B.R., Highley, E. and Banks, H.J., ed., Fumigation and Controlled Atmosphere Storage of Grain: proceedings of an international conference, Singapore, 14–18 February 1989. ACIAR Proceedings No. 25, 166–173.

Misra, U.K., Bhargava, S.K., Nag, D., Kidwai, M.M. and Lal, M.M. 1988. Occupational phosphine exposure in Indian Workers. Toxicology Letters, 42, 257–263.

Muthu, M., Majumder, S.K. and Parpia, H.A.B. 1973. Detector for phosphine at permissible levels in air. Journal of Agricultural Food Chemistry, 21, 184–186.

Philip Morris 1991. Managing insect pests of processed tobacco. Philip Morris, Switzerland. PME Method No. 752, 21 p.

Rajen, V.M. 1990. Aluminium phosphide analysis as per IS: 6438–1980 a review. Agrosynth Chemicals, Bangalore. 9 p.

Rajendran, S. 1989. Fumigant resistance: problems and its implications in the control of stored product insects in India. Pesticide Research Journal, 1, 111–115.

Rao, T.D.P. 1992. Tobacco: scope for enhanced quality. The Hindu - Agricultural Survey 1992. 83–85, The Hindu, Madras.

Sadana, B.K., Hira, C.K., Kanwar, J.K., Mann, S.K. and Sharma, K.K. 1988. Effect of income and education on prevalence of grain storage practices in Ludhiana district of Punjab. Bulletin of Grain Technolnology, 26, 191–196.

Shivanna, C.S. 1990. Handling, transport and storage of food grains. In: National symposium on newer dimensions in integrated pest management. Central Food Technological Research Institute, Mysore, 26 April 1990.

Sukprakarn, C., Nilpanit, P., Ataviriyasook, K., Budhasamai, K., Khouchaimala, L., Pvomsatit, B., Annis, P.C. and van S. Graver, J.E. 1986. Phosphine treatment for long term storage of milled rice. In: Naewbanij, J.O. ed., Advances in Grain Post-harvest Technology Generation and Utilization: Proceedings of the 11th ASEAN Technical Seminar on Grain Post-harvest Technology, Kuala Lumpur, Malaysia, August 1988, 221–225.

Taylor, R.W.D. and Halliday, D. 1986. The geographical spread of resistance to phosphine by coleopterous pests of stored products. Proceedings, British Crop Protection Conference—Pests and Diseases, Brighton, 1986. 607–613.

Thakre, B.D., Bansode, P.C. and Sone Lal, 1988. Use pattern of aluminium phosphide as grain fumigant at farm level in Madhya Pradesh. Bulletin of Grain Technology, 26, 220–223.

A new phosphine releasing product

C. Reichmuth*

Abstract

Phosphine (PH_3) for pest control can now be produced from new sachets containing aluminium phosphide. Additional ingredients delay the early generation of the gas for about 30 minutes. This delay allows time for distribution without the risk of inhalation by workers before leaving the treated area.

The new formulation ensures that the product can be safely transported and stored in cans or packages without free PH_3 in the enclosed airspace. Heavy metal containing mixtures to absorb free PH_3 are no longer necessary. The initially sorbed PH_3 is totally evolved later on, when moisture activates the distributed phosphide product and the usual generating reaction begins.

Mixtures of zeolites with magnesium phosphide products may combine the advantage of slow initial release with later accelerated and complete phosphine generation.

The principle chemistry of the new mixture is described and results presented on the differences between phosphine release from various formulations including the new product.

Introduction

Phosphine has been used since the early thirties in various fields of pest control, and particularly for disinfestation of grain in bags or bulk. Millions of tons of grain undergo a treatment each year. Pure phosphine has never been used for large-scale commercial use due to its flammability and the high risk of explosion. Freyberg in Germany 'tamed' this very active gaseous substance by formulating solid compounds containing phosphides and additional compounds from which phosphine is generated very slowly (Anon. 1934). One of the common methods to synthesise phosphine for insect and rodent pest control on sites is by hydrolysis of phosphides like AlP, Mg_3P_2 or Ca_3P_2:

Metal phosphide + Water → Phosphine ↑ + Metal hydroxide (1)

The exothermic reaction (1) may be more or less spontaneous and rapid, depending on the metal in the selected phosphide. Without any modifying agent, the development of phosphine occurs so quickly that within seconds the gas concentration in air locally around the phosphide exceeds the self ignition point of > 2% v/v. In addition to that, the rate of phosphine production is proportional to temperature and the amount of water which has access to the phosphide. These basic rules were known for a long time (Freyberg and Friemel 1965). The keys to the new formulation by Freyberg were the addition of a chemical ('protective compound') to the phosphide and a way of preparing the formulation under pressure to reduce the speed of the reaction and generation of

phosphine so that it was lower than the diffusion of phosphine from the formulation:

[Metal phosphide (57%) + protective compound (43%)) + Water → Phosphine ↑ + Metal hydroxide (2)

The protective compound, which amounts to 43 % by weight in most of the commercial formulations as indicated with formula (2), consists for instance of a mixture of paraffin and ammonium carbamate. This improves the properties of phosphide during manufacture to press powder, tablets or pellets and also develops CO_2 and NH_3 as inerting and warning gases. In 'strips' and 'plates' of Phostoxin[1] other chemicals are included together with magnesium phosphide to reduce the speed of reaction and release of phosphine (2).

Many authors have described dependance of the kinetics of reaction (2) on water, temperature, and the type of formulation (e.g. Banks 1991, Feuersenger 1955, Meuser et al. 1977 a, b, Mori and Kawamoto 1974, Rosebrook 1972).

All commercially available phosphide products have to pass safety examinations for inflammability and speed of reaction. One of the disadvantages of all products still is the phosphine gas content of the cans and canisters containing the formulation. This is set free immediately on opening the packing prior to use. Residual moisture present when the freshly produced formulations are packed into the cans or canisters in the factory is sufficient for production of this phosphine. Additionally, when applying and distributing the products after opening the pack, workers are in danger of inhaling the small amounts of phosphine which are produced at a low rate immediately moist air has access to the formulations.

A recent new approach is the development of formulations which prevent this type of early production of phosphine. This paper describes the advantage of the new Gas Ex-B bag[1] (Kapp and Moog 1991, 1992) in comparison with other normally used bags and pellets.

Materials and Methods

Description of the essential features of the new formulation

Phosphine generation can be prevented by addition of chemicals which tend to react with water before it diffuses into the phosphide. After long search, zeolites have been determined to be sufficiently hygroscopic to trap moisture (3) and prevent the reaction of phosphides to phosphine. Moreover, they are sorptive enough to trap traces of developing phosphine in the packing during the early stage of reaction (4) and (5), until they are saturated with the gas:

[(Metal phosphide + protective compound) + dry Zeolite] + Water → [(Metal phosphide + protective compound) + wet Zeolite] (3)

Instead of reaction (3), another step seems possible and likely to avoid the diffusion of phosphine from the product:

---

* Federal Biological Research Centre for Agriculture and Forestry, Institute for Stored Product Protection Königin-Luise-Straße 19, 14195 Berlin, Germany.

[1] Commercial products of Detia Degesch, Laudenbach, Germany.

[(Metal phosphide + protective compound) + dry Zeolite] + Water → [(Metal phosphide + protective compound) + (dry Zeolite + phosphine) + Metal hydroxide]   (4)

After the saturation of the zeolite with water, the generation of phosphine from the product begins:

[(Metal phosphide + protective compound) + wet Zeolite] + Water → Phosphine ↑ + [Metal hydroxide + wet Zeolite] (5)

The amount of zeolite together with the phosphide formulation in reaction (3), (4), and (5) determines the sorptive capacity for water. Some tiny amounts of phosphine which may develop are immediately sorbed by the zeolite (4). Later on in the process of the reaction, when the zeolite is saturated with water, further moisture diffuses to the phosphide and supports the generation of phosphine as described by reaction (5). When most of the phosphide has been transformed into phosphine and this gas has diffused away, any sorbed phosphine efficiently desorbes again from the zeolite. The new bags are made of Tyvek®, a material which has similar properties like Gore-Tex®, which reduces the speed of water transport to the phosphide formulation compared to the paper bags of the Detia Gas Ex- B product.

Comparison of different formulations

The release rate of different formulations was determined by Reichmuth (1981) by observing the build up of phosphine concentrations in a fumigation chamber. The first 30 minutes after opening the packs were especially investigated. The chamber was kept at constant temperature of 22°C and relative humidity of approximately 62%. Formulation, either in a new bag or an old conventional bag (each with 34 g product) or 56 pellets of Detia Gas Ex-P were exposed in a volume of 2.8 m^3. A number of replicate exposures were carried out.

A Miran infrared analyser operating continuously at a wave length of 9.5 μm was used to record the gas concentration. The analytical parameters were: about 3.5 volume of the cell, an optical path length of 21.75 m, slit 1 mm, response time 40 seconds, with high gain. Gas chromatography (AFID, Noack et al. 1978) was used to determine phosphine concentration for calibrating the Miran.

Results

Figure 1 shows data and fitted curves for release of phosphine from old conventional and new bags of Detia Gas Ex-B. The old bags started much earlier to develop gas though initially at a low rate. After about 15 minutes the rate increased significantly. During the experiments with the new Tyvek-bag, phosphine could be detected for the first time only after 30 minutes (detection limit of about 0.15 mg/m^3). For the regression of the concentration data. A function to describe the increase in concentration with time was fitted to the full set of data (Figure 2). Except for the first few minutes after opening the packings, a log-normal function of the form

$$y = a + b * \exp\{-0.5 * [\ln(x/c)/d]^2\}$$

described the different concentration characteristics very closely ($r^2 = 0.996$ or better).

It became obvious that the zeolites reduce the release rate of phosphine during the first 30 minutes of the initial fumigation period by factor of three or more. The regression curves in Figure 2 demonstrate the pronounced differences of the speed of phosphine generation for these formulations. After several hours, the total amount of released phosphine of the pellets and Tyvek bags was similar, between pellets and Tyvek bags as shown in Figure 4. The inhibition of the initial release rate can clearly be seen. Old and new bags evolved the phosphine markedly more slowly than pellets, with Tyvek bags giving the lowest rate (Figure 4).

In the normal practice of fumigation, the actual concentration of phosphine in air is the critical factor for worker safety. From Figure 5 it can be seen that the phosphine concentration remained in the chamber below 1 ppm for more than 30 minutes, and for 12 minutes below 0.1 ppm, the TLV (Threshold Limit Value) in many countries (Anon. 1988).

Discussion

Even competitors in the market of phosphine releasing products admit that the new Detia-Degesh formulation results in packs which are nearly free of gas on opening. This has been noticed also by the author.

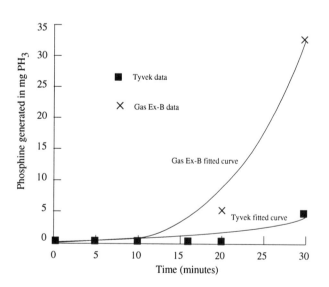

Fig. 1.   Comparison of the phosphine release in mg at 20°C and 63% r.h. in a gas tight fumigation chamber between old conventional and new Detia Gas Ex-B bags for the first 30 minutes after opening the canister

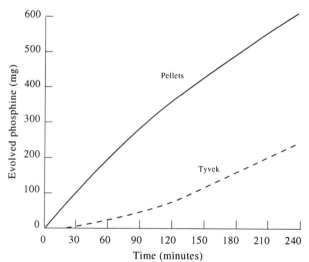

Fig. 2.   Comparison of the generation of phosphine in mg at 20°C and 62.5% r.h. and 64% r.h., respectively, between new Gas Ex-B (Tyvek) bags and Gas Ex-P pellets for 4 hours

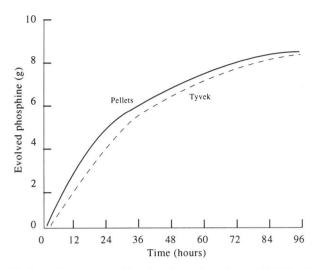

Fig. 3. Comparison of the phosphine release in mg at 20°C and 62.5% r.h. and 64 % r.h., respectively, between new Gas Ex-B (Tyvek) bags and Gas Ex-P pellets for 4 days

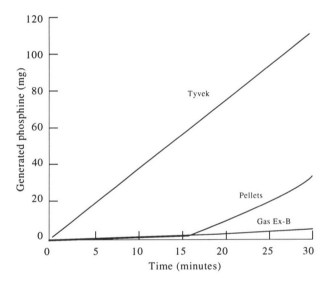

Fig. 4. Initial phosphine release in mg from new and old conventional Gas Ex-B bags and Gas Ex-P pellets within the first 30 minutes at 20°C and 62.5% r.h., and 64% r.h., respectively

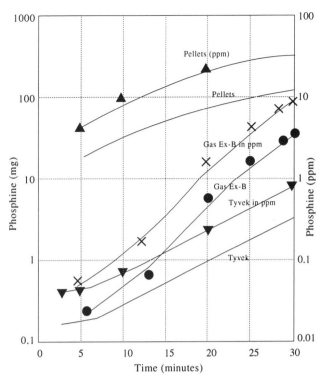

Fig. 5. Build up of phosphine concentrations in ppm from different formulations during the first 30 minutes of exposure

The new formulation tries to improve the idea of a light powder which can safely be transported, distributed and applied before the generation of the effective gas starts by hydrolysis with ambient water vapour, leaving behind harmless residues.

Acknowledgments

The author thanks Miss B. Hennig and Mr. G. Schmidt for splendid technical support. Mr. Schmidt also helped to prepare the graphs. Detia-Degesch supplied the products and gave financial support to Miss Hennig. Dr.C. Adler commented on the manuscript.

References

Anonymous 1934. title German Patent No. 698721 (November 7).

Anonymous 1988. Phosphine and selected metal phosphides. Environmental Health Criteria No. 73. WHO, Geneva, 100 p.

Banks, H.J. 1991. Influence of water and temperature on release of phosphine from aluminium phosphide-containing formulations. Journal of Syored Products Research, 27, 41–56.

Chakrabarti, B., Mills, K.A., Bell, C.H., Wonter-Smith, T. and Clifton, A.L. 1992. In: Fleurat-Lessard, F. and Ducom, P., ed., Proceedings of the Fifth International Working Conference on Stored-product Protection, Bordeaux, France, September 1990, 2, 775–784.

Feuersenger, M. 1955. Lebensmittelhygienische Fragen der Kornkäferbekämpfung mit Phosphorwasserstoff. Deutsche Lebensmittel Rundschau, 51, 293–296.

Freyberg, W. and Friemel, W. 1965. Verfahren zur Bekämpfung von Schädlingen [Procedure for pest control]. German Patent No. 1143053, 3p.

Kapp, W. and Moog, A. 1991. Method and means for preventing or delaying undesired phosphine levels. United States Patent, No. 5,015,475.

Kapp, W. and Moog, A. 1992. Verfahren zur Phosphinregulierung, Schädlingsbekämpfungsmittel und dessen Verwendung [Procedure to regulate phosphine, control pest agent and its application]. European Patent, No. 0 342 471 B, 119 p.

The presented results were obtained at a dosage of 1 bag/ 2.8 m³ at 20°C and 60% r.h. in an empty fumigation chamber. This dosage lies above that usual in many fumigations, but still in the range of some practical applications. The transport of this type of phosphine formulation is safer than of usual products, because of the reduced risk of ignition and explosion in case of any accidental breakage. In many countries and at least in Germany, worker safety plays a predominant role in the transport and use of fumigants in pest control. This convenient new formulation can potentially be dispensed without the risk of inhalation of phosphine for more than half an hour after opening the packs.

Other approaches try to overcome the problem of release of phosphine by applying the gas as a mixture with carbon dioxide from steel cylinders (Chakrabarti et al. 1992, Winks and Ryan 1992). These heavy weights are not easy to handle in many situations. The use of cylinder gas has the advantage of being interruptable at any time but it remains restricted to applicators with a fair degree of engineering skill and needs continuous observation.

Meuser, F., Rajani, C. and Reimers, H. 1977a. Bestimmung der Hydrolysegeschwindigkeit von pelletierten Metallphosphiden zur Getreidebegasung [Determination of the speed of hydrolysis of pelleted metal phosphides for grain fumigation]. Mühle + Mischfuttertechnik, 114, 423–426.

Meuser, F., Rajani, C. and Reimers, H. 1977b. Rate of hydrolysis of pelleted metal phosphides in grain fumigation. Milling Feed and Fertilizer, 160, (10) 15–16, 18, (11) 27–28, 34.

Mori, T. and Kawamoto, N. 1966. Studies on the properties and effect of fumigant, aluminium phosphide. Research Bulletin Japan Plant Protection Service, 12, 28–30.

Reichmuth, Ch. 1981. Inbetriebnahme der Begasungsstation des Instituts für Vorratsschutz der Biologischen Bundesanstalt für Land- und Forstwirtschaft in Berlin-Dahlem [The new fumigation laboratory of the Institute for Stored Product Protection of the Federal Biological Research Centre for Agriculture and Forestry in Berlin-Dahlem]. Nachrichtenblatt des Deutschen Pflanzenschutzdienstes, 33, 161–165.

Reichmuth, Ch. 1992. New techniques in fumigation research today. In: Fleurat-Lessard, F. and Ducom, P., ed., Proceedings of the Fifth International Working Conference on Stored-product Protection, Bordeaux, France, September 1990, 2, 709–725.

Rosebrook, D. 1972. Evaluation of phosphine preparation Detia Gas Ex.-B. Midwest Research Institute Project No. 3502–C. 38 p.

Tateya, A., Saeki, S. and Kawamoto, N. 1974. Effect of temperature and humidity on the decomposition of aluminium phosphides. Research Bulletin Japan Plant Protection Service, 12, 28–30.

Winks, R.G. and Ryan, R. 1992. Recent developments in the fumigation of grain with phosphine. In: Fleurat-Lessard, F. and Ducom, P., ed., Proceedings of the Fifth International Working Conference on Stored-product Protection, Bordeaux, France, September 1990, 2, 935–943.

Uptake of phosphine by stored-product pest insects during fumigation

C. Reichmuth*

Abstract

Phosphine (PH_3) is still the most important fumigant in grain storage world wide. Pest control of insects in large bulks of harvested agricultural products is often carried out with phosphine released from special formulations or even from cylinders.

A better knowledge of the mode of action enables the development of appropriate application of control procedures. In this process, the comparison of the efficacy of exposure to slowly increasing and decreasing phosphine concentrations with constant concentrations led consequently to the development of SIROFLO® by CSIRO Australia.

This study was carried out at the Stored Grain Research Laboratory of the Division of Entomology of CSIRO and describes the uptake of phosphine by *Sitophilus granarius, Tribolium castaneum, Rhyzopertha dominica,* and *Trogoderma variabile* including susceptible as well as phosphine-resistant or diapausing larvae. Two special fumigation chambers with pairs of Geiger tubes were constructed and used to establish uptake simultaneously during exposure to radio labelled PH_3. In addition, undecomposed PH_3 was determined as residue in fumigated insects with a special micro method.

The results show that the linear uptake rate decreases from living but already immobilised insects to deadly poisoned insects. The occurrence of change in slope can be linked to the time of a lethal effect of phosphine. Remarkably, resistant insects take up phosphine at a much lower rate. The total amount of incorporated phosphine leading to death is smaller or equal in resistant insects compared to susceptible strains. But even the resistant insects show a pronounced change in uptake rate after the lethal exposure period. Resistant insects remain active much longer when fumigated (concentration > 0.3 mg PH_3/L) and can clearly be distinguished from susceptible strains which are immobilised after a few minutes.

Introduction

In the last 60 years one chemical stands out for stored-product pest control: hydrogen phosphide or phosphine (PH_3). Experiments have been carried out to understand the mode of action of this fumigant (for example Banks 1975; Price and Mills 1988; Bolter and Chefurka 1990). One part of such experiments in toxicology consists in research in the speed of incorporation and investigations of the metabolic rate at which the poison is metabolised (Bond et al. 1969; Robinson 1969; Price 1984; Price and Mills 1988). This kind of research has been supported by surprising findings (Reichmuth 1986),

that changing concentrations of phosphine, as occurring in practice, are much less effective than constant concentrations at a low level when comparing the *CT*-(concentration × time) products. Winks (1992) transferred this knowledge into the practice of fumigation as the SIROFLO technique, where a constant low amount of phosphine is continuously purged through a bulk of grain in silo bins. How can this property of phosphine of increased efficacy at low concentrations be explained? This study tries to highlight the uptake behaviour of phosphine in treated insects of different species and strains during fumigation at constant and varying concentrations in comparison with chemical reaction of the incorporated gas into other non-gaseous compounds.

Materials and Methods

Preparation of radioactively labelled phosphine

Magnesium powder, 'cold' phosphoric acid and labelled ortho-phosphoric acid, $H_3{}^{32}PO_4$, were mixed together with water and some dilute hydrochloric acid in a clear quartz vessel. The mixture is dried gently under a stream of dry CO_2 over night. The CO_2 also removes residual oxygen. On heating the dry mixture, the exothermic fusion leads to radiolabelled magnesium phosphide. After cooling down, the phosphide is treated with 5% sulphuric acid giving an immediate production of phosphine. The phosphine is collected in a gas burette, with the CO_2 present later being removed by absorption with KOH solution.

Determination of phosphine uptake with radiolabelled gas

Phosphine uptake at 25°C by fumigated insects was determined on line by Geiger counting $^{32}PH_3$ in a specially constructed fumigation cell. The apparatus used for fumigation and radioactivity counting is shown in Figure 1. Insects, in batches of 10, were weighed and narcotised in the refrigerator prior to exposure into the fumigation cell FC_1. The phosphine concentration in the system and the required radioactivity was adjusted before the treatment of the insects started by opening clamp 5 and closing clamps 4 and 6. The fumigation of the insects commenced by opening clamps 4 and 6 and closing clamp 5. The counting device was set to zero. The cpm (counts per minute) values were recorded on paper together with the elapsed time. After correcting these values for the half life of ^{32}P (14.31 days), the uptake in ng PH_3 per 10 insects or per mg of insect could be calculated with the knowledge of the background counts, the phosphine concentration and the volume of the cell.

Determination of uptake by use of a new micro method

The tracer method described above does not distinguish between incorporated $^{32}PH_3$ and metabolised ^{32}P in any other form. Robinson and Bond (1970) have determined the eventual fate of incorporated $^{32}PH_3$ as in excreted hypophosphite

---

* Federal Biological Research Centre for Agriculture and Forestry, Institute for Stored Product Protection Königin-Luise-Straße 19, 14195 Berlin, Germany.

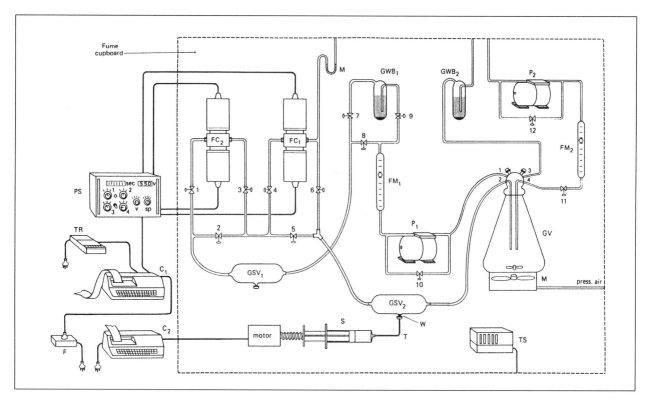

Fig. 1.   Apparatus and installation for monitoring phosphine uptake by insects at constant or varying concentrations: Fumigation chambers FC_1 and FC_2; during most of the experiments, FC_1 contains the insects; power supply (PS) and counting device (computer C_1) for the Geiger tubes; the voltage of the Geiger tube (550 V) can be adjusted with switch V for the 4 tubes 1, 2, 3, and 4; the elapsed time of the experiment is indicated in seconds or minutes depending on the switch between potentiometer 3 and 4; time reset button between potentiometer 1 and 2; sp for loudspeakers linked to the Geiger tubes; Filter F saves the computer from spikes derived from the main voltage supply; pump p_1 circulates the gas; flow adjustment at tube clamp 10, flow indication at flow meter FM_1, humidifying of the gas in gas washing bottle GWB_1 which contains a saturated NaCl solution and some crystalline NaCl at the bottom; regulation with clamps 7, 8, and 9; monitoring of the gas concentration at gas sampling vessel GSV_1 with syringe through the septum of this vessel; clamps 1 to 6 determine the actual path of the gas; manometer M shows the pressure in the system relative to ambient; the septum port in gas sampling bottle GSV_2 can be used to introduce gas slowly from a syringe S and tubing T; the plunger of the syringe is driven by a motor according to gas release characteristics which can be programmed into computer C_2; the tip of the syringe inside GSV_2 is covered with water W to avoid free diffusion of phosphine from the tip into the system; the gas in the gas mixing vessel GV with two septum ports for injection or withdrawing of gas is stirred by a magnet which is driven by another magnet outside on top of a rotor being moved by pressurised air to avoid electric heat transfer; for simulation of leakage port 3 and 4 of GV can be used to withdraw gas with pump p_2, regulated with clamp 12; flow indication at flow meter FM_2; the withdrawn gas is continuously replaced by fresh humidified (gas washing bottle GWB_2) air; the temperature of the installation is regulated constantly to 25±0.3°C using a thermosensor TS which is placed close to the apparatus inside a fume cupboard which is continuously sucking air outside the laboratory to install a negative pressure difference because of safety reasons; the leads of the electrical equipment are lead through holes in the walls of the cupboard.

and polyphosphate. To determine residual phosphine as such in fumigated insects, a micro headspace technique was developed, based on suggestions of Nowicki (1978) for the residue determination in treated foodstuffs. Figure 2 shows the apparatus for determination of residual phosphine released from fumigated insects. The insects are heated with concentrated KOH solution, which destroys the chitin membranes without significantly reacting with phosphine. This heating process transported most of the sorbed phosphine into the free air space of the vessel. After cooling, the released phosphine was determined with gas chromatography. Prior to phosphine determination fumigated insects were stored in liquid nitrogen to avoid losses of phosphine.

Test insects

All insects were taken from cultures of the Stored Grain Research Laboratory (SGRL), CSIRO Division of Entomology, Canberra.

Results

Experiments with radiolabelled phosphine

Figure 3 contains transformed cpm values transformed into phosphine content in ng PH_3/mg insect. *Sitophilus granarius* incorporated phosphine with $^{32}PH_3$ tracer differently at the four investigated concentrations of between 0.1 mg and 1 mg PH_3/L. With ascending concentration, the speed of uptake or the slope of the uptake rate increased correspondingly. The upper curve in Figure 3 starts bending after about 300 minutes of exposure, the second from top after about 400 minutes, the other second change slope within between 500 minutes and 1000 minutes with less pronounced change in rate. These times can be linked to lethal exposure at the given concentration. After about 30 minutes, all 10 weevils were immobilised at 0.95 mg PH_3/L. The place where the slope changed was not correlated with this immobilisation. The final slopes of the uptake curve for 0.53 mg PH_3/L and 0.26 mg PH_3/L do not correspond to the tendency of proportional reduction with

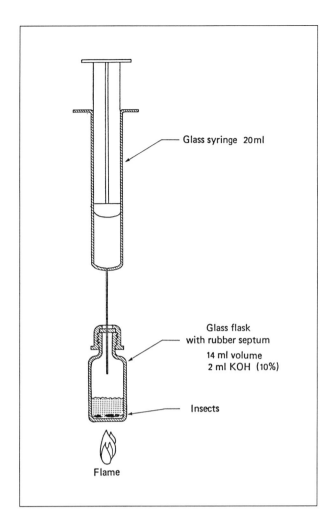

Fig. 2. Gas flask with variable volume: instrumentation to release phosphine from fumigated insects; the movable plunger of the syringe enables the expansion of the air-gas mixture in the glass flask when heated and contraction when cooling down again.

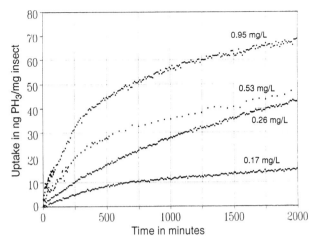

Fig. 3. Uptake in ng phosphine per mg insect of phosphine during 2000 minutes by *Sitophilus granarius* at constant concentrations of 0.17 mg/L (bottom line), 0.26 mg/L (second line from the bottom), 0.53 mg/L (second line from the top), and 0.95 mg/L (top line) at 25°C.

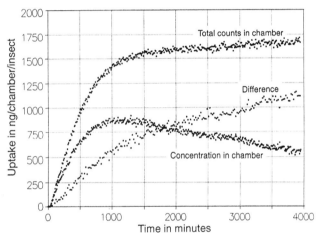

Fig. 4. Concentration of phosphine in ng per chamber (567µl) (line with maximum after 1000 minutes), or as total counts in chamber ng per 10 granary weevils (20.08 mg) (uppermost line). Line starting as the lowest line represents the difference between the two lines as ng per 10 granary weevils.

reduced concentration during exposure. These slopes of lines which result from uptake of mortally poisoned granary weevils are not parallel but fairly constant as are those of the curves during the beginning of exposure.

The response to varying concentrations of phosphine, as occurring in practice, are shown in Figure 4. The curve with the maximum value at 1000 minutes shows the gas concentration as measured in reference cell FC_2 (see Figure 1). The concentration is given in ng PH_3/567 µl, in order to have results comparable directly with uptake, which is given in ng PH_3/10 insects. The maximum concentration reached is 1.6 mg PH_3/L (Figure 5, top graph). The upper curve in Figure 4 contains combined data of concentration counts and uptake counts as measured in fumigation chamber FC_1 (see Figure 1). The subtraction of the concentration counts led to the typical uptake curve for 10 weevils of 20 mg weight in Figure 4 with the change in slope at about 750 minutes of exposure.

The difference in uptake of susceptible and resistant insects is shown for *Rhyzopertha dominica* (Figure 6) and *Tribolium castaneum* (Figure 7) at 1 mg PH_3/L. Even from the raw data in Figure 6, the difference is obvious. As in Figure 7, the resistant insects (TC_4P_{10} from the SGRL) take up only minute amounts of phosphine during exposure compared with susceptible strains (TC_4). Similar results were obtained with susceptible (RD_2) and resistant ($RD_{235}P_{10}$) *Rhyzopertha dominica* (Figure 6).

Whereas the susceptible insects lost weight from 16.89 mg to 14.49 mg (14 %) within 1525 minutes of exposure to phosphine, the weight loss in resistant beetles was determined only from 11.36 mg to 10.56 mg (8 %). This tendency was supported by other experiments.

The investigation of *Trogoderma variabile* (Tv) (strain CTV5 of SGRL) showed that diapausing larvae behave similarly to resistant insects in taking up less phosphine during exposure than nondiapausing Tv (Figure 8). According to Banks (pers. comm. 1984), diapausing larvae are defined here as larvae which have been separated from each other and not turned into pupa or adult during 5-7 weeks of insulation at 25°C. The results were obtained with batches of 5 larvae each and show the pronounced uptake of normal larvae compared to significantly reduced uptake by diapausing individuals.

The irregularities in Figure 8 may have been caused by mechanical disturbances during the experiment.

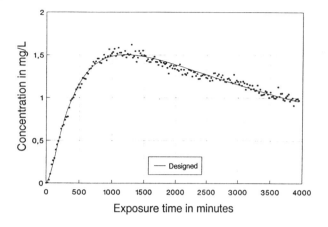

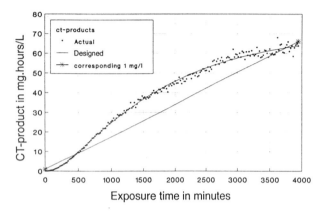

Fig. 5. Concentration profile corresponding to data in Figure 4, (upper graph) *CT*-product obtained under this profile and *CT*-product expected for constant phosphine concentration 1 mg/L as comparison (straight line, lower graph).

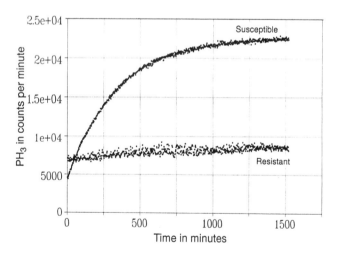

Fig. 6. Relative uptake in counts per minute versus time of susceptible and resistant adult *Rhyzopertha dominica* at 1 mg phosphine/L.

Experiments to determine phosphine micro chemically

After exposure of batches with 200 2–3 weeks old *Sitophilus granarius* to 1 mg PH₃/L, the content of undecomposed phosphine and the increase of difference in weight between untreated and fumigated insects was determined is a function of length of exposure (Figure 9). The two lines show fitted trends. Figure 10 and Figure 11 include results at gradually increasing phosphine concentrations. The concentration

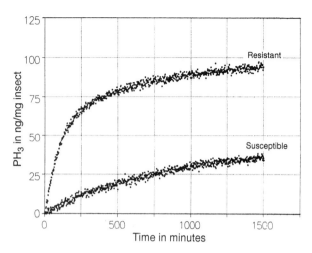

Fig. 7. Incorporation of phosphine in ng/mg insect by susceptible (bottom line) and resistant (top line) adult *Tribolium castaneum* during treatment with 1 mg/L, the amount of incorporated phosphine in susceptible insects after a lethal exposure time of 250 minutes corresponds roughly to the amount of incorporated phosphine in resistant insects after 750 minutes leading to death in this strain, the slope of both lines after 750 minutes is 0.0119 ng PH₃/mg insect/minute (upperline), and 0.0136 ng PH₃/mg insect/minute (lower line), respectively.

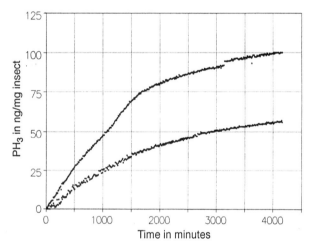

Fig. 8. Phosphine uptake in ng/mg insect of diapausing (lower line) and nondiapausing (upper line) larvae of *Trogoderma variabile* during exposure to 1 mg phosphine.

changes, observed time and percentage of immobilisation, content of phosphine, and mortality are indicated. Phosphine is still being incorporated after the insects have been immobilised and mortally poisoned. With the weight of 1 weevil of about 2 mg, the insects picked up about 70 pg phosphine within 5 hours at the given changing concentrations. There was a steep change in phosphine content in the insects after 2 hours of exposure, when immobilisation and mortal poisoning reached the 50% region (Fig. 11). Immobilisation did not necessarily lead to death at the same percentage as can be seen in Figure 11, where percentage of immobilisation reached about 90% after 3 hours of exposure with the same percentage of mortality requiring 4–5 hours.

Discussion

During the experiments with radiolabelled phosphine in the perspex fumigation chamber, I observed that resistant insects

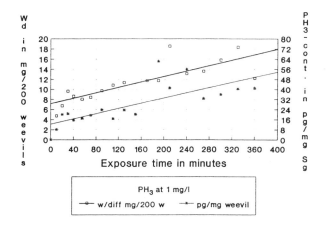

Fig. 9. Content of undecomposed phosphine (lower line) and difference in weight (upper line) following the exposure of *Sitophilus granarius* (14–21-day-old adults) to 1 mg/L at 26°C and 65% r.h.

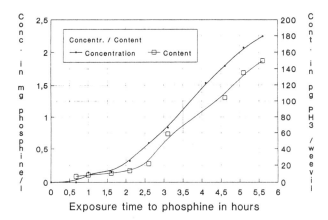

Fig. 10. Content of phosphine following the exposure of *Sitophilus granarius* to slowly rising concentrations at 25°C and 65% r.h.

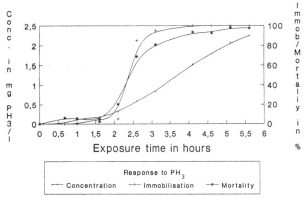

Fig. 11. Mortality and immobilisation in percent following the exposure of *Sitophilus granarius* to slowly rising concentrations at 25°C and 65% r.h.

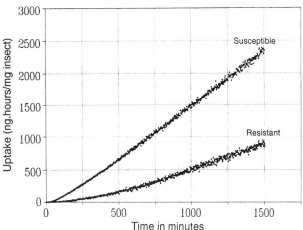

Fig. 12. Integral uptake in ng phosphine * hour per mg insect of resistant (lower line) and non-resistant (upper line) *Tribolium castaneum* at 1 mg/L phosphine.

have the ability to move around in the fumigated cell whereas the susceptible strains are immobilised after less than 1 hour at concentration above 0.5 mg/L. This finding has been developed into a rapid resistance test method (Reichmuth 1991 and 1992). At a given concentration, resistant insects pick up the gas at a rate equivalent to that for susceptible insects at lower phosphine concentrations (Figure 3, between 0.17 mg PH_3/L and 0.26 mg PH_3/L). The comparison of the CT-products for these strains of *Tribolium castaneum* (Winks 1984 and 1986) and the incorporated amounts of phosphine for control of susceptible TC_4 and the resistant TC_4P_{10} leads to the conclusion that resistant insects have at least the sensitivity to incorporated phosphine as susceptible ones. In Figure 7, the change in slope of the phosphine uptake occurs between 200 minutes and 300 minutes. Presumably, the incorporated amount of phosphine at this point is deadly. This can also be seen in Figure 12, which shows the integral of uptake over time for both lines from Figure 7. After 250 minutes, this integral is 250 ngh/mg insect in susceptible *Tribolium castaneum* (Tc). The latter value is reached in resistant Tc after about 750 minutes in Figure 12, again within the range of changing uptake rate in resistant Tc as shown in Figure 7 for the bottom line. The actual amount of incorporated phosphine seems to lead to death in all strains, with resistant or diapausing insects taking it up at a lower rate only.

The new micro method revealed, that, as in the other experiments with radiolabelled phosphine, the time of dramatic changes within the dying insects can be derived from uptake

data. Most of the incorporated phosphine seems to have been transformed into non gaseous compounds as in Robinson (1972), since only pg amounts of actual phosphine per mg insect could be detected (Figures 9 and 10), compared with amounts of some ng/mg insect which were incorporated according to the radiotracer experiments. The magnitude of the metabolic rate and its kinetics can be calculated from these figures.

The results in Figure 7 suggest that the final uptake rate of mortally poisoned insects is similar in resistant and non-resistant insects: 0.0119 ng PH_3/mg insect/minute for susceptible and 0.0135 ng PH_3/mg insect/minute for resistant *Tribolium castaneum* over the last 750 minutes of observation.

Results in Figures 6 and 8 appear to support this finding of constant uptake in dead insects at a given concentration. Figure 3 contains information that this rate may be concentration-dependant. On the other hand, the amount of incorporable phosphine should be limited by the chemistry of the insect body. This is saturated after some time and unable to react with more phosphine, provided that the body does not work like a catalyst for the oxidation of phosphine.

There is now a better understanding possible of the mode of action of phosphine in insect pest control. Phosphine uptake and lethal poisoning, especially in resistant insects, can be linked together. The question arises if, in the future, continuous observation necessary of insects during exposure to insecticides can be used to reduce the number of experiments and the time to determine the lethal conditions for particular fumigants. In the light of reduced funds for research, the technique described here could serve to save time and money in the course of development of a new compound like carbonyl sulphide (COS), for pest control. This material is in discussion as a replacement for the ozone depleting methyl bromide.

Acknowledgments

This study was only possible with the generous support of the German Ministry of Agriculture and CSIRO Division of Entomology. The former heads of the Stored Grain Research Laboratory (SGRL), Drs B. Champ and D. Evans and the former director of the Institute for Stored Product Protection, Dr R. Wohlgemuth, helped to organise my exchange visit with Dr. J. Desmarchelier (SGRL). I am most indebted to Dr R. Winks, who offered long time for scientific discussion and had the initial idea of the fumigation cell for continuous phosphine detection. The method of production of the radiolabelled phosphine was developed by Mr J. McKellar. Together with Mr C. Waterford, it was a pleasure to work in the field of phosphine analytical chemistry and biological efficacy. Miss Julie Gorman carried out the tedious micro analytical work. Her excellent recordings served to write this paper. Drs R. Winks, J. Banks, and J. Desmarchelier encouraged the work with their permanent readiness to discuss and interpret the results. Mrs A. Walton and Mrs S. Allen supplied the susceptible test insects. Drs R. Winks and J. Banks provided the resistant strains and the diapausing larvae. The 'friendly scientific climate' made my work at SGRL a success.

Finally, I thank my assistant, Mrs A. Paul, for patient work with thousands of data on activity counts. This work is not yet over.

References

Banks, H.J. 1975. The toxicity of phosphine to insects. Proceedings of the First International Working Conference on Stored-Product Entomology, Savannah, Georgia, USA, October 1974, 283–296.

Bell, C.H., Hole, B.D. and Evans, P.H. 1977. The occurrence of resistance to phosphine in adult and egg stages of strains of *Rhyzopertha dominica* (F.) (Coleoptera: Bostrichidae). Journal of Stored Products Research, 13, 91–94.

Bolter, C.J. and Chefurka, W. 1990. The effect of phosphine treatment on superoxide dismutase, catalase, and peroxidase in the granary weevil, *Sitophilus granarius*. Pesticide Biochemistry and Physiology, 36, 52–60.

Bond, E.J., Robinson, J.R. and Buckland, C.T. 1969. The toxic action of phosphine — absorption and symptoms of poisoning in insects. Journal of Stored Products Research, 5, 289–298.

Nowicki, T.W. 1978. Gas–liquid chromatography and flame photometric detection of phosphine in wheat. Journal of the Association of Official Analytical Chemists, 61, 829–836.

Price, N.R. 1981. A comparison of the uptake and metabolism of ^{32}P-radiolabelled phosphine in susceptible and resistant strains of the lesser grain borer (*Rhyzopertha dominica*). Comparative Biochemistry and Physiology, C 69, 129–131.

Price, N.R. 1984. Active exclusion of phosphine as a mechanism of resistance in *Rhyzopertha dominica* (F.) (Coleoptera: Bostrychidae). Journal of Stored Products Research, 20, 163–168.

Price, N.R. and Mills, K.A. 1988. The toxicity of phosphine to the immature stages of resistant and susceptible strains of some common stored product beetles, and implications for their control. Journal of Stored Products Research, 24, 51–58.

Reichmuth, Ch. 1986. The significance of changing concentrations in toxicity of phosphine. In: Proceedings of the GASGA Seminar on Fumigation Technology in Developing Countries, Tropical Development and Research Institute, Storage Department, Slough, March 1986, 88–98.

Reichmuth, CH. 1990. Toxic gas treatment responses of insect pests of stored products and impact on the environment. In: Champ, B.R., Highley, E. and Banks, H.J., ed., Fumigation and Controlled Atmosphere Storage of Grain: Proceedings of an international conference, Singapore, February 1989. ACIAR Proceedings No. 25, 56–69.

Reichmuth, Ch. 1991. A quick test to determine phosphine resistance in stored product insects. GASGA Newsletter, 15, 14–15.

Reichmuth, Ch. 1992. New techniques in fumigation research today. Proceedings of the Fifth International Working Conference on Stored-product Protection, Bordeaux, France, September 1990, 2, 709–725.

Reichmuth, Ch. 1992. Schnelltest zur Resistenzbestimmung gegenüber Phosphorwasserstoff bei vorratsschädlichen Insekten [Quicktest to determine phosphine resistance in stored product pest insects]. In: Mossakowski, ed., Proceedings of the European Congress of Entomology, Vienna, April 1991, 245–247.

Robinson, J.R. 1969. 33P: A superior radiotracer for phosphorus? International Journal of Applied Radiation and Isotopes. 20, 531–540.

Robinson, J.R. 1970. The toxic action of phosphine — studies with ^{32}PH$_3$; terminal residues in biological materials. Journal of Stored Products Research, 6, 133–146.

Robinson, J.R. 1972. Residues containing phosphorus following phosphine treatment: measurement by neutron activation. Journal of Stored Products Research, 8, 19–26.

Winks, R.G. 1984. The toxicity of phosphine to adults of *Tribolium castaneum* (Herbst): Time as a dosage factor. Journal of Stored Products Research, 20, 45–56.

Winks, R.G., and Waterford, C.J. 1986. The relationship between concentration and time in the toxicity of phosphine to adults of a resistant strain of *Tribolium castaneum* (Herbst). Journal of Stored Products Research, 22, 85–92.

Winks, R.G., and Ryan, R. 1992. Recent developments in the fumigation of grain with phosphine. Proceedings of the Fifth International Working Conference on Stored–product Protection, Bordeaux, France, September 1990, 2, 935–943.

Carbon dioxide under high pressure of 15 bar and 20 bar to control the eggs of the Indianmeal moth *Plodia interpunctella* (Hübner) (Lepidoptera: Pyralidae) as the most tolerant stage at 25°C

C. Reichmuth and R. Wohlgemuth*

Abstract

The combination of carbon dioxide and high pressure has been described and applied for stored-product pest control over the last ten years. The method requires autoclaves to hold pressures of about 20 to 40 bar. The greatest advantage compared to conventional control measures is the short lethal exposure period, in the range of minutes to a few hours depending on the pest species and stage of development.

The high cost of autoclaving restricts its use to high value products such as cocoa beans, nuts, almonds, drugs and spices, but this method could be adapted for grain in export and quarantine situations.

The present study describes in detail the efficacy of this new method on eggs of the Indianmeal moth *Plodia interpunctella*, the major insect pest in the German food industry. The experiments were carried out in the laboratory in a special chamber of 200 mL volume at controlled temperature of 25°C.

Very young eggs proved to be most tolerant, and 100% mortality was achieved within 40 minutes at 20 bar under carbon dioxide. 1 to 4-day-old eggs required about 20 minutes of exposure to 20 bar for complete kill.

Introduction

In the course of the extended study of the action of 'physically toxic' substances on the granary weevil by Ferguson and Pirie (1948), gases were applied at pressures much above atmospheric to attain the required lethal effects (Ferguson and Hawkins 1949). The authors pointed out very clearly that the actual concentration within the organism is the most important criterion to judge the toxicity of a substance. Most of the poisons proved to have a combined physical and chemical effect which was supported by narcosis of the organism. Unfortunately, the authors did not test carbon dioxide.

Mitsura et al. (1973) described treatments of the grain mite *Tyrophagus putrescentiae* with four different gases. For the first time, carbon dioxide was identified to be lethal within less than 60 minutes of exposure to all investigated stages within a pressure range of 6 to 26 bar. Dinitrogen monoxide, in contrast, caused only incomplete mortality, and hydrogen and nitrogen had very little effect.

The initial approach of Stahl and co-workers (Stahl and Rau 1985; Stahl et al. 1985) to control pest insects in drugs with pressurised carbon dioxide was prompted by the ban of ethylene oxide in Germany. During the search for alternative quick methods of disinfestation, among many gases tested, CO_2 under pressure showed the surprising property to kill all stages of the tested insects within less than 3 hours. Sometimes only minutes were needed. The group worked in the area of extraction of natural products with supercritical fluids and tried to control microorganisms with high pressure treatment (Rau 1985). Therefore, the step to apply carbon dioxide to insects was within range.

In the meantime, this method is common in Germany (Gerard et al. 1988a, b; Anon. 1989; Pohlen et al. 1989; Gerard et al. 1990; Corinth and Reichmuth 1991; Finkenzeller 1991; Reichmuth 1991, 1993; Rau 1993) and is registered as a stored–product protection procedure by several firms[1]. To be precise, it is the carbon dioxide under high pressure which is registered as a substance. Side effects on the quality of the treated products have been investigated and seem to be negligible (Pohlen et al. 1989). Still, the mode of action is not fully understood. It must be linked to the high solubility of carbon dioxide in water which enables a dramatic increase in uptake under high pressure. The question arises as to how much the rapidity of expansion after the pressurisation is responsible for a quick evaporation of the gas from the liquid leading to lesions of cell membranes (diver's disease). Gerard et al. (1988b) determined insect eggs to be the most tolerant stage, which supports the hypothesis: eggs have a low water content and very few cell membranes and are most stable as spheres.

This study reports on experiments on the efficacy of pressurised carbon dioxide against eggs of the Indianmeal moth *Plodia interpunctella*, one of the currently most important pest insects in the German food processing industry. This insect also causes losses in drugs and spices, and leads to various customer complaints.

---

* Federal Biological Research Centre for Agriculture and Forestry, Institute for Stored Product Protection, Königin–Luise–Strasse 19, 14195 Berlin, Germany.

[1] Carbo-kohlensäurewerke, 53557 Bad Hönningen, Germany.
Kohlensäure-werke R. buse GmbH &Co, 53557 Bad Hönningen, Germany.
Sauerstoffwerk F. Guttroff GmbH, 97877 Wertheim-Reichholzhofen, Germany.

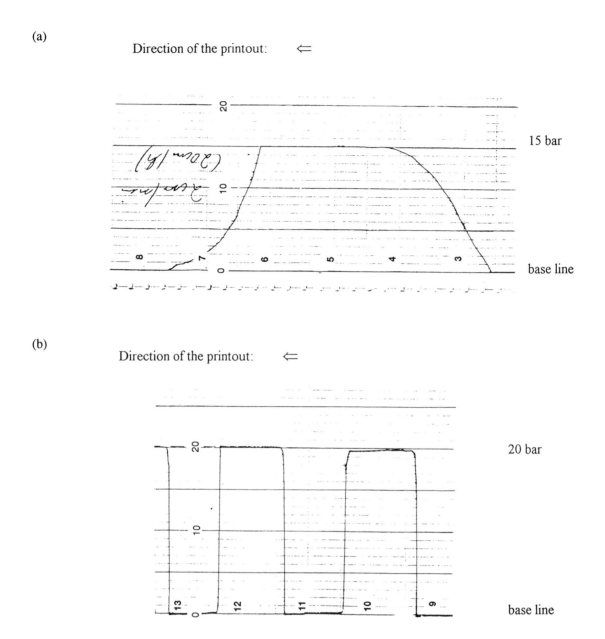

(a)

Direction of the printout: ⇐

15 bar

base line

(b)

Direction of the printout: ⇐

20 bar

base line

Fig. 1. Recorder printout of the speed of pressure build up and decay in the experimental CO_2 fumigation chamber, paper speed: 2 cm/ minute (a) low speed (s=slow) 1 minute, (b) high speed (q=quick) 1 second. Direction of printout ←

Materials and Methods

Insects

The eggs of *Plodia interpunctella* were taken from a strain which has been cultured at 25°C and 75% r.h. in the Institute for Stored Product Protection in Berlin, for more than 30 years. Adult moths laid their eggs during a sharply controlled time span of 22 hours or 2 hours, respectively. Egg age was adjusted by storing the freshly laid eggs for different times at 25°C.

The treated eggs were transferred into small glass rings of 13 mm diameter with nylon gauze-covered bottom and closed by a stop cock. The mesh of 250 µm allowed the hatching larvae to escape into the wheat bran outside the glass ring inside a Petri dish. One week later, dead eggs and husks of eggs which had been left behind by the hatched larvae were counted.

Fumigation procedure

After counting the eggs into small Petri dishes, batches of 4 with 100 eggs each were exposed to the pressurised carbon dioxide in a small laboratory pressure-tight chamber made of brass and of 400 mL volume. Water from a thermostat was pumped through copper tubes around the pressure cell to adjust the temperature at 25°C. The changes of pressure caused an increase to 32°C during pressurising and a decrease to about 18°C during depressurisation. These short-term temperature deviations, however, were readjusted to 25°C within seconds. Since the pure gas was taken from a cylinder, humidity was very low. The speed of pressure build up and decrease could be adjusted. Two types of pressure change were distinguished: quick pressure change within 1 second and slow pressure change within 1 minute to and from the required value (see Fig. 1). In both cases, only the exposure time at the final pressure was recorded. The final pressure and the speed of pressurisation and depressurisation were adjusted manually with valves and regulators before an experiment. By opening and closing of taps after introducing the eggs into the chamber, pressure was adjusted and recorded together with temperature. After the given exposure time aeration was started.

Pressure range tested

Previous experiments have shown that a pressure of 20 bars gives control of most of the insects within less than 2 hours (Prozell and Reichmuth 1990, 1991). Bearing in mind that each further bar requires the addition of an amount of carbon dioxide corresponding to the free volume in the chamber, the cost of treatment escalates rapidly with increasing pressure. To ensure the mechanical stability of the chamber for increasing pressure, the thickness of the walls of the chamber has to be increased. Safety and economy thus restrict the pressure range to less than about 40 bar. If sterilisation is the goal, higher pressures will be needed (Rau 1985). To use the chamber overnight or at weekends requires the determination of the lethal carbon dioxide pressure for longer exposure times.

Pressures of 15 and 20 bar were therefore selected to demonstrate some basic dependencies between exposure time, speed of pressure change, age of eggs, and mortality.

Results

Figure 2 contains information on the influence of the age of *Plodia interpunctella* eggs on the speed of control at 15 bar and 25°C . Quite clearly, the susceptibility is increasing with age from 1 to 4 days. The data in Figure 2(a) and (b) indicate that a slow change in pressure is more effective than the quick change. Because the quick change led to longer lethal exposure times, we identified the quick change as the experimental condition which resulted in the 'safest' determination of the exposure time needed to control eggs at a given age and pressure of CO_2. In other experiments not reported here quick pressure increase and slow pressure decrease, and converse conditions, were tested and found to support the quick change as the least effective (Reichmuth and Wohlgemuth, unpublished data). Comparing Figure 2 with corresponding data in Figure 3 shows the reduction in lethal exposure time from 30 to 10 minutes by pressure increase from 15 to 20 bar.

The mortality data in Figure 2 show pronounced scattering. This tendency was confirmed by data at 20 bar (Figure 3). Experiments were therefore continued at 15 bar with more precise differentiation between the ages. Young eggs had proven to be most tolerant. Accordingly, further treatments were carried out with 0–2, 2–4, 4–6, 6–8, and 22–24-hour-old eggs (see Figure 4). The scatter in mortality results was much reduced. Complete mortality of all tested eggs with an age of less than 8 hours occurred within 35 minutes of exposure. The acceleration of the control procedure by slower pressure changes can again be seen in Figure 4. Differences between the responses of these four different aged eggs were not very pronounced. Only the slightly older eggs in Figure 4(e) seemed to be more susceptible than the other tested young eggs. But they were more tolerant than the 1-day-old eggs in Figure 2(a).

Discussion

The egg stage of *Plodia interpunctella* is relatively tolerant to treatment with carbon dioxide under pressure compared with other stages (Gerard et al. 1988b). By differentiating the age of the eggs stepwise down to a Δ t of 2 hours, it could be shown that eggs less than 1 day old can survive exposure to CO_2

pressure of 15 bar for up to 40 minutes. The strongly changing susceptibility with age seems to be the reason for pronounced variations in response of eggs if they are not within well-defined age groups. If very young eggs are not included in the experiments to determine the necessary lethal exposure time, a recommendation of too-short exposure periods may result. The lethal effect is strongly dependent on the length of exposure and not on the speed of depressurisation. On the contrary, when pressure is slowly increased or decreased, the exposure to these slowly changing CO_2 pressures, increased above atmospheric, has already toxic effects. These seem to add to the effects caused at the final experimental pressure and reduce the lethal exposure period, compared with exposures when the pressure is very quickly adjusted within seconds. Temperature was not tested in this context, but it may play a role (Prozell and Reichmuth 1991). This supports the presumption that solubility in liquids, transport and reaction of carbon dioxide, and number of membranes which are temperature dependent, will all have a strong effect on the efficacy of pressurised CO_2 on insects.

References

Anon. 1989. Pest control with carbon dioxide. Food Technology, 52.

Anon. 1989. Schädlingsbekämpfung mit Kohlendioxid [Pest control with carbon dioxide]. Der praktische Schädlingsbekämpfer, 41, 200.

Corinth, H.–G. und Reichmuth, Ch. 1991. Verfahren und Einrichtung zum Entwesen von organischem Schüttgut [Procedure and installation to disinfest organic bulk material]. Patent of the Federal Republik of Germany, No. 39 30 470 [1992, new No. 04 17 430].

Ferguson, J. and Hawkins, S.W. 1949. Toxic action of some simple gases at high pressure. Nature, 164, 963–964.

Ferguson, J. and Pirie, H. 1948. The toxicity of vapours to the grain weevil. Annals of Applied Biology, 35, 532–550.

Finkenzeller, E. 1991. Verfahren und Einrichtung zum Entwesen von organischem Gut [Procedure and installation to disinfest organic produce]. European patent, No. 0 458 359 A 1, 6p.

Gerard, D., Kraus, J. und Quirin, K.–W. 1988a. Rückstandsfreie Druckentwesung mit natürlicher Kohlensäure [Residue free pressure disinfestation with natural carbon dioxide]. Gordian, 88, 90–94.

Gerard, D., Kraus, J., Quirin, K.–W. und Wohlgemuth, R. 1988b. Anwendung von Kohlendioxid (CO_2) unter Druck zur Bekämpfung vorratsschädlicher Insekten und Milben [Use of carbon dioxide (CO_2) to control stored product pest insects and mites]. Pharmazeutische Industrie, 50, 1298–1300.

Gerard, D., Kraus, J., Fröhlingsdorf, C.J. und Dallüge, A. 1990. Rückstandsfreier Vorratsschutz für Arznei–und Teedrogen [Residue free stored product protection for medical and tea drugs]. Deutsche Apotheker Zeitung, 130, 2014–2018.

Mitsura, A., Amano, R., and Tanabe, H. 1973. The acaricidal effects of compressed gas treatments on the grain mite *Tyrophagus putrescentiae*. Shokuhin Eisagaki–zasski, 14, 511–516.

Pohlen, W., Rau, G. und Finkenzeller, F. 1989. Erste praktische Erfahrungen mit einem Verfahren zur Druckentwesung mit Kohlendioxid [First practical experiences with a procedure to disinfest with pressurised carbon dioxide]. Pharmazeutische Industrie, 51, 917–918.

Prozell, S. und Reichmuth, Ch. 1990. Wirkung von Kohlendioxid unter Hochdruck auf den Kornkäfer *Sitophilus granarius* (L.) [Efficacy of carbon dioxide under pressure on the granary weevil *Sitophilus granarius* (L.)]. Mitteilungen der Deutschen Phytomedizinischen Gesellschaft, 20, 14.

Prozell, S. and Reichmuth, Ch. 1991. Response of the granary weevil *Sitophilus granarius* (L.) (Col.: Curculionidae) to controlled atmospheres under high pressure. In: Fleurat–Lessard, F., and Ducom, P., ed., Proceedings of the Fifth International Working Conference on Stored–product Protection, Bordeaux, France, September 1990, 2, 911–918.

Rau, G. 1993. Alternative Verfahren mit Gasen im Vorratsschutz [Alternative measures with inert atmospheres in stored product protection]. Deutsch Lebensmittel–Rundschau, 89, 216–219.

(a)

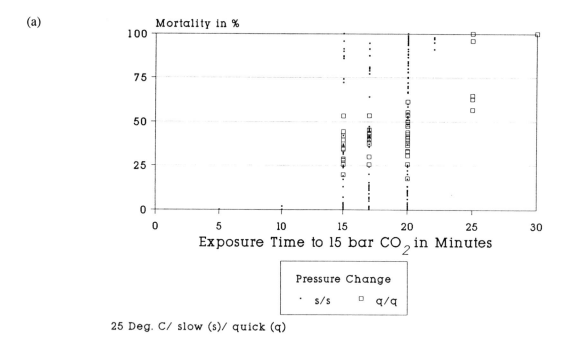

25 Deg. C/ slow (s)/ quick (q)

(b)

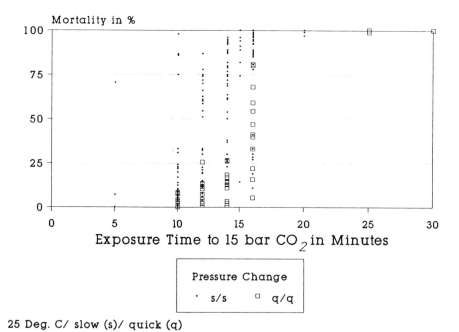

25 Deg. C/ slow (s)/ quick (q)

Fig. 2. Mortality results for the treatment of (a) 1 day- and (b) 2-day-old eggs of the Indianmeal moth *Plodia interpunctella* with 15 bar carbon dioxide at 25°C.

(c)

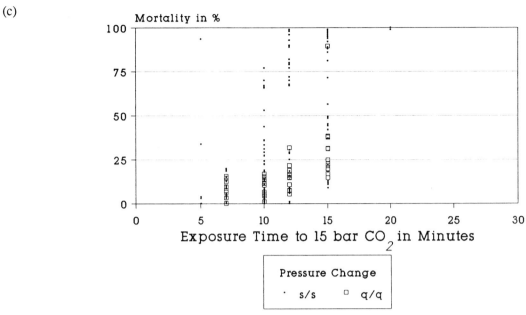

25 Deg. C/ slow (s)/ quick (q)

(d)

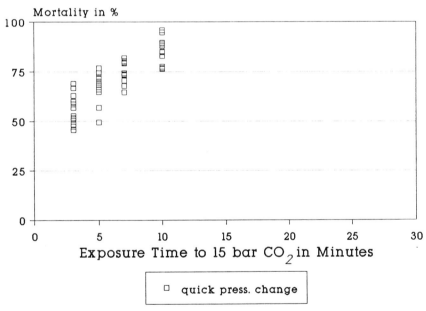

25 Deg. C

Fig. 2 – cont'd . Mortality results for the treatment of (c) 3 day- and (d) 4-day-old eggs of the Indianmeal moth *Plodia interpunctella* with 15 bar carbon dioxide at 25°C.

(a)

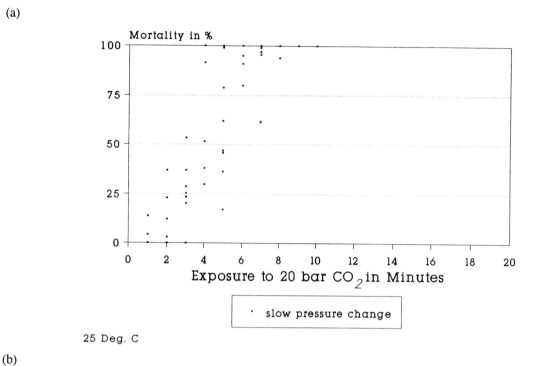

25 Deg. C

(b)

Fig. 3.   Mortality results for the treatment of (a) 1 day- and (b) 2-day-old eggs of the Indianmeal moth *Plodia interpunctella* with 20 bar carbon dioxide at 25°C.

(c)

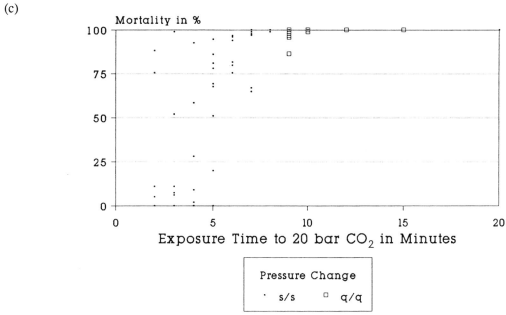

25 Deg. C/ slow (s)/ quick (q)

(d)

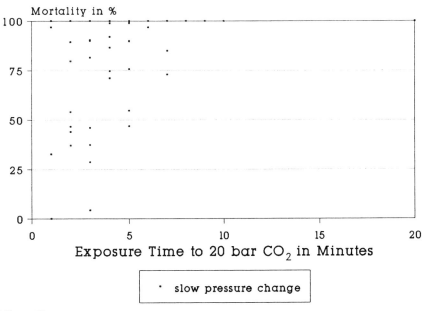

25 Deg. C

Fig. 3. – cont'd. Mortality results for the treatment of (c) 3 day- and (d) 4-day-old eggs of the Indianmeal moth *Plodia interpunctella* with 20 bar carbon dioxide at 25°C.

(a)

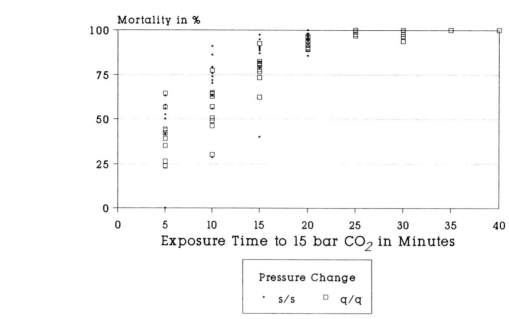

25 Deg. C/ slow (s)/ quick (q)

(b)

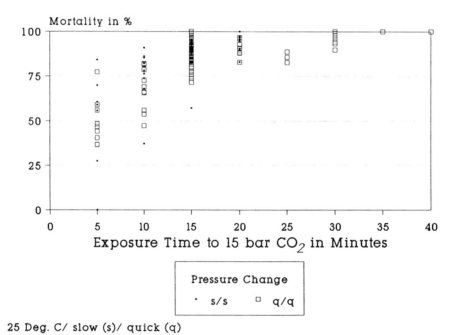

25 Deg. C/ slow (s)/ quick (q)

Fig. 4. Mortality results for the treatment of (a) 0–2 hour- and (b) 2–4 hour-old eggs of the Indianmeal moth *Plodia interpunctella* with 15 bar carbon dioxide at 25°C.

(c)

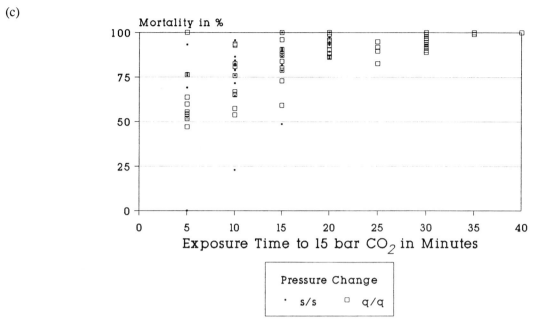

25 Deg. C/ slow (s)/ quick (q)

(d)

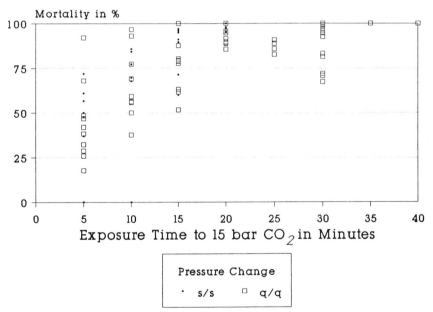

25 Deg. C/ slow (s)/ quick (q)

Fig. 4. – cont'd. Mortality results for the treatment of (c) 4–6 hour- and (d) 6–8 hour-old eggs of the Indianmeal moth *Plodia interpunctella* with 15 bar carbon dioxide at 25°C

(e)

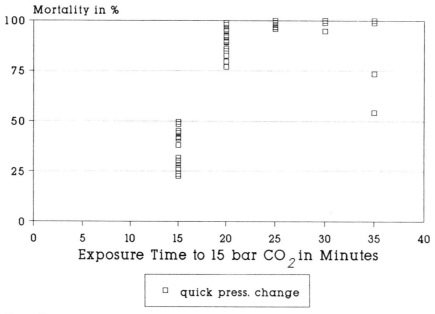

25 Deg. C

Fig. 4. – cont'd. Mortality results for the treatment of (e) 22–24-hour-old eggs of the Indianmeal moth *Plodia interpunctella* with 15 bar carbon dioxide at 25°C.

Rau, G. 1985. Die Anwendung von verdichtetem Kohlendioxid zur Qualitätsverbesserung von Drogen [Application of pressurised carbon dioxide to improve the quality of drugs]. Ph.D. Thesis, University of the Saarland, Saarbrücken, Germany, 100 p.

Reichmuth, Ch. 1991. New techniques in fumigation research today. In: Fleurat–Lessard, F., and Ducom, P., eds., Proceedings of the Fifth International Working Conference on Stored–product Protection, Bordeaux, France, September 1990, 2, 709–725.

Reichmuth, Ch. 1993. Vorratschutz: Entwesen mit Kohlendioxid [Stored product protection: disinfestation with carbon dioxide]. Die Mühle + Mischfuttertechnik, 130, 667–671.

Stahl, E. und Rau, G. 1985. Neues Verfahren zur Entwesung [A new method for disinfestation]. Anz. Schädlingskde., Pflanzenschutz, Umweltschutz, 58, 133–136.

Stahl, E., Rau, G. und Adolphi, H. 1985. Entwesung von Drogen durch Kohlendioxid–Druckbehandlung (PEX–Verfahren) [Disinfestation of drugs by pressure treatment with carbon dioxide (PEX procedure)]. Pharmazeutische Industrie, 47, 528–530.

Studies on the effect of carbon dioxide in insect treatment with phosphine

Y.L. Ren, *† I.G. O'Brien* and C.P. Whittle†

Abstract

Studies on the toxicity of mixtures of phosphine and carbon dioxide to adults of *Cryptolestes turcicus* (Grouvelle) showed that carbon dioxide enhanced the toxicity of phosphine. Studies on respiration of the tested insects showed an increase in phosphine consumption when carbon dioxide level increased. The optimum concentration of carbon dioxide for synergy with phosphine was in the range of 5–35% (v/v). The experiment also shows that the action of phosphine is stimulated through carbon dioxide stimulating respiration. However, once the concentration of carbon dioxide is above 35%, the toxicity of phosphine is gradually decreased because of the effect of narcosis.

Introduction

Phosphine has been widely used to control stored-grain insects. It is highly toxic to insects, relatively easy to apply, not appreciably phytotoxic, does not taint most commodities and leaves little residue. Phosphine plays an important role in integrated pest management, and it is still one of the most important methods, not only in eradication, but also, in China, in quarantine treatment. However, phosphine resistance of the lesser grain borer (*Rhyzopertha dominica* (F.)) and the flat grain beetle (*Cryptolestes turcicus* (Grouvelle)) has occurred in China (Table 1) and for some strains the resistance is extremely high. In fact it is not possible to determine LD50 values for these strains by fixed time (20-hour) exposure methods (e.g. Anon. 1975). Significant resistance to phosphine has been noted previously elsewhere (e.g. Tyler et al. 1983).

The present study was initiated to investigate possible methods for enhancing the efficacy of phosphine fumigations and, in particular, to study the effect of carbon dioxide on phosphine toxicity and the relationship between toxicity, respiration and fumigant uptake. *C. turcicus*, an important pest in China, was chosen as a representative insect.

The normal gases in the insect environment have some influence on the toxicity of fumigants to insects. Various techniques have been employed to alter the normal gases of the atmosphere to enhance the effectiveness of fumigants. In vacuum fumigation fumigants are found to penetrate commodities faster and to be more toxic to insects than in the normal atmosphere (Bond and Monro 1967). Carbon dioxide is often added with some fumigants to reduce flammability and explosive hazard and may increase the toxic effect of certain fumigants (Cotton and Young 1929). Carbon dioxide has also been used as a carrier to enable phosphine to penetrate to the bottom of the silo (Leesch 1990). Friedlander and Navarro (1979, 1983, 1984) have reported that carbon dioxide appears to have an effect on the metabolism of stored grain insects when under higher concentration. Liang Quan (1980) found that co-application in carbon dioxide increases the susceptibility of five strains of grain beetles to phosphine. Qou Shi-Jie (1980) reported that carbon dioxide potentiates the toxicity of phosphine against adults of *Tribolium castaneum* and *Sitophilus oryzae* at low concentration, e.g. at 30°C with carbon dioxide above 11.8% (v/v), 0.009 mg (PH_3)/L and an exposure time of 72 hours, all the stored grain insects could be killed, but no quantitative comparison was made with the phosphine without CO_2. Cotton (1930) investigated the use of carbon dioxide with fumigants and concluded that it might increase the toxic effects of fumigants by satisfying the sorptive properties in commodities when it is used with the fumigants. Cotton and Young (1929) and Cotton (1932) studied the uses of carbon dioxide to stimulate respiration which resulted in the increased toxicity of some fumigants. Jones (1938) and Aliniazee and Lindgren (1969) were able to show that the concentration of carbon dioxide was a factor in increasing the toxic effect of methyl bromide and other fumigants to the red flour beetle, *Tribolium castaneum* (Herbst). Kashi and Bond (1975) showed that concentrations of carbon dioxide ranging from 1 to 50% potentiated the toxic effects of phosphine on *Sitophilus granarius* (L.) and *Tribolium confusum* du Val. Desmarchelier (1984) showed that 25% carbon dioxide increased the toxicity of phosphine to some stored-product insects but not to the most tolerant stages. Rajendran and Muthu (1989) showed that carbon dioxide increased the toxicity of phosphine to adults of *S. granarius* and *T. confusum*. All of these studies have concentrated on the effects of carbon dioxide on the toxicity of fumigants to insects, largely in a qualitative sense.

Materials and Methods

Phosphine generation

Carbon dioxide and phosphine were generated and mixed in a fumehood in the device shown in Figure 1. The required mixture of gases was produced in the 500 mL fumigation chamber (B).

Phosphine was generated by mixing water and aluminium phosphide (E)

$$AlP + 3HOH \rightarrow Al(OH)_3 + PH_3$$

The phosphine–air mixture in generator (E) was transferred to the storage chamber (C) under pressure from (E), and saturated sodium chloride in (C) was forced into balance bottle (D2). The concentration of PH_3 for each batch was measured by gas chromatography.

* Faculty of Applied Science, University of Canberra, P.O. Box 1, Belconnen ACT 2616, Australia.
† Stored Grain Research Laboratory, Division of Entomology, CSIRO, GPO Box 1700, Canberra ACT 2601, Australia.

Preparation of apparatus

Test insects (20 mg, about 200 insects) were put into the fumigation chamber (B), stoppered and connected (J) to the rest of the apparatus. PH_3 was passed into (A), flushing tube 4, and discarded. In the same way, CO_2 (F) was used to flush tube 3, and discarded.

Mixing the gases

Because the required concentrations of CO_2 vary widely from 0 to 60% , and in order not to affect the test insects during the gas mixing operation, we employed a method of gas replacement under slight negative pressure by using the 500 mL hand pump (H). That is, with the tubes 2 and 5 connected, the cock (b) was opened and the pressure in the fumigation chamber (B) was reduced by not less than 76 mmHg, as measured by (I). CO_2 was then admitted to (B) by connecting tubes 3 and 5. This operation was repeated the required number of times (Table 2) to obtain approximately the necessary concentration of CO_2.

The data in Table 2 show the approximate relationship between the number of pump strokes and the final CO_2 concentration. The calculated volume of phosphine gas mixture for (B) is taken into (A) and CO_2 is added to achieve the required concentration. For example, when the required concentration of CO_2 is 30% (v/v), as indicated in Table 2, it is necessary to replace gases three times. The concentration of CO_2 was measured in the fumigation chamber (B) through (J), and according to the difference between target concentration and the determined value of CO_2 a calculated amount of CO_2 was added from the gas burette (A) together with the calcu-

lated required amount of PH_3. The mixed gases in the gas burette were admitted to the fumigation chamber (B) after one further pump stroke of equal volume. Cock (a) was then opened briefly to allow air to enter and restore atmospheric pressure in chamber (B). The final concentrations of CO_2 and PH_3 are determined by GC analysis and adjusted if necessary.

Determination of gas concentrations

Phosphine concentrations were determined using a Shimadzu instrument fitted with a flame photometric detector (FPD). Operating conditions were: 3 mm × 3 m glass column packed with 5% SE-30 (w/w) on 80–100 mesh Chromosorb 105, column temperature, 100°C; injection port and detector block temperatures, 110°C and 130°C. An Aerograph A-110-C model gas chromatograph was used for determination of carbon dioxide and oxygen. This instrument was equipped with a thermal conductivity detector (TCD) using a 4-filament hot wire with 3 mm × 1.5m stainless steel column packed with silica gel (100 mesh).

Treatment of insects

A phosphine-susceptible strain of flat grain beetle *C. turcicus* was collected in July 1984 from a grain storage in Sichuan province, China and cultured in the laboratory. Insects were reared at 30 ±1°C, 75 ± 2% r.h. Upon emergence, adults were transferred to 25°C, 70% r.h. and held at these conditions for at least 1 week before being used in experiments. All subsequent stages were conducted at 25°C and 70% r.h.

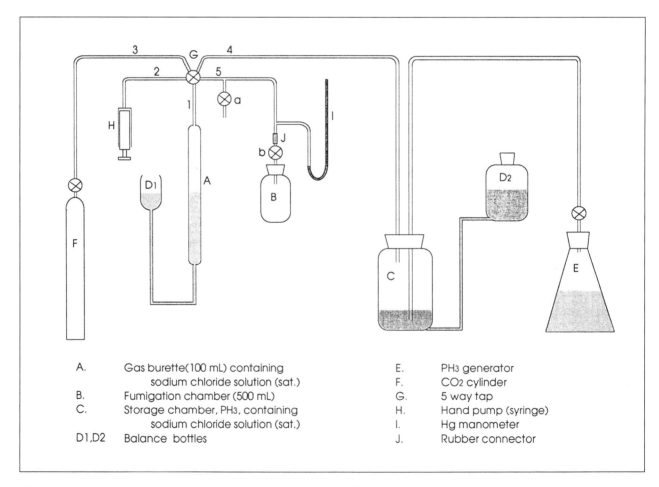

A.	Gas burette(100 mL) containing sodium chloride solution (sat.)	E.	PH3 generator
B.	Fumigation chamber (500 mL)	F.	CO2 cylinder
C.	Storage chamber, PH3, containing sodium chloride solution (sat.)	G.	5 way tap
		H.	Hand pump (syringe)
D1,D2	Balance bottles	I.	Hg manometer
		J.	Rubber connector

Fig. 1.   Apparatus for generating and mixing phosphine with carbon dioxide.

Batches of adults were exposed to each dosage (concentration × treatment time, mg.hour/L) and then transferred to petri dishes containing a thin layer of wholemeal wheat flour and stored under normal atmosphere. The end-point mortalities for individual treatment concentrations were determined over 14 days following termination of the exposure (Anon. 1989).

Measuring toxicity of phosphine under different carbon dioxide levels

Four batches of 50 adults were treated at each concentration of phosphine in 500 mL bottles under different concentrations of carbon dioxide (normal air, 5, 10, 15, 30, 35, 45 and 60%) for a 24-hour exposure period, following which they were transferred to petri dishes for the determination of mortalities. The phosphine concentrations used were 0.02, 0.05, 0.1, 0.2 and 0.5 mg/L. Data were analysed by probit analysis (Finney 1952). Standard errors of the slopes were in the range 0.16–0.74.

Measuring toxicity of CO_2 and of PH_3 in CO_2

Four batches of 50 adults were treated at each concentration of carbon dioxide in 500 mL bottles under different concentrations of phosphine for a 24-hour exposure period, following which they were transferred to petri dishes for the determination of mortalities. The carbon dioxide concentrations were: normal air, 3, 5, 10, 15, 19, 25, 28, 38, 50 and 58% (v/v); phosphine concentrations were 0, 0.071 and 0.104 mg/L for each carbon dioxide level.

Measuring oxygen consumption and intake of phosphine

The method used to determine oxygen consumption by *C. turcicus* exposed to carbon dioxide has been described by Aliniazee (1971). Three batches (20 mg each) of adults were treated in a 250 mL bottle at each concentration of carbon dioxide (normal atmosphere and 10, 20, 30, 40, 50, and 60%) and the same concentration of phosphine (0.104 mg/L) was added to each bottle. Gas samples were taken after 48 hours and the oxygen and phosphine levels were determined by gas chromatography. The consumptions were determined as O_2 mg/g(insect) and PH_3 μg/g(insect) per 24-hour period.

Results

The effect of carbon dioxide on intake of phosphine and oxygen by insects

Oxygen consumption in an atmosphere of 0.1 mg/L phosphine was found to increase with CO_2 concentration, and when insects were exposed to greater than 20% (v/v) carbon dioxide, oxygen consumption was more than double the normal consumption. The intake of phosphine by the insects was also more than double at that concentration of CO_2 (Fig. 2). However, with further increases in concentration of carbon dioxide, phosphine taken up by insects was increased slightly, but oxygen consumption was decreased in concentrations of CO_2 above 30% (see Fig. 2).

The effect of carbon dioxide on phosphine toxicity

We confirmed that phosphine toxicity to *C. turcicus* was influenced by the concentration of carbon dioxide used in the experiments. Phosphine toxicity increased with increasing concentration of carbon dioxide but decreased beyond 35% CO_2. The LC_{90} was achieved with lower concentrations of

phosphine as the CO_2 concentration increased. It was found that the concentration range of carbon dioxide for maximum enhancement of phosphine toxicity is 15–35% (see Fig. 3).

The mortality curve of *C. turcicus* is shown in Figure 4. Two concentrations of phosphine were used in the treatment of adult insects over a wide range of CO_2 concentrations (0–60%). We obtained a typical dosage–mortality curve under constant concentration of phosphine and increasing concentrations of carbon dioxide, and though carbon dioxide is

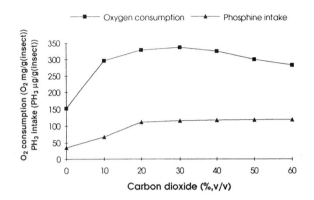

Fig. 2. Oxygen consumption and intake of phosphine (dosed at 0.1040.002 mg/L) by *C. turcicus* under carbon dioxide over 24 hours.

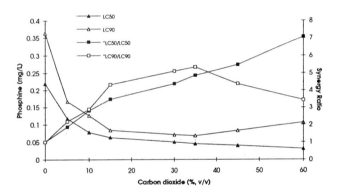

Fig. 3. The effect of carbon dioxide on the toxicity of phosphine (24 hours exposure). Toxicities (LC_{50}▲ and LC_{90}△) and Synergy ratios (*LC_{50}/LC_{50} ■ and *LC_{90}/LC_{90} □) vs CO_2 concentration. *LC_{50} and *LC_{90} values were determined in air, LC_{50} and LC_{90} with CO_2.

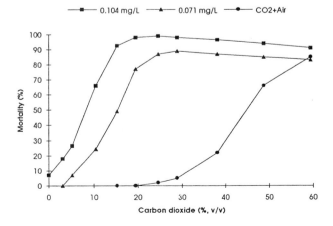

Fig. 4. Toxic effect of phosphine on *C. turcicus* adults (24 hours exposure) under different concentrations of carbon dioxide.

Table 1. Toxicity of phosphine to four representative resistant strains of *Rhyzopertha dominica* and *Cryptolestes turcicus*[a]

Strain[b]	Site of collection	Slope ± S.E.[c]	LD$_{50}$(95%CL) mg/L 20 hours, 25°C[c]	Resistance ratio
RdS	Susceptible strain	2.30 ± 0.63	0.008 (0.006–0.010)	1
Rd2	Nanping, Zhuhai	1.90 ± 0.12	0.141 (0.139–0.143)	18
Rd3	Nanping, Zhuhai	1.82 ± 0.04	0.068 (0.066–0.070)	8
Rd8	Guxiang, Chaozhou	2.26 ± 0.39	4.852 (4.850–4.854)	606
Rd10	Hudong, Lufeng	1.37 ± 0.08	0.182 (0.180–0.184)	23
CT	Hunan, Xiangtan	–	–	46

[a] Data from Guangdong Grain Storage Research Institute and determined by FAO, 20-hour exposure method.

[b] Rd = *R. dominica*, CT = *C. turcicuS*

[c] Plotted as per Finney (1952).

Table 2. The relationship between pump strokes and concentration of CO_2.

No. pump strokes	1	2	3	4	5	6	7	8
Concentration (%) of CO_2 (v/v)	10	19	27	34	41	47	52	57

clearly a mild fumigant (Fig. 4) it is also clear that the effect of carbon dioxide on the toxicity of phosphine is an apparent synergistic effect.

Discussion

As stated above, some researchers have reported that carbon dioxide can effectively increase toxicity of the fumigant PH$_3$ and, although they have discussed its possible action, a satisfactory explanation has not been provided. Nor has the lower efficacy in higher concentrations of CO_2 been shown previously. As can be seen in comparing Figures 2 and 4, there is a relationship between toxicity of phosphine and respiration of insects under different concentrations of carbon dioxide, both showing a maximum at 25–30% CO_2 for *C. turcicus* exposed to 0.1 mg/L phosphine. Comparison of mortalities in Figure 4 at two levels of phosphine show that while the optimum CO_2 concentration to produce a maximal kill (LC$_{90}$) at 0.07 mg/L is 30% (±5%), the optimum CO_2 for 0.1 mg/L phosphine (LC$_{99}$) is 20% (±5%). In CO_2 atmospheres above 30%, phosphine is less effective. It can be reasoned that the increase in phosphine toxicity with carbon dioxide is due to stimulation of insect respiration, as shown by increased oxygen consumption, causing increased phosphine interaction at the target site.

It is well known that the toxicity of phosphine to insects is dependent on respiration or oxygen consumption (Bond and Monro 1967). For example phosphine is not toxic to weevils treated at °C or in 100% nitrogen, i.e., under conditions where the weevils do not respire (Sato and Suwanai 1973). In this respect the observation of increasing oxygen consumption with increasing carbon dioxide concentration (Fig. 2) was not unexpected. However, the fact that the oxygen consumption peaks at about 30% CO_2 and the phosphine intake only increases slightly above 20% CO_2 requires explanation.

From Figure 4 it can be seen that at about 30% and above, CO_2 alone has an increasing lethal effect. It may be postulated that, although CO_2 causes an increase in oxygen consumption by stimulating respiration, above 30% the toxic effect causes a consequent lowering of oxygen consumption. The phosphine intake increases with CO_2 concentration (and hence with oxygen consumption) (Fig. 2) until the 20% CO_2 level, where the intake steadies at about 100 µg/g(insect) over 24 hours. Further increases in CO_2 afford only slight increases in phosphine uptake rate. It seems likely that the ability of the insect to absorb and detoxify phosphine is saturated at about

this level. Thus, phosphine would be expected to be most toxic with concentrations of CO_2 between 20 and 30%. In fact this is reflected in the LC$_{90}$ synergy ratio in Fig. 3 which shows that concentrations of CO_2 between 15 and 35% produce the greatest enhancement of the phosphine toxicity. From Figure 3 it can also be seen that although the LC$_{50}$ continues to fall with increasing CO_2 concentration, the LC$_{90}$ reaches a minimum at about 35%. This indicates a change in slope of the probit mortality–dosage lines with changing CO_2, possibly resulting from heterogeneity in the population of *C. turcicus* used in these experiments.

Some researchers have suggested that higher CO_2 concentrations may also assist fumigant intake by keeping the spiracles open (Bond and Monro 1967). On the other hand many compounds can penetrate the integument independently of the spiracles and the role of either spiracles or integument with respect to the toxicity of phosphine remains uncertain.

There is no apparent correlation between phosphine uptake and oxygen consumption above 30% CO_2. Increased uptake, even at a lower respiration rate, indicates that the mechanism of phosphine uptake by insects depends not only on respiration, but also on factors such as diffusion, and binding to components of insect tissue.

From these studies on the effect of carbon dioxide on the toxicity of phosphine, it is clear that there is an advantage in adding 15–35% CO_2. In fact for an LC$_{99}$ using 0.1 mg/L phosphine, the optimum CO_2 concentration may be near 20%. Within the 15–35% range it is apparent that CO_2 enhances the effect of the phosphine. The only advantage of co-applications of CO_2 above 35% may be the ability of CO_2 to promote a faster penetration of the PH$_3$ through the grain mass (Leesch 1990). These results may facilitate the determination of the cost effectiveness of adding CO_2 in phosphine fumigations. Carbon dioxide concentrations for controlled atmosphere applications are generally 40% or greater (Annis 1987). However, with combined phosphine and CO_2, lower levels of CO_2 are effective.

References

Aliniazee, M.T. 1971. The effect of carbon dioxide gas alone or in combinations on the mortality of *Tribolium castaneum* (Herbst) and *Tribolium confusum* du Val. (Coleoptera: Tenebrionidae). Journal of Stored Products Research, 7, 243–252.

Aliniazee, M.T., and Lindgren, D.L. 1969. Beetle and granary weevil at two different temperatures. Journal of Economic Entomology, 62, 904–906.

Annis, P.C. 1987. Towards rational controlled atmosphere dosage schedules: a review of current knowledge. In: Donahye, E. and Navarro, S., ed., Proceedings of the Fourth International Working

Conference on Stored-product Protection, Tel Aviv, Israel, September 1986, 128–148.

Anon. 1975. Recommended methods for the detection and measurement of resistance of agricultural pests to pesticides 16. Tentative method for adults of some major pest species of stored cereals, with methyl bromide and phosphine. FAO Plant Protection Bulletin, 23, 12–25.

Anon. 1989. Recommended methods for the detection and measurement of resistance of agricultural pests to pesticides. In: Suggested Recommendations for the Fumigation of Grain in the ASEAN Region, Part 1, Principles and general practice. Kuala Lumpur, ASEAN Food Handling Bureau, 105–118.

Bond, E.J. and Monro, H.A.U. 1967. The role of oxygen in the toxicity of fumigants to insects. Journal of Stored Products Research, 3, 295–310.

Cotton, R.T. 1930. Carbon dioxide as an aid in the fumigation of certain highly adsorptive commodities. Journal of Economic Entomology, 23, 231–233.

Cotton, R.T. 1932. The relation of respiratory metabolism of insects to their susceptibility to fumigants. Journal of Economic Entomology, 25, 1088–1103.

Cotton, R.T. and Young, H.D. 1929. The use of carbon dioxide to increase the insecticidal efficacy of fumigants. Proceedings of the Entomological Society of Washington, 31, 97–102.

Desmarchelier, J.M. 1984. Effect of carbon dioxide on the efficacy of phosphine against different stored product insects. Mitteilungen der Biologischen Bundesanstalt für Land- und Forst-wirtschaft, Berlin-Dahlem, Heft 220, 1–57.

Friedlander, A. and Navarro, S. 1979. The effect of controlled atmospheres on carbohydrate metabolism in the tissue of *Ephestia cautella* (Walker) pupae. Insect Biochemistry, 9, 79–83.

Friedlander, A. and Navarro, S. 1983. Effect of controlled atmospheres on the sorbitol pathway in *Ephestia cautella* (Walker) pupae. Experientia, 39, 744–746.

Friedlander, A. and Navarro, S. 1984. The effect of carbon dioxide on NADPH production in *Ephestia cautella* (Walker) pupae. Comparative Biochemistry and Physiology, 77B, 839–842.

Jones, R.M. 1938. Toxicity of fumigant–CO2 mixtures to the red flour beetle. Journal of Economic Entomology, 31, 98–309.

Kashi, K.P. and Bond, E.J. 1975. The toxic action of phosphine: role of carbon dioxide on the toxicity of phosphine to *Sitophilus granarius* (L.) and *Tribolium confusum* du Val. Journal of Stored Products Research, 11, 9–15.

Leesch, J.G. 1990. The effect of low concentration of gaseous carbon dioxide on the penetration of phosphine through wheat. In: Fleurat-Lessard, F. and Ducom, P., ed., Proceedings of the Fifth International Working Conference on Stored-product Protection, Bordeaux, France, September 1990, 857–864.

Liang Quan 1980. Studies on the synergy of phosphine activity in control of stored grain insects by controlled atmospheres. Grain Storage, No. 1, 1–11, 17 (in Chinese).

Qou Shi-Jie 1980. Investigation of the effect of oxygen and carbon dioxide on phosphine toxicity. 2nd China National Grain Storage Workshop, Collected Papers, Nan Ning, China, 79–89 (in Chinese).

Rajendran, S. and Muthu, M. 1989. The toxication of phosphine in the eggs of *Tribolium castaneum* Herbst (Coleoptera: Tenebrionidae). Journal of Stored Products Research, 25, 225–230.

Sato, K. and Suwanai, M. 1973. Studies on the characteristics of action of fumigants II. Entrance of hydrogen phosphide into weevil body under conditions of the failure to respire for the weevil. Botyu-Kagaku, 38, 213–216.

Tyler, P.S. Taylor, R.W. and Rees, D.P. 1983. Insect resistance to phosphine fumigation in food warehouses in Bangladesh. International Pest Control, 25, 10–13.

The impact of temperature, moisture content, grain quality and their interactions on changes in storage vessel atmospheres

R. Reuss,* K. Damcevski† and P.C. Annis†

Abstract

Knowledge of grain respiration is of interest in several areas of stored-products technology. In this study, insect-free wheat (moisture contents ranging from 9–13.2%) was sealed in storage vessels kept at temperatures ranging from 15–35°C. Changes in storage atmosphere were measured and rates of carbon dioxide production and oxygen consumption calculated. Carbon monoxide production was also measured.

Carbon dioxide production rates ranged from 1×10^{-5} to 450×10^{-5} mol/t/day. Oxygen consumption varied from 1×10^{-3} to 72×10^{-3} mol/t/day. Carbon dioxide production and oxygen consumption increased with temperature and relative humidity. Most variation in storage atmosphere composition could be modelled by a polynomial regression of log transformed rates against absolute humidity. Carbon monoxide levels observed ranged from 0 to 550 ppm. Higher temperatures led to higher concentrations of carbon monoxide across the range of moisture contents. Carbon dioxide and monoxide levels were closely correlated.

In a second experiment, changes in storage atmospheres of 8-year-old ASW grade wheat (moisture content 10.8%) were followed for 300 days. Carbon monoxide concentration reached 2363 ppm, Carbon dioxide and monoxide production rates were 3.3×10^{-4} and 1.35×10^{-4} mol/t/day. The oxygen consumption rate was 3.60×10^{-3} mol/t/day.

Introduction

In order to improve the efficacy of gaseous treatments, grain storage in sealed structures is becoming more common. This may lead to changes in storage atmospheres, which in turn may have important occupational health and safety implications. There is some possibility that these changes may also be helpful in maintaining the grain in an insect-free condition (Banks 1981; Navarro and Calderon 1980). In hermetic storage, atmosphere changes are due to metabolic activity of the grain, insects and fungi contained within a sealed storage environment. Oxygen (O_2) is lowered as it is used up by oxidative processes and carbon dioxide (CO_2), the end product of these processes, is elevated (Annis 1990; Banks 1981).

Bailey (1918) showed that carbon dioxide production of wheat increased with increasing moisture content. His measurements varied from 0.12 to 2.66 mol/t/day at 12.5–17.1% moisture content (m.c.) incubated at 37.8°C. He also found

that some varieties of wheat had higher respiration rates than others. He concluded that a temperature of 55°C was optimal for carbon dioxide production of wheat (7.21 mol/t/day at 17% m.c.).

Robertson et al. (1939) found that the respiration rate of wheat, oats and barley increased with increasing relative humidity (r.h.). For wheat they reported carbon dioxide production rates as low as 0.002 mol/t/day at a relative humidity of 57.6% (approximately 13% m.c.) and as high as 7.85 mol/t/day at 98% r.h. (>22% m.c.). Bailey (1940) reported that, when the logarithm of respiration rate of wheat was plotted against moisture content it approached a linear form, and proposed a simple model that predicted the amount of carbon dioxide produced by grain with moisture contents between 11 and 17%. He also derived formulae for the respiration of rice, barley and oats.

Ragai and Loomis (1954) studied the respiration of maize with moisture content of 14–24% and above at temperatures ranging from 8–30°C (carbon dioxide production at 30°C: 0.22 mol/t/day at 15% m.c. to 29.02 mol/t/day at 23% m.c.) They found that respiration depended partly on the storage atmosphere. They also treated grain with fungicides in an attempt to prevent carbon dioxide production by fungal respiration but observed little difference in the respiration of treated and untreated grain. Hummel et al. (1954) reported that mould-free grain showed low respiration rates, even at high moisture contents (15–31%), and at 30°C.

Bartholomew and Loomis (1967) examined carbon dioxide production of maize at 30°C. The grain used was viable or had been rendered non-viable by chemical treatments. The oxygen levels of the storage environments were controlled at 0, 21 and 100%. They found that carbon dioxide production rates of viable maize ranged from 0.001 mol/t/day at 2.4% m.c. and 0% oxygen to 0.008 mol/t/day at 12.6% m.c. after 10 days of storage. Non-viable maize produced 0.004 mol/t/day at 1.6% m.c. and 0% oxygen and a maximum of 0.006 mol/t/day at 13% m.c. and 21% oxygen. They reported that carbon dioxide production decreased with increasing storage time.

Carbon monoxide (CO) production of quiescent grain has only recently been recognised (Whittle et al. 1994). Even low concentrations of this gas may form a safety hazard. The threshold limit value (TLV) is set at 25 ppm (Anon 1992) and the short term exposure limit is 125 ppm. Many of the measurements reported by Whittle et al. (1994) exceeded these recommendations substantially.

The aims of the work described here are:

- to assist in modelling of head space concentrations of carbon dioxide, oxygen, and carbon monoxide.

- to develop a model that expresses the relationship between storage conditions, grain quality, and the composition of storage vessel atmospheres.

Results are expected to be useful in studies concerned with hermetic storage, aeration, and drying of wheat.

* University of Canberra, P.O. Box 1, Belconnen ACT 2616, Australia.
† Stored Grain Research Laboratory, CSIRO Division of Entomology, GPO Box 1700, Canberra ACT 2601, Australia.

Materials and Methods

The main observations were carried out on wheat from a single source (variety, Corella; grade, ASW; harvest, 1988). Moisture contents were determined according to ISO recommendation R 712-1968(E). Relative and absolute humidity were calculated on the basis of dewpoints measured with a condensation dewpoint hygrometer. Viability was measured according to methods suggested by the International Seed Testing Association (Anon. 1976). Fungal contamination was determined by direct plating wheat grains onto a suitable medium — dichloran 18% glycerol agar (Hocking 1991; Hocking and Pitt 1980).

Analysis of gas samples was carried out using a Fisher model 1200 Gas Partitioner with 80–100 mesh Columpak™ PQ (6.5 feet × 1/8 inch) and 60–80 mesh Molecular Sieve 13X (11 feet × 3/16 inch) columns in series. The conditions used were: carrier gas, helium at a flow rate of 30 mL/minute; oven temperature, 50°C. Concentrations were calculated on the basis of peak areas. Peak areas were calibrated periodically using a standard gas mixture with known carbon dioxide, oxygen, carbon monoxide and nitrogen composition.

The grain used had been stored in sealed screw-top jars each containing approximately 500 g of wheat. Moisture contents of these samples had been adjusted to four different levels before sealing. Storage temperatures were 15, 25 and 35°C. Accounting for all possible combinations of moisture content and temperatures, 12 jars were set up. From these jars, 200 g subsamples were obtained and placed in hermetically sealed flasks.

Before use, the grain samples were passed through a Börner divider to remove residual CO_2 and to ensure that subsamples were representative. The storage containers used were round-bottomed, 250 mL flasks sealed with a stopper containing a stopcock and a septum. These were designed to allow gas sampling through a septum without breaking the hermetic seal. The flasks were kept under the same conditions as the original samples. With replicates, 24 flasks were available for sampling. One of the flasks leaked during the experiment and a total of 23 yielded useable data.

Moisture content, humidity, viability and fungal contamination were measured at the beginning of the experiment. The atmospheres in the flask were sampled weekly, and in triplicate, over a period of 18 weeks. The volume of the sample taken was 1.2 mL which was replaced with dry air. Actual CO_2 and O_2 volumes in the flasks were computed from the free gas volume of the flasks and the measurements obtained from the gas partitioner. Carbon dioxide and O_2 volumes were calculated for all weekly measurements over the duration of the experiment.

Rates of gas production were calculated based on the averages of the replicate measurements. Regression analysis was carried out using *SAS* (Statistical Applications System, SAS Institute, 1987) and were based on the rates of CO_2 production and O_2 consumption.

An additional experiment was carried out using 8-year-old, ASW grade wheat. This had previously been stored in a 3 L sealed flask for 7 years. The moisture content and viability of this grain were measured as above. The wheat was split into 2 subsamples of approximately 1 kg each and resealed in 1 L flasks. The atmospheres in the flasks were sampled and analysed at approximately 4-week intervals over one year with each flask being sampled in triplicate. The volume of the samples taken was 1.2 mL which was replaced as detailed above.

Results

Moisture content, humidity, viability and fungal contamination

The moisture contents of samples at the beginning of the experiment are shown in Table 1. The maximum relative humidity measured was 60%. Absolute humidity (moisture content of the storage atmosphere) was in the range 3.6–23.4 g/m^3. Viability of wheat samples varied from 0 to 93.2%. Temperature and moisture content both had an impact on the viability of the samples. Viability decreased with rising temperature, reaching zero in samples of high moisture contents at 25°C. At 35°C most samples were no longer viable with the exception of the low moisture content samples (Table 1).

The percentage of grains contaminated by fungi ranged from 0 to 98%. Fungal contamination appeared to be dependent on moisture content and temperature. Contamination was high in samples whose moisture content ranged from 9–12% and that were kept at 15 and 25°C. Samples that were kept at 35°C showed low contamination except in the low moisture content range.

Table 1. Viability, fungal contamination, carbon dioxide production rate and oxygen consumption rate for wheat held in sealed vessels under a range of condition.

Temperature (°C)	Moisture content (%)	Relative humidity (%)	Absolute humidity (g/m^3)	Viability (%)	Fungal contamination	$CO_2 \times 10^{-5}$ (mol/t/d)	$O_2 \times 10^{-3}$ (mol/t/d)
15	9.0	27	3.6	90.8	98	1.0	7.500
	11.3	41	5.3	93.2	96	1.0	8.550
	11.6	48	6.1	78.4	92	1.0	7.250
	13.3	60	7.7	89.2	22	1.5	8.400
25	9.0	28	6.5	90.4	92	2.0	6.325
	10.9	44	10.0	83.6	44	3.5	6.075
	11.6	47	10.8	91.6	6	5.0	7.400
	13.1	58	13.4	2.8	0	9.5	1.008
35	8.9	27	10.7	80.4	12	5.5	7.850
	10.6	43	16.6	0	0	10.5	9.250
	11.3	44	17.6	0	0	140.0	12.450
	12.9	59	23.4	0	0	450.0	72.000

Carbon dioxide production and oxygen consumption

Carbon dioxide production rates ranged from 1×10^{-5} to 450×10^{-5} mol/t/day. As storage temperature increased, more CO_2 was produced. Similarly, increases in relative humidity appeared to lead to higher CO_2 production rates. Low viability occurred in samples with high CO_2 production rates. Samples with high fungal contamination had lower CO_2 production rates than those with low or no contamination (Table 1). Oxygen consumption rates varied from 1×10^{-3} to 72×10^{-3} mol/t/day. In general, oxygen consumption followed the trends of carbon dioxide production described above (Table 1).

Statistical analyses were performed using stepwise regression procedures. The most important predictors of log transformed carbon dioxide production and oxygen consumption were temperature, relative and absolute humidity and, to some extent, viability. A substantial amount of carbon dioxide production (adjusted $r^2 = 0.98$) could be described by a polynomial regression model ($F = 276.16$, df = 4,18, $p < 0.0001$) describing a curvilinear regression of log transformed carbon dioxide production rate against absolute humidity (Fig. 1):

$$\log(CO_2) = -3.79 - 0.6671AH + 0.1160AH^2 - 0.0070AH^3 + 0.0001AH^4 \qquad ... (1)$$

where AH is the absolute humidity in g/m^3 and carbon dioxide production rate is in mol/t/day.

Similarly, oxygen consumption was successfully modelled (adjusted $r^2 = 0.91$, $F = 74.81$, df = 3,19, $p < 0.0001$) using a cubic term and could be described as follows (Fig. 1):

$$\log(O_2) = -2.22 + 0.0433AH - 0.0061AH^2 - 0.0003AH^3 \qquad ... (2)$$

with the same units as in the carbon dioxide equation.

Carbon monoxide production

Carbon monoxide levels observed ranged from 0 to 550 ppm (mean = –1104, n = 1185), 64% of observations had no measurable amounts of carbon monoxide (< 250 ppm). The data were considered insufficient to calculate production rates or carry out detailed statistical analysis, but yielded some interesting trends. Concentrations between 250 and 300 ppm were found in 14% of all samples. Higher concentrations of carbon monoxide were not as commonly found and samples with carbon monoxide levels above 450 ppm comprised less than 4% of measurements taken.

The amount of carbon monoxide produced varied with temperature and moisture content. It appeared that higher temperatures led to higher concentrations of carbon monoxide across the range of moisture contents. Moisture content seemed to have less impact on carbon monoxide production than temperature (Fig. 2).

There was a strong correlation between the averaged CO_2 measurements and levels of carbon monoxide produced in the storage vessels (r = –0.72). When measurements of CO less than the limit of detection were dropped from the analysis, carbon dioxide became an excellent predictor of CO levels (r = 0.93). Oxygen consumption rate and CO concentration were negatively correlated, but the relationship was not very strong (r = –0.24 for all averaged measurements and r = –0.39 if zero measurements were disregarded).

Head space composition of 8-year-old wheat

The moisture content and viability of the wheat were 10.8 and 57.0%, respectively. After 21 days there was an almost linear rise in carbon dioxide and carbon monoxide concentrations with time. Oxygen concentration decreased almost linearly over the same period. The carbon monoxide concentration reached 2363 ppm in 299 days (Fig. 3).

The average CO_2 and CO production rates based on changes in gaseous composition after 21 days were 3.3×10^{-4} and 1.35×10^{-4} mol/t/day. The average O_2 consumption rate was 3.60×10^{-3} mol/t/day.

Discussion

Carbon dioxide production rates were quite low in all treatments with a maximum of 450×10^{-5} mol/t/day from samples stored at a relative humidity of 59% and 35°C. Oxygen consumption increased with temperature and humidity up to 72×10^{-3} mol/t/day at the maximum temperature and relative humidity used (35°C and 60% r.h.). Differences between oxygen consumption rates at low humidity were, however, minimal. Most of the variation in the composition of storage atmospheres could be described by a curvilinear regression using absolute humidity as a single factor (equations (1) and (2), Fig. 1). It appears that levels of carbon dioxide production and oxygen consumption depend mainly on the amount of moisture present in the storage atmosphere.

Samples stored at low humidity and temperatures had higher levels of fungal contamination then those kept at higher humidity and temperatures. None of the samples examined provided moisture levels (> 70% r.h.) suitable for significant fungal growth (Pitt and Hocking 1991). The contamination is therefore likely to be fungi producing viable spores and resting hyphae rather than actively growing. Fungal spores may be rendered sterile if exposed to 35°C over extended periods of time (Parrey and Pawsey 1984; Deacon 1984). Low moisture content may help preserve fungal spores under these conditions. In this respect, fungal spores may behave like seeds and lose viability, if they do not germinate, under conditions of high temperature and moisture content.

Viability of wheat was found to be dependent on temperature, moisture content and storage time as was expected from the literature (e.g. Villiers 1980). At 35°C most samples became non-viable during the 18 weeks of the experiment. The exception was samples kept at 9% m.c. which tolerated the temperature well. Samples which showed low viability had high carbon dioxide production and oxygen consumption rates. This may be due to the fact that low viability occurred with high humidity and storage temperatures. Viability is a factor which greatly depends on the amount of moisture present in the storage environment. The relationship between carbon dioxide production and viability may therefore be explained in the context of a regression of changes in storage atmospheres against absolute humidity.

Fungal contamination, which like viability was highest at low moisture contents and temperatures, was also negatively related to carbon dioxide production, possibly for similar reasons. In all cases fungal respiration could be discounted as a contributing factor to carbon dioxide production, as the most highly contaminated samples showed the lowest carbon dioxide production rates.

Carbon monoxide production appeared to be temperature dependent (Fig. 2). Carbon dioxide levels were a good indicator of carbon monoxide concentrations above the limit of detection (r = 0.92). Some 36% of measurements taken showed carbon monoxide levels higher than the short-term exposure limit, that is above 125 ppm.

A combination of the equations given earlier should make it possible to model the gaseous composition for atmospheres in fully sealed enclosures. Estimates can be made of the changes in the enclosed atmosphere of real storage structures by including terms that describe the degree of sealing of the enclosure.

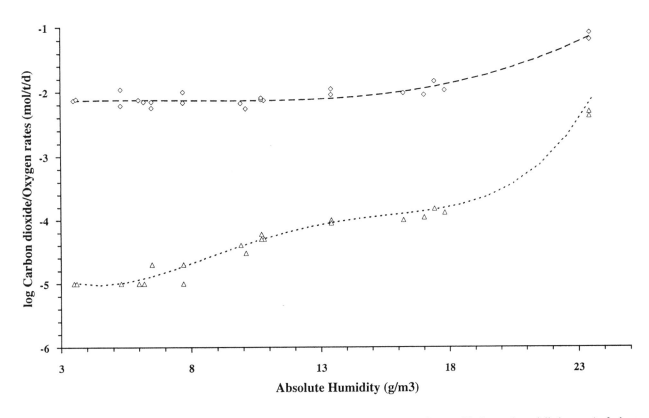

Fig. 1. Carbon dioxide production rate (Δ observed; modelled - - - -) and oxygen consumption rate (◇ observed; modelled – – – –) of wheat stored in sealed flasks. Data modelled using the polynomials of absolute humidity contained in equations (1) and (2).

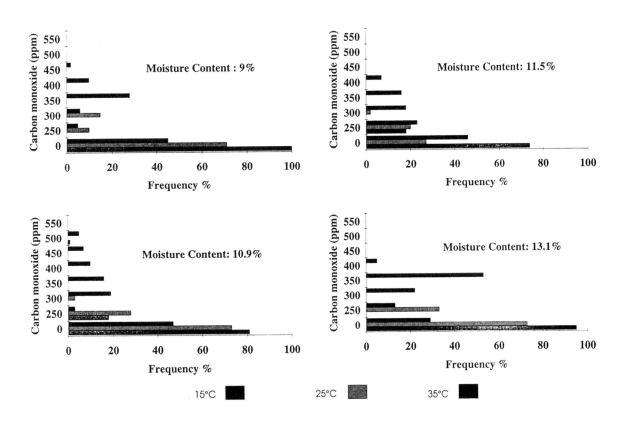

Fig. 2. Carbon monoxide content of storage atmospheres in wheat stored at various moisture contents and temperatures.

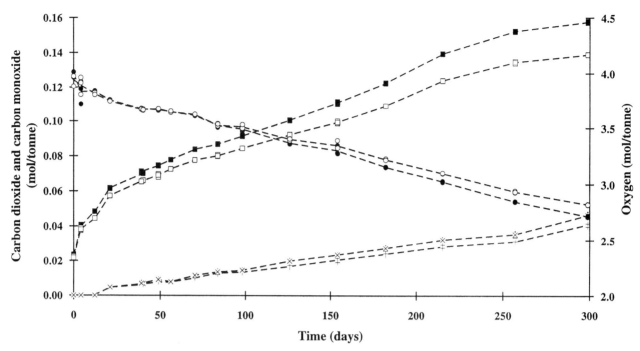

Fig. 3. Carbon dioxide (■ flask 1; ❑ flask 2), carbon monoxide (+ ×) and oxygen (● ○) concentrations in 8-year-old wheat stored in sealed flasks over a 300-day period. Symbols show individual observations for each of two flasks. The dashed lines connect the average values.

Conclusion

Carbon dioxide production and oxygen consumption rates of pest-free wheat stored at moisture contents of 9–13% at temperatures of 15–35°C were very low. Most variation in storage atmosphere composition could be explained by a curvilinear regression of log transformed rates against absolute humidity.

Long-term exposure of wheat to temperatures of 35°C appears to have fungicidal effects. At moisture contents likely to occur in wheat storage in Australia, fungal respiration is unlikely to contribute to modification of storage atmospheres.

Carbon monoxide levels occurring in the head space of wheat kept in sealed storage could present a hazard to workers that are required to enter storage facilities.

Acknowledgments

The authors wish to thank their colleagues at SGRL for all the help they were given. We also would like to thank M.L. Uhr from the University of Canberra for her assistance and patience. Facilities for this work were provided by the CSIRO Stored Grain Research Laboratory, Canberra.

References

Annis, P.C. 1990. Fumigation and CA as options for pest and quality control in stored grain. In: Champ, B.R., Highley, E. and Banks, H.J., ed., Fumigation and controlled atmosphere storage of grains. ACIAR Proceedings, No 25, 20–28.

Anon. 1976. International rules for seed testing. Seed Science and Technology, 4, 3–177.

Anon. 1992. Threshold limit values for chemical substances and physical agents and biological exposure indices (1992–93). Cincinnati, Ohio, American Conference of Government Industrial Hygienists, 15.

Bailey, C.H. 1918. Respiration of stored wheat. Journal of Agricultural Research, 11, 685–713.

Bailey, C.H. 1940. Respiration of cereal grain and flax seed. Plant Physiology, 15, 257–274.

Banks, H.J. 1981. Modified atmosphere and hermetic storage — effects on insect pests and the commodity. In: Champ, B.R. and Highley, E., ed, Proceedings of the Australian Development Assistance Course on the Preservation of Stored Cereals, Australia, August–October 1981, 2, 521–532.

Bartholomew, P.P. and Loomis, W.E. 1967. Carbon dioxide production by dry grains of *Zea mays*. Plant Physiology, 42, 120–124.

Deacon, J.W., 1984. Introduction to modern mycology. Oxford, Blackwell Scientific Publications, 103 p.

Hocking, A.D., 1991. Effects of fumigation and modified atmosphere storage on growth of fungi and production of mycotoxins in stored product. In: Champ, B.R., Highley, E., Hocking, A.D. and Pitt, J.I., ed., Fungi and mycotoxins in stored products: proceedings of an international conference, Bangkok, Thailand, April 1991. ACIAR Proceedings, No 36, 145–156.

Hocking, A.D. and Pitt, J.I. 1980. Dichloran–glycerol medium for enumeration of fungi from low moisture foods, Applied Environmental Microbiology, 39, 488–492.

Hummel, B.C.W., Cuendet., L.S., Christensen., C.M. and Geddes, W.F. 1954. Comparative changes in respiration, viability and chemical composition of mold free and mold contaminated wheat upon storage. Cereal Chemistry, 31, 143–149.

Navarro, S. and Calderon, M., 1980. Integrated approach to the use of controlled atmospheres for insect control in grain storage. In: Shejbal, J., ed., Controlled atmosphere storage of grains. Amsterdam, Elsevier, 53–78.

Parry, J.T. and Pawsey, R.K. 1984. Principles of microbiology for students of food technology. Cheltenham, U.K., Stanley Thornes Publishers Ltd, 39 p.

Pitt, J.I. and Hocking, A.D. 1991. Significance of fungi in stored products. In: Champ, B.R., Highley, E., Hocking, A.D. and Pitt, J.I., ed., Fungi and mycotoxins in stored products: proceedings of an international conference, Bangkok, Thailand, April 1991. ACIAR Proceedings, No 36, 16–21.

Ragai, H. and Loomis, W.E. 1954. Respiration of maize grains. Plant Physiology, 29, 49–55.

Robertson, D.W., Lute, A.M. and Gardner, R. 1939. Effect of relative humidity on viability, moisture content and respiration of wheat, oats and barley seed in storage. Journal of Agricultural Research, 59, 281–291.

SAS Institute 1987. SAS Applications guide, 1987 ed. Cary, NC. SAS Institute Inc., 252 p.

Villiers, T.A. 1980. Ultrastructural changes in seed dormancy and senescence. In: Thiman, K.V., ed., Senescence in plants. Boca Raton, Florida, CRC Press Inc., 39–66.

Whittle, C.P., Waterford, C.J., Annis, P.C. and Banks, H.J. 1994. The production and accumulation of carbon monoxide in stored grain, Journal of Stored Products Research, 30, 23–26.

Low-cost detector for the continuous monitoring of phosphine fumigation

R. Ryan*, S. Waddell*, P.W. Alexander†, K. Bowles†, L. Cherkson†, J. Morgan† and D. B. Hibbert†

Abstract

The general model $Ct = k$ describes the toxicity of phosphine when used as a fumigant in grain storage, where C is the concentration of phosphine, t is the exposure time, and k is a constant. This factor can be calculated only if data are available on the fate of phosphine gas after initial injection into grain storage silos. A simple hand-held phosphine monitor capable of continuous monitoring of phosphine levels is proposed. It could be installed in silos and used to monitor the phosphine levels over long periods with data-logging capability and connection to telemetry systems for remote viewing of real-time data. The monitor consists of a solid-state semiconductor sensor which has been shown to give a linear response to phosphine gas in the concentration range 0.1–300 ppm using standard gas dilution procedures to calibrate the sensor. The response of the sensor is rapid (2 second response to steady-state readings) and with little drift below 300 ppm phosphine. Exposures over 20 minutes showed little drift in the above range, but drifts occurred at 1000 ppm. An alarm can be fitted to indicate when limits reach predetermined concentrations, and feedback control of phosphine injection could be introduced. The monitor developed in this work is capable of continuous operation over long periods and can be controlled from an external computer with very rapid response times. It is of low cost and could be fitted into storage facilities to improve fumigation efficacy, cost efficiency and safe use of phosphine gas. Carbon monoxide produced during storage interferes only at concentrations above 200 ppm.

Introduction

Phosphine is an important gas for use in commercial agriculture. The major commercial use of phosphine is as a fumigant in grain storage. Traditionally, the phosphine was produced by simply throwing tablets of metallic phosphide into the grain. The tablets react with atmospheric moisture to form phosphine gas. There is now an interest to introduce phosphine gas directly from gas cylinders (2% phosphine in carbon dioxide) and there is a need to be able to monitor the level of phosphine in the storage areas to ensure the concentration is kept at an effective level. An appropriate sensor should be able to detect phosphine in the 100–200 ppm range for this application. Phosphine, however, is a highly toxic gas with a TLV (thresh-old limiting value) of 0.3 ppm. Therefore, there is an obvious need for a low-cost, reliable sensor to detect this gas at low concentrations for safety reasons.

Several gas sensors are currently available to detect phosphine. These include common badge types for personal exposure monitoring as well as photoionisation and infrared detectors. In addition, sensors sensitive to reducing gases, such as electrochemical types, Taguchi sensors (Figaro Engineering Inc. 1990; Alexander et al. 1993) and Langmuir-Blodgett film sensors may be applicable. Much work has been reported on gas sensors, particularly for the determination of O_2, CO_2, SO_2, H_2S, NH_3, CO, NO_2, HCl, H_2 and Cl_2 (Shurmer et al. 1987; Watson 1984; Mosely et al. 1991). The Taguchi types of tin-oxide semiconductor sensors are obvious candidates for phosphine sensors due to their high resistance to poisoning and low cost. The present paper describes a study of Taguchi sensors for continuous monitoring of phosphine and other potentially interfering gases such as hydrocarbons, CO, and CO_2.

Instrumentation

Preliminary studies were conducted with a Taguchi TGS825 sensor purchased from Figaro Engineering Inc. (Osaka, Japan). Single and multi-sensor studies were also conducted using a dual sensor gas analyser designed and manufactured at the Department of Chemistry, University of New South Wales, Australia (UNSW), and presently under a provisional patent by Unisearch Limited (Alexander et al. 1993). This analyser incorporated TGS813 and TGS822 sensors, purchased from Figaro Engineering Inc., attached to a printed circuit board via 6-pin connectors, and incorporating a small piston pump for aspiration of gas samples through the sensor cells. A glass T-piece was used to split the gas flow equally between two compartments containing the two sensors. The gas analyser was connected to a notebook computer via a 12-bit analog-to-digital converter for data logging on the computer. The system displays sensor voltages on an LCD with a switch to show the response of either sensor. Calibrations and calculations of gas concentrations detected by the sensors were performed on the notebook computer. The computer can also be used to control the monitor from a remote location.

Gas dilution apparatus

Gas dilutions for ethylene and phosphine calibrations were conducted using an Ecotech Model 8370 Gas Diluter on loan from The Commonwealth Industrial Gases (CIG). This is a portable system which allows mixing of two gases into precisely controlled gas flows to generate required concentration values. The system works by using a laminar flow device to force part of the gas stream through a section of heated capillary tube. The change of temperature (ΔT) between the ends of the capillary is a measure of heat absorbed by the gas. The amount of heat absorbed depends on the mass of the gas. Since mass is independent of temperature and pressure, the instrument is unaffected by temperature and pressure variations.

---

* The Commonwealth Industrial Gases Ltd, Chatswood, NSW 2067, Australia.
† Department of Analytical Chemistry, University of NSW, Kensington, NSW 2033, Australia.

Dilution gas (air) readings were given in L/minute and span gas (ethylene or phosphine) readings were given in mL/minute. The relationship for converting flow rates to concentrations is:

$$\text{Output concentration (ppm)} = \frac{\text{Input conc. (ppm) * span gas flow (mL / min.)}}{\text{Diluent gas flow (mL / min.) + span gas flow (mL / min.)}}$$

Ecotech quote the repeatability at 0.2% of full-scale readings for gas flows and ± 1% full scale for electrical accuracy and linearity.

Data acquisition and processing

The gas monitor unit developed at UNSW (Alexander et al. 1993) incorporates an LED readout and may be used as a stand-alone device or may be interfaced to a computer via an analog-to-digital converter (ADC). An RS232C serial output is incorporated for direct connection to computer and telemetry systems. The system has been programmed with the 'Think C' Compiler for Apple Macintosh® computers. In the multi-sensor studies the raw data may be displayed in real time in one window. Another split screen is used to display the calculated values for the two different gases, also in real time. The data can be stored as an ASCII text file and transferred subsequently to a scientific spreadsheet for data processing.

The data were processed using the IGOR® program. IGOR® stores the data as waves which may be displayed as graphs. Mathematical functions such as smoothing and integration are available, as well as provisions for labelling and annotating axes.

Reagents

The preliminary studies involved calibration and precision experiments with standards of ethyl acetate in water. The ethyl acetate was analytical grade and was purchased from Pronalys®. Single and multi-sensor studies involved calibration with phosphine and ethylene. Ethylene was obtained in a cylinder from CIG. Standards of 2 ppm and 1053 ppm ethylene in nitrogen gas were used. Phosphine was also obtained in cylinders from CIG. Standards of 300, 1000 and 2000 ppm were used.

Procedures for sensor calibrations

Ethyl acetate

The TGS825 sensor was calibrated with ethyl acetate in water at the following concentrations: 10, 50, 100, 200, 400, 600, 800 and 1000 ppm. The standards were made up in 100 mL volumetric flasks. The flasks were stoppered to allow ethyl acetate to accumulate in the headspace. The TGS825 sensor was allowed to warm up for 15 minutes before measurements, to ensure a stable baseline. Measurements were made by holding the inlet tube of the sensor apparatus at the top of the flask immediately after removing the lid. An Apple II computer programmed with a software package entitled MV/FILE was used to collect the data. As soon as the trace of the response on the computer screen peaked and started to drop, the sensor inlet tube was removed from the flask. The flask was then resealed, shaken, and allowed to stand to allow the headspace gases to re-equilibrate. The trace on the screen was allowed to return to the baseline before a repeat measurement was made. Multiple measurements of each standard were made. The standards were measured from low to high concentration to avoid the possibility of memory effects.

Precision was tested with the TGS825 by taking replicate measurements of the 800 ppm standard using the same sampling procedure stated above. The effects of response time, sampling time and positioning of the sensor inlet tube were tested by following the same procedure as above.

Calibration of the sensor response to phosphine and ethylene

TGS813 and TGS822 sensors were calibrated for various concentrations of phosphine. The phosphine from the standard gas cylinder of appropriate concentration was diluted with compressed air using the Ecotech Model 8370 Gas Diluter. The sensors were contained in the dual sensor gas analyser and could be calibrated simultaneously. The analyser and the dilution apparatus were turned on 15 minutes before measurements were to be taken, to ensure a stable baseline. Data points were acquired using a Macintosh® Powerbook computer. The compressed air was turned on and adjusted to the desired flow rate. The phosphine supply was then turned on at a very low flow rate. After the trace on the computer screen flattened out, the flow rate of the phosphine was increased in order to increase the concentration of the ethylene. This resulted in a series of stepwise increases in the voltage output. After the measurement with the maximum flow rate of phosphine, the flow supply was disconnected to allow the trace to return to the baseline. This procedure was repeated using a lower compressed air flow rate, allowing a series of measurements at higher phosphine concentrations. The above procedure was repeated for ethylene using the standardised gas cylinder containing 1053 ppm ethylene in nitrogen gas.

Results on the response of Taguchi sensors to phosphine

Figure 1(a) and 1(b) show the response for increasing concentrations of phosphine against a background of compressed air. Both sensors showed a significant response to the phosphine gas but higher response was achieved with the TGS822 sensor. Figure 1(c) shows that at higher concentrations of phosphine both sensors are subject to poisoning and do not give a steady-state response. At moderately high phosphine concentrations, it was found preferable to use the TGS813 rather than the TGS822 since it gave a lower response and was less prone to poisoning, as shown in Figure1(c).

Figure 2(a) shows the response of both sensors to a stream of CO_2 gas. This is important since phosphine gas is commonly diluted in cylinders with CO_2 gas. The TGS822 shows very low response to the CO_2 which makes it useful as a phosphine sensor. The response of the TGS813 is possibly due to CO_2 displacing O_2 at the surface of the sensor and thus changing the conductivity.

Calibration plots were prepared from the data in the 0–125 ppm (Fig. 2(b) and 2(c)) and 0–600 ppm ranges (Fig. 3(a) and 3(b)). At low concentrations (0–125 ppm), a linear response was obtained for the TGS822 with a correlation coefficient of 99.9%. The TGS813 gave a linear plot with correlation of 99.1%. The log–log plot of the TGS813 results gave marginally improved linearity with a correlation of 99.2%. The log–log plot for the TGS822 resulted in decreased linearity due to the effect of logarithms near zero. Over a wider concentration range, both sensors gave curved calibration plots. Curves of best fit were obtained with the following general equation:

Sensor voltage $= K0 + K1.\exp[-(\text{conc.} - K2)/K3]^2$

where K0, K1, K2 and K3 are constants. Log-log plots in this case resulted in improved linearity. These results were successfully fitted to the same form as above.

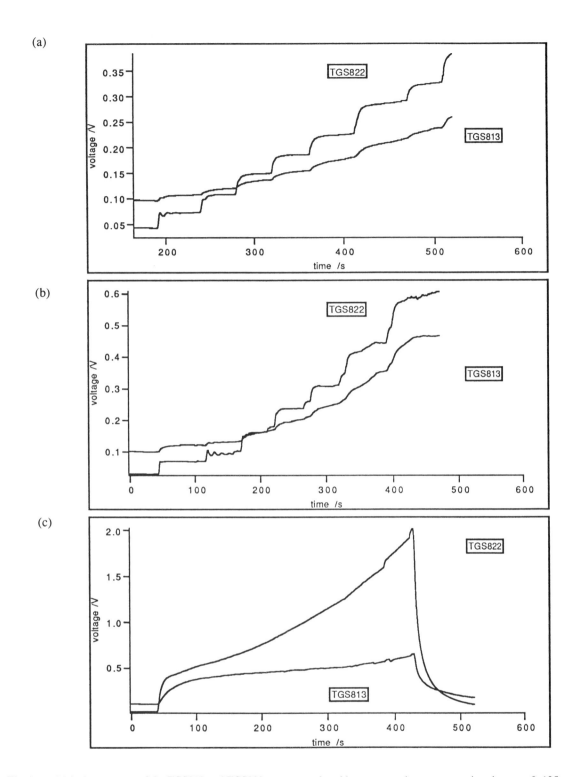

Fig. 1. (a) is the response of the TGS813 and TGS822 sensors to phosphine gas at various concentrations between 0–125 ppm. (b) is the response of the TGS813 and TGS822 sensors to phosphine gas at various concentrations between 0–600 ppm. The stepwise response is due to increases in the phosphine flow against a constant flow of diluent gas (compressed air). (c) is the response of the two sensors to phosphine at a higher concentration (1000 ppm).

Detection limits and sensitivity of each sensor

The sensitivity of the two sensors used indicated different response characteristics to phosphine, carbon monoxide and ethylene. The TGS813 responded well to ethylene but was less sensitive to phosphine. The TGS822 had a more sensitive response to phosphine but showed a definite effects of poisoning at higher phosphine concentrations. Conversely, the TGS822 was very insensitive to ethylene. Carbon monoxide showed very little response at either sensor when the concentration was below 200 ppm. Detection limits were calculated on the basis of the noise level observed for the background readings for each sensor. For the TGS813 sensor, the baseline noise level was 1 mV which was calculated to correspond to detection limits of 0.125 ppm for ethylene and 10.1 ppm for phosphine. For the TGS822 sensor, the response was insensitive to ethylene concentrations below 500 ppm, but gave a detection limit of 0.1 ppm for phosphine. It appears therefore

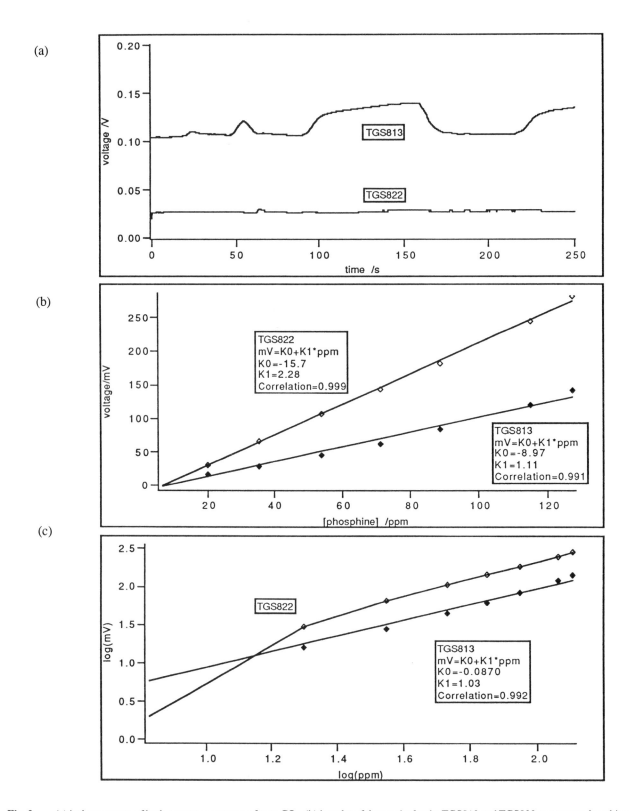

Fig. 2. (a) is the response of both sensors to a stream of pure CO_2. (b) is a plot of the results for the TGS813 and TGS822 sensors to phosphine at concentrations between 0–125 ppm. The line of best fit was calculated with constants as annotated. (c) is a log–log plot of the same data.

(a)

(b)

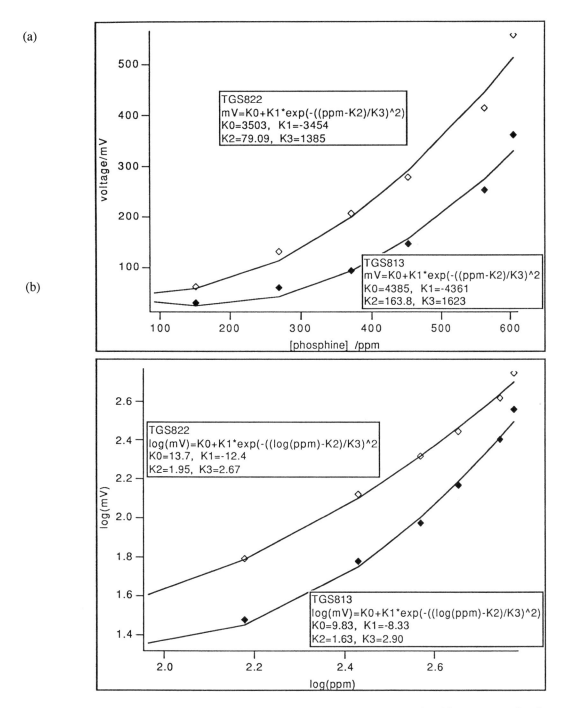

Fig. 3. (a) shows the calibration plots of the response of the TGS813 and TGS822 sensors to phosphine at concentrations between 0–600 ppm. The line of best fit was calculated with constants as annotated. (b) is a log–log plot of the same data.

that both sensors will be of use for measuring phosphine concentrations in grain storage, depending on the concentration of phosphine applied in the grain storage situation.

Conclusions

The monitor developed in this study has advantages over existing phosphine monitors for several reasons. It provides linear response in the range up to 300 ppm phosphine, it has a low detection limit, and rapid response times without poisoning problems below 300 ppm, and it is useful for data logging with computer control from a remote site.

References

Alexander, P.W., Hibbert, D.B., Cherkson, L. and Morgan, J.T. 1993. Australian Provisional Patent: Multi-sensor gas analyser. Unisearch Ltd, February 1993.

Figaro Engineering Inc. 1990. Figaro Gas Sensors, Product Document. Osaka, August 1990.

Mosely, P.T., Norris, V.J. and Williams, D. 1991. Techniques and mechanisms in gas sensing. Adam Hilger Series on Sensors, IOP Publishing.

Shurmer, H., Fard, A. Barker, J., Bartlett, P. Dodd, G. and Hayat, U. 1987. Physics in Technology, 18, 170–176.

Watson, J. 1984. Sensors and Actuators, 5(1), 29–42.

Controlled release of phosphine — an update

D. Schonstein,[*] W. Shore,[*] R. Ryan[†] and S. Waddell[†]

Abstract

The controlled release techniques possible with cylinders containing phosphine in carbon dioxide in minimising selection and controlling resistant strains were initially reported at the last International Working Conference on Stored-Products Protection at Bordeaux (Winks and Ryan, 1990). The benefit of controlled release has led to the growing international acceptance of Phosfume®, the patented non-flammable mixture of 2% wt. phosphine in carbon dioxide. The safe and accurate metering of the universal grain fumigant, phosphine, has revolutionised stored-product disinfestation. Innovations adopting this technology include;

- SIROFLO®, the flow-through fumigation application for phosphine in leaky silos developed by CSIRO;
- recirculation fumigation at export terminals developed by Western Australia Cooperative Bulk Handling (WACBH);
- low-dose (0.3 g/t) fumigation of pad storage by GRAINCO, Qld;
- long-term storage of peanuts in concrete silos by the Peanut Marketing Board (PMB), Qld;
- continuous low dose fumigation by the Ministry of Agriculture Fisheries and Food (MAFF), U.K.

CIG's provisional patented LoDose® regulator has the capability of dispensing a range of gas flow-rates from 0–100 L/minute at pressures as low as 0.1 kPa direct from the high pressure (5000 kPa) Phosfume® cylinder. The LoDose® phosphine dispensers have been proven in applications as diverse as bulk grain and on-farm fumigations.

Introduction

Australia currently enjoys a reputation for exporting grain that has one of the highest levels of insect freedom (Winks and Ryan 1993). This 'insect-free' status was achieved using residual pesticides, but during the last decade pesticide residues in grain have featured prominently as a concern of overseas customers. The market's demand for insect-free and pesticide residue-free food should ensure that residual pesticide grain protectants will be phased out.

The safe, accurate and controlled metering of the universal grain fumigant, phosphine, has revolutionised stored-product disinfestation. Gaseous phosphine [PH_3] fumigation is a proven cost-effective alternative to residual pesticides. Phosphine is an extremely flammable gas, and existing com-

mercial metallic phosphide formulations do not eliminate this hazard. The slow release of the phosphine from metallic phosphide formulations allows the phosphine to dissipate into the surrounding air and minimise explosive concentration levels. Phosfume™ is a patented (U.S. Patent No: 4889708) non-flammable mixture of phosphine (2% in weight) in carbon dioxide marketed by the international BOC Group, which can be easily, accurately and safely dispensed into all types of grain storage using gastight pipeline.

Innovations adopting this technology include:
- SIROFLO®
- recirculation fumigation
- pad/bunker fumigation
- continuous dose.

SIROFLO®

An exception to the gaseous grain fumigation's requirement of appropriate sealed storage is the CSIRO flow-through SIROFLO® fumigation technique. SIROFLO® is an innovative patented fumigation technique developed by Dr R.G. Winks of the CSIRO Stored Grain Research Laboratory, Canberra (Winks 1983). The traditional fumigant phosphine is used in a revolutionary way which makes it far more effective. The system is quite simple. A low concentration of phosphine gas from Phosfume™ cylinders is mixed with an air stream. The air–gas mixture is pumped into the base of the storage facility. The airstream produces a small positive pressure that causes the gas to spread evenly through the entire grain mass.

SIROFLO® obviates the need for expensive sealing ($3–5/t) and offers effective, essentially residue-free treatment at low cost; say 10–50 ¢/t depending on whether the treatment is short or long term plus $1/ t installation cost (Evans 1988).

CSIRO research suggests a solution to overcome the traditional problem of insect resistance in the case of phosphine by showing that phosphine-resistant strains succumb to the gas provided the exposure period is long enough. This can be achieved in gastight storage or by using the flow-through SIROFLO® fumigation technique (Winks and Ryan 1990).

For a minimum exposure period of 28 days, the recommended SIROFLO® minimum concentration is 10–22 ppm PH_3 with a continuous flow equivalent of approximately one volume change/day.

Developing Phosfume™ dispensing systems to provide the continuous long-term flow-through requirement of SIROFLO® involved an extended effort between CIG, CSIRO and the Bulk Grain Authorities. The equipment selection was complicated because Phosfume™ is dispensed as a liquid and early successful prototypes were later plagued by inherent impurity problems associated with phosphorus products. While phosphine's chemistry is not fully understood, it reacts with oxygen producing a polymer dust and oily phosphoric acid. Gas regulators and fine orifices in control valves are particularly vulnerable to the paste formed from these contaminants.

---

[*] CIG New Technology, 8/39 Herbert St, St Leonards, NSW 2065, Australia.

[†] The Commonwealth Industrial Gases Ltd, P.O. Box 288, Chatswood, NSW 2057, Australia.

CIG responded by implementing a program ensuring that oxygen was excluded from the Phosfume™ cylinders and developing the LoDose™ regulator. The initial response of analysing individual cylinder for oxygen contamination before filling is being superseded by the fitting of Minimum Pressure Relief [MPR] cylinder valves. The MPR valves used are identical to those proven in ultra high purity gas cylinders and exclude atmospheric air by maintaining a higher than ambient pressure of gas in the cylinder even if the cylinder valve is left open. The LoDose™ regulator was developed to accept liquid gas input and accommodate any polymer dust and phosphoric acid contaminants.

A schematic of the SIROFLO® technique using a LoDose™ regulator is shown in Figure 1. The high pressure liquid Phosfume™ mixture is vaporised to a low pressure gas and diluted ~1000 times in an air stream (i.e. ~26 ppm) before being dispensed into the base of the on-farm storage bin. The vaporisation and pressure reduction is achieved using the LoDose™ regulator. The LoDose™ regulator has the pressure reduction capability of ~50,000 i.e. it could reduce the 5000 kPa Phosfume™ to 0.1 kPa. The LoDose™ ambient-heat regulator shown in Figure 1 has a maximum continuous output flow of 3 L/minute (ie sufficient to treat a ~5000 m^3 storage using SIROFLO®).

The LoDose™ heater regulator shown in Figure 2 has a temperature controlled heater block inserted between the regulator's orifice seat and diaphragm controller. This regulator has a maximum continuous output of 25 L/minute (i.e. sufficient to treat a ~40,000 m^3 storage using SIROFLO®).

Both the LoDose™ ambient-heat and LoDose™ heater regulators have been successfully trailed by Bulk Grain Authorities dispensing Phosfume™ in SIROFLO® fumigation, and more than 200 units have been requested by the Bulk Grain Authorities to date. The LoDose™ heater regulator

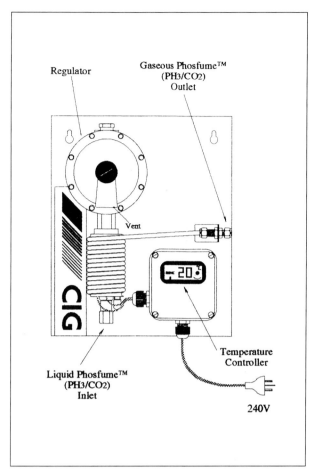

Fig. 2. LoDose™ heater regulator.

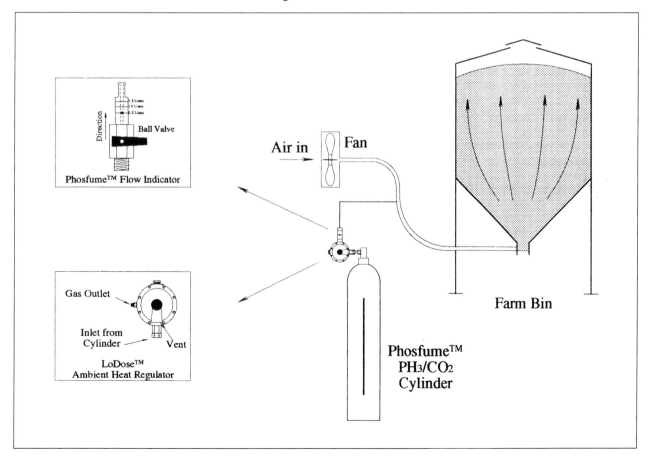

Fig. 1. SIROFLO® flow-through fumigation.

upgrade of the NSW GrainCorp SIROFLO® dispenser is shown in Figure 3.

SIROFLO® adoption by the Bulk Grain Authorities continues to grow and more than 6 million t of grain will be treated by this technique using Phosfume™ in 1994 (Winks and Ryan 1993).

Recirculation Fumigation

Cooperative Bulk Handling (CBH), the Western Australian grain handling authority, continues to use Phosfume™ at its ~1.5 Mt capacity Kwinana Grain Terminal. WACBH inject Phosfume™ into the forced air recirculation system of the 2200 t vertical storage (a total capacity of 912000 t is held in the 144 vertical cell which have individual dimensions of 11 m diameter and 30 m high).

The initial dispensing was achieved using a capillary tube restrictor to ensure the Phosfume™ was dispensed at a uniform rate over the time period required for the air blower to achieve one volume change. The LoDose™ heater regulator shown in Figure 2 is capable of uniformly dispensing the Phosfume™ dose required to treat the 2000 t vertical cells in the 5 hours required for one volume change. A schematic for a Phosfume™ recirculation system is shown in Figure 4.

CBH has access to low-cost metallic phosphide, but the 2 day saving in time using Phosfume™ is more critical than any increased costs over pellets [note: the cost of the Phosfume™ treatment is a low ~5¢ / t]. The additional benefits of eliminating occupational health and safety concerns associated with handling (pellets instantaneous generate PH_3 on exposure to atmospheric moisture) and disposal of spent pellets are an additional bonus (spent pellets can have ~5% unreacted metallic phosphide).

Western Australia commenced a program of sealing grain storage in the 1980s to enable grain to be fumigated with PH_3 effectively and today no residual pesticide grain protectants are used by the CBH.

Pad / Bunker Fumigation

Fumigation of pad or bunker storages using Phosfume™ has been perfected by GRAINCO, Qld (Ryan 1992). The tarpaulin-covered grain, typically ~30,000 t (Figure 5) is probed using a 9 mm diameter, 2 m stainless steel tube (with a 1.5 mm exit hole) connected to the Phosfume™ cylinder using a 30 metre length of teflon lined, stainless steel high pressure flexible hose. The dose rate of 0.3 g PH_3 / m^3 pioneered by GRAINCO has been recommended for addition to the Phosfume™ label. The innovative GRAINCO technique ensures quick release of phosphine, achieves peak phosphine concentration some four times that from metallic phosphide formulations, ensures rapid distribution (~140 ppm in 24 hours) and results in entomologically effective concentrations even after exposures of more than 3 months (daily loss of ~1.5 ppm in well constructed pad / bunker storage).

Continuous Dose

Requirement for low cost and 'nil electrics' fumigation equipment resulted in trials using LoDose™ regulators to dispense Phosfume™ in the fumigation of peanuts in long-term storage (Peanut Marketing Board [PMB], Kingaroy, Queensland) and fumigation of isolated on-farm grain storage (Blairgowrie Pastoral Company, Mendooran, NSW). The PMB used a LoDose™ heater regulator connected to the Phosfume™ cylinder supply and each silo (43 silos in group) was fitted with an individual low cost 0–1.5 L/minute Flow Indicator machined from clear acrylic tube and a low pressure ball valve (Fig. 1, inset). Measured phosphine level achieved in 'non-gastight' storage indicated entomologically effective concentrations are obtainable with continuous dosing. Results to date are encouraging enough to warrant additional trials.

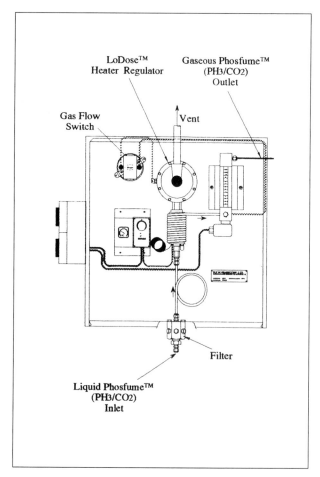

Fig. 3.   GrainCorp SIROFLO® upgrade.

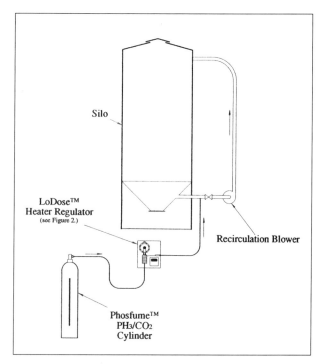

Fig. 4.   Phosfume™ recirculation system.

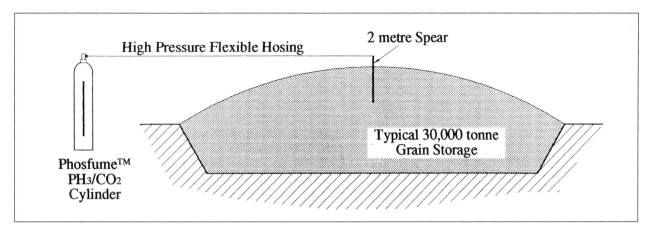

Fig. 5. Pad/bunker fumigation.

This technique using Phosfume™ was initially reported in a U.K. report (Bell et al. 1991) as a 'continuous flow system'. The difficulty of controlling gas flow experienced in the U.K. work was overcome using the LoDose™ regulators, although absence of appreciable back pressure required the inclusion of a restriction in the gas delivery line.

Conclusion

Phosfume™ capability to provide a controlled source of phosphine allows concentrations to be adjusted during a fumigation to compensate for unforseen air ingress. This adds an order of sophistication to grain fumigation. This, combined with the non-flammability and occupational health and safety benefits of Phosfume™, give significant advantages over alternatives. The new LoDose™ regulator provides long-term continuous flows required for techniques such as SIROFLO®. These techniques ensure that grain stored in non-gastight storages can be treated to achieve 'insect-free and residue-free' status and eliminate the need for traditional residual pesticide grain protectants.

The BOC Group Companies continues to develop a 'grain protection package' to include additional gases [eg carbon dioxide, nitrogen, Envirosol™ gas mixtures], gastight sealing of grain storage and dispensing equipment [e.g. SIROFLO®, non-cryogenic on-site nitrogen gas generators] to satisfy the need for cost effective 'residue-free' grain protection. With the deregulation of grain markets the BOC Group is focusing on the needs of on-farm fumigation.

References

Bell, C.H., Chakrabarti, B. and Mills K.A. 1991. New applications methods for the use of phosphine to disinfest bulk grain. Home Grown Cereals Association Project Report No. 41.

Evans, D.E. 1988. The control of stored grain pests: a changing scene. 38th Australian Cereal Chemistry Conference, 25–30 September 1988.

Ryan, R.F. 1992. Attaining insect-free and residue-free status in foods. Food Australia, 44(11), 556–557

Winks, R.G. 1983. Proc Aust Bulk Grain Handling Authority Stored Grain Protection Conference, Melbourne, Section 5,1–5.

Winks, R.G. and Ryan, R.F. 1990. Recent developments in the fumigation of grain with phosphine. 5th International Working Conference on Stored Product Protection, Bordeaux, 9–14 September 1990.

Winks, R.G. and Ryan, R.F. 1993. Some aspects of the history of chemical grain protection in Australia. Proceedings 43rd Australian Cereal Chemistry Conference, September 1988. 218–228.

A survey of phosphine and methyl bromide resistance in Malaysia

Z. Sulaiman, M. Rahim, M.E. Faridah and M. Rasal*

Abstract

Insects were collected from grain storages, flour mills and feed mills at 30 locations in Peninsular Malaysia during 1991–92. Laboratory bioassays using the FAO resistance testing method indicated that resistance to both methyl bromide and phosphine was widespread. Use of methyl bromide is not widespread in Malaysia but a large proportion of grain is fumigated prior to import. Phosphine is in common use in Malaysia.

Introduction

The earliest study on insect resistance to fumigants was reported by Champ and Dyte (1976). Their survey showed an increased tolerance of insects to fumigants, which subsequently led to worldwide reports of control failures resulting from resistance (Attia and Greening 1981; Borah and Chahal 1979; Taylor and Halliday 1986; Tyler et al. 1983; Zettler et al. 1989). In Malaysia, the first survey on resistance was carried out in 1985 (Rahim and Ong 1991). The results then indicated that some insects had shown resistance to malathion and methyl bromide. There was no indication that the test insects, *Sitophilus* spp., *Tribolium castaneum*, *Oryzaephilus surinamensis* and *Rhyzopertha dominica,* were resistant to phosphine although it is more widely used than methyl bromide to control insect infestation in stored paddy and milled rice. Methyl bromide is more often used on grain commodities before import into Malaysia, i.e. to meet the quarantine requirement.

During 1991-92 period, another survey was carried out. Samples of grains containing insects were collected from various Malaysian National Rice Board complexes/warehouses and private feed mills and paddy mills scattered all over Peninsular Malaysia. The objective of this study was to check on the extent of insect resistance to methyl bromide and phosphine in Malaysia food storages.

Methods and Materials

Four major stored-grain beetle pests were tested: *Sitophilus oryzae, S. zeamais, Tribolium castaneum* and *Rhyzopertha dominica.* Insects collected from various premises were cultured separately in fresh and sterilised media. The screening was carried out when the insects had established in the media and sufficient numbers of insects were available. Phosphine (generated from a commercial aluminium phosphide formulation) and methyl bromide were collected and injected using gastight syringes through rubber septum's in the lids of desiccators containing the test insects confined within glass rings, to give the required concentrations. The insects were exposed to various concentrations of methyl bromide (MeBr) and phosphine (PH_3) recommended by FAO for 5 and 20 hours (see Table 1), respectively (FAO 1975). After the exposure periods, the insects were transferred to sterilised culture media and held at 25°C and between 70-80% r.h. Assessments were carried out 14 days after treatment. For each test, between 40–50 insects and 3–4 replicates were used.

Results and Discussion

Table 2 shows the results obtained from the test carried out in the laboratory using both fumigants. The results obtained from the 1985 survey had indicated that the insect strains collected were not resistant to phosphine but to methyl bromide. In this study, however, the insects collected showed some levels of resistance towards both phosphine and methyl bromide, particularly *Tribolium castaneum* and *Sitophilus spp.* Thus, in order to overcome these insect resistance problems, there is a crucial need for researchers to identify new alternative fumigant(s) which are effective and at the same time they are safe and environmentally friendly.

Acknowledgment

The authors would like to thank the Director General of the Malaysian Agricultural Research and Development Institute (MARDI), and the Australian Centre for International Agricultural Research (ACIAR), for their helpful suggestions and assistance.

References

Attia, F.I. and Greening, H.G. 1981. Survey of resistance to phosphine in coleopterous pests of grain and stored products in New South Wales. General and Applied Entomology, 13, 93–97.

Borah, B and Chahal, B.S. 1979. Development of resistance in *Trogoderma granarium* Everts to phosphine in the Punjab. FAO Plant Protection Bulletin, 27, 77–80.

Champ, B. R. and Dyte, C.E. 1976. Report of the FAO global survey of pesticide susceptibility of stored grain pests. FAO Plant Production and Protection Series, No. 5. Rome, FAO, 297 p.

FAO 1975. Recommended methods for the detection and measurement of resistance of agricultural pests to pesticides. Tentative method for adults of some major beetle pests of stored cereals with methyl bromide and phosphine. FAO Method No. 16. FAO Plant Protection Bulletin, 23, 12–25.

Rahim, M. and Ong, S.H. 1991. Integrated use of pesticides in grain storage in the humid tropics. A completion report on ACIAR Project No. 8309/8311 and 8609.

* Food Technology Research Centre, MARDI, P.O. Box 12301, GPO, 50774 Kuala Lumpur, Malaysia.

Table 1. Insect species, fumigant, concentrations and exposure periods used in the test for resistance.

Insect species	Phosphine		Methyl bromide	
	Conc. (ppm)	Exposure period (hours)	Conc. (ppm)	Exposure period (hours)
Sitophilus oryzae	0.04	20	6	5
Sitophilus zeamais	0.04	20	6	5
Tribolium castaneum	0.04	20	12	5
Rhyzopertha dominica	0.03	20	7	5

Table 2. Insect strains which slow resistance towards methyl bromide and phosphine (PH_3).

Premises	*Tribolium castaneum*		*Sitophilus* spp.		*Rhyzopertha dominica*	
	Methyl bromide	PH_3	Methyl bromide	PH_3	Methyl bromide	PH_3
Complexes	4/7	3/10	4/9	1/9	0/4	4/4
Warehouses	2/4	1/5	1/3	0/4	0/1	–
Feedmills/flour mills	4/4	1/7	6/13	1/14	0/8	7/7

Taylor, R.W.D. and Halliday, D. 1986. The geographical spread of resistance to phosphine by coleopterous pests of stored products. In: Proceedings of 1986 British crop protection conference—pests and diseases, 17–20 November 1986, Brighton. British Crop Protection Council, 607–613.

Tyler, P.S., Taylor, R.W.D. and Rees, D.P. 1983. Insect resistance to phosphine fumigation in food warehouses in Bangladesh. International Pest Control, 25, 1–13, 21.

Zettler, J.L., Halliday, W.R. and Arthur, F.H. 1989. Phosphine resistance in insects infesting stored peanuts in the Southeastern United States. Journal of Economic Entomology, 82, 1508–1511.

Effectiveness of carbon dioxide under reduced pressure against some insects infesting dried fruit

L. Süss and D.P. Locatelli *[1]

Abstract

The effectiveness of CO_2 at different temperatures (20, 25 and 30°C) and exposure times (6, 9, 8, 24, 36, 48, 60 hours) in a vacuum autoclave (34.6–44 kPa) has been tested against the life stages (eggs, I and II instar larvae and mature larvae) of *Ephestia cautella* (Walk.), *Plodia interpunctella* (Herbst) and *Tribolium castaneum* Hbn. (eggs, larvae, pupae and adults).

Tests were carried out in a vacuum chamber ($3m^3$) and the samples were placed in the middle of 1 m^3 pallet of almonds.

Ninety-five and 99% mortalities were achieved after 55 and 64 hours at 20°C, 41 and 48 hours at 25°C, and 31 and 37 hours at 30°C. Pupae proved to be most resistant to the treatment. The I and II instar larvae of *Plodia interpunctella* and *Ephestia cautella* were less resistant than mature larvae. The LT_{99} at 20°C for the early and mature larvae of *Plodia interpunctella* are 40 and 64 hours, 27 and 48 hours at 25°C and 12 and 34 hours at 30°C respectively.

Introduction

Several species of insect are very common pests of dried fruit. These include, particularly, the moths *Ephestia cautella* (Walker), *E. elutella* Hubner, *E. figulilella* Greg. and *Plodia interpunctella* (Hübner), (Phycitidae) and the beetles, *Tribolium* spp. (Tenebrionidae) and *Oryzaephilus* spp. (Cucujidae). These polyphagous species live on a variety of foodstuffs ranging from flour to finished products, from tobacco to cocoa beans and almond meal. However, raisins, hazelnuts and almonds are most frequently infested.

Temperatures in warehouses are usually sufficient to permit the development of these species. They are often present in raw materials, but they can establish in processing premises and then infest finished products. For this reason it is important to control infestation of raw materials in storage.

Low temperature storage is usually used as a preventative measure against development of infestation as the activity and rate of reproduction of the insects is limited. High temperature treatment is used for products that can tolerate the treatment and, if there is a substantial infestation, fumigation at atmospheric pressure or under vacuum may be required (Bond, 1984).

Although there are fumigants currently available for treatment of dried fruit, there is an increasing preference for non-chemical methods that are intrinsically safe for humans. In the authors' opinion, it is better to develop alternative methods of pest control, such as controlled atmospheres (Jay et al. 1970; Gaunce et al. 1982; Soderstrom et al. 1991).

The present study was undertaken to see if a combination of factors: low pressure, high temperature and exposure to CO_2, could be used successfully to disinfest dried fruit.

Materials and Methods

The experiments were carried out on *E. cautella*, *P. interpunctella* and *T. castaneum*. Samples of the insects, at different stages of development, were treated with a high CO_2 atmosphere at low pressure inside a vacuum chamber (3 m^3). Their survival was observed. The chamber was the one previously used to assess control of rice pests (Locatelli and Daolio, 1993).

Test material was reared in incubators at $26 \pm 1°C$, $70 \pm 5\%$ r.h.

Samples of insects to be exposed were placed in gauze bags (0.2 mm mesh). For *T. castaneum*, samples of 20 adults, 20 mixed-age larvae or 20 pupae were used. For *P. interpuntella* and *E. cautella* samples of 20 first and second instar larvae, 20 mature larvae or 20 pupae were tested. Adults were not exposed as they are notoriously susceptible to the treatment.

Eggs of *T. castaneum* for testing were obtained by allowing 100 adults to lay for 5 days on 50 g samples of food substrate (wheat bran, whole wheat flour and wheat germ).

Eggs of the moths were obtained by allowing 50 gravid females to lay in a special plexiglass cylinder (15 cm diam., 40 cm high) fitted with a gauze base (18-mesh) through which the eggs could fall. The cylinders were held at $26 \pm 1°C$, $70 \pm 5\%$ r.h. on a 12 hours light: 12 hours dark cycle. Tests were carried out on 30–40-hour-old eggs, held under the laying conditions.

The test samples in the gauze bags were placed in the middle of 1 m^3 pallet loads of almonds held in jute sacks (50 kg). This is the form in which they are usually traded.

The exposure chamber was initially at 20°C, but a special heating system could be used to give the required temperature (20, 25 or 30°C) for each run. At the start of the exposure, the atmosphere within the chamber was modified by a vacuum pump to give an absolute pressure of between 34.6 and 37.3 kPa and by introduction of CO_2 to give a composition of 98% CO_2 by volume. At this composition the oxygen content was 0.5% or less. These conditions were maintained for test periods of 6 to 60 hours. At the end of the tests the internal pressure ranged from 38.7 to 44.0 kPa. It was necessary to maintain a high degree of vacuum throughout the entire exposure period to obtain the most rapid control.

Survival in the test samples was assessed immediately after completion of the exposure period and also 24 hours later, to check on delayed mortality. Surviving larvae and pupae were incubated until adult emergence. Each test was repeated four times, usually with four untreated controls. In some cases an estimate of control mortality was obtained from preliminary experiments. No appreciable mortality was recorded in untreated control batches of all larvae of *E. cautella* and *P. interpunctella*, and of larvae and adults of *T. castaneum*. Control survival in pupae of *P. interpunctella* in preliminary

* Istituto di Entomologia agraria - Università degli Studi, Via Celoria 2, I-20133 Milano, Italy.
[1] Research supported by National Research Council of Italy, Special Project RAISA, Sub-project N. 4 Paper N. 1297.

experiments was found to be 17 ± 1.0 and was assumed to be so for later tests.

Statistical analysis of the results was carried out using probit transformation to estimate lethal times for 95 and 99% mortality.

Results

Average survival times and standard deviations are given in Tables 1, 2, 3 for the different test species and stages, and different temperatures. Regression parameters from the probit analyses of these results and estimted LT_{95} and LT_{99} are given in Table 4. Goodness of fit and adequacy of the model are indicated by chi-square and significance levels (P).

The estimates for complete control (LT_{99}) can be used to indicate the most resistant stages and species to the treatment. Control regimens can be sent on the basis of these values and the best conditions for the foodstuff to be treated.

For *E. cautella*, the most resistant stages were the pupae and mature larvae. At 20°C, pupae and mature larvae required 63 and 58hours respectively. These times are much longer than required to kill first and second instar larvae (40 hours). At 25 and 30°C the LT_{99} of mature larvae and pupae were very similar, 45 and 43 hours and 31 hours each, respectively. First and second instar larvae and eggs were much more susceptible.

For *P. interpunctella*, mature larvae and pupae were also the most tolerant, with estimates of LT_{99} at 20, 25 and 30°C of 64,

60, 48, 48 and 34, 37 hours, respectively, and again, first and second instar larvae and eggs were more susceptible.

For *T. castaneum*, eggs and pupae were the most tolerant stages, with estimates of LT_{99} at 20, 25 and 30°C of 46, 46, 44, 44 and 17, 33 hours, respectively. The low value for the egg tolerance at 30°C is based on only two observations and has a large range of confidence limits.

There was no particular difference in the species in their response to the low pressure controlled atmosphere treatment. Eggs of the three species are the only stage to show markedly different tolerances.

Figures 1, 2 and 3 give the mortality data for *E. cautella*, *P. interpunctella* and *T. castaneum* as a plot of probit mortality against time. For most species and stages, the slope of the line doubles for increase in temperature from 20 to 30°C. Exceptions are the pupal stages of *P. interpunctella* and *T. castaneum*, mature larvae in *E. cautella* and adults of *T. castaneum*, all of which show lesser sensitivity to change in temperature.

Discussion

The probit transformation of the mortality data allows extrapolation to give the time required to give complete control, here taken to be the LT_{99}. It is also possible to extrapolate the LT_{99} expected to other temperatures, (Fig. 4) though this must be done continuously as the data values were only obtained for a narrow range of temperatures and the relation between LT_{99} and temperature may not be linear.

Table 1. Average survival (±SD) rate of different stages of *Ephestia cautella* (Walk.) and of control with the different temperatures and exposure times at chamber atmosphere of 98% CO_2 and $O_2 < 0.5\%$; absolute pressure 34.6–37.3 kPa.

Stage	Temp. (°C)		Exposure times (hours)						
			6	9	18	24	36	48	60
Eggs									
	20	treated	26 ± 4.2	21 ± 3.0	16 ± 1.7	10 ± 2.2	3.5 ± 2.6	0.2 ± 0.5	0
		control	41 ± 1.8	43 ± 3.9	42 ± 3.7	39 ± 2.2	38 ± 4.1	39 ± 3.3	31 ± 4.1
	25	treated	18 ± 2.6	12 ± 1.3	7.0 ± 0.8	1.3 ± 1.5	0		
		control	39 ± 2.1	41 ± 1.8	42 ± 3.4	36 ± 7.0	42 ± 3.9		
	30	treated	7.0 ± 2.6	6.0 ± 1.8	1.3 ± 1.5	0			
		control	41 ± 2.6	44 ± 3.7	41 ± 2.6	43 ± 4.2			
I and II instar larvae									
	20	treated	19 ± 1.0	18 ± 1.0	11 ± 1.0	6.5 ± 1.3	1.0 ± 1.4	0	
	25	treated	12 ± 1.3	7.5 ± 1.3	2.8 ± 1.0	0			
	30	treated	8.8 ± 1.0	5.3 ± 1.0	0				
Mature larvae									
	20	treated	20 ± 1.0	18 ± 0.5	16 ± 0.8	13 ± 1.0	7.3 ± 1.0	0.8 ± 1.0	0
	25	treated	19 ± 1.3	17 ± 1.3	13 ± 1.3	9.5 ± 1.3	0		
	30	treated	17 ± 1.0	11 ± 1.5	6.8 ± 1.0	1.0 ± 1.4	0		
Pupae									
	20	treated	17 ± 0.8	16 ± 0.6	13 ± 0.8	11 ± 0.8	8.5 ± 0.6	0.8 ± 1.5	0
		control	18 ± 0.8	18 ± 0.6	18 ± 1.0	18 ± 1.0	18 ± 1.0	18 ± 1.0	18 ± 0.8
	25	treated	16 ± 1.0	15 ± 1.0	12 ± 2.2	7.8 ± 1.9	0.8 ± 1.0	0	
		control	18 ± 1.0	18 ± 1.0	17 ± 1.0	17 ± 1.0	18 ± 0.8	17 ± 0.5	
	30	treated	16 ± 0.8	13 ± 1.3	6.3 ± 1.0	1.0 ± 1.2	0		
		control	18 ± 1.0	18 ± 0.8	16 ± 1.0	19 ± 0.6	18 ± 0.8		

Notes: The number or survivors in the control of I and II instar larvae and mature larvae was assumed to be 20, since no appreciable mortality was recorded in preliminary tests.

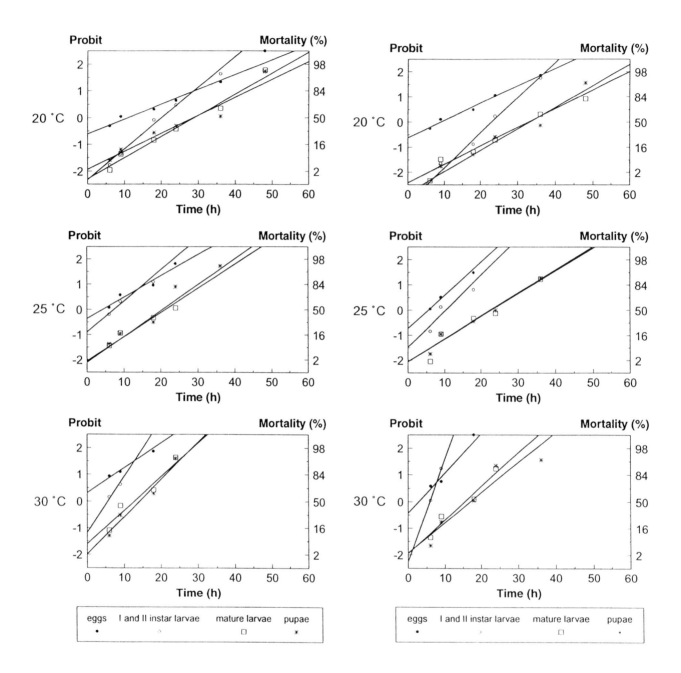

Fig.1. Mortality of *Ephestia cautella* (Walk.) eggs, I and II instars larvae, mature larvae and pupae with different temperatures (20, 25 and 30°C) and exposure times at chamber atmosphere 98% CO_2; $O_2 < 0.5\%$; absolute pressure 34.6–37.3 kPa.

Fig. 2. Mortality of *Plodia interpunctella* (Hbn.) eggs, I and II instars larvae, mature larvae and pupae with different temperatures (20, 25 and 30°C) and exposure times at chamber atmosphere 98% CO_2; $O_2 < 0.5\%$; absolute pressure 34.6–37.3 kPa.

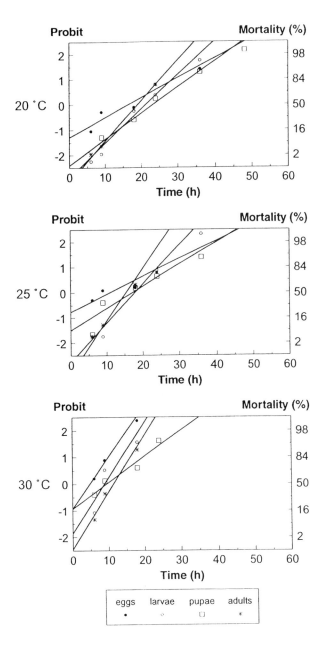

Fig. 3. Mortality of *Plodia interpunctella* (Hbn.) eggs, I and II instars larvae, mature larvae and pupae with different temperatures (20, 25 and 30°C) and exposure times at chamber atmosphere 98% CO_2; O_2< 0,5%; absolute pressure 34.6–37.3 kPa.

However, the general co clusion is that there is a need to adjust the time of treatment for the temperature of the foodstuff treated.

Table 5 gives the LT_{99} for the most tolerant stage of each of the three species tested for each of the three exposure temperatures. Complete kill (LT_{99}) of all stages of all the species was reached in 37 hours at 30°C and 64 hours at 20°C.

Our results indicate that carbon dioxide under reduced pressure is less toxic at lower temperatures, as shown by many authors for controlled atmospheres at ambient pressure. If the temperature is below the optimum for insect development, its metabolism is slower and thus less oxygen is required and the effect of the controlled atmosphere is lessened (Bailey and Banks 1980; Jay 1986).

The experiments described here show the different stages of the phycitid moths to be very tolerant at all temperatures compared with those of *T. castaneum*. Also, the mature larvae and pupae were found to be the most tolerant stages.

Generally, the last larval instar is reported to be the most tolerant stage to CO_2 (Jay 1984). It also has been observed with other species exposed to CO_2 that the pupae are very tolerant. Pupae, in the presence of unfavourable ambient conditions, can modify their metabolism to assist survival (Lindgren and Vincent 1970; Childs and Overby 1983).

The tolerance of *P. interpunctella* in the experiments described here was higher than observed previously (Locatelli and Daolio 1993). However, in the present case the test insects were placed in the middle of a pallet of bagged almonds where they were more protected. Under these conditions the rate of access of the CO_2 to the insects is slower and it takes a longer time to heat the whole mass of tested product to the test temperature.

Time required to obtain disinfestation with carbon dioxide under reduced pressure is considerably shorter than for normal controlled atmospheres (Annis 1987), but longer than required for vacuum fumigation with methyl bromide (Bond 1984).

Though the increase in temperature results in a higher mortality and could be useful in reducing the treatment period, some foodstuffs could be subject to changes in organoleptic quality and the use of increased temperature for reducing treatment time will be restricted to some products only. Undoubtedly, when assessing the utility of this method for other commodities, the possible organoleptic and quality changes will have to be considered. However, the use of this method is particularly appropriate, despite the technical difficulties, where the traditional disinfestation methods result in a high level of residues of the active material.

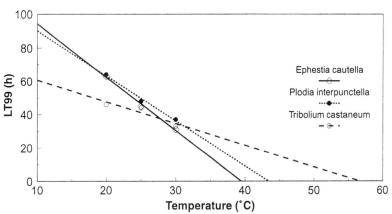

Fig. 4. Plot of the regression lines of LT_{99} and temperature for the *Ephestia cautella* (Walk.), *Plodia interpunctella* (Hbn.) and *Tribolium castaneum* Herbst.

Table 2. Average survival (±SD) rate of different stages of *Plodia interpunctella* (Hbn.) and of control with the different temperatures and exposure times at chamber atmosphere of 98% CO_2 and $O_2 < 0.5\%$; absolute pressure 34.6–37.3 kPa.

Stage	Temp. (°C)		Exposure times (hours)						
			6	9	18	24	36	48	60
Eggs									
	20	treated	23 ± 2.9	18 ± 2.2	12 ± 2.8	5.5 ± 1.3	1.3 ± 1.5	0	
		control	39 ± 3.9	39 ± 2.9	38 ± 1.7	38 ± 4.2	40 ± 2.9	40 ± 2.8	
	25	treated	17 ± 2.2	13 ± 2.1	2.8 ± 2.2	0			
		control	35 ± 4.6	42 ± 3.9	40 ± 2.9	40 ± 6.8			
	30	treated	11 ± 2.2	8.8 ± 1.7	0.3 ± 0.5	0			
		control	40 ± 3.3	39 ± 2.6	41 ± 1.7	41 ± 1.9			
I and II instar larvae									
	20	treated	20 ± 0	19 ± 1.2	16 ± 1.0	8.3 ± 1.7	0.8 ± 1.0	0	
	25	treated	16 ± 1.8	9.0 ± 0.8	4.3 ± 1.7	0			
	30	treated	9.5 ± 1.3	2.3 ± 1.5	0				
Mature larvae									
	20	treated	20 ± 0.5	19 ± 0.6	18 ± 0.6	15 ± 1.0	7.5 ± 0.6	3.5 ± 0.6	0
	25	treated	20 ± 0.6	17 ± 1.3	13 ± 1.3	11 ± 1.2	2.3 ± 1.3	0	
	30	treated	18 ± 0.5	14 ± 1.7	9.3 ± 1.0	2.3 ± 1.5	0		
Pupae									
	20	treated	17 ± 0.8	16 ± 1.0	15 ± 0.5	12 ± 1.0	9.3 ± 0.5	1.0 ± 2.0	0
	25	treated	16 ± 1.0	14 ± 0.8	11 ± 1.0	8.5 ± 1.3	1.8 ± 1.6	0	
	30	treated	16 ± 0.8	13 ± 1.3	8.3 ± 1.0	1.5 ± 1.7	1.0 ± 1.4	0	

Notes: The number or survivors in the control of I and II instar larvae and mature larvae was assumed to be 20, since no appreciable mortality was recorded in preliminary tests. The number of survivors in the control of pupae was measured only in the first part of the experiment and the same mortality was assumed as constant throughout the whole experiment. The number of survivors in this case was 17 ± 1.0.

Table 3. Average survival (±SD) rate of different stages of *Tribolium castaneum* (Herbst) and of control with the different temperatures and exposure times at chamber atmosphere of 98% CO_2 and $O_2 < 0.5\%$; absolute pressure 34.6–37.3 kPa.

Stage	Temp. (°C)		Exposure times (hours)						
			6	9	18	24	36	48	60
Eggs									
	20	treated	25 ± 4.2	19 ± 1.7	17 ± 1.7	6.0 ± 1.8	2.3 ± 1.7	0	
		control	29 ± 2.8	31 ± 3.4	31 ± 2.6	28 ± 2.8	28 ± 1.7	30 ± 1.3	
	25	treated	19 ± 2.5	13 ± 2.8	12 ± 2.5	6.8 ± 1.3	0		
		control	31 ± 4.0	29 ± 2.1	28 ± 2.2	32 ± 2.4	30 ± 1.7		
	30	treated	13 ± 2.8	5.5 ± 2.1	0.3 ± 0.5	0			
		control	32 ± 2.6	29 ± 2.2	29 ± 2.8	29 ± 2.2			
Larvae									
	20	treated	20 ± 0.5	20 ± 1.0	12 ± 2.2	7.0 ± 0.8	0.8 ± 1.0	0	
	25	treated	19 ± 1.0	19 ± 1.0	7.8 ± 1.7	3.8 ± 2.6	0.3 ± 0.5	0	
	30	treated	17 ± 1.0	6.0 ± 2.2	1.3 ± 1.5	0			
Pupae									
	20	treated	16 ± 1.3	14 ± 1.4	11 ± 1.8	6.0 ± 1.4	1.5 ± 1.9	0.3 ± 0.5	0
		control	16 ± 1.3	16 ± 1.3	15 ± 1.0	15 ± 0.8	16 ± 1.3	16 ± 1.0	16 ± 0.6
	25	treated	15 ± 1.3	9.8 ± 1.0	6.5 ± 1.3	4.0 ± 0.8	1.3 ± 1.5	0	
		control	15 ± 1.2	15 ± 0.9	15 ± 0.8	15 ± 1.0	16 ± 0.9	16 ± 1.0	
	30	treated	10 ± 0.8	7.0 ± 0.8	4.3 ± 1.0	0.8 ± 1.0	0		
		control	15 ± 1.3	15 ± 0.8	15 ± 1.0	15 ± 1.5	16 ± 0.6		
Adults									
	20	treated	20 ± 0.6	19 ± 0.8	11 ± 1.0	4.3 ± 1.0	0		
	25	treated	19 ± 0.5	18 ± 1.4	7.5 ± 1.9	0			
	30	treated	18 ± 0.5	13 ± 2.2	2.0 ± 2.5	0			

Notes: The number or survivors in the control of larvae and adults was assumed to be 20, since no appreciable mortality was recorded in preliminary tests.

Table 4 Regression parameters of the probit lines (intercept, slope and observed significance level of chi-square statistic, P) and lethal times to kill 95 and 99% (LT$_{95}$ and LT$_{99}$) estimated for different stages of *Ephestia cautella* (Walk.), *Plodia interpunctella* (Hbn.), and *Tribolium castaneum* (Herbst) with different temperatures at chamber atmosphere of 98% CO_2 and O_2 < 0.5%; absolute pressure 34.6–37.3 kPa.

Species	Stage	Temp. (°C)	Intercept	Slope ± SE	P	LT$_{95}$ (h) (95% CL)	LT$_{99}$ (h) (95% CL)
Ephestia cautella (Walk.)							
	Eggs						
		20	−0.61	0.06 ± 0.008	0.916	41 (35–50)	53 (44–67)
		25	−0.35	0.08 ± 0.017	0.854	24 (20–32)	32 (26–45)
		30	0.32	0.09 ± 0.031	0.993	14 (11–24)	22 (16–44)
	I and II instar larvae						
		20	−2.31	0.11 ± 0.018	0.989	34 (30–42)	40 (34–51)
		25	−0.89	0.12 ± 0.030	0.960	20 (17–30)	26 (2041)
		30	−1.17	0.21 ± 0.088	0.999	13 (10–39)	16 (12–56)
	Mature larvae						
		20	−2.29	0.08 ± 0.011	0.930	50 (44–59)	58 (51–71)
		25	−2.11	0.10 ± 0.017	0.685	36(31–45)	43 (36–55)
		30	−1.60	0.12 ± 0.024	0.739	26 (22–34)	31 (2642)
	Pupae						
		20	−1.93	0.07 ± 0.010	0.603	53 (46–65)	63 (54–79)
		25	−2.05	0.10 ± 0.016	0.903	38 (33–48)	45 (38–59)
		30	−1.98	0.14 ± 0.026	0.915	26 (22–34)	31 (26–41)
Plodia interpunctella (Hbn.)							
	Eggs						
		20	−0.62	0.07 ± 0.011	0.973	33 (28–41)	43 (36–55)
		25	−0.72	0.13 ± 0.027	0.986	18 (15–24)	23 (19–32)
		30	−0.41	0.15 ± 0.044	0.962	14 (11–22)	18 (14–33)
	I and II instar larvae						
		20	−3.22	0.14 ± 0.024	0.953	35 (31–42)	40 (3449)
		25	−1.47	0.14 ± 0.029	0.660	22 (18–29)	27 (22–37)
		30	−2.24	0.38 ± 0.150	.999	10 (8.5–22)	12 (9.6–30)
	Mature larvae						
		20	−2.41	0.07 ± 0.010	0.887	55 (49–65)	64 (56–67)
		25	−2.05	0.09 ± 0.014	0.760	41 (35–50)	48 (41–61)
		30	−1.94	0.13 ± 0.022	0.872	28 (24–36)	34 (2844)
	Pupae						
		20	−2.83	0.08 ± 0.013	0.726	52 (46–63)	60 (53–74)
		25	−2.05	0.09 ± 0.015	0.934	40 (35–51)	48 (40–62)
		30	−1.92	0.11 ± 0.020	0.495	31 (27–40)	37 (31–49)
Tribolium castaneum (Herbst)							
	Eggs						
		20	−1.29	0.08 ± 0.011	0.515	37 (32–46)	46 (39–58)
		25	−0.76	0.07 ± 0.012	0.367	34 (29–44)	44 (36–59)
		30	−0.92	0.19 ± 0.065	0.997	13 (11–24)	17 (13–35)
	Larvae						
		20	−2.88	0.14 ± 0.023	0.968	33 (2940)	38 (33–48)
		25	−2.75	0.15 ± 0.026	0.802	28 (25–34)	33 (28–41)
		30	−1.86	0.21 ± 0.049	0.345	17 (14–24)	20 (16–29)
	Pupae						
		20	−2.41	0.10 ± 0.018	0.924	39 (34–49)	46 (39–59)
		25	−1.50	0.09 ± 0.017	0.698	36 (30–48)	44 (36–60)
		30	−0.93	0.10 ± 0.025	0.960	26 (21–40)	33 (26–54)
	Adults						
		20	−3.05	0.16 ± 0.028	0.999	29 (26–36)	33 (2942)
		25	−3.20	0.21 ± 0.034	0.911	23 (21–28)	27 (23–33)
		30	−2.44	0.21 ± 0.040	0.979	19 (16–24)	22 (19–29)

Notes: SE= standard error; CL=confidence limits; h=hours

Table 5. Time needed to kill the 99% (LT$_{99}$) of individuals of the most resistant stage of the three species investigated, at different experimental temperatures in chamber atmosphere conditions of 98% CO_2 and O_2 < 0.5%; absolute pressure 34.6–37.3 kPa

Temp. (°C)	LT$_{99}$ (hours)		
	Ephestia cautella	*Plodia interpunctella*	*Tribolium castaneum*
20	63	64	46
25	45	48	44
30	31	37	33

References

Annis, P.C., 1987. Towards rational controlled atmosphere dosage schedules: a review of current knowledge. In: Donahaye, E. and Navarro, S., ed., Proceedings of the Fourth International Working Conference on Stored-product Protection, Tel-Aviv, Israel, September 1986, 128–142.

Bailey, S.W. and Banks, H.J. 1980. A review of recent studies of the effects of controlled atmospheres on stored product pest. In: Shejbal, J., ed., Controlled atmospheres storage of grains: proceedings of an international symposium, Castelgandolfo (Rome), May 1980. Elsevier, Amsterdam, 101–118.

Bond, E.J., 1984. Manual of fumigation for insect control. FAO Plant Production and Protection Paper 54, Rome, 232–237.

Childs, D.P. and Overby, J.E., 1983. Mortality of the cigarette beetle in high-carbon dioxide atmospheres. Journal of Economic Entomology, 76, 544–546.

Gaunce, A.P., Morgan, C.V.G. and Meheriuk, M. 1982. Control of tree fruit insects with modified atmospheres. In: Richardson, D.G. and Meheriuk, M., ed., Controlled Atmospheres for Storage and Transport of Perishable Agricultural Commodities. Timber Press, Beaverton, Ore. (USA), 383–390.

Jay, E.G., 1984. Imperfections in our current knowledge of insect biology as related to their response to controlled atmospheres. In: Ripp, B.E., ed., Controlled Atmosphere and Fumigation in Grain Storages: proceedings of an international Symposium, Perth (Australia) April 1983. Elsevier, Amsterdam, 493–508.

Jay, E.G., 1986. Factors affecting the use of carbon dioxide for treating raw and processed agricultural-products. In: GASGA Seminar on Fumigation Technology in Developing Countries. Tropical Development and Research Institute, London, 173–189.

Jay, E.G., Redlinger, L. M. and Laudani, H. 1970. The application and distribution of carbon dioxide in a peanut (groundnut) silo for insect control. Journal of Stored Products Research, 6, 247–254.

Lingren, D.L. and Vincent, L.E., 1970. Effect of atmospheric gases alone or in combination on the mortality of granary and rice weevils. Journal of Economic Entomology, 63, 1926–1929.

Locatelli ,D.P. and Daolio, E., 1993. Effectiveness of carbon dioxide under reduced pressure against some insects infesting packaged rice. Journal of Stored Products Research,, 29, 81–87.

Soderstrom, E.L., Brandl, D.G. and Mackey, B., 1991. Responses of *Cydia pomonella* (L.) (Lepidoptera: Tortricidae) adults and eggs to oxygen deficient or carbon dioxide enriched atmospheres. Journal of Stored Products Research, 27, 95–101.

Evolution of phosphine from aluminium phosphide formulations at various temperatures and humidities

Tan Xianchang*

Abstract

The rates of evolution of phosphine from aluminium phosphide tablets and sachets (produced in China, Shenyang Pesticide Factory) were determined in the laboratory in an air stream under controlled temperatures and humidities. Results showed that the maximum rate of evolution of phosphine was linearly related to the absolute humidity of the air. With absolute humidity moisture contents of air from 4.7 g/m^3 to 35.5 g/m^3 decomposition times ranged from 36 to 204 hours. The decomposition rate of the tablets was slightly faster than the sachets.

Introduction

Phosphine is an important fumigant for controlling insect pests of stored grain in China. Almost 85% of grain in state warehouses is fumigated with phosphine generated from aluminium phosphide every year. In actual practice, it was observed that aluminium phosphide tablets sometimes had not completely decomposed by the end of fumigation. Incomplete decomposition was noticed especially in short fumigations of 3–5 days exposure time and in dry grain.

Meuser et al. (1977) studied hydrolysis rates of aluminium phosphide in desiccators at two temperature and humidity combinations.

In this study, the decomposition rates of aluminium phosphide tablets and sachets were determined in the laboratory in an air stream under four temperature and humidity combinations, to provide basic data relevant to use of aluminium phosphide in fumigation practice.

Materials and Methods

Materials

The aluminium phosphide formulations used were produced at the Shenyang Pesticide Factory in China and contained 55% aluminium phosphide. Each tablet weighed 3.3 g and each sachet 33 g.

Methods

A special apparatus (Fig. 1) was assembled with the aim of passing air of constant humidity (± 5%) over the exposed tablets or sachets held at constant temperatures (±0.5°C). Dry air was produced by passing an airstream through calcium chloride tubes, and moist air by passing a second airstream through wash bottles filled with water. The flow rates were controlled by valves, and air of the desired humidity was obtained by varying the relative flow rates. The airstream of known constant humidity then passed through a flow meter and a wet and dry bulb hygrometer before it passed through the decomposition chamber. Every determination involved either 10 tablets or 1 sachet of the aluminium phosphide formulation. The air flow rate was 30 L/minute and this flow rate was calculated to provide sufficient moisture for complete reaction. The effluent gas was sampled using an automatic gas sampler. The phosphine concentration was determined using a colorimetric method in which the phosphine was absorbed in potassium permangamate solution and reacted with molybdenum blue (Boltz 1958). Formulations were considered to have decomposed completely when phosphine was no longer detected in the air stream.

Results and Discussion

The times for complete decomposition of both the tablet and sachet formulations under various temperatures and humidities are given in Table 1. The longest time was 204 hours for an air moisture content of 4.7 g/m^3 and the shortest was 36 hours for a moisture content of 35.5 g/m^3 and decomposition of tablets was slightly faster than for sachets. The formulation differences are unlikely to be significant in commercial fumigation practice.

The rates of evolution of phosphine at various times throughout the reaction are shown in Figure 2. Tablets reached the maximum rate slightly more rapidly than did sachets, and the maximum rates of evolution were slightly higher.

The maximum rates of evolution of phosphine for each temperature and relative humidity combination are listed in Table 2. Linear regression of these maximum rates with absolute humidities gave a correlation of coefficient $r = 0.963$, indicating a direct relationship between rate of evolution and absolute humidity. The data are consistent with those of Mori and Kawamoto (1977).

---

* Chengdu Grain Storage Research Institute, Ministry of Commerce, 95 Huapaifang, Chengdu, Sichuan 610031, China.

Table 1. Times for the complete decomposition of aluminum phosphide at various temperatures and relative humidities.

Temperature (°C)	Formulation	Relative humidity (%)							
		90		70		50		30	
		Decomposition time (hours)	Moisture content of air (g/m³)	Decomposition time (hours)	Moisture content of air (g/m³)	Decomposition time (hours)	Moisture content of air (g/m³)	Decomposition time (hours)	Moisture content of air (g/m³)
10	Tablet	102	8.5	143	6.6	198	4.7	—	—
	Sachet	106	—	138	—	204	—	—	—
20	Tablet	48	15.5	78	12	98	8.6	—	—
	Sachet	58	—	84	—	108	—	—	—
30	Tablet	35	27.9	40	21.7	72	15.5	108	9.3
	Sachet	42	—	48	—	78	—	120	—
40	Tablet	—	—	38	35.5	—	—	—	—
	Sachet	—	—	42	—	—	—	—	—

Table 2. Maximum rates of evolution of phosphine (g/hour) from 10 tablets or one sachet of aluminium phosphide at various combinations of temperature and relative humidity.

Temperature (°C)	Formulation	Relative humidity			
		90%	70%	50%	30%
10	Tablet	0.190	0.130	0.110	—
	Sachet	0.188	0.132	0.088	—
20	Tablet	0.346	0.268	0.190	—
	Sachet	0.350	0.270	0.172	—
30	Tablet	0.550	0.446	0.308	0.176
	Sachet	0.470	0.424	0.284	0.162
40	Tablet	—	0.736	—	—
	Sachet	—	0.552	—	—

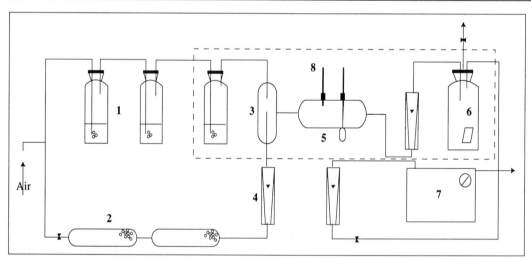

Fig. 1. Apparatus for determination of the evolution rate of phosphine from aluminium phosphide. (1) wash bottle, (2) calcium chloride tube, (3) gas mixer, (4) rotameter, (5) wet and drybulb hygrometer, (6) reaction vessel, (7) automatic gas sampler and (8) controlled temperature cabinet.

Conclusion

The rate of evolution of phosphine by aluminium phosphide is directly influenced by absolute humidity and was not directly influenced by temperature.

References

Boltz, D.F. 1958. Colorimetric determination of nonmetals. New York, Interscience Publishers, Inc., 32–36.

Meuser, F. Rajani, C. and Reimers, H. 1977. Rate of hydrolysis of pelleted metal phosphides in grain fumigations. Milling Feed and Fertiliser, 160, (10) 15–16, 18, (11) 27–28, 34.

Mori, T. and Kawamoto, N. 1977. Studies on the properties and effect of the fumigant aluminium phosphide. Research Bulletin of the Japan Plant Protection Service, 3, 24–25.

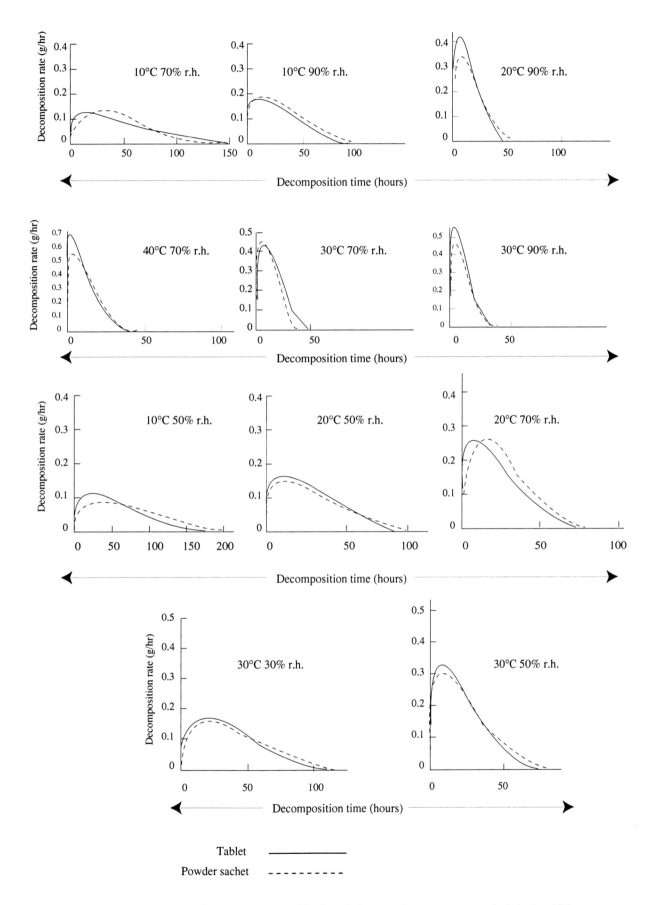

Fig. 2. Rate of evolution of phosphine from aluminium phosphide formulations at various temperatures and relative humidities.

Carbon dioxide fumigation of processed dried vine fruit (sultanas) in sealed stacks

C. Tarr,* S.J. Hilton,† J. van S. Graver† and P.R. Clingeleffer*

Abstract

Dried vine fruit, in Australia, is disinfested with ethyl formate and methyl bromide. Alternative disinfestation techniques for methyl bromide are being sought, and laboratory studies indicate carbon dioxide may be suitable. This paper describes two trials to test the feasibility of storing sultanas under sealed plastic membranes with an initial disinfestation using carbon dioxide.

The stacks were dosed at different rates, which gave 100% mortality of test insects (*Oryzaephilus surinamensis, O. mercator, Plodia interpunctella, Tribolium confusum* and *T. castaneum*) placed inside the cartons of fruit.

There was no evidence that condensation occurred during the periods of sealed storage, which lasted for 60 and 50 days, respectively. Organoleptic testing of samples, taken before and after the trials, indicated that sultana quality was unaffected by the treatment. The trials demonstrated that sealed stack storage has potential for long-term storage of large stacks of dried vine fruit.

Introduction

Dried vine fruit (currants, raisins/lexias, sultanas) is very susceptible to insect infestation during storage and transport. Over 20 insect species have been found infesting these products in Australia. The major pests are *Oryzaephilus surinamensis, O. mercator, Plodia interpunctella,* and various *Ephestia* spp. Other pests include *Tribolium castaneum, T. confusum, Carpophilus hemipterus* and *Drosophila melanogaster* (Tarr and Hilton, unpublished data).

The Australian dried fruit industry takes positive steps to prevent infestation of its product. Pest control methods applied include pre-packaging fumigation with ethyl formate and post-packaging fumigation with methyl bromide where stocks may be stored for prolonged periods. However, the industry seeks to reduce its use of chemicals that may leave undesirable residues. This objective, together with concern about environmental damage caused by methyl bromide, and potential restrictions on its production and availability, have led to investigation of alternative pest control measures. Attention has focused on modified atmospheres because of their minimal impact on the environment, greater worker safety, and because they leave no undesirable residues.

Initial studies have shown that sultanas stored in atmospheres up to 99% carbon dioxide (CO_2) for six months, produced no off-flavours or odours, and CO_2 had no deleterious effect on product colour, even at high temperatures (35°C).

All the major insect species infesting dried vine fruit in Australia are reported to be controlled by high CO_2 atmospheres (Annis 1987), except *Carpophilus hemipterus,* which has not been investigated. However, this gas has been used successfully to disinfest sultanas (van S. Graver and Hilton, unpublished data) in a sealed freight container using the method described by Banks (1988).

The commodity is usually packaged, after processing, in polyethylene-lined 14 kg cardboard cartons and stored in large stepped, 100–160 tonne stacks of tightly packed cartons (Figs 1 and 2). Carbon dioxide has the advantage that it can be used with little modification to current storage practice, by sealing carton-stacks of dried fruit in plastic membranes (Annis and van S Graver 1990). In situations where long-term storage (up to 18 months) is envisaged, the advantage conferred by storage within an insect-proof enclosure obviates the need for repeated treatments (Annis and van S. Graver 1987). The method also requires lower initial carbon dioxide gas concentrations than other modified atmospheres to obtain a successful treatment (Freidlander 1984).

The objective of the work reported here was to demonstrate that large stacks of packed sultanas could be disinfested with CO_2 and stored under sealed plastic membranes, without deleterious effects on the commodity.

Materials and Methods

Two trials were held in 1992. Each represented a different stacking configuration common in packing sheds and a different sealing system. The methodology used in both cases is described in full by Annis and van S. Graver (1990). After sealing, each stack was monitored for changes in CO_2 concentration and relative humidity. Bioassays were placed in each stack and removed after unsealing.

Trial 1

This trial was conducted from 1 April to 1 June 1992. One stack of 125 t of dried sultanas was used. The cartons, unitised on slip sheets (72 cartons per unit) and wrapped in stretch wrap (linear low density polyethylene), were stacked directly on the floor sheet. The stack was 10 m long, 6 m wide, and 3.6 m high (with a step at 2.1 m). To enhance gas distribution, a plenum was formed at floor level by laying a row of pallets along the central longitudinal axis of the stack (Fig. 1).

In this trial the cover sheet was sealed to the floor sheet with PVC solvent glue (Annis and van S. Graver 1990). The sheeted, sealed and pressure-tested stack was dosed with CO_2 delivered from a bulk tanker through a heat exchanger. Gaseous CO_2 was preferred to snow shooting because of the unknown effects of the freezing temperatures (−78.5°C) of CO_2 snow on the commodity and packaging.

Trial 2

This trial was conducted from 9 July to 19 September 1992. One stack of 139 t of dried sultanas was used. The stack con-

* CSIRO Division of Horticulture, PMB, Merbein Victoria 3505, Australia.
† Stored Grain Research Laboratory, CSIRO Division of Entomology, GPO Box 1700, Canberra ACT 2601, Australia.

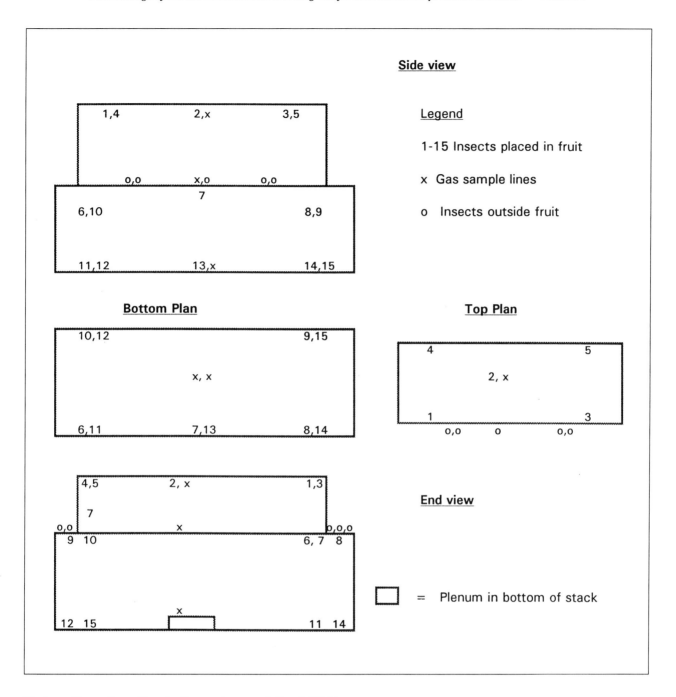

Fig. 1.   The positions of insect cultures and gas sample lines in Trial 1.

sisted of 9955 cartons stacked with four receding steps, each step one or two cartons wide. The cartons were hand stacked directly onto the floor sheet. To assist gas distribution, a 10–15 cm space, extending from the floor to the top of the stack, was left through the axis on both sides of the stack (Fig. 2).

The sheeted stack was sealed using a system of clamps and lengths of square-section pipe. Pressure applied by the clamps to the piping (laid along the periphery of the sheets), against the warehouse floor, held and sealed the sheets together.

The sheeted, sealed and pressure-tested stack was dosed with CO_2 delivered from a portable pallet tank through a heat exchanger. As in the previous trial, gaseous CO_2 was preferred to snow shooting.

Bioassays

Insect cultures for bioassay were placed in the cartons under 5–10 cm of fruit, with the inner polyethylene liner refolded and the cartons located at various positions in the stack (Figs 1

and 2). The insects, in 50 g of rearing medium, were held in chromed steel cages fitted with 60 gauge mesh gauze windows. The same species (*Oryzaephilus surinamensis, O. mercator, Tribolium castaneum, T. confusum* and *Plodia interpunctella*), were used in the both trials, with the exception of *Oryzaephilus mercator*, which was unavailable for the second trial.

The insects were reared on unprocessed dried vine fruit at 27°C and 60% r.h. at CSIRO Division of Entomology, Canberra, in a controlled temperature (CT) room. The day before each trial, 400 g whole cultures of each species containing all stages of development were divided into eight 50 g subsamples. Five subsamples were selected randomly as test (3), field control (1), and laboratory control (1), and the remainder discarded. The laboratory control for the first stack was held in the CT room, and for the second stack at the CSIRO Division of Horticulture, Merbein, in a CT room at 27°C and 50% r.h. for the duration of the trial.

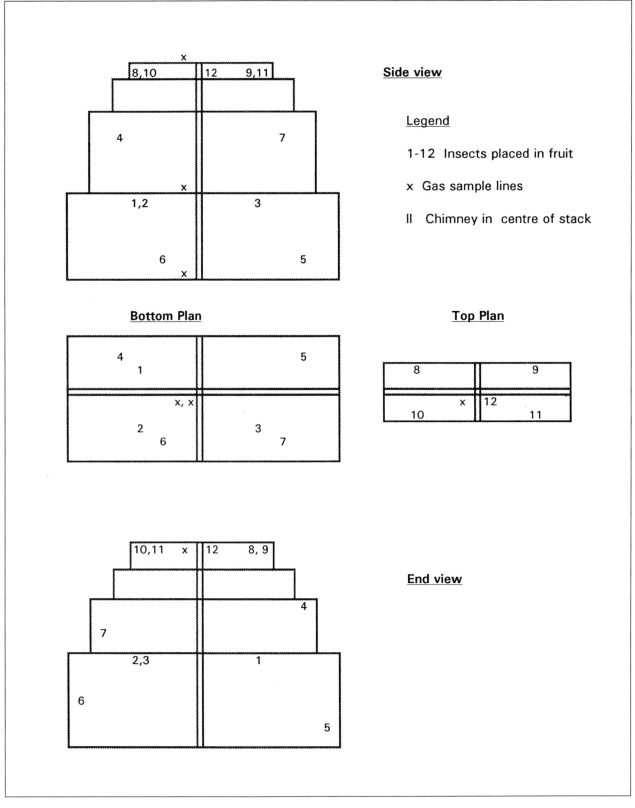

Fig. 2. The positions of insect cultures and gas sampling lines in Trial 2.

Test and field control samples were transported in an insulated box to the trial sites. Test samples were placed within the cartons in the stack, and field control samples taken to the CSIRO Division of Horticulture and held at ambient conditions for the remainder of the trial. These controls were not held at the site because of the risk from routine pest control measures that might have affected their value.

After 60 days, the first stack was opened and the insect samples collected and sent with field controls to CSIRO Division of Entomology for assessment. The second stack was opened after 50 days, and insect samples collected and taken for assessment to the CSIRO Division of Horticulture. Samples from the centre of the stack could not be removed for a week because of their inaccessibility. However, assessment and incubation were as with other test samples.

Assessment of test and control insects was made by hand sorting the samples and recording stages (adult, larvae or pupae) as alive or dead. Samples were reassessed after 30 days of incubation in the CT room to check for survival of immature stages in both cases.

Sample cultures of 200 g were also exposed in containers outside the dried fruit cartons (see Fig. 1) in the first trial. These were used as demonstration cultures at an industry meeting and were not incubated further.

Carbon dioxide concentrations and relative humidity in the enclosures

The stacks were dosed with CO_2 until gas concentrations at the top purge vent were greater than 80%. These dosages were 550 and 250 kg of CO_2/stack, respectively.

In both trials, initial determinations of the CO_2 concentration in the air purged from the stacks during dosing were made with a Riken gas interferometer.

Thereafter, for the duration of the storage period, CO_2 concentrations within the enclosures were monitored using Dräger CO_2 5%/A tubes (Cat. no. CH 20301). These tubes are able to measure CO_2 concentrations between 5 and 60%. Samples for this purpose were drawn through 2 mm internal diameter nylon piping from positions at the bottom, centre and top of the stack (Fig. 1).

In the first trial, the relative humidity within the enclosure was determined similarly, using Dräger water vapour (8101081) tubes (1–18 mg/L).

Quality assessment

In both trials fruit samples were taken for quality assessment before and after treatment. Samples were removed from the bottom, corners and sides of the stack. Taste testing of the samples taken from the stacks was conducted approximately one month after unsealing each stack and was based on a forced choice triangle test (ASTM 1968). The panel of 12 tasters was presented with four successive triangle tests. The tasters were required to identify the sample they believed to be different and state which taste, if any, they preferred. Comments on taints were encouraged as any unusual taste may have been attributable to the solvent glue used in the first trial, not the effect of CO_2 on the fruit.

Results and Discussion

Pressure tests

Both stacks successfully passed two pressure decay tests (Annis and van S. Graver 1990) before dosing with CO_2. Pressure decay halving times from 800 to 400 pa of 22, 24 and 34, 29 minutes were obtained in trials 1 and 2, respectively. These values are more than double the required standard, indicating that the membranes had been sealed effectively.

Carbon dioxide concentrations and relative humidity

Carbon dioxide concentrations obtained within the enclosures are displayed in Figure 3. These data were obtained using Dräger detector tubes, which have an upper limit of 60%. Thus, concentrations above this can only be interpolated. It can be seen that the requirement for CO_2 concentrations to be maintained above 35% at 15 days or longer (Banks et al. 1980; Annis and van S. Graver 1990) was successfully achieved in both trials.

The first stack was dosed with 550 kg of CO_2 (equivalent to 3.4 kg/t) with the concentration not falling below 35% after 60 days storage. This well-sealed stack would have been suitable for long-term storage.

The dose applied to the second stack was considerably lower — 250 kg of CO_2 (equivalent to 1.7 kg/t). This may be attributable to the spaces built along the axes of both sides of the stack. These were intended to assist the distribution of the gas inside the enclosure. However, a rapid rise in CO_2 concentration was detected during the purge, possibly because the gas was funnelled directly to the top of the enclosure. This led to a premature halt to dosing and was responsible for the lower gas concentrations achieved during the trial. Nonetheless, the CO_2 concentration was held above 35% for 21 days. Thus, both treatments were successful.

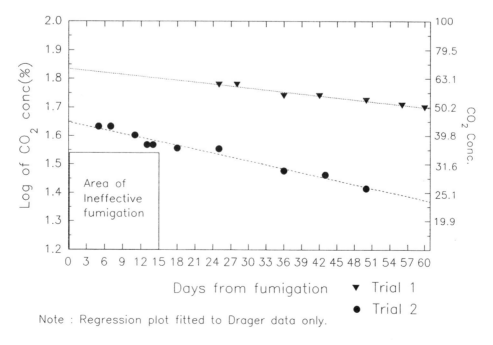

Fig. 3. Carbon dioxide concentrations in a sealed stack of sultanas following fumigation.

Water vapour readings were used as a guide to relative humidity changes in the first trial. The results are shown in Figure 4. It can be seen that water vapour changes within the enclosure resembled ambient water vapour changes. No moisture accumulation occurred during the trial, and since water vapour within the enclosure remained similar to external water vapour no moisture accumulation is expected to occur in future trials.

Bioassays

Insect mortalities in the first trial are given in Table 1. The treatment gave 100% mortality of all life stages of all cultures. Insect mortalities of the second trial were 100% for all adults, pupa and large larva present, but after the 30-day incubation a

few small, immature larvae were found (Table 2), giving total mortalities of between 100 and 89.9%. However, all cultures from the second trial had been incubated together and it was found that *O. surinamensis* had escaped through the 60 gauge mesh, with five adults from the controls found wandering amongst the samples. Subsequent reincubation of all immatures found in culture samples demonstrated that 'survivors' were all *O. surinamensis* regardless of the origin of the culture. Thus, it was concluded that the 'survivors' were offspring of the loose *O. surinamensis* or escapees from the controls.

We believe that a dose between the two used in these trials should produce desirable results with an economic advantage over the first trial's high dosage levels and a more certain insecticidal result than the second trial.

Table 1. Test insect mortalities in Trial 1.

Species	Adults	Larvae	Pupae	Adults	Larvae	Pupae	Total live insects	Mortality (%)
		Alive			Dead			
P. interpunctella								
Test	0	0	0	1	100	2	0	
Control	0	13	2	3	78	3	15	100
T. castaneum								
Test	0	0	0	34	96	0	0	
Control	16	34	0	6	5	0	50	100
T. confusum								
Test	0	0	0	183	276	2	0	
Control	205	118	0	8	0	0	323	100
O. mercator								
Test	0	0	0	423	19	0	0	
Control	327	45	7	48	1	0	379	100
O. surinamensis								
Test	0	0	0	317	20	1	0	
Control	352	32	3	45	0	0	387	100

Table 2. Test insect mortalities in Trial 2.

Species	Adults	Larvae	Pupae	Adults	Larvae	Pupae	Total live insects	Mortality (%)
		Alive			Dead			
P. interpunctella								
Test	0	0	0	0	77	7	0	
Control	0	43	0	0	76	9	43	100
T. castaneum								
Test	0	4	0	170	94	35	4[a]	
Control	189	562	10	137	0	2	761	99.47
T. confusum								
Test	0	2	0	47	5	0	2[a]	
Control	17	1	0	14	0	0	18	88.89
O. surinamensis								
Test	0	1	0	127	17	11	1[a]	
Control	108	88	1	54	4	0	197	99.49

[a]All larva found were early stage *O. surinamensis*, indicating contamination late in incubation.

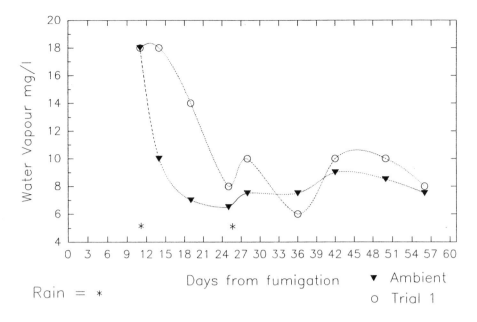

Fig. 4. Change in water vapour in a sealed stack of sultanas, compared with ambient water vapour.

Quality assessment

Fruit quality was unaffected by the treatment. The tasting panel could not differentiate between treated and untreated sultanas for either taste or taint. No change in colour was observed. Previous work by Tarr and Hilton (unpublished data) had shown that there was no significant deleterious effect on the colour of dried sultanas of a similar quality exposed at 15 or 35°C to high CO_2 concentrations for up to 6 months.

Conclusion

Disinfestation with CO_2, followed by storage under sealed plastic membranes, has potential application for dried vine fruit. The technique has a number of advantages over the current practice of sheet fumigation with methyl bromide. It eliminates a number of worker safety problems; particularly the hazards associated with working close to stacks while they are being fumigated and during the subsequent ventilation. Carbon dioxide has minimal impact on the environment and provides a residue-free treatment, which is increasingly advantageous in the markets for Australian dried fruit.

Acknowledgments

Thanks are due to the Dried Fruit Research and Development Council which, along with CSIRO, funded this research, and to the management of the Robinvale Producers Ltd and Mildura Cooperative Fruit Company Ltd for allowing us to use their products and premises for these trials. We are grateful to the staff at both sites for their enthusiastic assistance during the trials and to Maria Rosa and other CSIRO staff who provided technical support during the work.
The PVC enclosures and clamping system used to seal the sheets were manufactured by Commodity Storage Ltd, Riverstone, Sydney. Carbon dioxide was supplied by Liquid Air and CIG Australia. We thank the staff of these organisations for their support of our research.

References

Annis, P. C. 1987. Towards rational controlled atmosphere dosage schedules: a review of current knowledge. In: Donahaye, E. and Navarro, S., ed., Proceedings of the Fourth International Working Conference on Stored-product Protection, Tel Aviv, Israel, 21–26 September 1986. Bet Dagan, Israel, Permanent Committee, 128–148.

Annis, P.C. and van S. Graver, J. 1987. Sealed stacks as a component of an integrated commodity management system: a potential strategy for continued bag-stack storage in the ASEAN Region. CSIRO Division of Entomology Report No 42, 37–43.

Annis, P.C. and van S. Graver, J. 1990. Suggested recommendations for the fumigation of grain in the ASEAN Region. Part 2. Carbon dioxide fumigation of bag-stacks sealed in plastic enclosures: an operations manual, Kuala Lumpur, ASEAN Food Handling Bureau/Canberra, ACIAR, 58 p.

ASTM (American Society for Testing and Materials) 1968. Manual on sensory testing methods, sponsored by ASTM Committee E-18 on Sensory Evaluation of Materials and Products. Philadelphia, American Society for Testing and Materials, Special Technical Publication No 434, 77 p.

Banks, H.J. 1988. Disinfestation of durable foodstuffs in ISO containers using carbon dioxide. In: Ferrar, P., ed., Transport of fresh fruit and vegetables: proceedings of a workshop held at CSIRO Food Research Laboratory, North Ryde, Sydney, Australia, 5–6 February 1987. Canberra, ACIAR Proceedings No 23, 45–54.

Banks, H.J., Annis, P.C., Henning, R. and Wilson, A.D. 1980. Experimental and commercial applications of controlled atmosphere grain storage in Australia. In: Shejbal J., ed., Controlled atmosphere storage of grains: an international symposium held from 12–5 May 1980, at Castelgandolfo (Rome), Italy. Amsterdam, Elsevier, 207–224.

Freidlander, A. 1984. Biochemical reflections on a non-chemical control method. The effect of controlled atmosphere on the biochemical processes in stored products insects. In: Proceedings of the Third International Working Conference on Stored-product Entomology, 23–28 October 1983. Manhattan, Kansas, Kansas State University, 471–480.

The fumigation of bag-stacks with phosphine under gas-proof sheets using techniques to avoid the development of insect resistance

R.W.D. Taylor and A.H. Harris*

Abstract

The development of insect resistance to phosphine has been an increasing concern for the last decade. Until recently, it has been commonly assumed that, if resistance reached a level which made phosphine unusable, methyl bromide could, in many situations, be employed as an alternative. The listing of methyl bromide in 1992 as a depleter of ozone, suggests that this chemical may, eventually, be phased out for all but essential uses. With no other fumigants available, or likely to be introduced in the near future, the continued effectiveness of phosphine must now be considered imperative, and strategies introduced to avoid insect resistance.

The fumigation of bag-stacks, a common practice in developing countries, is often poorly carried out, leading to sub-lethal exposures to phosphine and potential insect resistance. Improvements in the methods employed, to ensure control of all stages of insect development, can be an important strategy in combating resistance. Trials with a technique of good quality fumigation of bag-stacks, under sheets, with particular attention paid to effective sealing at the floor level, demonstrated that 50% of the phosphine applied could be retained in stacks for at least five days. The introduction of routine fumigations of this standard could greatly assist in keeping the development of insect resistance to a minimum.

Introduction

In the last decade increasing concern has been expressed about the development of insect resistance to phosphine (Taylor 1989; Tyler et al. 1983; Winks 1986), and reports of several surveys of resistance have been published (Herron 1990; Sartori et al. 1990; Taylor 1989; White and Lambkin 1990; Zettler 1994). The recognition in 1992 (Watson et al. 1992) that methyl bromide could be an important contributor to depletion of the ozone layer has cast doubts on that chemical's continued availability as a fumigant. With no practical alternatives immediately available, the continued effectiveness of phosphine must now be regarded as imperative. The magnitude of phosphine resistance has generally been found to be greatest in developing countries, where standards of fumigation are often unsatisfactory and insects may be repeatedly exposed to low concentrations of fumigant (Mills 1983).

One of the commonest types of commodity fumigation in developing countries is the treatment of bag-stacks under gas-proof sheets and although, for such treatments, the original recommendation by phosphine manufacturers was a 72-hour exposure period, the advent of resistance has caused 5–7 day exposures to become standard practice in many countries.

Winks (1987) has described increased exposure periods as the key to controlling phosphine-resistant insects, and has suggested that exposure periods of 10 days may be necessary in some circumstances. The effective disinfestation of bag-stacks depends not only on using fumigation sheeting materials that are relatively impermeable to phosphine, but particularly on the adequate sealing of sheets, at floor level and where they are joined, to prevent gas leakage. The extension of fumigation exposures beyond the original 72-hour period will be of little value in some developing countries where, because of leakage due to poor sealing, a lethal concentration cannot be maintained. The routine adoption of well-sealed enclosures for fumigation must be regarded as critical in management strategies for phosphine resistance.

Authors differ about the recommended minimum lethal phosphine concentration to be used for the fumigation of stored commodities. Friendship et al. (1986) suggest that for a fumigation to be considered satisfactory a minimum concentration of 0.2 mg/1 (ca 150 ppm) phosphine should remain in the enclosure treated after five days. Van Graver and Annis (1994) recommend that a minimum phosphine concentration of 100 ppm be maintained for 7 days in well-conducted fumigations. monitoring of phosphine fumigations in developing countries has shown that gas retention in sheeted-stack fumigations frequently fails to meet either of the above recommendations. Figure 1 provides examples of phosphine gas retention in two fumigations of milled rice under sheets in Southeast Asia, and these are probably not atypical of many fumigations carried out in developing countries. Both treatments would probably fail the recommended regimes of gas retention indicated above.

Methods for Improving the Fumigation of Sheeted Stacks

The need arose recently to verify that the important beetle pest of maize, *Prostephanus truncatus* (Horn), could be effectively controlled by phosphine in sheeted-stack treatments in Africa. The ability to achieve complete control of the pest was considered particularly important since a related species, *Rhyzopertha dominica* (Fabricius), was one of the first insects to show resistance to phosphine in the field (Tyler et al. 1983). Also, because of the potential need to fumigate grain in Africa at high altitudes where low ambient temperatures may result in low grain temperatures, it was necessary to include one such location in the fumigation trial program.

The experimental program was carried out during the cooler months of August and September at three locations in Tanzania. The aim was to improve techniques of bag-stack fumigation in order to ensure maximum gas retention and minimise the opportunities for the selection of insects for resistance to phosphine. Duplicate 100 t stacks of maize were constructed on wooden pallets over good quality concrete floors in strategic grain reserve stores in Arusha, Dodoma and

* Natural Resources Institute, Central Avenue, Chatham Maritime, Chatham, Kent, ME4 4TB, United Kingdom.

Dar es Salaam, locations which have ambient average minimum temperatures during the cooler months, of 8, 11 and 18°C, respectively. During stack construction, nylon capillary tubes were placed near the top, middle, and bottom of stacks, in a vertical central line, to enable monitoring of phosphine concentrations during fumigation. A fourth sampling position was located peripherally, at floor level, beneath stacks, and a fifth was situated on the vertical stack surface midway between the base and the top of the stack. Temperature sensors, and insect cages containing all stages of *P. truncatus*, were located adjacent to all fumigant monitoring positions.

Laminated PVC sheets, having a supporting nylon scrim, and weighing approximately 360 g/m^2, were used in all fumigations. In laying the sheets, which were checked initially for holes or tears, particular care was taken to ensure that, where joins were necessary, overlapping of at least 1 m was provided at the sheet edges so that when these were folded together a gastight seal was formed. Considerable care was also taken to ensure that any grain spillage was removed before the fumigation sheet was positioned, and that there was a minimum of 1 m of sheet margin on the floor, around stacks, which was pulled tight to the corners to remove folds and channels through which fumigant could escape. To provide a good seal between the fumigation sheet and the floor, much larger sandsnakes were used to replace the narrow type commonly employed, which are often made from discarded fire hosepipe. The larger sandsnakes were fabricated locally from lightweight canvas and had a diameter of approximately 15 cm, which provided at least twice the weight and contact area on the fumigation sheet as those made from fire hosepipe. In order to ensure that good sealing to the floor was achieved, two sandsnakes were used at stack corners and where sheets were joined.

Aluminium phosphide tablets producing 1 g of fumigant were used in all fumigations, and were placed either on trays or in cotton bags (Arusha), on the ground beneath stacks. At Arusha, an application rate of 3 g phosphine/t was used where, although the ambient average daily minimum temperature was 12°C, a light insect infestation in some parts of stack resulted in a grain temperature varying from 22 to 26°C. At Dodoma and Dar es Salaam, a phosphine application rate of 2 g/t was used; there was little infestation in the maize stacks, with grain temperatures in the range 20–22°C and 25–26°C, respectively. The extent of gas retention in stacks and the

effectiveness of insect control were determined for both 5 and 7-day exposure periods. Phosphine concentrations were monitored in stacks by withdrawing samples at intervals through the nylon capillary tubing using a 50 mL gastight syringe, for direct injection into a Bedfont EC80 phosphine meter.

Results and Discussion

The average phosphine concentrations recorded in stacks, at the three locations, are given in Figures 2–4. The pattern of fumigant distribution within stacks over the first 24 hours was not identical, but phosphine became very evenly distributed in all stacks after approximately 40 hours and remained so until the end of the treatment periods. The distribution of phosphine in two of the fumigations is shown in the accompanying Tables 1 and 2 and is typical of all six stacks treated. The tables record the concentrations of gas at the different sampling positions during a 7-day fumigation at Dodoma (Table 1), and a 5-day fumigation at Arusha (Table 2). At Dodoma, the phosphine concentration at the end of the fumigation was slightly below 1.4 mg/L, having fallen from a maximum value of 2.0 mg/L during the preceding 5 days. Gas retention in all the stacks fumigated was sufficient to suggest that leakage between the fumigation sheets and the floor was minimal. Most of the leakage that occurred was probably through the fumigation sheets since the average fall in phosphine concentration, 7%/day, corresponded approximately to the known permeability of the type of sheet used.

A comparison of the maximum fumigant concentrations attained with the expected theoretical maximum values in stack treatments indicates the extremely good phosphine retention that occurred during the trials in Tanzania. At Dodoma, where a phosphine application rate of 2 g/t was used, and assuming a stowage factor for maize of 1.5 m^3/t, the theoretical maximum gas concentration with no allowance made for the volume occupied by the grain would have been of the order of 1.3 mg/L. However, allowing for the volume occupied by the grain, a maximum theoretical phosphine concentration of 2.0 mg/L or more was expected. The maximum concentration recorded was a little over 2 mg/L, and well within the expected range. At Arusha, where the application rate was 3 g phosphine/t, the evolution of fumigant was slower and the maximum concentration was not reached until 80

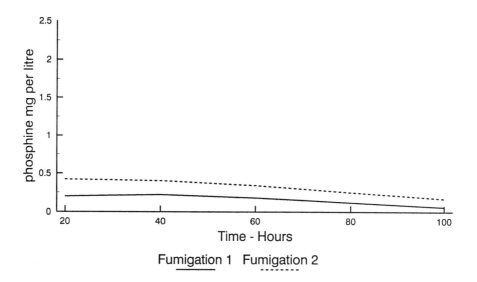

Fig. 1. Phosphine fumigations in Southeast Asia. The application rate is 2 g/ t of milled rice.

Table 1. Gas concentrations in mg/L recorded during a 7-day fumigation of maize at Dodoma, using phosphine applied at 2 g/t.

Sampling position	Time (hours)									
	18	25	42	49	66	74	90	114	138	162
Bottom centre stack	1.14	1.20	1.97	2.06	1.98	1.98	1.82	1.68	1.53	1.42
Middle centre stack	0.90	1.48	1.93	2.11	1.96	1.99	1.80	1.69	1.55	1.42
Top centre stack	1.30	1.54	2.02	2.11	1.95	2.01	1.83	1.69	1.57	1.45
Ground level under pallet	1.37	1.70	1.93	2.12	1.95	1.97	1.79	1.66	1.54	1.40
Stack surface midway – top of stack to floor	nd	1.88	2.05	2.17	1.99	2.01	1.84	1.70	1.58	1.46

nd= no data

Table 2. Gas concentrations in mg/L recorded during a 5-day fumigation of maize at Arusha, using phosphine applied at 3 g/t

Sampling position	Time (hours)								
	18	25	43	51	66	72	91	98	116
Bottom centre stack	0.70	1.03	1.66	1.83	2.15	2.29	2.33	2.40	2.33
Middle centre stack	0.67	0.99	1.65	1.86	2.15	2.29	2.33	2.40	2.33
Top centre stack	0.55	0.86	1.56	1.77	2.09	2.24	2.28	2.40	2.32
Ground level under pallet	0.67	1.01	1.64	1.88	2.17	2.39	2.34	2.40	2.33
Stack surface midway – top of stack to floor	0.75	1.09	1.64	1.91	2.18	2.37	2.34	2.40	2.35

hours. This may be attributed to the siting of the fumigant beneath the stack (in cotton bags) in direct contact with the concrete floor which was 1–2°C colder than the ambient store temperature, which fluctuated from 17 to 22°C. Retention of fumigant in stacks was particularly effective at Arusha, where only a small drop in phosphine concentration took place between the attainment of the maximum concentration and the end of the exposure period. All developmental stages of *P. truncatus* were effectively controlled in all the fumigations evaluated (Taylor and Harris 1994).

Using careful sheet placement and larger sandsnakes, the experimental program in Tanzania demonstrated that it is possible to retain at least 50% of the applied dose of phosphine in bag-stacks for 7 days or longer. This standard of fumigation was sufficiently good to suggest that, even at the lower ambient temperatures in Arusha (where grain stored for long periods might fall to 20°C), a phosphine application rate of 2 g/t would be expected to provide complete control of insect pests. Fumigations in which this level of gas retention is achieved should be attainable in routine practice at little extra cost, provided training and management inputs are properly applied. Monitoring of phosphine concentrations on a regular basis may, however, be necessary to ensure that gas retention is adequate in routine treatment programs. This measure would be fully justified as an aid to avoiding insect resistance to what could prove to be, in the future, the last remaining commodity fumigant.

References

Friendship, C.A.R., Halliday, D. and Harris, A.H. 1986. Factors causing development of resistance to phosphine by insect pests of stored produce. In: Howe, V., ed., GASGA Seminar on Fumigation Technology in Developing Countries, TDRI, Slough, England, March 1986, 141–149.

Herron, G.A. 1990. Resistance to grain protectants and phosphine in coleopterous pests of grain stored on farms in New South Wales. Journal of the Australian Entomological Society, 29, 183–189.

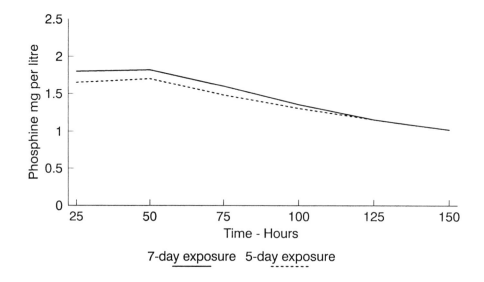

Fig. 2. Phosphine fumigation in Dar es Salaam. The application rate is 2 g/t of maize.

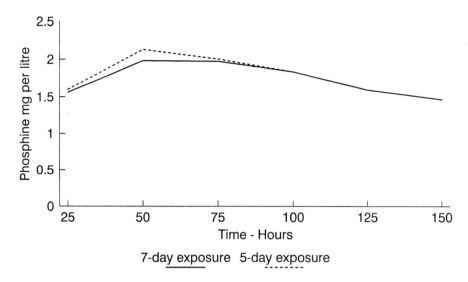

Fig. 3. Phosphine fumigation in Dodoma. The application rate is 2 g/t of maize.

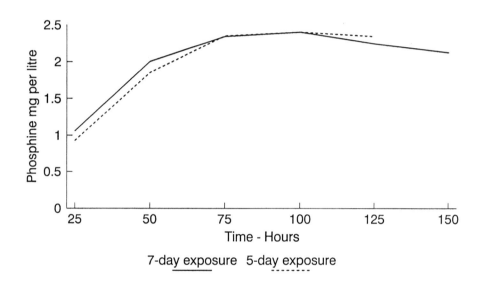

Fig. 4. Phosphine fumigation in Arusha. The application rate is 3 g/t of maize.

Mills, K. A. 1983. Resistance to the fumigant hydrogen phosphide in some stored product species associated with repeated inadequate treatments. Mitteilung-Deutschen Gesellschaft Fur Allgemeine Und Angewandte Entomolgie, 4, 98–101.

Sartori, M.R., Pacheco, I.A., and Vilar, R.M.G. 1990. Resistance to phosphine in stored grain insects in Brazil. In: Fleurat Lessard, F. and Ducom, P., ed., Proceedings of the Fifth International Working Conference on Stored-product Protection, Bordeaux, France, September 1990, II, 1041–1049.

Taylor, R.W.D. 1989. Phosphine a major grain fumigant at risk. International Pest Control, 31 (1), pp, 10–14.

Taylor, R.W.D. and Harris, A.H. 1994. Control of the larger grain borer, *Prostephanus truncatus* (Horn) (Coleoptera: Bostrichidae), in bagged maize by fumigation under gas proof sheets. FAO Plant Protection Bulletin, in press.

Tyler, P.S, Taylor, R.W.D. and Rees, D.P. 1983. Insect resistance to phosphine fumigation in food warehouses in Bangladesh. International Pest Control, 25, 10–13.

van S Graver, J. and Annis, P. 1994. Suggested recommendations for the fumigation of grain in the ASEAN region. Part 3. Phosphine fumigation of bag-stacks sealed in plastic enclosures: an operations manual. Kuala Lumpur, ASEAN Food Handling Bureau/ Canberra, Australian Centre for International Agricultural Research, 52.

Watson, R.T., Albritton, D.L., Anderson. S.O. and Lee- Bapty, S. 1992. Methyl bromide: its atmospheric science, technology, and economics. Synthesis report of the interim assessments of methyl bromide science, and technology and economics. UNEP, Ozone Secretariat, Nairobi, Kenya, June 1992.

White, G.G. and Lambkin, T.A. 1990. Baseline responses to phosphine and resistance status of stored-grain beetle pests in Queensland, Australia. Journal of Economic Entomology, 83 (5), 1738–1744.

Winks, R.G. 1986. The effect of phosphine on resistant strains. In: Howe, V., ed., GASGA Seminar on Fumigation Technology in Developing Countries, TDRI, Slough, England, March 1986, pp. 105–118.

Winks, R. G. 1987. Strategies for effective use of phosphine as a grain fumigant and the implication of resistance. In: Donahaye, E., and Navarro, S., ed., Proceedings of the Fourth International Conference on Stored-product Protection, Tel Aviv, Israel 1986, 335–334.

Zettler, J. L. 1994. Phosphine resistance in stored product insects. In: Navarro, S. and Donahaye, E., ed., Proceedings of an International Conference on Controlled Atmosphere and Fumigation in Grain Storage, Winnipeg, Canada, June 1992.

Effects of different speed of build up and decrease of pressure with carbon dioxide on the adults of the tobacco beetle *Lasioderma serricorne* (Fabricius) (Coleoptera : Anobiidae)

C. Ulrichs*

Abstract

The lethal effects of a sudden or slow increase and decrease of carbon dioxide pressure as a pest control technique were tested on adult tobacco beetles *Lasioderma serricorne*. The effect of exposure time and pressure change on the mortality was demonstrated. A quick build up and decay of the carbon dioxide pressure of 20 bar within 2 seconds resulted in higher mortalities than a slow increase and slow decrease within 2 minutes. When the pressure increase was slow and the decrease rapid, beetle mortality was higher, as when the increase was rapid and the decrease slow. It appeared that the decrease is more efficient than the increase for quickly obtaining mortality. The valance of increase and the valance of decrease were tested with 20 bar for 5 minutes and 25 bar for 4 minutes at 25°C.

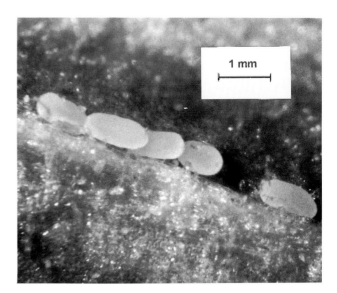

Fig. 1. Eggs of *Lasioderma serricorne* on a tobacco leaf

Introduction

Using techniques with pressurised fumigants in pest control is a new field in pest management. In order to apply the technique economically, it is important to determine its effects on the target organisms. Little is known on the susceptibility of the tobacco beetle to high pressures of carbon dioxide, apart from the studies of Gerard et al. (1988) and Stahl and Rau (1985). According to Gerard et al. (1988), the tobacco beetle is one of the most tolerant pest species towards pressurised carbon dioxide and was therefore chosen in this study.

Lasioderma serricorne causes problems not only on dried tobacco leaf but also on drugs and spices. Because of this economic importance, experiments were carried out to determine in more detail the possibility of controlling *Lasioderma serricorne* with carbon dioxide under high pressure as a possible alternative to conventional pest control.

Material and methods

Beetles

Tobacco beetles were reared on dried tobacco leaves and wheat bran including glucose, yeast, glycerine and water. The female insects laid eggs into folds of the tobacco leaves Fig. 1. Damage to the product is caused by the larvae which usually hatch after 10 days. The development of the beetles depends strongly on temperature and humidity. The optimum for larval development is 32.5°C and 70% r.h. Under these conditions development takes about 16 days (Heinze 1983). Test insects were reared under constant conditions at 25°C and 65–70% r.h. Larval development took about 40 days. The biology of *Lasioderma serricorne* is described in detail by Asworth (1993).

Treatment with carbon dioxide under high pressure

A small pressure chamber (Reichmuth and Wohlgemuth, these proceedings) was filled with carbon dioxide (Fig. 2). The volume of the chamber was 400 mL. The insects were inserted

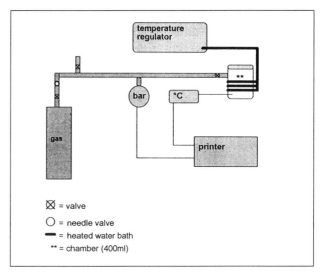

$\boxtimes$ = valve
$\bigcirc$ = needle valve
━ = heated water bath
** = chamber (400ml)

Fig. 2. Diagram of the apparatus used to treat insects with carbon dioxide under high pressure

---

* Federal Biological Research Centre for Agriculture and Forestry, Institute for Stored Product Protection Konigin–Luise–Straße 19, 14195 Berlin, Germany.

into the chamber in small glass vessels with a perforated plastic cap. Sensors were linked to a digital device and to a printer, recording changes in temperature and pressure. The temperature was controlled by a heated water bath.

The maximum pressure was 46 bar. The speed of building up or decreasing the pressure in the chamber was varied at constant temperature of 25°C.

- fast build–up of pressure : from 1 bar to 10 bar within 1 second
- slow build–up of pressure : from 1 bar to 10 bar within 60 seconds linear
- fast release of pressure : from 1 bar to 10 bar within 1 second
- slow release of pressure : from 1 bar to 10 bar within 60 seconds linear

Mortality tests

Fifty adult beetles were counted into a glass which was inserted into the pressure chamber. The pressure was built up at constant temperature. After the exposure time, the main valve was closed. Decompression was regulated with a needle valve. During the treatment, the beetles were exposed to maximum experimental pressure. The time for build up and decay of the pressure was not included in the exposure period. Three hours after treatment, the samples were bioassayed for survivors. Each test was carried out in four replicates. The control was kept at the same constant temperature without pressure.

Results

A fast build–up and decay of the pressure reduced the time needed to kill test insects. A comparison with slow pressure change is given in Figure 3. Slow increase and decrease of pressure and an exposure to 25 bar for 5 minutes resulted in 10% mortality. Mortality was 20% after 4 minutes at 25 bar. Higher mortalities of 40% and 45%, were achieved at the same pressure and exposure for 4 minutes by increasing the pressure quickly. Fast build-up and slow decay caused mortalities of 60% and 75%.

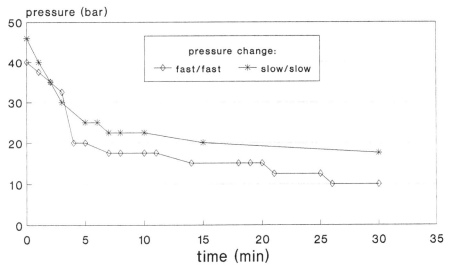

10 - 46 bar carbon dioxide

Fig. 3. Lethal dose for adult tobacco beetles at 25°C after treatment with carbon dioxide under pressure

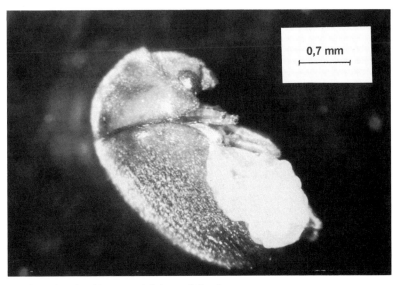

Fig. 4. Beetle with ruptured abdomen following pressure treatment

Independent of the speed of build-up of pressure, only fast decay led to physical disruption of the adult insects (Figure 4). When the beetles burst open, undestroyed eggs could be seen.

Discussion

The speed of increase and decrease of carbon dioxide pressure had a pronounced impact on the mortality. The more rapidly the changes in pressure occurred, the shorter was the time for complete kill of the adult insects. Carbon dioxide treatment at ambient pressure was required for 3 weeks to control the granary weevil *Sitophilus granarius* and the red flour beetle *Tribolium confusum* (Adler and Reichmuth 1989). A rapid build–up of pressure had less impact on mortality than rapid release of pressure (Fig. 5). Both rapid increase and rapid decrease led to higher mortalities than slow pressure increase and slow pressure decrease. In practice, the time for increasing and releasing the pressure is likely to be longer than under experimental conditions, because of the difference in size between a small laboratory autoclave and a large pressure chamber.

References

Adler, C. and Reichmuth, C.H. 1989. Zur Wirksamkeit von Kohlendioxid bzw. Stickstoff auf verschiedene vorratsschädliche Insekten in Stahl–Getreidesilozellen. Nachrichtenblatt des Deutschen Pflanzenschutzdienstes, 41, 177–183.

Asworth, J. R. 1993. The biology of *Lasioderma serricorne*. Journal of Stored Products Research, 29, 291–303.

Gerard, D., Kraus, J., Quirin, K.-W. and Wohlgemuth, R. 1988. Anwendung von Kohlendioxid unter Druck zur Bekämpfung vorratsschädlicher Insekten und Milben. Pharm. Ind., 50, 1298–1300.

Heinze, K. ed. 1983. Leitfaden der Schädlingsbekämpfung, Vorrats– und Materialschädlinge. IV Wissenschaftliche Verlagsgesellschaft mbH Stuttgart, 348 p.

Stahl, E. and Rau, G. 1985. Neues Verfahren zur Entwesung. Anzeiger für Schädlingskunde, Pflanzenschutz und Umweltschutz, 58, 133–136.

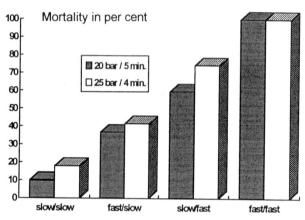

speed of pressure change (increase/decrease)

Fig. 5. Mortality of adult *Lasioderma serricorne* after treatment at different pressures and different speeds of pressure change (increase/decrease). Shaded columns, 20bar/5 minutes; open columns, 25 bar/4 minutes.

Response of the pea weevil *Bruchus pisorum* (L.) to phosphine

C.J. Waterford and R.G. Winks*

Abstract

Two field strains of *Bruchus pisorum* (L.) infesting peas were exposed to the fumigant phosphine at two fixed concentrations of 120 g/L and 240 g/L, and one decaying from 1.5 g/m^3, at 25 and 15°C. Treatment times were 5, 10, 14 and 21 days. Emergence after treatment was assessed at 2–3 weeks, 6–8 weeks and again at 12 months. Results indicate that further emergence of adults from fumigated peas occurred after 14 days exposure for all treatments in one strain and after 5 days exposure to 120 g/L for the other strain. All the emerged adults were dead when inspected. Dissection of subsamples of 100 peas indicated that survival was greatest in the pupal stage with increased survival in adult and larval stages at 15°C. Results indicate that to ensure control of all stages longer rather shorter exposures are desirable.

Introduction

Recommendations for phosphine fumigation of field peas for control of the pea weevil *Bruchis pisorum* (L.) in Australia are based largely on the biology of the pest and an assumption that its tolerance of phosphine is similar to that of some of the other stored product bruchids. The aim of this study was to examine the validity of these recommendations and to establish dosages needed to control the pea weevil in stored field peas.

Materials and Methods

Infested peas from two areas of Australia were used. One was supplied by the Department of Agriculture, Victoria and referred to as 'Strain A' while the other, 'Strain B', was supplied by the Department of Agriculture, South Australia. Both strains were from plots of untreated peas to ensure a high level of infestation and were machine harvested. Samples of each strain were placed in position on grids by means of contact adhesive, and X-rayed at 30 kV, 3 mA for 2 seconds. Examination of the radiographs indicated the level and stages of infestation present. The infested peas were divided and packed in porous paper parcels each weighing 800 g. These were placed into chambers (Fig. 1) through which conditioned air was passed. Both strains, heavily infested with all stages of the pea weevil *B. pisorum* (L.), were exposed to constant concentrations, 120 µg/L and 240 µg/L. Samples were exposed at two temperatures, 25 and 15°C, for periods of 5, 10, 14 and 21 days.

Constant concentrations were applied by means of mass-flow controllers, a diaphragm pump, and a cylinder of compressed gas containing 1 g/m^3 phosphine in nitrogen (Fig. 1). Air from the pump and phosphine from the gas cylinder were controlled through two mass-flow controllers so that when the output from each was blended the appropriate concentration was produced. The resultant gas mixture was passed through distilled water in a bubbler held at 15°C in a water bath. The saturated mixture was then warmed to 25°C thus lowering the relative humidity to 57%. This equates to 10–11% moisture content for peas. The gas was then passed to a manifold where it was divided into equal flows one for each chamber. The moisture content of the peas supplied was not measured.

A concentration of 1.5 g/m^3 phosphine was applied to a third set of samples and allowed to decay during the experiment in an attempt to simulate a field exposure. The samples were placed in a glass aspirator of 20 L volume (Fig. 2). Apart from absorption and chemical breakdown of phosphine, to simulate gas loss, gas was allowed to diffuse through two 6 mm holes at the top of the chamber. Further gas loss occurred when samples were removed from the chamber at the end of each exposure period. The phosphine was injected into the sealed chamber through a septum at the top and the gas mixed by means of a small magnetically driven fan at the base of the chamber for 5 minutes at the beginning of the exposure.

Concentrations applied were checked by means of gas chromatography using the response of a flame photometric detector compared with the response of prepared gas standards. These were prepared by means of volumetric dilution of a high concentration phosphine source analysed with a Gow-Mac gas density balance. Samples of peas were removed at the end of each exposure period, transferred from the porous paper parcels into glass jars, sealed with black filter paper and held at 25°C and 57% r.h. Each sample was sieved at 2 weeks or at the end of exposure, whichever was the longer, and again at 6–8 weeks and 12 months. Adults were assessed as alive or dead, numbers recorded, and adults removed. In addition, at 14 days or the end of exposure and again at 6–8 weeks a subsample of 100 peas was selected from each treatment sample and dissected to assess the effect on immature stages.

Results

The X-ray plates indicated that the two strains of weevil were at a slightly different stage of development. Strain B being further advanced in development and hence containing a much higher proportion of the pupal and adult stages. Conversely, Strain A revealed a higher proportion of early larval stages and no adults. This was confirmed by the dissection of the subsamples (Table 1) which showed no adults, 13 pupae and 23 larvae for Strain A, as against 3 adults, 18 pupae and only 8 larvae in Strain B. In addition, the sieving done at 14 days (Table 2) showed no adult emergence from the controls of Strain A, whereas in Strain B, 38 adults had emerged from the control sample.

---

* CSIRO Division of Entomology, GPO Box 1700, Canberra ACT 2601, Australia.

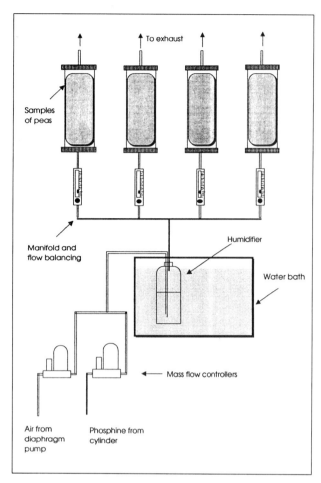

Fig. 1. Apparatus used to expose samples of infested peas to fixed concentrations of phosphine.

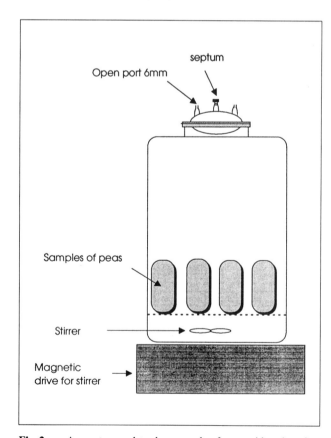

Fig. 2. Apparatus used to dose sample of peas with a decaying concentration of phosphine.

Table 1. Developmental stages of *B. pisorum* (L.) present in samples of 100 control peas dissected 14 days from start of exposure

Strain	Larvae		Pupae		Adults	
	Live	Dead	Live	Dead	Live	Dead
A	5	18	5	8	0	0
B	3	5	12	6	2	1

An estimate of the numbers of pea weevil exposed in each 800 g sample, was obtained from the radiographs and average weight of peas dissected. These data, combined with average control mortality, gave an indication of the number of live insects treated in each sample (Table 4). The sieving results, (Table 2.) which looked only at emergence, show no survival of emerged adults at any treatment for either strain, i.e. all emerged adults were dead when inspected. However, since all adults were removed after sieving and again at subsequent sievings it would appear that for Strain B further emergence occurred up to 14 days for all dosage combinations and for 21 days exposure at 120 μg/L and a treatment temperature of 25°C. Strain A appeared to be controlled by a 10-day exposure, there being further emergence of adults at 5-day exposures for this strain at 120 μg/L but none at 240 μg/L. It may be that this apparent emergence of adults is due to adults being killed just before emergence and, with desiccation, falling out on subsequent sieving, or that they were seriously affected by their exposure and died during, or on, emergence. However, the condition of stages present in the dissected material (Table 3) showed that survival within the pea was present up to 6 weeks after commencement of exposure. This was most apparent in the pupal stage. In addition, the reduced efficacy of phosphine at the lower temperature of 15°C is indicated by survival of the adult and larval stages.

The natural mortality of emerged adults was high for both strains, 13% for Strain A and 33% for Strain B. This mortality was on top of approximately 66% mortality for all stages present in the infested peas. This very high level of mortality during maturation of the pea weevil would seem to be caused by storage and handling alone. It is not possible to say what effect these conditions had on the results of this experiment.

Discussion

The data indicate that some survival of *B. pisorum* (L.) to a phosphine fumigation is possible after 14 days exposure. More importantly, it appears that this species follows the same pattern as other stored product pests in that the pupal stage appears to be the most tolerant stage of development. It would seem then that fumigating as early as possible in the life-cycle, would increase the probability of a successful fumigation and have the added advantage of minimising damage done to the peas by any infestation present. The emergence of a number of adults from Strain B between the seiving at 2 weeks and the subsequent examinations at 6–8 weeks and 12 months would suggest that some pupae survived and developed to adults. Since none of these emerged adults was found alive, whether they died because of natural mortality, handling stress induced by the examination of each sample or because of the chronic effects of phosphine cannot be said.

There are two possible reasons for the substantial difference in response by the two strains. Firstly, Strain A could be generally more susceptible to phosphine and hence at the dosages applied more easily controlled. Secondly, since Strain A was at an earlier stage of development, the predominant stage being fumigated would have been the larvae and this stage is presumably easier to control. However, there was a proportion

Table 2. Incubation results for field peas infested with *B. pisorum* (L.) exposed to phosphine for two fixed and one decaying concentration at two temperatures. All emerged adults in the treatments were dead.

Assessment	Exposure											
	5 days			10 days			14 days			21 days		
	2	6	52 weeks	2	6	52 weeks	2	6	52 weeks	2	6	52 weeks
120 µg/L 25°C												
Strain A	0	2	1	0	0	0	0	0	0	0	0	0
Strain B	21	1	5	17	2	7	24	1	3	12	0	1
240 µg/L 25°C												
Strain A	0	0	0	0	0	0	0	0	0	0	0	0
Strain B	15	2	2	11	0	2	14	0	2	20	0	0
120 µg/L 15°C												
Strain A	0	1	0	0	0	0	0	0	0	0	0	0
Strain B	24	2	2	17	2	8	0	0	2	11	0	0
240 µg/L 15°C												
Strain A	0	0	0	0	0	0	0	0	0	0	0	0
Strain B	14	0	4	25	1	5	23	0	2	16	0	0
1.5 g/m^3 25°C				decaying to 0.005 g/m^3 at 21days								
Strain A	0	0	0	0	0	0	0	0	0	0	0	0
Strain B	17	0	2	12	0	2	13	0	1	12	0	0
1.5g/m^3 15°C				decaying to 0.004 g/m^3 at 21days								
Strain A	0	0	0	0	0	0	0	0	0	0	0	0
Strain B	12	0	2	12	0	1	10	1	1	14	1	0

Table 3. Survival of internal stages of *B. pisorum* (L.) in peas exposed to phosphine at 120 and 240 µg/L and 1.5 g/m^3 decaying to 0.005 g/m^3 concentration at two temperatures. Samples of 100 peas from each treatment were dissected two or three weeks and six weeks after start of exposure.

Assessment	Larvae				Pupae				Adults			
					Exposure (days)							
	5	10	14	21	5	10	14	21	5	10	14	21
25°C												
120 µg/L												
2 weeks	0	0	0	0	5	1	2	2	0	0	0	0
6 weeks	0	0	0	0	0	1	0	0	0	0	0	0
240 µg/L												
2 weeks	0	0	0	0	6	0	1	0	0	0	0	0
6 weeks	1	0	0	0	1	1	0	0	0	0	0	0
1.5 g/m^3												
2 weeks	0	0	0	0	0	0	0	0	0	0	0	0
6 weeks	0	0	0	0	0	1	0	0	0	0	0	0
Total 25°C	1	0	0	0	12	4	2	2	0	0	0	0
15°C												
120 µg/L												
2 weeks	1	0	3	0	8	1	3	2	1	0	0	1
6 weeks	0	0	1	0	0	2	0	0	3	2	2	0
240 µg/L												
2 weeks	1	0	0	0	7	2	0	0	1	0	0	0
6 weeks	1	0	0	0	1	2	0	0	0	1	0	0
1.5 g/m^3												
2 weeks	0	1	0	0	0	1	0	0	0	0	0	0
6 weeks	0	2	0	0	0	1	0	0	1	0	0	0
Total 15°C	3	3	4	0	16	9	3	2	5	2	2	1

Table 4. Estimated number of pea weevils, *B. pisorum* (L.), treated in each 800 g sample of peas

Strain	No of peas X-rayed	% infested	Average weight of 100 peas	No. of insects in 800 g sample	
				Total	Alive
A	822	46	22.7g ± 1.0	1620	593
B	837	36	15.1 g ± 0.5	1800	498

of pupae present in both strains which suggests the first possibility could be the case. The implications of this are that a fairly broad range of tolerance to phosphine may exist in the field. How this range of tolerance arose is open to speculation. However, to ensure disinfestation of all strains present in the field the underlying need for extended exposure when fumigating with phosphine remains, regardless of what developmental stage is present, since the absence of the pupa cannot be guaranteed.

As far as recommendations for phosphine treatment of *B. pisorum* (L.) are concerned there is no evidence from the response of these two strains to support a reduction in time of exposure for a treatment. In addition, the use of gastight enclosures is essential to retain the gas, or a method of application of the fumigant, such as SIROFLO®, which sustains a lethal concentration for the required time. The emergence of one adult, at between 6 weeks and 12 months, after 21 days exposure may suggest an even longer exposure than 14 days would be desirable.

Acknowledgment

Thanks to Mr Greg Baker of the Department of Agriculture, South Australia and Mr Mark Smith of the Department of Agriculture and Rural Affairs, Victoria for the supply of infested peas used in this study.

Correlation between phosphine resistance and narcotic response in *Tribolium castaneum* (Herbst)

C.J. Waterford and R.G. Winks*

Abstract

Phosphine, at high concentrations, induces a narcotic response in insects similar to 'knockdown' from which they can recover if exposure is not excessive. For most species of stored-product pest investigated, this response has been associated with dramatic changes in tolerance of phosphine at high concentrations (up to 50 times). This study reports changes in resistance to phosphine over 30 generations of selection of a susceptible strain of *Tribolium castaneum*. The selection, based on narcotic response to brief exposures at high concentrations of phosphine, provided a method of identifying the target beetles without killing them. In addition, as the resistance level changed so did the narcotic response. The correlation between the changes in time to narcosis and time to death is discussed. The narcotic response to phosphine or 'knockdown' at high concentrations has been used as the basis of a quick method for indicating the presence of resistant field strains that exhibit this form of phosphine resistance.

Introduction

Phosphine is one of the few remaining fumigants available for the disinfestation of stored products now that methyl bromide is likely to be restricted in use because of its ozone depleting nature. However, increased tolerance of stored-product pests to phosphine in recent times (Tyler et al. 1983) has made the effective use of phosphine more difficult. Studies of phosphine resistance in *Rhyzopertha dominica* (F.) (Price 1984) show that one resistance mechanism is active exclusion where the insect actively keeps the fumigant away from susceptible sites. At high concentrations, phosphine also induces a narcotic effect from which insects can recover if the exposure is not excessive (Winks 1984). The response is similar to the knockdown response when insects are exposed to insecticides. The high concentration region above about 0.5 mg/L, where the narcotic response is most pronounced, is associated with significant changes in tolerance of phosphine of up to 50 times in *Tribolium castaneum* (Herbst) (Winks 1984). In addition, it has been shown that resistant strains of *Rhyzopertha dominica* (F.) take longer to succumb to the narcotic effect than do susceptible strains. On the basis of this behavioural response, a quick test to indicate the presence of resistance has been proposed (Reichmuth 1991). It was thought that narcosis was a form of protective mechanism, in that the insects when narcotised did not take up as much phosphine. However, preliminary work for this investigation showed that it was the insects that resisted narcosis or remained active which survived. Those that succumbed quickly were the first to die. This observation is consistent with a mechanism of active exclusion. This study examines the relationship between, and changes in, narcotic response times at higher concentrations and resistance levels at a lower concentration in selections of a susceptible strain of *T. castaneum*.

Materials and Methods

Origin and maintenance of insect material

Test insects used were a susceptible strain of *T. castaneum* (CTC_4) held in laboratory culture since collection in 1965 from a produce merchant's store in Brisbane, Australia, and selections cultured from this strain. Culturing and general handling techniques follow those described in Winks (1982).

Fumigation chambers

The phosphine exposure and selection were carried out in a purpose-built chamber (Fig. 1) in which the insects could be placed and the air conditioned. The chamber top could then be sealed and the enclosed space dosed with an appropriate volume of phosphine through the septum. The insects were observed through the glass top. At the end of exposure, phosphine was rapidly removed and the insects sorted for selection. Narcotic response times were determined in small plastic cell-culture flasks with optically clear sides. Response lines for strains were determined in a multi-chamber apparatus described in Winks and Waterford (1983).

Production and measurement of phosphine

Phosphine used to dose chambers was produced by hydrolysis of pellets of aluminium phosphide (Phostoxin®) according to a published method (Anon. 1975) and introduced into the chosen chamber using gastight syringes. The concentration of the phosphine source was determined by gas chromatography using the response of a Gowmac® gas density detector. The volume injected was calculated from the source concentration and the volume of the fumigation chamber.

Selection methods

All dosing and handling were carried out in a laboratory that was maintained at 25°C. One hundred adults were used for each selection. They where placed in the selection chamber fitted with a quick-release lid and a septum through which gas could be injected (Fig. 1). The insects were starved and conditioned in an incubator at 25°C, 57% r.h. overnight. The lid was placed on top and clamped shut sealing the chamber. Phosphine source, approximately 18 μL, sufficient to produce a narcotic concentration of 2 mg/L was injected through the septum. The gas was immediately stirred by repeated removal and injection of a quantity of the atmosphere within the chamber, using a 10 mL syringe. The insects were then

* Stored Grain Research Laboratory, CSIRO Division of Entomology, GPO Box 1700, Canberra ACT 2601, Australia.

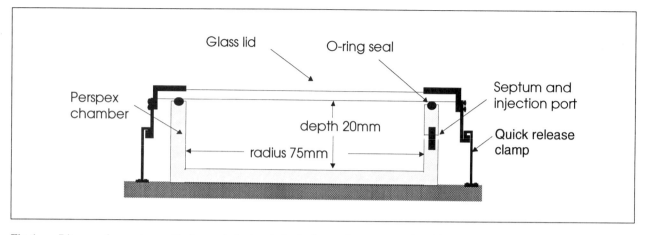

Fig. 1. Diagram of apparatus used to dose and select on the basis of narcotic response to phosphine.

observed through the glass top. Two selections were made based on the behavioural response of the insects. The first, in which the insects were observed until more than 50% had become immobile or narcotised, was termed narcotic tolerance. The top was removed and the phosphine fanned off. The 30 least narcotised or most active beetles were then removed into one recovery dish, using soft forceps, the remainder into another dish. The reverse was done for the narcotic-susceptible selection. The insects were observed until about 20% had succumbed or ceased to move. The top of the chamber was removed and the 30 most deeply narcotised removed into a recovery dish with the soft forceps. The remainder (active beetles) were placed into another dish. Time of exposure was recorded in both cases. The mortality was assessed at 7, 14 and 21 days to ensure end-point mortality was reached. The survivors of the 30 least narcotised were set up as parents of a strain designated CTC_4NR_1 where NR means narcotic resistant. The survivors of the 30 most narcotised were set up as parents of CTC_4NS_1 where NS means narcotic susceptible. The progeny of these two strains became the test insects for the next selection. The narcotic-resistant progeny were selected for 30 generations. The narcotic-susceptible progeny were selected for 10 generations.

Narcotic and mortality response assessments

The parent strain (CTC_4), the 7th, 10th and 30th generation for narcotic resistance (CTC_4NR_7, CTC_4NR_{10} and CTC_4NR_{30}) and narcotic susceptibility (CTC_4NS_{10}) were assessed to determine times to narcotic response following a modified method described in Winks (1984). Instead of groups of 10 being assessed periodically for the number responding, the time to narcosis was determined for a number of individual insects. This was done by placing individual insects into small flasks and injecting sufficient volume of phosphine to provide an atmosphere of 2 mg/L. The insects

were observed and the time to narcosis was recorded. Narcotic response time was estimated from linear regression analysis of the cumulative response of a number of individuals. Three strains (CTC_4, CTC_4NS_{10}, CTC_4NR_{10}) and the 20th generation for narcotic tolerance (CTC_4NR_{20}) were also assessed for mortality response when exposed to a fixed concentration of 0.1 mg/L phosphine for a range of exposure times in a multi-chamber apparatus described in Winks and Waterford (1983). Groups of 200 adults were used at each exposure time. End-point mortality response was determined from successive observations using the method recommended by Winks (1982). Results were analysed using the method of Finney (1971).

Results

Dosage estimates and parameters of probit regression equations fitted to end-point mortalities of adults of the tested selections are given in Table 1. Probit mortality lines are shown in Figure 2. These show a change in resistance to phosphine in opposite directions from the parent strain CTC_4 depending on the method of selection. Table 2 shows the estimates of times to narcosis for various selections and parameters of regressions of time to narcosis on exposure to 2 mg/L phosphine. The regression lines for times to narcosis (Fig. 3) show similar response changes in both directions from the parent strain, depending on the method of selection.

Discussion

The selection on the basis of narcotic tolerance led to a steady and significant change in both narcotic tolerance and phosphine resistance and the selection based on narcotic susceptibility led to a small decrease in both narcotic tolerance and resistance to phosphine. There was no evidence of a rapid

Table 1. Dosage estimates and parameters of regression of probit mortality on log dosage for adults of *Tribolium castaneum* exposed to 0.1 mg/L phosphine for various exposure times at 25°C, 57% r.h

Strain	LD$_{50}$ mg.hour/L	LD$_{99}$ mg.hour/L	Slope ±SE	Mean probit response (Y)	Heterogeneity	
					χ^2	d.f
CTC_4NS_{10}	0.229	0.378	10.7 ± 0.54	5.33	2.63	6
CTC_4	0.249	0.425	10.0 ± 0.48	5.26	2.83	7
CTC_4NR_{10}	0.965	2.59	5.43 ± 0.50	5.23	28.13	6
CTC_4NR_{20}	1.84	3.653	7.8 ± 0.41	5.01	1.18	4

Table 2. Times to 50 and 99% narcosis (NT_{50} and NT_{99}), and parameters of the regression of probit transformed times to narcosis for adults of *Tribolium castaneum* when exposed to 2 mg/L phosphine at 25°C, 57% r.h.

Strain	NT_{50} (minutes)	NT_{99} (minutes)	r^2	No. used (n)	Intercept (a)	Slope b $\pm$ SE
CTC_4NS_{10}	5.2	6.6	0.98	96	−11.19	22.5 ± 0.37
CTC_4	7.1	10.9	0.96	88	−5.56	12.4 ± 0.27
CTC_4NR_7	7.8	14.2	.93	96	−3.0	8.9 ± 0.24
CTC_4NR_{10}	25.8	86.7	0.91	94	−1.2	4.42 ± 0.14
CTC_4NR_{30}	263	420	0.79	19	−2.33	11.43 ± 1.4

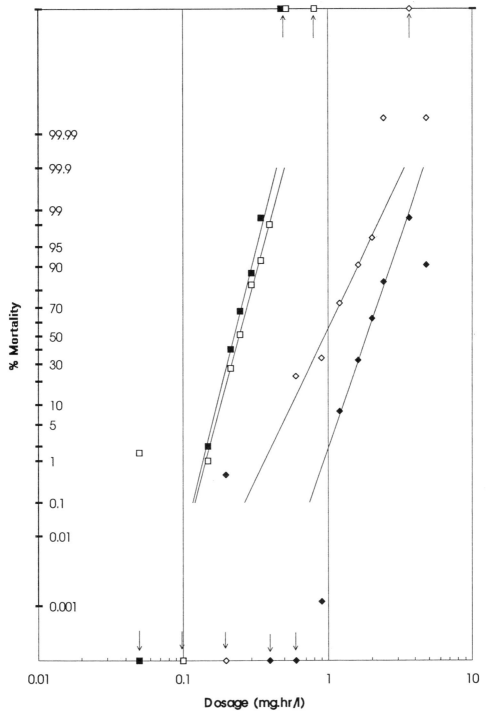

Fig. 2. Probit lines fitted to end-point mortalities for narcotic tolerant strains CTC_4NR_{10} (◇—◇) and CTC_4NR_{20} (◆—◆), a narcotic-susceptible strain, CTC_4NS_{10} (■—■) and the parent strain CTC_4 (□—□) after exposure to 0.1 mg/L phosphine for a range of exposure times.

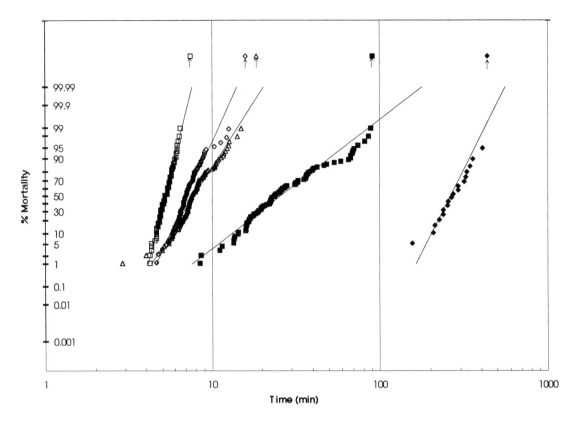

Fig. 3. Response of individual insects showing time to narcosis for the selected strains: narcotic-resistant CTC_4NR_7 ($\triangle$—$\triangle$), CTC_4NR_{10} ($\blacksquare$—$\blacksquare$) and CTC_4NR_{30} ($\blacklozenge$—$\blacklozenge$), narcotic-susceptible CTC_4NS_{10} ($\square$—$\square$) and the parent CTC_4 ($\diamond$—$\diamond$). The lines are regressions through the probit-transformed cumulative responses of each strain.

shift in resistance or narcotic tolerance, just a gradual change in the slopes of the response lines, first a decrease in slope and then an increase. When the exposure times for narcotic tolerance selections are graphed against generation number (Fig. 4) the rate of change seemed slow at first, then seemed to increase sharply after about the 10th generation with some suggestion of slowing down after the fifteenth.

The times to death when exposed to 0.1 mg/L phosphine, of the parent and selected strains of *T. castaneum,* compared with times to narcosis when exposed to 2 mg/L phosphine fall along a straight line when the time to mortality is log transformed (Fig. 5).

A question could be posed: is this a specific correlation or could it be a general one for all species? Equivalent points for

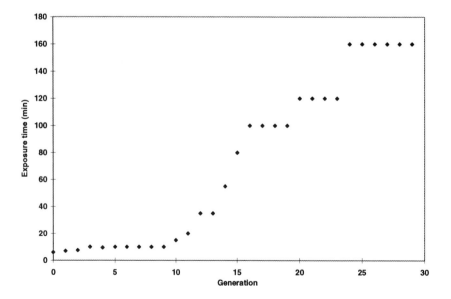

Fig. 4. Exposure times used to achieve approximately > 50% narcosis for the purpose of selecting the 30% least narcotised for 30 generations of selection.

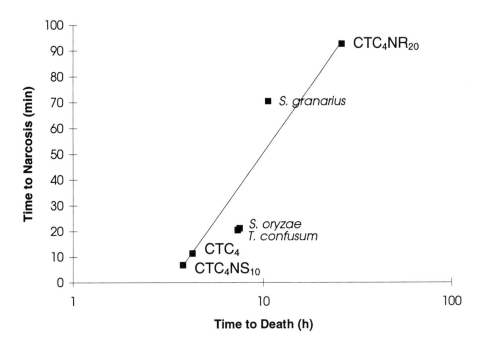

Fig. 5. Time to death (log scale) at 0.1 mg/L phosphine versus times to narcosis at 2 mg/L for narcotic-tolerant (CTC_4NR_{10}) and narcotic-susceptible (CTC_4NS_{10}) strains with the parent strain (CTC_4) □—□ 99% response levels. Equivalent points for three other stored-product species are added for comparison.

other species are plotted and they do fall generally along the same line. The position of these points is not inconsistent with a general correlation between narcotic response and lethal response. However, more research is needed.

If active exclusion is the general resistance mechanism for phosphine, and increased narcotic tolerance is a means of quickly determining the presence and efficiency of the mechanism, then techniques based on assessing the narcotic response, as proposed by Reichmuth (1991), may be a robust means of rapidly detecting strains with this form of resistance before a fumigation. With not much more effort the level of resistance could be determined in less than a day. This information could be useful in deciding what treatment should be applied to an infested commodity. Techniques based on the narcotic response could also be a useful tool for rapid screening of field strains possessing this type of resistance, without killing the parents — thus allowing a more detailed assessment of particularly resistant strains.

However, there may be other mechanisms of resistance possible that would not be detected by this response. The only way to assess these is by determining the full lethal response of the strain from graded response lines. In addition, the observation is in adults and says nothing about the relative response of other stages which may be more important in determining the significance of resistance to phosphine in the field.

References

Anon. 1975. Recommended methods for the detection and measurement of resistance of agricultural pests to pesticides. 16. Tentative method for adults of some major species of stored cereals, with methyl bromide and phosphine. Rome, FAO Plant Protection Bulletin, 23, 12–25.

Finney, D.J. 1971. Probit analysis, 3rd ed. Cambridge, Cambridge University Press.

Price, N.R. 1984. Active exclusion of phosphine as a mechanism of resistance in *Rhyzopertha dominica* (F.) (Coleoptera: Bostrichidae). Journal of Stored Products Research, 20, 163–168.

Reichmuth, C. 1991. A quick test to determine phosphine resistance in stored-products research. GASGA Newsletter, No. 15, 14–15.

Tyler, P.S., Taylor, R.W. and Rees, D.P. 1983. Insect resistance to phosphine fumigation in food warehouses in Bangladesh. International Pest Control, 25, 10–13, 21.

Winks, R.G. 1982. The toxicity of phosphine to adults of *Tribolium castaneum* (Herbst.): time as a response factor. Journal of Stored Products Research, 18, 159–169.

Winks, R.G. 1984. The toxicity of phosphine to adults of *Tribolium castaneum* (Herbst.): time as a dosage factor. Journal of Stored Products Research, 20, 45–56.

Winks, R.G. and Waterford, C.J. 1983. Multi-chamber apparatus for laboratory studies with gases. Laboratory Practice, 32, 62–65.

New aluminium phosphide formulations for controlled generation of phosphine

C.J. Waterford, C.P. Whittle and R.G. Winks*

Abstract

New formulations of phosphine-generating products have been developed and tested for use with on-site phosphine generators. The formulations are inherently safer during storage and transport before use than current commercial formulations. They are well suited to controlled production of phosphine for continuous application technology such as SIROFLO®. However, when used in appropriate generators they allow for controlled phosphine production that is useful in a range of fumigation practice. Patent applications have been lodged on the inventions described in this paper.

Introduction

Phosphine is the most widely used fumigant for disinfesting stored grain. Indeed, it could soon be the only fumigant available for this purpose in many countries. The other commonly used fumigant, methyl bromide, now recognised as a potent ozone depletor, is most likely to be banned or highly restricted by the year 2000. Unless other fumigants are developed and approved, phosphine will become the principal fumigant gas.

Phosphine is a colourless gas, which is odourless when pure, but the technical product has an odour of garlic. The gas can be generated on-site by the hydrolysis of aluminium, magnesium or zinc phosphides. Aluminium phosphide (AlP) is the most commonly used to generate phosphine for disinfestation of stored products. Phosphide as tablets, pellets, or as a powder in sachets, slowly reacts with atmospheric moisture to release phosphine. This reaction has been used in several different methods of application. Thus:

- tablets of AlP can be added directly to the commodity;
- tablets of AlP can be placed in removable dispensing containers over the stored product; and/or
- a 'blanket' containing sachets of AlP can be placed over the stored product.

Although these methods, when properly applied, can be very effective, there are some disadvantages, including the following.

- Tablets added directly to the grain leave a residue in the grain, consisting largely of aluminium hydroxide, but with some unreacted phosphide (Bruce et al. 1962; Vardell et al. 1973; Banks 1987).
- In storages without recirculation equipment, distribution and levels of phosphine cannot be controlled once the dose of AlP has been applied.

- Application rates are sometimes increased excessively in the hope this will overcome the reasons for gas loss and maintain levels in leaky storages.
- The most efficient use requires a well-sealed storage structure.

Direct application of phosphine as a gas (rather than by generation from AlP) for fumigation has advantages and methods have been developed recently to exploit these. The introduction of Phosfume® gas (2% phosphine in carbon dioxide) and the SIROFLO® (Winks 1988) and SIROFUME® technologies are examples. SIROFLO® is a technique where a low concentration of phosphine is passed through stored grain for sufficient time to ensure that all life stages of an insect infestation are exposed to a lethal dose (concentration × time). SIROFUME® is a technique where the level of phosphine in the headspace of a sealed store is monitored and periodically topped up to replace gas lost. For these purposes phosphine can be obtained in cylinders, usually diluted in nitrogen or carbon dioxide. However, there are circumstances where use of cylinders is neither desirable nor warranted. For example, use of phosphine gas for SIROFLO® applications in on-farm storages may be better serviced by a phosphine generator. Prototype generators have been constructed (Banks and Waterford 1991) which use commercially available AlP tablets or pellets as a source for the phosphine. While the feasibility of using such devices for continuous phosphine production has been demonstrated, we had concerns about the safety of using currently available formulations of tablets or pellets in this application. Thus, it was decided to investigate alternative formulations of AlP better suited for use in a phosphine generator.

Initial experiments indicated that admixture of AlP with water-immiscible carriers, such as paraffin oil and paraffin wax, produce extrudable pastes or solid formulations. Such formulations are relatively stable in air but produce phosphine when added to water. This paper reports studies to investigate the properties of these formulations, and trials that were used to test the more promising formulations in prototype generators.

Materials and Methods

The AlP used was technical grade, granular 86% AlP (Detia Germany 1988). A fine powder, suitable for making formulations, was made by grinding AlP in an electric blender in air. The phosphide was sieved with a 250 micron mesh in a well-ventilated fumehood after blending for about 30 seconds. The portion that did not pass through the sieve was returned to the blender and topped up with more AlP for further grinding. Using this technique a 1 kg pack of technical grade AlP yielded 950 g (95%) of phosphide with particle size < 250 microns. The formulations produced from this powder were homogeneous, of a smooth consistency and of uniform reactivity.

All subsequent preparation of formulations was also carried out in a well-ventilated fume hood and it was not necessary to provide an inert atmosphere for any of these procedures.

* Stored Grain Research Laboratory, CSIRO Division of Entomology, GPO Box 1700, Canberra ACT 2601, Australia.

However, with larger quantities, provision for a blanket of inert gas may be prudent.

The following properties were considered to be important with respect to suitable carriers for the AlP. They should be:

- non-reactive to AlP
- non-flammable
- sufficiently viscose to prevent the phosphide settling out
- of low volatility
- stable in the temperature range −40 to +60°C
- non-reactive with packaging material and unable to diffuse through it
- non-reactive with metals such as stainless steel and aluminium
- non-water absorbing.

Paraffin oils and waxes possess many of these properties.

Paste formulations

Paraffin oil admixed with ground AlP powder in the ratio of 73% AlP/27% paraffin oil produced a paste which could be readily extruded from a syringe or canister with an orifice > 3 mm. The basic formulation was modified to eliminate settling or separation by substituting a proportion of the paraffin oil in the carrier with petroleum jelly. The modifications included, 50% paraffin oil–50% petroleum jelly, and petroleum jelly alone. These formulations increased the viscosity of the carrier and in turn the viscosity of the final paste mixture.

Solid formulation

A solid formulation (27% wax, 73% powder) was made by suspending finely divided phosphide in molten paraffin (BDH Ltd, paraffin wax with ceresin, congealing point ca 60°C), and allowing the mixture to cool. The wax was heated to about 90°C so that the powder could be admixed before the wax congealed.

Evolution of phosphine from formulations in water

Samples of phosphide formulation (ca 1.0 g) were weighed into a 150 mL flask fitted with a connection to a gas burette. The flask was flushed with nitrogen then 100 mL of reactant, usually water, was added and the flask sealed. The reaction vessel was held in a water bath to maintain a constant temperature of either 23 or 40°C (±1°C). Evolution of phosphine was followed by recording gas burette readings with time. Measurements were made for the first 30 minutes without stirring the mixture. Thereafter, the mixture was agitated using a magnetic stirrer. For comparison unformulated AlP powder, ground to the same particle size, was also used.

Additives chosen from a variety of readily available detergents and surfactants were assessed as possible means to further control the release of phosphine from formulations and included:

- Teepol® — Shell, household detergent
- BP Comprox® — British Petroleum (BP), general purpose detergent
- PEG 600 — ICI, polyethylene glycol, surfactant
- Heptane — May & Baker, laboratory solvent
- Dishwashing powder — alkaline non-foaming detergent.

The materials sold as Teepol® and Comprox® are mixtures of anionic and non-ionic detergents and are subject to changes in formulation depending on availability of materials. In the present work the Teepol® was newly acquired material whereas the Comprox® was taken from a drum of material that had been in store for some years.

Evolution of phosphine from paste formulations in air

Phosphide paste formulations were placed in shallow containers and the exposed surface area measured. A container of paste was placed in a glass vessel fitted with an inlet line providing air of known humidity at 100 mL/minute, and an outlet line fitted with a sampling septum. The humidifier was a glass gas-washing bottle containing either water or an appropriate water/glycerol solution. The reaction vessel and the humidifier could be maintained separately at any given temperature from 23 to 40°C (Fig. 1). Phosphine evolution was calculated from concentrations measured at intervals by gas chromatographic analysis (FPD) of the outlet gas stream.

Conditions were changed to test the effects on the evolution of phosphine from the paste formulation of:

- exposed surface area,
- humidity, and
- temperature.

Analysis of phosphide residues and evolved phosphine gas

Spent formulation samples were recovered from reactions where the paste was stirred with water for 48 hours and then allowed to stand for 24 hours. Samples were analysed for unreacted phosphide by a modification of the method of Bruce et al. (1962) using mercuric chloride to trap released phosphine. A known amount of phosphine gas and a sample taken from a spent sachet containing AlP were analysed for comparison. The latter was taken from a Detia AlP 'blanket'.

Gas samples were taken at random from the phosphine generated from the paste formulations and analysed on a gas chromatograph fitted with a gas density balance.

Phosphine production from prototype generators

A prototype generator was constructed to evaluate selected formulations at rates sufficient to fumigate a 2000 t bin of grain. For a bin of this size, SIROFLO® protocol requires a supply of phosphine of about 1.5 g/hour. The major components of the generator are shown in Figure 2. They include:

- a 20 L reaction vessel with a stirrer
- ability for the headspace of the reaction vessel to be flushed with air or an inert gas
- ability to pass this same gas through the water to produce extra agitation
- a port for the introduction of the paste
- a port to extract phosphine produced and any gas used for stirring or flushing
- a delivery mechanism for the formulation.

The phosphine generated was swept from the headspace with air from a small diaphragm pump. It was piped into a measured air flow of a standard SIROFL® system fitted to a 40 t farm bin filled with wheat. The concentration of phosphine produced in the air flow was measured with a CiTicel® phosphine sensor, and the output recorded on a chart recorder. The response of the sensor was calibrated periodically by using an analysed phosphine gas standard (20 ppm from The Commonwealth Industrial Gases Ltd, Australia).

A second generator (Fig. 3) was constructed to evaluate the generation of phosphine from an exposed surface area of formulation.

A prototype dispenser (Fig. 4) was used to evaluate the solid formulation, where plugs (4.5 × 2 cm) were dispensed, one at a time, from the bottom of an upright canister containing a total of 12 plugs.

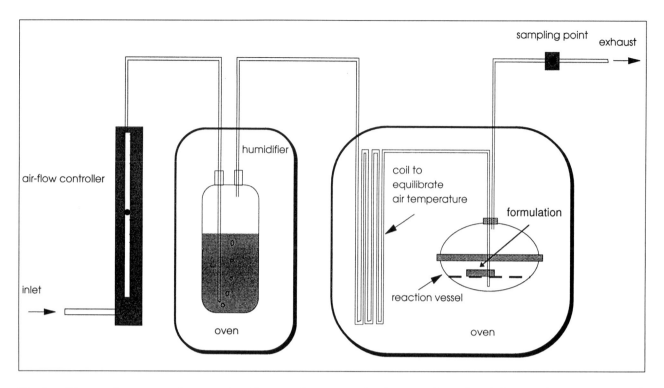

Fig. 1. Diagram of apparatus used to assess phosphine output from a range of surface areas of paste formulations at various temperatures and relative humidities.

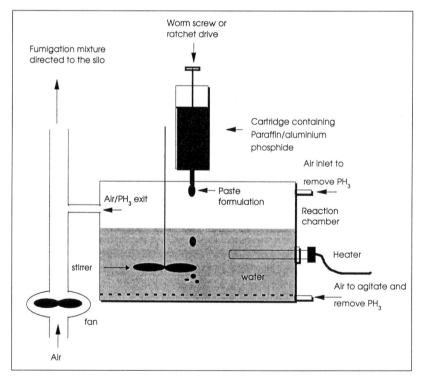

Fig. 2. Diagram of phosphine generator used to assess phosphine-generating paste formulations.

Results

Formulations in water

The results from unstirred reactions are presented in Figure 5. The reaction rate of the unstirred formulation in water is very much less than that observed for the powder, even when the reaction mixture was heated to 40°C. The reaction rate of all formulations in water when not stirred was very low com-pared with the reaction rate of the unformulated powder. With one notable exception, addition of materials intended to aid the dispersal of the formulation in water had no effect on unstirred reactions (Fig. 6). The addition of Teepol® to the water promoted a marked increase in the reaction rate, in strik-ing contrast to other detergents, surfactants and solvents, which did not.

When the paste formulations were continuously stirred in water the reaction rates in general were observed to be signifi-

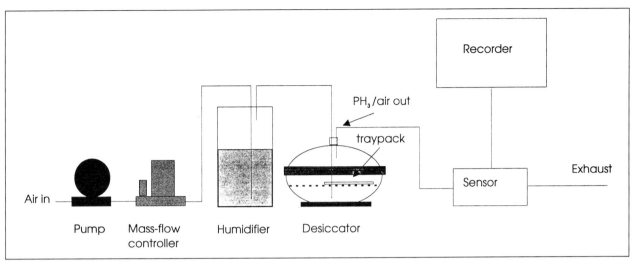

Fig. 3. Diagram of apparatus used to assess production of phosphine from tray packs using paste formulations based on 20 and 27% paraffin with the balance aluminium phosphide.

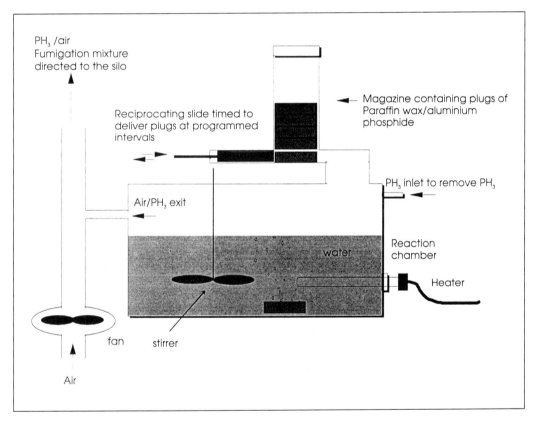

Fig. 4. Schematic of phosphine generator fitted with a mechanism to deliver wax plugs of a phosphine-generating formulation.

cantly faster than the unstirred reactions. Raising the temperature to 40°C increased the reaction rate (Fig. 7) and increasing the viscosity of the formulation reduced the reaction rate (Fig. 8) but, with one exception, the changes were not large. The behaviour of the solid preparation made with paraffin wax differed markedly in two respects. Firstly, this formulation was not greatly affected by stirring and secondly, the reaction rate over the period of digestion was found to be remarkably steady. A 5 mm × 12 mm sample of solid formulation retained its form but became thinner with time during the course of the reaction.

Addition of dispersal agents to the stirred reactions gave unexpected results. Heptane or 3% Comprox®, the latter a non-foaming detergent, caused a significant decrease in the reaction rate for the stirred mixture. Teepol® caused an increased reaction for the unstirred mixture which was not altered by subsequent stirring. None of the remaining additives was effective in significantly increasing the reaction rate over that observed in water alone.

Paste formulations in air

The effect of change of humidity and temperature on the release rate of phosphine from exposed formulation is shown in Figure 9. Initially the air was at 23°C with a relative humidity of >90%. After 24 hours the temperature of the reaction vessel was raised to 40°C while the humidifier was maintained at 23°C, which kept the absolute moisture content of the air constant. Apart from a minor excursion due to disturbance of the apparatus, the rate of evolution of phosphine was not altered by the increase in temperature. However, the evolution rate doubled when the humidifier was also heated to 40°C, restoring the relative humidity to >90%, thus increasing the absolute moisture content of the air passing over the paste.

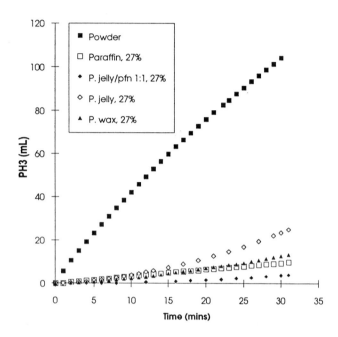

Fig. 5. Phosphine released from aluminium phosphide formulations (1.0 g) in unstirred water, compared with release from aluminium phosphide powder. Formulations were made from aluminium phosphide powder and 27% paraffin oil, 27% of a 1:1 petroleum jelly/paraffin oil mixture, 27% petroleum jelly, and 27% paraffin wax.

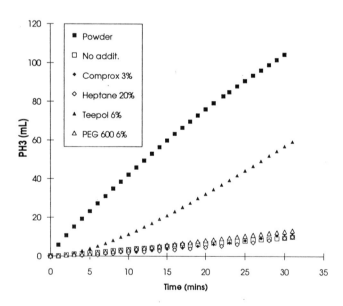

Fig. 6. Phosphine released from 73/27 aluminium phosphide/paraffin oil formulation (1.0 g) when placed in unstirred water with various additives, compared with unformulated aluminium phosphide powder in water and the formulation with no additives.

Analysis of phosphide residues and evolved phosphine gas

The results of phosphide analysis are presented in Table 1. The amount of unreacted phosphide in the used formulation samples was found to be very low, even in a freshly obtained sample. By contrast the sachet sample contained 4–5% unreacted phosphide, a typical result for preparations that use exposure to air to generate phosphine (Banks 1987).

Table 2 shows the composition of the gas produced by reaction of the formulations with water. The small amounts of air that were present in the samples can be attributed to diffusion during sampling and the water over which the gas was stored. The methane could have resulted from the reaction of aluminium carbide with water. The method used to produce technical grade AlP is known to produce small amounts of the carbide as a contaminant.

Table 1. Phosphine residue analyses

Substrate	Condition	Residue analysis[a]
Phosphine (control)[b]	Gas 0.61 mg	0.63 mg
Celphos® sachet	Used	4.2–4.6%[c]
AlP/paraffin 73/27	10 weeks after reaction inwater	10 ppm[c]
AlP/paraffin 73/27	Stirred in water 48 hours, stand 24 hours, N_2 4 hours	28 ppm[c]

[a] Bruce et al. (1962)
[b] Known amount of phosphine injected into analytical apparatus and analysed as if released from a substrate.
[c] Expressed as AlP remaining in spent product (w/w)

Table 2. Concentration of phosphine in samples of gas produced by the reaction of paste with water.

Component	Concentration (mole%)	Retention time (minutes)[a]
Air[b]	2.7	1.35
	2.3	1.41
	2.4	1.40
	2.36	1.39
Methane[c]	1.27	1.69
	1.28	1.75
	1.22	1.75
	1.39	1.73
Phosphine	96.07	4.11
	96.42	4.18
	96.39	4.15
	96.25	4.12

[a] Analysed by GC using a Gowmac Gas Density Balance detector on a Tracor MT150 GC fitted with a 6 ft × 1/4 inch stainless steel column packed with 80–100 mesh Porapak Q.
[b] The air comes partly from diffusion during sampling and partly from the water over which the gas was stored.
[c] Aluminium carbide in the technical grade AlP is the source for the small percentage of methane.

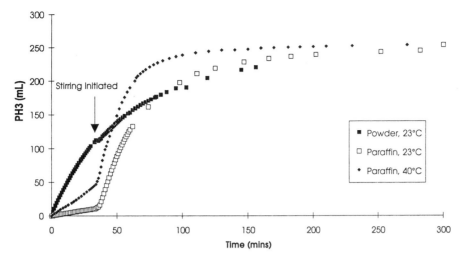

Fig. 7. Phosphine released by 73/27 aluminium phosphide/paraffin oil formulation (1.0 g) in water at 23 and 40°C, compared with aluminium phosphide powder. Stirring commenced at 31 minutes.

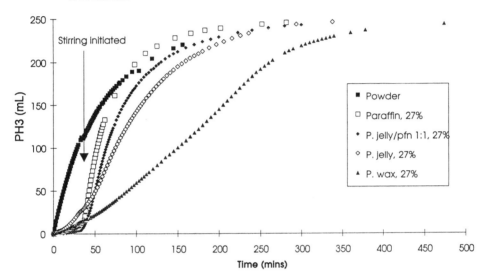

Fig. 8. Phosphine released from formulations (1.0 g) in water with stirring after 31 minutes, compared with release from aluminium phosphide powder. Formulations were made from 27% paraffin oil, 27% of a 1:1 mixture of petroleum jelly and paraffin oil, 27% petroleum jelly, and 27% paraffin wax.

Performance of paste formulations in a prototype generator

The phosphine production rate for a typical trial is presented in Figure 10. In most cases the reaction rate came to an equilibrium and then declined when addition of the formulation ceased. Generally, the reaction commenced more rapidly when detergent was included with the water.

Performance of paste formulations when used as 'tray-packs'

One novel aspect of the work suggested a method of generating phosphine by exposing a thin layer of formulation to moist air in what has been loosely termed a tray-pack. Figure 11 shows evolution of phosphine from paste when used as a tray-pack where the formulation contained 20% paraffin. The output tracked the ambient temperature and hence the available moisture provided by the humidifier held at ambient conditions. The output of phosphine ranged from about 100–300 µg/hour/cm² for the exposed surface area. At this rate the exposed surface area of a tray-pack required to provide 1 g/hour output would be

about 0.5 m². However, the increased volume of the reaction products of the formulation absorbed some of the paraffin oil, which together produced an increased surface area and reaction rate towards the end of the time over which phosphine was produced. Some means of steadily reducing the available surface area during the reaction would prevent this.

Performance of solid formulation in prototype generators

A typical production profile for the solid formulation from a plug of 4.4 cm radius and 1.8 cm depth, with all surfaces available for reaction, is shown in Figure 12. The plug was suspended in the reaction vessel of the generator. The water, containing 5% Teepol®, was heated to approximately 28°C. This single addition produced phosphine at an average rate of approximately 1 g/hour over 38 hours.

By increasing in the depth of formulation and restricting the reaction to one surface only phosphine production was extended to 7 days (Fig. 13). The plugs had a radius of 2.7 cm and were 4.5 cm deep. The plug was removed and examined on two occasions during the trial in an attempt to correlate production rates with the progress and nature of the digestion of the

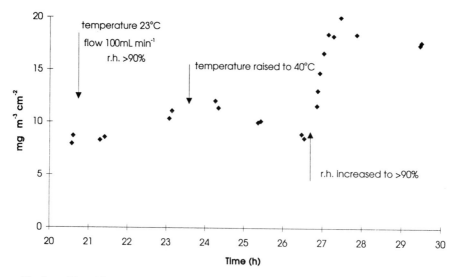

Fig. 9. Phosphine released from the surface of a 73/27 aluminium phosphide/paraffin oil paste formulation on exposure to air at differing temperatures and humidities.

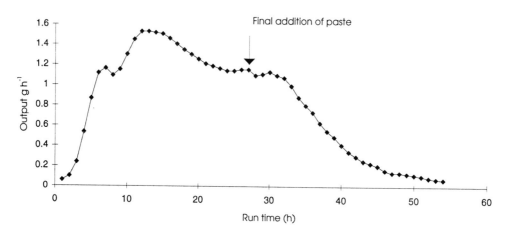

Fig. 10. Phosphine output from 73/27 aluminium phosphide/petroleum jelly paste formulation. Nominal addition rate of 6 g paste/hour. No air stirring.

plug. On the first inspection, the face of the plug was distorted down one side, the effect of which was to increase the surface area available for reaction. This increased surface area would explain the level of reaction rate just before the inspection. A large sliver of spent formulation above the reactive face, which had split off the plug, may have caused the higher reaction rate some hours earlier by doubling of the reactive surface. The plug was replaced in the generator and the reaction continued. The second inspection towards the end of the reaction indicated that the sloughing off or splitting from the face of the plug had continued. Passage of water to the unreacted face of the plug and the escape of phosphine gas may have been slowed by the increasing depth of reacted residue towards the end. This may explain the rise and fall of the rate about the expected rate for this evaluation.

Assessed also were four identical plugs (36 g each, 73% AlP) in which the base and sides of each plug were coated with an additional layer of wax (1–2 mm) so that only the top surface remained reactive. One was reacted with unstirred water at a constant temperature of 30°C. The generated phosphine was extracted with air and monitored as previously described. The sample produced phosphine over a period of 45 hours.

The second was reacted at ambient temperature (12–25°C). The recorder trace from this trial was very noisy, the average reaction rate was generally lower and the plug took about 10 hours longer than the first plug to digest. The third was reacted in the same way as the second, with the concentration recorded at a point within the grain in order to assess the damping effect that the grain would have on the pulses of phosphine production. The recorder trace showed that the time to complete digestion of the plug was similar to the second plug and significant damping of the concentration had occurred over the reaction period once the gas had passed into the grain.

The fourth was left exposed to air of about 40% r.h. Signs of a slow reaction with the surface changing from greenish to grey were apparent after a day. The reaction proceeded slowly over several months producing a 'growth' of reacted material on the surface. However, the surfaces coated with the layer of extra wax did not react.

A dispenser (Fig. 4) with the capacity to add plugs to a generator at a predetermined rate was also evaluated. The rate chosen was one plug a day. Phosphine was produced for 8 days. Conditions were changed during the trial to assess the effect on output. The first two plugs were reacted at ambient

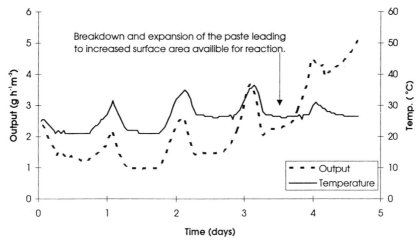

Fig. 11. Phosphine output from a tray pack of 80/20 aluminium phosphide/paraffin oil exposed to humidified air at ambient temperatures.

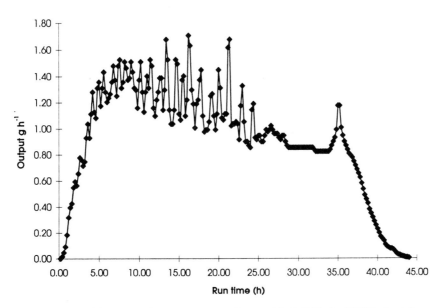

Fig. 12. Phosphine output from a generator using one block (150 g) of 73/27 aluminium phosphide/paraffin wax formulation.

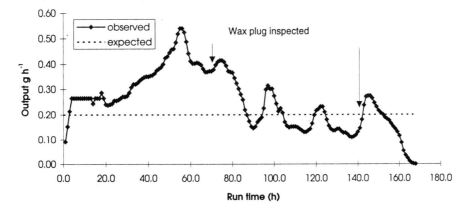

Fig. 13. Generator output from a 73/27 aluminium phosphide/paraffin wax block (140 g) held in an open glass vessel. The vessel containing the wax block was removed for inspection at the times indicated.

temperature without any additional heating. The concentration produced in the constant air flow showed a slow start and uneven reaction rates as the temperature of the water changed with ambient conditions. When the water was maintained at 32°C the plugs were reacted in less than 24 hours. When the temperature was reduced to 22°C the output became more uniform. Changes in reaction rate were presumably caused by changing viscosity of the paraffin wax with temperature. In this type of generator the production rate of phosphine can be controlled by exposed surface area of the plug and the rate of addition of plugs to the reaction vessel.

Discussion

These laboratory investigations showed that AlP can be formulated as an extrudable paste or a solid formulation. Although the formulations may not be miscible with water, reaction of the AlP with water to produce phosphine can still occur.

The paste which was less viscous showed a tendency to settle out or separate with time. The more viscous preparations did not suffer from this problem but were more difficult to extrude from a dispenser. Extrusion pressure depends on the viscosity of the preparation and the size of the outlet orifice. Selection of a suitable extrudable paste formulation will therefore depend, in part, on the design of the dispensing mechanism.

The solid formulation based on paraffin wax was more stable and easier to handle, and the phosphine output was generally more controllable.

Based on inherent reaction rates, the formulations incorporating a paraffin as carrier were found to be less hazardous than phosphide tablets, pellets, or blankets when exposed to air. Reaction with water (liquid or vapour) takes place only at the exposed surface of the formulation and the reaction is slow with no increase in temperature or danger of spontaneous combustion. Deliberate attempts to promote ignition or violent reaction were unsuccessful. For this reason it is anticipated, that these formulations will present less hazard during transport and storage than the currently available commercial preparations of AlP when accidentally exposed to air. It was observed that reaction in air was less likely to proceed to completion with the less viscous pastes. For pastes containing a higher content of paraffin oil the reaction essentially ceased once the surface layer had reacted. Conversely, for a paste containing only 20% paraffin oil and formulations where solid wax was the carrier, reaction in air continued slowly to completion.

Reactivity of the phosphide formulations in water was generally found to be low without agitation. For many preparations, less than 4% of the total potential phosphine was released after being in water at 23°C for 30 minutes. By comparison, under the same conditions, >50% of unformulated AlP powder reacted. Stirring the water and heating increased the reaction rate. Unexpectedly, most attempts to accelerate the reaction rate by addition of solvents or detergents were unsuccessful. Some of the additives inhibited the reaction. A notable exception was the detergent, Teepol®, which promoted rapid reaction even when unstirred.

The reactivity of AlP in paraffin wax was also unexpected. Tablets or plugs of this formulation were found to react in water steadily at the surface of the preparation. This behaviour, which produced a relatively constant reaction rate, indicated that such a formulation would be suitable for use in a phosphine generator.

It was anticipated that tests on reactivity in air would provide information relevant to safety and product packaging considerations. This aim was achieved and the investigation

led to another useful discovery. It was found that phosphine production from a fixed surface area of formulation, exposed in an air-flow of constant relative humidity, remained the same for many days. The surface area required to produce say 1 g/hour indicated that a generator using the formulation as a thin layer in a tray was feasible. This discovery has been incorporated in a patent application (Waterford et al. 1993).

Selected paste formulations tested in a prototype generator fitted with a motor driven piston for extruding paste formulation from a canister provided a supply of phosphine up to the required 1.5 g/hour. The design of the prototype generator had provision for an inert gas to sweep phosphine from the generator. In practice, this facility was unnecessary and air was used. Phosphine was constantly in excess of the flammability limit in the reaction vessel before being diluted into the inlet airstream of the silo. Even when positive steps were taken to promote ignition, spontaneous combustion or polymerisation of the phosphine, none of these phenomena occurred.

The expended paste formulations appear to pose no disposal hazard as the reaction proceeds to completion within the generators. After exposure to air, exhausted AlP pellets contain up to 5% of unreacted phosphide occluded in the residue of aluminium hydroxide (Banks 1987). This was not the case for paste formulations reacted with water where the residues were found to contain little (ppm level), if any, unreacted phosphide.

Tray-packs of phosphide paste are a novel possibility for use in a phosphine generator and have the advantage of simplicity. The results obtained with tray-packs are sufficient to demonstrate the potential of this technique. However, further work is needed to determine the best formulation for this application. Initial results indicated, that the formulation used in traypacks may require less carrier than that required for pastes designed for extrusion from a canister. This is a matter for further investigation.

The most promising results were obtained from the solid formulation made from AlP and paraffin wax as plugs or tablets. This performed very reliably in a phosphine generator and appears to be the preferred option for further development.

Conclusion

It is entirely feasible to safely produce sufficient phosphine for the purpose of fumigating large bulks of stored products by controlling the way water and AlP interact, including the addition of formulations of AlP to an excess of water. These discoveries have led to the development of a number of novel methods for generating phosphine as well as those detailed in this paper. As previously indicated the novel aspects arising from these and other studies are covered by patent applications (Banks and Waterford 1991; Waterford et al. 1993). These aspects include:

* phosphine generation by addition of phosphide formulation to an excess of water
* pellet dispensers for use in a phosphine generator
* the use of permeable membranes for controlling flow of water to phosphide
* paste and wax formulations of phosphide and reaction of these materials in water and in air.

Acknowledgment

This work was supported financially, in part, by a grant from Roussel Uclaf.

References

Banks, H.J. 1987. A continuous, cumulative procedure for analysis of phosphine. Journal of Stored Products Research, 23, 213–221.

Banks, H.J. and Waterford, C.J. 1991. Method and apparatus for generating phosphine, and safety system for phosphine generators. Patent Application No. PCT/AU91/00264.

Bruce, R.B., Robbins, A.J. and Tuft, T. O. 1962. Phosphine residues from Phostoxin-treated grain. Agriculture and Food Chemistry, 10, 18–21.

Vardel, H.H., Cagle, A., and Cooper, E. 1973. Phosphine residues on soybeans fumigated with aluminium phosphide. Journal of Economic Entomology, 66, 800–801.

Waterford, C.J., Whittle, C.P. and Winks, R.G. 1993. Formulations, method and apparatus for the controlled generation of phosphine. Patent Application No. PCT/AU93/00270.

Winks, R.G. 1988. SIROFLO®, a new grain protection method. Stored Grain Australia, 1, 4–5.

Fumigation of dried vine fruit for export

P. Williams[*]

Abstract

Methyl bromide is currently used extensively for quarantine fumigations and for fumigating dried fruit prior to export. Recent studies have been conducted to develop a modern code of practice for the use of this fumigant by the dried fruit industry. This work has led to improved fumigation techniques and safety procedures. It has also demonstrated that it can take up to 12 days to completely ventilate large stacks (200 t) of boxed sultanas so that concentrations of methyl bromide are at or below the occupational exposure standard or threshold limit value (TLV) of 5 ppm.

The construction of large stacks can be modified by splitting stacks and introducing ducts to improve ventilation but, in the absence of direct monitoring, 10 days of ventilation are necessary before stacks are broken down. This necessitates a change in regulations for ventilation.

There are now concerns about the ozone depleting properties of methyl bromide, and the phasing out of its use has been proposed under the Montreal Protocol. Improved codes of practice for the use of methyl bromide will only meet short- term requirements for industry, and there is an urgent need to develop alternative treatments for control of insects in export produce.

Introduction

Methyl bromide is a fumigant which can kill insect pests rapidly (within a few hours) and is used extensively for quarantine and export fumigations. The dried fruit industry in Victoria, Australia uses methyl bromide for fumigating stacks of cartons of processed fruit to control insect infestations in fruit for export. Fruit is fumigated approximately every month and three days before export in accordance with the Dried Fruit (Export) Orders, using as a standard for fumigation the National Health and Medical Research Council (NH&MRC) 'Code of Practice for the Fumigation of Dried Fruit with Methyl Bromide'(1971). This document has provided the industry with useful operational guidelines, but it has become outdated since it uses Imperial units and contains insufficient information on bromide residues, personal protection and safe ventilation procedures. This paper is concerned with obtaining information required to modernise the code of practice. It is also concerned with more general implications of some of the findings, e.g. in relation to worker safety.

Before the current study, the only detailed data on fumigations of large stacks of processed sultanas (174 t) were in reports on two fumigations conducted by the Department of

Primary Industries and Energy (DPIE). The reports expressed concern that workers required to break down stacks a few hours after removal of fumigation sheets could be exposed to unacceptable levels of fumigant. The use of fan-forced aeration was endorsed, since no fumigant was detected after 22 hours of fan forced ventilation of a 174 t stack. However, it was stated that 'due to the limit of accuracy of the gas analyser used, 'not detected' should be interpreted as meaning that there was no more than 100 ppm present'.

More precise information on methyl bromide concentrations during ventilation was required for a modern code of practice because, the short-term exposure standard, or short term-exposure limit (STEL), for methyl bromide is 15 ppm and the occupational exposure standard or threshold limit value (TLV) for prolonged exposure is 5 ppm.

Materials and Methods

Methyl bromide fumigations of 200 t stacks of 15 kg boxes of sultanas were monitored at the Mildura Co-operative Fruit Company Ltd packing sheds at Red Cliffs and Irymple, Victoria. The sultana boxes were of corrugated cardboard with polythene bag liners. Fumigations all lasted 24 hours with fumigant dosages ranging from 18 to 48 g/m^3. The stacks were built on aluminium sheeting (ACI Vapastop) placed over the concrete floor to reduce gas leakage.

Each stack was built with flexible ducting laid around its base in such a way that it could be connected to an extraction fan on conclusion of the fumigation. At first, fumigations were carried out on solid stacks of boxes, but for later fumigations stacks were split into two 100 t stacks about 50 mm apart. In the split stack fumigations, T-pieces were introduced connecting the ducting around the stacks with weldmesh ducts, with cross sectional areas the same as the end of a sultana box ($0.0459 m^2$), built into the bases of stacks. Each duct was linked to a chimney, with a cross-sectional area the same as the top of a sultana box ($0.108 m^2$), running vertically through the stack (Fig. 1).

The gas-proof sheets used to cover the stacks were made of nylon fabric, proofed on both sides with PVC and ca. 0.37 mm thick. The sheets were tailored to fit over metal frames fitted with wheels. Sheets were raised to allow the frame to be wheeled over a stack and were then lowered to cover the stack and surrounding ducting. Sand snakes were used to hold the edges of the sheeting down onto the Vapastop sheets at the base of the stack to form a seal. Piping to deliver methyl bromide to the stack was fitted to the top of the frame and down one side, where a gas fitting, running through the fumigation sheet, was connected to a methyl bromide vaporiser for introduction of the fumigant. Before fumigation commenced, a risk area was delineated with ropes and warning signs 15 m from the stack in accordance with AS 2476 (ASA 1981).

Initially, fumigation staff were equipped with canister respirators, overalls and gloves in accordance with the NH&MRC Code of Practice. Following a review of safety procedures, they were equipped with self-contained breathing apparatus, the compressed air cylinders for which were carried on a

---

[*] Victorian Institute for Dryland Agriculture, c/- State Chemistry Laboratory, Department of Agriculture, 5 Macarthur Street, East Melbourne, Victoria 3002, Australia.

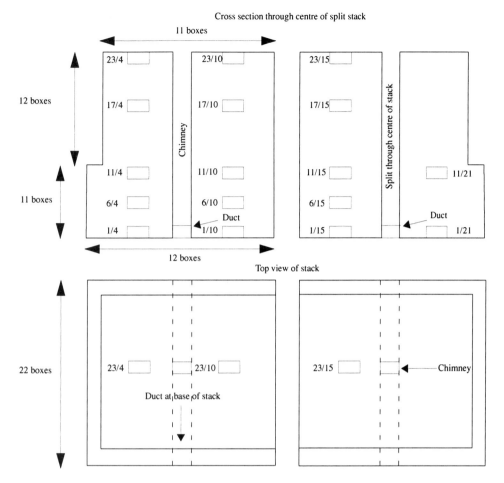

Fig. 1. Diagram of stack of 15 kg boxes of sultanas showing a split stack with basal ducts and chimneys to improve ventilation following fumigation. Boxes identified by vertical/horizontal numbers. All boxes had external gas lines; boxes 1/4, 6/4, 11/4, 17/4, 23/4, 11/10 and 17/10 had internal gas lines; boxes 1/4, 6/4, 11/4, 17/4, 23/4, 11/10, 23/10, 17/15 and 23/15 had external thermocouples; and boxes 1/4, 11/4 and 23/4 had insect cages.

trolley. Flexible air lines linked the cylinders to full face respirator masks. The cylinders were fitted with audible warning devices to indicate when the air level was low.

Methyl bromide concentrations were determined using either a gas chromatograph (GC) fitted with an electron capture detector, or by gas detector tubes (Drager). The GC was fitted with a J&W Scientific DB-624 fused silica column 30 m × 0.53 mm ID with a film thickness of 0.3μm; operating temperatures were oven 50°C, injection port 250°C and detector 300°C. Gas sampling lines were placed at different locations within stacks both outside and within boxes of sultanas. Gas samples for analysis were taken from the lines during fumigation and ventilation periods. Stack and ambient temperatures were recorded using an electronic data logger (Datataker) attached to T-type thermocouples located with some of the gas lines and outside stacks. In most fumigations, insect bioassay cages were inserted into some of the boxes of sultanas and into ducts at the bases of stacks (Fig. 1). The insect species used were *Tribolium castaneum* (Herbst) (Tenebrionidae, Coleoptera), *Oryzaephilus surinamensis* (L.) (Silvanidae, Coleoptera), and *Plodia interpunctella* (Hubner) (Pyralidae, Lepidoptera).

To commence ventilation, an extraction fan was connected to a chimney pipe through which gas was exhausted above the packing shed roof. The intake of the fan was connected to the ducting around the stack. The fan was switched on and the fumigation sheet was raised on the side opposite the fan. This enabled the fan to draw fresh air through the stack to flush out the methyl bromide. After ca. 20 hours, the extraction fan was

switched off and disconnected. The sand snakes were removed and the fumigation sheet was either removed completely, or the sides were raised, and the stack was subjected to fan-forced ventilation from either one large or two smaller fan units for a further 24 hours after which the fumigation sheet was removed, if it was still in place, and the stack was left under natural ventilation.

After fan-forced ventilation was completed, the atmosphere in the shed in the vicinity of the stack was tested for methyl bromide by a fumigator wearing a respirator. A halide lamp was used as a check for significant contamination and if there was no indication of methyl bromide then gas detector tubes were used in the immediate vicinity of the stack to confirm that concentrations in the area were at or below the TLV of 5 ppm.

Results and Discussion

The NH&MRC Code of Practice (NH+MRC 1971) recommends the use of full facepiece canister respirators in accordance with the Australian Standard Z18 of 1968. However, under the current standard, Australian Standard AS 1716–1991 'Respiratory protective devices' (ASA 1991), the maximum gas concentration for which canister filters can be recommended is 50 times the TLV, i.e. 250 ppm for methyl bromide. Also, WorkCover Authority of N.S.W. (1991) recommends that canister filter respirators should never be used for protection against highly toxic gases such as methyl bromide. Consequently, it was recommended to the industry that fumigators be equipped with self-contained breathing equipment. Industrial response was hastened by a ban on use of canisters (which incorporated a phase-in period) imposed by trade unions.

As previously reported (Williams and Henderson 1993), a solid 200 t stack with a temperature of ca. 21°C was fumigated with 18 g/m^3 of methyl bromide. During ventilation concentrations of ca. 100 ppm of methyl bromide were detected at the base of the stack 8 days after ventilation commenced and it took nearly 12 days before concentrations throughout the stack were reduced to 5 ppm or less. This result prompted the use of split stacks and the introduction of ducts and chimneys to facilitate ventilation (Fig. 1).

The results of monitoring methyl bromide concentrations during ventilation of a split stack with ducts and chimneys and a stack temperature of ca. 19°C are given in Table 1. Despite the improved ventilation it still took 9.5 days before methyl bromide concentrations were at or below 5 ppm at the base of the stack. The zero or low concentrations of methyl bromide recorded after 7 days ventilation in the outer and upper regions of the stack indicate the potential for lowering methyl bromide concentrations more rapidly by reducing stack size. However, reducing stack size means that the floor space required to house the fruit must be increased and this presents most packing sheds with major storage problems.

Location of a stack in a well-ventilated part of a packing shed, e.g. with open doors on either side of the stack, and high temperatures, can aid removal of methyl bromide. A 200 t split stack in a well-ventilated location was fumigated with 24 g/m^3 of methyl bromide when the ambient temperature was 40°C. It took only 6 days of ventilation before methyl bromide concentrations were reduced to 5 ppm or less.

The caged insects used in all of the trials were dead when examined. Many insects caged in unfumigated boxes of sultanas were alive and live insects were also found amongst the sultanas in some of these boxes. Thus, it appeared that the fumigations were successfully controlling infestations.

Gas detector tube readings for methyl bromide taken immediately around stacks (including readings taken with detector tubes touching the boxes) on completion of fan-forced ventilation never exceeded the occupational exposure standard of 5 ppm, in accordance with DPIE findings. However, analysis by GC and/or detector tubes of gas samples from within stacks at this time always showed methyl bromide concentrations in excess of both the TLV and STEL, demonstrating the danger of relying on the DPIE findings (Table 1).

Information gained in this study indicates the importance of checking fumigant concentrations within bulks of produce after an initial ventilation period, to ensure that concentrations do not exceed the occupational exposure standard before people are required to handle the produce.

There are now concerns about the ozone-depleting properties of methyl bromide, and the phasing out of its use has been proposed under the Montreal Protocol. Consequently this study will only meet short-term requirements for industry. Work on developing and testing alternative fumigants and fumigant mixtures for control of insect pests in a variety of products is planned to meet future requirements. The Stored Grain Research Laboratory, Division of Entomology CSIRO, and the Division of Horticulture CSIRO, have already conducted some trials on the use of carbon dioxide as an alternative treatment for the dried fruit industry. The carbon dioxide treatments proved successful when stacks of fruit were enclosed in tailor made gas-proof covers, which enabled toxic concentrations of carbon dioxide to be maintained for a minimum 15 day treatment period (Tarr et al., these proceedings).

Conclusions

Monitoring ventilation of large stacks of dried fruit showed that, on completion of fan-forced ventilation, concentrations of methyl bromide in the immediate vicinity of stacks were below the TLV of 5 ppm and that consequently it was safe to work within the area. However, concentrations of methyl bromide persisting within the stacks made it unsafe to break them down. In all circumstances, there was a need to extend the current maximum 2-day period allowed before breaking down of stacks. In general, a 12-day ventilation period would be needed for a solid 200 t stack and a 10-day period for a split stack with ducts and chimneys. Only if a stack was fitted with gas sampling lines to enable methyl bromide concentrations to be monitored would it be reasonable to take advantage of favourable conditions to reduce the ventilation period required.

It is important that organisations undertaking fumigations of produce conduct some monitoring to check that their fumigations comply with current health and safety standards as well as existing fumigation schedules. It is also important that fumigators be equipped with appropriate protective respirators and clothing that meets current standards.

Table 1. Concentration of methyl bromide (ppm) in a 200 t split stack in boxes of sultanas at different times after commencement of ventilation. Ventilation of both sub-stacks was aided by a chimney and basal duct as in Figure 1.

Vertical location in stack		1.5 days				2.5 days				7 days			
Top	Box 23	52	35	30		36	37	26		0	0	0	
	Box 17	51	50	86		63	63	45		0	0	0	
Middle	Box 10	128	850	212	288	91	330	118	119	0	8	2	0
	Box 6	356	1497	418		139	666	171		0	3	5	
Base	Box 1	1059	2118		2163	476	1091		861	18	75	35	60

Vertical location in stack		8 days				8.5 days				9.5 days			
Top	Box 23	0	0	0									
	Box 17	0	0	0									
Middle	Box 10	0	3	0	0	0							
	Box 6	0	12	0		2					2		
Base	Box 1	10	35	16	30	3	7	0	5	0	4	0	3
Horizontal location in stack		Box 4	Box 10	Box 15	Box 21	Box 4	Box 10	Box 15	Box 21	Box 4	Box 10	Box 15	Box 21

Acknowledgments

The author thanks Mr D.I. Allen (State Chemistry Laboratory, Victoria), and Mr A.P. Henderson and Ms J. Lupton (Institute of Plant Sciences, Victoria), for their assistance with this study, and the Mildura Cooperative Fruit Company Ltd for providing facilities and assistance. The work was carried out with financial support from the Dried Fruits Research and Development Council.

References

ASA (Australian Standards Association) 1981. AS 2476 General fumigation procedures. North Sydney, Standards Association of Australia, 12 p.

ASA. 1991. AS 1716 Respiratory protective devices. North Sydney, Standards Association of Australia, 72 p.

NH+MRC (National Health and Medical Research Council) 1971. Methyl bromide fumigation: Fumigation of dried fruit (Code of practice for the fumigation of dried fruit with methyl bromide). 10 p.

Williams, P. and Henderson, A. P. 1993. Methyl bromide fumigation of dried vine fruits. In: Corey, S., Dahl, D. and Milne, W., ed., Pest control and sustainable agriculture, Canberra 1992. 489-491.

WorkCover Authority of New South Wales. 1991. List of approved respiratory protective devices. Lithgow, Australia, Industrial Printing Co., 40 p.

Phosphine fumigation of stored field peas for insect control

P. Williams* and C. P. Whittle†

Abstract

The grain industry has on occasions experienced fumigation failures when fumigating stored field peas with phosphine to control pea weevil. In relation to this problem the present study has examined the susceptibility of the pea weevil to phosphine and the extent to which phosphine is sorbed and desorbed by field peas. Rates were not affected by changes in variety, condition or source of the peas. Fumigations with 1400 ppm phosphine in 80% filled containers showed that >200 ppm phosphine remained after 21 days. Studies of susceptibility of the developmental stages of pea weevil *Bruchus pisorum* (L.) established that relatively low concentrations of phosphine applied in sealed containers are sufficient to kill all stages of the beetle. No beetles survived in treatments with 30 ppm or more of phosphine. Absorbed phosphine desorbs readily from peas. In aired samples, phosphine levels were found to fall from above 12 ppb to less than 0.1 ppb in two days.

The results indicate that problems with commercial fumigations are most likely to arise from failure to maintain adequate gas tightness rather than excessive sorption of phosphine by peas or a tolerance of phosphine by pea weevils.

Introduction

In spring, adult pea weevils, *Bruchus pisorum* (L.), fly into crops of field peas, *Pisum sativum* L., attracted by the flowers. They feed on pollen in the flowers before laying eggs on the pea pods. The larvae which hatch from the eggs bore through the pods into pea seeds where they develop to the adult stage. At harvest time in summer all stages of pea weevil can be found in the pods. Most adults emerge from the peas within the first 2 months of storage, but some adults can remain within undisturbed seed until the following spring (Comery and Chaffey 1987). The pea weevil does not reproduce in stored peas.

Control of the pea weevil requires effective treatment of both larvae and pupae within the peas as well as of the adult weevils. Small quantities of peas sometimes contaminate other grains and if live insects are present in the peas, export may be delayed. In addition, the adults are active fliers and can create problems at export terminals by flying from contami-

nated peas onto previously treated peas and other grains that are being loaded onto ships. Pea weevils must be controlled in peas for export from Australia in order to meet the nil tolerance for live insects in export grain required under the *Export (Grain) Regulations* of the *Customs Act 1901–1971*.

Contact insecticides are considered unsuitable for control of the weevils in stored grain because the insecticidal effects are largely restricted to the emerged adults. Some are so active that they will leave the grain before being killed. The grain industry relies on fumigation shortly after storages are filled to minimise damage by developing weevils and limit the dispersion of adults. Methyl bromide and phosphine are the only fumigants currently registered for this use in Australia. However, methyl bromide has been implicated as a potent ozone depletor and its use will be limited in accordance with the Montreal Protocol. Further controls over those already agreed to are anticipated.

There are differences in phosphine fumigation strategies employed for control of pea weevil. The recommended dosage, 2 g/t, may be applied as a single dose at the start of fumigation, or as two applications each of 1 g/t, the first at the start of the fumigation period and the second half-way through. The two-stage dosing attempts to counteract losses of phosphine (from sorption or leakage) and ensure that effective concentrations of the fumigant are still present at the end of the fumigation. Recommended minimum fumigation times vary from 21 to 28 days. In the 1980s shorter fumigation times were used but there is now evidence that this could permit survival of some weevils (Williams 1990; Waterford and Winks, these proceedings). An alternative method of treatment, SIROFLO®, in which low concentrations of phosphine are maintained by automatic gas introduction during the fumigation period (Winks 1993), is now sometimes used.

Reported failures of some phosphine fumigations are of concern to industry, particularly because the use of the alternative and effective fumigant methyl bromide is likely to be restricted in the future. High tolerance of phosphine by some life stages of the pea weevil and high sorption of phosphine by peas under some conditions have been considered to be possible causes of fumigation failures. This study is concerned with examining these factors.

Materials and Methods

Field peas cultivar (cv.) Dun were grown at the Wimmera Research Station, Dooen, Victoria and left untreated with insecticide before harvest so that they could become heavily infested with pea weevils. Peas harvested from the outer edges of the planted area, where the weevils are most numerous (Comery and Chaffey 1987), were segregated for use in laboratory fumigation studies with immature pea weevils. Peas harvested from the centre of the crop were used for phosphine sorption studies. Samples of the Dun cultivar were collected from Walpeup, Dooen, Murtoa, Woomelang and Donald in Victoria during the 1992 and/or 1993 harvests. Samples of the Dinkum cultivar were collected from Walpeup in 1993 for use in the sorption studies.

* Victorian Institute for Dryland Agriculture, c/o State Chemistry Laboratory, Department of Agriculture, 5 Macarthur Street, East Melbourne, Victoria, 3002, Australia.
† Stored Grain Research Laboratory, CSIRO Division of Entomology, GPO Box 1700, Canberra, A.C.T. 2601, Australia.

Pea weevil fumigations

Peas containing immature pea weevils at different stages of development were fumigated in sealed 4.5 L jars in the laboratory. Phosphine was injected with a syringe into the jars via rubber septa in the screwtop lids. Jar-to-lid joints were sealed with silicone rubber to ensure gastightness. In each jar the peas (250 g) were loaded into a heavy duty wire gauze cage mounted on a perforated plastic container attached to the base of the jar. Within the plastic container was a magnetic stirrer. Different concentrations of phosphine were achieved in the jars by injecting different volumes from a standard cylinder containing 10000 ppm (ca 14 g/m^3) of phosphine in nitrogen. Before introducing a volume of phosphine into a jar a syringe was used to remove a similar volume of air from the jar. This avoided increasing the pressure within the jar to above that of the surrounding atmosphere. Immediately after introduction of phosphine the gas within a jar was mixed using the magnetic stirrer.

In each fumigation experiment a minimum of four replicate jars were used for the untreated controls and for each phosphine concentration. The latter ranged from 0.01 to 1.4 g/m^3 (10–1000 ppm). The jars were held in a dark room or incubator at a temperature of 23–25°C during the fumigation periods which ranged from 21 to 28 days.

After fumigation the jars were opened and ventilated under a fume hood. The open tops were covered with terylene voile held in place by elastic bands to prevent insects from leaving or entering jars during ventilation. After ventilation, pea samples, at least 100 peas per jar, were dissected and examined for pea weevils. Dead and live pea weevils were counted and discarded and the remaining peas were incubated at 27°C (60–70% r.h.) and examined for emerged adults at monthly intervals for three months.

Phosphine sorption and desorption studies

Sorption of phosphine by field peas was studied using 100 mL glass vials fitted with Mininert® valve lids to provide gas sampling ports. Each vial was loaded with ca 62 g of peas; either whole peas or a mixture of whole and damaged peas were used. The filling ratio was ca 80%. The vial was sealed with the valve lid, the valve was opened and a syringe needle was inserted through the septum to allow air pressure within the vial to equilibrate with that outside. A gas syringe was used to remove 8 mL of air from the vial, then 8 mL of a phosphine–nitrogen mixture from a 10000 ppm (ca 14 g/m^3) phosphine standard cylinder was injected into the vial (restoring the pressure equilibrium) and the valve was closed. Assuming a true density of 1.48 g/mL for the peas, the free air space was ca 58 mL and the addition of the phosphine should result in a concentration of ca 2 g/m^3 of phosphine (ca 1400 ppm) in the free air space within the vial. The true density of the peas was assumed to be 1.48 g/mL on the basis of studies of the true densities of grains, including field peas from Dooen, Victoria (J. Cassells and H.J. Banks, pers. comm., 1992).

Five replicate vials were set up for each batch of peas to be assessed for phosphine sorption. The peas were held at a constant temperature of 23°C with moisture contents at particular values in the range 9.8–12.2% w.b. for at least 2 weeks before being used in experiments.

Moisture content determinations were made using an air oven method in which a small sample of peas was ground and ca 20 g was placed in a moisture determination tin which was weighed before and after the sample was dried to constant weight in an oven at 105°C. Samples were placed in the oven overnight and the following morning were transferred to a desiccator to cool before being weighed so that the moisture loss could be calculated.

The phosphine sorbed by the peas in the vials was monitored by analysing the phosphine concentrations in samples of gas from the vials during 21 days. Each vial was shaken to mix the gas within immediately before a gas sample was taken for analysis.

An hour after phosphine was injected into a set of replicate vials of peas, a 10 L sample from each vial was injected into a gas chromatograph (GC) to analyse phosphine concentrations. The GC was fitted with a flame photometric detector and a J&W Scientific DB-5 fused silica column 16 m 0.25 mm i.d. with a film of 5% phenyl methyl polysiloxane of 0.25 μm thickness. Operating temperatures were: oven 50°C, injection port 150°C and detector 200°C. Samples were then taken at 2-hourly intervals for the first 8 hours and thereafter at 1–3-day intervals. When the vials were not being sampled they were kept in dark rooms or incubators at 23°C.

Phosphine desorption from fumigated peas was monitored by successive measurements of phosphine concentrations in the peas using the Brockwell (1978) method of phosphine residue analysis. This method, which was described for the analysis of phosphine in wheat, uses grinding to release sorbed phosphine into an enclosed headspace. The released phosphine is then determined by gas chromatographic analysis. The method was tested for peas by placing a known weight of peas in the analytical apparatus and introducing a known amount of phosphine. The peas were then ground and the phosphine gas in the headspace determined. This process was repeated for different amounts of phosphine. Phosphine recovery was 75% and linear in the range 0.1–40 ppb. Undamaged peas (cv. Dun from Donald, 250 g) were fumigated by treatment with 1.5 g/m^3 of phosphine in an air stirred container for 24 hours. After this time the phosphine was removed and the peas divided into 30 g samples in petri dishes and placed in a fume hood. Individual samples were analysed for phosphine by the Brockwell method at various times up to 28 hours. Peas were also treated with a constant flow of phosphine at 0.1 g/m^3 for 48 hours and then analysed as above.

Freshly fumigated (1.5 g/m^3) peas were allowed to stand in air for 0.5 hours and then 30 g samples (solid volume, 20 mL) were placed in glass bottles (120 mL) sealed with Mininert® valves. After 95 hours the headspace atmospheres were analysed for phosphine gas (by GC), and the vessels were opened and the peas analysed for residual phosphine by the Brockwell method. For the purposes of comparison both results were expressed as parts per billion with respect to the weight of peas in the vessel.

Results and Discussion

Pea weevil fumigations

About 6% of the peas used in the fumigation experiments were infested, ca 97 peas in each fumigation jar. The distribution of developmental stages in these peas was 21% larvae, 5% pupae and 74% adults. Most of the pupae and adults were alive at the start of the fumigations, but up to 85% of the larvae were dead. First and second instar larvae accounted for most of the dead. In fumigations in which dosages of phosphine have ranged from 0.04 to 1.4 g/m^3 (30–1000 ppm) all fumigated pea weevils were killed, whereas most insects in unfumigated control jars remained alive. The dead insects in the fumigated jars included many newly developed adults in the process of emerging from within the peas. The duration of the fumigations may have been sufficient to allow pupae, which are believed to be the most phosphine tolerant life stage, to develop into adults, which are more susceptible. Consequently, some emerged and emerging adults may have been pupae at the start of a fumigation. At concentrations of phosphine below 0.04 g/m^3 some weevils survived. Further experiments at phosphine levels below 0.04 g/m^3 are currently in progress.

Phosphine sorption and desorption studies

Sorption of phosphine by field peas is neither rapid nor extensive when compared with wheat and many other commodities (Banks 1993). At a constant temperature (23°C) and after an initial 8 hours, when sorption is more rapid, the amount of phosphine (C) absorbed with time (t) can be represented by the relationship $\log C = a + kt$, where a and k are constants. A comparison of data for fumigations carried out at constant temperature (Table 1) showed that there was little variation in the value of the slope (k) and also that there were no obvious effects arising from changes in the condition or source of the peas. This is further illustrated in Figure 1, in which results of phosphine sorption experiments with whole field peas of cv. Dinkum (12.09% m.c.) from Walpeup and whole and damaged field peas of cv. Dun (9.90% m.c.) from Dooen are shown. Over 21 days the phosphine concentration in the headspace was found to fall by less than one order of magnitude. Sealed containers, filled to 80% capacity with peas and fumigated with 2 g/m^3 phosphine, were found to contain better than 0.28 g/m^3 (200 ppm) phosphine after 21 days. This is well in excess of the minimum initial dose of 0.04 g/m^3 (30 ppm) required for complete control.

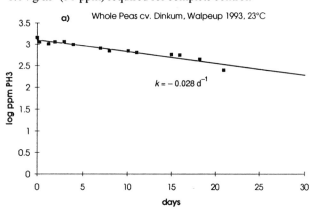

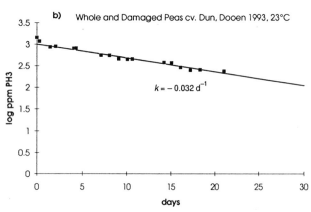

Figure 2 shows the release of sorbed phosphine from peas after fumigation at 1.5 g/m^3. The greater part of the phosphine desorbs within 8 hours of commencing aeration and after about 2 days the amount of phosphine in the peas is below detectable limits (0.1 ppb). A similar result was obtained for peas which had been fumigated at 0.1 g/m^3. However, in this case the initial amount of sorbed phosphine in the peas was only 2.5 ppb. When freshly fumigated peas containing 15 ppb of sorbed phosphine were placed in a sealed container and allowed to stand, the phosphine equilibrated between the headspace and the peas. Where the volume of the headspace was 6 times that of the peas and after 95 hours standing, the equivalent of 14 ppb phosphine was detected in the headspace of the container but no phosphine could be detected in the peas. These results indicate that when sorbed phosphine equilibrates between the atmosphere and the peas, the desorption of the phosphine is rapid and, after several air exchanges, the amount of sorbed phosphine in the peas can be expected to be below detectable limits (<0.1 ppb).

Conclusions

The laboratory experiments in which pea weevil larvae, pupae and adults were exposed to phosphine in sealed containers showed that relatively low initial dosages of ca 0.04 g m^3 could kill the weevils given an adequate exposure period of 21 days at 23–25°C. The sorption studies showed that field peas do not sorb excessive amounts of phosphine, and the desorption experiments demonstrated that the sorbed phosphine is readily desorbed leaving no detectable phosphine residue in the peas.

The pea weevil does not breed in stored peas and if insects that survive a fumigation are to breed, they must leave the storage and fly to fields where pea crops are to be planted. There they mix with other pea weevils which have remained in the fields and have never been exposed to phosphine. The opportunity for concentration of resistant genes and selection of highly resistant populations is thus much reduced compared with that for insects that breed continuously in stored commodities where populations may be exposed many times without dilution by susceptible individuals.

These results indicate that there should be no problems in controlling pea weevils in stored peas using a single dose of phosphine applied at the currently recommended rate of 1.5 g/m^3 (2 g/t) and a minimum fumigation time of 21 days at

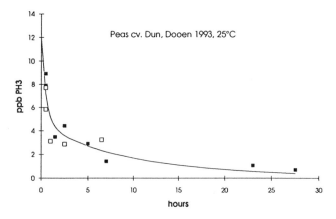

Fig. 1.   Phosphine sorption by a) whole field peas cv. Dinkum (12.09% m.c.) from Walpeup and b) mixed whole and damaged peas cv. Dun (9.90% m.c.) from Dooen. Sorption measured by phosphine loss from headspace.

Fig. 2.   Phosphine levels measured in peas at various times after fumigation at 1.5 g/m^3 for 24 hours. Data from two experiments fitted to a curve expected for a diffusion controlled loss from a spherical particle (Banks 1986).

Table 1. Comparison of slopes from regression of log concentration against time for phosphine sorption by various samples of field peas fumigated with 2 g/m^3 at 23°C over 21 days.

Year	Location	Cultivar	Condition	Moisture content (% w.b.)	Slope (d^{-1})
1993	Dooen	Dun	whole and damaged	9.90	-0.032
1993	Dooen	Dun	whole	9.84	-0.029
1992	Donald	mixed	whole and damaged	11.01	-0.028
1992	Woomelang	mixed	whole and damaged	11.32	-0.029
1992	Murtoa	Dun	whole and damaged	9.87	-0.029
1993	Walpeup	Dun	whole	12.15	-0.031
1993	Walpeup	Dinkum	whole	12.09	-0.028

the storage temperatures normally experienced after harvest (usually 20–30°C). There is no record of phosphine resistant strains of pea weevil occurring, and the development of strains capable of withstanding a well conducted fumigation is considered unlikely. It appears that failing to ensure or maintain the necessary gastightness in storages under fumigation (Banks and Annis 1984) is the most likely cause of fumigation failures.

Acknowledgments

The authors thank Mr A.P. Henderson, and Ms J. Lupton, Institute of Plant Sciences, Victoria, and Ms M. Connell, State Chemistry Laboratory, Victoria, for their assistance with this study. Mr C.J. Waterford, CSIRO Division of Entomology, is also thanked for helpful discussions. The work was carried out with financial support from the Grains Research and Development Corporation.

References

Banks, H.J. 1986. Sorption and desorption of fumigants on grains: mathematical descriptions. In: Champ, B.R. and Highley, E., ed., Pesticides and Humid Tropical Grain Storage Systems. Proceedings of an international seminar, Manila, Philippines, 1985. ACIAR Proceedings No. 14, 179–193.

Banks, H.J. 1993. Uptake and release of fumigants by grain: sorption/desorption phenomena. In: Navarro, S. and Donahye, E., ed., Proceedings of an International Conference on Controlled Atmosphere and Fumigation in Grain Storages, Winnipeg, Canada, June 1992. Jerusalem, Caspit Press Ltd, 241–260.

Banks, H.J. and Annis, P.C. 1984. Importance of processes of natural ventilation to fumigation and controlled atmosphere storage. In: Ripp, B.E., Banks, H.J., Bond, E.J., Calverley, D.J., Jay, E.G. and Navarro, S., ed., Controlled atmosphere and fumigation in grain storages. Amsterdam, Elsevier, 299–323.

Brockwell, C.A. 1978. Determination of phosphine in wheat by headspace gas–liquid chromatography. Journal of Agricultural and Food Chemistry, 25, 962–964.

Comery, J. and Chaffey, R. 1987. Pea weevil. Department of Agriculture and Rural Affairs, Victoria, Agnote, 3 p (Order No. 3735/87 Agdex 166/622).

Williams, P. 1990. Fumigation of stored field peas — current strategies. Proceedings National Pea Weevil Workshop, Melbourne, 1990. Department of Agriculture and Rural Affairs, Victoria, Conference Proceedings No. 38, 67–71. Agdex 166/622.

Winks, R.G. 1993. The development of SIROFLO in Australia. In: Navarro, S. and Donahye, E., ed., Proceedings of an International Conference on Controlled Atmosphere and Fumigation in Grain Storages, Winnipeg, Canada, June 1992. Jerusalem, Caspit Press Ltd, 399–410.

Measurement of resistance to grain fumigants with particular reference to phosphine

R.G. Winks and E.A. Hyne*

Abstract

The dosage of a fumigant is a function of the concentration to which an insect is exposed and the duration of exposure to that concentration. Since any measure of resistance is usually a comparison of dosages that elicit a similar response, it follows that fumigant resistance may be influenced by the choice of variables used to measure or characterise resistance. When the ratio of the coefficients of the dosage variables (the toxicity index) of the strains being compared is equal, the magnitude of resistance will be constant over a range of dosages. The converse is true when the ratios are not equal. In addition, when the ratios are greater than, or less than one, the measure of resistance will depend on the choice of the independent variable on which the comparison is based. With phosphine it has been shown that the toxicity index varies between strains and is usually less than one. This can have a significant effect on the interpretation of resistance data and the implications of such resistance in practical fumigations.

This paper describes the underlying principles that influence the detection and measurement of fumigant resistance and describes results of experiments that examine the differences associated with the choice of the independent variable on which dosage comparisons are made. Resistance factors obtained for two strains of *Rhyzopertha dominica* varied ×5 and ×13 in one strain and between ×10 and ×31 in another strain depending on whether the comparisons were based on estimates of LC_{99} from a 20-hour test or on estimates of LT_{99} from a test at 0.05 mg/L. By contrast, resistance factors, derived from times to population extinction of mixed age cultures of the same two strains were ×1.4 and ×2.4 at 0.05 mg/L and ×1.4 and ×1.4 at 0.1 mg/L.

The results of this study question whether any of the reported resistances to phosphine pose a threat to the continued usefulness of this fumigant, except in fumigation enclosures that are inherently leaky and as such, unsuitable for fumigation. They also raise questions about the methods used for detecting and measuring resistance.

Introduction

Resistance to phosphine was first observed in stored-product insects over 30 years ago as a cross resistance following selections of *Sitophilus granarius* with methyl bromide (Monro et al. 1961). Moreover, the FAO survey of resistance in stored-product pests (Champ and Dyte 1976), in which widespread resistance to phosphine was detected, is now almost 20 years old. Clearly, phosphine resistance has been present in insect populations around the world for a long time and over the last decade some quite high levels of resistance have been

* Stored Grain Research Laboratory, CSIRO Division of Entomology, GPO Box 1700, Canberra ACT 2601, Australia.

reported. Nevertheless, phosphine is still an effective fumigant and is likely to remain so for some years to come.

At the time of the FAO survey, in 1973, it was decided to include testing for resistance to the fumigants methyl bromide and phosphine. For that purpose a method that one of us (Winks) had developed was employed. This method was later adopted as the FAO test method for fumigant resistance (Anon. 1975). The method uses modified laboratory desiccators as fumigation chambers and is based on the exposure of adult insects to selected concentrations for a fixed exposure period. The exposure period nominated was 20 hours, which was chosen to accommodate a 24-hour testing routine by a single person when many samples had to be tested, i.e. allowing time before testing commenced to terminate tests started on the previous day and to prepare insects for that day's dosing. A 24-hour test period was considered unsuitable for such a requirement. A fixed exposure period was chosen as the basis of the method, partly for convenience and partly because it was the basis of apparatus and techniques that were simpler and cheaper than those required for tests based on fixed concentrations. Moreover, many earlier studies on fumigants, particularly methyl bromide, had been based on fixed exposure periods.

The FAO test method has been employed in a number of laboratories over the years and has proved to be an effective method for detecting fumigant resistance. A number of laboratories have pursued studies of resistance to phosphine using methods derived from the FAO method. Largely, the methods employed have simply extended the exposure times to periods more suited to some of the strains being tested (e.g. Mills 1983). While the FAO test method has proved to be a valuable tool for detecting resistance to fumigants, its limitations, particularly in the case of phosphine, need to be recognised. Perhaps its greatest limitation in this context lies in its inability to provide a meaningful estimate of the level of resistance. This is partly because of the choice of fixed exposure periods as the basis of the method and partly because of the intrinsic characteristics of the toxicity of phosphine, i.e. the unequal contribution of concentration and time in the toxic action of phosphine. However, at the time the method was developed, the relationship between concentration and time in the toxicity of phosphine to insects was not fully understood.

In this paper we report results of experiments that highlight the intrinsic characteristics of fumigant resistance using strains of *Rhyzopertha dominica* that are resistant to phosphine and, in so doing, demonstrate one of the limitations of the FAO method.

Theoretical considerations

The dosage of a fumigant is the product of the concentration (*C*) and time (*t*) of exposure. If these dosage variables acted equally we would have a simple relationship of the form $C \times t = k$, where *k* is a constant describing the dosage required for a specified level of response, e.g. mortality. This equation is sometimes referred to as Haber's Rule. Although this relationship might be a convenient description for calculating dosages in some field situations, it does not provide an acceptable description of the toxicity of a fumigant, or any other dose/response relationship in which the dosage is comprised of more than one variable. Indeed, in most cases, Haber's Rule is the exception to the rule. However, it would seem that Haber's Rule (Haber 1924) has been largely misquoted for a good many years. In comparing the toxicity of a number of gases with potential use in chemical warfare, he proposed that 'the smaller the product of concentration and time producing a fatal response in the laboratory animal, the more poisonous the chemical weapon'. This, however, does not imply that the product of concentration and time for a specific level of response for a given gas is constant.

The more general form of the relationship between concentration and time is $C^n t = k$ (Winks 1984) in which the value of n, the 'toxicity index', provides a measure of the relative importance of concentration and time to the dosage. The toxicity index is a measure of the slope of the regression of time-to-a-given-response-level (e.g. LT$_{99}$) over a range of fixed concentrations and described by the equation:

$$\log t = \log k + n \log C$$

When $n = 1$ concentration and time contribute equally to the effectiveness of the dosage. When n is less than one, time is the more significant component of dosage and when n is greater than 1 the concentration has greater effect.

Apart from being significant in terms of the influence of the dosage variables on a given dosage, the value of n also influences the significance of resistance measurements. When n is greater than or less than 1, the choice of the fixed dosage variable, concentration or time, will influence the magnitude of a resistance factor determined for the same strain. If n is less than 1, experiments in which exposure time is the fixed component of dosage will produce a higher resistance factor than when concentration is the fixed component of dosage. The converse is true when n is greater than 1. This may be demonstrated by considering a range of say LT$_{99}$ values that could be expected from a range of fixed concentrations for a resistant strain and a susceptible strain both of which have a value of n of 0.5. Hypothetical data for such strains are plotted in Figure 1.

From the equations to these lines it may be deduced that the hypothetical resistant strain is 10 times resistant when LT$_{99}$s are compared, i.e. when concentration is the fixed component of dosage, but is 100 times resistant when LC$_{99}$s are compared, i.e. when exposure time is the fixed component of dosage, e.g., a comparison of the LC$_{99}$s (or LD$_{99}$) from an experiment based on a range of concentrations at a fixed exposure period of 20 hours (or any other fixed time).

When the values of n are different between two strains, a simple comparison (i.e. one in which the resistance level remains constant over the range chosen) no longer applies. In this case the level of resistance will not only depend on the choice of the fixed component of dosage, but will also depend on the particular value of the fixed component. This is illustrated in the data of Figure 2 in which the value of n of the resistant strain is 0.4 and that in the susceptible strain is 0.95.

From the data of Figure 2 it may be deduced that, if a fixed exposure period was used, the resistance factor would vary depending on the length of the exposure period chosen (Table 1).

It may also be deduced that if a fixed concentration was used as the basis of a test the resistance factor would vary depending on the concentration chosen (Table 2).

Clearly, when strains being compared have different values of n, large differences in the measured resistance factor can be obtained depending on the choice of the fixed component of dosage. The principal significance of these differences lies in their interpretation. Frequently they are interpreted to indicate the likelihood of failure of the fumigant in question to control the resistant strain, or the extent by which the dosage needs to be increased to control the resistant strain. With only a single

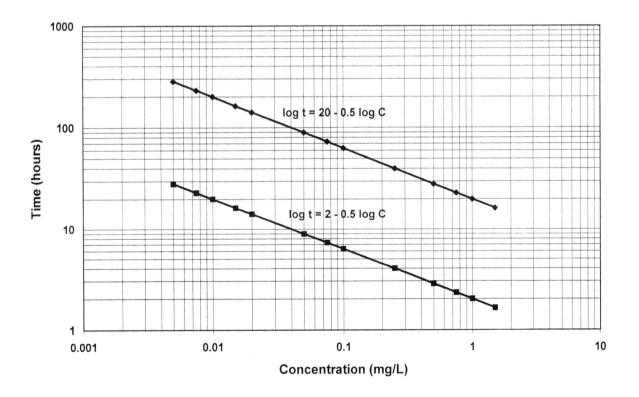

Fig. 1.   Comparison of two strains with the same slopes of the regressions of LT$_{99}$ over a range of fixed concentrations.

Table 1. Resistance factors derived for fixed exposure periods from the data of Figure 2.

Fixed exposure time (hours)	Strain 1 (mg/L)	Strain 2 (mg/L)	R factor
6	20.29	0.073	277
20	1.00	0.021	49
48	0.11	0.008	14
96	0.02	0.004	5

Table 2. Resistance factors derived from different fixed concentrations from the data of Figure 2.

Concentration (mg/L)	Strain 1 (hours)	Strain 2 (hours)	R factor
0.01	126.2	39.7	3
0.05	66.3	8.6	8
0.1	50.2	4.5	11
0.5	26.3	1.0	27
1	20.0	0.5	40

fixed concentration or single fixed exposure time being used as the basis of comparison between two such strains, gross errors of judgement could be made.

While the data of Figures 1 and 2 represent specific response levels such as the LD_{99}, a more general description, embracing all response levels, is provided by the equation:

$$Y = a + b_1 X_1 + b_2 X_2$$

where Y is the probit response, X_1 is log concentration (C), X_2 is log time (t) and b_1 and b_2 are the coefficients of log C and log t, respectively. This equation describes a plane representing the probit response to a range of dosages obtained by

various combinations of the dosage variables, concentration and time (Finney 1971). For such a plane, in which an interaction term is not significant, a simple relationship between concentration and time may be derived in the form $C^n t = k$ as follows:

$$(Y - a)/b_2 = (b_1/b_2)\log C + \log t$$

taking antilogs:

$$k = C^n t$$

where $n = b_1/b_2$.

For n to equal 1, b_1 must equal b_2. In addition, for the values of n from two strains to be the same, the ratio b_1/b_2, for each strain, must be the same. Clearly, both propositions are unlikely to be true except in a few cases *as convenient approximations*. Therefore, resistance factors, derived in accordance with the FAO test method, and those that have been reported in the literature, (e.g. Mills 1983; Zettler 1990) are unlikely to provide any real guide to their practical significance. Their only value lies in determining some measure of relative resistance between strains.

To examine the foregoing theory, experiments were conducted with resistant and susceptible strains of *R. dominica* using various concentrations and times and then comparing these results with comparisons based on times to population extinction of mixed-age cultures.

Materials and Methods

Three experiments were conducted, two with adults and the third with mixed-age cultures.

Experiment 1: Adult insects were exposed for a fixed time (20 hours) to various concentrations of phosphine.

Experiment 2: Adults were exposed to a fixed concentration of phosphine (50 μg/L) for various exposure periods.

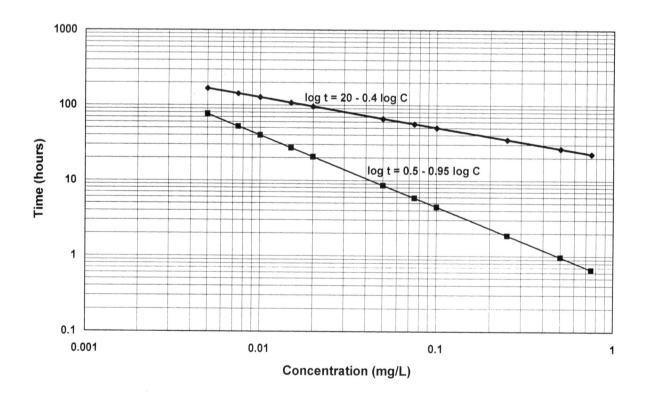

Fig. 2. Comparison of two strains with different slopes of the regressions of LT_{99} over a range of fixed concentrations.

Experiment 3: Mixed-age cultures were exposed to a fixed concentration of phosphine (50 μg/L) for a range of exposure times.

Origin and maintenance of insect material

Three strains of *R. dominica* were used in all experiments:
1. A phosphine-susceptible strain, CRD2, cultured in the laboratory since 1967 from stock obtained from the Pest Infestation Control Laboratory (U.K.).
2. A phosphine-resistant strain, CRD235P10, derived from laboratory selection with phosphine of 10 generations of a strain from Borivli, India 1972 that was detected as resistant to phosphine during the FAO survey of resistance in stored-product insects (Champ and Dyte 1976).
3. CRD316, a phosphine-resistant strain collected in 1989 from a grain store at Trangie, New South Wales.

The insects were reared on whole wheat containing some whole wheat flour. Before use, the wheat was conditioned to 12% m.c. and sterilised by heating to 60°C. Cultures were established by placing 200 adults on 30 g of flour in glass jars sealed with filter paper tops. After 7 days the adults were removed and 400 g of whole wheat was added to the jar. This method produced an F_1 generation without any F_2 adults at 6 weeks. The age of the adult insects ranged from 2 to 4 weeks at the time of the experiments. Mixed-age cultures were established by placing 300 adults on 1000 g of wheat and 200 g of whole wheat flour in 2 L jars. The wheat and flour were arranged in layers in the jar. The adults were not removed from these cultures. This method produces a culture in which every stage of the life cycle is present. The mixed-age cultures were 6 weeks old at the time of the experiment. Samples from these cultures were examined using X-rays to determine the relative abundance of the various developmental stages and to ensure that there were sufficient pupae and late instar larvae present to give a reasonable expectation that pupae of all ages were present at the start of experiments. In most species, the pupa is the most tolerant stage to phosphine and young pupae, the most tolerant of this stage. All insect cultures were reared at 30°C, 70% r.h. and insects used in experiments were conditioned at 25°C for at least 1 week before use.

Fumigation chambers

Experiment 1

The fumigation chambers used were modified 2.5 L desiccators described in Anon. (1975). Insects were confined in the desiccators within glass rings 2.5 cm high and 2.4 cm inner diameter. The glass rings had stainless steel mesh glued to one end and a stainless steel mesh lid was attached to the other end with a rubber band to prevent insects from escaping. The glass cages were assigned to each desiccator randomly. For all insect strains, two replicates of 50 insects were used for each dose and four replicates used for the control.

Experiment 2

A multi-chamber apparatus, from which insects can be removed at selected times following exposure to a constant phosphine concentration, was used (2nd apparatus, Winks and Waterford 1983). There was no recirculation and gas was supplied by mass flow controllers as outlined below. Insects were confined to glass rings and assigned to chambers in the same manner as in Experiment 1.

Experiment 3

Each culture was divided into two 3.1 L perspex flow-through fumigation chambers (Fig. 3). Each chamber was fitted with perspex screw top lids (1) and made gastight with

11.5 cm diameter neoprene O-rings. The lids were lined with stainless steel mesh to prevent insects and dust escaping through the inlet (2) and outlet (3). Six chambers were connected in parallel and supplied with a constant, flow-through concentration of phosphine.

Fumigation procedures

All insects were placed into the fumigation chambers 24 hours before the start of the experiments.

Experiment 1

A source of phosphine was prepared according to Anon. (1975). The required dose of phosphine for each desiccator was achieved by injecting calculated volumes of the phosphine source using gas tight syringes.

Experiments 2 and 3

The multi-chamber apparatus and the flow-through fumigation chambers were supplied with a constant concentration of phosphine. This was achieved by blending compressed air and phosphine from cylinders of phosphine in nitrogen using two Brooks mass flow controllers (5850E series). The diluted phosphine was humidified to 57%, by passing it through distilled water maintained at 15°C before being warmed to the laboratory temperature of 25°C.

The flow of the phosphine/air mixture to the multi-chamber apparatus was at a rate of 2 L/minute. In Experiment 3 the total flow of the phosphine-air mixture from the mass flow controllers was 200 mL/minute. The flow into each of the six fumigation chambers was regulated using Brooks flowmeters.

Throughout each experiment phosphine concentrations were monitored using a gas chromatograph fitted with a flame

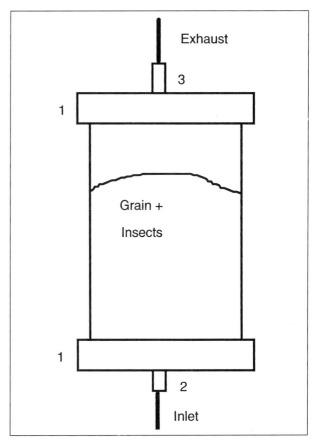

Fig. 3.   Flow-through fumigation chamber of apparatus used to expose mixed-age cultures of insects to a constant concentration of phosphine.

photometric detector. The flame photometric detector was calibrated using samples of a gas mixture, the concentration of which was determined using a gas density balance. The response of a gas density balance is predictable from the molecular weights of the components of the sample.

Post-fumigation procedures

Experiments 1 and 2

At the end of the exposure periods adult insects were removed from the chambers and transferred to 9 cm polystyrene Petri dishes containing a layer of wholemeal flour. During the post treatment holding period, the insects were held at 25°C, 67% r.h. Mortality assessments were made at 7, 14 and 21 days following the end of exposure. These periods were chosen to ensure that a mortality end-point was reached in each of the treatments (Winks 1982).

Experiment 3

At intervals of 24 hours for the experiment at 50 mg/L and of 2–3 days at 100 mg/L, samples of approximately 200 g were taken from the whole cultures and placed in glass jars the tops of which were black filter paper sealed around the edges with paraffin wax. These samples were incubated at 25°C, 60% r.h. and assessed for the presence of live adults by sieving. Assessments were made 24 hours after exposure and then at 8 weeks and 16 weeks later. From these observations, the time to population extinction (TPE) was determined as the first exposure time that produced no survival of insects at 8 weeks or 16 weeks, providing subsequent exposure periods also produced no survival.

Results

Using the methods described in Anon. (1975), and in accordance with Experiment 1 above, discriminating dosage tests were conducted on the three strains. The mortality response of each strain at the dosages chosen is given in Table 3. It is clear from these results that both strains CRD316 and CRD235P10 would be diagnosed as resistant according to this test and the mortalities obtained were either at or below the lowest mortality recorded during the FAO survey of phosphine resistance in 1973. Moreover, from the data of Mills (1983) the strain CRD235P10 would appear to be similar to the Bangladesh strain at the discriminating dosage, judged from its response at 0.8 mg.hours/L.

In accordance with experiments 1 and 2 the response of these strains was determined to a graded series of concentrations at a fixed exposure period of 20 hours and to a graded series of exposure times at a fixed concentration of 0.05 mg/L. Probit lines were fitted to the data and resistance factors determined from comparisons of both the LD_{50} and LD_{99} (Table 4). There was a marked difference in the measured level of resistance depending on the method chosen.

Probit planes fitted to data from a number of experiments in accordance with the protocol of Experiment 2 for CRD2 and CRD235P10 and over a range of fixed concentrations of 0.0025 to 0.05 mg/L for CRD2, and from 0.05 to 0.35 mg/L for CRD235P10, were also examined. The probit planes satisfied the simple model in which the interaction between concentration and time was not significant. The parameters of these planes in which Y is the probit response, are given in the following equations:

CRD2 $Y = 5.96 + 6.86 \log C + 10.88 \log t$
CRD235P10 $Y = -4.89 + 9.40 \log C + 13.14 \log t$

From these planes, estimates of the LC_{99} and the LT_{99} were derived and resistance factors calculated at 0.05 mg/L and at 20 hours. At 0.05 mg/L the resistance factor was ×8 while at 20 hours the resistance factor was ×22 (Fig. 4).

Table 3. The response of strains of *Rhyzopertha dominica* to dosages of phosphine at or above the discriminating dosage for this species (Anon. 1975).

Concentration (mg/L)	Exposure time (hours)	Mortality response (%)		
		CRD2	CRD316	CRD235P10
0.03[a]	20	100	3	NT[b]
0.04	20	100	NT	1
0.052	20	100	47	NT
0.075	20	100	NT	3
0.093	20	100	86	NT

[a] FAO test discriminating dosage
[b] NT = not tested

Table 4. Resistance factors derived from the response of adults of strains of *Rhyzopertha dominica* exposed to phosphine over a graded series of concentrations for 20 hours and over a graded series of times at 0.05 mg/L

	CRD2 (mg. hours/L)	CRD316 (mg. hours/L)	Resistance factor	CRD235P10 (mg.hours/L)	Resistance factor
Fixed exposure time of 20 hours					
LD_{50}	0.11	1.71	16	2.93	28
LD_{99}	0.19	2.50	13	5.77	31
slope	9.29	14.15		7.91	
Fixed concentration of 0.05 mg/L					
LD_{50}	0.31	1.1	4	2.46	8
LD_{99}	0.52	2.33	5	5.13	10
slope	10.54	7.13		7.26	

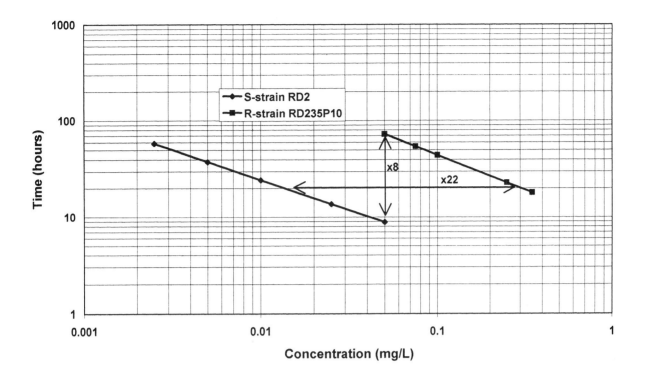

Fig. 4. The relationship between concentration and time in the LD_{99} values derived from probit planes fitted to mortality response data for adults of a susceptible strain (CRD2) and a resistant strain (CRD235P10) of *Rhyzopertha dominica* exposed to a range of fixed concentrations of phosphine for various exposure times at 25°C, 60% r.h.

The values of the toxicity index for these two strains were 0.63 for CRD2 and 0.72 for CRD235P10. Thus the resistance factor does not change greatly over a wide range of concentrations and decreases with increasing concentration. With a resistant strain of *Tribolium castaneum* the opposite change was found and the resistant factor decreased with decreasing concentration (Winks and Waterford 1986).

Resistance factors derived from mixed-age cultures exposed to 0.05 mg/L are given in Table 5 together with results from a similar experiment at 0.1 mg/L. The times to population extinction (TPE) are based on emergence in samples up to 16 weeks from the end of the exposure periods. TPEs are recorded as the earliest exposure period from which there was no emergence in samples providing that this was also true in samples from longer exposure periods.

At 0.05 mg/L TPEs ranged from 5 to 7 days for the three strains and the resistance factors for the two strains based on TPE was ×1.4 for CRD316 and ×2.4 for CRD235P10. A comparison with similar data obtained at 0.1 mg/L showed that a doubling of the concentration made no discernible difference in strains CRD2 and CRD316 but did reduce the TPE and the resistance factor in CRD235P10.

Discussion

The results of these experiments with different strains of *R. dominica* demonstrate that estimates of the level of resistance can vary widely depending on the method chosen. This is particularly so in terms of the choice of the independent variable of dosage. In the experiments of this study with adults, there was an approximate 3-fold difference in the estimated levels of resistance at the LD_{99} and about a 4-fold difference at the LD_{50} level. If inferences had been drawn from the results with adults about the practical difficulties of controlling such strains, such inferences would have been seriously at odds with the reality of the situation. For example, a resistance factor of ×31 obtained from the 20-hour test with CRD235P10 might have suggested that this strain would not have been controlled with normal dosages of phosphine and that an alternative method of control may have been needed. In reality, a dosage of 1.5 g/m^3, which is a common recommendation, would have achieved control.

The various quick resistance tests based on the correlation between the narcotic response and resistance (Reichmuth 1991; Bell et al., these proceedings; Waterford and Winks, these proceedings) also have the potential to exaggerate the level of resistance if they are used for anything more than

Table 5. Times to population extinction (TPE) and derived resistance factors following exposure of mixed-age cultures of three strains of *Rhyzopertha dominica* to constant concentrations of phosphine at 25°C and 60% r.h.

Strain	TPE (days)	Resistance factor	TPE (days)	Resistance factor
		0.05 mg/L		0.1 mg/L
CRD2	5		5	
CRD316	7	1.4	7	1.4
CRD235P10	12	2.4	7	1.4

detecting the presence or absence of resistance in a strain. These tests can be quite misleading if they are used to estimate resistance levels. Moreover, they are based on an, as yet, unproven correlation between resistance and the narcotic response and in addition, have the potential to fail to detect resistance arising from a mechanism that is not related to the narcotic response.

Tests like the FAO method and any derivatives based on simple increases of the exposure periods should only be used as means to detect resistance. When comparisons with the reference (susceptible) strain are made, inferences about the practical significance of such comparisons should be resisted.

Conclusion

The results of this study demonstrate that, with phosphine, where the value of the toxicity index *n* is likely to be less than one:

- any resistance factor derived from experiments in which fixed exposure periods are used with a range of concentrations, *will exaggerate the level of resistance.*
- any resistance factor based on comparisons at, for example, only one fixed concentration, *will be inadequate and could be quite misleading.*
- any resistance factor derived from adults only, *should not be used as the basis for adjusting dosage rates or exposure times in the field.*
- any resistance level estimated from a quick test method, *should not be used as the basis for adjusting application rates or exposure periods in practice.*

While times to population extinction provide estimates of levels of resistance that give reasonable guides to the practical significance of such resistance, experiments can take up to 6 months to complete.

References

Anon. 1975. Recommended methods for the detection and measurement of resistance of agricultural pests to pesticides. 16 Tentative method for adults of some major species of stored cereals, with methyl bromide and phosphine. FAO Plant Protection Bulletin, 23, 12–25.

Champ, B.R. and Dyte, C.E. 1976. Report of the FAO global survey of pesticide susceptibility of stored grain pests. Rome, Food and Agriculture Organization of the United Nations, FAO Plant Production and Protection Series, No. 5, 297 p.

Finney, D.J. 1971, Probit analysis, 3rd ed. Cambridge, Cambridge University Press.

Haber, F. 1924. Fünf vorträge aus den jahren 1920–1923 über die darstellung des ammoniaks aus stickstoff und wasserstoff die chemie im kriege das zeitalter der chemie neue arbeitsweisen zur geschichte des gaskrieges. Berlin, Julius Springer, 23–92.

Mills, K.A. 1983. Resistance to the fumigant hydrogen phosphide in some stored-product species associated with repeated inadequate treatments. Mitteilungen der Deutschen Gesellschaft für Allgemeine und Angewandte Entomologie, 4, 98–101.

Monro, H.A.U., Musgrave, A.J. and Upitis E. 1961. Induced tolerance of stored-product beetles to methyl bromide. Annals of Applied Biology 49, 373–377.

Reichmuth, C. 1991. A quick test to determine phosphine resistance in stored product insects. GASGA Newsletter, 15, 14–15.

Winks, R.G. 1982. The toxicity of phosphine to adults of *Tribolium castaneum* (Herbst): time as a response factor. Journal of Stored Products Research, 18, 159–169.

Winks, R.G. 1984. The toxicity of phosphine to adults of *Tribolium castaneum* (Herbst): time as a dosage factor. Journal of Stored Products Research, 20, 45–46.

Winks, R.G. and Waterford, C.J. 1983. Multi-chamber apparatus for laboratory studies with gases. Laboratory Practice, 32, 62–65.

Winks, R.G. and Waterford, C.J. 1986. The relationship between concentration and time in the toxicity of phosphine to adults of a resistant strain of *Tribolium castaneum* (Herbst). Journal of Stored Products Research, 22, 85–92.

Zettler, J.L. 1990. Phosphine resistance in stored product insects in the United States. Proceedings.of the Fifth International Working Conference on Stored-product Protection, Bordeaux, France, September 1990, 1075–1081.

Control of the common clothes moth *Tineola bisselliella* (Hummel) (Lepidoptera: Tineidae) and other museum pests with nitrogen

A. Wudtke[*] and C. Reichmuth[*]

Abstract

Many products, including wool, fur and bird feathers, are attacked by the common clothes moth, *Tineola bisselliella*. The pest may cause high economic losses in warehouses, museums and households. Thorough prophylaxis and control is of special importance for museum exhibits and other valuable objects. This paper reports on studies of the use of nitrogen as an alternative to traditional chemical means of control of museum textile pests.

Because of their higher susceptibility, the lethal exposure period for moths is presumably shorter than for beetles. From mortality data in stored-product protection it can be presumed that pupae will be most tolerant, followed by eggs. In this study, eggs of *Tineola* were investigated for their susceptibility to high nitrogen and low oxygen atmospheres at 25°C and 32°C. Practical control measures with pure nitrogen have been carried out in collections of cultural institutions and museums. Some results are presented in this paper.

Introduction

The common clothes moth *Tineola bisselliella* (Hummel) (Lepidoptera: Tineidae) and other museum pests are established worldwide and cause damage amounting to millions of dollars (Parker 1990). Especially in museums, due to optimal temperature and moisture for the development of the pests, these losses are severe. Unique artefacts destroyed cannot be replaced. Because of the high economic and cultural value of artefacts and their sensitivity to certain treatments, it is no longer recommended or acceptable to use toxic chemicals, including gases, which tend to react with the artefacts. In Germany and other countries, there are now strong restrictions for the use of ethylene oxide (EO) (Gilberg 1991). Moreover, the health of visitors to museums can be negatively affected by insecticides. Methyl bromide is very effective but it reacts with sensitive surfaces of paintings (Reichmuth et al. 1991). EO reacts with water to form ethylene glycol, which reacts as an aggressive solvent of the exhibits (Florian 1987).

Nitrogen as an inert gas does not react with most other chemical substances. Because of this property it can be used as a replacement gas for life-supporting oxygen, for optimal control of museum pests.

This paper reports on experiments with nitrogen and the eggs of the common clothes moth and summarises some field experiments which have successfully been carried out in German museums to control pest insects.

Material and Methods

Laboratory and field experiments were performed with museum pests, including the common clothes moth as the most important insect. The investigations and the engineering were supported by a team with wide recent experience in the field of stored-product protection (Adler and Reichmuth 1989; Gilberg 1989; Reichmuth 1987; Tunç et al. 1982). These experiences and useful information from the literature could be transferred to the application of nitrogen for museum pest control (Koestler 1992; Gilberg 1989; Pinninger 1992).

Laboratory experiments

The treatments with nitrogen were carried out in 550 mL-Dressel flasks. Several bottles were linked together with tubing. The required gas mixture was prepared by use of manometric instrument (SETARM, Rampé à Gaz) and steel cylinders which were evacuated prior to mixing. For treatment of caged insects in the bottles, the gas was introduced into the system of linked flasks from these cylinders. Between the gas cylinder and the first bottle another Dreschel flask was introduced which contained a saturated $NaNO_2$ solution to humidify the dry nitrogen to 65% r.h. To adjust the flow, a rate metre with a range of 0–20 L/hour was used.

The last bottle in the row was connected with an oxygen analyser (TORAY LF–750), which measured the oxygen content (precision of O_2 determination ± 0.015 vol.-%).

For breeding, the insect cultures were held at 25°C and 65% r.h. in a constant temperature room. In the course of the laboratory experiments, eggs of different age were tested. To obtain the eggs, adults were placed on felt and removed after 6–12 hours (Titschack 1922, 1926). After ageing at 25°C, the 4 and 5-day-old eggs were placed in special constructed cages with 120 µm mesh screen on one side and fine tissue on the other side. Experimental temperatures were 25°C and 32°C. After the treatment, the hatch rate of the eggs was determined for four days. The higher temperature was expected to lead to higher mortality or faster control (Valentin and Preusser 1990). Every test was accompanied by one untreated control and another control treated only with pressurised air at the same flow rate.

Field experiments

The following treatments were tested.
1. Wrapping the exhibits in PVC-or nylon-laminates and welding together the seams before N_2-treatment. This application can be used anywhere, especially when gastight chambers or rooms are not available and the objects are too brittle to be moved.

* Federal Biological Research Centre for Agriculture and Forestry, Institute for Stored Product Protection, Königin–Luise–Str. 19, 14195 Berlin, Germany.

In a museum in Berlin, a nitrogen treatment was carried out in a 30 m³ commercially available plastic tent. The fumigation process was undertaken as described above. However, the firm involved did not measure oxygen levels and there was no control engineering to adjust for constant low oxygen content.

In a cultural institute in Potsdam, an antique cupboard and other valuable cultural pieces were treated in a bag made of polyethylene. During the two experiments, nitrogen was introduced into the bag up to a pressure difference of 20–50 Pa.

2. Modifying existing fumigation and vacuum chambers constructed for use with toxic gases such as EO, for application of controlled atmospheres, especially in museums.

In the Hamburg Ethnic Museum, two experiments at 20°C and 65% r.h. with nitrogen were performed in a 12 m³ vacuum chamber. Little modification of fittings and door gaskets was necessary (Reichmuth et al. 1994).

In the course of these trials, control of insects was tested not only in textile exhibits but also in paintings and in test blocks of wood. The chamber was filled as completely as possible to reduce the free space and thereby the N_2 consumption.

The oxygen content in the chamber and the amount of nitrogen needed were recorded (Fig. 1). At the beginning of the fumigation, four cylinders of nitrogen were used to replace most of the oxygen. Another four cylinders were successively used during the experiment. When the pressure difference between the inside and outside of the chamber dropped below 5 Pa, a magnetic switch automatically opened a valve in the connecting tube to the cylinder to keep the nitrogen content within the chamber at a high level.

A second trial in this museum was carried out 5 months later under similar conditions. Together with samples of the common clothes moth, some dermestids and anobiids were distributed equally inside the chamber as test insects.

3. Treatment of cultural or moth-infested objects in systems such as the fumigation 'bubble' from Rentokill[1] or B&G Equipment[1]

4. Fumigation of gastight rooms or whole buildings, such as the herbarium of the Botanical Museum, Berlin.

Results

Laboratory results

These first nitrogen experiments with *Tineola bisselliella* were carried out with eggs of different ages. In comparison to the control group with more than 95% hatch, 100% mortality was achieved after 3 days of exposure to nitrogen at 25°C and 32°C. Oxygen concentration in the test atmosphere was 1.85%.

In a second series of experiments with 2% oxygen in nitrogen at 25°C (Fig. 2), 5-day-old eggs were slightly more susceptible than 4-day-old eggs under the experimental conditions. Complete mortality was achieved after 8 days.

Field experiment results

1. After the treatment of the antique wood cupboard, despite some leaks in the polyethylene foil, all introduced wood pests were found to be dead after 6 weeks of exposure.

2. The aim of the experiments in the chamber was to reduce the oxygen content below 2% during the treatment. Because the gaskets of the door of the chamber were constructed for use under vacuum, the tightness during the overpressure of 5

Pascal (0.000005 bar) was not satisfactory. Due to leakage and ingress of oxygen, the nitrogen content dropped to 96.5%.

During the whole experiment of more than 4 weeks, the average oxygen content could be adjusted to 1.5% which was sufficient to kill most of the introduced pest insects (*Attagenus smirnovi*, *Trogoderma angustum*, and *Anthrenus verbasci*) After the first experiment, only some cigarette beetles, *Lasioderma serricorne*, and one of the hide beetles, *Dermestes maculatus*, survived the treatment.

Some other wood pests exposed (*Anobium punctatum* De Geer, *Hylotrupes bajulus* L. and *Lyctus brunneus* Stephens) did not survive (Reichmuth et al. 1994).

The nitrogen consumption during the first experiment with 8 nitrogen cylinders at 12 m³, was rather high and affected by undetected leaks. The consumption of the gas is described in Figure 1.

The first 4 bottles were used solely to replace the oxygen, and the last bottle was empty after a short period due to a faulty connection of tubing.

During the experiment, the highest divergence of room temperature from 20°C was ±1°C which resulted from bad climate control of the room containing the chamber. The gas was adjusted to normal room temperature by use of a long copper tube. The relative humidity within the chamber varied widely during the first experiment. Therefore, conditioned silica gel was used during the second trial to stabilise the humidity.

3. The experiment with the commercial tent was cancelled after 8 weeks because of the high oxygen content despite the use of high amounts of nitrogen. All insects survived.

Discussion

Laboratory experiments

The first experiment with slightly less than 2% oxygen was successful within 3 days at both 25 and 32°C.

The lethal exposure time of several days with oxygen contents of 2% lies within the range of results with eggs of stored-product pest moths (Reichmuth 1987). There are not enough data for statistical analysis. On the other hand, the tendency of older eggs to die earlier was confirmed by results at both sublethal exposure periods.

Fields trials

Especially for the treatment in plastics tents, described under 2, gastightness is the crucial condition and must necessarily be installed. Otherwise, the gas consumption will be too high and the treatment too expensive and, moreover, ineffective. An electronic regulating device is helpful and saves time.

The process of filling bags, bubbles, or tents with the exhibits should be as effective as possible, to reduce the empty space.

During the experiments with nitrogen in the vacuum chamber, too much gas was used due to the untight gasket.

The presented trials were performed with commercial 50 L pressurised gas cylinders. In future, other means of supply of nitrogen may be possible, such as pressure swing machines with semipermeable membranes. Krabiell et al. (1992) describe the performance of such a machine producing 150 m³ N_2/hour. In many cases, nitrogen-application was criticised because of the long exposure times needed. In a museum situation, time is mostly not a restriction.

The length of time needed for effective treatment is also dependent on the thickness of the infested material which has to be penetrated by nitrogen to replace the oxygen. Normally, thick wood needs more time for uniform distribution than textiles.

---

[1]Quotation of commercial names and trade marks implies no endorsement or recommendation of products.

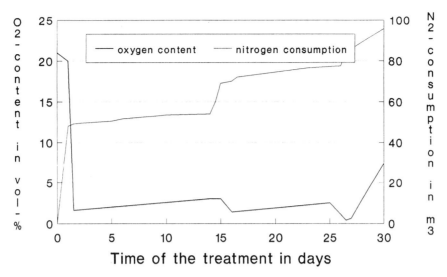

Fig. 1. Nitrogen application in the former vacuum fumigation chamber of the Hamburg Ethnic Museum. The nitrogen consumption and the oxygen content achieved are indicated. Pronounced changes in the slope of the lines correspond to change of the nitrogen cylinders. A total of 93 m^3 of nitrogen was used.

Fig. 2. Mortality of eggs of different age of the common clothes moth after treatment with 2% oxygen in nitrogen at 25°C.

In the field experiment reported, exposure of 4 weeks at 25°C and 65% moisture was sufficient to kill all the pests tested. For the control of wood pests, exposure times of more than 4 weeks are generally necessary (Reichmuth 1993).

For households, it may be convenient, to treat objects such as carpets or artefacts with nitrogen in plastic bags for pest control (Wudtke 1994). Oxygen-consuming substances such as AGELESS® are available to add to the bags to create low oxygen content atmospheres (Gilberg 1990). The cheapest plastic for this purpose is PVC. To evaluate the oxygen-containing air at the beginning of the procedure, only a vacuum cleaner is necessary.

All of the methods of N$_2$ application presented are practicable and relatively easy to apply, provided sufficient gastightness can be achieved. When calculating the costs of nitrogen treatments and comparing them with use of conventional insecticide treatments, it is necessary to consider not only the economic costs but also the ecological benefits. In the latter context, nitrogen treatment is clearly ahead of treatments with toxic gases.

References

Adler, C., and Reichmuth, Ch. 1989. Zur Wirksamkeit von Kohlendioxid bzw. Stickstoff auf verschiedene Insekten in Stahlsilozellen (Efficacy of carbon dioxide and nitrogen on different insects in steel silo bins). Nachrichtenblatt des Deutschen Pflanzenschutzdienstes, 41, 177–183.

Florian, M.–L. 1987. The effects on artefact materials of the fumigant ethylene oxide and freezing used in insect control. ICOM Committee for Conservation. In: Proceedings of the 8th Triennial Meeting, Sydney, Australia 6–11 September 1987, 199–208.

Gilberg, M. 1989. Inert atmosphere fumigation of museum objects. Studies in Conservation, 34, 80–84.

Gilberg, M. 1990. Inert atmosphere disinfestation using AGELESS® oxygen scavenger. ICOM Committee for Conservation. In: Proceedings of the 9th Triennial Meeting, Dresden, German Democratic Republic 26–31 August 1990, 2, 812–816.

Gilberg, M. 1991. The effects of low oxygen atmospheres on museum pests. Studies in Conservation 36, 93–98.

Koestler, R.J. 1992. Practical application of nitrogen and argon fumigation procedures for insect control in museum objects. In: 2nd International Conference on Biodeterioration of Cultural Property, Yokohama, Japan, 5–8 October 1992, 96–98.

Krabiell, K., Fokken, A. and Gester, R. 1992. Luftzerlegung. Frischer Wind durch neue Techniken (Separation of the components of air. Fresh wind with new techniques). Gas aktuell, 44, 14–19.

Pinninger, D.B. 1992. New developments in the detection and control of insects which damage museum collections. Biodeterioration Abstracts, 2, 125–130.

Parker, T.A. 1990. Clothes moths. In: Mallis, A., ed., Handbook of pest control, 346–375.

Reichmuth, Ch. 1987. Low oxygen content to control stored product insects. In: Donahaye, E. and Navarro, S., ed., Proceedings of the 4th International Working Conference of Stored–product Protection, Tel Aviv, Israel, September 1986, 194–207.

Reichmuth, Ch. Unger, A. and Unger, W. 1991. Stickstoff zur Bekämpfung holzzerstörender Insekten in Kunstwerken (Nitrogen to control wood destroying insects in artefacts). Restauro, 4, 246–251.

Reichmuth, Ch., Unger, A., Unger, W., Blasum, G., Piening, H., Rohde–Hehr, P., Plarre, R., Pöschko, M. and Wudtke, A. 1994. Nitrogen flow fumigation for the preservation of wood, textiles and other organic material from insect damage. In: Donahaye, E. and Navarro, S., ed., Proceedings of the International Conference on Controlled Atmospheres and Fumigation in Grain Storages, Winnipeg, Canada, 11–13 June, 1993. Jerusalem, Caspit Press Ltd.

Titschack, E. 1992. Beiträge zu einer Monographie der Kleidermotte *Tineola biselliella* HUM.(Monograph of the common clothes moth *Tineola biselliella* HUM.) Zeitschrift für Technische Biologie, 19, 168 p.

Titschack, E. 1926. Untersuchungen über das Wachstum, den Nahrungsverbrauch und die Eierzeugung.II. *Tineola Bisselliella* Hum. (On the growth, feeding behaviour, and egg production of the common clothes moth). Zeitschrift für wissenscheaftliche Zoologie, 128, 509–569.

Tunç, I., Reichmuth, Ch. and Wohlgemuth, R. 1982. A test technique to study the effects of controlled atmospheres on stored product pests, Zeitschrift für Angewandte Entomologie, 93, 493–496.

Valentin, N. and Preusser, F. 1990. Insect control by inert gases in museums, archives and libraries. Restaurator, 11, 22–33.

Wudtke, A. 1994. Alternative Methoden zur Bekämpfung von Museumsschädlingen mit inerten Gasen am Beispiel der Kleidermotte *Tineola bisselliella* (Hum.) (Studies on alternative control of Tineola bisselliella (Hum.) with carbon dioxide or nitrogen). Anzeiger für Schädlingskunde, Pflanzenschutz und Umweltschultz, in press.

Fumigation and Controlled Atmospheres — Session Summary

Conveners: N.R. Price, C. Waterford and P. Annis

The biggest single issue in the last four years has undoubtedly been the implication of methyl bromide in ozone depletion. It has been significant in terms of our positive approach to this, that there have been no papers on methyl bromide per se at this conference.

Rather the emphasis has been on making the most of the existing alternative, phosphine. In the light of concerns on health and safety and environmental issues, scientists were urged to collect data on the environmental fate of phosphine to bolster the data in support of this useful fumigant.

New methods of formulation and application (mixtures with carbon dioxide and in combination with heat, and constant flow systems) have been strongly represented.

In the light of this need to nurture phosphine, there was much discussion on the practical significance of resistance and the need to distinguish between detection of the resistance gene and the significance in terms of field dosages. The importance of rapid detection as an early warning was acknowledged, but the need for a test which reflected the field significance was stressed.

In the same context the need for improved education and training in good fumigation practice, as perhaps the single most important factor in the continued effective use of phosphine, was discussed.

Alternative fumigants have emerged at this meeting. Carbonyl sulphide is developing as an exciting prospect for the future and the re-evaluation of other existing alternatives, such as methyl isothiocyanate, offer promise. Indeed this may rekindle interest in the international scientific community in the search for yet more alternative chemicals.

Returning to the theme of protecting useful fumigants, a general agreement emerged that:

1. Government and regulatory authorities, in co-operation with manufacturers and distributors, provide accurate information for the effective training of fumigators.

2. Adoption of effective fumigation practice be fostered by:

   • appropriate extension by government;

   • effective training for fumigators including regulation;

   • identification of incentives that ensure fumigations are carried out in the best possible way, including; effective preparation of enclosures using pressure test standards and; concentration monitoring during the fumigation to ensure minimum effective concentrations are exceeded.

Controlled atmospheres are in many ways a mature technology with a long research history but they have not been widely used. Nevertheless, there is now an upsurge in interest in CA use as the conventional fumigants become harder to use, mainly on environmental grounds. This increased interest was shown by the number of papers on old methods revisited, and on novel methods of application. Of special interest was the use of high pressure carbon dioxide to disinfest small batches of commodity. This technique, if developed to full commercial standards, could replace methyl bromide for small-scale fumigations.

Several papers on biogeneration and hermetic storage aimed to address problems associated with the more toxic fumigants. The renewed interest in these ancient storage technologies is a reminder that all that is old is not yet ready for discarding. One of the papers on dry grain respiration showed that, despite its long history, all research aspects of this topic have not been exhausted.

There were several papers on the effects of CA on a range of storage pests, contributing to the steadily increasing knowledge base on quantitative aspects of CA toxicology, which are the basis for determining controlled atmosphere dosage regimes for specific circumstances.

Certain CA topics clearly need further work and discussion, and it was recommended that these be brought to the attention of the organisers of the next controlled atmosphere and fumigation conference, to be held in Cyprus in 1996. They included the following:

- Standards required for CA treatment, especially pressure testing
- Revisions to recommended CA dosage regimes, with particular reference to specific sets of conditions, such as species present, commodity treated, extreme temperatures and other modifying factors
- Studies on the economics of controlled atmospheres.

Engineering

Developments in silo design for the safe and efficient storage and handling of grain

A. W. Roberts*

Abstract

This paper presents an overview of some recent developments in the technology of bulk solids handling as it relates to the grain industry. The paper focuses on silo and discharge equipment design, emphasising the need to understand the relevant properties of the grain and how these relate to silo geometry and discharge flow pattern to generate the load patterns exhibited in the silo walls. With the emphasis on grain conditioning procedures, such as aeration, to control grain quality, it is important to also take note of the influence of any such grain conditioning on the bulk storage and flow properties of the grain. Any change in the flow properties of the grain can influence, sometimes significantly, the pressures and loadings occurring in silo walls. The matter of structural integrity and safety of the silo and the efficiency and reliability of the discharge equipment, such as feeders used to control discharge rate, are of paramount importance.

The paper includes a brief review of discharge flow patterns in bins and silos and presents an overview of the complex pressure loading patterns in silo walls. Single outlet, multiple outlet symmetric silos are discussed as well as silos with eccentric loading due to either non-symmetric loading, or silos with eccentric discharge openings. Brief mention is made of the relevant flow properties such as angles of internal friction, bulk strength and friction angles between the grain and silo walls. Particular mention is made of the influence of grain moisture content variations, interstitial air, and temperature variations on the pressures and hence stresses occurring in silo walls.

Methods of controlling loads in tall, single and multi-outlet silos are discussed. Particular mention is made of the use of anti-dynamic tubes to control the flow pattern and, hence, control the magnitude of the wall pressures generated during discharge.

A recurring problem in bins and silos is the phenomenon of 'silo-quaking', a term used to describe pulsating loads which may be experienced during discharge. If the frequency of the flow pulsations is in harmony with any of the natural frequencies of the structure, severe dynamic loads and stresses can occur. The various mechanisms of 'silo quaking' will be outlined and methodologies to determine the magnitude and distributions of the dynamic loads will be given.

The paper will be illustrated by results from tests performed on pilot scale model silos, as well as by case studies drawn from field experience.

design of safe and functional storage and handling plant which ensures efficient and reliable operation. A full understanding of modern technology of silo and handling plant design, which takes account of the relevant physical properties of the granular materials under the varying environmental conditions during storage, is essential.

Over the past three decades much progress has been made in the theory and practice of bulk solids handling. Test procedures for determining the strength and flow properties of bulk solids have been developed and analytical methods have been established to aid the design of bulk solids storage and discharge equipment. Much progress has also been made in the understanding of the complex nature of loading conditions that may be generated in walls of storage bins and silos. There are now reliable theories and associated design procedures that may be applied, with confidence, to the solutions of practical silo design problems. Yet despite this progress, there are still a number of uncertainties particularly in relation to changes in environmental conditions in grain silos, notably those due to temperature and moisture as well as those due to unpredictable flow patterns. The latter often arises in the case of tall silos and in silos with multi or eccentric outlets.

The purpose of this paper is to present an overview of the present state of knowledge associated with silo design for the safe storage of bulk granular materials. The factors influencing the loads in the walls of silos are many and varied and, as is often the case, the loads are a combination of several interacting effects. The paper, focuses on silo wall loads indicating how these are influenced by the discharge flow pattern. The complexity of loading patterns due to eccentric and multiple outlets is mentioned. The effects of moisture content on grain swelling is discussed with particular reference to the significant increase in silo wall loads that may occur. In addition, the influence on wall loads of pressures due to aeration and thermal expansion and contraction of silos is indicated. The use of anti-dynamic tubes to control the wall pressures during discharge of grain in tall single and multi-outlet silos is outlined. Finally the phenomenon of 'silo quaking', a term used to describe pulsating loads which may be experienced during discharge is discussed. If the frequency of the flow pulsations is in harmony with any of the natural frequencies of the structure, severe dynamic loads and stresses can occur. The various mechanisms of 'silo quaking' are outlined and methodologies to determine the magnitude and distributions of the dynamic loads is given.

Introduction

Bulk materials handling operations are a key function in agricultural grain production. The capital costs of storage and handling plant are quite substantial, so too are the operating and running costs. With the emphasis on grain quality, it is not only important that attention be given to the maintenance of appropriate grain conditioning during storage, but also to the

Handling Plant Design—Basic Concepts

As background information to the study of grain storage and handling, the basic concepts of handling plant design for the general class of bulk materials is briefly reviewed.

General remarks

The procedures for the design of handling plant, such as storage bins and silos, feeders and chutes are well established and follow the four basic steps:

---

* Institute for Bulk Materials Handling Research, The University of Newcastle, NSW 2038, Australia

(i) Determination of the strength and flow properties of the bulk solids for the worst likely flow conditions expected to occur in practice.

(ii) Determination of the bin, stockpile, feeder or chute geometry to give the desired capacity, to provide a flow pattern with acceptable characteristics and to ensure that discharge is reliable and predictable

(iii) Estimation of the loadings on the bin and hopper walls and on the feeders and chutes under operating conditions.

(iv) Design and detailing of the handling plant including the structure and equipment.

The general theory pertaining to gravity flow of bulk solids and associated design procedures are fully documented (Jenike 1961, 1964; Arnold 1982; Roberts 1988). For the purpose of the present discussion, the salient aspects of the general philosophy are briefly reviewed.

Modes of flow in bins and silos of symmetrical geometry.

Following the definition by Jenike (Jenike 1961, 1964) the two principal modes of flow are mass-flow and funnel-flow. These are illustrated in Figure 1.

In mass-flow, the bulk solid is in motion at every point within the bin whenever material is drawn from the outlet. There is flow of bulk solid along the walls of the cylinder (the upper parallel section of the bin) and the hopper (the lower tapered section of the bin). Mass-flow guarantees complete discharge of the bin contents at predictable flow rates. It is a 'first-in, first-out' flow pattern; when properly designed, a mass-flow bin can re-mix the bulk solid during discharge should the solid become segregated upon filling of the bin. Mass-flow requires steep, smooth hopper surfaces and no abrupt transitions or in-flowing valleys.

Mass-flow bins are classified according to the hopper shape and associated flow pattern. The two main hopper types are conical hoppers which operate with axi-symmetric flow and wedged-shaped or chisel-shaped hoppers in which plane-flow occurs. In plane-flow bins, the hopper half-angle α will usually be, on average, approximately 8° to 10° larger than the corresponding value for axi-symmetric bins with conical hoppers. Therefore, they offer larger storage capacity for the

same head room than the axi-symmetric bin, but this advantage is somewhat offset by the long slotted opening which can give rise to feeding problems. The transition hopper, which has plane-flow sides and conical ends, offers a more acceptable opening slot length. Pyramid shaped hoppers, while simple to manufacture, are undesirable in view of build-up of material that is likely to occur in the sharp corners or in-flowing valleys. This may be overcome by fitting triangular-shaped gusset plates in the valleys.

Funnel-flow occurs when the hopper is not steeply sloped and the walls of the hopper are not smooth enough. In this case, the bulk solid sloughs off the top surface and falls through the vertical flow channel that forms above the opening. Flow is generally erratic and gives rise to segregation problems. In the case of cohesive bulk solids, flow will continue until the level of the bulk solid in the bin drops an amount H_D equal to the draw-down. At this level, the bulk strength of the contained material is sufficient to sustain a stable rathole of diameter D_f as illustrated in Figure 1b. Once the level defined by H_D is reached, there is no further flow and the material below this level represents 'dead' storage. This is a major disadvantage of funnel-flow. For complete discharge, the bin opening needs to be at least equal to the critical rathole dimension determined at the bottom of the bin corresponding to the bulk strength at this level. However, for many cohesive bulk solids and for the normal consolidation heads occurring in practice, ratholes measuring several metres are often determined. This makes funnel-flow impracticable. Funnel-flow has the advantage of providing wear protection of the bin walls, since the material flows against stationary material. However it is a 'first-in last-out' flow pattern which is unsatisfactory for bulk solids that degrade with time. It is also unsatisfactory for fine bulk solids of low permeability. Such materials may aerate during discharge through the flow channel and this can give rise to flooding problems or uncontrolled discharge.

In the case of cohesive bulk solids, the disadvantages of funnel-flow may be overcome by the use of expanded-flow, as illustrated in Figure 2. This combines the wall protection of funnel-flow with the reliable discharge of mass-flow. Expanded-flow is ideal where large tonnages of bulk solid are

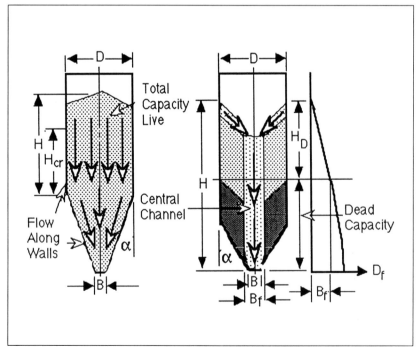

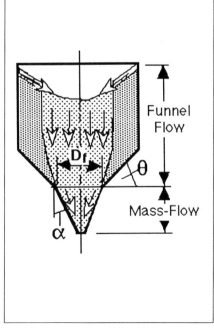

Fig. 1   Modes of flow (a) Mass-flow; (b) Funnel-flow

Fig. 2.   Expanded flow

to be stored. For complete discharge, the dimension at the transition of the funnel-flow and mass-flow sections must be at least equal to the critical rathole dimension at that level. Expanded-flow bins are particularly suitable for storing large quantities of bulk solids while maintaining acceptable head heights. The concept of expanded-flow may be used to advantage in the case of bins or bunkers with multiple outlets.

Generally speaking, symmetric bin shapes provide the best performance. Asymmetric shapes often lead to segregation problems with free flowing materials of different particle sizes and makes the prediction of wall loads very much more difficult.

Free-flowing granular materials

Since most bulk grains at low moisture contents are generally free flowing, they have no bulk strength. Hence funnel flow, as illustrated in Figure 3, is a viable option. During discharge, an 'effective transition' (ET) forms and this defines the lower region of the flow channel during discharge. Once the head drops during the last stages of discharge, the grain will eventually settle at its final repose angle as indicated as indicated in Figure 3a. However, the grain retained in the 'dead' region of the silo may degrade with time and contaminate fresh grain loaded into the silo. For this region, the silo should be self emptying, which may be accomplished by using a hopper bottom as illustrated in Figure 3b.

Even though funnel-flow silos may be used to store and handle grain, the disadvantages of a 'first-in, last-out' flow pattern which could lead to a reduction in grain quality, needs to be noted. For this reason, the advantages of mass-flow should not be overlooked even though a steeper hopper is required as in Figure 1a. Certainly, if the grain is at a higher moisture content, it will exhibit cohesive strength. In this case mass-flow is definitely the preferred option.

Mass-flow and funnel-flow limits for symmetrical bins

Established theory due to Jenike

The mass-flow and funnel-flow limits have been defined by Jenike on the assumption that a radial stress field exists in the hopper (Jenike 1961, 1964). These limits are well known and have been used extensively and successfully in bin and silo design. The limits for axi-symmetric or conical hoppers and hoppers of plane-symmetry depend on the hopper half-angle α, the effective angle of internal friction δ and the wall friction

angle ϕ. Once the wall friction angle and effective angle of internal friction have been determined by laboratory tests, the hopper half-angle may be determined. The bounds for conical and plane-flow hoppers are plotted for three values of δ in Figure 4. In functional form,

$$\alpha = f(\phi, \delta) \qquad (1)$$

In the case of conical or axi-symmetric hoppers, it is recommended that the half-angle be chosen to be 3° less than the limiting value. For plane-flow, the bounds between mass and funnel-flow are much less critical than for conical hoppers. In plane-flow hoppers, much larger hopper half angles are possible which means that the discharging bulk solid will undergo a significant change in direction as it moves from the cylinder to the hopper. For plane-flow, the design limit may be selected; if the transition of the hopper and cylinder is sufficiently radiused so that the possibility for material to build-up by adhesion is significantly reduced, then a half-angle 3° to 4° larger than the limit may be chosen.

Typically for wheat, average values of the friction angles are $\delta = 30°$, $\phi = 20°$ for polished stainless steel or polished mild steel and $\phi = 25°$ for mild steel plate as rolled. For mass-flow in conical or axi-symmetric hoppers, the required hopper half-angles (Figure 1(a)) are $\alpha_c = 25°$ for the polished surfaces and $\alpha_c = 17°$ for the rolled mild steel. In the case of plane-flow, wedged-shaped mass-flow hoppers, the corresponding angles are $\alpha_p = 35°$ and $\alpha_p = 27°$ respectively.

Modification to mass-flow limits—more recent research

Since in the work of Jenike, flow in a hopper is based on the radial stress field theory, no account is taken of the influence of the surcharge head due to the cylinder on the flow pattern developed, particularly in the region of the transition. It is been known for some time that complete mass-flow in a hopper is influenced by the cylinder surcharge head. For instance, there is a minimum level H_{cr} which is required to enforce mass-flow in the hopper (Thompson 1984). For the mass-flow bin of Figure 1a, this height ranges from approximately $0.75\,D$ to $1.0\,D$.

More recent research has shown that the mass-flow and funnel-flow limits require further explanation and refinement. For instance, Jenike published a new theory to improve the prediction of funnel-flow; this led to new limits for funnel-flow which give rise to larger values of the hopper half-angle than previously predicted, particularly for high values of the wall friction angle. In the earlier theory, the boundary between mass-flow and funnel-flow was based on the condition that the

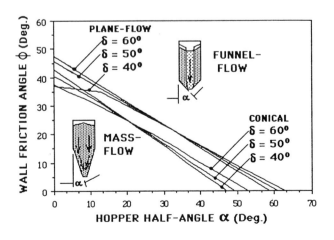

Fig. 3.   Funnel-flow for free-flowing granular material   **Fig. 4**   Limits for mass-flow for conical and plane-flow channels

stresses along the centre line of the hopper became zero. In the revised theory the flow boundary is based on the condition that the velocity becomes zero at the wall.

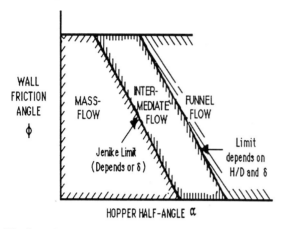

Fig. 5. Flow regimes for plane-flow hopper defined by Benink (1989)

In a comprehensive study of flow in silos, Benink (1989) has identified three flow regimes, mass-flow, funnel-flow and an intermediate flow as illustrated in Figure 5. Whereas the radial stress theory ignores the surcharge head, Benink has shown that the surcharge head has a significant influence on the flow pattern generated. He derived a fundamental relationship for H_{cr} in terms of the various bulk solid and hopper geometrical parameters, notably the $H{:}D$ ratio of the cylinder and the effective angle of internal friction δ. Benink developed a new theory, namely the 'arc theory', to quantify the boundaries for the three flow regimes. This theory predicts the critical height H_{cr} at which the flow changes.

Intermediate-flow is illustrated in Figure 6. Grain flows more quickly in the central flow channel. The nature of the flow is such that pulsating loads may be induced. The characteristics of pulsating dynamic loads is discussed in Section 8.

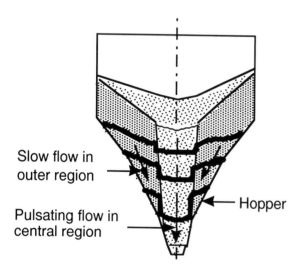

Slow flow in outer region

Pulsating flow in central region

Hopper

Fig. 6. Intermediate flow illustrating pulsating flow regime.

Hopper opening dimension for mass-flow

The principles embodied in mass-flow hopper design are illustrated in Figure 7. Having selected the basic shape, that is, whether the hopper is to be axi-symmetric (conical) or plane-flow (wedge-shaped), the hopper half angle α is determined

on the basis of the wall friction angle ϕ. It then remains to determine the required dimension of the channel to ensure that arching does not occur and the required flow rate is achieved. In most cases, the critical dimension for flow occurs at the outlet.

The parameter used to describe the strength of a bulk solid is the Flow Function which is abbreviated as '*FF*'. The *FF* represents the variation of the unconfined yield strength σ_c as a function of the major consolidation pressure σ_1 and is illustrated in Figure 7a. For free-flowing grain, $\sigma_c = 0$ and the FF graph coincides with the horizontal axis. Referring to Figure 7b, the stress acting in the arch is represented by $\overline{\sigma}_1$ and is related to the major consolidation pressure σ_1 by the 'flow factor' ff, a flow channel parameter, as follows:

$$\overline{\sigma}_1 = \frac{\sigma_1}{ff} \qquad (2)$$

For the critical condition, the minimum opening dimension B occurs when $-\sigma_1 = \sigma_c$ and is given by

$$B_{min} = \frac{\overline{\sigma}_1 H(\alpha)}{\rho g} \qquad (3)$$

where $H(\alpha)$ = Arch thickness and hopper geometry factor parameter based on Hopper Half Angle (Jenije 1961, 1964; Arnold 1982; Roberts 1988).

ρ = Bulk Density

g = acceleration due to gravity

In practice, the actual opening dimension B is made larger than B_{min} in order to achieve the desired flow rate. For the enlarged opening, the actual Flow Factor ff_a applies and the major consolidating pressure at the outlet increases to the value corresponding to the actual opening dimension B.

Silo Wall Loads

In bin and silo design, the prediction of wall loads continues to be a subject of some considerable complexity. In view of its obvious importance it is a subject that has attracted a good deal of research effort. A brief selection of published research which deals with the subject of wall pressures is given in Arnold (1982); Roberts and Ooms (1983); Ooms and Roberts (1985); Roberts (1988a,b); Rombach and Eibl (1989); Ooi and Rotter (1989); and Wu (1990).

Despite the widely varying approaches to the analysis of bin wall loads, it is clear that the loads are directly related to the flow pattern developed in the bin or silo. The flow pattern which a mass-flow bin exhibits is reasonably easy to predict and is reproducible. However, in funnel-flow bins the flow pattern is more difficult to ascertain, especially if the bin has multiple outlet points, the loading of the bin is not central and/or the bulk solid is prone to segregation. Unless there are compelling reasons to do otherwise, bin shapes should be kept simple and symmetric.

The subject of bin loads has been addressed in several design codes notably the recent Australian Standard AS3774-1990, 'Loads on Bulk Solids Containers' (SAA 1990). The earlier codes concentrated on funnel-flow bins, but the later codes cover both mass-flow and funnel-flow bins. In particular, AS 3774-1990, is very comprehensive in the range of loading conditions included and in the types of bins considered.

Symmetric mass-flow bins

The stress fields and normal wall pressures occurring in mass-flow bins for the initial filling and flow conditions are shown in Figure 8. When a bin is initially filled from the empty condition, a peaked stress field occurs as in Figure 8a; the major principal pressure is almost vertical. When flow occurs, the stress field in the hopper switches to an arch stress field, the switch travelling up the hopper becoming locked in at the transition as in Figure 8b. In the arched stress field, the load is transmitted to the wall of the hopper with the major principal stress acting more in a horizontal direction.

Above the hopper, that is in the cylinder, the peaked stress field remains, although imperfections in the cylinder wall which give rise to localised flow convergences cause over-pressures to occur in the cylinder. Imperfections in bin walls, which give rise to over-pressures, may be due to manufacturing and/or constructional details such as weld projections or plate shrinkages in the case of steel bins or deformation of form work in the case of concrete bins. Jenike has used strain-energy methods to analyse these over-pressures during flow in the cylinder. In view of the difficulty of using the Jenike strain energy method, design codes generally employ over-pressure

factors to account for flow conditions in the cylinder. This is indicated by the upper bound p_n curve for the cylinder in Figure 8b.

It is to be noted that when the bin discharges and the flow is stopped, the stress fields and corresponding pressures shown in Figure 8b will remain. The stress field does not revert to that of Figure 8a; this only occurs if the bin is completely emptied and then filled again.

Wall pressures in symmetric funnel-flow bins

In the case of symmetrical funnel-flow bins or silos, the effective transition previously defined, controls the flow channel in the lower region of the bin as indicated in Figure 9. Above the effective transition the flow occurs along the walls similar to that occurring in a tall mass-flow bin. An over-pressure occurs at the effective transition similar to the switch

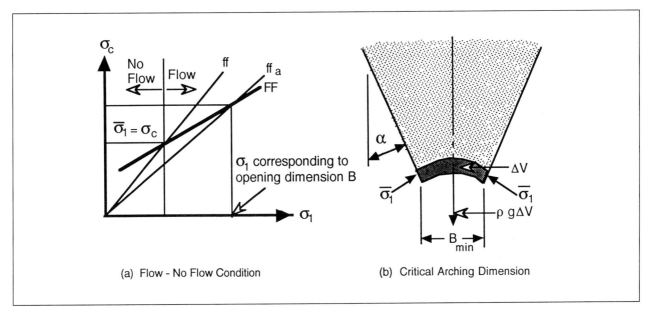

(a) Flow - No Flow Condition

(b) Critical Arching Dimension

Fig. 7.   Determination hopper opening dimension for mass-flow.

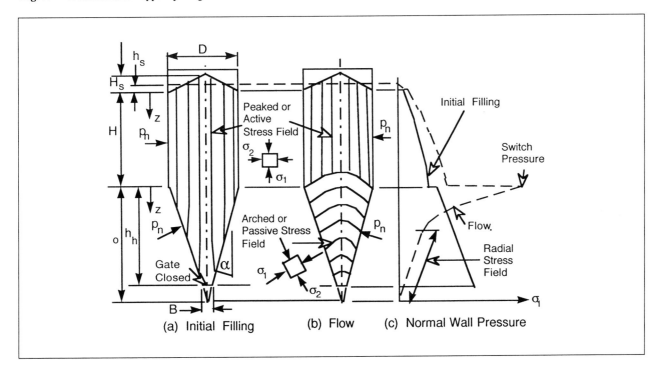

(a) Initial Filling   (b) Flow   (c) Normal Wall Pressure

Fig. 8.   Pressures acting in mass-flow bins.

pressure in mass-flow, although the magnitude, whilst significant, may not be as severe as the mass-flow switch pressure. Since the effective transition is not stationary but tends to move during the discharge process, the procedure for design involves defining the lowest position at which the effective transition is likely to occur. In the upper cylindrical section of the bin above the effective transition, imperfections in the walls can give rise to increased pressures. The flow pressures in funnel-flow bins, depicted in Figure 9, are normally computed by the application of over-pressure factors on Janssen. The procedures are given in AS 3774 (SAA 1990).

Multiple outlet bins and silos and with eccentric discharge silos

Multiple outlet silos and silos with an eccentric discharge point give rise to complex loading patterns as a result of the flow channels that develop above the outlets. Referring to Figure 10, the pressures acting on the wall in the region of the outlet are of lower magnitude than the pressures that act in the section of the bin or silo where there is no flow. Consequently bending moments are induced in the walls leading to a distortion of the silo's symmetrical cylindrical shape. Apart from the bending stresses which are superimposed on the average hoop stresses at each location, the distorted shape of the silo may give rise to buckling problems.

The subject of eccentric loads in silos is addressed in a comprehensive way in AS3774 (SAA 1990). The loading patterns illustrated in Figure 10 are based on this Standard. It is to be noted that where there are several symmetrically located outlets, the conservative approach to design is to assume that the possibility exists for only one of the outlets to operate at a particular instant.

Moisture Variations in Grain

Free expansion due to moisture increase

Grain, such as wheat, when subjected to increases in moisture, if not completely constrained, will undergo an increase in volume. Changes in moisture content can occur during aeration of grain in bulk storage when the humidity of the circulating air increases. The amount of grain swelling is illustrated in Table 1. This shows the effects of wheat grains after prolonged exposure to high humidity conditions, the results being based on samples conditioned in a humidity cabinet. As indicated the moisture content increased from 9.54% to 20.07% after 4 days, the corresponding increase in volume due to swelling being 17.8%. For moisture changes of lower order the degree of swelling is correspondingly lower.

Constrained expansion due to moisture increase

In a storage bin situation the volume increase due to swelling is constrained. To gauge the possible volume expansion under such constrained conditions, tests were conducted on wheat contained in cylindrical cells with rigid bottoms and side walls with a loaded piston being located on the top of the contained samples, the pistons being free to move in the vertical direction. The moisture content of the grain in the various cells was varied and movement of the piston due to swelling of the grain over varying periods was noted. A typical set of results is shown in Figure 11 which applies to a pressure of 28 kPa. The results indicate, for example, that an increase of moisture content from 9% to approximately 23%, the corresponding expansion of the bulk grain is 1.2%. Under normal conditions of storage moisture variations of this magnitude are unlikely. However the results indicate the order of swelling that may occur during constrained storage; on the basis of such results the increased loading on the walls of silos may be estimated.

Loads in Grain Silos due to Grain Moisture Increase

Possible loading conditions

The loading conditions that may occur are shown in Figures 12 and 13. When grain is subject to swelling, it tends to expand upward as well as outward, the upward movement being resisted by wall friction. In the limiting condition the wall friction acts downward as indicated in Figure 13. The wall friction may not be fully mobilised and the resistance at the wall to upward expansion may be less than for the limiting condition.

Assuming the pressure is uniform across the cross-section, the following differential equation arises

$$\frac{dp_n}{dz} \pm \frac{4\mu K_j}{D} p_n = \gamma K_j \tag{4}$$

where

+sign applies to the Janssen case of Figure 12

−sign applies to the swelling case of Figure 13

μ = coefficient of friction as the wall

p_v = vertical pressure

p_n = lateral pressure

D = silo diameter

γ = bulk specific weight of wheat

Table 1 Free expansion of humidified wheat

Sample	Storage time (hours)	Moisture content (w.b.)		Volume increase (%)
		Initial (%)	Final (%)	
1	1.0	6.32	7.38	0.62
2	3.75	4.91	8.65	4.74
3	5.0	10.83	13.83	6.05
4	24.0	11.16	16.58	14.52
5	29.0	9.51	19.26	16.40
6	95.0	9.54	20.07	17.79

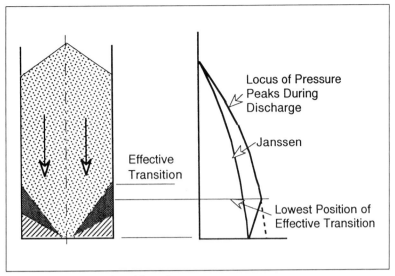

Fig. 9.   Pressures in funnel flow silo.

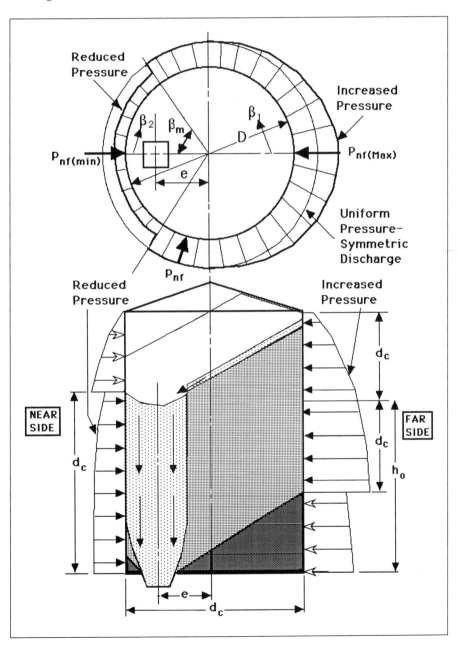

Fig. 10.   Circumferential pressure variation due to operation of one eccentric outlet.

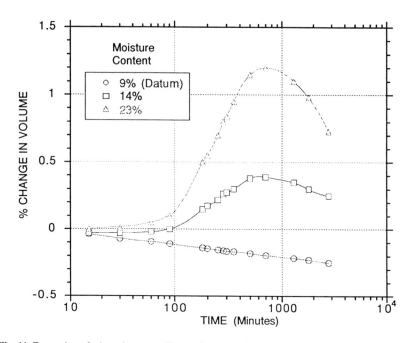

Fig. 11. Expansion of wheat due to swelling under constrained storage pressure applied: 28 kPa.

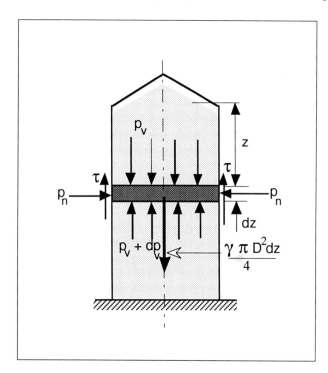

Fig. 12. Janssen pressure due to static loading.

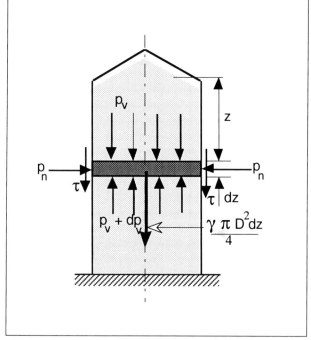

Fig. 13. Effect of upward swelling producing reverse friction forces at wall

Janssen Equation

Putting the + sign into (4) and solving yields the Janssen equation

$$p_n = \frac{\gamma D}{4\mu}\left(1 - e^{-4\mu K_j z/D}\right) \qquad (5)$$

and

$$p_v = \frac{p_n}{K_j} \qquad (6)$$

For silo design, where the loading is symmetrical, it is usual to apply an over-pressure factor to the pressure p_n of equation (5).

'Piston' equation for grain swelling

Grain swelling may occur if moist air is circulated through stored grain via aeration ducts as illustrated in Figure 14. Physically the effect of grain swelling is likened to a piston trying to push the column of grain upward (Roberts 1988). Putting the minus sign into equation (4) and solving yields

$$P_n = \frac{\gamma D}{4\mu}\left(e^{4\mu K_j z/D} - 1\right) \qquad (7)$$

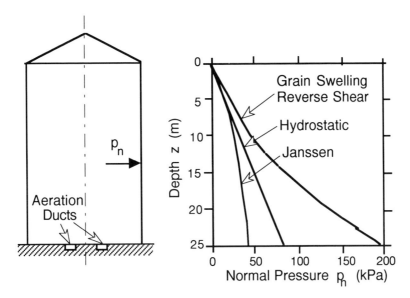

Fig. 14. Wall pressure distributions in 15 m diameter by 25 m high wheat silo.

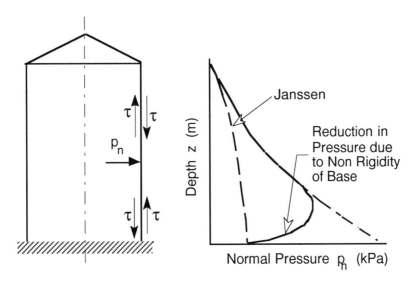

Fig. 15. Reduction in wall pressures in lower region of silo walls.

Equation (7) assumes the wall friction is fully mobilised. An intermediate case of interest occurs when the wall resistance reduces to zero which implies that the vertical wall support just equals the upward force at the wall due to swelling. In this case the vertical pressure p_v is equal to the hydrostatic pressure

$$p_v = \gamma z \tag{8}$$

and
$$p_n = K_j \gamma z \tag{9}$$

The graphs depicting the wall pressures, based on equations (5), (7) and (8) are shown In Figure 14.

The large magnitude of the pressures given by equation (7) other than for the Janssen pressure given by equation (5) is clearly evident. The pressure curves in Figure 14 apply to a 15 m diameter by 25 m tall wheat silo and have been plotted for a constant value of $K_j = 0.4$. It is possible that K_j will be higher for the grain swelling case, increasing the pressures beyond those plotted.

Some general comments

There seems no doubt that an increase in moisture content of the grain can cause a significant increase in the silo wall pressures. The key to this behaviour is the direction and magnitude of the friction force at the wall. When the friction force acts downward, a 'piston effect' is produced in the silo where the forces in the silo wall resisting upward expansion of the grain can become quite considerable. This would be particularly the case for tall silos where the height is several times the diameter.

The extent to which the frictional forces at the wall change in magnitude and direction over the full height of the silo are virtually impossible to predict. If the floor of the silo is not entirely rigid, then the frictional forces in the wall in the lower region of the silo would act upward and the pressures exerted on the silo walls in this region would reduce. This is depicted in Figure 15.

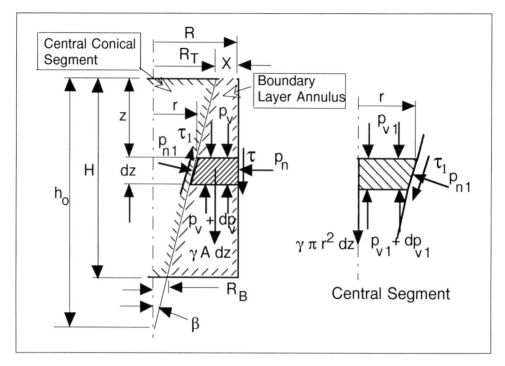

Fig. 16. Silo pressure model showing boundary layer annulus to account for grain swelling.

Simplified theory for estimating silo wall pressures due to grain swelling

It is evident that the large size of actual silos would be such that pressure distribution at any cross section would not be uniform. Furthermore, there is every reason to believe that the moisture content of the stored grain would not be uniformly distributed throughout the mass; the geometry and location of the air supply ducts suggest this would be the case.

In reality the behaviour of the grain during swelling is likely to produce a boundary layer effect. The model used by Roberts (1988a,b) to analyse this case is illustrated in Figure 16. A conical rupture surface is assumed, this surface being sloped at an angle β to the vertical and intersecting the top surface of the silo at a radius R_T. It is assumed that the conical central portion of the grain mass is able to expand upwardly due to grain swelling and induce pressures in the outer, wedge-shaped annulus comprising the boundary layer. The outer boundary layer annulus is constrained by the silo walls; as the thickness of this annulus increases, the greater become the proportion of the grain mass which must be constrained by the silo walls and hence the greater become the pressures on the walls.

The differential equations for the grain pressures have been derived and numerical solutions obtained(Roberts 1988a). The silo in this case is 15 m diameter by 25 m high and stores wheat. Some typical results for various R_B/R_T ratios and various values of X (Figure 16) being shown in Figure 17. As indicated, the presence of the boundary annulus significantly influences the wall pressures. In general, as the dimension X increases, then the closer does the pressure distribution approach that depicted in Figure 14, and the greater becomes the pressure. For comparison purposes the Janssen curve and the pressure curve for reverse friction or shear of Figure 14 are also included in Figure 17.

The change in pressure with change in dimension X are shown more clearly in Figure 18. At the base of the silo, that is at $z = 25$ m, the pressure increases quite significantly as X increases. At the depth $z = 20$ m the increase in pressure is less pronounced, while at the depth $z = 10$ m, there is little variation in pressure with increase in X. Based on the assumptions

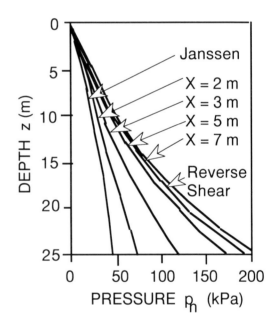

Fig. 17. Influence of boundary layer on silo wall. Pressure distributions, $R_B/R_T = 0.5$.

made in the foregoing analysis, the results indicate that the maximum pressure at the base of the silo with zero boundary layer is close to 200 kPa. With the presence of a boundary layer the pressures acting at the wall are reduced considerably. The peak pressure with zero boundary layer is 4.2 times the Janssen pressure while with a boundary layer thickness of say $X = 2$ m, the wall pressure at the base of the silo is approximately 1.5 times the Janssen pressure. These values are based on the pressure ratio $K = 0.4$ and $R_B/R_T = 0.5$. It is to be noted that based on experiments to examine grain swelling, and allowing for the relative elasticity of the wheat and the steel silo shell, the estimated increase in pressure is of the order of 2 to 1.

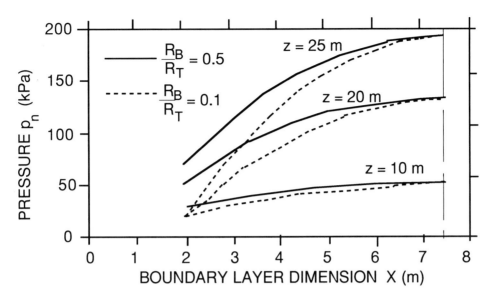

Fig. 18. Variation of silo wall pressures with boundary layer. Dimension at selected depths.

While the foregoing analysis is an approximation, it does provide an indication of the influence of a boundary layer in tempering the wall pressures generated as a result of grain swelling.

Other Factors Influencing Bin and Silo Loadings

In addition to moisture, other environmental factors may influence bin and silo wall loadings. Two of these factors, namely aeration and temperature effects, are briefly discussed.

Silo aeration— effect of air pressure

Pressures due to aeration must be added to the pressures generated by the bulk material itself. Jenkyn (1978) considered likely packaging configurations during aeration and concluded that body-centred cubes would occur most frequently. These produce a maximum increase in value

occupied by the material of approximately 25% with a corresponding decrease in density. Jenkyn recommends a design pressure given by

$$P_{des} = P_h + P_a \qquad (10)$$

where

$$P_h = \frac{0.8\gamma D}{4\mu}\left(1 - e^{-4\mu K_o h/D}\right) \qquad (11)$$

γ = Bulk specific weight
D = Silo diameter
μ = Coefficient of friction
h = Head
K_o = Lateral pressure coefficient.
Under aeration conditions K_o may approach unity.
The aeration pressure is given by

$$P_a = \frac{\Delta_p h}{h_o} \qquad (12)$$

where Δ_p = Maximum aeration pressure
h_o = Effective maximum material depth.

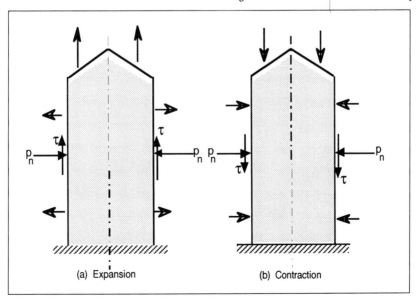

Fig.19. Expansion and contraction of silo.

The foregoing assumes symmetrically located air injection points and a uniform air distribution. In practice, variations in grain packing densities and variations in moisture throughout the bulk grain mass could give rise to non-uniform pressure distributions in eccentric loadings.

Temperature variations

The variations methods use to analyse the wall pressure in steel silos generally assume a rigid silo. This is not the case. Daily, weekly and monthly variations in atmospheric temperatures cause the shell of a silo to expand and contract with some settling of stored material during expansion with little recovery, if any, during contraction. This can be a problem in the case of silos storing bulk materials over long periods of time without discharge.

When the silo expands due to temperature increase, it does so by expanding radially outward and upward. This allows the grain to settle. When the silo contracts it does so by contracting radially inwards and downward, the latter effect causing the friction force at the wall to switch from the initially upward direction to the downward direction, as depicted in Figure 19. This has the effect of significantly increasing the pressure in a similar way to that of increased moisture. Since the temperature cycles on a daily basis, as well as decreasing on a longer-term basis, the combined effect of the progressive settling of the grain and the action of the reversed friction force at the walls could have a most significant influence on the structural safety of the silo.

Measurements of daily temperature fluctuations and pressures generated in walls of steel silos have indicated quite clearly a significant increase in pressures during the night period when temperatures were lower than during the day. Apart from the daily temperature fluctuations, the overall changes from summer to winter can have a significant effect. For example, in a 15 m diameter steel silo, a temperature increase of 20° could cause an expansion in the order of 3 mm. If, hypothetically, the silo contained wheat and was completely rigid so that the contraction of the silo were prevented as the temperature dropped, an increase in wall pressure of 18 kPa would occur, this being 41% of the maximum static wall pressure. However, the elasticity of the wheat relative to steel needs to be taken into account. The modulus of elasticity of wheat varies significantly from one wheat type to another [16]. Based on the properties of individual grains, a typical elasticity modules for wheat is $E_W = 2.07 \times 10^9$ kPa gives

$(E_W/E_S) = 0.01 = 1\%$

Using this ratio, one daily expansion and contraction could produce a residual increase in wall pressure of 0.1 kPa. While this is in itself insignificant, the cumulative effect of expansion and contraction over a period of days and weeks is significant. For example, extrapolation of the data for a two-month period indicates a cumulative increase in pressure of 11 kPa or 25% of the maximum static wall pressure. This is an upper bound value since the modules E_W for bulk wheat is less than that given above. Furthermore the analysis is greatly simplified; the actual process of expansion and contraction coupled with the effect of cyclical loadings on the grain and influence of moisture variations is an extremely complex problem to analyse. However the quoted example indicates the possible order of magnitude of increase in wall pressure due to repeated expansion and contraction.

It must be noted that the foregoing discussion assumes uniform temperatures around the periphery of the silo. This will not occur in practice; there will be differential expansion owing to the position of the sun relative to the silo walls. The loading pattern will be more complex and the resulting pressure changes around the periphery will not be uniform.

Use of Anti-dynamic Tubes to Control Silo Pressures

In tall grain silos, the effective transition occurs low down the silo walls; as a result, mass-flow of grain with flow along the walls occurs over a substantial height of the silo above the effective transition. The effect is to cause dynamic pressures to be generated, these pressures being in the order of two to three times the static pressures generated after the silo is filled from the empty condition.

As shown by Reimbert (see Thompson 1984), it is possible, by the use of an anti-dynamic tube, to control the flow pattern so that funnel-flow always occurs without flow along the walls. In this way, the wall pressures never exceed the static values. Reimbert's anti-dynamic tube, placed centrally in a symmetrical silo, extends almost the whole height and has a series of holes or ports to allow grain to enter the tube at various levels.

A variation of the Reimbert tube is the tremmie tube which has no holes in the walls and extends slightly less than half the height of the silo. Research using this type of anti-dynamic tube was conducted at the University of Newcastle, Australia (Ooms and Roberts 1985; Roberts 1988b). Figure 20 shows, schematically, the 1.2 m diameter by 3.5 m tall model flat bottom test silo and a sample set of test results for the normal wall pressures. Load cells fitted in the wall of the silo enabled these measurements to be taken.

The work was initiated in order to provide a simple and low cost solution to controlling the pressures in a number of badly cracked concrete grain silos approximately ten times the scale. In effect, the tremmie tube divides the tall funnel-flow silo into two squat silos in series. The top half of the silo discharges first followed by the bottom part once the level drops below the top of the tremmie tube and the tube empties. Ports in the bottom of the tube allow grain to flow laterally to the silo outlet.

The design of the bottom ports and tube sizing in relation to the silo outlet dimension are important in order to promote automatic choking of the lateral flow at the bottom until the tube empties. No valves are necessary. The arrangement ensures that at no time does the effective transition intersect the walls of the silo and hence the pressures never exceed the values corresponding to the static or initial filling condition. This is illustrated in the test results of Figure 20.

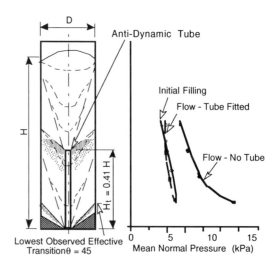

Fig. 20. Model silo studies using anti-dynamic tube.

The anti-dynamic tube described above has been successfully used to control the flow in silos having eccentric and multiple discharge points. Typical applications are shown in Figure 21. Without the tube in place, the walls of such silos are subject to significant bending stresses in addition to the hoop stresses.

It is to be noted that the anti-dynamic tube described here is suitable only for free flowing, cohesionless bulk solids such as grain. They should not be used for cohesive bulk solids. It should also be noted that the draw-down loads on anti-dynamic tubes can be quite high. The computation of these loads and design of location rods has been discussed in Ooms and Roberts (1985) and Roberts (1988b).

Pulsating Loads in Bins – 'Silo Quaking'

As is often the case, the solution of one problem which leads to an improvement in plant performance exposes other problems which require further research and development. In terms of silo loads, a re-occurring problem is the phenomenon of 'silo quaking', a term used to describe pulsating loads which may be experienced during discharge. If the frequency of the flow pulsations is in harmony with any of the natural frequencies of the structure, severe dynamic loads and stresses can occur. The problem of 'silo quaking' is discussed in Roberts et al (1991); Roberts (1993,1994); and Craig and Roberts (1994). The discussion that follows provides an overview of the 'silo quaking' problem as it relates to mass-flow, funnel-flow, expanded-flow and intermediate-flow bins.

The silo quaking problem

Tall mass-flow and funnel-flow bins

The flow characteristics in mass-flow bins are illustrated in Figure 22. As discussed in **Free flowing granular materials**, there is a minimum level H_{cr} which is required to enforce

1 - Stage 1 Discharge; 2 - Stage 2 Discharge; 3 - Stage 3 Discharge

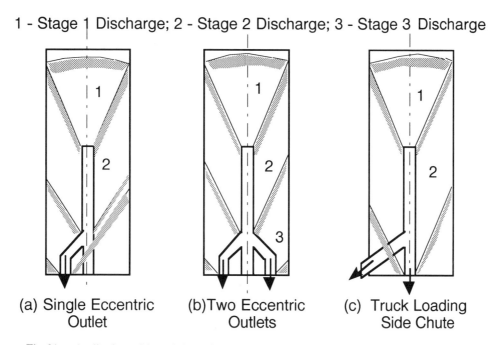

(a) Single Eccentric Outlet

(b)Two Eccentric Outlets

(c) Truck Loading Side Chute

Fig. 21.   Applications of the anti-dynamic tube for eccentric discharge.

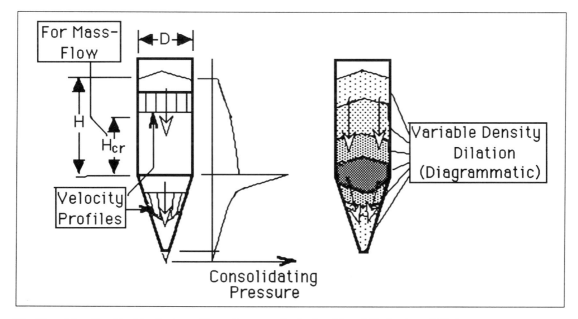

Fig. 22.   Mass-flow bin, (a) velocity profiles and pressure distribution; (b) variable density and dilation.

mass-flow in the hopper. Typically, this height ranges from approximately $0.75\,D$ to $1.0\,D$. As the material flows, it dilates leading to variations in density from the static condition. This is depicted pictorially in Figure 22b. With $H > H_{cr}$, the flow in the cylinder is uniform or 'plug-like' over the cross-section, with flow along the walls. In the region of the transition, the flow starts to converge due to the influence of the hopper and the velocity profile is no longer uniform. The velocity profile is further developed in the hopper as shown. As the flow pressures generate in the hopper, further dilation of the bulk solid occurs. As a result of the dilation, it is possible that the vertical supporting pressures decrease slightly reducing the support given to the plug of bulk solid in the cylinder. This causes the plug to drop momentarily giving rise to a load pulse. The cycle is then repeated.

A similar action to that described above may occur in tall funnel-flow bins or silos where the effective transition intersects the wall in the lower region of the silo. As a result, there is flow along the walls of a substantial mass of bulk solid above the effective transition.

Funnel-flow and expanded-flow bins

During funnel-flow in bins of squat proportions, where there is no flow along the walls, as depicted in Figure 23, dilation of the bulk solid occurs as it expands in the flow channel. As a result some reduction in the radial support given to the stationary material may occur. If the hopper is fairly steeply sloped, say ($\theta \geq \delta$), then the stationary mass may slip momentarily causing the pressure in the flow channel to increase as a result of the 'squeezing' action rather like that of a collet used to clamp a rod of steel in a lathe. The cycle then repeats, the flow being characterised by a 'ratchet' type effect.

Expanded-flow bins are commonly used to store bulk solids in large tonnages. Pulsating loads, illustrated in Figure 24, can

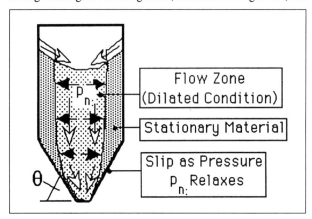

Fig. 23.   Funnel-flow bin.

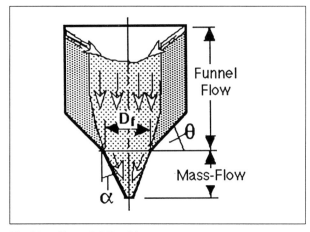

Fig. 24.   Expanded-flow bin.

occur in such bins, particularly if the slope angle θ of the transition is too steep. Pulsating loads can also be experienced in intermediate-flow bins of the type shown in Figure 6. The central flow channel leads to more rapid flow in this region giving rise to pulsating flow due to the 'slip stick' nature of the friction generated at the boundary of the central and outer flow channels.

Pilot scale mass-flow bin handling wheat

Studies of pulsating flow have been conducted in the laboratories of The University of Newcastle using a mass-flow test bin (Roberts et al. 1991). The bin is illustrated in Figure 25. It is 1.2 metre diameter with a 3.5 metre tall cylindrical section constructed of mild steel; the conical mass-flow hopper is constructed of stainless steel. The bin is fitted with load cells which measure, simultaneously, both normal pressure and shear stress.

By way of illustration, typical test results using wheat are presented. Fourteen load cells, located as shown in Figure 25, are capable of measuring both normal pressure and wall shear stress. For the purpose of the present discussion on load pulses, the loads acting in the cylinder are examined. A series of tests were conducted for different stored heads in which the normal pressures and shear stresses at several locations were recorded during filling, during undisturbed storage and during emptying. Referring to Figure 25, the locations at which the measurements are reported are 5, 7, 10, 12, 13 and 14. Three head heights of the wheat in the cylinder were examined:

H = 3.44 m	2.41 m	1.1 m
H/D = 2.9	2.0	0.9

Pressures pulsations during settling after filling

A dynamic 'slip stick' effect was observed in both the normal pressure and shear records during filling and during settling after filling. By way of example Figure 26, shows sample records for location 14 for the ratio $H/D = 2.9$. A dynamic 'slip stick' effect is depicted in both the normal pressure and shear records during filling; this effect continues on after the bin is filled, the effect being most pronounced in the shear stress record where the pulses are most evident. It is quite clear that during undisturbed storage, the stored mass approaches, asymptotically, its critical state consolidation condition by a pulse type settling action. The phenomena depicted in Figure 26 was observed at all locations, with the pulses at each location occurring at the same time intervals. Records of the undisturbed settling were taken over prolonged time periods. The same effect was observed for the three H/D ratios examined. As the settling time increased, the pulse period also increased in an exponential manner as illustrated in Figure 27. This shows that the settling characteristics for the two cases $H/D = 2.9$ and $H/D = 2.0$ are virtually the same.

Pressures during emptying

Figure 28 shows for $H/D = 2.9$, the normal wall pressures and shear stresses at locations 5 and 14. At location 5, the dynamic wall pressure and shear stress have each taken approximately one minute to reach their maximum values. At this location, which is near the hopper transition, the long rise time is no doubt associated with the time taken to establish the arched stress field in the hopper. The influence of a pulsating flow effect is depicted in the records for location 5, as for all other locations, even though the amplitude of the pulsations in the normal pressure and shear stress at location 5 is small compared with the average values. The amplitude increased relative to the average values at higher locations in the bin. For instance, as Figure 28b shows, at location 14, the pulsations

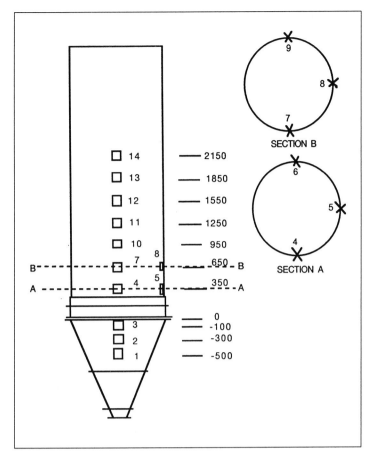

Fig. 25. Mass-flow test bin.

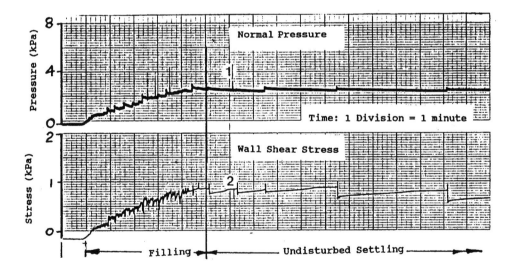

Fig. 26. Normal pressure and shear stress records for wheat at location 14 of test bin

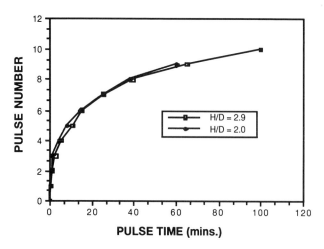

Fig. 27. Load pulsations during undisturbed storage of wheat in test bin.

(a)

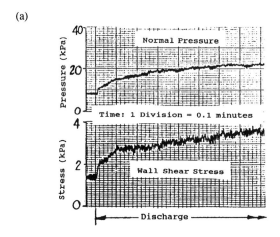

(b)

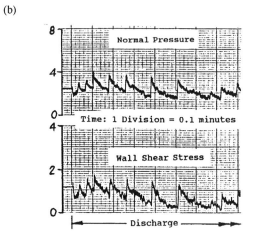

Fig. 28. Pressures and shear stresses for wheat during emptying $H/D = 2.9$. (a) Results for location 5 (b) Results for location 14.

are quite pronounced, the period of the pulsations averaging approximately 5 seconds.

Effect of surcharge head

Similar dynamic effects were experienced for the case of $H/D = 2.0$, but when the head was reduced to $H/D = 0.9$, there was no evidence of any pulsations during discharge. This con-

Perimeter $P = (2)^{1-m} (\pi)^m (L)^{1-m} (D)^{m+1}$ (18)

firms the conclusion that self excited pulsating flows in mass-flow bins which give rise to the 'silo quaking' phenomenon are associated with surcharge heads greater than a minimum value of, say, $H/D = 1.0$.

Dynamic loads in tall mass and funnel-flow bins

Roberts(1993) developed a simplified theory to determine the magnitudes of the dynamic loads due to 'silo quaking'. The loading conditions are depicted in Figure 29.

Shock loads

It is postulated that due to 'silo quaking' shock loads are imposed on a plane at some location in the cylinder wall. While there may well be several shock planes, the case of a single shock plane occurring at a height H_{cr} in the cylinder is considered as the worst possibility as far as loading is concerned. Above this plane, the bulk solid is considered to behave as a plug which imparts step loads due to 'slip-stick' motion as may arise from variations in density and in wall friction, as the friction changes from limiting or static to kinetic values. The behaviour is illustrated in Figure 29.

After each shock occurs, the dynamic wave will decay with time according to the bulk elastic properties and damping characteristics of the consolidated bulk solid. It is assumed that the amplitude of the shock pressure Δp_{wy} normal to the wall will decay exponentially with distance measured on each side of the shock plane. That is,

$$\Delta p_{wy} = \Delta p_{wo} e^{-\mu K y/R} \qquad (13)$$

Where the maximum increment Δp_{wo} in lateral wall pressure is given by

$$\Delta p_{wo} = \frac{F_D}{A} K \qquad (14)$$

Equation (13) may be used to estimate the peak dynamic pressures due to silo quaking. Δp_{wy} is the additional wall pressure applied to the initial or static wall pressure p_{ni} as indicated in Figure 29 (b). As $y \to h$, it is assumed that Δp_{wy} 'tails off' to approach the flow pressure at the top surface in the cylinder.

F_D is the total shock load and acts over the cross-sectional area A. F_D is given by

$$F_D = k_d \left\{ \frac{W}{h} \left[1 - e^{-\mu K h/R} \right] \left[\frac{R}{\mu K} - h_s \right] + W_s \right\} \qquad (15)$$

The two weight components are

$$W = \rho P R h \qquad (16)$$

and $W_s = \rho P R h_s$ (17)

where

$\gamma = \rho g$ = bulk specific weight, kN
ρ = bulk density, t/m^3
P = perimeter of flow channel, m
R = effective radius, m
h = head, m
h_s = surcharge head,
k_d = Dynamic load factor

Effective radius $R = \dfrac{D}{2(1+m)}$ (19)

Surcharge head $h_s = \dfrac{H_s}{m+2}$ (20)

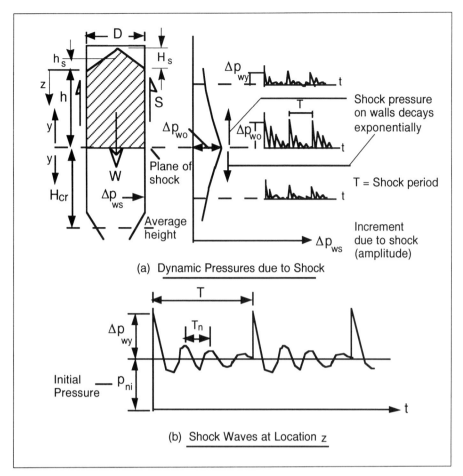

Fig. 29. Dynamic loads induced in silo.

where
$m = 0$ for plane-flow
L = length of flow channel,
$m = 1$ for axi-symmetric flow
D = diameter or width of flow channel
H_s = actual surcharge head
h_s = Effective surcharge head

If the load F_D is suddenly applied at each pulse period, then $k_d = 1.0$. Under severe conditions, such as when the moisture content of the bulk solid is high, the 'slip stick' effect is likely to be more pronounced. In this case $k_d > 1.0$.

Wave motion—brief discussion

The natural frequencies of the stored material are given by

$$f_i = \frac{i\lambda}{4h} \quad i = 1,3,5..... \tag{21}$$

where λ = wave velocity such that

$$\lambda = \sqrt{\frac{E}{\rho}} \tag{22}$$

E = elastic modulus of bulk solid
ρ = bulk density

The wave velocity for the bulk solid will depend on the properties of the bulk solid including the consolidation conditions and moisture content. It has been determined, for example, that for a certain dry bulk solid material, λ ranges from 250 to 400 m/second. Assuming $\lambda = 300$ m/second, then the natural frequencies will be $f_i = \frac{75}{h}\, \frac{225}{h}\, \frac{375}{h}...$(Hz) for i = 1,3,5.. respectively. For instance if h = 10 m, this gives $f_i =$ 7.5, 22.5, 37.5... Hz. With low damping, the fundamental

period $T_n = 0.133$ seconds; the period will increase as damping increases.

Effect of reducing head due to discharge

It is assumed that the location of the shock plane remains substantially at the critical height H_{cr} during discharge. Therefore, as the level in the bin drops, h decreases and the amplitudes of the shock pressure normal to the wall, Δp_o and Δp_s, also decrease. When the level drops to the critical height H_{cr}, $h = 0$ and the shock pulses virtually disappear. The number of cycles of dynamic load due to 'silo quaking' during *each* discharge is obtained from the equation

Number of pulses per discharge cycle =

$$\frac{\text{Time to discharge to critical height } H_{cr}}{\text{Pulse period T}}$$

It is also noted from equation (17) that, as h (Figure 8) decreases, the natural frequencies of the stored bulk solid will increase.

Period of pulses

As shown by Roberts (1993), the pulse period may be estimated from the following relationship

$$T = \left(t_o + \frac{v}{a}\right) + \sqrt{\frac{v}{a}\left(2t_o + \frac{v}{a}\right)} \tag{23}$$

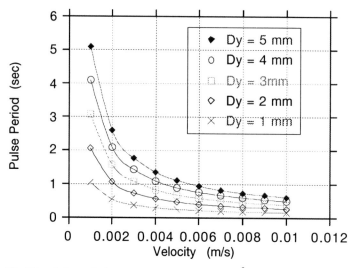

Fig. 30. Pulse period versus velocity a = 1 m/second^2

where

$$a = \frac{F_D g}{W_T} \qquad (24)$$

$$v = \frac{Q}{\rho A} \qquad (25)$$

$$A = \left(\frac{\pi}{4}\right)^m (L)^{(1-m)} (D)^{(m+1)} \qquad (26)$$

$$t_o = \frac{\Delta\varepsilon_y}{v} \qquad (27)$$

The variables given by equations (23) to (27) are defined as
a = acceleration of upper mass during pulse motion, m/second^2

v = average velocity of bulk solid in the cylinder during discharge, m/second

Q = discharge rate, kg/s

A = cross-sectional area of cylindrical section of bin

F_D = dynamic force defined by equation (10)

W_T = total weight of upper mass, N

t_o = time for motion of upper mass to be initiated

$\Delta\varepsilon_y$ = expansion of consolidated mass in vertical direction due to dilation

The parameter $\Delta\varepsilon_y$ is difficult to determine precisely. It is dependent on properties of the bulk material including the degree of consolidation and the particle size distribution. The acceleration given by equation (24) will always be less than the gravitational acceleration, since $F_D < W_T$. For the acceleration a = 1 m/second^2, which is a typical value, the variation of the expected pulse period T with the average velocity v for various values of $\Delta\varepsilon_y \equiv Dy$ is shown in Figure 30. This indicates the velocity dependency characteristic of the pulse period. As indicated, the pulse period is relatively insensitive to average velocity for velocities above 0.004

Example—Pilot scale wheat silo of Figure 25

To illustrate the application of the theories presented, reference is made to the experimental silo investigation described in Section 3. The following data apply:

D = 1.2 m

H = 3.44m

$H_S = 0$

$\mu = \tan 30^0 = 0.577$

$K = 0.4$

$H_{cr} = 1.4$ m (assumed)

$h = 1.5$

$Q = 5$ t/hR =

$F(D/4) = 0.3$ m

$\rho = 0.85$ t/m^3

Amplitude of Normal Pressure

Equation (16) : W = 18.9 kN

Equation (15): F_D = 9.6 kN

Equation (14): Δp_{wo} = 3.4 kPa

In order to compare the predicted and measured results for the load cell location 14 of Figure 25, equation (13) is used noting that y = 2.15 - 1.2 ≈ 1.0 m. This gives Δp_{wy} = 2.06. From Figure 28, the measured value of Δp_{yo} ≈ 1.9 kPa which is in reasonably close agreement with the predicted value.

Pulse period

Assuming that $\Delta\varepsilon$ ranges from 5 to 7 mm for the wheat, using equations (19) to (23), the estimated pulse periods are, respectively, T = 3.5 seconds and T = 4.9 seconds. From Figure 28, the actual pulse period ranges from 2 to 7.5 seconds averaging around 5 seconds. On this basis it is concluded that equation (19) is a reasonable predictor of the pulse period.

Squat funnel, intermediate and expanded flow bins

The effective dynamic mass of bulk solid in the central flow channel of the hopper section is the total mass in this section less the frictional support due to shear at the boundaries between the central and outer flow channels. This is depicted in Figure 31.

Here the analysis is similar to that presented for tall bins, except in this case the flow pulsations are generated in the lower hopper section. The pulsating mass is as indicated in Figure 11. Equations (15) to (20) apply in this case with the following substitutions:

$D_f = D$ = width or diameter of flow channel and

μ = coefficient of internal friction of bulk solid

Normally $\tan \phi_t \geq \mu \geq \sin \delta$ (26)

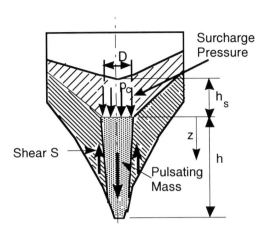

Fig. 31. Effective mass in pulsating flow.

where ϕ_t = static angle of internal friction and δ = effective angle of internal friction

Silo quaking problem—two case study examples

Field observations confirm the foregoing description of the mechanisms of 'silo quaking' (Roberts et al. 1991; Roberts 1993, 1994). For illustration purposes, two case study examples of pulsating flow in bins are presented. While the examples refer to coal bins, the observations could be equally applicable to grain.

Tall silos with chisel-shaped hoppers

Severe silo quaking problems were experienced in an installation of three, 12700 t, reinforced concrete coal silos. The silos, which are shown schematically in Figure 32, stand 58 m above the ground; the height above the base is 52 m, the internal diameter is 21.4 m and the wall thickness is 0.3 m. Adjacent silos are connected by concrete along the vertical lines of intersection. The silos have chisel-shaped, plane-flow hoppers lined with carbon steel plate, the half angle α being 40°, this being the angle with respect to the vertical. The hopper splits into three outlets, the lower hoppers being lined with stainless steel. The maximum discharge rate is 2700 t/hour through vibratory feeders. The reported period of the shocks was 3 –5 seconds, the shock loading being most severe when the silos were substantially full. After several years of use, severe damage started to occur, with sections of concrete being dislodged, particularly at the junction of adjacent silos.

Computation of shock loads

The shock plane is assumed to be located at a height of 26 m above the base of the hopper. The head of coal above the shock plane is $h = 22$ m. Using the procedures presented in Section 4, the following values are calculated:

$W = 77626$ kN

Equation (15):$F_D = 50130$ kN

Equation (14):$\Delta p_{wo} = 56$ kPa

Equation (13) has been used to compute the locus of the shock pressure amplitude, the results being illustrated in Figure 33. Also shown are the static and flow pressures. In view of the curved transition of the chisel-shaped hopper with the cylinder, the pressure profiles will vary around the periphery of the bin. Figure 13 applies to the side of the bin where the chisel-shaped hopper has its highest intersection point with the cylinder wall.

Estimation of pulse period

The application of equations (19) to (23) give the following predicted values of the pulse period:

For discharge rate $Q = 2700$ t/hour; $T = 2.4$ seconds

These values, which are based on $\Delta\varepsilon = 5$mm for the coal, are in general agreement with the approximate values of 3–5 seconds as reported for the actual silo.

Vibration of structure

A critical factor in the operation of the silos is the influence of the dynamic characteristics of the overall structure. Noting that the silos are supported on columns on a base, which, in turn, is supported on piles, a simplified dynamic model of the silos is shown schematically in Figure 34. There will be vertical and lateral stiffness due to the columns and piles as well as vertical and lateral stiffness due to the concrete connecting adjacent silos. In view of the significant variation in the silo mass from the full to the empty condition, there is a significant variation in the natural frequencies.

The natural frequency for the fundamental mode is given by

$$\omega_n = \sqrt{\frac{k}{M}}$$

Noting the variation in mass, the natural frequency of each silo will change as follows:

$$\frac{\omega_n full}{\omega_n empty} = 0.57$$

The natural frequency of the full silo is only 0.57 of the empty silo. While the various stiffness values of the structures are not known, it is possible that the load pulse frequency of 3 to 5 seconds could excite one of the silo modes. As each silo fills and empties, there will be different mass contents and, hence, different frequencies for adjacent silos. As a result, there will be dynamic coupling between adjacent silos which could impose significant loads on the concrete connecting these silos.

The modes of vibration, while complex, would involve a combination of vertical and sideways swaying motion, the latter induced by non-symmetrical loadings of coal in the silos as well as variations in ground stiffness in the zone of the supporting piles.

Multi-outlet bins

Silo-quaking problems have been known to occur in bins with multiple outlets. By way of illustration, consider the large coal bin shown in Figure 35. The bin has seven outlets, six around an outer pitch circle and one located centrally. The hopper geometries provide for reliable flow permitting complete discharge of the bin contents. Coal was discharged by means of seven vibratory feeders onto a centrally located conveyor belt. When the bin was full or near full, severe shock loads were observed at approximately 3 second intervals during discharge. The discharge rate from each feeder was in the order of 300 t/hour. When the level in the bin had dropped to approximately half the height, the shock loads had diminished significantly. With all the outlets operating, the effective transition was well down towards the bottom of the bin walls and the critical head H_m was of the same order as the bin diameter and greater than D_F. Substantial flow occurred along the walls, and since the reclaim hoppers were at a critical slope for mass and funnel-flow as determined by flow property tests, the conditions were right for severe 'silo quaking' to occur.

Confirmation of the mechanism of silo quaking was obtained in field trials conducted on the bin. In one series of tests the three feeders along the centre line parallel with the reclaim conveyor were operated, while the four outer feeders were not operated.

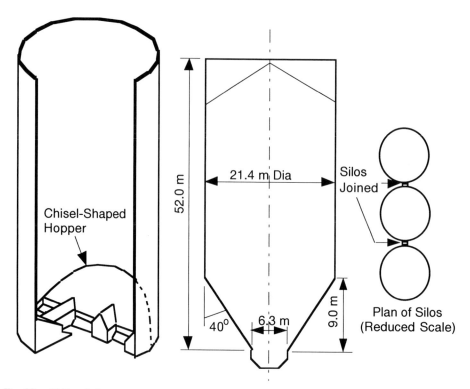

Fig. 32. Tall coal silos

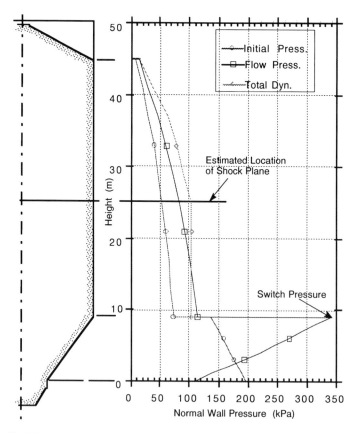

Fig. 33. Pressure distributions for bin.

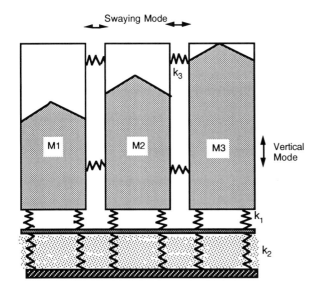

Fig. 34. Simplified dynamic model of silos.

This induced funnel-flow in a wedge-shaped pattern as indicated in Figure 35, with the effective transition occurring well up the bin walls, that is $H_m < H_{cr} (=D_F)$ or $H_m << D$. The same was true when only the central feeder (Fdr. 1) was operated; in this case the stationary material in the bin formed a conical shape. Under these conditions, the motion down the walls was greatly restricted and, as a result, the load pulsations were barely perceptible.

In a second set of trials, the three central feeders were left stationary, while the four outer feeders were operated. This gave rise to the triangular prism shaped dead region in the central region, with substantial mass-flow along the walls. The load pulsations were just as severe in this case as was the case with all feeders operating. Dynamic strain measurements were made using strain gauges mounted on selected support columns. When the bin was full (or near full), the measured dynamic strains with $H_m >> H_{cr}$ were in the order of 4 times

the strains measured when the flow pattern was controlled so that $H_m < H_{cr}$.

Concluding Remarks

This paper has presented an overview of some recent developments in the technology of bulk solids handling as it relates to the grain industry. The paper has focused on silo design, emphasising the need to understand the relevant properties of the grain and how these relate to silo geometry and discharge flow patterns to generate the load patterns exhibited in the silo walls. While the loading patterns in the case of symmetrical silos are well defined, in the case of multi-outlet silos and silos with eccentric discharge the loadings are much more complex leading to bending and buckling stresses in the walls.

With the emphasis on grain conditioning procedures, such as aeration, to control grain quality, it is important to also take note of the influence of any such grain conditioning on the bulk storage and flow properties of the grain. In this respect, increased moisture content which produces grain swelling can cause substantial increases in silo wall loads. Daily and seasonal fluctuations in temperature which cause expansion and contraction of silo walls, particularly in the case of steel silos, can also cause substantial silo wall pressure fluctuations leading to increased loads during periods of contraction and structural fatigue.

Methods of controlling loads in tall, single and multi-outlet silos have been discussed. Particular mention has been made of the use of anti-dynamic tubes to control the flow pattern and, hence, control the magnitude of the wall pressures generated during discharge.

A re-occurring problem in bins and silos is the phenomenon of 'silo quaking', a term used to describe pulsating loads which may be experienced during discharge. The various mechanisms of 'silo quaking' have been outlined and methodologies to determine the magnitude and distributions of the dynamic loads have been reviewed. If the frequencies of the flow pulsations are in harmony with any of the natural frequencies of the structure, severe dynamic loads and stresses can occur.

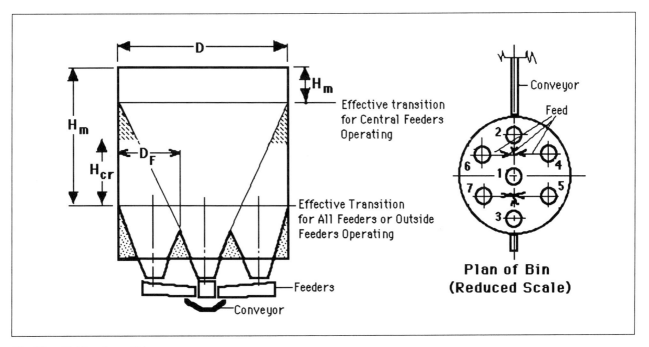

Fig. 35. Multi-outlet coal bin.

It is quite evident that, in recent years, significant advances have been made in research and development associated with bulk handling systems for granular materials. It is gratifying to acknowledge the increasing industrial awareness and acceptance throughout the world and particularly in Australia of modern bulk materials handling testing and plant design procedures. It is also noted that areas for ongoing research are continuing to be identified; provided Universities and research establishments are encouraged to pursue this research, then the future efficiency of bulk materials handling as a key function industry and agriculture is assured.

References

Arnold, P.C. and Roberts, A.W., 1966. Stress distributions in loaded wheat grains, Journal of Agricultural Engineering Research, 11, 1.

Arnold, P.C., McLean, A.G. and Roberts, A.W. 1982. Bulk solids: storage, flow and handling. The University of Newcastle Research Associates (TUNRA), Australia.

Benink, E.J. 1989. Flow and stress analysis of cohesionless bulk materials in silos related to codes. Doctoral Thesis, The University of Twente, Enschede, The Netherlands.

Jenike, A.W. A Theory of flow of particulate solids in converging and diverging channels based on a conical yield function. Powder Tech., 50, 229–236.

Jenike, A.W. 1961. Gravity flow of bulk solids. The Univ. of Utah, Engineering Experiment Station, Bulletin 108, USA

Jenike, A.W. 1964 Storage and Flow of Solids. The Univ. of Utah, Engineering Experiment Station, Bulletin 123, USA.

Jenkyn, R.T., 1978. Calculation of material pressures for the design of silos. Proceedings Institution. Civil Engineers, 2, 65.

Ooi, J.Y. and Rotter, J.M. 1989. Elastic and plastic predictions of storage pressures in conical hoppers. In: Third International Conference on Bulk Materials Storage, Handling and Transportation. Newcastle, Australia, The Institution of Engineers, 203–207.

Ooms, M. and Roberts, A.W. 1985. The reduction and control of flow pressures in cracked grain silos. Bulk Solids Handling, (5), 5, 1009–1016.

Roberts, A.W. 1988a. Modern concepts in the design and engineering of bulk solids handling systems. The University of Newcastle, Australia, TUNRA Bulk Solids Research.

Roberts, A.W. 1988b. Some aspects of grain silo wall pressure research—influence of moisture content on loads generated and control of pressures in tall multi–outlet silos. In: Proceedings Conference, Chicago, USA, May 11–24.

Roberts, A.W. 1993. Mechanics of self excited dynamic loads in bins and silos. Oslo, Norway. Proceedings of the 2nd Reliable Flow on Particulate Solids Symposium 983–1004.

Roberts, A.W and Ooms, M. 1983. Wall loads in large steel and concrete bins and silos due to eccentric and other factors. In: Proceedings 2nd International Conference on the Design of Silo for Strength and Flow, Powder Advisory Centre, U.K. 151–170.

Roberts, A.W., Scott, O.J. and Wiche, S.J. 1991. Silo quaking. A pulsating load problem during discharge in bins and silos. Proceedings Bulk 2000, The Institution of Mechanical Engineers, U.K., 7.12.

Rombach, G. and Eibl, J. 1989. Numerical simulation of filling and discharging processes in silos. In: Third International Conference on Bulk Materials Storage, Handling and Transportation. The Institution of Engineers, Newcastle, Australia, 48–52.

Standards Association of Australia (SAA) 1990. Australian Standard AS3774–1990, Load on bulk solids containers

Thomson F.M. 1984. Storage of particulate solids. Van Nostrand, Handbook on Powder Science and Technology, 9.

Wu, Y.H. 1990. Static and dynamic analysis of the flow of bulk materials through silo. PhD. Thesis, The University of Wollongong, Australia.

Temperature studies on steel silos in North Africa

El H. Bartali*

Abstract

The control of physical parameters is a useful means of evaluating the storage quality of grains and legumes. A study was undertaken to assess the performance of storage systems of cereals and legumes in North African climate situations. Bulk storage of three commodities: soft wheat, barley and fava beans in vertical and horizontal silos were investigated. The study involved a multidiscilplinary team and was lead by the Department of Agricultural Engineering, Rabat, Morocco. Four storage units made of different construction materials were investigated and included: one upright 44 t steel unit filled with soft wheat; one 45 t horizontal unit with corrugated aluminium sheet walls containing barley; one 15 t capacity clay straw, naturally ventilated and filled with soft wheat; and a reed silo containing 1.5 t of fava beans. All the units were equiped with a network of temperature and relative humidity sensors that allowed measurement at various time intervals of this data. The temperature and r.h. data were utilised to decide the periods of mechanical ventilation of the steel vertical unit and the aluminium horizontal silo. Ventilation was applied using either ambient or refrigerated air. Sampling was carried out at regular time intervals to assess changes in physical properties and rate loss in terms of storage period. The insect population was monitored using insect traps placed at specific locations in the storage units. The paper presents an analysis of the results obtained.

Introduction

Adequate grain storage is meant to preserve the initial quality and other characteristics the grain had before being stored. North African countries are developing production of grain, a staple food, with the objective of achieving self-sufficiency and reducing imports. Steel silos are being progressively and increasingly used beside other appropriate storage structures by farmers and state controlled sector in order to upgrade grain storage capacity and reduce losses. Given the generally hot climate in the region, stored grain is threatened by excess heating in steel silos, particularly in summer if appropriate aeration is not performed. Ambient air ventilation consists in applying frequent doses to prevent the start of grain heating, or to cool grain by successive steps whenever ambient air temperature is favourable. This requires continuous monitoring of temperature and relative humidity of grain and ambient air to maintain quality of stored product.

* Agricultural Engineering Department, Agronomic and Veterinary Institute Hassan II, B.P. 6202, Rabat, Morocco.

Materials and Methods

An investigation on the behaviour of grain stored in a steel silo was carried out in Meknes, central of Morocco, known for its continental climate. The galvanised steel silo is 44 t capacity, 3.70 m high and 4.3 m in diameter with a conical roof (Fig. 1). The silo is equipped with a fan with 1.3 m^3/second airflow, 1.5 Kw power and total pressure of 55 mm of water gauge. Grain handling is carried out mechanically by an auger equipped for a 6 t/hour filling and emptying rate. The study involved storing soft wheat for one year from January 1989 to December 1990 and monitoring temperature and relative humidity in specific locations of grain and air.

A computer run network of temperature and humidity sensors hooked to data loggers enabled hourly measurements and storage of data. A total of 13 temperature sensors and 1 relative humidity sensor was placed within the grain and on the structure (Fig.2). The silo is equipped with sampling perforations placed at bottom, mid height and top that made it possible to analyse grain quality at regular intervals, 45 days or 90 days, and whenever necessary during aeration.

Investigation was conducted in an experiment station on grain storage, located in an extension station. Research funds were provided by the Moroccan Ministry of Agriculture.

Aeration of grain was performed mostly at night. Continuous monitoring of grain and air temperatures determined when grain ventilation was necessary and possible, what level of grain cooling could be reached and what is the gain with respect to insect control.

Results and Discussion

Temperature variations

Temperature variations recorded at the centre and outside of the silo over a period of 10 months are presented in Figure 3. Maximum air temperatures are shown to reach upper forties in summer. Minimum temperatures reached a few degrees below zero in winter. Grain temperature variations at the centre of the silo are reduced due to thermal inertia of grain and fluctuate between maximum and minimum air temperatures.

Grain aeration

Over 9 months of storage, 8 aeration interventions took place summarised in Table 1 and Figure 4.

Since most ventilation operations took place at night, grain and air characteristics, particularly temperature and relative humidity, tend to vary. There is a need to make several measurements of the variables and evaluate night means. Therefore, intervals between data measurements and recordings were lowered to 20 minutes during ventilation.

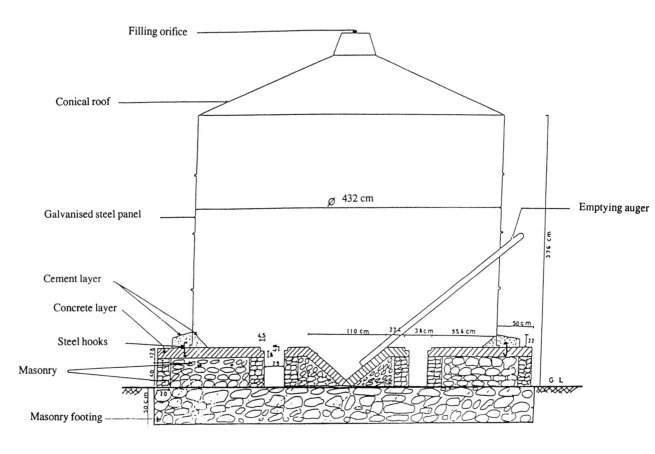

Fig. 1.   Steel silo, 44 t capacity.

Table 1.   Calendar of aerations and air and grain characteristics.

Ventilation doses	Start	Finish	Duration (hours)	Relative humidity (r.h.) of cooling air (%)	Equilibrium (r.h.) of grain
28/2–1/3/1990	28/2 at 10 pm	1/3 at 6.30 am	8 hours 30 minutes	81	55
14/2-18/3	14/3 at 10 pm	18/3 at 7.00am	27 hours and 30 minutes	79.6	53
22–23/3	22/3 at 12.30 am	23/3 at 6.40 am	11 hours and 30 minutes	80	56
4/4	4/4 at 2.00 am	4/4 at 4.00 am	2 hours	83	54
6–10/9	6/9 at 12.30 am	6/9 at 4.45 am	9 hours and 45 minutes	65	51
27–29/7	27/7 at 8.35 pm	29/7 at 5.00 am	12 hours	45	56
1/11	10.45 pm	6.10 am	7 hours and 25 minutes	73	51
4–9/10	4/10 at 10.40 pm	9/10 at 6.00 am	27 hours	68	52

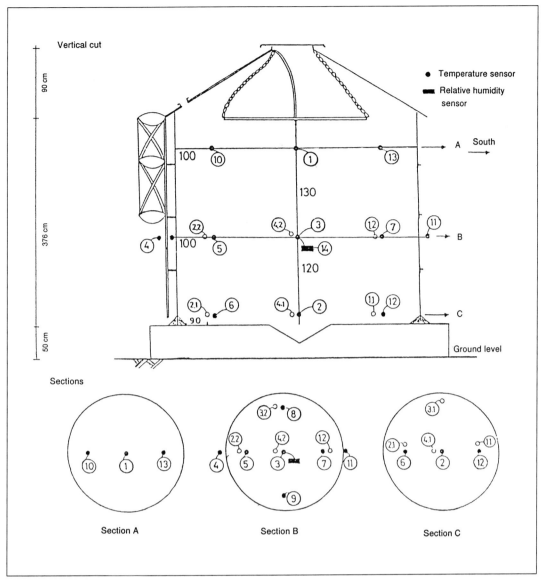

Fig. 2. Locations of temperature and relative humidity sensors in 44 t experimental silo. Measurements are in cm.

Evolution of grain temperature during ventilation

Grain is a poor heat-conducting material. Its temperature changes slowly when ventilation is not applied. However, outside air temperature can undergo rapid and large changes. The storage period outlined in Figure 4 may be divided into three parts.

In the first part, both air and grain temperatures decrease over a 2-month period starting from filling. Grain remains at a temperature close to air temperature. No ventilation was necessary given the safe conservation conditions provided by the low temperatures, as indicated by the conservation diagram.

A second part, that ends in April, is characterised by frequent ventilation interventions due to:

• increases in grain temperature reaching values beyond recommended safe storage figures of the diagram; and

• limited time where air temperature values allow grain cooling, approximately 6 hours per day.

Ventilation has made it possible to decrease grain temperature to 14°C.

A third part, from April–October, corresponds to a progressive increase in grain temperature and more rapidly than in the second part. As indicated in Figure 5, relatively high air temperatures did not allow ventilation at the beginning of this

period. During this period a sharp decrease in temperature took place particularly during the eighth ventilation dose where temperature was halved. Analyses carried out by entomologists showed that during this period, development of storage insect species both in number and diversity has increased with temperature.

Detail of a ventilation dose

In order to make detailed observations on the ventilation process, Figures 5 and 6 depict the variation of temperature during ventilation time, along two vertical axes, in the centre and one near the wall of silo.

The eighth ventilation operation that took place in October was selected for this purpose. A high temperature difference between air and grain was noticed –14°C. This temperature difference that happened after a sharp decrease in air temperature in October, could create condensation problems. Such problems were avoided by conducting ventilation over longer periods of time.

At each ventilation, blown air cools the lower layers of grain at the bottom of the silo (sensors #6 and #2) (Figs 5 and 6). A wave of cooling air develops and progresses slowly along air

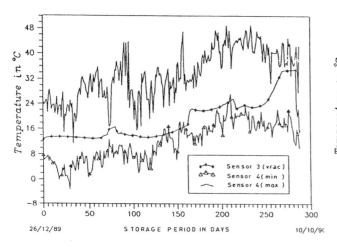

Fig. 3. Maximum and minimum temperatures at the centre of the bulk. See Figure 2 for locations of sensors.

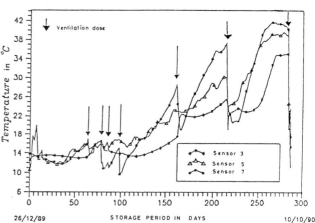

Fig. 4. Effect of ventilation on temperature. See Figure 2 for locations of sensors.

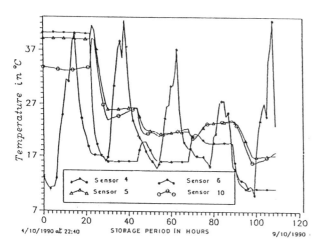

Fig. 5. Sidewall temperature and ventilation doses. See Figure 2 for locations of sensors.

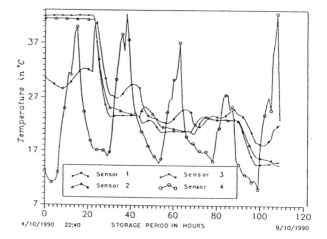

Fig. 6. Central axis temperatures. See Figure 2 for locations of sensors.

stream. All grain layers are then successively cooled from the bottom to the top of the silo, as shown in Figure 6. Lower layers of grain where sensor #6 is placed cools first, followed by middle height layer with sensor #5 and last the upper layer monitored by sensor #10. This upper layer is influenced by changes in ambient air temperature.

Temperatures recorded by sensor #6, located near ventilation conduit, approaches cooling air temperature, after each ventilation whereas temperatures of upper layers often remain higher than external air temperature.

Effect of relative humidity of cooling air

For a given relative humidity of interstitial air within grain mass and a given temperature, there exists an equilibrium moisture content of grain, outside ventilation period.

Table 1 shows relative humidities of cooling air and grain for each ventilation. Relative humidity of cooling air is measured by a psychromèter and of the grain by a relative humidity sensor placed at the centre of grain silo.

Values of the ratio of relative humidity of grain to relative humidity of air are lower than 1, indicating that an under-cooling of grain takes place.

Conclusions

This study conducted on temperature distribution in aerated steel grain bins leads to the conclusion that in hot temperate climates such as the one in continental Morocco, grain aeration is justified and recommended. Although grain moisture content at harvest period is in the range 10–12 %, the grain mass needs be cooled in order to be safely stored. Such cooling by aeration is possible since climatic conditions in this area indicate that 6 hours per day are favourable to carry out ventilation of grain.

The following pattern could be followed for ventialtion of grain stored in a steel bin in the area:

- First cooling of grain whose initial temperature at harvest time is high and ranges from 30–40°C will allow a reduction in grain temperature down to 22°C in July–August.
- Second cooling carried out in October–November will decrease grain temperature to 14°C.
- Third cooling undertaken in December–January will make it possible to reduce grain temperature to approximately 10°C. High thermal inertia of grain will maintain this temperature in spring.

References

Bachraoui, Z. et El Alami, N., 1991. Contribution au Suivi et à l'Evaluation des Performances de Cinq Systèmes de Stockage des Céréales dans le Saiss. Mémoire de fin d'étude génie rural, IAV Hassan II, Rabat Maroc, Janv. 1991

Bartali, H. 1990. Situation de la Post Récolte en Afrique du Nord, Invited key note paper. Comptes rendus du séminaire international sur la post récolte en Afrique, AUPELF–UREF, Abidjan, Côte d'Ivoire, Fev. 1990

Bartali, H. 1990. Storage structures and equipment invited key note paper. Proceedings of the Fourth International Conference on Stored–Product Protection, Bordeaux, France, 5–8 September, 1990.

Bartali, H. 1992. Monitoring and evaluation of grain storage systems. International symposium on stored grain ecosystems, organised by Department of Agricultural Engineering, University of Manitoba, Winnipeg, Canada, 7–11 June, 1992.

Bartali, H. and Hatfield, F., 1990 Decreasing ambient temperature in cylindrical silos. Journal of the American Concrete Institute, 87(1).

Gers 1989. La Station Expérimentale sur le Stockage des Céréales du Saiss, Projet Maitrise des techniques de stockage des céréales, DPV, IAV Hassan II Rabat Maroc, Rapport No.1, Nov. 1989.

Gers 1991. Stockage Adéquat des Céréales, journée d'étude organisée par IAV Hassan II, projet Maitrise des Techniques de Stockage des Céréales, DPV, Rapport No. 4, Juillet 1991.

Lasseran, J.C. 1990. La Ventilation des Grains: Conditions d'utilisation en Climat Tempéré Chaud. Séminaire de la 2 Section de la Commission Internationale du Génie Rural sur les Structures de Stockage des céréales, des légumineuses et de leurs dérivés. Edit. Bartali, Rabat, Maroc, 28–30 Nov. 1990

Development of a programmable aeration controller

P. A. Gibbs*

Abstract

This paper describes the development of a remotely accessible (via modem) PC-based system for the control and monitoring of aeration systems. The system comprises a PC and interfacing hardware, modem, weather station and sensors in the grain. This system offers more flexible and efficient control of aeration, reduced maintenance costs and more efficient supervision by regional offices of storage sites which are often difficult to access. As a research tool, the system has similar benefits. In particular, access to data and system status via the telephone network saves time and travel expenses, and reduces the dependence on local operators for collecting information. The programmability of the system ensures that it can be easily adapted to different control strategies and new applications as these are developed.

Information gathered in the first two years of trials of the system will be used to improve control algorithms and to verify mathematical models of heat and mass transfer in grain stores under aeration. These trials have already demonstrated the feasibility and usefulness of such remotely controlled systems.

Introduction

There has been little recent work on aeration controllers in Australia. Existing systems have only a single mode of operation, whereas microprocessor technology offers greater flexibility and improved performance.

The PMCAM (programmable microprocessor control and monitoring system), also known in the control industry as a SCADA (supervisory control and data acquisition) system, is a multipurpose computer-based data control system which can be programmed to control aeration systems, log system information and allow remote control and monitoring and remote collection of data.

The failure of unattended systems is not always apparent on casual inspection. With a microprocessor controller we can avoid this problem by checking for system faults, looking for conflicting behaviour, using numerical models of aeration in bulks to control aeration for maximum cooling, automatically switching aeration off at the optimum time, and even aerating without cooling when some ventilation is desirable.

By avoiding the need to travel to remote sites to collect data we can save considerable time and money. With control and supervision from a central location, we reduce the need for training operators at each site and can instead concentrate on training regional operators in more detail. We also have

---

* Stored Grain Research Laboratory, CSIRO Division of Entomology, GPO Box 1700, Canberra, ACT 2601, Australia.

greater flexibility since we can modify the control program at any time and download it over the telephone network.

Hardware

Figure 1 gives the layout of the PMCAM hardware. The components inside a standard mini-tower computer case are: 80386SX motherboard with 1MB RAM, low capacity hard disc (40 or 80MB), 3.5" floppy disc drive, video card, serial card, HDD and FDD controller card and an analog-to-digital and digital I/O card. Components mounted inside the main PMCAM enclosure, but outside the PC case are: the external modem connected to the serial port, interfacing and multiplexing board with built in watchdog timer, and one analog input card for each set of 16 channels of analog input, connected to the interface card.

Nominally the motherboard should be at least an 80286 with 1 Mb RAM. A serial I/O and disc controller card controls the floppy disc and hard disc drives, and provides the interface for serial communication. For this application we need a low-cost modem, and this usually means medium speed. Both 1200 and 2400 baud must be available as some country phone systems are very noisy and should not be run faster than 1200 baud. Low cost is essential because of the risk of lightning strikes and the need to reduce overall costs.

We are currently using a 12 bit analog-to-digital converter card with 16 analog input, 2 analog output, 16 digital input and 16 digital output channels. The analog interface card was designed simply to allow easy connection of sensor wires to the analog input cable. The multiplexer (MUX) card is used to expand the number of analog channels available from 16 to a maximum of 1216. A watchdog timer card reboots the computer if the control software fails to send a signal at regular intervals. This provides more reliable operation.

The components outside the PMCAM enclosure are a weather station and the grain temperature and perhaps moisture sensors. The weather station normally contains two dry-bulb temperature sensors and one relative humidity sensor. It can also house a rain sensor. Typically, the relative humidity sensor will be a dual temperature/r.h. sensor.

The first system constructed, installed at a large site in South Australia, used existing thermocouple sensor cables and a requirement for RTD (resistance thermometer device) sensors in the duct. We currently use semiconductor temperature sensors rather than thermocouples or RTDs for their simple interfacing requirements, and lower overall cost. Thermocouple cable is approximately five times the cost of copper wire. Instrumenting a 20 m high silo with thermcouple wire would incur a considerable wiring cost.

The phone connection may be either a mobile phone with modem interface or a direct connection to a fixed line into the site. The 1993 season was the first time we used mobile phones, and they were a mixed blessing. Although cheaper than a permanent phone line in particular sites where the telephone line is far from the silos, reception is not always good.

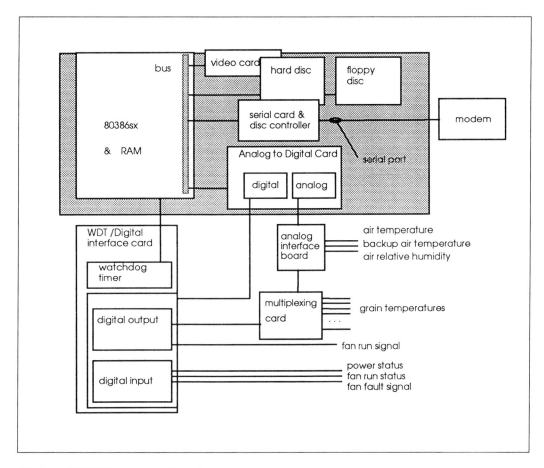

Fig. 1. PMCAM hardware configuration.

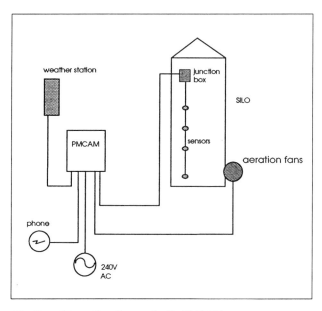

Fig. 2. Connection diagram for the PMCAM.

Discussion

Initial field trials were undertaken at three sites around south-eastern Australia.

Currently, we have 17 PMCAMs installed in 5 States from Western Australia to Queensland, aerating or monitoring the storage of grain in silos ranging from 100 to 15 000 t capacity.

We have tested about four different hardware configurations for the PMCAMs. One important finding, although based on only four machines, was that the reliability of the expensive industrial grade motherboards was not as good as expected for the price. Of three initial industrial grade motherboards with battery backed RAM discs, one failed in the first year. For the second season of trials we installed an additional PMCAM with a standard commercial grade motherboard and hard disc drive at less than half the price and had satisfactory performance. We currently use standard rather than industrial grade equipment. Of the 17 systems installed to date we have had motherboard failures in only two of the standard machines.

Our field experience has shown us that a backup air temperature sensor should be used with each PMCAM so the PMCAM can function in the event of the main dry-bulb temperature sensor in the weather station failing.

The most serious problem we have had with the systems is the reliability of relative humidity sensors, a critical factor if we are to use wet-bulb temperature control. The cost of commercially available wet-bulb sensors is of the order of $1400. This is excessive considering that it will almost double the cost of the system. Three different makes of relative humidity sensors used in the field from 1990 to 1992 were unreliable. Indeed, one of the temperature sensors was also unreliable, and prompted us to install a backup, simple, temperature sensor in each site as a fail-safe device.

In late 1993 we tested two each of three other relative humidity sensors. The best performance, based on accuracy and repeatability, was given by the Vaisala combined temperature and relative humidity sensor at a cost of approximately $700. Although this was twice the previous price for relative humidity sensors, we have installed them in the bulk of our new sites and to date have had good, reliable performance.

Trials at one site in South Australia with 10 cells (5 fans) under aeration have allowed us to run four different aeration control algorithms at the same time at one site. The strategies were dry- and wet-bulb set point, and dry- and wet-bulb time proportioning control strategies [see Desmarchelier and Wilson (1994) for an explanation of these strategies]. The use of the PMCAM has firstly made it possible to have the one control system running all four of these algorithms (a very significant leap forward in flexibility) and secondly to monitor the conditions in the grain.

At this site each cell is fitted with a platinum resistance thermometer in the aeration duct. Since the system uses suction, this sensor gives an indication of the grain temperature in the bin. Trials over two years have shown that this duct air temperature gives a good indication of the time of passage of the first cooling front through the grain (Fig. 3).

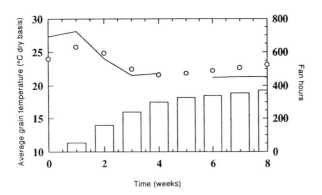

Fig. 3. Duct temperature as a rough guide to the time at which the first cooling front has passed. The open circles represent average grain temperature, while the lines are the duct temperature and the bar chart shows the fan hours.

Costs

Hardware costs are currently just under $2000 per system excluding weather station, sensors, sensor wiring, and the installation cost to connect to power and telephone. Our current weather station configuration costs approximately $1200. Temperature sensors and other wiring add another $1000 to $2000. Table 1 shows the trend of PMCAM cost over the duration of the project. Currently, the capital cost of storage capacity for just the PMCAM on its own without sensors and installation, etc., is 14 cents/t for a 15000 t horizontal store and 96 cents/t for a 2000 t vertical store. If more than one vertical cell is controlled then the capital cost is reduced.

In 1992 we achieved energy consumption rates of between 0.5 and 0.8 kWh/t (at 14 cents/kWh this is equivalent to 7 cents/t and 11.2 cents/t, respectively), for final average grain

dry-bulb temperatures of between 21 and 23°C from initial temperatures in the 28 to 30°C range.

In 1993 we achieved a final average grain dry-bulb temperature of close to 14°C from around the same starting temperature. This final temperature is slightly lower than the 15°C dry-bulb temperature limit recommended in order to ensure that later fumigation is effective (Australian recommendations for PH_3). Energy consumptions for the most aerated cell in South Australia, and the shed in New South Wales were 1.7 and 1.5 kWh/t (23.8 cents/t and 21.0 cents/t), respectively. The lower temperatures reached in 1993 naturally resulted in an increased energy cost compared with the 1992 trials. Nevertheless, these energy costs are less than expected for aeration systems. For example, costs of 50–100 cents/t are often attributed to aeration. We over-aerated slightly in the horizontal shed in 1993 and yet we still obtained energy costs less than half that of this value.

The observed energy consumption figures are in line with those of Navarro and Calderon 1982, but less than old Australian reports of 2.5–4.0 kWh/t (Sutherland 1968). Further reductions in energy consumption of at least 10% should be possible by careful tuning of the control algorithm, and by incorporating models to predict the time of passage of the cooling front, thereby allowing the fans to be switched off before excessive aeration occurs. These modifications are made possible because the PMCAM control program can be rewritten very easily.

The usefulness of the PMCAM for research purposes can be demonstrated by the use of temperature data collected by the system in a 15000 t capacity horizontal store. Contour plots of the data show the passage of the cooling fronts on a half-hourly basis, and provide a very detailed picture of the progress of the aeration. These data are also being used in conjunction with moisture measurements to map the distribution of moisture around the aeration ducts.

For the future one would not expect the present form of PMCAM to be used in medium to large sites because PLC (industrial programmable logic control) systems are more rugged, more likely to be in place for controlling other aspects of the existing machinery, and because the software becomes so complicated that it should be left for the experts in industrial control systems. The use of PMCAMs at smaller sites will affect the way larger sites are controlled. The results derived from operating smaller sites with the PMCAMs and optimising the control strategies can be incorporated into the software controlling the larger sites.

The PMCAM has shown itself to be a useful tool for the control and monitoring of aeration. Such systems will be of great value in the control and or monitoring of many other processes.

Using the PMCAM we have demonstrated that aeration is a much cheaper option than previously believed in Australia.

Table 1. Capital cost of PMCAMs over the term of the project ($Aust). The cost of sensors is not included, since the requirements depend on the site layout and number of sensors for the particular research involved. The cost of software is also not included.

PMCAM number	Year	Cells	Tonnes per cell	Capital cost ($)	Capital cost per tonne of capacity ($/t)
1	1990	10	1400	8410	0.60
2	1991	1	2200	4030	1.83
3	1991	1	2200	3375	1.53
4	1992	1	15000	2110	0.14
5	1993	1	2000	1920	0.96
6	1993	1	14000	1920	0.14

The flexibility and cost saving nature of PMCAMs makes them an excellent research tool, and, in its more appropriate forms linked into PLC systems, will prove to be an excellent way of controlling larger-scale processes.

Acknowledgments

The funding of the project by the Grains Research and Development Corporation is gratefully acknowledged. Without the support and assistance of the Bulk handling Organisations the field trials would not have been possible. The design, construction and testing of the in house PMCAM components were very ably performed by Paul Elkerbout.

References

Desmarchelier, J. M. and Wilson, S. G. 1994. A guide to grain aeration. Melbourne, CSIRO Publications Unit, in press.

Navarro, S. and Calderon, M. 1982. Aeration of grain in subtropical climates. FAO Agricultural Services Bulletin, 52, 119 p.

Sutherland, J. W. 1968. Control of insects in a wheat store with an experimental aeration system. Journal of Agricultural Engineering Research, 13(3), 210–219.

Observations on large-scale outdoor maize storage in jute and woven polypropylene sacks in Zimbabwe

L. Kennedy and A.D. Devereau*

Abstract

Sacks made from woven polypropylene are replacing jute sacks for commodity storage in developing countries. In sub-Saharan Africa this has coincided with an increase in stackburn, a condition in which maize becomes discoloured during storage, losing commercial and possibly nutritional value. This has occurred mainly in Zimbabwe but also in Ghana and Malawi. Investigations have shown that stackburn is due to chemical changes in the grain induced by high temperatures during storage; it may in part be due to non-enzymic browning. In Zimbabwe insect metabolic activity may be a possible cause of heating.

The possible link between these high temperatures and the adoption of woven polypropylene (wpp) sacks was investigated by monitoring identical 1100 t stacks of jute and wpp sacks in Zimbabwe. Heating attributed to insects occurred throughout each bagstack, causing temperature rises to 40°C, but the effects were dissimilar in each bagstack. The jute bagstack, especially in the lower half, cooled for a period when fumigated but the wpp bagstack showed much less or no reaction. The wpp bagstack, especially at mid height, heated at a higher rate than the jute stack. These factors caused the temperature to be generally cooler for longer periods in the jute bagstack than in the wpp bagstack.

Introduction

Jute and woven polypropylene sacks

Heavy duty sacks made of woven jute are widely used for the storage and transport of agricultural commodities. They are cheap, tough, flexible, have high tensile strength and have good handling and stacking properties (Coveney 1969). Sacks with a capacity of 60–90 kg are typically used. Other natural fibres, e.g. sisal, may also be used for sacks, but the amounts involved are small in comparison.

Woven polypropylene (wpp) sack manufacture was developed in Japan in the late 1960s and was quickly adopted in Europe, South Africa, Australia and North America (Paine 1991). High density polypropylene film is extruded and slit into narrow tapes which are heated, stretched and woven into sacks, usually of a plain weave. These sacks are lighter than jute but relatively stronger.

* Natural Resources Institute, Central Avenue, Chatham Maritime, Kent, United Kingdom, ME4 5NJ.

Wpp sacks have replaced jute sacks in a number of African countries for commodity storage. For example, in Zambia in 1988, 10 million natural fibre sacks were made in comparison to 60 million synthetic sacks (FAO 1992). They are produced locally using imported raw materials and are cheaper than imported jute sacks. Little work appears to have been done to investigate the effects of the new sacks on the quality of commodities stored in them.

Stackburn

Large quantities of maize, up to 1.2 Mt, are stored annually in Zimbabwe (Tyler 1992). One of the main storage methods is to build large outdoor stacks from sacks of maize, covering them with tarpaulins and sometimes plastic sheets, a technique called cover-and-plinth or CAP storage. These bagstacks may hold up to 5000 t of maize and may be left for one year or more.

Wpp sacks have largely replaced jute sacks for the storage and handling of maize in Zimbabwe. During recent years, coinciding with the introduction of wpp sacks, maize throughout the centre of some bagstacks has been found to be discoloured from a light tan through to a dark red-brown colour after storage. The discoloration, termed stackburn, results in the affected maize being downgraded with attendant financial losses. Instances of stackburn have been reported in Ghana and Malawi, also in maize stored in wpp stacks (Tyler 1992).

Investigations by NRI have so far identified the discolouration as partly due to non-enzymic browning, or the Maillard reaction. Further work on the nature of the chemical reactions is under way. The extent and rapidity of browning was found to be closely linked to the temperature and moisture content (m.c.) of the maize. Field trials using wpp bagstacks in Zimbabwe showed that heating was occurring in bagstacks during storage, raising the temperature in the interiors to approximately 40°C over a prolonged period. It seemed likely that these prolonged high temperatures were a major factor in the discoloration process.

Causes of heating within stored grain include the following:
a. mould respiration, releasing heat, which occurs when the moisture content of grain is higher than a critical limit and mould growth occurs;
b. insect respiration, again releasing heat, caused by heavy insect infestations in the grain;
c. grain respiration, releasing heat;
d. rises in ambient temperature.

In all cases the grain acts as a thermal insulator, retarding the escape of heat from the grain in cases a, b and c so that a temperature rise results, while inhibiting the transmission of heat into the grain in case d; only long-term (seasonal) changes in ambient temperature affect the interiors of stored grain.

Preliminary investigations in Zimbabwe showed that the moisture content of the maize during bagstack storage was insufficient for mould growth, less than 14.25%, and no moulding was observed. Analysis of samples from the bagstack showed that some insects, mainly *Tribolium* spp. and *Sitophilus* spp., were present at many positions in the bagstack at the beginning of the investigation; after 12 months they were distributed throughout the bagstack, often at high densities. The peak temperatures observed, around 40°C, were considerably above those on the stack surfaces, ruling out heating from ambient temperature rises. Aerobic grain respiration heating was thought to play only a minor role because of the high rate of heating observed in comparison to the low rates of heating attributed to aerobic respiration by Christensen and Sauer (1982) and Gough (1985), though the potential role of anaerobic respiration was not clear. Insect heating may therefore be a likely cause of the high temperatures which in turn led to the discoloration of the maize.

The preliminary trials suggested the mechanism of the discoloration but did not throw any light on the possible link between the change from jute to wpp sacks and stackburn. A further trial, described in this paper, was therefore undertaken to compare the storage conditions within otherwise identical wpp and jute bagstacks.

Method

The trial was undertaken in Zimbabwe, at the Grain Marketing Board (GMB) depot at Concession, approximately 40 km north of Harare. Two bagstacks of approximately 1100 t of white dent maize were built during August 1991 on adjacent earth plinths covered with wooden poles on tarpaulins. One bagstack was built of 21226 new wpp sacks of 50 kg capacity, the other of 12662 re-used jute sacks of 90 kg capacity. The maize for both stacks was the current season's harvest. That for the wpp stack was of mixed commercial and communal origin, transferred from a nearby depot, while that in the jute stack was of commercial origin only, transferred into jute sacks from a bulk silo at Concession. The bagstacks were approximately 16 m long, 14 m wide and 5.75 m high to eaves. Fumigations were carried out by the GMB on day 13 after building started (wpp only), then on days 84, 170 and 246, by covering the stacks with airtight sheets and applying methyl bromide gas for two days. The trial was concluded prematurely after 7 months since the maize was required to offset shortages caused by the drought which developed in southern Africa in early 1992.

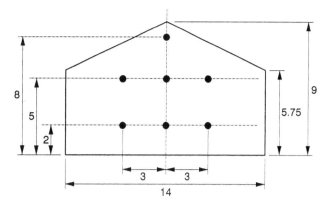

Fig. 2. Cross-section of bagstack showing sensor positions (dimensions in metres).

Each bagstack was instrumented identically during construction. Figures 1 and 2 show the position of thermocouples in the bagstacks. The majority were in seven vertical planes spaced evenly along the length of each stack, labelled A to G in Figure 1. Each plane contained seven thermocouples placed 2 m, 5 m and 8 m from the base of the stack, shown in Figure 2, forming three horizontal planes through the bagstack. Thermocouples were also placed between the tarpaulin and sack surfaces on the outside faces of the bagstacks and underneath the bottom layer. The thermocouples were connected to automatic dataloggers which recorded temperatures every 2 hours.

Changes in moisture content were recorded every 2 weeks using a total of 13 moisture sensors (Gough 1980) positioned next to thermocouples at the bases, on one face and at the centre of each vertical plane in each bagstack.

Nine microphones were also placed in each stack next to thermocouples on vertical planes C, D and E. At the time of this trial, NRI was developing an acoustic detection technique for insect infestations, and had produced a prototype system which was used here. It was subsequently confirmed in the laboratory that exposure to methyl bromide (or phosphine) does not affect the response of the microphones. The microphones were monitored and recorded every 2 weeks.

Maize samples taken from thermocouple positions at the start and finish of the trial were analysed for moisture content by the ISO routine method (ISO 1980) and assessed for insect infestation by the GMB. Note that all moisture content data are expressed on a wet weight basis. Meteorological data were taken from the nearest recording station.

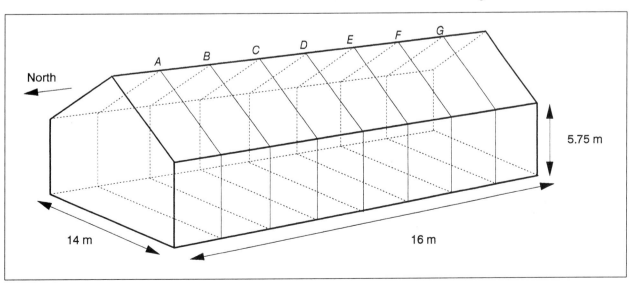

Fig. 1. Bagstack showing sensor planes (not to scale).

Results

Ambient conditions

Daily average maximum and minimum ambient temperatures for the trial period are shown in Figure 3. These were similar to 15-year averages for Harare.

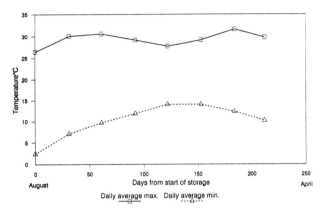

Fig. 3.   Daily average ambient temperatures.

The jute bagstack

Temperatures in the bagstack interior were remarkably consistent between vertical planes. Planes B and F are largely typical of the whole stack and are shown in Figures 4 and 5 (reduced to 24 hour averages) with results for equivalent wpp positions. On the faces of the jute bagstack temperatures were higher than the average ambient at the start of the trial and tended to rise, reaching 35°C in some cases. At the base the temperatures rose by approximately 5°C from 20°C.

The lower level temperature sensors (2 m from the bagstack base) showed temperatures rising at approximately 0.3°C/day from the start of the trial until the day 84 fumigation, after

which temperatures fell sharply for a period of 30–60 days. Heating then resumed. At some positions temperatures rose to around 40°C by day 130 and stayed approximately constant until the fumigation on day 170, after which they fell. At other positions, where cooling had been greater after the day 84 fumigation, the subsequent heating was slower and temperatures did not reach 40°C before the day 170 fumigation, after which they fell sharply.

The mid-level sensors (5 m from the base) showed much lower but steadily accelerating temperature rises at first and did not cool after the day 84 fumigation. All positions reached 40°C by around day 120, stayed approximately constant until the day 170 fumigation, then fell sharply, though some positions started to cool before the fumigation.

At the top of the bagstack (8 m from the base) temperatures rose from approximately 25°C to 30°C between days 0 and 160, falling slightly after the day 84 fumigation. All positions rose sharply to 40°C after day 160. A few positions cooled after the day 170 fumigation.

The critical moisture content for the maize, above which mould will grow, was determined as 14.75% at 27°C and approximately 14.25% at 37°C in earlier trials. Moisture content determinations for samples from the jute bagstack were in the range 11.4–13.0% at the start of the trial and 10.5–13.6% after. During the trial the moisture content sensors at the mid-level interior positions showed a rise to day 130 then a fall. The highest peak was at 14.2% moisture content, most others were between 13.5–13.7% moisture content. Positions at the base mostly stayed level though one rose to above the critical moisture content level.

The microphones detected insects at many of the monitored positions between the start of the trial and the fumigation at day 84. There were no detections 4 days after the day 84 fumigation but they gradually returned until the day 170 fumigation, after which they again stopped except at one position. Inspection of samples taken from the bagstack at the end of the trial revealed *Sitophilus* spp. at most positions and *Tribolium* spp. at all positions.

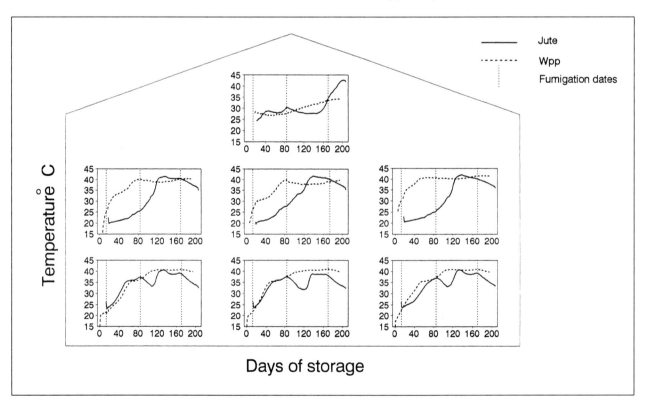

Fig. 4.   Plane B temperatures, jute and wpp bagstacks.

The wpp bagstack

Temperatures on the base and faces of the wpp stack were again slightly above ambient and rose during the trial in a very similar manner to those in the jute stack. In the interior of the stack the temperature records were again very consistent for each vertical plane. Results for planes B and F are shown in Figures 4 and 5 with those for the jute bagstack.

About half of the lower level temperature sensors (2 m from the bagstack base) showed rapid initial temperatures rises which were very similar to those at equivalent positions in the jute bagstack. These positions reached a steady limit of approximately 40°C at around day 100. At the other positions, towards the south and east of the stack, the heating rate was lower; temperatures were still rising at the end of the trial, reaching only 30°C in some cases. All positions showed steady but slow cooling or a reduction in heating rate after the day 170 fumigation, and this continued until the end of the trial. Only a few positions showed any cooling after the day 13 and 84 fumigations.

The mid-level sensors (5 m from the base) showed temperatures rising very rapidly to 40°C by day 100 then staying steady until the end of the trial. In many positions there was limited cooling after the day 84 fumigation but not after the day 170 fumigation. A few positions towards the east and around the mid-length of this level heated at a very much lower rate, not reaching 40°C by the end of the trial.

At the top of the bagstack, temperatures behaved in a similar way to those in the jute stack, rising from approximately 25°C to 30°C, though they showed no cooling after fumigation. The most significant difference was the complete absence of any temperature rise at the end of the trial.

Moisture content determinations were in the range 11.1–14.2% before the trial and 10.5–13.3% after. Sensors at the mid-level showed no rises during the trial in contrast to the slight rises in the jute bagstack. Base positions did show rises but only above the critical moisture for mould growth at one position.

Microphone results were very similar to those from the jute stack, showing insects present at all monitoring positions at various times throughout the trial, with detections generally stopping after fumigations, though there was one instance of a detection immediately after one fumigation. Inspection of samples taken from the bagstack at the end of the trial revealed a similar situation to the jute stack, i.e. widespread infestations of *Sitophilus* spp. and *Tribolium* spp.

Discussion

The similarities between the two stacks may be considered first. Heating to above ambient temperatures occurred over long periods throughout the interior of both stacks, especially at the lower and mid levels. At many positions the temperature reached 40°C then remained approximately steady. Temperature records were remarkably similar over similar levels in each stack, with the exception of the lower wpp level which showed two distinct heating rates. The higher of these rates, seen at half of the lower level positions, was very similar to heating rates at equivalent jute bagstack positions. Positions at the edges of the bagstacks heated but were cooler than the interiors. Moisture contents were below critical levels for mould growth for each stack throughout the trial with only minor exceptions.

As noted earlier, the main causes of heating inside a stack of grain are mould respiration, insect respiration, grain respiration and seasonal ambient temperature changes. As moisture contents were below critical limits in each stack, mould heating could not be regarded as playing a significant part in any heating that occurred. Aerobic grain respiration heating may also be ruled out for the same reasons given in the introduction. Temperatures inside each stack were without exception higher that those on the surfaces of the stacks and outside the bagstacks, ruling out seasonal ambient effects. This leaves heat from insect respiration as the likely cause of the temperature rises. The microphone results show that there

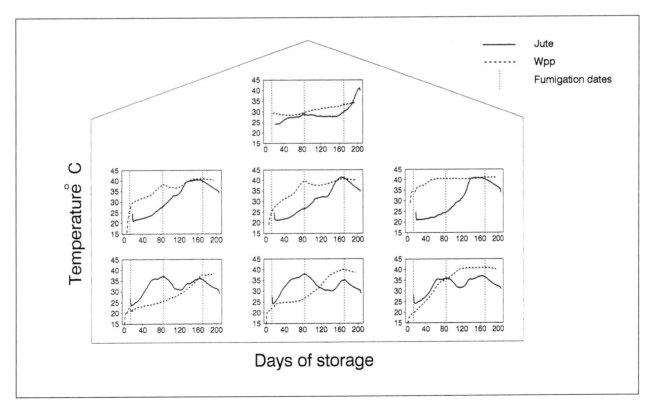

Fig. 5. Plane F temperatures, jute and wpp bagstacks.

were insects present throughout the trials in each stack, and sample analyses show widespread insect infestations at the end of the trials. Other evidence supporting this conclusion is the temperature limit of 40°C seen in each stack and the cooling which occurred after fumigation to varying degrees in each stack. The lethal temperatures for *Sitophilus* spp. and *Tribolium* spp., insects which were found in greatest number in the sample analyses, are reported by various researchers (Howe 1962a, Reddy 1950, Gonen 1977) as 35–40°C.

Howe (1962b) proposed a process by which insect heating in grain may progress. An insect infestation in the grain respires and produces heat, causing a rise in temperature which in turn increases the reproductive and metabolic rate of the insects. The infestation thus grows at an accelerating rate, releasing heat at an increasing rate and resulting in greater temperature rises. This autocatalytic process continues, with ever-increasing temperature rises, until the temperature becomes unfavourable for the insects and they either migrate or die. This hypothesis was substantiated with field and laboratory observations of insect heating, which showed rapid increases in insect populations accompanied by rapid temperature rises to a limit of approximately 40°C.

The heating observed in the present trials follows many of the patterns described by Howe (1962b), supporting the conclusion that insect respiration is a likely cause of heating in both bagstacks. The principal difference in behaviour was the relative uniformity of heating seen throughout the bagstack layers; Howe (1962b) observed nuclei of heating known as hotspots.

This leads to a consideration of the differences between the two bagstacks. The jute bagstack cooled after both of its fumigations, mainly in the lower levels, while the wpp stack remained almost unaltered, or cooled at a much lower rate. In about half of the lower wpp positions, temperatures rose at a lower rate than equivalent jute positions. Conversely, temperatures at mid-level positions in the jute stack rose, at least initially, at a very much lower rate than equivalent wpp positions. At the top level of the jute stack there was a sharp rise in temperature at the end of the trial that did not occur in the wpp stack.

Possible reasons for the differences in behaviour after fumigations between the two bagstacks, assuming that insect heating occurred, include the following:
- fumigations were more effective at controlling infestations in the jute stack than in the wpp stack;
- fumigations were effective in both stacks but the jute stack was able to cool while the wpp was not, i.e. the ability to cool could be dependent on the sack material or bagstack geometry.

The first reason is to be favoured since both stacks showed the ability to cool on different occasions, though higher rates were seen in the jute stack. Note, however, that the jute stack cooled significantly only at the lower level, for a relatively short time, and at the middle level there was cooling after the last fumigation but not after earlier ones. The conclusions that may be drawn are that fumigations were of variable effectiveness, were at best only partially successful but were more effective in the jute stack than in the wpp stack. This may indicate differences in the way in which the fumigant gas penetrates the sacks. Nevertheless, temperature differences caused by the thermal effects of the different sack materials or the methods of construction of the two stacks cannot be ruled out totally.

The lower rate of cooling in about half of the lower wpp positions occurred only towards one end of the stack. Differences in the amount of solar radiation falling on the different faces of the bagstack could not have caused this difference as both the wpp and jute bagstacks received equal amounts of

solar radiation but the effect was apparent only in the wpp bagstack. Differences in thermal behaviour of the jute and wpp sack materials or in the methods of stack construction are unlikely to be significant in this case as only half of the lower wpp positions heated at the lower rate, the other half matched the heating rate in the jute stack. Differences in the degree of insect infestation in the wpp sacks might have contributed to the temperature differences. The maize in the wpp stack was of mixed commercial and communal sources and was likely therefore to have had widely varying initial insect concentrations. Although fumigated at the start of the trial, the evidence from this work suggests that this may not have been as effective as supposed. The reversed situation at the mid-level, with the jute stack heating more slowly than the wpp stack, again supports differences in insect infestations.

The sudden rise in temperature at the top of the jute stack is difficult to explain. There was no equivalent rise at the top level of the wpp bagstack. Before the rise, the temperatures and temperature patterns at the top of the wpp and jute bagstacks were very similar; they were cooler than the lower positions, which would be expected by their closeness to the stack exteriors. The sudden rise in temperature in the jute bagstack may have been caused by a sudden increase in insect numbers, though the occurrence of the rise all along the stack length would not be expected, and the positions were close to where the fumigant gas was applied to the bagstack. The rise was too sharp to have been caused by heat rising from lower levels.

Conclusions

The following conclusions are drawn:
- heating occurred in each stack and may be attributable to insect respiration
- fumigations varied in effect and were at best only partially effective in stopping heating
- fumigations were consistently more effective in the jute stack than in the wpp stack, causing some positions to be cooler for longer periods in the jute stack than in the wpp stack
- differences in heating rate between the two stacks could not be positively attributed to differences in sack materials or the method of stack construction, although it is not possible to discount these factors.

Acknowledgments

The assistance of Mr Stanley Nosenga, Mr Noah Kutukwa and Mr Paul Zimbizi of Zimbabwe's Grain Marketing Board is gratefully acknowledged. We also acknowledge the help given by colleagues at NRI. This work was funded through the Natural Resources and Environment Department of the U.K.'s Overseas Development Administration.

References

Christensen, C.M. and Sauer, D.B. 1982. Microflora. In: Christensen, C.M., ed., Storage of cereal grains and their products. Minnesota, American Association of Cereal Chemists, Inc., 219–240.

Coveney, R.D. 1969. Sacks for the storage of food grains. Tropical Stored Product Information, 17, 3–19.

FAO (Food and Agriculture Organisation of the United Nations) 1992. Problems of natural fibre bag industries in some African countries. FAO, Committee on Commodity Problems, Intergovernmental Group on Jute, Kenaf and Allied Fibres, CCP: JU 92/7.

Gonen, M. 1977. Susceptibility of *Sitophilus granarius* and *S. oryzae* (Coleoptera: Curculionidae) to high temperatures after exposure to supra–optimal temperatures. Entomologia Experimentalis et Applicata, 21, 243–248.

Gough, M.C. 1980. Evaluation of a remote moisture sensor for bulk grain. Journal of Agricultural Engineering Research, 25, 339–344.

Gough, M.C. 1985. Physical changes in large-scale hermetic grain storage. Journal of Agricultural Engineering Research, 31, 55–65.

Howe, R.W. 1962a. Observations on the rate of growth and disruption of moulting in the larvae and pupae of *Tribolium castaneum* (Herbst)(Coleoptera: Tenebrionidae) at sub-optimal temperatures. Entomologia Experimentalis et Applicata, 5, 211–222.

Howe, R.W. 1962b. A study of the heating of stored grain caused by insects. Annals of Applied Biology, 50, 137–158.

ISO (International Organisation for Standardisation) 1980. International Standard ISO 6540–1980(E).

Paine, F.A., ed. 1991. The packaging user's handbook. Glasgow, Blackie, 596 p.

Reddy, D.B. 1950. Ecological studies of the rice weevil. Journal of Economic Entomology, 43, 203–206.

Tyler, P.S. 1992. Heat and discolouration of bagged maize. World Grain, September, 14–16.

Quality enhancement of stored grain by improved design and management of aeration

J-C. Lasseran*, G. Niquet* and F. Fleurat-Lessard†

Abstract

Aeration is the right preservation technique for all grains in order to control all quality criteria and reduce running costs in storage facilities if associated with a temperature monitoring system. Very often warehouses are poorly equipped and/or badly managed with regard to aeration because of a lack of practical knowledge leading to preconceived ideas. The air feeding network (pipes and ducts) has to be properly designed to minimise pressure drops for an appropriate air flow rate (5–15 m^3/hour/m^3). The fan (centrifugal in most cases) must be selected to satisfactorily work on any species of grain (rape, barley, wheat, rice, maize, sunflower, soybean etc). The working mode of the fan (blowing or sucking) is of prime importance; a rise in static pressure of 850 Pa will generate a 1°C increase in temperature. Usual static pressures range from 2–8 kPa. So, apart from bins or stores of small height, it seems technically wiser to have a suction fan aspirating the fresh air at the top of the bin and throwing out the used air at ground level in order to avoid compression and obtain better benefit from cooler periods. Grain has to be aerated several nights in succession during 4–10 favourable hours per day, i.e. on both sides of the daily minimum, the fan being controlled by a temperature regulator. Since this procedure has been questioned experiments were carried out during the 3 last years, in close cooperation with the INRA Research Centre of Bordeaux, in a medium size metallic bin (500 t, height 16 m). They have shown that under temperature climates 2 or 3 aeration periods, from summer to autumn, are sufficient to cool the grain from 30–5°C or less. Afterwards, grain temperature remains stable during winter, and slowly rises in the spring. Tests applied to grain stored for one year showed an excellent hygienic condition, no marked local rewetting even when aerating with a high relative humidity, no chemical change, a slight increase in the technological criteria of quality, and last but not least, a total disinfection in the case of an initial pest infestation: this means chemicals are not needed, a great advantage for store operators (no insecticide expenditure), millers and consumers (no risk of residues). These observations have been confirmed by checks made on wheat stored in large round concrete bins (diameter 9 m, height 40 m, capacity 2000 t).

Introduction

The optimisation of an aeration system rests in the first place on well designed installations (Lasseran 1988). The air feeding network (pipes and ducts) has to be properly calculated to minimise pressure drops for an appropriate airflow rate (5–15 m^3/hour/m^3). The fan (centrifugal in most cases)

must be selected to satisfactorily work on any species of grain. The choice of ducts is also an important question (Lasseran 1993).

The objective of this paper is to take another look at regular aeration (no chilling unit involved) in order to preserve all quality criteria of grain by improving the design and use of the technique. Furthermore, so as to take into account international trade regulations on chemical residues, and meet consumer requirements for sound and healthy foods, the paper will stress the limitation of insecticides, which is a worldwide concern (Cuperus et al. 1993).

Grains harvested or dried at commercial moisture levels (14–15 % w.b.) and at temperatures as high as 20–35°C cannot be preserved during a long-term storage period if they are not cooled to at least below 12°C. Otherwise, they are liable to be damaged and/or tainted by (1) storage microflora (mainly moulds) and possibly mycotoxins, (2) insects and mites, the most dangerous being the granary weevil, *Sitophilus granarius*, (3) enzymatic reactions leading to dry matter losses, decrease in viability, lowering in breadmaking qualities, etc.

Recent findings (Lasseran et al. 1990) on cold survival of *Sitophilus granarius* at 7°C have shown that all stages (eggs, larvae I, larvae II, prepupae, pupae and adults) are killed after 2 months exposure; larvae III and larvae IV are more cold-resistant, as indicated in Figure 1, but die after a 3-month exposure. As these experiments were carried out without acclimation to cold, it seems wiser, in practice, to work from a 4–5°C lethal level.

With regard to aeration, a high relative humidity of the cooling air is not a major inconvenience: the risk of locally rewetting the grain is very limited if the difference in temperature between grain and air is greater than 5°C. Thus, aeration at night or early in the morning, when the ambient air temperature is lower, is more efficient and cheaper (electricity night rate).

When a fan's static pressure exceeds 2 kPa (e.g. wheat aerated in a bin higher than 15 m, with an airflow rate of 10 m^3/hour/m^3), air temperature is increased by 2°C, for 3 kPa by 3°C, etc. When blowing air through the grain mass, the fan compresses and heats the cooling air: this disadvantage is suppressed when air is sucked from top to bottom of the bin by a fan at ground level. This method of operation permits the maximum cooling potential of the ambient air, which is important in subtropical or mediterranean type climates. Contrary to non-experimentally based assertions, the small perforations in the sheet-steel of air ducts do not get blocked up by grain particles when aspirating the air.

Cooling of the grain can be divided into 2 or 3 aeration cycles, from harvest season in summer up to the end of autumn. The goal is to rapidly reach the 12°C level in the grain mass to prevent insect reproduction, and then to continue to cool to near 8–10°C, possibly 4°C, to kill any living pests. Each cooling cycle is also divided in several steps cumulating the coldest hours of various days in succession. With an airflow rate of 10 m^3/hour/m^3 of grain, the total cumulative time of aeration is about 100–120 hours, e.g. 12 to 15 nights at the rate of 8 'favourable hours'/day (Fig. 2).

* ITCF, Department of Agro-Industrial Quality, F-91720 Boigneville, France
† INRA, Laboratory of Stored-Product Entomology, F-33883, Villenave D'Ornon, France.

Materials and Methods

A 600 t metallic storage bin (Fig. 3), 16 m high, horizontal cross-section 6.42 × 6.42 m, filled with freshly harvested wheat, equipped with a 6.5 kW centrifugal fan (airflow rate: 7 m³/hour/m³) controlled by a thermostat for automatic running when ambient temperature is lower than the setting point, has been used for a long-term storage experiment. A fan sucked in fresh air at the top of bin and emitted the used air at ground level, since the static pressure was 3.3 kPa. The storage time was 300 days (mid-August 1992 to mid-June 1993).

A bin temperature monitoring system (5 cables, 6 sensors in each), a computer and a printer were used to measure and record grain temperatures throughout the storage period. A set of 12 small grain cages (0.5 dm³) infested with all stages of *S. granarius* (25 adults and 100 hidden forms) were fixed on two cables, in the centre and a corner of the bin.

After the first cooling cycle is achieved (thermostat setting point 18°C), the threshold is adjusted on a new setting point (12°C) for the second cycle, and finally 2°C for the third cycle (Fig. 4). In practice such a low setting is not needed, 5°C is sufficient, but the aim was (1) to verify the influence of cold on insect mortality or survival, (2) study grain natural reheating in the spingtime.

Results and Discussion

Grain temperature during the storage period (Fig. 5)

Grain is rapidly cooled to 17°C in summer after harvest due to the important diurnal-nocturnal difference in temperature (10–12°C on average, Fig. 2). Then, to 8.5°C in early autumn,

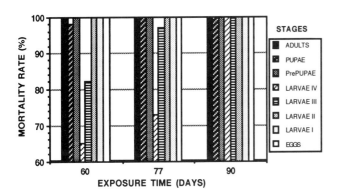

Fig. 1. Mortality rate of *Sitophilus granarius* at 7°C (without acclimation) as a function of exposure time.

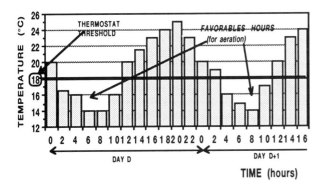

Fig. 2. Typical summer weather (Orléans, France) and 'favourable hours' for aeration

finally to –2°C just before winter. After a 300-day storage time, grain temperature rises slightly to +2°C in the centre of the bin (3.2 m from walls), and more markedly to 14°C near the edge (0.6 m from walls).

Grain moisture content

No significant overall change was observed: 12.4%wb at bin loading, 12.8% at emptying. A light local rewetting was noted on the grain surface only, with a maximum of 16.8% at 0 to 0.3 m depth, then a fast decreasing gradient was seen to 1.50 m.

Microflora (bacteria and moulds) (Table 1)

A decrease in the total number of bacteria and moulds was noted: this disinfection is an important advantage for the milling industry since the International Commission on Microbiological Specifications for Foods recommends the limit of wholesomeness at 10^4 mould propagules per gram. Furthermore, no mycotoxins were detected in the samples taken when emptying the bin.

Table 1. Evolution of microflora

Microflora	Bacteria Nb/g	Moulds Nb/g
Loading	1.2 106	12.9 103
Unloading	0.3 106	5.25 103

Insects (*Sitophilus granarius*, granary weevil)

The rate of mortality of the insects confined in the cages was 100% at the centre of the bin and 98.8% at 0.60 m from the walls. Surprisingly, the total number of dead insects in the cages after the experiment ranged from 25 to 60 individuals, though they were initially 125, and potentially 250 after a 60-day storage period before the temperature dropped below 10°C: clearly, the first cooling cycle which brought the grain temperature to 17°C in two weeks markedly disrupted the hatching of the larvae.

No insects, dead or alive, were found in the rest of the grain. This suggests that moderate cold as it was the case at the beginning of the trial is sufficient enough to prevent insects from invading sound grain, and that aeration is more effective than chemical insecticides in controlling insects.

Breadmaking quality of wheat (Table 2)

No significant change was observed between bin loading and unloading with regard to various conventional tests: Zeleny, Chopin alveogram, French breadmaking mark, Hagberg falling number.

Table 2. Evolution of breadmaking quality

Breadmaking quality of wheat	Zeleny (mL)	Hagberg (seconds)	French breadmaking mark/300
Loading (17/08/92)	49	356	178
Unloading (14/06/93)	45	441	179

Cost of aeration

The specific electricity consumption measured for each aeration cycle is 2.5 kWh/t, leading to the practical ratio of 5 kWh/t for 2 cycles. As night (off-peak) electricity is cheaper,

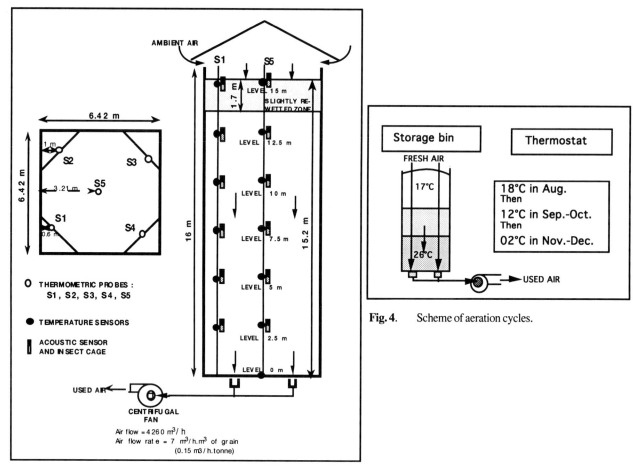

Fig. 3.   Diagram of the experimental bin used for aeration trials

Fig. 4.   Scheme of aeration cycles.

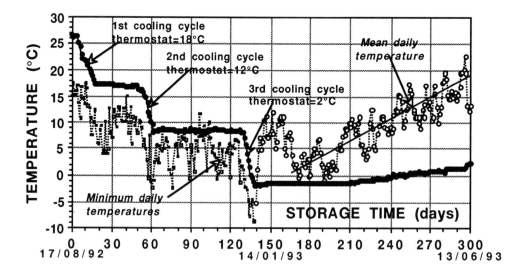

Fig. 5.   Aeration of wheat in the experimental bin. Grain temperature in the centre of the bin. Minimum (period of cooling cycles) and mean (reheating period) daily temperatures

the operating cost of aeration in the case of the trial was 0.1 US$/t; that is to say less than with insecticides for an equal duration (300 days): 0.3 to 0.7 US$/t.

Conclusions

In temperate climates, aeration is the best technique to preserve grain whatever the quality criterion may be.

It has to be undertaken when the ambient temperature is close to the daily minimum, e.g. mostly in the night, and preferably with the fan sucking the air through the grain if the static pressure exceeds 2 kPa. There is no need for a chilling unit. A simple and properly adjusted thermostat is sufficient to provide automatic control and permit grain to cool from 30°C or more, to 4°C in 3 aeration cycles within 100 days. When grain is initially sound, 2 aeration cycles are sufficient to achieve 8–10°C and ensure safe storage for at least one year. These prescriptions have been confirmed by observations made on wheat stored in large round concrete bins (diameter 9 m, height 40 m, capacity 2000 t).

A properly designed and managed aeration facility decreases the operating cost of storage because (1) insecticide treatments are not needed and (2) the electricity tariff is cheaper at night, when the fans are used.

The slight rewetting of the grain (on top surface when sucking, above the ducts when blowing) is not a major drawback: a small 'turning over' of the grain in the storage bin

eliminates the damp layers. Well cooled grain, naturally reheats very slowly in the springtime.

With regard to sanitation, mould growth is inhibited, pests are totally controlled without chemicals,which removes the risk of residues when selling, processing and consuming the grain.

References

Cuperus, C. Noyes, R.T. Fargo, W.S. et al., 1993. Reducing pesticide use in wheat postharvest systems. Cereal Food World, 38 (4), 199–203.

Lasseran, J.C. 1988. The preservation of grains and the measurement of grain temperature in storage bins (silo-thermometry). In: Multon, J.L., ed., Preservation and storage of grains and seeds. Paris, Lavasier.

Lasseran, J.C. 1993. Improvement of aeration duct system an appropriate management to control the sanitary quality of grain (in Spanish). FAO Study, Viale delle Terme di Caracalla, 00100 Rome, Italy.

Lasseran, J.C. and Fleurat-Lessard, F. 1990. Aeration of grain with ambient or artificially cooled air: a technique to control weavils in temperate climates. In: Fleurat-Lessard, F. and Ducom, P., ed., Proceedings of the 5th International Working Conference on Stored-product Protection, Bordeaux, France, September 1990.

Chilled aeration and storage of U.S. crops — a review

D.E. Maier*

Abstract

The contamination of grain with insects, insect fragments, fungi and mycotoxins is one of the major concerns of the U.S. grain industry. In the past, pesticides have been used to achieve therapeutic control. However, insects are developing resistance, consumers are concerned about toxins and pesticide residues, many pesticides are being withdrawn from use because of high costs to register or develop new ones, and environmental considerations have made the production of certain pesticides illegal. Therefore, alternative, preferably non-chemical, methods of effective pest control are needed in the postharvest grain storage environment. One alternative, aeration cooling, can be used to lower the carrying capacity of the stored-grain ecosystem for insects. However, under certain climatic conditions and in various geographic locations, ambient aeration cannot completely inhibit insect activity and preserve grain quality. When grain temperatures cannot be sufficiently reduced, chilled aeration is a technically viable and economically feasible grain-conditioning alternative.

Grain chilling is a non-traditional, non-chemical treatment technology of bulk stored commodities to prevent spoilage and stored-product insects. Grain is cooled using a mobile refrigeration system that controls both the temperature and relative humidity of the aeration air. Grain chilling was first applied in the United States in the late 1950s, and subsequently investigated in the southwest and midwest for the storage of high-moisture grains. However, the technology was never commercialised. In 1988, researchers at Michigan State University began the first serious evaluation of chilled grain aeration in more than 20 years. The success of those tests led to the development and field testing of a new grain chiller developed by Purdue University in cooperation with AAG Manufacturing (Milwaukee, WI), which has resulted in the commercialisation and sale of the first U.S.-built grain chillers.

This paper reviews the trials that established the technical feasibility and biological desirability of chilled aeration and storage of grains in the U.S. It also presents the development and optimisation of the first generation U.S. grain chiller; field testing and computer simulation were used to optimise the system. The current status of utilisation of grain chilling and conditioning technology in the U.S. grain industry will be presented.

Introduction

A stored grain bulk is a man-made ecological system in which deterioration is an on-going process that results from interactions among physical, chemical and biological variables.

Damage by insects, fungi, heating and sprouting causes millions of dollars in economic losses to U.S. grain farmers, handlers and processors every year. Although quality of harvested grain can never be improved with storage time, the rate of deterioration can be slowed with an integrated postharvest system management approach that combines engineering, biological, and economic principles. Sanitation of harvesting, handling, and processing equipment and storage structures is essential. Drying grain reduces its spoilage potential, and cleaning it before binning removes fines, filth, damaged kernels, and foreign material. Proper aeration practices maintain low grain temperatures, and the application of a residual pesticide may reduce insect damage.

Protecting grain with a pesticide is recommended for grains stored in the southern and south-eastern U.S. for any length of time, while it is recommended only for long-term (over one-season) storage in the midwest. As grain is marketed and moves through various facilities, the identity of a lot is lost and additional pesticide treatments occur. Thus, the number of pesticide applications can increase, and potentially toxic residues may accumulate in the final food and feed products. Recent reports from the mid-western United States indicate that the lesser grain borer, *Rhyzopertha dominica* (F.) may be developing resistance to chlorphyrifos-methyl (Reldan) (Beeman and Wright 1990; Zettler and Cuperus 1990). As a result, this pest has been removed from the label. The label for pirimiphos-methyl (Actellic) now specifies that it will only suppress, not control *R. dominica* (F.) if it is applied at the maximum recommended rate. As additional insects develop resistance to currently available pesticides, greater emphasis must be placed on alternative methods for protecting our food supply.

Alternative pest control techniques must focus on limiting insect reproduction and growth by controlling temperature and moisture content in the grain bulk, and by utilising multiple tools such as insect sampling with pheromone and pitfall traps, risk-benefit analysis of control strategies, plant resistance, and biological methods (Hagstrum and Flinn 1992). Maintaining low temperature- and moisture-levels in bulk-stored grain was identified in a major study on 'Enhancing the quality of U.S. grain for international trade' (U.S. Congress 1989) as the main way to preserve grain quality, and to prevent damage from moulds and insects. One technology that has been successfully utilised to preserve grain quality during storage is grain chilling. It permits the short- to long-term storage management of grain independent of the ambient conditions. The chilled aeration of grain has been applied commercially in over 50 countries during the past 30 years (H. Brunner, pers. comm., 1990), but only recently in the United States. It is estimated that annually over 20 million tonnes of grain are cooled with grain chilling systems. Grain chilling is accepted as a grain conditioning technology in much of Western Europe, while currently most units appear to be sold in Southeast Asia. In the 1960s grain chillers were primarily used as a means of preserving high moisture grain. Today, the primary application for grain

---

* Department of Agricultural Engineering, Purdue University, West Lafayette, Indiana, 47907–1146, USA.

chilling is in preserving the keeping quality of dry grain through chilled aeration and non-chemical pest management.

Commonly, grain is stored in bins, silos and flat storages that either do not allow cooling because of a lack of aeration systems, or allow cooling to within several degrees of the minimum ambient temperature using conventional aeration systems. In contrast, grain chilling is defined as the cooling of grain independent of the minimum ambient temperature by using a refrigerated air system. In a grain chilling system, ambient air is ducted over a bank of refrigeration coils in order to decrease the air temperature (Fig. 1). Because dry grain will absorb moisture from wet air, the air is reheated a few degrees to reduce the relative humidity to 60-75%. The amount of reheating and the final air temperature are adjustable by the operator to achieve the desired aeration conditions. Once the grain has been initially cooled, only occasional rechilling for short time periods is required to maintain chilled storage conditions due to the insulating properties of the grain itself. The ability to control both the bin inlet temperature and relative humidity is desirable for conditioning stored grain and maximising its end-use processing value.

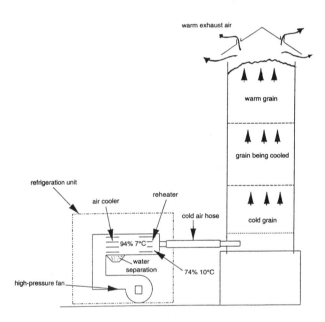

warm exhaust air

warm grain

grain being cooled

refrigeration unit

air cooler   reheater

cold air hose   cold grain

94% 7°C

water separation

74% 10°C

high-pressure fan

Fig. 1. Schematic of the grain chilling process. The binned grain is cooled independent of the minimum ambient temperature by using air conditioned to operator-selected temperature and relative humidity levels.

Historical Development of Grain Chilling

The change-over from traditional harvesting methods to high-volume combining of grain crops after World War II caused a rapid expansion in the use of heated-air drying to preserve grains (Hall 1980). However, in many installations the limited capacity of grain dryers soon created bottlenecks (Saul and Lind 1958). In Europe the search for an alternative preservation method led to the application of refrigerated aeration with saturated air (Burrell 1974). Lowering the temperature of wet grain to less than 10°C within 24 hours of harvest allowed for the 2–3 week risk-free storage of small grains up to 20% moisture, and of maize up to 35% moisture (Heidt 1963).

According to Reimann (1927) the idea of cooling grain artificially was first proposed by a German engineer in 1917. The concept seemed impractical and expensive at the time. According to Burrell (1974), the use of a refrigeration system to dry grain from 20% moisture to 16% was proposed in France by Leroy in 1950. As early as 1958 a cold-air drying system was sold in Germany (Escher-Wyss 1960). These closed-cycle batch systems consisted of a heat pump to dehumidify and cool the exhaust air from the top of the grain pile. The cold, dry air was forced back into the bottom of the bulk. In 1961 the Escher–Wyss company of Lindau, Germany began with the production of commercial grain chillers (Heidt 1963), which consisted of a cold-air fan, evaporator coil, compressor, condenser and cooling fan. The chillers had cooling capacities of 50 t of grain/day.

Munday (1965) described commercial grain chillers manufactured in England primarily designed for on-farm use. By 1970 grain chillers incorporated automatic cold air regulation and a reheater after the evaporator to control the relative humidity of the chilling air automatically (Sulzer-Escher Wyss 1970). Reheating is generally accomplished with waste heat from the condenser. In the newly developed American chiller, however, a separate glycol loop returns heat to the chilled air after the evaporator (Maier et al. 1993a). This approach reduces the cooling load on the evaporator and minimises the use of refrigerant. Most chiller manufacturers use motorised dampers to control the airflow across the evaporator, and thus provide a constant cold-air temperature into the grain bin. Only recently has a chiller been introduced with a variable-frequency blower drive to optimise airflow performance (IKZ 1993).

There are at least seven major commercial manufacturers of grain chillers worldwide: 1) Series 'Goldsaat' by Fritz Döring Co. (Prüm, Germany); 2) Series 'Grain Cooler' by PM-Luft Co. (Kvänum, Sweden); 3) Series 'Granifrigor' by Sulzer-Escher Wyss Co. (Lindau, Germany) (Fig. 2); 4) Series 'DUK' by Uniblock Zanotti Co. (Suzzara, Italy); 5) Series

Table 1. Summary of the main design parameters of the largest commercial grain chillers of each of four manufacturers (Fritz Döring, Prüm, Germany; Sulzer-Escher Wyss, Lindau, Germany; PM-Luft, Kvänum, Sweden; Uniblock Zanotti, Suzzara, Italy; AAG Manufacturing, Milwaukee, Wisconsin, USA.).

	Goldsaat	Granifrigor	Grain	DUK	Chill'd Aire
	GK 480 NDI	KK400	Cooler 8000	100	GTC 3000
Chilling capacity [t/day]					
	400	335	350	350	350
Airflow at 2000 Pa [m³/hour]					
	16750	16300	16250	17000	16900
Evaporator capacity [kW]					
	128	107	107	110	115
Connected load [kW]					
	N/A	54	55	55	55

'RM Grain Cooling Units' by MacBea Co. (Parkdale, Australia); 6) Series 'LK Grain Coolers' by IKZ (Zwickau, Germany); and 7) Series 'Chill'd Aire' by AAG Manufacturing (Milwaukee, WI). The largest chilling units of the leading manufacturers are similar in performance and capacity (Table 1). The compressor capacities range from 107 to 130 kW. The fans provide airflow rates of about 16500 m³/hour at 2000 Pa of static counter pressure. The average grain chilling capacity is about 350 t/day.

Fig. 2. Commercial grain chiller Granifrigor KK400 manufactured by Sulzer-Escher Wyss, Lindau, Germany.

In his comprehensive state-of-the-art review, Burrell (1982) summarised one of the main uses of grain chilling as the protection of grain from insect infestation by storing cooler grain. He concluded that 'refrigeration has so far had little worldwide impact.' Although grain chilling may so far have had little impact on the grain industry, Maier (1992) has summarised the wealth of application experiences that has been reported from around the world, i.e., Germany (Heidt 1963; Heidt and Bolling 1965; Brunner 1985), Great Britain (Burrell 1964; Sullivan and Sebestyen 1973; McLean and Barlett 1991), Australia (Sutherland et al. 1970; Hunter and Taylor 1980; Sutherland 1986), Israel (Navarro et al. 1973; Donahaye 1974; Ben-Efrain et al. 1985), Italy (Baldo and Brunner 1983; Finassi 1987; Bissaro and Bertoni 1989 et al.), France (Lasseran and Fleurat-Lessart 1988; Berhaut et al. 1988), Spain (Clapers 1970; Torres 1983, 1985; Rius 1987), Latin America (Malaga 1973; Cunille 1988; Moreira et al. 1993), and Southeast Asia (Tuckett 1982; Brunner 1985; Chek 1989; Maier et al. 1993c). The application experiences in the United States are the main focus of this review.

Chilling of High-Moisture U.S. Grains

Sorghum chilling in Texas in 1959–68

In the United States conditioned-air storage to maintain grain quality found its first application in 1959 at a commercial elevator in Texas (McCune 1962). The objectives of using conditioned-air were (1) to reduce moisture losses due to ambient aeration in Texas, (2) to control insect infestation by lowering grain temperatures, and (3) to store grain at higher moisture contents for feeding purposes. The initial field test employed a 26.4 kW evaporator coil. The results showed that conditioned-

air grain storage had good potential, and that the grain temperatures could be maintained below 10°C. The field work was followed by detailed laboratory tests to determine the design factors, such as equipment size, operating parameters, appropriate moisture contents, and allowable storage times (McCune et al. 1963). In later work, Person et al. (1966) and Sorensen et al. (1967) described the thermodynamic considerations necessary to design chilled grain storage systems. Haile and Sorenson (1968) determined the effect of respiration heat of sorghum on the cooling load requirements for conditioned-air storage systems in Texas.

Maize chilling in Illinois 1965–66

Shove (1966) proposed a process called 'dehydrofrigidation' to dry grain at low temperatures in an insulated dome structure using a refrigeration system. The first phase consisted of cooling the shelled maize from its harvest temperature to –1 to 10°C within 24 hours. Thirty-five kW of refrigeration were considered sufficient to cool a daily load of 102–122 t in the corn belt of the United States. The experimental test encountered several problems, including moulding of the wet grain in the top layers, and inadequate controls to properly condition the air. However, refinement of the process was not pursued further.

Maize chilling in Indiana 1966–70

Ambient and refrigerated aeration in the United States was further investigated by Tuite et al. (1970). Maize at moisture contents of 18–22% was evaluated with respect to changes in moisture content, temperature, mould population and fat acidity. Tests were conducted between 1965 and 1969 with field-harvested shelled maize in 6.5 t insulated bins. The treatments included continuous aeration, intermittent aeration, and refrigerated aeration with and without recirculation at airflow rates of 0.5–1.0 m³/minute. Continuous ambient aeration was sufficient if the maximum initial maize moisture content was not higher than 21% and was reduced below 16% by April. The refrigerated aeration system had several design shortcomings, including the lack of automatic de-icing of the evaporator coil. However, the system did cool the maize faster in the fall, maintained the temperatures closer to the desired level, and maintained lower temperatures for 4–6 weeks longer in the spring than the ambient aeration system.

Maize chilling in Nebraska 1967–69

Thompson et al. (1969) compared the performance of temperature control systems for the storage of high-moisture maize at 23.3–24.8%. The main objective of cooling was to maintain quality until drying, feeding, or processing. One bin was equipped with a standard ambient aeration system, and the second bin with a 5.3 kW refrigeration system. Both systems delivered airflow rates of 0.4 m³/minute. Since the tests were conducted between November and April, no significant performance differences between the systems were observed. It was concluded that under midwestern conditions a system with continuous ambient aeration performs better than a more complicated mechanical refrigeration system when storing high-moisture maize. However, the refrigerated aeration system controlled the grain temperature and moisture content in the bulk better than the conventional system.

The temporary storage of high-moisture grain in the United States continued to be of interest to researchers over the next 20 years (Thompson 1972; Stewart 1975; Felkel 1978; Friday et al. 1989; Strohshine and Yang 1990), but, no additional work on chilling high-moisture grains has been reported.

Chilling of Low-Moisture U.S. Grains

Maize and wheat chilling in Michigan in 1988–91

Maier (1992) utilised field tests and simulation techniques for an in-depth analysis of the chilled aeration and storage of low-moisture maize and wheat in the midwestern U.S. The simulation model of the chilled aeration and storage system consists of three parts: (1) the chilled aeration model, (2) the grain chiller model, and (3) the chilled grain storage model. The single-stage refrigeration cycle of the commercial grain chiller used in the study is modelled with steady-state external energy balances across the condenser, reheater and evaporator. The model developed accurately predicts performance in terms of the flow rate, temperature and relative humidity of the air into and out of the chiller, the evaporating and condensing refrigerant temperatures and capacities of the chiller refrigeration cycle, and the electrical power consumption under transient conditions. The chiller model was successfully integrated into a time-dependent grain aeration and storage model for upright, steel silos that simulates the heat and mass transfer with two coupled differential equations in the grain bulk during the aerated period, and the heat transfer with a two-dimensional conduction equation during the non-aerated period (Maier and Bakker-Arkema, 1994). The effect of the aeration and storage conditions on the deterioration of stored grain due to respiration was evaluated using the concept of dry matter loss. Experimental data collected at a commercial elevator for a three-year period were used to validate the systems model (Bakker-Arkema et al. 1989; Maier et al. 1989).

Fall cool-down of maize

The simulated cool-down of a 597-t bin of maize is compared in Figure 3 for the 1988 through 1990 seasons. The temperature profiles and the lengths of the cool-down cycles are obviously influenced by the year-to-year variation of the ambient conditions. In October of 1988 it required 195 hours to chill the maize from 17°C to below 7°C, compared with 180 hours in 1990, and 162 hours in 1989. Associated with longer cooling times are higher energy costs. In addition, managing the cool-down is critical since the bins are filled in rapid sequence during the fall harvest, and sufficient cooling capacity has to be available to keep up with the harvest rate. The predicted power consumption is 2296 kWh in 1988, 1842 kWh in 1989, and 2157 kWh in 1990, resulting in efficiencies of 0.38 kWh/t/°C, 0.31 kWh/t/°C, and 0.36 kWh/t/°C, respectively, during the three seasons.

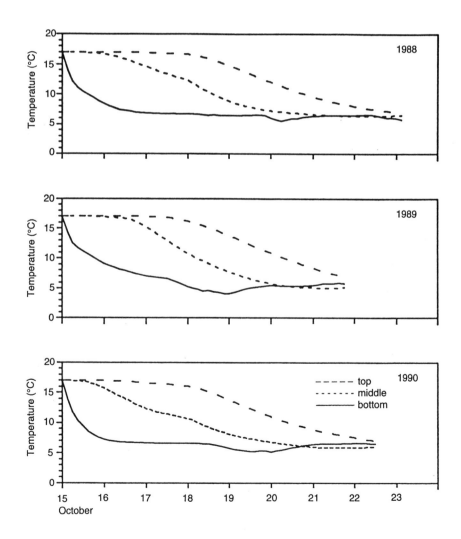

Fig. 3.   Simulated chilled aeration of a 597-t bin of maize in mid-October for the 1988, 1989 and 1990 seasons in Michigan. Initial maize temperature 17°C and moisture content 16.5%. Top: – – –; Middle - - -; Bottom ——.

Summer cool-down of wheat

The simulated cool-down of a 579-t bin of wheat is compared in Figure 4 for the 1988 through 1991 seasons. Given the same grain conditions, bin dimensions and chiller settings, the cooling times vary by as much as 3 days. The fastest and slowest cool-downs are predicted in back-to-back years. In July of 1988 it requires 207 hours to chill the wheat from 30°C to below 15°C, compared with 270 hours in 1989, 225 hours in 1990, and 258 hours in 1991. The predicted power consumption is 2632 kWh in 1988, 3517 kWh in 1989, 2854 kWh in 1990, and 3335 kWh in 1991. This results in cooling efficiencies of 0.30 kWh/t/°C, 0.40 kWh/t/°C, 0.33 kWh/t/°C, and 0.38 kWh/t/°C during the four seasons, respectively.

The simulated performance of the grain chiller is shown in Figure 5 for the 1989 cool-down. Given the cyclical behaviour of the ambient temperature and relative humidity, the chiller opens and closes the air throttle to maintain the bin inlet air conditions at a constant 14°C (13°C at the chiller plus 1°C

heat gain in the duct) and 67% r.h. The variable airflow is a unique characteristic of a chilled aeration system; ambient aeration maintains a constant airflow but has variable air inlet conditions.

Chilled versus ambient versus no aeration

In the northern United States ambient temperatures drop low enough in the late fall (and winter) to reduce temperatures in stored grain to safe levels. Even in the southern United States ambient temperatures usually decrease sufficiently in the late fall or early winter to achieve stored grain temperatures below 10–15°C with conventional ambient aeration systems. In other parts of the country, such as the Great Plains, grain storage is frequently practiced in bins without aeration systems entirely. There the crop is stored after the harvest and left to cool as a function of the weather conditions. Thus, for chilled aeration to be adopted as an alternative technology, it must be biologically desirable, and have additional economic and environmental benefits. The following summary of results

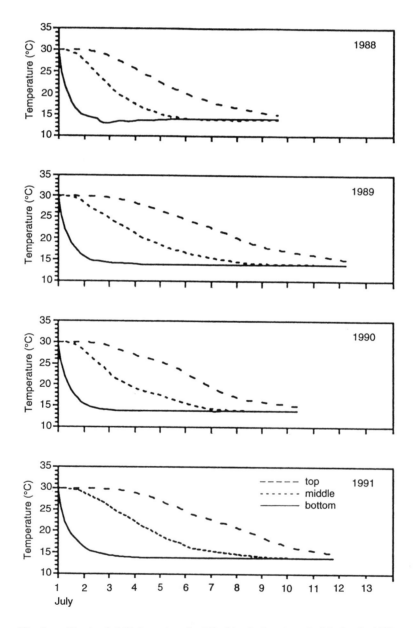

Fig. 4.   Simulated chilled aeration of a 579-t bin of wheat in early July for the 1988, 1989, 1990 and 1991 seasons in Michigan. Initial wheat temperature 30°C and moisture content 14%. Top: – – –; Middle - - -; Bottom ——

demonstrates its benefits under Michigan conditions, and makes the technology at least as desirable in the warmer climates of the southern United States.

Maize

Table 2 summarises the comparison of chilled, continuous ambient and no aeration in a bin containing 597 t of maize stored between 15 October 1989 and 14 October 1990 in mid-Michigan. The initial grain temperature is 17°C, and the initial moisture content is 16.5%. The minimum bulk temperature in the non-aerated maize reaches 7.5°C after about 5 months of storage; after 11 months the bulk temperature reaches 17°C again. The ambient aeration of the maize to 7°C at an airflow rate of 0.1 m³/minute/t takes 4 times longer than cooling the bulk to 10°C. The reason for this significant difference is a period of warmer weather during which the fan rewarmed the grain before it could reach 7°C. Chilling the grain to 7°C and 10°C is accomplished within about 1 week in both cases.

The average dry matter loss of the maize after 12-month storage is 37% less for chilled aeration to 7°C than for ambient cooling, and 68% better than for no aeration. Chilled and ambient aeration to 10°C are equally successful in minimising dry matter loss. The critical dry matter loss limit of 0.5%, which indicates a loss in grade, is reached after 10 months of non-aerated storage. Without close manual supervision, or an automatic aeration controller to selectively cool the bin, the temperature and moisture content may cycle significantly during the cool-down with ambient air. The effect of cycling appears to be more damaging to the long-term grain quality than simply storing at a 3°C higher grain temperature.

Obviously, the costs of chilled versus ambient aeration have to be taken into account to make an economic recommendation of the preferred system. The operating costs for one-time chilling of a 597-t maize bin under mid-Michigan conditions immediately after harvest to 10°C are about double per tonne compared with aerating with ambient air using a comparable 5.5 kW fan. Chilled and ambient aeration to 7°C require the same amount of energy. The comparison also indicates that one-time chilling to a low of 7°C in the fall may be sufficient with respect to the quality preservation of maize during the 12-month storage under Michigan conditions.

Wheat

Table 3 compares the chilled, continuous ambient and no aeration in a bin containing 579 t of wheat stored between 1 July 1989 and 30 June 1990 in mid-Michigan. The initial grain temperature is 30°C, and the initial moisture content is 14%. The lowest bulk temperature in the non-aerated wheat is 13°C after about 10 months of storage. Continuous ambient aeration to 10°C at 0.1 m³/minute/t takes 1.5 times longer than cooling the bulk to 15°C. Chilling the wheat to either 10 or 15°C is accomplished within about 1 week in each case. The average dry matter loss of the wheat after 12-month storage is 63–67% lower for chilled aeration than for ambient cooling, and 68–71% better than for no aeration. The operating costs for one-time chilling after summer harvest to 10°C are about 31% higher compared with chilling to 15°C. Aerating continuously with ambient air can be 5–7 times more expensive than chilling once.

Rice chilling in Louisiana and Michigan in 1991

In a field test utilising a German-built chiller, 86 t of rice were chilled from 30°C to less than 22°C while ambient temperatures and relative humidities were as high as 38°C and 90%, respectively. The chilled rice was successfully shipped to a midwestern processing plant by rail in the middle of the summer without any fumigation application. Normally, each rail car is fumigated before sealing and remains under fumigation during shipment. The rice was rechilled and stored without quality deterioration for several months before end-use processing. The trial was conducted by Michigan State University, and the detailed results remain confidential.

Table 2. Comparison of chilled, ambient and no aeration of a 597-t maize bin between 15 October 1989 and 14 October 1990. Initial grain temperature 17°C and moisture 16.5%.

	Cool-down time for bulk to reach		DM loss (%) after cooling and 12 months storage[a]		Moisture content (%) after cooling and 12 months storage[a]		Specific energy (kWh/t) for cooling to	
	7°C	10°C	7°C	10°C	7°C	10°C	7°C	10°C
No aeration	5 months[b]	2.5 months	0.641	0.64	16.51	16.5	–	--
Ambient	4 wks	1 wk	0.52	0.41	16.1	16.2	3.1	1.3
Chilled	1 wk	1 wk	0.38	0.41	16.1	16.2	3.1	2.6

[a] Assuming no moisture changes during storage
[b] Lowest bulk temperature reached was 7.5°C

Table 3. Comparison of chilled, ambient and no aeration of a 579-t wheat bin between 1 July 1989 and 30 June 1990. Initial grain temperature 30°C and moisture 14%.

	Cool-down time for bulk to reach		DM loss (%) after cooling and 12 months storage[a]		Moisture content (%) after cooling and 12 months storage[a]		Specific energy (kWh/t) for cooling to	
	10°C	15°C	10°C	15°C	10°C	15°C	10°C	15°C
No aeration	10 months[b]	8 months	1.1[b]	1.1	14.01	14.0	–	–
Ambient	4.5 months	3 months	0.96	0.95	11.0	11.3	65	40
Chilled	1 wk	1 wk	0.32	0.35	13.2	13.3	8.9	6.1

[a] Assuming no moisture changes during storage
[b] Lowest bulk temperature reached was 13°C

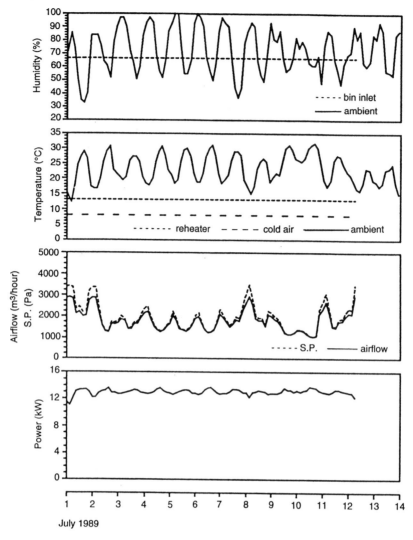

Fig. 5. Simulated performance of a German-built grain chiller cooling the 579-t bin of wheat in July 1989. Bin inlet air temperature 14°C and 67% r.h.

Rice and maize chilling in Texas

Conventional and chilled aeration of rough rice and food maize were investigated for summer conditions (August-October) along the humid gulf coast of Texas (Maier et al. 1992). Of four ambient aeration strategies — continuous; fan on from 10 a.m.. to 4 p.m.; fan on from 10 p.m. to 4 a.m.; fan on whenever the relative humidity is below 75% — continuous aeration performed best because it resulted in the smallest moisture losses and gradients in the commercially-sized silos. Chilled aeration was able to minimise moisture losses and maintained average grain temperatures below 15°C while bulk temperatures in conventionally aerated rice and maize remained between 21° and 38°C. Significant cycling of the grain temperatures and moisture contents was observed in the aerated bins where several warming and cooling cycles moved through the grain before final storage temperatures were reached in the fall. One-time chilled aeration of maize immediately after harvest to 15°C was completed within 120–160 hours of cooling, while maintaining the moisture content uniformly at 13%.

Wheat chilling in Kansas in 1991

Hellemar (1993) reported the successful chilling of a 2500 t concrete silo of wheat at a commercial elevator in Kansas during the summer of 1991. Initial grain temperatures were 32–35°C and 14.5% moisture. Chilling to 15–17°C was completed within 144 hours. After 4 months of storage the wheat was shipped by rail. The moistures and temperatures remained unchanged, and no insects were detected. In comparison, an additional 2500 t silo of wheat was stored at the same initial temperatures but at only 12–13% moisture. During the 4-month storage period the wheat was turned once and fumigated twice in the silo. Because of insect activity the rail car had to be fumigated at the time of shipment. The wheat temperature was 29°C and the moisture 10–11% at the time of shipping. Treatment costs were estimated at $0.20/t for chilling, and $0.80/t for turning and fumigating. In addition, approximately 56 t of wheat were lost in the non-chilled silo.

Maize chilling in Indiana in 1992

During the summer of 1992, 80 t of maize were chilled and maintained at temperatures below 13°C. Repeated chilling cycles were administered while optimising the performance of a prototype U.S. grain chiller developed by Purdue University researchers in cooperation with AAG Manufacturing, Milwaukee, WI (Maier et al. 1993a) (Fig. 6). At the end of the trial, the maize was sold as No. 1 grade to a local wet milling plant. The performance characteristics of the prototype chiller are summarised in Table 4. They are compared to the Granifrigor KK140 (Sulzer-Escher Wyss, Lindau, Germany), which was previously used by Michigan State University researchers. Five test runs were conducted (Table 5). In runs 1 and 2 the chiller operated with the initial factory settings. Runs 3 and 4 were conducted after raising the compressor pressure limit to increase refrigeration capacity. Before run 5 the air throttle controller and a faulty glycol valve were replaced. The optimisation of the outlet chiller air conditions during the five tests is illustrated using data from runs 1 and 5. The ambient inlet and chilled air temperatures and relative humidities during run 1 are shown in Figure 7. [A power failure of 3 hours occurred 36 hours into the test.] The average chilled air outlet temperature and relative humidity were 0.6°C and 6.5 percentage points above the set points, respectively (Table 6). The airflow rate

Fig. 6. Prototype grain chiller developed jointly by Purdue University and AAG Manufacturing (Milwaukee, Wisconsin, USA) operating during the optimisation test runs in 1992.

from the chiller into the bin ranged from a low of 1783 m³/hour to a high of about 2377 m³/hour. The average airflow was 2094 m³/hour with a standard deviation of 319 m³/hour. The air throttle reacted to increasing and decreasing ambient refrigeration loads of the evaporator coil. However, the reaction of the controller was somewhat abrupt. Some hunting of the feedback controller occurred during high temperature periods, which contributed to increased chilled air conditions after the reheater coil. Closer examination of the subsequent test runs revealed that the off-the-shelf feedback air throttle controller did not have sufficient capability to interact with the variable load on the reheat cycle. Additionally, a faulty glycol flow control valve was detected.

The ambient and chilled temperatures and relative humidities during run 5 are shown in Figure 8. Although the outlet air temperature was about 1.8°C above the set point, the relative humidity was within 3 percentage points of the expected LARH (Table 6). The effect of the corrected glycol valve was significant. The heat gain of the air across the reheater coil was the highest for any of the test runs, and accounted for the improved control over the relative humidity of the chilled air. A significant pre-cooling effect reduced the cooling load on the evaporator, which was highest in terms of temperature reduction compared to previous runs. The sought-after control over the chilled air conditions by regulating both the airflow given a fixed refrigeration capacity, and the load on the evaporator coil through pre-cooling the air is one of the distinguishing features of the Purdue-AAG chiller compared with other commercial units

The results of the experimental test runs clearly indicated that the Purdue-AAG prototype grain chiller operated exceptionally well. The refrigeration capacity of the unit was maximised, and control over the chilled air temperature and relative humidity was improved. Variability of the relative humidity was reduced from 6.5–8.8 percentage points to

Table 4. Specifications of the Purdue prototype grain chiller versus a similar-sized commercial unit.

	Purdue prototype[a]	Granfrigor KK140[b]
Evaporator capacity (kW)	26.4	32.7
Airflow at 2000 Pa (m³/hour)	2300	4400
Chilling capacity (t/day)	53[c]	100
Connected load (kW)	9.3	13.0

[a]Manufactured by AAG Manufacturing, Milwaukee, WI
[b] Manufactured by Sulzer-Escher Wyss, Lindau, Germany
[c] Estimated based on airflow-chilling capacity

Table 5. Experimental test runs conducted with the Purdue prototype grain chiller, leaving air temperature (LAT) settings at the evaporator and reheater coils, and estimated leaving air relative humidity (LARH) at the chiller outlet.

Run ID	Duration	Evaporator LAT	Reheater LAT	Expected LARH[a]
1[b]	5/19 - 5/23/92	7.2°C	12.8°C	70%
2[b]	5/27 - 6/05/92	4.4°C	7.2°C	80%
3[c]	6/19 - 6/26/92	7.2°C	12.8°C	70%
4[c]	6/08 - 6/12/92	4.4°C	7.2°C	80%
5[d]	6/29 - 7/06/92	7.2°C	12.8°C	70%

[a] Calculated assuming saturation at the evaporator coil
[b] Factory settings
[c] After raising compressor pressure limit
[d] After replacing the air throttle controller and faulty glycol valve

2.1–3.0 points during the tests. Outlet temperatures were achieved within 2°C of the set point. Successful testing of this prototype has resulted in the commercialisation of the first U.S.-built grain chillers.

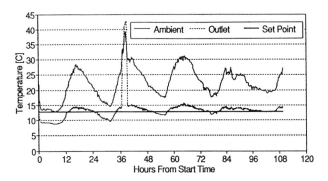

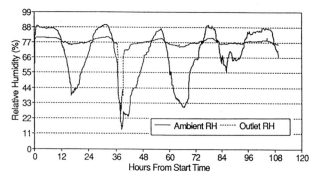

Fig. 7. Inlet and outlet temperatures and relative humidities during the first test run of the Purdue-AAG prototype grain chiller in the summer of 1992. Top: ambient/outlet temperature vs. time. Bottom: ambient/outlet r.h. vs. time. Runs started 5/19 11: 50 am and ended 5/23 at 12:10 pm.

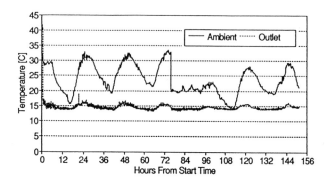

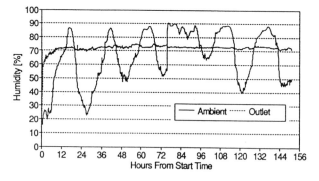

Fig. 8. Inlet and outlet temperatures and relative humidities during the final test run of the Purdue-AAG prototype grain chiller in the summer of 1992. Top: ambient/outlet temperature vs. time. Bottom: ambient/outlet r.h. vs. time. Runs started 6/29 1: 03 pm and ended 7/6 at 10:50 am.

Wheat chilling in Carrolton, Texas in 1992

Approximately 2250 t of 12% moisture wheat were chilled in a concrete silo from an initial temperature of 23–24°C to 15–16°C in the middle of June at a commercial elevator in Texas (Hellemar 1993). The wheat was stored through the end of December, and quality was maintained without any additional pest management treatments. The trial was conducted utilising a Swedish-made grain chiller.

Maize chilling in Indiana since 1992

Four 6-t bins were filled in July 1992 with No. 2 yellow maize and intentionally infested with adult *Sitophilus oryzae* (L.) (rice weevils) (Adams et al. 1993). Four different pest management techniques have been employed: fumigation, continuous/controlled ambient aeration, no aeration, and intermittent chilling. Temperatures, moisture contents, mould development and insect development have been measured intermittently. The study is on-going at Purdue University, and its preliminary results are reported elsewhere in these proceedings.

Rice chilling in California since 1993

Of primary concern to the rice miller is the head yield, or the amount of unbroken kernels. Broken kernels are worth only approximately 50% that of whole rice kernels. It is estimated that an average-sized U. S. rice miller can increase returns by $262 500 with a 1% decrease in broken kernels (Spadaro et al. 1980). Milled rice is extremely sensitive to changes in air temperature and relative humidity. When rice is exposed to high or low humidities, it is likely to fissure, which destroys its processing value. Once the rice is milled it must be slowly cooled to a suitable storage temperature (Lan and Kunze 1992). If ambient weather conditions are not favourable for ambient aeration, controllers must be used to operate the fans only when air conditions permit. The aeration process must be accomplished gradually and within a relatively narrow band of relative humidity in order to prevent (or minimise) fissuring of the rice kernels due to excessive adsorption (or desorption) of moisture. This approach can result in prolonged storage times and delayed shipment of rice. Humidity controllers were used by one commercial rice processor to operate aeration fans whenever ambient conditions allowed. However, daily and seasonal changes in the ambient weather conditions limited the usefulness of this system, and made operation unpredictable. This approach also revealed that ambient aeration did not always prevent fissuring. The only viable solution appeared to be to utilise a conditioning system that controls both the temperature and the relative humidity of the aeration air independent of climatic conditions.

The Chill'd Aire Aeration and Conditioning System developed jointly by Purdue University and AAG Manufacturing (Milwaukee, Wisconsin) was used to conduct a field test to verify the benefits of this chilling system for the conditioning of milled rice (Maier et al. 1993b). At a commercial rice processing facility in California, 175 t of milled rice were chilled in a concrete silo from an initial temperature of 26–29°C to 11–12°C within 48 hours (Fig. 9). [The unit automatically shut down for a brief period due to cool ambient conditions.] The movement of the cooling front through the rice is obvious from the temperature readings of the thermocouples located at 3.6, 5.4, and 7.2 m. The cooling front reached the top of the pile after about 27 hours (i.e., 21 hours of operating time). Airflow during the trial averaged 1872 m^3/h. Control over the temperature and relative humidity of the air entering the storage silo within 0.5°C and 0.2% of set

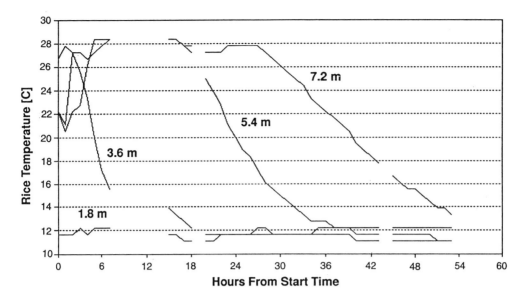

Fig. 9. Temperatures of milled rice conditioned with the Chill'd Aire Aeration and Conditioning System (AAG Manufacturing, Milwaukee, Wisconsin, USA) operating at a commercial rice processing facility in California in 1993.

point, respectively, was critical to minimise damage to the rice. The rice was used for further processing and was of superior quality to milled rice aerated with ambient air. The results were confirmed in a second test. Because the air used to chill the rice is temperature and humidity controlled, conditioning can be accomplished in a predictable time period, cost-effective manner, and independent of weather conditions. This application has been scaled-up and is the first commercial implementation of the grain chilling technology in the United States.

Future plans

Currently, field tests for research and demonstration purposes are continuing and additional ones are planned in the southern, south-eastern and mid-western United States. The primary focus will be on specialty and food-grade crops such as popcorn, white and yellow food maize, rice and wheat. Additionally, the potential benefits of chilled aeration and storage as a non-chemical pest management technology will be further quantified.

Summary

The effectiveness of chilled aeration for the optimisation of end-use processing quality by conditioning bulk grain, and its potential as a non-chemical pest control technology, have been investigated and demonstrated for the past 6 years for a variety of crops and geographic locations in the United States.

Commercialisation of a U.S.-built grain chiller and the first sale of any grain chiller in the U.S. represent the beginning of what is expected to lead to a more broad-scale adoption and utilisation of chilled aeration in the next few years. The need for the adoption of this particular postharvest technology has only recently been highlighted in the report 'Alternatives to methyl bromide: assessment of research needs and priorities', published by the U.S. Department of Agriculture, which identified chilled aeration of unprocessed grains as one of only four immediately available alternative technologies.

References

Adams, W.H. 1993. Comparison of Stored-Insect Management Techniques. Paper No. 93–6513. ASAE, St. Joseph, MI.

Bakker-Arkema, F.W., Maier, D.E., Mühlbauer, W. and Brunner, H. 1989. Initial cool-down in chilled maize storage. pp. 2261–2266. In: Agricultural Engineering, Vol. 4. Dodd, V.A. and Grace, P.A., (eds.) Balkema Inc., Rotterdam, The Netherlands.

Baldo, R. and Brunner, H. 1983. Relieving the dryer through chilling (In German). Sulzer- Escher Wyss reprint from Die Mühle + Mischfuttertechnik. (31).

Beeman, R.W. and Wright, V.F. 1990. Monitoring for resistance to chlorphyrifos-methyl, pirimiphos methyl, and malathion in Kansas populations of stored-product insects. J. Kan. Entomol. Soc. 63, 385–392.

Ben-Efrain, A., Lisker, N., and Henis, Y. 1985. Spontaneous heating and the damage it causes to commercially stored soybeans in Israel. Journal of Stored Products Research, 21, 179–187.

Berhaut, P., Desnos, G., Franquet, V., Lasseran, J.C., Niquet, G., Poichotte, J.L. 1988. Study of different aeration cooling techniques

Table 6. Summary of the average ambient (in) and chilled (out) air conditions with standard deviations during the five test runs of the Purdue prototype grain chiller.

Run ID	T_{in} (°C)	r.h. $_{in}$ (%)	T_{out} (°C)	r.h. $_{out}$ (%)
1	22.2 ± 5.1	68.4 ± 19.1	13.4 ± 3.9	76.5 ± 6.5
2	17.6 ± 5.3	60.7 ± 25.2	9.0 ± 1.2	75.9 ± 8.8
3	16.9 ± 6.0	64.9 ± 19.8	11.1 ± 1.7	72.8 ± 6.9
4	21.2 ± 4.8	45.1 ± 17.3	10.5 ± 1.5	77.1 ± 2.1
5	23.2 ± 4.9	66.7 ± 19.0	14.6 ± 1.8	72.2 ± 3.0

— Silo Champagne Cereales de Coolus (In French). Report of the Institut Technique des Cereales et des Fourrages, Boigneville, France.

Bissaro, F. and Bertoni, R. 1989. Refrigerated preservation of rice (In Italian). Freddo. 43(6):697–701.

Brunner, H. 1985. Why, when and how to chill grain (In German). Sulzer-Escher Wyss reprint from Die Mühle + Mischfuttertechnik. (26).

Burrell, N.J. 1964. Hot news on cold air. Farmer and Stockbreeder. 3 November 1964.

Burrell, N.J. 1974. Chilling. In: Christensen, C.M. (ed.) Storage of Cereal Grains and Their Products. AACC, St. Paul, MN.

Burrell, N.J. 1982. Refrigeration. In: Christensen, C.M. (ed.) Storage of Cereal Grains and their Products. AACC, St. Paul, MN.

Chek, T.I. 1989. Application of paddy cooling technique in Malaysian rice industry. Workshop on Grain Drying and Bulk Handling and Storage Systems in ASEAN, Pitsanuloke, Thailand, 17–20 October 1989.

Clapers, R. 1970. Rice Cooling. Internal Report. Sulzer-Escher Wyss, Lindau, Germany.

Cunille, A. 1988. Sasetru experience with pellets. Internal Report. Sulzer-Escher Wyss, Lindau, Germany.

Donahaye, E., Navarro, S. and Calderon, M. 1974. Studies on aeration with refrigerated air – III. Chilling of wheat with a modified chilling unit. Journal of Stored Products Research, 10, 1, 1-8.

Escher Wyss. 1960. User note for grain drying with the Escher Wyss cooling unit (In German). Escher Wyss, Lindau, Germany.

Felkel, C.A. 1978. Storability of high moisture corn when stored at simulated post-harvest temperatures at three locations in Indiana. MS-thesis. Purdue University, W. Lafayette, IN.

Finassi, A. 1987. Recent progress in rice cultivation in Italy. FAO Int. Com. Rep. 36(1):5–7.

Friday, D., Tuite, J. and Stroshine, R. 1989. Effect of hybrid and physical damage on mould development and carbon dioxide production during storage of high-moisture shelled corn. Cereal Chemistry, 66(5), 422–426.

Haile, D.G. and Sorensen, J.W., Jr. 1968. Effect of respiration heat of sorghum grain on design of conditioned-air storage systems. ASAE Transactions. 11(3):335–338.

Hagstrum, D.W. and Flinn, P.W. 1992. Integrated pest management of stored-grain insects. pp. 535–562. In: Storage of cereal grains and their products. D.B. Sauer (ed.) AACC, St. Paul, MN.

Hall, C.W. 1980. Drying and Storage of Agricultural Crops. AVI Publ. Co. Westport, CT.

Heidt, H. 1963. Grain chilling, a valuable aid in combine harvesting (In German). Die Muhle. 100(39):417–418.

Heidt, H. and Bolling, H. 1965. Cooling of kernels (In German). Escher Wyss reprint from Die Mühle. 102(10, 11, 12).

Hellemar, J. 1993. The big chill: a grain handling alternative. Proceed. GEAPS Exchange '93. GEAPS, Minneapolis, MN.

Hunter, A.J. and Taylor, P.A. 1980. Refrigerated aeration for the preservation of bulk grain. Journal of Stored Products Research 16, 123–131.

IKZ 1993. Sales brochure 'Getreidekühler'. Industriekühlung GmbH Zwickau, Zwickau, Germany.

Lan, Y. and Kunze, O.R. 1992. Fissures related to stress distribution in milled rice. Paper No. 92–6553. ASAE, St. Joseph, MI.

Lasseran, J.C. and Fleurat-Lessart, F. 1988. Ventilation de refroidissement et refrigeration comme technique de lutte contre les insectes. Journee Technique du GLCG, 2. February 1988, France.

Maier, D.E., Kelley, R.E., Bakker-Arkema, F.W. and Brunner, H. 1989. Longterm chilled grain storage. Paper No. 89-6541. ASAE, St. Joseph, MI.

Maier, D.E. 1992. The chilled aeration and storage of cereal grains. Ph.D. thesis, Michigan State University, E. Lansing, MI.

Maier, D.E., Bakker-Arkema, F.W. and Moreira, R.G. 1992. Comparison of conventional and chilled aeration of grains under Texas conditions. Applied Engineering in Agriculture, 8(5), 661–667.

Maier, D.E., Rulon, R.A., Guiffre, A.F. and Wilson, C.J. 1993a. Development of a new commercial grain chiller. Paper No. 93–6514. ASAE, St. Joseph, MI.

Maier, D.E., Adams, W. H., Hudson, P. and Guiffre, A.A. 1993b. Chilled conditioning of milled rice. Paper No. 93–6515. ASAE, St. Joseph, MI.

Maier, D.E., Bakker-Arkema, F.W. and Ilangantileke, S.G. 1993c. Ambient and chilled paddy aeration under Thai conditions. Agricultural Engineering Journal, 2(1,2), 15–33.

Maier, D.E. and Bakker-Arkema, F.W. 1994. Chilled Corn Aeration and Storage in Michigan. ASAE Transactions, in press.

Malaga, G. 1973. Experiences with cold storage with Granifrigor KK60. Internal Report. Sulzer-Escher-Wyss, Lindau, Germany.

McCune, W.E. 1962. Conditioned Air for Maintaining Quality of Stored Grain. Paper presented at Southeastern Electric Exchange General Sales Conference, New Orleans, LA, November 7–9 1962.

McCune, W.E., Person, N.K. Jr. and Sorensen, J.W. Jr. 1963. Conditioned air storage of grain. ASAE Transactions. 6(3):186–189.

McLean, K.A. and Barlett, D. 1991. Chilled preservation of barley (In German). Die Mühle + Mischfuttertechnik. 128(18), 225–227.

Moreira, R.G., Maier, D.E. and Dalpasquale, V.A. 1993. Aeration of grains using natural and chilled air (In Spanish). FAO Conference for Grain Drying and Storage. Canela, Rio Grande do Sul, Brazil 19–23 October 1993.

Munday, G.D. 1965. Refrigerated Grain Storage. Journal and Proceedings of the Institution of Agricultural Engineers. 21(2):65–74.

Navarro, S., Donahaye, E. and Calderon, M. 1973. Studies on aeration with refrigerated air - I. Chilling of wheat in a concrete elevator. Journal of Stored Products Research, 9, 253–259.

Person, N.K. Jr., Sorenson, J.W. Jr. and McCune, W.E. 1966. Thermodynamic considerations in designing controlled storage environments for bulk grain. ASAE Transactions. 520–523.

Reimann, E. 1927. The influence of temperature on grain storage (In German). Die Mühle. (12):367–368

Rius, J. (1987). Experiences with storage of paddy using Grainifrigors in Spain. Sulser-Escher Wyss, Lindau, Germany.

Saul, R.A. and Lind, E.F. 1958. Maximum Time for Safe Drying of Grain With Unheated Air. ASAE Transactions. 2:29–33.

Shove, G.C. 1966. Application of dehydrofrigidation to shelled corn conditioning. Paper No. 66–351. ASAE, St. Joseph, MI.

Sorensen, J.W. Jr., Person, N.K. Jr., McCune, W.E. and Hobgood, P. 1967. Design methods for controlled-environment storage of grain. ASAE Transactions. 10(3):366–369.

Spadaro, J.J., Matthews, J. and Wadsworth, J.I. 1980. Milling. pp. 360–402. In: Rice: Production and utilisation, B. S. Luh, (ed.) The AVI Publishing Company, Inc. Westport, CT.

Stewart, J.A. 1975. Moisture migration during storage of preserved, high moisture grains. ASAE Transactions. 18(2):387–393, 400.

Stroshine, R.L. and Yang, X. 1990. Effects of hybrid and grain damage on estimated dry matter loss for high-moisture shelled corn. ASAE Transactions. 33(4):1291–1298.

Sullivan, S.O. and Sebestyen, E.J. 1973. Problems of grain preservation in storage facilities. Flour and Animal Feed Milling. May 1973.

Sulzer-Escher Wyss. 1970. Kernel cooling unit Granifrigor (In German). Sales brochure. Sulzer-Escher Wyss, Lindau, Germany.

Sutherland, J.W., Pescod, D., Airah, M. and Griffiths, H.J. 1970. Refrigeration of Bulk Stored Wheat. Australian Refrigeration, Air Conditioning and Heat 24(8), 30–34, 43–45.

Sutherland, J.W. 1986. Grain Aeration in Australia. pp.206–218. In: Champ, B.R. and Highley, E. (eds.) Preserving grain quality by aeration and in-store drying: Proceedings of an international seminar, Kuala Lumpur, Malaysia, 9–10 October 1985. ACIAR Proceedings No. 15.

Thompson, T.L., Villa, L.G. and Cross, O.E. 1969. Simulated and experimental performance of temperature control systems for chilled high-moisture grain storage. Paper No. 69–856. ASAE, St Joseph, MI.

Thompson, T.L. 1972. Temporary storage of high-moisture shelled corn using continuous aeration. ASAE Transactions. 15:333–337.

Torres, J.R. 1983. Storage of paddy with the Granifrigor System (In German). Internal Report. Sulzer-Escher Wyss, Lindau, Germany.

Torres, J.R. 1985. Experiences with paddy storage using Granifrigor grain chillers (In German). Internal Report. Sulzer-Escher Wyss, Lindau, Germany.

Tuckett, G.J. 1982. Letter to Sulzer-Escher Wyss, Lindau, Germany.

Tuite, J., Foster, G.H. and Thompson, R.A. 1970. Moisture limits for storage of corn aerated with natural and refrigerated air. Paper No. 70–306. ASAE, St. Joseph, MI.

U.S. Congress. 1989. Enhancing the Quality of U.S. Grain for International Trade. Office of Technology Assessment, Washington, DC.

Zetter, J.L. and Cuperus, G.W. 1990. Pesticide resistance of *Tribolium castaneum* (Coleoptera: Tenebrionidae) and *Rhyzopertha dominica* (Coleoptera: Bostrichidae) in stored wheat. Journal of Economic Entomology, 83, 1677–1681.

Pest management of stored maize using chilled aeration — a mid-west United States perspective

L. J. Mason*, D.E. Maier†, W.H. Adams† and J. L. Obermeyer*

Abstract

Management of stored-grain pests such as insects and fungi requires an integrated system approach that combines engineering, biological and economic principles. Chilled aeration is a non-traditional, non-chemical preservation technology for storage of cereal grains. Although it has been used effectively throughout the world to prevent and control insect infestations in stored grains, chemical pesticides are still the primary method utilised in the United States. The process of chilled aeration followed by low temperature storage is a technically feasible alternative to pesticide application for controlling stored-grain insects in the mid-western region of the United States.

This paper presents a preliminary summary of a comprehensive study in progress under midwestern United States conditions that includes physical, biological, and economic factors. Maize in four 250 bushel bins has been stored since July 1992, and was intentionally infested with insects. Four different management techniques have been employed as treatments (i.e. fumigation, chilled aeration, ambient aeration, and no aeration). Temperature, moisture content, mould development and insect population levels have been measured throughout the year. Although the study is not yet complete, preliminary results favour chilled aeration as a pest management strategy for optimum grain quality preservation.

Introduction

A stored-grain bulk is an ecological system created by humans in which deterioration results from interactions among physical, chemical, and biological variables, such as temperature, moisture, carbon dioxide, oxygen, microorganisms, insects, mites, rodents, birds, geographical location and storage structure. Among these variables, insects create numerous quality problems, and cause substantial economic losses (estimated at over $12 million in Indiana, USA in 1990). Preventing insect infestations of stored commodities requires an integrated system management approach that combines engineering, biological, and economic principles.

At a recent USDA workshop on alternatives for methyl bromide, (29 June-1 July 1993, Crystal City, Virginia) ambient and chilled aeration of stored grains, pulses, milled products and animal feeds was rated as one of only four high priority research areas. Although aeration is an existing technology, its utilisation as a preventative pest management tool has not been fully investigated to date. Other high priority research areas designated were monitoring, decision-support systems, and modified atmospheres. The consensus among workshop participants was that no direct substitute for methyl bromide exists that is as effective and fast-acting, nor is the development of such a therapeutic tool expected. Thus, we believe that research efforts should focus on preventative pest management techniques, such as aeration.

Insects and fungi create numerous quality problems in stored grain. Preventing insect infestation and mycotoxin occurrence begins on the farm with sanitation in storage facilities and grain cleaning. Application of a grain-protecting pesticide treatment is currently recommended for long-term (over one-season) storage. As grain is marketed and moves through various facilities, the identity of a lot is lost and additional pesticide treatments may occur. Thus, the number of pesticide applications can increase and toxic residues may accumulate in the final food or feed product.

Recent reports from the mid-western United States indicate that the lesser grain borer, *Rhyzopertha dominica* (F.), may be developing resistance to chlorphyrifos-methyl (Reldan) (Beeman and Wright 1990; Zettler and Cuperus 1990). As a result, this pest has been removed from the label. The label for pirimiphos-methyl (Actellic) now specifies that it will only suppress, not control *R. dominica* (F.). As additional insects develop resistance to currently available pesticides, greater emphasis must be placed on alternative methods of protecting our food supply including pheromone and pitfall traps, managing grain temperature and moisture content to limit insect reproduction and growth through aeration cooling, plant resistance, and biological control. Integrated pest management (IPM) is one strategy that can replace or reduce the use of synthetic chemicals as the primary control method. It involves insect sampling, risk-benefit analysis, aeration management, and the use of several control strategies (Hagstrum and Flinn 1992).

Insects are very sensitive to temperature changes in their environment (Mullen and Arbogast 1984). These changes in turn can influence population growth by reducing rates of development, survivorship, and age-specific fecundity (Hagstrum and Throne 1989). Most stored-product insects will not lay eggs below 15°C, and eggs laid above 15°C will not hatch below that temperature (Sinha and Watters 1985). Larval development is slowed in the range of 5-10°C and may result in death after a very short period of time (Sinha and Watters 1985). Conversely, stored-product insects thrive at approximately 29°C, and after 80 days of storage at or above 21°C any grain lot is likely to reach the economic threshold of insects if no protective measures are taken. Aeration cooling of grain as an insect control measure has potential for several reasons. In temperate climates the rate of growth of an insect population can be slowed by aeration with cool ambient air (Cuperus et al. 1986, 1990). Additionally, if low temperatures are attained rapidly and are maintained long enough, insects may be killed. Chilling to 3–4°C of barley infested with *Sitophilus granarius* (L.) prevented the development of a

---

* Department of Entomology, 1158 Entomology Hall, Purdue University, West Lafayette, Indiana, USA 47907–1158.
† Agricultural Engineering Department, 1146 Agricultural Engineering, Purdue University, West Lafayette, Indiana, USA 47907–1146.

severe infestation, with approximately 97% control (Burrell 1967). Another reason for utilising aeration cooling is that lower temperatures if applied quickly prevents insects from acclimatising and insect population growth is slowed. Desmarchelier et al. (1979) examined the influence of chilling fumigated grain. Populations that were not cooled recovered to detection levels after just 10 weeks, while those subjected to a fast cooling did not achieve a detection level until week 34.

Evans (1979, 1983) examined the effects of thermal acclamation and humidity on the survivorship of several species of beetles. There was considerable interaction between temperature and humidity; as well as considerable variation between species. In another study, Fields (1990) demonstrated that if cooling occurred rapidly during the fall when insects had not been previously exposed to cool temperatures, all adult rusty grain beetles (*Cryptolestes ferrugineus*) (Stephens) were killed. Conversely, if cooling occurred mid-winter, there was a 60% survival rate. Thus, the rapid application of cold temperatures in warm grain appears to be an effective pest management tool.

Many experts do not recommend freezing maize during aeration. However, at the 1992 NC-151 (Marketing and Delivery of Quality Cereals and Oilseeds) annual meeting it was noted that an elevator in northern Minnesota killed insects in grain by aerating with air at –18°C (DeJean 1992). Related strategies which may hold promise include cycling the grain between warm and cold conditions over a period of several days. The additional stress placed on the insects may result in killing of larger percentages of adults and larvae. Under certain conditions, aeration cannot completely inhibit insect activity even in temperate climates because it cannot prevent the development of grain temperatures that are optimum for stored grain insects (i.e. 21–29°C). Such temperatures occur in recently harvested wheat and carry-over shelled maize stored in the summer. For such situations, chilled aeration may be utilised to maintain quality grain without the use of chemical protectants as compared to ambient aeration or no aeration (Maier and Bakker-Arkema 1993).

Although numerous aeration cooling studies have been conducted in the past, none has been published for maize stored in the mid-west corn belt region of the U.S. In addition, few studies have presented the potential synergistic effect between insects and fungi. The principal investigators are aware of pest management practices in numerous commercial popcorn and food-grade maize handling facilities that are implemented on a calendar basis. In most cases the aeration fan is operated manually, and its operation is generally neglected after fall cool-down is complete. In other cases, fumigation treatments occur once a month during the spring and summer whether insects are detected in the grain or not. In other cases, grain protectants are sprayed onto inbound grain originally binned on farms and delivered to the handling facility throughout the season. It is apparent that even these food-grade grain operators, who are generally considered more sophisticated than commercial grain handler, lack understanding of the potential of aeration as an alternative pest management strategy.

Agricultural engineers and entomologists at Purdue University are cooperating to examine the engineering and biological aspects of four different storage management techniques and their impact on stored-grain insects. The techniques examined include fumigation, chilled aeration, continuous/controlled aeration with ambient air, and undisturbed storage. The research trials are designed to evaluate 'real world' storage situations. The research objectives are to:

- compare the practices of fumigation, chilled aeration, continuous/ controlled aeration with ambient air, and undisturbed storage to manage stored-grain insects;

- add to the database of integrated pest management (IPM) knowledge and use this knowledge to investigate preventative and therapeutic management strategies; and
- aid in the development of computer models to simulate aeration, storage and pest management techniques more efficiently.

Only data from the first objective will be presented in this paper.

Information generated in this research could be used in stored-grain management decision-support systems such as the Stored Grain Advisor (SGA) being developed by the United States Grain Marketing Research Laboratory, USGMRL (Hagstrum and Flinn 1992), or the Stored-Corn Management Instruction System being developed at Purdue University. Computer-based management programs provide the flexibility and capability needed to implement a comprehensive IPM program. They can estimate insect densities and potential spoilage, and recommend alternative control measures, such as chilled aeration, or operate aeration fans more intelligently than current control schemes.

Materials and Methods

On 28–29 July 1992 four round metal bins were each filled with approximately 6.5 t of maize. The maize was acquired from a commercial elevator and had been in storage since autumn 1991. The average initial moisture content of the No. 2 maize was 14% wet basis. No visible insect or fungal damage was apparent. The bins are 2.2 m (7 ft) in diameter and were filled to approximately 2.2 m (7 ft) depth.

The maize in all bins has been continuously monitored since 30 July 1992. However, the storage period can be divided into two distinctly different trials. The first trial ran from 28 July 1992 until 16 June 1993. The second trial began on 16 June 1993 and is in progress. Insects were introduced into all bins to establish uniform initial populations. During the first trial the insect population failed to become fully established. The reason for this failure has not been fully determined except that initial infestation may have been too low and development too slow to reach detectable levels. Insects were reintroduced during the summer of 1993, and have reached detectable levels. Data for the second trial only will be reported in this paper.

Bins were assigned one of four treatments: chilled aeration, fumigation, ambient aeration, and undisturbed. Chilled aeration was accomplished through the use of a prototype grain chiller developed jointly by Purdue University and AAG Manufacturing (Milwaukee, Wisconsin, USA). A small 0.0125 hp centrifugal fan has been used for ambient aeration. It was sized to deliver 0.1 m^3/minute/t. From the beginning of the storage period to the autumn of 1993 the fan was operated continuously. In the autumn of 1993 control over the aeration fan was changed by installing a Sentry PAK (Sentry Technologies, Chanhassen, Minnesota) automatic aeration controller. The unit is operated in storage mode and activates the fan intermittently to maintain grain temperature near the 21-day average ambient temperature.

The grain is monitored for temperature, moisture content, insect activity, and fungal growth. Temperatures are taken hourly with an automatic data acquisition system. Thermocouples are placed in five locations throughout the bin (three centre and two perimeter). Centre thermocouples are vertically located 0.6 m apart starting 0.6 m from the bottom (bottom, middle and top centre, respectively). Perimeter thermocouples are located 15-25 cm from the southern side wall and 0.6 and 1.5 m from the bottom (bottom and top perimeter, respectively). Maize samples are taken approximately each month to determine fungal growth and moisture contents.

Fungal growth is evaluated by using a serial dilution method (Marks 1992). The moisture content is determined using the drying oven method (ASAE standard S352.2).

Insect populations are monitored with probe traps. Five probe traps are placed in each bin just below the grain surface on the north, south, east, west sides and centre of the bins. During the first trial the probes were inserted on a weekly basis and left in for 3 days. During trial 2 the bins are monitored continuously. Trap catches are collected on a weekly basis and the number and species of insects recorded. The longer sampling time in trial 2 has allowed the acquisition of a more representative insect population. During the winter months traps are checked only once per month due to lower insect activity.

In the first trial the bins were infested with *Sitophilus oryzae* (L.) (rice weevil) at a rate of 250 insects per bin. After the completion approximately of one complete life cycle, the treatments were applied beginning on 24 August 1992. The fumigated bin was treated with phosphine (Phostoxin; Degesch) at the recommended label rate. After the recommended period of fumigation the bin was aerated. The fans in both the fumigated bin and the non-fumigated ambient aeration bin were continuously operated until the end of October 1992. At that time ambient temperatures were determined to be low enough to maintain grain temperatures through the winter. The ambient aeration bin was turned on again on 5 May 1993. For the chilled aeration treatment, 10°C air was moved through the grain mass for a 24-hour period on 23 August 1992. The bin was chilled three more times on 11 September 1992, 16 October 1992 and 11 May 1993. The control bin was left undisturbed for the entire period.

On 16 June 1993 three new populations of insect species were introduced into the bins. The species included *S. zeamais* (Motsch.), *S. oryzae* (L.) and *Tribolium confusum* (Jacquelin du Val) (maize weevil, rice weevil and confused flour beetle, respectively). The rate of infestation was 250 insects/species per bin. The chilled bin was cooled with 7°C air on 27 September 1993. The fumigated bin was treated on 24 September 1993 with phosphine, same as the first trial. In August 1993 fungal growth was detected on the surface of the grain in the chilled and undisturbed bins. To assess the extent of visible fungal growth and grain heating, the maize in each bin was turned in late September. This process included removing the maize from each bin, cleaning any mouldy grain from any of the surfaces of the bins and returning the grain to the bin. Due to the unloading, thorough mixing of each batch of maize occurred. Visible inspection and sampling of the maize showed it to be in good condition, and thus the trial was continued.

Results and Discussion

The results of trial one proved to be inconclusive due to the lack of detectable insect levels. Rice weevil populations failed to become established throughout the trial period. It was not possible to distinguish between the treatment effects because of the limited numbers of insects collected. However, during this trial the grain temperature control due to ambient, chilled and no aeration performed as expected (Fig. 1). The temperature of the control bin displays a steady decrease into the winter months. When compared with ambient temperatures, it follows the same pattern with a small time delay. The ambient aeration bin shows a great deal of fluctuation. This is due to the daily temperature changes. Temperatures in the chilled bin show a number of steps. These are the chilling cycles. The sudden drops and slow increases are very pronounced. The grain temperature at the centre of the four bins decreased from an initial 25°C to about 8°C by 5 December 1992. [Note: The temperature in the fumigated bin is not shown but it displayed the same pattern as the aerated bin during this period.]

The second trial is proving to be more successful. The introduction of new insects contributed substantially to the establishment of populations. Figures 2–5 illustrate the total number of insects trapped per bin per species during each sampling. Internal infesters (*Sitophilus* spp. (rice and maize weevils)) detected in each bin are shown in Figure 2 between 14 May through 25 December 1993. For most of the summer the populations were essentially non-detectable. The sudden increase in populations by October seem to indicate an incubation period of about three months for the population to establish. Observations in October indicated populations to be largest in the fumigated and untreated (control) bins. Clearly, the eradication treatment with phosphine fumigation was not effective. Additionally, fumigation has no residual effect and the potential for reinfestation exists. Populations were again high in the fumigation bin by 15 October 1993. Controlled and chilled aeration appear to maintain grain temperatures sufficiently low to suppress *Sitophilus* spp. populations during warmer weather. However, intermittent chilling for short duration shows slightly poorer treatment effects, and without additional chilling populations increased in November. By mid-December, populations declined to below detectable levels.

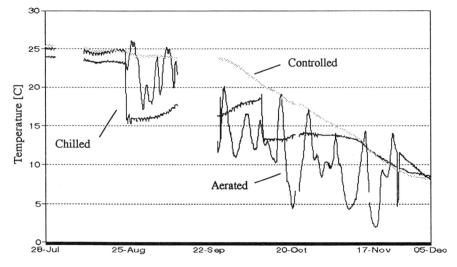

Fig. 1. Grain temperatures during summer, autumn and winter 1992.

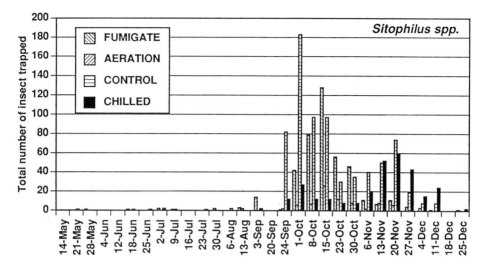

Fig. 2. Total number of *Sitophilus* spp. (rice and maize weevils) collected in probe traps during trial 2.

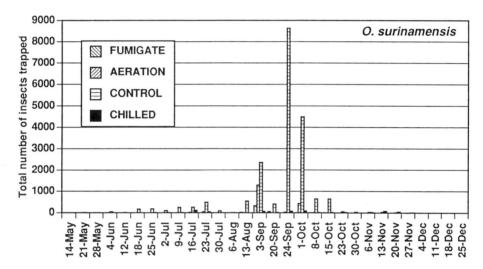

Fig. 3. Total number of *Oryzaephilus surinamensis* (sawtoothed grain beetle) collected in probe traps during trial 2.

The external feeders detected in large numbers include *Oryzaephilus surinamensis* (L.) (sawtoothed grain beetle) and *T. confusum* (confused flour beetle) (Figs 3 and 4). The control bin shows a gradual increase in sawtoothed grain beetle (Fig. 3) and confused flour beetle (Fig. 4) populations through the summer and into September when populations exploded (>8000/bin sawtoothed grain beetle and >4500/bin confused flour beetle). This increase corresponds to the steady increase in grain temperature throughout the summer. The substantial drop in the population in early October corresponds to the rotation of the bins and a decrease in grain temperature. The fumigation treatment shows a considerable decrease in the confused flour beetle population within 2 weeks of treatment from nearly 300 insects on 20 September to approximately 50 on 15 October (Fig. 4). The reason the population remained high on 1 October was probably due to the fact that the insects had 4 days to get trapped before fumigation occurred on 24 September; and because populations were on the increase. The bin under ambient aeration showed little change in population levels over the sampling period. An increase in confused flour beetles can be seen in the chilled bin during the 20 September 1993 trapping. However, population levels were considerably lower than in the control. Population decreased from 544 insects to a total of 10 insects in the chilled bin once the bin

was chilled on 27 September 1993. Populations remained low throughout the rest of the 1993 season.

Fungal growth in all the bins resulted in large numbers of fungus-feeding insects. The primary fungi found were *Fusarium monilifourne*, *Aspergillus flavus* and *A. glaucus*. The most prominent fungus/feeding insect was *Typhaea stercorea* (hairy fungus beetle). Sampling results are summarised in Figure 5. The control bin showed a gradual increase of *T. stercorea* until decreasing significantly through August. Population rebounded in September and October and finally dropped of near zero in November. The ambient aerated and chilled bins both show substantial increases through August with populations exceeding 4500/bin per week in the aerated bin and 3500/bin in the chilled bin. Subsequently, *T. stercorea* numbers in the aerated, chilled and control bins decreased while the population in the fumigated bin increased. This probably indicates insect movement between bins due to excessive population densities. The fumigation treatment successfully controlled the increase in *T. stercorea* at the end of September. The chilled aeration treatment on 27 September 1993, reduced insect populations from 500 in September to zero after 1 October. Low temperatures in late autumn helped to maintain smaller populations in all bins.

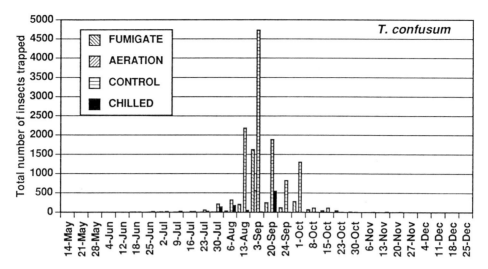

Fig. 4. Total number of *Tribolium confusum* (confused flour beetle) collected in probe traps during trial 2.

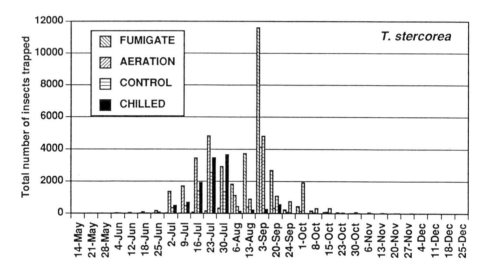

Fig. 5. Total number of *Typhaea stercorea* (hairy fungus beetle) collected in probe traps during trial 2.

The current trial will continue through the spring of 1994. This will provide a full year of data to see how the insects react to the treatments and the change of season. Preliminary results indicate that ambient and chilled aeration have beneficial therapeutic effects on insect populations in mid-west stored maize. Optimising the timing of chilling will be investigated further and will be part of the trial in year three. Storing maize without the availability of temperature control through aeration appears clearly undesirable.

The condition of the grain is such that the trial will not likely continue through the summer of 1994. However, a third trial is planned which will use 1993 maize currently in storage at Purdue University. The information gathered during these trials will be the basis for laboratory studies on the effects of environmental conditions on insect development and provide data needed for computer modelling of insect population dynamics in the stored-grain ecosystem.

References

Beeman, R.W. and Wright, V.F. 1990. Monitoring for resistance to chlorphyrifos-methyl, pirimiphos-methyl, and malathion in Kansas populations of stored-product insects. Journal of the Kansas Entomological Society, 63:385–392.

Burrell, N.J. 1967. Grain cooling studies - II: Effect of aeration on infested grain bulks. Journal of Stored Products Research, 3(2), 145–154.

Cuperus, G.W., Noyes, R.T. Fargo, W.S. Clary, B.L. Arnold, D.C. and Anderson, K. 1990. Management practices in a high-risk, stored wheat system in Oklahoma. American Entomologist, 36, 129–134.

Cuperus, G.W., Prickett, C.K. Bloome, P.D. and J.T. Pitts 1986. Insect populations in aerated and unaerated wheat in Oklahoma. Journal of the Kansas Entomological Society, 59, 620–627.

Desmarchelier, J.M., Bengston, M. Evans, D.E. Heather, N.W. and White, G. 1979. Combining temperature and moisture manipulation with the use of grain protectants. In: Evans, D.E. ed., Australian contributions to the symposium on the protection of grain against insect damage during storage, Moscow, 92–98.

DeJean, J. 1992. Continental Grain Company, Chicago Division. Marketing and Maintenance of Quality in Market Channels. NC-151 Annual Meeting. 12 February 1992, Chicago, IL.

Evans, D.W. 1979. The effect of thermal acclamation and relative humidity on the oxygen consumption of three *Sitophilus* species. Journal of Stored Products Research, 15, 87–93.

Evans, D.W. 1983. The influence of relative humidity and thermal acclamation on the survival of adult grain beetles in cooled grain. Journal of Stored Product Research, 19, 173–180.

Fields, P.G. 1990. Cold-hardiness of field and laboratory acclimated *Cryptolestes ferrugineus*. Proceedings Fifth International Working

Conference on Stored-product Protection, Bordeaux, France. 9–14 September 1990.

Hagstrum, D.W. and Flinn, P.W. 1992. Integrated pest management of storedgrain insects. In: Sauer, D.B. ed., Storage of cereal grains and their products. AACC St. Paul, MN. pp 535–562.

Hagstrum, D.W. and Throne, J.E. 1989. Predictability of stored-wheat insect population trends from life history traits. Environmental Entomology, 18, 660–664.

Maier, D.E. and Bakker-Arkema, F.W. 1993. Analysis of chilled corn storage in the midwestern U.S.A. ASAE Transactions.

Marks, B.P. 1992. Effects of storage history and hybrid on storability of shelled corn. M.S. thesis, Purdue University, West Lafayette, IN.

Mullen, M.A. and Arbogast, R.T. 1984. Low temperatures to control storedproduct insects. In: F.J. Baur ed., Insect management for food storage and processing, 257–264.

Sinha, R.N. and Watters, F.L. 1985. Physical and chemical control of storedproduct insects. In: Sinha, R.N. and Watters, F.L. ed., Insect pests of flour mills, grain elevators, and feed mills and their control, 57–74.

Zettler, J.L. and Cuperus, G.W. 1990. Pesticide resistance in *Tribolium castaneum* (Coleoptera: Tenebrionidae) and *Rhyzopertha dominica* (Coleoptera: Bostrichidae) in stored wheat. Journal of Economic Entomology, 83, 1677–1681.

Mobile drive-over hoppers and stackers for filling and emptying grain bunkers

F.M. Miller*

Abstract

F.A. Miller and Son Pty Ltd (FAMSON) has developed specific machinery for the receival of grain from road transport and transferring the grain to 30 metre wide bunkers. The system is known as a FAMSON D.O.H.S. (Drive Over Hopper and Stacker).

The machinery is portable, can handle wheat at 300 t/hour, is environmentally clean, does not damage the product and loads bunkers from the side so that no trucks drive on the prepared pad The equipment is then shifted inside the bunker and used to outload in conjunction with front-end loaders. Using the hopper as a surge bin and loading trucks on the roadway outside the bunker gives excellent outload rates up to 400 t/hour.

The equipment has proven efficient and cost effective during extensive use over the last three harvests in South Australia, where there were 2 Mt in bunker storage at the end of the 1993–94 harvest. The versatility of the equipment has expanded its use to shed filling and outloading, direct rail wagon loading and grain transfer.

History of Development

F.A. Miller & Son Pty Ltd (FAMSON), has a history in materials handling solutions, particularly grain, and have developed a range of air-supported, 'belt-in-tube' conveyors since 1980.

South Australian Cooperative Bulk Handling (SACBH), the bulk handling authority for South Australia, had until the late 1980s used grain throwers to bunker grain. The inherent problems with this method were extreme dust generation, which caused local governments to ban the use of this equipment near towns, and high double handling costs, as only specialist contractor trucks can be used to load grain into throwers. These trucks had to reverse onto the bunker pad creating a serious safety hazard because the thrower works inside the bunker area. A thrower can handle up to 200 t/hour but, because of truck changeover, this was not averaged over a day's work.

FAMSON was approached by SACBH to design a better system. In 1989 a design for filling 20 m wide bunkers was agreed upon and a machine built. Because of the heavy harvest, the prototype was used at several sites and handled approximately 40000 t of wheat and barley in the first harvest. The equipment performed above expectations. From the experience gained with this prototype, design criteria were set, resulting in the first of the current style equipment being built in 1990 to fill 30 m wide bunkers.

---

* F.A. Miller & Son Pty Ltd, 51 Chandos Terrace, Lameroo, South Australia, 5302, Australia.

The final 1990 design had to incorporate many improvements to bring bunker storage to a viable system, not merely a method of unloading the permanent storage system as an emergency method of handling a bumper crop. Specific requirements met by the FAMSON drive over hoppers and stackers (DOHS) system included:

- Environmentally friendly — a minimum of dust and of disruption to the area by noise.
- Complete safety — all reversing of delivery trucks to be eliminated.
- Minimum staff to operate.
- Compatible with permanent existing silo system so growers' trucks, after sampling and weighing, can be directed to permanent or bunker storage.
- Completely self contained and portable, so units can be relocated as harvest requirements of storage become apparent.
- Able to receive from all types of road vehicles — farmers tray-top trucks with side emptying bins through to road trains.
- Cost effective to operate and simple to maintain.
- Capable of filling 30 m wide bunkers with an approximate capacity of 100 t of wheat per metre.
- Able to receive at rates up to 300 t/hour.
- Vehicles to be kept off prepared bunker pad.
- Able to be used in the bunker unloading process.

All these design criteria have been met and are standard features of the FAMSON DOHS system.

Description of drive-over hopper

The drive-over receival hopper is 8 m long, 3.6 m wide and 0.85 m high with 4 ramps to allow trucks to pass over the unit. (Fig. 1). A lid folds back from the centre to form side bulkheads in operation.

The bottom of the hopper is shaped in 4 'V' sections to take 4 ribbon augers which run along the 8 m length and bring the grain to the centre of the hopper. Built into the base of the hopper below the auger level is a transverse box section tunnel with a sliding valve in the top surface. Into this tunnel is fitted a conveyor belt which moves the grain out from under the hopper and delivers it to the stacker.

The unit is powered by a quiet pack diesel over hydraulic power pack operating through a solenoid controlled valve bank. The power pack is bolted to the hopper in operation. The power pack forms part of the total assembly and moves as part of it. The operator has a panel of switches to control the functions of the hopper. The unit has a hydraulic jacking system to raise it and to lower the wheels to move along the bunker.

The entry and exit ramps and sliding gang valve are operated by hydraulic rams. The four ribbon augers in the hopper base and the transfer conveyor are driven by hydraulic motors.

Fig. 1. Mobile drive over hoppers and stackers for filling and emptying grain bunker storage

Description of the stacker

The mobile stacker (Fig. 2) consists of an air-supported 'belt-in-tube' conveyor 25 m long mounted on a mobile frame. The unit has a 16 m cantilever overhang in front of the frame and is raised and lowered using telescopic hydraulic rams.

An air supported 'belt-in-tube' conveyor (Fig. 3) was chosen for the high capacity at steep angles. The unit operates at 25° when filling a bunker. The conveyor has very high integral strength in the tubular design for minimum weight to allow cantilevering.

The frame mounts on truck wheels on the main axle. These wheels can be fitted to tow the unit in the normal lengthwise manner or at 90o to the tube for movement along the bunker wall. The double rear castors work in either direction. The frame is counter-balanced with concrete ballast blocks.

Power is from a quiet pack diesel engine over hydraulic power pack operating through a solenoid controlled valve bank. The operator has a panel of switches to control the stacker functions. The electrical controls are on a 30 m flex to allow the operator to be located at the hopper for inload and at the delivery point for outload.

The delivery of grain at the discharge is controlled by a multi-directional outlet giving movement across and along the stack. The outlet is operated by hydraulic rams (Fig. 4).

Mobility of system

Once assembled both the hopper and stacker are towed into position with a tractor or front-end loader.

The hopper can be stripped down by removing the transfer conveyor, the four ramps, the wheels and operator walkways to load onto a flat top semi-trailer. The stacker has the belt released and the tube assembly folds 6 m from the discharge end to reduce length for transfer. The main wheels and castor assemblies are removed and the unit fits on to an extendable trailer (Fig. 5). To dismantle and load on to road transport takes

Fig. 2.   The mobile stacker

approximately 6 hours with experienced personnel. Reassembly takes about the same time.

Method of operation

Once in place at the side of a bunker, the system is operated by one main operator and an assistant.

Trucks drive onto the hopper and discharge as directed. The sliding gang valve is opened to control the gravity flow on to the transfer conveyor which feeds the main stacker. As the level in the hopper drops, the ribbon augers are engaged to maintain the feed rate onto the transfer conveyor. During this period the next truck moves on to the hopper and begins discharge to maintain the maximum output rate.

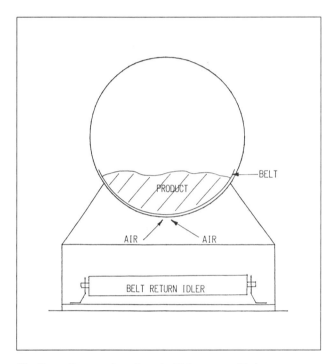

Fig. 3. Cross-section of a typical 'air supported' belt conveyor.

Once the stack is built to the required height using the stacker outlet to fill along the bunker as far as possible, the stacker and hopper are towed forward until the stacker outlet feeds backwards to give the best heap without extra moves. The movement along the stack takes approximately 10 minutes.

At the end of the day, or in bad weather, the lid folds down to cover the hopper.

Method of operation — unloading a bunker

Unloading a bunker is accomplished by setting up the hopper and stacker inside the bunker walls so that the stacker discharge is over the bulkheads above the side road in line with the area occupied by the hopper when loading the bunker. In this way trucks drive on the prepared highway alongside the bunker to be filled. No reversing is needed.

The hopper is modified by removing the four ramps and fitting bulkheads across the ends to form a deeper unit. The hopper is then set up to feed to the stacker. Front end loaders are used to fill the hopper and can build up a reserve while trucks are moving or returning from delivery.

The operator stands on a tower near the delivery end of the stacker where he can direct the truck driver and see the grain in the truck. The electrical controls for the hopper and stacker are located in the operator's tower so that he can control the hopper gang valve and stacker outlet directions to load the trucks.

As the bunker is unloaded, so the system is moved along to stay close to the grain stack for the front-end loaders.

Conclusion

The DOHS system, as developed, has proven to be a very efficient and cost-effective system for the rapid handling of grain in to and out of bunker systems. Sixteen units are currently in operation. More are being considered.

The versatility of the equipment lends it to many other uses. Units are now being used to load and unload long storage sheds, and for direct transfer of grain from road transport to rail wagons for bulk rail direct shipping.

Fig. 4. Multi-directional outlet.

Fig. 5.   Extendable trailer carrying tube assembly.

At a bunker site in Cummins, South Australia a system for receiving from bulk rail to store in bunkers has been developed. The rail line runs over a pit in which two hoppers are placed to receive from the bottom discharge rail wagons. The grain is transferred to road trucks which, in turn, deliver to the bunkers under construction. Trains up to 1200 t capacity are unloaded in under 4 hours, allowing three trains per 12-hour day.

Using controlled aeration for insect and mould management in the south-western United States

R.T. Noyes*, G.W. Cuperus†, and P.Kenkel§

Abstract

The use of controlled aeration systems as an integrated pest management (IPM) tool has been studied continuously in Oklahoma since the mid-1980s. Field research and demonstrations in storage bins and silos at farms and commercial elevators were used to test the principles of grain and headspace air temperature manipulation to minimise insect populations. This work was an extension of research studies by Evans and other Australians in the 1970s and early 1980s, adapted to Oklahoma's high risk storage environment for hard red winter wheat.

An extension fact sheet on use of aeration for insect management in stored grain was developed in 1987 (revised in 1990) by leaders of the OSU Stored Grain IPM Committee, formalising a step-by-step management procedure for use in pest management. This process uses an inexpensive ($US400-$US1000) custommade aeration controller to operate aeration blowers on grain silos or tanks at preset temperatures. An elapsed time hour meter accumulates blower(s) operating time to provide aeration management operating time data. Aeration controllers usually pay for themselves in less than one year.

Aeration systems ranging from one blower per small tank or silo, to large elevator storage complexes with as many as 24 blowers on two large steel tanks operated by one aeration controller were developed. Electro-mechanical aeration controllers and electrical circuits are illustrated with component costs in U.S. dollars.

Stored-Grain Pest Management

Stored-grain pest management is the organised, long–term approach to maintaining the quality of the grain, minimising chemical control inputs, and preserving the integrity of the grain storage system. To implement an effective management program and integrate control practices, operators must understand the ecology of the storage system. Through this understanding, techniques can be integrated into grain storage systems to prevent or minimise losses. These management techniques must focus on factors that influence storability, including:

- grain temperatures throughout the mass;
- grain moisture and seed interstice relative humidity;
- grain condition (peaked, fines and foreign material);
- storage structure and conditions;
- storage timing;
- aeration timing; and
- sources of grain insects and moulds.

Grain temperature — a pest management tool

Grain temperature is a major stored–grain management tool. Insects and moulds are regulated by temperature. Harvest temperatures vary widely for grain crops across the U.S. In northern states, grain is generally harvested later, can be stored at higher moisture levels, and can be cooled much sooner after harvest than grain in central and southern locations (Table 1).

Insects and moulds are poikilothermic — the higher the temperature (within limits), the greater the development. Most grain insect activity begins to slow above 35°C (95°F). Thus, the ability to maintain grain temperatures at or above 35°C for considerable periods of time is a useful grain insect management tool. Also, since most grain insect and mould activity is greatly reduced at grain temperatures below 15°C (60°F), rapidly cooling grain from the 30 to 35°C range to about 15°C or lower, provides an added grain insect management option. Thus, planned temperature reductions by controlled aeration can significantly reduce insect and mould populations (Fig. 1).

Grain moisture — storage and marketing problems

Grain moisture is another critical factor that regulates storability. The higher the levels of grain moisture, the greater the potential for high populations of stored–grain insects and moulds. Table 1 illustrates safe grain storage moisture levels for southern, central, and northern U.S. storage regions. As shown by these data, grain is at higher risk in southern states than in central and northern states due to longer periods of warm temperatures and higher relative humidity between harvest and aeration cooling. So, recommended storage moisture levels are lower in the South.

During aeration, grain moisture content is reduced by an estimated 1/3 to 1/2 percent during one fall aeration cooling cycle, and 1/4 to 1/3 percent during one winter cooling cycle. Insects and moulds can be managed by strategically using aeration to lower and equalise grain temperature. Fall grain cooling is critical in eliminating moisture migration and reducing insect and mould risks.

Effects of Temperature and Moisture on Insect Populations

Stored–grain insects and moulds are like other biological organisms—the longer the amount of time under favourable conditions, the more they develop. Figure 2 gives data projections by Hagstrum and Flinn (1990) showing expected effects of grain moisture and grain temperature on insect populations in grain aerated by selected target dates. This figure illustrates

* Biosystems & Agricultural Engineering Department, 224 Agricultural Hall, Oklahoma State University, Stillwater, Oklahoma, USA 74078-0469.
† Department of Entomology, 127 Noble Research Center, Oklahoma State University, Stillwater, Oklahoma, USA 74078-0464.
§ Department of Agricultural Economics, 512 Agricultural Hall, Oklahoma State University, Stillwater, Oklahoma, USA 74078-0505.

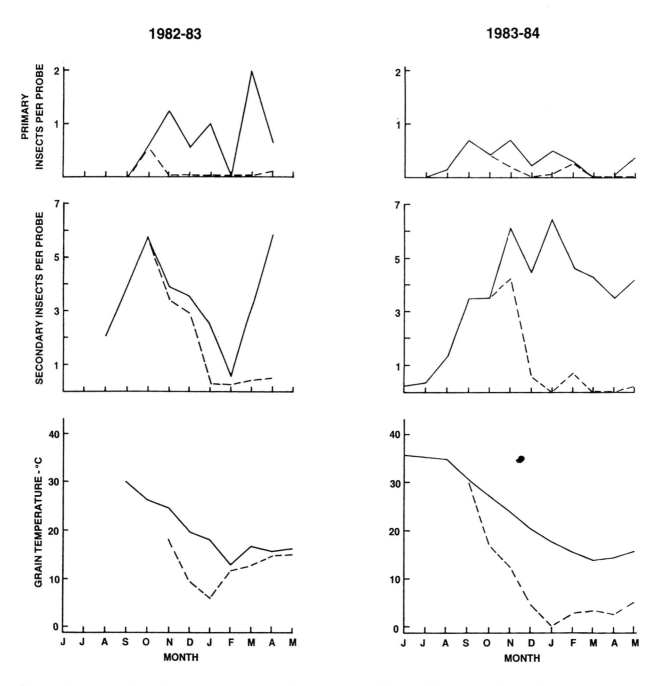

Fig. 1. Comparison of aerated to unaerated wheat storage effects on insect populations: solid line, unaerated; broken line, aerated.

predicted relationships between two grain temperatures and two moisture levels with insect frequencies when aeration is completed by 1 October.

These projections closely model field experience during the past decade in southern high plains wheat storages. This prediction system involves exponential population development. If temperature and grain moisture levels are favourable, stored–grain insects and moulds will increase in an exponential (non–linear) fashion. Managers must be aware of the risk based on the time the product has been stored when grain temperatures and moisture levels are conducive to growth.

Table 1 lists safe storage moistures for 10 grain and seed crops in the U.S. by general latitude location. Table 2 lists the approximate time that maize (shelled corn in the U.S.) can be expected to remain in storage before it deteriorates (loses 1% dry matter) enough to change at least one market grade level

based on storage moisture and temperature of the grain in storage. Recommended airflow rates by crop for several moisture levels are listed in Table 3.

Storage Time vs. Moisture and Temperature

Grain moisture and temperature interact significantly; both must be interactively managed. Table 2 lists estimates of allowable time that maize or shelled corn can be stored in aerated storage at various moisture contents and grain temperatures before the product loses 1% dry matter, reducing maize one market grade or more, depending on other grading factors in a sample.

Table 2 does not account for significant differences in hybrid seed varieties, variations in moisture tester or meter accuracy, and other important management conditions, such

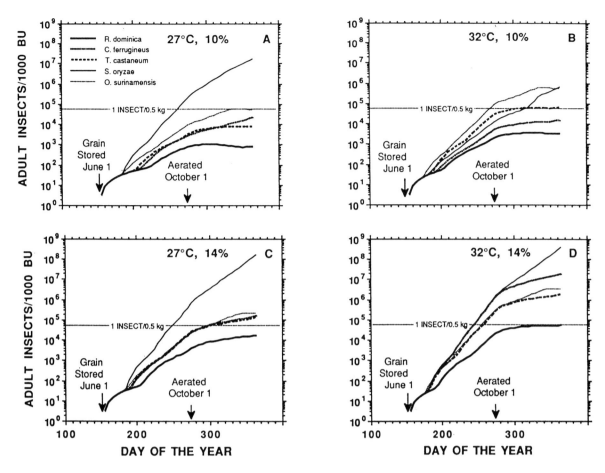

Fig. 2 Predicted effects of aeration temperatures and grain moisture levels on the growth of five stored-grain insect species with aeration completed on October 1.

as solar radiation on the side of storages, that can radically change localised grain temperature and cause spoilage due to rapid mould development when the grain is held in storage in a very moist condition. That is why aeration is necessary when holding grain above the safe storage levels listed in Table 1.

Moisture Migration

Grain that contains acceptable moisture levels and uniform temperatures can be stored safely. Maintaining uniform grain temperatures requires close management or thermally insulated storage. When grain is stored at safe moisture levels but is not aerated properly, uneven grain temperatures can cause movement of moisture through the grain mass, called moisture migration.

In cold weather, temperatures in the outer two to three feet (sides and top) of grain cool much faster than grain near the centre. Cold dense air settles by gravity through the outer grain to the floor, moves inward near the bottom of the storage. As air warms and becomes lighter, it expands and rises in the centre grain (Fig. 3), and its relative humidity drops.

For each 11°C (20°F) rise in air temperature, the percent relative humidity is cut in half. Air at 4.4°C (40°F) with 80% r.h. will drop to 40% r.h. when it warms to 15.6°C (60°F); this air is below the equilibrium relative humidity of most dry grain and will absorb moisture from the grain. When air moves through grain and warms to 11°C (20°F), it is typically below grain equilibrium moisture and absorbs moisture. As warm moist air reaches cold surface, it condenses moisture on the cold grain. Periodic warming of headspace air activates

moulds, causing grain to crust and seal over. Moisture migration, also called top crusting (Fig. 3), can occur even in safe grain moistures of 9–11% when grain is not properly managed and aerated.

Controlled Aeration — Stored Grain Pest Management Tool

Aeration is used to manage grain temperature by cooling grain to uniform temperature levels in the fall and winter. The general principles of controlled aeration for grain received at safe storage moistures are as follows: 'Harvest heat' should be left in the grain as a barrier to insect entry during the warm summer months. Insect populations will increase in the early fall when grain surface temperatures begin to lower. (Populations may build earlier if grain is cooled during the summer to temperatures that are ideal for insect reproduction and growth.) When a strong (3–5 day) weather front develops with air temperatures that will allow cooling grain to 13–18.5°C (55–60°F), aeration systems should be operated. The objective is to quickly cool the grain mass across the optimum range of insect feeding and breeding, 21–32°C (70–90°F), to temperatures below 13–18.5°C (55–60°F).

These ideal aeration weather patterns in Oklahoma and the south–west U.S. are the result of major fall air temperature changes. Figure 4 shows the average amounts of 55°F available in Oklahoma during the months of October–November based on 30–year weather data. Cold weather fronts begin arriving in Oklahoma and the south–west during mid to late September. Aeration systems with medium airflow, 0.15–0.30 m³/minute/m³ (0.20–0.40 cfm/bu Table 6), can cool the entire

Table 1. Maximum moisture contents for aerated grain storage

Grain type and storage time	Maximum moisture content for safe storage (percent wet basis)		
	South	Central	North
Shelled maize and sorghum			
Sold as #2 grain by spring	14	15	15
Stored 6 to 12 months	12	13	13
Stored more than 1 year	12	13	13
Soybeans			
Sold by spring	13	14	14
Stored 6 to 12 months	12	12	13
Stored more than 1 year	11	11	12
Wheat, oats, barley, rice			
Stored up to 6 months	12	13	14
Stored 6 to 12 months	11	12	13
Stored more than 1 year	10	11	12
Sunflower			
Stored up to 6 months	10	10	10
Stored 6 to 12 months	9	9	9
Stored more than 1 year	8	8	8
Flaxseed			
Stored up to 6 months	9	9	9
Stored more than 6 months	7	7	7
Edible beans			
Stored up to 6 months	13	14	15
Stored 6 to 12 months	12	13	14
Stored more than 1 year	10	11	12

Notes: 1. Values for good quality, clean grain and aerated storage. 2. Reduce one percent for poor quality grain, such as grain damaged by blight, drought, etc. Reduce each entry by two percent for nonaerated storage.
Adapted from Anon. (1987).

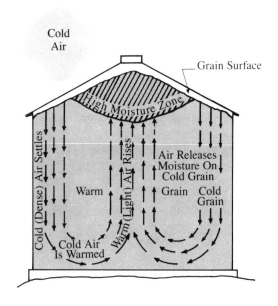

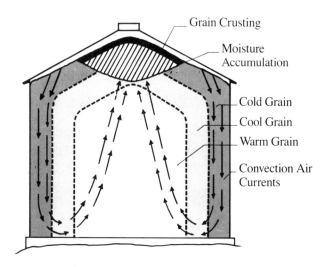

Fig. 3. Example of moisture migration or top crusting in summer-harvested grain stored several months without aeration.

grain mass in 36–72 hours. Thus, an operator can take advantage of those 3–5 day periods of cold air to gain earlier control of insects by rapidly cooling the grain. But, even if only 50–60% of the grain is cooled during the first cold front, cooling may be completed with the next cold weather in the next week or two, and the entire storage is placed in safer insect and mould conditions.

During one aeration cycle, the grain temperature will typically be reduced from 26.7–32.2°C (80–90°F) to 12.8–18.3°C (55–65°F), a reduction of 8–19°C (15–35°F). Grain that is not cooled in early to mid–fall (mid–September through early November) will develop various stages of moisture migration or top crusting. Early fall aeration eliminates most if not all of this temperature differential. Data needed to manage an aeration system are listed in Tables 1–6.

Aeration controller thermostats monitor ambient air temperatures and start the blower(s) automatically at preset temperatures that are ideal for grain cooling. The controller also stops the blower(s) if the air temperature rises above the thermostats preset upper temperature limit. It will precisely repeat these blower start/stop conditions at any time and as many times as the outside air temperature moves below or above the thermostat setting. The aeration controller makes sure that the cooling air is at least as cool or colder than the thermostat setting before the blower will operate.

An hour meter accumulates the hours of operation each time the blower runs. This provides the grain manager with the

exact amount of aeration time since aeration is started. Based on blower airflow rates, and the data from Table 6, managers can predict the amount of aeration time needed to finish a cooling front. As the hour meter approaches the calculated shut off time, the grain mass temperatures should be checked to see if the cooling front has moved through the grain. Aeration can be used to manipulate grain temperatures for insect management if a properly designed aeration controller is used for precise cooling of the grain.

Aeration System Management Problems

Data from throughout the U.S. indicate that many grain managers do not operate aeration systems properly. Mistakes include waiting too late in the fall to begin aeration, not operating aeration blowers under optimum temperature conditions, missing favourable weather pattern air temperatures, running blowers continuously night and day in temperatures that rewarm the grain, failing to operate blowers the required number hours to move a cooling front completely through the grain mass, or operating blowers excessively and removing excess grain moisture (reducing market weight). Timing and duration of early and mid-fall cooling fronts are difficult to predict. Thus, manual operation of aeration systems is difficult if not impossible to manage satisfactorily.

Aeration Controller Trials in the South-western U.S.

In Oklahoma, grain harvested in June is typically stored at safe storage moisture levels at temperature levels of 35–38°C (95–100°F), and often reaches storage temperatures of 40.5°C (105°F) or more. Dry wheat (11–12% moisture, wet basis) will store satisfactorily at 40.5–43.3°C (105–110°F). If aeration is used to cool wheat during the summer, the wheat is placed in a condition that favours insect infestation. Steel tank storage headspace day-time temperatures typically range between 45 and 57°C (113–135°F). During nights the headspace air varies between 29.5 and 35°C (85–95°F). Thus, steel tank surface grain temperatures cycle between these temperature extremes, a thermal environment that is not conducive to insect entry, but does cause protectants to degrade. Concrete silo sidewalls and roofs provide a thermal lag, so headspace temperatures change much more slowly in concrete than steel structures.

For Oklahoma, central and north Texas and southern Kansas, aeration systems should be prepared for operation by early September. National weather reports should be monitored closely to determine when the first usable cold weather will enter the storage region. From Kansas to south Texas, thermostats require different settings. Weather fronts of at least 3–5 days duration with temperatures in the 15–21°C (60–70°F) range can be used effectively. For these temperatures, upper limit thermostats should be set to switch when air temperatures drop to 15–18°C (60–65°F) for Oklahoma and Kansas; in central to south Texas, settings of 19–21°C (65–70°F) may be necessary to capitalise on early favourable cooling weather.

As the actual temperatures in a cooling front are determined, the controller thermostat should be set on the upper acceptable limit to achieve maximum cooling. If night temperatures drop to 12–13°C (53–55°F) and day–time peak temperatures are 22–24°C (72–75°F), consider setting the thermostat to minimise off time. In Oklahoma, early fall cold weather fronts typically last 3–5 days. In some years, several fronts may come during September a few days apart, but in other years, there may be 2–3 weeks before the second cold weather arrives. Figure 4 shows the available hours of 13°C (55°F) temperatures available in October and November, based on 30-year U.S. Weather Bureau data for Oklahoma. About 90% more hours are available each month at 15.5°C (60°F) than at 13°C. So, instead of setting the controller at 15.5°C (60°F), the ideal upper limit, grain managers may gain 50–75% more fan time by operating the blowers at temperatures 2–3°C higher just to get the overall cooling front completed before a cold weather front recedes.

Airflow Rates vs. Cooling Time

Completing the aeration cooling cycle in minimum time is very important for insect and mould management. The highest aeration airflow rates that can be justified economically should be used. Typical commercial aeration airflow rates are 0.04–0.08 m^3/minute/m^3 (0.05–0.1 cfm/bu). At these rates, completing one cooling cycle in the fall typically requires 150–300 hours at 0.08 at 0.04 m^3/minute/m^3. Blowers that deliver 0.20–0.40 m^3/minute/m^3 (0.25–0.5 cfm/bu) will reduce cooling time to about 40–80 hours. The 0.20 m^3/minute/m^3 flow rate would allow the grain manager to cool a storage in 80 hours, or just over 3 days of continuous operation. Tables 4 and 5 list approximate aeration blower power (kW) per 100 m^3 (2823 bushels) at specific airflow rates for a range of storage depths for wheat, maize, grain sorghum, soybeans and other grain crops. Table 6 lists approximate cooling hours required for aeration at specific airflow rates.

It costs more money to go fast. Increasing from 0.08 to 0.20 m^3/minute/m^3 involves an increase of 2.5 times the airflow, but about 6 times the power. Pressure (and thus power) increases as a function of the square of the airflow rate increase. Electric power delivery wiring and controls may also require expensive modifications. Local power company engineers should be consulted.

The cost of insect damage may easily offset a significant increase in power costs. Changes in blower size and power should be evaluated carefully. One replacement cost strategy for a storage facility with a variety of storage and blower sizes is to move large blowers to smaller storages to increase the airflow, then purchase and install bigger blowers on the larger storage.

If all storage units can be completely cooled with the first cold weather by using 0.20 m^3/minute/m^3 airflow rate rather than partially cooling it using 0.08 m^3/minute/m^3 airflow, the manager does not have to wait several days or weeks to complete cooling. Control of insect populations is established early. When the entire grain mass is cooled rapidly, from warm to cold temperatures in 2–5 days, insects will die in

Table 2. Estimated storage durations for shelled maize held at specific moistures and temperatures during which one percent dry matter loss occurs[a]

Temperature		Days				
°C	°F	14	15.5	17	18.5b	20b
		Moisture content (wet basis)				
10.0	[50]	256	128	64	32	16
15.6	[60]	128	64	32	16	8
21.1	[70]	64	32	16	8	4
26.7	[80]	32	16	8	4	2
32.2	[90]b	16	8	4	2	1
37.8	[100]b	8	4	2	1	0

Source: H. J. Raney et al. (1987).
[a]Conditions during which one percent dry matter loss occurs and quality is reduced by one market grade.
[b]Continuous aeration is required at moisture levels of 18 to 20 percent with grain and/or air temperatures above 80 °F.

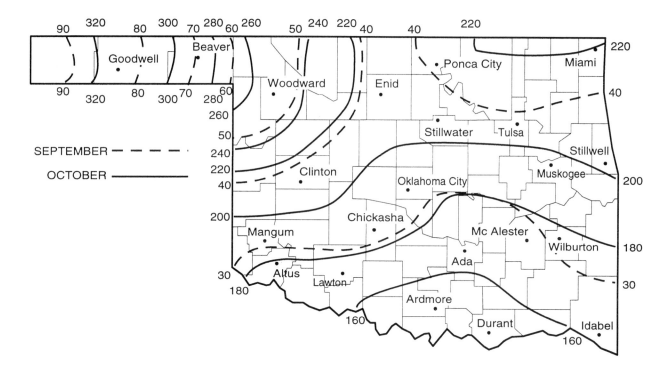

Fig. 4.   Hours of 12.8°C (55°C) weather available annually in Oklahoma during September and October (30 year weather data).

Table 3. Recommended minimum airflow rates for aeration

Crop	Moisture content	Cfm/bu range
Shelled maize,	14 percent and below	1/10 to 1/4
Sorghum	15 to 16 percent	1/4 to 1/2
	18 percent +	1/2 to 1
Wheat, Oats,	13 percent and below	1/10 to 1/4
Barley, Rice	14 to 16 percent	1/4 to 1/2
	17 percent +	1/2 to 1
Soybeans	10 to 11 percent	1/10 to 1/4
	12 to 13 percent	1/4 to 1/2
	14 percent +	1/2 to 1
Sunflowers	8 to 9 percent	1/10 to 1/4
	10 to 11 percent	1/4 to 1/2
	12 to 13 percent	1/2 to 1

Source: Stored Grain Management Handbook, Agricultural Engineering Department, Kansas State University, Manhattan, Kansas, USA.

about 2 weeks (Khan 1990). Then, not only has fumigation been replaced by grain temperature manipulation and management, but aeration for mould control has been completed, placing the grain in a safer condition against *moisture migration* and mould.

OSU Aeration Controller Designs

Aeration controllers have been developed at Oklahoma State University (OSU) for a wide range of grain storage facilities in State. These systems range from small corrugated steel 100–250 t (3500–10000 bushel) tanks with one vane axial blower per tank on farms, to a commercial grain facility with two 8000 t (30000 bu) steel tanks with 12 blowers per tank and another elevator with four 2250 t (80000 bu) concrete silos with 3 blowers per silo, or 12 blowers on one controller.

The aeration controller on the two 8000 t steel tanks operated eight 6 kW blowers mounted on base aeration ducts and four 2.5 kW roof exhausters on each tank, or a total of 24 blowers. Each tank has a separate control circuit panel for 12 blowers, but all 24 blowers operate from one set of thermostats in the *master* control box. Either tank can be operated independently of the other, or both can be operated simultaneously. Blowers are grouped with two 6 kW and one 2.5 kW blower operated as a unit by each time delay relay, with four groups of blowers on each tank. Blower groups are started at 7 second intervals; the entire system is on-line in 56 seconds.

All controllers use recording hour meters to log the continuous hours of blower operation time. Figure 5 shows an aeration controller used on commercial storage structures with multiple large power blowers. Each controller has a control selector switch for manual, off, automatic operation. The manual selector switch position is used for blower maintenance and service work, or to test various functions of the aeration system. It operates the blower motors directly. The automatic switch position sets the circuit for automatic thermostat on/off blower control, based on the set-point temperature. The hour meter records the time the blowers run, regardless of whether manually or automatically controlled.

Some aeration controller circuits use both high and low temperature thermostats. Two thermostats are used to limit the range of cold air temperatures forced into the grain, such as an upper limit setting of 15.5°C (60°F) and a lower limit setting of 4.4°C (40°F), to control grain temperatures precisely within an 11°C (20°F) band. Either high or low temperature switch opening will interrupt blower operation.

Figure 5 illustrates a typical commercial grain facility controller which uses only a high or upper temperature limit thermostat. This system was designed to operate multiple blowers on two grain tanks where the operator wanted separate hour meters for each set of blowers. Two relays and one time delay relay control the blower groups in this installation. In multiple blower systems, time delay relays are used to

Table 4. Approximate blower power, kW/100 m³ [HP/1000 bushels]-- wheat/sorghum.

Grain depth		Airflow rate m³/minute/m³ [(cfm/bu)]											
m	[ft]	0.8	[1]	0.6	[3/4]	0.4	[1/2]	0.2	[1/4]	0.08	[1/10]	0.04	[1/20]
4.6	[15]	3.5	[1.65]	2.3	[1.11]	1.0	[.47]	.23	[.11]	.08	[.040]	–	–
6.1	[20]	6.3	[2.99]	4.0	[1.91]	1.66	[.79]	.42	[.20]	.10	[.050]	.04	[.020]
7.6	[25]	14.3	[6.80]	7.0	[3.33]	2.65	[1.26]	.63	[.30]	.14	[.065]	.05	[.024]
9.1	[30]	20.0	[9.50]	11.0	[5.22]	4.04	[1.92]	.95	[.45]	.17	[.080]	.06	[.029]
10.7	[35]	29.2	[13.88]	14.9	[7.08]	5.88	[2.79]	1.37	[.65]	.21	[.10]	.07	[.034]
12.2	[40]	–	–	20.0	[9.51]	7.75	[3.68]	1.83	[.87]	.30	[.14]	.08	[.040]
13.7	[45]	–	–	–	–	10.87	[5.16]	2.44	[1.16]	.34	[.16]	.10	[.048]
15.2	[50]	–	–	–	–	13.33	[6.33]	2.82	[1.34]	.40	[.19]	.12	[.057]
18.3	[60]	–	–	–	–	20.1	[9.55]	4.33	[2.06]	.59	[.28]	.16	[.076]
21.3	[70]	–	–	–	–	–	–	5.94	[2.82]	.82	[.39]	.20	[.096]
24.4	[80]	–	–	–	–	–	–	7.67	[3.64]	1.05	[.50]	.27	[.13]
27.4	[90]	–	–	–	–	–	–	10.43	[4.95]	1.39	[.66]	.36	[.17]
30.5	[100]	–	–	–	–	–	–	–	–	1.66	[.79]	.42	[.20]

Sources: Anon. (1962, 1988),

Table 5. Approximate blower power kW/100 m3 [hp/1000 bushels] — maize/soybeans.

Grain depth		Airflow rate m³/minute/m³ [(cfm/bu)]											
m	[ft]	0.8	[1]	0.6	[3/4]	0.4	[1/2]	0.2	[1/4]	0.08	[1/10]	0.04	[1/20]
4.6	[15]	1.3	[0.61]	0.65	[0.31]	0.27	[.13]	0.08	[.04]	0.042	[.020]	–	–
6.1	[20]	2.5	[1.20]	1.2	[0.57]	0.5	[.24]	0.13	[.06]	0.048	[.023]	–	–
7.6	[25]	5.3	[2.50]	2.0	[0.95]	0.8	[.39]	0.21	[.10]	0.059	[.028]	0.021	[.010]
9.1	[30]	8.0	[3.80]	3.2	[1.54]	1.2	[.58]	0.29	[.14]	0.070	[.033]	0.023	[.011]
10.7	[35]	11.6	[5.50]	4.6	[2.20]	1.8	[.84]	0.42	[.20]	0.080	[.038]	0.027	[.013]
12.2	[40]	–	–	6.5	[3.10]	2.3	[1.11]	0.53	[.25]	0.090	[.043]	0.032	[.015]
13.7	[45]	–	–	–	–	3.3	[1.55]	0.67	[.32]	0.110	[.052]	0.036	[.017]
15.2	[50]	–	–	–	–	4.0	[1.90]	0.86	[.41]	0.135	[.064]	0.040	[.019]
18.3	[60]	–	–	–	–	6.0	[2.86]	1.3	[.61]	0.204	[.097]	0.048	[.023]
21.3	[70]	–	–	–	–	–	–	1.9	[.90]	0.284	[.135]	0.063	[.030]
24.4	[80]	–	–	–	–	–	–	2.6	[1.25]	0.379	[.18]	0.080	[.038]
27.4	[90]	–	–	–	–	–	–	3.5	[1.65]	0.484	[.23]	0.097	[.046]
30.5	[100]	–	–	–	–	–	–	–	–	0.653	[.31]	0.114	[.054]

Sources: Anon. (1962, 1985).

sequence the startup of all blowers or groups of blowers at approximately 6–8 second intervals.

An electrical circuit schematic diagram for two simple controllers is illustrated in Figure 6. A more complex controller circuit diagram is shown in Figure 7. Both circuits use a basic 'ladder' diagram for wiring and circuit analysis. All aeration controllers are set up so the original blower motor starter manual operating switches can still be used for controlling individual blowers when the aeration controller is set to the off position.

A typical component parts list for an aeration control with time delay relays that can operate 4 blower motors or groups of motors is shown in Table 7. Prices of the materials for the controller are 1990 US$. After building one or two units, an experienced electrical technician can assemble, wire and test these boxes in 1–2 days, depending on the complexity of the system and diagram (Noyes, et al. 1992a,b).

Electromechanical aeration controllers built from standard off-the-shelf components are reliable, easy to troubleshoot and repair, and are relatively inexpensive. Aeration controllers usually pay for themselves in less than one year. An electri-cian's labour costs US$30/hour for 8 hours assembly time, or US$240 labour per controller, parts are US$318 (Table 7) and it costs US$120 for installation: that's US$678 for a unit that will operate all blowers reliably and accurately in a 1–3 tank complex. Compared to computerised controllers that cost US$3500 without installation, and require considerable learning to operate effectively, the OSU electro-mechanical controller is a good investment. If a relay, timer, thermostat, switch or hour meter fails, pull it out and plug in a new one, purchased locally. Computers and microprocessors are wonderful for complex processes. Aeration management need not be a complex process.

OSU's Aeration Management System

Aeration is part of the overall stored-grain management plan. Developing an aeration controller system to operate aeration blowers without including other 'best management storage practices' does not constitute an effective insect control program. Individual grain management systems require different levels and amounts of management input and time.

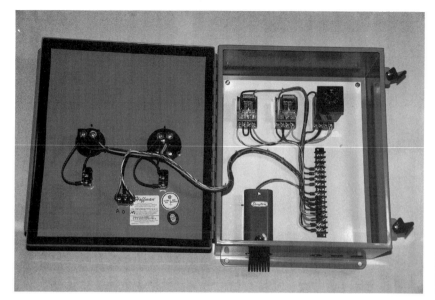

Fig. 5.   Typical electro-mechanical aeration controller used for commercial elevators.

Producers and elevator operators must develop stored-grain management strategies depending on their location, facility, product, and harvest time (Noyes et al. 1989a,b; 1990).

Stored grain aeration management strategies include the following:

• planned aeration using automatic aeration control—an aeration controller using one thermostat system (both high and low temperature settings on some systems) operating at preset temperatures controls all blowers on multiple tanks or silos in sequence

• maintaining grain temperatures above or below the optimum insect feeding and breeding range of 21–32°C (70–90°F)

• high airflow — using higher than the normal airflow rates of 0.08 m³/minute/m³ (0.1 cfm/bu) allows shorter aeration times (see Table 6); instead of 120 hours for a complete cooling cycle at 0.08 m³/minute/m³ (0.1 cfm/bu), using 0.20–0.40 m³/minute/m³ (0.2–0.5 cfm/bu) would reduce cooling time to 25–60 hours

• maximum use of available favourable cool or cold weather systems, using high airflow rates

• sealing tank, bin or silo base openings — keeps cold air from draining out of grain mass and pulling in warm air, and keeps insects from entering at base; grain managers can monitor insects in the grain storage headspace

• roof blowers/vents open except when fumigating — allows air exchange that minimises humid air buildup in headspace and roof condensation

• housekeeping/cleanup—thoroughly clean bin aeration ducts and unload auger trenches (where insects thrive on grain dust and foreign material) as well as other areas of storage

• empty tank or silo pesticide spray and/or fumigate— to remove initial insect populations before fresh harvested grain is stored; very important if aeration ducts and unload augers are not thoroughly cleaned or vacuumed

• grain cleaning—removes grain dust and fines that insects thrive on, and improves aeration; clean grain provides less insect attraction. Aeration is much more productive — blower operating times are typically 25–50 % less for clean grain than grain with significant levels of trash and foreign matter

• grain spreading/leveling, or pulling core of fines from silo or tank by running centre unload conveyor — improves management inspection and sampling, aeration (eliminates core of fines and foreign material under fill point), grain temperature maintenance, fumigation

Table 6.   Airflow rate, m³/minute/m³ vs. cooling time, hours[a]

Season	Low aeration (hours)		Medium aeration (hours)				High aeration (hours)		
	0.04	0.08	0.16	0.24	0.32	0.4	0.5	0.6	0.8
Summer	180	90	45	30	24	18	15	12	9
Fall	240	120	60	40	30	25	20	15	12
Winter/Spring	300	150	75	50	40	30	25	20	15

[a] Assumes clean grain at safe storage moisture. Grain that is peaked and has foreign material concentrated under the fill point(s), cooling may require 50% additional time or more.

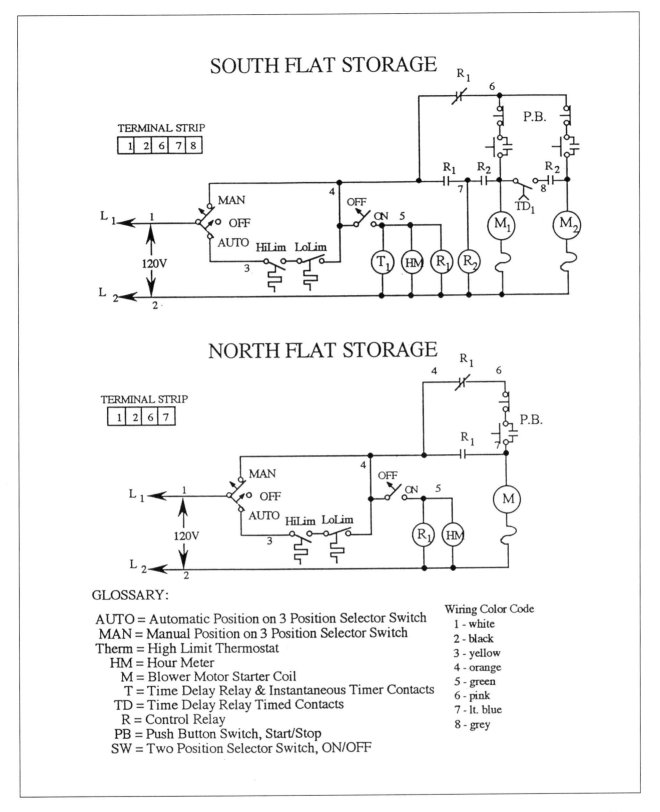

Fig. 6. Wiring schematic diagram for aeration controller circuits using high and low temperature thermostats to control one or more blower motors.

- temperature cable thermocouple reading/recording — provides periodic grain mass thermal profile, a valuable management asset
- grain monitoring/sampling — provides grain moisture and temperature, and insect and mould management data
- fumigation — is a backup system if grain has infestation at or above economic population thresholds and weather for grain cooling is not available; storage prepared for

controlled aeration management by sealing roof vents and blowers, and fill points is ready for fumigation.

When properly coordinated, these management practices will help maintain grain quality, reduce inputs, and preserve grain quality. To effectively store grain, it must be managed from a total systems approach. Stored grain should be inspected regularly. Key management components including grain moisture and temperature, insect and mould population

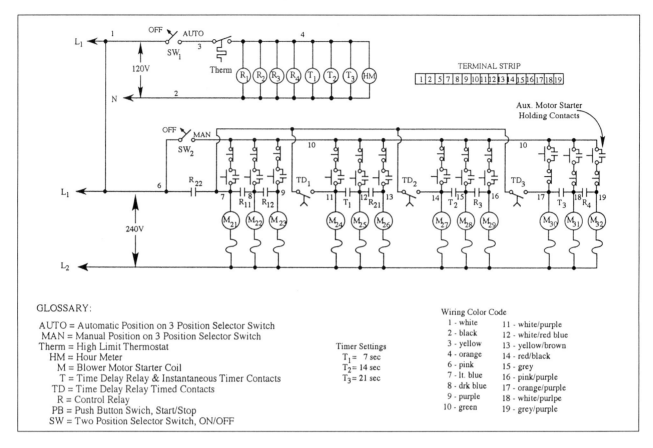

Fig. 7. Wiring schematic diagram for aeration controller circuits using a high temperature thermostat to control 12 blower motors.

levels, and the storage structure should be checked. The following sections discuss selected grain management strategies.

Aeration Management Data

Data from throughout the U.S. indicate that many producers and grain managers do not operate aeration systems properly and do not run their blowers the correct number of hours. This may be because they do not have the correct management data at their disposal. The use of automatic aeration controllers with hour meters to optimise aeration blower time should be a priority stored-grain management strategy. Grain should be aerated during major seasonal air temperature changes in the fall and winter to prevent moisture migration. Some of the data needed to manage an aeration system properly are listed in Tables 1–6. Seal Storage Base and Sidewall Openings
Seal all tank and silo base and sidewall openings, including aeration blowers, augers, slide gate push rods, U-trough covers, foundation cracks, missing bolt holes and sidewall doors or access ports. This sealing process must be extremely thorough. Insects can and will enter any small access opening. Sealing the base restricts insect access to the top of the structure where it can be more easily monitored. Sealing auger and blower or blower inlet or outlet openings will prevent cold air from leaking out of the storage and warm convection air currents from moving up through the storage, removing grain moisture. Use professional fumigation sealing materials. Seal for non-leak fumigation, then leave storage openings sealed except when in use or when cleaning.

Note: Do not seal roof aeration exhaust or inlet vents except for fumigation. The storage head space must have free air movement to minimise humid air buildup and roof condensation.

Level vs. Peaked Grain Surfaces

Level grain is much easier to manage than peaked grain. Peaked grain, shown in Figure 8, is difficult to manage because of several problems:

- peaked grain is very difficult to cool; after cooling, peaked grain rewarms rapidly — temperatures can not be controlled
- at least 25–50% more aeration time is required to cool peaked grain compared with storages with grain surfaces that are level or slightly rounded
- grain protectants deteriorate more rapidly in hot, peaked grain
- peaked grain has a much larger surface area than level grain (visualise the outer surface of a cone vs. the base area of the cone); grain rewarms rapidly in peaks due to warm head space temperatures, which allows insect and mould populations to accelerate
- fumigation of peaked grains is more difficult and generally not as effective
- grain that is peaked usually has a core of foreign material down the centre of the grain mass; this core is difficult to cool, and attracts and harbours insects populations.

An effective method for cleaning out centre concentrations of fines and trash from peaked grain is to unload the centre grain from each 0.6–1.2m layer as tanks or silos are being filled or loaded. After the final fill layer, remove the peak grain to an inverted cone approximately half the bin diameter across the top (Fig. 9), then level the grain or leave the shallow depression. Even when a tank or silo is completely filled before removing grain, running the centre unload conveyor to pull the peak out will remove a core of grain about 25–50 cm in diameter that contains a high concentration of grain fines. This action will open the centre and loosen the grain mass, allowing

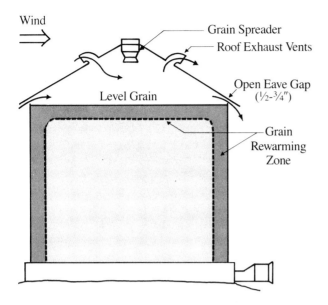

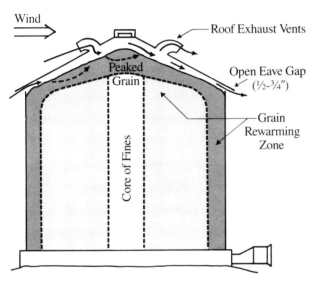

Fig. 8. Peaked grain vs. level grain surface in storage bins.

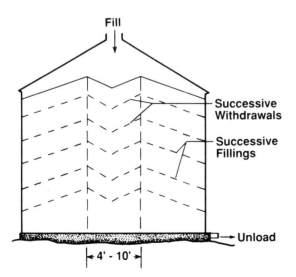

Fig. 9. Withdrawing layers of grain during filling to remove most fines and foreign material from a 1–2 m centre core.

improved air penetration of the centre of the grain mass, especially with the peak depth removed.

To determine when the peak grain has been pulled out, place confetti or newspaper shreds in the grain at the top and scatter them in continuously from the peak about half way down the slope. Observing when the confetti or pieces of paper stop coming out will indicate when the grain surface cone is about half the storage tank diameter.

Monitoring and Sampling

Monitor grain and sample for insects every 3–4 weeks throughout the storage period. Fumigate if insects exceed economic threshold levels — population levels of insects that will likely cause significant economic losses if not treated. If a storage has an area of warm grain that is infested, complete fumigation may be required if grain cannot be turned. Deep cup probes or vacuum samplers can be used to determine the extent of insect infestations.

Probe traps are recommended and can be used to monitor populations and make treatment decisions. Economic thresholds for probe traps need to integrate grain temperature, insect species and market destination criteria. (Fargo et al. 1989, 1994). Close monitoring during late August–early September allows managers to determine whether aeration may occur soon enough or if they may need to fumigate.

Aeration with cold air can help control insect populations if detected early before they reach economic threshold populations. However, insects generally infest peaked grain and cores of fines or wet areas below fill points where moisture condensation has occurred. These areas are often dense and sometimes crusted over with mould which restricts or blocks forced air movement.

Summary

Controlled aeration is a major management tool for elevators and grain storage facilities in the south-west United States. Many of the grain management principles discussed in this paper can be applied in other areas of the world, especially in regions where climatic conditions are similar. Principles such as sealing storage to keep insects from entering storage bases and using head space temperatures and retaining 'harvest heat' can be used to deter insect infestation as long as grain is stored at safe moisture levels and is monitored for pest problems.

References

Anon. 1985. Grain Storage Management Workshop Handbook, Farmland Industries, Inc., Kansas City, Missouri, USA.

Anon. 1962. Aeration in country elevators. Marketing Research Report No. 178, Agricultural Marketing Service, United States Department of Agriculture, Washington, D.C., USA.

Anon. 1987. Managing dry grain in storage, Publication No. AED-20, Midwest Plan Service, Iowa State University, Ames, Iowa.

Cuperus, G. W., C. K. Prickett, R. D. Bloome, and J. T. Pitts. 1986. Insect populations in aerated and unaerated grain storage in Oklahoma. Journal of the Kansas Entomological Society, 59:620–627, Kansas State University, Manhattan.

Cuperus, G. W., R. Higgins, and E. Williams. 1989. Integrated pest management, In: 1989 Oklahoma grain elevator workshop manual, Circular E-881, Cooperative Extension Service, Oklahoma State University, Stillwater.

Fargo, W.S., G.W. Cuperus, E.L. Bonjour, W.E. Burkholder & R.T. Noyes. 1994. Influence of probe trap type and attractants on the capture of four stored-grain coleoptera. Journal of Stored Products Research (In press).

Fargo, W.S., D.R. Epperly, G.W. Cuperus, B.L. Clary & R.T. Noyes. 1989. Influence of temperature and duration on the trap capture of four stored grain insect species. Journal of Economic Entomology. 82: 970–973.

Hagstrum, D. W., and P. W. Flinn. 1990. Simulations comparing insect species differences in response to wheat storage conditions and management practices, Journal of Economic Entomology, 83:2469–75.

Harner, J. P., and R. T. Noyes. 1988. The principles of grain aeration, Section 6, In: Stored grain management, Kansas Cooperative Extension Service, Kansas State University, Manhattan.

Khan, N. I. 1990. The effects of various temperature regimes and cooling rates on the mortality and reproductive abilities on two stored grain insect species. Thesis, Oklahoma State University.

Noyes, R. T., D. R. Epperly, B. L. Clary and G. W. Cuperus. 1992. Elevator managers summary report—Oklahoma wheat elevator electrical energy reduction demonstration project, OSU Grant No. 3881 OIL/SECP 88, Oklahoma Cooperative Extension Service, Oklahoma State University, Stillwater, 28 p.

Noyes, R. T. 1991. Aeration of Texas coastal region grain storage: critical management considerations, In: Proceedings, South Texas grain quality conference, Corpus Christi, Texas, May 22–23, 32p.

Noyes, R. T. 1990. Aeration of grain storage. In: Proceedings, 3rd national stored grain pest management training conference, Kansas City, Missouri, Oct. 20–25.

Noyes, R. T. (ed.), and B. L. Clary. 1989 a. Aeration of commercial grain storage, In: 1989 Oklahoma grain elevator workshop manual, Circular E-881, Cooperative Extension Service, Oklahoma State University, 135 p.

Noyes, R. T., B. L. Clary, H. A. Cloud, and B. A. McKenzie. 1989 b. Aeration of stored wheat, In: Wheat pest management handbook, USDA Extension Service, Washington, D.C., 68p.

Noyes, R. T., B. L. Clary, and G. W. Cuperus. 1990. Maintaining quality of stored grain by aeration. OSU Extension Facts No. 1100, Oklahoma Cooperative Extension Service, Oklahoma State University, Stillwater, 4p.

Raney, H. G., et al. 1987. Management of on-farm stored grain, Cooperative Extension Service, University of Kentucky, Lexington, 98p.

Closed loop fumigation systems in the south-western United States

R.T. Noyes* and P. Kenkel†

Abstract

'Closed Loop Fumigation' [CLF] was developed and used in the early 1920s for methyl bromide fumigation in the U. S. and other major grain producing areas. A 'new' low power closed loop fumigation process using low volume blowers and duct systems was developed for commercial phosphine fumigation use in the late 1970s by James Cook. He patented the process in 1980.

CLF installations were installed in wheat storage structures at Kansas and Oklahoma elevators in the mid-to-late 1980s and data were recorded. CLF systems were installed at Oklahoma elevators in 1993 as part of a demonstration project funded by the Oklahoma Department of Commerce. Design and economic data have been developed on some CLF systems installations and operations.

New design methods for installing plumbing and blowers for increased operating flexibility were developed to reduce installation and operating costs. Multiple tanks have been connected to one blower and blowers are moved from site to site to minimise capital investment. Preliminary results have been documented on operating procedures and costs.

Benefits of CLF systems are: 1. reduced worker exposure; 2. quicker fumigation response time; 3. lower operating cost for fumigants through reduced labour and possible lower fumigant requirement; 4. reduced regulation compliance costs; and 5. potentially better fumigation efficacy with the same management expertise.

Background

Closed loop fumigation (CLF) was originally known as a recirculation process developed for methyl bromide fumigation in the U.S. and other major grain-producing areas. Reports cite recirculation of methyl bromide as early as the 1920s.

A new low power, low volume closed loop gas recirculation fumigation process using centrifugal blowers and duct systems was developed in the late 1970s for commercial fumigation using phosphine tablets, pellets, packets, or strips. James S. Cook of Houston, Texas received U.S. Patent No.4,200,657 on this process on 29 April 1980 (Cook 1980).

CLF Benefits

Although initial installation costs of CLF systems are substantial, financial payback can be relatively short, and elevator worker satisfaction is enhanced. Compared with conventional probe and tarp fumigation methods, management of CLF fumigation is simplified due to: (1) improved timing for multiple tank fumigations; (2) reduced labour; (3) reduced housekeeping; (4) less grain damage losses and operating expenses from handling compared with fumigation while turning; (5) reduced fumigant cost; and (6) faster fumigation response, application, and purge timing.

CLF Fumigation Procedures

In CLF, phosphine pellets are preferred over tablets because of quicker, more uniform gas release. Pellets can be placed on the grain surface, probed into the grain 10–20cm, or placed in a shallow layer (not over 2 pellets or tablets deep) on a board, tarp, metal or plastic sheet on the grain for ash recovery. Pellets or tablets enclosed in packets, blankets or strips can be used to recover phosphine dust. After placing or probing the phosphine pellets and sealing the structure, the CLF blower is turned on. It pulls the gas mixture from the storage head space through a small diameter (7–15 cm) duct, pipe or hose into the suction side of the blower, then pushes the gas through a duct into the base of the storage, forcing it up through the grain back to the storage bin headspace. The blower is usually operated 5–8 days, depending on grain and weather conditions.

Well designed CLF systems mix the gas and air thoroughly. In a tightly sealed silo or grain tank, fumigant use may be reduced to 50–75% of amounts (or minimum label rates in some cases) used in conventional probe and tarp methods. CLF should not be used on poorly sealed storages.

With the CLF system in place, response time to begin fumigation of a large number of silos or storage tanks requiring several days preparation and application time by a group of workers can be reduced to a few hours using two or three people. Fumigant pellet placement for all storage units can be made the same day. By fumigating all storage units at the same time, reinfestation of fumigated storages by insects migrating from adjacent non-fumigated tanks or silos is eliminated. Fumigant reductions of 25–50% are reported with improved results, due to continuous recirculation and uniform distribution. In some cases, operators find they are able to reduce dosages to minimum label rates, depending on the facility.

Gas Distribution Designs

Typical pipe sizes are 10 cm internal diameter connected to the inlet and outlet of a 0.0746 kW (0.10 hp) blower, and 12.7–15.25 cm, suction and pressure piping for a 0.375 kW (0.5 hp) blower. At piping junction tees or crosses, 12.7 cm pipes may be used for lateral conduits requiring reduced gas flows. A 0.0746 KW blower will produce about 5.5–6.0 m^3/minute (about 190–210 cfm) while the 0.375 kW blower will deliver 20–23 m^3/minute (700–800 cfm). Air velocities are so low in grain that there is little flow resistance from the grain. Tube, hose or pipe sidewall flow resistance produces most of the blower pressure loss.

---

* BioSystems and Agricultural Engineering Department, Oklahoma State University, Stillwater, Oklahoma, USA 74078–0469.

† Department of Agricultural Economics, Oklahoma State University, Stillwater, Oklahoma, USA 74078–0505.

Concrete silos

In 4.5–7.6 m inside diameter concrete or steel silos, an open bottom pipe or tube that discharges at the bottom of the vertical sidewall or extends downslope to the centre of the silo is the typical design. Figure 1 shows three pipe and blower configurations in one-way sloped or cone hopper bottom silos. In 9.0–15.0 m diameter concrete silos, 2, 3, or 4 pressure pipes spaced evenly around the tank or silo perimeter manifolded to a single blower provides better gas distribution. In hopper bottom tanks, a single pipe run down the slope to within 0.6–1m of the bottom of the hopper provides excellent uniformity as the hopper acts as a gas distribution funnel.

Figure 2 illustrates the piping setup for CLF blowers to vent phosphine gas from the storage. If explosion-proof motors, switches, wiring conduits, and controls are used that meet electrical codes, the entire piping system can be mounted inside the silo with the blower mounted on roof or wall brackets, or inside elevator head houses with the motor mounted on the floor. If explosion-proof components are not used, all motors, switches, junction boxes, and wiring must be mounted outside the bin or elevator head house in open air. Suction pipes are not required for blowers mounted in the headspaces unless blowers are used for venting Fig. 2.

Grain pressures in 30–40 m grain depths place great stress on piping mounted inside silos, so fastening pipes securely to inside walls is critical. Pipe mounted to inside walls should be securely fastened at 1 m intervals. If an interior ladder is available, mounting the duct against the wall and ladder brackets or side rails provides construction convenience and structural stability.

Warning. PVC pipe should not be used inside silos, due to static electricity generated by sliding grain on plastic pipe, unless each pipe is well grounded by a knowledgeable electrician.

Continuous grounding must be installed in PVC plastic pipe to continually discharge static voltage to a positive ground such as metal water piping or concrete reinforcing bars.

Plastic PVC pipe is a popular duct material for use on the outside of storages due to light weight, chemical resistance, low cost, and ease of fabrication and assembly. If an outside blower is used for purging, disconnect the suction hose at the blower inlet, open the roof hatch and turn the blower on.

For external CLF blower assembly, the suction pipe must be extended through the silo roof into the head space (Fig. 2). The pressure pipe can be installed through the roof and along the inside wall to the base [e.g. secured to the ladder siderail down the wall], or it can be installed down the outside of the tank and enter the grain inside the tank at the base of the wall. The blower can be installed near the base with a long suction pipe and a short pressure pipe, or on top of the silos with a short suction line and a long pressure line. If aeration systems are involved, the external pressure pipe can be extended through the aeration blower manifold, or can be manifolded from one main line down the side of the tank to two or more aeration blower transitions.

Steel tanks

On 20–40 m diameter tanks with multiple aeration blower and transition positions, a CLF piping system design using two or three small blowers with short pressure pipe runs and small suction piping may be simpler and less expensive than trying to use one large blower with extensive larger piping or hose systems. Getting uniform gas distribution is more difficult in large diameter tanks than it is in tall silos, where the silo diameter is much smaller than the grain depth (Fig. 3).

Based on Cook's patent (Cook 1980), recirculation systems should be designed to provide a gas flow range of 0.0024–0.0064 m^3/minute/m^3(0.003–0.008 cfm/bu).

Higher gas flow rates of 0.004–0.0064 m^3/minute /m^3(0.005–0.008 cfm/bu) will offset poor distribution duct patterns and accelerate getting lethal gas levels to all parts of the storage. Figure 4 shows the CLF system modified so blowers are used for venting the gas when fumigation time has been completed. Tanks with aeration systems should use aeration blowers to vent the fumigant gases. Immediately after venting is completed, operators should reseal aeration blower openings to keep insects from reinfesting the storage at the base level.

On a 10600 m^3(300000 bushel) welded steel tank in Kansas City, two 0.062 kW (1/12 hp) centrifugal blowers were plumbed to 15 cm (6 inch) PVC suction pipes fitted to openings half way up the roof slope. The outlet of each of the blowers was plumbed by flexible hose to 10 cm (4 inch) pressure pipes manifolded to three aeration blower duct transitions per blower, placed symmetrically around the tank base. Figures 3 and 4 illustrate the layout concept for welded steel tanks.

Construction piping options

Construction costs vary widely based on the difficulty of installing piping through roofs and securing the pipes to the sidewalls internally. External pipe mounting on steel tanks is relatively simple as brackets can be spaced 2.4–3.6 m (8–12 feet) apart. Self-tapping or threading bolts can be used to fasten pipe mounting brackets on corrugated steel tanks with thinner steel walls. Pneumatic conveyor piping brackets may be used for CLF piping. Galvanized angles or channels with standard U-bolts or formed all-thread U-bolts make a good anchoring scheme. Piping half-bands welded to angles for welding or bolting to bins or silos are another alternative. Used aluminum irrigation pipe may be available at salvage prices in some areas for CLF systems use; 10, 12.7 and 15 cm diameter aluminum tubing or pipe make good CLF gas flow ducts and irrigation tube connectors make suitable joints pipe.

Sealing structures

Sealing bin or silo openings is a key factor in CLF system operation. Welded steel and concrete tanks are usually sealed better than bolted steel tanks unless the bolted tanks were well caulked. Roof to sidewall air gaps and the space between roof panel ridges and fill rings are critical sealing areas in corrugated steel tanks. Exposed roof panel ends under the fill ring flashing collects grain dust and makes a natural breeding place for insects. These openings should be sealed with a foam sealer. For standard bolted tanks without intensive caulking, recirculation airflow rates should be higher than for welded steel or concrete tanks, in the 0.0056–0.008 m^3/minute/m^3 (0.007–0.010 cfm/bu) range.

One key to CLF is sealed structures that hold an adequate gas level to maintain 150–200 ppm for at least 48 hours so gas can penetrate kernels and kill insect larvae and eggs. Commercial fumigators claim a complete kill of all insect stages is possible at 125–150 ppm if held for 3–4 days. Low grain/air temperatures cause slow gas release which can extend CLF blower operation from 5–8 days, which makes sealing critical. If tanks are well sealed with good aeration ducts, CLF systems with 0.373–0.0746 kW blowers will work satisfactorily on 8800–10600 m^3 (250000–300000 bu.) tanks. The 0.373 kW blower can provide adequate gas flow for two large tanks (Noyes 1993).

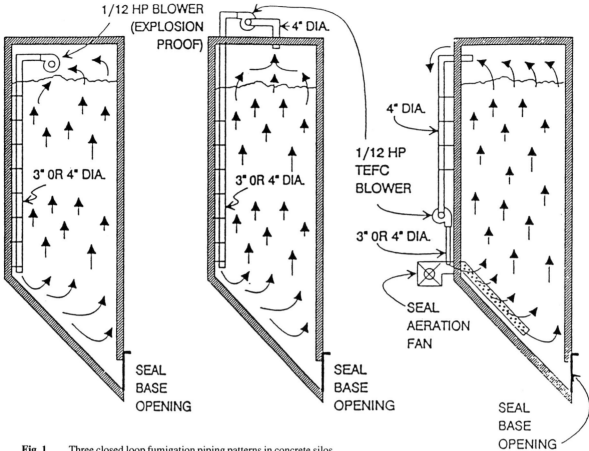

Fig. 1.   Three closed loop fumigation piping patterns in concrete silos.

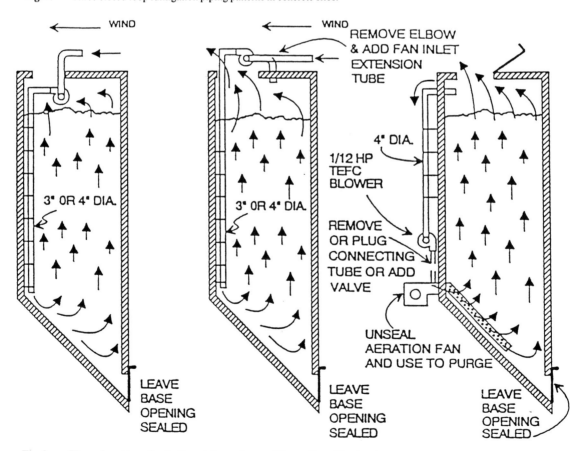

Fig. 2.   Three closed loop fumigation piping patterns set for venting of fumigant gases.

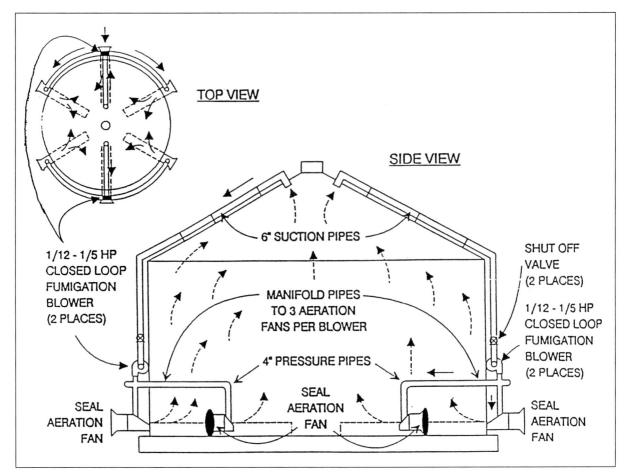

Fig. 3.   Large steel tank with closed loop fumigation pipes connected to aeration ducts.

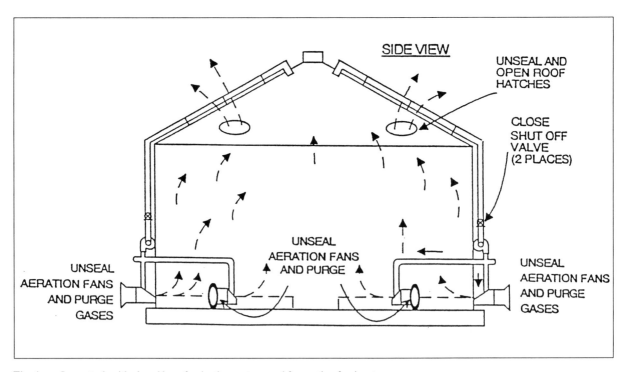

Fig. 4.   Large tank with closed loop fumigation system used for venting fumigant gas.

Corrugated steel tanks lose gas continually during fumigation if sidewall seams and eave gaps, roof vents, aeration blowers, unload conveyors, sidewall entry doors, and bolt holes are not well sealed. With gas leaks, too much fumigant may be lost, especially if high winds cause serious in-flow of air and leakage or out-flow of air and gas, diluting gas levels. New bolted steel tanks to be used with CLF should be carefully sealed for gas leaks at wall/roof joints and eaves during construction.

Distribution of the gas is also very important. The key is that the base duct system must provide fairly uniform distribution. The lower the gas flow rate, the more critical the uniformity of gas distribution becomes.

Blower Specifications

Blowers used for phosphine gas handling should be manufactured from materials that are resistant to chemical deterioration. Aluminum or plastic wheels and housings are preferred because they are also spark resistant. Steel blower wheels and housings should be coated with epoxy or other tough, spark resistant materials.

Gas flow rates range from 0.0016–0.008 m^3/minute/m^3 (0.002–0.010 cfm/bu) to provide a total air change every 50–200 minutes or about 6–24 changes per day. According to Cook's patent (1980), 0.00056 m^3/minute/m^3 (0.0007 cfm/bu), a 0.0746 kW blower delivering 5.95 m^3/minute in a 10,620 m^3 tank (210 cfm in a 300000 bushel tank), is a satisfactory air flow. Normal aeration at 0.08 m^3/minute/m^3 (0.1 cfm/bu) moves one air change in a full bin in 5 minutes, 20 times faster than a CLF blower delivering 0.004 m^3/minute/m^3 (0.005 cfm/bu), which requires 100 minutes.

Low gas flow rates of 0.0016–0.004 m^3/minute/m^3 [0.002–0.005cfm/bu] with air exchange times of 250–100 minutes [4.16–1.67 hours/cycle] are quite low relative to tank or silo volume.

A basic closed loop fumigation system blower and duct design for a single 6 m × 30–40 m concrete silo is a 0.0622–0.0746 kW centrifugal blower with a 10 cm inlet and outlet that operates at 250–500 Pa (1/2 inch water column, [W.C.]) static pressure at a rated airflow of 160–190 m^3/minute, or about 0.004–0.008 m^3/minute/m^3 through 700–1800 m^3 silos and tanks; this would provide one air exchange every 1–2 hours or 12–24 air changes per day.

In concrete silos where internal vents are difficult to seal, combining the silos into a larger storage volume may simplify the installation and reduce installation expense. Suction and pressure pipes from multiple silos can be manifolded to one larger blower to simplify operation and reduce control costs.

For 1770–3540 m^3 concrete silos (50000–100000 bushels), 0.125–0.250 kW blowers delivering 7–10 m^3 (250–350 cfm) at 250–375 Pa (1–1.5 inches W.C.) combined suction and positive static pressures are recommended. Steel tanks with 3500–14000 m^3 (100000–400000 bushels) volume may use one 0.188–0.377 kW blower or use two lower hp blowers, depending on the layout of the aeration system and the piping required to get good gas distribution.

Gas Level Monitoring

To develop valuable historical use date, monitor gas concentration levels daily at key locations in the storage throughout the fumigant recirculation period during the first time that a new CLF system is used. Records of previous gas level readings provide valuable management data. Operators should probably use high label rates (at least 75% of maximum dosages) during the first application of the new system, to make sure gas levels are adequate. If initial gas

readings are sufficiently high, application levels may be reduced by stages during future fumigations. If satisfactory gas levels and kill results warrant, application can be reduced to minimum label rates.

However, if fumigant levels drop below minimum recommended levels during the first 24–48 hours of operation due to gas leakage, more fumigant may have to be added. If fumigant must be added, it should be done by workers wearing appropriate personal protective equipment, or by pouring pellets through a small roof opening using a long probe tube which can be moved to distribute the pellets across the grain surface. Workers should stand up wind of the pour spout and use caution.

Purging phosphine with standard aeration blowers can be done in a few hours. If CLF blowers are used, at least 1–3 days of continuous blower operation should be used. Exhaust air should be sampled from tank head spaces through tubes run through bin wall openings to make sure the air inside is safe for personnel entry. Appropriate personal protection equipment should be used when remote sampling tubes are not used and gas levels are sampled directly from silo or tank head spaces or from the exhaust air.

Venting Fumigant with CLF Blowers

Once fumigation is complete, tanks or silos must be adequately aerated before workers can re-enter storage units or grain can be shipped. For storages with aeration, unseal roof vents and use aeration blowers to vent gas, then reseal blowers to block reinfestation of the silo or tank at the base level.

Warning. Keep all personnel away from open CLF system exhaust tubes or vents when using aeration or gas recirculation systems to purge tanks or silos of fumigant to avoid phosphine gas poisoning. Exhaust tubes should be ducted outside away from personnel areas.

When using CLF blowers for purging silos or tanks, open vents or hatches must be located away from the fresh air supply of the blowers. The fresh air supply may need to be ducted to the blower from several feet away so exhaust air is not recirculated. Prevailing winds need to be considered and tall standpipes may need to be used to avoid dilution of fresh air. If blowers are inside the storage as shown in Figure 2, the blower air supply must be controlled from outside to avoid the need for self-contained breathing apparatus equipped personnel to change the piping before venting or purging the fumigant. (Noyes 1993) On external mounted blower and piping systems where the tank or silo has no aeration system, remove the suction return pipe and open roof exhaust vents or doors and operate the blower to purge the tank.

For storages without aeration, use CLF blowers for forced air purging (Figs 2 and 4). Storages can be purged much faster and to much lower gas levels using CLF blowers than by conventional gravity draft venting. Storage bases can remain sealed, minimizing insect reinfestation.(Noyes et al. 1989a,b) CLF blowers that provide 0.0016–0.0064 m^3/minute/m^3 (0.002–0.008 cfm/bu) airflow should be operated continuously for 1–3 days when venting storages because of non-uniform air distribution. Operate aeration systems rated at 0.04–0.4 m^3/minute/m^3 (0.05–0.5 cfm/bu) 3–5 hours for venting.

Regardless of the method of venting the gas, monitor the air quality or gas level in each storage structure with appropriate gas sampling equipment, preferably through remote sampling tubes, before entering the storage. Air samples must be taken at entrances and in work areas of the storage and recorded to confirm that the fumigant has been satisfactorily purged. Workspace air samples should be taken and data recorded before workers resume normal re–entry, to ensure safe concentration levels of phosphine gas below the 0.225 ppm (0.3 ppm × 0.75 — phosphine gas monitoring tube accuracy is ± 25%) for phosphines.

Field Test Results

In well sealed storage structures, closed loop systems require less fumigant than conventional probe and tarp fumigation methods. Field data from closed loop systems in operations in Texas, Oklahoma, and Kansas are listed in Tables 1 and 2 (Noyes et al. 1989a,b) Table 1 is a comparison of type and quantity of phosphine in small sealed volume tests using 0.19 m^3 (50 gallon) barrels. Table 2 is an analysis of the standard fumigation process (non recirculation) compared with closed loop fumigation on a 10 600 m^3 (300 000 bushel) welded steel tank. Uniformity was greatly improved throughout the grain mass with CLF.

Installation and Operation Economics

For steel tanks, preliminary construction and installation cost estimates are about US\$2500–US\$3000 for a 7000–10000m^3 steel tank, or about US\$ 0.30–0.45/m^3, or about US\$0.40–0.60/t. Because concrete silo volumes are smaller and the structures taller, installation cost per unit volume is substantially higher than for steel. Cost estimates for CLF systems in concrete silos are US\$750–1000 for a 400–600 m^3 concrete silo, or about US\$1.20–1.50/m^3, or about US\$1.60–2.00/t. (Noyes et al. 1989a,b)

On a unit volume basis, it costs about 2–4 times as much to install CLF in concrete silos as in steel tanks. Silo costs estimates are high because each is treated as an individual storage unit with its own plumbing, blower and controls. However, if several silos are plumbed together into combined volumes that are similar in size to individual steel tanks where fewer, larger powered blowers and controls are used, installation costs may be reduced to a level more competitive with large steel tanks. One large vertical standpipe from the top of silos down the outside to ground level serving 6–8t concrete silos should be more economical than individual pipes (Kenkel et al. 1993, 1994)

Also, if an elevator with concrete silos is currently turning grain one additional time to fumigate concrete silos with automatic dispensers, CLF savings will also include the shrink (from grain dust and moisture losses) and conveyor operating costs, plus additional labour associated with turning.

Table 1. Phosphine comparison tests [raw data from field tests]

Day	Date	Time	Temperature		Hours	ppm
		Test 1 — 1 Pellet Phostoxin/0.189 m^3 (50 gallon) barrel				
Fri.	10-02-87	3:30 p.m.			0	0
	10-02-87	4:30 p.m.	18.3°C	65 °F	1	45-50
Mon.	10-05-87	9:30 a.m.	14.4 °C	58°F	66	excess of 50
Wed.	10-14-87	9:30 a.m.	13.9°C	57°F	282	excess of 50
		Received high range tubes Friday morning				
Fri.	10-16-87	1:30 p.m.	15.0°C	59°F	334	450
		Friday 4:45 pull plug from hose				
Mon.	10-19-87	8:30 a.m.	11.7°C	53°F	401	450
		Forced out with air				
		Test 2 — 1 Pellet Fumitoxin/0.189 m^3 (50 gallon) barrel				
Mon.	10-19-87	10:00 a.m.	12.2 °C	54• F	0	0
	10-19-87	2:00 p.m.	10.6°C	51°F	4	175
	10-19-87	4:45 p.m.	11.1°C	52°F	6.75	510
	10-19-87	7:00 p.m.	11.1°C	52°F	9	525
	10-19-87	10:00 p.m.	10.6°C	51°F	12	505
Tue.	10-20-87	8:00 a.m.	12.2°C	54°F	22	600
	10-20-87	4:00 p.m.	13.9°C	57°F	30	550
		31 hours 5:00 p.m. remove both caps				
Wed.	10-21-87	8:30 a.m.	11.1°C	52°F	46 1/2	0
		Test 3 — 2 Pellets Fumitoxin/0.189 m^3 (50 gallon) barrel				
Wed.	10-21-87	1:45 p.m.	13.3°C	56°F	0	0
	10-21-87	2:15 p.m.	13.3°C	56°F	1/2	21
	10-21-87	4:45 p.m.	16.7°C	62°F	3	100
Thur.	10-22-87	7:45 a.m.	13.3°C	56°F	18	1190
	10-22-87	9:45 a.m.	13.3°C	56°F	20	1200
	10-22-87	11:45 a.m.	15.6 °C	60°F	22	1080
		(Different tubes)				
	10-22-87	4:45 p.m.	18.9°C	66°F	27	900-1050
Wed.	10-28-87	1:45 p.m.	18.9°C	66°F	168	1080-1200

Source:Mike Stringer, Mgr., Fairfax Elevator, Union Equity Cooperative Exchange, Kansas City, Kansas, 1988.

Table 2. Comparison of recirculation vs. non-recirculation on 10 600 m^3 (300 000 bu.) welded steel tank.

	No recirculation—ppm[a]							
	10/27 19 hrs	10/27 25 hrs (windy)	10/28 43 hrs	10/28 49 hrs (2 days)	10/29 67 hrs	10/29 73 hrs (3 days)	10/30 93 hrs (4 days)	11/2 162 hrs (7 days)
Headspace	300	325	175	400	500	475	525	325
3 m	20	0	15	0	10	20	75	400
6 m	0	0	0	0	0	0	20	375
9 m	125	100	200	175	200	200	250	325
12 m	275	400	425	400	450	450	575	300
Fan	400	175	625	500	2100	1125	1000	1000
	Recirculation—ppm[b]							
	10/27 19 hrs	10/27 25 hrs (1 day)	10/28 43 hrs (windy)	10/28 49 hrs (2 days)	10/29 67 hrs	10/29 73 hrs (3 days)	10/30 93 hrs (fans off) (4 days)	11/2 162 hrs (7 days)
Headspace	200	425	675	800	1000	950	1050	350
3 m	125	125	650	500	850	775	825	650
6 m	190	275	575	600	850	825	875	625
9 m	210	275	600	650	900	875	925	525
12 m	300	425	450	750	950	950	1100	310
Fan	300	400	700	700	950	925	900	150

[a]30 flasks—24 probed into surface, 6 in aeration ducts.
[b]30 flasks broadcast on the surface.

Preliminary estimates indicate that CLF systems installed in steel tanks will pay back in 6–8 years if the system results in lower fumigant usage. While it may take 20–30 years for fumigant savings to equal the cost of installing a CLF system in concrete silos, costs will be paid back in less than 4 years if CLF eliminates an additional grain turning operation. Private application using CLF may also replace the cost of commercial application with some savings to the elevator. Additional benefits of timeliness, labour reduction, easier management and safety are subjective factors that are difficult to place a monetary value on, but do provide significant capital advantages.

CLF operating costs are extremely low. Costs to run 0.062–0.125 kW (1/12–1/6 hp) blowers are about the cost of electricity for a 75–150 w light bulb — less than 2 cents/hour — about 1.5–3.0 kWH/day, or US$0.15–30/day per blower at US$0.10/kWH. A 0.187–0.375 kW (1/4–/2 hp) blower costs about the same to operate as a 200 or 400 w bulb — 2–4 cents/hour or US$0.50–1.00/day

The electrical expense to fumigate and purge a 6 m × 30 m (20 ft by 100 ft) silo with a 0.075 kW (1/10 hp) blower operated 6–10 days for fumigation and purging is about US$1.50–US$2.00. An 8850 m^3 (250 000 bu) bin with a 0.187 kW (1/4 hp) blower costs US$3.50–US$10.00 to operate for 6–10 days.

References

Cook, J.S. 1980. Low airflow fumigation method. U.S. Patent No. 4200,657. P. O. Box 5421, Houston, Texas 77021. Issued 29 April. [Note: This patented closed loop fumigation process, known as the 'J–System' was purchased from Cook by Degesch America, Inc., P. O. Box 116, Weyers Cave, Virginia 24486].

Kenkel, P. and Noyes, R.T. 1993. Costs and benefits of installing closed loop fumigation systems in commercial elevators. Stillwater, Oklahoma State University, Cooperative Extension Service, OSU Fact Sheet No. 219, 4 p.

Kenkel, P., Noyes, R.T., Cuperus, G.W. and Criswell, J.T. 1994. Updated estimates of the costs and benefits of closed loop fumigation systems: field results from an Oklahoma country elevator. In : Proceedings 1994 Texas High Plains Grain Elevator Workshop, Amarillo, TX, Texas A&M University Extension Center, 27 January 1994, 8 p.

Noyes, R.T. 1993. Closed loop fumigation system: design and management. Presented at Indiana Stored Grain Pest Management Workshop, Purdue University, Lafayette, Indiana, 9 September 1993, 18 p.

Noyes, R.T., Clary, B.L and Stringer, M.E. 1989a. Closed loop fumigation. In: 1989 Proceedings, Fumigation Workshop, Oklahoma Cooperative Extension Service. Stillwater, Oklahoma State University, Circular E–888, 103–112.

Noyes, R.T., Stringer, M.E. and Clary, B.L. 1989b. Closed loop fumigation. In: 1989 Oklahoma Grain Elevator Workshop Manual, Oklahoma Cooperative Extension Service. Stillwater, Oklahoma State University, Circular E–881, 7 p.

Engineering input in the design of on-farm storage in India

B.D Shukla and K.K Singh[*]

Abstract

The present production of foodgrains in India is 176 million t against an estimated need of 220 million t by the end of this century. Hence, the production in future must increase to meet the need of the ever growing population of the country. Nearly two–thirds of the foodgrains produced are retained by the farmers for their personal use. Hence, the majority of the total produce is handled and stored by the farmers in about 6 000 000 villages in various types of stores. Traditional storage structures are still practiced to a great extent in these villages, which have a wide variation in socioeconomic conditions of the farmers. It is a herculean task to cover all the villages and all the farmers for improving their storage facilities through the Government's efforts, but during the last decade, scientifically designed improved storage structures have been adopted in rural India and the avoidable quantitative and qualitative losses have been reduced. Engineering input, like design data, foundation design, construction materials, construction or fabrication technology, material balance, insect-pests infestation technique, etc., have been used in design of appropriate structures, keeping in mind the socioeconomic condition of the farmers. Based on the construction materials and placement, these structures can been classified as: (i) indoor metallic bins, (ii) indoor non-metallic bins, (iii) outdoor metallic bins, and (iv) outdoor non-metallic bins. The essential design factors of each bin type have been described in terms of its capacity, strength, material used, location, handling of grain, and quality of grain and economics.

Introduction

The expected world's population by the Year 2000 is about 5000 million. In India, the population is expected to reach 1000 million. How to feed these people is the concern of everyone. Hence, in a country like India, agricultural production must increase to a target of about 220 million tonnes by the end of this century.

Immediately after harvest, the foodgrains undergo several types of physiological changes which are governed by the environmental conditions. They must be protected from spoilage and pillage, and they must be preserved under conditions which are not prejudicial to their qualities and do not render them unfit for human consumption. Therefore, suitable transportation and storage of grain are essential.

In India, nearly two-thirds of the foodgrains produced are retained by the farmers for their personal use. Hence, the majority of total production is handled and stored by the farmers in about 6 000 000 villages, The farmers in rural areas use traditional stores made of mud, plant stem, straws, stones and several types of other local materials. These storage structures have several drawbacks (Shukla and Patil 1988) which ultimately affect the quality of grains and reduce the storage life.

In a developing country like India, it is a herculean task to cover all the villages and all the farmers for improving their storage facilities through the Government's efforts, but during the last decade, scientifically designed, improved storage structures have been adopted in rural India and the avoidable qualitative and quantitative losses have been reduced. Engineering inputs like design data, foundation design, construction materials, construction and fabrication technologies, material balance, insect pest infestation technique, etc. have been used in design of such structures, keeping in view the socioeconomic conditions of farmers. These structures have found a place in rural areas and have been able to replace the traditional and age-old technology of storing the foodgrains. This paper briefly highlights the engineering aspects of design and development of farm storage of small capacity to meet the need of farmers who are responsible for preserving a huge quantity of the country's produce.

Classification of Structures

In terms of the engineering design, the storage structures on farm level can be classified by two means: (i) indoor and outdoor structures and, (ii) metallic and non-metallic structures (Raman 1988, 1990).

Selection of materials suitable to local climatic conditions is the main basis in the design of these structures. Because India is a vast country with varied resources and climatic conditions, a single design cannot cater for the requirements of farmers of different regions.

However, in general, the metallic structures are designed using materials like galvanised plain sheets, galvanised corrugated sheets and aluminium sheets, while the non-metallic structures are designed using materials like bricks, cement, wood, stone slab, and to some extent with plant stems. The reinforced cement concrete (RCC) structures are designed using steel as well as cement.

Indoor metallic structures

These structures are of small capacity and used by individual farmers. They are of the following types:

Domestic bins

These bins have been designed for capacities ranging from 0.3 –2.75 t with 24 or 22 gauge galvanised plain sheets. The thickness of the sheet depends upon the capacity of the bin. The outlet of these bins is designed according to the rate of consumption of stored materials (grains) and the properties of the stored commodity. For example, in the structure designed for storing wheat, the outlet is inclined as the quantity of grain withdrawn each time is usually not large. The inlet is reasona-

---

[*] Central Institute of Agricultural Engineering, Nabibagh, Berasia Road, Bhopal–462 018, India.

bly big for easy loading. The larger capacity bins such as 1 t and 0.62 t is designed in two sections, joined by a metallic belt and mild steel clamps for easy transportation. Even though main body of the bins is fabricated with 22 or 24 gauge galvanised plain sheet, it becomes necessary to use thicker gauge materials for other components such as inlet and outlet surrounds, covers, etc. For these components 20, 18 and 16 gauge galvanised plain sheets are used depending on the load, number of handlings, etc. In high capacities, two sections of the bins are tightened by means of metallic belt and clamps with the help of nuts and bolts. All the mild steel components have been painted properly. A locking facility has also been provided for inlets and outlets.

Urban bins

These bins are designed either circular or rectangular in shape in capacities ranging from 0.09–0.3 t. Since these bins are small, there is no need of providing outlets. The body of the bin is designed using 24 gauge galvanised plain sheets. With a little more strengthening of joints, either with mild steel flats and mild steel angles, some of the designs can also be made multipurpose, so that they can be used as a bed or seat. Castor wheels can be fitted to the structures so that they can be moved from place to place.

Spiral lock seam tube bins

Light weight, machine-made metal tubing is used for the construction of this bin. The tubes are usually made of 24 gauge galvanised plain sheet strips spirally wound, with locked seams for greater strength and rigidity. They are mainly used for ducting and are available in different diameters: 700 mm and 900 mm diameter are found to be most convenient for the design of domestic bins. By fixing top and bottom plates, and inlets and outlets, these tubes are made indoor storage structures.

Welded wire mesh bins

For high moisture grain not infected with microorganisms, this type of bin is useful. The design of this bin is simple: a 75 × 25 mm welded wire mesh bolted to a metallic base supported by steel legs. To ensure air circulation which is essential to bring down the moisture content of the grain, hessian cloth lining inside the mesh is provided. A $4m^3$ size structure is the most economical capacity of this type bin. The roof can be designed with 16 gauge mild steel sheet or rubberised cloth. All the components made of mild steel should be painted properly.

Indoor non-metallic structures

These structures are designed to meet the need of farmers with the help of locally available materials either for direct application or in modified forms.

Pucca kothi

Pucca kothi is a storage structure made of burnt bricks. To protect against rodent invasion and to provide strength, 2–3 walls are sometimes constructed. The floor and roof are constructed using reinforced bricks and stone slabs and the floor is elevated. Bins have a capacity of about 1 t and wooden and steel inlets and outlets respectively.

Cavity wall bins

Unburnt sun-dried bricks are used for the construction with an air-gap of 100 mm to provide insulation, and wood and burnt bricks are used for the construction of roof and floor respectively. These storage structures are provided with metallic outlets with covers and inlet covers of wood or other materials.

Plywood bins

Rectangular, circular and square sheets of 4mm plywood are glued with synthetic resin and hessian cloth and then nailed together according to the design. The plywood flooring is elevated and a skirting is provided at the bottom with stone chips and cement to prevent rodent attack.. The capacity of this structure is 1 t.

Outdoor metallic structures

High capacity outdoor structures are required to store grain for longer periods. Engineering inputs have helped to minimise cost, provide strength, and protect grain from insect pests and rodents. Design features of some popular structures are described below.

Flat and hopper bottom module bins

Flat and hopper bottom bins of different designs are used. Capacities vary from 2–2.5 t. In the modular design, bins having fixed diameter and variable capacities can be constructed by varying the height of the bin. Both galvanised iron sheet (GI) of 20 gauge and aluminium sheet of 15 gauge are used for construction of these bins.

Flat bottom bins are placed on steel legs or brick pillars. Inlets and outlets with locking arrangements are provided, as well as ladder rests and loading platforms. The roof and wall stiffeners are designed either of 18 gauge galvanised sheet or 16 gauge aluminium sheet depending on the body of the bin. Reinforced circular steel concrete is added to the base and roof, and all joints are sealed with cotton cord and thick white paint. Bitumen compounds are used at the joints of base and roof sheets, and nuts and bolts are also provided with washers to make them airtight.

Outdoor flat bin

Bin capacity is 3.2 t using 22 gauge galvanised plain sheets. The base is made of brick masonry and cement. Wall and roof stiffeners are used, with locking inlet and outlets. Bitumen compounds are used at the joints of base and roof sheets, and nuts and bolts are also provided with washers to make them airtight.

Corrugated galvanised sheet bin

A 10 t bin was designed with 24 gauge galvanised corrugated sheets. Special machinery is necessary for curving the corrugated sheets. No wall stiffeners are necessary as the corrugated sheet itself is strong enough to bear the load. The base is designed with 16 gauge mild steel sheet supported by brick masonry column. In these bins instead of roof stiffeners, trusses are used for supporting the roof and these trusses are bolted to the bin wall by means of cleats. Suitable outlets and inlets are provided with locking arrangements. A ladder is provided for easy loading. Proper painting is recommended for mild steel components.

Galvanised plain sheet with adjustable height

In this bin, the sheets are used breadthwise instead of lengthwise so that the height of the bin can be altered without changing the base and roof. This type of bin is designed for capacities of approximately 4 t by using 3 galvanised plain sheets and 4 galvanised plain sheets per layer respectively. The 22 or 24 gauge galvanised plain sheets are used depending upon the capacity. Wall stiffeners are not necessary as there is a join of sheets all round the bin and roof inlets with a locking arrangement as in the case of the module bin. The brick masonry base is economical for this kind of design. A ladder rest and platform are also provided. Proper painting is recommended to mild steel components.

High moisture grain storage bin

This design is able to provide safe storage conditions for storing grain of higher moisture content (as high as 22% moisture content in the case of paddy). The basic design consists of two concentric shells with an annular space in between. The inner shell is of perforated sheet and the outlet shell is designed with 22 or 24 gauge galvanised plain sheet. The base sheet is perforated to facilitate air movement. Ventilation holes (75 mm apart) at the periphery of wall sheets enable hot air to escape. The roof consisted of 22 or 20 gauge galvanised plain sheets supported by trusses made of mild steel as in the galvanised corrugated sheet bin. Suitable outlets and inlets with locking arrangements are provided. An air control gate is incorporated in the design so that air can enter from the sides, thus increasing air movement in the bin. Ladder rests and ladder are also included in the design. Proper precautions are necessary if the bin is to be designed for small capacities (3 t). Larger capacities need more perforated pipes inside for proper air circulation. Extra heating of atmospheric air if necessary, is supplemented by a suitable solar reflector.

Composite bin

The bin consists of both wood and steel. Wooden columns provide support to the galvanised steel wall. Because of the support afforded by the wooden columns, 22 gauge sheets are used. With this technique, bins ranging from 3–14.5 t can be constructed. No wall stiffeners are necessary and wooden beams are used for support in the roof. Suitable outlets and inlets with locking arrangements, ladder-rest and ladder are also provided. A brick masonry base is suitable and all mild steel parts are properly painted.

Outdoor non-metallic structures

The following types of non metallic structures are used:

Reinforced cement concrete ring bin

This type of bin is designed either with locally available reinforced cement rings or rings cast in situ. Rings are cast with 1:2:4 cement, sand, and aggregate mixtures. Cement and sand mortar at a ratio of 1:2 is recommended for joining the 300 mm rings. The brick masonry base is designed with a reinforced cement concrete roof, and metallic inlet and outlets are provided. The outlet cover is lined with wooden inserts/discs to prevent condensation at inlet. Proper water proofing treatment is given for preventing seepage of water into the structure. The structure is designed for various capacities.

Reinforced brick bin

Designed for various capacities, this bin is made of reinforced steel bricks. The bin is constructed of two reinforced brick walls, with an inner layer of polyethylene. The roof is of reinforced cement concrete, and the inlet and outlets are similar to that of the reinforced cement concrete ring bin.

Hollow block bin

Hollow cement concrete blocks, in a ratio of 1:6 cement and sand mixture are being used to construct storage structures. The remaining features are similar to that of reinforced cement concrete bins or reinforced brick bins.

The Economics of Engineering Input

In modern design there has been a great saving of steel and cement (Shukla and Patil 1990). For example, in warehouse design, mild steel has been replaced by cold twisted deformed steel. The number of compartments are also reduced. The angle iron trusses are replaced with tubular trusses as they are lighter in weight compared to steel trusses. The greatest advantage in this design is that no intermediate columns are provided to hold the structure as single open roof serves the purpose. This gives maximum utilisation of space without any obstruction. A comparison of consumption of steel trusses of different design is given in Table 1.

Table 1. Steel requirement in different designs of trusses used in storage structures.

Type of truss	Weight (t)	Steel (kg/m^2)	Reduction in steel use (%)
Structural steel truss	50	16.70	–
Tubular truss with tubular purlin	36	12.00	28
Tubular truss with cold roll from section purlin	31	10.40	38

Scientific design has reduced the use of steel and cement by 60% in the construction of these structures. The use of high tensile steel for reinforcement and tubular trusses and cold rolled sections for roof purlins has brought down the demand for steel. Cement use has also been reduced due to application of under pile foundation and the elimination of base concrete for floors. A comparison of use of cement and steel consumption in construction of storage structures is given in Table 2.

Table 2. Requirements of cement and steel in different types of warehouse design.

Type of design	Cement (kg/m^2)	Steel (kg/m^2)
RCC flat roof warehouse	193.50	38.00
Warehouse with structural steel designs	139.00	41.50
Conventional warehouse with tubular truss	99.50	23.00
Conventional warehouse with tubular truss and design on ultimate load theory	94.50	18.50
Conventional warehouse by adopting revised design of floor	74.00	18.50
Conventional warehouse with re-designed truss with cold rolled from sections of purlins	74.00	17.00

Reduction of space in alleyways is accomplished by eliminating intermediate covered columns and providing large single spare roof trusses. Space requirements for alleyways in different types of designs are shown in Table 3. Approximate costs of construction of different designs are given in Table 4.

Strategies to meet the need

The basic objectives are to meet the storage requirements of rural producers and consumers so that losses can be reduced, enabling producers and consumers better access to the market.

Table 3. Space requirements for alleyways in different designs of warehouses.

Type of warehouse	Alleyways (%)
RCC roof warehouse	37
Tashspan warehouse	30
Single-span warehouse	25
Modified single-span warehouse	24

Table 4. Cost of construction per tonne in different designs of warehouses.

Design	Capacity (t)	Cost of construction/t	
		Indian rupees	US$
RCC flat roof	5000	1600	54
Twin span warehouse	5000 t	1500	50
Conventional warehouse	5000 t	1200	40
Modified conventional warehouse	5000 t	1000	34
Grain warehouse	650 t	1100	37
Rural warehouse	200 t	1400	47

Decentralised facilities available through cooperatives have to be part of a large network, but should not be too large to cater for the small hinter land. Godowns at village level have a capacity of 100 t, with godowns built at secondary/terminal markets having capacities ranging between 250–500 t.

In order to maintain the viability of investment in godowns, multiple activities have to be undertaken. In a country like India, modernisation of storage is a lengthy process, and bulk storage systems must be encouraged. Constant research is underway to improve the system and design of Indian storage conditions, as well as extension of the improvements to users.

Conclusions

Engineering input in the design of scientific storage structures has played a key role in providing appropriate and economical storage systems on-farm as well as in rural areas of India.

Acknowledgement

The authors are thankful to Dr R.S. Devnani, Director CIAE Bhopal for providing facilities for the preparation of this paper and Mr Achuthan N.M. for typing the manuscript.

References

Raman C.P. 1988. Improved storage structures. In: Girish, G.K. and Goyal R.K. ed., Proceedings of a regional workshop on on-farm storage, Hapur, India, 29 February–12 March 1988.

Raman C.P. 1990. Bulk storage for storage of foodgrains. In: Girish G.K. and Kumar, A. ed., Proceedings of a regional workshop on warehouse management of stored foodgrains, New Delhi, India, 19 March–6 April.

Shukla B.D. and Patil R.T. 1988. Overview of grain drying and storage problems in India. Australia, GASGA Executive Seminar Series No. 2, 7–27.

Shukla B.D. and Patil R.T. 1990. Recent developments in engineering of foodgrain storage in India. In: Lessard, F. and Ducom, P. ed., Proceedings of the 5th International Working Conference on Stored-product Protection, Bordeaux, France, 9–14 September 1990.

Design chart for in-store maize drying under tropical climates

S. Soponronnarit, P. Wongvirojtana and A. Nathakaranakule*

Abstract

Product quality, drying capacity and energy consumption have to be taken into consideration in drying system design. Long periods while in-store drying maize may cause product deterioration due to successive fungi, aflatoxin and dry matter loss. Determination of air flow rate and depth of grain is usually a problem due to the time-consuming calculation process. The objective of this study is to demonstrate the development of a design chart for in-store maize drying under tropical climates. A near equilibrium drying model was developed that included the effects of heat and moisture from respiration to improve accuracy. The model was finalised by being incorporating a submodel for aflatoxin development. Simulation results given various conditions were used to construct four quadrant charts. Minimum specific airflow rates (m^3/minute/m^3 of grain) which corresponded to a dry matter loss of 0.5 % were derived. It was observed that corresponding aflatoxin concentration were lower than 50 ppb. The minimum specific airflow rate increased with initial moisture content but decreased with bed thickness. Other critical values such as pressure drop, specific energy consumption and drying time could also be obtained from the charts. These charts are expected to be useful for design engineers to reduce calculation time and to help select the appropriate design and operating parameters for in-store maize drying.

Introduction

Methods of grain drying may include continuous flow and fixed bed. In-store drying is one method of fixed bed drying without grain movement during drying and storage. Grain is dried slowly with ambient air or air at a temperature a few degrees higher than ambient. Design of in-store drying includes selection of items such as thickness of grain bed, airflow rate, size of fan and air duct, and drying time. In addition, energy consumption, dry matter loss due to respiration, and toxic substances produced during drying and storage have to be considered. These values depend on initial moisture content of grain and ambient air parameters. Complex mathematical models and computer programs as well as a trained personnel are needed for the design. It also takes time to investigate appropriate design and operating parameters. A design chart furnished with adequate information would be useful.

Mathematical grain drying models can be classified as non-equilibrium, i.e. no equilibrium between product and drying air (Brooker et al. 1974), near-equilibrium, which assumes thermal equilibrium between product and drying air (Bakker Arkema et al. 1977; Thompson et al. 1968) and equilibrium

models. The second model predicts a slightly faster drying than the first, but requires much less computer time. The last model is appropriate only to slow drying with ambient air and can provide only the position not the shape of drying zone. In in-store drying, energy input to the drying system is for the fan only. Therefore, calculation of air temperature rise while flowing across the fan and air passage is very important because it significantly affects the drying rate (Soponronnarit 1988). In some cases, especially drying of high moisture grain, heat and moisture liberated by respiration should be included in the model because this significantly improves the accuracy of the model (Soponronnarit and Chinsakolthanakorn 1990).

The main parameters of grain quality are moisture content and other physical properties such as degree of breakage, colour, and foreign matter. In the case of maize, aflatoxin is also considered, but we can find no equations describing development of aflatoxin.

The objective of this paper is to develop a mathematical model able to describe drying of high moisture maize under hot, humid climates. Also, design charts are developed from simulation results.

Materials and Method

Mathematical drying model

The development of a drying model accommodating equations covering drying rate, energy and mass conservation, enthalpy change of air while flowing across a fan, dry matter loss and heat and moisture liberated from respiration process was presented by Soponronnarit (1988) and Soponronnarit and Chinsakolthanakorn (1990). Details are as follows :

1. Energy conservation for a thin layer

$$c_a T_o + (2502 + c_v T_o) W_o + R c_{pw} T P_o = \\ c_a T_f + (2502 + c_v T_f) W_f + R c_{pw} T_f \qquad (1)$$

where
C = specific heat, kJ/kg°C
T = air temperature,°C
TP = grain temperature,°C
W = absolute humidity, kg H$_2$O/kg dry air
R = ratio of dry grain mass to dry air mass, kg dry matter/kg dry air
and subscripts
a = dry airp
w = wet grain
f = after drying
v = water vapour
o = before drying

2. Mass conservation for a thin layer

$$W_f - W_o = (M_o - M_f) R \qquad (2)$$

---

* School of Energy and Materials, King Mongkut's Institute of Technology Thonburi Suksawat 48 Rd., Bangkok 10140, Thailand.

where
M = grain moisture content, decimal dry-basis.

3. Thin layer drying

The thin layer drying equation developed by Westerman et al. (1973) was manipulated. It is valid for temperature and relative humidity ranges of 38-71°C and 10–60%, respectively.

4. Energy balance at the fan

$$\dot{m}(c_a + c_v W_o)\Delta T_f = \dot{m}P / \rho_a \eta_f \tag{3}$$

where
$\dot{m}$ = mass flow rate of dry air, kg / second
ΔT_f= temperature rise by the fan, °C
ρ_a= density of dry air, kg/m^3
P = total pressure Pa
η_f= fan efficiency, decimal

5. Dry matter loss, heat and moisture from respiration

The equations of dry matter loss developed by Thompson (1972) and Steele (1967) were manipulated. Heat and moisture liberated from respiration was calculated by the following equations (Soponronnarit and Chinsakolthanakorn 1990).

$$\Delta T_h = 15778 \, DML/c_{pw} \tag{4}$$
$$\Delta M_h = 0.6 \, DML \tag{5}$$

where
ΔT_h = temperature rise of grain, °C
ΔM_h = moisture rise of grain, decimal dry-basis
DML = dry matter loss, decimal

In order to make the calculation possible, it is necessary to employ state equations of moist air. In this paper, the equations developed by Wilhelm (1976) were manipulated.

6. Development of aflatoxin

Aflatoxin is developed by fungi under favourable conditions. Experiments on the development of aflatoxin in maize stored in plastic bags of about 10 kg under room conditions in Bangkok were conducted by Kawashima et al. (1990). It was found that aflatoxin concentration increased with initial moisture content of grain and storage time. The development of equations describing the development of aflatoxin and employing these data will be explained in the next section.

Experiment

To verify the accuracy of the mathematical model, eight test runs of maize drying were conducted in a bin with a diameter of 0.75 m. The height of the grain bed was 1.4 m. Grain moisture content was measured at 20 cm interval by using air oven at 103°C for 72 hours. Temperature was also measured at the same positions, using type K thermocouples (±1°C) connected to a data logger. Air velocity was measured by a hot wire anemometer.

Initiative for constructing design charts

A design chart was constructed to overcome the inconveniences mentioned in the introduction to this paper. For this study, design charts for in-store maize drying under tropical climates were developed. It was assumed that grain maintained its quality if dry matter loss during drying was less than 0.5%. Therefore, the dry matter loss was kept at 0.5% for each computer simulation run. Final moisture content was assumed to be 14% wet-basis. As a result, the airflow rate corresponding to the dry matter loss of 0.5% was the minimum value for maintaining grain quality. In addition, system pressure drop

was calculated (must be less than 1500 Pa) by assuming that it was 1.5 times the pressure drop in the grain bed. The pressure drop data for popcorn developed by Shedd (1953) were manipulated for this calculation. In addition, drying time and specific energy consumption were computed. These values depended on the thickness of grain bed and initial grain moisture content. Simulation results were then used to construct four quadrant design charts.

Results and Discussion

Equation of aflatoxin

Figure 1 shows the result of curve fitting using empirical equations which can be written as follows:

$$AFB_1 = A_1 + A_2 t + A_3 t^2 + A_4 t^3$$

where

$$A_1 = -3031.360 + 248.7328 M_w - 4.917356(M_w)^2 \tag{6α}$$

$$A_2 = 2512.535205.0159 M_w + 4.022257(M_w)^2$$

$$A_3 = 544.2637 + 44.03857 M_w 0.8539792(M_w)^2$$

$$A_4 = 22.918111.856915 M_w + 0.03609485(M_w)^2$$

and
AFB_1= aflatoxin concentration (B_1), ppb
t = storage time, day
M_w = moisture content, % wet-basis

The above equation is valid for moisture contents ranging from 20.7 to 28.9% wet basis and appropriate only for hot ambient air (about 25–35°C). If the value of AFB_1 or $d(AFB_1)/dt$ from equation (6a) was negative, the value of $d(AFB_1)/dt$ was made zero. In case of moisture content below 20.7% wet-basis,

$$AFB_1 = 0 \tag{6b}$$

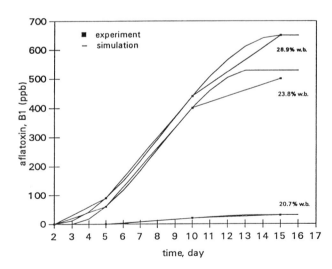

Fig. 1. Evolution of simulated and measured aflatoxin concentration.

Comparison between experimental and simulated results

Figure 2 shows comparative results of experiment and simulation. The latter can be divided into simulation with and without heat and moisture from respiration. It was found that the accuracy of the model was improved when heat from respiration was included especially at the top layer which was dried latest. In case of lower initial grain moisture content (not

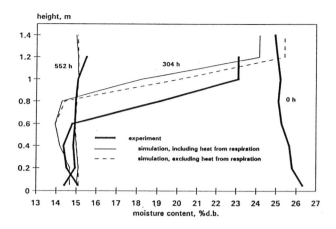

Fig. 2.   Simulated and measured moisture profiles, test no. 4, December 1991.

shown in this paper), there was no difference between the two cases. This was due to lower respiration in drier grain.

Design chart

Figure 3 shows the simulation results for continuous ventilation, using ambient air conditions of September 1988, which is the worst month during the 5-year period, in four quadrant chart. Final moisture content was 14% wet-basis and dry matter loss was 0.5%. It was found that aflatoxin concentration was lower than 50 ppb in all simulation runs presented in the chart. Minimum airflow rate increased along with initial grain moisture content, but decreased with bed depth. Other important values such as pressure drop, drying time and energy consumption, are also evident. Figure 4 shows the sim-

ulation results using ambient air condition of September 1992 which is the best month during the 5-year period. It was similar to the previous case except that energy consumption increased along with initial moisture content. In the former case, energy consumption increased with initial moisture content at low level, to be reversed at high moisture levels. At higher initial moisture content, temperature rise by the fan was higher due to higher airflow rate and thus improved drying potential significantly especially for poor weather as in the former case.

Figures 5 and 6 show the simulation results of intermittent ventilation corresponding to Figures 3 and 4, respectively. They are similar to Figure 4. Minimum airflow rates were a little higher but energy consumption decreased significantly especially for the case of poor weather. Figures 7 and 8 show the evolution of corresponding ambient temperature and relative humidity.

Conclusion

The following conclusions can be drawn from this paper:

1. The mathematical model was accurate and was improved when heat and moisture liberated from respiration were included, especially for the case of high initial grain moisture content.

2. The design charts may be useful to designers of in-store maize drying under hot, humid climates. They help select appropriate design and operating parameters.

3. From the design charts, it may be concluded that in-store maize drying under hot, humid climates is feasible provided that initial grain moisture content is not so high as to necessitate reducing grain depth to the extent that storage may not be economical.

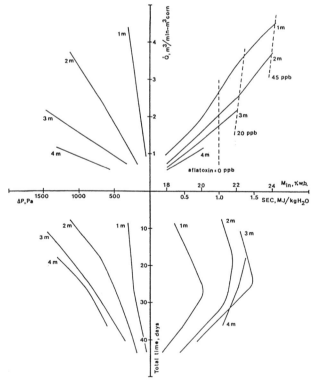

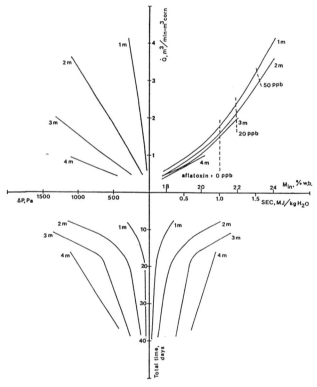

Fig. 3.   Design chart for in-store maize drying (continuous ventilation), ambient mean temperature = 28.2°C, ambient relative humidity = 81.5%, dry matter loss = 0.5%. M_{in} = initial moisture content; Q = airflow rate for 0.5% dry matter loss; ΔP = total pressure drop; SEC = specific energy consumption.

Fig. 4.   Design chart for in-store maize drying (continuous ventilation), ambient mean temperature = 28.4°C, ambient relative humidity = 77.8%, dry matter loss = 0.5%. M_{in} = initial moisture content; Q = airflow rate for 0.5% dry matter loss; ΔP = total pressure drop; SEC = specific energy consumption.

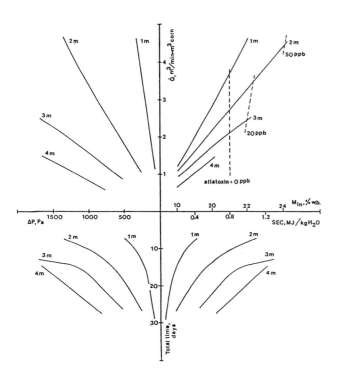

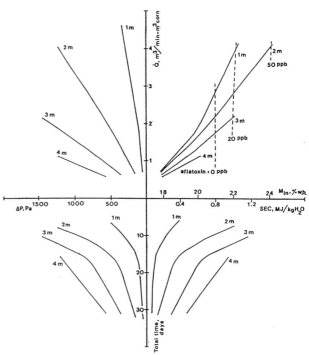

Fig. 5. Design chart for in-store maize drying (intermittent ventilation), inlet relative humidity below 75%, ambient mean temperature = 28.2°C, ambient relative humidity = 81.5%, dry matter loss = 0.5%. M_{in} = initial moisture content; Q = airflow rate for 0.5% dry matter loss; ΔP = total pressure drop; SEC = specific energy consumption.

Fig. 6. Design chart for in-store maize drying (intermittent ventilation), inlet relative humidity below 75%, ambient mean temperature = 28.4°C, ambient relative humidity = 77.8%, dry matter loss = 0.5%. M_{in} = initial moisture content; Q = airflow rate for 0.5% dry matter loss; ΔP = total pressure drop; SEC = specific energy consumption.

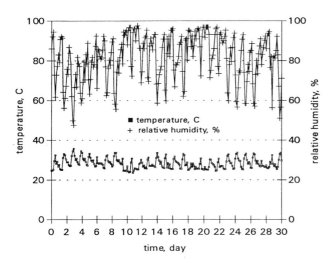

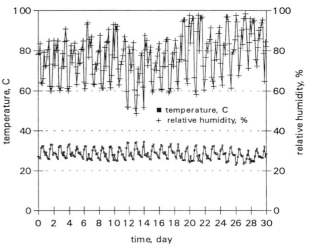

Fig. 7. Evolution of ambient air temperature and relative humidity, September 1988.

Fig. 7. Evolution of ambient air temperature and relative humidity, September 1992.

Acknowledgment

The authors express their sincere thanks to the Australian Centre for International Agricultural Research for financial support to this project.

References

Bakker-Arkema, F.W., Becker, S. and Brooker, D.B. 1977. Feasibility of solar energy grain drying in Missouri and Michigan. Paper presented at the 1977 annual meeting, ASAE, St. Joseph, Michigan.

Brooker, D.B., Bakker-Arkema, F.W. and Hall, C.W. 1974. Drying cereal grains. AVI, 265 p.

Kawashima, K., Siriacha, P., Kawasugi, S., Saito, M., Okazaki, H., Tonboon-ek, P., Manabe, M. and Buangsuwon, D. 1990. Studies on quality preservation of maize by the prevention of aflatoxin contamination in Thailand, Part II. Tropical Agriculture Research Centre and Department of Agriculture, Ministry of Agriculture and Cooperatives, Thailand.

Shedd, C.K. 1953. Resistance of grains and seeds to air flow, Agricultural Engineering, 34, 616–619.

Soponronnarit, S. 1988. Energy model of grain drying system. ASEAN Journal on Science and Technology for Development, 5 (2), 43–68.

Soponronnarit, S. and Chinsakolthanakorn, S. 1990. Effect of heat and water from respiration on drying rate and energy consumption. ASEAN Journal on Science and Technology for Development, 7 (2), 65–83.

Steele, J.L. 1967. Deterioration of damaged shelled corn as measured by carbon dioxide production, Unpublished Ph.D. Thesis, Iowa State University of Science and Technology.

Thompson, T. L. 1972. Temporary storage of high moisture shelled corn using continuous aeration, Transactions of the American Society of Agricultural Engineers, 15, 333–337.

Thompson, T.L., Peart, R.M. and Foster, G.H. 1968. Mathematical simulation of corn drying — a new model, Transactions of the American Society of Agricultural Engineers, 11(4), 582–586.

Westerman, P.W., White, G.M. and Ross, I.J. 1973. Relative humidity effect on the high temperature drying of shelled corn, Transactions of the American Society of Agricultural Engineers, 16, 1136–1139.

Wilhelm, L.R. 1976. Numerical calculation of psychrometric properties in SI units. Transactions of the American Society of Agricultural Engineers, 19 (2), 318–321, 325.

Advances in research on in-store drying

G.S. Srzednicki and R.H. Driscoll *

Abstract

In-store drying (near ambient air drying) has been used in temperate climates by the grain industry since about 1950. Current research has adapted this technique to the humid tropics, by means of computer simulation based on long-term climatic records, combined with application of automatic controllers.

This paper focuses on its application to crops prone to rapid quality deterioration, such as paddy and maize. Quality models are discussed for the dynamics of aflatoxin build-up in storage. Drying strategies are presented which take into account quality parameters for each crop.

The effects of modification of the traditional design (such as grain stirrers), and the interfacing of in-store drying with the post-harvest chain, are shown to be logistically and economically feasible.

Introduction

The main objective of drying of farm produce is to reduce the water activity of a product from its harvest level to a safe level for extended storage. After drying, the rate of deterioration due to respiration of the grain, and to insect, chemical, bacterial and fungal activity (all of which are functions of temperature and humidity), should be minimised, resulting in maintenance of the quality of the stored product.

Grains, which are some of the most important crops worldwide, can be dried in a variety of ways. The traditional method for centuries has been sundrying on the ground. The ground may be covered, bare, the compacted ground of a farmyard, or in more recent times a sealed road or the concrete clad surface of a basketball ground. The increase of grain production due to agronomic research and the mechanisation of agriculture has led to development of various techniques of mechanical drying. One major factor leading towards adoption of mechanical drying has been the increased use of large capacity bulk–handling equipment for grain. The systems which are currently in use range from high-temperature, high-capacity, single-stage, continuous-flow dryers with high energy inputs to various types of in-bulk drying systems, operating at high or low temperature, with lower energy inputs but longer drying times. There are several types of in-bulk dryers (McLean 1980) such as:

- warehouses with in-floor or above-floor aeration ducts
- ventilated silos with perforated floors and vertical air-flow (round or square)
- ventilated silos with separate inlet and exhaust ducts

- ventilated silos with vertical ducts and radial air-flow.

In addition to single-stage systems, combination or two-stage systems have been devised in order to take into account the different drying rates of grain at different moisture contents. Typically this involves high-temperature, high-speed drying in order to reduce the moisture content from its harvest level to give a product water activity of around 0.85. For rice and maize, this corresponds to a moisture content of about 18% (wet basis). This moisture, being concentrated near the surface of the grain, can be removed more easily. After completion of the first stage of drying, grain is transferred into a storage bin, where it is cooled and dried using lower air temperatures. The second stage involves removal of moisture from the centre, and so is diffusion controlled leading to a reduced drying rate. This drying stage is called the 'falling rate period'.

In-store drying is synonymous to low-temperature in-bin drying. It may be used where grain remains in store until milled or exported, or where drying is seen as the primary purpose of the equipment, with the grain being removed to another bin for aerated storage. The advantage of the latter is increase in throughput and reduction of capital cost per unit dried. This situation may arise at trader level where fast turnover of grain is required so that fresh stocks may be purchased.

The main advantages of two-stage drying are (Morey et al. 1981):
- reduced energy requirements;
- increased drying system capacity; and
- improved grain quality.

The reduced energy requirement over conventional drying technology is due to the increased air efficiency compared with continuous-flow dryers, so that less heat is vented to the atmosphere. The second point is related to the capacity of the first stage 'fast' dryer, since discharge of the grain at a higher moisture content before cooling will free the dryer for the next load of high moisture grain, which is where continuous dryers are more efficient. The third point relates to the relaxation time given the grain during second-stage drying, which allows moisture gradients within the grain to relax, preventing the outer layer of the grain from being overdried and hence made brittle and susceptible to cracking.

Conventional Systems

In-store drying has been practised in the USA since the 1950s among maize farmers in the North Central Region and rice farmers in Texas (Do Sup Chung et al. 1986). However, the technique was used mainly on a trial-and-error basis and occasionally led to spoilage of grain due to extremely low airflow rates combined with too-high drying temperatures.

Systematic studies on in-store drying began in the early 1970s when Thompson (1972) developed a mathematical model predicting changes in grain temperature, moisture content and dry matter deterioration, taking into account factors such as heat transfer through the walls of the bin, respiration of the grain mass, and conditioning of the grain through

---

* Department of Food Science and Technology, University of New South Wales, P.O. Box 1, Kensington NSW 2033, Australia.

continuous aeration. This model was called the 'near equilibrium model'. The method of analysis was to construct a heat and mass balance across a single thin layer of material. A deep bed of grain was considered as multiple layers of grain, with a stream of air perpendicular to each layer. The initial work was for maize, the primary commercial target in the USA at the time. Farmers were required to reduce the maize moisture quickly to a level of 15.5%. Maize at too high or too low a moisture attracted a penalty. In addition, maize at too high a moisture content ran the risk of deterioration due to moulds or insects. Computer simulations were run based on this model, using weather data (temperature and relative humidity) covering long periods of time from different locations of the USA, in order to study the drying process in each place.

The technique was soon extended to other crops. In temperate climates, particularly in the USA (Barrett et al. 1981, on wheat) the U.K. (Smith and Bailey 1983, on barley) and Canada (Muir et al. 1991, on rapeseed), significant research was done on in-store drying of wheat and barley respectively. Airflow rates and weather records were studied in order to study and design in-store dryers.

Recent Research on In-Store Drying in the Tropics

Although weather conditions in humid tropical climates are less favourable than temperate climates for in-store drying, due to high ambient temperatures and relative humidities, research on the use of aeration in combination drying had began as early as the sixties (Calderwood 1966). Initially, aeration was used for cooling paddy previously dried in a high temperature dryer, by aerating a mass of grain with ambient air in a holding bin. In the late 1970s researchers in a number of subtropical and tropical countries started studying conditions for successful adoption of in-store drying of paddy, later also researching its application to crops such as maize, peanuts and soybeans. They extended the techniques initially used in temperate climates, namely optimisation by application of drying models, to tropical conditions. In order to compensate for the higher daily relative humidities, higher airflow rates and additional heat were included in the study (Adamczak et al. 1986; Driscoll and Srzednicki 1991; and Driscoll et al. 1989).

Considerable research into in-store drying of paddy was conducted in Australia from the 1970s onwards. The reason for this research was a dramatic increase in rice production in the recently created Murrumbidgee Irrigation Area in southwestern New South Wales. High quality standards and the need for very competitive pricing forced the rice processors to look for the most cost-effective method for drying and storing freshly harvested paddy (Bramall 1986). The collaboration between researchers from the Commonwealth Scientific and Industrial Research Organisation (CSIRO), and later from the University of New South Wales, who studied the fundamentals of deep-bed drying of granular solids, resulted in the development of improved strategies for in-store drying in Australia. The following conditions were determined as essential for successful drying (Bramall 1986):

- segregation of procured paddy according to moisture contents
- monitoring of grain moisture and temperature on a regular basis
- use of low speed fans
- use of aeration strategies taking into account daily fluctuations in weather conditions.

Originally, small-scale 100 t capacity radially aerated bins were used. They have gradually been replaced by bins aerated from the bottom through on-floor or in-floor ductings. The sheds currently used in the Murrumbidgee Irrigation Area have a capacity of 3000-5000 t when fully loaded, i.e. when the grain bed is 7-12 m in height. New storage bins have been installed for 'fast' drying where the grain bed is limited to 2-3 m (Semple 1988).

A series of collaborative research projects supported by the Australian Centre for International Agricultural Research (ACIAR) was launched in the early 1980s on in-store drying in the humid tropics. These projects involved Australian and international research organisations; the University of New South Wales, the Philippine National Postharvest Institute for Research and Extension (NAPHIRE), the Thai King Mongkut's Institute of Technology Thonburi (KMITT) and the Malaysian Agricultural Research and Development Institute (MARDI).

The main objectives of the research were:
- to determine thermophysical data for the main grain crops, with the aim of using these data to design drying systems for these crops;
- to investigate first-stage drying options for areas where a two-stage drying strategy was required;
- to provide appropriate technology for complete drying systems for main crops, especially paddy and maize, in the humid tropics; and
- to study the effects of various drying strategies on quality of the crop.

The results of this research have been presented at seminars and published in the literature. The following are the main outcomes from the projects:
- A very comprehensive set of weather data covering at least 10 years has been collected from various locations in Malaysia, Thailand, Indonesia and Australia (and to a lesser extent in the Philippines).
- Thermophysical data comprising bulk and true density, porosity, angle of repose, coefficient of static friction, specific heat, equilibrium moisture contents at various temperature and relative humidity levels, and thin-layer drying rates have been determined for a range of commercially important varieties of paddy and maize. Thermophysical data have also been collected for soybeans, peanuts and mungbeans.
- Baseline data regarding the structure of the grain industry in the collaborating countries in Southeast Asia have been collected. These include the geographic distribution of crops, cropping calendars, quantities procured daily by processors, and storage and milling capacities.
- A computer drying simulation model based on the thermophysical data has been developed Driscoll (1986). The model is based on thermodynamic equilibrium between air and grain during the drying process as described by Sutherland (1984). The simulation model includes grains such as paddy (Australian and Asian varieties), maize, peanuts, soybeans, mungbeans, barley and other products. Different strategies can be simulated for in-store drying, among them constant aeration, relative humidity control, time control and modulated burner control. The model makes provision for options such as recirculation of air, stirring of grain, dehumidification, and heat losses through walls.
- A great many simulations have been performed in order to assess the feasibility of in-store drying under the climatic conditions prevailing on selected sites in the main grain-growing regions of the collaborating countries in Southeast Asia. As a result of the analysis of the computer simulations it has been established that two-stage drying with in-store as a second stage is feasible under the conditions of the humid tropics.

- Confirmation of the computer model predictions was achieved by means of firstly pilot plant studies, using 1-5 t of paddy, conducted in Australia and in Asia, and secondly industrial scale experiments with up to 500 t, conducted at government grain complexes or privately owned rice mills in Southeast Asia. These experiments also demonstrated that paddy could be dried successfully during the wet season down to safe storage or milling level within the required period of 13–23 days, depending on the chosen strategy, the initial grain condition and the bed depth.
- Programmable controllers were added to the dryers. The most appropriate drying strategies for each location were programmed into these controllers, using parameters derived from computer simulation.

Quality Considerations

There are different quality standards for different grain crops, so that it is not possible to have a general quality model. However, some quality criteria are applicable to a wide range of grains; for example deterioration due to respiration or the action of certain microorganisms. Deterioration due to respiration is usually called dry matter loss, and can be determined by the amount of carbon dioxide produced, as described by Steele et al. (1969). The extent of fungal activity can be defined in a variety of ways; for example in terms of plate counts or by chemical analyses of constituents of metabolites. Some researchers (Seitz et al. 1979; Schwadorf and Muller 1989) have suggested using measurement of the amount of ergosterol present in the grain as an indicator of fungal growth. Ergosterol is a sterol found in the cell membranes of the fungi, as well as in the grain itself. Changes in the amount of ergosterol present after harvest are indications of mould activity on the grain.

Among the quality parameters of greatest economic importance for paddy is head rice yield, the proportion of head rice (rice which is 3/4 kernel size or larger) in the total amount of milled rice. Fissuring during postharvest handling of paddy will decrease the head rice yield. The mechanism of fissuring is associated with readsorbtion of moisture by grain that has already been dried (Kunze and Prasad 1978). Hence, the head rice yield is closely related to the drying strategy.

Although fissuring is also important to maize quality, a more significant quality parameter is the aflatoxin content. Aflatoxins are a secondary metabolite formed by spoilage fungi of the genus *Aspergillus*, particularly *A. parasiticus* and *A. flavus*. Aflatoxins belong to the group of difuranocoumarin compounds (Bhat 1991). They cause concern because of their potential carcinogenic effects on humans, and, in livestock production, because they lead to a significant decrease in growth rate and increase mortality among poultry and pigs. Aflatoxins are produced in a range of produce, but especially in maize, peanuts, copra, tree nuts and milk.

Improvement in quality of the dried product may often decide whether a particular drying technique is to be adopted or not. This can be shown in an example taken from a recently completed study involving in-store drying in a rice-producing co-operative in Thailand using increase in head rice as the primary quality parameter (see Table 1).

As far as aflatoxin control in maize is concerned, earlier work in the USA has shown that, when using low temperature drying, grain above 17% m.c., stored at temperatures between 13–41°C should not be exposed to relative humidities above 85% for more than 48 hours (Ross et al. 1979). Since the equilibrium relative humidity increases with temperature for a given moisture content, any delay in drying under tropical conditions combined with an increase in temperature can result in moisture adsorbtion and increased risk of aflatoxin

Table 1. Financial analysis of in-store drying facility at Chachoengsao agricultural cooperative (Soponronnarit et al. 1993).

Basic data			
Storage capacity	100 tonnes		
Establishment costs	148 536 Baht		
Drying time	218.2 hours		
No of fans	1		
Fan power	11.7 kW		
Fan operating time	179.1 hours		
Electricity cost	1.65 Baht/kWh		

Assuming an increase in head yield by 5%: Internal rate of return			
Price increase in Baht/ tonne	One drying/annum		Two dryings/annum
50	7.3%		18.2%
100	11.6%		23.9%
150	14.4%		29.5%

Payback period			
Price increase in Baht/tonne		Interest	
	12%	15%	20%
50	i) >15	i) >15	i) >15
	ii) 5.2	ii) 6.3	ii) 10.5
100	i)12.9	i) >15	i) >15
	ii) 3.6	ii) 4.0	ii) 5.1
150	i) 7.8	i) 10.9	i)>15
	ii) 2.7	ii) 3.0	ii) 3.5

i) One drying per annum
ii) Two dryings per annum

formation. Recent work conducted in Thailand by Wongviroj-tana et al. (1993) has shown that, with careful monitoring of the quality of incoming grain and the choice of an appropriate drying strategy, it is possible to control the aflatoxin build-up by using in-store drying technology, as shown in Table 2. However, it is feasible only for bin loading moisture contents below 19% and for low initial aflatoxin levels.

Improvement of Conventional Techniques

Conventional systems, which use fans and ducts to distribute air through the grain bed, use outside air that is conditioned to the required average relative humidity by preheating. The drying process can be controlled manually by a skilled operator systematically monitoring ambient air conditions and grain moisture content, provided he has a good understanding of the in-store drying process, or by means of a micropro-cessor-based controller. The first option (manual control) provides only limited accuracy, is labour intensive and needs a trained operator. The microprocessor-based controllers can be operated continuously, if necessary using a battery back-up, and are directly linked to temperature and relative humidity sensors. They execute commands from their program algo-rithms, which provide the intelligence for the dryer. New installations above a certain critical size will opt for automatic controllers. An automatic control system should include a possible choice of drying strategies; for example continuous aeration, relative humidity control, time control or modulated burner control. The relative humidity control option offers the possibility of integrating information from the drying model, as well as the measured grain temperature and so to continu-ously adjust the set-point during the drying process, resulting in a considerable saving in airflow, heater power and overall cost (Ryniecki and Nellist 1991a,b).

As previously mentioned, the drying model developed at the University of New South Wales includes recirculation of the exit air, an option which may improve the efficiency of a dryer, as described by Driscoll and Intong (1991). The advantage of recirculating a proportion of the exit air is basically of benefit to products with slow drying rates or using high airflow rates.

Since an in-store dryer is inherently efficient by means of the submergence of the drying front within the grain mass during most of the drying, recirculation offers little to an in-store grain drying facility until the leading edge of the drying

front reaches the top of the bed, which is the last 10–20% of the drying time of a grain batch, whereafter there will be increased advantage as the exit relative humidity drops. Thus, the recirculation rate should be zero initially and then raised to 90% once the drying front broaches the top of the bed. This means the recirculation hardware (ducts and dampers) is useful only for a small percentage of the total drying time. To offset this limited advantage in energy use, recirculation will require additional capital costs, will increase pressure drop (requiring a larger, more expensive and power hungry fan), and may increase the difficulties of loading and unloading the bins. The researchers found an interesting possible advantage, which was to increase the speed of drying by increasing the amount of burner heat. If done without recirculation, the inlet grain layer would be overdried. However, by using exit air to condition the inlet air, this could be prevented while still retaining a higher air temperature. Mixing exit air with the inlet air allowed a faster drying front to be propagated through the bed. Research is continuing into whether this effect can be used to economic advantage, as the issue becomes fairly complex.

Stirring is performed with vertical augers which bring grain from the bottom layers of the stack to the surface and vice versa. It is a useful tool to prevent overdrying of the bottom layers which are the first ones to be in contact with drying air. Wilcke and Bern (1986a,b) studied the effects of stirring on dryer performance in a number of maize batches over two years. They found advantages in terms of reducing the pressure drop initially, reduction in drying time, electric energy use and prevention of overdrying. However, they found that stirring was eventually increasing airflow resistance by producing small amounts of fines and pushing them towards the plenum chamber at the bottom of the bin. Stirring is repre-sented in the simulation by mixing the grain from all layers at set intervals.

The computer simulation developed at the University of New South Wales was used by the present authors to study the effects of stirring on paddy under typical harvesting condi-tions for Malaysia and Thailand. The ambient conditions during that period are shown in Figures 1 and 2. The drying strategy was based on relative humidity control, with ambient air being used within 60–75% limits and additional heat being supplied from 75–95%. The maximum temperature rise (burner at full capacity) was set at 5°C and the airflow rate was

Table 2 Effects of in-store drying on final aflatoxin content under Thai conditions in a 1.4 m bed (Wongvirojtana et al. 1993).

Basic data	Experiment					
	i) Cont. aeration	ii) Cont. aeration	iii) Cont. aeration	iv) Cont. aeration	v) Time control (12 hours on)	vi) Time control (12 hours on)
Initial moisture content in % wb	19.1	25.5	20.3	20.8	17.3	18.0
Airflow rate in m³/minutes/m³	3.0	4.5	1.5	2.5	4.0	2.5
			Ambient conditions			
			Relative humidity in %			
Average	85.0	83.0	75.8	87.7	76	82.9
Range	62–98	58–98	43–98	70–98	40–98	60–93
			Temperature in °C (average)			
	28.0	28.0	25.9	27.0	27.0	26.7
Drying time (hours)	144	216	360	264	132	244
			Aflatoxin content in ppb			
Before drying	0	16	51	48	119	274
After drying	0	666	61	282	69	187

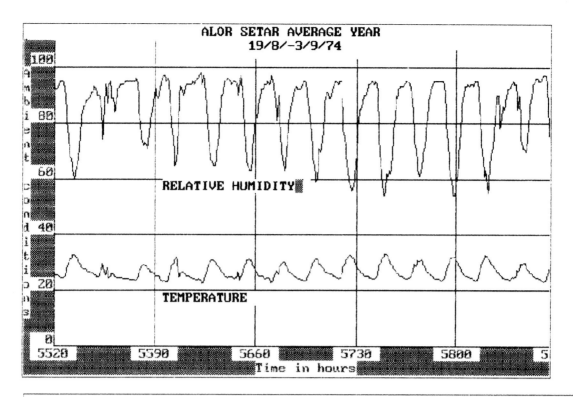

A) Average year

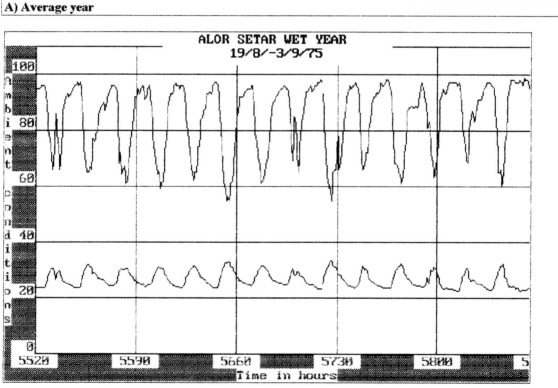

B) Wet year

Fig. 1.   Weather pattern during wet season in Northeast Malaysia. A) Average year; B) Wet year.

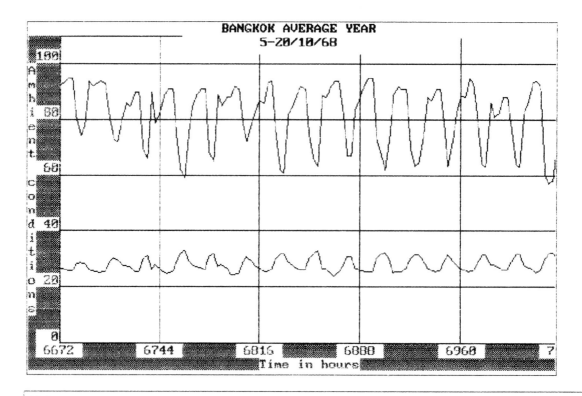

A) Average year

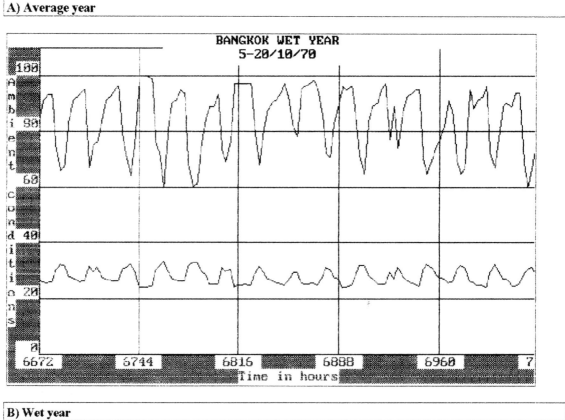

B) Wet year

Fig. 2   Weather pattern during wet season in Central Thailand.

set at 8.6 mL/minute. Under these conditions, the effects on the moisture gradient within a 3 m bed are shown in Figure 3. It appears that stirring has a dramatic effect in reducing the top-to-bottom moisture differential in the grain bed. The effect is noticeable in average as well as in wet years but appears more pronounced in average years.

Conclusions

In concluding this paper, we can say that the recent research on in-store drying has explored various ways of fine-tuning the technique in order to optimise the use of ambient air for drying of grain. Computer simulations have been used to investigate possible applications in the humid tropics. The technique has proven feasible under tropical conditions, but within stricter operating limits than for temperate climates. More consideration is currently being given to the quality of the final product, but additional systematic studies are required for a fuller understanding of the mechanisms of product deterioration, especially related to the formation of mycotoxins. Improvement of conventional methods, namely through inclusion of recirculation and stirring, are proving to have a favourable effect on energy cost and product quality. Yet it is unlikely that recirculation for in-store drying will ever be economically effective. Microprocessor-based control systems are an essential tool for optimising the drying systems. Future directions in the development of in-store drying for Australian conditions may well be along the lines of model-based control, where sufficient fundamental data on the drying properties of the product have been provided in order to develop accurate process algorithms.

References

Adamczak, T., Loo Kau Fa, Driscoll R.H. and Mochtar M. 1986. Preventing paddy backlogs using two stage drying. In: De Mesa, B.M. ed., Grain protection in postharvest systems: Proceedings of the Ninth ASEAN Technical Seminar on Grain Postharvest Technology, Singapore, 26–29 August 1986, 83–98.

Barrett, J.R., Okos, M.R. and Stevens, J.B. 1981. Simulation of low temperature wheat drying. Transactions of the American Society of Agricultural Engineers, 24 (4), 1042–1048.

Bhat, R.V. 1991. Aflatoxins: successes and failures of three decades of research. In: Champ, B.R., Highley, E., Hocking, A.D. and Pitt, J.I. ed., Fungi and mycotoxins in stored products: proceedings of an international conference held at Bangkok, Thailand, 23–26 April 1991. ACIAR Proceedings No. 36, 80–85.

Bramall, L.D. 1986. Paddy drying in Australia. In: Champ, B.R. and Highley, E. ed., Preserving grain quality by aeration and in–store drying: proceedings of an international seminar held at Kuala Lumpur, Malaysia, 9–11 October 1985. ACIAR Proceedings No. 15, 219–223.

Calderwood, D.L. 1966. Use of aeration to aid rice drying. Transactions of the American Society of Agricultural Engineers, 9 (6), 893–895.

Do Sup Chung, Boma Kanuyoso, Erickson, L. and Chong-Ho Lee. 1986. Grain aeration and in-store drying in the USA. In: Champ, B.R. and Highley, E. ed., Preserving grain quality by aeration and in-store drying: proceedings of an international seminar held at Kuala Lumpur, Malaysia, 9–11 October 1985. ACIAR Proceedings No. 15, 224–238.

Driscoll, R. H. 1986. The application of psychrometrics to grain aeration. In: Champ, B.R. and Highley, E. ed., Preserving grain quality by aeration and in-store drying: proceedings of an international seminar held at Kuala Lumpur, Malaysia, 9–11 October 1985. ACIAR Proceedings No. 15, 67–80.

Driscoll R.H., Adamczak T. and Samsudin A. 1989. Control options for in-store drying. In: De Mesa, B.M. ed., Grain postharvest systems: Proceedings of the Tenth ASEAN Technical Seminar on Grain Postharvest Technology, Bangkok, Thailand 19–21 August 1987, 44–51.

Driscoll, R. H. and Intong, C.L. 1991. Modelling of recirculation for through-drying of granular products. Food Australia, 43 (10), 456–458.

Driscoll R.H. and Srzednicki G.S. 1991. Design of complete drying systems under Southeast Asian climates. In: Naewbanij, J.O. ed., Grain postharvest research and development: priorities for the nineties. Proceedings of the Twelfth ASEAN Seminar on Grain Postharvest Technology, Surabaya, Indonesia, 29–31 August 1989, 3–11.

Kunze, O.R. and Prasad, S. 1978. Grain fissuring potentials in harvesting and drying of rice. Transactions of the American Society of Agricultural Engineers, 21 (2), 361–366.

McLean, K.A. 1980. Drying and storing combinable crops. Ipswich, Suffolk (U.K.), Farming Press Ltd, 281 p.

Morey, V.R., Gustafson, R.J. and Cloud, H.A. 1981. Combination high-temperature, ambient-air drying. Transactions of the American Society of Agricultural Engineers, 24 (2), 509–512.

Muir, W.E., Sinha, R.N., Zhang, Q. and Tuma, D. 1991. Near-ambient drying of canola. Transactions of the American Society of Agricultural Engineers, 34 (5), 2079–2084.

Ross, I.J., Loewer, O.J. and White, G. M. 1979. Potential for aflatoxin development in low temperature drying systems. Transactions of the American Society of Agricultural Engineers, 22 (6) 1439–1443.

Ryniecki, A. and Nellist, M.E. 1991a. Optimisation of control systems for near-ambient grain drying: part 1, the optimization procedure. Journal of Agricultural Engineering Research, 48, 1–17.

Ryniecki, A. and Nellist, M.E. 1991b. Optimization of control systems for near-ambient grain drying: part 2, The optimizing simulations. Journal of Agricultural Engineering Research, 48 19–35.

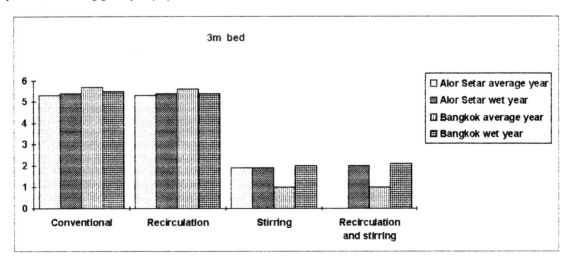

Fig. 3   Effects of recirculation and stirring on top-to-bottom moisture differential in paddy

Schwadorf, K. and Muller, H.M. 1989. Determination of ergosterol in cereals, mixed feed components, and mixed feeds by liquid chromatography. Journal of the Association of Official Analytical Chemists, 72, (3), 457–462.

Seitz, L.M., Sauer, D.B., Burroughs, R. and Mohr, H.E. 1979. Ergosterol as a measure of fungal growth. Phytopathology, 69, 1202–1203.

Semple, R.L. 1988. Post harvest technology: storage development and application in developing Asian countries. Annex 1: Deniliquin Storage/Development NSW Australia. In: Semple, R.L., Hicks, P.A., Lozare, J.V. and Castermans, A., ed., Grain storage systems in selected Asian countries: proceedings of the workshop/study tour of grain storage systems, Zheijang province, People's Republic of China, May 1988. Bangkok, REAPASIA, 9–60.

Smith, E.A. and Bailey, P.H. 1983. Simulation of near-ambient grain drying. II: Control strategies for drying barley in Northern Britain. Journal of Agricultural Engineering Research, 28, 301–317.

Soponronnarit. S., Chalidapongse, P., Nathakaranakule, A., Prachayavarakorn, S., and Intrachantra, S. 1993. Full report on design and testing of in-store drying system at the agricultural co-operative in Chachoengsao. Phase I. KMITT, School of Energy and Materials, Bangkok, Thailand (unpublished).

Steele, J.L., Saul, R.A. and Hukill, W.V. 1969. Deterioration of shelled corn as measured by carbon dioxide production. Transactions of the American Society of Agricultural Engineers, 12 (5), 685–689.

Sutherland, J.W. 1984. The design and prediction of performance of batch driers for paddy rice. International Journal for Development Technology, 2, 119–129.

Thomson, T.L. 1972. Temporary storage of high moisture shelled corn using continuous aeration. Transactions of the American Society of Agricultural Engineers, 15 (2), 333–337.

Wilcke, W.F. and Bern, C.J. 1986a. Natural-air corn drying with stirring: I. Physical properties effects. Transactions of the American Society of Agricultural Engineers, 29 (3), 854–859.

Wilcke, W.F. and Bern, C.J. 1986b. Natural-air corn drying with stirring: II. Dryer performance. Transactions of the American Society of Agricultural Engineers, 29 (3), 860–867.

Wongvirojtana, P., Soponronnarit, S. and Nathakaranakule, A. 1993. Feasibility study of in-store corn drying under tropical climates. In: Naewbanij, J.O., Manilay, A.A. and Frio, A.S. ed., Increasing handling, processing and marketing efficiency in the grain postharvest system: Proceedings of the Sixteenth ASEAN Seminar on Grain Postharvest Technology, Phuket, Thailand, 24–26 August 1993. 265–283.

Modelling heat and mass transfer phenomena in bulk stored grains

G.R. Thorpe*

Abstract

Markets for durable produce, such as food grains, continue to impose increasingly stringent requirements for high quality commodities This is reflected in demands for zero or very low levels of chemical pesticide residues in grains, very low levels of mycotoxins, high head yield of paddy, a high malting yield of barley and so on. As a result, the design and operation of grain storage systems must be constantly reviewed. Mathematical modelling of the stored grains ecosystem offers the possibility of devising methods of manipulating microclimate within a bulk of grain to ensure that desirable properties of the grain can be preserved during storage. This paper describes some of the recent developments in the formulation of the equations that govern heat and moisture transfer in stored grains. As a result of the work, it has proved possible to calculate from first principles rate processes that occur in bulk-stored grains. A rational method of choosing the simplest, yet accurate, mathematical models is described and results from the work suggest that thermal equilibrium between grains kernels and the intergranular air may be assumed in most commercial applications involving aeration or drying. An analysis of heat and moisture transfer in bulks of high moisture content respiring grains identifies terms that account for the fate of mass and energy associated with the oxidation of the grain substrate. A method of calculating heat and moisture transfer in grain stores of any shape is outlined.

Introduction

Increasingly strict requirements continue to be imposed on the quality of food grains presented onto domestic and international markets. For example, consumers demand low or zero levels of chemical pesticide residues in grains, very low concentrations of mycotoxins, a high milling yield of rice, a high malting yield of barley and so on. In tropical countries, losses resulting from consumption by insects, mites and moulds continue to be unacceptably high. An improved understanding of the stored grains ecosystem is a key to solving these problems. This anthropogenic ecosystem is much simpler than many that occur naturally in that its biological and physical diversity is relatively limited, and the system may be biologically isolated from its wider environment. Furthermore, the microchemical within a grain store can be manipulated by drying or cooling the commodity, or fumigating it with a gas such as phosphene that is toxic to insects.

The devising of strategies to manipulate the microclimates within bulks of stored grains to meet the demands of the market is clearly a multidisciplinary activity that requires

inputs from stored products entomologists, toxicologists, mycologists, mathematicians, engineers and so on. This paper outlines some of the recent advances in modelling physical aspects of the stored grains ecosystem, and discusses some areas that require further development. Topics discussed include the establishment of equations that govern heat and mass transfer in bulk stored grains on a firmer foundation than has been traditional. Once the equations have been derived they must be solved subject to boundary conditions that reflect the behaviour of commercial grain stores. For example, fumigant gases released in the headspace of a silo permeate through the grain bulk, and before we can estimate their rate of dispersion we must be able to define the boundary conditions at the interface between the grains and the air in the headspace. It is pointed out in this paper that there is a research focus on solving the equations subject to boundary conditions that reflect the types of grain store found in practice, and this will allow for systems optimisation. Commercial computer software is now available that enables heat, mass and momentum phenomena to be fairly readily simulated. This software awaits exploitation by the community of post-harvest technologists.

Some Practical Considerations

The local temperature, grain moisture content and composition of the intergranular atmosphere generally dominate biological phenomena that occur in a bulk of stored grains. It is the manipulation of these variables that is available to managers of stored products systems. For example, grain may be cooled by forcing through it cool air. If every region of the grain bulk is to benefit by adopting this strategy the system must be designed so that the air flow rate in every region is sufficiently high to ensure that the grain is cooled sufficiently quickly. At the peripheries of the store where the ambient temperature may prevent the grain from cooling it is desirable to investigate the likelihood of grain drying in these regions. These considerations imply that we must choose the most appropriate location and size of aeration ducts and type of aeration fan required to force air through the grain at the required rate. It is also important to ensure the temperature and humidity of the air used to ventilate the grain are suitable to achieve the required degree of grain cooling. This implies some selection and control of the operation of the aeration system.

Before these commercial considerations can be addressed we require engineering equations that can be used to reliably design storage systems. Such equations are based on the physics of heat and mass transfer and fluid flow. In addition, it is vital that the impact of manipulating the design variables on the biological environment within stored grains can be predicted.

The Fundamental Equations

Most analyses of the heat, mass and momentum transfer phenomena that occur in bulk stored grains are based on the so-called continuum approach, as outlined by Bejan (1984). This

---

* Department of Civil and Building Engineering, Victoria University of Technology, P.O. Box 14428, Melbourne Mail Centre, Melbourne, Australia 3000.

terminology arises from the fact that the bulk of grain is considered to be a continuum analogous to a single phase fluid or solid. Properties such as density and composition are deemed to vary continuously in space, whereas in reality they suffer discontinuities at the grain/air surfaces. Such an approach does not account for the fact that heat, mass and momentum transfer in stored grains are more reliably described in terms of phenomena that occur on the length scales of the grain kernels and intergranular pores. For example, heat conduction in the intergranular air is governed by Fourier's law, namely

$$\mathbf{q}_\gamma = k_\gamma \nabla T_\gamma \qquad (1)$$

where $\mathbf{q}_\gamma$ is the heat flux, k_γ is the thermal conductivity of air and ∇T_γ is the temperature gradient in the pores. The flow of the air between the grain kernels is governed by the Navier-Stokes equation, which for steady flows may be expressed as

$$\nabla p_\gamma = \rho_\gamma \mathbf{g} + \mu_\gamma \nabla^2 \mathbf{v}_\gamma \qquad (2)$$

Equations 1 and 2 are constitutive equations of classical continuum mechanics, and they are well established and of wide applicability. We may also write similar point equations that govern heat and moisture transfer in the grain kernels, and although these may assume the forms of Fourier's and Fick's laws they are only broad descriptions of the phenomena. This is because unlike the intergranular air the grain kernels themselves do not constitute a continuum, but they are a microporous system. However, experimental evidence (Crapiste et al. 1988) suggests that such continuum descriptions are applicable to microporous media such as grain kernels.

Equations 1 and 2 apply to phenomena that occur on the length scales of the grain kernels and intergranular spaces, whereas designers and managers of grain storage systems are concerned with the overall behaviour of a grain store. Engineering design equations need to be expressed in terms of variables such as temperature, grain moisture content, fumigant concentration that vary on the length scale of the macroscopic system. Such variables consist of averages of the temperatures and moisture contents of the grain kernels and air in a representative elementary volume within the grain bulk, as depicted in Figure 1.

Provided the radius of the elementary volume is large compared with the grain kernels and small with respect to the size of the grain bulk the averaged quantities vary smoothly with distance, and their time constants are large compared with those associated with the small length scales. The advantages of the method of volume averaging include:

• Details of the small scale phenomena may be retained in the analysis, and this leads to the possibility of estimating rate coefficients such as the effective diffusivity of moisture through grain bulks (Thorpe et al. 1991a,b) and thermal conductivity (Nozad et al. 1985a,b) from first principles without recourse to experimentation.

• A more detailed analysis of the grain/air/water system enables stored grains technologists to select the simplest mathematical models that are consistent with the required accuracy of calculation (Thorpe and Whitaker 1992a,b).

• The gaining of greater insights into the physical processes that occur in grain bulks. For example, the approach has been used by Thorpe (1993) to identify in greater detail than previously the mass and energy transfers that occur in bulks of respiring grains.

• The possibility of invoking theorems to prove the uniqueness of the engineering design equations derived from the fundamental assumptions.

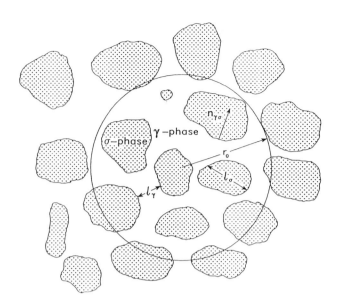

Fig. 1. A representative elementary region of the grain bulk

The Method of Volume Averaging

Carbonell and Whitaker (1984) provide a comprehensive and general description of the method of volume averaging. Its application to specific problems in grain storage engineering are given by Thorpe et al. (1991a,b), Thorpe and Whitaker (1992a,b) and Thorpe (1993). In this paper the emphasis is on the results of the analyses rather than on the methodology. However, it is necessary to present a few key definitions and theorems in order to clarify the discussion. The phase average of a quantity Γ_ω is denoted by $\langle \Gamma_\omega \rangle$ and it is defined by:

$$\langle \Gamma_\omega \rangle = \frac{1}{V} \int_{V_\omega} \Gamma_\omega dV \qquad (3)$$

where V_ω is the volume of the ω–phase in representative region. The intrinsic phase average Γ_ω of is defined by:

$$\langle \Gamma_\omega \rangle^\omega = \frac{1}{V_\omega} \int_{V_\omega} \Gamma_\omega dV \qquad (4)$$

and if Γ_ω refers to grain moisture content, say, it is the its intrinsic phase average moisture content that is likely to be measured.

When we take the volume average of Equation 1, say, we obtain the average of a Laplacian which is of limited practical use, whereas the Laplacian of an averaged quantity would be a far more useful quantity. The latter may be obtained from the former by successive applications of the volume averaging theorem proved independently by Anderson and Jackson (1967), Marle (1967), Slattery (1967) and Whitaker (1967) and it may be stated as:

$$\langle \nabla \circ \Gamma_\omega \rangle = \nabla \circ \langle \Gamma_\omega \rangle + \frac{1}{V} \int_{A_{\omega\alpha}} \mathbf{n}_{\omega\alpha} \circ \Gamma_\omega dA \qquad (5)$$

in which $\circ$ represents a tensor multiplication operator of any order, i.e. Γ_ω may be a scalar, vector or tensor of any order. It is useful to make use of the spatial decomposition proposed by Gray (1975), namely

$$\Gamma_\omega = \langle \Gamma_\omega \rangle^\omega + \tilde{\Gamma} \qquad (6)$$

The point equations that govern heat, mass and momentum transfer in the grain kernels and intergranular pores may be expressed in terms of volume averaged quantities and spatial deviations from them. Carbonell and Whitaker (1984) have discussed the length-scale constraints that must be satisfied when applying the method of volume averaging. To ensure that averaged quantities vary smoothly with spatial distance the length scales of the grain kernels and intergranular pores, l_γ and l_σ (which are about the same) must be much smaller that tne radius, r_o, of the averaging volume. In addition, the radius of the averaging volume must be much smaller than the length scale, L, over which significant changes in the averaged variables occur. These constraints may be stated as

$$l_\gamma, l_\sigma \ll r_o \ll L \qquad (7)$$

If the equations that govern the spatial deviations can be formulated and solved we are able to completely define the behaviour of the heat, mass and momentum transfer phenomena that occur in a bulk of grain and other porous media as is evident from the work of Nozad et al. (1985a,b), Ochoa et al. (1986), Thorpe et al. (1991a,b) and Whitaker (1986). This is one of the strengths of the method of volume averaging.

The Effective Diffusion Coefficient in Stored Grains

Moisture diffuses through bulk grains as a result of vapour pressure gradients that arise because of gradients in grain moisture content and/or temperature. An early attempt by Pixton and Griffith (1971) to quantify the rate of moisture diffusion in stored grains made use of the grain moisture content as the driving potential, but this approach lacks generality because the vapour pressure of the interstitial moisture is a non-linear function of grain moisture content and temperature. This results in the effective diffusivity of moisture also being a function of these two variables. Thorpe (1981,1982) recognised that the diffusion of moisture through grains depended almost exclusively on vapour pressure gradients and he used the experimental data of Griffith (1964) and Pixton and Griffith (1971) to calculate the effective diffusivity of moisture as 0.212 that of the diffusivity of moisture vapour in free air. Later, Thorpe et al. (1991a,b) exploited the method of volume averaging to calculate from first principles the effective diffusivity of moisture content in bulk wheat to be 0.24 that of the diffusivity in free air, i.e. in very close agreement with the one obtained experimentally.

The analysis presented by Thorpe et al. (1991a,b) appears to be quite formidable, but it is based on the simply stated equations that govern moisture conservation in the intergranular air and the grain kernels. The equation governing moisture transfer in the intergranular air is:

$$\frac{\partial C_\gamma}{\partial t} = \nabla \cdot \left(\mathcal{B}_\gamma \nabla C_\gamma \right) \qquad \text{in } V_\gamma \qquad (8)$$

in which C_γ and $\mathcal{B}_\gamma$ represent the concentration of moisture vapour in tne intergranular air and the molecular diffusion coefficient in of water vapour in air respectively. At the interface of the grain kernels and the intergranular air we have

$$-\mathbf{n}_{\gamma\sigma} \cdot \mathcal{B}_\gamma \nabla C_\gamma = -\mathbf{n}_{\gamma\sigma} \cdot \mathcal{B}_\sigma \nabla C_\sigma \qquad \text{at } A_{\gamma\sigma} \qquad (9)$$

where C_σ and $\mathcal{B}_\sigma$ represent the concentration of water in the grain kerneis and the diffusion coefficient of moisture in the solid phase, namely the grain kernels. A unit normal directed from the γ-phase, the intergranular air, to the σ-phase is denoted by $\mathbf{n}_{\gamma\sigma}$. The concentration, C_σ of the mois-

ture on the grain surface is related to that in the air at the surface and the temperature $T_{\gamma\sigma}$ by a general isosteric relationship, thus

$$C_\sigma = \mathcal{F}(C_\gamma, T_{\gamma\sigma}) \qquad \text{at } A_{\gamma\sigma} \qquad (10)$$

Moisture transfer in the grain kernels is governed by

$$\frac{\partial C_\sigma}{\partial t} = \nabla \cdot \left(\mathcal{B}_\sigma \nabla C_\sigma \right) \qquad \text{in } V_\sigma \qquad (11)$$

The statement of the mathematical problem is completed by stating the boundary conditions at the periphery of the system shown in figure 1 as

$$C_\gamma = F(\mathbf{r}, t) \qquad \text{at } A_{\gamma e} \qquad (12)$$

and

$$C_\sigma = G(\mathbf{r}, t) \qquad \text{at } A_{\sigma e} \qquad (13)$$

where $A_{\gamma e}$ and $A_{\sigma e}$ are the areas of the γ- and σ-phases at the periphery of the region under consideration.

Equations 8–13 describe moisture transfer in bulk stored grains, but as they stand they are of little practical value to a grains storage technologist concerned with managing or designing grain stores. This is because they apply only on length scales associated with the grain kernels and the intergranular pores, whereas in practice we are concerned with phenomena that occur on much larger scales, such as the height of a silo. However, the equations contain some information on the rates of mass transfer in the intergranular air and the grain kernels since the diffusion coefficient of moisture through air is widely reported, see Wexler (1965) for example. Furthermore, the diffusion coefficient of moisture in grains has been given by Becker and Sallans (1971) or it may be calculated from drying rate constants such as that presented by O'Callaghan et al. (1971) using the method of Jury (1967). From these constitutive equations of continuum mechanics Thorpe et al. (1991a) show how to apply the volume averaging theorem, expressed as equation 5, and the definition of the spatial deviation, equation 6, to derive a mass transfer equation expressed in terms of a volume averaged concentration, spatial deviations of concentration and empirical rate coefficients. It may be written as

$$\left\{ \varepsilon_\sigma \frac{\partial \mathcal{F}}{\partial \{C\}} + \varepsilon_\gamma \right\} \frac{\partial \{C\}}{\partial t} = \nabla \cdot \left\{ \varepsilon_\gamma D_\gamma \left[\nabla \{C\} + \frac{1}{V_\gamma} \int_{A_{\gamma\sigma}} \mathbf{n}_{\gamma\sigma} \tilde{C}_\gamma dA \right] \right\}$$
$$+ \nabla \cdot \left\{ \varepsilon_\sigma D_\sigma \left[\frac{\partial \mathcal{F}}{\partial \{C\}} \nabla \{C\} + \frac{1}{V_\sigma} \int_{A_{\sigma\gamma}} \mathbf{n}_{\sigma\gamma} \tilde{C}_\sigma dA + \frac{\partial \mathcal{F}}{\partial \langle T \rangle} \nabla \langle T \rangle \right] \right\} - \varepsilon_\sigma \frac{\partial \mathcal{F}}{\partial \langle T \rangle} \frac{\partial \langle T \rangle}{\partial t} \qquad (14)$$

in which $\{C\}$ is an equilibrium phase weighted concentration defined as

$$\{C\} = \varepsilon_\sigma \mathcal{F}^{-1}\left(\langle C_\sigma \rangle^\sigma, \langle T \rangle \right) + \varepsilon_\gamma \langle C_\gamma \rangle^\gamma \qquad (15)$$

and where $\mathcal{F}^{-1}$ is the inverse of the isosteric equation and it my be defineu as

$$C_\gamma = \mathcal{F}^{-1}\left(C_\sigma, T_{\gamma\sigma} \right) \qquad (16)$$

and $\langle T \rangle$ is a phase weighted volume average temperature defined as

$$\langle T \rangle = \varepsilon_\gamma \langle T_\gamma \rangle^\gamma + \varepsilon_\sigma \langle T_\sigma \rangle^\sigma \qquad (17)$$

Note that in equation 15 we have used volume averaged arguments of $\mathcal{F}^{-1}$ which Whitaker (1987,1988) has shown to be permissible provided the length scale constraint, 7, is satisfied.

The next task is to find expressions for the spatial deviations $\tilde{C}_\gamma$ and $\tilde{C}_\sigma$ as they occur in a bed of grain. Before this can be done we need to specify the geometry of the grain/air interface. Given that the grain kernels settle in a grain bulk in a random order, this cannot be done precisely, but some average or idealised geometry might be used. Thorpe et al. (1991a,b) assumed that the grains constitute a spatially periodic porous medium, as illustrated in Figure 2. The equations that govern the behaviour of $\tilde{C}_\gamma$ and $\tilde{C}_\sigma$ are difficult to formulate and solve even for the system depicted in Figure 2, and the geometry used is depicted in Figure 3. This has the advantage of realising analytical solutions of the equations that govern $\tilde{C}_\gamma$ and $\tilde{C}_\sigma$, and these may be substituted into equation 14 to arrive at an equation that governs the diffusive moisture transfer of moisture through bulk grains. Thorpe et al. (1991b) show that the effective diffusivity, D_{eff}, of moisture through bulk stored grains is given by the expression

$$D_{eff} = D_\gamma \left\{ \frac{2\kappa - (\kappa - 1)\varepsilon_\gamma}{2 + (\kappa - 1)\varepsilon_\gamma} \right\} \Big/ \left\{ \varepsilon_\sigma \frac{\partial \mathcal{F}}{\partial \{C\}} + \varepsilon_\gamma \right\} \qquad (18)$$

where

$$\kappa = \frac{\partial \mathcal{F}}{\partial \{C\}} \frac{\mathcal{B}_\sigma}{\mathcal{B}_\gamma} \qquad (19)$$

Local Mass and Thermal Equilibrium in Bulks of Grain

Optimisation studies of grain storage systems, such as low temperature aeration, often require that many simulations be carried out to help identify the optimum design or operating conditions. For this reason the mathematical models that are used should be as simple, yet as accurate, as possible. This remains important even in this era of readily accessible and rapidly increasing computer power. This is because the models of grain stores apply to increasingly complicated and commercially important geometries, as opposed to the relatively simple heat and moisture transfer models in one-dimensional systems on which the techniques were developed.

Mathematical models of heat and moisture transfer in bulks of grain span a spectrum of complexity. The most complicated identify four independent variables in time and space, namely the temperatures and the moisture contents of the intergranular air and grain kernels. The next lower tier of complexity of models assumes that the air and grains are in local thermal equilibrium, but there is a finite resistance to mass transfer between the grain kernels and the intergranular air. This results in there being three dependent variables at each point in space and time, namely one temperature and a moisture content in each of the two phases present. On the third tier are the simplest models such as those presented by Sutherland et al. (1971) and Ingram (1979) in which both local mass and thermal equilibrium are assumed throughout the bulk. The formulation of such models in one-dimension allows analytical solutions to be obtained. These are useful for illustrating the passage of temperature and moisture waves through bulks of grain, and they are numerically the most efficient to solve, but their analytical solutions are very restrictive in the initial and boundary conditions that can be accommodated.

A rational means must be found for determining the most appropriate model for a given set of conditions, and this is the motivation of the work of Thorpe and Whitaker (1992a,b). The analysis entails introducing macroscopic deviations of temperature and moisture content. Whitaker (1991) defines macroscopic temperature deviations $\hat{T}_\gamma$ and $\hat{T}_\sigma$ as

$$\langle T_\gamma \rangle^\gamma = \langle T \rangle + \hat{T}_\gamma \qquad (20)$$

and

$$\langle T_\sigma \rangle^\sigma = \langle T \rangle + \hat{T}_\sigma \qquad (21)$$

and when the sorption isotherm of the grains is non-linear Thorpe et al (1991a) define macroscopic concentration deviations as

$$\langle C_\gamma \rangle^\gamma = \{C\} + \hat{C}_\gamma \qquad (22)$$

and

$$\langle C_\sigma \rangle^\sigma = \mathcal{F}(\{C\}, \langle T \rangle) + \hat{C}_\sigma \qquad (23)$$

It can be noted from equations 20 to 23 that when local thermodynamic equilibrium occurs the macroscopic deviations

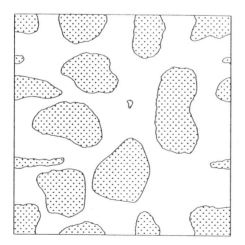

Fig. 2. A representation of a bulk of grains that displays spatial periodicity.

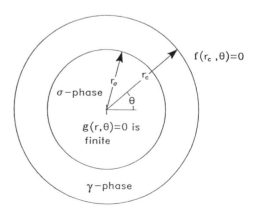

Fig. 3. A simplified unit cell of a spatially periodic porous medium.

are zero Additionally, it is readily deduced that the macroscopic spatial deviations are intimately related to differences in intrinsic averaged quantities in the two phases, i.e.

$$\hat{T}_\gamma = \varepsilon_\sigma \left(\langle T_\gamma \rangle^\gamma - \langle T_\sigma \rangle^\sigma \right) \qquad (24)$$

and

$$\hat{C}_\gamma = \varepsilon_\sigma \left[\langle C_\gamma \rangle^\gamma - \mathcal{F}^{-1} \left(\langle C_\sigma \rangle^\sigma , \langle T \rangle \right) \right] \qquad (25)$$

The work of Thorpe and Whitaker (1992a) suggests that the thermal energy balance in a ventilated bed of non-respiring grains may be expressed in the terms

$$(26)$$

$$\{ \varepsilon_\gamma (C_\gamma c_p) + \varepsilon_\sigma (C_\sigma c_p) \} \frac{\partial \langle T \rangle}{\partial t} + \varepsilon_\gamma (C_\gamma c_p) \langle v_\gamma \rangle \cdot \nabla \langle T \rangle + \varepsilon_\gamma (C_\gamma c_p) \nabla \cdot \langle \hat{v}_\gamma \hat{T}_\gamma \rangle + \varepsilon_\sigma \langle h, \dot{m} \rangle$$
$$= \nabla \cdot \left[\varepsilon_\gamma k_\gamma \left\{ \nabla \langle T \rangle + \frac{1}{V_\gamma} \int_{A_{\gamma\sigma}} \mathbf{n}_{\gamma\sigma} \hat{T}_\gamma dA \right\} \right] + \nabla \cdot \left[\varepsilon_\sigma k_\sigma \left\{ \nabla \langle T \rangle + \frac{1}{V_\sigma} \int_{A_{\gamma\sigma}} \mathbf{n}_{\sigma\gamma} \hat{T}_\sigma dA \right\} \right]$$
$$c_{p,l} \langle \dot{m} \rangle \hat{T}_\gamma - \varepsilon_\gamma (C c_p)_\gamma \frac{\partial \hat{T}_\gamma}{\partial t} - \varepsilon_\gamma (C_\gamma c_p) \langle v_\gamma \rangle \cdot \nabla \hat{T}_\gamma + \nabla \cdot (\varepsilon_\gamma k_\gamma \nabla \hat{T}_\sigma)$$
$$\varepsilon_\sigma (C c_p)_\sigma \frac{\partial \hat{T}_\sigma}{\partial t} + \nabla \cdot (\varepsilon_\sigma k_\sigma \nabla \hat{T}_\sigma)$$

As intimated above, thermodynamic equilibrium is approached as the macroscopic spatial deviations become small. But small compared with what? Whitaker (1991) suggests that they should be considered small compared with the thermal conduction terms in equation 26, which are usually the smallest in magnitude. This implies that in beds of ventilated grains the following constraints must be satisfied

$$\varepsilon_\gamma (C c_p)_\gamma \frac{\partial \hat{T}_\gamma}{\partial t} + \varepsilon_\sigma (C c_p)_\sigma \frac{\partial \hat{T}_\sigma}{\partial t} \ll \nabla \cdot \left[\left(\varepsilon_\gamma k_\gamma + \varepsilon_\sigma k_\sigma \right) \nabla \langle T \rangle \right] \qquad (27)$$

$$\varepsilon_\gamma (C c_p)_\gamma \langle v_\gamma \rangle^\gamma \cdot \nabla \hat{T}_\gamma \ll \nabla \cdot \left[\left(\varepsilon_\gamma k_\gamma + \varepsilon_\sigma k_\sigma \right) \nabla \langle T \rangle \right] \qquad (28)$$

$$\nabla \cdot \left[\varepsilon_\gamma k_\gamma \nabla \hat{T}_\gamma + \varepsilon_\sigma k_\sigma \nabla \hat{T}_\sigma \right] \ll \nabla \cdot \left[\left(\varepsilon_\gamma k_\gamma + \varepsilon_\sigma k_\sigma \right) \nabla \langle T \rangle \right] \qquad (29)$$

and

$$\left(c_p \right)_l \langle \dot{m} \rangle^\gamma \cdot \nabla \hat{T}_\gamma \ll \nabla \cdot \left[\left(\varepsilon_\gamma k_\gamma + \varepsilon_\sigma k_\sigma \right) \nabla \langle T \rangle \right] \qquad (30)$$

Thorpe and Whitaker (1992b) show that the thermal energy equations for each of the solid and fluid phases may be combined to yield estimates of the macroscopic deviations, and that these may be used in the restrictions 27 to 30. Similar reasoning may be applied to develop restrictions that must be satisfied if local mass equilibrium is to occur between the grain kernels and the intergranular air.

As a result of the work of Thorpe and Whitaker (1992a,b) it is possible to show that in practical applications of grain drying or aeration local thermal equilibrium between the kernels and the air may be assumed to exist. This applies to the drying of small grains, such as canola or large tree nuts such as walnuts. The reason for this apparent anomaly is that the width of a drying wave, say, is very narrow when the grains being dried are small, hence temperatures change relatively rapidly but thermal equilibrium is maintained because of the smallness of the particles. When commodities that consist of large particles are being dried the drying wave is wide and temperatures change more slowly in the drying wave, and because of this, thermal equilibrium is maintained between the phases. A consequence of this finding is that mathematical models of

heat and moisture transfer processes in bulk stored grains can usually be formulated on the assumption of local thermal equilibrium between the grains and the air. It should be noted that moisture equilibrium cannot generally be assumed, particularly in grain dryers, which is as one would expect. In aeration systems, mass equilibrium is approached as cooling and wetting waves widen when they pass through the grain bulk.

Heat and Mass Transfer in Respiring Bulks of Grains

When grains are stored with moisture contents that result in the relative humidity of the intergranular exceeding about 70% there is a likelihood of mould activity. This gives rise to the production of heat and moisture, and in unaerated grain bulks particularly this may result in moisture migration, thus causing more spoilage of grain. Although some analyses of heat and moisture transfer in respiring bulks of grains have been presented (Thompson, 1971) they fail to fully account for the requirements of mass and energy conservation. For example, when grain respires the carbohydrate substrate of the grain kernels is oxidised. In the process heat and moisture are liberated. As well as these sources of energy and mass, the moisture associated with the disappearing substrate must be accounted for and the surface energy that binds the moisture to the substrate must also be considered. A convenient attack on this problem is described by Thorpe (1993,1994). It commences with statements of the mass and energy conservation equations for the reactive and inert components of the intergranular air and the grain kernels. The boundary conditions at the air/grain kernel interface are stated and the analysis proceeds to develop mass and energy conservation equations expressed in terms of a volume averaged temperature and grain and air moisture contents. The outcomes of the work are expressions for heat and moisture transfer in ventilated beds of respiring grains for two extreme cases - one in which the bed bridges as the grain substrate is consumed and the other in which the bed of grains is deemed to slump. In the bridging bed case the moisture balance may be written

$$\left\{ \varepsilon_\gamma + \varepsilon_\sigma \frac{\partial \mathcal{F}}{\partial \langle (\rho_l)_\gamma \rangle^\gamma} \right\} \frac{\partial \langle (\rho_l)_\gamma \rangle^\gamma}{\partial t} + \left\{ \mathcal{F} \left(\langle (\rho_l)_\gamma , \langle T \rangle \right) - \langle (\rho_l)_\gamma \rangle^\gamma \right\} \frac{\partial \varepsilon_\sigma}{\partial t}$$
$$+ \varepsilon_\sigma \frac{\partial \mathcal{F}}{\partial \langle T \rangle} \frac{\partial \langle T \rangle}{\partial t} + \nabla \cdot \langle (\rho_l)_\gamma \rangle^\gamma \langle v_\gamma \rangle$$
$$= \mathbf{D}_{eff} \cdot \nabla \nabla \langle (\rho_l)_\gamma \rangle^\gamma + a_v \langle (r_l) \rangle_{\gamma\sigma} \qquad (31)$$

In equation 31, $\langle (\rho_l)_\gamma \rangle^\gamma$ is the concentration of water vapour in the intergranular atmosphere. The concentration of water in the grains, $\langle (\rho_l)_\sigma \rangle^\sigma$ (another way of expressing grain moisture content), is related to the concentration of moisture vapour in the air, $\langle (\rho_l)_\gamma \rangle^\gamma$ and temperature, $\langle T \rangle$, by a general isosteric function expressed in terms of spatially averaged variables, thus

$$\langle (\rho_l)_\sigma \rangle^\sigma = \mathcal{F} \left(\langle (\rho_l)_\gamma \rangle^\gamma , \langle T \rangle \right) \qquad (32)$$

The features that distinguish equation 31 from those that govern moisture transfer in grains that are not respiring are the presence of the terms

$$\left\{ \mathcal{F}\left(\langle(\rho_I)_\gamma,\langle T\rangle\rangle\right) - \left\langle(\rho_I)_\gamma\right\rangle^\gamma\right\}\frac{\partial\varepsilon_\sigma}{\partial t}$$

and $a_V\langle(r_1)\rangle_{\gamma\sigma}$. The former represents changes in moisture concentrations as the grains substrate is consumed by respiration. In particular

$$\mathcal{F}\left(\langle(\rho_I)_\gamma,\langle T\rangle\rangle\right)\frac{\partial\varepsilon_\sigma}{\partial t}$$

accounts for the moisture bound to the disappearing substrate and $\left\langle(\rho_I)_\gamma\right\rangle^\gamma\frac{\partial\varepsilon_\sigma}{\partial t}$ arises from the fact that in a bridging bed the intergranular spaces are increasing and they are filled, in part, by moisture vapour. The second term, $a_v\langle(r_1)\rangle_{\gamma\sigma}$, is the rate of production of moisture per unit volume of the grain bulk, and it arises from the oxidation of the grain substrate.

An equation that governs thermal energy conservation in a respiring bulk of grains has been derived by Thorpe (1993) to be

$$\left\{\varepsilon_\gamma\sum_{i=1}^4(c_i)_\gamma\left\langle(\rho_i)_\gamma\right\rangle^\gamma + \varepsilon_\gamma c_{v2}\left\langle(\rho_I)_\gamma\right\rangle^\gamma + \varepsilon_\sigma\sum_{i=1}^2(c_i)_\sigma\left\langle(\rho_i)_\sigma\right\rangle^\sigma\right\}\frac{\partial\langle T\rangle}{\partial t}$$

$$+\varepsilon_\sigma c_{v2}f_2\left(\left\langle(\rho_I)_\sigma\right\rangle^\sigma\right)\frac{\partial\langle T\rangle}{\partial t} + \left\{\sum_{i=1}^4(c_i)_\gamma\left\langle(\rho_i)_\gamma\right\rangle^\gamma\langle v_\gamma\rangle\right\}\cdot\nabla\langle T\rangle$$

$$-\varepsilon_\sigma h_s\frac{\partial\left\langle(\rho_I)_\sigma\right\rangle^\sigma}{\partial t} - (\rho_2)_\sigma\frac{\partial\varepsilon_\sigma}{\partial t}\int_0^W h_s dW$$

$$= K_{e\!f\!f}:\nabla\nabla\langle T\rangle + a_v\left\langle(r_2)_\sigma\right\rangle_{\gamma\sigma}H_o - a_v\langle(r_1)\rangle_{\gamma\sigma}\langle h_v\rangle^\gamma \quad (33)$$

The features that distinguish equation 33 from those that do not include respiration are the terms

$$(\rho_2)_\sigma\frac{\partial\varepsilon_\sigma}{\partial t}\int_0^W h_s dW$$

and

$$a_v\left\langle(r_2)_\sigma\right\rangle_{\gamma\sigma}H_o - a_v\langle(r_1)\rangle_{\gamma\sigma}\langle h_v\rangle^\gamma$$

The first term accounts for the energy associated with the water bound to the grain substrate that disappears as a result of oxidation. The latter group of terms arises from the heat of respiration, and the term $a_v\langle(r_1)\rangle_{\gamma\sigma}\langle h_v\rangle^\gamma$ is associated with the fact that the thermal energy equation is expressed in, amongst other variables, air moisture content and it represents a correction to the heat of respiration which is based on liquid water being formed.

It should be noted that the term

$$\varepsilon_\sigma c_{v2}f_2\left(\left\langle(\rho_I)_\sigma\right\rangle^\sigma\right)\frac{\partial\langle T\rangle}{\partial t}$$

should arise in analyses of heat and moisture transfer in both respiring and non-respiring grains, although in this context it appears to have been overlooked by previous authors. Close and Banks (1972) point out its existence in an analysis of coupled heat and mass transfer in silica gel, and it occurs because the integral heat of wetting of the grains is a function of temperature. The constant c_{v2} arises in an expression that

relates the latent heat of vaporisation, h_v, of free water and temperature, i.e.

$$h_v = 2502 - c_{v2}T \quad (34)$$

and $f_2\left(\left\langle(\rho_I)_\sigma\right\rangle^\sigma\right)$ is a function of the grain moisture content and it may be derived from sorption isotherms as demonstrated by Thorpe et al (1990). The function may be expressed in terms of simple polynomials.

Equations 31 to 33 have been programmed (Thorpe, 1994) for numerical solution by computer, and they were used to assess the likely success of aerating bulk stored grains in the humid tropics.

Solution Procedures

Having established the equations that govern heat and moisture transfer within bulks of grains they must be solved to satisfy the initial and boundary conditions. The initial conditions correspond to the grain temperature and moisture content distributions at the start of the storage period. The boundary conditions on the intergranular air usually imply that the walls and floor of the store are impermeable to gases. Temperatures are usually imposed on the walls and roof of the store and these may be estimated from climatic and solar radiation data, and the floors of grain stores are usually considered to be adiabatic, that is they are perfectly thermally insulated. A procedure for solving the equations that govern heat and mass transfer in three-dimensional bulks of grains is presented by Singh et al (1993b)

A problem of considerable commercial importance is the design of bunker stores for the long term storage of grains. The grains are often stored with initial temperatures exceeding 30°C and as the weather after harvest becomes progressively colder, thermal gradients are set up in the grain bulk and these promote natural convection currents. Warm air from the centre of the bulk rises to the peak where its relative humidity increases, thus increasing the moisture content of the grain in this region. The grain in the vicinity of the peak can become so wet that it is completely spoiled. It is important to investigate design and operational procedures that can overcome these difficulties, and mathematical modelling can be a useful adjunct to this process. Several mathematical models have been developed to predict the rate of migration of heat and moisture resulting from natural convection currents in bunker stored grains. The heat, mass and momentum transfer equations are cast in terms of derived variables, that is a stream function in two-dimensional flows and a vector potential in three-dimensional flows. The resulting equations are solved by discretising them, and solving them on a grid of orthogonal mesh points. In the studies of Nguyen (1987) and Freer et al (1990) the nodes coincide with the physical boundary of the grain store. More recently, Singh and Thorpe (1993a,b) have adopted a numerical scheme whereby the arbitrary physical shapes of grain stores can be transformed (or mapped) into simple shapes such as squares for two-dimensional flows and rectangular parallelepipeds for three-dimensional flows. The basic idea of mapping two-dimensional flow is shown in figure 4. Every point in the physical domain corresponds to a point in the computational domain, in this case a square. The points in the physical domain have (x,y) coordinates and the coordinates in the computational domain are (ξ,η). By the chain rule of differentiation we can express differentials in the physical domain in terms of those in the computational domain, thus

$$\frac{\partial T}{\partial x} = \frac{\partial T}{\partial \xi}\frac{\partial \xi}{\partial x} + \frac{\partial T}{\partial \eta}\frac{\partial \eta}{\partial x} \qquad (35)$$

and

$$\frac{\partial T}{\partial y} = \frac{\partial T}{\partial \xi}\frac{\partial \xi}{\partial y} + \frac{\partial T}{\partial \eta}\frac{\partial \eta}{\partial y} \qquad (36)$$

Terms such as $\partial \xi /\partial x$ and $\partial \eta /\partial y$ arise from the geometry of the grain store and they can be easily calculated from algebraic formulae or numerically. As Singh and Thorpe (1993a) demonstrate, analogous expressions may be developed for higher derivatives. These expressions may be substituted into the governing heat and moisture transfer equations and solved in the computational domain in (ξ,η) coordinates. The thermal energy balance, equation 33, is thus expressed

$$\left\{\varepsilon_{\tau}\sum_{i=1}^{4}(c_i)_{\tau}\big\langle(\rho_i)_{\tau}\big\rangle^{\tau} + \varepsilon_{\tau}c_{v2}\big\langle(\rho_i)_{\tau}\big\rangle^{\tau} + \varepsilon_{\sigma}\sum_{i=1}^{2}(c_i)_{\sigma}\big\langle(\rho_i)_{\sigma}\big\rangle^{\sigma}\right\}\frac{\partial\langle T\rangle}{\partial t}$$
$$+\varepsilon_{\sigma}c_{v2}f_2\big(\langle(\rho_i)_{\sigma}\rangle^{\sigma}\big)\frac{\partial\langle T\rangle}{\partial t} + \sum_{i=1}^{4}(c_i)_{\tau}\big\langle(\rho_i)_{\tau}\big\rangle^{\tau}\left[u_{\tau}\left\{\alpha_i\frac{\partial T}{\partial \xi} + \alpha_2\frac{\partial T}{\partial \eta}\right\}\right.$$
$$+v_{\tau}\left\{\alpha_3\frac{\partial T}{\partial \xi} + \alpha_4\frac{\partial T}{\partial \eta}\right\}\right] - \varepsilon_{\sigma}h_s\frac{\partial\langle(\rho_i)_{\sigma}\rangle^{\sigma}}{\partial t} - (\rho_2)_{\sigma}\frac{\partial\varepsilon_{\sigma}}{\partial t}\int_0^W h_s dW$$
$$= \beta_1\frac{\partial^2 T}{\partial \xi^2} + \beta_2\frac{\partial^2 T}{\partial \eta^2} + \beta_3\frac{\partial^2 T}{\partial \xi\partial \eta} + \beta_4\frac{\partial T}{\partial \xi} + \beta_5\frac{\partial T}{\partial \eta} + a_v\big\langle(r_2)_{\sigma}\big\rangle_{\gamma\sigma} H_\sigma - a_v\big\langle(r_i)\big\rangle_{\gamma\sigma}\langle h_v\rangle^{\tau} \qquad (37)$$

in which

$$\alpha_i = \frac{\partial \xi}{\partial x}, \quad \alpha_2 = \frac{\partial \eta}{\partial x}, \quad \alpha_3 = \frac{\partial \xi}{\partial y}, \quad \alpha_4 = \frac{\partial \xi}{\partial y} \qquad (38a)$$

$$\beta_1 = \alpha_1^2 + \alpha_3^2, \beta_2 = \alpha_2^2 + \alpha_4^2, \beta_3 = 2(\alpha_1\alpha_2 + \alpha_3\alpha_4), \qquad (38b)$$

$$\beta_4 = 2\left(\frac{\partial^2 \xi}{\partial x^2} + \frac{\partial^2 \xi}{\partial y^2}\right), \quad \beta_5 = 2\left(\frac{\partial^2 \eta}{\partial x^2} + \frac{\partial^2 \eta}{\partial y^2}\right) \qquad (38c)$$

The important feature of equation 37 is that it can be solved using the same procedures as Nguyen (1987) and Freer et al. (1990), but the shape of the grain store can be arbitrary.

Future Developments

Modelling transport phenomena in the headspace

To date, mathematical models of grain stores have not treated the heat, mass and momentum transfer phenomena that occur in the headspace of grain stored in the same detail as the phenomena in the grain bulk. Reasons for this are the complex nature of the turbulent fluid flow in the headspace, and the as yet incompletely resolved physics of the interaction of the air in the headspace and the grain bulk. Singh et al. (1993a) and Singh et al. (1994) have studied laminar flows in fluids overlaying porous media using two formulations of the boundary conditions at the fluid/porous medium interface, namely the Beavers-Joseph (Beavers and Joseph 1967) conditions and the Brinkman (1947) conditions. This research needs to be extended to account for the turbulent nature of the flow in the headspaces of grain stores, and the logical next step is to exploit the methods used by Gatheri et al. (1993) who have developed mathematical descriptions of turbulent flows in enclosures.

The use of commercial software packages

Fundamental studies of computational fluid dynamics carried out some two decades ago are now giving rise to powerful and fairly user-friendly commercial software packages. The packages are used in industries such as aerospace, chemi-

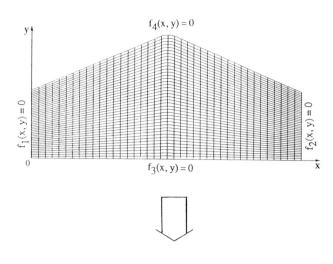

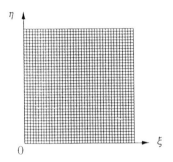

Fig. 15. Mapping the cross-section of a bulk of grain into an orthogonal computational domain.

cal processing, automotive design and so on to predict the performance of components such as cooling towers, turbomachinery, furnaces, air conditioning systems and quite recently in grain storage applications. The packages, such as PHOENICS produced by CHAM (Proprietary information), can be used to discretise the differential equations that govern heat, mass and momentum transfer in stored grains, and in particular the coordinate system can be manipulated to fit the shape a grain store, say. Such packages clearly offer considerable scope for grain storage technologists to solve very complicated problems.

Acknowledgment

The author is grateful to Associate Professor I. D. G. Mackie, Head, Department of Civil and Building Engineering, for encouraging this work to be carried out.

Nomenclature

a_v	Area per unit volume of bed, 1/m	T_ω	Temperature of ω-phase, K.
c_i	Specific heat of ith species, J/kgK.	u	Horizontal component of velocity, m/s
c_{v2}	Constant in expression for latent heat of evaopration of free water, J/kgK	$\mathbf{u}_i$	Diffusion velocity of ith species, m/s.
$\{C\}$	Phase weighted equilibrium concentration, kg/m^3	v	Vertical component of velocity, m/s.
		$\mathbf{v}_i$	Velocity of ith species, m/s.
C_ω	Concentration of species ω-phase, kg/m^3	V	Volume of representative elementary region, m^3.
$\mathcal{B}_\omega$	Binary diffusion coefficient, m^2/s	V_ω	Volume of ω-phase in representative elementary region, m^3.
$cw1,...,cw5$	Empirical constants in the isotherm equation.	$\mathbf{w}$	Velocity of surface of grain kernels being consumed by respiration, m/s
$\mathbf{D}_{eff}$	Effective diffusivity, m^2/s	W	Moisture content of grain (dry basis), kg/kg.
$\mathcal{F}$	Sorption isotherm		
f_2	An empirical function of moisture content.	x,y	Cartesian coordinates, m.
H_o	Heat of oxidation of grain substrate, J/kg	*Greek symbols*	
h_s	Heat of sorption of moisture on grain substrate, J/kg	α	Refers to α-phase
h_v	Latent keat of vaporization of water, J/kg.	$\alpha_1,.....,\alpha_4$	Mesh transformation functions.
$\mathbf{K}_{eff}$	Effective thermal dispersivity tensor, J/kg/K	$\beta_1,.....,\beta_5$	Mesh transformation functions.
k_ω	Thermal conductivity of ω-phase, J/m/K.	γ	Refers to γ-phase
l_ω	Length scale of ω-phase, m.	Γ_ω	A quantity defined in the ω-phase.
L	Length scale of macroscopic system, m.	$\langle \Gamma_\omega \rangle$	Superficial volume average of Γ_ω.
$\dot{m}$	Rate of moisture transfer, kg/s/m^3	$\langle \Gamma_\omega \rangle^\omega$	Intrinsic volume average of Γ_ω.
$\mathbf{n}_{\alpha\omega}$	Unit normal in the direction $\alpha-\omega$	$\langle \Gamma \rangle_{\alpha\omega}$	Area average of Γ at the $\alpha-\omega$ surface.
p_o	Grain-specific constant in isotherm equation, Pa.	$\tilde{\Gamma}_\omega$	Spatial deviation of Γ_ω from $\langle \Gamma_\omega \rangle^\omega$.
p_s	Saturation pressure of free water, Pa.	$\hat{\Gamma}$	Macroscopic deviation of $\langle \Gamma_\omega \rangle^\omega$ from
p_ω	Pressure in ω-phase, Pa.		$\langle \Gamma \rangle$
$\mathbf{q}_\omega$	Heat flux through ω-phase, W/m^2.	ε_ω	Void fraction of ω-phase.
r_i	Rate of production or disappearance of ith species per unit area, kg/s/m^2	κ	Defined by equation 19.
r_o	Radius of representative elementary volume, m.	μ	Viscosity, kg/m/s.
$\mathbf{r}$	A position vector, m.	ξ,η	Coordinates of the computational domain.
		ρ_i	Density of ith species, kg/m^3.
t	Time, s	σ	Refers to σ-phase
$\langle T \rangle$	Phase weighted average temperature, K.	ω	Refers to ω-phase
T^o	Reference temperature, K.		

References

Anderson, T.B. and Jackson, R. 1967. A fluid mechanical description of fluidised beds. Industrial Engineeing and Chemistry Fundamentals, 6, 527–539.

Beavers G. and Joseph, D.D. 1967. Boundary conditions at a naturally permeable wall. Journal of Fluid Mechanics, 110, 197–207.

Becker, H.A. and Sallans, H.R. 1971. Drying wheat in a spouted bed. On the continuous moisture diffusion controoled drying of solid particles in a well-mixed, isothermal bed. Chemical Engineering Society, 13, 97–112.

Bejan, A. 1984. Convection heat transfer. New York, Wiley.

Brinkman, H.C. 1947. A calculation of the viscous force exerted by a flowing fluid on a dense swarm of particles. Applied Science Research, A1, 27–34.

Carbonell, R.G. and Whitaker, S. 1984. Heat and mass transfer in porous media In: Bear, J. and Corapciolglu, M. Y., eds., Mechanics of fluids in porous media, Nijhof, Brussels, 121–198.

CHAM (Proprietary information) What is PHOENICS? Concentration Heat and Momentum Limited, London, UK.

Close, D.J. and Banks, P. J. 1972. Coupled equilibrium heat and single adsorbate transfer in fluid flow through a porous medium – II. Predictions for silica gel-air using characteristic charts. Journal of Chemical Engineering Science, 27(5),1157–1169.

Crapiste, G.H., Whitaker, S. and Rotstein, E. 1988. Drying of cellular material I. A mass transfer theory. Chemical Engineering Science, 43, 2919–2928.

Freer, M.W., Siebenmorgan, T.J., Couvillion, R.J. and Loewer, O.J. 1990. Modeling temperature and moisture content changes in bunker-stored rice. Transactions ASAE, 33, 211–220.

Gatheri, F.K., Reizes, J.A., Leonardi, E. and de Vahl Davis, G. 1993. The use of variable false transient parameters for the solution of natural convection problems. Fifth Australasian Heat and Mass Transfer Conference, Brisbane, 6–9th December, 1993.

Gray, W.G. 1975. A derivation of the equation for multiphase transport. Chemical Engineering Science, 30, 229–233.

Griffith, H.J. 1964. Bulk storage of grain: a summary of factors governing control of deterioration. Melbourne, Australia, CSIRO Division of Mechanical Engineering, Report ED8.

Ingram, G.W. 1979. Solution of grain cooling and drying problems. Journal of Agricultural Engineering Research, 24, 219–232.

Jury, S.H. 1967. An improved vesion of the state equation for molecular diffusion in a dispersed phase. AIChEJournal, 13,1124–1126.

Marle, C.M. 1967. Ecoulements monophasiques en milieu poreux. Rev. Fr. Petr., 2, 327–356.

Nguyen, T.V. 1987. Natural convection in stored grains—a simulation study. Drying Technology, 5, 541–600.

Nozad, I, Carbonnel, R.G. and Whitaker, S. 1985a. Heat conduction in multiphase systems— I. Theory and experiment for two phase systems. Chemical Engineering Science, 40, 843–855.

Nozad, I, Carbonnel, R.G. and Whitaker, S. 1985b. Heat conduction in multiphase systems—II. Experimental method and results for three phase systems. Chemical Engineering Science, 40, 857–863.

O'Callaghan, J.R., Menzies, D.J. and Bailey, P.H. 1971. Digital simulation of agricultural drier performance. Journal of Agricultural Engineering Research, 16, 223–244.

Ochoa, J.A. Stroeve, P. and Whitaker, S. 1986. Diffusion and reaction in cellular media. Chemical Engineering Science, 40, 943–855.

Pixton, S.W. and Griffith, H.J. 1971. Diffusion of moisture through grain. Journal of Stored Products Research, 7, 133–152.

Singh, A.K and Thorpe, G.R. 1993a. A solution procedure for three-dimensional free convective flows in peaked bulks of grain. Journal of Stored Products Research, 28 (3), 221–235.

Singh, A.K and Thorpe, G.R. 1993b. Application of a grid generation technique to the numerical modelling of heat and moisture movement in peaked bulks of grains. Journal of Food Processing Engineering, 16 (2), 127–145.

Singh, A.K., Leonardi, E and Thorpe, G.R. 1993a. Three-dimensional free convection in a confined fluid overlying a porous layer. Journal of Heat Transfer, 115 (3), 631–638..

Singh, A.K., Leonardi, E and Thorpe, G.R. 1993b. A solution procedure for the equations that govern three-dimensional free convection in bulk stored grains. Transactions ASAE, 36(4), 1159–1173.

Singh, A.K., Moore, G.A. and Thorpe, G.R. 1994. Effect of the ratio of the depths of fluid and porous layers on free convective flows in tall rectanglar cavities. 12th National Heat and Mass Transfer Conference, January 5–7, Bombay, India.

Slattery, J.M. 1967. Flow of viscoelastice fluids through porous media. AIChE Journal, 13,1066–1071.

Sutherland, J.W., Banks, P.J. and Griffith, H.J. 1971. Equilibrium heat and moisture transfer in air flow through grain. Journal of Agricultural Engineering Research, 16, 368–386.

Thompson, T.L. 1972. Temporary storage of high-moisture sheeled corn using continuous aeration. Transactions ASAE, 15, 333–337.

Thorpe, G.R. 1981. Moisture diffusion through bulk stored grain. Journal of Stored Products Research, 17, 39–42.

Thorpe, G.R. 1982. Moisture diffusion through bulk grain subjected to a temperature gradient. Journal of Stored Products Research, 18, 9–12.

Thorpe, G.R. 1993. Heat and mass transfer in ventilated bulks of respiring porous hygroscopic media. Fifth Australasian Conference on Heat and Mass Transfer, December 6–9, Brisbane, 1993.

Thorpe, G.R. 1994. Heat and moisture transfer in ventilated bulks of respiring grains—A theoretical analysis and technological application. Report, Department of Civil and Building Engineering, Victoria University of Technology, Melbourne, Australia.

Thorpe, G.R. and Whitaker, S. 1992a. Local mass and thermal equilibria in ventilated grain bulks. Part I The development of heat and mass conservation equations. Journal of Stored Products Research, 28, 15–27.

Thorpe, G.R. and Whitaker, S. 1992b. Local mass and thermal equilibria in ventilated grain bulks. Part II The development of constraints. Journal of Stored Products Research, 28, 29–54.

Thorpe, G.R., Ochoa, J.A. and Whitaker, S. 1991a. The diffusion of moisture in food grains. I The development of a mass transfer equation. Journal of Stored Products Research, 27, 1–9.

Thorpe, G.R., Ochoa, J.A. and Whitaker, S. 1991b. The diffusion of moisture in food grains. II Estimation of the effective thermal diffusivity. Journal of Stored Products Research, 27, 11–30.

Thorpe, G.R., Stokes, A.N. and Wilson, S.G. 1990. The integral heats of wetting of food grains, Journal of Agricultural Engineering Research, 46, 71–76.

Wexler, A. 1965. Humidity and moisture. In: Wexler, A., ed., Measurement and Control in Science and Industry, 1, Reinhold, New York.

Whitaker, S. 1967. Diffusion and dispersion in porous media. AIChE Journal, 13, 420–427.

Whitaker, S. 1986. Flow in porous media I: A theoretical derivation of Darcy's law, Transport in Porous Media, 1, 3–25.

Whitaker, S. 1987. The role of the volume averaged temperature in the analysis of non–isothermal, multiphase transport phenomena. Chemical Engineering Communication, 58, 171–183.

Whitaker, S. 1988. Coments and corrections concerning the volume-averaged temperature and its spatial deviation. Chemical Engineering Communication, 70, 15–18.

Whitaker, S. 1991. Some improved estimates for the principle of local thermal equilibrium. Industrial Engineering and Chemistry Research 30, 983–997.

Grain aeration system controlled by computer

Wu Zidan and Li FuJan*

Abstract

A mathematical model based on experiment work was constructed to predict the effect on grain aeration of changing physical parameters. The variation of these parameters was observed in order to define the most effective way to control aeration. With this information, computer controlled aeration systems have been possible.

Introduction

In China, grain aeration is one grain storage technology which has been extensively applied in recent years. How to control the aeration process scientifically and efficiently is a very important subject. In the past, setting temperature, humidity or time alone was often used for aeration control. However, this is not very efficient or suitable in many areas in China where there are great changes in air temperature, high humidity and different grain types.

Over the past few years, we have been researching and experimenting with using computers to control aeration. By building a mathematical simulation to imitate the changes of each parameter in the aeration process, the best time for aeration can be calculated and selected by computer. Equipment for grain aeration control by computer manufactured in Tianjin city in 1990 has been been put into use extensively in China. This paper will discuss some theoretical and practical problems associated with this research.

Grain Equilibrium Absolute Humidity and the Conditions of Aeration Control

In controlling the grain aeration process, the main problem is to correctly select a group of atmosphere parameters, such as the upper and lower limits of temperature, humidity and conditions of dew point, etc. This group of parameters must satisfy the following requirements:

First, it must satisfy the specific aim of aeration, for example, it must select completely opposite conditions of humidity in two different kinds of aeration: reducing moisture content or regulating grain quality (increasing moisture content of grain slightly in order to improve grain milling quality).

Second, it must satisfy not only the efficiency of aeration, but also the time required for aeration, which is the key to reducing energy use and costs.

---

* State Administration of Grain Reserve, Ministry of Internal Trade, Beijing, People's Republic of China.

Third, it must avoid adverse effects, for example, if you want to lower temperature, you must protect the grain from gaining moisture.

Fourth, it must ensure the security of aeration, in particular, protecting the grain from dewing.

For all of these conditions to be satisfied at once, it is necessary to find the best equilibrium point for many factors.

The first part of this process concerns the relationship between the moisture content of the grain and the humidity of the air. Figure 1 is an equal temperature curve of the moisture content of grain and the relative equilibrium humidity. It is a 'S' curve which is composed of an absorption curve and an opposite absorption curve. Its shape and location is dependent on the temperature and species of the grain. Because in most cases of aeration, the grain is in a state of opposite absorption, and in reality the r.h. of the atmosphere is not less than 20%, the mathematical simulations use the curve of opposite absorption and only similate the part of the curve where r.h. exceeds 20%. Regression analysis of the relation between the moisture content and the equilibrium humidity was determined for five grain species: wheat, maize, paddy, rice and soya bean.

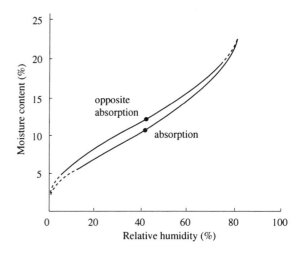

Fig. 1. The equal temperature curve of the moisture content of grain and the relative equilibrium humidity.

These were then incorporated in a mathematical simulation (MS) which represents the relation between grain moisture content and the equilibrium humidity. It is:

The equilibrium relative humidity (r.h.);

r.h.(%) = e(K1X+Y+A)/B1

The equilibrium absolute humidity (AH);

Ps2(mmHg) = e(K2X+Y+A)/B2

The absolute humidity of atmosphere;

Ps1(mmHg) = e(K3X+Z+C)/B2

The temperature of dew point;

T(°C) = D/p+E

Where:

A, B1, B2, C, D, E are constants;

K1, K2, K3 is relevant to grain species;
X is relevant to grain temperature;
Y is relevant to grain moisture;
Z is relevant to r.h.;
p is relevant to the equilibrium absolute humidity.

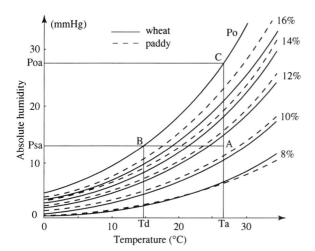

Fig. 2. The curve graph of grain equilibrium absolute humidity.

Using experimental data, the MS then produced the curves for grain equilibrium absolute humidity (Fig. 2). In Figure 2 the ordinate is the absolute humidity (unit: mmHg), the curve Po is the curve of saturation absolute humidity of atmosphere under one atmospheric pressure, the rest are curves of the equilibrium absolute humidity of grain, showing the variation of the equilibrium absolute humidity of grain under different conditions of temperature and moisture content. The project of point A on the ordinate and abscissa presents the absolute humidity, Psa, and temperature, Ta, of this point. The projection of points B and C on the ordinate and abscissa of the curve, represents the temperature of dew point Td and the saturation absolute humidity Poa of point A; the radio of Psa/Poa is r.h.(%) of point A. If the moisture content which is on the curve of grain equilibrium absolute humidity across point A is W%, the Psa, r.h.a and Td respectively represent the grain equilibrium absolute humidity, the equilibrium r.h. and grain temperature of dew point of point A.

If the equilibrium absolute humidity of the different grain species is different under the condition of same temperature and moisture content, then from Figure 2, it can be seen that the homologous points of dew and equilibrium r.h. are also different. Therefore the difference in grain species must be considered in the aeration process.

The following examples illustrate the relationship between these different parameters and the condition of aeration control.

Example 1. Judging whether a grain depot can be ventilated for cooling given the following conditions: the wheat temperature is 30°C, the moisture content is 11.5%, the atmosphere temperature is 20°C, r.h. is 80%.

When the moisture content of the wheat is 11.5%, temperature (Ta) is 30°C, its homologous point is A, then from Figure 3:
the equilibrium absolute humidity: Pa = 16.4 mmHg
the saturation absolute humidity: Poa = 31.6 mmHg
the equilibrium r.h.: r.h.a = Pa/Poa = 51.9%
When atmosphere temperature Tb = 20°C,
the saturation absolute humidity of atmosphere:
Pbb = 17.3 mmHg,

the absolute humidity of atmosphere:
Psb = 17.3 × 80% = 13.9 mmHg.

Comparing point A with B: the r.h. of point B is higher than point A, but its absolute humidity is lower than point A, showing that the moisture of grain will not increase during aeration, so the condition of humidity in aeration can be fulfilled. Technology Regulations of Grain Aeration (Commercial Ministry, PRC) state that the difference in temperature between the grain and atmosphere in the beginning of ventilation cannot be less than 8°C, which means that in this case, the upper limit of atmosphere temperature is 30°C–8°C = 22°C (line CD). In this example then, the air temperature is lowered to 22°C, and this satisfied the requirements of the conditions concerning temperature and so the grain depot may be ventilated.

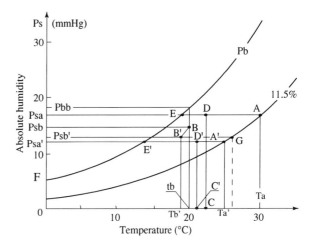

Fig. 3. The analysis of the wheat aeration for reducing temperature. The fold line CDEF gives a border conditions of ventilation, which looks like a 'window'. If the state of atmosphere is within the 'window', the ventilation is permitted, otherwise, it is not.

Example 2. If the temperature of wheat in example 1 after ventilation is lowered to 25°C, the moisture has not changed, the atmosphere temperature is lowered to 19°C, and the r.h. of atmosphere is lowered to 77%, should aeration be continued or not?

The Technology Regulations of Grain Aeration further state that the difference in temperature required to stop ventilation is 4°C. That is to say, the upper limit of atmosphere temperature is 21°C (line C'D', Fig. 3), thus the temperature of atmosphere is still suitable for aeration.

Because of the reduction of grain temperature, the state of the grain moved from point A to A' (Fig. 3), equilibrium AH reduced to 11.7 mmHg; and the state of atmosphere moved from B to B', AH reduced from 13.9 mmHg to 12.5 mmHg. Now the 'window' is C'D'E'F, point B' has already moved away from the 'window'. So if the aeration is continued, the humidity of the grain will increase. This means that the aeration cannot be continued.

From the examples above mentioned it is clear to see the dynamic interation of the parameters in aeration. Now, we can conclude that aeration control is not reliable if constant temperature and humidity set points for the atmosphere condition are used to control ventilation. This is because even though the temperature and humidity set points of the atmosphere condition are suitable for ventilation at the beginning, they may not be satisfactory through the whole process of ventilation.

Table 1. The criteria for determining the possibility of ventilation.

The aim of aeration	The condition of aeration control
Reduce temperature	Ps1<Ps2
	Beginning: t2 – t1 > 8°C
	In sub-tropic areas: t2 – t1 > 6°C)
	During process: t2 – t1 > 4°C
	In sub-tropic areas: t2 – t1> 3°C)
Reduce moisture	Ps1 < Ps21
	t2 > td1
Regulate grain quality	Ps1 > Ps22
	t2 > td1

Where:

t1 — atmosphere temperature;
t2 — grain temperature;
td1 — the temperature of atmosphere dew point;
Ps1 — the AH of atmosphere;
Ps2 — the grain AH where grain temperature is t2;
Ps21 — the grain AH which reduces the moisture of grain by one percent, and the temperature of grain is equal to that of the atmosphere t1;
Ps22 — the grain AH which increases the moisture of the grain by 2.5%, and the temperature of grain is equal to that of atmosphere t1.

Definition of Conditions of Aeration Control

Table 1 outlines the main conditions for control of aeration.

Definition of the temperature limits for cooling aeration

Considering the efficiency of aeration and the opportunity of ventilation, the difference in temperature between the atmosphere and grain must not be less than 8°C at the beginning of ventilation and 4°C during ventilation in all areas except the sub-tropics in China. Because the annual differences in temperature are less in the sub–tropical areas of China, efficiency of ventilation must be partly sacrificed for the sake of ensuring enough opportunity to ventilate. So the differences in temperature should be 6°C in the beginning and 3°C during ventilation.

Definition of the humidity limits for aeration concerned with reducing moisture and regulating grain quality

For reducing moisture, the humidity condition for aeration is the grain equilibrium absolute humidity that reduces the moisture of the grain by 1%, while the temperature of the grain is equal to that of the atmosphere. For regulating grain quality, the humidity condition is the equilibrium absolute humidity that increases the moisture content of the grain by 2%, while the temperature of grain is equal to that of atmosphere. The equilibrium absolute humidity used here is an opposite absorption curve graph, and the homologous moisture value on the absorption curve is generally 2–2.5% higher than that on the opposite absorption curve under the same humidity level. Therefore, we need to add 2.5% moisture to grain moisture as the difference in moisture to compensate the difference between two curves and ensure that the humidity is increased effectively in the process of aeration for regulating grain quality.

Concerning the condition of dewpoint

There are two types of condensation in commodity ventilation: One is called 'internal dew', formed from water vapour in the stack, internally condensed on cool air when the temperature of the atmosphere is lower than the dewpoint of the stack. The other is 'exterior dew', formed from water vapour in the air condensed onto cool grain when the temperature of the grain is lower than the dewpoint temperature. The 'internal dew' has slight effect on ventilation in practice, because the dewing will stop immediately once the hot humid air inside the stack is displaced by large amounts of cool dry air from the outside. Because the moisture content of the grain can be changed by the air drawn into the stack, the 'exterior dew' is important. In order to avoid the occurrence of 'exterior dew', ventilation should not proceed when the temperature of the grain is lower than the dewpoint of the atmosphere.

The Principle and Practice of Automatic Computer Monitoring of Grain Ventilation

An automatic computer monitor for ventilation of crops has been produced in TianJin. Its characteristics are as follows:

1. Using the mathematic model similar to the equilibrium AH curve of commodities, it can simulate all the variations of equilibrium AH of different species or moisture content of crops which change with grain temperature. Moreover, it can convert the r.h. and AH of atmosphere, dewpoint of grain and that of atmosphere at the same time.

2. According to different purposes of ventilation, it can open a 'window' for ventilation with the ideal border conditions. That is, by a decision control procedure, it can define the reasonable ventilating temperature and humidity conditions automatically, then can judge whether the ventilation is permitted or not. Moreover, by using the data concerning atmosphere and grain conditions collected by its sensors, it can rectify the 'window' conditions automatically and constantly, so the efficiency and security in the whole process of ventilation is ensured.

The automatic control installation can be used conveniently. Once you install the basic parameters of ventilation purpose, commodity type and moisture content on the keyboard, the equipment can substitute for skilled staff to control the process of ventilation automatically. Moreover, it can forecast and limit the possibility of forming dew. In terms of energy efficiency, it is generally 20–50% better than that of personal control.

Engineering — Session Summary

Conveners: J. Ford and D.E. Maier

The Engineering Session consisted of an all day session that included 15 oral paper presentations and over 20 posters, which covered the following areas of interest:

I. Pest prevention and management through aeration (7 oral papers)

II. Modelling and analysis of storage and drying (4 oral papers)

III. Design and management of structure, and equipment (4 oral papers)

The stage for the session was set by the keynote address of Dr. Allan Roberts, which reviewed succinctly the challenges facing engineers when designing silo bulk handling facilities. The presentation was well received by the predominantly non-engineering audience. The engineering session enjoyed a good attendance with an audience high of 60. Participants took full advantage of the 30–45 minute discussion periods following each of the three topic areas.

Highlights of the aeration paper included U.S. work on chilled aeration and conditioning. Specifically, as it is beginning to be implemented for some of the higher value commodities for both pest prevention and product quality preservation in the U.S. grain industry. British research highlighted the effectiveness of ambient aeration cooling as part of an integrated storage strategy. British, Australian, Moroccan, Chinese and U.S. work on the utilisation and development of fan control strategies illustrated the fact that a proper understanding of aeration management is critical for automatic systems to succeed. Much of the discussion centred around whether aeration control strategies should be based on dry bulb or wet bulb temperatures, or relative humidity. Although no consensus developed, agreement did exist that aeration strategies must take into account regional if not local climatic conditions for aeration to be a successful tool in an integrated bulk storage management system.

Australian research on programmable fan controllers pointed the way to the future where computer-based systems will provide managers remote, on-line and real-time decision support, as well as extensive flexibility to optimise strategies. During the discussion period it was pointed out that a prototype commercial system was recently introduced in the U.S. grain industry. In the future, aeration cooling using ambient and/or chilled air will become every storage managers number one preventive pest management tool. Engineers, entomologists and mycologists need to cooperate in promoting this alternative, non-chemical product protection technology as the primary line of defence against quality deterioration!

Another key to improved product quality is a better understanding of the influence of various parameters on the drying and storage of bulk commodities. Because of the complexity of the physical, chemical and biological ecosystem at hand, numerical modelling provides a powerful tool for analysis of optimum strategies and limiting constraints. Work from Australia, China and Thailand highlighted the effectiveness and powerfulness of modelling. The discussion period pointed out the continuing need to refine the numerical approach, and the trade offs between numerical accuracy and field applicability.

Finally, an engineering session would not be complete without looking at the design of the equipment and structures that are necessary to make drying, storage and handling of bulk crops possible. Work from India presented an update on design improvements of on-farm storage's in that vast country. British research highlighted product quality differences observed in large-scale outdoor sack storage's in Zimbabwe when different bagging materials were used. U.S. work presented a summary on the design of dosed-loop fumigation systems in concrete and steel storage structures at commercial elevators in the U.S. wheat belt. Australian engineering ingenuity was demonstrated with the design of a new high-capacity stacker for filling and emptying of large-scale grain bunkers.

Unfortunately, this particular Australian presentation represented the only contribution by an industry representative in the Engineering session. In the discussion sessions it was emphasised that future session organisers and participants should make a more concerted effort to increase industry participation in oral and poster presentations. Engineers working outside the academic and research community need to be invited to become involved, and hopefully provide an important reality check as to feasibility, costs, economic incentives, appropriateness and transferability of technology. This would add an invaluable dimension to any working conference!

Last but not least, it must be pointed out that engineers and engineering solutions were essentially part of all sessions at this conference. Thus, future conferences should consider using a session title less broad than 'Engineering'. This would hopefully eliminate some confusion as to submitting a paper to what session, and attract a broader range of papers from which to select oral presentations from. Finally, discussions among the participants of the engineering session suggested the following three engineering-related topics for the next conference:

(1) Sealing procedures and technology for new and existing structural designs of storage and processing facilities.

(2) Advances in sensor technology used for monitoring and detecting gases, dusts, noise, odours, product quality, and product damage due to insects or fungi.

(3) Preventive stored commodity management techniques including the utilisation and control of aeration cooling.

Sampling and Trapping

The use of sex pheromones to control *Ephestia kuehniella* Zeller (Mediterranean flour moth) in flour mills by mass trapping and attracticide (lure and kill) methods

P. Trematerra*

Abstract

Since the identification of the components of the sex pheromone of *Ephestia kuehniella* Zeller in the 1970s, considerable progress has been made in the use of this pheromone for control. This paper reports the results obtained in flour mills using mass trapping and attracticide (lure and kill) methods. The mass trapping method when applied in two large flour mills, using a multifunnel trap every 260–280 m^3 baited with 2 mg of TDA (Z9, E12-14Ac, with a daily release of 13 µg) removed a large number of males from the mills. This prevented the expected increase in the population of *E. kuehniella* which remained at a constant low level. Subsequently, the number of larvae and damage to products in all departments were reduced. The number of fragments revealed in filth tests was also low. Control of *E. kuehniella* males by the attracticide method, using laminar dispensers (2 × 2 cm), baited with 2 mg of TDA and 5 mg of cypermethrin, showed that the combined action of the pheromone and the insecticide achieved good results. The success using attracticide applications every 220–280 m^3 was encouraging. The sublethal and lethal effects of attracticide formulations were found to be more intense if associated with a silhouette of inverted triangular shaped forms as a 'sign stimulus'. The control achieved with these two methods made the usual second fumigation of the mills unnecessary.

The employment of pheromones for the control of pest populations in mills may come into more widespread use if made price competitive with conventional insecticides. A benefit of the attracticide method is that it allows a broad spectrum of insecticides to be used selectively, thus preventing the death of beneficial insects, which may occur when conventional insecticides are used. The use of synthetic pheromones may lead to a drastic reduction of chemical treatments with subsequent economic and quality advantages. Goods may be protected from possible pesticide residues and thus improve the image of the company's products.

Introduction

In the Mediterranean region *Ephestia kuehniella* Zeller (Lepidoptera: Phycitidae) is a major pest of flour mills. It is generally present in Italian mills all year round with fluctuations that peak in June and also between August and September. The pheromone of *E. kuehniella* was identified in the 1970s as a blend of Z9E12-14Ac (TDA), Z9E12-14OH (TDO) and Z9-14Ac (TA) (Brady et al. 1971a,b; Brady 1973; Kuwahara and Casida 1973; Sower et al. 1974). Since then, the main component, TDA, has been used for monitoring *E. kuehniella* in many parts of the world. To pass from simple monitoring of storage phycitid moths to more sophisticated

uses of their pheromone such as mass trapping, attracticide (lure and kill) and even mating disruption, it is often necessary to use the exact doses released of their pheromone components (Bommer and Reichmuth 1980). Taking into consideration the work of Read and Haines (1976), Mankin et al. (1980), Levinson and Buchelos (1981), Süss and Trematerra (1982), Trematerra and Rossi Porzio (1982), Levinson and Hoppe (1983), Hodges et al. (1984), Burkholder and Ma (1985), the activity of TDA and TA was investigated.

Using rubber septa it was found that *E. cautella* (Walker) responded well to a mixture of TDA + TA at doses of 10 ± 5 µg (100 ng released daily); *E. kuehniella* to TDA at doses of 2000 µg (13000 ng released daily) and *Plodia interpunctella* (Hübner) to a mixture of TDA + TA at doses of 1 ± 0.5 µg (10 ng released daily) (Süss and Trematerra 1985, 1986; Trematerra 1986).

Research has been carried out with the aim of reducing or finding alternatives to traditional chemical treatments, which often amount to two fumigations and several other limited insecticide treatments per year (Fig. 1). Direct control of infestations of *E. kuehniella* in flour mills by pheromones has been reported: mass trapping, by Levinson and Buchelos (1981), Trematerra and Battaini (1987), and Trematerra (1988, 1990); the possibility of the attracticide method, or lure and kill (an insecticide–pheromone combination), by Trematerra and Capizzi (1987, 1991); and mating disruption, by Trematerra and Capizzi (1987).

This paper reviews the studies undertaken in Italy from 1986 to 1993 on the possibility of controlling *E. kuehniella* infestations by sex pheromones using mass trapping and attracticide methods.

Studies on Mass Trapping Method

Preliminary observations in two different flour mills

Integrated control of *E. kuehniella* can be achieved in flour mills by mass trapping and the limited use of insecticides (Trematerra and Battaini 1987). Experiments were performed in two different unheated flour mills, one a small traditional kind, the other a large processing plant with modern machinery.

In this preliminary study, funnel traps (mastrap type, produced by G. Donegani Institute, Novara, Italy) (Fig. 2) with rubber dispensers baited with 2 mg of TDA (release rate of 13 µg daily) were used. The traps were placed in the most infested parts of the processing area of the traditional mill, covering an area of 567 m^3, while in the industrial mill covering 1560 m^3. Using the pheromone dispersion model proposed by Mankin et al. (1980) and consideration of the work of Süss and Trematerra (1985), two traps were placed in the former and six traps in the latter (in both areas one trap per 260–280m^3 was used). The traps were placed 2–2.5 m from the floor and 3.5–4 m from the walls.

The pipe unions were opened in each mill during temporary stoppages in the processing, to enable the pheromone to attract

* Dipartimento di Scienze Animali Vegetali e dell'Ambiente, University of Molise, Campobasso, Italy.

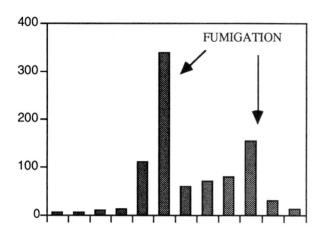

Fig. 1. Dynamic population of *Ephestia kuehniella* males in a fumigated flour mill.

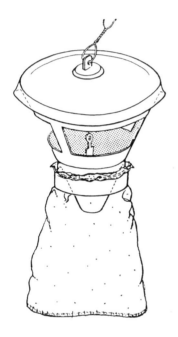

Fig. 2. The funnel trap used in the experiments.

moths hiding in the machinery. Trap captures and residual infestation of adults on the walls and machinery were recorded every week. Experiments were performed from April to October, when the moths were most numerous. As a control, observations were made in a different mill using pheromone traps for simple monitoring. The rubber dispensers used in the experiments were replaced at three-month intervals, when they became ineffective. The population fluctuations of *E. kuehniella* in the untreated flour mill are shown in Figure 3.

The results from the traditional mill (Fig. 4) show a marked increase in *E. kuehniella* during the first two weeks of June, until it was fumigated. This fumigation had a limited effect on the population. The 3402 captures in the traps prevented any further increase in the population, which remained limited and constant during the following months.

The results from the experiments in the industrial mill are reported in Figure 5. They show that the pheromone traps attracted a large number of *E. kuehniella* (8825 males), higher on average than trapped in the traditional mill.

A comparison of the percentage of males trapped to those free in the environments showed the mass trapping method to have a capture effectiveness over 94%. During the trial, monitoring traps trapped fewer *E. kuehniella* than normally found during months when moth development was unfavourable.

Of interest was the small number of females present in the residual infestation (14 specimens in the traditional mill and 388 in the industrial mill). This suggests their possible migration to adjacent rooms not permeated with pheromone. With regard to this, observations in other areas of the mills showed an abnormal abundance of females (over 75% of all moths present) compared with the control mill, where female presence was 50–55%.

Application of the method in a large flour mill

These experiments were performed in an unheated flour mill situated near Parma, in northern Italy. The mill was large (about 15000 m^3) and produced approximately 100 t/day of flour from spring wheat, *Triticum aestivum* L. (Trematerra 1988).

Following the findings of Trematerra and Battaini (1987), 56 funnel traps baited with 2 mg of TDA were placed in the mill, with a further four traps outside the mill. These outside traps were placed near loading equipment which was frequently flour covered. Trap capture and residual infestation of adults on the walls and machinery were recorded between 9 and 10.00 am every two weeks from May to November.

The monthly mean temperatures of the mill varied between 18 and 30°C from May to November, which allowed the continuous development of *E. kuehniella* for seven to eight months of the year.

The total fumigation with methyl bromide carried out in the last week of April in the flour mill led to a rapid decrease of the number of moths caught in the pheromone traps. Suppression of the number of moths caught in traps lasted only two to three weeks. The treatment was only partially successful, and rapid recolonisation of the mill occurred by insects present on the outside wall of the building, as evident from the peak catches of *E. kuehniella* in outdoor traps.

Trap captures in the mill are reported in Figure 6, and residual infestations of adults on the walls and machinery in Figure 7. Pheromone traps generally attracted a high number of male moths (about 15600). A total of 13777 *E. kuehniella* males was trapped in the mill, whereas 1813 were trapped in the four traps situated outside the mill (Fig. 8).

The continuous presence of the traps in the mill resulted in a population reduction of 95–97%. The use of this method led to a reduction in chemical use within the mill, with a few limited treatments needed, plus one rather than the usual two fumigations required.

The catches of males varied according to the weather, being relatively high at the end of May, gradually decreasing during June and early July but increasing between July and August, and then abruptly declining from September to November. This confirmed the results observed in the preliminary tests in the smaller mills.

The prolonged presence of the traps in the flour mill, particularly during periods when climatic conditions were favourable for moth development led to a drastic reduction of infestation to the level of 'insectistasis' as defined by Levinson (1983). As observed above, the number of females present in the mills indicated an extremely low percentage which was repeatedly found throughout the tests. The pheromone present in the treated environment may have induced female moths to leave the mill for other areas more suitable for reproduction, or the absence of males may have stimulated dispersal. High concentrations of synthetic sex pheromone caused an

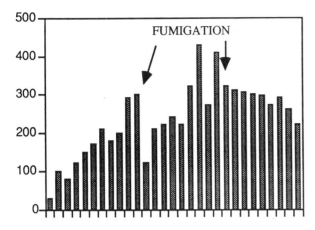

Fig. 3 Dynamic population of *Ephestia kuehniella* in the untreated flour mill.

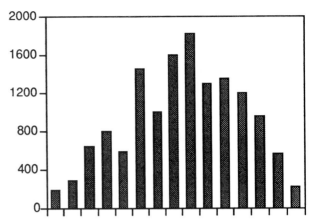

Fig. 6. Catches of *Ephestia kuehniella* males in the flour mill.

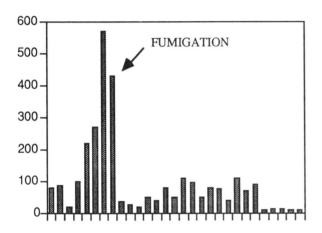

Fig. 4. Weekly catches of male *Ephestia kuehniella* in the environment of the traditional flour mill.

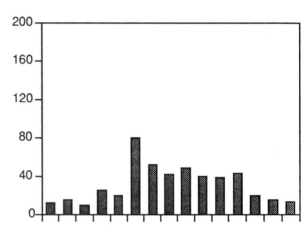

Fig. 7. Residual infestation of adults of *Ephestia kuehniella* in the flour mill.

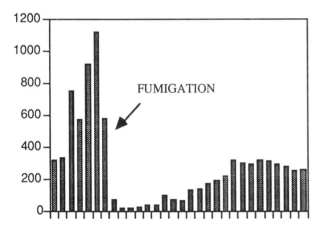

Fig. 5. Weekly catches of male *Ephestia kuehniella* in the environment of the industrial flour mill.

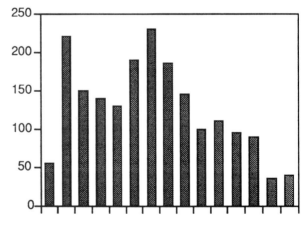

Fig. 8. Catches of *Ephestia kuehniella* males outside of the flour mill.

increased flight activity amongst females, which may have increased dispersal.

Three years of mass trapping

The experiment was performed over three years in an unheated flour mill situated in Pianura Padana, northern Italy. The mill consisted of a large building of 20000 m^3 in which about 125 t/day of flour were produced from spring wheat. Funnel traps with 2 mg of TDA (daily release of 13 µg) were used in the test. The dispensers remained effective for about 2.5–3 months and were then replaced. Following the work of Trematerra and Battaini (1987) and Trematerra (1988), 67 traps were placed inside the mill and five outside. Trap capture and residual infestation of adults on the walls and machinery were recorded every two weeks. The success of the trial was noted by observing the fluctuations of larval presence in machinery and insect fragments in the flour produced by the mill, and determined by filth tests.

Control observations were made in a different mill, situated in the same area, using pheromone funnel traps for simple monitoring. Mean trap catches reflecting the moth population in the control mill are reported in Figure 9.

In 1987 the mill was fumigated with methyl bromide in the last week of April and again between August–September. This resulted in a rapid decrease of the number of moths trapped in the pheromone treated mill, but for only two to three weeks. The fumigations were only partially successful, with rapid recolonisation of the mill, possibly from insects present on the outside walls of the building or from insects surviving the fumigation. This was seen in the peak catches of *E. kuehniella* in traps, and similar observations were made in 1988 and 1989. Total trap captures in the mill during the three years of the trials are reported in Figure 10, with the trap captures from traps situated outside the mill shown in Figure 11.

The traps removed males from *E. kuehniella* populations preventing an increase in the residual population. The effectiveness of mass trapping, gauged by the percentage of males trapped as compared with those free in the environment in the course of the experiments, was about 90–95%. The prolonged presence of the traps in the flour mill, particularly during periods when the moths were able to breed, led to a reduction throughout the mill, including areas where no processing occurs, such as the stock yards and sales office.

the investigation showed that moths that are harmful to food supplies are found outdoors. The well-founded assumption that they might re-infest appropriate goods should be taken into consideration with regard to storage, especially since it has hitherto been taken for granted that an infestation can arise only from infested products.

The control program considered here made a second fumigation (in August–September) of the mill unnecessary in both 1988 and 1989. The use of the pheromone was accompanied by careful cleaning, particularly of the machinery, which further reduced the possibility of moth reproduction.

There was also a consequent economic and qualitative advantage arising from the protection of the products from pesticide residues and an enhanced reputation for the company.

The impressive reduction in the population density of the moth raises the question of whether insectistasis can be achieved in flour mills by mass trapping alone. Extrapolation of the data suggests that using pheromone traps for longer periods should result in a further reduction of the population density of *E. kuehniella*. However, it was not possible to eliminate the infestation, or even to reduce the level to insectistasis, if trapping was not accompanied by careful cleaning, particularly in corners and inside the machinery where moths were able to reproduce undisturbed.

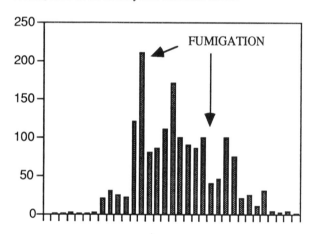

Fig. 9. Dynamic population of *Ephestia kuehniella* males in the control flour mill.

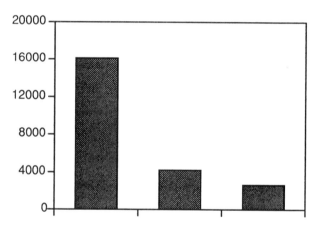

Fig. 10. Total captures inside the flour mill during three years of mass trapping.

The pheromone traps attracted a high number of male moths, with a total of 16108 trapped in the mill during 1987, 4158 in 1988, and 2639 in 1989. It is interesting that 1813 males were also trapped in the five traps situated outside the mill during 1987, 871 in 1988, and 670 in 1989. The continuous presence of traps both inside and outside the mill caused a marked decrease in the *E. kuehniella* population during the second and the third year, resulting in levels 26% and 16%, respectively, of those recorded during the first year.

In the same trial, high levels of the infestation occurred in rooms where flour or bran silos were present. In these areas the moths were able to find considerable residues of flour and other goods. During the three years of experiments, moths were observed to be more abundant on floors where it was easier for the gravid female to enter from outside the mill and colonise the environment (Fig. 12).

Investigations on the appearance of moths of the genera *Ephestia* and *Plodia* outside warehouses and food processing factories were reported by Trematerra (1988). The results of

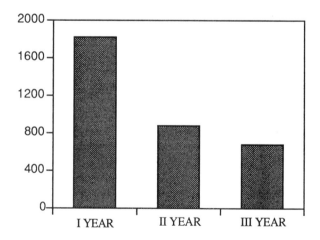

Fig. 11. Total captures outside the flour mill during three years of mass trapping.

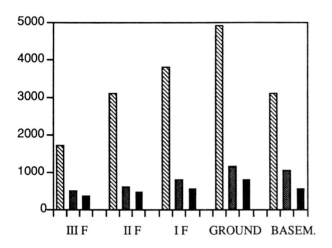

Fig. 12. Dynamic population of *Ephestia kuehniella* males in different floors of the flour mill during three years of mass trapping.

Assuming a highly efficient trapping system and trapping regime have been established, the problem still remains of accurately assessing the effects of the treatment. With regard to the experiments reported above, the residual infestation of adults throughout the three years of the trial was determined as well as larval presence in machinery. Fragments of moths in the flour were determined by monthly filth tests.

Regular inspection of catches in pheromone traps and in rooms revealed the periodic appearance of a summer generation of *P. interpunctella* and *Sitotroga cerealella* (Olivier), but their population density was low. The presence of *Tribolium castaneum* (Herbst) and other Coleoptera in the mill in August was controlled by localised insecticide treatments.

As observed before in the other mills controlled by mass trapping, the small number of *E. kuehniella* females present in the residual infestation was of special interest. This number, compared with that found in the environment, indicated an extremely low percentage which was repeatedly found throughout the tests.

Furthermore, electroantennogram tests on virgin females and on mated females revealed a positive response of mated females to the TDA. This response maybe explained as a disturbed response to the presence of the pheromone (Trematerra and Capizzi 1991), although the biological significance is not fully understood.

Studies on the Attracticide Method

Preliminary applications

The method involves combining insecticide on one side of the laminar dispenser with the attractant effect of the pheromone.

The preliminary study was carried out in a large mill by Trematerra and Capizzi (1987). The technique of the mass-trapping trials (Trematerra and Battaini 1987) was applied, but instead of funnel traps, laminar dispensers (2×2 cm) were used baited with 2 mg of TDA and treated with 10 mg of cypermethrin. The dispensers were placed on the walls and machinery 1.8–2.0 m from the floor, with one dispenser for every 220–280 m^3.

The effectiveness of this method was compared with mass trapping which was used to protect another area of the same mill (Fig. 13).

The residual infestation of adults on the walls, machinery and in traps was recorded weekly. Experiments were per-

formed from April to May of the following year (Fig. 14). The laminar dispensers and the natural rubber dispensers in the traps were replaced every 2.5–3 months.

The attracticide method eliminated over 90% of the males, with a stabilised residual infestation at acceptable levels. The success of this method may be ascribed to the males touching the attraction source again and again (Fig. 15). The prelimi-

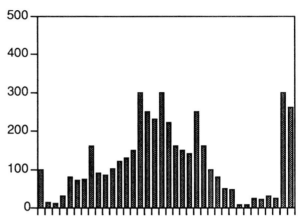

Fig. 13 Dynamic population of *Ephestia kuehniella* during a year of mass trapping.

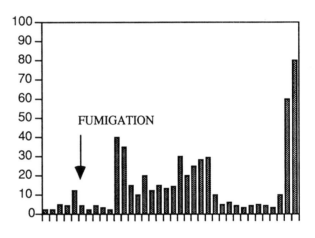

Fig. 14. Residual infestation in the environment treated by mass trapping.

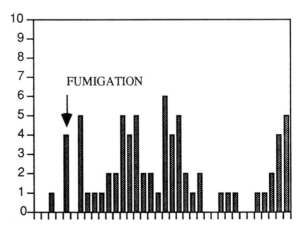

Fig. 15. Residual infestation in the environment treated by the attracticide method.

nary data showed that this method and mass trapping have the same degree of control of *E. kuehniella*.

Control of *E. kuehniella* in an entire mill is currently in progress. In this case the sublethal and lethal effects of attracticide formulations were combined with a silhouette of inverted triangular forms as a 'sign stimulus'. These forms were used in conjunction with the pheromone and cypermethrin (5 mg) applied on the surface of the laminar dispenser. Using the experimental plan reported above, a long-term program was undertaken in central Italy in a large mill that produced flour from *T. aestivum* and from *T. durum* L. After two years the attracticide method removed a high percentage of males from the mill and prevented an increase in the population of *E. kuehniella*, which remained limited and constant during every month of the year. The numbers of larvae and damage to products in all departments were diminished and the fragments revealed in the filth test were also negligible.

Studies on effectiveness

The effectiveness of the attracticide formulations depends initially upon the interaction between the pheromone call and the visual effect of the location of the laminar dispenser. Then, after contact with the treated layer, effectiveness is dependent upon the insecticide effects which modify (at the sublethal level) the behaviour or induce mortality.

With regard to this, Trematerra and Capizzi (1991) performed behavioural tests involving olfactometer, electroantennograms and insecticide activity in order to clearly determine the importance of these factors in the attracticide method. In field tests they checked the measure of control of *E. kuehniella* in practice.

In the olfactometer tests the percentage of *E. kuehniella* males responding to pheromone–insecticide dispenser was 80–90%, confirming that cypermethrin had little influence on their sexual behaviour, as reported for *Pectiniphora gossypiella* (Saunders) by Haynes et al. (1986). These tests also showed no difference in the distribution of virgin females in the presence or absence of TDA. Importantly, gravid females significantly increased their mobility and distribution in the presence of the pheromone. This suggests that gravid *E. kuehniella* can perceive their own pheromone, as reported for other Lepidoptera by Mitchell et al. (1972), Birch (1977), Palaniswamy and Seabrook (1978). This may result in a repellent effect, thus inducing dispersion.

In the electroantennogram tests, males, mated females and virgin females revealed different responses to control and pheromone tests. The average response of males to the stimuli was quantified at 0.42 mV, while the control was 0.17 mV. In mated females the response was 0.24 mV and 0.12 mV, whilst in virgin females it was 0.10 mV and 0.21 mV, respectively. In *E. kuehniella* males, the positive response to the pheromone is expected but in the mated female it may be explained as a disturbed response to the presence of pheromone substance.

Gould (1984) reported certain pyrethroid insecticides to be repellent to some insects, but Trematerra and Capizzi (1991) found no evidence that cypermethrin was responsible for any statistically significant reduction in source contact when comparing control using the dispenser with pheromone alone or when in competition with virgin females. Cypermethrin showed a slight reduction in source contact, but this did not result in any significant change in the behavioural sequence. Males that contacted attracticide dispensers showed symptoms of sublethal poisoning that was only manifested in their ability to perform the behavioural sequence involved in locating the source of pheromone. Percentage mortality during the experiments is reported in Figure 16. It can be seen that the level of mortality of males in the cages was not significantly

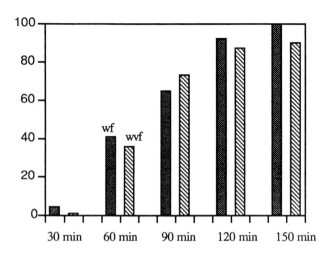

Fig. 16. Percentage mortality of males during the tests performed in big cages; without females (wf), with virgin females (wvf).

affected by the simultaneous presence of females. In this case the females survived for a long time. We observed that 56% of males were killed in 30 minutes without mating; 12% in 50 minutes after a behavioural sequence, 18% in 50 minutes after a courtship sequence and only 14% were able to mate with the female. These effects in the field may be more intense than in our experiments because of the high population density.

The possible interaction between optical and pheromone stimuli was studied by recording choices of *E. kuehniella*, males and females of nine different figures varying in shape and position. The experiments were carried out in two unheated mills: one was small, traditional, and wooden; the other a large processing plant, recently built, with advanced machinery. Light brown cardboard figures of approximately 25 cm^2 were placed on a rectangular sticky panel (85 cm x 70 cm) divided into nine sectors. The insects which landed around the sides of these figures were counted and sexed every three days. Experiments were repeated six times, randomising the position of the figures.

The preferential flight of *E. kuehniella* towards the light-brown cardboard figures resulted in the attraction of a significantly higher number of males to inverted triangular forms resembling the female than to any of the others. The mean catch of 106.25 and 103.375, was higher than the catches on the other figures (Table 1).

Control of *E. kuehniella* males in mills by the attracticide method showed that the combined action of TDA and cypermethrin spread on the surface of a laminar dispenser positioned on the inverted triangular figure achieved good results (Figs 17–18).

According to Traynier (1968), male *E. kuehniella* in the presence of the sex pheromone attempt to mate with objects resembling females. In our case the orientation preferences for the silhouette of inverted triangular forms can thus be regarded as a sign stimulus attracting *E. kuehniella* males to a configuration which resembles females.

Levinson and Buchelos (1981) found that male *Ephestia* spp. do not fly to a source of TDA in complete darkness, whereas they readily flew towards the pheromone at dim light. This suggests that in *E. kuehniella* optical stimulation of the retina is important for taking off. Previously, Kennedy and Marsh (1974) described the optomotor reaction in *P. interpunctella*, *E. cautella* and *E. kuehniella* males to a striped pattern on the ground as an anemotactic guide to the appropriate pheromone stimulus. Interaction between optical and

Table 1 Interaction between optical and pheromone stimuli, comparative responses of *Ephestia kuehniella* to various figures. Means followed by the same letter do not differ significantly. Duncan's multiple range test: P<0.05, capital letters; P> 0.01, small letters.

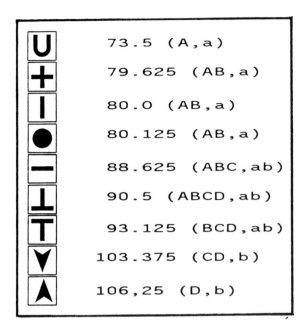

U	73.5 (A,a)
+	79.625 (AB,a)
I	80.0 (AB,a)
●	80.125 (AB,a)
—	88.625 (ABC,ab)
T	90.5 (ABCD,ab)
T	93.125 (BCD,ab)
▼	103.375 (CD,b)
▲	106,25 (D,b)

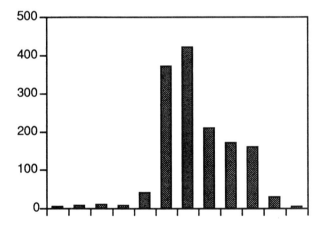

Fig. 17. Dynamic population of *Ephestia kuehniella* in untreated flour mill.

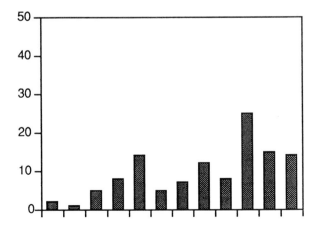

Fig. 18. Dynamic population of *Ephestia kuehniella* in flour mill protected by the attracticide method.

pheromone stimuli in other insects of stored products, such as *E. cautella* and *P. interpunctella*, was also reported by Levinson and Hoppe (1983).

Leos-Martinez (1989) used Lindgren funnel traps without any attractant to determine the daily flight periodicity of *Ahasverus advena* (Waltl), *Cryptolestes ferrugineus* (Stephens) *Oryzaephilus surinamensis* (L.), *Rhyzopertha dominica* (F.), *T. castaneum*, *E. cautella*, *P. interpunctella* and *S. cerealella*. Trematerra and Daolio (1990), using unbaited multifunnel traps, caught *C. ferrugineus*, *R. dominica*, *S. oryzae* and *T. castaneum*. With the same trap more species of Coleoptera and Lepidoptera were collected in a flour mill (Trematerra et al., these proceedings). Quartey and Coaker (1992) undertook studies to improve trap design and found that *E. cautella* responded visually to shapes of different colour and size. They showed that *E. cautella* were attracted to darker coloured vertical rectangles on a white background and to brown and white stripes.

The success of the laminated pheromone dispensers for *E. kuehniella* control in field applications is most encouraging. The sublethal and lethal effects of attracticide formulations was found to be more intense if associated with a silhouette of inverted triangular form. In this case, the combined action of TDA and cypermethrin applied on the surface of the laminar dispenser achieved results similar to those obtained by mass trapping.

The use of pheromones to control pest populations through the attracticide method may become more popular and successful if it is made price competitive with conventional insecticide treatments and lower doses of pheromones are used than those applied in mass trapping or in mating disruption methods. Another benefit is that the attracticide technique allows a broad spectrum of insecticides to be used selectively, preventing the deaths of beneficial insects which may occur with conventional applications of insecticides.

Conclusions

Considerable progress has been made in recent years in controlling *E. kuehniella* with pheromones by mass trapping and by using the attracticide method. Mating disruption trials have been also carried out for species of *Ephestia* in warehouses by permeation with TDA. The isolated environment within such large stores certainly provides a situation where immigration is virtually excluded. The relatively high population densities that occur with such enclosures do not, however, suggest this approach to be promising (Barrer 1976; Haines and Read 1977). However, with *E. kuehniella* few studies have been undertaken. In Italy, mating disruption using laminar dispensers distributing 0.2 mg of TDA/m^3, has resulted in a considerable reduction of an infestation in a traditional flour mill over a three-year period. The number of insects in the funnel traps fell from thousands trapped during the first year to a few males during the third year of mating disruption (Trematerra and Capizzi 1987). Further data are necessary to optimise this control method especially in order to reduce the quantity of pheromones used and thus the risk of residues in food.

Better control of *E. kuehniella* in flour mills can be obtained by the use of mass trapping, attracticide or mating disruption methods, together with careful cleaning of the rooms, particularly in the corners and, above all, inside machinery. This procedure eliminates the possibility of insect reproduction in areas where food is present.

In the case of complex infestations where more than one species of moth or beetle is involved, treatments which use only pheromones become complex and are not effective. Integrated treatments which use different means, including chemicals, achieve better results. An important role should be assigned to all prevention techniques which effectively keep infested goods out of food-processing plants, mills, storehouses, etc.

Insectistasis can be readily achieved by continuous supervision of environments by pheromone traps, in combination with a limited number of curative measures appropriately timed.

The use of pheromones is therefore one of several modern techniques that show promise in controlling stored-product insects. Their use may lead to a drastic reduction of chemical treatments with consequent economic and qualitative advantages. Goods may be protected from possible pesticide residues and hence make a company's products more attractive to consumers.

References

Barrer, P.M. 1976. The influence of delayed mating on the reproduction of *Ephestia cautella* (Walker) (Lepidoptera: Phycitidae). Journal of Stored Products Research, 12, 165–169.

Birch, M.C. 1977 Responses of both sexes to *Trichoplusia ni* (Lepidoptera: Noctuidae) to virgin females and to synthetic pheromone. Ecological Entomology, 2, 99–104.

Bommer, H. and Reichmuth, Ch. 1980. Pheromone der vorratsschaedlichen Motteh (Phycitinae, speziell Mehlmotte *Ephestia kuehniella* Zeller) in der biologischen Schaedlingsbekaempfung. Mitteilungen aus der Biologischen Bundesanstalt für Land-und Forstwirtschaft Berlin-Dahlem, 198, 1–114.

Brady, U.E. 1973. Isolation, identification and stimulatory activity of the second component of the sex pheromone system (complex) of female almond moth, *Cadra cautella*. Life Science, 13, 227–235.

Brady, U.E., Nordlund, D.A. and Daley, R.C. 1971a. The sex stimulant of the Mediterranean flour moth *Anagasta kuehniella*. Journal of Georgia Entomological Society, 6, 215–217.

Brady, U.E., Tumlinson, J.H., Brownlee, R.B. and Silverstein, R.M. 1971b. Sex stimulant and attractant in the Almond Moth. Science, 171, 802–804.

Burkholder, W.E. and Ma, M. 1985. Pheromones for monitoring and control of stored-product insects. Annual Review of Entomology, 30, 257–272.

Gould, F. 1984. Role of behavior in the evolution of insect adaptation to insecticides and resistant host plants. Bulletin of the Entomological Society of America, 30, 34–41.

Haines, C.P. and Read, J.S. 1977. The effect of synthetic female sex pheromones on fertilization in warehouse population of *Ephestia cautella* (Walker) (Lepidoptera, Phycitidae). Report L 45, Tropical Product Institute, London: 10pp.

Haynes, K.F., Li, W.G. and Baker, T.C. 1986. Control of pink bollworm moth (Lepidoptera: Gelechiidae) with insecticides and pheromones (attracticide): lethal and sublethal effects. Journal of Economic Entomology, 79, 1466–1471.

Hodges, R.J., Benton, F.P., Hall, D.R. and dos Santos Serodio, R. 1984. Control of *Ephestia cautella* (Walker) (Lepidoptera, Phycitidae) by synthetic sex pheromones in the laboratory and store. Journal of Stored Products Research, 20, 191–197.

Kennedy, J.S. and Marsh, D. 1974. Pheromone-regulated anemotaxis in flying moths. Science, 184, 999–1001.

Kuwahara, Y. and Casida, J.E. 1973. Quantitative analysis of the sex pheromone of several phycitid moths by electron capture gas chromatography. Agriculture and Biological Chemistry, 37, 681–684.

Leos-Martinez, J. 1989. Periodicidad diaria de vuelo de insectos de productos almacanados. XX Congreso National de Entomologia. Oaxtepec, Morelos, Mexico, 91–108.

Levinson, H.Z. 1983. Integrated manipulation of storage pests involving insectistasis. Mitteilungen der Deutschen Gesellschaft für Allgemein angewandte Entomologie, 4, 101–103.

Levinson, H.Z. and Buchelos, C. Th. 1981. Surveillance of storage moth species (Pyralidae, Gelechiidae) in a flour mill by adhesive traps with notes on the pheromone-mediated flight behaviour of male moths. Zeitschrift für angewandte Entomologie, 92, 233–251.

Levinson, H.Z. and Hoppe, T. 1983. Preferential flight of *Plodia interpunctella* and *Cadra cautella* (Phycitinae) toward figures of definitive shape and position with notes on the interaction between optical and pheromone stimuli. Zeitschrift für angewandte Entomologie, 96, 491–500.

Mankin, R.W., Vick, R.W., Mayer, M.S., Coffelt, J.A. and Callagan, P.S. 1980. Model for dispersal of vapors in open and confined spaces: applications to sex pheromone trapping in a warehouse. Journal of Chemical Ecology, 6, 929–960.

Mitchell, E.R., Webb, J.C., and Hines, R.W. 1972. Capture of male and female cabbage loopers in field traps baited with synthetic sex pheromone. Environmental Entomology, 1, 525–526.

Palaniswamy, P. and Seabrook, W.D. 1978. Behavioral responses of the female eastern spruce budworm, *Choristoneura fumiferana* (Lepidoptera: Tortricidae) to the sex pheromone of her own species. Journal of Chemical Ecology, 4, 649–655.

Quartey, G.K. and Coaker, T.H. 1992. The development of an improved model trap for monitoring *Ephestia cautella*. Entomologia Experimentalis et Applicata, 64, 293–301.

Read, J.S. and Haines, C.P. 1976. The function of the female sex pheromones of *Ephestia cautella* (Walker) (Lepidoptera: Phycitidae). Journal of Stored Products Research, 12, 49–53.

Sower, L.L., Vick, K.W. and Tumlinson, J.H. 1974. (Z, E)-9,12-tetradecadien-1-ol: a chemical released by female *Plodia interpunctella* that inhibits the sex pheromone response of male *Cadra cautella*. Environmental Entomology, 3, 120–122.

Süss, L. and Trematerra, P. 1982. Valutazioni in laboratorio ed osservazioni in magazzino dell'attività del feromone sessuale dei Lepidotteri Ficitidi infestanti le derrat. Atti Giornate Fitopatologiche, 3, 330–340.

Süss, L. and Trematerra, P. 1985. Valutazione dell'attività di TDA a diverse concentrazioni nel controllo de *Ephestia kuehniella* Zeller (Lepidoptera: Phycitidae). Tecnica molitoria, 10, 821–829.

Süss, L. and Trematerra, P. 1986. Control of some Lepidoptera Phycitidae infesting stored-products with synthetic sex pheromone in Italy. In: Donahaye, E., and Navarro, S., ed., Proceedings of the Fourth International Working Conference on Stored-product Protection, Tel-Aviv, Israel, September 1986, 606–611.

Traynier, R.M. 1968. Sex attraction in the Mediterranean flour moth, *Anagasta kuehniella*: location of the female by the male. Canadian Entomologist, 100, 5–10.

Trematerra, P. 1986. Valutazione dell'attività di richiamo di due componenti il fermone sessuale femminile di *Plodia interpunctella* Hübner. Atti Giornate Fitopatologiche, 1, 187–194.

Trematerra, P. 1988. Suppression of *Ephestia kuehniella* Zeller by using a mass trapping method. Tecnica molitoria, 18, 865–869.

Trematerra, P. 1990. Population dynamic of *Ephestia kuehniella* Zeller on flour mill: three years of mass trapping. In: Fleurat-Lessard, F., and Ducom, P., ed., Proceedings of the Fifth International Working Conference on Stored-product Protection, Bordeaux, France, September 1990, 3, 1435–1443.

Trematerra, P. and Battaini, F. 1987. Control of *Ephestia kuehniella* Zeller by mass trapping. Journal of Applied Entomology, 104, 336–340.

Trematerra, P. and Capizzi, A. 1987. Esperienze di controllo delle infestazioni di *Ephestia kuehniella* Zeller nei mulini mediante feromoni. In: Domenichini, G., ed., Atti IV Simposio, Difesa antiparassitaria nelle industrie alimentari, Piacenza, 511–518.

Trematerra, P. and Capizzi, A. 1991. Attracticide method in the control of *Ephestia kuehniella* Zeller: studies on effectiveness. Journal of Applied Entomology, 111, 451–456.

Trematerra, P. and Daolio, E. 1990. Capture of *Rhyzopertha dominica* (F.) (Col., Bostrichidae) with Dominicalure multifunnel traps and considerations on trapping of non target species: *Sitophylus oryzae* (L.), *Cryptolestes ferrugineus* (Stephens) and *Colidium castaneum* (Herbst). Journal of Applied Entomology, 110, 275–280.

Trematerra, P. and Rossi Porzio, R. 1982. Risposte olfattometriche ed elettroantennogtrafiche di alcuni Lepidotteri Ficitidi al TDA, componente del feromone sessuale. In: Domenichini, G., ed., Atti III Simposio, La difesa antiparassitaria nelle industrie alimentari e la protezione degli alimenti, Piacenza, 601–614.

Grain storage in a small-farm ecosystem: Angoumois grain moth movement and management

R.J. Barney and P.A. Weston*

Abstract

Sitotroga cerealella (Olivier), the Angoumois grain moth (AGM), is found worldwide, both preharvest and in postharvest storage of grains such as wheat, rice, maize (corn), millet, and sorghum. Many other non-crop, plant hosts have been found, and others probably exist. In addition to having flexible nutritional requirements, AGM is highly mobile and a primary coloniser. These attributes allow it to survive in a variety of habitats and move from patch to patch as conditions change.

More than 40 000 Angoumois grain moths were captured during the two plus years: 35 713 in pheromone traps and 5203 in bin-vent traps. AGM activity at the bins and at the greatest distance from the bins (550 m) preceded moth activity at maize fields.

Introduction

The Angoumois grain moth (AGM), *Sitotroga cerealella* (Olivier), has been recognised as a grain pest for over 250 years. It is found worldwide in both preharvest and postharvest storage of such grains as wheat, rice, maize (corn), millet, and sorghum. Many other non-crop, wild-plant hosts have been found (Joubert 1966), and others probably remain to be discovered. Moths have even been found in a forest about 5 km from the nearest storage facilities or fields (Cogburn and Vick 1981).

In addition to having flexible nutritional requirements, AGM is highly mobile and a primary coloniser. These attributes allow AGM to survive in a variety of habitats and move from patch to patch as conditions change. Simmons and Ellington (1933) hypothesised that in Maryland, AGM move from grain storages to ripening wheat fields in summer. Stockel (1971) reported that in southwest France, AGM produce three generations per year: the first two in grain storage or wild grasses and the third on ears of maize in the field.

The Angoumois grain moth was first recognised in Kentucky as a pest of wheat in 1852. Recent survey work in Kentucky has revealed AGM as the most significant moth pest of on-farm, stored, shelled maize. Central Kentucky, located in the Inner Bluegrass, is a rural area of gently rolling hills underlain by limestone. Although the Bluegrass is known for its horses, burley tobacco and bourbon, Kentucky on-farm, grain-storage capacity averages 190 million bushels (Kentucky Agricultural Statistics 1992). Kentucky ranks forth in the nation in number of farms (91 000 farms averaging 63 ha),

signifying the preponderance of small, family farms in the state.

The abundance and spatial distribution of insects in agricultural landscapes is seldom static, with populations typically moving through multiple habitats over the course of a season (Landis 1994). From a grain moth perspective, a typical small farm is a patchwork of prime habitats (maize fields and storage bins) and less stable or less desirable habitats (wild/alternative hosts, livestock feed, grain spills). From a grain grower/manager perspective, a farm is a patchwork of prime manageable, revenue-producing habitats (maize fields and storage bins), manageable, but nonrevenue-producing habitats (wild/alternative hosts, livestock feed, grain spills), and unmanageable, nonrevenue-producing, off-farm habitats.

The purpose of this study was to monitor AGM movement among these habitats on a small, central Kentucky farm. The objectives were: (1) to determine temporal patterns of AGM movement, (2) to determine spatial patterns of AGM movement, and (3) to determine if knowledge of these patterns has management implications.

Materials and Methods

The experiment was conducted at the Kentucky State University Research Farm, Franklin County, KY in 1991, 1992 and 1993. The 83-ha farm is primarily used for vegetable plot studies and livestock grazing. A stored-grain, bin complex is located near the centre of the farm and plots of maize were grown at various locations on the farm each year.

At the initiation of the experiment (July 1991), shelled maize was stored in a bin complex, consisting of two 1000-bu bins and 12 300-bu bins. All bins held 'DeKalb 689' hybrid, with the large bins containing maize harvested in November 1988 (three-year old) and the smaller bins containing maize harvested in November 1989 (two-year old). Maize in the small bins remained in place throughout the experiment while the large bins were refilled every other year.

The trapping regime consisted of two parts: monitoring AGM in and around the storage structures with sticky traps in the top vents of the bins, and monitoring AGM activity across the farm and near maize plots with pheromone traps arranged in a circular grid, radiating throughout the farm with the storage complex as the centre.

The first series of traps ('bin-vent traps') were 76 × 127 mm (611 × 1211) yellow, double-sided, sticky traps (Olson Products, Medina, OH). The traps were suspended with clamps from wire grates within the vents. Each trap (n = 10) was exposed on both sides, thereby having one side facing the inside of the bin and the other facing away. Traps were monitored from 27 September to 8 November 1991 in eight bins. The same bins were monitored from 15 May to 13 November 1992 and from 31 March to 3 November 1993.

The second series of traps were delta-style, sticky-board insert, AGM pheromone traps (Insects Limited, Indianapolis, IN). Traps were arranged in concentric rings of 65, 175 and 400 m from the bins. Traps were attached to wooden stakes at a height of 1.3 m and located at the four cardinal compass

---

* Atwood Research Facility, Kentucky State University, Frankfort, Kentucky 40601 USA.

points, next to maize plots, and other locations. Traps (n = 17) were monitored weekly from 23 August to 8 November 1991. In 1992 paired traps (n = 34) were located at 65 400 and 550 m to provide replication and cover a greater portion of the farm. Traps were sampled from 22 May to 13 November 1992 and from 30 April to 12 November 1993.

All traps were monitored weekly, and inserts were replaced as necessary. Insects were counted without removing them from the sticky surface.

Results

Greater than 40 000 Angoumois grain moths were captured during the two plus years: 35 713 in pheromone traps and 5203 in bin-vent traps. The pheromone-baited traps were much more efficient at catching moths than passive, non-baited sticky traps.

AGM activity at the bin-vents was greatest in July during 1992, with activity decreasing each month until frost (Fig. 1A). In 1993 the activity showed the same general distribution, but peak activity was delayed until August.

Male moth activity across the farm, as measured by pheromone trapping, was very heavy during September and October each year (Fig. 1B). Field activity was concentrated in and around maize plots from August to November.

Results from trapping at different distances and directions from the bins were inconclusive. Activity at 550 m was very

high (>70 AGM per trap) in July 1992 and August 1993; this is the same time that activity peaked at the bins.

Discussion

A study of planting date influence on preharvest infestation of maize by AGM in central Kentucky (Weston et al. 1993) outlined some management options based on ovipositional behaviour. Combined with the trapping data in the present study, the following scenario in central Kentucky is suggested:

May–June	moths in storage become active
July–August	large moth exodus from bins
	high activity at 550 m, off-farm, wild hosts
August–September	activity centred at maize fields
September–October	oviposition in field
October–November	harvest/binning of infested grain.

Temperature changes between years will shift events to earlier or later in the season.

Many insects require one type of habitat for overwintering and one or more additional habitats for feeding, mating, resting, and oviposition. Patches in close spatial association frequently constitute a source/sink relationship. For the Angoumois grain moth, storage, fields, and wild hosts can each serve as both source and sink, depending on the time of year. The association of these habitat patches in both time and space is a critical feature of the functioning of an agricultural landscape (Landis 1994). These aspects must be considered when designing an ecologically-sound, arthropod management system.

References

Cogburn, R.R. and Vick, K.W. 1981. Distribution of Angoumois grain moth, almond moth, and Indian meal moth in rice fields and rice storages in Texas as indicated by pheromone-baited adhesive traps. Environmental Entomology, 10, 1003–1007.

Joubert, P.C. 1966. Field infestations of stored-product insects in South Africa. Journal of Stored Products Research, 2, 159–161.

Kentucky Agricultural Statistics 1992. Kentucky Agricultural Statistics Service, Louisville, KY.

Landis, D.A. 1994. Arthropod sampling in agricultural landscapes: ecological considerations. In: Pedigo, L.P. and Buntin, G.D. ed., Handbook of sampling methods for arthropods in agriculture. Boca Raton, CRC Press, 15–31.

Simmons, P. and Ellington, G.W. 1933. Life history of the Angoumois grain moth in Maryland. United States Department of Agriculture Technical Bulletin, 351, 1–34.

Stockel, J. 1971. Utilisation du piegeage sexuel pour lletude du deplacement de l'alucite *Sitotroga cerealella* (Lepidoptera: Gelechiidae) vers les cultures des mais. Entomologia Experimentalis et Applicata, 14, 39–56.

Weston, P.A., Barney, R.J. and Sedlacek, J.D. 1993. Planting date influences preharvest infestation of dent corn by Angoumois grain moth (Lepidoptera: Gelechiidae). Journal of Economic Entomology, 86, 174–180.

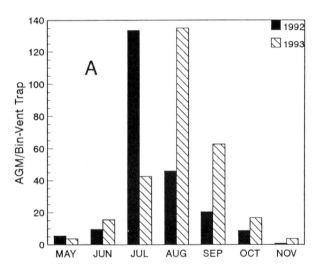

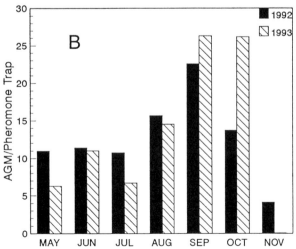

Fig. 1. Trap catches of angoumois grain moth. (A) bin-vent trap; and (B) pheromone trap.

The use of multiple trapping methods to assess population size: an evaluation

J.H. Brower, L. Smith and E.P. Wileyto*

Abstract

A biological control test in six 1500 bushel metal bins of shelled maize provided an opportunity to compare the accuracy of several trapping methods in determining population numbers of three storage pests and their parasites. The species sampled were the maize weevil, *Sitophilus zeamais*, the red flour beetle, *Tribolium castaneum*, and the Indianmeal moth, *Plodia interpunctella*. Trapping methods included vacuum sampling and probe (pitfall) trapping at 15 points throughout the grain mass, corrugated cardboard refuge traps at the grain surface and pheromone flight traps using a mark–release–recapture method, The results of each of these methods will be compared at different population densities.

Introduction

It is now generally accepted that for the detection of insect populations in bulk grain, traps are much more effective than conventional sampling techniques (Wilkin 1991). This is particularly true for low density or very aggregated populations of insects in large bulks of stored grain. The fact that traps can remain in place for considerable periods of time is probably their major advantage. The greatest disadvantage is that insect trap catch is not easily translatable into meaningful population estimates (Hagstrum et al. 1991). The number of insects caught in a trap is very dependent on the size and behaviour of each species and on a number of environmental variables (Cuperus et al. 1991). The most useful type of trap for detecting insects in bulk grain is the pitfall type of probe trap that is available in a number of modifications (White et al. 1991). These traps are a very sensitive indicator of the presence of insect infestation in bulk grain if they are placed in a number of locations throughout the grain bulk and left for several to many days. Although the estimation of absolute population numbers from probe trap catches may be imprecise, the comparison of the effectiveness of various treatments in similar conditions can be relatively accurate if the number of traps is adequate. In this paper we will discuss the use of correlated vacuum samples and grain probe catches and moth flight traps to assess the effectiveness of an integrated biological control regimen in bins of stored maize.

* Stored-Product Insects Research and Development Laboratory, USDA, ARS, 3401 Edwin Street, Savannah, Georgia 31405 USA.

Methods and Materials

In order to conduct a test of biological control in stored shelled maize, six 1500 corrugated metal bins were constructed. Bins were circular, 2.75 m in diameter × 4.57 m in height, resting on a flat concrete foundation. The roof of each bin had a centre opening of 0.9 m for grain loading, an access door near the periphery, and three screened air vents for aeration. Bins were equipped with raised perforated metal drying floors, and 0.46 m vane axial fans with 2.0 hp motors. Each bin also had a permanently installed horizontal under-bin unloading auger with a dedicated 3/4 hp motor. All bin seams, joints and small openings were sealed during bin construction and again after bin completion or after bin filling. Bins were instrumented with thermocouples at nine locations in the grain mass and one in the headspace above the grain. Temperature measurements were recorded hourly from each location.

About 1025 bu of shelled maize (Pioneer 3320) from the 1992 crop was loaded into each of the bins between 13 October and 26 October 1992 and then levelled. A number of random samples were taken from the grain stream during the loading of each bin, and the moisture content and infestation levels were determined. Because live insects were detected, all bins were sealed, openings were taped and the bins were fumigated with methyl bromide in January and again in February. Bins were sampled and refumigated if any live insects were detected, until no live insects were detected in any of the bins.

The bins were aerated after each fumigation and for additional periods in order to dry the maize down to a grain moisture content of less than 15% so that the grain would keep for a year without spoilage (Mills 1992). Bins were aerated 18–23 December 1992, 22–25 and 28–29 January and 2 February for a total aeration time of 110 hours.

Bins were sampled at five-week intervals using a Probe-A-Vac® pneumatic grain sampler. Fifteen 4-L samples were aspirated from each bin on each sampling date. The surface of the grain in each bin was divided into five quadrats of equal area (ca 2 m^2) with one round area in the centre and four truncated wedges aligned with the four compass points (N,E,S,W). The grain was then divided into three layers: bottom (0–1.1 m), middle (1.1–2.2 m), and top (2.2–3.3 m). On each sampling date a spot was chosen at random on each surface quadrat, marked with a surveyor's flag, and using the vacuum sampler a 4-L sample was aspirated from each of the three layers to yield 15 samples per bin. These samples were immediately placed in labelled jars, returned to the laboratory and sifted to remove all of the living and dead insects. The insects were then separated from the debris, identified, counted and discarded, and the debris was returned to the samples. The samples were then incubated for four weeks at 30 ± °C and about 70% r.h., and the samples were sifted at two and four weeks to count any insects that emerged during the four-week incubation period. Samples were then frozen and discarded, and jars were cleaned before the next sampling date. Each sample was briefly placed in a Motomco 919 Automatic Moisture Tester® to determine the grain moisture content.

The day following the vacuum sampling, Trappit® grain probe traps were placed in each of the 15 locations from which vacuum samples had been taken. These traps are clear acrylic tubes (370 mm × 27 mm dia.) with a large number of small holes through which the insects pass before they fall down a funnel into a specimen tube enclosed in the bottom of the trap. These traps were left in place for five days and then retrieved using a cord tied to the top of each trap. The traps were removed to the laboratory, where the contents of each was removed, separated into species and all insects both alive and dead were counted. Samples were collected at five-week intervals for the duration of the test.

In addition, mark-recapture flight traps were developed to monitor population levels of the Indianmeal moth, *Plodia interpunctella* (Hübner). Marking was accomplished by marking stations, otherwise identical to sticky traps in structure and lure but modified to mark and release adults back to the population. One trap and one marking station were placed into each bin, suspended on a line about 1.5 m above the grain surface, and traps were retrieved and replaced at regular intervals. Moths caught in the pheromone baited sticky traps were assayed with a shortwave ultraviolet light to determine if they were unmarked or marked with a fluorescent powder from the marking station. The moth population could then be estimated using formulas described by Wileyto et al. (1994) and Wileyto in these proceedings.

Bins disinfested by fumigation were reinfested intentionally by releasing known numbers of pest insects into each bin on a set schedule. Three pest species were released on February 11 1993: 500 unsexed adults of *Sitophilus zeamais* Motschulsky, 100 unsexed adults of *Tribolium castaneum* (Herbst), and 25 pairs of *P. interpunctella* per bin. At five-week intervals throughout the test, smaller numbers of the three pest species were added to each bin to simulate natural immigration. Numbers used per bin were 50 *S. zeamais*, 10 *T. castaneum* and five pairs of *P. interpunctella*, and these were introduced four times.

Bins were assigned randomly to either a check treatment or to a biological control treatment. Pests were added to all six bins, but the biological control agents were added only to the three biological control bins. To establish a population of the predatory warehouse pirate bug, *Xylocoris flavipes* Reuter, 200 adult bugs were released weekly starting on February 18 1993, and continuing through the test period. Starting on March 11, 1993, and weekly thereafter, 1000 adult *Anisopteromalus calandrae* (Howard) (a parasite of *Sitophilus* weevils) and 200 adult *Bracon hebetor* Say (a parasite of pyralid moths) were released into the three designated bins.

Results

Unfortunately the test was late getting started because of technical problems, and the grain temperatures were low when the test started. Just before the start of releases on February 4 1993, the average grain temperatures were very close to 8°C in the six bins. Temperatures warmed about 4.5°C per month, reaching 30°C by the start of July.

In the southeastern United States, maize is harvested in the early fall at a fairly high moisture content, and typically dried to some extent before storage. The maize received for this test had already been cleaned and dried, but the moisture content was still 15.5%. This is above the recommended moisture content for long-term storage of maize (Mills 1992) and the maize was further dried using the aeration fans on the bins. Bins were aerated during three periods: 18–23 December 1992 for 50 hours; 22–24 January 1993 for 42 hours; and 28

January –2 February 1993 for 18 hours. Thus, aeration totaled 110 hours and average grain moisture content was reduced to 14.0% (Fig. 1). Maize moisture content varied only slightly throughout the course of the test (Fig. 1).

Fifteen vacuum samples (4 L) were used to assess the absolute density of insects in each bin and to compare this to the results from the trap catches. Beetle populations were compared, primarily between the vacuum samples and the pitfall probe trap samples. A comparison of results from these two methods for the maize weevil, *S. zeamais*, (Fig. 2) showed very similar population abundance and nearly identical population trends. Early in the year while grain temperatures were low, both techniques indicated low population numbers until about the middle of May. At that time both methods indicated that maize weevil populations started to increase through July. Slight decreases occurred by August, perhaps because of high populations of the parasitoid, *A. calandrae*, in both check and treatment bins by that time. In general, a five-day sampling period for the pitfall probe traps yielded slightly higher numbers of maize weevils than did the vacuum samples (Fig. 2). Overall, the population trends for *S. zeamais* in check bins and biological control bins were very similar, but there was a tendency for more weevils to be present in the check bins. In general, weevil populations remained low in all bins during the course of this test.

Tribolium castaneum populations exhibited different growth curves than the ones for *S. zeamais* (Fig. 3). No *T. castaneum* were detected before May 19, 1993, but their populations increased very rapidly to a high at the end of the test. The pitfall probe traps proved to be a more sensitive monitoring method for *T. castaneum* than the vacuum probe samples (Fig. 3). Although population curves were similar using both methods, the pitfall traps always caught more *Tribolium* adults than were contained in 4 L samples of grain. This difference was accentuated later in the year as grain temperatures increased to over 30°C. In general, both sampling techniques showed that there were fewer adult *T. castaneum* in the biological control bins than in the check bins.

Flat grain beetles, *Cryptolestes* spp., were not introduced into the test bins of this experiment, but several nearby sources of these insects were present. Although bins were tightly sealed, eventually they were all infested by *Cryptolestes* spp. These small active pests were first detected by the pitfall probe traps in March (Fig. 4) and it was not until July when populations had increased significantly that they were found in the vacuum samples. In all cases, the pitfall traps showed significantly more *Cryptolestes* than the vacuum samples. As with *Tribolium*, the greatest differences between the two methods were at the end of the test when temperatures were greatest. No parasitoid was added to control *Cryptolestes* spp. but *X. flavipes* is a very effective predator of these pests (Brower and Press 1992). The vacuum samples showed more *Cryptolestes* adults in the check bins than in the biological control treatment bins, but the pitfall traps did not show any consistent difference between the two treatments (Fig. 4).

Most parasitoids and the predator, *X. flavipes*, are small and fragile and were very poorly represented in vacuum samples of grain. When host populations are low, as in our test, very few parasitoids emerge from incubated grain samples. However, the pitfall probe traps proved to be very good at sampling the small active parasitoids and predators used in this experiment. In fact, the pitfall traps not only documented the spread of the maize weevil parasitoid, *A. calandrae*, into the check bins (Fig. 5) but they also showed that the parasitoid increased rapidly on the larger host populations present in the check bins and actually exceeded the number of parasitoids in the release bins. After this peak of parasitoid density was reached on July 23, 1993 (Fig. 5), the abundance of the host

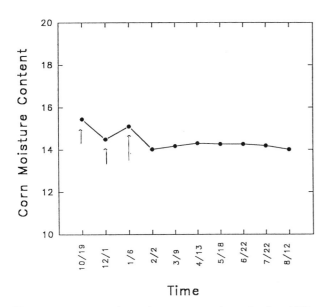

Fig. 1. Average grain moisture content from October 1992 –August 1993. Points are the average for the six bins with 15 moisture determinations per bin. Arrows indicate periods of aeration.

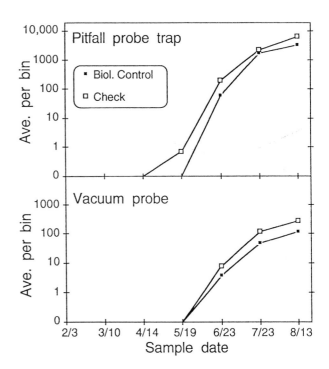

Fig. 3. Number of adult *Tribolium* spp. in pitfall traps and probe samples. Points are averages of total number of insects in three bins as determined by 15 samples/bin

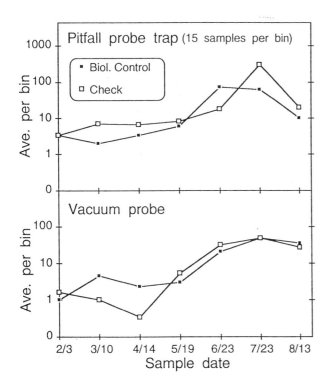

Fig. 2. Comparison of number of adult *S. zeamais* obtained by two different sampling methods. Each point represents the average number of insects in three bins as determined from 15 samples/bin.

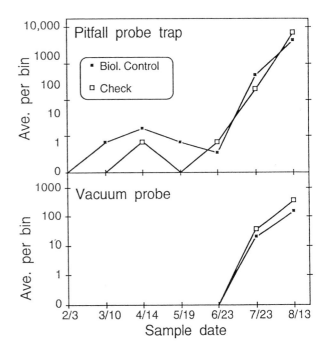

Fig. 4. Number of adult *Cryptolestes* spp. in pitfall traps and probe samples. Points are averages of total number of insects in three bins as determined by 15 samples/bin.

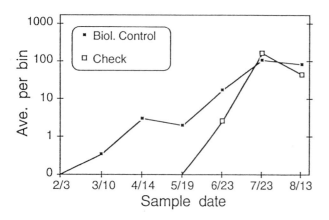

Fig. 5. Number of live parasitoids (*Anisopteromalus calandrae*) recorded in pitfall probe traps. Points represent averages of three bins with 15 samples/bin.

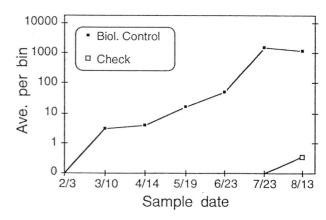

Fig. 6. Number of live predators (*Xylocoris flavipes*) recovered in pitfall probe traps. Points represent averages of three bins with 15 samples/bin.

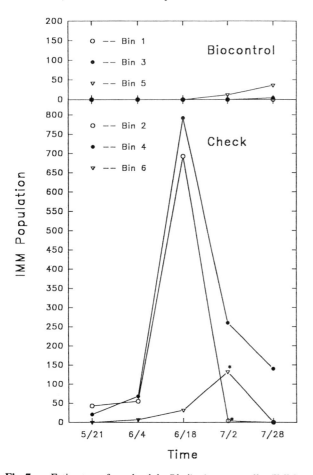

Fig. 7. Estimates of total adult *Plodia interpunctella* (IMM) populations in each of six bins (three treatment and three check) as calculated from mark-release-recapture data (see text for description of technique).

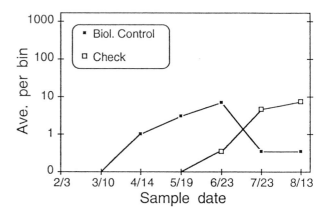

Fig. 8. Number of live *Bracon hebetor* recovered in pitfall probe traps in both biological control and treatment bins.

decrease the effectiveness of this predator. In any case, this demonstrated that pitfall traps are an effective sampling tool for *X. flavipes* and probably also a good collecting technique for this species.

Standard grain sampling techniques are very poor for estimating moth populations, especially the pyralid pests such as *Plodia* sp. and *Ephestia* spp. This proved to be true in our case since both vacuum samples and probe pitfall traps indicated few to no adult *P. interpunctella* present. A *Plodia* larval sampling technique using corrugated cardboard rolls within 0.5 m metal cylinders inserted into the grain as larval refuge and pupation sites on the grain surface also proved to be negative. However, the pheromone baited flight traps used in the mark–release–recapture technique proved well suited to determining moth population abundance. Using estimating techniques described by Wileyto (these proceedings), the total moth population in each bin was calculated for a number of sampling dates and is shown in Figure 7. *Plodia interpunctella* populations in the biological control bins remained very small throughout the course of this test. In contrast, the adult populations in the check bins increased to over 125 in one bin and to 700–800 in the other two bins (Fig. 7). *Plodia interpunctella* populations peaked in mid-June and then declined precipitously to zero in two of the three bins by late July. Coincidentally with this abrupt decline in abundance was the presence and rapid increase in a specific parasitoid population.

populations in the check bins declined (Fig. 2). This is probably the first use of pitfall grain probe traps to monitor parasitoid population abundance. In addition, these traps caught significant numbers of the predator, *X. flavipes* (Fig. 6). This bug is typically apterous or brachypterous and does not fly, and so it did not appear in check bins until August 13th and then in very low numbers. In contrast, it appeared in traps in the release bins early in the test and in increasing numbers through most of the test period (Fig. 6). In fact, such large numbers were caught during the last two months of the test (about 1000/bin) that it was feared this removal might

Bracon hebetor is a parasitoid of the pyralid moths associated with stored products and, released into the treatment bins, significant populations are present during the period from April to June (Fig. 8). Few hosts were present in treatment bins during this period but large host populations were present in adjacent check bins (Fig. 7). Visual observations and probe trap catches showed that *B. hebetor* invaded the three check bins during May and June and these populations increased rapidly to outnumber the populations in the treatment bins that still were receiving weekly releases of *B. hebetor* (Fig. 8). The abrupt decline in moth abundance apparently resulted from the actions of these parasitoid populations.

Discussion

The determination of insect abundance in large bulks of grain has always been a difficult problem but one of great importance. Direct sampling methods such as grain triers and vacuum sampling are difficult, time consuming, and sometimes impractical, but the only way to get an absolute estimate of population density. To avoid some of the above problems, many types of insect traps have recently been developed. These are often much more effective at detecting low density insect populations than direct samples, and under some conditions they may catch large numbers of insects. However, environmental conditions greatly affect trap efficiency in most cases (Cuperus et al. 1991), and relating trap catch to absolute population density has proved to be very difficult (Barak and Harein 1982, Lippert and Hagstrum 1987). Thus, the present experiment was designed to sample 15 locations in each of six grain bins having different numbers of insects, by both vacuum samples and probe pitfall traps and to compare these results. This is the first report of those efforts.

Trends of pest populations were very similar using the two different techniques; however, the indication of absolute population size was not well correlated. Species that are more sedentary such as the maize weevil showed a much closer correspondence between the two methods than did highly mobile species such as the flat grain beetles. Environmental conditions seemed to affect the movement of highly mobile species more than they affected more sedentary ones. This study confirmed findings of others (Barak and Harein 1982; Lippert and Hagstrum 1987) that traps are a more sensitive detection method for low density insect populations. This study is perhaps the first to demonstrate the utility of pitfall probe traps for the detection and estimation of relative abundance of biological control agents. It appears that because effective beneficials are usually highly mobile, they are caught by pitfall traps very efficiently.

Differences in insect pest abundance between check bins and biological control bins were modest but usually demonstrable in this first year of the test. However, sampling and trapping techniques provided enough information to suggest improvements that might increase the degree of control attainable, and some of these changes have now been implemented.

Acknowledgments

Special thanks go to Wen Biran, Visiting Scientist from the PRC, for help in all phases of this study. Thanks also to R.T. Arbogast, J.E. Baker, J.G. Leesch, J.E. Throne and D.K. Weaver for helpful discussions, and to the technicians who did most of the work, J.A. Barron, A.M. Davis, R.D. Jones, P.L. Lang, R.A. Potter, J.L. Robertson, S.E. Sing and J.L. Thorpe.

References

Barak, A.V. and Harein, P.K. 1982. Trap detection of stored grain insects in farm-stored, shelled corn. Journal of Economic Entomology 75, 108–111.

Brower, J.H. and Press, J.W. 1992. Suppression of residual populations of stored-product pests in empty corn bins by releasing the predator, *Xylocoris flavipes* (Reuter). Biological Control 2(1), 66–72.

Cuperus, G.W., Fargo, W.S., Flinn, P.W. and Hagstrum, D.W. 1991. Variables affecting capture of stored-grain insects in probe traps. Journal Kansas Entomological Society 63, 486–489.

Hagstrum, D.W., Flinn, P.W., Subramanyam, Bh., Keever, D.W. and Cuperus, G.W. 1991. Interpretation of trap catch for detection and estimation of stored-product insect populations. Journal of the Kansas Entomological Society 63, 500–505.

Lippert, G.E. and Hagstrum, D.W. 1987. Detection of estimation of insect populations in bulk stored wheat with probe traps. Journal Economic Entomology 80, 601–604.

Mills, J.T. 1992. Safe storage guidelines for grains and their products. Postharvest News and Information 3, 111–115.

White, N.D.G., Arbogast, R.T., Fields, P.G., Hillmann, R.C., Loschiavo, S.R., Subramanyam, Bh., Throne, J.E. and Wright, V.F. 1991. The development and use of pitfall and probe traps for capturing insects in stored grain. Journal of the Kansas Entomological Society 63, 506–525.

Wileyto, E.P., Ewens, W.J. and Mullen, M.A. 1994. Markov/recapture population estimates: a tool for improving interpretation of trapping experiments. Ecology, in press.

Wilkin, D.R. 1991. Detection of insects in bulk grain. Journal of the Kansas Entomological Society 63, 554–558.

The use of a managed bulk of grain for the evaluation of PC, pitfall beaker, insect probe and WBII probe traps for trapping *Sitophilus granarius, Oryzaephilus surinamensis* and *Cryptolestes ferrugineus*

P.M. Cogan and M.E. Wakefield*

Abstract

A 100 tonne flat-store of wheat has been used to evaluate PC (surface and buried), pitfall beaker, insect probe and to a lesser extent WBII probe traps for trapping *Sitophilus granarius, Oryzaephilus surinamensis* and *Cryptolestes ferrugineus*. Each of three trials evaluated the traps against one species over an eight-week period. *S. granarius, O. surinamensis* and *C. ferrugineus* were seeded into the bulk at 1.0, 0.3 and 0.75 per kg, respectively. The WBII traps were evaluated against *O. surinamensis* (for the last two weeks of the trial) and *C. ferrugineus* only. 'Tinytalk' temperature data recorders were used to record ambient temperatures, those at the grain surface at 5 cm depth and for trials with *O. surinamensis* and *C. ferrugineus* temperatures on and within traps. The temperature and trap records showed the trials to be conducted close to the movement threshold for each species. The traps throughout the eight-week trial periods trapped only 0.3% *S. granarius*, 2.75% *O. surinamensis* and 0.9% *C. ferrugineus* from those released. The surface PC and pitfall beaker traps were the most effective for trapping *S. granarius,* both PC traps for *O. surinamensis* and the WBII trap for trapping *C. ferrugineus*. The surface PC trap showed a close correlation of trap catch with the PC buried trap for all three species. Both of these traps also correlated well with maximum temperatures recorded on the surface of the grain during each week of the *O. surinamensis* trial. The three trials showed the importance of considering temperature in trapping programs and further demonstrate the importance of trap type and position with regard to trapping different grain beetle pests.

traps—one placed upon the surface and one buried—in order to detect all the storage beetle species found in U.K. grain stores. However, those storekeepers that use traps tend to use one type either designed for surface trapping or one for detecting insects located beneath the grain surface (Prickett and Muggleton 1991). For this reason the PC trap was developed (Cogan et al. 1990), one trap performing both functions.

Following the success of the introduction of traps into commercial stores, storekeepers now have some experience of traps and interpreting trap catch. Interpretation varies from store to store but the awareness of traps has led many storekeepers to implement remedial measures when any trap shows insect presence. This usually takes the form of localised surface treatments with insecticide. Evaluation of traps in such stores has thus become extremely difficult.

For this reason our approach has been to set up a representative grain store for the evaluation of traps, including the PC trap. The major grain pests of the U.K., *Sitophilus granarius, Oryzaephilus surinamensis* and *Cryptolestes ferrugineus*, have been used to evaluate PC, pitfall beaker (Cogan and Wakefield 1987), insect probe (Burkholder 1984) and to a lesser extent, the WBII (Burkholder 1988) traps.

Traps are most useful and provide the earliest warning of damage if they trap at low infestation levels. Considerable warning is also obtained if traps are able to trap at low temperatures prior to the grain heating to the breeding temperatures of the beetle pests. Insect movement for storage beetles is considered to start above 5°C (Field 1992) and breeding above 12°C (Howe 1965). Commercial stores in the U.K. often maintain their grain below 12°C for much of the year. For the trials reported here, the grain was maintained at a temperature sufficiently low for each species so as to provide information on low temperature trapping.

Introduction

The use of traps for the detection of insect pests in large bulks of grain has been shown to be far better than the use of sampling techniques (Cogan and Wakefield 1987). The initial evaluation of detection methods for beetle pests in grain was undertaken by the authors in flat (floor) stores in both farm and commercial stores. Commercial stores were found to be more uniform in storage practices and provided a more reliable experimental site. As the intended end use of the grain was more predictable, experiments could be planned with some reliability.

Commercial storekeepers in the U.K. have implemented trapping as a means for the detection of beetle pests. Cogan and Wakefield (1987) recommended the use of two

Method

The site housing the experiment was a shed approximately 90 × 60 m with a 15 m height to the roof girders.

The grain walling consisted of 1.5 × 1.75 m wide panels bolted to supports to form a barrier 3 m in height. The back wall of the store was built to a width of 7.8 m and the two sides projected forwards 11.2 m (see Fig. 1a). There was no wall at the front.

A central aeration duct approximately 0.45 m diam connected to an aeration fan was positioned at the rear of the store and extended through to within 1 m of the front of the grain slope. The duct was sealed at the front end.

One-hundred tons of pesticide-free feed wheat were added to the store so that for a distance of 5 m from the back of the store the grain was level and at a depth of 2 m. This area represented 59% of the grain volume and 41% of the exposed surface area. For the rest of the bulk (7.8 × 6.5 m) the grain sloped down to the floor at the front (Fig.1b).

* MAFF, Central Science Laboratory, London Road, Slough, Berks, SL3 7HJ, U.K.

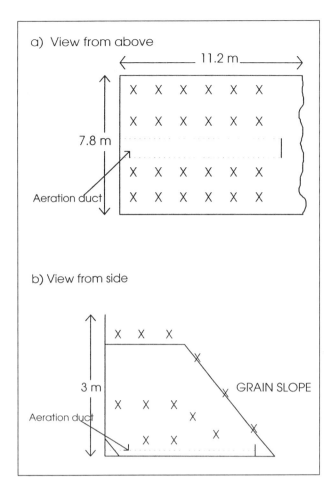

Fig. 1. Plan of 100 t floor store. X = trapping position.

Duckboards were placed centrally from the front to the rear of the store. A second path of duckboards was placed across the back of the store approximately 1 m from the rear grain walling.

One week after delivery of the grain for the first trial, a 1 kW electric fan heater was connected to the intake of the aeration fan. After one week the heater was increased to 2 kW. After a further week excessive condensation made it necessary to stop heating the grain.

For the second trial it was not necessary to heat the grain, but for the final trial heating using 2 kW was undertaken throughout the period.

Addition of insects

A vacuum sampler attached to a 2 cm diameter tube was used to remove a core of grain down to the bottom of the bulk. Adult insects were added to the grain bulk by pouring into the tube in batches of 500 to prescribed depths within the grain. After pouring in the insects at the lowest depth, 2 m, the tube was then removed to 1m depth before pouring in the next batch of insects. This was repeated at 0.5 m and 0.25 m. The distribution of batches was made so that they were added at regular intervals in a grid pattern across the bulk. The number of batches necessarily varied with the depth of the grain and number of insects of each species.

First trial

A total of 100 batches of 500 *S. granarius* (a laboratory insecticide-susceptible population, in culture for more than 20 years) were added to provide a density of 0.5/kg.

After eight weeks, and very few insects recorded in the traps, a further 50000 *S. granarius* were added to make a total of 100000 beetles in the grain, i.e. 1.0/kg.

Second trial

O. surinamensis (a laboratory insecticide-susceptible population, in culture for six years) were added at 0.3/kg. A total of 60 batches of 500 insects were put into the grain (30000 or 0.3/kg).

Third trial

A total of 75000 (0.75/kg) *C. ferrugineus* (a standard insecticide susceptible laboratory population, in culture for more than 10 years) were added.

Disinfestation between trials

After the first trial the grain was fumigated using methyl bromide. The 100 t of feed wheat was then turned. This was achieved by removing the grain from the store using a tractor fitted with a 1 t bucket and once the store was empty returning the grain back as for the first trial. For the final trial the wheat was disinfested using carbon dioxide from burner gas. The wheat was turned and returned to the store as for the second trial.

Monitoring the bulk

Samples of the wheat were taken after delivery and analysed for insecticide contamination. No insecticides were detected.

At each visit, spot temperatures were recorded using a portable temperature probe. Twelve locations across the bulk were recorded at each visit, taking readings at approximately 5 cm and at 1.5 m depths. Ambient temperature within the shed was also recorded.

'Tinytalk' temperature data recorders were used to record at two-hour intervals the temperature of the grain in the surface layer at approximately 5 cm depth. The 'Tinytalks' were sited approximately 1 m from the rear wall, one in the middle and one on either side of the bulk. Additional 'Tinytalks' were added during the second and third trials so that two trapping positions were monitored, one at the back of the bulk and one in the middle of the bulk. 'Tinytalks' were placed at 5 cm depth in the grain, on the PC trap top on the grain surface at the centre and within a pitfall beaker located at the back of the bulk.

For analysis, only the maximum temperatures recorded each week by the 'Tinytalks' were considered, as trap capture is largely dependent upon temperatures that allow insect activity. Temperatures within the bulk throughout the trials were not sufficiently low for a long enough period to cause insect mortality.

Trap location

Traps were positioned at 24 locations (Fig. 1). At each location there were two PC traps; one with its lid placed upon the surface (PC) and one with its lid approximately 5 cm below the grain surface (PC buried). Also at each location a pitfall beaker trap and an insect probe trap were inserted into the grain. Traps were initially positioned 1 m apart, one at each cardinal point. At each position the traps were assigned randomly but keeping the PC and PC buried traps opposite each other. Where depth permitted (Fig. 1), insect probe traps were also inserted to depths of 1 m (20 locations) and 2 m (four locations). These insect probe traps were buried near to the centre of each trapping location.

At each weekly visit the traps were examined and the catches recorded. The traps were then rotated to occupy the cardinal position 90 degrees either clockwise (even numbered trap positions) or anti-clockwise (odd numbered positions).

For the last two weeks of the second trial and throughout the final trial WBII traps were included. WBII traps were placed randomly between the traps on every other trap location and moved as described above.

Each trial lasted 10 weeks from the initial addition of the insects. Traps were placed in position one week after addition of the insects, thus providing eight weeks of trap catch results. Only 11 *S. granarius* were trapped in the first part (0.5/kg) of the first trial and therefore trap catches were recorded for a further eight weeks after introduction of the second batch of insects. Results for the second part only (1.0/kg) are considered in this paper.

One week after introduction of each batch of 50000 *S. granarius*, and inclusion of *O. surinamensis*, gravity spear samples were taken at 2 m (where possible), 1 m and 0.5 m. On these occasions, 1 kg samples were obtained and sieved at 12 locations, roughly corresponding with the temperature recording positions. Spear samples were not taken for *C. ferrugineus*.

A total of 32 bait bags (Pinniger 1975) were positioned at 1.5 m intervals at approximately 2 m from the outside edge of the grain walling. These were checked at weekly intervals for escapes from the experiment. A further metre outside the bait bags, 1 m-wide band of chlorpyrifos-methyl emulsifiable concentrate was sprayed at field dose to prevent any insects that had escaped from leaving the building.

Trap types were compared using Mann-Whitney ranking and trap capture relationships with temperature and trap types (and positions) were investigated.

Results

The wheat was delivered at a moisture content of 15.5%. For the second and third trials the moisture content of the bulk of the grain was approximately 16%, with the surface ranging between 16 and 19%.

Wheat was delivered for the first trial at a temperature of approximately 8°C. Heating raised the grain temperature to a 10.5°C mean (range 6.8–13.0°C) at the surface and 10.8°C (9.2–12.3°C) at 1.5 m in the bulk, before addition of the first batch of insects. By the time the second batch of insects was added the temperature had fallen to 3°C at the surface (range 2.9–3.4°C) and 6.4°C at 1.5 m (4.6–7.7°C).

For the second trial the initial grain temperature averaged 13.7°C (range 13.2–14.0°C) on the surface and 15.1°C (13.5–16.1°C) at a depth of 1.5 m in the bulk, at the time of the addition of the insects.

For the final trial the average temperatures at the surface were 9.3°C (range 7.8–12.3°C) and at 1.5 m, 12.9°C (10.7–16.5°C), at the time of the addition of the insects.

Results of the trap catches for each trap type along with the number of traps that were positive, i.e. the number of traps which recorded insects present, are presented in Tables 1–5.

For all three trials a high correlation was found between the trap catches for the PC and PC buried trap with r = 0.75 for *S. granarius*, r = 0.95 for *O. surinamensis* and r = 0.90 for *C. ferrugineus*.

First trial (*S. granarius*)

Maximum spot temperatures for the surface (5 cm depth) rose from 3.4°C at the time of the addition of the second batch of insects, to 9.0°C two weeks later. For the rest of the trial, maximum spot temperatures remained between 6.1°C and

Table 1. Number of *S. granarius* recorded in traps over an eight-week trial period from 100000 released into a 100 t grain bulk.

Week	P.C traps			Insect probe traps		
	Pitfall	Surface	Buried	Surface	1 m	2 m
1	11	12	15	3	0	0
2	44	63	17	0	1	0
3	15	15	1	0	0	0
4	13	3	0	0	0	0
5	10	14	0	0	0	0
6	10	11	5	0	0	0
7	9	7	2	0	0	0
8	4	3	0	1	0	0
Total	116	128	40	4	1	0

9.8°C. Maximum spot temperatures at 1.5 m, rose from 7.7°C to 8.1°C then remained between 5.9°C and 8.3°C for the rest of the trial.

The 'Tinytalk' surface maximum temperature readings for the second period (1.0/kg) are presented in Figure 2. Surface temperatures remained below 9.5°C for the first five weeks of the trial, rising to 12°C for the final three weeks.

All trap types, except for the 2 m insect probe traps, trapped *S. granarius* but only two were trapped in the 1 m probes (Table 1). A total of 300 *S. granarius* (0.3%) were trapped from the 100000 released into the grain.

During the trial, the greatest number trapped in one week was 125 (43% of those trapped). This occurred during the second week (Table 1).

Comparing the effectiveness of the traps with regard to their ability to detect *S. granarius*, the PC and pitfall beaker were significantly better (p < 0.05), i.e. more per trap and more positive traps, than the PC buried and the insect probe traps (Tables 4 and 5).

No *S. granarius* were found in the bait bags and only one was found in the gravity spear samples.

Second trial (*O. surinamensis*)

Surface spot readings fell from a mean of 13.1°C (range 12.5–13.1°C) in the second week to 7.0°C (6.1–7.6°C) at the end of the trial. During the fourth week only, the temperature rose to give a mean of 13.9°C (13.5–14.6°C).

Spot temperatures at 1.5m were at their maximum on the second week with a mean of 15.4°C (range 14.8–15.7°C). The mean fell gradually throughout the trial to 12.7°C (10.4–13.8°C) at the end of the trial.

'Tinytalk' temperatures recorded on the PC trap-top fell from a maximum of 16.9°C during week two to 8.4°C by week six, continuing with this temperature as a maximum until the end of the trial. In the surface layer (5 cm depth) the maximum temperature fell from 15.2°C in week one to 7.8°C by week seven (Fig. 2).

A total of 990 *O. surinamensis* (2.75%) were trapped from the 30000 released into the grain (Table 2). The greatest number trapped in one week was 268 (27% of those trapped), during the first week of trapping. Numbers trapped decreased except for week six and for the final week (Table 2).

All trap types trapped *O. surinamensis* (Table 2). Although only three were recorded in the 2 m insect probe traps, 138 were recorded from the 1 m insect probe traps, with 49 of these recorded during the last week.

Comparing the effectiveness of the traps with regard to their ability to detect *O. surinamensis,* i.e. positive traps (Table 5),

Table 2. Number of *O. surinamensis* recorded in traps over an eight-week trial period from 30000 released into a 100 t grain bulk.

		PC trap		WB11	Insect probe traps		
Week	Pitfall	Surface	Buried		Surface	1 m	2 m
1	13	92	91	-	42	29	1
2	4	59	91	-	20	17	1
3	1	45	66	-	4	1	0
4	7	29	62	-	7	9	1
5	2	16	30	-	3	9	0
6	4	20	24	-	8	14	0
7	5	10	17	12	4	10	0
8	4	22	26	8	1	49	0
Total	40	293	407	20	89	138	3

Table 3. Number of *C. ferrugineus* recorded in traps over an eight-week trial period from 75000 released into a 100 t grain bulk.

		PC traps		WB11	Insect probe traps		
Week	Pitfall	Surface	Buried		Surface	1 m	2 m
1	2	13	7	16	6	32	3
2	0	0	3	66	12	4	1
3	40	1	3	22	16	30	8
4	1	1	2	19	12	12	4
5	0	0	1	16	11	4	2
6	12	6	8	15	5	2	11
7	4	6	6	3	4	16	0
8	8	37	78	21	30	21	8
Total	67	64	108	178	96	121	37

Table 4. Number per trap during the eight-week trapping periods. [a]Extrapolated from two weeks' data.

		PC trap		WB11	Insect probe traps		
	Pitfall	Surface	Buried		Surface	1 m	2 m
S. granarius	5.0	5.4	1.7	-	0.2	0.1	0
O. surinamensis	1.7	12.2	17.0	7.3[a]	3.7	6.9	0.8
C. ferrugineus	2.8	2.7	4.5	16.2	4.0	6.1	9.3

[a]Extrapolated from two weeks' data.

Table 5. Number of traps recording insects (positive traps) per trapping position during the eight-week trapping period. * Extrapolated from two weeks' data.

		PC traps		WB11	Insect probe traps		
	Pitfall	Surface	Buried		Surface	1 m	2 m
S. granarius	2.4	2.1	0.6	-	0.1	0.1	0
O. surinamensis	1.5	3.8	4.9	4.0*	2.1	2.4	0.8
C. ferrugineus	0.8	0.8	1.1	2.8	2.0	1.9	2.5

both surface and buried PC traps proved significantly better (p < 0.05) than the probes and the pitfall beaker. The PC buried traps proved significantly better (p < 0.05) than the 1 and 2 m insect probe traps.

There was a high correlation between the trap catches for PC and pitfall beaker traps (r = 0.97) and to a lesser extent with PC buried and surface insect probe traps (r = 0.77). Maximum temperatures recorded by the 'Tinytalk' temperature data recorders on the trap-top correlated closely (both r > 0.8) with PC and PC buried trap catch.

Only three *O. surinamensis* were found in the bait bags and none were found in the gravity spear samples.

Third trial (*C. ferrugineus*)

The ambient temperature throughout the *C. ferrugineus* trial was between 3.4°C and 8.1°C. Spot temperatures fluctuated widely at the beginning of the trial with the surface mean 12.3°C (range 9.3–13.6°C) and 1.5 m mean 16.5°C (range 10.3–25.6°C).

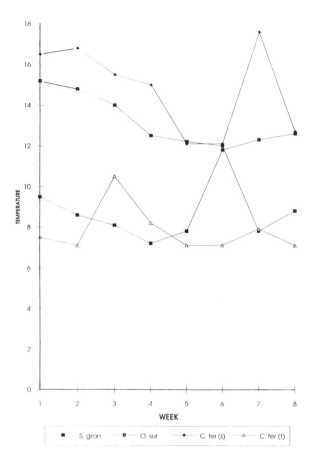

Fig. 2. Weekly maximum temperatures at surface layer, s—surface, t—trap top.

Temperatures were lower by week two, with the surface mean at 9.1°C (6.1–14.6°C) and 1.5 m mean 10.7°C (7.4–13.6°C).

Temperatures were more consistent on the surface of the bulk for the rest of the trial, at approximately 8°C. However, the 1.5 m spot temperature means fluctuated from a 15.3°C (week 3) to 9.5 °C (week 5) and to 12.8°C by the end of the trial.

The 'Tinytalk' temperature recorder buried 5 cm into the surface layer at the back of the bulk showed a gradual drift in maximum temperature from 16.8°C in the second week to 12°C by week five, then up to 17.6°C in the penultimate week (Fig. 2). A similar pattern was observed in the centre surface layer trap but the 'Tinytalk' positioned upon the trap top recorded between 5°C and 9°C lower each week compared with those in the surface layer. Temperatures on the trap top were between 7.1°C and 8.2°C except for the third week when it was 10.5°C (Fig. 2). At the front of the bulk, on the sloping grain, temperatures fell from 8°C (maximum) to 6°C over the same period and then rose to 8°C in the last week.

The 'Tinytalk' placed in the pitfall beaker trap in the surface layer fluctuated wildly from 15°C in the second week to 7°C by week five and then back to 14°C by the final week.

Condensation and damping were evident on the surface of the bulk after five weeks of the trial.

A total of 671 *C. ferrugineus* (0.89%) were trapped from the 75 000 released. In the final week 203 (30% of the 671) were trapped, the highest weekly catch. None were trapped in the traps positioned at the front of the grain slope.

All trap types trapped *C. ferrugineus*. Thirty-seven (6% of trap captures) were recorded in the 2 m insect probe traps and

only 131 (20%) were recorded in the surface traps (PC and pitfall beaker combined). Over 70% of the total PC buried trap catch was recorded during the last week (Table 1).

Comparing the effectiveness of the traps with regard to their ability to detect *C. ferrugineus* (Tables 4 and 5), the WBII trap was significantly (p < 0.05) better than the PC, pitfall beaker or PC buried.

The 1 m insect probe trap catch showed a high correlation with the ambient and also the surface spot temperatures (both r > 0.83).

No *C. ferrugineus* were found in the bait bags surrounding the bulk.

Discussion

The store was set up to represent a typical 100 t farm store. The layout appears to be representative of the U.K., but no farm store would be monitored so heavily.

The ability to heat the grain using fan heaters connected to the aeration fan had some success but condensation in the first trial became a problem so the heating had to be abandoned. Condensation and damping of the grain also became a problem after five weeks in the final trial.

Temperatures achieved throughout the trials were sufficiently low as to provide information on low temperature trapping.

The lack of insects found in the spear samples is unsurprising, with only 12 kg of samples taken and a density of 0.5 and 0.3/kg for *S. granarius* and *O. surinamensis*, respectively. This lack of recovery of the two species further demonstrates how unwise it is to rely on spear samples for detecting insects. The insects had not left the grain bulk during the trials or they would have been detected by the bait bags. It must be assumed that the insects were therefore in the bulk but not detected except as determined by the trap catches. The percentage trap catches for *O. surinamensis* (2.75%) and *C. ferrugineus* (0.9%) were lower than those recovered (8.6 and 2.9 %, respectively) in the bin trial by Wilkin et al. (1990). For *S. granarius* the recovery was the same at 0.3%. It must be noted that the initial temperature of the grain in the Wilkin et al. trial was far higher than the trials presented here and that the more mobile species (*O. surinamensis* and *C. ferrugineus*) would be expected to find their way to the traps more readily than the less mobile *S. granarius*.

Although *S. granarius* is the most cold hardy storage beetle in the U.K. (Howe 1965) it is unlikely that it moves much below 9°C. Wilkin et al. (1990) trapping the same laboratory strain of *S. granarius* used in these trials, found that pitfall beaker trap catch ceased at temperatures below 6.5°C and only occasionally were *S. granarius* trapped below 12°C. In the experiment reported here, some of the *S. granarius* found in the pitfall beaker traps may have fallen in involuntarily through movement of the grain, but those in the PC surface traps must have been trapped and therefore indicate insect movement throughout the eight weeks of the trial.

The low temperatures recorded by both the 'Tinytalk' temperature recorders and the spot readings for the period when the *S. granarius* were present at 1.0/kg do not explain the relatively high numbers trapped. It is important to note that it is the maximum temperatures, as seen in the 'Tinytalk' records, which should be considered when viewing the possibility of movement. The Tinytalk readings were at two-hour intervals and therefore reflect far more accurately the temperature changes in the trials than do the weekly spot readings.

For the whole of the first trial, spot surface temperatures remained below 10°C and at 1.5 m below 8°C. The 'Tinytalk' recordings agree with the surface spot recordings and show the surface layer maximum temperature at 9.5°C on the first week but after that date not above 8°C until week six. By this time most of the *S. granarius* that were trapped in the trial had been trapped.

The *S. granarius* were a laboratory population and as such might be considered to be less likely to move at low temperatures than field populations. The trap catches and temperature data show that *S. granarius* are able to move and be trapped at temperatures below 10 °C , particularly on the grain surface. An explanation for their movement might lie in the position of the 'Tinytalk' recorders, which for the first trial were buried 5 cm below the grain surface. Temperatures may have fluctuated more widely closer to the grain/air interface as occurred in the other trials and may have reached such temperatures as to allow movement.

As was expected from both laboratory and field trials (Cogan and Wakefield 1985, 1987), the insect probe traps performed poorly against *S. granarius*. A total of just five insects were recorded during the trial.

Although more *S. granarius* were trapped by the PC surface traps (Table 4), the pitfall beaker trap recorded more positive trap catches (Table 5). Surprisingly, the PC buried traps trapped 12% of all the *S. granarius* trapped.

The PC trap catch was possibly aided by the low temperatures as the *S. granarius* would have supposedly been less able to 'dangle' i.e. hang by just two legs into the trap hole as described in Wakefield and Cogan (1993) and then withdraw from the trap hole to avoid capture. This may help explain the lack of correlation between increase in temperature and trap catch reported for *Rhyzopertha dominica*, *Tribolium castaneum* and *S. oryzae* (Fargo et al. 1989). Traps do not catch all insects encountering the trap. Some may 'dangle' in the trap hole then withdraw. Trap catch falls with temperature decrease due to fewer insect / trap encounters over a period of time. As temperatures fall, a point is reached where trap catch will increase due to the inability of the insects to escape from an encounter with the trap hole, resulting in a greater success rate for the trap.

For the *O. surinamensis* trial, temperatures were higher than in the first trial. Throughout the trial weekly maximum temperatures fell at the surface from 17°C at the start down to less than 9°C at the end. These temperatures correlated well with the catch found in the PC surface and buried traps. This relationship between temperature and trap catch should greatly assist with the calibration of the PC traps when relating *O. surinamensis* infestation with trap catch.

The surface trap catches, i.e. in PC and pitfall beaker, also correlated well, although the PC trap (surface and buried) proved to be the most effective trap type used for *O. surinamensis* detection.

The high number recorded in the 1 m insect probe traps during the last week may well reflect migration of this species down into the bulk as temperatures at the surface lowered during the final two weeks.

The temperatures in the *C. ferrugineus* trial showed the greatest fluctuations but at the surface layer the temperature remained above 12°C throughout the period. This was above the temperature at which traps caught this species in the trials conducted by Wilkin et al. (1990). Temperatures on the surface traps as determined by the 'Tinytalk' recorder on the PC trap top showed the importance of recording at the trap/air interface. The surface layer was found to be between 5°C and 9°C above this trap top temperature where maximum temperatures were 7.1°C for four weeks and only rose above 10°C during one week. This rise to 10°C coincided with the largest trap catch for the pitfall beaker traps with 60% of their total caught in the one week.

The high trap capture in the last week of the *C. ferrugineus* trial is harder to explain but may have been due to the sudden

5°C increase in maximum temperatures during the penultimate week. The lowering by 5°C in the final week may have resulted in many beetles migrating into the bulk of the grain, having moved to the surface when the temperature was more favourable during the preceding week. This is supported by the trap catch of the PC buried and surface insect probe traps both of which had their largest catch in the final week. A difficulty with this hypothesis is that the PC trap on the grain surface also recorded its largest catch during the final week when the temperature at the trap was at a maximum of 7.1°C.

Cuperus et al. (1990) considered that aggregation of species such as *C. ferrugineus* makes trap catch sensitive to trap placement. Watters (1969) and Loschiavo (1983) found *C. ferrugineus* movement to be affected by moisture and temperature. Trap catch also increases significantly as grain temperature increases (Loschiavo and Smith 1986; Fargo et al. 1989). This trial considered temperature alone but we only found a correlation between catch in the probe traps at 1 m depth with ambient and spot surface temperatures. The correlation with ambient may have been due to the influence of the aeration fan for much of the trial; the temperature of the heated air being largely determined by the temperature of the ambient air drawn into the fan.

The number of insects trapped may in future be of importance for accurately determining the population size within the grain bulk. However, as traps currently catch a small percentage of those present, as found in this trial, the number of positive traps is probably of more importance. Positive traps indicate those traps which consistently catch the pest species and therefore give a measure of trap sensitivity. In this trial the difference between the results for trap catch numbers and positive trap numbers was small for *S. granarius* but for the other two species, the differences between the traps was noticeably less when positive traps were considered.

The questions raised by these trials indicate that closer examination of data from such trials is necessary. Temperature recorder data needs to be directly related to individual trap catch. This is particularly relevant if traps are to be calibrated to relate infestation or risk of infestation with trap catch For this relationship to be explored it would appear necessary to increase the density of insects used in such trials in order to provide sufficient trapping data for analysis.

Acknowledgments

We should like to thank The Pesticide Safety Directorate, MAFF for funding of this work, all colleagues at CSL, Slough who helped with the trials and John Chambers for critical review of the manuscript.

References

Burkholder W.E. 1984. Stored product insect behaviour and pheromone studies. Keys to successful monitoring and trapping. In: Proceedings of the Third International Working Conference on Stored-product Protection, Manhatten, Kansas, USA. Kansas State University Press, 20–23.

Burkholder W.E. 1988. Some new lures, traps and sampling techniques for monitoring stored product insects. In : Proceedings of the XVIII International Congress of Entomology, Vancouver, Canada, 444 pp.

Cogan, P.M. and Wakefield, M.E. 1985. Detection of insects in large bulks of grain. MAFF, Storage Pests Department Report No 14. MAFF, UK 22 pp.

Cogan, P.M. and Wakefield, M.E. 1987. Further developments in traps used to detect low-level infestations of beetle pests in bulk stored grain. Proceedings of the BCPC Symposium, Reading, U.K. (BCPC Monograph No. 37). Ed T.J. Lawson ,161–167.

Cogan, P.M., Wakefield, M.E. and Pinniger, D.P. 1990. PC, A novel and inexpensive trap for the detection of beetle pests at low densities in bulk grain. In: Proceedings of the Fifth International Working

Conference on Stored-product Protection, Bordeaux, France, September 1990, 2, 1321–1330.

Cuperus, G.W., Fargo, W.S., Flinn, P.W. and Hagstrum, D.W. 1990. Variables affecting capture of Stored-grain insects in probe traps. Journal of the Kansas Entomological Society 63, 4, 486–489.

Fargo, W.S., Epperly, D., Cuperus, G.W., Noyes, R.T. and Clary, B.L. 1989. Influence of temperature and duration on the trap capture of stored-product insect species. Journal of Economic Entomology 82, 970–973.

Field, P. 1992. The control of stored-product insects and mites with extreme temperatures. Journal of Stored Products Research 28, 2, 89–118.

Howe, R.W. 1965. A summary of estimates of optimal and minimal conditions for population increase of some stored product insects. Journal of Stored Products Research 1, 177–184.

Loschiavo, S.R. 1983. Distribution of the rusty grain beetle (Coleoptera) in columns of wheat stored dry or with localised high moisture content. Journal of Economic Entomology 76, 881–884.

Loschiavo, S.R., and Smith, L.B. 1986. Population fluctuations of the rusty grain beetle *Cryptolestes ferrugineus* (Coleoptera: Cucujidae), monitored with insect traps in wheat stored in a steel granary. Canadian Entomology 118, 641–647.

Pinniger, D.B. 1975. A bait trap technique for assessment of stored-product beetle populations. Tropical Stored Product Information 19, 9, 9–11.

Prickett, A.J. and Muggleton, J. 1991. Commercial grain stores 1988/1989—England and Wales—Pest incidence and storage practice—part 2. HGCA Project Report No 29 Home-Grown Cereals Authority 119 pp.

Watters, F.L. 1969. The locomotor activity of *Cryptolestes ferrugineus* (Stephens) (Coleoptera: Cucujidae) in wheat. Canadian Journal of Zoology 47, 1177–1117.

Wilkin, D.R., Armitage, D.M., Cogan, P.M. and Thomas, K.P. 1990. Integrated pest control strategy for grain. HGCA Report No 24 Home-Grown Cereals Authority 87 pp.

New trends in stored-grain infestation detection inside storage bins for permanent infestation risk monitoring

F. Fleurat-Lessard*, A.-J. Andrieu† and D.R. Wilkin§

Abstract

Accurate and reliable detection and monitoring of insects in grain bulks is an essential part of commercial trading or research into pest management systems. Current methods of monitoring grain quality during storage have limitations, several of them sufficiently serious to prevent long-term risk assessment, a fundamental requirement of proper pest management.

New approaches in pest detection will help to change this situation. Mechanical collection and automated sieving of samples should be available in the near future to assess insect infestation in moving grain.

Integrated pest management strategies have been much improved by installation of in-bin acoustic sensors which automatically alert to the need for insect control. These systems can reduce the chances of an infestation going undetected at a lower cost than permanent chemical control.

The information provided from such monitoring tools could be linked to an expert-system that would play an active role both for risk assessment and decision-making during grain storage.

Introduction

Stored grain is susceptible to attack and damage by insects and much effort has to be expended on pest control and prevention. Monitoring of insects in stored grain can be the most costly part of grain storage management. A wide range of approaches to pest management has been attempted (Flinn and Hagstrum 1991; Subramanyam et al. 1991; Wilkin et al. 1991; White 1992) but all rely on a reliable estimate of insect numbers, which can be doubtful or impossible at very low density levels (i.e. less than 1 insect/kg of grain). Nevertheless, estimating insect numbers is also an essential part of judging the need for control measures and, therefore, a means of optimising the economics of pest control. The ability to estimate populations becomes even more important when more sophisticated pest management systems are attempted.

Methods Already Used for Insect Detection in Samples of Grain

Despite the clear need for accurate determination of insect populations in grain, the method most widely used by commerce, the assessment on samples withdrawn from the grain mass, is crude, labour intensive and insensitive. Some limitations of sampling have been revealed in pilot scale trials (Wilkin and Fleurat-Lessard 1991) through experiments in static grain bulks.

The collection and removal of samples was used for insect detection in grain bulks. Dead adult insects were added to the grain during bin loading operations at three rates: 0.2 insects/kg, 1 insect/kg and 5 insects/kg for each species. The results showed that, despite the collection of a larger number of samples than recommended in sampling standard, and of large quantities of grain, detection was only reliable at the highest population density. This work provides a clear illustration of the restricted value of collection and examination of samples to detect and estimate insects in grain even if they are dead and motionless.

The collection of a larger number of samples or greater quantities of grain per unit under examination is not usually done in grain transactions even if some tentatives are under experimental extension. When insects are alive, we have demonstrated the poor value of such means of assessment of insect density in the case of aggregative patterns of dispersion or of migration inside the grain bulk during the storage period (Fleurat-Lessard and Poisson 1982).

On the other hand, mechanical sampling allows larger quantities of grain to be obtained, with better representativeness of the samples when good sampling rules are followed. Usually, such machines work by sucking a sample of grain from the lorry, passing grain and air through a cyclone and delivering the grain to the laboratory. Work by Hurburgh et al. (1985) indicate that vacuum sampling of grains and fluid-lift conveying of samples could introduce several biases to the results of insect contamination rates (for example over-estimates of lighter dead insects, and the lethal effects of mechanical impacts during conveying). More complicated devices (i.e. cross-flow samplers) have been used for quality assessment of grain during loading of boats in different countries (Friedrich 1981; Gy 1982, 1983). Nevertheless, this method is labour intensive and these systems have been progressively abandoned. The intrinsic value of mechanical samplers in pest detection is a much debated question. Often, the plethora of samples represent a limitation for the analysis operations which follow sampling. However, automatic seeving devices for impurities grading or insect recovery in large samples (10 kg) are available (for example 'insectomat' in U.K.) and could be used for checking insect presence and numbers at a density as low as 1 insect per 5 kg of grain.

Among the promising insect-detection techniques under investigation is the detection and measurement of insect by-products, such as uric acid or CO_2 release rate in a sample of grain, or directly in a storage bin, and the use of various radiation-based methods. The latter include X-rays (ISO 1987), reflected or transmitted near infrared (Pinniger et al. 1986), or nuclear magnetic resonance (Chambers et al. 1984). All these methods are designed to work with samples of grain, often of a very limited size. The difficulties surrounding the use of samples have been detailed earlier and the same limitations will apply to any method based on samples. Even if they are

* INRA, Laboratoire des Insectes des Denrées Stockées, B.P. 81, 33883 Villenave d'Ornon Cédex, France.
† INRA, Laboratoire d'acoustique, Domaine de Vilvert, 78353 Jouy en Josas, France.
§ Hathersage, 39 Denham Lane, Chalfont St Peters, Gerrards Cross, SL9 0EP Bucks., United Kingdom.

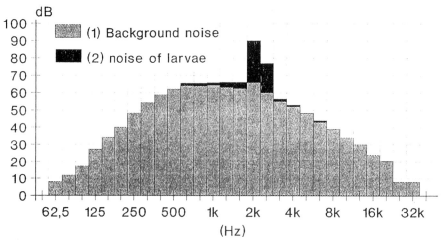

Fig. 1. Noise spectra in uninfested grain (1) and after addition of infested kernels with *Sitophilus granarius* larvae (2). (Without filter for extreme frequencies.) (With frequency filtering system.)

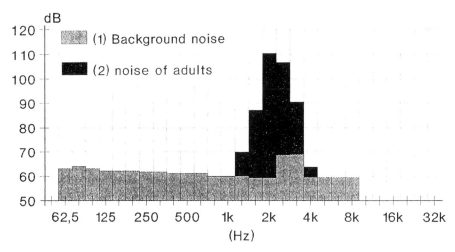

Fig. 2. Noise spectra with INRA acoustic probe in infested grain (1) and with *Sitophilus granarius* adult (2). (With frequency filtering system.)

very precise, the accuracy of the detection is depends on the reliability of the sample.

Results Obtained with Traps and Acoustic Probes

The solution to permanent monitoring of insect risks is 'in-bin detection': if insects can be monitored *in situ* it is possible that some of the limitations associated with sampling can be overcome. Trapping and acoustic detection have obvious potential for use in this way. Pitfall traps or perforated probe traps sunk into the surface or at a small depth in multi-thousand tonne bulks of grain are extensively used in spite of the absence of a scientific principals base. Results published by Loschiavo (1975), Burkholder (1983), E.J. Wright (pers. comm. 1990), Cogan and Wakefield (1987) and Wilkin (1991), confirmed that either pitfall or probe traps, or a combination of the two, were much more effective in detecting insects in static bulks of grain than conventional sampling methods. Research on trapping seems to have concentrated on enhancing the effectiveness of an already effective method by developing improved traps (Cogan et al. 1991) and by adding pheromone or food attractants to existing traps (Chambers 1987; Burkholder and Ma 1985). This work has undoubtedly been successful but fails to address the problem of interpretation of

trapping results in terms of estimating insect population densities. But it is the first step in the early warning of insect presence which does not need any sophisticated material or training.

Acoustic detection relies on the sounds produced by insects as they move and feed within a grain bulk (Fleurat-Lessard and Andrieu 1986; Hagstrum et al. 1988; Andrieu and Fleurat-Lessard 1990; Flinn and Hagstrum 1991). These sounds are readily detected with relatively simple equipment but separating insect and background noise can be difficult and identification of the specific sound spectrum is extremely hard inside a grain bulk (Figs 1 and 2). Nevertheless, acoustical sensors can provide, without disturbing the grain bulk when taking a sample, density estimates that are in good agreement with estimates obtained through exhaustive systematic sampling in all locations of a grain bin (Figs 3 and 4). In addition, acoustic detection reveals the hidden infestation by the larvae of weevils and borers, which is a substantial advantage for early warning. Most of the recent development in acoustic detection of insects has concentrated on methods of analysing the sound spectrum of insect noises and developing computer-based analysis programs that will identify the insect species through their sound spectrum. This approach has a great potential for incorporation in an automated insect monitoring system in situ, directly installed inside grain bins (Fig. 5).

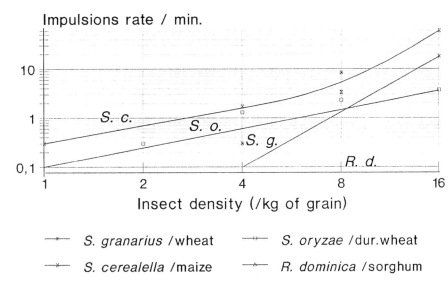

Fig. 3. Significant noise impulsions resulting from feeding activity of insect larvae in different cereal grain. (1 kg grain samples, temperature 20°C).

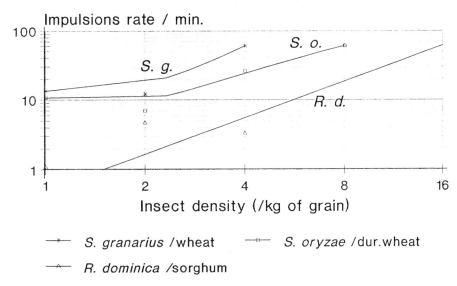

Fig. 4. Significant noise impulsions resulting from feeding activity of moving adults in different cereal grain.

The Importance of New Insect Detection Devices

Validation of early-detection devices is currently in the pre-development phase in Europe. Either probe traps or pitfall traps and acoustic probes are used in large grain storage facilities to test them in full-scale conditions and for long-term storage of grain.

For instance, an acoustic probe and recording system was used last year in France inside a large bin (650 m³) filled with 500 t of wheat (Fig. 6). The system gave permanent information about insect activity during all the storage period. These records were of primary importance because storage was driven with a cooling aeration technique (Lasseran and Fleurat-Lessard 1991). With insect activity deep inside the grain bulk, acoustic detection will depend on temperature (Fig. 7).

It was demonstrated in France (Fleurat-Lessard and Andrieu 1986) and later in USA (Hagstrum et al. 1988, 1990) that there is a good correlation between insect noise and population density of the granary weevil. Much more reliable than pittfall or probe traps, the acoustic detection in situ, with acoustic probes installed inside bins, allows forecasting of dangerous density levels of primary noxious insect species about three or four months before reaching the critical threshold (Andrieu and Fleurat-Lessard 1984, unpublished data).

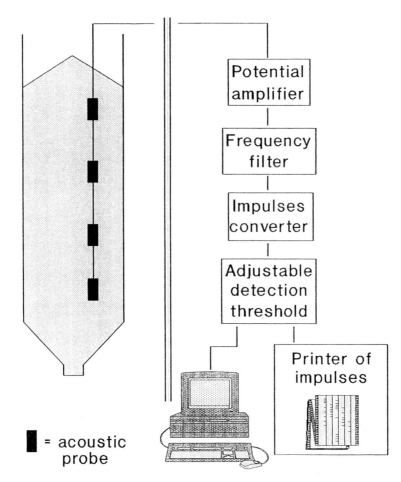

Fig 5.   Detection of insects inside bins of grain with acoustical probes and a signal
treatment system.

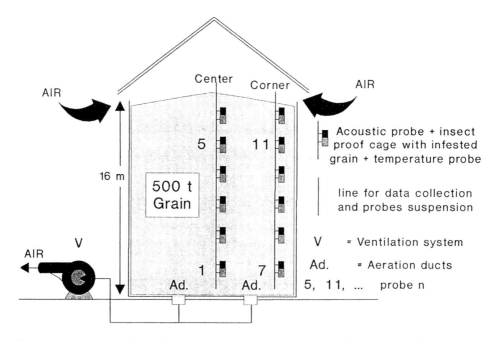

Fig. 6.   Experimental bin for full-scale experimentation on insect activity during cooling of grain,
recorded by acoustic probes.

With acoustical assessment of population density, insect population dynamics is accurately predictable, using only physical conditions criteria, provided that the main insect pest species are accurately characterised for biotic potential and population dynamics in different biotic situations such as fluctuations of intrinsic rate of increase on different cereal species or varieties, and temperature threshold for development (Birch 1953; Hardman 1978; Hagstrum and Throne 1989).

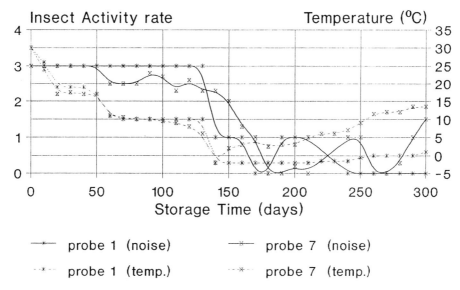

Fig. 7. Relation between insect activity recorded by acoustic probes and temperature in the grain bulk.

Further mathematical modelling for insect demography is not a constraint provided that the main physico-chemical parameters inside the grain bulk are known.

The research approach today tends to combine simulations obtained in several submodels (temperature and moisture migration related to climatic changes, ventilation or meteorological database; residual level of pesticide residues using pattern of insecticide degradation with time, for instance). Only a very small part of this modelling is useful for integration into management strategies for insect infestation in static grain bulks. The ability to forecast population trends for providing as soon as possible prophylactic and remedial advice is only possible if the lowest level of density or activity of insect is detectable in the grain bulk. Most modelling tentatives are based on population density levels much higher than the tolerance threshold, or only use fictive density rates for simulation. All the expert systems under study for grain pest management during storage need an early warning system such as acoustic probes (Flinn and Muir 1992).

Conclusion

Accurate and reliable detection of insects in grain is an essential part of commercial trading or research into pest management systems, especially through the actual development of expert systems designed for the monitoring of grain quality during all the conservation period (Ndiaye 1994). Current methods have serious limitations, several of which are sufficiently serious to prevent long-term risk assessment, a fundamental requirement of proper pest management.

New approaches to pest detection will help to change this. Work on mechanical collection and automated sieving of samples is under way, and commercial equipment to assess insect infestation on samples collected in moving grain should be available in the near future. Integrated pest management strategies have a great deal to commend them in the field of static bulk grain storage, particularly when long-term storage is being attempted. Permanently installed acoustic sensors, which automatically indicate the need for insect control, can improve pest management by reducing the chances of an infestation going undetected. Ultimately, the results from such monitoring could be linked to an expert system that would play an active role both for risk assessment and decision-making during grain storage.

References

Andrieu, A.J. and Fleurat-Lessard, F. 1990. Insect noise in stored foodstuffs. Compact Disk, INRA Ed., Versailles, France.

Andrieu, A.J., and Fleurat-Lessard, F. 1984. Réalisation et expérimentation d'un prototype d'équipement de détection acoustique automatique d'insectes dans une cellule de stockage de blé, Confidential report, 53pp (unpublished data).

Birch, L.C. 1953. Experimental background to the study of the distribution and abundance of insects. The influence of temperature, moisture and food on the innate capacity for increase of three grain beetles. Ecology 34, 698–711.

Burkholder, W.E. 1983. Stored-product insect behaviour and pheromone studies: keys to successful monitoring and trapping. In : Proceedings of third International Working Conference on Stored-Product Entomology Manhattan KS, USA. October, 1983, 20–33.

Burkholder, W.E., and Ma. M. 1985. Pheromones for monitoring and control of stored-product insects. Annual Revue of Entomology 30, 257–272.

Chambers, J. 1987. Recent developments in techniques for the detection of insect pests of stored products, 151–160. In: Lawson, T.J. Stored Product Pest Control. BCPC Monograph 37, Reading U.K., March 1987, 151–160.

Chambers, J., Mc Kevitt, N.J. and Stubbs, M.R. 1984. Nuclear magnetic resonance spectroscopy for studying the development and detection of the grain weevil, *Sitophilus granarius (L.)(*Col: Curculionidae), within wheat kernels. Bulletin of Entomological Research 74, 707–724.

Cogan, P.M., Wakefield, M.E. and Pinniger, D.B. 1991. PC, a novel and inexpensive trap for the detection of beetle pests at low densities in bulk grain, In: Fleurat-Lessard, F. and Ducom, P., ed., Proceedings of Fifth International Working Conference on Stored-Product Protection, Bordeaux, France, September 1990, 2, 1321–1329.

Cogan, P.M. and Wakefield, M.E. 1987. Further developments in traps used to detect low level infestations of beetle pests in stored grain. In: Lawson, T.J. Stored Product Pest Control, BCPC Monograph 37, Reading, U.K., March 1987, 161–168.

Fleurat-Lessard, F. and Poisson, J. 1982. Evolution biologique et physique d'un stock de blé tendre en présence d'une infestation par le charançon des grains: *Sitophilus granarius* L. (Col. Curculionidae) et mesure des pertes pendant le stockage, 83–121.

La conservation des céréales en France, Actions Thématiques Programmées INRA, INRA Editions, Versailles, France, 83–121.

Fleurat-Lessard, F. and Andrieu, A.J. 1986. Development of a rapid method to determine insect infestation in grain bins with electro-acoustic devices. In: Donahaye, E.J. and Navarro, S., ed., Proceedings of Fourth International Working Conference on Stored-Product Protection, Tel-Aviv, Israel, September 1986, 643.

Flinn, P.W. and Hagstrum, D.W. 1991. An expert system for managing insect pests of stored grain, In: Fleurat-Lessard, F. and Ducom, P., ed., Proceedings of Fifth International Working Conference on Stored-Product Protection, Bordeaux, France, September 1990, 3, 2011–2018.

Flinn, P.W. and Muir W.E. (1992). Expert system concept. In: Jayas, D.S., White, N.D.G., Muir W.E. and Sinha, R.N., ed., Proceedings International Symposium on Stored Grain Ecosystems, Winnipeg, Canada, June 1992.

Friedrich, W. 1981. Le prélèvement automatique d'échantillons: un moyen de détecter la présence de substances indésirables réparties de manière inégale. Industries des Céréales, 9, 37–40.

Gy, P. 1982. Sampling of particulate materials. Theory and practice. Elsevier, Amsterdam, 432 pp.

Gy, P. 1983. Sampling errors: they may deprive the analytical results of any meaning. Analysis 11, 413–440.

Hagstrum, D.W., Webb, J.C. and Vick, K.W. 1988. Acoustical detection and estimation of *Rhyzopertha dominica* larval populations in stored wheat. Florida Entomologist 71, 441–447.

Hagstrum, D.W., Webb, J.C. and Vick, K.W. 1990. Acoustical monitoring of *Rhyzopertha dominica* (F.)(Col.: Bostrichidae) populations in stored wheat. Journal of Economic Entomology 83, 625–628.

Hagstrum, D.W. and Throne, J.E. 1989. Prediction of stored-wheat insect population trends from life history traits. Environmental Entomology 18, 660–664.

Hardman, J.M. 1978. A logistic model simulating environmental changes associated with the growth of populations of rice weevils, *Sitophilus oryzae*, reared in small cells of wheat. Journal of Applied Ecology 15, 65–87.

Hurburgh, C.R., Bern, C.J. and Cox, D.F. 1979. Evaluating grain probing devices and procedures. Summer Meeting of ASAE and CSAE, Winnipeg, Canada, June 1979.

I.S.O. 1987. Cereals and pulses: determination of hidden insect infestation - parts I - IV, standard 6639.

Lasseran, J.C. and Fleurat-Lessard, F. 1991. Aeration of grain with ambiant or artificially cooled air: a technique to control weevils in temperate climates. In: Fleurat-Lessard, F. and Ducom, P., ed., Proceedings of Fifth International Working Conference on Stored-Product Protection, Bordeaux, France, September 1990, 2, 1221–1231.

Loschiavo, S.R. 1975. Field tests of devices to detect insects in different kinds of grain storage. Canadian Entomologist 107, 385–389.

Ndiaye, A. 1994. Le (futur) système de pilotage des stocks de grain par un système expert informatisé. In: compte-rendu Journée Technique GLCG, Paris, February 1994.

Pinniger, D.B., Chambers, J. and Cogan, P.M. 1986. New approaches to the detection of pests in grain. In: Flannigan, B. Spoilage and mycotoxins of cereals and other stored products. C.A.B. International, Slough, U.K., 7–12.

Subramanyam, Bh., Hagstrum, D.W. and Harein, P.K. 1991. Upper and lower temperature thresholds for development of six stored-product beetles. In: Fleurat-Lessard, F. and Ducom, P., ed., Proceedings of fifth International Working Conference on Stored-Product Protection, Bordeaux, France, September 1990, 3, 2029–2038.

Wilkin, D.R., 1991. Detection of insects in bulk grain. Journal of Kansas Entomology Society (in press).

Wilkin, D.R. and Fleurat-Lessard, F., 1991. The detection of insects in grain using conventional sampling spears, In: Fleurat-Lessard, F. and Ducom, P., ed., Proceedings of fifth International Working Conference on Stored-Product Protection, Bordeaux, France, September 1990, 3, 1455–1454.

Wilkin, D.R., Mumford, J.D. and Norton, G. 1991. The role of expert systems in current and future grain protection. In: Fleurat-Lessard, F. and Ducom, P. Proceedings of fifth International Working Conference on Stored-Product Protection, Bordeaux, France, September 1990, 3, 2039–2047.

White, N.D.G. 1994. Insects, mites and insecticides in stored grain ecosystems. In : Jayas, D.S., White, N.D.G., Muir, W.E. and Sinha, R.N., ed., Proceedings of International Symposium on Stored Grain Ecosystems, Winnipeg, Canada, June 1992, in press.

Acoustical monitoring of stored-grain insects: an automated system[1]

D.W. Hagstrum[*], P.W. Flinn[*] and D.Shuman[†]

Abstract

An automated system for monitoring insect populations in stored grain with acoustical sensors was tested in six bins storing 65 to 110 t of newly-harvested wheat on four farms in Kansas during 1992 and 1993. During both years, sounds were detected more frequently as insect density increased during the storage period and the acoustical sensors detected insects 16 to 31 days earlier than grain trier samples.

The number of times that acoustical sensors detected insects was correlated with insect densities in grain samples over the range of 0.5 to 7.5 insects per kilogram of grain. Acoustical detection increased by one each time insect density increased by 0.305 insects per kilogram of grain. The correlation between acoustical detection and insect density will enable us to estimate insect density without taking grain samples.

Insects were detected at only 5 to 15 of the 56 sensor locations at which grain samples were taken. These locations were generally near the grain surface in the centre of the bin. Acoustical sensors at these locations should provide the most effective insect monitoring in farm bins.

Introduction

Research on the use of acoustical sensors to monitor stored-grain insect populations has been reviewed (Hagstrum 1991). Acoustical sensors were first used to detect insect larvae feeding inside kernels of grain. The most accurate system for detecting larvae inside kernels in large grain samples, as a method for grain grading, counts the number of locations in a kilogram of wheat at which insect sounds are detected (Shuman et al. 1993). Another approach has been to probe or install cables with acoustical sensors instead of taking grain samples. An automated system that uses acoustical sensors on cables to monitor stored-grain insect populations has been developed and shown to be effective in laboratory tests (Hagstrum et al. 1991).

We report here the results of the first two years of a three-year field study which compares the effectiveness of this automated system in monitoring insect populations with that of grain trier samples.

Materials and Methods

Our automated insect detection system was field tested in six bins storing 65 to 110 t of newly-harvested wheat on four farms in Kansas during 1992 and 1993. Seven flexible cables were installed vertically in the grain mass along a transect across the diameter of the bin. The cables to either side of the centre cable were 30 cm from centre and the remaining cables were 60 cm from the nearest cable. The 3 m of each cable that were in the grain had 20 sensors (MuRata PKM28-2AO, Smyrna, Georgia) 15 cm apart. The cables were 21 m long and connected to electronic instrumentation located in a small trailer next to the bins. The instrumentation monitored sensors in two bins. The signal from sensors was amplified 10 000 times (Bruel and Kjaer Model 2610, Marlborough, Massachusetts), filtered (Krohn-Hite Model 3700 variable band-pass filter, Avon, Massachusetts), and the number of voltage spikes above a 0.22 volt trigger level during a 10-second interval were counted (Hewlett-Packard universal counter Model 5316A, Wichita, Kansas). An IBM-compatible computer stored the data, reset the counter after making the count for each sensor and controlled switching (Hewlett-Packard switch /test unit Model 3235) to the next sensor. The system reads each sensor for 10 seconds 27 times per day during the first three to four months of storage.

Grain samples were taken every two weeks from in front of each sensor in the top 1.2 m of grain with a compartmented grain trier. The insects in each grain sample were counted and these numbers were compared with the average number of 10 second intervals per day in which sounds were detected during the five-day period prior to sampling.

Results

Insect densities were highest near the top of the grain mass in the centre. In this region, the number of intervals in which insect sounds were detected began to increase on day 39, but the first insect was not detected in grain samples until day 70, 31 days later (Fig. 1). The insect species present in this bin included *Rhyzopertha dominica* (F.), *Tribolium castaneum* (Herbst) and *Sitophilus oryzae* (L.). On days 39, 44, 48, 54 and 57 and from day 67 to day 78, there were significantly fewer (p < 0.05) 10-second intervals without sounds at the sensor locations where grain trier samples confirmed the presence of insects than at the locations where they did not (Table 1). For the days in Figure 1 that are not shown in Table 1, differences were not significant between day 15 and day 34, and were significant between day 79 and day 125.

In a second bin, the number of intervals with insect sounds began to increase by day 44 and insects were first detected in grain samples on day 60, 16 days later. The shorter time between first acoustical detection and first detection with grain samples in the second bin may be due to only *R. dominica* being found in this bin. This species is more difficult to detect with acoustical sensors (Hagstrum and Flinn 1993). In the second bin after day 44, the numbers of intervals with insect sounds were consistently significantly different

---

[1] This article reports the results of research only. Mention of a proprietary product or pesticide does not constitute an endorsement or a recommendation for its use by USDA.

[*] U.S. Grain Marketing Research Laboratory, Agricultural Research Service, United States Department of Agriculture, 1515 College Avenue, Manhattan, Kansas 66502, USA.

[†] Insect Attractants, Behaviour and Basic Biology Laboratory, Agricultural Research Service, United States Department of Agriculture, P.O. Box 14565, Gainesville, Florida 32604, USA.

Table 1. Statistical comparison of detection of sounds at locations where insects were found in grain samples to that at other locations.

Day of storage	Number of intervals without sounds				t	Probability
	Insects present in grain samples		Insects absent from grain samples			
	Mean	SD	Mean	SD		
35	22.0	5.6	23.5	1.0	0.70	0.51463
36	23.0	2.0	23.9	0.9	1.09	0.32503
37	24.0	2.0	24.6	2.3	0.52	0.62305
38	22.2	1.5	23.0	0.9	1.23	0.27214
39	22.6	1.7	25.2	0.7	3.81	0.01249
40	20.8	1.8	22.0	1.7	1.25	0.26501
41	22.5	2.5	25.3	0.5	2.96	0.03143
42	22.8	2.4	23.9	1.5	1.05	0.34096
43	23.3	3.5	25.5	0.5	1.50	0.19428
44	18.6	5.4	25.7	0.0	3.22	0.02345
45	21.3	4.3	23.3	3.7	0.70	0.51290
46	23.4	1.3	23.8	1.5	0.37	0.72532
47	24.0	1.7	23.9	2.2	−0.05	0.96446
48	22.0	1.7	24.7	1.1	3.26	0.02251
49	20.4	5.7	22.2	1.4	0.71	0.50887
53	18.8	8.8	25.6	0.6	1.88	0.11891
54	17.9	6.3	25.6	0.5	3.24	0.02300
55	18.8	6.8	24.7	1.0	2.23	0.07574
56	15.7	6.8	20.7	1.7	1.88	0.11890
57	16.0	9.0	25.1	1.7	2.57	0.04982
58	15.8	6.0	22.9	5.1	2.07	0.09275
59	15.4	7.9	24.2	3.3	2.54	0.05203
60	15.6	6.9	22.5	4.5	2.06	0.09412
61	11.9	7.6	19.1	6.4	1.71	0.14811
62	12.9	8.4	20.8	7.2	1.69	0.15097
63	11.7	5.4	19.6	6.5	1.72	0.14570
66	12.2	8.7	20.6	5.7	1.87	0.12060
67	10.7	8.0	20.3	3.8	2.66	0.04497
68	9.3	6.0	21.8	3.6	4.36	0.00732
69	8.0	5.9	19.7	3.7	4.10	0.00936
70	8.7	5.2	21.8	3.1	4.93	0.00435
71	6.7	4.8	20.4	4.2	4.62	0.00572
72	3.6	3.3	17.5	5.2	5.36	0.00305
73	1.9	1.3	17.2	2.7	12.36	0.00006
74	2.7	1.6	19.6	4.4	8.61	0.00035
75	2.2	1.4	19.0	3.5	10.40	0.00014
76	1.7	0.5	18.9	5.0	8.37	0.00040
77	3.9	1.7	18.6	6.1	5.55	0.00262
78	4.3	3.4	18.8	4.2	4.88	0.00454

between the sensor locations with and those without insects in grain samples.

The number of times that acoustical sensors detected insects was correlated with the number of insects in grain samples over the range of 0.5–7.5 insects/kg grain (Fig. 2). Acoustical detection increased by one each time insect density increased by 0.305 insects/kg grain and the intercept of the regression equation was not significantly different from zero (p=0.96). The correlation (r^2=0.53) between acoustical detection and insect density will enable us to estimate insect density with the automated insect monitoring system. The confidence intervals indicate that insect density can generally be estimated within plus or minus one or two insects.

In four bins, the number of sensor locations at which insects were found ranged from 5 to 15 of the 56 sensor locations at which grain samples were taken. Acoustical sensors at these locations should provide the most effective insect monitoring in farm bins.

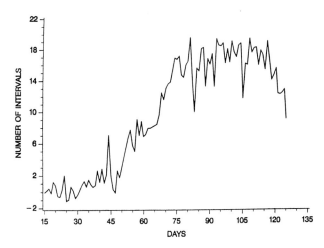

Fig. 1. Increase during the wheat storage period in the number of time intervals in which insects sounds are detected. The number of intervals with insect sounds is calculated by subtracting the number of intervals without sounds at locations where insects are present in grain samples from the number of intervals without sounds at locations where insects are absent from grain samples.

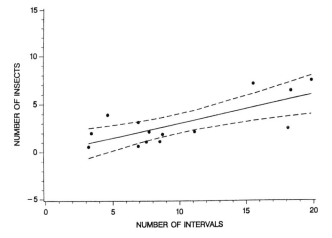

Fig. 2. Correlation between the number of insects (y) and the number of time intervals (x) in which insect sounds are detected where y=0.305x–0.048 with r^2=0.53. The dashed lines show the 95% confidence limits.

Discussion

The automatic monitoring system for stored-grain insects which estimates insect densities from the number of times that insect sounds are detected will reduce the labour required for insect monitoring and will improve pest management by pro-

viding more accurate and up-to-date information about insect infestations. Automation is needed because large numbers of samples are generally needed to detect stored-grain insects (Hagstrum and Flinn 1992). The lower the insect density the larger the number of samples needed to be 95% confident of detection. With standard sampling procedure, only one sample per 27 t of grain is needed with a mean insect density of 6 insects/kg grain but >100 samples are needed with a density of 0.02 insects/kg grain. Because insects have a high population growth rate, in addition to sampling thoroughly, it is also necessary to sample frequently to be sure that infestations are discovered before insect populations reach unacceptable levels (Hagstrum and Flinn 1990).

Although the automated system was tested on farms to make scaling up from laboratory studies with 0.135 t lots of wheat more practical, it is more likely to be used at elevators. With the automated system, a computer in the main office could provide a list of bins that will need insect control. Insect population growth models can be used to forecast which bins will need insect control in one, two or three months (Hagstrum and Flinn 1990). This information could be useful in deciding which grain to sell first or in making sure that grain with different insect infestation levels is not combined to fill an order. By networking computers, we could follow a lot of grain as it moved through the marketing system and control insects at the most appropriate time.

References

Hagstrum, D.W. 1991. Automated acoustical detection of stored-grain insects and its potential in reducing insect problems. In: Fleurat-Lessard, F and Ducom, P., ed., Proceedings of the Fifth Interational. Working Conference on Stored-Product Protection. Bordeaux, France, 1341–1349.

Hagstrum, D.W. and Flinn, P.W. 1990. Simulations comparing insect species differences in response to wheat storage conditions and management practices. Journal of Economic Entomology 83, 2469–2475.

Hagstrum, D.W. and Flinn, P.W. 1992. Integrated pest management of stored-grain insects. In: Sauer, D.B. ed., Storage of cereal grains and their products. American Association of Cereal Chemists, St Paul, Minnesota, 535–562.

Hagstrum, D.W. and Flinn, P.W. 1993. Comparison of acoustical detection of several species of stored-grain beetles (Coleoptera: Curculionidae, Tenebrionidae, Bostrichidae, Cucujidae) over a range of temperatures. Journal of Economic Entomology 86, 1271–1278.

Hagstrum, D.W., Vick, K.W. and Flinn, P.W. 1991. Automatic monitoring of *Tribolium castaneum* populations in stored wheat with computerised acoustical detection system. Journal of Economic Entomology 84, 1604–1608.

Shuman, D., Coffelt, J.A., Vick, K.W. and Mankin, R.W. 1993. Quantitative acoustical detection of larvae feeding inside kernels of grain. Journal of Economic Entomology 86, 933–938.

Responses of *Tribolium castaneum* to different pheromone lures and traps in the laboratory

A. Hussain*, T.W. Phillips† and M.T. AliNiazee§

Abstract

Three different pheromone lure designs and three different trap designs were tested with *Tribolium castaneum* for behavioural response in the laboratory. Two-choice pitfall bioassays were used for lure studies in which beetles oriented to one of two holes in the floor, below which were placed stimulus or control materials. The test arena for trap studies was a large tray with one layer of wheat grains on the floor. Traps were placed randomly within the arena. All lures were least effective when new due to high release rate of pheromone. The membrane/reservoir design elicited very low beetle responses when newly removed from the package, but after five to six weeks of aging/aeration these baits elicited high response (50–60%). The rubber septum was attractive in weeks one to three but declined in activity thereafter. Laminated baits were consistently attractive from one to three weeks (60–65% response). Laminated baits were thus used for trap comparisons. Trapping studies compared corrugated cardboard/pitfall, circular inclined ramp pitfall, and rectangular inclined ramp pitfall trap designs. The rectangular inclined ramp pitfall trap was most effective and caught 25% of released beetles compared to 10% and 4% in cardboard and circular traps, respectively.

Introduction

Pheromones are effective monitoring tools for a number of stored-product pests (Burkholder and Ma 1985). Pheromone-based monitoring and detection requires that synthetic pheromone must be formulated into a controlled release device and used in a trap that is effective, durable, and serviceable. Optimum designs for both the trap and the controlled release device are primary objectives for pest management (Barak et al. 1990). Several designs for controlled release pheromone dispensers have been used for different insect species, each one giving characteristic release for specific pheromones (McDonough 1991). Traps of various designs have been developed for stored-product insects, but their efficacy for specific insect species has not been fully investigated.

Despite the early identification of 4,8-dimethyldecanal (DMD) as the aggregation pheromone for *Tribolium castaneum* (Suzuki 1980), limited work has been reported on the use of the pheromone in trapping systems. Barak and Burkholder (1984) describe a trap made of corrugated cardboard with an oil-filled pitfall cup that was effective for several

beetle species. They show that *T. castaneum* was caught in this trap, and eventually this design was patented and commercialised (Barak and Burkholder 1984). Controlled release pheromone lures provided with the cardboard traps are simply rubber septa impregnated with synthetic pheromone. Mullen (1992) reported a new trap design for *T. castaneum* and compared it with the cardboard trap and two other designs. The objectives of this study were to (a) evaluate the activity and longevity of various controlled dimethyldecanal release devices that employed three different release designs, and (b) test the effectiveness of several trap designs.

Methods and Materials

Insects

A laboratory colony of *Tribolium castaneum*, established in 1990 from beetles collected on a farm in Dane County, Wisconsin, was used in all experiments. Beetles were reared on a mixture of whole wheat flour and brewer's yeast (95:5, v:v) in a growth chamber maintained at $27 \pm 1°C$, 60% r.h. and a photoperiod of 16:8 (L:D) hours. Parent beetles were sifted from cultures one week after inoculation, and new adult progeny were removed for bioassay five to seven days after emergence.

Lure formulations and bioassay

Commercially produced lures described below were all formulated with the 4(R), 8(R,S) isomeric blend of DMD at >90% purity. The three general lure designs studied differed in their formulation and mechanism of release. These designs will be referred to as 'septum', 'membrane', and 'laminate' lures throughout this report. Septum lures were provided by Trécé Inc., Salinas, CA, and were composed simply of a red rubber septum (sleeve stopper type), 9.1×18.8 mm, onto which a solution of DMD had been applied (loading rate not provided). The pheromone is thus soaked into the rubber and is released slowly over time as a function of the rubber matrix. Membrane lures were from Consep Membranes Inc., Bend, OR, and contained DMD in a reservoir between an impermeable backing material and a plastic membrane through which the pheromone evaporated slowly as a function of membrane characteristics. Membrane lures were flattened rectangular devices that had a circular releasing surface and a reservoir loaded with 8.0 mg of DMD. Laminate lures, provided by Hercon Environmental Co., Emigsburg, PA, were a 'sandwich' design in which 1.0 mg of DMD was formulated into a PVC reservoir and placed between an impermeable Mylar bottom and a permeable PVC top piece. Pheromone release from the laminate lure is primarily through the top permeable layer, and its rate is a function of film thickness, film composition, and total lure dimension (i.e. area of release surface). Replicate samples of the lure designs were subjected to bioassay for activity at different ages (one to two-week intervals) following their removal from the packages. Maximum ages of lures tested were six weeks for septum lures, 12 weeks for

---

* Department of Entomology, University of Agriculture, Faisalabad, Pakistan.

† USDA ARS, Department of Entomology, University of Wisconsin, Madison, WI USA 53706.

§ Department of Entomology, Oregon State University, Corvallis, OR USA 97331.

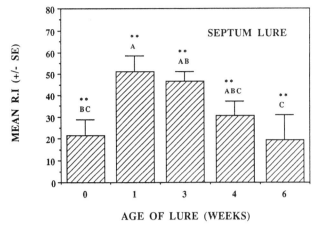

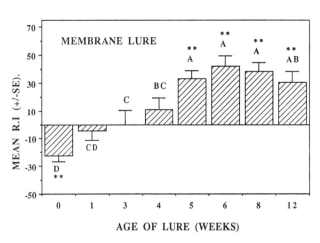

Fig. 1. Response of *T. castaneum* to rubber septum lures in two-choice pitfall bioassays. Histogram shows mean response index and standard error (N=10). The response index (RI) was calculated as RI = (T–C/Tot) × 100, for which T is the number responding to the treatment, C is the number responding to the control, and Tot is the total number of insects released. Mean RIs followed by different letters are significantly different (P<0.05, ANOVA and LSD). Significant difference between treatment and control within each age class are indicated as ** (P<0.01).

Fig. 2. Response of *T. castaneum* to membrane lures in a series of two-choice pitfall bioassay. Histograms show mean response index (RI, see above) and standard error (N=10). Mean RIs followed by different letters are significantly different (P<0.05, ANOVA and LSD). Significant difference between treatment and control within each age class are indicated as ** (P<0.01).

membrane lures, and 14 weeks for laminate lures. Individually identified lures were aged in a fume hood at room temperature between bioassays. A two-choice pitfall bioassay, in which beetles oriented to one of two holes in the floor of a steel can arena below which were placed stimulus or control materials (described by Phillips et al. 1993), was used to evaluate all lures in this study. For all bioassays the test lure was placed in the treatment dish and the control dish was left empty. Twenty adult *T. castaneum* were released in each can arena and given two hours to respond; ten replicates of each lure and age class were deployed.

Quantification of DMD released from laminate lures

Release of pheromone from laminate lures over time was determined by collecting volatiles from lures and subjecting these to quantitative GC analysis. Aeration chambers, collection procedures, and quantitative GC analyses were the same as those reported for live beetles by Hussain et al. (these proceedings). Five lures were aerated for one hour immediately after removal from their packages. After aeration the lures were aged in a fume hood and analysed again.

Trap designs

Three different trap designs were obtained from commercial suppliers, and two modifications of one of the designs was made, yielding the following five designs.

1. Storgard traps from Trécé Inc. (Salinas, CA), were made of corrugated cardboard and are 9 cm on a side when folded. Three out of four folds are punched out to fit in a small cup. Corrugations are oriented diagonally across each section and most flutes lead to the cup that contains oil (described by Barak and Burkholder 1984).

2. The modified storgard used the fundamental storgard, but had a bigger cup of 4 cm square and 1 cm high. Additionally, the orientation of the corrugations was changed from diagonal to perpendicular in relation to the trap sides, and the direction of corrugations was alternated for adjacent layers.

3. A second modification of the above traps, the rounded storgard trap, consisted of four circular, 4.5 cm diameter plates of corrugated cardboard. All four plates had 4 cm square punchouts. The two bottom circular plates fit the 4 cm square × 1 cm high cup and were mounted on a circular cardboard of the same size. The other two were used as a cover and punchout openings led into the cup. The two top and two bottom corrugated plates were glued together to make two main top and bottom pieces. The circular corrugated plates were placed on each other so that orientation of the corrugations was perpendicular.

4. The Fuji Trap (Japan Tobacco Co.) was a rectangular ramp-and-pitfall design, approximately 4×10×1 cm, constructed primarily from hard paper. Responding insects enter ramps on each end of the trap, climb the ramp to the edge of a 4×4×1 plastic cup. The opening for the cup is covered with cloth netting, onto which beetles crawl, but from which they eventually lose footing and fall onto an oil-soaked pad. The whole trap is covered with a plastic rectangular sleeve that provides support to hang a lure.

5. The Trécé 'Flit-Trak' (pronounced 'flight-track') trap was also a ramp-and-pitfall design and consisted of a wide-mouth plastic cup and an octagonal paper lid and base. The cup is of an inverted cone shape that is 10 cm diam. at the bottom, and from which a ramp rises at a 50° angle to a 5 cm diameter top. The ramp had a rough surface to facilitate insect crawling whereas the top edge was smooth so that insects would fall into the cup. The lure was glued to the upper portion of the lid.

Trap bioassays

All trap experiments were conducted in a 92×92×9 cm steel tray with one layer of whole wheat grains in the arena. A layer of grains provided footing as well as a natural environment to the responding beetles. The sides of the trays were coated with liquid Teflon to prevent insect escape. All trap experiments used laminate lures that had been aged 9–12 days in a fume hood. All pitfall traps contained 0.5 mL oil in their reservoirs. The oil was provided by Trécé Inc. and was a mixture of grain and mineral oils. Trécé oil presumably enhanced the attraction of beetles and served to kill the trapped beetles by suffocation. Only two types of traps were tested at a time, and each was placed randomly in the arena a

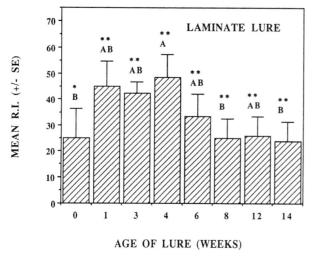

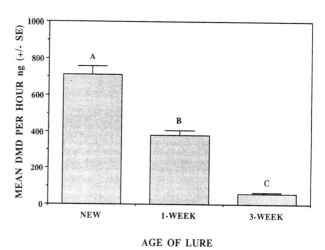

Fig. 3. Response of *T. castaneum* to laminate lures in a two-choice pitfall bioassay. Histograms show mean response index (RI, see above) and standard error (N=10). Mean RIs followed by different letters are significantly different (P<0.05, ANOVA and LSD). Significant difference between treatment and control within each age class are indicated as ** (P<0.01) and * (P<0.05).

Fig. 4. Amounts of pheromone released by laminate lures at different ages. Volatiles from lures were collected on super-Q and analysed with quantitative GC. Mean release rates with different letters are significantly different (P<0.05, ANOVA and LSD).

minimum of 15 cm from a side and 30 cm from another trap. Eight replicates of each pair-wise comparison were performed. One-week-old, mixed sex beetles were counted and kept without food for 30 minutes before the experiments. An inverted glass funnel of 2 cm diam. was placed in the centre of the tray arena and the test beetles were released into it. Fifty insects were kept confined under the inverted glass funnel for 15 minutes and then were released into the arena by lifting the glass funnel. After releasing the beetles in the arena the trays were covered with a screen to prevent any insect escape by flight. Bioassays were conducted for 20 hours in complete darkness at 27°C and 60 ± 10% r.h., after which the number of beetles caught in each trap was determined.

Results and Discussion

All commercial lure designs elicited significant attraction by *T. castaneum* under certain conditions, but response varied greatly with design and age of the lure. All of the lures exhibited low activity when first removed from their packages (0 week), and typically elicited maximum response at one or more weeks of age. Rubber septum lures (Fig. 1) showed highest activity at one week of age, and yielded similar responses at three and four weeks of age. Significantly lower responses to rubber septum lures were recorded at six weeks of age, and these were similar to the initial low response to new lures. Responses to the membrane lures were characteristically low when lures were first taken from the packages, and then showed increased activity with age. The membrane lure design tested here (Fig. 2) significantly repelled beetles when new and elicited no significant responses on weeks one, three and four, but were significantly attractive during later test periods. Membrane lures possessing different release characteristics were evaluated by Hussain (1993) and elicited responses similar to those reported here. Responses of *T. castaneum* to laminate lures revealed similar patterns over time as those found for membrane and septum lures (Fig. 3).

All ages of the laminate lure design tested (labelled 'slow' by the manufacturer) elicited significant attraction of *T. castaneum* compared to blank controls. Lowest responses to laminate lures were at week 0 and on weeks 6–14, and highest responses occurred during weeks 1–4 (Fig. 3). Additional designs of laminate lures examined by Hussain (1993) yielded less desirable response characteristics than the designs tested here. Amounts of DMD released over time from laminate lures used here were determined by GC analysis of collected volatiles (Fig. 4). It is clear that the new lures (0 week) that elicited low behavioural response were releasing much higher levels of pheromone than the older lures that were attractive to beetles.

Responses of *T. castaneum* to controlled release formulations of DMD are similar to what was observed for response to different doses of synthetic pheromone on filter paper discs (Hussain 1993). For that dose/response study, attraction was typically low at low doses, increased to some optimum level with increased doses, then decreased sharply at very high doses. The lure studies presented here reflect the same pattern as that for dose/response, but in reverse. Brand new lures (0 week) were presumably releasing at very high levels (e.g. Fig. 4), and thus were repellent, neutral, or slightly attractive. As release rate decreases with time the level of pheromone approaches an optimum level and attraction is highest. As release rate slows even more with age, attractive response diminishes. Loss of activity from DMD lures in these studies could also be caused by oxidation of the aldehyde group on the pheromone molecule. The ideal controlled release device should emit pheromone at a constant rate over a specified period of time (Roelofs 1979). Chemical analyses of volatiles from the laminate lure designs (Fig. 4) indicates release rate was not stable over the three-week test period, and behavioural responses of beetles reflect this variation in pheromone release.

A series of two-choice experiments were conducted with the five different traps in large tray arenas (Table 1). Initial experiments compared the original storgard design with the modified storgard design (bigger square cup, corrugations perpendicular to sides) and determined that the modified design captured significantly more beetles. A subsequent experiment (Expt 2, Table 1) determined that there was no difference in response of beetles to the square modified storgard and the rounded storgard. The Flit-Trak traps and Fuji traps each caught significantly more beetles than the Rounded Stor-

Table 1. Responses of *Tribolium castaneum* to various trap designs in large tray arena bioassays.

Exp. #	Traps tested	Mean % caught (±SE)[a]	Difference[b]
1	Storgard vs	11.50 + 2.38	**
	modified storgard	26.00 + 2.85	
2	Modified storgard	20.50 + 3.46	NS
	vs rounded storgard	24.50 + 3.89	
3	Flit-Trak	23.00 + 2.93	*
	vs rounded storgard	11.50 + 1.68	
4	Fuji	26.00 + 4.44	**
	vs rounded storgard	15.50 + 2.35	
5	Fuji	25.50 + 3.44	**
	vs Flit-Trak	12.50 + 2.60	

[a]Per cent of beetles caught in each trap after release of 50 beetles.
[b]Differences between traps determined by Student's t-tests on arcsin transformed percentages; **, P<0.01; *, P<0.05; NS, P>0.05; N=8.

gard traps (Table 1, Expts 3 and 4). When Fuji and Flit-Trak traps were compared in the same experiments the Fuji trap captured significantly more beetles (Table 1, Expt 5). Thus, of the two ramp style traps tested, it appears that the Fuji trap, which employs a straight cardboard ramp at each end, is more effective at trapping *T. castaneum* than the Flit-Trak trap, which employs a circular plastic ramp.

It is clear from these studies that formulation and release rate of pheromone is very important for response of *T. castaneum*. Most commercial pheromone lures tested elicited poor responses when first removed from their packages, but yielded positive responses by beetles after some period of aging. This 'burst' effect of slow release pheromone devices is commonly noted (McDonough 1991). A large amount of pheromone is presumably released early in the life of the device as the transfer matrix becomes saturated with pheromone and release equilibrates. Our studies also indicate that trap design can be very important in capture of *T. castaneum*. The success of the ramp-pitfall designs may relate directly to a natural tendency of beetles to be negatively geotropic (Mullen 1992). Additionally, ramp traps may allow for more unimpeded release of pheromone from the lure compared to cardboard traps, in which there is not a direct path for escape of the pheromone from the lure to the outside of the trap. These studies provide a framework for future studies that should address efficacy of lures and traps in field situations.

Acknowledgments

We appreciate reviews of the manuscript by Robert J. Bartelt, USDA ARS in Peoria, IL, and Joel K. Phillips, USDA ARS in Madison, WI. Research was partially funded by a training grant from USAID. We appreciate travel support for AH provided by Insects Limited Inc., Indianapolis, IN, and Consep Membranes Inc., Bend, OR.

References

Barak, A. V. and Burkholder, W. E. 1984. A versatile and effective trap for detecting and monitoring stored-product Coleoptera. Agriculture, Ecosystems and Environment 12, 207–208.

Barak, A. V., Burkholder, W. E. and Faustini, D. L . 1990. Factors effecting the design of traps for stored-products insects. Journal of the Kansas Entomological Society 63, 466–485.

Burkholder, W.E. and Ma, M. 1985. Pheromone for monitoring and control of stored product insects. Annual Review of Entomology 30, 257–272.

Hussain, A. 1993. Chemical ecology of *Tribolium castaneum* Herbst (Coleoptera: Tenebrionidae): factors affecting biology and application of pheromone. Ph. D. Dissertation. Corvallis, Oregon, Oregon State University, 119 pp.

Hussain, A., Phillips, T.W., Mayhew, T.J. and AliNiazee, M.T. These Proceedings. Pheromone biology and factors affecting its production in *Tribolium castaneum*.

McDonough, D.L. 1991. Controlled release of insect sex pheromone from a natural rubber substrate. In: Hedin, P.A. ed., Naturally occurring pest bioregulators. ACS Symposium Series 449, 106–125.

Mullen, M.A. 1992. Development of a pheromone trap for monitoring *Tribolium castaneum*. Journal of Stored Products Research 28, 245–249.

Phillips, T.W., X.-L. Jiang, W.E. Burkholder, J.K. Phillips, and H. Tran-Quoc. 1993. Behavioral responses to food volatiles by ecologically different stored-product Coleoptera, *Sitophilus oryzae* (Curculionidae) and *Tribolium castaneum* (Tenebrionidae) Journal of Chemical Ecology 19, 723–734.

Roelofs, W.L. 1979. Establishing efficacy of sex attractants and disruptants for insect control. Entomological Society of America 97 pp.

Suzuki, T. 1980. 4, 8-Dimethyldecanal: the aggregation pheromone of the flour beetles, *Tribolium castaneum* and *T. confusum* (Coleoptera: Tenebrionidae). Agricultural Biology and Chemistry 44, 2519–2520.

Response of *Prostephanus truncatus* and *Teretriosoma nigrescens* to pheromone-baited flight traps

G.E. Key*, B.J. Tigar[†], E. Flores-Sanchez[†], and M. Vazquez-Arista[†]

Abstract

The behaviour of *Prostephanus truncatus* (Horn) was observed in the vicinity of a source of synthetic aggregation-pheromone. These observations are discussed in relation to trap design and use. Pheromone-baited sticky traps, in various configurations, and funnel traps were evaluated for their efficiency at catching both *P. truncatus* and its predator, *Teretriosoma nigrescens* Lewis under field conditions in Mexico. The funnel trap caught most *P. truncatus* and *T. nigrescens* while delta traps caught least. However, using unfolded delta traps as flat, sticky sheets increased their catch and when two traps were deployed to make a double-sided sticky trap the number of *P. truncatus* caught was similar to that in the funnel traps.

Single-sided sticky traps were also used to investigate the effect of height of traps and of surrounding vegetation on catch. Both *P. truncatus* and *T. nigrescens* were caught in traps placed at heights of 0 m, 1 m and 2 m with a tendency towards higher catches of both species at 0 m and 1 m, which was significant when surrounding vegetation was also 0–1 m high.

Introduction

Prostephanus truncatus (Horn) (Coleoptera:Bostrichidae) is a primary pest of farm-stored maize and causes sporadic but locally important damage in Central America and Mexico (Mesoamerica) (Böye 1988; Giles and Leon 1975; Hoppe 1986). It was introduced accidentally into east and west Africa where it has become a serious pest to both stored maize and dried cassava (Golob and Hodges 1982; Hodges et al. 1985). *Teretriosoma nigrescens* Lewis (Coleoptera:Histeridae) is a predator of *P. truncatus* and occurs in close association with it in Mesoamerica and shows a kairomonal response to its aggregation pheromone (Böye et al. 1988; Rees et al. 1990). The predator has recently been released into both west and east Africa (GTZ unpublished; NRI unpublished).

Pheromone trapping is the main technique available for monitoring the distribution and relative abundance of *P. truncatus* and *T. nigrescens* but there is little information available to help interpret trap catches. The most commonly used traps are of the 'delta' type, which for convenience are normally hung between 1 and 2 m above the ground at the approximate height of maize cobs (Dendy et al. 1989a; Rees et al. 1990). Dendy et al. (1989a) concluded that *P. truncatus*

flies directly to the pheromone and does not attempt to walk on the sticky surface. However, observations made while trapping *P. truncatus* in Mexico indicated that they occur on the vegetation surrounding the trap and on the outer surfaces of the trap itself. The aims of the trials reported here were to examine the relative efficiency of different trap designs and their deployment in relation to the behavioural response of insects, so that recommendations may be made for monitoring both species in the field.

Methods and Materials

All the trials were undertaken the vicinity of the Irapuato unit of the Centro de Investigaciones y Estudios Avanzados de I.P.N. (CINVESTAV) in Guanajuata state, central Mexico. Polythene pheromone vials (supplied by the Natural Resources Institute (NRI), Chatham, U.K.) measuring 24 mm $\times$ 8 mm and impregnated with 2 mg of Trunc-call 1 and 2 in the ratio of 1:1 by weight were used throughout.

All data were analysed using the multivariate analysis of variance (ANOVA) and multiple regression analysis.

Behavioural response to pheromones

A landing target constructed from a sheet of white card (1.0 $\times$ 1.4 m) was marked with a 10 cm^2 grid and hung vertically from an external wall, 15 cm above the ground. A 5 cm^2 square marked the centre to which a pheromone vial was attached.

For three consecutive days the target was monitored at intervals from 16.20–20.17 hours, for insect arrivals. For each insect the following were noted: position of landing, the time of landing and the time to reach the centre or to take off. On 13 occasions insects were removed because too many were present to permit accurate tracking. Records of less than one minute were given a value of 0. 5 minutes. Weather conditions were uniformly warm and still during the observations.

Influence of trap design on catch

In a series of three trials the following trap types were tested.
1. Delta I. Pink bollworm delta traps, with a sticky surface of approximately 270 cm^2 (supplied by Agrisense Ltd., Pontypridd, U.K.) were suspended from a single tie at the top centre.
2. Delta II. As for delta I but suspended from additional ties at both ends to prevent rotation in the wind.
3. Single. A single-sided sticky trap made by unfolding a delta I trap, the sheet was suspended lengthways and weighted below.
4. Double. Similar to the single trap except made by fixing two unfolded delta I traps back to back.
5. Cylinder. Consisted of a delta I trap folded to form a cylinder with the sticky side facing outwards, hung vertically and secured at one side to prevent spinning.

* Manchester Metropolitan University, Biological Sciences, John Dalton Building, Chester Street, Manchester, MI 5GD, U.K.
† CINVESTAV/NRI Larger Grain Borer Project, Centro de Investigaciones y Estudios Avanzados de I.P.N., Apartado Postal 629, Irapuato, Gto, Mexico.

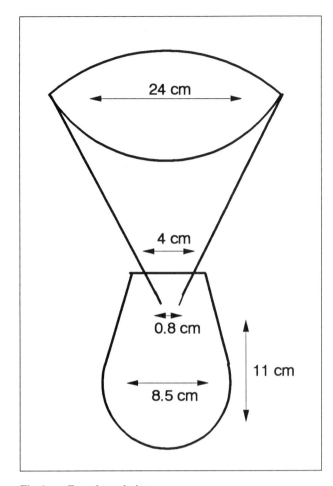

Fig. 1.   Funnel trap design.

6. Funnel. Similar to that described by Dendy et al. (1989b) but having a plastic, barrel-shaped collecting vessel into which the funnel is inserted (Fig. 1). The funnel was constructed from a rolled acetate sheet, held in place with a metal rivet, with an elliptical shaped opening having the front edge lower than the back to improve access to flying beetles. A piece of tissue paper was inserted into the collecting vessel to provide a foot-hold for captured insects and to deter them from boring out of the plastic container.

Each trap was baited with a standard pheromone vial, already described; pheromones were tied to the traps to prevent loss and stuck to the centre panel of the delta I, single and double traps, or hung inside the funnel at the back.

First trial

All traps except delta II were tested in this trial. Four replicates of each design were suspended at a height of 1 m from trees and bushes, 100 m apart, around four fields some 55 km from CINVESTAV. At the time of the trial, asparagus, sorghum and maize were being cultivated at the site. Initial trap sequence was randomised and traps were rotated by one position each day. Pheromone vials were dedicated to trap position and were not rotated with the traps. This was intended to obviate any effects of position and variation in the pheromone vials. Catch was recorded daily for four consecutive days. Delta I, single and double traps were replaced by fresh traps each day, while funnel traps were emptied and re-hung with a clean piece of tissue paper.

Second trial

Similar to the first trial, but undertaken over five consecutive days in a field of immature sorghum adjoining

CINVESTAV and only using delta I and delta II traps hung at a height of 1 m above the ground from cane tripods placed equi-distant (100 m) around the perimeter of the field.

Third trial

Similar to the second trial but using four replicates of delta II, single and cylinder traps.

Influence of trap height on catch

Three pheromone-baited, single-sided sticky traps (see earlier) were hung from the same vertical pole with the base of the trap at 0 m, 1, or 2 m up the pole. Traps were changed daily for four or five days with the pheromone remaining at the same position throughout. Four replicates were established in fields as follows:

1. A maize field with the crop at the 'milky' stage, plants about 3 m tall and cobs occurring at 1 to 2 m. Replicates were placed 100 m apart.
2. A field of maturing sorghum with panicles at between 1 and 1.25 m. Replicates were placed 150 m apart.
3. A field of grassy turf, approximately 5–10 cm high, where the nearest tall objects, e.g. bushes, trees and walls, were at least 5 m from the traps. Replicates were placed about 60 m apart.

Results

Behavioural response to pheromones

There were 114 landings of *P. truncatus* and two of *T. nigrescens* on the landing target. *P. truncatus* reached the centre on 24 occasions but took off before reaching the centre 77 times.

Mean time to reach the centre of the target was 7.48 ($\pm$7.97) minutes ($n = 21$), with a mean of 6.57 ($\pm$1.85) squares crossed. Mean time between landing and take-off was 2.02 ($\pm$2.4) minutes ($n = 76$), with a mean of 1.58 ($\pm$2.2) squares crossed. Data was extremely variable as shown by the standard deviations.

Only 2.1% of landings on the original target occurred within 20 cm of the centre with 85% at a distance of 50 and 70 cm. Many beetles landed along one side of the board (52.1% of landings) or along the lower edge (20.8% of landings).

General observations of *P. truncatus*, responding to the pheromone lures, showed that insects approaching the card frequently dashed themselves against it. Before landing insects tended to hover close to the card surface below the level of the pheromone for several seconds and then either landed directly or with an upward swoop. Those that approached the pheromone took a more or less straight path towards it, although in some cases the approach was interspersed with periods of inactivity. This suggests that even this close to the pheromone there was a concentration gradient for them to follow. On the two occasions when *T. nigrescens* was recorded, the insects walked directly to the pheromone, arriving within one minute of the first observation.

Influence of trap design on catch

Numbers of both species caught were found to be highly significant for trap type ($F(3,139) = 6.68$, $p < 0.001$ and $F(3,139) = 23.73$, $p < 0.0001$ for *P. truncatus* and *T. nigrescens* respectively). Figures 2 and 3 show the mean numbers of *P. truncatus* and *T. nigrescens* caught, respectively. Double and funnel traps caught the most *P. truncatus* and delta I (standard delta) the least. There was no significant difference between

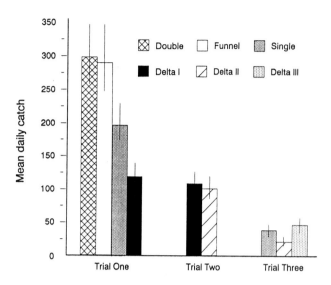

Fig. 2. Influence of trap design on catch of *Prostephanus truncatus* in three separate trials. Bars show means ±SE. For an explanation of trap design see text

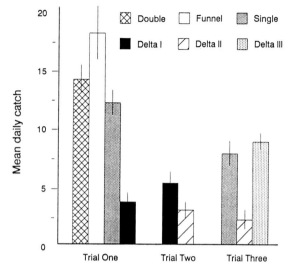

Fig. 3. Influence of trap design on catch of *Teretriosoma nigrescens* in three separate trials. Bars show means ±SE. For an explanation of trap design see text.

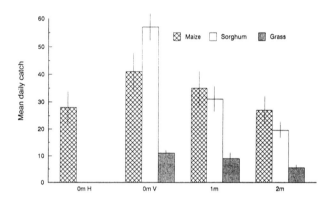

Fig. 4. Influence of trap height in various environments on catch of *Prostephanus truncatus*. Bars show means ±SE. H = horizontally and V = vertically-placed traps.

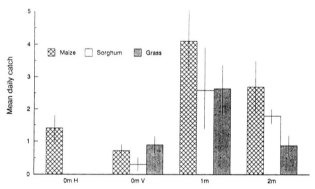

Fig. 5. Influence of trap height in various environments on catch of *Prostephanus truncatus*. Bars show means ±SE. H = horizontally and V = vertically-placed traps.

numbers of *P. truncatus* caught in delta I and delta II. Cylinder traps obtained catches similar to those of single traps.

Variation in catch was greater between than within days indicating that environmental conditions are an important factor influencing catch. In the third trial, mean numbers of both species were low in comparison to the first and second trials.

The double-sided sticky traps have a tendency to rotate and beetles may be caught on both sides, possibly influencing catch. Mean catch of *P. truncatus* was 137.9 (±133.4) from the side with a pheromone and 66.9 (±37.7) from that without a pheromone, and for *T. nigrescens* was 11.3 (±8.4) and 6.7 (±4.9) respectively.

Influence of height on catch

Figures 4 and 5 show catches of *P. truncatus* and *T. nigrescens* respectively at different heights in the three sites. A large proportion of catch variation is likely to have been caused by changes in environmental conditions throughout the experiment. Height of trap was significant for numbers of *P. truncatus* in sorghum ($F_{(2,45)}$ = 22.02, p<0.001) and

grassland ($F_{(2,57)}$ = 5.85, p < 0.01) but not in maize. In both sorghum and grassland, greatest catch was achieved at 0 m and least at 2 m with a similar trend in maize.

Although far fewer *T. nigrescens* were caught than *P. truncatus*, height of trap was also significant in all three trials ($F_{(3,60)}$ = 6.05, p < 0.01; $F_{(2,45)}$ = 8.45, p < 0.001; $F_{(2,57)}$ = 4.73, p < 0.05 in maize, sorghum and grassland respectively). Most were caught at 1 m in all three crops. Days of high catch of *T. nigrescens* did not necessarily coincide with those of *P. truncatus*.

Discussion

Behavioural observations

Observational data differ from those of Dendy et al. (1989a), in that insects were not found to fly directly to the pheromone source. The tendency for insects to land within 1 m of the pheromone and walk towards it explains those seen on vegetation in the vicinity of traps. However, catch was found to be higher when surrounding vegetation was at trap level, suggesting that repeated landing and taking-off in the vicinity of pheromone sources occurs.

The insects caught by the delta I and II traps were usually concentrated at the two entrances of such traps, indicating that insects land on the external surface and walk into the traps, although some may alight immediately on entering. A more complex Lingren funnel was used by Leos-Martinez et al. (1986) to capture the closely related bostrichid, *Rhyzopertha dominica* (F.), and was found to be superior to various other sticky traps, including a delta trap. This type of trap presents a lot of edges for landing which may influence its success.

A white card was used to provide a good contrast and to allow easy observation of the beetles. The sticky, external surfaces of the sticky traps were also white; however, the funnel trap had a clear funnel and a yellow base and the delta traps had a bright pink external surface. It is possible that colour and search image cues influence the results for both species, and the attraction of some scolytids and platypodids may be increased by using red traps.

Trap design

In all the trials described here, funnel traps consistently caught more, and standard delta traps fewer, beetles than all other traps. This differs from a trial carried out in Tanzania where the results indicated that funnel and delta designs caught similar numbers of *P. truncatus* (Dendy et al. 1989b). This could be due to a number of factors, including trap design, placement, trapping period and behavioural differences of the separate populations. The success of funnel traps may result from attracting insects to fly directly into the acetate funnel where they hit the sides and so fall into the collecting vessel. Traps with external sticky surfaces are more likely to catch those insects which dash themselves against the surface in the vicinity of the pheromone, compared with delta traps which only catch those which either land or walk into the internal sticky surface.

Trap height

In the sorghum and grassland trials maximum catch for both species was achieved in sticky traps near ground level, between 0 and 1 m. In the maize field catch was more varied, although more insects were caught between 0 and 1 m than 2 m, suggesting that this is the preferred height regardless of surrounding vegetation. For *P. truncatus* this may be explained by the tendency of insects to land on surrounding vegetation in the vicinity of traps so that in the maize more landing surfaces were available at all trap heights.

Relation between *P. truncatus* and *T. nigrescens*

No relation in trap catch was found between numbers of *T. nigrescens* and *P. truncatus*. In the trials on trap design, days when high numbers of *T. nigrescens* were caught did not correspond to those for *P. truncatus*. It is possible that their responses to concentration and ratio of components or environmental conditions are different. Observations in the field and laboratory indicate that *T. nigrescens* is more agile and active than *P. truncatus* which may affect the speed with which each species responds. Occasional peaks in the number of *T. nigrescens* have been observed during other trials in Mexico (National Resources Institute 1990) which may last between one and three days, during which time ratios of *T. nigrescens* to *P. truncatus* ranged from 1:9 to 1:35. The former is similar to ratios found in established laboratory populations of *T. nigrescens* (1:9–11) (Leliveldt and Laborious 1990) and

T. nigrescens caught from farm stores in Costa Rica (1:7.5) (Böye 1988).

Practical application

Observations indicated that a large, sticky surface would be a very efficient trap, but this would not be practical for a large-scale monitoring program. The delta I design used here, although less efficient, is readily available and proved easy to handle in the field with reasonable resistance to weathering. The funnel traps were custom-built and the materials from which they were constructed were readily available and made them inherently stronger than the delta traps. Once constructed they were quick to place in the field and had the advantage of catching live beetles of both species, although they tended to fill with water during rain, drowning the insects. It may be possible to improve their design by attaching a rain guard where live insects are required and by using a glass collection vessel to prevent boring. Rain did not appear to affect numbers caught by funnel traps nor did it appear to wash trapped insects off the exposed sticky surfaces of the single, double, or cylinder traps. However, the card from which these latter traps were constructed tends to disintegrate following repeated heavy rain, causing them to break and fall, particularly the single traps.

The difference in sensitivity between trap types has implications for both future experimental trials and in the monitoring of field populations. This is particularly so in Africa where monitoring for *P. truncatus* forms part of the containment and control programs of the pest, and where it is necessary to estimate the spread and abundance of *T. nigrescens* introduced into east and west Africa.

Acknowledgments

The authors would like to thank the director, Dr Ariel Alvares, for the use of the facilities at CINVESTAV; NRI Chatham for support; Dr Patrick Osborne for help with the design of the funnel traps; and Ricardo Vargas for technical assistance. We are also grateful to Arq. Gustavo Garcia, Dr Jorge Molina and La Escuela de Agronomia de la Universidad de Guanajuata for the use of their field sites. This work is part of an IPM initiative, Larger Grain Borer Program for Africa, of the Overseas Development Administration of the U.K. Government.

References

Natural Resources Institute 1990. Second Quarterly Report, IPMI Larger Grain Borer Control Program, Irapuato, Mexico. Natural Resources Institute, Chatham, U.K.

Böye, J. 1988. Autokologische untersuchengen zum verhalten des grossen kornbohrers *Prostephanus truncatus* (Horn) (Coleoptera; Bostrichidae) in Costa Rica. Doktorarbiet, Institut fur Phytopathologie der Christian-Albrechts-Universitat, Kiel. xiii + 195pp. (In German with English summary).

Böye, J., Burde, S., Keil, H., Laborius, G.A. and Schulz, F.A. 1988. The possibilities for biological integrated control of the larger grain borer *(Prostephanus truncatus* (Horn)) in Africa. In: G.G.M. Schulten and A.J. Toets eds, Workshop on the containment and control of the Larger Grain Borer. Arusha, Tanzania. Ministry of Agriculture and Livestock Development, Tanzania, with assistance from Food Agriculture Organisation of the United Nations, Rome, Italy. Report B Technical papers presented at workshop, 110–139.

Dendy J., Dobie, P., Saldi, J.A., Smith S.L. and Uronu, B. 1989a. Trapping the Larger Grain Borer *Prostephanus truncatus* in maize fields using synthetic pheromones. Entomologia Experimentalis et Applicata 50, 241–244.

Dendy, J., Dobie, P., Saidi, J.A. and Sherman, C. 1989b. The design of traps for monitoring the presence of *Prostephanus truncatus* in maize fields. Journal of Stored Products Research 25, 187–191.

Dick, K.M. 1990. Biological control of the larger grain borer in Africa: a component of an integrated pest management strategy. In:

Markham, R.H. and Herren, H.R. ed., Biological control of the larger grain borer. proceedings of an IITA/FAO Coordination Meeting. Cotonou, Republic of Benin, 2–3 June 1989, 5–15.

Giles, P.H. and Leon, O.S. 1975. Investigating the problems in farm–stored maize. In: Proceedings of the First Working Conference on Stored-Product Entomology. Savannah, USA, 1974, 68–76.

Golob, P. and Hodges, R.J. 1982. Study of an outbreak of *Prostephanus truncatus* (Horn) in Tanzania. Tropical Products Institute, Report G164. Typewritten, 23 p.

Hodges, R.J. 1986. The biology and control of *Prostephanus truncatus* (Horn) (Coleoptera:Bostrichidae) a destructive pest with an increasing range. Journal of Stored Products Research 21, 1–14.

Hodges, R.J., Meik, J. and Demon, H. 1985. Infestation of dried cassava *(Manihot esculenta* Crantz) by *Prostephanus truncatus* (Horn) (Coleoptera:Bostrichidae). Journal of Stored Products Research 21, 73–77.

Hoppe, T. 1986. Storage insects of basic food grains in Honduras. Tropical Science 26, 2538.

Leliveldt, B. and Laborius, G.A. 1990. Effectiveness and specificity of the antagonist *Teretriosoma nigrescens* Lewis (Col:Histeridae) on the larger grain borer *Prostephanus truncatus* (Horn) (Col:Bostrichidae). In: Markham, R.H. and Herren H.R., ed., Biological control of the larger grain borer. Proceedings of an IITA/FAO Coordination Meeting, Cotonou, Republic of Benin, 2–3 June 1989, 87–102.

Leos-Martinez J., Granovsky, T.A., Williams, H.J., Vinson S.B. and Burkholder, W.E. 1986. Estimation of aerial density of the lesser grain borer *Rhyzopertha dominica* (Coleoptera: Bostrichidae) in a warehouse using dominicalure traps. Journal of Economic Entomology 79, 1134–1138.

McFarlane, J.A. 1988. Pest management strategies for *Prostephanus truncatus* (Horn) (Coleoptera:Bostrichidae) as a pest of stored maize grain: present status and prospects. Tropical Pest Management 34, 21–32.

Rees, D.P. 1985. The life history of *Teretriosoma nigrescens* Lewis (Coleoptera:Histeridae) and its ability to suppress populations of *Prostephanus truncatus* (Horn) (Coleoptera:Bostrichidae). Journal of Stored Products Research 21, 115–118.

Rees, D.P. 1987. Laboratory studies on predation by *Teretriosoma nigrescens* Lewis (Coleoptera:Histeridae) on *Prostephanus truncatus* Horn (Coleoptera:Bostrichidae) infesting maize cobs in the presence of other maize pests. Journal of Stored Products Research 23, 191–195.

Rees, D.P., Rodriguez Rivera, R. and Herrara Rodriguez, F.J. 1990. Observations on the ecology of *Teretriosoma nigrescens* Lewis (Coleoptera:Histeridae) and its prey *Prostephanus truncatus* (Horn) (Coleoptera:Bostrichidae) in the Yucatan peninsular, Mexico. Tropical Science 30, 153–165.

Watters, F.L. 1984. Biology and control of *Prostephanus truncatus*. Proceedings of the GASCA Workshop on the larger grain borer, *Prostephanus truncatus*, 24–25 February 1983, TPI, Slough; Publ. GTZ, Eschborn, 49–61.

Development of immunoassays for quantitative detection of insects in stored products

G. B. Kitto*, F.A. Quinn* and W.E. Burkholder†

Abstract

Immunoassay procedures for the detection of insects and insect remains have been developed for several types of stored products. These tests, which are based on enzyme-linked immunosorbent assay (ELISA) procedures, couple excellent reproducibility with ease of use. One such assay is for use with whole grain and milled grain products. The test detects all major grain insect pests and gives a positive and linear response to larval and pupal life stages and adult insects. The assay is applicable to a wide variety of types of grain and milled grain products, including wheat, rice, barley, oats, and maize. A similar type of assay procedure has also been developed for the analysis of insect contamation of a broad variety of spices. Additional ELISA tests have been developed which are species specific. These provide information about the relative degrees of infestation by insect species that are either beneficial or harmful. Such species-specific assays are available for the granary weevil, *Sitophilus granarius*, and the kaphra beetle *Trogoderma granarium*, and for *Laelius pedatus*, *Bracon hebetor*, *Trichogramma pretiosum* and *Xylocoris flavipes*, which are used for biological control.

Introduction

The amount of insect contamination present in a stored-product sample provides an indication not just of the amount of contamination present at the time of testing, but also gives an indication of the quality of sanitation to which the material has been subjected in past storage and handling. Such testing additionally serves as a guide to the amount of insect infestation likely to be encountered in future storage and transport. For these reasons it is important to have test procedures which can reliably and quantitatively estimate the amount of insect material present in stored food product materials. It would be particularly convenient to have a test that could be applied across a very broad range of stored-food products and which was capable of detecting all the major stored-product insect species. It would also be desirable to have a single test which could be applied to both processed and unprocessed food products. Unfortunately, the array of tests presently available for detecting insect contamination are far from meeting these requirements. Most of these tests involve visual inspection elements, are highly variable and quite subjective. For example, the visual examination for live insects in whole grain

for export provides no direct information about the degree of internal infestation. Yet, if present, these internally developing insects are likely to cause very serious problems during transportation and storage. Similarly, the assessment of insect damaged kernels only provides an indication that insects were present at some time in the past and again gives no reliable indication of the number of internally developing insects (Russell 1988).

With processed food products, such as flour, the primary test for insect contamination is the insect fragment count (U.S. Food and Drug Administration 1988). This assay too, is fraught with difficulties, since it evaluates only the number of insect fragments present without regard to their size. Thus, it is only an indirect measure of the mass of insect material present. Moreover, the test itself is highly variable (Kurtz 1965) and is highly dependent upon the type and life stage of the insects present in the food before processing. It has been demonstrated that the dead insects yield many more fragments than do live ones, and that dead adult insects can contribute as much as 50 times as many fragments as a dead larvae (Sachdeva 1978). While the insect fragment test has been a long-term standard for food analysis, it is worth noting that the test requires lengthy technician training, is time consuming and relatively costly.

It was against this background that we set out to develop modern biochemical assays, based on the immunological recognition of insect material in stored products. The aim was to develop assay procedures that combine a high degree of accuracy and repeatability with the requisite degree of sensitivity. Another goal was to devise tests that could be used with minimal training and at a variety of stored-food product handling and storage sites and which would be of low cost. One general type of procedure which has the potential of attaining these goals is the enzyme-linked immunosorbent assay (ELISA). This type of assay is used extensively in clinical diagnostics and is seeing increasing use in agriculture fields, such as the testing for pesticides (Mei et al. 1991) and aflatoxins (Vanderlaan et al. 1988).

One common feature required of all tests for insect contamination of stored-food products is that they be capable of detecting minute quantities of insect material in the presence of very large amounts of background material. The high selectivity of immunological assays makes them well-suited to this purpose. An overview of the immunoassay procedures that we have developed for insects in stored food products is shown in Figure 1. A measured sample of the foodstuff is briefly blended with an extraction fluid, in a common household blender, to solubilise any insect material present. An aliquot of this sample is then used in the ELISA test. Typically this test is carried out in plastic multi-well strips or plates. After the sequential addition of several reagents to the wells, colour develops in proportion to the amount of insect material present in the sample. The amount of colour in the wells is measured in a colorimetric ELISA reader and the data can either be printed out directly or stored for further analysis and manipulation in a computer database.

* Department of Chemistry and Biochemistry, The University of Texas at Austin, Austin, Texas 78712, USA.

† USDA/ARS, Department of Entomology, University of Wisconsin, Madison, WI 53706, USA.

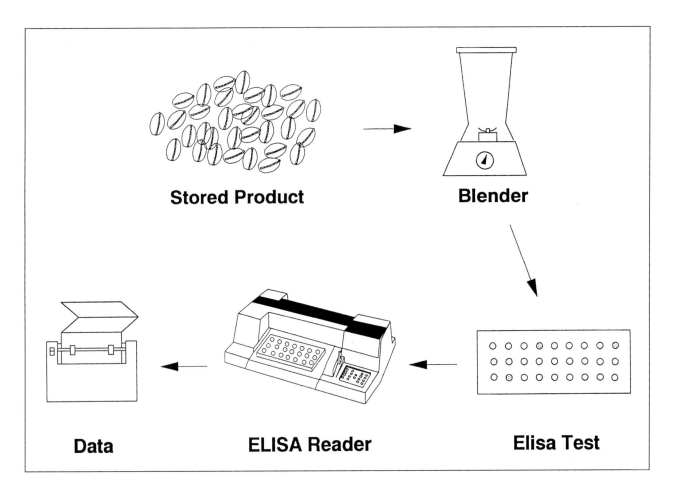

Fig. 1. An overview of the immunoassay procedure for the detection of insects, or insect remains, in stored products.

Development of Immunoassays for Insects in Whole and Milled Grain

For detecting grain insects by immunoassays, a suitable insect material to detect (antigen) had to be selected. Such an antigen would ideally be present in all of the common grain insects in substantial quantities and in all the major life stages, be easy to extract from whole grain and processed grain samples and differ little from insect to insect. The insect muscle protein myosin was chosen for immunoassay development since it meets the above criteria. This protein is present in all insect life stages except eggs, differs little in structure between all the common grain insect pests and is readily extracted using high salt solutions.

The general features of the ELISA assay that we developed using myosin as the test material are shown in Figure 2. Briefly, the walls of the wells of a plastic microtiter plate are precoated with rabbit polyclonal antibodies. When an extract of the grain or flour is added, any insect material present is selectively bound to the antibody coating the wells. After extraneous plant material is washed away, a second antibody, which is conjugated to an enzyme, is added to the wells. This second antibody also binds selectively to any insect myosin material present. After rinsing, substrate for the bound enzyme is added and colour subsequently develops in proportion to the amount of myosin present. The amount of myosin detected correlates well to the mass of insect material present. The mass of insect material present in turn correlates well to the number of insects present. A typical assay using this procedure is shown in Figure 3. In this case, samples of clean grain were spiked with the indicated number of granary weevils. The data indicate not only the linearity of the assay, in

this case showing a coefficient of variation $r = 0.97$, but also the sensitivity of the assay which is readily capable of detecting as little as one granary weevil per 50 g sample of grain. The myosin ELISA assay has been shown to give positive linear results with all of the grain insect pests listed in Table 1, which represents a majority of the insects found in stored-grain products. The assay has been tested with all of the major wheat varieties as well as with barley, oats, rice, maize, soybeans, and sorghum and has been found to give a linear response to insect contamination in all cases (Kitto 1991; Quinn et al. 1992). Some slight variations in extraction procedures are required for specific types of grain. The assay has been shown to be highly reproducible, as indicated in Figure 4 which shows multiple replicate samples tested on a single day. In this case, the average reading was 0.975 with a standard derivation of 0.02.

The high degree of reproducibility of the myosin immunoassay was also established by assaying samples of grain of varying levels of insect contamination, obtained from grain mills across the United States, and assaying them repeatedly over several days. An example of this type of ruggedness testing of the assay is shown in Figure 5, using a series of hard red winter wheat samples which were assayed either two or three times with a high degree of reproducibility. The ability of the immunoassay to accurately assess the degree of internal infestation was also examined, in collaborative trials carried out with the U.S. Department of Agriculture. In these studies, which were carried out in a blind fashion, an excellent correlation was observed between the number of internally infesting granary weevils present, as determined by x-ray analysis, and the response in the immunoassay procedure (Schatzki et al. 1993).

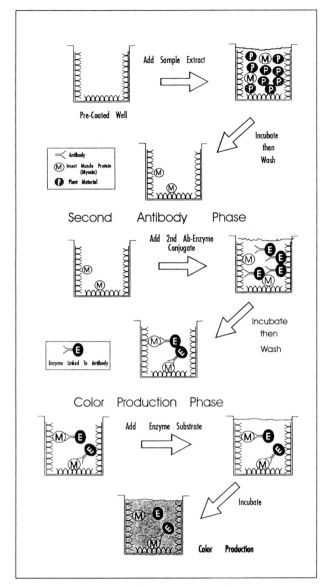

Fig. 2. The myosin sandwich ELISA procedure for quantitative detection of grain insects in stored products.

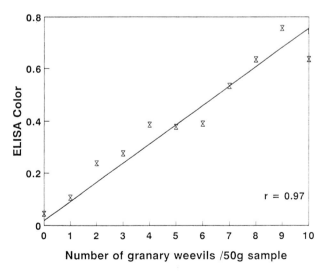

Fig. 3. Correlation between the amount of colour produced in the myosin sandwich ELISA (absorbance at 414 nm) and the number of granary weevils present in 50 g samples of hard red winter wheat. Points represent the average of three determinations.

r = 0.97

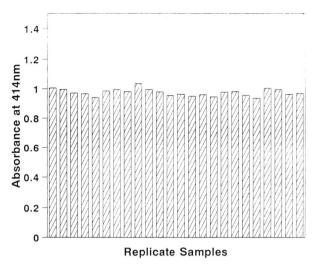

Fig. 4. Replicate samples assayed using the insect myosin sandwich ELISA procedure.

Table 1. Grain pests detectable using ELISA assay

Sitophilus zeamais	Maize weevil
Sitophilus granarius	Granary weevil
Sitophilus oryzae	Rice weevil
Trogoderma variabile	Warehouse beetle
Trogoderma glabrum	glabrous cabinet beetle
Tribolium castaneum	red flour beetle
Tribolium confusum	confused flour beetle
Oryzaephilus mercator	merchant grain beetle
Oryzaephilus surinamensis	saw toothed grain beetle
Rhyzopertha dominica	lesser grain borer
Prostephanus truncatus	large grain borer
Lasioderma serricorne	cigarette beetle
Stegobium paniceum	drugstore beetle
Callosobruchus maculatus	cowpea weevil
Attagenus megatoma	black carpet beetle
Alphitobius diaperinus	lesser mealworm
Plodia interpunctella	Indianmeal moth

Of particular significance is the fact that the immunoassay procedure can be applied equally well to either whole grain or milled grain materials. As shown in Figure 6, the total amount of insect material found in the flour, shorts, and bran fractions of a sample of milled wheat totalled the amount of insect material present in the original whole grain sample. This means that the immunoassay procedure can be used to accurately predict, by analysis of incoming whole grain, what amount of insect material will be present in the finished product. It should be noted that with the immunoassay procedure there is no difficulty in quantifying the amount of insect material present in the shorts and bran fractions which are extremely difficult to assay reliably using insect fragment counts.

Because of the accuracy and reliability of the immunoassay procedure and its ability to predict flour quality from whole grain analysis, this allows for the test to be used to quantitatively blend grain samples of differing levels of insect contamination to produce finished products of a desired

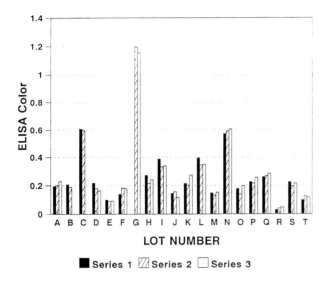

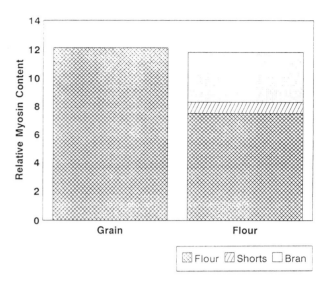

Fig. 5. Reproducibility of the myosin insect ELISA procedure using mill supplied grain samples with varying degrees of insect contamination. Samples of the grain were extracted and assayed separately either two or three times over a five-day period.

Fig. 6. The amount of insect myosin present in whole grain (hard red winter wheat) and in the flour, shorts and bran fractions milled from this grain, as assayed by the myosin ELISA procedure.

quality level. An example of the use of the myosin immunoassay for such blending is shown in Figure 7.

The microwell procedure described above is well suited for use in small laboratories, requires only about two days training, and is capable of evaluating approximately 20 samples in a two-hour period using our current methodologies. It would also be very useful if a simpler and quicker test was available for on-site testing of incoming grain and outgoing product at shipping docks and for checking the quality of export goods. For these reasons, we are engaged in developing a rapid, test-strip assay which can give quantitative results in a 15 minutes. An enhanced sensitivity immunoassay, using a biotin/streptavidin ELISA procedure, and which offers an approximately ten-fold enhancement over the regular ELISA is available for situations where detection of very low levels of insect contamination is necessary (I. Heller and G.B. Kitto, unpublished results).

Development of a Species-Specific Grain Insect Immunoassay

In addition to having available rapid tests which can accurately measure the total degree of insect contamination in stored-food products, it would be useful to have available species-specific tests which could provide information about the relative degrees of infestation by different insect pests. This type of information could be useful in selecting the most appropriate type of pesticide to be used on the stored product, or for providing feedback to the growers for more effective control regimes. As a model for this type of approach, a monoclonal antibody-based procedure has been developed for the quantitative detection of *Sitophilus granarius* (Chen and Kitto 1993). As illustrated by the example shown in Figure 8, this species-specific assay can accurately determine the degree of infestation caused by the granary weevil in a mixed insect population. In this particular case, a hard red winter wheat sample was spiked with the indicated number and variety of insects. The total degree of insect infestation was measured using the insect myosin ELISA, while the contribution of granary weevils to this total was estimated by the monoclonal antibody assay. The *S. granarius*-specific ELISA very accurately reflects the

degree of granary weevil infestation. A similar species-specific assay has also been developed for the stored-grain pest *Trogoderma granarium*, the kaphra beetle (Stuart et al. 1994). This assay is capable of readily distinguishing the adults, pupae and larvae of this species from six other *Trogoderma* species and from other common grain pests.

Development of Immunoassays for Beneficial Insects

The protection of stored-food products against insect damage is clearly of major importance, yet the number of chemical agents for this purpose is very restrictive and insect resistance to these chemical insecticides is increasing. An alternative to the use of chemical pesticides is biological control using beneficial insects. Such beneficial insects can either kill deleterious insects by eating these pest insects directly, or more indirectly by infesting the egg or larvae of the pest species. A number of small-scale tests have shown the effectiveness of this biological control approach (Brower 1988; Keever et al. 1986). For such biological control methods to be used on a wide scale for stored-crop protection, much must be learned about their efficacy, the timing of insect release and about insect dispersal. Current assays for beneficial insects, like those for pest species, depend upon time consuming and expensive sieving and visual tests that require skilled personnel. A rapid and effective means for quantifying the number and spatial distribution of beneficial insects in stored products is highly desirable. In addition it would also be very helpful to have available rapid quantitative tests which could monitor the effectiveness of product cleaning in getting rid of any remaining beneficial insects (which are usually external feeders) when the product is processed.

Any test for beneficial insects necessarily requires a high degree of specificity. Advantage has been taken of the exquisite selectivity of monoclonal antibodies in our initial work on the development of species specific assays for beneficial insects. A monoclonal antibody based system has been developed for the parasitic wasps *Bracon hebetor* and *Laelius pedatus*, effective control agents for grain insect pests (Stuart and Burkholder 1991). These immunoassay systems have the

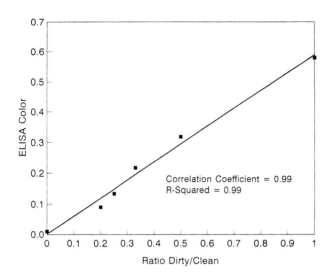

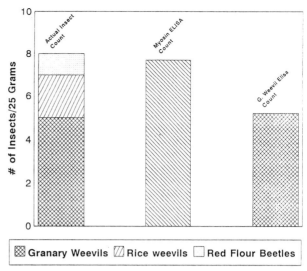

Granary Weevils **Rice weevils** **Red Flour Beetles**

Fig. 7. Assessment of the application of the insect myosin ELISA for blending grain. Clean hard red winter wheat was spiked with granary weevils and ground. This spiked dirty wheat was mixed with ground clean grain in varying proportions to give the equivalent of 0–2.5 granary weevils per 50 g sample.

Fig. 8. Use of a species-specific ELISA to determine the proportion of granary weevils in a mixed insect population in hard red winter wheat. A sample of clean grain was spiked with the indicated number of insects. The total insect infestation was assessed by the myosin ELISA and the granary weevil content was assessed using a monoclonal antibody-based species-specific ELISA.

desired selectivity, cross-reacting only with the beneficial species and not with a range of grain insect pests. Similar monoclonal antibody-based immunoassay procedures have been developed in our laboratories for other beneficial insects, including the parasitic wasp *Trichogramma pretiosum* and the pirate bug *Xylocoris flavipes*.

Application of Insect Immunoassays to Other Stored Products

Because a variety of stored-food products harbor the same types of insect pests as grains, it was anticipated that the insect myosin immunoassay procedure might also be applicable to a wide range of food products. An extensive series of tests was therefore carried out to assess the applicability of this test to a broad spectrum of spices, nuts, and dried fruits. These studies included both an examination of whether the ELISA procedure might be affected by materials present in these food products and whether the assay procedure was of the requisite sensitivity and linearity. In the case of spices, the same type of extraction employed with grains was found to be satisfactory for analysis of these flavouring materials. A wide array of different spices of both the leafy and nut-like varieties proved amenable to insect analysis by the ELISA procedure, giving a linear and sensitive response. In a few types of spices, including star anise and lavender, however, inhibition of the immunoassay was observed. Preliminary investigations into this problem revealed that the diminution in signal is likely due to non-specific binding of the antibody by plant materials present in these spices and experimentation is underway to circumvent this difficulty.

A similar situation was found when a variety of nuts and dried fruits were tested for applicability of the insect ELISA assay. A large range of these food materials, including products as diverse as cashews and dried peaches, were perfectly suited to the myosin immunoassay, but a few have so far proven recalcitrant. Mites and aphids are also important destructive agents in several types of stored products. Prelimi-

nary investigations have shown that the myosin ELISA assay is also capable of measuring the degree of infestation by these organisms, including the detection of such commercially important species as tobacco mites.

Summary

The new immunologically based assays for insect contamination in stored products bring a much higher degree of objectivity and quantitation to the analytical process, since they eliminate most of the subjectivity present in the visually based tests that are currently being used. Not only do the ELISA assays provide for a highly reliable and repeatable inspection process, they also require little training time and are of moderate cost. At the present time, the ELISA tests which have been most thoroughly tested and documented require the use of a modestly equipped laboratory. If large numbers of samples need to be analysed, then the ELISA procedure can readily be automated, as has been done for clinical chemistry ELISA tests.

In many circumstances it would be highly desirable to have as an option extremely rapid test procedures through which the results could be obtained in a matter of minutes. This type of assay would be invaluable for evaluating incoming loads of stored product and for spot checking the samples for the export market. The test-strip assay that we have under development appears to be very well suited for this type of approach.

The use of these immunoassay techniques for measuring insect contamination in stored products does raise regulatory questions. The immunoassays detect the mass of a particular type of insect material in a sample. The detected material can be insect myosin, or other proteins in the case of the monoclonal antibody-based tests. The mass of this insect component has very definitely been shown to correlate with the total mass of insect material present in the samples, and this mass in turn, correlates very well with the number of insects present (see Figure 3 and Schatzki et al. 1993). Thus, the immunoassay procedures can be used with no difficulty as a direct replacement for present assays which relate to numbers of insects in a sample. On the other hand, for processed stored products, such as wheat flour, where the primary current analytical procedure is the insect fragment count, a new type of regulatory unit will be needed for use with the immunoassays. This is because the insect fragment count simply measures the number of fragments present without regard to the actual mass of insect material present. Over the broad range, there is reasonable agreement between the ELISA procedure and insect fragment counts, but at low levels of contamination the correspondence between the two types of assays is weaker. This lack of correlation is due almost entirely to the variability of the insect fragment assay, as determined by direct comparisons of the two techniques using samples of known insect content (G.B. Kitto, P. Behrens and I. Heller, unpublished data). From studies based on the amount of insect test material present in different insect life stages and a knowledge of the length of time spent in each life stage, it is possible to define an 'average insect unit' for the immunoassay procedures (Schatzki et al. 1993). It is proposed that such a unit could be used for setting appropriate regulatory standards.

Acknowledgments

This work was supported in part by funds from the U.S. Department of Agriculture, the Millers' National Federation, the American Spice Trade Association, the Foundation for Research, and Research Applications Inc. We are grateful to the following individuals for their contributions to these studies: Jim Bair, Dr. Alan Barak, Dr. Patricia Behrens, Wen-Min Chen, Susan Fullilove, Steve Grissamore, Jason Hale, Ilana Heller, Dr. Donald Koeltzow, Dr. Vera Krischick, Jim Lemburg, Paul Proske, Kathy Richter, David Shedd, Dr. Tom Schatzki, Janet Stevenson, Dr. Melissa Stuart, Sabrina Trieff, and Pei Wang. We also thank Drs Dean Appling and Jon Robertus for their helpful comments on the manuscript.

References

Brower, J.W. 1988. Population suppression of the almond moth and the Indianmeal moth by release of *Trichogramma pretiosum*. Journal of Economic Entomology 81, 944–948.

Chen, W-M and Kitto G.B. 1993. Species-specific immunoassay for *Sitophilus granarius* in wheat. Food and Agricultural Immunology 5, 165–175

Keever, D.W., Millen, M.A., Press, J.W. and Arbogast, R.T. 1986. Augmentation of natural enemies for suppressing two major insect pests in stored farmers stock peanuts. Environmental Entomology 15, 767–770.

Kitto, G.B. 1991 A new biochemical technique for quantitating insect infestation in grain. Association of Operative Millers' Bulletin, 5835–5838.

Kurtz, O'D. 1965. Comparison of AACC and AOAC methods for extraneous materials in flour. Journal of the AOAC 48, 554–558

Mei, J.V., Yin, C-M., Carpino, L.A. and Ferguson, B.S. 1991 Hapten synthesis and development of immunoassays for methoprene. Journal of Agricultural Food Chemistry 39, 2083–2090.

Russell, G.E. 1988. Evaluation of four analytical methods to detect weevils in wheat: granary weevil, *Sitophilus granarius* in soft white wheat. Journal of Food Protection 51(7), 547–533.

Sachdeva, A.S. 1978. Effect of infestation stage, form and treatment on fragment count in flour. M.S. Thesis, Kansas State University.

Schatzki, T.F., Wilson, E.K., Kitto, G.B., Behrens, P.Q. and Heller, I. 1993. 'Determination of hidden *Sitophilus granarius* (Coleoptera: Curculionidae) in wheat by myosin ELISA,' Journal of Economic Entomology 86, 1584–1589.

Stuart, M.K., Barak, A.V. and Burkholder, W.E. 1994. Immunological identification of *Trogoderma granarium* Everts (Coleoptera: Dermestidae). Journal of Stored Products Research (in press).

Stuart, M.K. and Burkholder, W.E. 1991. Monoclonal antibodies specific for *Laelius pedatus* (Bethylidae) and *Bracon hebetor* (Braconidae), two hymenopterous parisitoids of stored-product pests. Biological Control 1, 302–308.

Vanderlaan, M., Watkins, B.E. and Stanker, L. 1988. Environmental monitoring by immunoassay. Environment, Science and Technology 22, 247–254.

Development of pheromone-baited insect traps

M.A. Mullen*

Abstract

An initial step in the development of an effective insect trap is to study the traps currently in use. In this paper the efficiency of several commercially available pheromone traps in capturing adult *Tribolium castaneum* under warehouse conditions is examined. To illustrate the development of an effective trap, the evolution of the Savannah trap from a tennis-ball canister to the final commercial product is discussed.

Introduction

Early detection of pest populations with pheromone-baited traps is one way to reduce our dependence on pesticides. Having identified and synthesised many pheromones, we can now concentrate on the design and development of effective traps to make use of these pheromones. In stored-product environments it is essential that we develop traps that are effective, sensitive, and selective.

The development of effective insect traps must bring together a variety of disciplines. The developer must be, as expected, an entomologist with a strong background in insect behaviour, a video specialist, and a design engineer.

DeCoursey (1931) reported the development of an efficient trap for *Tribolium confusum* Jacquelin du Val. The traps contained no pheromone and consisted of corrugated paper containing various food baits. Flour was often used as the bait and although it was effective as an attractant, it provided no means of killing the insects. In general, food baits do not attract over long distances. Incorporation of insect pheromones has improved the effectiveness of traps. Multilayered corrugated paperboard traps for trapping stored–product insects were designed by Burkholder (1976), Barak and Burkholder (1976, 1985), and Williams et al. (1981). These traps were similar to the DeCoursey trap in that they were constructed of corrugated cardboard and took advantage of insect behaviour by providing a place to hide. The traps incorporated a food attractant, a pheromone bait, and a killing agent. Barak and Burkholder (1985) improved the traps by replacing the insecticide with a plastic pitfall containing an oil bait which served as both a food attractant and a killing agent. A modification of this trap was made by Barak (1989) as an effective trap for *Trogoderma granarium* Everts.

Other factors must be considered when developing traps. It was reported by Barak and Burkholder (1985) and Mullen (1992) that some insect species are hesitant about walking onto a sticky surface. In the development of the trap described

by Mullen (1992), as well as other trap types, some glues repelled insects or were not strong enough to capture them. Other considerations are dust resistance, an ability to discourage humans or animals from disturbing the trap, and placement of the traps in such a way that normal warehouse traffic and commodity movement do not affect performance.

Development of the Savannah Trap

The development of the trap initially known as the Savannah trap was an evolution of design. From previous studies of other traps it was determined that an effective trap for *Tribolium* sp. would take into account several aspects of pest behaviour. Because no single pheromone trap would be effective for all species, a compromise between convenience and efficiency is essential. An initial step in trap development was the evaluation of commercially available traps in detecting various Coleoptera in storage environments. This study led to the development of a new trap that incorporated some of the features of previous traps as well as innovations to help eliminate some of their limitations.

The Storgard® trap is a widely used trap that uses a pheromone and oil bait to lure insects into the corrugated paper trap and hopefully into an oil filled pit in the centre. This trap consists of four layers of corrugation with only one layer leading to the oil pit. The result was a trap that attracted the insects, but allowed them to escape. The trap simply provides a dark place for the insects to hide. The Trappit® trap and its improvement, the Window® trap, are constructed of corrugated plastic with a clear window that makes it easy to count the trapped insects. However, close observation with a video camera showed that the beetles entered the trap but hesitated to walk out on the sticky surface. Improvement of the Window® trap eliminated the corrugated plastic and substituted a paperboard ramp. Also studied was the Allure® trap, a box trap which requires insects to walk up a ramp and fall over a sharp edge onto a sticky surface. The Fuji Tribo® trap includes a ramp with a net material suspended over an oil filled pit. All of these traps have design features that needed to be incorporated into a new trap design. The Storgard® trap had a dark place for insects to hide as well as an oil pit that provided a place for a food attractant and a killing agent. The Trappit®, Window®, and Tribo® trap provide an easy way to count the trapped insects, but they are not reusable. The Allure® trap had a long ramp for the insects to climb as well as a sticky surface to trap the insects. The later Window® trap, the Tribo® trap, and the Storgard® trap can only be entered from the ends and insects were often found along the sides of the trap with no way to enter.

As an example of the development of a successful trap, I will trace the steps and considerations made during the design of the Savannah trap. During the evolution of the Savannah trap several designs were developed and tested before the final design was selected. All were of the pitfall type. The earliest was constructed from layers of 1/8" posterboard. Each layer was smaller than the next, creating a pyramid with a series of steps. At the top and in the centre a recessed plastic cup served as the pitfall. The plastic cup contained a small amount of

*USDA ARS Stored-Product Insects Research and Development Laboratory, 3401 Edwin Street, Savannah, Georgia USA 31405, USA.

wheat–oat oil that acted both as an attractant and as a killing agent. A pheromone–impregnated rubber septum was placed on a small piece of aluminium foil in the centre of the cup. Early tests on the preliminary trap design were conducted in painted wooden chambers 1.2 m square. Chambers were covered with a wooden frame top covered with cloth screen to prevent the insects from escaping but still allow air movement.

Further modifications of the basic design resulted in a trap constructed from Plexiglas®. These round traps were approximately 6 cm in diameter and 1 cm high and were machined so that the 2 cm sides sloped upward at an angle of 35 degrees. At its apex the slope was reversed so that a downward slope of about 45 degrees led to a pit. The pit was constructed from a 2 cm stainless steel cup that was fitted into the trap through a hole that was machined from the base through the top. The longer outer slope was roughened to make a surface for the insects to crawl up, while the shorter inner slope was polished to induce the insects to slide into the pit once they crossed over the apex of the trap. As in the previous design, the pit was filled with an oil bait.

The final experimental trap was constructed from a plastic tennis-ball canister (Fig. 1A). The top 2.5 cm of the can was removed with a band saw. Five 2.5 cm slits were cut in the top part of the canister (Fig. 1B, 1) to form openings through which the insects could enter the trap, and to provide legs to support the top. The concave bottom of the canister was inverted and a 3.0 cm hole (2) was drilled through the centre. As with the Plexiglas® trap, the upward slope was roughened to allow the insects to crawl up and the shorter, steeper downward slope was left smooth to induce insects to slide into the trap. The canister top with the legs (3) was then fastened to the bottom with epoxy glue and pushed down so that the entire trap was about 2.5 cm tall and 6 cm in diameter. The trap was designed so that the canister lid also served as the top of the trap. A hole was made in the centre of the lid to accept the pheromone-impregnated rubber septum (4). This arrangement allowed part of the septum to protrude from the top of the lid as well as into the centre of the trap. Another lid was used as the bottom of the trap (5). This lid was trimmed so that only the minimal vertical surface remained for the insect to climb. The trapping surface at the bottom of the trap was made from a piece of index card cut into a 6 cm circle. The disk was coated with Tangletrap® and was held in place with double stick tape. A fibre disk (6) 2 × 6 mm was with the oil bait and placed on a pin pushed through the top of the trap.

The early cardboard pyramid traps were tested in the painted plywood boxes previously described. The traps were placed into the centre of the chamber and 50 one-week old *T. castaneum* (Herbst) adults were placed in each chamber. Trap catch was recorded after 24 hours. The corrugated paperboard Storgard® flour beetle traps were used as standards for comparison.

Additional tests were conducted in a 19 322 m^3 warehouse located at the U.S. Naval Supply Centre in Jacksonville, FL. USA. This series of tests compared the Plexiglas® trap, the Savannah trap, the Storgard® trap, and the Trappit® trap. Traps were placed on the floor near supports and around the walls. Tests were run for 16 weeks. After the first eight weeks the Plexiglas® trap was omitted and the PT 6 Allure® trap was added. Eleven traps of each type were used. The Savannah traps used the Storgard® rubber septa lure. Initially, 1650 two-week old adult *T. castaneum* adults were released every other week from 36 release points scattered around the warehouse. Insects were counted and removed from the traps before each release. Lures and traps were replaced every eight weeks, resulting in two replications over time.

Similar tests were conducted in smaller rooms at the Savannah laboratory. These tests were conducted in 15 × 5.6 × 1.7 m rooms. These tests compared the Savannah, Storgard®, and the Trappit® traps. Two traps of each type were placed around the perimeter of the room. The Storgard® septum lure was used in the Savannah trap. One thousand two-week-old unsexed *T. castaneum* adults were released at various points throughout the exposure room. Counts were made at 24-hour intervals for 72 hours. When possible, captured insects were removed from the traps after counting. Because of its construction, captured insects could not be removed from the Trappit® trap. Only insects that were actually entrapped in the trap were counted. Insects that were in the trap but were not entrapped were returned to the exposure room. At the end of each test the room was cleaned to remove insects that remained between replications. Traps were moved to different locations in the room between replications. Lures and traps were replaced monthly. Each test was replicated 16 times.

The Savannah trap was tested to see if it would capture other stored-product species. It was tested against the maize weevil, *Sitophilus zeamais* Motschulsky, in a small shed which had contained maize. The shed was completely emptied of all maize residues, swept, and the seams were caulked. No pheromone was used and the only attractant was an oily bait made from ground maize extracted with pentane. In other tests the

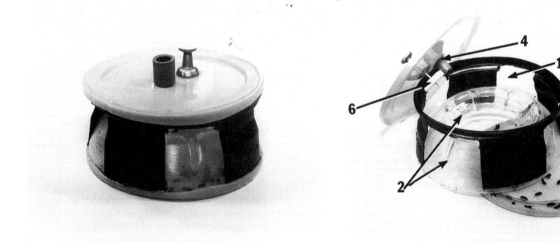

Fig. 1. Left: the completely assembled Savannah trap (support legs have been painted for contrast in photograph). Right: exploded view of the trap showing: (1) entrance slits, (2) concave bottom assembly with 3.0 cm opening, (3) top assembly with support legs, (4) pheromone-impregnated rubber septa, (5) base with sticky bottom, (6) fibre bait pad.

trap was baited with four types of lures for the cigarette beetle, *Lasioderma serricorne* (Fab.). These were the Serrico® lure (Fuji Flavor LTD, Tokyo, Japan), Lasiolure® (der B.A.T. Cigarettenfabriken, GmbH), and the Trécé® lure at 10.0 mg and 1.0 mg loads (Trécé, Salinas, CA). Tests were conducted in 15×15.6 m$\times$1.7 m exposure rooms at the Savannah laboratory. Traps were placed in the four corners of the rooms and 250 one-week-old unsexed adults were released every Tuesday and Friday. Counts were made every three days and the rooms were cleaned between releases to remove uncaptured insects. A total of 16 replications was completed using the same lure.

Data were analysed by the GLM procedure of the Statistical Analysis System (SAS Institute 1987).

After 24 hours most of the *T. castaneum* released in the chambers were captured in both cardboard pyramid trap and the Storgard® trap. Although more beetles entered the Storgard® trap, only a few were actually captured in the oil filled pit, indicating that the beetles were attracted to the trap but did not enter the pit. In additional tests, beetles entered unbaited traps almost as readily as baited ones, indicating that the beetles were seeking harbourage. This phenomenon occurred repeatedly in small chamber tests.

The results of the warehouse tests using the Savannah trap, the Storgard® trap, the Trappit® trap, the Allure® trap and the Plexiglas® trap are presented in Figure 2. The Allure® trap was tested for eight weeks and the Trappit® and Storgard® traps for 16 weeks (two replications). The Plexiglas® trap was effective but because the trap was not covered, it was fouled with filth and raided by rodents, and insect numbers were often estimated by counting insect fragments. Data for this trap are not included. The data presented are the mean number of insects trapped at each count. The Savannah trap was more effective than all traps tested and both the Savannah trap and the Storgard® were more effective than either the Allure® or the Trappit® (P>0.001).

In the small warehouse tests the Savannah trap was compared with the Storgard® trap and the Trappit® trap (Fig. 3). In all replications the Savannah trap captured more *T. castaneum* adults than either of the other traps (P>0.0001). The number of insects trapped by the Storgard® and Trappit® traps did not differ significantly. However, the Storgard® trap attracted almost as many insects as did the Savannah trap, the difference being that the Storgard® trap did not entrap the insects and the insects could leave the trap once they entered. Because both traps used the same rubber septum lure impregnated with R,R and R,S of 4,8-dimethyl decanal, the difference in trap catch can be attributed only to trap efficiency. Using only a food lure made from a pentane extract of

dried maize, three Savannah traps captured a total of 450 maize weevils in 10 days in an empty warehouse. Cigarette beetle adults were caught when the trap was baited with the Lasio®, Serrico®, and Trécé® lures. Although no statistical difference was found the trap was equally effective when tested with various lures for the cigarette beetle.

The final trap design is shown in Figure 4 and was the result of consultation with a design engineer and a mold maker. The design is such that it can be made in the most economical way possible. It is vacuum molded from one sheet of plastic. This allows several traps to be made at one time and the mold is considerably cheaper than an injection mold. The surface of the mold produces the rough outer climbing surface as well as the smooth surface needed in the cup to prevent escape. The outer cover of the trap is made from a coated paperboard. This coating not only provides a printing surface, but prevents curling. Holes are punched in the top to hold the pheromone lures.

The Savannah trap meets all the requirements of an effective trap as stated by Barak and Burkholder (1985). It is versatile in that it is effective against a wide variety of species. It is made of plastic and is durable, with replaceable parts. Unlike the Storgard® trap, it is practically escape-proof. Trap capture can be increased by removing obstructions near the pheromone dispenser (Tingle and Mitchell 1979). The design of the Savannah trap allows the pheromone dispenser to protrude through the top, thus providing an unobstructed pathway for the pheromone to disperse. Incorporation of an appropriate bait enhances the effect of the pheromone by providing a short distance attractant. This is especially important in a warehouse environment where there is little air movement to effect pheromone dispersal (Mankin et al. 1980). Molecules of pheromone disperse slowly in a closed environment and under some conditions they stick to absorbent surfaces which can then act as secondary pheromone emitters. This makes it difficult for insects to locate traps over long distances and as a result they tend to be attracted to the nearest trap (Vick et al. 1990). Proper use of pheromone traps will allow warehouse workers to locate patch infestations of insects that often occur in stored-product environments. Early detection of infestation will provide the warehouse workers more latitude in choosing the most effective control procedure.

The Savannah trap is now patented and is being marketed as the FLIT–TRAK M^2® by Trécé of Salinas, CA.

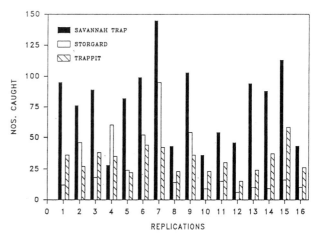

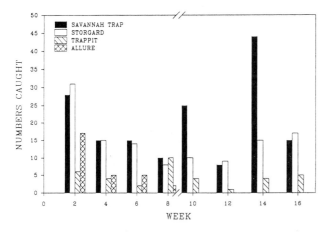

Fig. 2. Large warehouse test comparing four *T. castaneum* traps; lures and traps were replaced every eight weeks.

Fig. 3. Small warehouse test comparing the Savannah trap, the Storgard® trap, and the Trappit® trap for *T. castaneum*. Lures and traps were replaced monthly; each test used two traps and was replicated 16 times.

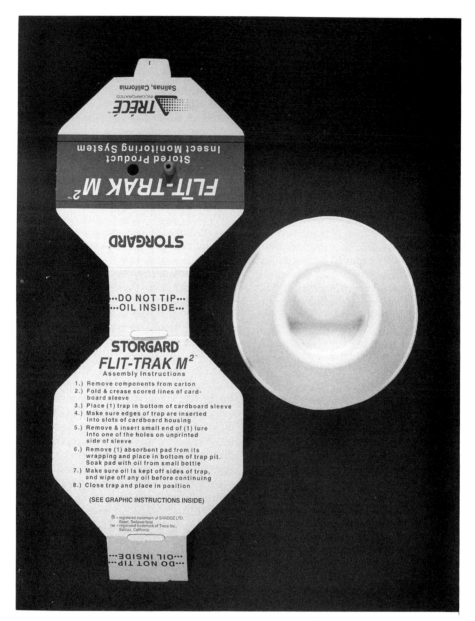

Fig. 4.   The commercial product known as the FLIT–TRAK M^2® as produced by Trécé of Salinas, CA, USA.

References

Barak, A.V. 1989. Development of a new trap to detect and monitor Khapra beetle (Coleoptera: Dermestidae). Journal of Economic Entomology, 82, 1470–1777.

Barak, A.V. and Burkholder, W.E. 1976. Trapping studies with dermestid sex pheromones. Environmental Entomology, 5, 111–114.

Barak, A.V. and Burkholder, W.E. 1985. A versatile and effective trap for detecting and monitoring stored-product Coleoptera. Agriculture, Ecosystem and Environment, 12, 207–208.

Burkholder, W.E. 1976. Applications of pheromones for manipulating insect pests of stored-products. In: Kono, T. and Ishii, T., ed ., Insect pheromones and their applications. Nagota and Tokyo, Japan Protection Association, 111–112.

DeCoursey, J.D. 1931. A method of trapping the confused flour beetle, *Tribolium confusum* DuVal. Journal of Economic Entomology 24, 1079–1081.

Mankin, R.W., Vick, K.W., Mayer, M.S., Coffelt, J.A. and Callahan, P.S. 1980. Dispersal of vapors in open and confined spaces: applications to sex pheromone trapping in a warehouse. Journal of Chemical Ecology, 6, 929–950.

Mullen, M.A. 1992. Development of a pheromone trap for monitoring *Tribolium castaneum*. Journal of Stored Products Research, 28, 245–249.

SAS Institute 1987. SAS/STAT Guide for personal computers, version 6 edition. SAS Institute, Inc, Cary, NC.

Tingle, F.C. and Mitchell, E.R. 1979. *Spodoptera frugiperda*: factors affecting pheromone trap catches in corn and peanuts. Environmental Entomology, 8, 989–992.

Vick, K.W., Mankin, R.W., Cogburn, R.R., Mullen, M.A., Throne, J.E., Wright, V.F. and Cline, L.D. 1990. Review of pheromone-baited sticky traps for detection of stored-product insects. Journal of the Kansas Entomological Society, 63, 526–532.

Williams, H.J., Silverstein, R.M., Burkholder, W.E. and Khorramshahi, A. 1981. Dominicalure 1 and 2: Components of aggregation pheromone from male lesser grain borer, *Rhyzopertha dominica* (F.). Journal of Chemical Ecology, 7, 759–780.

Effect of single and multiple species release on the capture of *P. interpunctella* and *C. cautella* in pheromone-baited traps

M.A. Mullen*

Abstract

The effect of the presence of the Indianmeal moth, *Plodia interpunctella* (Hübner), on the capture of the almond moth, *Cadra cautella* (Walker), in pheromone-baited traps was evaluated. It was found that when both sexes of each species were released into a large warehouse simultaneously, the capture of male *C. cautella* was reduced to 5.5 ± 0.8 in traps baited with a two-component lure as compared with 11.4 ± 2.2 when only *C. cautella* were released. This decrease in response may have been caused by an inhibitory substance produced by the female *P. interpunctella* that affects the response of male *C. cautella* to the pheromone, or to confusion caused by an increase in the amount of pheromone present. The reduced response of *C. cautella* must be considered when using pheromone-baited traps to estimate population levels when both species are present.

Introduction

The use of pheromone-baited traps to monitor stored-product insects is an important part of a pest management program (Ahmad 1987; Vick et al. 1986). The Indianmeal moth, *Plodia interpunctella* (Hübner), and the almond moth, *Cadra cautella* (Walker), are serious pests of stored food in warehouses in the United States and throughout the world. Both share the same major sex pheromone component (Z,E)-9,12-tetradecdien-1-ol-acetate (Brady and Nordlund 1971; Brady et al. 1971; Kuwahara et al. 1971 a, b).

Cogburn and Vick (1981) monitored population trends of Angoumois grain moth, *Sitotroga cerealla* (Olivier), in traps baited with (Z,E)-7,11- hexadecadien-ol acetate and also *C. cautella* and *P. interpunctella* in pheromone-baited traps containing (Z,E)-9,12-tetradecadien-1-ol acetate. Levinson and Buchelos (1981) and Sifner and Zdarek (1982) demonstrated the feasibility of monitoring multiple pest species of pyralid moths in traps baited with species specific pheromones. In some cases, closely related species can be attracted to a single pheromone. However, Mullen et al. (1991) found the presence of male and female *P. interpunctella* in equal numbers with *C. cautella* reduced capture of *C. cautella* males in traps baited with the same pheromone to less than 1% of the number released. At the same time 39% of the *P. interpunctella* males were recaptured. We speculated that this reduction in capture was caused by an inhibitory influence by the female *P. interpunctella*. Thus, while the value of moth pheromones for monitoring populations has been demonstrated, additional

work on the composition of the pheromone is needed (Chambers 1990).

In this study we attempted to determine if the presence of female *P. interpunctella* inhibited the capture of male *C. cautella*. We also tested the effectiveness of a specific pheromone blend on the capture of male *C. cautella*. This information will be used as part of a study to develop an effective pest monitoring system for use in warehouses containing both packaged and unpackaged foods susceptible to infestation by these two pests.

Materials and Methods

Test facility

All lures were tested in a $22\,100$ m^3 ($7 \times 61 \times 51$ m) warehouse. The roof of the warehouse had three arches with skylights extending along the entire length. Because of the arches the ceiling height varied from 4.6 to 7.3 m. Six small windows and three overhead doors were spaced evenly over each 51 m length. The windows were never opened and only one door was used during the test. At the centre of each 61-m width were large doors that led to other warehouse space. One door was opened during the day and both were closed at night. The warehouse was used for storing furniture and office items. The traffic in the warehouse was generally low but varied from week to week. Lights were often left on at night by warehouse workers.

Traps and lures

The rubber septa lures (Trécé Inc., Salinas, CA USA) contained either (Z,E) 9,12-Tetradecadien-1-ol-acetate as a single component for *P. interpunctella* or a two-component pheromone containing 85% (Z,E) 9,12-Tetradecadien-1-ol-acetate and 15% Z-9-Tetradecadien-1-ol-acetate for *C. cautella*. Each lure type was loaded with 1 mg of pheromone. The trap used was the Pherocon 1C wing trap (Trécé Inc.). Traps and lures were replaced every three weeks at the beginning of each new test.

Trap arrangement

Twenty-four traps were placed in the warehouse so that a density of approximately one trap per 920 m^3 was achieved. Each trap was suspended at a height of 4 m. Traps were baited with pheromone lures in the following sequence *P. interpunctella* (one-component), *C. cautella* (two-component) and blank. This was repeated so that there were 8 one-component, 8 two-component, and 8 unbaited traps.

Insects

The insects were reared on artificial moth medium (Boles and Marske 1966) at 27 ± 2°C and at 60 ± 10% r.h. with a 12:12(L:D) hour photoperiod. Pupae were collected in corru-

* U. S. Department of Agriculture, Agricultural Research Service, Stored-Product Insects Research and Development Laboratory, Savannah, Georgia USA.

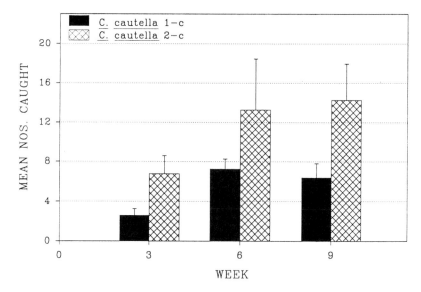

Fig. 1. Mean (±SEM) catch of released *C. cautella* in traps baited with either a one component (1-C) or a two-component pheromone (2-C). The data are totals for three replicates and are separated by lure type.

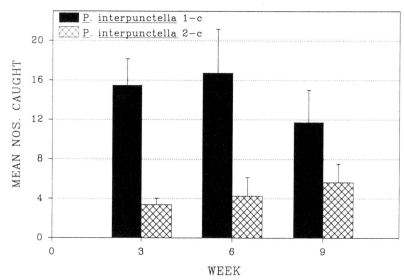

Fig. 2. Mean catch (±SEM) of released *P. interpunctella* in traps baited with either a single component lure (1-C) or a two-component pheromone (2-C). The totals for the three replicates are separated by lure type.

gated paper rolls. For each test 1000 unsexed pupae were removed from the rolls, separated into groups of approximately 28 each, and released at 36 locations in the warehouse. The sex ratio was assumed to be 1:1 as found by Mullen et al. (1991). Adult emergence from the released pupae ranged from 96–100% throughout the tests as determined by counting exuvia.

Experimental procedure

During week one, *C. cautella* pupae only were released from 36 locations in the warehouse. After three weeks, trap catch for each type of lure was recorded. Lures and traps were replaced and the procedure was repeated for a total of three replications. The same procedure was repeated for the *P. interpunctella*. For the final test a total of 1000 *P. interpunctella* and 1000 *C. cautella* were released from 36 locations with approximately 28 of each species at each location. The trap configuration was the same as in the previous tests and the trapped moths were counted and separated by species and lure

type. Temperature and relative humidity were monitored using a hygrotheromograph The temperature averaged 26°C and the humidity ranged from 40–90% for the duration of this test. Data for the capture by lure type for single species releases were analyzed using chi-square as an index of dispersion and the data for the combined species releases was analysed using a Spearman Rank correlation (SAS Institute 1987).

Results

Cadra cautella were captured in the traps baited with the two-component *C. cautella* lure and in traps with the one-component *P. interpunctella* lure. The number of moths of either species caught in the blank (control) traps averaged less than one moth per trap per week and was not considered significant. *Cadra cautella* were captured in greater numbers in all traps baited with the two-component lure with a mean (±SEM) of 11.4 ± 2.2 per trap compared to 5.4 ± 0.7 ($\chi^2 = 70.0$; $P > 0.00001$) (Fig. 1) in traps baited with the commer-

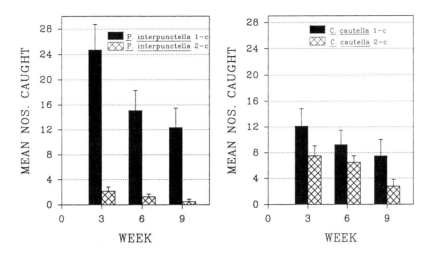

Fig. 3. Trap catch when both *C. cautella* and *P. interpunctella* released simultaneously in a large warehouse. (a) Totals for insects attracted and captured in traps baited with the *P. interpunctella* one-component lure. (b) Totals for insects attracted and captured in traps baited with the *C. cautella* two-component lure.

cially available single component formulation sold for both *P. interpunctella* and *C. cautella*. These results suggest that the two-part formulation is more effective for attracting and capturing male *C. cautella* than the single component pheromone now marketed.

More *P. interpunctella* were captured in traps baited with one-component *P. interpunctella* pheromone than with the two-component *C. cautella* pheromone. An average of 14.7 ± 2.0 males were trapped in the *P. interpunctella*-baited traps and 4.4 ± 0.9 in the two-component *C. cautella*-baited traps ($\chi^2 = 20.92$; $P > 0.00001$) (Fig. 2). From the data it appears that the best lure for *P. interpunctella* is the single component lure.

When both species were released simultaneously the capture of *C. cautella* in was significantly reduced in traps baited with either lure. An average of 1.3 ± 0.3 *C. cautella* were captured with one-component lures and 5.5 ± 0.8 in traps with a two-component lure. This is a reduction from 5.4 ± 0.7 and 11.4 ± 2.2 in one-component and two-component baited traps, respectively, when only *C. cautella* were released. This compares to 17.4 ± 2.2 *P. interpunctella* in one-component baited traps and 9.6 ± 1.4 in two- component baited traps (Fig. 3). The lack of a positive and significant correlation coefficient ($R^2 = -0.24329$; $P = 0.0957$) between lure types is consistent with the hypothesis that the reduction in the number of male *C. cautella* captured was the result of confusion caused by the presence of the *P. interpunctella* females and not the pheromone bait.

Discussion

The presence of female *P. interpunctella* may have inhibited the attraction and capture of male *C. cautella*. Nakajima (1970) observed that male *P. interpunctella* were highly responsive to extracts of female *C. cautella*. Although no reason was given, male *C. cautella* were much less responsive to extracts from female *P. interpunctella*. However, Coffelt and Vick (1987) found that the synthetic pheromone in commercially produced lures does not exactly mimic that produced by the female. Mullen et al. (1991) reported that when both species were released simultaneously, commercially produced *P. interpunctella* lures caught less than 1% of

the male *C. cautella* compared to 39% of the male *P. interpunctella* in the same study. Ahmad (1987) reported the capture of male *C. cautella* using the single component pheromone when *P. interpunctella* were not present.

Sower et al. (1974a) reported that the responses by males of both *P. interpunctella* and *C. cautella* became habituated to the naturally produced sex pheromone. Roelofs and Comeau (1970) suggested that this habituation may be due to the presence of structurally similar components that interfere with specific sex pheromone binding sites on the antennal sensilla of males. The suppression of *C. cautella* may be due to the presence of an inhibitor produced by the female *P. interpunctella* as reported by Ganyard and Brady (1971) and later identified as (Z,E)-9,12-tetradecadien-1-ol by Sower et al. (1974b) and Krasnoff et al. (1984). Because *P. interpunctella* were present in equal numbers, the females may have produced enough of this inhibitor to affect the response of the male *C. cautella*. Suppression was evident although the two component pheromone specific for the *C. cautella* was present. Alternatively, because both species share the same major pheromone component, the apparent decrease in response by *C. cautella* may have been the result of confusion resulting from the presence of pheromone producing female *P. interpunctella*. Even though cross mating is not known to occur, female *P. interpunctella* may have been considered by the male *C. cautella* as mates, or the increased amount of pheromone present made it more difficult to find the appropriate trap.

The single component *P. interpunctella* lure is very effective for attracting and capturing male *P. interpunctella*. However, when both species are present, a reduction in the capture of *C. cautella* can be expected. This reduction could be especially important in studies that attempt to estimate actual populations based on trap catch alone.

References

Ahmad, T.R. 1987. Effect of pheromone trap design and placement on capture of almond moth, *Cadra cautella* (Lepidoptera: Pyralidae). Journal of Economic Entomology 80, 897–900.

Boles, H.P. and Marske, F.O. 1966. Lepidoptera infesting stored products. In: C.N. Smith ed., Insect colonization and mass production. Academic Press, New York, 255–70.

Brady, U.E. and Nordlund, D.A. 1971. Cis-9, trans-12-tetradecadien-yl acetate in the female tobacco moth, *Ephestia elutella* (Hübner) and evidence for an additional component of the sex pheromone. Life Sciences 10 (part II), 797–801.

Brady, U.E., Tumlinson, J.H.,. Brownlee, R.G. and Silverstein., R.M. 1971. Sex stimulant and attractant in the Indianmeal moth and in the Almond moth. Science 171, 802–804.

Chambers, John 1990. Overview of stored product insect pheromones and food attractants. Journal of the Kansas Entomological Society 63, 490–499.

Coffelt, J.A. and Vick, K.W. 1987. Sex pheromones of *Ephestia cautella* (Walker) (Lepidoptera: Pyralidae): influence of mating on pheromone titer and release rate. Journal of Stored Products Research 23, 119–123.

Cogburn, R.R. and Vick, K.W. 1981. Distribution of angoumois grain moth, almond moth, and Indianmeal moth in rice fields and storages in Texas as indicated by pheromone baited adhesive traps. Environmental Entomology 10, 1003–1007.

Ganyard, M. C.,Jr. and Brady, U.E. 1971. Inhibition of attraction and cross attraction by interspecific sex pheromone communication in Lepidoptera. Nature 234, 415–416.

Krasnoff, S.B., Vick, K.W. and Coffelt, J.A. 1984. (Z,E)–9, 12-tetradecadien 1-ol: A component of the sex pheromone of *Ephestia elutella* (Hübner) (Lepidoptera: Pyralidae). Environmental Entomology 13, 765-767.

Kuwahara, Y., Hara, H., Ishii, S. and Fukami, H. 1971a. The sex pheromone of the Mediterranean flour moth. Agricultural and Biological Chemistry 35, 447–448.

Kuwahara, Y., Kitamura, C., Takahaski, S., Hara, H. Ishii, S. and Fukami, H. 1971b. Sex pheromone of the almond moth and the Indian meal moth: Cis–9, trans-12-tetradecadienyl acetate. Science 171, 801–802.

Levinson, H.Z., and Buchelos, C.Th. 1981. Surveillance of storage moth species (Pyralidae: Gelechidae) in a flour mill by adhesive traps with notes on the pheromone mediated flight behavior of male moths. Zeitschrift Fur Angewandte Entomologie 92, 233–251.

Mullen, M.A., Highland, H.A. and Arthur, F.H. 1991. Efficiency and longevity of two commercial sex pheromone lures for the Indianmeal moth and the almond moth. (Lepidoptera: Pyralidae). Journal Entomological Science 26, 64–68.

Nakajima, Minoru 1970. Studies on the sex pheromone of the stores grain moths. . In: D.L. Wood, R.M. Silverstein, and M. Nakajima ed., Control of insect behavior by natural products. Academic Press, New York, 209–217.

Roelofs, W.L. and Comeau, A. 1970. Sex pheromone perception: Synergists and inhibitors for the red banded leaf roller attractant. Journal Insect Physiology 17, 435–438.

SAS Institute 1987. SAS/STAT guide for personal computers, version 6 edition. SAS Institute, Inc., Cary, NC.

Sifner, F. and Zdarek, J. 1982. Monitoring of stored food moths (Lepidoptera: Pyralidae) in Czechoslovakia by means of pheromone traps. ActaEnt. Bohemoslov. 79, 112–122.

Sower, L.L., Vick, K.W. and Ball, K.A. 1974a. Perception of olfactory stimuli that inhibit the responses of male phycidid moths to sex pheromones. Environmental Entomology 3, 277–279.

Sower, L.L., Vick, K.W. and Tumlinson, J.H. 1974b. (Z,E)–9,12-tetradecadien-1-ol: A chemical released by female *Plodia interpunctella* that inhibits the sex pheromone response of male *Cadra cautella*. Environmental Entomology 3, 120–122.

Vick, K.W., Koehler, P.G. and Neal, J.J. 1986. Incidence of stored-product phycitine moths in food distribution warehouses as determined by sex pheromone baited traps. Journal Economic Entomology 79, 936–939.

Monitoring field populations of *Lyctocoris campestris*, a predator of stored-grain insects: assessment of different trap designs

M.N. Parajulee* and T.W. Phillips†

Abstract

A year-round sampling of field populations of *Lyctocoris campestris* was conducted in Madison, Wisconsin, USA during 1992–93. Three different trapping methods were used to monitor relative populations: pitfall probe, cardboard, and sticky (flight) traps. One absolute sampling technique was also employed by using a 1140 mL cup sampler. Probe traps caught adult bugs throughout the year, but other traps were unable to detect bugs during times of low density. Probe traps caught significantly more bugs than cardboard and flight traps at most times throughout the trapping season. Absolute sampling catches were low, making comparison between absolute and relative sampling difficult. During early summer, flight traps caught more bugs than cardboard traps, while the opposite was true in early fall. The reversed response to cardboard and flight traps through the season might be due largely to environmental factors. Flight activity was higher because of relatively higher temperatures in the storage and surroundings during summer, thus flight trap catches were higher. A lower temperature during fall drove bugs inside the grain mass, increasing cardboard trap catches. Earlier work, based on probe trap catches, reported that *L. campestris* showed a highly female-biased sex ratio in the field as opposed to a 1:1 sex ratio in laboratory colonies. The present study found that the estimate of sex ratios in the field can differ with trapping methods. Examination of environmental factors and behaviour of bugs in the grain suggested that relative sampling methods are unreliable for sex-ratio studies in the field, and an absolute sampling method must be used to detect the true field sex ratio of *L. campestris*. Understanding this type of seasonal pattern across different traps could be valuable in monitoring predator populations during biological control operations.

Introduction

The predacious bug *Lyctocoris campestris* (F.) (Heteroptera: Anthocoridae) is commonly found in stored grains and other stored-product habitats. *L. campestris* is a generalist predator of stored-product insects that has been shown to be a potential biological control agent for these pests (Parajulee and Phillips 1992a, b). Because of the growing interest in using natural enemies in stored-product pest management systems and the realisation of the predatory potential of *L. campestris,* detailed biological studies have been undertaken in recent years. Parajulee and Phillips (1992b) first described the rearing

techniques and the laboratory biology of *L. campestris*. Laboratory studies have shown that *L. campestris* can use a wide range of prey species of any stage and size that it can overpower (Parajulee and Phillips 1993). Parajulee et al. (1994) report a detailed account of *L. campestris* predatory response to different prey species and habitat structures. Our earlier work (Parajulee and Phillips 1992b) reported that *L. campestris* prefer mouldy and moist microenvironments over dry, and that the sex ratio is highly female biased in the field.

For the efficient use of natural enemies in pest management programs it is important that a thorough knowledge of the biologies, including distribution, abundance, population structure, and seasonality of both pests and natural enemies be known. Therefore, a two-year trapping study was initiated to document the field ecology of *L. campestris* in flat storages of shelled maize in Wisconsin. Specific objectives of this study were: 1) to monitor the seasonal patterns of *L. campestris* in the field; and 2) to examine the effectiveness of different trapping methods to monitor *L. campestris* population dynamics.

Materials and Methods

Experimental site and duration

A year-round sampling study of *L. campestris* was conducted for two consecutive years in 1992 and 1993 in a privately owned flat storage of shelled maize located 15 km west of the University of Wisconsin, Madison, Wisconsin, USA, hereinafter called Storage I. The sampling study was carried out mainly by trapping, except that there was one form of absolute sampling (see below). Storage I was a metal building (20 m × 40 m) on a cement slab with a buttressed 5 m tall inner wooden wall. Maize was harvested in October and December of 1991 and 1992, respectively, and thermally dried to a moisture content of 12–14% before storage. The surface of the storage was divided into 70 equal plots (2.7 m × 3.7 m) and we sampled alternating plots (up to 35) in a systematic manner. A nine-month (July 1992 to March 1993) sampling study was also conducted in another flat storage of shelled maize owned by the same farmer located 13 km west of Madison campus, hereinafter called Storage II. Storage II was a wooden building, approximately 20 m × 10 m, with maize filled to a height of 3 to 4 m. Storage II contained two to five-year-old maize in the lower layers, but had an approximately 1 m upper layer of maize harvested in 1991.

Trap types

We used three different relative sampling and one absolute sampling methods to monitor the *L. campestris* population in the field. Both relative and absolute sampling methods were deployed during 1992, while only relative sampling methods were used during 1993 in Storage I. Probe traps only were deployed in Storage II. Traps used in our study are described below.

* Department of Entomology, University of Wisconsin-Madison, WI 53706, USA.

† USDA ARS, Stored-Product Insects Research Unit, Department of Entomology, University of Wisconsin-Madison, WI 53706, USA.

Probe traps

We used WB-II pitfall probe traps (Trece Inc., Salinas, CA) in our entire trapping study. Probe traps were 45.0 × 3.3 cm plastic cylinders, of which the top 30 cm was constructed of a mesh (3.5 mm × 2 mm openings) and the bottom 15 cm consisted of a removable, tapered tip that concealed a small inner funnel with a lower opening of 8 mm. Probe traps employ a pitfall design to passively capture insects that move through grain (Barker et al. 1990).

Fight traps

Flight traps were made of 480 mL yellow plastic drinking cups (Fabr-Kal Corp., Kalamazoo, Michigan) the outer surface of which was uniformly covered with the sticky material Tangletrap® (The Tanglefoot Co., Grand Rapids, MI). Cups were then inverted and an S-shaped hook was inserted at the bottom surface of the cup as a hanging device.

Cardboard traps

Cardboard traps were simply single pieces of single-faced corrugated cardboard (18 cm × 13 cm) inserted into a paper envelope (19 cm × 14 cm) with top and bottom ends open. The trap was inserted vertically in the grain mass and insects were able to crawl through either end of the envelope and hide in the corrugations of the trap.

Cup sampler

A 1140 mL cup sampler was used to obtain an absolute sample from each experimental plot. The cup sampler is a cylindrical device (7.5 cm ID × 26 cm height) of which the bottom end is pointed so that the device can be inserted into the grain mass; when the handle is pulled up, the top cover opens and a grain sample falls into the cylinder.

Sampling procedures

Each of 35 experimental plots in Storage I received a probe trap that was inserted vertically into the maize so that the top edge was 1–2 cm above the surface, a pair of flight traps (one 10 cm and the other 1 m above the grain surface) to monitor flying insects, and a cardboard trap inserted just below the grain surface. All these traps were deployed in May of 1992 and checked every week until November of 1993. Samples from probe traps were emptied and the same traps were replaced, but cardboard and flight traps were replaced with new traps every week. *L. campestris* caught in each trap type and plot were sorted by sex and stages and were recorded. The grain temperature at each experimental plot was recorded at 20 cm depth by using a digital thermometer (Fisher Scientific, Pittsburgh, PA). Two grain samples (absolute samples) were taken from each experimental plot each week during 1992 by using a 1140 mL cup sampler. One cup sample was taken as a column of 0–30 cm from the grain surface, and the second sample was drawn as another column of 30–60 cm from the surface, each randomly selected within a plot. Maize samples were returned to the laboratory and all insects were separated from the maize by sifting. Moisture content was also determined for each maize sample every other week by using a Steinlite digital moisture meter (Seedburo Equipment Company, Chicago, IL). *L. campestris* were sorted and identified by sex and stage, and counted as before. Surveying was continued through the winter of 1993, but the frequency of sampling was every two to four weeks. In April 1993, a weekly sampling scheme was resumed and continued through November, 1993 when the owner removed the 1992 crop from the storage and added the 1993 crop. A daily high and low air temperatures and relative humidities (r.h.) were recorded at 1 m above the grain near the centre of the storage building using

a mini hygrothermograph (Cole-Parmer Instrument Company, Chicago, Ill), and calculated as mean daily temperatures and relative humidities. We then calculated weekly mean temperatures and relative humidities for each sampling date. Similarly, mean grain temperature and grain moisture for the entire site were calculated for each sampling date. There were only four pitfall probe traps deployed in Storage II throughout the sampling period. *L. campestris* caught in each of four probe traps were sorted by sex and stages and were recorded as before. Grain temperatures were also recorded in Storage II on a weekly basis.

Data analysis

Storage I

Because of low bug counts early and late in each year, we used data from 15 consecutive weeks, July 1 to October 14, 1992 and June 30 to October 13, 1993, for analysis. Absolute grain sampling data from 1992 only were used for the sex ratio analysis. During 1992 there were 19 experimental plots that remained in the storage throughout our 15-week experimental period. However, the number of plots in the storage were variable during 1993 because of the continuous removal of maize by the owner.

Storage II

Probe trap data from the entire nine-month trapping period were used for the analysis of sex ratio and seasonal trend.

Seasonal patterns of *L. campestris* adults were shown by plotting observed counts against sampling dates. The efficiency of cardboard, probe and flight traps in catching insects was examined by comparing the proportion of each trap type catching bugs for each sampling date during 1992 (protected Fisher's exact test; SAS Institute 1989).

Sex ratios (percentage of males) of adult *L. campestris* from Storage I were calculated for each sampling date and trap type across all the experimental plots. Significant departures from a sex ratio of 0.5 were then calculated for each sampling date and trap type (G-test; Sokal and Rohlf 1981). Because of very low counts in cardboard and flight traps during 1993, sex ratios were calculated and compared by summing counts across 15 weeks for these two trap types. Similarly, probe trap bug counts from Storage II were also summed across nine months for sex ratio analysis.

Results

Seasonal patterns

Adult *L. campestris* were first trapped in Storage I during second week of May in both 1992 and 1993 sampling years. However, total counts in all trap types were very low, with no bugs in cardboard traps, < 0.2 bugs/trap/week in probe traps, and < 0.009 bugs/plot/week in flight traps until June 30. Hence, we report statistical inferences based on 15-week sampling periods for both sampling years in Storage I. During 1992 more bugs were caught on the flight traps early in the season (June–July) and low numbers were caught in the cardboard traps. As the season progressed, numbers of bugs caught in the cardboard traps increased while the opposite was true for the flight traps (Fig. 1). Overall, probe traps caught more bugs on most sampling dates than the other two trap types (G-test; P<0.001; Sokal and Rohlf 1981), but the seasonal trend was two directional as opposed to the trends depicted by flight and cardboard traps. Probe traps caught low numbers in July and increased steadily for almost half of the sampling season during 1992, then showed a maximum abundance of 154 bugs on August 12, and finally declined steadily as the season pro-

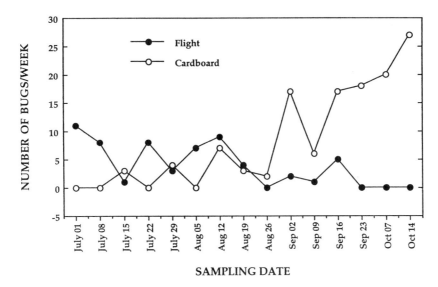

Fig. 1. Seasonal abundance of *L. campestris* monitored by flight and cardboard traps in Storage I of shelled maize from July 1 to October 14, 1992.

gressed (Fig. 2). The probe catch abundance during 1993 was much lower and more fluctuating than in 1992 (Fig. 3). A continuous habitat disturbance during 1993 was probably the cause for low counts of bugs in both flight and cardboard traps and no apparent seasonal trend was detected. The seasonal trend found in Storage II (Fig. 4) was similar to the trend found in Storage I in 1992 (Fig. 2), particularly during July to October. Bug abundance peaked in about mid to late August (Fig. 4) in Storage II. Despite a very low sample size (only two probe traps in 1993) we collected *L. campestris* adults from Storage II during most of the winter of 1993 until we terminated the trapping study in that storage in March, 1993. Since absolute sampling during 1992 in Storage I found low counts (mean of 3.33 bugs/week) of *L. campestris*, a seasonal pattern was not apparent. Low counts in grain samples also made comparison between absolute and relative sampling difficult.

Effectiveness of traps

The probe trap was the most effective design in capturing *L. campestris* on most sampling dates (Table 1). Proportions of traps that contain bugs are in agreement with the seasonal pattern we observed before. For instance, the proportion of flight traps containing bugs declined considerably as the season progressed and no flight trap detected any bugs after mid-September. Similarly, the proportion of cardboard traps containing bugs consistently increased throughout the sampling season as the grain temperature declined (Table 1, Fig. 5). Except in a few cases, the proportion of probe traps containing bugs were >0.5 and the efficiency of traps did not vary much with the sampling season. As late as mid-October, 60% of the probe traps captured bugs. It is interesting to note that during and after mid-September the proportion of probe traps containing bugs did not differ from that of cardboard traps.

Table 1. Proportions of traps that captured bugs per sampling date for each of three relative sampling methods in Storage I, 1992 (N= 19 traps).

Sampling date	Cardboard	Probe	Flight
July 1	0.105a	0.474b	0.368ab
July 8	0a	0.737c	0.316b
July 15	0.158a	0.632b	0.053a
July 22	0a	0.632b	0.053a
July 29	0.211a	0.632b	0.053a
Aug 5	0.053a	0.632b	0.158a
Aug 12	0.263a	0.737b	0.158a
Aug 19	0.158a	0.579b	0.158a
Aug 26	0.158a	0.684b	0a
Sep 2	0.316a	0.737c	0.053b
Sep 9	0.263a	0.579b	0.053a
Sep 16	0.474a	0.579a	0.053b
Sep 23	0.474a	0.474a	0b
Oct 7	0.571a	0.214ab	0b
Oct 14	0.500a	0.643a	0b

Proportions with different letters in each row are significantly different (protected Fisher's Exact Test; SAS Institute 1989).

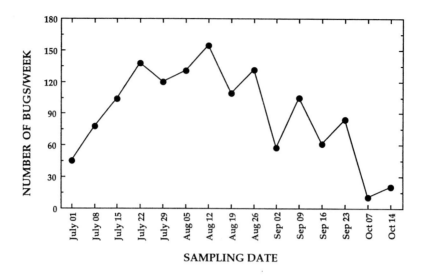

Fig. 2.   Total number of *L. campestris* caught per week in probe traps in Storage I of shelled maize from July 1 to October 14, 1992

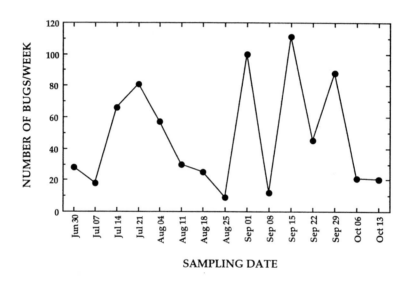

Fig. 3.   Total number of *L. campestris* caught per week in probe traps in Storage I of shelled maize from June 30 to October 13, 1993.

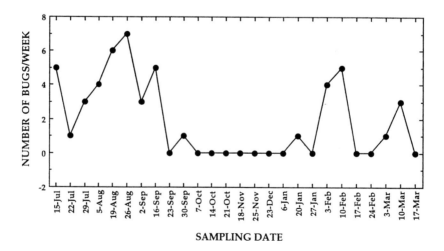

Fig. 4.   Total number of *L. campestris* caught per week in probe traps in Storage II of shelled maize from July 15, 1992 to March 17, 1993

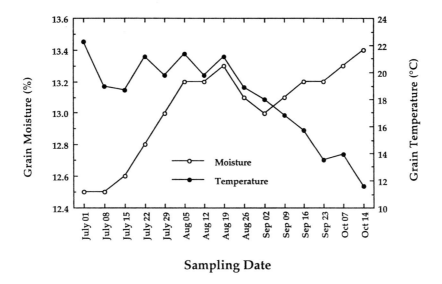

Fig. 5.   Weekly record of mean grain moisture content (%) and mean grain temperature (°C) across all plots in Storage I of shelled maize from July 1 to October 14, 1992.

Hence, late in the season both probe and cardboard traps were equally efficient in detecting *L. campestris*. However, the number of bugs caught in cardboard traps was much lower than the bugs captured in probe traps.

Sex ratio

Earlier field work with *L. campestris* (Parajulee and Phillips 1992b) and the present study found a highly female-biased sex ratio in probe traps, but these data may not reflect the true population sex ratio. For both sampling years, probe traps caught significantly more females than males (G-test; P<0.05; Sokal and Rohlf 1981) in Storage I, with sex ratios of O. 18 and 0.21 during 1992 and 1993, respectively. Probe traps also showed a highly female-biased sex ratio (0.20) in Storage II (G-test; P<0.05; Sokal and Rohlf 1981). Flight traps showed the sex ratios of 0.36 and 0.27 during 1992 and 1993, respectively, but these sex ratios showed no deviation from 0.5 (G-test; P>0.05; Sokal and Rohlf 1981). Similarly, cardboard traps showed no deviation in the sex ratios from 0.5 during 1992 (0.46), but showed a female-biased sex ratio (0.13) during 1993. Absolute sampling of grain during 1992 also showed an essentially equal sex ratio (0.52), and no seasonal variation was detected in the sex ratio by absolute sampling (P>O. 10, G -test for 18 individual sampling dates, June 24 to October 21).

Discussion

Seasonal patterns

The estimates of population abundance of *L. campestris* during 1992 from all three trap types were influenced by one or more environmental factors. Correlation analysis (Table 2) showed that all the environmental factors monitored in our study, except relative humidity, were strongly correlated with sampling dates. Similarly, all three trap types showed a significant correlation between sampling dates and trap catches, indicating a seasonal effect on observed *L. campestris* abundance. Correlation analysis (Table 2) showed that grain and air temperatures and grain moistures were important factors affecting bug abundance in cardboard traps. A high correlation between cardboard trap catches and temperatures (Table 2) suggests that temperature is very important in determining bug abundance in cardboard traps. The lower temperatures (both air and grain) might have led the bugs to move into the grain subsurface to seek refuge in the cardboard traps. Similarly, a moist environment, associated with cooler temperatures, could be favourable for *L. campestris* survival as well as oviposition. On the other hand, at higher temperatures (in early and mid-season sampling) bugs may have experienced higher surface mobility and increased flight

Table 2   Correlation coefficients among trap types and environmental variables for 1992 in Storage I.

	Week	Moisture	Grain temperature	Air temperature	r.h.	Cardboard trap	Flight trap
Grain moisture	O. 84						
Grain temperature	−0.85	−0.56					
Air temperature	−0.77	−0.48	0.87				
Relative humidity	0.05[a]	−0.31[a]	−0.22[a]	−0.02[a]			
Cardboard trap	0.86	0.62	−0.89	−0.89	0.06[a]		
Flight trap	−0.69	−0.47	0.70	0.60	−0.02	−0.59	
Probe trap	−0.44	−0.07[a]	0.62	0.80	−0.38[a]	−0.69	0.31[a]

[a]Correlation coefficients are not significantly different from zero (PROC CORR; SAS Institute, 1989; N = 15C).

activity, thus leading to a decrease in cardboard trap catches and an increase in the flight trap catches (Fig. 1, Table 2). The negative correlation between flight trap and cardboard trap catches is reflected in the opposite trends of these two traps in correlations with all environmental factors (Table 2). Bug abundance in probe traps also was influenced by both grain and air temperatures. The higher temperature might have enhanced the mobility of bugs in the grain leading to more bugs falling into the probes. Although grain moisture was not correlated with probe trap catches in this study, in earlier work we found that *L. campestris* was caught in probe traps in high moisture mouldy areas of the grain compared to drier areas (Parajulee and Phillips l992b). High moisture in 'hot spots' is a direct result of zones of fungal activity that harbour many prey species. Overall, grain moisture in the current study increased through the season as grain temperature decreased. Thus, increased grain moisture may be a consequence of fungal activity in certain areas or simply from a decrease in temperature throughout a grain mass, and bug activity (measured by flight and probe traps) is more directly related to temperature.

Effectiveness of traps

A suitable sampling method must be devised to accurately predict predator population levels before augmentation for a cost-effective biological control program (e.g. Shipp et al. 1992). Flight traps appear to be effective in early season and cardboard traps in late season in monitoring *L. campestris* field populations. Probe traps seem to be most effective in detecting *L. campestris* population throughout the sampling season. Absolute sampling was labour intensive, yet it could not detect low populations of *L. campestris*. Hence, we recommend using probe traps in monitoring *L. campestris* field populations. Performance of relative sampling methods is influenced by the behaviour of insects being sampled, habitat structure, environmental factors, and the effectiveness of traps in catching insects. Thus, the results based on relative sampling must be interpreted with caution. The timing and duration of the activity period of insects depend on the life history characteristics of the species in question (Wolda 1988). Hence, we would typically expect the same seasonal pattern if all the traps had a similar performance. Thus, differences in trap performance could lead to misinterpretation of seasonal pattern of the species if only one type of trap were used.

Sex ratio

Seasonal activity patterns and differences in sex ratios may reflect different behavioural responses of bugs to trap designs. Probe traps consistently showed a female-biased sex ratio of *L. campestris* field populations for the entire sampling studies. If only probe traps were used in field sampling, one might consider an apparent female-biased sex ratio. However, the other two relative sampling methods, flight traps and cardboard traps, showed essentially equal sex ratios. No deviation in the sex ratios in flight traps suggests that both sexes are similar in flight activity. Cardboard traps caught bugs mostly during the second half of the sampling season during periods of higher grain moisture and lower air and grain temperatures, which may have caused L. *campestris* to seek refuge. Once *L. campestris* find refuge, they hide until they have to leave to search for prey. We have observed in the field that one of the prey of *L. campestris*, the Indianmeal moth larvae, tend to seek such refuge as cardboard traps for pupation. Hence, *L. campestris* may not have to leave cardboard traps if they encounter their prey in the trap. The equal sex ratio revealed by cardboard traps must be attributed to the similar behaviour of both male and female *L. campestris* in seeking refuge. Laboratory studies have consistently shown equal sex ratios under various rearing conditions (Parajulee and Phillips 1992, 1993). More importantly, absolute sampling of grain during 1992 in Storage I showed an essentially equal sex ratio (0.52), and no seasonal variation was detected in the sex ratio by absolute sampling. Absolute samplings were independent of insect mobility or any environmental factor because the samples were taken instantly along with their habitat substrate. An equal sex ratio in these grain samples confirmed that males and females exist in equal numbers in the field. Therefore, the trap type and the time of sampling can affect the sample count and consequently the sex ratio.

Acknowledgments

We thank Erik V. Nordheim, Departments of Statistics and Forestry, University of Wisconsin-Madison, for providing statistical advice in analysing data. We gratefully acknowledge Rosa A. Franqui for her contribution in designing and deploying cardboard and flight traps for both sampling years, and Peter Edelman for technical assistance. We greatly appreciate reviews of the manuscript by Joel K. Phillips (USDA ARS, Madison, WI), Rosa A. Franqui and Todd J. Mayhew (both University of Wisconsin-Madison). Financial support for travel was provided from a research support agreement between USDA ARS and the University of Wisconsin, and from Insects Limited Inc., Indianapolis, David Mueller, President. This article reports the results of research only. Mention of a proprietary product does not constitute an endorsement or a recommendation for its use by USDA. All programs and services of the U.S. Department of Agriculture are offered on a nondiscriminatory basis without regard to race, colour, national origin, religion, sex, age, marital status, or handicap.

References

Barak, A.V., Burkholder, W.E. and Faustini, D.L. 1990. Factors affecting the design of traps for stored-product insects. Journal of the Kansas Entomological Society, 63, 466–485.

Parajulee, M.N. and Phillips, T.W. 1992a. Biology and field activity of *Lyctocoris campestris* (Heteroptera: Anthocoridae), a predator of stored-product insects. In: Jayas, D.S., White, N.D.G., Muir, W.E. and Sinha, R.N. ed., Extended Abstracts: International Symposium on Stored-Grain Ecosystems, Department of Agricultural Engineering, University of Manitoba, Winnipeg, Manitoba, Canada, June 1992, 10–13.

Parajulee, M.N. and Phillips, T.W. 1992b. Laboratory rearing and field observations of *Lyctocoris campestris* (Heteroptera: Anthocoridae), a predator of stored-product insects. Annals of the Entomological Society of America 85, 736–743.

Parajulee, M.N. and Phillips, T.W. 1993. Effect of prey species on development and reproduction of the predator *Lyctocoris campestris* (Heteroptera: Anthocoridae). Environmental Entomology 22, 1035–1042.

Parajulee, M.N., Phillips, T.W. and Hogg, D.B. 1994. Functional response of *Lyctocoris campestris* (F.) adults: effects of predator sex, prey type, and experimental habitat. Biological Control 4, 80–87.

Poole, R.W. 1974. An introduction to quantitative ecology. McGraw-Hill, New York, 532 p.

SAS Institute 1989. SAS users guide: statistics, 2nd ed. SAS Institute, Cary, NC.

Shipp, J.L., Zariffa, N. and Ferguson, G. 1992. Spatial patterns of and sampling methods for *Orius* spp. (Hemiptera: Anthocoridae) on greenhouse sweet pepper. Canadian Entomologist 124, 887–894.

Sokal, R.R. and Rohlf, F.J. 1981. Biometry, The Principles and Practice of Statistics in Biological Research. W.H. Freeman and Co., New York, 859 p.

Wolda, H. 1988. Insect seasonality: why? Annual Review of Ecology and Systematics 19, 1–18.

Comparison between two methods of insect sampling in stored wheat

P.R.V.S. Pereira, F.A. Lazzari, S.M.N. Lazzari and A.A. Almeida*

Abstract

Sampling techniques for insect detection on bulk-stored grain can provide meaningful information for management, marketing, and research. Thus, efficient, accurate, and economic techniques will reduce the number of treatments, levels of residues, and risks to the operator and, consequently, better quality products will be obtained. Unbaited cylindrical plastic probe traps and grain sampling with a sieve were used to detect insects in stored wheat, in southern Brazil. The samples were taken from a large capacity silo, every fifteen days, during seven months. The total number of insects collected with the traps was significantly higher (21 604 individuals) than that obtained by the grain-sampling (1326), at p = 0.01. However, relative efficiency and accuracy, besides the differences between the techniques, must be recognised when interpreting the data. Seven insect species were detected with the probe traps and six with the sieve. The flat and rusty grain beetles, *Cryptolestes* spp., were the most abundant species detected by both methods, representing 97.68% in the traps and 83.26% in grain sampling. Based on the occurrence and middle dominance of Palma (1975), only *Cryptolestes* spp. were classified as 'common' species (in the probe traps), while the others were either 'intermediate' or 'rare'. The lesser grain borer, *Rhyzopertha dominica*, and the rice and maize weevils, *Sitophilus* spp., were poorly collected by both methods due to their behaviour (distribution patterns and mobility). The main advantages of the probe traps over the sieve were: easier handling; continuous action, not requiring the constant presence of the operator; and detection of most species even at low infestation levels. The combination of the probe trap and the sieve data can give basic information about the species and their relative abundance for the stored-wheat management program in Brazil.

Introduction

A good insect sampling technique that is repetitive, accurate and less dependent on the operator is required for stored-grain pest management and marketing decisions, and also for research purposes. Based on a good monitoring program, control measures can be taken at the appropriate population level, reducing significantly the losses and costs.

It has been observed (Subramanyam and Harein 1989) that unbaited probe traps are more sensitive than grain sampling devices in detecting adults of various insect species in barley.

The number of insects captured in probe traps provides a relative measure of insect activity in grain, but does not esti-

mate abundance of insects in the grain area sampled (Subramanyam et al. 1989). The number of adults captured in traps may vary depending on the insect species, trap duration, and grain temperature (Fargo et al. 1989; Subramanyam and Harein 1989; White et al. 1990).

The most common species caught in probe traps were the flat and rusty grain beetles, *Cryptolestes* spp., the red flour beetle, *Tribolium castaneum*, and the fungus feeder groups (Subramanyam and Harein 1989, 1990; White et al. 1990). Some species, such as the lesser grain borer, *Rhyzopertha dominica* and the maize and rice weevils *Sitophilus* spp. are not easily detected with probe traps, but, pheromones, escape-proof products, and other sampling devices may be combined with the probe trap technique to increase catches (White et al. 1990).

In planning a trapping program, one of the first considerations is the estimation of trap efficiency so that the number of insects caught can be converted to absolute insect density. Regression equations have been used (White et al. 1990) for calculating trap efficiency over a range of environmental conditions for several different stored-grain beetles. Also, trapping methods should be standardised and trapping data combined with grain-sample data for a more complete view of the kinds of insects that are present in the grain (White et al. 1990). Also, one must recognise the differences between techniques when comparing data.

The only stored-insect monitoring technique adopted in Brazil is the standard grain sampling sieve. Thus, the objective of this study was to compare the standard device to the probe trap technique to show the advantages of the traps for stored-wheat insect detection and for determining the relative occurrence and middle dominance of species.

Materials and Methods

The study was carried out in a silo in southern Brazil, with 1200 t of wheat from the 1991 crop. Twelve cylindrical probe traps, made with perforated plastic tubes of 30 cm long and 2.0 cm internal diameter, with 2.8 mm round holes sloping downwards (Subramanyam and Harein 1990; Subramanyam et al. 1989), were installed in the centre and periphery of the silo, according to the cardinal points, at a depth of 20.0 cm. The traps were checked every 15 days for seven months, and the insects removed and frozen for later counting and identification.

For grain sampling, a sieve 24.0 × 24.0 cm with 2.8 cm high edges and a mesh of 2.8 mm round holes, was used to take samples from about the same places and depth, and at the same dates as the probe traps. The insects obtained in the tray were taken to the laboratory for counting and identification.

A simple statistical variance analysis was applied to test the significance of the difference between the catches by the two methods.

The abundance of the specimens was related to the occurrence and middle dominance of species according to the criteria proposed by Palma (1975). The occurrence of each insect species was calculated by dividing the number of

* Universidade Federal do Paraná, Departamento de Zoologia –Entomologia, Caixa Postal 19020; 81531-970 Curitiba, Paraná, Brazil.

samples where a given species occurred by the total number of samples multiplied by 100. Thus:

    0.0–25.0% = accidental species;
    25.1–50.0% = accessory species;
    50.1–100.0% = constant species.

The middle dominance was calculated dividing the number of individuals of a given species by the total number of individuals multiplied by 100. Thus:

    0.0–2.5% = accidental species;
    2.6–5.0% = accessory species;
    5.1–100.0% = dominant species.

These two classifications were grouped into:

Common species
    Constant and dominant.
Intermediate species
    Constant and accessory;
    Constant and accidental;
    Accessory and accidental;
    Accessory and dominant;
    Accessory and accessory;
    Dominant and accidental.
Rare species
    Accidental and accidental.

Results

Table 1 presents the number of insects captured by the cylindrical probe traps and by the grain sample sieve during a period of 207 days. In all sampling days, the probe traps captured more insects (94%) than the sieve sampling (6%).

Table 2 shows, by the variance analysis between the two sampling methods, that the number of insects sampled with the probe traps was significantly higher (p=0.01) than the number of insects captured by the sieve.

The insect species collected with the traps and the sieve were about the same, but the occurrence of most species was higher in the probe traps (Table 3). Based on the occurrence and middle dominance, the species were categorised according to the Palma's classification, as shown on Table 3.

The flat and rusty grain beetles, *Cryptolestes* spp., were the most abundant species detected in both devices. They were considered 'common' species only in the probe traps, but 'intermediate' in the sieve. All other species were only 'intermediate' or 'rare' in both devices, by Palma's general classification (Table 3). Analysing the occurrence of the species, which represents the relative frequency of each species in the samples, it can be observed that *Cryptolestes* spp. were 'constant' (above 50.1%); *Typhaea stercorea*, the hairy fungus beetle, 'accessory' (25.1%–50.0%); and the others only 'accidental' (below 25.0%) in the traps. Using the sieve technique, *Cryptolestes* spp. and *Sitophilus* spp. were classified as 'constant' and all the others 'accidental'.

The middle dominance, which means the number of a given species related to the total number of individuals, showed differences for the same species when comparing both techniques (Table 3). In the probe traps, only *Cryptolestes* spp. were placed as 'dominant' species (above 5.1%), while all the

Table 2. Variance analysis comparing probe trap and sieve catches of insects in stored wheat.

Source of variation	DF	TS	AS	F
Methods	1	16957.928	16957.928	9.64**
Residue	22	36060.658	1639.121	
Total	23	53018.586		

**Significantly different at p=0.01.

others were 'accidental' (below 2.5%). The species of *Cryptolestes* and *Sitophilus* were 'dominant' in the sieve.

Figure 1 shows the percentage of insects (total species) captured by the probe traps and sieve over the seven-month period. It can be noticed that the traps captured the insects in a consistent way, independent of the infestation level. The conventional sieving technique detected a noticable number of insects only when the infestation was relatively high.

Discussion

Insect sampling with probe traps was significantly superior to the standard grain-sieving technique: it proved to be easier to handle; it was more efficient in capturing stored-wheat insects either at high or low infestation levels; it captured a greater number of individuals of most species; the insects were captured alive and could be used for other research tests; and the continuous catching did not require the constant presence of the operator. However, the differences between the two techniques must be considered, because the traps were left inside the grain mass and removed only for recovering the insects, while the sieve was applied only during the grain sampling period.

The sieve method did not present consistent results; even during the periods of high infestation (as detected by the traps), the number of insects sampled was much less than that caught with the probe traps.

Pheromone baits can be placed inside the traps for detection of certain insect species, such as *R. dominica* and *Sitophilus* spp., which were, in this experiment, poorly collected with the probe traps, and a little more efficiently by the grain sieving.

It is known (Fargo et al. 1989; Subramanyam and Harein 1989; White et al. 1990) that the number of insects captured in the traps are related to species behaviour (aggregation, mobility), temperature, moisture, duration of trapping period, location, and number of traps, as well as insect density. Aggregation, low mobility, and probably low infestation levels could explain the low catches of *Rhyzopertha dominica*. On the other hand, *Sitophilus* spp. have high mobility and stay alive for long periods inside the chamber, thus they might climb and escape from the traps. The use of Fluon® or a killing agent in the collecting chamber can overcome the problem of *Sitophilus* spp. escaping from the traps (White et al. 1990). These species were also collected in low numbers in unbaited traps by other authors (Subramanyam and Harein

Table 1. Number of insects captured in stored wheat by probe traps and by grain sieving.

Methods					Sampling days						
	1	20	30	45	102	120	135	151	192	207	Total
Probe traps (%)	250	160	45	107	38	193	3244	14352	856	2359	21604
	97	83	47	63	79	89	98	97	80	84	94
Sieve	8	32	55	63	10	22	51	421	212	452	1326
(%)	3	17	53	37	21	11	2	3	20	16	6

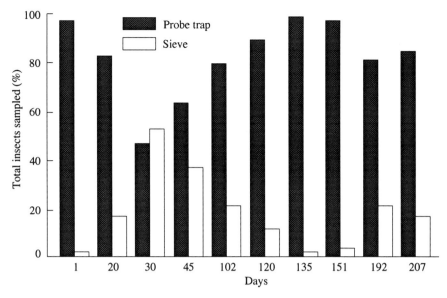

Fig. 1.   Comparison between two methods for sampling stored wheat insects over 207 days.

Table 3.   Total number of individuals, occurrence, middle dominance, and Palma's classification of insect species sampled with probe traps and sieve. (C = common, R = rare, I = intermediate).

Insect species	Probe trap				Sieve			
	Nº total individ.	Occurrence (%)	Middle domin.(%)	Palma's general classif.	Nº total individ.	Occurrence (%)	Middle domin.(%)	Palma's general classif.
Ahasverus advena (Col: Cucujidae)	25	9.2	0.12	R	53	5.8	3.99	I
Cryptolestes spp. (Col: Cucujidae)	21102	70.0	97.68	C	1104	40.0	83.26	I
Alphitobius spp. (Col: Tenebrionidae)	75	5.8	0.35	R	0	0.0	0.0	R
Tribolium castaneum (Col: Tenebrionidae)	9	4.2	0.04	R	1	0.8	0.08	R
Sitophilus spp. (Col: Curculionidae)	45	11.7	0.21	R	148	40.8	11.16	I
Rhyzopertha dominica (Col: Bostrichidae)	5	2.5	0.02	R	09	5.8	0.68	R
Typhaea stercorea (Col: Mycetophagidae)	343	29.2	1.58	I	11	3.3	0.83	R

1989; White et al. 1990), but pheromone lures proved to increase the catches.

In conclusion, despite the advantages of the probe traps, the use of data from the traps to estimate population density is difficult. However, insect catches can be converted to absolute insect densities by calculating trap efficiency for each species under a range of environmental conditions. The combination of probe trap and sieve data can give a more complete view of the insect species and their infestation levels in bulk-stored grain. Meaningful ecological information such as distribution, occurrence, abundance, dominance and diversity can be obtained by using the probe trap technique. Also, by the kind and relative numbers of the detected species, one can determine the conditions of storage and have a guide for management decisions. More studies are required in this subject, in Brazil.

References

Fargo, W.S., Epperly, D., Cuperus, G.W., Clary, B.C. and Noyes, R. 1989. Effect of temperature and duration of trapping on four stored grain insect species. Journal of Economic Entomology 82(3), 970–973.

Hagstrum, D.W. and Flinn, P.W. 1992. Integrated pest management of stored-grain insects. In: Sauer, D.B. ed., Storage of cereal grains and their products. American Association of Cereal Chemists, St. Paul, 535–562.

Hagstrum, D.W., Flinn, P.W., Subramanyam, Bh., Keever, D.W. and Cuperus, G.W. 1990. Interpretation of trap catch for detection and estimation of stored-product insect populations. Journal of the Kansas Entomological Society 63(4), 500–505.

Madrid, F.J., White, N.D.G. and Loschiavo, S.R. 1990. Insects in stored cereals and their association with farming practices in southern Manitoba. The Canadian Entomologist 122, 515–523.

Palma, S. 1975. Contribución al estudio de los sifonofors encontrados frente a la costa de Valparaiso. Aspectos ecologicos. In : II Simposio Latinoamericano sobre Oceanografia Biologica, Universidad d'Orient, Venezuela, 2, 119–133.

Subramanyam, Bh. and Harein, P.K. 1989. Insects infesting barley stored on farms in Minnesota. Journal of Economic Entomology 82(6),1817–1824.

Subramanyam, Bh. and Harein, P.K. 1990. Accuracies and sample sizes associated with estimating densities of adult beetles (Coleoptera) caught in probe traps in stored barley. Journal of Economic Entomology 83, 1102–1109.

Subramanyam, Bh., Harein, P.K. and Cutkomp, L.K. 1989. Field tests with probe traps for sampling adult insects infesting farm-stored grain. Journal of Agricultural Entomology 6(1), 9–21.

White, N.D.G., Arbogast, R.T., Fields, P.G., Hillmann, R.C., Loschiavo, S.R., Subramanyam, Bh. and Wright, V.F. 1990. The development and use of pitfall and probe traps for capturing insects in stored grain. Journal of the Kansas Entomological Society 63(4), 506–525.

Using pheromones for location and suppression of phycitid moths and cigarette beetles in Hawaii—a five-year summary

L.H. Pierce*

Abstract

This paper summarises observations during a five-year period to evaluate the efficacy of various pheromone-enhanced techniques for the location and suppression of phycitid moths and cigarette beetles, *Lasioderma serricorne* (Fabricius), in field situations in Hawaiian food facilities.

Location of infested products was accomplished using a fixed grid triangulation technique based on the inverse relationship between the number of captures per trap and the relative proximity of the infestation. This technique was successfully employed to locate hidden infestations of cigarette beetles and phycitid moths in various food facilities.

Pheromone mass trapping was found to be a cost-effective alternative to general (ULD) fogging for the suppression of phycitid moths. Increasing the number of pheromone moth traps in a food warehouse while reducing the frequency of general fogging progressively suppressed the phycitid moth captures during a three-year period.

In two food warehouses, pheromone mass trapping suppressed Indianmeal moth and cigarette beetle populations with and without regular pesticide treatments. Captures of these insects declined more rapidly in the regularly treated warehouse.

Pheromone enhanced mortality (PEM) techniques employ pheromone lures in conjunction with pyrethrin fog and/or cyfluthrin to suppress populations of stored-product insects. Several PEM techniques were employed to suppress populations of: 1. cigarette beetles in a bakery using time-mist foggers (battery operated aerosol); 2. phycitid moths in a grain elevator using spot applications of cyfluthrin; and 3. both phycitid moths and cigarette beetles in a feed warehouse using a combination of time-mist foggers and cyfluthrin.

Introduction

For the past five years, Food Protection Services has studied and evaluated a variety of pheromone enhanced techniques to locate, suppress and control phycitid moths and cigarette beetles in field situations in Hawaii. The following paper summarises long-term field observations in various Hawaiian food and feed facilities. Most of this work stemmed from a need to solve specific insect infestation problems. As data from each facility was collected and analysed, comparisons were made and the following series of semi-controlled field studies evolved.

* Food Protection Services, 95–715 Hinalii Street, Mililani, Hawaii 96789, USA.

Location of Infestation using Fixed Grid Triangulation

Continuous monitoring of large food storage facilities was implemented using large numbers of pheromone traps deployed in fixed grid patterns at 4 to 15 m intervals. The fixed grid technique facilitated early detection, location and removal of insect sources before the fugitive insects had a chance to migrate to other infestable products. This technique was employed at each of the facilities discussed in this paper.

Using the fixed grid triangulation technique, the layout of the pheromone trapping grid is not changed from week to week. Figure 1 shows a representative facility diagram with pheromone trap positions. The distance between traps remains constant and the lures in the traps are all the same age. Circles are drawn on the facility trapping diagram around each of the three trap sites having the highest number of captures in an active area. The relative sizes of the circles are estimated by converting the number of captures at each site to their respective percentage share of the combined total captures at the three sites. Then, each percentage share is subtracted from 100. The resultant numbers approximate the relative distances from the infestation and the radiuses of the triangulation circles drawn on the diagram.

Example: the number of captures at site 5 = 76 insects, site 6 = 41 and site 2 = 13. The total number of captures at these three sites equals 130 insects. Therefore, the percent share of captures at each of the sites is: 58%, 32% and 10% respectively. Subtract each of the percentage values from 100. The resulting numbers (42, 68, and 90) have a ratio of 4:7:9 and approximate the relative distances of each trap from the source of the insects. These numbers are used as radiuses of the circles drawn on the trapping diagram. The area where the three circles intersect indicates the approximate location of the infestation.

While this technique has limitations, it has been successful in locating both cigarette beetle, *Lasioderma serricorne* (Fabricius) and phycitid moth infestations in food warehouses.

Pheromone Trapping as an Alternative to General Fogging

The cost effectiveness of pheromone mass trapping as an alternative to general (ULD) fogging was investigated in a food warehouse. Phycitid moth populations had persisted in spite of a relatively costly fogging program in the warehouse. It was found that general fogging at three to four-week intervals had little effect on the number of moth captures in the warehouse. Fogging was phased out over a three-year period as shown in Figure 2.

A pheromone moth trapping grid was set up in the ca 42 500 m^3, warehouse. The initial low level monitoring at ca 1 trap/8500 m^3 was steadily increased to mass trapping levels of ca 1 trap/530 m^3. The number of weekly moth captures in the facility decreased from 61 to 27 as the level of pheromone trapping was increased as shown in Figure 3. The annual cost of the moth mass trapping program was about one-third the $6000 annual cost of the phased out fogging program as shown in Figure 4.

Pheromone Mass Trapping for Population Suppression

Two infested food warehouses were studied to determine the effects of increasing levels of pheromone trapping on Indian-meal moths, *Plodia interpunctella* (Hubner) and cigarette beetles, *Lasioderma serricorne* (Fabricius). In warehouse HW, standard chemical pest control measures were regularly employed including: weekly ULD fogging, time-mist fogging, fumigation, removal of infested products, and residual spraying around infested areas. In cohort warehouse HT, approximately half the size, limited ULD fogging and spraying were done initially, but stopped. Periodically, dry pet food items were fumigated in containers.

Warehouse HW had a volume of ca 79000 cm^3 and the trapping density for moths was increased 12-fold from N=30 to N = 372 (ca 1 trap/2630 m^3 to 1 trap/210 m^3). The number of cigarette beetle traps increased from N=30 to N= 61 (ca 1 trap/2630 m^3 to 1 trap/1300 m^3) during the 2.25 years of observation.

The average weekly moth captures decreased ca 96% from an average of 512 during the first quarter of monitoring to 18 during the last quarter. Cigarette beetle captures decreased ca 99% from 573 to 7 during these same periods. Figure 5 shows cigarette beetle average weekly capture data compared with the number of traps deployed by quarter. Figure 6 shows Indianmeal moth average weekly capture data and Figure 7 presents the same data on a weekly basis.

At the conclusion of this 120-week mass trapping program, a comprehensive inspection of every food item in the warehouse revealed no products which were actively infested by either insect. The two species which had been pheromone mass-trapped were no longer found and no captures were recorded in the pheromone traps during the last weeks of observation. In contrast, a number of products were still infested by all of the other stored-product insects which were present at the beginning of the trapping program. These principally included: red flour beetles *(Tribolium castaneum)*, drugstore beetles *(Stegobium paniceum)*, merchant grain beetles *(Oryzaephilus mercator)*, red-legged ham beetles *(Necrobia rufipes)*, and *Trogoderma* sp.

In warehouse HT, the number of cigarette beetles and Indianmeal moths also declined as the level of trapping in the ca 33000 m^3 facility was increased from ca 1 trap/3300 to ca 1 trap/330 m^3 as shown in Figures 8 and 9. Although the mass trapped insects were not completely eliminated in this warehouse, they were suppressed. The average weekly number of cigarette beetle captures decreased 81% from 73 during the first year to 14 during the fifth year of observation. The average weekly moth captures increased initially as the number of traps was increased, but then decreased 75% from eight during the third year to two during the fifth year.

Pheromone Enhanced Mortality

Pheromone enhanced mortality (PEM) techniques employ pheromone traps and/or lures in conjunction with minute amounts of pyrethrin fog and/or cyfluthrin to suppress populations of stored-product insects. Several PEM techniques were successfully employed to suppress phycitid moth and cigarette beetle populations.

Pheromones and Time-mist Foggers

The long-term effects of pheromone trapping with and without continuous time-mist fogging on cigarette beetle populations in two infested bakeries, DA and HC, were observed over a five-year period. Bakery DA had a volume of ca 13900 m^3 and bakery HC had a volume of ca 14300 m^3. Both bakeries had previously been ULD fogged in futile attempts to control the infestations.

The level of pheromone trapping in each facility was slowly increased from ca 1 trap/1410 m^3 to mass trapping levels of ca 1 trap/240 m^3. Several types of battery-operated time-mist foggers were deployed adjacent to the pheromone traps along the heavily infested back wall area in bakery DA. Each fogger dispensed ca 52 mg of 0.975–1.80% pyrethrin aerosol every 7.5–15 minutes. The products and food equipment were covered with plastic sheeting before the foggers were activated each night during non-production hours.

Weekly ULD fogging was discontinued in bakery DA during the first year after the time-mist foggers were installed. The number of time-mist fogger units deployed ranged from 10–24. ULD fogging was started and stopped several times for three to four month periods during the first three years in order to assess the separate and combined effects of both types of fogging. No ULD fogging was done in bakery DA during the last two years of observation.

The average weekly number of cigarette beetles captured in bakery DA with time-mist fogging declined steadily over the five years from 464 during the first year to 121 during the fifth year as shown in Figure 10.

Bakery HC was pheromone trapped, but no regular ULD fogging was conducted in the facility. The average weekly number of captures in the bakery HC remained relatively constant once the maximum trapping levels were reached as shown in Figure 11.

A total of 60 pheromone traps were deployed in each bakery during the last two years of observation. The cost of the pheromone trapping program (N=60) was ca $180/month. The additional cost of time-mist fogging in bakery DA was ca $180 month. The cost of the previous ULD fogging program in bakery DA was ca $400/month. The use of pheromone trapping in combination with time-mist fogging cost slightly less and was more effective than weekly ULD fogging.

Pheromones and Cyfluthrin

Short-term, rapid suppression of fugitive phycitid moths was achieved in a grain elevator by placing pheromone lures on 20 cyfluthrin treated spots (ca 1 m^2) on the outside of the silo cones in the elevator base and on 10 spots on the corrugated walls in the top section of the elevator. During the initial 21-week observation period, this technique was employed twice for two-week periods when the moth populations in the elevator began to increase. The number of fugitive moths seen and pheromone trap captures in the elevator declined dramatically, but rebounded after a few weeks when the treatment was stopped (Fig. 12).

A total of 28548 pycitid moths were captured in the elevator during the first 21 weeks of observation. Of these, 15731 moths (55%) were captured in the traps (N=20) at the top of the elevator and 12817 moths (45%) were captured in the traps in the base (N=40).

Long-term effects of this suppression technique were investigated during the subsequent 22-week observation period. Moth lures were placed on the inside of each of the 40 silo top hatch covers. The inside surface of the hatch covers were spot treated with 1% cyfluthrin wettable powder. The hatch covers were dusted off and resprayed approximately every two

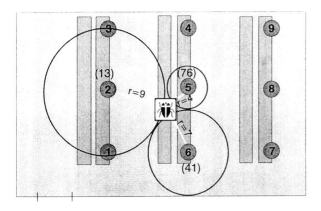

Fig. 1. Fixed grid triangulation technique.

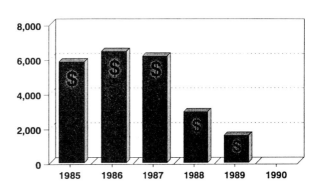

Fig. 2. ULD fogging cost in warehouse HM, 1985–90.

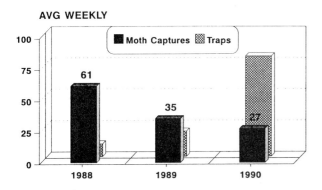

Fig. 3. Pheromone trapping effects on moth captures in warehouse HM.

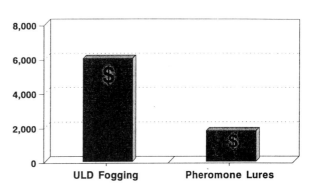

Fig. 4. ULD fogging vs pheromone costs in warehouse HM.

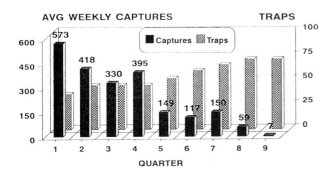

Fig. 5. Pheromone trapping effects on cigarette beetle captures in warehouse HW.

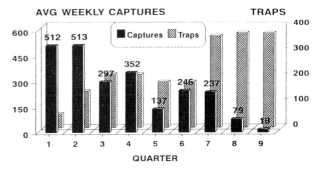

Fig. 6. Pheromone trapping effects on Indianmeal moth captures in warehouse HW (quarterly basis).

weeks. No additional chemical treatments were done in the base of the elevator. The arrays of pheromone traps in both the elevator top and base were monitored weekly.

The number of moth captures declined from 28 548 during the first 21 weeks to 9557 during the last 22 weeks of continuous treatment at the top of the elevator. Moth captures in the top section of the elevator declined 84% from 15 731 during the first 21 weeks to 2535 during the last 22 weeks and remained suppressed during treatment. There were 12 817 moths captured in the elevator base during the first 21 weeks and an additional 7022 captured in this untreated area during

the second 22 weeks. Moth captures in the base rebounded initially, but subsequently declined 45% compared to the previous 21-week period.

This technique proved to be highly effective for both immediate and long-term suppression of fugitive moths in the grain elevator. The percent reduction in moth captures in the treated top section of the elevator was double that of the untreated bottom section (Fig. 13).

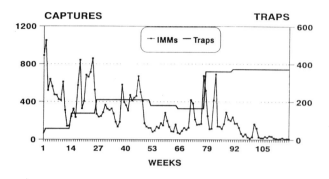

Fig. 7. Pheromone trapping effects on Indianmeal moth captures in warehouse HW (weekly basis).

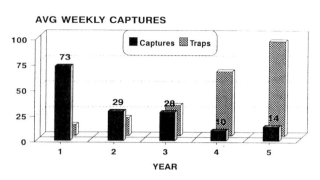

Fig. 8. Pheromone trapping effects on cigarette beetle captures in warehouse HT.

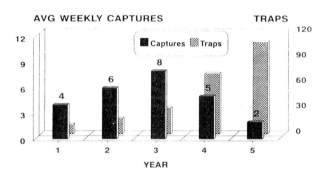

Fig. 9. Pheromone trapping effects on Indianmeal moth captures in warehouse HT.

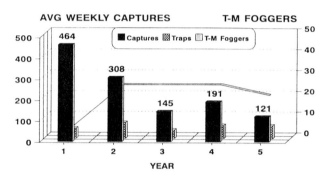

Fig. 10. Pheromone trapping and time-mist fogging effects on cigarette beetle captures in bakery DA.

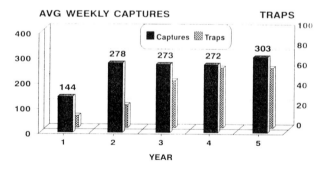

Fig. 11. Pheromone trapping effects on cigarette beetle captures in bakery HC.

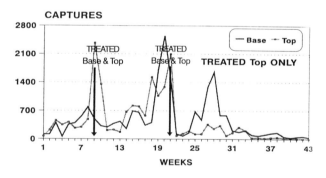

Fig. 12. Pheromone and cyfluthrin effects on phycitid moth captures in a grain elevator (weekly basis).

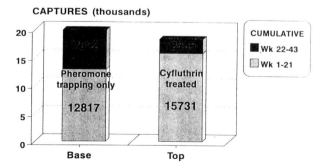

Fig. 13. Pheromone and cyfluthrin effects on phycitid moth captures in a grain elevator (by location in elevator).

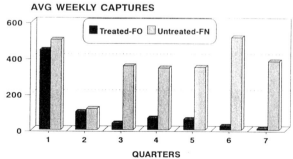

Fig. 14. Comparison of cigarette beetle captures in treated and untreated warehouses FO and FN.

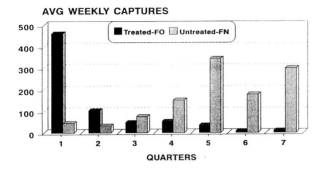

AVG WEEKLY CAPTURES

Fig. 15. Comparison of phycitid moth captures in treated and untreated warehouses FO and FN.

Pheromone Mass-trapping with Time-mist Foggers and Cyfluthrin

The combined effects of pheromone attractants in combination with time-mist fogging and spot treatment were evaluated during two-year observation period in two feed warehouses. Both warehouses were heavily infested with a variety of stored-product insects including phycitid moths and cigarette beetles.

Time-mist foggers were set up inside the treated warehouse FO along the front wall behind pheromone trapping arrays. Wall sections behind each of the trapping arrays were spot treated with 1% cyfluthrin wettable powder. The time-mist foggers operated continuously and the front wall spots were cleaned and resprayed monthly.

Average weekly cigarette beetle captures in the treated warehouse FO declined from 444 to 7 during the seven quarters of PEM treatment. Average weekly moth captures dropped from 461 during the first quarter to 10 during the last quarter of observation. Capture rates for these same two insects in untreated warehouse FN remained the same or increased during this same period (Figs 14 and 15).

Pheromone enhanced mortality resulting from the use of pheromone attractants in conjunction with the minute amounts of pyrethrin and cyfluthrin was highly effective in suppressing both fugitive phycitid moths and cigarette beetle in this feed warehouse.

References

Burkholder, W.E and Ma, M. 1985. Pheromones for monitoring and control of stored-product insects. Annual Review of Entomology 30, 257–72.

Haines, C.P and Read, J.S. 1977. The effects of synthetic female sex pheromones on fertilization in warehouse population of *Ephestia cautella* (Walker) (Lepidoptera:Phycitidae). Rapport 45, 10.

Hodges, R.J., Benton, F.P, Hall, D.R. and Serodio, R. dos S. 1984. Control *of Ephestia cautella* (Walker) (Lepidoptera:Phycitidae) by synthetic sex pheromones in the laboratory and store. Journal of Stored Products Research 20, 4, 191–197.

Levinson, A.R., and Buchelos, C.Th. 1988. Population dynamics of *Lasioderma serricorne* F. in tobacco stores with and without insecticidal treatments: a three year survey by pheromone and unbaited traps. Journal of Applied Entomology 106, 201–211.

Levinson, H.Z. and Levinson, A.R. 1987. Pheromone biology of the tobacco beetle *(Lasioderma serricorne* F., Anobiidae) with notes on the pheromone antagonism between 4S,6S,7S- and 4S,6S,7R-serricornin. Journal of Applied Entomology 103, 3, 217–240.

Mankin, R.W., Vick, K.W., Mayer, M.S., Coffelt, J.A. and Callahan, P.S. 1980. Models for dispersal of vapors in open and confined spaces: applications to sex pheromones trapping in a warehouse. Journal of Chemical Ecology 6, 5, 929–950.

Mueller, D., Pierce, L., Benezet, H. and Krischik, V. 1991. Practical application of pheromone traps in food and tobacco industry. Journal of the Kansas Entomological Society 63, 4, 549–553.

Mueller, D. and Pierce, L. 1992. Stored Product Protection in Paradise. Pest Control 60, 6, 34–36.

Pierce, L. and Okumura, G. 1992. Mass trapping with allure pheromone products. Pest Management Quarterly 11, 1, 6.

Previtt, P.F., Benton, F.P., Hall, D.R., Hodges, R.J. and Serodio, R. dos S. 1989. Suppression of mating in *Ephestia cautella* (Walker) (Lepidoptera:Phycitidae) using microencapsulated formulations of synthetic sex pheromone. Journal of Stored Products Research 25, 3, 147–154.

Sower, L.L. and Whitmer, G.P. 1977. Population growth and mating success of Indianmeal moths and almond moths in the presence of synthetic sex pheromone. Environmental Entomology 6, 1, 17–20.

Suss, L. and Trematerra, P. 1986. Control of some Phycitidae infesting stored-products with synthetic sex pheromone in Italy. Proceedings of the 4th International Working Conference on Stored-Product Protection, Tel-Aviv, Israel, 1986, 606–611.

Trematerra, P. and Battaini, F. 1987. Control of *Ephestia kuehniella* Zeller by mass-trapping. Journal of Applied Entomology 104, 336–340.

Trematerra, P. 1988. Suppression of *Ephestia kuehniella* Zeller by Using a Mass-Trapping Method. Tecnica Molitoria, 18, 865–869.

Trematerra, P. 1991. Population dynamic of *Ephestia kuehniella* Zeller in flour mill: three years of mass-trapping. Proceedings of the 5th International Working Conference on Stored-Product Protection, Bordeaux, France, September 1990, III, 1435–443.

Trematerra, P. and Capizzi, A. 1991. Attracticide method in the control of *Ephestia kuehniella* Zeller: studies on effectiveness. Journal of Applied Entomology 111, 451–456.

Improved early detection of internal infestation by flotation using product-adapted salt solutions

K. Richter and P. Tchalale*

Abstract

The success of stored-pest control strategies depends upon the effectiveness of the methods used for detecting and monitoring. Although many methods have been developed, the early detection of insects feeding inside grain kernels remains difficult. In particular, a method which is effective, rapid, accurate and easy to handle is needed by farmers, grain dealers and millers in developing countries.

Usually, the flotation method is used with standard salt solutions or water. It was found that a salt solution with product-adapted density improves the accuracy of detection (up to 100% of the real infestation) because grain varieties and cultivars differed much more than expected in kernal density. Moreover, adapted solutions permitted a much earlier detection of the hidden infestation after oviposition.

The optimum density of the solution was found to be 12–13% below the average density of sound grain. At this value errors and misinterpretations were minimised.

Using this method, a 75% detection level was reached 10 days after oviposition compared with 22 days with standard salt solutions (example: triticale/*Sitophilus oryzae* (L.)). The data showed a clear correlation (r = +0.964**) between close adaptation of density of the flotation medium and earliest possible detection.

The method is proposed for the inspection of samples of cereals and grain legumes after harvest, before storage or export, when the sieving method gives negative results for pests causing internal infestation, or when circumstances dictate that the internal feeders need to be detected.

Introduction

Detecting internal insect infestation in cereal and legume grains is important in the prevention of damage during storage, the minimisation of insect fragments in the milling process, and the maintenance of seed quality parameters.

If adults have been removed from the kernels by cleaning or threshing after having laid eggs into grains during transit, storage or in the field under favourable climatic conditions, grain internally infested by larvae, pupae or even newly hatched adults may be mistaken for uninfested grain. A very low population density of adults may lead to a very high level of internal infestation over four to eight weeks by the steady egg-laying process before the hatching of the F_1-generation (Fig. 1).

Tests for the detection of insects feeding inside the kernels include presumptive tests such as staining (Goossens 1949), flotation (White 1956) and cracking flotation separation (Harris et al. 1952), sound detection (Shuman et al. 1993; Vick et al. 1988), CO_2-determination (Bruce et al. 1982), NMR-inspection (Pinniger et al. 1986) and direct visualisation of internal infestation by X-ray analysis (Milner et al. 1950). Most of these methods are expensive, time consuming or require sophisticated equipment..

Although many methods have been developed, the early detection of insects feeding inside the kernel remains difficult. The need for a method which is effective, rapid, accurate and easy to handle is evident for farmers, grain dealers and millers, particularly in developing countries.

The present report suggests the use of product-adapted salt solutions to separate internally infested grain from noninfested grain. This method meets the demand for effectiveness, rapidity, accuracy and simplicity.

Material and Methods

The rice weevil *Sitophilus oryzae* (L.), and the Brazil bean weevil, *Zabrotes subfasciatus* (Boh.), were used as the test species in these experiments because their larvae feed exclusively within seeds.

The rice weevil and Brazilian bean weevil were reared on bread wheat and triticale, and on phaseolus beans, respectively.

Developmental stages (from eggs to adults) of known age were obtained by allowing adults to oviposit on uninfested seeds for 48 hours (rice weevil) and four days (Brazilian bean weevil).

Adults were then removed and individual kernels were checked microscopically to identify kernels containing eggs. Selected seeds were then incubated at 26 ±1°C and 65 ± 5% r.h. in darkness. This procedure was replicated in intervals of one to three days to obtain grain infested by larvae at all developmental stages. When the adults began to emerge from the oldest sample, all repetitions of the same series were examined by flotation. All separated grains were then inspected for the presence of internal infestation and for the identification of developmental stages.

Product-adapted salt solutions were prepared as follows: After determining the mean density of the grain lot (healthy), we adjusted the highest density for the salt solution to 12–13% of this value (0.55 kg/L $NaNO_3$ for density = 1.24 g/cm^3). Lower density steps were investigated by using saturated and unsaturated NaCl-solution (density = 1.19 and 1.04 g/cm^3), and water.

Results

In all experiments, the recovered percentage of infested grain in relation to the development stage of the larva is well

---

* University of Leipzig, Institute of Tropical Agriculture, Fichtestrasse 28, D-04275 Leipzig, Germany.

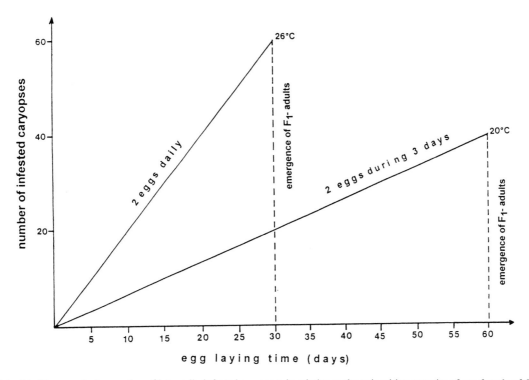

Fig. 1. Calculated increase in the number of internally infested caryopses in relation to the oviposition capacity of one female of *Sitophilus granarius* under different temperatures, using data from Jacobi (1982) and Swatonik (1975a, 1975b).

described by the curves calculated as logistic exponential functions.

The data obtained by flotation with the media in question (salt solutions and water) were significantly different at the 1% level (bars not shown in the figures).

Using the optimum adapted salt solution with a density = 1.24 g/cm³ for triticale and wheat, a 75% detection level could be obtained 10 days and 16 days after oviposition, respectively (Figs 2 and 3). First records of hidden infestation were made possible five and six days after oviposition for the cultivars tested.

In relation to these results, the maximum detection level for triticale and wheat using water averaged 49.5% and 30.0% and for the low-density media 62.4% and 36,7%, respectively.

Moreover, it is worth noting that even a small reduction in density (from 1.24–1.19 g/cm³) may drastically reduce the percentage of detectibility from 75% to 55% and the point of earliest traceable infestation from 10 to 22 days in case of triticale (Fig. 2). The analogous results for wheat were 16.5 and 22.5 days (Fig. 3).

To emphasise the relationship between the early detection of infestation in the product and density of the product, a rela-

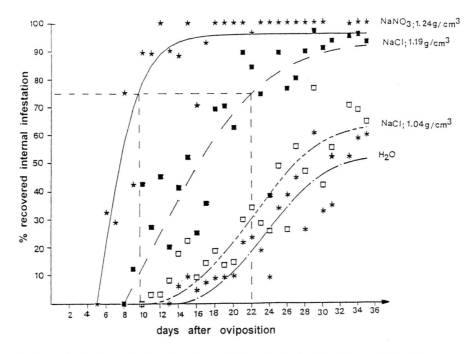

Fig. 2. The effect of various salt solutions on the detection of internal infestation by flotation (*S. oryzae/* triticale).

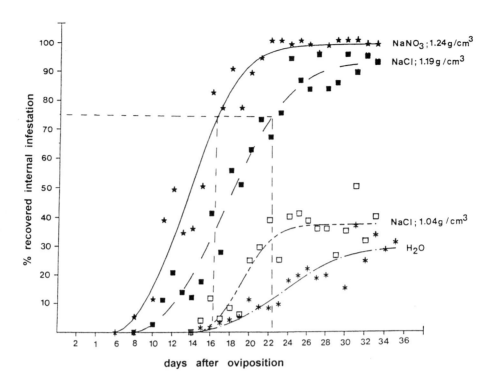

Fig. 3. The effect of various salt solutions on the detection of internal infestation by flotation (*S. oryzae*/ wheat).

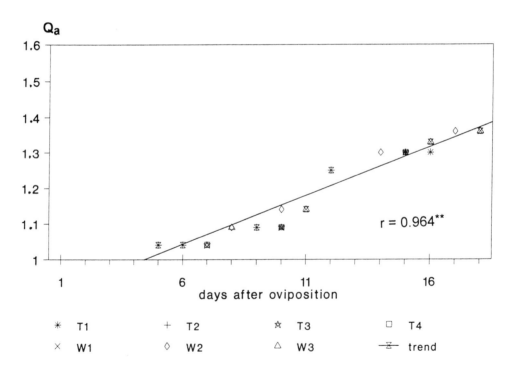

Fig. 4. Values for possible early detection in dependence of adaptation quotient Qa (T = triticale; W = wheat).

tionship for the adaption of the density of flotation medium to the mean density of healthy product was calculated, the quotient of adaption Q_a:

$$Q_a = \frac{\text{Mean density of product}}{\text{Density of salt solution}}$$

Using this equation, we could find a strong, significant correlation between time after oviposition and adaptation ratio, independent of the product tested (Fig. 4).

Because of the lower feeding percentage in relation to the weight of the kernel, infestation by the Brazilian bean weevil

was traceable only at older developmental stages. The point for earliest traceable infestation was found to be nine to ten days after oviposition, with one larva per kernel.

Discussion

Usually the flotation method is performed using standard salt solutions of a density = 1.19 g/cm^3, fixed salt solutions or plain water (Apt 1950; Peng and Morallo-Rejesus 1987; White 1956). We found that a salt solution with product-adapted density improves the accuracy of detection to up to

100% of the real infestation, because the density of grains of varieties and even cultivars differs much more than expected. Moreover, adapted solutions allow a much earlier detection of the hidden infestation after oviposition.

For preparing such flotation media, the adaption quotient Q_a will be a good measure. The closer the Q_a value to 1, the better the adaptation will be. The optimum density of the solution was found to range between 12 and 13% below average density of sound grain, corresponding to an Q_a value of 1.04. At this value errors and misinterpretations are minimised.

Using this method, a 75% detection level will be reached by 10 days after oviposition compared to 22 days with the standard salt solution (example: Triticale/*Sitophilus oryzae* L.). The data show a clear correlation ($r = +0.964$) between close adaptation of density of the flotation medium and earliest possible detection.

Comparing the data obtained by flotation with those from X-ray inspection or NMR-technique, it becomes evident that the detectibility starts approximately at the same time, around six to ten days after oviposition for both rice weevil and Brazilian bean weevil (Pinniger et al. 1986; Spirina 1982; Milner et al. 1950).

For further investigation, e.g. to distinguish species or developmental stages, the separated kernels can be checked by cutting or X-ray inspection.

The method is proposed for the inspection of samples of cereals and grain legumes after harvest, prior to storage, and before exportation when the sieving method gives negative results for pests or when situations dictate that the internal feeders must be detected.

References

Apt, A.C. 1950. A method for detecting hidden infestation in wheat. Milling Production 15, 1–5.

Bruce, W.A., Street, M.W., Semper, R.C. and Fulk, D. 1982. Detection of hidden insect infestations in wheat by infrared carbon dioxide gas analysis. USDA, Advances in Agricultural Technology, South. Series 26.

Goossens, H.I. 1949. A method for staining insect egg plugs in wheat. Cereal Chemists 26, 419–420.

Harris, K.L., Nicholson, J.F., Randolph, L.K. and Trawick, J.L. 1952. An investigation of insect and rodent contamination of wheat and wheat flour. Journal of Association of Official Agricultural Chemists 35, 115–158.

Jacobi, H. 1982. Getreidebearbeitung und-lagerung. Landwirtschaftsverlag Berlin.

Milner, M., Lee, M.R. and Katz, R. 1950. Application of X-ray technique to the detection of internal insect infestation of grain. Journal of Economic Entomology 43, 933–935.

Peng, W.K. and Morallo-Rejesus, B. 1987. Grain storage insects. In: Rice Seed Health. Proceedings of the International Workshop, March 16–20, 1987. Los Baños, Phillipines.

Pinniger, D.B., Chambers, J. and Cogan, P.M. 1986. New approaches to the detection of pests in grain. International Biodeterioration Supplements 22, 1986.

Shuman, D., Coffelt, J.A., Vick, K.W. and Mankin, R.W. 1993. Quantitative acoustical detection of larvae feeding inside kernels of grain. Journal of Economic Entomology 86, 933–938.

Spirina, T.S. 1982. Zernovki-opasnye vrediteli Zapasov [The bruchids—dangerous pests on stored products]. Zascita Rastenii, Moskva, 9, 40–46.

Swatonek, F. 1975a. Untersuchungen zur Biologie des Kornkäfers, *Sitophilus granarius* (L.) I. Teil. Bodenkultur 26, 278–290.

Swatonek, F. 1975b. Untersuchungen zur Biologie des Kornkäfers, *Sitophilus granarius* (L.) II. Teil. Bodenkultur 26, 379–398.

Vick, K.W., Webb, J.C., Weaver, B.A. and Litzkow, C. 1988. Sound detection of stored-product insects that feed inside kernels of grain. Journal of Economic Entomology 81, 1489–1493.

White, G.D. 1956. Studies on separation of weevil infested from noninfested wheat by flotation. USDA, AMS 1, 10 pp.

The use of various insect traps for studying psocid populations

R. Roesli and R. Jones*

Abstract

The experiment reported here aimed to find the most reliable and cost-effective method of obtaining estimates of population densities of psocids, especially Liposcelidae, a group of small insects commonly found in very large numbers in humid tropical grain storage.

Sampling and trapping were done at silos (8 –10 m diameter) belonging to feed-milling companies in southern and northern Queensland. A 0.70 m compartmentalised spear sampler and three different insect traps, namely probe, pitfall and bait bag traps, were compared. Up to 68 spear samples were usually taken before, during or after trapping. Depending on the size of the silo, five or six of each type of trap were placed in the top 70 cm of the grain, and left to trap insects for 4 hours.

There were good correlations between density estimates obtained from trapping and spear sampling. When population numbers were very low, pitfall traps were best at detecting psocids. To estimate mean abundances, probe traps were the most reliable, and good estimates of mean density were obtained with two traps per silo.

Introduction

Psocids, especially *Liposcelis* spp., are small, soft bodied insects commonly found in very large numbers in humid tropical grain storage. These occurrences are often associated with grains regularly treated with insecticides and fumigants (Haines and Pranata 1982; Pranata et al. 1983). Until recently, psocids in storage were not thought to be pests. They had only been considered as an annoyance to the worker when found in large numbers, or as indicators for damp or bad storage conditions. However, McFarlene (1982), using small samples of milled rice, found that significant weight loss occurred due to heavy infestation of *Liposcelis bostrichophilus* Badonnel..

Studies that generate more information on the ecology and the biology of psocid in stores are needed to develop successful control measures. Such studies rely on the development of easy sampling or trapping methods that can give accurate population estimates.

Storage insect populations have usually been estimated by a spear sampling method. However, Hodges et al. (1985) suggest that large numbers of samples must be taken to obtain an accurate population estimate. Taking large numbers of spear samples is difficult, very time consuming and laborious, and the separation of insects from the grain samples presents a further problem.

This study aimed to find a psocid trapping method that will enable estimates of psocid population densities to be obtained reliably and easily.

Materials and Methods

Location and commodity

Sampling and trapping were carried out in silos (8 to 10 m diameter) belonging to feed milling companies of southern and northern Queensland. Three silos of a feed milling company located in Ayr, one silo at Clifton and three silos at Warwick were chosen for sampling and trapping. The types of grains that were sampled from each silo are listed in Table 1.

Table 1. Lists of silos, types of grain and numbers of sampling points of the experiment

Silo	Grain types	Number of sampling points
Ayr1	Sorghum	68
Ayr2	Sorghum	62
Ayr3	Barley	68
Clifton	Maize	40
Warwick 1	Oat	40
Warwick 2	Maize	40
Warwick 3	Oat	22

Spear sampling

The spear sampler used was about 70 cm long and originally designed for use with polypropylene bags. It was manually modified to a compartmentalised spear sampler with five compartments and an aluminium outer sleeve.

Samples were taken before (mostly), during or after the trapping was carried out. All samples were taken from the top 70 cm of the grain bulk. Numbers of sample points varied according to the size of silo, availability of time and sampling equipment during sampling, and infestation level (Table 1).

Using five funnels of 80 mm diameter supported by a wooden stand, grain samples obtained were removed separately from each of the five spear compartments into 30 mL plastic vials.

Separation of insects from the grain samples was done, soon after the samples arrived and were weighed in the laboratory, using extraction methods developed by Leong and Ho (1990). Insects were then preserved in 5 mL plastic vials, and psocids were counted using a low-power microscope.

Psocid trapping

Three insect traps, namely probe, pitfall and bait bag traps, were used in this experiment (Fig. 1.). The probe trap was a 370 mm × 27 mm diameter trap resembling the spear sampler, but with holes along the upper section and a glass tube inside

* Zoology Department, James Cook University of North Queensland, Townsville, Q 4811 Australia.

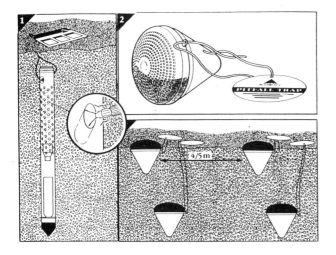

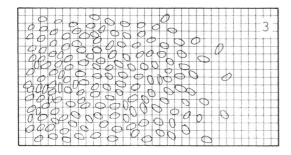

Fig. 1. Probe trap and pitfall trap (1 and 2, reproduced from commercial leaflet with permission of AgriSense), and bait bag trap (3).

the trap. The pitfall trap was a 95 mm × 125 mm, cone-shaped trap with catching holes through the top of the trap. The bait bag trap was a sealed envelope (8 × 16 cm) of netlon plastic mesh (2.00 mm apertures) containing 60 g of brown rice.

Six, for bigger silos like Ayr1, Ayr2 and Ayr3, or five, for the rest of the silos, of each type of trap were placed in the top 70 cm of the grain and left to trap insects for four hours.

At the end of trapping period, insects trapped in probe and pitfall traps were directly removed into 5 mL vials and preserved. Bait bag traps were removed from the grain, put into clickseal bags and sealed; insects were then removed from the trap by shaking the plastic bag approximately 20 times. The trap was then taken out of the bag, and insects were removed into 5 mL vials. Psocids were counted later under the microscope.

Data analyses

In order to find the correlation between the spear sampling result and the trap catches, simple linear regression analyses were done to the mean psocid numbers per trap and the psocid mean numbers per 100 g estimated from spear samples.

To evaluate the numbers of trap sample replicates needed to obtain reliable correlations with mean spear sample densities, analyses were repeated with randomly chosen subsets of the trap catches. This procedure was repeated 10 times for each subset size (1, 2, 3, 4, or 5 trap samples). The mean, standard error and range of the resultant R^2 value were then plotted against the number of trap samples per subset.

Results and Discussion

At least three species of *Liposcelis*, i.e. *L. bostrichophilus*, *L. entomophilus* (Enderlein) and *L. paetus* Pearman and one non-*Liposcelis* species were recognised from the samples and trap catches.

The results of the spear sampling are presented in Table 2. The highest psocid population was found in Ayr 1 silo, and the lowest in Warwick 3 silo.

The results of each trapping method are presented in Table 3. The highest psocid catches of all traps were also found in Ayr1 silo, and the lowest in Warwick 3 silo. The rank order of psocid mean numbers from probe trap catches (Table 3) is much the same as that of psocid numbers per spear compartment (Table 2).

The results (Tables 2 and 3) show that at very low psocid infestation, i.e. at Warwick 3 situation, most sampling and trapping methods, except pitfall trap failed to trace their presence. Pitfall traps may therefore be the most reliable method of detecting early infestations or reinfestations of psocids.

The regression analyses of trapping against sampling method are shown in Figures 2 and 3. There is significant correlation for all trap types between the mean trap catch and density estimates obtained from spear samples. However, the probe trap gave the best and most reliable correlation with

Table 2. The means and standard error of psocid numbers per spear compartment and per 100 g.

Silo	Psocids/compartment	Psocids/100 g
Ayr1	1.6030 ± 0.163	22.208 ± 2.249
Ayr2	0.119 ± 0.021	2.169 ± 0.375
Warwick 1	0.050 ± 0.015	3.554 ± 1.286
Ayr3	0.012 ± 0.006	0.203 ± 0.101
Clifton	0.010 ± 0.007	0.317 ± 0.224
Warwick 2	0.005 ± 0.005	0.100 ± 0.100
Warwick 3	0	0

Table 3. Mean numbers of trap catches and the standard errors of probe, pitfall and bait bag traps.

Silo	Probe	Pitfall	Bait bag
Ayr1	778.33 ± 673.97	168.83 ± 40.02	10.33 ± 7.33
Ayr2	27.33 ± 6.11	53.67 ± 18.33	2.67 ± 0.61
Warwick 1	12.40 ± 7.28	46.20 ± 42.00	1.40 ± 0.93
Ayr3	0.33 ± 0.33	1.00 ± 0.52	0.67 ± 0.67
Clifton	0.20 ± 0.20	2.20 ± 0.49	0.80 ± 0.48
Warwick 2	0.20 ± 0.20	4.40 ± 0.75	0.20 ± 0.20
Warwick 3	0.00 ± 0.00	0.60 ± 0.24	0.00 ± 0.00

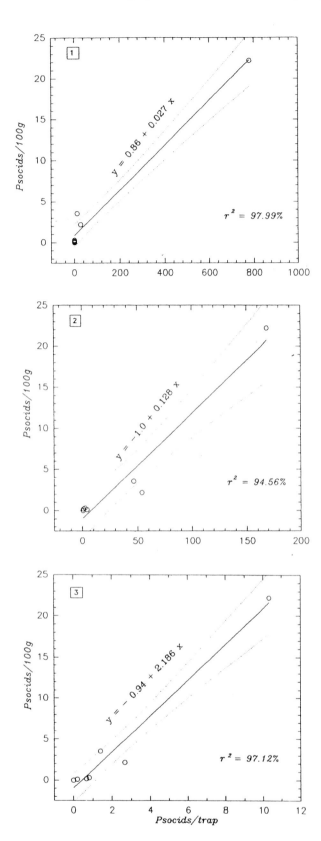

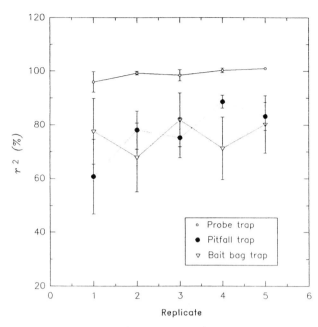

Fig. 3. The average R^2 (%) values of randomly chosen subsets of trap catches against replicate numbers of trap.

Conclusion

There are good correlations between density estimates obtained from trapping and from spear sampling, and therefore traps can be used to estimate psocid population densities.

When population numbers are very low (at early infestation), pitfall traps do best at detecting psocids.

To estimate mean abundances, probe trap are the most reliable, and good estimates of mean density may be obtained with sample sizes of two traps per silo.

Acknowledgments

The authors would like to thank Dr. R. Hodges, NRI the U.K., for his assistance; Mr Paul Cogan, MAFF, the U.K., and AgriSense the U.K., for providing free samples of traps; and the feed milling companies in Ayr, Clifton and Warwick for allowing us to sample the silos.

References

Haines, C.P. and Pranata, R.I. 1982. Survey of insects and arachnids associated with stored products in some parts of Java. In: Teter, N.C. and Frio, A.S. ed., Progress in Grain Protection. Proceedings of the 5th Annual Workshop on Grains Post-Harvest Technology. Southeast Asia Cooperative Post-Harvest Research and Development Program, Laguna, Philippines, 17–48.

Hodges. R.J., Halid, H., Rees, D.P., Meik, J. and Sarjono, J. 1985. Insect traps as an aid to pest management in milled rice stores. Journal of Stored Products Research 21 (4), 215–229.

Leong, E.C.W. and Ho, S.H. 1990. Techniques in culturing and handling of Liposcelis entomophilus (Enderlein) (Psocoptera: Liposcelidae). Journal of Stored Products Research 26 (2), 67–70.

McFarlene, J.A. 1982. Damage to milled rice by psocid. Tropical Stored Products Information 44, 3–10.

Pranata, R.I., Haines, C.P., Roesli, R. and Sunjaya. 1983. Dust admixture treatment with permethrin for the protection of rough rice and milled rice. In: Tetter, N.C., Semple, R.l. and Frio, A.S. ed.,. Maintaining Good Grain Quality. Puncak Pass, Bogor, Indonesia, South-East Asia Cooperative Post-Harvest Research and Development Program and ASEAN Food Handling Bureau (Proceeding of the sixth Annual Workshop on Grains Post-Harvest Technology), 132–136.

Fig. 2. Correlations between the mean trap catch of probe trap (1), pitfall trap (2), and bait bag trap (3); and density estimates obtained from spear samples.

spear sample densities. Even when only 1 trap was used, the average R^2 values exceed 90%. Little or no improvement in the R^2 magnitude or reliability was achieved by increasing the number of traps per silo above two.

Trapping stored-product insects using an unbaited multifunnel trap

P. Trematerra, G. Rotundo and A. De Cristofaro*

Abstract

Unbaited multifunnel traps have been used in a warehouse where different cereals were stored and two rooms of a flour mill where bran and cracked kernel waste were present.

In the cereal warehouse only Coleoptera were trapped, whereas in the flour mill Coleoptera, Diptera and Lepidoptera were trapped.

During a six-week trial in the cereal warehouse, the multifunnel traps caught adults of *Cryptolestes ferrugineus* (Stephens), *Rhyzopertha dominica* (F.), *Sitophilus oryzae* (L.) and *Tribolium castaneum* (Herbst).

The species found in the flour mill were different, more numerous and dependent upon location. In the room storing bran, the most frequent species were the moths *Ephestia kuehniella* Zell., *Pyralis farinalis* (L.) and *Nemapogon granellus* (L.), and the beetles *R. dominica* and *T. castaneum*. In the room storing the cracked kernel waste Lepidoptera were predominant with *E. kuehniella*, *N. granellus*, *Hofmannophila pseudospretella* (Staint.) and *P. farinalis* present.

The silhouette of the multifunnel trap (shaped like a tree trunk) may attract these species because the shape resembles their original food source.

A comparison made with other commercial pheromone traps showed the versatility of the multifunnel trap in complex infestations where various species of flying beetles and moths are present. The multifunnel trap in such situations may provide a better means of the detection of pests, especially when used as part of a control program.

Introduction

Trematerra and Girgenti (1988) noted in studies on the behaviour of the Lesser grain borer *Rhyzopertha dominica* (F.) in a grain store that the multifunnel traps they used also captured *Cryptolestes ferrugineus* (Stephens), *Sitophilus oryzae* (L.) and *Tribolium castaneum* (Herbst).

This led to the investigation undertaken in a cereal warehouse (Trematerra and Daolio 1990) of the role of synthetic dominicalure on nontarget species, the effects of silhouette on trapping efficiency and the influence of trap position in different lit areas of the store. It was found that for *R. dominica* the orientation to the traps could be mediated by chemical and visual cues. Also *C. ferrugineus*, *S. oryzae* and *T. castaneum* did not respond to dominicalure but visual cues may have played a major role in the orientation of the beetles to the multifunnel traps.

*Dipartimento di Scienze Animali Vegetali e dell'Ambiente, University of Molise, Via Cavour 50, I - 86100 Campobasso, Italy.

The effectiveness of the multifunnel trap in a warehouse and flour mill was used to test the supposition that this response is a relict of an ancient behaviour pattern from times in which the *C. ferrugineus*, *R. dominica*, *S. oryzae* and *T. castaneum* had not adapted to storage conditions.

Materials and Methods

The exercise was undertaken in a cereal warehouse and in two rooms of a flour mill where bran and cracked kernel wastes were present. Both places were unheated.

Using three multifunnel traps, preliminary trials were carried-out for six weeks during June–August, in the cereal warehouse. Subsequently two traps were used to observe the effectiveness of the traps for nine months in two locations of a flour mill.

The multifunnel trap was 80 cm high, with a 30 cm diameter and consisted of 9 vertically aligned white funnels. The traps were placed in the middle of each location and positioned 30–60 cm from the floor (Fig. 1).

The multifunnel traps were compared with the effectiveness of other commercial traps baited with pheromones of some moths and beetles infesting stored products and with the same traps without baits. The following substances have been used (Z,E)-9,12-tetradecadien-1-yl acetate (or TDA) for phycitid moths and (Z, E)-7,11-hexadecadien-1-yl acetate (or HDA) for *S. cerealella,* and pheromones of *R. dominica* (Dominicalure 1+2), *Sitophilus* spp. (sitophinone) and *Tribolium* spp. (dimetyldecanal). The traps were inspected weekly.

Results and Discussion

Numerous male and female insects were found in the unbaited multifunnel traps. Many flying Lepidoptera and Coleoptera were trapped both in the warehouse and in the flour mill.

Only the Coleoptera *C. ferrugineus*, *R. dominica*, *S. oryzae* and *T. castaneum* were trapped in the cereal warehouse. This contrasts with the flour mill where different orders: Coleoptera, Diptera and Lepidoptera were trapped.

In the cereal warehouse during the six-week trial the multifunnel trap caught 11 250 adults of *C. ferrugineus*, 3183 of *R. dominica*, 423 of *S. oryzae* and 2157 of *T. castaneum* (Table 1). In the flour mill the species found were more numerous and differentiated according to the location. In the bran room, the most frequent species among the moths were *E. kuehniella* , *P. farinalis* and *N. granellus* , and among the beetles *R. dominica* and *T. castaneum* (Table 2). In the room where cracked kernel waste was present the Lepidoptera were predominant with *E. kuehniella, N. granellus, H. pseudospretella* and *P. farinalis* (Table 3).

Data for the comparison with the commercial pheromone traps used at the same time and in the same locations are reported in Table 2 for Lepidoptera and Table 3 for Coleoptera.

The pheromone traps of Lepidoptera were more effective compared to the multifunnel traps, but they may be used only

Table 1.   Insects trapped by multifunnel traps in the cereal warehouse

	Species				
Weeks	*C. ferrugineus*	*R. dominica*	*S. oryzae*	*T. castaneum*	Total
I	6400	553	151	728	7832
II	1200	664	36	139	2039
III	300	509	20	122	951
IV	1050	714	61	403	2228
V	1500	519	98	467	2584
VI	800	224	57	298	1379
Total	11250	3183	423	2157	17013

Table 2.   Insects trapped in the room of flour mill with bran

	Traps				
Species	Multi funnel	TDA[a]	HDA[a]	Control[a]	Total
Lepidoptera					
E. elutella	-	4	-	-	4
E. kuehniella	56	197	3	2	258
H. pseudospretella	26	1	-	-	27
N. granellus	55	9	8	1	73
P. farinalis	15	1	-	-	16
P. interpunctella	2	18	-	-	20
S. cerealella	3	1	14	-	18
Coleoptera					
T. castaneum	-	-	1	-	1

[a]Wing trap produced by G. Donegani Institute, Novara, Italy.

Table 3.   Insects trapped in the room of flour mill with cracked kernel waste

	Traps					
Species	Multifunnel	*Rhyzopertha* pheromone[a]	*Sitophilus* pheromone[b]	*Tribolium* pheromone[c]	Control[d]	Total
Lepidoptera						
E. kuehniella	75	-	-	-	8	83
H. pseudospretella	4	1	1	1	1	8
N. granellus	26	-	-	-	-	26
P. farinalis	41	3	3	3	1	51
P. interpunctella	1	1	-	1	-	3
S. cerealella	-	-	1	-	1	2
Coleoptera						
A. unicolor	4	3	6	4	6	23
C. ferrugineus	12	85	114	68	71	350
R. dominica	686	917	144	43	76	1886
S. oryzae	9	33	51	29	43	165
T. castaneum	56	103	126	132	127	544
T. mauritanicus	1	21	23	24	25	104
T. molitor	-	2	-	1	1	4
Diptera						
C. pipiens	1	-	-	-	-	1
M. domestica	1	-	-	-	-	1

[a]Cardboard trap baited with 2.5 mg + 2.5 mg of dominicalure 1+2.
[b]Cardboard trap baited with 1 mg of sitophinone.
[c]Wyatt-Wynn trap baited with 2 mg of dimethyldecanal.
[d]Cardboard trap unbaited.

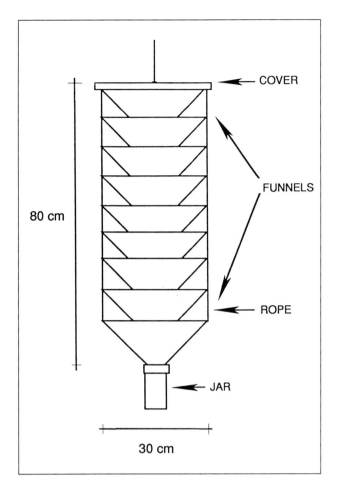

Fig. 1   Multifunnel trap used in the experiments.

the Lepidoptera:Phycitidae by Lorenz (1937) and Levinson and Hoppe (1983), in *Pectinophora gossypiella* Saunders by Farkas et al. (1974), in *Lymantria monacha* (L.) by Schneider (1981), in *Coleophora laricella* (Hübner) by Witzgall and Preisner (1984).

A review on factors affecting the design of traps for stored-product insects was reported by Barak et al. (1990).

Kennedy and Marsh (1974) described optomotor reactions in *E. cautella*, *E. kuehniella* and *P. interpunctella* males to a stripe pattern on the ground as an anemotactic guide to the appropriate pheromone stimulus. Levinson and Buchelos (1981) found that *Ephestia* males did not fly to a source of TDA in complete darkness, while they readily flew toward the pheromone at dim light, suggesting that optical stimulation of the retina is important for taking off.

Levinson and Hoppe (1983) showed that male *E. cautella* and *P. interpunctella* are visually attracted to certain shapes, preferably towards vertical rectangles. Leos-Martinez (1989) used Lindgren funnel traps without any attractant to determine the daily flight periodicity of *Ahasverus advena* (Waltl), *C. ferrugineus*, *O. surinamensis*, *R. dominica*, *T. castaneum*, *E. cautella*, *P. interpunctella* and *S. cerealella*. Trematerra and Capizzi (1991) reported that in the presence of sex pheromone, males of the Mediterranean flour moth attempt to mate with objects resembling females. Quartey and Coaker (1992) improved trap design by studying visual responses of *E. cautella* to rectangles of different colour and size and to brown and white stripes. They showed that moths were attracted to darker coloured vertical rectangles on a white background and to brown and white stripes. Buchelos and Levinson (1993) reported the efficacy of multisurface traps with and without pheromone addition, for monitoring and mass trapping of *Lasioderma serricorne* F.

Conclusions

Before mating, most male or female storage insects respond to their pheromone in combination with supplementary key stimuli. The sequence of sensory stimuli can be used in the design of attractant traps to be employed in the manipulation of storage pests. Trap design should be based on a thorough knowledge of the insect's behaviour and ecology. In this respect the preferential flight of many insects toward unbaited multifunnel traps is vital for developing trapping strategies. Accurate studies for estimating pest population levels are a necessity in order to maximise the efficiency of control procedures.

Trap design for monitoring storage moths has generally been empirical although consideration has been given to flight behaviour and to the form of the pheromone plume. Other factors may also affect trap efficiency such as its shape and colour, location of entry ports and ability to retain the insect after it enters the trap. The type of lure, release rate and plume dimension influence the active space of the plume and the insect's response to the lure. Trap positioning, environmental factors and trap type can also affect insect behaviour.

Ideally, traps should detect low levels of infestations which would otherwise go unnoticed and operate across a range of conditions to reflect accurately population changes in the target species.

In the case of complex infestations, where more than one species of flying insect is involved, the use of the multifunnel trap may be effective and may also reduce the costs of sampling and monitoring.

for the control of phycitid moths. Similarly the traps baited with HDA captured mainly *S. cerealella*. By comparison, the unbaited multifunnel traps collected a large numbers of *H. pseudospretella*, *N. granellus* and *P. farinalis*.

The multifunnel traps located in the cracked kernel waste room collected many species of Lepidoptera and Coleoptera. In the commercial beetle traps, nontarget species were also trapped, probably attracted by refuges. This was also observed in the unbaited cardboard traps used as control. The multifunnel trap caught more species of Coleoptera and with its high capacity there was no need to change any part of the trap during the trapping period. In the complex infestations with more than one species of moths and flying beetles involved, the multifunnel trap should provide better monitoring of all species and thus enable better pest management.

The success of the multifunnel trap directly relates to the flying insect, which depends upon factors influencing the initiation of flight such as population density, starvation, age and time of day on initiation of flight of adult insects, and by the capacity of the insects to fly towards the silhouette of the trap (McNeil 1991; Barrer et al. 1993).

It is possible that the species captured in a large number showed an attraction towards the silhouette of the multifunnel traps. This trap does resemble a tree trunk which may have been their original food source. Trunks of various trees may have provided the 'sign stimulus' for settling, calling and mating.

Similar behaviour was observed for some Coleoptera:Scolitydae by Vité and Bakke (1979), Borden et al. (1982), Lindgren et al. (1983), Birch (1984), Niemeyer (1985), McLean et al. (1987), Chénier and Philogène (1989) and for

References

Barak, A.V., Burkholder, W.E. and Faustini, D.L. 1990. Factors affecting the design of traps for stored-product insects. Journal of the Kansas Entomological Society 63, 4, 466–485.

Barrer, P.M., Starick, N.T., Morton, R. and Wright, E.J. 1993. Factors influencing initiation of flight by *Rhyzopertha dominica* (F.) (Coleoptera: Bostrichidae). Journal of Stored Products Research 29, 1–5.

Birch, M.C. 1984. Aggregation in bark beetles. In: Bell, W. and Cardé, R.T. eds, Chemical Ecology of Insects. London, Chapman and Hall, 331–353.

Borden, J.H., King, C.J., Lindgren, B.S., Chong, L., Gray, D.R., Oehlschlager, A.C., Slessor, K.N. and Pierce, H.D. 1982. Variation in response of *Trypodendron lineatum* from two continents to semiochemicals and trap form. Environmental Entomology 11, 403–408.

Buchelos, C.Th. and Levinson, A. 1993. Efficacy of multisurface traps and Lasiotraps with and without pheromone addition, for monitoring and mass-trapping of *Lasioderma serricorne* F. (Col., Anobiidae) in insecticide-free tobacco stores. Journal of Applied Entomology 116, 440–448.

Chénier, J.V.R. and Philogène, B.J.R. 1989. Evaluation of three trap designs for the capture of conifer-feeding beetles and other forest Coleoptera. Canadian Entomologist 121, 159–167.

Farkas, S.R., Shorey, H.H. and Gaston, L.K. 1974. Sex pheromone of Lepidoptera. Influence of pheromone concentration and visual cues on aerial odour-trial following by males of *Pectinophora gossypiella*. Annals of the Entomological Society of America 67, 633–638.

Kennedy, J.S. and Marsh, D. 1974. Pheromone-regulated anemotaxis in flying moths. Science 184, 999–1001.

Leos-Martinez, J. 1989. Periodicidad diaria de vuelo de insectos de productos almacenados. XX Congreso National de Entomologia. Oaxtepec, Morelos, Mexico, 91–108.

Levinson, H.Z. and Buchelos, C.Th. 1981. Surveillance of storage moth species (Pyralidae, Gelechidae) in a flour mill by adhesive traps with notes on the pheromone-mediated flight behaviour of male moths. Zeitschrift für angewandte Entomologie 92, 233–251.

Levinson, H. Z. and Hoppe, T. 1983. Preferential flight of *Plodia interpunctella* and *Cadra cautella* (Phycitinae) toward figures of definite shape and position with notes on the interaction between optical and pheromone stimuli. Zeitschrift für angewandte Entomologie 96, 491–500.

Lindgren, B.S., Borden J.H., Chong, L., Friskie, M. and Orr D.B. 1983. Factors influencing the efficiency of pheromone-baited traps for three species of Ambrosia beetles (Coleoptera: Scolytidae). Canadian Entomologist 115, 303–313.

Lorenz, K. 1937. Uber die Bildung des Instinktbegriffes. Naturwissenschaften 25, 289–300; 307–318; 324–331.

McLean, J.A., Bakke, A. and Niemeyer, H. 1987. An evaluation of three traps and two lures for the Ambrosia beetle *Trypodendron lineatum* (Oliv.) (Coleoptera: Scolytidae) in Canada, Norway, and West Germany. Canadian Entomologist 119, 273–280.

McNeil, J.N. 1991. Behavioural ecology of pheromone-mediated communication in moths and its importance in the use of pheromone traps. Annual Review of Entomology 36, 407–430.

Niemeyer, H. 1985. Field response of *Ips typographus* L. (Col., Scolytidae) to different trap structures and white black flight barriers. Zeitschrift für angewandte Entomologie 99, 44–51.

Quartey, G.K. and Coaker, T.H. 1992. The development of an improved model trap for monitoring *Ephesta cautella*. Entomologia Experimentalis et Applicata 64, 293–301.

Schneider, I. 1981. Freiland-und Laboruntersuchungen über das Verhalten maennlicher Nonnenfalter (*Lymantria monacha*) gegenüber Disparlur. Mitteilungen deutsch Gesellschaft Allgemein Angewandte Entomologie 2, 343–347.

Traynier, R.M. 1968. Sex attraction in Mediterranean flour moth, *Anagasta kuehniella*: location of the female by the male. Canadian Entomologist 100, 5–10.

Trematerra, P. and Girgenti, P. 1988. Etologia di *Rhyzopertha dominica* (F.) in presenza del feromone di aggregazione. In: Süss, L. and Trematerra, P. ed., La conservazione dei cereali: realtà ed esperienze a confronto. Milano, Mo.Ed.Co., 181–193.

Trematerra, P. and Daolio, E. 1990. Capture of *Rhyzopertha dominica* (F.) (Col., Bostrichidae) with Dominicalure multifunnel traps and considerations on trapping of not target species: *Sitophylus oryzae* (L.), *Cryptolestes ferrugineus* (Stephens) and *Colidium castaneum* (Herbst). Journal of Applied Entomology 110, 275–280.

Trematerra, P. and Capizzi, A. 1991. Attracticide method in the control of *Ephestia kuehniella* Zeller: studies on effectiveness. Journal of Applied Entomology 111, 451–456.

Vité, J.P. and Bakke, A. 1979. Synergism by an Ambrosia beetle. Naturwissenschaften 66, 528–529.

Witzgall, P. and Preisner, E. 1984. Behavioural responses of *Coleophora laricella* male moths to synthetic sex attractant, (Z)-5-decenol, in field. Zeitschrift für angewandte Entomologie 98, 15–33.

The potential of insect self-marking for the interpretation of trap catch

E.P. Wileyto*

Abstract

Self-marking is an approach used to make absolute estimates of insect populations using trap data. It is relatively labour free because traps and marking stations (traps modified to mark and release insects) are placed in the population simultaneously, and the user walks away. Dynamic models describing insect movements into marking stations and traps are used to determine the numbers of marked and unmarked insects expected to be found in the trap upon return. These models have been expanded to allow the estimation of parameters related to population growth, in addition to population size. The statistical methods used to infer population size, and other population parameters, are straightforward and used extensively in ecological research. This paper reviews the self-marking models and their application.

Introduction

Knowing the size of an insect population can be very important. For someone who manages a grain storage facility, this quantity may represent the information needed to make a decision about treatment. For someone else, determining population size may be an integral part of ongoing research. Regardless of need, populations are seldom estimated, except in a relative way, because of two problems. First, trapping alone gives only relative numbers or indices, not absolute estimates of size or density. Information about 'capture probability' or 'trapping efficiency' is needed to make the leap from a relative to an absolute measure, but it is not generally available. The only hope has been to relate trap efficiency statistically to the environment where the trap is placed, so that knowing enough about weather, and the items competing with traps for an insect's attention would allow us to infer capture probability (Andow 1984; Shore and McLean 1988; Hagstrum et al. 1988; Hagstrum et al. 1988). The approach is too complex; it incorporates many errors, first in describing the relationship between environment and capture probability, and second in assessing the state of the insect's current environment. The approach also demands that a manager track many aspects of the environment in order to feed the model and get the answer. Work in the field is best kept simple.

Traditional methods of population estimation, such as mark–recapture and removal, can be used to estimate not only population size, but also rates of recruitment and mortality (see Seber 1982; White et al. 1982; Nichols 1992). The problem is that these methods are too labour intensive to be used routinely for insect monitoring, and often do not perform

well when used for insect research. For mark–recapture methods, the initial marking phase may require that many thousands of insects be captured, marked, and returned to the population in good condition. Even then, getting few or no recaptures is a common occurrence. Removal methods would seem to be made for use with trapping, since they work by repeating the trapping efforts, and modelling the degree to which prior trapping depresses later trapping (White et al. 1982). Yet, the return can be small for the labour, requiring two or three sequential trap observations just to estimate population size.

The answer comes in part from using the techniques of statistical inference that are behind the traditional methods of population estimation. However, we bring them to bear against a unique class of marking and trapping models built on the assumption that the insects can do the work for us (Wileyto et al. 1994). For trapping, the observer may use any of the common trap designs on the market today, or design one for special use. Marking is accomplished by marking stations, otherwise identical to traps in structure and bait, but modified to mark and release insects back to the population. Traps and marking stations are placed in the population simultaneously, and when the traps are retrieved, they contain samples of marked and unmarked insects. The difficult work of interpreting results is kept in the laboratory, on the computer. Technology in the field is kept comparatively crude and easy to use.

Self-marking methods, like the traditional methods of population estimation, can be used to obtain estimates, not only for population size, but also for other parameters such as adult recruitment rates. The approaches are also alike in that they each consist of two parts: 1) a deterministic portion describing what is expected in the trap; and 2) a probabilistic portion, there because real trap results are, in part, a result of chance. This paper will emphasise the deterministic side. I will show how the self-marking scheme is modelled to produce expected results for a trap, and discuss the variety of additions that can be made to the model to invade further the lives of our favourite insects. I will also discuss how well they perform. Finally, I will mention some specific applications of recruitment estimates that have been articulated recently.

Estimation Models

Most methods of population estimation assume that the observer cannot find and count everyone (Seber 1982). Rather, a portion of the population is caught, leaving an unknown number at large. However, by having those caught split up into different categories, according to a specific probability model which is related to population size, an estimate can be made (Seber 1982). For instance, the mark–recapture method captures and marks a known number of animals before returning them to the population. That number, divided by the total population, is the capture probability for marked individuals during the next capture round. Likewise, the removal method assumes a constant (per capita) capture probability for each of several sequential trapping episodes, while

---

* USDA-ARS Stored Product Insects Research and Development laboratory, 3401 Edwin Street, Savannah, GA 31405 USA.

the population is diminished by a known number over each episode.

For either of the traditional methods, capture probabilities represent what is expected to be seen on average. The actual trapping results occur in part by chance. Closed population models of mark and recapture assume a hypergeometric sampling distribution, while removal methods assume a multinomial distribution. From there it is easy to use maximum likelihood techniques to obtain simultaneous estimates of population size and capture probability. (For information on these methods, see Seber 1982; Nichols 1992).

The self-marking approach, likewise, captures a portion of the population (Wileyto et al. 1994). The relationships between capture probabilities for marked and unmarked individuals are determined from the assumption that traps and marking stations are equally attractive, and operate at the same rate. Figure 1 is a schematic representing the flow of insects from the unmarked population (F) into the trap (C, R), under the original sampling scheme. (Category M indicates marked yet free insects.) Marking stations are used at a 1:1 ratio with traps, so that both per capita marking and trapping rates are equal to λ. The trajectories for all categories are obtained by solving for a set of differential equations:

$$E(F)' = -2\lambda F$$
$$E(C)' = \lambda F$$
$$E(M)' = \lambda F - \lambda M \quad (1)$$
$$E(R)' = \lambda M$$

Then, after substituting p for $e^{-\lambda t}$, we can specify the probabilities of being in category C, R, or of not being in the trap:

unobserved	captured, unmarked	captured, marked
p,	$1/2(1-p^2)$,	$1/2(1-p)^2$

This set of multinomial probabilities allows the use of maximum likelihood techniques to find a population estimate $\left(\hat{N}\right)$ simultaneously with the parameter p, given by:

$$\hat{N} = \frac{(C+R)^2}{2R} \quad (3)$$

which has a positive bias when the number of recaptures is small. (Bias indicates that on average, the estimate misses the true population size. Bias is virtually corrected by adopting the following version with slight modification in the denominator (see Wileyto et al.1994):

$$\hat{N} = \frac{(C+R)^2}{2(R+1)} \quad (4)$$

The approach can be improved by using a double marking scheme, which yields increased degrees of freedom. For example, suppose that for every trap, we use two red marking stations and one blue marking station. Insects can show up in the trap as unmarked (C), or red (R), or blue (Q), or red and blue (category S is doubly marked from visiting both types of marking station). Figure 2 shows a generalisation of this scheme, which uses α red and β blue marking stations. (Using unspecified ratios allows complete flexibility in sampling scheme specification after the experiment is done.) Again, solving simultaneously a set of differential equations provides the trajectory for each of the categories in the sampling scheme, and specifies multinomial probabilities to be used in estimation. Estimation then relies on numerical methods (Pollard 1977). Further, the added degrees of freedom may be used to test the fit of the model, using a chi-squared or other goodness-of-fit test.

The closed population models mentioned above have been tested thoroughly by simulating the sampling distributions of the population estimate over a wide range of parameter values,

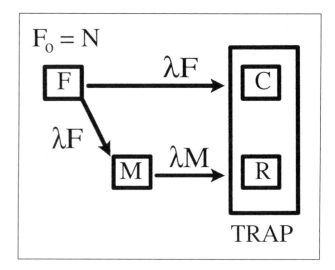

Fig. 1. The original sampling scheme. Traps and marking stations are available simultaneously, and per capita marking rates and trapping rates are both assumed to equal λ. The model projects the number of marked and unmarked insects expected in the traps.

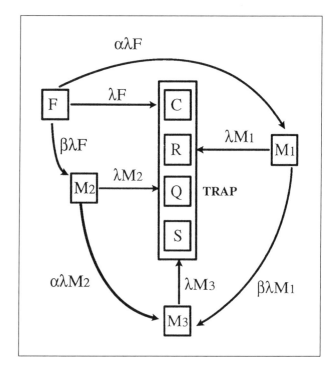

Fig. 2. The double marking scheme. For every trap, there are α red marking stations and β blue marking stations. Insects may arrive in a trap as unmarked, red, blue, or doubly marked. This doubles the degrees of freedom for the model.

and they work well. Generally, the estimate is biased (on average, misses the true value for population size) when the expected number of recaptures is small (Stuart and Ord 1991). Bias is easily corrected using a technique called the jackknife (except for the original 1:1 scheme, which is corrected analytically); the jackknife is a systematic resampling technique which empirically replicates the sampling distribution of the estimate (Efron 1982; Hinkley 1983; Potvin and Roff 1993). The corrected estimate is virtually unbiased unless the expected number of recaptures falls below five (Wileyto, unpublished).

Confidence limits may be obtained either as two standard-error limits, or by using the profile-likelihood technique. The two standard-error technique uses the variance estimate that comes from the information matrix that is used in the numerical estimation routine (Mood et al. 1974; Lehmann 1983). The profile-likelihood technique uses the likelihood ratio to determine confidence limits (Venzon and Moolgavkar 1988; Arnason et al. 1991). The profile-likelihood approach performed better than the two standard-error approach, producing as short an interval as two standard errors, but including the true value of the population size much closer to 95% of the time.

Adding to the Model

Degrees of freedom may be added to the sampling scheme by borrowing the approach used in the removal method, that of revisiting the trap to make sequential observations. Figure 3 shows a simplified version of the double mark sampling scheme, except that the traps are inspected a second (or third) time. This nets four degrees of freedom for each observation, for totals of eight or twelve. The sequential observations with many degrees of freedom are needed for the next step which is adding other ecological processes to the trapping model. Figure 3 also shows the simplest addition that can be made: recruitment of adults to the free and unmarked class (F). These could come from outside of the population, or from the larval age classes. The recruitment process could contain any process equation that makes sense, and is useful to the modeller. I have begun work with just two models. Model 1 (Fig. 3) works under the assumption that recruitment is constant over the sampling period, and proportional to the adult population by some factor b. This requires the estimation of b, in addition to the usual parameters of initial population size and time. Model 2 (Fig. 3) works under the assumption that recruitment is changing through the sampling period and uses a quadratic function of time to profile that variable recruitment. This requires the estimation of three additional parameters: b_1, b_2, and b_3.

It should be noted here that the trap categories are no longer represented by a multinomial probability model. However, the actual number in each category is represented by a Poisson distribution, where the mean is the expected count based upon solving the differential equations that arise from Figure 3.

Performance of the recruitment estimates must also be evaluated by simulation. Parameter b from model A could often be evaluated from a single visit to a trap, although two visits worked more consistently. Confidence limits from a single visit were too broad to be useful. Using two visits, b was biased for small expected numbers of recaptures. Bias adjustment by jackknifing has not yet been tested. Again, the profile-likelihood approach was found to be superior to the two standard-error approach.

Model B has not yet been tested, though I have made trial estimates, using a sampling scheme with three observations. The parameters b_2 and b_3 appear to have very tight confidence limits compared to b_1 or to b from model A.

Applications

I would like to touch briefly upon one of the potential applications for recruitment models. The quadratic function estimated will allow inferences to be made about the age distribution of the underlying larval population. Insects that have colonised a stored commodity often go through a series of boom and bust cycles before the age distribution becomes stable. Although augmented biological control for stored commodities is only now being investigated, we already know that timing of parasitoid release can be very important, especially if only a single release is to be made (L. Smith, pers. comm.). The quadratic model can tell us whether the age classes required by the parasitoid are on an upswing, downswing, or if stationary, whether they are on a crest or in a trough.

Conclusion

Self-marking offers a simple approach to estimating parameters of insect populations. It relies on the construction of dynamic models representing insects arriving at marking stations and traps. Those models might include only changes due to marking and trapping in a closed population, or they may incorporate constructs from ecological models representing change or structure in the population. Applications include inference of population age structure to aid in timing of treatment.

Acknowledgments

Research was conducted as part of the Agricultural Research Service project on Population Ecology and Predictive Models for Packaged Commodities, CRIS No. 6605–43640–018–00D, M.A. Mullen senior investigator. I thank M.A. Mullen for his ideas which set the project in motion, and I thank W.J. Ewens for being my mentor and for introducing me to estimation problems. I would also like to thank R.T. Arbogast, M.A. Mullen, L. Smith, D.K. Weaver for their comments on earlier versions of this manuscript.

References

Andow, D.A. 1984. Estimating relative population density of many species from trap data. Applied Entomology and Zoology 19, 391–392.

Arnason, A.N., Schwarz, C.J. and Gerrard, J.M. 1991. Estimating closed population size and number of marked animals from sighting data. Journal of Wildlife Management 55, 716–730.

Efron, B. 1982. The jackknife, the bootstrap and other resampling plans. (Regional Conference Series in Applied Mathematics). Philadelphia, PA: Society for Industrial and Applied Mathematics.

Hagstrum, D.W., Flinn, P.W., Subramanyam, B., Keever, D.W. and Cuperus, G.W. 1990. Interpretation of trap catch for detection and estimation of stored product insect populations. Journal of the Kansas Entomological Society 63, 500–505.

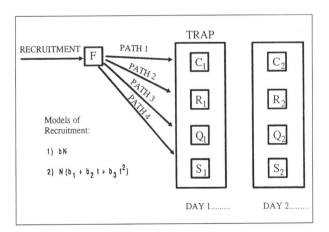

Fig. 3. Another view of the double marking scheme. The paths into the trap are simplified, but degrees of freedom are expanded by having the observer visit the traps more than once. The added degrees of freedom are useful for estimating recruitment to the adult population, the arrow at the left. Thus far, recruitment has been modelled as a constant rate (model 1), or as a quadratic profile over time (model 2).

Hagstrum, D.W., Meagher, R.L. and Smith, L.B. 1988. Sampling statistics and detection or estimation of diverse populations of stored product insects. Environmental Entomology 17, 377–380.

Hinkley, D.V. 1983. Jackknife methods. Encyclopedia of statistical sciences 4, 280–7.

Lehmann, E.L. 1983. Theory of point estimation. (Probability and mathematical statistics series). New York, NY: Wiley.

Mood, A.M., Graybill, F.A., and Boes, D.C. 1974. Introduction to the theory of statistics. (3rd Edition). New York, New York: McGraw-Hill Book Co.

Nichols, J. D. 1992. Capture–recapture models. Bioscience 42, 94–102.

Pollard, J.H. 1977. A handbook of numerical and statistical techniques. Cambridge, U.K., Cambridge University Press.

Potvin, C. and Roff, D.A. 1993. Distribution free and robust statistical methods: viable alternatives to parametric statistics? Ecology 74, 1617–1628.

Seber, G.A.F. 1982. The estimation of animal abundance. (Second Edition). New York, NY, MacMillan.

Shore, T.L., and McLean, J.A. 1988. The use of mark recapture to evaluate a pheromone based mass trapping program for ambrosia beetles in a sawmill. Canadian Journal of Forest Research 18, 1113–1117.

Stuart, A. and Ord, J.K. 1991. Kendall's advanced theory of statistics. (5th edition). Volume 2. New York, NY: Oxford University Press.

Venzon, D.J., and Moolgavkar, S.H. 1988. A method for computing profile-likelihood based confidence intervals. Applied Statistics 37, 87–94.

White, G.C., Anderson, D.R., Burnham, K.P. and Otis, D.L. 1982. Capture-recapture and removal methods for sampling closed populations. Springfield, VA: National Technical Information Service, U.S. Government Printing Office.

Wileyto, E. P., Ewens, W. J. and Mullen, M. A. Markov/recapture population estimates: a tool for improving interpretation of trapping experiments. Ecology, in press.

The statistical interpretation of insect self-marking and trapping

E.P. Wileyto*

Abstract

Insect self-marking offers a new approach for estimating capture probabilities, hence population size, from trap catch data. The novel aspect of the method is that it uses dynamic models of insect movement to generate categorical probability models of trap observations. Hence, the number of marked and unmarked individuals in a trap can be used to estimate the population size, using well known statistical techniques. This paper reviews, step-by-step, how to get from a dynamic self-marking model to estimates of population parameters and confidence limits. It also provides guidance on how to test the performance of any home-grown trapping model. This should help the reader understand and use the accompanying software (the program SELFMARK, demonstated in the trapping session), and it should help those who wish to develop their own self-marking models for specific purposes.

Introduction

Self-marking methods of population estimation (Wileyto et al. 1994) were developed especially for use with insect populations and trapping data. They work by allowing insects to mark themselves at marking stations (traps modified for mark and release) prior to arrival in a trap. The approach is technically related to the conventional methods of population estimation, which are mark–recapture and removal. However, the manual labour and impracticalities involved with using traditional methods on insect populations are replaced by simple dynamic models which describe the movement of insects from the free, unmarked population into the traps. Such models may also incorporate constructs from ecological models which reflect changes in population size and structure, enabling the user to address problems more interesting than population size alone.

The sampling schemes are very flexible, allowing the user to use any number of traps and marking stations. The original scheme specified a 1:1 ratio between marking stations and traps. Later versions incorporated double marking, where two different marking stations result in four different marking categories arriving in the trap (i.e. unmarked, red, blue, and doubly marked). The ratios between marking stations and traps were left unspecified in the later version, so that the user could incorporate exactly what was done. If more information is needed, for smaller confidence intervals or estimates of other ecological parameters, the traps can be visited by the user more than once, providing eight or twelve degrees of freedom for two or three visits, respectively.

* USDA-ARS, Stored Product Insects Research and Development Laboratory, 3401 Edwin Street Savannah, GA 31405 USA.

The approach is split logically into two facets. The first is deterministic. It uses differential equations to determine what is expected in the traps, depending upon the initial population size and time elapsed in trapping. This expectation may be in the form of capture probabilities for closed populations (those which presume to have a fixed number of trials, and have no loss or gain, apart from trapping, during the sampling period). The expectation may also be a number, related to initial population size, and population changes, when the population is not assumed to be closed.

The deterministic trapping model tells us what to expect if we know the parameters of the experiment, but it does not tell us exactly what we will see. In order to deal with uncertainties, and to make sensible estimates, we need the other facet, a probability model which reflects the role that chance plays in determining trap results. The deterministic parts of self-marking models have been described elsewhere (Wileyto, these proceedings). This paper deals with the probability and estimation portions of the model. It is presented to help the user understand the accompanying software demonstration, but also to give enough information so that users will be able to construct self-marking models, to make estimates for their own purposes.

Adding Chance to Fate

Closed and open population models incorporate different probability distributions. The closed population model specifies fixed capture probabilities for each marking category, and has a fixed number of trials which is equal to the population size, so that it reasonably meets the assumptions of the multinomial probability distribution (Stuart and Ord 1992a).

Suppose that there are three categories of events. Event one happens with probability p, event 2 with probability q, and event 3 with probability r. ($p+q+r=1$) The probability of obtaining event 1 i times, event 2 j times, and event 3 k times, out of M trials ($M=i+j+k$), without regard to the order, is

$$\Pr(i,j,k) = \frac{M!}{i!\,j!\,k!}\, p^i q^j r^k$$

(see Stuart and Ord 1992a). From Wileyto et al. (1994) we obtain probabilities of three trapping events:

unobserved	captured, unmarked	captured, marked	
p,	$1/2(1-p^2)$,	$1/2(1-p)^2$	(1)

The N population members make up N trials, and the above events occur $N–C–R$, C, and R times, respectively. By substitution, the probability of obtaining those event totals, without regard to order is:

$$\Pr(N-C-R,C,R) =$$
$$\frac{N!}{(N-C-R)!\,C!\,R!}\left(\frac{1}{2}\right) p^{N-C-R}(1-p^2)^C(1-p)^{2R} \quad (2)$$

Open populations cannot be represented by a fixed number of trials, because they are composed of an unknown number of individuals, each of which is present and exposed to the risks of trapping for unknown portions of the sampling period.

Arrival at a trap in any of the marking categories meets the assumptions of a Poisson distribution (see DeGroot 1975, Chapter 5, Theorem 1, regarding sums of random variables drawn from different Poisson distributions). This is convenient because the Poisson distribution has a single parameter, the mean or expected number of events (λ), and the outcome from the deterministic model is the expected number in each category. According to the Poisson distribution, the probability of seeing i events, when the expected number is , λ is given by:

$$Pr(i) = \frac{\lambda^i}{e^\lambda i!}$$

Using the capture probability (Eqn 1) as an example, the expected number of marked and captured individuals is

$$E(R) = \frac{N}{2}(1-p)^2$$

By substitution, the probability of obtaining exactly R recaptures in the Poisson model is

$$Pr(R) = \frac{\left(\frac{N}{2}(1-p)^2\right)^R}{e^{\left(\frac{N}{2}(1-p)^2\right)}R!} \tag{3}$$

Estimating Population Size and Other Parameters

Estimates for the parameters of any trapping model, and confidence limits, are easily obtained using the method of maximum likelihood (see Pollard 1977 for a review of methods). Maximum likelihood estimates are known to have many desirable properties: they are asymptotically unbiased and normally distributed, with a minimum variance (Stuart and Ord 1992b).

In order to use the method, one must first write down the likelihood function representing the model of interest. The term likelihood is synonymous with joint probability. For our closed population trapping model, it represents the probability of obtaining C unmarked and R marked insects in the trap, and $N–C–R$ insects which escape capture. For the multinomial model, likelihood (L) is identical to the expression for probability (Eqn 2):

$$L = \frac{N!}{(N-C-R)!C!R!}\left(\frac{1}{2}\right)^{C+R} p^{N-C-R}(1-p^2)^C(1-p)^{2R}$$

For a Poisson model with marking categories of unmarked, red, blue, doubly marked, with trap counts of C, R, Q, and S respectively, the likelihood would be:

$$L = Pr(C) \bullet Pr(R) \bullet Pr(Q) \bullet Pr(S)$$

The distinction between likelihood and joint probability is a logical one made by R.A. Fisher (Stuart and Ord 1992b). When referring to probability, parameters of the model are fixed, and the events or data are allowed to vary. When referring to likelihood, data are fixed, and the parameters are allowed to vary. Strictly speaking, events are probable while parameters are likely.

Likelihood functions are roughly bell shaped, though often skewed. Maximum likelihood techniques make an estimate by choosing the most likely value for a parameter, often using the simpler form of the log of the likelihood function. For the closed population model, the parameters N and p must be estimated simultaneously. The solution for $\hat{N}$ (trials) from multinomial capture data is determined from the likelihood (L) by setting $L(N,p)=L(N–1,p)$, and is well known as:

$$\hat{N} = \frac{\text{Total caught} \{C + R + ...\}}{1 - \hat{p}} \tag{4}$$

(See Seber 1982; Burnham et al. 1987.) The solution of $\hat{p}$ comes from setting

$$\frac{\partial \ln L}{\partial p} = 0$$

where $\ln L$ is the log of the likelihood function, and this can be done either analytically or numerically. (See Pollard (1977) for a review of numerical methods.)

The solution for $\hat{N}$, the initial population size in the open model, is structurally similar to that of the closed model (Eqn 4), except that the denominator is now the expected total rate of capture per capita (per initial individual):

$$\hat{N} = \frac{\text{Total caught} \{C + R+...\}}{\frac{E(C)}{\hat{N}} + \frac{E(R)}{\hat{N}}+...}$$

Bias of the Estimate

Though maximum likelihood estimates are asymptotically unbiased, they are often biased for smaller samples. Bias is defined as the difference between the mean value of the estimate of a parameter and the true value of the parameter:

$$\text{Bias} = E(\hat{N}) - N$$

The mean of $\hat{N}$ cannot be obtained from the multinomial probability distribution unless $\hat{N}$ has a simple analytical form. However, the mean can be approximated by simulation, as shown extensively, later in this paper.

Estimation and reduction of bias for an individual sample may be accomplished through the jackknife technique (Potvin and Roff 1993; Hinkley 1983; Efron 1982). The technique uses a systematic resampling of K observations by sequentially dropping one observation at a time, to produce K samples of $K–1$, and an empirical probability distribution for the estimate. The estimate of bias is then

$$\text{BIAS} = -\frac{K-1}{K}\sum_K(\hat{N}_K - \hat{N}_{K-1})$$

$$= -\frac{K-1}{K}\left[C(\hat{N}_C - \hat{N}_{C-1}) + R(\hat{N}_R - \hat{N}_{R-1})+...\right]$$

since $\{C+R + ...\} = K$.

Efron (1982) suggests a corresponding correction in which each of the K observations is added (duplicated) in turn, to produce a slightly different estimate of bias:

$$\text{BIAS} = \frac{K+1}{K}\sum_K(\hat{N}_K - \hat{N}_{K+1})$$

$$= \frac{K+1}{K}\left[C(\hat{N}_C - \hat{N}_{C+1}) + R(\hat{N}_R - \hat{N}_{R+1})+...\right] \tag{5}$$

This estimate of bias has an advantage when the probability of capture in several categories is small. We cannot generate estimates using the Markov-recapture technique if too many categories have zero entries, and having an entry of one in several categories would then disable the standard jackknife method. The method described above (Eqn 5) is not disabled by such data. This latter approach is incorporated in the SELF-MARK program.

Confidence Intervals

Two approaches may be used for determining confidence limits for the estimate of N. The first approach is through the theoretical approximation of variances of the estimates, which are obtained from the inverse of the information matrix (Mood et al. 1974; Lehmann 1983), found by taking the second derivatives of the log-likelihood functions. These second order partials are evaluated using the maximum likelihood estimates for all parameters. An approximate 95% confidence interval for population size N may then be obtained as

$$\text{C.I.}(95)_{\hat{N}} = \hat{N} \pm 2\sqrt{V\hat{A}R(\hat{N})}$$

The two standard-error approach at times produces a confidence interval which misses the true parameter value more often than it should. A more recent approach is the profile likelihood interval (see Venzon and Moolgavkar 1988; Arnason et al. 1991; Lebreton et al. 1992). Using the closed population model as an example, the approach bases the upper and lower limits on the likelihood ratio

$$LR = L(\hat{N},\hat{p})(L(N,p^*))$$

where p^* is the maximum likelihood estimate of p, given N). Since $2\,Ln\,(LR)$ is distributed approximately as χ^2 with one degree of freedom, limits occur where

$$-2\,Ln(LR) = 3.84$$

the 95% significance value for a two sided chi-square test. Profile-likelihood intervals tend to be as broad as the corresponding two standard-error interval, although the location differs, and the interval may be asymmetric about the estimate (Lebreton et al. 1992). (For a high performance algorithm that calculates confidence limits by profile likelihood, see Venzon and Moolgavkar 1988.)

Testing the Performance of Estimates and Confidence Intervals

Self-marking models, like all other novel statistical models, must be tested before they are ready for public consumption. I would like to mention the procedures for testing the estimates and confidence intervals before closing.

First, the parameter estimates. The property of greatest interest is bias. When an estimate is biased, on average it misses the true value (e.g. the true population size) in a particular direction. In order to determine the bias of an estimate, we must know the sampling distribution of that estimate. Whenever the estimate or sampling distribution does not have a simple analytical form, the distribution must be simulated. This is easily done by assuming parameter values, generating random trap catches on the computer, calculating the resulting parameter estimates, and finally, the average. If an estimate is found to be biased, then we might employ the jackknife resampling scheme to correct for the bias. If this is done, we must then work out the bias of the corrected estimate using the same approach.

It is also important to determine how well the confidence limits perform. Here, there are two criteria. First, a confidence interval should be as short as possible. There is a theoretical limit to how short an interval can be. The variance calculated using the information matrix, with known parameter values, is really a lower limit to the variance. You generally get close, on average, to an interval length which reflects this variance by using the two standard-error approach. The profile likelihood interval also gets close to this value. Testing is done by evaluating the average interval length, again by simulating the sampling distribution of interval lengths using random trap data.

The second goal is that a confidence interval should cover what it proports to cover. That is, a 95% confidence interval for population size should include the true size 95% of the time. Testing is done, again, by simulating random trap numbers. Here is where the profile likelihood interval usually surpasses the two standard-error interval. Two standard errors will often fall short for some parameter combinations, covering 80 to 90%, rather than 95%, whereas the profile likelihood interval will remain close to 95%. It should be noted that the two standard-error interval may perform well under some combinations of parameters, and is computationally much more simple.

Random trap numbers may be generated under a variety of circumstances, which vary depending upon the context of the test. The dynamic model behind self-marking and trapping is an abstraction, a simplification of what happens in the real world. We generally avoid making the trapping model increasingly complex to deal with the real world. Rather, we examine how well an estimate will do in spite of the simplification. One type of test which is always done assumes that the random trap catches are generated in exactly the way expressed in the construction of the model, and reflected in the joint probability distribution (Eqns 2 and 3). (For convenience, let us refer to these as type 1 random numbers.) Any bias found here is due to the nature of the estimate, and not to any real world intervention. This is also the sort of bias that can be helped by using the jackknife resampling scheme described earlier (Eqn 5).

Another common test generates random trap catches from a probability model other than the original (type 2 random numbers). For instance, we might use a closed population estimate on data generated by an open population model. This allows us to determine the impact of specific violations of assumptions, violations that we might be likely to see when employing the method in the real world.

Finally, a less common test generates random trap catches against the backdrop of a realistic population model that might include spatial structure, age structure, weather, and many ecological or behavioural factors (type 3). This is most appropriate for examining the ability of the estimate to improve our decision-making in the real world; we might want to know how often we would make the right or wrong decisions about treatment with and without the model.

Regardless of the circumstances under which the random trap numbers are generated, tests for bias and performance characteristics of confidence intervals should be performed before a procedure is put into routine use. Tests using type 1 random numbers should always be run. Tests using type 2 numbers should usually be done, because there are normally some common violations of assumptions. Type 3 random numbers will only be useful when inspecting the performance of decision rules.

Conclusion

The self-marking and trapping approach to population estimation uses a set of differential equations representing insect visitation to marking stations and traps, in order to generate a model of the numbers expected in the trap by category. The estimation of population size, and other parameters, results from a straightforward application of estimation techniques used commonly for population estimation in wildlife ecology and fisheries. The maximum-likelihood technique works well for estimation, although estimates may be biased under some circumstances. Bias may be adjusted using the jackknife resampling technique. For novel models of marking and trapping, extensive testing should be performed before they are put into general use, to be certain that we understand the behaviour of the estimate and confidence interval.

Acknowledgments

Research was conducted as part of the Agricultural Research Service project on Population Ecology and Predictive Models for Packaged Commodities, CRIS No. 6605–43640–018–00D, M.A. Mullen senior investigator. I thank M.A. Mullen for his ideas which set the project in motion, and I thank W.J. Ewens for being my mentor and for introducing me to estimation problems. I also thank R.T. Arbogast, M.A. Mullen, L. Smith, and D.K. Weaver for their comments on earlier versions of this manuscript.

References

Arnason, A.N., Schwarz, C.J. and Gerrard, J.M. 1991. Estimating closed population size and number of marked animals from sighting data. Journal of Wildlife Management, 55, 716–730.

Burnham, K.P., Anderson, D.R., White, G.C., Brownie, C. and Pollack, K.H. 1987. Design and analysis for fish survival experiments based on release-recapture. American Fisheries Society Monograph 5.

DeGroot, M.H. 1975. Probability and statistics. Reading, MA, Addison-Wesley Publishing Co.

Efron, B. 1982. The jackknife, the bootstrap and other resampling plans. (Regional conference series in applied mathematics). Philadelphia, PA, Society for Industrial and Applied Mathematics.

Hinkley, D.V. 1983. Jackknife methods. Encyclopedia of statistical sciences, 4, 280–7.

Lebreton, J., Burnham, K.P., Colbert, J. and Anderson, D.R. 1992. Modelling survival and testing biological hypotheses using marked animals: A unified approach with case studies. Ecological Monographs 62, 67–118.

Lehmann, E.L. 1983. Theory of point estimation. (Probability and mathematical statistics series). New York, Wiley.

Mood, A.M., Graybill, F.A., and Boes, D.C. 1974. Introduction to the theory of statistics. (3rd edition). New York, McGraw-Hill Book Co.

Pollard, J.H. 1977. A handbook of numerical and statistical techniques. Cambridge, U.K., Cambridge University Press.

Potvin, C. and Roff, D.A. 1993. Distribution free and robust statistical methods: viable alternatives to parametric statistics? Ecology, 74, 1617–1628.

Seber, G.A.F. 1982. The estimation of animal abundance. (2nd edition). New York, MacMillan.

Stuart, A. and Ord, J.K. 1991. Kendall's advanced theory of statistics. (5th edition). Volume 1. New York, Oxford University Press.

Stuart, A. and Ord, J.K. 1991. Kendall's advanced theory of statistics. (5th edition). Volume 2. New York, Oxford University Press.

Venzon, D.J., and Moolgavkar, S.H. 1988. A method for computing profile-likelihood based confidence intervals. Applied Statistics 37, 87–94.

Wileyto, E.P., Ewens, W.J. and Mullen, M.A. Markov/recapture population estimates: a tool for improving interpretation of trapping experiments, Ecology, in press.

The detection of insects in grain during transit—an assessment of the problem and the development of a practical solution

D.R. Wilkin*, D. Catchpole† and S. Catchpole†

Abstract

Pest detection in grain, particularly during transit, is of great commercial importance. Most current methods will not detect insects at low population densities and often cannot estimate levels of infestation accurately. This limits management options and encourages the prophylactic use of pesticides.

Much research continues to be directed at improving methods of detection. However, the results so far either offer very limited improvements or require much more development before they are suitable for commercial application. Following a thorough review of the available options, it was concluded that some method of separating insects rapidly from large (5 kg or more) samples of grain, offered the most practical route to improved detection.

Prototype machines using sieves or aspiration to separate insects from grain were built and tested. The best results were obtained with a flat-bed, reciprocating sieve and development was concentrated on this design. The rate of flow of grain through the machine had a profound effect on the recovery of insects but the angle of the sieve and the amplitude of shaking had much less effect. The optimised prototype sieve was able to give consistent recoveries of two important species of grain pest close to 100%. Tests showed that this unit could process 10 kg of wheat in 1.8 minutes, recovering almost 100% of check insects in the grain. It proved capable of detecting infestations at population densities of 0.2 insects/kg and also gave reliable estimates of population densities.

Introduction

Rapid decisions must be made about the quality of individual lots or loads of grain, particularly during transit. The consistency and accuracy of such decisions will affect the price paid for the grain and will also influence the perceived value of the grain to the buyer. Often, the detection of insects will result in the rejection of a parcel of grain and will certainly affect the attitude of the buyer towards further storage or the need for treatment before storage. More importantly, failure to detect insects may lead to a load of infested grain being added to an otherwise uninfested bulk.

Currently, the majority of rapid assessments are based on the examination of a sample or samples. There are no internationally recognised methods, although several national bodies have well documented techniques for internal use. All seem to be based on the collection and examination of relatively small samples of grain (0.5–1.0 kg).

* 39 Denham Lane, Chalfont St Peter, Gerrards Cross, Bucks., SL9 0EP, U.K..

† Samplex, Gaymers Way, North Walsham, Norfolk, NR28 0AN, U.K.

Work by Cogan and Wakefield (1987) suggested that, in small, 50 kg containers of grain, the likelihood of detecting insects by the collection of samples was a simple factor of the amount of grain removed and population density. However, Wilkin and Fleurat-Lessard (1990), working with artificially contaminated 20-t bulks of wheat, found that the removal and examination of samples will only detect insects with a high degree of probability when the number/kg exceeds five. The chance of detection was related to the population density and the number of samples or weight of grain that was examined. For example, only one 250 g sample in 80 contained an insect when the infestation rate was 0.2 insects/kg but rose to about 1 in 2 when the rate was 5 insects/kg. The basic conclusion of the work was that the standard approach to insect detection by sampling was extremely insensitive and was likely to miss substantial populations of insects, as well as underestimating the population density.

The aim of the work reported in this paper was to investigate the options for rapid detection of insects in grain and then, if possible, to develop a simple machine that would offer substantial improvements over current methods.

Current Approaches to Detecting Insects in Samples

It is widely recognised that examination of samples is an essential part of detecting insects in bulks of grain during transit and, to some extent, during storage. Therefore, the sampling technique is also likely to influence the effectiveness of detection (Hurburgh 1983; Wilkin 1991) but this was considered to be beyond the scope of this investigation. However, having obtained an appropriate sample, the effectiveness of the method of assessment will also be of great importance, both in terms of its ability to detect insects and its commercial acceptability.

Methods of detecting insects in samples of grain can be divided into two parts: novel approaches and mechanical separation.

Novel methods

There have been a number of research projects aimed at investigating methods of detecting insects in samples of grain. Several of these date back many years, but as yet, none seem to have been developed to a stage of commercial acceptability. Some of the more effective approaches are described below.

Detection of carbon dioxide

Insects respire and one of the by-products of respiration is carbon dioxide. Measurement of the carbon dioxide emissions from a grain sample can give an indication of the presence of insects and the level of infestation. The principles of the method were described by Howe and Oxley (1952), together with some of the advantages and disadvantages. Modern equipment offers the possibility of relatively rapid assessment of samples but only at the expense of costly and complicated apparatus (Bruce and Street 1974). It is also difficult to make the method work reliably with grain having a moisture content of about 15% or more, as the carbon dioxide produced by the grain may mask infestation. A final disadvantage is that at least five minutes per sample is required for an assessment.

Detection of uric acid

Uric acid is another by-product of insect metabolism which can be detected in a sample of grain (Wehling et al. 1984). However, this method requires expensive analytical apparatus and the results do not necessarily give an indication of current infestation. Uric acid is a relatively stable compound and may have accumulated in the sample over a period of time before analysis. Also the uric acid detected in the grain may have been produced by harmless, stray insects rather than damaging pests.

X-ray analysis of grain samples

Samples of grain can be examined by using X-rays and insects, particularly those hidden within grains, will show up (Fesus 1972). This method is used commercially in France by flour mills to examine samples of incoming grain. However, it requires expensive equipment, is relatively slow (about 30 minutes/sample) and will only work with very small samples of grain. Future development with equipment and perhaps image analysis technology could render this approach viable but this seems unlikely within the next five years.

Acoustic detection

Insects produce sound when moving and feeding in grain and this can be detected with sensitive equipment. Much research has been directed at this technique (Hagstrom 1990), with particular emphasis being placed on the use of microprocessors to differentiate between the sounds of insects and background noise. The general approach shows great promise but seems to be still some way from commercial application.

Nuclear magnetic resonance or near infra-red spectroscopy

Spectrographic examination of grain samples is already widely used in the grain industry to determine a number of quality parameters. Chambers et al. (1984) and Wilkin et al. (1988) demonstrated the potential of these techniques to detect insects and mites. However, there are some fundamental disadvantages associated with current analytical equipment, which limits the methods to working with small samples of grain (100 g). A major breakthrough in either the equipment or its method of use will be required before these methods can be considered as having potential for commercial use.

Extraction of insects with Berlese-type equipment

Live insects can be persuaded to leave a grain sample by the application of dry heat. The Canadian Grain Commission uses this technique to check some samples of grain collected during an annual audit process. Smith (1977) demonstrated that the method was effective with dry grain (<14%) but recovered less than half the insects from grain with a moisture content of 16%. It uses relatively simple equipment and could work with large samples of grain, but is unsuitable for rapid assessments.

Mechanical separation

Despite the work reported above, sieving a sample of grain to remove insects, or some other form of physical separation, is by far the most widely used method and would appear to offer the greatest potential for the speedy development of an improved approach to detection. Surprisingly, little research

has been carried out on this topic, perhaps because of its apparent simplicity.

Sieving relatively small (250 g–1 kg) samples is by far the most widely used method of detecting insects in grain, both in the U.K. and on a worldwide basis. The Canadian Grain Commission, the Australian Wheat Board and the U.S. Federal Grain Inspection Service all rely on sieving numerous, relatively small samples of grain to detect insects. None of these bodies quotes any scientific research to support the method of sieving that is recommended, although methods of sample collection are carefully prescribed.

Unpublished work by Goodman (1979) suggested that manual sieving was not entirely reliable in removing insects from samples of grain. It is extremely surprising that there does not appear to be any published data on this topic, despite the widespread, international use of the technique.

Sieving of larger samples has been recognised as a method of improving the detection of insects in grain. White (1983) describes an inclined plane sieve for use with 25 kg samples. However, tests of the efficacy of this simple device suggested that up to seven passes down the sieve were needed to recover all the test insects seeded into a sample. This would indicate that a simple sieve may not provide the complete answer to the problem of detection, irrespective of the size of samples that are used.

Summary

The following conclusions can be drawn from the above review of available methods:

- Insect detection is inextricably linked to the amount of grain that is examined, regardless of the method used. However, commercial constraints limit the amount of time that can be spent examining samples.
- The commercial options for the rapid assessment of large samples of grain for the presence of insects are limited.
- Much recent and current research on the topic seems to be aimed at developing methods that are not suitable for the rapid assessment of large samples or are still some way from commercial application. In addition, some of the methods will require the use of expensive equipment.
- Mechanical separation (sieving or aspiration) seemed to offer the best option for rapid detection of insects in large samples of grain.

Investigations and Development Work

Assessment of manual sieving

Despite the widespread use of manual sieving, there are no data to confirm that it is an effective method of removing insects from grain. Therefore, a small experiment was done to assess the efficiency of manual sieving in removing insects from samples of grain under controlled conditions. In general, this was based on the procedures used to assess grain offered for sale into the U.K.

Materials

English wheat of unknown origin and a moisture content of about 13% was used as the substrate. This grain was divided into 500 g batches and live, adult *Oryzaephilus surinamensis* or *Sitophilus granarius* of unknown age and sex, were collected from laboratory cultures and added to the samples. These samples were then sieved over a 200 mm diameter sieve of the type recommended by the Intervention Authorities, with a 2 mm stainless steel mesh.

Methods

Insects were added to the batches of grain at rates of 1/sample and 5/sample and each species was assessed separately. The insects were allowed between 5 and 10 minutes to disperse before sieving commenced. In all but one case, five replicate samples were used with each density and species. Exceptionally, in one case 10 replicates were used.

Each infested sample was tipped into the sieve and shaken by the same operator for 30 seconds. The sievings in the receiver were examined and the number of insects recorded. Batches of grain to which no insects had been added were assessed in the same way to confirm that the grain was uninfested.

Results

Complete recovery of *S. granarius* was obtained at both rates of inclusion. Complete recovery of *O. surinamensis* was obtained at 5/sample and in 9 out of 10 replicates at 1/sample.

Conclusions

Given conditions as described, manual sieving should detect the presence of either of the two species in a sample without difficulty. It should also allow the population density within the sample to be estimated with a high degree of precision. Therefore, any new device must aim to offer a similar level of effectiveness.

Production of prototypes

When options for prototype machines were considered, certain constraints had to be applied to limit the range of designs that could be considered. These constraints included, likely cost of the final machine, ease of manufacture and, most importantly, ease of use. Three options were considered: a rotating drum sieve (Fig. 1); a zig-zag aspirated separator (Fig. 2); and an eccentric, reciprocating table sieve (Fig. 3). Working prototypes of all three machines were built and tested.

Testing of prototypes

Initially, all prototypes were tested for their ability to remove fine material from 10 kg batches of grain. The two sieve based separators gave broadly similar results but the aspirated separator removed almost twice as much material. This was because it was also removing some small and light grains that were not separated by the sieves.

For the second series of tests, 50 dead adult *S. granarius* were added to the 10 kg batches of wheat that had been cleaned. About 50% of the insects were recovered using the table sieve, none were recovered with the rotating drum sieve and about 25% with the aspirated separator. However, with the latter machine the insects were recovered as fragments.

Some time was spent investigating the performance of each machine and assessing the effects of modifications. The poor performance of the rotary sieve appeared to be related to the relatively small area of sieve that was in contact with the grain at any one time. Using very slow rates of throughput and internal baffles allowed this unit to recover some insects but never as many as the other sieve. Therefore, the rotating drum sieve was abandoned.

The number of insects recovered by the aspirated separator could be increased by careful adjustment of the rate of airflow but some of the dead insects were still being damaged by passage through the machine. It is possible that live insects would not have been damaged to the same extent. Difficulties were also encountered in adjusting the airflow and obtaining consistency between samples. It was, therefore, decided to abandon this approach.

Work on the reciprocating sieve showed that recovery could be improved by reducing the depth of grain flowing over the sieve, and up to 75% of dead *S. granarius* could be separated from the grain in a single pass. Further development work was concentrated on this machine.

Modifications to the reciprocating sieve

The rate of flow of the grain through the sieve and the depth of grain on the sieve plate affected the recovery of insects, with slower, thinner flows giving best results. However, even at the slowest flow rates, recovery of dead insects was not as good as obtained with the hand-operated sieve. Amplitude of shaking and inclination of the sieve plane had relatively little influence on the rate of recovery.

Initially, the machine used a 2 mm, round hole sieve plate and the area of these holes was smaller than the area of the

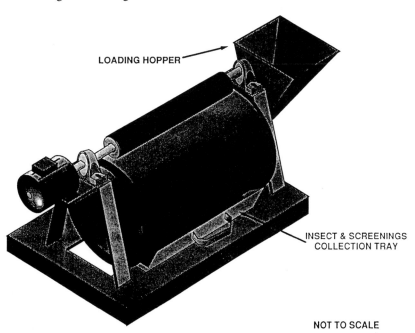

LOADING HOPPER

INSECT & SCREENINGS COLLECTION TRAY

NOT TO SCALE

Fig. 1.　Rotating drum sieve.

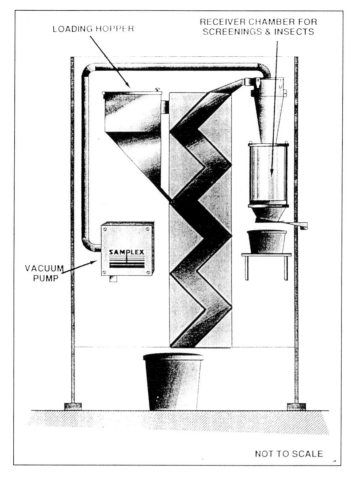

Fig. 2. Zig-zag aspirated separator

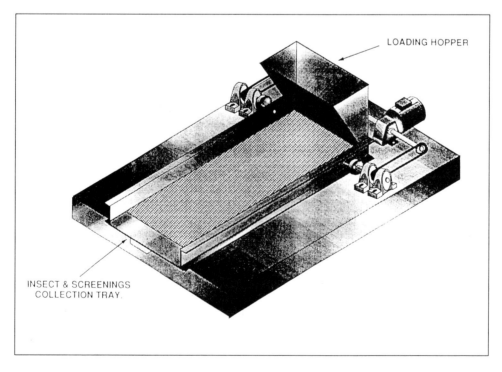

Fig. 3. Eccentric reciprocating table sieve.

apertures in a 2 mm mesh sieve. For mechanical reasons, meshes are less satisfactory than perforated plates in mechanised sieves. Therefore, a sieve plate with a 2.5 mm round hole was substituted. This produced an immediate improvement in recovery to almost 100% and allowed the rate of throughput to be increased, giving a time to sieve 10 kg of less than 2 minutes.

A full assessment of this machine was then done using live insects. Ten live adult *S. granarius* or *O. surinamensis* were added to 5 kg samples of wheat and each sample was passed through the machine three times. Three replicate samples were tested for each species. Between 80 and 90% of the *S. granarius* were recovered on the first pass and 100% were recovered by the third pass. Complete recovery of *O. surinamensis* was achieved on the first pass. A further, replicated test with *S. granarius* was carried out with two insects added to each 5 kg sample. Complete recovery was achieved on the first pass.

A final test was carried out in which the 10 adults of both species were added to 5 kg samples of wheat and allowed 24 hours before the samples were sieved. Between 80 and 100% recovery of both species was obtained on the first pass. However, total recovery was not achieved in every case, even after three passes through the machine, and it is possible that some insects had escaped from the grain.

Development and Testing of a Pre-production Prototype

The prototype machines had been built to test the principles involved, but did not address the needs of operating under practical conditions, nor did they incorporate all of the features which were likely to improve the ability of a sieve to recover insects. Any practical machine must be easy to use and safe, as well as being effective. Ease of use must include features such as ease of loading, simplicity of operation, ease of cleaning and low noise level.

A machine was constructed, incorporating the basic features of the prototype, flat-bed sieve, but taking account of the requirements for safety, ease of use and incorporating features that were likely to optimise efficiency. The machine was constructed in stainless steel, although a final production version may use other materials (Fig. 4). The main sieve was a flat plate with 2.5 mm holes (Fig. 5), although an option was included to incorporate two sieve plates of different meshes. The sieves were easily detachable for cleaning (Fig. 6). The insects and sievings were collected in a removable receiver (Fig. 7).

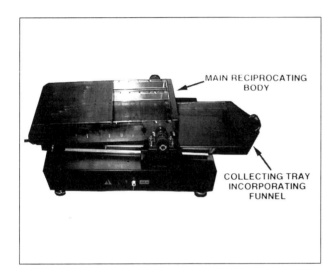

Fig. 4.   Pre-production prototype reciprocating sieve.

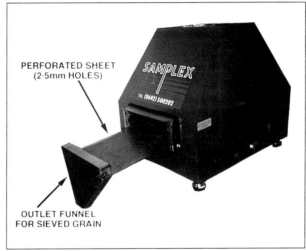

Fig. 5.   Pre-production prototype reciprocating sieve, showing flat plate with 25 mm holes.

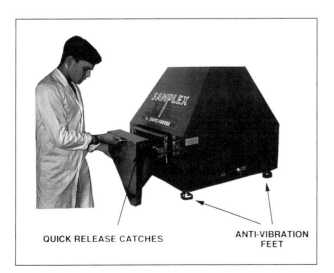

Fig. 6.   .Pre-production prototype reciprocating sieve, showing detachable sieves.

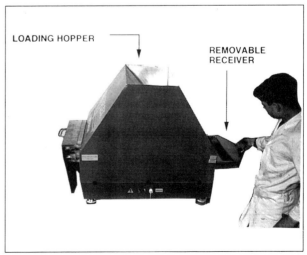

Fig. 7.   Pre-production prototype reciprocating sieve, showing removable receiver.

Table 1. The mean recoveries of insects from 5 kg samples of wheat or barley using the final prototype reciprocating sieve. Each test result is the mean of three replicates.

Test no.	Conditions of test	Percent recovered on first pass	
		S. granarius	*O. surinamensis*
1	10 live insects/5 kg wheat	97	_[a]
2	As above, insects left on grain for 5 minutes before sieving	100	_[a]
3	1 live insect/5 kg wheat	100	_[a]
4	10 live insects/5 kg wheat	_[a]	100
5	1 live insect/5 kg wheat	_[a]	100
6	100 live insects/5 kg	_[a]	97
7	5 live insects of each species/5 kg of barley	107[b]	173[b]

[a]No insects of this species were included in the test.

[b]The barley was used 'as supplied' by a merchant and was already infested with both species of insect used in the tests. Assessments of the grain without additional insects confirmed that the results in the table reflected the inherent infestation in the barley plus the test insects.

A detailed test program was undertaken to confirm that this machine would extract insects from grain at least as well as the first prototype and also to confirm that it was easy to use.

Test procedures

English wheat of unknown origin was used for most of the tests but English feed barley was used in one case. Initial tests were with dead insects but then live insects were substituted. Adult *O. surinamensis* or *S. granarius* taken from laboratory cultures were used. Five-kg samples were used for all tests and each sample was passed through the machine up to three times or until all insects were recovered. For each test, three replicate batches were used. For initial tests, the flow rate through the machine was 5 kg/30 seconds. Subsequently, the rate was reduced to 5 kg/50 seconds.

Results

The initial tests with dead insects and a rapid flow rate gave a mean recovery rate for 10 dead *S. granarius* of 87% on the first pass. A further 10% was recovered during the second pass and the final 3% of the insects were recovered on the third pass. When the tests were repeated with the slower flow rate, 100% of the *S. granarius* were recovered on the first pass. The rates of recovery at the slower flow rate are summarised in Table 1.

The recovery rate of insects in the above tests was always high and the insects appeared undamaged and were still alive after extraction. When the recovery was less than 100%, the missing insects were almost always recovered by passing the grain through the machine a second time. Stripping down and cleaning the machine failed to yield any insects lodged in the sieve or hopper system.

A further series of tests were done using grain containing about 2% screenings. This did not affect the recovery rate but did slow the final assessment as the insects were somewhat more difficult to spot in the receiver.

Conclusions

An assessment of the options available for the rapid detection of insects in grain during transit, suggests that whatever method is used, it must be capable of dealing with multi-kilogram samples if the low population densities are to be detected or estimated accurately. The mechanical separation of insects from samples of grain was chosen as the most appropriate method for the development of a machine that

could be produced and offered for sale within a limited time period.

The pre-production prototype, flat-bed, reciprocating sieve that was developed during this project is able to process large samples of grain rapidly (10 kg of wheat/1.8 minutes). It is able to detect reliably two important species of grain pest at population densities of 0.2 insects/kg. This is comparable to the population densities detected by trapping methods used during storage and is greatly superior to the level of 2–5 insects/kg detectable with current, commercial sampling and sieving methods. The test results from this machine also show that it was possible to estimate the population density to a high degree of accuracy. The pre-production prototype appears to meet all requirements for ease of use and safety, and would appear to be well suited for use in a commercial grain laboratory.

Acknowledgments

This work was supported by a grant from the Home-Grown Cereals Authority of the U.K. The authors are grateful for the assistance with the technical review provided by Francis Fleurat-Lessard of INRA, Bordeaux.

References

Bruce, W.A. and Street, M.W. 1974. Infrared carbon dioxide detection of hidden insects. Journal of Georgia Entomological Society 9(4), 260–265.

Chambers, J., McKevitt, N.I. and Stubbs, M.R. 1984. Nuclear magnetic resonance spectroscopy for studying the development and detection of the grain weevil, *Sitophilus granarius*, within wheat. Bulletin of Entomological Research 74, 707–724.

Cogan, P.M. and Wakefield, M.E. 1987. Further developments in traps used to detect low level infestation of beetle pests in stored grain. British Crop Protection Council Monograph 37, 161–168.

Fesus, J. 1972. Detection and estimation of internal pest infestation in seeds by the application of X-ray techniques. OEPP/EPPO Bulletin 3, 65–76.

Goodman, A. 1979. The removal of insects from grain by sieving. Unpublished data in the possession of D.R. Wilkin.

Hagstrom, D.W. 1990. Automated acoustical detection of stored grain insects and its potential in reducing insect problems. In: Fleurat-Lessard, F. and Ducom, P. ed., Proceedings of the Fifth International Working Conference on Stored-Product Protection, Bordeaux, France, September 1990, II, 1341–1350.

Howe, R.W. and Oxley, T.A. 1952. Detection of insects by their carbon dioxide production. Her Majesty's Stationary Office, London.

Hurburgh, C.R. 1983. Grain testing and shrinkage. Proceedings of grain storage and conditioning conference, Iowa State University, Ames Iowa.

Smith, L.B. 1977. Efficiency of Berlese-Tullgren funnels for the removal of the rusty grain beetle, *Cryptolestes ferrugineus*, from wheat. Canadian Entomologist 109, 503–509.

Wehling, R.L., Wetzel, D.L. and Pederson, J.R. 1984. Stored wheat insect infestation related to uric acid as determined by liquid chromatography. Journal of Association of Official Analytical Chemists 67(3), 644–647.

White, G.G. 1983. A modified inclined sieve for separation of insects from wheat. Journal of Stored Products Research 19(2), 89–91.

Wilkin, D.R. 1991. An assessment of methods of sampling bulk grain. Project Report No. 34, Home-Grown Cereals Authority, HGCA, London.

Wilkin, D.R., Cowe, I.A., Thind, B., McNicol, B. and Cuthbertson, D.C. 1986. The detection and measurement of mite infestation in animal feed using near infrared reflectance. Journal of Agricultural Science (Cambridge) 107, 439–448.

Wilkin, D.R. and Fleurat-Lessard, F. 1990. The detection of insects in grain using conventional sampling spears. In: Fleurat-Lessard, F. and Ducom, P. ed., Proceedings of the Fifth International Working Conference on Stored-Product Protection, Bordeaux, France, September 1990, III, 1445–1455.

Trapping *Trogoderma variabile* (Coleoptera: Dermestidae): a comparison of traps and techniques for adult and larval monitoring

E.J. Wright* and R.L. Delves†

Abstract

As part of a large Australian research program on the warehouse beetle, *Trogoderma variabile*, various trap designs and trapping methods were tested for larvae and adults of this species. Due to the cryptic habits of the larvae, detection is very difficult and an effective larval trap would be very useful as a monitoring tool. Three designs of larval traps were tested in a heavily-infested building. None of them performed well. Trapping of adult males using pheromone-baited sticky traps was far more successful. A comparative study of three brands of commercial pheromone lures was conducted, with one brand giving a higher catch than the others. For one brand the effect of ageing on the lure was examined and the results suggested that more frequent replacements were required than was recommended by the manufacturer. The importance of vertical positioning of traps on performance was studied on the outside of an infested grain storage structure where traps were placed between 0 and 4.2 m above ground. The highest numbers were caught less than 0.3 m above ground. Between 1.5 and 4.2 m, catch numbers were similar, suggesting that trap height has a minimal effect on catch within this range.

Introduction

Trogoderma variabile Ballion, the warehouse beetle, is a relatively new pest in Australia, having been present only since the late 1970s (Hartley and Greening 1983). It is a serious pest of storage environments, especially of packaged processed products that are not quickly consumed. It is an important pest for quarantine reasons in Australia because of its resemblance to *Trogoderma granarium*, a species that does not occur in Australia and against which there are many international quarantine barriers (Wright 1993).

A large research program has recently been completed on the biology and control of this species in Australia. To be able to monitor populations for ecological studies and to evaluate effectiveness of control treatments, it was necessary to have reliable methods for detection and estimation of populations of larvae and adults. Here we report on the testing of a number of trap designs and trapping methods.

The larvae of *T. variabile* are exceedingly cryptic, and when in diapause move relatively little and seldom from their refugia. The standard method of detecting larvae of *T. variabile* is by very thorough inspection, particularly in the very difficult and inaccessible locations within a structure. This is challenging and requires a significant amount of experience,

so a reliable trap would be very useful. There are a various types of traps available that claim to capture larvae. We tested two of these, along with a very simple and inexpensive cardboard trap that has been used by Wright (1991) with much success for other species of stored-product insects.

The adults of *T. variabile* are known to be good fliers and the males respond to a sex pheromone, (Z)-14-methyl-8-hexadecenal, produced by the female (Cross et al. 1977). During the unsuccessful warehouse beetle eradication program in New South Wales in 1977–1981, an attempt was made to use pheromone lures in traps but no adult males were caught, probably because the main flight period had passed (Hartley and Greening 1983). The potential for use of pheromone-baited traps was so great that pheromone traps were tried again, this time with considerable success. Here we report on comparisons of three types of pheromone lures and a field trial to determine the optimum height above ground for setting the pheromone trap.

Experimental Methods and Results

Larval traps

The traps

1. Consep box trap (Consep Membranes, Inc., 213 Southwest Columbia, P.O. Box 6059, Bend, Oregon 97708, USA). This is a sticky trap designed to catch adults and larvae, primarily for use inside buildings. It uses a proprietary lure containing both sex pheromone and unspecified food attractants. This trap measures $15 \times 8.5 \times 5$ cm, with flaps that fold inwards to provide the opening to the trap and ramps for the larvae to walk up. The interior is completely coated with a film of glue.
2. Khapra beetle trap (Barak 1989). This trap is designed to be attached to walls, and attracts adults using a pheromone lure, and larvae using wheat germ or sesame oil. The insects are captured by drowning in the reservoir of wheat germ oil, housed in a piece of corrugated cardboard which is enclosed in a cardboard jacket that is attached to a vertical surface.
3. Cardboard flour trap. This home-made cardboard trap (Wright 1991) contains a small amount of whole wheat flour as a food attractant for larvae. It acts as a harbourage for the insects and does not kill them. It is a single piece of corrugated cardboard (10×15 cm) that has been sealed along one long side with cellulose tape and filled with 0.4 g of whole wheat flour as a feeding attractant for the larvae.

Site and trapping details

The trapping trials were divided into two periods. During period 1 (14 August –17 October 1990) 8 khapra beetle traps, 11 box traps and 11 cardboard traps were set at various locations in an office in Griffith, New South Wales known to be infested with *T. variabile*. During period 2 (17 October 1990–4 October 1991), the number of box and cardboard traps was reduced to 3, but all 8 khapra beetle traps were retained.

* Stored Grain Research Laboratory, CSIRO Division of Entomology, GPO Box 1700, Canberra ACT 2601, Australia.
† P.O. Box 228, Hanwood New South Wales 2680, Australia.

The box and cardboard traps were placed together in pairs to enable a direct comparison between the two types of traps that rested on horizontal surfaces. All traps were checked and the insects were removed and identified weekly. When an exuvium was found in a trap without any larvae, it was assumed that the larva had entered the trap for long enough to moult but had then escaped. The larval exuvium is sufficient for species identification, so the positive identification of a *T. variabile* exuvium was considered a larval detection.

Results

Very few insects were captured over the 14 months of the test, so no inferences about seasonal activity are possible. Totals only are presented (Table 1).

The box trap caught an average of 0.003 insects/trap day and retained no larvae at all, although 1 exuvium was found. Because the lure contained pheromone, it did catch a few adults, but overall, this trap was quite uninformative.

The cardboard trap caught only 4 larvae, giving a capture rate of 0.002 insects/trap day.

The khapra beetle trap caught both adults and larvae, with a capture rate of 0.01 insects/trap day. When considering larval catches only, the rate of capture was 0.005/trap day.

Discussion

None of the traps could be considered effective or useful. Over the time when very few larvae were caught by the traps, larvae were found often on the floors and furniture by office staff. Nevertheless, of the traps tested, the khapra beetle trap was best for detection of the presence of *T. variabile*, because it caught both adults and larvae. From the lack of catches of larvae by the box trap, we conclude that the food attractant in the lure used in this trap was ineffective. The cardboard trap was the best for larvae alone, although no larvae were trapped in the second, year-long period of the test. This trap was by far the least expensive of the three.

Pheromone traps for adults: comparison of commercial lures

Three commercial pheromone lures were tested, made by: Consep Membranes, Inc. (213 Southwest Columbia, P.O. Box 6059, Bend, Oregon 97708, USA); Trécé Incorporated (635 South Sanborn Road, Suite 17, P.O. Box 5267, Salinas, CA 93915, USA); and Agrisense-BCS Limited (Treforest Industrial Estate, Pontypridd, Mid Glamorgan, CF37 5SU, U.K.). The Consep lure was a round disk with an adhesive backing and an advertised life of 16 weeks. The Trécé lure was a rubber septum impregnated with pheromone, with a suggested replacement schedule of 3–4 weeks. The Agrisense lure was a thin plastic square about 1×1 cm, with an advertised lifespan of 3 months.

The first lures used in 1990 were from Consep, starting in August 1990. The lures from Trécé became available a few months later, and were tested against the Consep lures (experiment 1).

The Agrisense lure became available in Australia at a competitive price in late 1991, with a longer advertised lifespan than the Trécé. The Trécé and Agrisense lures were compared in experiment 2.

The trap used for the adult monitoring was a Delta-style trap. This is a home-made sticky trap constructed from 2-litre milk cartons (R.A. Vickers, unpublished data).

Experiment 1: Comparison of old and new Consep lures and new Trécé lures

Methods

The Consep lure had been in use for 12 weeks at the start of the trial. At the office site used for the larval traps study, three sets of three traps were placed in trees around the outside of the office. Within each set of traps there was 1 lure of each kind. Every four days, the catch was collected and the traps rotated one position within each set. This was repeated 6 times between 20 November and 26 December 1990. The lures were not changed during this trial.

Results

The aged Consep lure had definitely lost activity (Fig. 1) after 12 weeks in the field. There were differences in catches between the new Consep and Trécé lures for the first 2 weeks of the test. These data suggest that output of the Consep lure was different and more favourable than that of the Trécé lure at the start, but that it functioned very much like the Trécé lure

Table 1. Numbers of *Trogoderma variabile* caught in 3 larval traps in an infested office building in Griffith, New South Wales. Period 1 is 15 August–17 October 1990, Period 2 is 17 October 1990–4 October 1991. A=adults, L=larva, E=larval exuvium.

| | Horizontal traps | | | | Vertical traps | | |
| | Box trap | | Cardboard trap | | | Khapra beetle trap | |
Trap location	Period 1	Period 2	Period 1	Period 2	Trap location	Period 1	Period
1	0	–	0	–	1	2L	3A, 1L
2	0	–	0	–	2	1A	2A 5L 1E
3	0	–	0	–	3	0	4A
4	0	–	0	–	4	1L	6A
5	0	–	1L	–	5	1L	4A, 1L
6	0	2A	1L	0	6	0	3A
7	0	1A	0	0	7	0	3A
8	0	–	1L	–	8	0	2A
9	1A	–	1L	–			
10	0	1A, 1E	0	0			
11	0	–	0	–			
Totals	1A	4A, 1E	4L	0		1A, 4L	27A, 7L, 1E

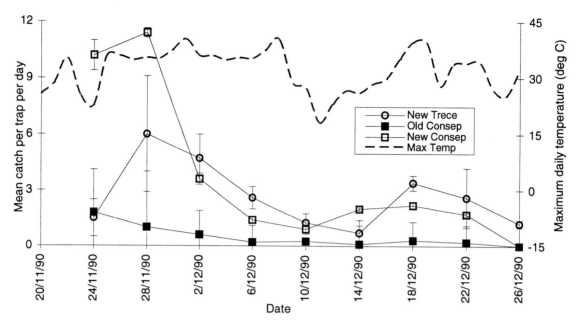

Fig. 1. The trap catches of the aged and new Consep lures and new Trécé lures for *Trogoderma variabile*, with the maximum daily temperatures. Error bars represent ±1 standard error, n = 3.

after 2 weeks. Because the same catch pattern is reflected by all baits, it suggests that the trap catches were indicative of the population through the period of the trial. The daily maximum temperatures (Fig. 1) throughout the period were suitable for flight of *T. variabile* (unpublished data) so the decline of catches represents the real decline of the adult population at this time of the year.

Discussion

The Consep lure was advertised to last for 16 weeks at a constant 25°C. During the 12 weeks it had been out in the field, the average maximum daily temperature was only 21.4°C. Thus it appears that the effective lifespan of the lure in Australia was much less than expected, given the temperatures to which it was exposed. The difference in catch rate of the new lures emphasises the inherent differences in release rates of different types. The difference in lures is more than just the amount released, but also the pattern of release.

Experiment 2: Comparison of Trécé lure vs Agrisense lure

Methods

This trial compared the Trécé lure with the Agrisense lure. The procedure was similar to the preceding test, except that 6 pairs of traps were used around the office site. The catch was collected and the trap positions changed every 4 days between 9 March–21 April 1992, for a total of 10 trapping periods. Lures were replaced on 16 April at trap positions 1, 2 and 4.

Results

The traps with the Agrisense lure caught more *T. variabile* than did those with the Trécé lure on 8 of 10 trapping periods (Table 2). Also, within pairs, the Agrisense traps caught the same or more *T. variabile* than the Trécé trap in 48 of 59 pairs (1 trap lost one trapping period). Changing the lure in half of the trap positions did not obviously affect trap catch in the

Table 2. Trap catches of *Trogoderma variabile* for the comparison between the Agrisense (A) and Trécé (T) lures, using paired traps at 6 locations outside an infested office building in Griffith, NSW, 9 March–21 April 1992

Collection date	Trap locations												Totals	
	1		2		3		4		5		6			
	A	T	A	T	A	T	A	T	A	T	A	T	A	T
13/3/92	6	3	5	2	7	2	1	0	8	1	3	0	30	8
17/3/92	15	3	0	2	4	2	7	12	3	0	2	4	31	23
22/3/92	6	6	0	0	1	0	5	0	0	1	1	0	13	7
26/3/92	6	1	0	0	0	0	2	1	2	1	0	0	10	3
31/3/92	0	0	0	1	0	0	1	1	0	1	0	1	1	4
4/4/92	5	4	0	1	1	0	–	1	1	0	1	0	8	6
8/4/92	5	1	1	1	1	0	7	0	2	5	0	0	16	7
12/4/92	2	6	0	0	0	1	3	1	2	2	0	0	7	10
16/4/92	1	1	0	0	0	1	1	0	0	0	0	0	2	2
21/4/92	10	3	1	0	0	0	0	0	0	0	0	0	11	3
Totals	56	28	7	7	14	6	27	16	18	11	7	5	129	73

next trapping period, in part because there were few beetles flying so late in the season.

Discussion

The Agrisense lure was shown to be the more effective one, especially in the first 4 trapping periods. It would be the lure of choice for a program where detection of low numbers of insects was required.

Height of trap

Methods

The trap used in this trial was the Delta-style trap used previously, with the Trécé lure, which was replaced every 4 weeks.

Two trap locations were chosen at a grain storage facility at Emery, New South Wales at either end of a large grain shed, where there was a suitable vertical post from which to hang traps. Initially traps were hung at 0.3, 1.5, 2.4, 3.3 and 4.2 m above ground on 16 December 1992. The insects were removed from the traps weekly, except on the first occasion when they were removed after 2 weeks. After 4 trapping periods, an additional trap was set just above the ground at each location. The trapping continued for 13 more weeks until the end of the flight season.

Results

The numbers caught were converted to the proportion of insects caught at that location in each trap, for each trapping period. These data were then averaged to give the mean proportion caught at each height (Fig. 2). As virtually no insects were caught in the last 4 trapping periods (23 March–21 April 1992) because the flight season had ended, these data were omitted from the analysis. It was clear from the first trapping periods that there was a much higher catch at the 0.3 m height (Fig. 2a), prompting the additional trap set just at ground level. The trend continued (Fig. 2b), with an average of 70% of the insects caught at or below 0.3 m. The exact height above the ground, up to 1.5 m had a very large effect on the catch.

Discussion

It is clear that maximum trap catch is obtained with the trap very close to the ground. If detection of very low populations is desired, then traps should be placed close to the ground.

In certain kinds of ecological studies, it more important to have catches comparable between locations than to catch the maximum number of insects. During seasonal flight activity studies, traps at different sites were placed between 1.5 and 2.5 m above ground level, for operational reasons. The results of this height trial show that the catch is fairly constant within this range, which means that the data for these different trap sites are comparable, at least in terms of height effects.

Conclusions

None of the larval traps tested was of any use and there appears to be no substitute for the experienced inspector for detecting larval infestations.

All of the pheromone lures were effective at attracting adult male *T. variabile* although there were some minor differences. The catches in traps baited with Consep and Trécé pheromone lures were similar, except that catches for Consep lures were much higher than for Trécé lures for the first 2 weeks. A comparison of new Trécé and Agrisense lures showed that almost twice as many insects were caught with the Agrisense lure, and this lure would be a better choice when detection of very low populations was important. The effective life of a lure depends on many factors, including temperature, wind and exposure to ultraviolet light. The effective life of the lure indicated by the manufacturer is generally based on constant and moderate conditions. If it is intended to use the lure under extreme conditions, it is important to test the effective life under those conditions. Even so, in one of the trials reported here, when daily temperatures were far from extreme, the effective life of the Consep lure was shown to be much less than the advertised 16 weeks. As a result of these trials, the lures were changed every 4 weeks in ecological studies on *T. variabile* to ensure as much consistency as possible between locations and seasons.

Maximum catches of adult males were achieved by pheromone traps very close to the ground, so during detection programs for, for example, quarantine, they should be set as low as possible. However, when comparable data between trapping locations is needed for ecological work and it is not possible to set traps on the ground, catches from traps set between 1.5 and 2.5 meters high are sufficiently high and uniform to be useful.

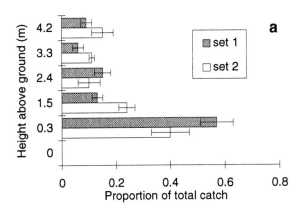
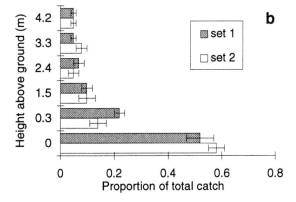

Fig. 2. The results of the experiment investigating the effect of height on pheromone trap catch, Emery, New South Wales. a, Results for trapping period 16 December 1991–21 January 1993 with no trap at ground level. b, Results for trapping period 21 January–23 March 1992 including a trap at ground level. Error bars represent ±1 standard error.

Acknowledgments

We thank Dr Alan Barak (USDA, APHIS, PPQ, Hoboken) and Consep Membranes, Inc. for donating their traps and lures for testing. R. Vickers and D.P. Rees made constructive suggestions on the manuscript. This research was supported by a grant from the Rural Industries Research and Development Corporation and the New South Wales rice industry.

References

Barak, A.V. 1989. Development of a new trap to detect and monitor Khapra beetle (Coleoptera: Dermestidae). Journal of Economic Entomology, 82, 1470–1477.

Cross, J.H., Byler, R.C., Silverstein, R.M., Greenblatt, R.E., Gorman, J.E. and Burkholder, W.E. 1977. Sex pheromone components and calling behavior of the female dermestid beetle, *Trogoderma variabile* Ballion (Coleoptera: Dermestidae). Journal of Chemical Ecology, 3, 115–125.

Hartley, D.J. and Greening, H.G. 1983. Review of warehouse beetle control programme. A report to the Standing Committee on Agriculture. Department of Agriculture New South Wales. vi+43 p.

Wright, E.J. 1991. A trapping method to evaluate efficacy of a structural treatment in empty silos. In: Lessard, F. and Ducom, P., ed., Proceedings of the Fifth International Working Conference on Stored-product Protection, Bordeaux, France, September 1990, 3, 1455–1462.

Wright, E.J. 1993. *Trogoderma variabile* (Coleoptera: Dermestidae) in Australia. In: Corey, S.A., Dall, D.J. and Milne, W.M., ed., Pest control and sustainable agriculture. Melbourne, CSIRO Publications, 373–375.

Sampling and Trapping —
Session and Workshop Summary

Conveners: E. Jane Wright and Paul Cogan

Sampling and trapping insects generated considerable interest with a keynote address and 22 contributed papers presented during the conference and a full-day workshop that was attended by about 70 people.

Insect traps baited with pheromones have been used for many years to monitor populations of stored-product insects in a variety of situations. Our contributed paper and workshop discussions clearly showed that we are moving forward both in terms of interpretation of trap catch but also in the use of pheromones and traps in IPM and active control of pests. Our keynote speaker, Dr Pasquale Trematerra, spoke on his ground-breaking work on the use of pheromones in integrated pest management of moth pests in commercial flour mills. This topic was expanded upon during the workshop with a presentation on the long term population suppression of phycitid moths and *Lasioderma serricorne* in food warehouses in Hawaii. Further discussion identified a need for more well-controlled but realistic trials of these techniques to realise their potential.

Another set of papers concerned the interpretation of trap catch results and included several papers that compared different trapping methods and one that concerned the difficulties of interpreting the direction of movement of moths from trap catches. We heard of important steps towards the calibration of trap catches and the new self-marking system and associated analysis of mark-recapture data was particularly interesting.

Detection of insects remains a topic of considerable interest and we heard about the development of immunoassays to detect insect contamination of grain, milled grain products and spices, an improved flotation method for detecting internal feeders, and a sieving system suitable for use with grain in transit. Since the last conference, the use of automated acoustical detection systems has advanced considerably from small experimental bulks to larger bulks in both France and the USA.

We were reminded of the diversity of behaviours in our pest complex with a number of papers concerning the special considerations and trapping methods for different species including; *Trogoderma variabile*, *Prostephanus truncatus*, *Lyctocoris campestris*, *Rhyzopertha dominica*, *Tribolium castaneum*, *Plodia interpunctella*, *Ephestia cautella*, and Psocoptera.

During the workshop we considered the issue of trap design and were treated to the insider's view of development of a new commercial trap. Our commercial colleagues pointed out that every manufacturer wants to sell a simple product that is easy to use, does everything and is disposable and urged the researchers to make the design simple and consider alternative uses of the design.

Given that the purity of pheromone in lures can markedly affect trap catch, the idea that manufacturers should state the purity of their pheromone lures was discussed at some length in the workshop. The commercial view was that this was unlikely to be considered at present.

We addressed the issue of standardisation of trapping research methods during the workshop with a panel discussion. There was general agreement that a minimum level of reporting of the trapping conditions and data presentation would be desirable for published papers so that the published data will be interpretable and useable by others. The session conveners have undertaken to prepare a note for publication in appropriate journals outlining these standards. The hope is that the quality of published papers on trapping will be improved and that researchers will have a guide to assist them in designing new trapping studies.

Insect Biology

Pheromones of stored-product insects: current status and future perspectives

T.W. Phillips*

Abstract

Pheromones have been isolated and identified from over 30 species of stored-product beetles and moths during the past 25 years. Existing gaps in our basic knowledge include: 1) discovery of new compounds in a few groups; 2) refined knowledge of chemical signals, response behaviours; and slow-release delivery systems; 3) development of cost-effective chemical syntheses on commercial scales; 4) identification of food attractants or pheromone synergists; and 5) identification of attractants for female moths. Application of new knowledge and improvements on existing pheromone-based technologies are expected. The most common application presently is with pheromone-baited traps for survey, detection, and monitoring, primarily of pyralid moths, cigarette beetles, and dermestids in processed food situations. Improvements in traps will occur in trap design, lure formulation, and data handling. Other applications of pheromones have been suggested or researched, but few have undergone extensive testing. Mass trapping of male moths is generally thought ineffective for control, but optimisation could result in biologically significant capture of males; mass trapping beetles with aggregation pheromones captures females and should be effective in control. Mating disruption was demonstrated many years ago and more work should be done on implementation. The 'attracticide' method holds numerous opportunities for pest suppression, especially as new and improved attractants are developed and more effective pesticides and pathogens for dissemination become available. Biorational alternatives are urgently needed to replace ineffective or unsafe methods of traditional pest control, and pheromone-based technologies can provide many solutions.

Introduction

The first stored-product insect pheromone was identified nearly 30 years ago from the black carpet beetle, *Attagenus unicolor* (=*megatoma*) (Silverstein et al. 1967), and since that time many advances in pheromones of storage pests have resulted in valuable tools for pest management (Burkholder 1990). From the hundreds of insects recorded from stored commodities and foods, pheromones for many of the common species have been investigated and synthetic formulations of pheromones are available for some of the most prevalent pests. Consequently, pheromone-based monitoring systems have been developed that facilitate early pest detection and

integrated pest management for overall reduction of pesticide use and increased product quality (e.g. Mueller and Pierce 1992). These trapping programs are mostly centred around detection of phycitine moths, cigarette beetles, dermestid beetles and other pests of processed foods in finished product warehouses and factories (Faustini et al.1991).

Research is still required on some of the important warehouse pests and on pests of raw grain in order to develop effective monitoring systems with their pheromones. Sensitivity and levels of detection can be improved with more thorough knowledge of insect behaviour, semiochemistry, and redesigned trapping and formulation technology. Suppression methods employing pheromones were proposed decades ago, but their practical development in food storage systems has yet to come. In this brief review I will summarise the existing knowledge on available stored-product insect pheromones, I will identify research needs that could enhance pheromone-based monitoring in specific pest management systems, and I will review proposed strategies for using pheromones to suppress insect populations.

Pheromones Available for Use

Table 1 is a list of stored-product insects for which pheromones have been chemically identified. This list is similar to those published earlier (e.g. Burkholder and Ma 1985; Burkholder 1990), but with some recently reported entries. Of the 38 species listed, 50% have synthetic pheromones commercially available for use as formulated lures for traps. These 19 species include the most serious insect pests of stored grains and stored-food products worldwide. Some pheromones are active for multiple species of pests, thus enhancing their versatility as monitoring tools. For example, males of the first five moth species listed under Pyralidae can be trapped using the same compound, (*Z, E*)-9,12-tetradecadienyl-acetate. Some pheromones have been identified for stored-product mites (e.g. Kuwahara et al. 1982), but they have not been developed commercially and will not be considered here. In general, much of the basic research of discovery and identification has been done on pheromones of the major stored-product pests, hence refined understanding and new applications of pheromone systems provide the current challenges to researchers and pest managers.

Research Needs

New chemistry

Pheromone identification and synthesis is required for some species of important stored-product pests. One important species that remains to be fully studied is the cowpea weevil, *Callosobruchus maculatus*, a serious pest of stored legumes that uses a female-produced sex pheromone (Qi and Burkholder 1981). The sex pheromone for a closely related species, *C. analis*, was recently identified (Cork et al. 1991), but work is apparently still in progress on the pheromone for

* USDA ARS Stored-Product Insect Research Unit, Department of Entomology, University of Wisconsin, Madison, Wisconsin, USA 53706. This article reports results of research only. Mention of a proprietary product does not constitute and endorsement or a recommendation for its use by USDA.

Table 1.   Species of stored-product insects for which pheromones have been identified, and information on commercial availability.

Taxa/Species	Sex producing	Commercial availability	Reference[a]
Coleoptera (beetles)			
Family Anobiidae			
Stegobium paniceum	Female	Yes	Kuwahara et al. 1978
Lasioderma serricorne	Female	Yes	Chuman et al. 1983
Family Bostrichidae			
Rhyzopertha dominica	Male	Yes	Williams et al. 1981
Prostephanus truncatus	Male	Yes	Hodges et al. 1984
Family Bruchidae			
Acanthoscelides obtectus	Male	No	Horler 1970
Callosobruchus chinensis	Female	No	Tanaka et al. 1981
C. analis	Female	No	Cork et al. 1991
Family Cucujidae			
Ahasverus advena	Male and Female	No	Pierce et al. 1991b
Cathartus quadricollis	Male	No	Johnston and Oehlschlager 1986
Cryptolestes ferrugineus	Male	No	Wong et al. 1983
C. pusillus	Male	No	Millar et al. 1985a
C. turcicus	Male	No	Millar et al. 1985b
Oryzaephilus surinamensis	Male	No	Pierce et al. 1985
O. mercator	Male	No	Pierce et al. 1985
Family Curculionidae			
Sitophilus granarius	Male	No	Phillips et al. 1987
S. oryzae	Male	No	Schmuff et al. 1984
S. zeamais	Male	No	Schmuff et al. 1984
Family Dermestidae			
Attagenus unicolor	Female	Yes	Silverstein et al. 1967
A. brunneus	Female	No	Fukui et al. 1977
Anthrenus flavipes	Female	Yes	Fukui et al. 1974
A. sarnicus	Female	No	Finnegan and Chambers 1993
A. verbasci	Female	Yes	Kuwahara and Nakamura 1985
Dermestes maculatus	Male	No	Levinson et al. 1978
Trogoderma glabrum	Female	Yes	Yarger et al. 1975
T. granarium	Female	Yes	Cross et al. 1976
T. variabile	Female	Yes	Cross et al. 1976
Family Tenebrionidae			
Tribolium castaneum	Male	Yes	Suzuki 1980
T. confusum	Male	Yes	Suzuki 1980
Tenebrio molitor	Female	No	Tanaka et al. 1986
Lepidoptera (moths)			
Family Gelichiidae			
Sitotroga cerealella	Female	Yes	Vick et al. 1974
Family Pyralidae			
Cadra cautella	Female	Yes	Brady 1973
C. figulilella	Female	Yes	Brady and Daly 1972
Ephestia elutella	Female	Yes	Brady and Nordlund 1971
E. kuehniella	Female	Yes	Brady et al. 1971a
Plodia interpunctella	Female	Yes	Brady et al. 1971b
Pyralis farinalis	Female	No	Landolt and Curtis 1982
Amyelois transitella	Female	No	Coffelt et al. 1979
Family Tineidae			
Tineola bisselliella	Female	Yes	Yamaoka et al. 1985

[a]References are those giving the chemical identification or analysis of pheromones; more literature may exist on biology and application (see Burkholder 1990).

C. maculatus. Pheromones for cucujid beetles in the genera *Cryptolestes* and *Oryzaephilus* have been identified (Table 1), but the lack of an efficient large-scale chemical synthesis of the pheromones has prevented their commercial application. Additionally, some pheromones have been identified and could be synthesised commercially, but they have not been put to use because data on field activity are lacking or their use in a pest management scheme has not been developed (e.g. *Anthrenus* spp., *Callosobruchus* spp., and *Sitophilus* spp.).

Formulation and delivery systems

The performance of certain commercially available pheromone systems could be improved with changes in chemical formulation or delivery systems. A pheromone-based trapping system is being used for *Tribolium* flour beetles, but effectiveness of traps in detecting pest populations is equivocal (David Mueller, pers. comm.). Males of *T. castaneum* and *T. confusum* produce 4,8-dimethyldecanal, a molecule that can exist as four distinct optical isomers or enantiomers. A complete mixture of all four enantiomers yield fairly low activity in baited traps (Javer et al. 1990), but the pure 4R, 8R isomer gives optimal response (Levinson and Mori 1983). Hence suppliers of *Tribolium* pheromone may produce sub-optimum lures if they do not use synthetic pheromone with the proper enantiomeric composition.

Formulation of *Tribolium* pheromone into slow-release devices is also very important for eliciting optimum response of beetles. Work in my laboratory has investigated the response of *T. castaneum* to different pheromone lure formulations using a two-choice pitfall bioassay (see Phillips et al. 1993 for details of bioassay). Commercial lures of different ages were tested against blank controls (see Hussain 1993 for complete listing of lures tested). Examples of variation in response to two different lure designs are given in Figure 1. Membrane lures, in which liquid pheromone was held in an impermeable reservoir and evaporated through a semi-permeable membrane on one surface, were repellent to *T. castaneum* when new, and did not elicit significant attraction until five weeks of age (Fig. 1a). However, laminate lures, in which the pheromone was formulated into a plastic matrix and then released through a semi-permeable membrane, were attractive when initially removed from the package, and remained active for the 14-week course of the study (Fig. 1b). Differences in activity of these lures over time is likely to be due to the amount of pheromone being released at a given time. Other work has shown that high doses of 4,8-dimethyldecanal, like those probably coming from new membrane lures (Fig. 1), are repellent to *T. castaneum* in laboratory bioassays (Hussain 1993). Proper release rate, and the formulation required to achieve it, are therefore very important for effective pheromone lures for *Tribolium*. We also know that the aldehyde group on the pheromone molecule reacts with air over time to form the carboxylic acid, which is inactive. *Tribolium* pheromone, as with other unstable pheromones, should probably be protected with an antioxidant while in the slow release device.

Minor components

Despite successful use of some pheromones in traps, the pheromone system of the pest insect in question may not be fully understood. Minor components may exist that could improve trap performance, particularly under low population densities. A single component pheromone, (Z, E)-9, 12-tetradecadienyl acetate (referred to as ZETA), is available for the Indianmeal moth, *Plodia interpunctella*, and four other species of stored-product Pyralidae (Table 1). The corresponding alcohol to ZETA, (Z, E)-9, 12-tetradecadienol

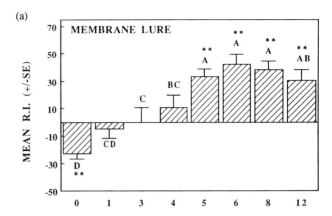

(a)

MEMBRANE LURE

MEAN R.I. (+/-SE)

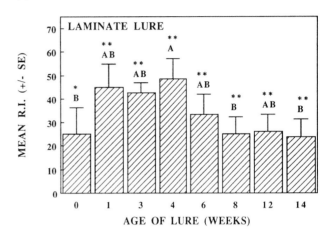

(b)

LAMINATE LURE

MEAN R.I. (+/- SE)

AGE OF LURE (WEEKS)

Fig. 1. Responses of *Tribolium castaneum* to pheromone lures of different ages in a two-choice pitfall bioassay. Histograms show means and standard errors for response index (R.I.), calculated as [(T-C)/total]x100, for which 'T' is the number responding to the treatment (a lure), 'C' is the number responding to the control (blank), and 'total' refers to the total number of beetles released (usually 20 one-week-old adults of mixed sex). Means for response to lures of different ages with different letters are significantly different (P<0.05, ANOVA and Fishers PLSD test), and stars indicate significant differences in response to treatment and control within an experiment (*, P< 0.05, **, P< 0.01, Student's t-test). (a) Membrane lure; (b) Laminate lure.

(ZETOH), is produced by female *P. interpunctella* and it also enhances responses of male moths when combined with ZETA (Vick et al. 1981; Phillips unpublished). However, ZETOH has the opposite effect of inhibiting responses by the almond moth, *Cadra cautella* (Sower et al. 1974), hence the material is not added to commercial lures intended for multiple species. I studied the responses of male *P. interpunctella* natural and synthetic pheromone sources in a wind tunnel (Fig. 2). When a small cage containing a virgin female *P. interpunctella* was presented upwind, all male moths tested flew upwind and contacted the cage. A one-week-old pheromone lure (rubber septum) in the wind tunnel elicited upwind flight in male moths, but very few males actually contacted the pheromone source; a similar response occurred when 1.0 mg of ZETA was applied to filter paper and presented. When 100 ng of ZETA on filter paper was tested in the wind tunnel I observed flight and contact responses of males similar to those elicited by virgin females. However, this dose of 100 ng represents about 10 times the amounts produced by a virgin female (Sower and Fish 1975). When 10 ng of ZETA was tested, a

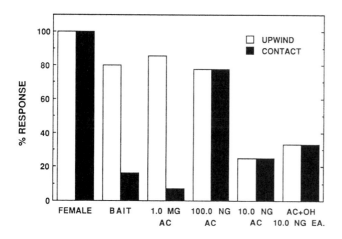

Fig. 2. Responses of male *Plodia interpunctella* to natural and synthetic sources of sex pheromones in a wind tunnel bioassay. Histograms represent the percentage of females (n=10–20) to: 1) take flight and initiate characteristic upwind orientation; and 2) to contact the source for a minimum of 10 seconds. Sources of attraction listed along the bottom of the graph are: 1) a single virgin female, one day old, in a small cage; 2) a one-week old pheromone lure (Trécé Inc.) containing ZETA in an open 60 mm glass Petri dish; 3) 1.0 mg of ZETA (referred to as AC for acetate) applied to filter paper in a dish; 4) 100.0 ng of ZETA on paper in a dish; 5) 10.0 ng of ZETA on paper in a dish; and 6) 10.0 ng each of ZETA and ZETOH on paper in a dish. Individual males, one-day old, were released downwind and given 3 minutes to respond.

biologically relevant dose, both flight and contact responses were very low; addition of ZETOH at 10 ng had no effect (Fig. 2). Thus synthetic pheromone components released at levels similar to one or a few virgin females failed to elicit attraction in *P. interpunctella*. Male moths obviously respond from a distance to high levels of ZETA coming from commercial lures, but moth capture in traps is probably dependent on relatively high population levels. Analysis of volatile chemicals produced by female *P. interpunctella* revealed individual variation in the number and amounts of compounds produced (Fig. 3). In addition to ZETA and ZETOH, which may vary quantitatively between females, several other unidentified compounds were detected. It is clear from behavioral studies (Fig. 2) that ZETA alone cannot account for pheromone-mediated mate finding in *P. interpunctella*, and potential exists for discovery of new pheromones that may yield improved pest management tools.

Female attractants

Sex pheromones produced by female insects attract only males. Traps baited with synthetic sex pheromones can only monitor populations of males and impact on overall population increase is limited since females are not affected. Work is underway in various laboratories to develop attractants for female moths of agricultural importance (e.g. Mitchell et al. 1990; Landolt 1989). Work in my laboratory has focused on attractants and oviposition stimulants for the Indianmeal moth (Phillips and Strand 1994). In close-range oviposition experiments, mated female *P. interpunctella* oviposited on dishes containing cornmeal-based rearing medium. While investigating the potential for epideictic (spacing or dispersal) pheromones, we found that females preferred to oviposit on cheesecloth patches that had been contaminated by conspecific wandering fifth instar larvae. A similar oviposition response was reported by Corbet (1973) for the Mediterranean

flour moth, *Ephestia kuehniella*. With *P. interpunctella* we found that a combination of food plus larval secretions elicited a greater oviposition response than either food alone or larval secretions alone. In wind tunnel bioassays female moths flew upwind only when food was present, and contacts by females were highest to the combination of food plus larval secretions (Phillips and Strand 1994). We are currently isolating the material that elicits oviposition in *P. interpunctella* in hopes of identifying new semiochemicals that can be used to manipulate behaviour of female moths.

Food odours

Our results with a combination of food odours and insect produced volatiles in *P. interpunctella* reflect a recurring theme on the importance of food and food volatiles in the pheromone systems of insects. Several examples are known from stored-product insects in which food is very important for pheromone production and food volatiles serve as attractants or synergists to pheromones. Grain oils are known to be attractive to stored-grain insects (e.g. Nara et al. 1981), as are volatiles from fresh grains and from fungus-infected grains (e.g. Pierce et al. 1990; 1991a). Recently we compared the responses of two ecologically different beetle species, *T. castaneum* and *S. oryzae*, to a collection of grain oils and found that the two species responded quite differently to the same oils (Phillips et al. 1993). In the same study we also showed that male *T. castaneum* responded to food volatiles at levels higher than females, and that responses of both sexes to synthetic pheromone was increased by the addition of food volatiles. Hussain (1993) found that male *T. castaneum* produced pheromone only when feeding on cracked wheat, and that pheromone production ceased when beetles were removed from food. A scenario involving food and mate-finding can be proposed for *T. castaneum*: male beetles locate food sources at which they feed, produce pheromone, and attract females for mating and reproduction. Walgenbach et al. (1987) had similar results with *S. zeamais*, in which weevils responded stronger to a combination of pheromone and food volatiles compared to pheromone or food separately. Synergism of host plant volatiles with beetle-produced pheromones is a common phenomenon in scolytid bark beetles (reviewed in Raffa et al. 1993), and a synergistic role for plant volatiles in pheromone systems in various insect groups is now being realised (Dickens et al. 1990). Potential certainly exists to improving existing trapping technology for stored-product insects by addition of food volatiles to increase insect response to pheromone lures.

Applications

Monitoring and detection

The predominant application of pheromones of stored-product insects is as tools for monitoring and detection of pest populations. Pheromone-baited traps are used extensively to monitor populations of pyralid moths, cigarette beetles (*Lasioderma serricorne*), and dermestid warehouse beetles (*Trogoderma* spp.) in processed food warehouses and factories. Typical recommendations are for a grid-work of traps to be established in a structure and monitored at regular intervals for the capture of insects. If one or more traps in a particular area captures more insects than others, then an increased density of traps should be deployed in that area in order to localise the source of the pests (Burkholder 1990). Monitoring of insects infesting raw grains has relied mostly on probe traps that do not require pheromone lures, although synthetic phe-

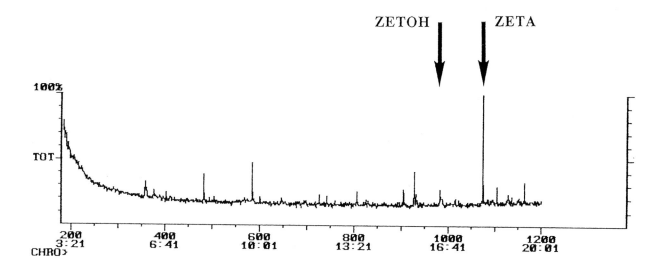

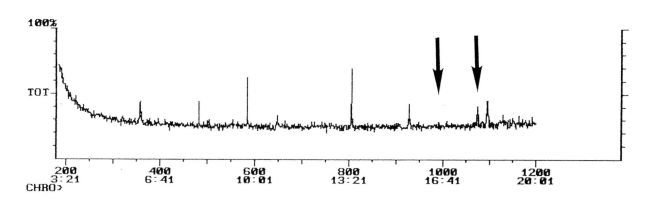

Fig. 3.   Total ion chromatograms (GC) of volatiles collected from two different one-day old *P. interpunctella* for 16 hours through the scotophase. Arrows indicate the retention times for ZETOH and ZETA. The x-axis is the retention time in seconds and minutes:second, and the y-axis is ion current relative to the largest peak. Volatiles were collected on Super-Q polymer, extracted with hexane, and analysed on a Schimadzu GC-14A with a 30 m DB-1 column (J & W Scientific) coupled to a Finnegan 800 series Ion Trap Detector.

romones for the grain borers *Rhyzopertha dominica* and *Prostephanus truncatus* have been used to study the geographic spread of these pests within regions of new range extension (e.g. Fields et al. 1993; Hodges et al. 1984).

New developments in the use of pheromone-baited traps could improve their application in pest management schemes. New trap designs are being developed and some show promise in improving the detection limits with certain insects (e.g. Mullen 1992; Quartey and Coaker 1992). Once traps and lures are optimised, new computer-based methods to organise and interpret data from traps will greatly enhance the value of the data and facilitate informed decision-making based on trap catch (e.g. Insects Limited 1994). Warehouse managers could then maintain long-term databases on insect activity in their facilities and easily detect changing trends in pest numbers in a relatively short time. In addition to estimating relative numbers of insects by observing changes in trap catch over time, pest managers may also employ new techniques to estimate absolute numbers of insects in a facility. Wileyto et al. (1994) have developed a trapping-marking–re-trapping method to estimate population sizes of insects using pheromone traps. In a validation experiment of their mark-recapture model, Wileyto et al. (1994) estimated sizes of known populations within statistically acceptable levels of confidence. Population estimation may provide a more realistic numerical

value that can be used with greater confidence than simply uncalibrated trap catches for making pest management decisions.

Suppression

Although population monitoring is the primary use of pheromones in agriculture, the idea of actually suppressing insect populations with pheromones has been proposed from the very beginning of pheromone research. Mass trapping with pheromones, or simply catching as many insects as possible to reduce overall numbers, has immediate appeal but has received limited success. In the case of female-produced sex pheromones, those used most commonly in storage systems, only males are trapped. Hence any attempt to suppress the population by trapping males would require a sufficient number of males to be trapped so that nearly all females would go unmated. Theoretical considerations of mass trapping males take into account the density of males in the population and the potential number of matings a male is able to secure in a lifetime (Lanier 1990; Roelofs et al. 1970). If a male can mate with ten females in his lifetime, as is the case for the Indianmeal moth (Brower 1975), then up to 90% of the male population can be trapped and not have any effect on mated females or the subsequent larval generation. Under high population levels the rate of female encounter would be high and mass trapping more difficult to achieve. However, under low population levels males would locate females less frequently and intensive trapping could conceivably reduce male populations to biologically significant levels. Proper experiments of mass trapping are difficult to conduct due to inadequate controls or poor replication, although various studies have reported successful suppression of storage pests by mass trapping males with sex pheromones (e.g. Mueller and Pierce 1992; Suss and Trematerra 1986). Mass trapping both sexes of a population using aggregation pheromones should be intuitively more effective than mass trapping just females. Aggregation pheromones produced by males are known from several beetle species that infest stored products (Table 1), but no studies have been conducted to suppress populations of these insects by mass trapping.

Mating disruption is a potentially effective control measure in which the air in the area containing the pest population is permeated with sex pheromone. Although the mechanism of mating disruption is incompletely understood, it seems that males fail to mate with females due to either a physiological affect on pheromone perception or due to a behavioural inability to locate and contact a real female amongst all the synthetic sources of pheromones (Cardé 1990). The limitations and theoretical bases of mating disruption are similar to those for mass trapping of males: a substantial proportion of the male population must fail to locate females, and success is more likely under relatively low population levels. Mating disruption has been successfully applied to Lepidoptera in field crops (e.g. the pink bollworm on cotton) and has been commercialised in such cases. Although several successful experiments have been reported using mating disruption with stored-product pyralid moths (e.g. Prevett et al. 1989; Sower and Whitmer 1977), no commercial applications of this method are currently available. Clearly mating disruption is a potentially effective pheromone-based control method for storage moths that requires further consideration.

Attracticide

One final application of pheromones for insect pest suppression in stored-products is the 'attracticide' or 'lure-and-kill' method. The attracticide concept involves using a pheromone or other attractive semiochemical to lure insects to an area or a specific point source whereby they contact a toxicant or are destroyed or disabled by other means. Attracticidal control is analogous in a way to mass trapping, although many more insects are affected because the attracticide is broadcast over a large area and the killing effect is not limited to individual traps. Trap cropping represents a variation of the attracticidal method in agriculture whereby the target insects are lured to a plot of crop plants (or one or more trees in the case of bark beetles), and the crop is subsequently destroyed. The more common concept of an attracticide involves the formulation of an insecticide with a feeding stimulant and a long-range attractant, either in liquid droplets or in some encapsulated or solid matrix, that can be applied evenly over large areas. Target insects orient to the formulation where they feed on or contact the insecticide and then die. In agriculture the attracticide method has been used with success against tephritid fruit flies (Cunningham et al. 1990) and cotton boll weevils (Ridgeway et al. 1990). In stored-products there are many promising results on the use of the attracticide concept in flour mills and warehouses (Trematerra, these proceedings). One variation on the attracticide concept is for the dissemination of pathogenic organisms into the pest population. Instead of an insecticide, the attractive source could contain inoculum of an insect pathogen that could be spread among contacting insects and then

amplified in the next generation to bring about an epizootic. Suppression of experimental dermestid beetle populations using a protozoan pathogen and pheromone-based dissemination was reported by Shapas et al. (1977). Current work on microbial insect pathogens and biotechnology could conceivably produce inoculum of maximal virulence and high environmental tolerance that would be applicable to the attracticide method.

Conclusions

The practice of stored-product protection is facing serious challenges worldwide as important issues arise regarding food safety, government regulation, and biological pressure from insects that become increasingly more difficult to control. In the United States, for example, one fumigant and two residual protectants are available for use on raw grain, and finished food warehouses and factories have very limited options for chemical pest control. Insect populations are rapidly evolving resistance to the few chemicals being used, and the general public is calling for limited use of pesticides on food while also approaching a zero-tolerance attitude toward insect contamination in agricultural commodities. Safe and effective alternatives to conventional pesticides are needed, and potential exists for development of pheromone-based methods. Basic research on the isolation and identification of pheromone compounds flourished during the 1970s and 1980s, and consequently pheromones are known from most of the major pest insects in stored-products. It is time now to move forward and apply existing knowledge of pheromones and insect behavior into effective biorational methods to manage insect pests in stored-product systems.

Acknowledgments

I am very grateful to the organisers of the biology section, Barry Longstaff and Jonathan Donahaye, for inviting me to make this presentation, and I thank Jane Wright for her encouragement and organisational input with the conference. John Hutchinson provided technical assistance on wind tunnel bioassays. I appreciate reviews of the manuscript by Wendell Burkholder, Joel Phillips, and Rosa Franqui. I am indebted to David Mueller for providing helpful information on commercial use of pheromones. I want to make a special note of thanks and appreciation to Wendell Burkholder for his encouragement and for giving me the opportunity to study stored-product insects.

References

Brady, U. E. 1973. Isolation, identification, and stimulatory activity of a second component of the sex pheromone system (complex) of the female almond moth, *Cadra cautella* (Walker). Life Science, 13, 227–235.

Brady, U. E. and Daly, R. C. 1972. Identification of a sex pheromone from the female raison moth, *Cadra figulilella*. Annals fo the Entomological Society of America. 65, 1356–1358.

Brady, U. E., and Nordlund, D. A. 1971. *cis*-9, *trans*-12-Tetradecadien-1-yl acetate in the female tobacco moth *Ephestia elutella* (Hübner) and evidnec for an addtional copmponent of the sex pheromone. Life Science, 10, 797–801.

Brady, U. E., Nordlund, D. A. and Daley, R. C. 1971a. The sex stimulant of the Mediterranean flour moth *Anaghsta kuehniella*. Journal of the Georgia Entomological Society, 6, 215–217.

Brady, U. E., Tumlinson, J. H., Brownlee, R. G. and Silverstein, R. M. 1971b. Sex stimulant and attractant in the Indian meal moth and in the almond moth. Science, 171, 802–803.

Brower, J. H. 1975. *Plodia interpunctella*: effect of sex ratio on reproductivity. Annals of the Entomological Society of America, 68, 847–851.

Burkholder, W. E. 1990. Practical use of pheromones and other attractants for stored-product insects. In: Ridgeway, R. L., Silverstein, R. M., and Inscoe, M. N., ed., Behavior-modifying chemicals for insect management, applications of pheromones and other attractants, New York, Marcel Dekker, Inc., 497–516.

Burkholder, W. E. and Ma, M. 1985. Pheromones for monitoring and control of stored-product insects. Annual Review of Entomology, 30, 257–272.

Cardé, R. T. 1990. Principles of mating disprution. In: Ridgeway, R. L., Silverstein, R. M., and Inscoe, M. N., ed., Behavior-modifying chemicals for insect management, applications of pheromones and other attractants, New York, Marcel Dekker, Inc.,47–71.

Chuman, T., Mochizuki, K., Kato, K, Ono, M. and Okubo, A. 1983. Serricorone and serricorole, new sex pheromone components of cigarette beetle. Agricultural and Biological Chemistry, 47, 1413–1415.

Coffelt, J. A., Vick, K. W., Sonnet, P. E. and Doolittle, R. E. 1979. Isolation, Idnetification, and synthesis of a female sex pheromone of the navel orangeworm, *Amyelois transitella*. Journal of Chemical Ecology, 5, 955–966.

Corbet, S.A., 1973. Oviposition pheromone in larval mandibular glands of *Ephestia kuehniella*. Nature. 243: 537–538.

Cork, A., Hall, D. R., Blaney, W. M. and Simmonds, M. S. J. 1991. Identification of a component of teh female sex pheromone of *Callosobruchus analis* (Coleoptera: Bruchidae). Tetrahedron Letters, 32, 129–132.

Cross, J. H., Byler, R. C., Cassidy, R. F. Jr., Silverstein, R. M., Greenblatt, R. E., Burkholder, W. E., Levinson, A. R. and Levinson, H. Z. 1976. Porapak-Q collection of pheromone components and isolation of (Z)- and (E)-14-methyl-8-hexadecanal, potent sex attracting components, from the frass of four species of *Trogoderma* (Coleoptera: Dermestidae). Journal of Chemical Ecology, 2, 457–468.

Cunningham, R. T., Kobayashi, R. M. and Miyashita, D. H. 1990. The male lures of tephritid fruit flies. In: Ridgeway, R. L., Silverstein, R. M., and Inscoe, M. N., ed., Behavior-modifying chemicals for insect management, applications of pheromones and other attractants, New York, Marcel Dekker, Inc., 255–267.

Dickens, J.C., E.B. Jang, D. M. Light & A.R. Alford, 1990. Enhancement of insect pheromone responses by green leaf volatiles. Naturwiss. 77:29–31.

Faustini. D. L., Barak, A. V., Burkholder, W. E. and Leos-Martinez, J. 1991. Combination-type trapping for monitoring stored-product insects-a review. Journal of the Kansas Entomological Society, 63, 539–547.

Fields, P. G., VanLoon, J., Dolinski, M. G., Harris, J. L., and Burkholder, W. E. 1993. The distribution of *Rhyzopertha dominica* (F.) in western Canada. Canadian Entomologist, 125, 317–328.

Finnegan, D. E. and Chambers, J. 1993. Identification of the sex pheromone of the guernsey carpet beetle, *Anthrenus sarnicus* Mroczkowski (Coleoptera: Dermestidae). Jounal of Chemical Ecology, 19, 971–983.

Fukui, H., Matsumura, F., Barak, A. V. and Burkholder, W. E. 1977. Isolation and identification of a major sex attractant component of *Attagenus elongatulus* Casey (Coleoptera: Dermestidae). Journal of Chemical Ecology, 3, 541–550.

Fukui, H., Matsumura, F., Ma, M. C. and Burkholder, W. E. 1974. Identification of the sex pheromone of the furniture carpet beetle, *Anthrenus flavipes* LeConte. Tetrahedron Letters, 40, 3563–3566.

Hodges, R. J., Cork, A., and Hall, D. R. 1984. Aggregation pheromones for monitoring the greater grain borer, *Prostephanus truncatus*. British Crop Protection Conference-Pests and Diseases, Brighton, pp. 255–259.

Horler, D. F. 1970. (-)-Methyl-n-tetradeca-*trans*-2, 4, 5, trienoate, an allenic ester produced by the male dried bean beetle, *Acanthoscelides obtectus* (Say). Journal of the Chemical Society C, 6, 859–862.

Hussain, A. 1993. Chemical ecology of *Tribolium castaneum* Herbst (Coleoptera: Tenebrionidae): factors affecting biology and application of pheromone. Ph.D. thesis, Oregon State University, Corvallis, 118 p.

Insects Limited, Inc. 1994. Pest monitoring software, Ver. 3.0. Insects Limited, Inc. Indianapolis, Indiana.

Javer, A., Borden, J. H., Pierce, H. D. and Pierce, A. M. 1990. Evaluation of pheromone-baited traps for monitoring of cucujid and

tenebrionid beetles in stored grain. Journal of Economic Entomology, 83, 268–272.

Johnston, B. D. and Oehlschlager, A. C. 1986. Synthesis of the aggregation pheromone of the square-necked grain beetle *Cathartus quadricollis*. Journal of Organic Chemistry, 51, 760–767.

Kuwahara, Y, Fukami, H., Howard, R., Ishii, S., Matsumura, F. and Burkholder, W. E. 1978. Chemical studies on the Anobiidae: sex pheromone of the drugstore beetle, *Stegobium paniceum* (L.) (Coleoptera). Tetrahedron, 34, 1769–1774.

Kuwahara, Y. and Nakamura, S. 1985. (Z)-5- and (E)-5-Undecenoic acid: identification of the sex pheromone of the varied carpet beetle, *Anthrenus verbasci* L. (Coleoptera: Dermestidae). Applied Entomology and Zoology, 20, 354–356.

Kuwahara, Y., Thi My Yen, L., Tominaga, Y., Matsumoto, K., Wada, Y. 1982. 1, 3, 5, 7-Tetramethyldecyl formate, Lardolure: aggregation pheromone of the acarid mite, *Lardoglyphus konoi* (Sasa et Asanuma). Agricultural and Biological Chemistry, 46, 2283.

Landolt, P.J., 1989. Attraction of the cabbage looper to host plants and host odor in the laboratory. Entomol. Exp. Appl. 53: 117–124.

Landolt, P. J. and Curtis, C. E. 1982. Interspecific sexual attraction between *Pyralis farinalis* L. and *Amyelois transitella* (Walker). Journal of the Kansas Entomological Society, 55, 248–252.

Lanier, G. N. 1990. Principles of attraction-annihilation: mass trapping and other means. In: Ridgeway, R. L., Silverstein, R. M., and Inscoe, M. N., ed., Behavior-modifying chemicals for insect management, applications of pheromones and other attractants, New York, Marcel Dekker, Inc., 25–45.

Levinson, H. Z., Levinson, A. R., Jen, T.-L., Williams, J. L. D., Kahn, G., and Francke, W. 1978. Production site, partial composition, and olfactory perception of a pheromone in the male hide beetle. Naturwissenschaften, 65, 543–545.

Levinson, H. Z. and Mori, K. 1983. Chirality determines pheromone activity for flour beetles. Naturwissenschaften, 70, 190–192.

Millar, J. G., Pierce, H. D. Jr., Pierce, A. M., Oehlschlager, A. C., Borden, J. H. and Barak, A. V. 1985a. Aggregation pheromones of the flat grain beetle, *Cryptolestes pusillus* (Coleoptera: Cucujidae). Journal of Chemical Ecology, 11, 1053–1070.

Millar, J. G., Pierce, H. D. Jr., Pierce, A. M., Oehlschlager, A. C. and Borden, J. H. 1985b. Aggregation pheromones of the grain beetle, *Cryptolestes turcicus* (Coleoptera: Cucujidae). Journal of Chemical Ecology, 11, 1071–1081.

Mitchell, E.R. Tingle, F.C. and Heath, R.R. 1990. Ovipositional response of three *Heliothis* species (Lepidoptera: Noctuidae) to allelochemicals from cultivated and wild host plants. Journal of Chemical Ecology, 16, 1817–1827.

Mueller, D. and Pierce, L. 1992. Stored product protection in paradise. Pest Control, June, pp. 34–36.

Mullen, M. A. 1992. Development of a pheromone trap for monitoring *Tribolium castaneum*. Journal of Stored Product Research, 28, 245–249.

Nara, J.M., Lindsay, R.C. and Burkholder, W.E. 1981. Analysis of volatile compounds in wheat germ oil responsible for an aggregation response in *Trogoderma glabrum* larvae. Agricultural and Food Chemistry, 29, 68–72.

Phillips, J. K., Miller, S. P. F., Anderson, J. F., Fales, H. M. and Burkholder, W. E. 1987. The chemical identification of the granary weevil aggregation pheromone. Tetrahedron Letters, 28, 6145–6146.

Phillips, T.W., Jiang, X.-L., Burkholder, W.E., Phillips, J.K and Tran-Quoc, H. 1993. Behavioral responses to food volatiles by ecologically different stored-product Coleoptera, *Sitophilus oryzae* (Curculionidae) and *Tribolium castaneum* (Tenebrionidae) Journal of Chemical Ecology, 19, 723–734.

Phillips, T. W., Strand, M. R. 1994. Larval secretions and food odors affect orientation in female Indinameal moths, *Plodia interpunctella*. Entomologia Experimentalis et Applicata, 71, 185–193.

Pierce, A. M., Pierce, H. D., Jr., Oehlschlager, A. C., and Borden, J. H. 1985. Macrolide aggregation pheromones in *Oryzaephilus surinamensis* and *O. mercator*. Journal of Agricultural and Food Chemicstry. 33, 848–852.

Pierce, A. M., Pierce, H. D., Oehlschlager, A. C., and Borden, J. H. 1990. Attraction of *Oryzaephilus surinamensis* (L.) and

Oryzaephilus mercator (Fauvel) (Coleoptera: Cucujidae) to some common volatiles of food. Journal of Chemical Ecology, 16, 465–475.

Pierce, A.M., Pierce, H.D, Borden, J.H. and Oehlschlager, A.C.. 1991a. Fungal volatiles: semiochemicals for stored-product beetles (Coleoptera: Cucujidae). Journal of Chemical Ecology, 17, 581–597.

Pierce, A. M., Pierce, H. D., Oehlschlager, A. C., and Borden, J. H. 1991b. 1-Octen-3-ol, attractive semiochemical for the foreign grain beetle, *Ahasverus advena* (Waltl) (Coleoptera: Cucujidae). Journal of Chemical Ecology, 17, 567–580.

Prevett, P. F., Benton, F. P., Hall, D. R., Hodges, R. J. and Dos Santos Serodio, R. 1989. Suppression of mating in *Ehestia cautella* (Walker) (Lepidoptera: Phycitidae) using microencapsulated formulations of synthetic sex pheromone. Journal of Stored-Products Research, 25, 147–154.

Qi, Y.T. and Burkholder, W. E. 1981. Sex pheromone biology and behavior of the cowpea weevil *Callosobruchus maculatus* (Coleoptera: Bruchicae). Journal of Chemical Ecology, 8, 527–534.

Quartey, G. K. and Coaker, T. H. 1992. The development of an improved model trap for monitoring *Ephestia cautella*. Entomologia Experimentalis et Applicata, 64, 293–301.

Raffa, K. F., Phillips, T. W., Salom, S. M. 1993. Strategies and mechanisms of host colonization by bark beetles. In: Schowalter, T. D., and Filip, G. M., ed., Beetle-pathogen interactions in conifer forests, London, Academic Press, pp. 103–128.

Ridgeway, R. L., Inscoe, M. N. and Dickerson, W. A. 1990. Role of the boll weevil pheromone in pest management. In: Ridgeway, R. L., Silverstein, R. M., and Inscoe, M. N., ed., Behavior-modifying chemicals for insect management, applications of pheromones and other attractants, New York, Marcel Dekker, Inc., 437–471.

Roelofs, W. L., Glass, E. H., Tette, J. and Comeau, A. 1970. Sex pheromone trapping for red-banded leaf roller control: theoretical and actual. Journal of Economic Entomology, 63, 1162–1167.

Schmuff, N., Phillips, J. K., Burkholder, W. E., Fales, H. M., Chen, C., Roller, P., and Ma, M. 1984. The chemical identification of the rice and maize weevil aggregation pheromones. Tetrahedron Letters, 25, 1533–1534.

Shapas, T. J., Burkholder, W. E. and Boush, G. M. 1977. Population suppression of *Trogoderma glabrum* by using pheromone luring for protozoan pathogen dissemination. Journal of Economic Entomology, 70, 469–474.

Silverstein, R. M., Rodin, J. O., Burkholder, W. E. and Gorman, J. E. 1967. Sex attractant of the black carpet beetle. Science, 157, 85–87.

Sower, L. L. and Fish, J. C. 1975. Rate of release of the sex pheromone of the female Indian meal moth. Environmental Entomology, 4, 168–169.

Sower, L. L. and Whitmer, G. P. 1977. Population growth and mating success of Indian meal moths and almond moths in the presence of synthetic sex pheromone. Environmental Entomology, 6, 17–20.

Sower, L. L., Vick, K. W. and J. H. Tumlinson. 1974. (Z,E)–9, 12-Tetradecadien-1-ol: a chemical released by female *Plodia interpunctella* that inhibits the sex pheromone response of male *Cadra cautella*. Environmental Entomology, 3, 120–122.

Süss, L. and Trematerra, P. 1986. Control of some Lepidoptera Phycitidae infesting stored-products with synthetic sex pheromone in Italy. Donahaye, E. and Navarro, S, ed. Proceedings of the Fourth International Working Conference on Stored-Product Protection, Tel Aviv, Israel, September 1986, 606–611.

Suzuki, T. 1980. 4,8-Dimethyldecanal: the aggregation pheromone of the flour beetles, *Tribolium castaneum* and *T. confusum* (Coleoptera: Tenebrionidae). Agricultural and Biological Chemistry, 44, 2519–2520.

Tanaka, Y., Honda, H., Ohsawa, K. and Yamamoto, I. 1986. A sex attractant of the yellow mealworm beetle, *Tenebrio molitor* L., and its role in the mating behavior. Journal of Pesticide Science, 11, 49–55.

Tanaka, K., Ohsawa, K., Honda, M. and Yamamoto, I. 1981. Copulation release pheromone, erectin, from the Azuki bean weevil, *Callosobruchus chinensis* L. Journal of Pesticide Science, 7, 535–537.

Vick, K. W., Coffelt, J. A., Mankin, R. W. and Soderstrom, E. L. 1981. Recent developments in the use of pheromones to monitor *Plodia interpunctella* and *Ephestia cautella*. In: Mitchell, E. R., ed., Management of insect pests with semiochemicals, concepts and practice, New York, Plenum Press, 19–28.

Vick, K. W., Su, H. C. F., Sower, L. L., Mahany, P. G. and Drummond, P. C. 1974. (Z,E)-7, 11-Hexadecadien-1-ol acetate: the sex pheromone of the Angoumois grain moth, *Sitotroga cerealella*. Experientia, 30, 17–18.

Walgenbach, C.A. Burkholder, W.E., Curtis, M.J. and Khan, Z.A. 1987. Laboratory trapping studies with *Sitophilus zeamais* (Coleoptera: Curculionidae). Journal of Economic Entomology, 80, 763–767.

Wileyto, E. P., Ewens, W. J. and Mullen, M. A. 1994. Markov-recapture population estimates: a tool for improving interpretation of trapping experiments. Ecology, In Press.

Williams, H. J., Silverstein, R. M., Burkholder, W. E. and Khorramshahi, A. 1981. Dominicalure 1 and 2: components of aggregation pheromone from male lesser grain borer *Rhyzopertha dominica* (F.). Journal of Chemical Ecology, 7, 759–780.

Wong, J, W., Verigin, V., Oehlschlager, A. C., Borden J. H., Pierce, H. D. Jr., Pierce, A. M. and Chong, L. 1983 Isolation and identification of two macrolide pheromones from the flat grain beetle, *Cryptolestes ferrugineus* (Coleoptera: Cucujidae). Journal of Chemical Ecology, 9, 451–474.

Yarger, R. G., Silverstein, R. M. and Burkholder, W. E. 1975. Sex pheromone of the female dermestid beetle *Trogoderma glabrum* (Herbst). Journal of Chemical Ecology, 1, 323–334.

Yamaoka, R., Shiraishi, T., Ueno, T., Kuwahara, Y., and Fukami, H. 1985. Structural elucidation of Koiginal I and II, the sex pheromones of the webbing clothes moth, using capillary GLC/MS. Mass Spectroscopy, 33, 189.

Studies on the relative susceptibility of varieties and germplasm lines of sesame to infestation by *Tribolium castaneum* (Herbst) (Coleoptera: Tenebrionidae)

R.K. Murali Baskaran, M.S. Venugopal and C.V. Sivakumar*

Abstract

Growth and development of red flour beetle *Tribolium castaneum* (Herbst) on wheat flour and nine varieties and three germplasm lines of sesame was studied at 30±2°C under laboratory conditions. Development from first instar larvae to adult was fastest on sesame line ES 22 and slowest on TM V6 taking 37.9±0.6 and 57.1 ±0.2 days respectively. Survival to adulthood ranged from 81.7% on wheat flour, 61.8% on ES 22 to 32.5% on TM V6.

Weight loss due to insect activity was higher on white-seeded lines TMV 2, SVPR 1 and ES 12 than on black, or brown, seeded varieties. It was concluded that, except for ES 22 and SI 250, susceptibility to infestation by *T. castaneum* was mainly attributed to the oil content and colour of the testa.

Introduction

Sesame seeds and its cake in storage are attacked by many storage pests causing losses to weight of produce and its germinability (Mahadevan 1988). Sesame seeds are attacked by the rice moth (*Corcyra cephalonica* (Stainton)), the red flour beetle (*Tribolium castaneum* Herbst), the khapra beetle (*Trogoderma granarium* (Herbst)) and the almond moth (*Cadra cautella* (Wlk.)) in India and Brazil; *Cryptolestes pusillus* Schon and *Oryzaephilus mercator* (Fauvel) in Africa; and *Tribolium confusum* Duv. in Uganda and Africa. Among the different storage pests, *T. castaneum* and *C. cephalonica* are cosmopolitan pests, causing additional damage (the latter by way of webbing) (Srivastava 1964; Mookherjee and Bose 1967). Different varieties of seeds show varying susceptibility to storage insects infesting them.

In Tamil Nadu, nine varieties and three germplasms of sesame have been developed which include varieties resistant to *Antigastra catalaunalis* (ES 22, ES 12 and SI 250) as well as high yielding varieties (TMV 3, TMV 2 and SVPR 1). Since the relative susceptibility of these varieties and germplasms to stored product insects had not been investigated, the present study was undertaken to determine the susceptibility of nine varieties and three germplasms to infestation by the beetle, *Tribolium castaneum*, an important storage pest of sesame in Tamil Nadu.

Materials and Methods

Sesame varieties, TMV 1, 2, 3, 4, 5 and 6, CO 1, SVPR 1 and Paiyur 1 and three germplasms ES 22, 12 and SI 250, harvested during the rainy season 1992, were obtained from the Regional Research station, Vridhachalam and Paiyur and Cottom Research Station Srivilliputtur, Tamil Nadu. Varieties, TMV 2, SVPR 1 and ES 12 were white seeded and the others were uniformly brown or black in colour and were not generally distinguishable from one another.

Culture of *T. castaneum* was maintained on the whole wheat flour with 5% yeast powder at 30 ± 2°C. The eggs of the beetle were extracted as per the methodology given in Figure 1.

Growth parameters of *T. castaneum* on varieties and germplasms of sesame.

Sound seeds of sesame were disinfested in an oven at 55°C for 2 hours before use. Fifty one day old eggs were placed in 250 mL conical flasks containing 30 g of whole seeds, and their growth and development compared with that in wheat flour. For proper aeration, the mouth was plugged with absorbant cotton. Each seed material was replicated thrice.

Observations on larval and pupal periods, weight/pupa, per cent pupation, average developmental period, weight/adult and per cent adult emergence were made.

Indices for varieties and germplasms of sesame

Various indices — larval–pupal index, pupal weight index and adult weight index (Deshmukh et al. 1982), adult emergence index (Tripathi et al. 1982), developmental index (Prasad and Bhattacharya 1975), general growth index (Pant and Dang 1969) and Howes growth index (Howe 1971) — were computed by using wheat flour as the standard.

Finally, a suitability index was calculated by taking the average of the above seven indices.

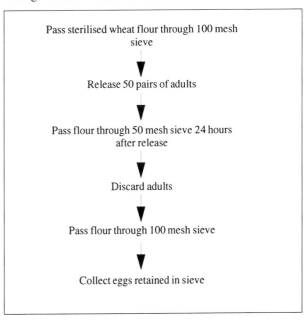

Fig. 1. Flow chart for extracting eggs of *T. castaneum*.

* Department of Agricultural Entomology, Agricultural College and Research Institute, Madurai-625104, Tamilnadu, India.

Sex ratio

Emerged adults were sexed based on the presence of a shallow, oval pit with erect golden yellow hairs on the venter of the fore femur of the male (Hinton 1942).

Losses and viability

Loss due to infestation was estimated by weighing the developed adult during feeding activity, and the viability of seeds was determined by performing a germination test at the end of the completion of the first generation.

Results

Out of nine varieties and three sesame lines tested, ES 22, TMV 2, SI 250, ES 12, TMV 5 and SVPR 1 were highly susceptible to *T. castaneum* while TMV 6 and CO 1 were least susceptible.

Growth and development of *T. castaneum*

The larval development was fastest in ES 22, lasting 25.70 ± 0.44 days and slowest in TMV 6, lasting 43.45 ± 0.38 days (Table 1), which in turn influenced the average developmental period. The developmental period of *T. castaneum* on various seed materials of sesame ranged from 37.90 ± 0.59 to 57.10 ± 0.25 days, which differed significantly among samples. White seeded sesame TMV 2, ES 12 and SVPR 1 had fairly low developmental period, ranging from 38.55 ± 0.81 to 40.60 ± 0.71 days while wheat flour recorded the developmental period of 34.85 ± 0.81 days. There was not much difference among different varieties and germplasms in the pupal period, however, it ranged from 6.95 ± 0.04 to 8.20 ± 0.03 days.

The pupa and adult gained the maximum weight when the grubs were reared on wheat flour (3.12 ± 0.65 mg/pupa and 3.81 0.73 mg/adult). In susceptible seed materials, weights ranged from 2.15 ± 0.48 to 2.25 ± 0.73 mg/pupa and 2.75 ± 0.52 to 2.98 ± 0.72 mg/adult, while in other seed materials, the range was 1.78 ± 0.77 to 1.99 ± 0.38 mg/pupa and 2.13 ± 0.51 to 2.73 ± 0.34 mg/adult.

The lowest pupation was recorded on TMV 6 (46.3%) which was significantly different from other seed materials, followed by CO 1 (53.4%), Paiyur 1 (62.7%) and TMV 3 (63.7%). The highest per cent pupation was recorded on TMV (72.7%) which was on par with ES 22 (72.1%), ES 12 (72.1%) and SI 250 (71.9%).

Adult emergence ranged from 32.5% in TMV 6 to 61.8% in ES 22. The adult emergence obtained on TMV 6 differed significantly from other seed materials. On the other hand, Paiyur 1, TMV 5, TMV 4 and TMV 1 did not differ significantly. Highest adult emergence of 81.7% was obtained on wheat flour (Table 1).

Indices for varieties and germplasms of sesame

All indices of growth and development of *T. castaneum* indicated that ES 22, TMV 2, SI 250, ES 12, TMV 5 and SVPR 1 were highly susceptible to infestation, whereas TMV 6 and CO 1 were resistant to feeding.

Howes growth index on various seed materials ranged from 0.026 and 0.047, being lowest on TMV 6 and highest on ES 22 (Table 2). Suitability index ranged from 0.686 to 1.135, and was greater than 1 on wheat flour and ES 22, while others which were less preferred by *T. castaneum* recorded less than 1.

Sex ratio

Unsuitability of grains favoured males, which outnumbered females on TMV 6 (2.10:1), CO 1 (2.00.1), TMV 1, TMV 3, TMV 4 and Paiyur 1 (1.50:1), SVPR (1.30:1), TMV 5 and ES 12 (1.12:1), while in others females dominated (Table 2).

Losses and viability of seeds

The highest loss in weight (12.9, 12.8 and 12.8%) was sustained by white seeded sesame TMV 2, SVPR 1 and ES 12 while the lowest (5.2%) was recorded on TMV 6 which was significantly different from other seed materials. After completion of the first generation, the viability of seeds ranged from 41.7 to 66.7% on various varieties and germplasms of sesame (Table 3).

Discussion

T. castaneum showed a preference for ES 22, TMV 2, ES 12 and SI 250 over other seed materials of sesame for feeding. The varieties and germplasms of sesame showing short developmental period were preferred and are therefore more likely to be infested under normal conditions of storage. TMV 6 was resistant to infestation with the highest developmental period of 57.10 ± 0.25 days. Maximum development period was also recorded on N 32 (54.2 days) and shortest of 42.6 days on JT 7 (Sachan and Chandel 1991). The insects reared on groundnut germplasm took less time than those reared on sesame (Singh 1985).

Adult emergence was hindered in resistant varieties, TMV 6 and CO 1, which were significantly different from each other. Appreciably better adult emergence was recorded on sesame, 68.99% on whole seeds and 88.47% on crushed seeds (Urs and Mookherjee 1966) In the present study, it seems probable that the sesame testa provided the main resistance to infestation in the apparently resistant varieties TMV 6 and CO 1. Williams and Mills (1980) also reported that undamaged pericarp is a factor in resistance of sorghum varieties to damage by weevils.

Weight of the adult ranged from 2.13 ± 0.51 to 2.98 ± 0.72 mg/adult in different sesame seed materials. In contrast to this, Sachan and Chandel (1991) and Singh (1985) recorded the maximum adult weight of 0.51 mg/adult and 0.51 mg/adult on JT 7 and PT 303 respectively.

The values of Howes growth index ranged from 0.026 to 0.046 and was the highest in wheat flour (0.055). The present study is supported by Sachan and Chandel (1991) who recorded Howes growth index, ranging from 0.048 to 0.085 with 0.658 in standard Murag–1. The values for the suitability index agree with the findings of Sachan and Chandel (1991) who recorded maximum and minimum of 1.094 and 0.777 for C 6 and TMV 3 respectively, whereas, in the present study, it was 1.135 and 0.686.

Interestingly, the suitability index had an influence on the sex ratio and the susceptible host favoured females, which outnumbered males. The resistant varieties TMV 6 and CO 1 were found favourable to male, recording 2.10:1 and 2.00:1 male:female. This result supports the findings of Sachan and Chandel (1991) who recorded more numbers of males on N 32 and TMV 3 and an equal number of males and females on TC 25, whereas in the present study, no variety or germplasm recorded a balanced sex ratio.

White seeded sesame TMV 2, ES 12 and SVPR 1 were highly susceptible to infestation when compared to brown or black seed, recording 12.9, 12.8 and 12.8% weight loss respectively (Table 3). White seeded varieties are rich in oil content which in turn become more susceptible to the attack of

Table 1. Growth parameters of *T. castaneum* on sesame varieties and germplasms

	Larval period (days)	Pupal period (days)	Weight/pupa (mg)	Pupa (%)	Average developmental period (days)	Weight /adult (days)	Adult (%)
TMV 1	31.97±0.85	7.60±0.04	1.81±0.50	(53.70)cde	44.65±0.31	2.58±0.55	(40.69)d
TMV 2	26.10±1.18	6.90±0.08	2.15±0.48	(58.52)b	38.55±0.81	2.86±0.81	(48.85)c
TMV 3	31.76±0.93	7.65±0.03	1.99±0.38	(52.95)e	44.60±0.28	2.55±0.75	(39.67)de
TMV 4	32.05±0.40	7.45±0.06	1.93±0.48	(54.95)cd	45.95±0.97	2.53±0.84	(40.51)d
TMV 5	32.15±0.88	7.60±0.08	1.95±0.75	(53.35)de	44.80±0.83	2.73±0.34	(40.68)d
TMV 6	43.45±0.38	8.20±0.03	1.78±0.77	(42.88)g	57.10±0.25	2.13±0.51	(34.76)f
CO 1	41.00±0.73	7.70±0.05	1.81±0.81	(46.95)f	53.70±0.89	2.34±0.78	(37.34)e
Paiyur 1	34.77±0.51	7.70±0.05	1.88±0.73	(52.36)e	47.30±0.55	2.65±0.58	(40.69)d
SVPR 1	28.76±1.31	7.15±0.08	2.01±0.70	(55.49)c	40.60±0.71	2.85±0.63	(48.65)c
ES 22	25.70±0.44	6.95±0.04	2.24±0.51	(58.08)b	37.90±0.59	2.95±0.53	(51.83)b
ES 12	26.60±0.97	7.20±0.03	2.25±0.73	(58.12)b	38.60±0.75	2.75±0.52	(50.30)bc
SI 250	26.85±1.36	6.90±0.05	2.23±0.81	(57.99)b	38.75±0.33	2.98±0.72	(50.22)bc
Wheat flour	22.15±0.57	6.45±0.08	3.12±0.65	(67.11)a	34.85±0.81	3.81±0.73	(64.78)a
CD (P=0.05)				1.87**			2.33**

Notes: Figures in parentheses are arcsin values.
    Means followed by same letters are not significantly different (P=0.05).
    ** Significant at 5 and 1% level.

Table 2. Indices for the development of *T. castaneum* on sesame varieties and germplasms.

Variety/ germplasm	Larval-pupal index[a]	Pupal weight index[b]	Adult weight[c]	Adult emergence[d]	Developmental index[e]	General growth index[f]	Howe's growth index[g]	Suitability index[h]	Sex ratio Male:Female
TMV 1	0.723	0.580	0.677	0.52	1.281	2.030	0.036	0.835	1.50:1
TMV 2	0.867	0.689	0.751	0.69	1.106	2.785	0.045	0.990	0.95:1
TMV 3	0.725	0.638	0.669	0.50	1.279	2.006	0.036	0.836	1.50:1
TMV 4	0.724	0.619	0.664	0.52	1.319	2.090	0.035	0.853	1.50:1
TMV 5	0.719	0.625	0.717	0.52	1.286	2.003	0.036	0.964	1.25:1
TMV 6	0.554	0.571	0.559	0.39	1.638	1.066	0.026	0.686	2.10:1
Co 1	0.685	0.580	0.614	0.45	1.541	1.302	0.029	0.746	2.00:1
Paiyur 1	0.673	0.603	0.696	0.52	1.357	1.803	0.034	0.812	1.50:1
SVPR 1	0.796	0.644	0.748	0.69	1.165	2.361	0.043	0.921	1.30:1
ES 22	0.876	0.718	0.774	0.76	1.088	2.805	0.047	1.135	0.77:1
ES 12	0.846	0.721	0.721	0.72	1.108	2.711	0.046	0.972	1.12:1
SI 250	0.847	0.715	0.782	0.72	1.112	2.678	0.046	0.985	0.99:1
Wheat flour	–	–	–	–	–	3.828	0.055	1.941	0.75:1

[a]Larval pupal Index $\dfrac{\text{Average larval period on wheat flour (days)}}{\text{Average larval period on test material (days)}}$ $\dfrac{\text{Average pupal period on wheat flour (days)}}{\text{Average pupal period on test material (days)}}$

[b]Pupal weight index $\dfrac{\text{Average pupal weight on test material (mg)}}{\text{Average pupal weight on wheat flour (mg)}}$

[c]Adult weight index $\dfrac{\text{Average adult weight on test material (mg)}}{\text{Average adult weight on wheat flour (mg)}}$

[d]Adult emergence index $\dfrac{\text{Adult emergence on test material (\%)}}{\text{Adult emergence on wheat flour(\%)}}$

[e]Developmental index $\dfrac{\text{Average developmental period on test material (days)}}{\text{Average developmental period on wheat flour(days)}}$

[f]General growth index $\dfrac{\text{Population(\%)}}{\text{Average larval period (days)}}$

[g]Howe's growth index $\dfrac{\text{log adult emergence (\%)}}{\text{Average developmental period (days)}}$

[h]Suitability index $\dfrac{\text{Sum of all indices on test material}}{7}$

Table 3. Reaction of sesame varieties and germplasms to *T. castaneum*

Variety/germplasm	Weight loss (%)	Germination (%)
TMV 1	10.9 (19.32)e	50.9 (45.51)de
TMV 2	12.9 (21.01)g	43.2 (41.07)gh
TMV 3	11.8 (20.09)f	50.9 (45.52)de
TMV 4	10.2 (18.58)d	50.7 (45.39)e
TMV 5	8.2 (16.64)c	52.6 (46.47)d
TMV 6	5.2 (13.12)a	66.7 (54.76)a
CO 1	8.0 (16.43)c	61.9 (51.89)b
Paiyur 1	7.1 (15.39)b	60.4 (48.36)c
SVPR 1	12.8 (20.92)g	47.8 (43.74)f
ES 22	13.5 (21.52)f	42.1 (40.57)g
ES 12	12.8 (20.92)g	41.7 (40.19)gh
SI 250	12.2 (20.40)f	40.6 (39.55)h
CD (P = 0.05)	0.41**	0.93**

Notes: Figures in parentheses are arcsin values.
Means followed by same letters are not significantly different (P=0.05).
**Significant at 5 and 1% level..

T. castaneum (Directorate of Oilseeds Research 1988). Germination of the seeds was reduced considerably due to the destruction of the germ portion of seeds. The viability of different varieties and germplasms ranged from 41.7 to 66.7%.

Acknowledgments

The authors are grateful to the Regional Research Station, Vridhachalam and Paiyur and Cotton Research Station, Srivilliputtur for the supply of sesame varieties and germplasms.

References

Directorate of Oilseeds Research 1988. Annual report XXXIV kharif workshop on oilseeds. Rajendranagar, Hyderabad, Andhra Pradesh.

Deshmukh, P.D., Rathore, Y.S. and Bhattacharya, A.K. 1982. Effect of temperature on the growth and development of *Diacrisia obliqua* (Walker) on five host plants. Indian Journal of Entomology 44, 21–33.

Hinton, H.E. 1942. Secondary sexual characters of *Tribolium*. Nature, 149(2), 500–501.

Howe R.W. 1971. A parameter for expressing the suitability of an environment for insect development. Journal of Stored Products Research, 7, 63–65.

Mahadevan, N.R. (1988) Screening sesamum entries for resistance to leaf webber *Antigastra catalaunalis* Duphoncel. Madras Agricultural Journal, 75, 225.

Mookherjee, P.B. and Bose B.N. 1967. Some insecticide trials for the protection of til seed in storage. Indian Journal of Entomology, 29, 103–104.

Pant N.C. and Dang K. 1969. Food values of several stored commodities in the development of *Tribolium castaneum* Herbst. Indian Journal of Entomology, 31, 147–151.

Prasad J. and Bhattacharya A.K. 1975. Growth and development of *Spodoptera littoralis* (Boisd.) (Lepidoptera: Noctuidae) on several plants. Zeitschrift angewande Entomologie, 79, 34–48.

Sachan, G.C. and Chandel S.K. 1991. Relative susceptibility of some sesame germplasms to *Tribolium castaneum* (Herbst) Bulletin of Grain Technology, 29(1), 31–34.

Singh, M.N. 1985. Developmental behaviour of *Lasioderma serricorne* (F.), *Tribolium castaneum* (Herbst), *Ephestia cautella* (Wlk.) and *Corcyra cephalonica* (Stnt.) on toria and rai germplasms M.Sc. (Ag.) thesis, G.B. Pant University of Agriculture and Technology, Patnagar.

Srivastava A.S. 1964. Entomological research during the last ten days in the section of Entomologists to Government of U.P., Kanpur (India) Res. Mem. Uttar Pradesh No. 3.

Tripathi A.K., Bhattacharya, A.K. and Verma S.K. 1982. Development behaviour of *Mythimna separata* (Walker) on some monocotyledonous plants. Indian Journal of Entomology, 44(4), 355–367.

Urs K.C.D. and Mookherjee P.B. 1966. Development of *Tribolium castaneum* (Herbst) on the whole and crushed seeds of sesame Indian Journal of Entomology, 28, 234–240.

Williams J.O. and Mills R.B. 1980. Influence of mechanical damage and repeated infestation of sorghum on its resistance to *Sitophilus oryzae* (L.) (Coleoptera: Curculionidae). Journal of Stored Products Research, 16, 51–53.

A comparison of the demography of four major stored grain coleopteran pest species and its implications for pest management

S. J. Beckett, B. C. Longstaff and D. E. Evans*

Abstract

Detailed demographic data are now available for four of the major stored grain pests, *Sitophilus oryzae, Rhyzopertha dominica, Oryzaephilus surinamensis* and *Tribolium castaneum*, over a comprehensive range of temperature and relative humidity combinations.

This paper presents a between species comparison of the major demographic components of age specific survival and fertility, immature development period and finite rate of increase. The variation between species, evident in the data, is reflected in their environmental limits, as well as in the optimum and suboptimum conditions for reproduction and population growth.

The implications of these findings for pest management, especially with regard to setting limits for grain receival moisture content and for aeration strategies, are discussed.

Introduction

In Australia, the current status of four of the major stored grain coleopteran pest species, *Sitophilus oryzae, Rhyzopertha dominica, Oryzaephilus surinamensis* and *Tribolium castaneum*, is in part a reflection of the changing history of pest management practices this century.

Prior to the introduction of malathion between 1960 and 1965 (Murray 1979), *S. oryzae* was a major problem in Australia. The insect caused significant storage losses and was frequently identified in grain exports. From 1917 to 1919, for example, numerous outbreaks in Victoria, South Australia and New South Wales reached what were considered to be plague proportions (Winterbottom 1922). However, with the systematic application of malathion, a dramatic decline in the incidence of the pest occurred. Since then, the low levels achieved have been maintained with new generations of pesticides and, in general, it now ranks a poor fourth in the frequency of reported infestations at central storages (Bulk Grains Queensland Ltd (BGQ) 1991; Co-operative Bulk Handling Western Australia Ltd (CBH WA) 1991; Grain Elevators Board of Victoria (GEB Vic.) 1991; NSW Grain Corporation Ltd (GrainCorp) 1991; South Australian Co-operative Bulk Handling Ltd (SA-CBH) 1991).

R. dominica, on the other hand, has consistently been a major stored grain pest. It was the repeated occurrence of malathion resistant strains of *R. dominica*, rather than *S. oryzae*, that led to the urgent development of alternative pesticides in the 1970s (Murray 1979). The majority of phosphine-resistant strains reported so far have also been *R. dominica* (Caddick 1990). Today, it is consistently the most frequent

cause of infestations at central storages and rejections at point of export (BGQ 1991; CBH WA 1991; GEB Vic. 1991; GrainCorp 1991; SA-CBH 1991; Australian Wheat Board (AWB) 1988).

Traditionally, *O. surinamensis* and *T. castaneum* have not been considered major threats to grain exports in Australia. In the past, outbreaks of *O. surinamensis* have been relatively uncommon. However, over the past decade, the incidence of this insect has increased significantly (BGQ 1991; CBH WA 1991; GEB Vic. 1991; GrainCorp 1991; SA-CBH 1991; AWB 1988). It is now a key pest in some areas and is becoming a threat to grain exports. A major contributing factor appears to have been the development of pesticide resistance (Allen et al. 1988; insect resistance to grain protectants in the Victorian central grain handling system 1988; Sproul and Emery 1988; Wallbank 1988).

On the other hand, the reported incidence of *T. castaneum* has generally been consistent and though it has been relatively easy to treat, outbreaks of this pest have been quite frequent. For this reason, the cost of control makes it economically important (BGQ 1991; CBH WA 1991; GEB Vic. 1991; GrainCorp 1991; SA-CBH 1991; AWB 1988).

Experience has shown that significant changes in pest management or grain storage have the potential of altering the relative status of stored grain pests. The decline in importance of *S. oryzae*, for example, coincided with the conversion from bag to bulk storage in the 1950s and, as previously mentioned, the introduction of residual chemical protectants in the 1960s. Today, pest management relies less on the application of protectants and more on alternatives such as phosphine fumigation and aeration. If *S. oryzae,* which already is relatively tolerant to phosphine, were to develop major resistance to that fumigant, it has the potential to become a serious pest once again. The general move away from the residual treatment of grain, as well as higher receival limits for grain moisture at central storages, could further compound the problem.

To understand the extent of a potential problem, such as the one outlined above, it is essential to have detailed demographic data of the major pest species over a comprehensive range of temperature and moisture combinations. With such information, predictions can be made as to the potential threat of a given mixed infestation in a defined environmental condition and the likely dominant species. Furthermore, by knowing those conditions close to the theoretical development and the theoretical population growth thresholds for each species, steps can be taken which will avoid a major infestation of one or more of these species.

Detailed demographic data are now available for four major pest species on wheat, over a comprehensive range of temperature and relative humidity combinations (Longstaff, unpublished data; Beckett and Evans 1994; White 1987; Longstaff and Evans 1983), enabling a comparison to be made of the effects of environment on the population growth of each species. This paper reviews the data for the main demographic parameters and represents them in a directly comparable format. It discusses the population dynamics of each species

---

* Stored Grain Research Laboratory, CSIRO Division of Entomology, GPO Box 1700, Canberra ACT 2601, Australia.

and explores the possible implications for pest management in the light of the different demographies exhibited.

The Demographic Data

The range of available demographic data varies for each species. Data for *S. oryzae* exist at temperatures from 15 to 32.5°C and at relative humidities from 20% to 70% r.h. (Longstaff and Evans 1983); data for *R. dominica* exist from 15 to 39°C for the same range of relative humidities (Longstaff, unpublished data); data for *O. surinamensis* range from 20 to 35°C and from 30 to 70% r.h. (Beckett and Evans 1994); and finally, data for *T. castaneum* generally cover a range from 20 to 40°C and from 35 to 70% r.h. (White 1987). Although these four studies were carried out under broadly similar environmental conditions, specific temperature and relative humidity combinations often differ between them. Therefore, it has frequently been necessary to interpolate between points so that a direct comparison can be made.

The effect of temperature on a given demographic parameter has been considered over the full range of available data for each species. On the other hand, consideration of the effect of relative humidity has been limited to four equally spaced conditions over the best range for direct comparison. These are 35%, 45%, 55%, and 65% r.h. For *S. oryzae*, it was necessary to convert measurements of percentage moisture content to the equivalent relative humidity (Wilson 1990).

The data for *S. oryzae*, *R. dominica* and *O. surinamensis* have been gained for optimum rearing conditions. Both *S. oryzae* and *R. dominica* were reared on whole wheat (var. Olympic), while *O. surinamensis* was reared on kibbled wheat (var. Corella). On the other hand, *T. castaneum* was reared on partly broken whole wheat (var. Oxley), which though not optimum for population growth, reflects a realistic grain storage environment for the insect (White 1987). All four species were reared in a pesticide-free environment.

The description of the rate of population growth has been standardised in terms of the weekly finite rate of increase (λ) (Birch 1948), which is the natural antilog of the weekly infinite rate of increase (r_m). Actual and interpolated data are used for all species, except in the case of *T. castaneum*, where all data are interpolated from a three dimensional graph (White 1987).

To gain information on the possible variation in demographic strategy employed by the four species, data for the net rate of reproduction (R_0) and adult survival (l_x) have been compared after an 8-week period. Eight weeks was chosen as the maximum observation period common to all species and conditions. As 95% of the value of λ at a given condition is accounted for by the time parental females reach an age equivalent to their immature development period (Birch 1948, Figure 3), the presented values of R_0 are close to those required to achieve the actual values of, at temperatures roughly above 21°C for *S. oryzae* and *O. surinamensis*, and 24°C for *R. dominica* and *T. castaneum*. The value of R_0 depends on the product of a female's fertility and her ability to survive. By considering adult survival, an indication of its impact on the net rate of reproduction can be determined.

Furthermore, the relative combined effects that differences in the net rate of reproduction and the immature development period (D) have on the value of λ have also been considered. The relationship can be described in simple terms by the expression $r_m = \log_e R_0/T_c$, where T_c (the cohort generation time) is the sum of the mean age of the female at oviposition time) and the immature development period (Evans 1982). Where models have been developed to described fertility and

survival, as in the case of *S. oryzae* and *T. castaneum*, these have been used to extract the required data.

The relationship between the weekly finite rate of increase (λ), temperature and relative humidity for all species is displayed in Figure 1. In general, the rate of population growth increases as temperature increases at a given relative humidity until an optimum temperature is reached. That temperature and the maximum population growth rate achieved differ for each species. In all situations, populations subjected to drier conditions grow slower than those subjected to more moist conditions. Above the optimum temperature for population growth, the rate of growth declines rapidly.

However, at 35% r.h. values of λ for *S. oryzae* are below 1 and decline with increasing temperature, indicating negative population growth. In the case of *T. castaneum*, White (1987) was unable to observe immature development at temperatures at 35% r.h. However, this model predicts a very small positive population growth between 25 and 35°C ($\lambda = 1.01$ to 1.11). Because of this discrepancy, values of λ at 35% r.h. have not been considered in this study. On the other hand, under similar circumstances positive population growth rates start to increase at 19 and 20°C for *R. dominica* and *O. surinamensis*, with a maximum value of λ (1.63) achieved by *R. dominica* at 30 and 33°C, and by *O. surinamensis* (1.78) at 30°C. Rates of growth for both species are similar as temperature increases to 24°C. Then, from 25 to 32°C the dry conditions favour *O. surinamensis* more so than *R. dominica*. However, as temperature increases further the population growth rate for *O. surinamensis* quickly decreases and reaches zero ($\lambda = 1$) by 35°C. The value of λ for *R. dominica* starts to decrease at about the same temperature but not as rapidly, with a λ value of 1 being reached between 37 and 38°C.

The general relationship of initial increase in population growth rate with increased temperature occurs for *S. oryzae* at the higher relative humidities considered (45%, 55%, and 65% r.h.), with maximum values of λ (1.77, 1.93, and 2.03) achieved at 27°C. For the other species, maximum values of λ (1.94, 2.21, and 2.39) are achieved by *R. dominica* at 33°C, *O. surinamensis* (1.95, 2.13, and 2.29) at 32°C, and *T. castaneum* (1.64, 2.05, and 2.31) at 35°C.

As temperature increases to 27°C at the higher relative humidities, the value of λ for *S. oryzae* is consistently greater than that for the other species. However, at 30°C at 45% r.h. the value of λ is less than that for *R. dominica* and *O. surinamensis*, and at 30°C at 55% and 65% r.h. it is also less than that for *T. castaneum*. On the other hand, as temperature increases to 32.5°C at these relative humidities, λ values for *R. dominica* and *O. surinamensis* are similar. Above this temperature, the population growth rate for *O. surinamensis* decreases faster than that for *R. dominica*. By contrast, values of λ for *T. castaneum* are the lowest for all species over the majority of temperatures/humidity combinations. However, in general, the relative difference in the size of the value of λ at a given temperature decreases as relative humidity increases so that the rate of population growth at a given temperature at 65% r.h. is very similar to that for *R. dominica* and *O. surinamensis*. Moreover, *T. castaneum* reaches its optimum population growth rate at a higher temperature (35°C) at relative humidities above 35% r.h than all other species. At 37°C, λ values for *T. castaneum* at these relative humidities are greater than those for *R. dominica* and it is still achieving positive population growth rates at 39°C/55% r.h. and 40°C/65% r.h.

Since initially the relationship between the rate of population growth and temperature for a given relative humidity is more or less linear, theoretical thresholds of population growth have been calculated for the four species (Table 1). Apart from negative population growth for *S. oryzae* at temperatures at 35% r.h., the thresholds of population growth are

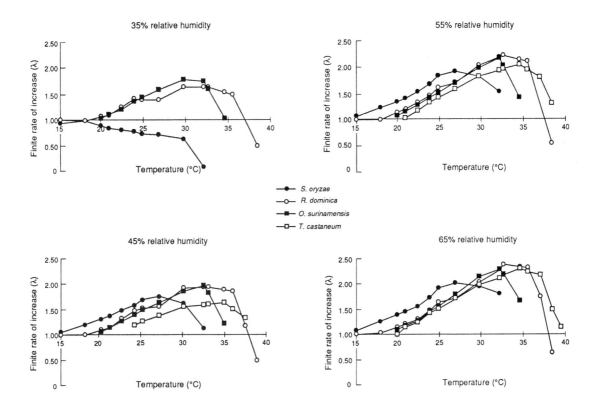

Fig. 1. The weekly finite rate of population increase (λ) for four stored-grain coleopteran species over a range of temperature (°C) and relative humidity (r.h.) combinations.

Table 1. The threshold of weekly finite rate of reproduction (λ) for four stored-grain coleopteran species in terms of wet bulb (T_{wb}) and dry bulb (T_{db}) temperature (°C) over a range of relative humidities (r.h).

% r.h.	T_{wb} (°C)	r^2	°T_{db} (°C)	r^2		°T_{wb} (°C)	r^2	T_{db} (°C)	r^2	n
	Sitophilus oryzae					*Rhyzopertha dominica*				
35.0	-	-	-	-	-	9.3	0.9274	17.3	0.9287	11
45.0	8.8	0.9892	14.7	0.9905	8	10.9	0.9532	17.4	0.9545	11
55.0	10.2	0.9778	14.9	0.9785	8	12.3	0.9722	17.5	0.9728	11
65.0	11.2	0.9766	14.8	0.9782	8	13.8	0.9659	17.7	0.9652	11
	Oyrzaephilus surinamensis					*Tribolium castaneum*				
35.0	11.0	0.9959	19.3	0.9963	7	-	-	-	-	-
45.0	12.1	0.9911	18.9	0.9920	8	13.5	0.9946	20.7	0.9970	4
55.0	13.6	0.9989	19.1	0.9989	8	14.7	0.9958	20.3	0.9969	6
65.0	15.2	0.9978	19.3	0.9975	8	15.5	0.9969	19.8	0.9974	7

little affected by the different relative humidities considered (*S. oryzae*: 14.8°C, SD 0.08; *R. dominica*: 17.5°C, SD 0.15; *O. surinamensis*: 19.2°C, SD 0.17; *T. castaneum*: 20.3°C, SD 0.37).

Values of R_0 for each species, i.e. the sum of adult female progeny per original female over an 8-week period, are displayed in Figure 2. The relationship between R_0, temperature and relative humidity, is similar to that of λ with values increasing as temperature increases until an optimum is reached. Beyond this optimum the values for R_0 quickly decrease. Values at a given temperature are also generally lower at drier conditions than at more moist conditions. Maximum values at temperatures ranging from dry to moist

conditions range from 74.6 at 24°C/45% r.h. to 131.7 at 24°C/65% r.h. for *S. oryzae*, 158.2 at 24°C/ 35% r.h. to 316.9 at 35°C/65% r.h. for *R. dominica*, 41.5 at 30°C/35% r.h. to 117.0 at 32.5°C/65% r.h. for *O. surinamensis*, and 3.5 at 30°C/35% r.h. to 245.1 at 35°C/ 65% r.h. for *T. castaneum* (values for R_0 at 35% r.h. for *T. castaneum* are predicted rather than actual).

In general, after an 8-week period, as temperature increases at 35% r.h., adult survival (l_x) starts to decrease below 80% at 24°C for *S. oryzae*, 30°C for *R. dominica*, 35°C for *O. surinamensis* and 32°C for *T. castaneum*. However, over the range of temperatures up to 33°C at relative humidities above 35% r.h., survival for all species, apart from *O. surinamensis*, is not less

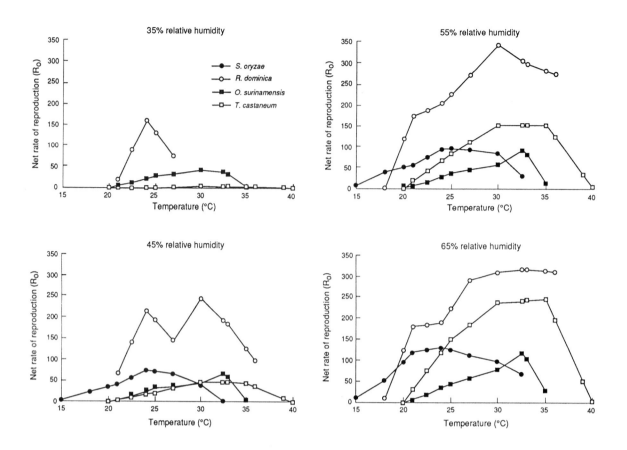

Fig. 2. The net rate of reproduction (R_o) after an 8-week period, for four stored grain coleopteran species over a range of temperatures (°C) and relative humidity (r.h.) combinations.

than 80%. Survival for *O. surinamensis* on the other hand drops slightly between 24 and 30°C at 35% r.h. and 27 and 30°C at 45% and 55% r.h., but remains above 65%. Values of l_x are not available above 33°C for *S. oryzae,* but are available for *R. dominica* up to 39°C. Therefore, it can be seen that values of l_x for *R. dominica* only drop below 60% at 35°C/35% r.h. and 39°C at the higher relative humidities. At these latter conditions adult mortality suddenly becomes more or less complete. Data are not available for *O. surinamensis* above 35°C, but survival is not less than 50% at this temperature at 35% r.h. and not less than 60% at the higher relative humidities. Values of l_x for *T. castaneum* decrease rapidly at 35°C/35% r.h. and at 39°C at the higher relative humidities. However, at 39°C, survival is still between 10 and 25%, but mortality more or less complete at 40°C. Therefore, for most conditions, there are high levels of adult survival for all species after 8 weeks, suggesting relatively similar limited effects on the final values of R_0. Low survival after 8 weeks, which has a major impact on the value of R_0, is only apparent at very suboptimum conditions specific for each species.

The data for immature development (*D*) are displayed in terms of a rate, $1/D$, in Figure 3. Like values of λ and R_0 there is a similar relationship between development period, temperature and relative humidity, with development periods becoming increasingly shorter as temperature increases, until an optimum is reached. Furthermore, development periods are longer at drier conditions than at more moist conditions. At all temperatures at 35% r.h. both *S. oryzae* and *T. castaneum* are unable to develop. However, for *R. dominica* the optimum development temperature at 35% r.h. is 33°C (42.3*D*) and for *O. surinamensis* it is 32.5°C (25.2*D*). The optimum development temperature for *S. oryzae* is 30°C at the higher relative

humidities (42.0, 38.5, and 35.7*D*), for *R. dominica* it is 33°C (33.7, 30.3, and 29.2*D*), for *O. surinamensis* they are 32.5°C at 45% and 55% r.h., and 35°C at 65% r.h. (23.7, 22.5, and 21.4*D*), and for *T. castaneum* it is 35°C (30.6, 25.4, and 22.3*D*).

Values for $1/D$ have also been used to determine the theoretical development thresholds and thermal constants for each species at the four relative humidities (Table 2). Apart from the inability of *S. oryzae* and *T. castaneum* to develop at 35% r.h., the theoretical development thresholds for each species are little affected by the difference in the relative humidities considered (*S. oryzae*: 11.4°C, SD 0.27; *R. dominica*: 16.6°C, SD 0.52; *O. surinamensis*: 15.3°C, SD 0.27; *T. castaneum*: 16.8°C, SD 0.44). On the other hand, thermal constants range from 703 day degrees at 45% r.h. to 609 at 65% r.h. for *S. oryzae*, 730 day degrees at 35% r.h. to 483 at 65% r.h. for *R. dominica*, 393 day degrees at 35% r.h. to 327 at 65% r.h. for *O. surinamensis*, and 548 day degrees at 45% r.h. to 372 at 65% r.h. for *T. castaneum*.

Discussion

Species differ in the range of environmental conditions at which they show positive population growth and in those conditions at which population growth is at an optimum.

In general, the demography of *S. oryzae* is unsuited to higher temperatures at lower humidities, but is favoured above the other species at cooler more moist conditions. On the other hand, *R. dominica* thrives more so than any other species over the greatest range of temperature and relative humidity combinations. This is especially true at higher temperatures over all

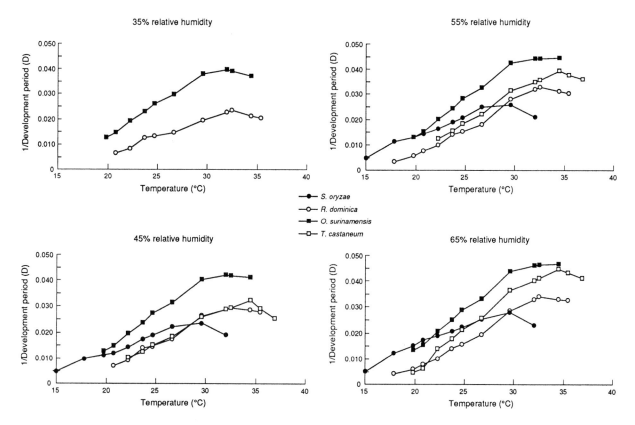

Fig. 3. The net rate of immature development (1/D) for four stored grain coleopteran species over a range of temperature (°C) and relative humidity (r.h.) combinations.

Table 2. The theoretical development thresholds (°C) and thermal constants (day degrees) for four stored-grain coleopteran species over a range of relative humidities (r.h.)

% r.h.	Development threshold	Thermal constant	r^2	Development threshold	Thermal constant	r^2
	Sitophilus oryzae			*Rhyzopertha dominica*		
35.0	-	-	-	15.7	729.8	0.9832
45.0	11.6	702.8	0.9847	17.1	521.9	0.9808
55.0	11.7	624.3	0.9887	16.8	494.8	0.9854
65.0	11.1	609.4	0.9620	16.7	482.7	0.9890
	Oryzaephilus surinamensis			*Tribolium castaneum*		
35.0	14.9	393.1	0.9989	-	-	-
45.0	15.3	360.9	0.9983	16.6	547.7	0.9867
55.0	15.5	338.7	0.9985	16.5	458.1	0.9871
65.0	15.6	327.1	0.9994	17.5	371.5	0.9798

the relative humidities considered. *O. surinamensis* is relatively successful at moderate temperatures at all the relative humidities, but it is more successful than the other species at the drier end of the humidity range. By contrast, the population growth of *T. castaneum* is relatively weak under cool and dry conditions. However, as temperature increases at higher relative humidities, population growth rate improve considerably. At the upper range of viable temperatures at humid conditions, *T. castaneum* becomes more successful than even *R. dominica* and is the only species capable of population growth above 38°C.

Relative to the other species, the finite rate of reproduction for *R. dominica* can in part be described by slower immature developmental rates (1/D) (Fig. 2) and higher levels of fertility (R_0) (Fig. 3). On the other hand, *S. oryzae* and *O. surinamensis* generally exhibit rapid immature development and short generations rather than high fertility to achieve similar or greater rates of population growth. For example, at 20°C at relative

humidities above 35% r.h., *R. dominica* has a greater net rate of reproduction than *S. oryzae*. However, due mainly to more rapid immature development, *S. oryzae* maintains a greater finite rate of reproduction as temperature increases to 27°C. This strategy is put to even greater use by *O. surinamensis* over practically the entire range of environmental conditions. Its fertility is consistently far less than that of the other species at even the most optimum of conditions. However, low values of R_0 are compensated for by low values of D. In this way, *O. surinamensis* maintains values of similar to or greater than those for the other species. As temperature increases, values of D and R_0 for *T. castaneum* are more affected by levels of relative humidity than they are for the other species. For example, at most temperatures at 45% r.h. values of D for *T. castaneum* are generally similar to those for *R. dominica*. However, at temperatures above 27°C at 65% r.h., only *O. surinamensis* has lower values of D. Values of R_0 at a given temperature also increase dramatically with increased relative humidity so that above 24°C at 65% r.h. *T. castaneum* is second only to *R. dominica* in levels of fertility. The combination of quicker immature development and higher fertility means that at most temperatures at 65% r.h., *T. castaneum* has similar values for λ to *R. dominica*.

Population growth, wet bulb temperature and implications for pest management

There are potentially important consequences of these results for the use of cooling by aeration as a pest management strategy, especially in an increasingly pesticide-free environment. The insects of most concern must first be *S. oryzae* and second, *R. dominica*. *S. oryzae* has the lowest theoretical thresholds of development and population growth (11.4°C and 14.8°C) (Tables 1 and 2), and at relatively low temperatures such as 18°C, the value of λ is quite significant. Although the degree of relative humidity is not very important (apart from dry conditions, i.e. 35% r.h.), the value of λ does vary from 1.20 [doubling time (*t*) = 3.8 wks] at 45% r.h. to 1.28 (*t* = 2.8 wks) at 65% r.h. (Fig. 1). *R. dominica* is not quite so cold tolerant (development threshold = 16.6°C, population growth threshold = 17.5°C). However, at 20°C the value of λ varies from 1.06 (*t* = 11.9 wks) at 35% r.h. to 1.15 (*t* = 5 wks) at 65% r.h. Furthermore, it must be remembered that *O. surinamensis*, although it has a higher population growth threshold (19.2°C), is relatively successful at cool dry conditions and has λ values similar to *R. dominica* at just 21°C at humidities as low as 35% r.h.

If increased grain moisture receival limits are to be considered, it is important to be aware of the increasing effect that relative humidity has on the population growth rate of all species as temperatures increase. This is especially the case for *S. oryzae* at cooler conditions. For example, at 22°C values of λ vary from about 0.83 at 35% r.h. [10.1% equilibrium grain moisture content (emc)] to 1.56 at 65% r.h. (13.0% emc). Population growth rates for *T. castaneum* also respond in a similar dramatic manner where at 22°C/35% r.h. population growth is at best negligible but at 22°C/65% r.h., λ = 1.24.

Using limited available data, Desmarchelier (1988) showed that rates of population growth in several stored grain pests, including the four species considered here, are linearly related to wet bulb temperature (T_{wb}) over temperatures ranging from somewhat above the theoretical threshold of population growth to the optimum for population growth. However, with the availability of considerably more information, it has been possible to re-examine this relationship in more detail and to quantify the degree of accuracy achieved by this rule of thumb.

Desmarchelier (1988) used data pooled from a range of temperature and moisture contents to show that a linear relationship exists between λ and T_{wb} for all species at and below the optimum wet bulb temperature for population growth. However, when plots of λ against T_{wb} at each relative humidity are considered, what was a single line has clearly become a family of lines (Table 1). Using these data to calculate the theoretical threshold for population growth (λ = 1) in terms of T_{wb}, the value is found to vary with relative humidity for each species (*S. oryzae* and *T. castaneum*: from 45% to 65% r.h., variation of 2.4 and 2.0 °C. *R. dominica* & *O. surinamensis*: from 35% to 65% r.h., variation of 4.5 and 4.2°C). This is problematic in terms of setting aeration temperatures because it suggests that, though a population at a higher relative humidity may be controlled at a higher T_{wb}, significant rates of population growth occur at the same T_{wb} at lower relative humidities. In the same way, if a lower T_{wb}, equivalent to the population growth threshold at a lower relative humidity were used, unnecessarily high energy inputs

would then be used to control insects at higher relative humidities (Beckett and Evans 1994).

By contrast, the theoretical threshold for population growth over the same range of relative humidities, in terms of dry bulb temperature, varies considerably less (*S. oryzae* and *T. castaneum*: from 45% to 65% r.h., variation of 0.2 and 0.9 °C. *R. dominica* & *O. surinamensis*: from 35% to 65% r.h., variation of 0.4 and 0.4°C) (Table 1). This is due to the fact that the effect of relative humidity on rate of population growth becomes less as temperature decreases towards the minimum theoretical threshold, where it is almost insignificant, as has been demonstrated in the case of *O. surinamensis* (Beckett and Evans 1994).

Also of concern is that, on occasion, different dry bulb temperature/relative humidity combinations can, in some cases, give very different values for λ even though they translate as the same wet bulb temperature. For example, for *S. oryzae*, 25°C/55% r.h. and 30°C/35% r.h. both give approximately 15.5°C T_{wb}, but λ values of 1.84 and 0.86 are recorded at each condition respectively. For *R. dominica*, 30°C/55% r.h. and 36°C/35% r.h. both give 22.8°C T_{wb}, but in this case values of 2.03 and 1.51 are recorded at each condition respectively. Similar examples occur for *O. surinamensis* and *T. castaneum*.

To make accurate predictions of population growth rate for a given species, it would seem more prudent to consider dry bulb temperature and relative humidity rather than wet bulb temperature. For this purpose, contour diagrams of λ values set at intervals of 0.5, incorporating the entire range of the available data for each species, are presented in Figure 4. In this way, if for a particular reason total population control is not considered necessary, aeration conditions can be determined to give the level of control required. In terms of population doubling time (*t*), where *t* = $\log_e (2.0)/\log_e (\lambda)$, an insect population halves in one week if λ = 0.5, is static if λ = 1, doubles in 1.7 weeks if λ = 1.5, and doubles in one week if λ = 2.

Finally, as immature development decreases with increasingly favourable environmental conditions, the likelihood of selection for pesticide resistance increases for all species. However, from the cursory look at the demographic strategies employed by each of the four species, it can be seen that due to its reliance on rapid immature development rather than on high levels of fertility, *O. surinamensis* has a selective advantage for pesticide resistance over the other species. This is especially the case over the moderate temperature range. The increasingly frequent occurrence of resistant strains in Australia over recent years (Allen et al. 1988; Insect resistance to grain protectants in the Victorian central grain handling distriact 1988; Sproul and Emery 1988; Wallbank 1988) may in part be due to this factor. As temperature increases at high relative humidity, immature development also becomes increasingly rapid for *T. castaneum* relative to other species. So at these conditions, it too is more likely to select for pesticide resistance.

References

Allen, P., Henry, K. and Kelly, B. 1988. Insect trace-back system in South Australia. Australian Wheat Board Pest Control Conference, Melbourne, March 1988. Section 3.2.3, 5 pp. Australian Wheat Board, Melbourne.

Australian Wheat Board. 1988 General review of insect infestation, patterns, rejections at terminals and market complaints for insect infestations and pesticide residues 1984–1987. Australian Wheat Board Pest Control Conference, Melbourne, March 1988. Section 1.3. 19 pp. Australian Wheat Board, Melbourne.

Australia Wheat Board. 1988. Insect resistance to grain protectants in the Victorian central grain handling system. Australian Wheat Board Pest Control Conference, Melbourne, March 1988. Section 3.2.1, 7 pp. Australian Wheat Board, Melbourne.

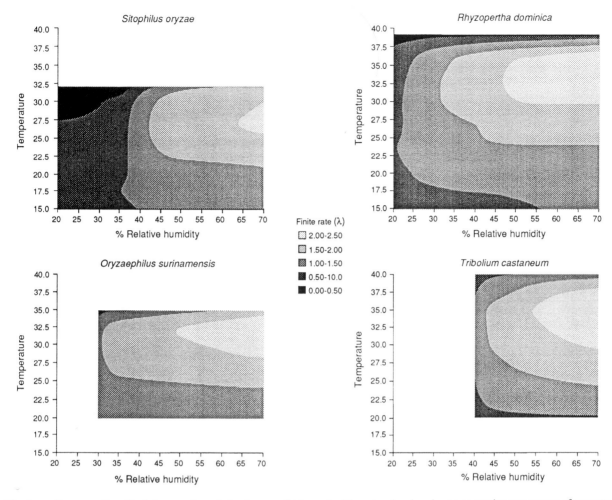

Fig. 4. Contours of weekly finite rate of population increase (λ) for each of four stored grain coleopteran species over a range of temperature (°C) and relative humidity (r.h.) combinations. (λ) values below one for *T. castaneum* have been derived from White's (1987) model.

Beckett, S.J. and Evans, D.E. 1994. The demography of *Oryzaephilus surinamensis* (L.) (Coleoptera: Silvanidae) on kibbled wheat. Journal of Stored Products Research, in press.

Birch, L.C. 1948. The intrinsic rate of natural increase of an insect population. Journal of Animal Ecology, 17, 15–26.

Bulk Grains Queensland Ltd. 1991. The nature, distribution & frequency of infestations: A BGQ situation report. Grain Industry Pest Control Conference, Canberra, March 1991. Section 1.1.1, 4 pp. CSIRO Division of Entomology, Canberra.

Caddick, L.P. 1990. Survey on the occurrence of insecticide and fumigant resistance in central grains handling systems 1988–1990. AWB/GRDC Workshop on insect control failures and pesticide resistance, Melbourne, December 1990. Australian Wheat Board, Melbourne.

Co-operative Bulk Handling Western Australia Ltd. 1991. Co-operative Bulk Handling Limited, Situation report — Infestation levels. Grain Industry Pest Control Conference, Canberra, March 1991. Section 1.1.2, 6 pp. CSIRO Division of Entomology, Canberra.

Desmarchelier, J. M. 1988. The relationship between wet-bulb temperature and the intrinsic rate of increase of eight species of stored-product Coleoptera. Journal of Stored Products Research, 24, 107–113.

Evans, D.E. 1982. The influence of temperature and grain moisture content on the intrinsic rate of increase of *Sitophilus oryzae* (L.) (Coleoptera: Curculionidae). Journal of Stored Products Research, 18, 55–66.

Grain Elevators Board of Victoria. 1991. Grain Elevators Board Situation report — Infestation levels. Grain Industry Pest Control Conference, Canberra, March 1991. Section 1.1.5, 3 pp. CSIRO Division of Entomology, Canberra.

Longstaff, B. C. and Evans, D. E. 1983. The demography of the rice weevil, *Sitophilus oryzae* (L.) (Coleoptera: Curculionidae),

submodels of age-specific survivorship and fecundity. Bulletin of Entomological Research, 73, 333–344.

Murray, W.J. 1979. Infestation patterns in the Australian wheat industry during the past two decades—immediate short term solutions and a prospective for the future. Australian Contributions to the Symposium on the Protection of Grain against Insect Damage during storage, Moscow 1978. Ed. D.E. Evans. Canberra: CSIRO Division of Entomology.

NSW Grain Corporation Ltd. 1991. GrainCorp Situation report—Infestation levels. Grain Industry Pest Control Conference, Canberra, March 1991. Section 1.1.3, 1 pp. CSIRO Division of Entomology, Canberra.

South Australian Co-operative Bulk Handling Ltd. 1991. South Australian Co-operative Bulk Handling Limited Situation report—Infestation levels. Grain Industry Pest Control Conference, Canberra, March 1991. Section 1.1.4, 9 pp. CSIRO Division of Entomology, Canberra.

Sproul, A.N. and Emery, R.N. 1988. Insecticide resistance in grain insects. Australian Wheat Board Pest Control Conference, Melbourne, March 1988. Section 3.2.4, 5 pp. Australian Wheat Board, Melbourne.

Wallbank, B. E. 1988. Resistance to organophosphates in *Oryzaephilus surinamensis* in New South Wales. Australian Wheat Board Pest Control Conference, Melbourne, March 1988. Section 3.2.2, 3 pp. Australian Wheat Board, Melbourne.

White, G.G. 1987. Effects of temperature and humidity on the rust-red flour beetle, *Tribolium castaneum* (Herbst) (Coleoptera: Tenebrionidae), in wheat grain. Australian Journal of Zoology, 35, 43–59.

Wilson, S.G. 1990. ASEC: A PC program to determine air seed equilibrium conditions. CSIRO Division of Entomology, Stored Grain Research Laboratory Publication. 1 pp.

Winterbottom, D.C. 1922. Weevil in Wheat and Storage of Grain in Bags. Government Printer, Adelaide.

A new host of the groundnut seed beetle, *Caryedon serratus* (Ol.) in Israel

M. Calderon*

Abstract

Caryedon serratus (Ol.), well known as a serious pest in stored groundnuts, was first reported in Israel, in 1962, as *Caryedon gonagra* (F.). Its hosts then were seeds of two desert *Acacia* species. Since then this bruchid has been reported in more host plants, the most recent one being seeds of the ornamental tree, *Bauhinia variegata* (Caesalpiniaceae). A short account of its life cycle in the different host plants is given.

Introduction

One of the first comprehensive studies of the groundnut seed beetle (*Caryedon gonagra* (F.), a previous synonym of *C. serratus*), was done by Davey (1958), and observations on its biology were reported by Prevett (1966).

Caryedon serratus was first reported in Israel by Calderon (1959) as *Caryedon fuscus* (Goeze) which was identified by Southgate and Pope (1958) as *C. gonagra* (Fabricus). However, in 1966, the above types were found by Decelle to be synonyms of *Caryedon serratus* (Ol.), which is today the accepted name for the groundnut seed beetle.

This species was extensively studied for many years by P.F. Prevett, and important information on its origin, distribution and hosts (Prevett 1967) and notes on its biology (Prevett 1966) were reported. The distribution of this species (discussed in detail in Davey 1958) indicates that its origin was Asia, and that it breeds in the seeds of *Tamarindus indica* (Cesalpiniaceae) and other legumes. Through international trade *C. serratus* spread and was established in Africa, also in unshelled groundnuts, in which it became a primary pest. It should be noted, however, that this species was never found in groundnuts in Israel. On the other hand, *C. serratus* was found on other host plants in this country, and in 1976, was identified and described by B.J. Southgate as *Caryedon serratus* (Ol.) subsp. *palaestinicus*. (Southgate 1976). It is probable that the species we found in *B.variegata*, may also belong to the above mentioned subspecies. However, according to its identification, obtained from Prof. J. Decelle in 1990 (per. comm.) it is referred to in this paper as *C. serratus* (Ol.).

Host recorded in Israel

Acacia tortilis complex (Mimosaceae)

C.serratus (Ol.) was first recorded and described in Israel as *Caryedon fuscus* (Goeze) in 1959 (Caryedon 1959). It was found developing in pods of *Acacia tortilis* complex, later identified as *Acacia tortilis* (Forsk.) Hayne subsp. *tortilis*

* The Agricultural Research Organization, Department of Stored Products, The Volcani Center, P.O. Box 6, Bet Dagan, Israel.

(Hayne) Brenan, and *A. tortilis* (Forsk.) Hayne subsp. *radiana* (Savi) Brenan (Halevi 1974). The *Caryedon* specimen and the acacia pods were collected in the arid south of Israel and the Dead Sea depression.

An account of the behaviour of *C. serratus* as a pest of the above mentioned *Acaciae* was given by E. Donahaye et al. in 1966. It should be noted that *Acaciae spirocarpa*, given in the above mentioned paper was wrongly identified and the correct denomination should be *Acacia t. radiana*.

The specimens were collected in the seed pods of the above desert *Acaciae* in the arid south of the country. The pods were heavily attacked by *C. serratus* and by other bruchids belonging to the genus *Bruchidius*. The ripe seed pods which had fallen on the ground were so heavily damaged that this may represent a limiting factor to the spread of these acacia species in that area.

The adults lay their eggs on the seed pods developing on the trees or on dry pods still found on the sandy ground. The developed larvae, within the pods, pupate in oval silk cocoons, which are found stuck on the dry pods or loose on the ground. The mature larvae may remain in cocoons for long periods of time, depending on the ambient conditions. Adults emerge, mate and lay eggs when the environment is favourable. The time from the forming of the cocoon cell, until the emergence of the adults, may take from three to seven months. The extended period during which the insects remain in the pupal cocoons indicated that this species spends the winter months in a state of dormancy. It is estimated that the insect develops during the year about two overlapping generations.

The authors provide a list of *Hymenoptera* and *Diptera* parasites found in the dry pods collected from the ground, which, however, may be associated with other bruchid species.

Prosopis farcta (Banks et Sol.) MacBride (Mimosaceae)

This thorny shrub is commonly found in uncultivated fields of heavier soil, in many parts of the country and it is considered an undesirable weed. Its pods are oblong egg-shaped and their integument thick and spongy. The plant is not higher than 2 m.

C. serratus developing in the seeds of *P. farcta* were found and studied by Belinsky (1976). The distribution of *C. serratus* in this country, its development and reproductive biology in seeds of this new host were studied in laboratory conditions by Belinsky and Kugler (1978). In this work *C. serratus* was referred to as *C. serratus* subsp. *palaesinicus* as identified by B.J. Southgate 1976.

Adults start emerging toward spring and lay eggs on fruits from the previous year still found on the plants. Larvae penetrate into the fruit and try to reach the seeds in which they develop, moult and remain until the end of the fourth instar and then leave the seed and the fruit and pupate in their typical silk cocoons. Some larvae pupate inside the pod near its surface but in general the mature larvae pupate inside the pod near its surface but in general the mature larvae leave the fruit to pupate. The second generation lay eggs on the new green pods that are just ripening and the development of larvae continues. The authors estimate that there are at least three generations throughout the year and the larvae or the prepupae pass the winter in a dormant state. Laboratory trials carried out by the authors showed that at 25°C and 50% r.h. females laid significantly more eggs than at 30°C and 70% r.h. The longevity of females was significantly greater than males under both conditions. Belinsky and Kugler (1978) investigated also the host preference of this species for *Prosopis farcta* or *Arachis hypogea* (groundnuts). This topic is discussed later in the paper.

The new host *Bauhinia variegata* (Caesalpiniaceae)

Several species of the genus *Bauhinia* have been recorded as hosts of the groundnut seed beetle (Davey 1958; Yvon de Luca 1962; Prevett 1967); however, *B. variegata* is not amongst them. *B. variegata* is a beautiful ornamental tree which may be found in gardens in many parts of Israel. According to the New York Botanical Garden Encyclopedia (1981) this *Bauhinia* is native of India, South East Asia and China, and today is a favourite ornamental tree in Florida, Hawaii and other warm regions. It is a medium-sized, stiff-branched tree with heart shaped leaves and beautiful white flower clusters. Seed pods become dark brown when ripe and are up to 20–30 cm long. *C. serratus*, breeding on pods and seeds of this tree, was found a few years ago and in 1990 was sent for final identification to Prof. J. Decelle. During the last two years the following notes on its biology were observed:

The tree starts flowering at the end of April and the beginning of May, and the green pods commence forming and elongating. During June the fruits are ripening, their colour becomes yellowish-brown to dark brown. Ripe pods are collected under laboratory conditions in jars. During August the larvae pupating in the oval silk cocoons can be seen at the bottom of the jar, and towards the end of August the first adults emerge. It is estimated therefore that the development of the first generation takes, in nature, about 2 months.

The adults continue laying eggs on the pods kept in jars, and young larvae penetrate the pods and develop in the seeds, causing extensive damage. Adults of the second generation (under laboratory conditions) appear in large numbers during November and December with plenty of pupal cocoons stuck on the pods or on the bottom of the jar. Adults in the jars were seen during all of December and January. It is supposed, therefore, that this insect develops at least two generations during the year, while the larvae or the pupae of the second (or the third) generation spend the winter in dormancy, either in dry pods which have fallen on the ground or in other debris, until the forming of new pods on the trees. This assumption, however, is not yet proved.

Discussion

C. serratus is important in Israel because it is known as a destructive pest of stored groundnuts in Africa and in other parts of the world. Groundnuts (*Arachis hypogea*) are grown locally on quite a large scale and a considerable part of the harvest is exported. The groundnut storage places are not far from areas in which the natural hosts of *C. serratus* (especially in the case of *B. variegata*) are found. Therefore, Calderon et al. (1967) investigated the possibility of *C. serratus* developing in local groundnuts, when transferred from its natural host *Acacia tortilis*. The results obtained showed clearly that *C. serratus* is capable of developing in the locally stored groundnuts. This point was more thoroughly studied by Belinsky and Kugler (1978). They examined the host preference and the development of this species when reared in *P. farcta* versus in groundnuts. Results showed that the beetles developed much

more successfully on their natural host, *P. farcta*, than in groundnuts, and when given a choice the females would oviposit only on *P. farcta*. The authors deduction is that although this species can develop on groundnuts it will not become a pest of groundnuts in Israel in places where *P. farcta* grows.

C. serratus adults grown in *B. variegata* pods were also transferred in jars containing shelled and unshelled groundnuts. Adults oviposited in August and developed at least two generations until winter under laboratory conditions, while live adults, pupal cocoons and perforated groundnuts are seen in the jars at the beginning of February. However, no preference tests were carried out in this case.

It should be noted again that *C. serratus* found in the first two hosts were identified by B.J. Southgate as *C. serratus* subsp. *palaestiniscus*. It is very probable that our specimens from *B. variegata* belong also to the same subspecies. The preference of this species for its natural host, and its relatively less successful development in groundnuts, may be due to characteristics pertaining to the new subspecies. However, the fact that the insects can develop freely in groundnuts indicates the possibility that it may become a pest of stored groundnuts in Israel if established in a groundnut storage place at a considerable distance from its natural host.

References

Belinsky, A. 1976. A survey of entomofauna of *Prosopis farcta* (Mimosaceae) in Israel, with special reference to *Caryedon serratus* (Coleoptera, Bruchidae), M.Sc. thesis, Tel Aviv University, 82 pp.

Belinsky, A. and Kugler, J. 1978. Observations on the biology and host preference of *Caryedon serratus palaestinicus* (Coleoptera, Bruchidae) in Israel.

Calderon, M. 1959. The Bruchidae of Israel. Ph.D. thesis, the Hebrew University of Jerusalem. (in Hebrew with English summary).

Calderon, M., Donahaye, E. and Navarro, S. 1967. The life cycle of the Groundnut Seed Beetle, *Caryedon serratus* (Ol.) in Israel. Israel Journal of Agricultural Research, 17, 145–148.

Davey, P.M. 1958. The Groundnut Bruchid *Caryedon gonagra* (F.) Bulletin of Entomological Research, 49, 385–404.

Decelle, J. 1966. *Bruchus serratus* Ol. 1790, espece-type du genre *Caryedon* Schoenherr, 1823. Revue of Zoology and Botany in Africa, 74, 1–2.

Donahaye, E., Navarro, S. and Calderon, M. 1966. Observations on the life cycle of *Caryedon gonagra* (F.) on its natural host in Israel, *Acacia spirocarpa*, Tropical Science, 8, 85–89.

Halevy, G. 1974. Effect of gazelles and seed beetles (Bruchidae) on germination and establishment of *Acacia* species. Israel Journal of Botany, 23, 120–126.

Prevett, P.F. 1966. Observation on biology in the genus *Caryedon* Schonherr (Coleoptera, Bruchidae) in Northern Nigeria, with a list of associated parasitic Hymenoptera. Proceedings of the Royal Entomological Society of London (A), 41, 9–16.

Prevett, P.F. 1967. Notes on the biology, food plants and distribution of Nigerian Bruchidae (Coleoptera), with particular reference to the northern region. Bulletin of the Entomological Society of Nigeria, vol 1, 3–6.

Prevett, P.F. 1967. The field occurrence of *Caryedon serratus* (Ol.), the Groundnut Seed Beetle (Coleoptera, Bruchidae), in Uganda. Journal of Stored Products Research, vol. 3, 267–268.

Southgate, B.J. 1976. A new subspecies of *Caryedon* (Goleoptera: Bruchidae), from the Middle East. Israel Journal of Zoology, 25: 194–198.

Yvon de Luca 1962. Contribution aux Bruchidae (Coleopteres) d'Algerie. Leurs hotes—leurs parasites—leurs stations. Memoires de la Societe d'Histoire Naturelle de l'Afrique du Nord, No. 7, 115 pp.

Dynamics and expansion of populations of stored product beetles

Z. Ciesielska*

Abstract

Dynamics and activity of populations of *Oryzaephilus surinamensis* (L.), *Sitophilus granarius* (L.), *Sitophilus oryzae* (L.) and *Rhyzopertha dominica* (F.) were investigated. Experiments were conducted in different environmental conditions, in single species cultures and in interspecific competition, in conditions of unidirectional emigration to jars having a glycerine trap and two-directional migration into a jars with grain. All the experiments were carried out at 27°C and ca 70% r.h., in 5–10 and 20–25 repetition.

It was shown that the level and changes in population number are influenced by environmental conditions and the population and interspecific relationships. One of the ways of regulating the population number was through change of sex ratio. The populations of all the investigated species showed high tendencies to migration and dispersal, with indices of migration from 60–80%. Migration was especially high in the initial period of attack of new habitat and in interspecific competition.

Introduction

The dynamics of an ecological system (including a population) can be forecast on the basis of the attributes of its elements (individuals) or on the basis of characteristics of the populations and environmental factors which condition the activity of the population. Attributes of individuals alone cannot predict the functioning of a whole ecosytem; investigations must be focused on population characteristics and chosen ecological factors. (Ciesielska 1978, 1985; Lacey et al. 1992; Loschiavo and Smith 1986; White and Sinha 1980; Sinha 1984).

Data concerning the sex structure of populations of most animals show that the primary sex ratio is close to 1 (Deevey 1947; Kalela and Oksala 1966; Lampio 1965). No influence on the sex of an individual with a chromosomal type of sex determination is possible. Ecological factors can indirectly influence the primary sex ratio in a population by selecting genotypes with tendency to producing offspring of a determined sex predominance (Petrusewicz 1965). However, the sex structure of a population can also be influenced by different migration of males and females after the period of reproduction thus leading to division and formation of groups of different sex ratio.

There is a number of aspects of ecological importance concerning investigations on sex structure of populations. Apart from the possibility of determining the reproductive potential by investigating the directions of changes of sex ratio, the course of dynamics of the population and its activity can also be predicted. From an experimental point of view, populations of grain beetles are very suitable for such model investigations.

Dynamics and activity of populations of *Oryzaephilus surinamensis* (L.), *Sitophilus granarius* (L.), *Sitophilus oryzae* (L.) and *Rhyzopertha dominica* (F.) are described. Observations concerned population density, age and sex structure, interspecific competition, and such environmental factors as food, its lack or distribution and environmental conditions enabling migration. Apart from population dynamics, the tendency to migration and dispersion was analysed to estimate population activity.

The tendency for migration and dispersal was displayed by the insects looking for new habitats and food. These processes intensify with progressing grain exploitation and increase of population density. Based upon preliminary investigations (Ciesielska 1992) a hypothesis was forwarded that age and sex structure exert considerable influence on the migration process.

Material and Methods

Populations of *O. surinamensis*, *S. granarius*, *S. oryzae* and *R. dominica* were studied in single-species and competitive two-species systems. In experiments on migration, *R. dominica* was studied in competition but without any migration possibility. In some of the preliminary experiments, special jars were used, which made the expansion of *R. dominica* possible.

The series of experiments concerned:

Dynamics of populations in various environmental and population conditions, as: different food and different depths of its layer. 2 variants of initial density: 20 individuals (optimal) and 100 individuals (overdensity) for 40 g grain. Long-term experiments concerning the relation between population numbers and sex ratio in single species cultures and in interspecific competition.

Migration and dispersal tendencies of populations in single species systems and interspecific competition. Open jars with grain infested with initial populations in one- and two-species systems (initial density 40 individuals for 40 g of grain) were placed in bigger tightly closed jars. Two experimental variants were applied:

1) unidirectional emigration to jars without any food having a glycerine trap,

2) two-directional migration to jars with an analogical grain content.

Experiments were checked every 30 days according to a previously described method of carrying out long term experiments (Ciesielska 1978). All experiments were conducted at 27°C and r.h. ca 70%. Experiments concerning population dynamics under various environmental conditions were replicated 10 times, the others 20–25 times. Results were based on mean data. The following were calculated: standard deviation, indices of variability, percentage indices of migration and sex ratio. Models of the course of processes were based on statistical analysis of results.

---

* Institute of Biology, Dept. of Ecology and Environment Protection, Pedagogical University, Podbrzezie 3, 31–054 Kraków, Poland

Results

Population numbers within *S. granarius*, *S. oryzae*, *R. dominica* and *O. surinamensis* are influenced by environmental conditions, intraspecific and interspecific interactions. The decrease in the population number may occur because of an increased mortality rate, decreased oviposition rate and/or decreased developmental rate. Another way of regulating the population number is to change the sex ratio.

A leading factor regulating population number is the different male and female mortality rates at various stages of development of individuals and the whole population. Changes in sex ratio were observed within groups of individuals either surviving or dying at juvenile stages; in conditions of various thickness of the layer of the substrate from favourable conditions to undesirable; in aging populations; with food gradually running out; and finally in interspecific competition (Figs 1,2; Tables 1–3). A wide range of factors facilitated drawing more general conclusions.

The sex structure of a population is not a permanent feature, although the primary structure is generally balanced. It is formed only in subsequent periods of a population's development. The secondary sex structure is correlated with some factors both of the environment and the populations working together. Also at juvenile stages a balanced sex ratio is maintained with the tendency to female dominance. Such a pattern is assumed to be natural to the population of granary beetles. The dominance of females is the reason for the high biological activity and reproductiveness of these species.

A more interesting problem seems to be the development of the secondary sex ratio under conditions of interspecific competition. (Figs 1,2; Tables 1–3). At the final stage of population development in the two species culture where the competition results in replacement of *O. surinamensis* by *R. dominica*, a typical feedback between the fluctuation of the population growth curve and sex ratio occurs (Fig. 2). In the case of *O. surinamensis* and *S. granarius*, however, at the initial stage of the increase of the number of individuals in a population, interactions between these two species in a given set of environmental conditions can be classified as proto-cooperation. In this structure, sex ratios approximate 1, with the tendency towards female dominance for *S. granarius* and male dominance for *O. surinamensis*. At subsequent stages of the population growth the relation between these species converts into a typical competition. It can be stated then, that the process of replacing the feebler population as a result of competition begins with the change of its sex structure.

Earlier research showed that the thickness of the layer of the grain has a considerable importance for *S. granarius* development (Ciesielska 1978), with developmental rate slowed when the substrate layer is too thin. The low number of individuals in the grain from 0.5 to 1.0 cm thick is correlated with dominance of males. After approximately 90 days, the sex ratio ranges from 1.3 to 1.4. However, in a culture where the thickness of the layer of the grain is between 3 and 4 cm the sex ratio does not exceed 0.8 in analogous period of time (Fig. 1, Tab. 3). This indicates the dominance of females.

Summing up, on the basis of this set of experiments, the following can be stated. Under normal conditions, deterioration of living conditions causes the decrease in the population number, which is usually preceded by a change in the population's sex structure towards the dominance of males (sex ratio>1). The increase in sex ratio is caused by dying out of females (in the group of dying-out individuals, the sex ratio is <1). The dominance of female individuals within the population is followed by periods of increase in the population number. Thus, the increase results not only from an increased

reproductive rate, which refers to the intensity of reproduction of respective females.

Changes in the sex ratio of a population, occurring due to differential rates of mortality and survival of individuals of opposite sexes, is one way of regulating the population number, especially when the population faces extinction as a consequence of worsening environmental conditions. Deterioration of environmental conditions causes the decrease in a population and also makes the process of change in the sex structure of a population faster. The amount of females decreases. The model of the process is shown in Figure 3.

Activity of population is not only the ability to increase its numbers. It is also reflected by the tendency to disperse. The calculated indices of migration show a high migration and dispersion tendency in all the investigated species. Their values vary according to environmental conditions and system (1- or 2-species) from 40 to over 80 %. Observations on uni-directional migrations to traps showed a high migration activity even during the initial period of infesting the grain. From the populations of *O. surinamensis*, *S. granarius* and *S. oryzae* about 60% of individuals emigrated within the time period from 30 to 90 days. The rest of the population left the initial habitat within the successive 30–60 days. In 60% of replicates all adults of *S. granarius* and *O. surinamensis* left the initial substrate within the first 60 days. In the other experiments the population numbers remained at a very low level for another 90 days.

In 2-species competitive systems migration activity was higher. After 30 days about 80% of individuals left the initial population. This applied to both systems in which the two competitive populations could leave the culture jar as well as the system where only one species was able to emigrate (*O. surinamensis* with *R. dominica*). From the 60th day the population number stabilised at a low level. In 70% of replicates all individuals of populations in 2-species competition left the initial habitat. (Table 4).

Natural environmental conditions were simulated where continuous two-directional migrations (with the possibility to return to the initial habitat) were possible. In this series of experiments too the results showed predominance of emigration processes over immigration ones. The populations numbers of *O. surinamensis* and *S. granarius* in the outer jars was always greater. However, migration activity changes in time. Two periods of peak activity were distinguished: the first is the initial period of inhabiting the grain, and the second after a time lapse of about 180–200 days as a result of grain exploitation and overdensity of population (Table 5).

The population *S. oryzae* showed a tendency to uniform dispersion. This was shown by the emigration index within range of about 50% maintained during the whole investigation period.

Mortality of the initial populations *O. surinamensis* and *S. granarius* was higher as a rule than of the groups of emigrants. In single species cultures, mortality indices in groups of emigrants ranged from 6–15%, and in initial populations they are about twice as high (Table 5).

Analogous results were observed within competitive interspecific systems, at a higher mortality, however. In emigrating groups the mortality was about 15–17% and in initial populations about 60%.

In populations of *S. oryzae*, the relation between migration and mortality was also different. The tendency to uniform dispersion expressed by an equalised migration index was reflected in a uniform course of mortality. In single species cultures, mortality indices in migrating groups fluctuated on average 3.7% and in initial populations 4.1%. In competition systems, mortality indices in groups of migrating individuals

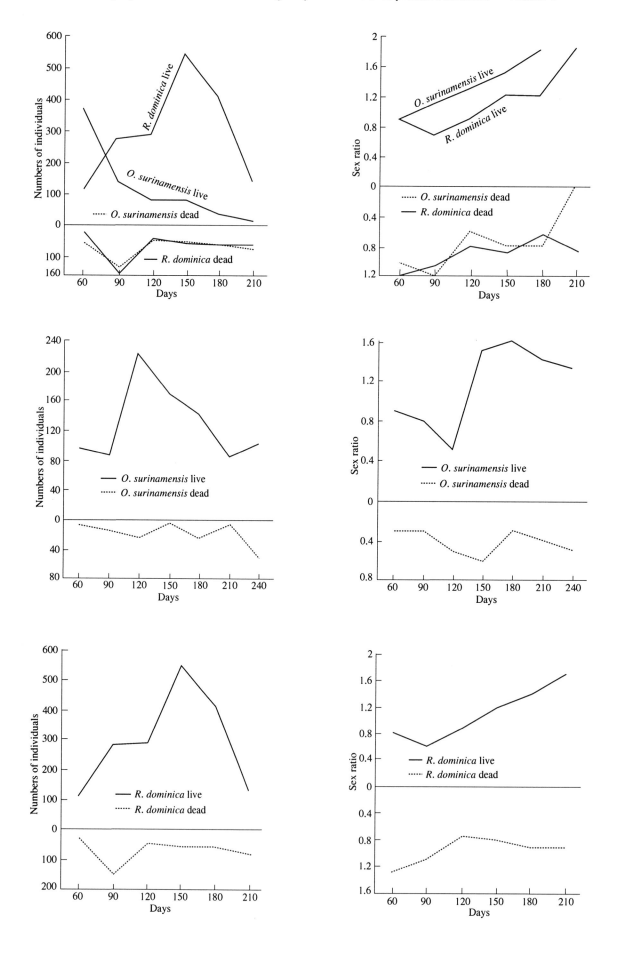

Fig. 1.   Changes of numbers of individuals and sex ratio in populations of *O. surinamensis* and *R. dominica* in interspecific competition and in single cultures.

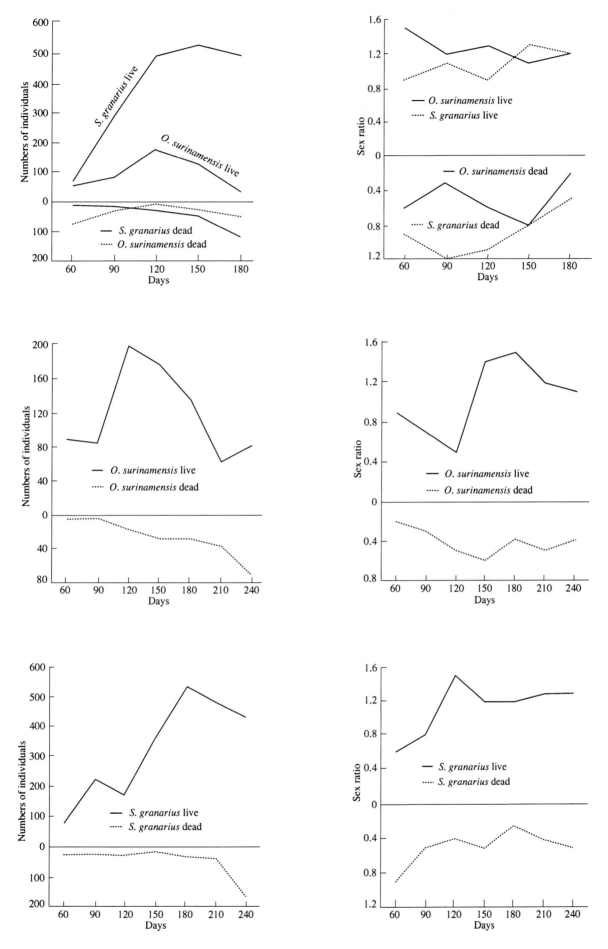

Fig. 2. Changes of numbers of individuals and sex ratio in populations of *S. granarius* and *O. surinamensis* in interspecific competition and in single species cultures.

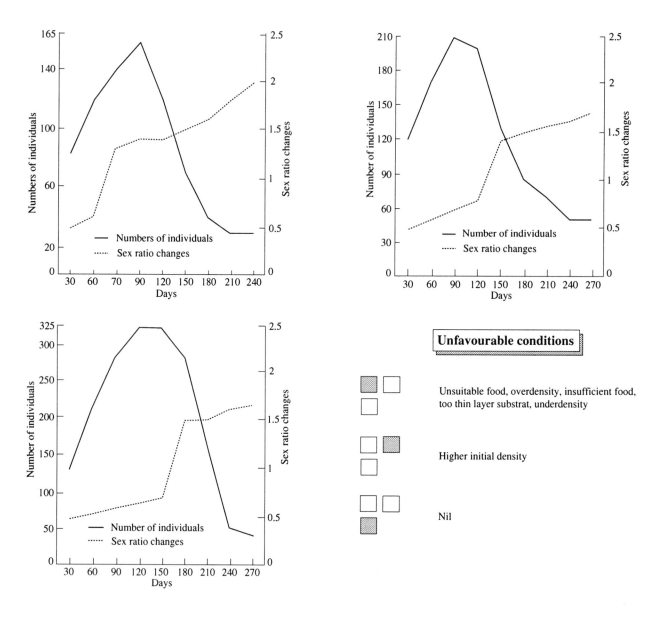

Fig. 3. Sec ratio changes in populations in various environmental conditions.

was on average 40.5%, whereas, in initial populations it was 37.5%.

Two models of migration processes were observed. The first occurred when more active individuals emigrated (maybe younger ones) characterised by a low death rate. In this group emigration processes predominate over immigration processes. The second occurred where uniform dispersion took place, not distinguishing in the population more or less active groups of a higher or lower mortality (Table 5).

The sex ratio of the population is a measure of the group activity and consequently of its value for the development and expansion of the population. Preliminary investigations showed a higher emigration tendency of females (Ciesielska 1992). Continued long term investigations confirmed this result.

The females of the investigated species demonstrated a higher migration activity, confirmed by sex ratio values less than 1 in groups of emigrating adults (Table 5). Highest migration activity was observed in females in the initial period of infestation of the grain (days 30–90). In this time the sex ratio values are from 0.4 to 0.8. The second period with a predominance of females within migrating groups was noticed after

the lapse of 180 to 200 days. In this time overdensity of the population and increased grain exploitation took place. The tendency to migrate by *R. dominica* populations was especially high. Previous experiments indicated that females of *R. dominica* are twice as active as males (Table 5).

In competitive systems the activity of females is also higher. It was especially distinct in the situation where *O. surinamensis* was in competition with *R. dominica*. Females of *O. surinamensis* `escaped' from this competitive system extremely actively.

In summary, it should be stated that migration activity of the investigated species led to formation of aggregations or to uniform dispersion. Continuous infestation of stored grain by populations of granary beetles results without doubt from their exceptional tendency to migrate and disperse. Migration is as much a property of a species as a reaction to changing environmental–food conditions and intraspecific and interspecific competition. Various factors condition the migration processes: behavioural–physiological, properties of a species and populations and environmental factors.

Table 1. Sex ratio (males/females) in populations cultivated in various population and environmental conditions

Species	Days	Favourable conditions	Higher initial density	Lack of food, competition	Depth of wheat layer	
					> 3 cm	ca 0.5 cm
O. surinamensis	0-60	0.6	0.7	1.04		
	60-90	0.9	0.9	1.2		
	90-120	0.4	1.2	1.4		
	120-180	1.2	1.3	1.46		
	180-220	1.2		1.1		
	220-		1.3			
R. dominica	0-60	0.53	0.7	0.8		
	60-90	0.69	1.1	1.2		
	90-120	1.1	1.32	1.32		
	120-180	1.3	1.25	1.25		
	180-220	1.09	1.1	1.1		
	220-	1.2				
S. granarius	0-60	1.1	0.7	0.6	0.7	1.6
	60-90	0.7	1.1	0.9	0.9	2.3
	90-120	0.9	0.7	1.4	0.7	1.7
	120-180	1.1	1.3	1.1	1.1	1.5
	180-220	1.5	1.3	1.3	1.1	1.4
	220-250			1.4	1.0	1.2
	250-			1.6		
S. oryzae	0-60	1.0	0.9	1.0		
	60-90	1.1	0.8	0.9		
	90-120	0.9	1.1	1.2		
	120-180	1.1	1.1	0.13		
	180-220	1.1	1.2	1.7		
	220-		1.3	1.6		

Discussion

Both reproduction and mortality are closely connected with the sex structure of the population, usually expressed as the sex ratio. In the majority of species, the sex ratio is approximately 1 during the initial phases of population development. In groups of adults deviations from this ratio can occur differently in different age classes, for example, in *Perdix perdix* or *Lophotortyx californica* proportion of males increases in older classes (Lack 1954). In mammals, sex ratios are close to 1 in embryonic and juvenile stages, with a tendency to deviation from this proportion in adults (Lampio 1965, Kalela and Oksala 1966, Gliwicz 1990). Sex structure of populations can undergo modifications by differentiation of the growth rate or maturity of individuals of one sex.

Predominance of individuals of one sex influences population dynamics. Variability of sex ratio can also be considered from the point of view of influence of certain factors on the rate of mortality of individuals of different sexes. Deviations of the sex ratio from 1 are often the consequence of different survival of males and females as well as different seasonal migrational ability (Deevey 1947, Petrusewicz 1965). Dependence of the sex ratio on density is encountered but this is not the rule. Under low density conditions the predominance of males leads to a decrease of population numbers (Ciesielska 1985, 1988).

Laboratory populations of stored product insects are especially convenient for investigations constituting the basis both for theoretical considerations and practical applications (White and Sinha 1980; Sinha 1984; Loschiavo and Smith 1986; Lacey et al. 1992; Cofie-Agblor et al. 1992). The first part of investigations, showed that the main factor regulating population numbers is differential survival of males and females at various ages and periods of population development (Ciesielska 1978, 1988). Based upon very numerous experiments and replicates, changes of the sex ratio were analysed within dying-out and surviving groups of individuals, in juvenile stages, in conditions of various densities, at lack of food and its overabundance, in favourable and unfavourable conditions of food distribution (depth of the substrate layer), and in ageing populations connected with gradually decreasing of food.

A multidirectional analysis of the problem permits some general conclusions. Sex structure of a population is not its constant property, although the primary structure is equal as a rule. Changes occur only at later developmental stages of the population. The secondary sex structure is correlated with environmental and population factors acting jointly.

When environmental conditions are poor, the decrease of population numbers is preceded by a change in sex structure tending to favour males. Increase of the value of the sex ratio (>1) takes place by means of increase of mortality of females. A higher participation of females in a population (sex ratio <1) preceded periods of increase of population numbers. Thus, it is the effect not only of an increase of the reproductive index value determining the biotic potential of a population but also of the sex structure. In all the investigated populations of grain beetles not endangered with unfavourable environmental conditions, a tendency to a higher proportion of females prevails.

Table 2. Sex ratio (males/females) among adults dying out in populations cultivated in various environmental conditions

Species	Days	Favourable conditions	Higher initial density	Lack of food, competition	Depth of wheat layer	
					> 3 cm	ca 0.5 cm
O. surinamensis	0-60	0.5	0.4	1.33		
	60- 90	0.3	1.1	1.12		
	90-120	0.6	0.6	0.42		
	120-180	0.5	0.7	0.69		
	180-220	0.4	0.8	0.34		
	220-	0.4	0.7			
R. dominica	0-60	0.3	0.33	1.25		
	60-90	90	0.4	0.9		
	90-120	0.3	0.09	1.3		
	120-180	0.8	0.62	0.7		
	180-220	0.6				
	220-					
S. granarius	0-60	1.0	0.4	0.9	0.6	0.3
	60- 90	0.5	0.8	1.7	0.5	1.3
	90-120	0.7	0.8	1.2	0.4	1.6
	120-180	0.8	0.6	1.1	0.8	1.0
	180-220			1.0	0.6	3.3
	220-250			0.8	0.4	1.7
	250-1.0			1.0		
S. oryzae	0-60	0.3	0.4	0.9		
	60-90	0.5	0.5	0.8		
	90-120	1.3	0.6	1.0		
	120-180	1.6	0.3	1.1		
	180-220	0.4	0.8	1.0		
	220-	0.9				

Table 3. Sex ratio (males/females) in *S. granarius* populations in various initial density and depth of layer of wheat

Adults	Depth of wheat layer							
	ca 0.5 cm				ca. 3.5 cm			
	Initial density				Initial density			
	100 ind	S.E.	20 ind	S.E.	100 ind	S.E.	20 ind	S.E.
Live	1.85	0.43	1.25	0.81	0.80	0.02	0.83	0.27
Dead	0.70	0.25	1.35	1.12	0.76	0.11	0.60	1.20
Total	1.227	0.67	1.30	0.97	0.78	0.60	0.61	0.68

Directional change of the sex ratio occurs by differential mortality and survival of individuals of different sexes and is one of the ways of regulating the population numbers during times when the existence of the population is endangered due to deterioration of environmental conditions.

On the other hand, a higher activity of females expressed by their higher proportion in groups of emigrants assures population emigration success. In populations of grain beetles this is enhanced by a lower mortality within groups of emigrating individuals.

Dispersal behaviour of animals has been investigated under natural and laboratory conditions (Surtees 1965, Lidicker 1975). Investigations on the emigrating behaviour of natural populations of Microtinae (Gliwicz 1988, 1990) showed that mortality in groups of migrants is higher. Differences in relation to data obtained for populations of grain beetles do not only result from differences of taxonomy groups and laboratory investigations. Investigations of Gliwicz concern small

mammals whose high mortality takes place mainly during searching for new areas. Whereas, after finding such areas the reproductive success is often higher than in initial populations. Hence, the problem whether the migration behaviour brings profit only to migrants or to the whole populations is still disputable. It seems that emigration improves conditions also to these individuals which do not leave the donor habitat by decreasing population density. Tschinkel (1981) suggested that larvae of *Zophobas atratus* showed an increasing tendency to overdispersal as the density of the population increased.

In conclusion, it should be stated that migration behaviour developed in response to both physiological and ecological factors. Hence, the course of the process is the result of their influence. The established high tendency to migration and dispersion in populations of the investigated species of granary beetles and especially high migration activity of females is a direct cause of their continuous expansion.

Table 4. Indices of migration (means only presented)

Days	O. surinamensis		S. granarius		S. oryzae	
	Single species	In competition	Single species	In competition	Single species	In competition
Undirectional emigration to traps						
30–90	59	78	62	81	56	68
90–150	40	26	35	14	38	23
Two-directional migrations						
30–90	78	88	81	72	58	68
90–150	59	64	72	69	48	51
150-210	64	65	67	59	48	57
210+	84	78	79	81	58	61

Table 5. Sex ratio (males/females) and mortality

Days	O. surinamensis		S. granarius		S. oryzae	
	Initial population	Migration groups	Initial population	Migration groups	Initial population	Migration groups
Single species culture: sex ratio						
30–90	1.4	0.4	1.1	0.7	0.7	0.9
90–150	1.6	0.8	0.9	0.5	0.7	0.7
150-210	1.4	1.0	1.1	0.9	0.8	0.9
210+	2.1	0.8	1.4	0.6	1.4	0.8
Mortality %						
30–90	23.9	6.2	13.0	6.8	1.9	2.9
90–150	32.6	3.7	15.9	11.8	6.2	7.6
150-210	18.7	10.6	31.8	9.7	5.3	2.4
210+	36.0	14.7	37.1	12.9	3.3	2.1
In competition: sex ratio						
30–90	1.5	0.5	1.2	0.8	1.0	0.8
90–150	2.0	0.9	1.1	0.6	0.9	1.0
150-210	1.2	0.6	0.9	1.0	1.1	0.6
210+	1.4	1.1	1.3	0.9	0.9	0.8
Mortality (%)						
30–90	64.4	1.2	14.2	13.3	24.0	43.3
90–150	55.0	13.8	12.5	10.8	39.7	56.3
150-210	84.1	17.6	35.3	22.6	34.7	37.9
210+	53.8	28.2	41.2	27.1	42.2	33.0

References

Ciesielska, Z. 1978. Interactions among populations of granary beetles (*Sitophilus granarius* L., *Rhyzopertha dominica* F. and *Oryzaephilus surinamensis* L.). Polish Ecological Studies, 4, 5–44.

Ciesielska, Z. 1978. The relationship between sex ratio and populations dynamics of *Sitophilus granarius* L. (Col. Curculionidae) and *Oryzaephilus surinamensis* in various environmental conditions. Mitteilungen Deutschen Gesellschaft für Allgemeine und Angewandte Entomologie, 4, 272–274.

Ciesielska, Z. 1988. Dynamics of granary beetles populations and changes of sex ratio: Simulation model approaches. Proceedings XVIII International Congress of Entomology, Vancouver, Canada, July 1988, 454.

Ciesielska, Z. 1992. Tendencies to migration in granary beetles populations. In: Jayas, D.S. et al., ed., International Symposium on Stored-Grain Ecosystems, Winnipeg, Canada, June 1992.

Cofie-Agblor, R., Muir, W.E., and Sinha, R.N. 1992. Heat production of adult grain beetles in stored wheat. In: Jayas, D.S. et al., ed., International Symposium on Stored-Grain Ecosystems, Winnipeg, Canada, June 1992.

Deevey, E.S. 1947. Life tables for natural populations of animals. Quarterly Review of Biology, 22.

Gliwicz, J. 1988. Seasonal dispersal in non-cyclic populations of *Clethironomys glaerolus* and *Apodemus flavicollis*. Acta Theriologica 33, 263–272.

Gliwicz, J. 1990. First born, their dispersal and vole cycles. Oecologia, 83, 519–522.

Kalela, O., Oksala, T. 1966. Sex ratio in the wood leming, *Myopus schisticolor* (Lilljeb) in nature and in captivity. Annales Universitatis Turkuensis, 37.

Lacey, J., Ramakrishna, N., and Smith, J. 1992. Interactions between water activity, temperature, and different species on colonisation of grain and mycotoxin formation. In: Jayas, D.S. et al., ed., International Symposium on Stored-Grain Ecosystems, Winnipeg, Canada, June 1992.

Lack, D. 1954. The natural regulation of animal numbers. Oxford University Press, 403p.

Lampio, T. 1965. Sex ratio and the factors contributing to them in the squirrel *Sciurus vulgaris* in Finland. Finnish Game Research 25.

Lidicker, W.Z. 1975. The role of dispersal in the demography of small mammals. IBP Small Mammals, Their Productivity and Population Dynamics, 103–128.

Loschiavo, S.R., and Smith, L.B. 1986. Population fluctuations of the rusty grain beetle, *Cryptolestes ferrugineus* (Coleoptera:

Cucujidae), monitored with insect traps in wheat stored in a steel granary. Canadian Entomologist, 118, 641–647.

Petrusewicz, K. 1965. Some regularities in male and female numerical dynamics in mice populations. Acta Theriologica, 4.

Sinha R.N. 1984. Effects of weevil (Coleoptera: Curculionidae) infestation on abiotic and biotic quality of stored wheat. Journal of Economic Entomology, 77, 1483–1488.

Surtees, G. 1965. Laboratory studies on dispersion behaviour of adult beetles in grain. Journal of Applied Ecology, 2, 71–80.

Tschinkel, W.R. 1981. Larval dispersal and cannibalism in a natural population of *Zophobas atratus* (Coleoptera:Tenebrionidae). Animal Behaviour, 29, 990–996.

White, N.D.G., and Sinha R.N. 1980. Principal component analysis of interrelations in stored-wheat ecosystems infested with multiple species of insects. Researches on Population Ecology, 22, 33–50.

Bioassays with bruchid beetles: problems and (some) solutions

P.F. Credland*

Abstract

Up to 20 species of seed beetle (Coleoptera: Bruchidae) have been described as pests of stored legume seeds. The most important, in an international context, are members of the genera *Callosobruchus*, *Acanthoscelides* and *Zabrotes*. Species of *Bruchus* are more frequently encountered in the field and pose less of a problem in storage situations.

There have been attempts in several organisations to breed cultivars of the primary legume crops susceptible to bruchid attack which provide some measure of resistance to the beetles. There are also numerous papers especially, but not exclusively, from developing countries describing surveys of local cultivars and selections for bruchid resistance. If any reports of bruchid resistance either among traditional varieties or new products of genetic engineering are to be worthwhile and compatible, then it is essential that the bioassays provide reliable information which is unambiguous and from which reliable conclusions can be drawn. Regrettably, this has not always been the case. What are the problems in designing effective bioassays using bruchids? Can they be overcome? If so, how?

Bruchid Beetles as Pests

There are about 1300 species of seed beetle in the family Bruchidae. Of these about 20 are recognised as being pests, usually in stored legume seeds and especially in developing countries (Southgate 1979). Four species are of cosmopolitan importance. These are *Callosobruchus maculatus*, *C. chinensis*, *Acanthoscelides obtectus* and *Zabrotes subfasciatus*. Other species of *Callosobruchus* including *C. analis*, *C. rhodesianus* and *C. subinnotatus* constitute a secondary group of storage pests, and *Bruchus pisorum*, *B. rufimanus* and *Bruchidius atrolineatus* are important as pests in the field and early stages of storage (Southgate 1978).

Most experimental work has concentrated on the four species of international importance since all of these species are responsible for significant losses from stores of seed held by subsistence farmers for food, marketing and as a source of seed for the next crop. Because of the nature of the situations in which these losses occur, accurate data on the scale of the damage are hard to find but frequently indicate that it exceeds 40% (Caswell 1981; Cardona and Karel 1990).

Even in countries, such as Australia, with access to conventional fumigant and chemical insecticides, the success of treatments has not always been established, and work on bruchid control is still in progress (e.g. Daglish et al. 1993).

Control Measures for Bruchids

The essence of any control measure is to reduce biological fitness (Begon et al. 1990) which in this context, is usually

measured simply as increased mortality. Actually it is not especially important which fitness parameters are affected. For example, increased larval mortality or decreased fecundity would both contribute to control of a pest population. Any means of decreasing fitness compatible with environmental safety, where the environment is taken to include people and their domestic stock, would be potentially acceptable.

In the context of their predominant importance as pests in developing countries, bruchid control is not a simple matter. Chemicals require a recurring expenditure and, for their safe use, appropriate levels of education. Biological control involving the application of predators, parasites or microbiological agents demands initial expenditure and, often, appropriate subsequent monitoring (Dent 1991). Neither funding nor expertise are always available. Cultural control is usually embodied within conventional traditional practices and may not readily be improved. Consequently, there is a major incentive to identify and use 'resistant' cultivars of the crops which may, in the longer term, provide a relatively cheap additional protection against losses. In some cases the identification of such cultivars has simply been based on a survey of crops already in use in different areas, and in others has involved systematic screening of a large germplasm bank and subsequent plant breeding (Singh 1978), the costs being borne largely, though not exclusively, by aid agencies. One assumes that the objective is to provide a self-sustaining and hence inexpensive control, although this assumption has already been questioned since insect resistance to the resistant factor(s) has already been demonstrated (Dick and Credland 1986b).

Bruchid Beetles as Experimental Subjects in Bioassays

Whether studies of bruchids are for purely academic or for more practical reasons, much of the work demands a series of effective bioassays. However, the literature on bruchid control is composed of numerous inconsistent reports which frequently cannot be compared with each other in any worthwhile way. The explanation for this situation lies in the diversity of bioassays which have been employed and, sometimes, inadequate appreciation of the variables which must be controlled. If progress is to be made in bruchid control and results are to be applicable on a global rather than local scale, then an understanding of the problems, and ideally standardisation of protocol, must be more widespread than is currently found.

A Brief and Simplified Theoretical Background

A bioassay is simply an alternative term for an experiment in which biological material is the test subject. There is almost always a comparative element in the experimental design; the effect of a treatment is determined by comparing the behaviour of the living material in the presence and absence of the treatment or by comparing the effects of a range of treatments of which at least one is presumed to be stable and

* Department of Biology, Royal Holloway, University of London, Egham Hill, Egham, Surrey, TW20 0EX, U.K.

'known'. In conventional terminology, some form of 'control' is an essential feature of the experimental design.

The bare essentials of any worthwhile bioassay are, like those of any other experiment, that it could be repeated by another scientist in another location. This presupposes that the original bioassay has stipulated variables and the other components are precisely defined and fixed. The only variable should be the treatment under investigation. The physical conditions of the assay should be regulated in an appropriate way and the subjects described appropriately.

In practice biological experiments can seldom, if ever, conform to this ideal because of the nature of the subjects and, sometimes, of the treatments. Biological material rarely approaches uniformity. Assays undertaken on clones or highly inbred strains of animals exhibiting minimal genetic variation and maintained under similar, strictly controlled conditions may be useful in some contexts, but the inherent variation between individuals is thereby deliberately minimised although it is a feature of almost all biological material. The variation of response is frequently as important as a repeated response elicited in a number of replicates. Bioassays should reveal this variability among the subjects as well as a 'normal' response. Furthermore, results determined in a series of tests on a single population, collection or culture of insects are almost invariably ascribed to the species although we know that in many cases there are major differences between populations (Diehl and Bush 1984). Here is a paradox; we demand that bioassays are repeatable but we know that our subjects, the test organisms, are inherently variable.

It is possible to apply materials to an insect in a very precise dose and measure many chemical parameters with great precision, although sometimes with less accuracy than is suggested (see Robertson and Preisler (1992) for an analysis and explanation of these problems). Thus, bioassays involving treatments with purified chemicals are *relatively* easy to standardise compared with research in which biological material constitutes the treatment as well as the subject. Where one is interested in the response of one sort of biological material to another, bioassays are much more complex and interpretation correspondingly more difficult.

The design of bioassays must include consideration of the physical conditions, the subjects, and the treatment. These three elements will now be considered separately, although this is an artificial division and there are complex interactions between them. No attempt is made to deal with the statistical basis of experimental design since Robertson and Preisler (1992) have already provided a comprehensive treatment of this issue in the context of pesticide bioassays with arthropods in general. The following sections are concerned only with bioassays involving pest species of bruchid in laboratory situations, particularly with reference to the determination of resistance among their host seeds, although the general principles are universally applicable.

Physical Conditions

It has been known for at least 30 years that many environmental variables affect the fitness of bruchids. Howe and Currie (1964) showed that different temperatures and relative humidities had a very significant effect on the fecundity, developmental speed and longevity of several of the most important bruchid species. Southgate (1964) pointed out that *C. rhodesianus* was typically found in cooler areas than *C. maculatus*, and *Zabrotes subfasciatus* is a pest of *Phaseolus vulgaris* at lower altitudes (which are warmer and more humid) being replaced by *A. obtectus* at higher altitudes (Cardona and Karel 1990). One should therefore consider whether the performance of these pairs of species should be

compared under the same physical conditions or under those in which they are normally found. Lighting regimes are often less important but in some cases are critical; *Bruchidius atrolineatus* will effectively stop breeding altogether in certain lighting regimes as it exhibits a facultative reproductive diapause (Lenga et al. 1991).

Some bruchids limit oviposition in the absence of an adequate supply of hosts. Females of some populations of *Callosobruchus maculatus* withhold eggs if provided with 10 seeds and lay significantly more on 40 or 100 (Credland 1986). Thus, realised fecundity (Mitchell 1990) is reduced although potential fecundity is maintained.

Some, perhaps all, bruchids will oviposit on a range of inappropriate substrates for reasons which are not fully understood. Many bruchids will attach eggs to the walls or floor of Petri dishes or glass tubes in experimental situations. This can occur even if surplus seeds which bear no eggs but appear indistinguishable from others bearing eggs are readily available. Therefore it is easy to underestimate fecundity and impossible to estimate the significance of a treatment applied to the beetles using this parameter without consideration of this potential problem.

In the case of *A. obtectus*, the proportion of eggs attached to seeds is immensely variable but no more than 50% are usually attached and many are easily dislodged (Quentin et al. 1991; D. Parsons and C. Moss, unpublished data). Although it is well known that larvae of this species move among seeds it is inevitable that fecundity measurements demand additional care. The variable fecundity figures quoted for this species may well be due to the omission of eggs not attached to seeds.

The Subjects

Inter and intraspecific variation among bruchids

Different species of bruchid typically occupy different habitats. Each species characteristically infests a different host although there is now a considerable measure of overlap among the major pest species (Southgate 1978). *A. obtectus* and *Z. subfasciatus* are usually found in stores of common bean, *P. vulgaris*, with which they are associated in South and Central America where they originated (Cardona and Karel 1990). Both species have now been found on beans and cowpeas, *Vigna unguiculata*, in Africa (Meik and Dobie 1986; Olubayo 1993). *C.maculatus*, the cowpea seed beetle, probably originated on cowpeas in West Africa but can now be found on chickpea (*Cicer arietinum*), adzuki bean (*Vigna angularis*), lentils (*Lens culinaris*) and numerous other legume seeds. *C. chinensis*, the adzuki bean weevil, probably evolved in Southeast Asia in association with a wild progenitor of the adzuki bean (Yoshida et al. 1986), but has now spread onto numerous alternative hosts. Thus there are major differences in host utilisation and although the pest species are more polyphagous than their wild relatives, there are limits to the host range of each species (Credland 1990); for example, *C. maculatus* cannot reproduce on *P. vulgaris* (Janzen et al. 1976).

The fact that Southgate (1964) could identify *C. rhodesianus* as more typical of cooler areas than *C. maculatuss*, epitomises the differences in physical conditions to which even closely related species are adapted. Therefore if one is comparing species, the physical conditions employed are of paramount importance. One needs to decide if the objective of an assay is to compare performance under a single or multiple set of conditions, or to determine the limit of each species' performance under optimal or marginal conditions.

Interspecific differences are probably less of a concern, since they are more readily recognised, than intraspecific differences. The very fact that the species have different names indicates a measure of difference even if it is not always defined. Intraspecific differences are much more difficult to quantify and are frequently not even recognised, although there has been a greater appreciation of the problem in the last few years (Credland 1990). Fujii (1968, 1969) showed that differences existed among populations of bruchid species nearly 30 years ago, but the practical implications have only become apparent more recently (e.g. Credland 1990; Credland and Dendy 1992). In these and other studies major differences are expressed between geographical biotypes (Diehl and Bush 1984) in parameters such as fecundity and developmental speed.

One of the simplest demonstrations of the differences between populations can be provided by looking at susceptibility to an insecticide (in one laboratory), since physical conditions and treatment are identical in each assay. Exposing adults of five populations of *C. maculatus*, maintained without exposure to any insecticide for at least 150 generations, to malathion immediately exposes variation (Fig. 1). It is therefore not surprising that the capacity to survive in so-called 'resistant' seeds should also be exhibited (Dick and Credland 1986a; Giga et al. 1993). Separate populations respond differently in the same bioassay, although the scale of the differences vary with the nature of the assay. It should also be obvious that averaging results of assays in different laboratories, probably using different biotypes, has no biological meaning, being unrepresentative of the species or, perhaps, any part thereof.

Bioassays conducted on a single biotype do not represent the response of that bruchid species to a treatment.

Individual variation

Hidden within the previous argument is the assumption that individual insects within the population differ, and that differences can be overcome with appropriate statistical tests or manipulations (Robertson and Preisler 1992). This is probably the case in many situations where appropriate attention is applied to other parameters of the bioassay, such as the physical conditions, but the extent of individual variation is frequently extreme and unpredictable.

One major problem is the incidence of polymorphism in *C. maculatus* (Utida 1981) which is primarily induced by elevated seed water content (Sano-Fujii 1984) but also has a genetic basis (Sano-Fujii 1986; Messina 1987). A comparable problem may occur in other species (Credland 1990). Most authors have worked in conditions which ensure production of the normal (nonflight) form which reproduces immediately after adult emergence and has a higher fecundity than the active or flight form (Utida 1972). The extent of individual variation under tightly controlled conditions can be illustrated by reference to recent studies on *Acanthoscelides obtectus* (Parsons and Credland, in prep.). In my laboratory this species exhibits greater individual variation within a population, under regulated physical conditions, than any other bruchid on which we have worked. For example, most bruchids which have been studied exhibit a predictable pattern of oviposition activity under specified conditions. All the biotypes of *C. maculatus* which have been studied lay most of their eggs on the first day after emergence under optimal or near-optimal physical conditions (Dick and Credland 1984). The number of eggs laid on each succeeding day declines so that virtually all have been laid by the sixth day after emergence (see Credland and Wright 1989, for details). In the case of *Zabrotes subfasciatus* most eggs are usually laid on day 2 (Howe and Currie 1964; Meik and Dobie 1986). However, in the case of *A. obtectus* there is immense variation and although an overall pattern such as that in Figure 2a can be obtained, it is a composite of numerous diverse patterns such as those in Figure 2b which represent the daily oviposition of six individual females. Full sibs may display all these and other patterns, regardless of that displayed by their mother or paternal aunts. Whether the patterns have a genetic basis or represent the product of complex environmental interactions is unknown.

In many cases, it can be shown that statistically significant variation in parameters such as adult weight or fecundity has an environmental basis and, hence, can be manipulated by the physical conditions to which animals were exposed before or during the bioassay. A simple example of the situation where inattention to detail can produce wide fluctuations in parameters frequently used in bioassays, is the association between

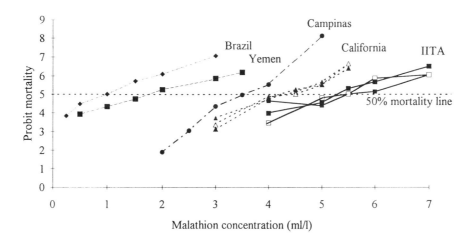

Fig. 1. Probit mortality of five populations of *Callosobruchus maculatus* exposed to a commercial emulsifiable concentrate of malathion. Each point represents the mean of 9 replicates of 10 adult beetles which were confined on 5 cm² of filter paper to which the specified concentration had been applied. Mortality was assessed after 10 hours exposure at 25°C. Three entirely separate assays were undertaken for the California and IITA populations, and one for each of the others. 1 mL/L malathion corresponds to exposure to 0.0031 mL of technically pure active material.

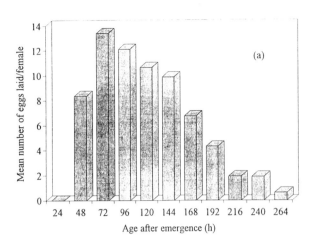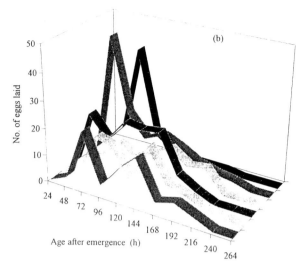

Fig. 2. Daily oviposition by female *Acanthoscelides obtectus* reared at a uniform density and provided with four fresh, conditioned seeds per day. Females were allowed to mate during the initial 24 hours and put on beans after this period. (a) Mean results from 101 individual females. (b) Results for six individual females to show the immense variation which occurs even under rigorously standardised conditions.

larval density within a seed, the number and weight of adult insects which emerge and the fecundity of the females among them (Credland et al. 1986; Desroches 1990).

It is now well established that increasing larval density within a seed leads, not surprisingly, to a greater number of adults emerging in many populations of some bruchids such as *C.maculatus* and *C. chinensis*. However, the increase is not linear and competitive interactions produce a plateau representing the maximum number of individuals which can successfully complete their development in a seed of given size (Bellows 1982a, b; Smith and Lessells 1985). The larvae of some populations, such as the South India biotype of *C. maculatus*, are highly competitive and it is rare for more than a single adult to emerge from any seed (Mitchell 1991); this represents an interesting extreme case of the differences noted among populations by Dick and Credland (1984). There are also interspecific differences since all recorded *C. analis* larvae appear to be competitive and it is rare for more than a single adult to emerge from any seed (Toquenaga and Fujii 1990). Therefore, survival of individual larvae is a factor not only of their own viability in the context of their host and physical conditions, but also of the number of other larvae within the seed that they occupy. Where multiple larvae do survive in a single seed, then their weight is often significantly smaller as the larval density increases (e.g. Credland et al. 1986). Thus, increasing larval survival can result in higher densities and smaller body weights, the former suggesting an increase in fitness and the latter, a decrease. However, neither may reveal anything about the chemistry or physical attributes of the seed since the results can be a consequence of space and individual behaviour. Female fecundity in *C. maculatus* increases with body weight and there is no independent role of larval density (Colegrave 1993). Therefore, like body weight, fecundity of insects is markedly affected by larval density.

These examples illustrate that experimental protocols which fail to regulate larval density can produce a mass of data which are so interwoven that it is virtually impossible to make a sensible interpretation of their meaning. Nevertheless, perhaps the most common assays have involved introducing 'x' insects to 'y' seeds and measuring precisely those parameters of individual insects which are subject to modification by the unregulated larval density.

Treatments

If the physical conditions and protocol of the bioassay and the inherent variation in the living material, the bruchid subject, can be recognised, controlled and their effects thereby limited, the remaining variable(s) is the treatment applied.

In those cases where the treatment is 'simple', the assay, for example, of a predetermined dose of a pure chemical formulated in a defined way and applied in a standard manner, there are no problems which differ from those which impinge on routine assays undertaken in numerous laboratories every day. Concerns over purity, degradation, losses during application, the definition of incapacity or death in a practical context, do not require any abnormal attention in the context of bruchid beetles. The same criteria for rigour which apply to all insecticide assays should be employed (Robertson and Preisler 1992).

The problems are much greater in assays which involve botanical material and especially whole seeds. The search for bruchid resistance in seeds has always depended on, and continues to depend on, exposing seeds to beetles and observing parameters such as oviposition, development speed and larval survival (CIAT 1986). Subsequently reports usually cite a germplasm bank or cultivar name or number and regard this as a constant in the experimental analysis. Unfortunately, this is inappropriate and can sometimes be misleading.

Perhaps the simplest problem to understand is the source of the material. In the great majority of cases, the bioassays are not conducted by the collector or breeder of the seed. If the seed has been collected from the field, it is imperative that the material is homogeneous, and not a heterogeneous mix of seeds, ideally from an individual plant and certainly from a single stand or area. Frequently it is unknown if this is the case and certainly it is rarely stated in the published accounts. While different plant species may possess characteristic cocktails of allelochemicals (Liener 1982), it is well known that the relative concentrations of the components of the cocktails and the absolute levels of each component can vary for both genetic and environmental reasons (Zangerl and Berenbaum 1993). For example, chemical composition may vary with soil characteristics and exposure.

Whilst it is unreal to imagine that complete analyses of each seed tested or even a representative sample could be undertaken in every bioassay, a simple account of the source of the seed, its consistency in appearance and, if it were bred, its pedigree would be an invaluable adjunct to the data presented.

In addition to any chemical differences between seeds, the physical differences between accessions or cultivars can have far reaching consequences in terms of resistance measurement. Size alone can determine the number of bruchids which emerge, so that comparative assays in which seed number is constant can be misleading if larval density is not controlled. Conversely, if seed weight is kept constant then the number of seeds may vary considerably. Some bruchids distribute their eggs uniformly among the available seeds and only a single adult emerges from each seed (Thanthianga and Mitchell 1990). Seed number will then be more important than seed weight in determining the number of progeny completing development.

The physical characteristics of the seed testa can determine acceptability for oviposition but may not be related to the anti-biotic nature of the seed (Messina and Renwick 1985). Nwanze et al. (1975) showed that rough seeds were less acceptable to *C. maculatus* than smooth ones. If no account is taken of differential oviposition, then it would be easy to attribute resistance to seeds, measured as adult emergence, on the basis of their acceptability for oviposition. (While it is true that reduced acceptability could potentially be used to contribute to resistance, it is important to be able to segregate behavioural and physiological resistance at an early stage in screening.)

Seed shape is seldom mentioned in bioassays with bruchids but is critically important. For example, some wild accessions of *Phaseolus vulgaris* are flattened and reniform but mungbeans (*Vigna radiata*) are almost spherical. The larvae of *Acanthoscelides* move from one seed to another before penetrating the testa. They need to be able to bridge a gap somewhat shorter than their bodylength to provide a brace to maintain their position whilst excavating the early stages of a penetration hole. Rounded seeds generally provide fewer such points (Quentin 1991) and any movement of the seed disrupts boring and can prevent penetration, so providing a physical means of their control (Quentin et al. 1991). In many assays, especially if the timing of penetration is to be measured precisely, so requiring regular examination of the seeds, then the movement of the seeds within their containers is enough to inhibit penetration and false resistance may be recorded (C. Moss, per. comm. 1993).

Among other seed differences which may influence the outcome of bioassays but have seldom been considered methodically, are the hardness of the testa (Thiery 1984), its softening on exposure to air with different water content, its age (since some of its constituent chemicals may degrade thereby producing or destroying toxins), and the means of its storage which can affect any or all of these parameters. Furthermore, the effects of any microorganisms which may be associated with the seed, especially in damp conditions, need to be considered.

Some Pragmatic Solutions

Many of the problems which have been mentioned in the context of bruchids are found and have probably been solved in work with many other pests. Unfortunately these solutions, if they exist, have not permeated very far into the world of bruchid biology and perhaps this provides some explanation of why many basic problems remain and why so much work stands in isolation and cannot be integrated into more helpful solutions of the important practical problems which face us today (Singh 1990). In a database of over 1000 publications, excluding purely taxonomic works, on bruchids accumulated over the last 12 years, there is not a single paper which provides

genuine, reliable *data* on losses that they cause in the field or store anywhere in the world. There are numerous assertions, estimates and guesses. Literally dozens of papers deal with 'resistance' in numerous hosts, cultivars and accessions, but there is no pattern among the data and it is impossible to compare almost any one paper with any other. The preceding sections explain, in part, how this latter situation has arisen. Why it has arisen is more contentious but may have much to do with the availability of literature in those areas where most of the bruchid problems occur and the understandable demand for quick local solutions rather than an aspiration to broaden the basis of our fundamental knowledge. Nevertheless, the consequence is that much work which could be important is overlooked and there can be no doubt that a wider dissemination of knowledge and experience would be valuable. The scientific community will be more receptive when the work has a more structured basis and it is apparent that the data have been collected in a rigorous way which excludes or at least recognises the variables to which reference has been made. What is required? What practical steps can be taken to alleviate if not solve the problems?

In terms of the physical conditions, it should be a minimum requirement that conditions are defined and controlled. Ambient conditions permit such variation in insect performance (fitness) that very limited value can be ascribed to data obtained under such conditions. The bioassay could hardly be repeated in the same laboratory under the stability of some tropical conditions, and certainly never be copied elsewhere. Temperature, relative humidity and the lighting regime need to be defined with means and the range. It is important that they are actually measured rather than set since, for example, even an open dish of water which is continually replenished will not maintain humidities above about 30% in some incubators (P. Credland, unpublished data). Depending on the objective of the study and the species concerned, there is no reason why these three variables should be constant throughout the day; some diurnal cycling may be desirable or essential but requires definition. The actual values which are preset will usually be determined in the context of either those experienced by the species in its normal local habitat, those which produce maximum fitness (if they have been defined) or those under which it competes with other species. An explanation of the values selected, in the preceding terms, would always be desirable.

The capacity of many bruchids to lay on 'inappropriate' surfaces needs to be taken into account. It can be minimised or avoided altogether by enclosing seeds in containers lined with coarse emery paper on which oviposition has never been seen in, literally, thousands of individual cases (Dick and Credland 1984). If smooth glass or plastic containers are used, then checks for oviposition on these surfaces are essential. These considerations do not apply to *Acanthoscelides* in which the larvae are motile and can move among seeds but care must be taken in counting the eggs.

The subjects of the experiments, the insects, must be described in unambiguous terms. If it is intended that the results of the assays should be applicable to a species, then recognition of 'the biotype problem' is a minimal requirement (Diehl and Bush 1984). No single population can represent the diversity expressed by an entire cosmopolitan species (Credland 1990); this is not because the species has no defining characteristics but is because responses, both terminal and intermediate, may include steps or make metabolic demands on characteristics which do show intraspecific variation. Therefore, while a purely local problem needs to be approached with reference to its own endemic bruchid population, one cannot extrapolate from these results to a different locality or to the species as a whole.

Preceding any bioassay it is imperative that the insects to be employed are reared at a density of one insect per seed to minimise variation due to polymorphism and standardise their size and potential fecundity (Mitchell 1990). However, it is important that the range of responses as well as the 'typical' response of the population is recognised. Environmentally induced variation can arise even in regulated physical conditions as explained previously. The adult density on seeds to be tested and larval density within the seeds need careful regulation since both affect many of the parameters usually used in bioassays (e.g. Padgham et al. 1992; Schoonhoven et al. 1983; Simmonds et al. 1989). With unlimited seed, maintaining a larval density of one per seed is to be preferred since this excludes any variation in competitiveness within the species. Furthermore, infestation with single larvae can be established in very small seeds which are too small to allow the production of two adults for physical reasons. The uniformity in size among adults which develop individually is greater than among adults developing at variable and unknown densities, permitting more precise comparisons between hosts to be made (B. Tran and P. Credland, unpublished data). Individual variation among the progeny can then be accommodated by the use of statistical methods of the kind explained by Robertson and Preisler (1992).

If larval density is to be regulated, one obvious sequitur is that oviposition and subsequent assays need to be conducted independently. If this is not done then increasing egg number may lead to an increase in density (hatched eggs/seed), thence smaller, less fecund adults, and, possibly, more rapid development (Credland et al. 1986). Thus a change in one parameter can lead to correlated changes in others, producing spurious or misleading results. The combined, accurate, measurement of oviposition on the seed and subsequent measurement of suitability for development can only compromise each other to the extent that neither measurement has any value. Bearing in mind that many bruchids will oviposit on an enormous range of surfaces, if seed is limited it is generally preferable to measure aspects of development and mortality with precision. Except in the case of *A. obtectus*, some oviposition on the seed must occur for this experiment to be undertaken and therefore some degree of acceptability has been demonstrated.

The final factor to be considered is the nature of the seed itself. In the determination of any differential resistance among seeds, the quantity of homogeneous seed which is available usually delimits the experimental protocol. The size of mature seeds is a characteristic of the cultivar or wild form, not of the species. Red kidney beans may weigh 500 mg and be 17 mm long compared with some wild accessions of the same species, *Phaseolus vulgaris*, whose seeds weigh only 30 mg and are 5 mm long (C. Moss, unpublished data). It is impossible to generalise about the treatment of this variable. For example, if larvae are not competitive and the individual seeds are large then it may be possible by removal of excess eggs and ensuring separation of eggs on each seed to exploit the seed size by allowing a small, known number of larvae to penetrate and, perhaps develop. (Single infestation is still to be preferred, if possible.) Conversely, to undertake assays with more than one egg on some of the very small, flattened seeds of wild *P. vulgaris* is wasteful since no more than one adult will develop for size reasons alone. The proportion of eggs producing an adult can be manipulated in such situations by simply changing the egg: seed ratio. Seed shape also needs to be considered; if seeds are spherical then oviposition by those species which glue their eggs to the seeds and penetration by larvae of *Acanthoscelides* are more difficult. These problems are minimised if more than one seed is used to provide additional contact points and the seed(s) fit closely into the container. If seeds are in short supply, clean glass beads of comparable size to the seeds may provide a suitable alternative.

Seed conditioning is an essential prerequisite. With large (5–6 mm) seeds such as cowpeas, it takes at least 2 weeks in the experimental conditions for the moisture content to reach equilibrium (Dick and Credland 1984). Since seed water content can determine the morph of the adult *C. maculatus*, which are characterised by different fecundities (Sano-Fujii 1984; Ouedraogo et al. 1991), the importance of this period cannot be overestimated. Measurements of seed water content should be made routinely. Seed age is a parameter which has not been investigated systematically but is a potential problem. Since the purpose of most experiments is to investigate seeds that will typically be stored for less than a year, naturally dried, mature seeds harvested within the past year should be used when possible.

Conclusion

The object of this contribution was to try and expose the problems in undertaking measurement of bruchid resistance in seeds. The difficulties can be divided into three areas, although interactions between them result in making a clean division impossible. Some of the problems can be overcome by careful consideration of experimental design including pretreatment of the test organisms and the biological material (seeds) under investigation. By unifying design of bioassays, or at least recognising the problems to which attention has been drawn and providing adequate descriptions of conditions and data, far greater interchange of information and more rapid progress will be made. The difficulties which have been highlighted do not constitute a reason for hopelessness or despair; they can all be overcome or mediated. It is a fact that biotypical differences may ultimately prevent the pronouncement of *simple* answers referable to a whole species, but if the problem is appreciated then satisfactory progress can be made with even the most cosmopolitan pest species. Total uniformity in experimental design may be too much to expect and perhaps undesirable in some respects, but a widespread appreciation that experiments should be repeatable and that accounts should provide all the information to enable this to happen would be an invaluable advance.

Acknowledgments

I am indebted to Dr Frans van Alebeek and Dr Alan Gange whose critical comments on a previous draft were immensely valuable, and to my students and assistants whose efforts have provided much of the material for this paper. Mike Profit and Deborah Parsons provided Figures 1 and 2 respectively from their unpublished work.

References

Begon, M., Harper, J.L. and Townsend, C.R. 1990. Ecology (Second edition). Oxford, Blackwell, 945 p.

Bellows, T.S. Jr. 1982a. Analytical models for laboratory populations of *Callosobruchus chinensis* and *C. maculatus* (Coleoptera, Bruchidae). Journal of Animal Ecology, 51, 263–287.

Bellows, T.S. Jr. 1982b. Simulation models for laboratory populations of *Callosobruchus chinensis* and *C. maculatus*. Journal of Animal Ecology, 51, 597–623.

Cardona, C. and Karel, A.K. 1990. Key insects and other invertebrate pests of beans. In: Singh, S.R. ed., Insect pests of tropical food legumes. Chichester, John Wiley and Sons, 157–191.

Caswell, G.H. 1981. Damage to stored cowpea in the northern part of Nigeria. Samaru Journal of Agricultural Research, 1, 11–19.

CIAT (Centro Internacional do Agricultura Tropical). 1986. Main insect pests of stored beans and their control. A study guide.

Colegrave, N. 1993. Does larval competition affect fecundity independently of its effect on adult weight? Ecological Entomology, 18, 275–277.

Credland, P.F. 1986. Effect of host availability on reproductive performance in *Callosobruchus maculatus* (F.) (Coleoptera: Bruchidae). Journal of Stored Products Research, 22, 49–54.

Credland, P.F. 1990. Biotype variation and host change in bruchids: causes and effects in the evolution of bruchid pests. In: Fujii, K., Gatehouse, A.M.R., Mitchell, R. Johnson, C.D. and Yoshida, Y., ed., Bruchids and legumes: economics, ecology and coevolution, Dordrecht, Kluwer Academic Publishers, 271–287.

Credland, P.F. and Dendy, J. 1992. Comparison of seed consumption and the practical use of insect weight in determining effects of host seed on the Mexican bean weevil, *Zabrotes subfasciatus* (Boh.). Journal of Stored Products Research, 28, 225–234.

Credland, P.F. and Wright, A.W. 1989. Factors affecting female fecundity in the cowpea seed beetle, *Callosobruchus maculatus* (Coleoptera: Bruchidae). Journal of Stored Products Research, 25, 125–136.

Credland, P.F., Dick, K.M. and Wright, A.W. 1986. Relationships between larval density, adult size and egg production in the cowpea seed beetle *Callosobruchus maculatus*. Ecological Entomology, 11, 41–50.

Daglish, G.J., Hall, E.A., Zorzetto, M.J., Lambkin, T.M. and Erbacher, J.M. 1993. Evaluation of protectants for control of *Acanthoscelides obtectus* (Say) (Coleoptera: Bruchidae) in navybeans (*Phaseolus vulgaris* (L.)). Journal of Stored Products Research, 29, 215–219.

Dent, D. 1991. Insect pest management. Wallingford, CAB International, 604 p.

Desroches, P. 1990. Influence de la densite larvaire intracotyledonaire sur les capacites reproductrices d'Acanthoscelides *obtectus* Say (Coleoptera: Bruchidae). Annales de la Societe entomologique de France, 26, 93–102.

Dick, K.M. and Credland, P.F. 1984. Egg production and development of three strains of *Callosobruchus maculatus* (F) (Coleoptera : Bruchidae). Journal of Stored Products Research, 20, 221–227.

Dick, K.M. and Credland, P.F. 1986a. Variation in the response of *Callosobruchus maculatus* (F.) to a resistant variety of cowpea. Journal of Stored Products Research, 22, 43–48.

Dick, K.M. and Credland, P.F. 1986b. Changes in the response of *Callosobruchus maculatus* (Coleoptera: Bruchidae) to a resistant variety of cowpea. Journal of Stored Products Research, 22, 227–233.

Diehl, S.R. and Bush, G.L. 1984. An evolutionary and applied perspective of insect biotypes. Annual Review of Entomology, 29, 471–504.

Fujii, K. 1968. Studies on interspecies competition between the azuki bean weevil and the Southern cowpea weevil. III. Some characteristics of strains of two species. Researches in Population Ecology, 10, 87–98.

Fujii, K. 1969. Studies on the interspecies competition between the azuki bean weevil and the Southern cowpea weevil. IV. Competition between strains. Researches in Population Ecology, 11, 84–91.

Giga, D.P., Kadzere, I and Canhao, J. 1993. Bionomics of four strains of *Callosobruchus rhodesianus* (Pic) (Coleoptera, Bruchidae) infesting different food legumes. Journal of Stored Products Research, 29, 19–26.

Howe, R.W. and Currie, J.E. 1964. Some laboratory observations on the rates of development, mortality and oviposition of several species of Bruchidae breeding in stored pulses. Bulletin of Entomological Research, 55, 437–477.

Janzen, D.H., Juster, H.B. and Liener, I.E. 1976. Insecticidal action of the phytohemagglutinin in black beans on a bruchid beetle. Science N.Y., 192, 795–6

Lenga, A., Thibeaudeau, C. and Huignard, J. 1991. Influence of thermoperiod and photoperiod on reproductive diapause in *Bruchidius atrolineatus* (Pic) (Coleoptera, Bruchidae). Physiological Entomology, 16, 295–303.

Liener, I.E. 1982. Toxic constituents in legumes. In: Arora, S.K. ed., Chemistry and biochemistry of legumes, New Delhi, Oxford and IBH Publ. Co., 217–257 .

Meik, J. and Dobie, P. 1986. The ability of *Zabrotes subfasciatus* to attack cowpeas. Entomologia experimentalis et applicata, 42, 151–158.

Messina, F.J. 1987. Genetic contribution to the dispersal polymorphism of the cowpea weevil (Coleoptera: Bruchidae). Annals of the Entomological Society of America, 80, 12–16.

Messina, F.J. and Renwick, J.A.A. 1985. Resistance to *Callosobruchus maculatus* (Coleoptera: Bruchidae) in selected cowpea lines. Environmental Entomology, 14, 868–872.

Mitchell, R. 1990. Behavioural ecology of *Callosobruchus maculatus*. In: Fujii, K., Gatehouse, A.M.R., Mitchell, R. Johnson, C.D. and Yoshida, Y. ed., Bruchids and legumes: economics, ecology and coevolution, Dordrecht, Kluwer Academic Publishers, 317–330.

Mitchell, R. 1991. The traits of a biotype of *Callosobruchus maculatus* (F.) (Coleoptera: Bruchidae) from South India. Journal of Stored Products Research, 27, 221–224.

Nwanze, K., Horber, E. and C. Pitts 1975. Evidence of ovipositional preference of *Callosobruchus maculatus* (F.) for cowpea varieties. Environmental Entomology, 4, 409–412.

Olubayo, F. 1993. A study of storage bruchids (Coleoptera: Bruchidae) infesting cowpeas (*Vigna unguiculata* (l.) Walp.) in Kenya and their control. Ph.D. thesis, University of Newcastle-upon-Tyne, England.

Ouedraogo, P.A., Monge, J.P. and Huignard, J. 1991. Importance of temperature and seed water content on the induction of imaginal polymorphism in *Callosobruchus maculatus*. Entomologia experimentalis et applicata, 59, 59–66.

Padgham, J., Pike, V., Dick, K. and Cardona, C. 1992. Resistance of a common bean (*Phaseolus vulgaris* L.) cultivar to post-harvest infestation by *Zabrotes subfasciatus* (Boheman) (Coleoptera: Bruchidae). I. Laboratory tests. Tropical Pest Management, 38, 167–172.

Quentin, M.E. 1991. Host-colonisation behaviours of the bean weevil, *Acanthoscelides obtectus* (Say), in stored beans. Ph.D. thesis, Michigan State University, USA.

Quentin, M.E., Spencer, J.L. and Miller, J.R. 1991. Bean tumbling as a control measure for the common bean weevil, *Acanthoscelides obtectus*. Entomologia experimentalis et applicata, 60, 105–109.

Robertson, J.L. and Preisler, H.K. 1992. Pesticide bioassays with arthropods. CRC Press, Boca Raton. 127 pp.

Sano-Fujii, I. 1984. Effect of bean water content on the production of the active form of *Callosobruchus maculatus* (F.) (Coleoptera: Bruchidae). Journal of Stored Products Research, 20, 153–161.

Sano-Fujii, I. 1986. The genetic basis of the production of the active form of *Callosobruchus maculatus* (F.) (Coleoptera: Bruchidae). Journal of Stored Products Research, 22, 115–124.

Schoonhoven, A.V., Cardona, C. and Valor, J. 1983. Resistance to the bean weevil and the Mexican bean weevil (Coleoptera: Bruchidae) in noncultivated common bean accessions. Journal of Economic Entomology, 76, 1255–1259.

Simmonds, M.S.J., Blaney, W.M. and Birch, A.N.E. 1989. Legume seeds: the defences of wild and cultivated species of *Phaseolus* against attack by bruchid beetles. Annals of Botany, 63, 177–184.

Singh, S.R. 1978. Resistance to pests of cowpea in Nigeria. In: Singh, S.R., van Emden, H.F. and Taylor, T.A. ed., Pests of grain legumes: ecology and control. London, Academic Press, 267–279.

Singh, S.R. ed. 1990. Insect pests of tropical food legumes. Chichester, John Wiley and Sons, 451 pp.

Smith, R.H. and Lessells, C.M. 1985. Oviposition, ovicide and larval competition in granivorous insects. In: Sibly, R.M. and Smith, R.H. ed., Behavioural ecology; ecological consequences of adaptive behaviour, Oxford, Blackwell Scientific Publication, 423–448.

Southgate, B.J. 1964. Distribution and hosts of certain Bruchidae in Africa. Tropical Stored Product Information, 7, 277–279.

Southgate, B.J. 1978. The importance of the Bruchidae as pests of grain legumes, their distribution and control. In: Singh, S.R., Van Emden, H.F. and Taylor, T.A. ed., Pests of grain legumes: ecology and control, London, Academic Press, 219–229.

Southgate, B.J. 1979. Biology of the Bruchidae. Annual Review of Entomology, 24, 449–473.

Thanthianga, C. and Mitchell, R. 1990. The fecundity and oviposition behaviour of a South Indian strain of *Callosobruchus maculatus*. Entomologia experimentalis et applicata, 57, 133–142.

Thiery, D. 1984. Hardness of some fabaceous seedcoats in relation to larval penetration by *Acanthoscelides obtectus* (Say) (Coleoptera: Bruchidae). Journal of stored Products Research, 20, 177–181.

Toquenaga, Y. and Fujii, K. 1990. Contest and scramble competition in two bruchid species, *Callosobruchus analis* and *C. phaseoli* (Coleoptera, Bruchidae). 1. Larval competition curves and interference mechanisms. Researches on Population Ecology, 32, 349–363.

Utida, S. 1972. Density dependent polymorphism in the adult of *Callosobruchus maculatus* (Coleoptera, Bruchidae). Journal of Stored Products Research, 8, 111–126.

Utida, S. 1981. Polymorphism and phase dimorphism in *Callosobruchus*. In: Labeyrie, V. ed., The ecology of bruchids attacking legumes (pulses). The Hague, Dr W. Junk Publishers, 143–147.

Yoshida, Y., Igarashi, H. and Shinoda, K. 1986. Life history of *Callosobruchus chinensis* (L.) (Coleoptera, Bruchidae). In: Donahaye, E. and Navarro, S. ed., Proceedings of the 4th International Working Conference on Stored Product Protection, Tel Aviv, Israel, September 1986, 471–477.

Zangerl, A.R. and Berenbaum, M.R. 1993. Plant chemistry, insect adaptations to plant chemistry, and host plant utilisation patterns. Ecology, 74, 47–54.

The distribution and PCR-based fingerprints of *Rhyzopertha dominica* (F.) in Canada

P.G. Fields[*] and T.W. Phillips[†]

Abstract

The first record of *Rhyzopertha dominica* in North America is in 1861. Since then it has become a major pest in the USA and is found occasionally in Canada. We sought answers to the following questions: What is the extent of its distribution in Canada? What are the sources of infestation? What is its genetic background?

Pheromone traps detected *R. dominica* mainly in Manitoba, though it was trapped in the other western provinces. It was not caught in southern Ontario or Quebec. In Manitoba the total numbers caught from 9 traps were 2538 (1990), 4226 (1991), 723 (1992), and 289(1993). Despite extensive sampling, *R. dominica* was rarely detected in grain held in storage.

Genetic differences among *R. dominica* populations were investigated using the PCR-based fingerprinting method of Randomly Amplified Polymorphic DNA (RAPD-PCR). Surveyed genotypes from Canada were shared by many distantly separated populations in the USA, suggesting that gene flow and migration are widespread.

Introduction

Rhyzopertha dominica, the lesser grain borer (Coleoptera: Bostrichidae), is a major pest of stored cereals. It is thought to have originated from the Indian subcontinent, but now has a cosmopolitan distribution (Potter 1935). *R. dominica* was found in the USA in 1861 (Leconte 1862), with widespread distribution occurring by the 1920s (Back and Cotton 1922). *R. dominica* has been found in grain stored on farms in all USA states bordering the Canadian prairie provinces (Storey et al. 1983; Subramanyam and Harein 1989).

There are occasional records of *R. dominica* occurring in Canada since the late 1800s (Fields et al. 1993). However, several extensive surveys of insects in grain stored on farms in the Canadian prairies have failed to detect any *R. dominica* (Liscombe and Watters 1962; Loschiavo 1975; Smith and Barker 1987; Madrid et al. 1990). Each year the Canadian Grain Commission conducts Berlese funnel extractions for stored-grain insects in over 100000 samples of prairie grain from primary elevators (grain received directly from the farm), rail cars and terminal elevators (grain destined for export). From 1975 to 1993 only nine *R. dominica* were found, and all of these since 1989 (J. Van Loon, unpublished data).

Occasionally *R. dominica* has been found in feed mills (Steinbach, Manitoba 1981; Aldergrove, BC 1987; Lyster, Quebec 1987; P.G. Fields, unpublished data). As feed mills are

heated throughout the winter and sometimes import feed-grade corn from the USA (Fields et al. 1993), we believe that corn infested with *R. dominica* from the USA could be a source of infestation for other stocks of grain in Canada. In 1990 over 600000 t of grain was imported from the USA into Canada; in 1991, less than 300000 t. In both years over 90% of the grain was corn. Assuming the infestation levels are similar to what they were in the late 1970s (Storey et al. 1982), then Canada would import over a million *R. dominica* each year (Fields et al. 1993).

The purpose of this study is to determine the distribution of this species across Canada and to determine genetic relationships among Canadian populations and between Canadian and U.S. populations. These data should be useful in testing three hypotheses: importation of infested grain, windborne migration (e.g. Smith and MacKay 1989) or an established Canadian population, put forth to explain the increased incidence of *R. dominica* in western Canada.

Materials and Methods

Pheromone trapping

Methods were similar to those of Fields et al. (1993). Eight-unit Lindgren funnel traps (Phero Tech Inc., Delta, BC, Canada) (Lindgren 1983; Cogburn et al. 1984; Leos-Martinez et al. 1986, 1987), were baited with *R. dominica* aggregation pheromone (Williams et al. 1981) (10 µL of dominicalure 1, 97% pure and 10 µL dominicalure 2, 95% pure) in 1990 and 1991 or with commercially available pheromone loaded onto rubber septa (Trécé Ltd., Salinas, USA) in 1992 and 1993.

In Manitoba, traps were placed within 5 m of the grain storage structures at three dairy farms, three primary elevators and three feed mills. Outside Manitoba, traps were placed near large grain terminals or transfer elevators. Each week the collection jars at the bottom of each trap were emptied and refilled with a solution (2 L water, 40 mL 6% sodium hypochlorite to slow microbial growth, 2 mL liquid soap to reduce the surface tension, and 500 mL ethylene glycol to reduce evaporation). In 1990 and 1991 pheromones were changed weekly, in 1992 and 1993 pheromones were changed every 4 weeks.

Daily maximum and minimum temperatures were collected from a Stevenson screen at Glenlea, Manitoba. For the summer of 1990, the temperatures at Glenlea were compared to 3 other locations in Manitoba (Altona, Plum Coulee, Niverville) closer to the other pheromone traps. No significant difference was found (ANOVA, p=0.61) between the four locations, therefore only one location was used.

In 1990, pheromone traps (eight-unit traps) were placed 100 and 1000 m north, south, east and west of two grain elevators located at Dominion City, Manitoba. Traps (four-unit traps) were also placed inside each elevator at the receiving door, loading to railcar door, the bottom of the bucket elevator and

---

[*] Agriculture and Agri-Food Canada, Research Station, 195 Dafoe Road, Winnipeg, Manitoba, Canada, R3T 2M9.

[†] United States Department of Agriculture, Agriculture Research Service, Department of Entomology, University of Wisconsin, Madison, Wisconsin, 53706, USA.

the top of the bucket elevator. Insects were collected weekly from 8 August to 10 October.

PCR fingerprinting

We used the polymerase chain reaction (PCR) to analyse randomly amplified polymorphic DNA (referred to as RAPD-PCR), to study genetic diversity within and among populations of *R. dominica*. RAPD-PCR was introduced by Williams et al. (1990) as a means of identifying large numbers of heritable genetic markers, and since has been applied to population studies (Hadrys et al. 1992). RAPD-PCR uses a single randomly designed 10-base primer that typically anneals to a number of complimentary sequences in the genomes of most organisms. Only a single primer is added to the PCR reaction, and products are amplified only when priming sites occur sufficiently close to one another (generally less than 2 kb) in an inverted orientation. RAPD products are heritable dominant genetic markers. Polymorphism in RAPD-PCR products, as evidenced by the presence or absence of a particular band, results from either a mutation in the sequence at one of the priming sites, or a change (duplication or deletion) in the intervening sequence yielding a fragment of different size.

Insects were collected using pheromone-baited multiple funnel traps with wheat in the collection jar. Live specimens were frozen at –80°C and held at –45°C until extraction. RAPD-PCR was conducted using techniques modified from Williams et al. 1990. Genomic DNA was extracted from frozen individual *R. dominica* by grinding in a Tris-HCl EDTA (disodium ethylene diamine tetraacetate) buffer containing SDS (sodium dodecyl sulfate). The preparation was incubated at 70°C for 30 minutes, proteins extracted with KOH, precipitated with isopropanol, washed in ethanol, dried, and resuspended in 10 μL water. The extract was diluted 10 fold again to yield enough material for over 100 PCR experiments. For PCR, 1 μL of dilute sample DNA, about 1–10 ng, was combined with 200 ng of a 10-base primer, 1.25 μL of Taq polymerase, '10x' buffer provided by the manufacturer for the Taq polymerase, 10 mM of $MgCl_2$ for DNA stability and 2 mM of dNTPs (a mix of adenine, guanine cytosine and thymine for primer extension). Total reaction volume in a sterile tube was 23 μL. An over-abundance of primer relative to sample DNA is important to ensure that primers have a high probability of annealing to available complimentary sites anywhere on the genome before genomic DNA reanneals in the cooling cycle. Multiple reaction tubes were set up in a thermal cycling machine. Our temperature program was 1) denaturation at 94°C for an initial 3 minutes; 2) annealing at 36°C; 3) extension at 72°C for 1 minute; 4) back to 94°C for 1 minute to begin a new cycle, for a total of 45 cycles. PCR products were separated by electrophoresis on a 3% agarose gel at 90 VAC for about 1.5 hours, and visualised under UV light following ethidium bromide staining. Frequency of RAPD products (bands on gels) were computed for each population.

Results and Discussion

Pheromone trapping

As in 1990 and 1991 (Fields et al. 1993), *R. dominica* was found across western Canada in 1992. Although over nine traps were placed in eastern Canada, no *R. dominica* were caught (Fig. 1). In 1993 the only *R. dominica* found outside of Manitoba were near grain terminals in Vancouver (Fig. 2).

There were no differences between the three site types, farms, elevators or feed mills in any of the four years (Table 1). This suggests that imported grain from the USA is not the sole source of the *R. dominica* found in Canada. Furthermore, southern Ontario and Quebec import grain from the USA, but no *R. dominica* were found in 1992 in these areas.

Intensive sampling around two elevators in southern Manitoba showed that the closer traps were placed to grain elevators the more *R. dominica* were caught (Fig. 3). However, eight traps placed inside the elevators caught only 41 insects, and all of these from the loading and receiving doors. We can think of two explanations for this. One, *R. dominica* are in the elevator but are rarely detected because they react much differently to pheromone baited traps in the confined space of an elevator. Two, the source of the *R. dominica* was not the elevators, and insects were trapped closer to the elevators because they were attracted to elevators because of the grain volatiles (Dowdy et al. 1993). This second explanation supports the windborne origin hypothesis.

Low temperatures may explain why fewer insects were caught in 1992 and 1993 than 1991 (Table 1, Fig. 4) No insects were caught when the mean daily maximum temperature for the week was below 18°C. There was an exponential increase in the number of *R. dominica* caught as the temperature increased (Fig. 4). It is doubtful that the change in pheromone delivery systems in 1992 would account for the reduction in *R. dominica* caught. We showed earlier that there was no difference in insects caught using the two systems (Fields et al. 1993).

PCR fingerprinting

Twenty populations of *R. dominica* from six geographic regions in North America are currently being studied, and data from eight populations are reported here. Ten different random primers were studied for utility in providing markers. To date, two gave consistently interpretable banding patterns

Table 1. The total number of *Rhyzopertha dominica* caught between late May and late August in pheromone baited flight traps outside various storage facilities in Manitoba Canada (n=3 per site type, mean ± standard error of the mean), and the monthly mean temperature for June, July and August from Glenlea, Manitoba.

| Year | Site type[a] | | | | |
	Farms	Elevators	Feed mills	Totals[b]	Temperature (°C)
1990	458 ± 220	159 ± 68	229 ± 198	282 ± 100[ab]	19 ± 1
1991	403 ± 125	448 ± 237	521 ± 305	470 ± 119[a]	21 ± 1
1992	68 ± 41	94 ± 39	80 ± 77	80 ± 28[b]	15 ± 0.2
1993	42 ± 35	40 ± 21	14 ± 8	32 ± 13[b]	13 ± 1

[a]No significant differences between site types (ANOVA p=0.90).

[b]Significant differences between years are indicated by different letters (ANOVA p=0.006, Duncans MRT p<0.05).

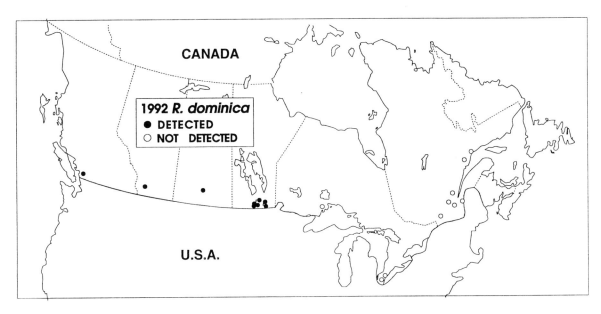

Fig. 1. The places that *Rhyzopertha dominica* were caught in pheromone-baited fight traps placed outside grain storage facilities in 1992: closed symbols had insects, open symbols did not.

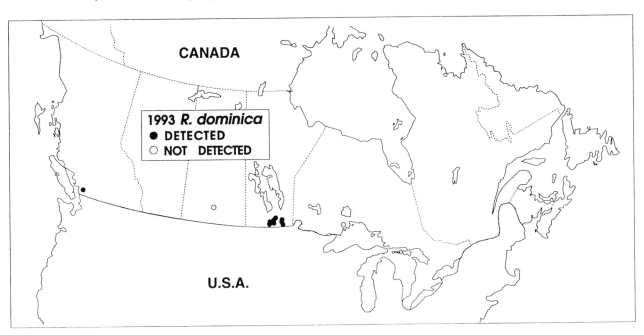

Fig. 2. The places that *Rhyzopertha dominica* were caught in pheromone-baited fight traps placed outside grain storage facilities in 1993: closed symbols had insects, open symbols did not.

(Fig. 5) and revealed polymorphism within and among populations. These two primers had the following sequences (5' to 3'): no. 160, CGATTCAGAG; no. 192, GCAAGTCACT.

Of the RAPD products observed following amplification with different primers, none were unique to any one *R. dominica* population and all occurred in each populatiosome level (Fig. 6). The only exception to this being the absence of 192/400 in the Coaldale Alberta population. Statistical analysis of the data is precluded by small sample sizes, but some relationships can be inferred at this time. Populations that differ substantially in band frequency are less closely related than those with similar band frequencies, and probably have not experienced gene flow recently. Populations fixed for presence or absence of a band possess less genetic diversity than those polymorphic at that band site (Hadrys et al. 1992). Thus our sample from Coaldale, Alberta is fixed for the presence of 160/700 and the absence of 192/400, which suggests that a genetic bottleneck occurred with the founding

of this population by a few genetically similar individuals. If such a founder event was not followed by subsequent immigration of genetically variant individuals, then genetic diversity would be expected to remain low (Wright 1978). A recent founder event at the Coaldale site would be consistent with its status as the most remotely collected (far from commercial centres and established populations) *R. dominica* population in this study. More data on the frequency of genetic markers in *R. dominica* populations is needed to clarify genetic relationships among populations and discuss models of range expansion.

In summary, none of the three hypotheses concerning the origins of *R. dominica* were strongly supported or discounted. It is likely that grain imported from the USA contains some *R. dominica*, but we did not trap more *R. dominica* outside feed mills than other grain handling facilities. This strongly suggests that imported grain is not the sole source of insects. The windborne hypothesis is supported by the intensive

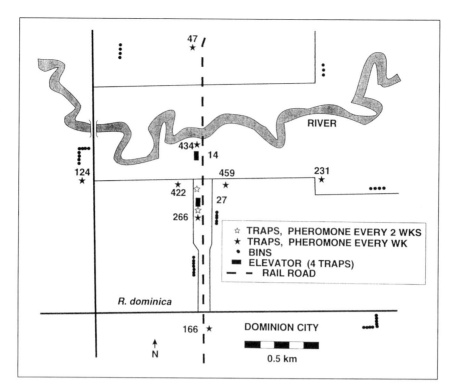

Fig. 3. The total number of *Rhyzopertha dominica* caught from 8 August to 10 October 1990 using pheromone-baited flight traps. There were four traps placed inside each of the two elevators.

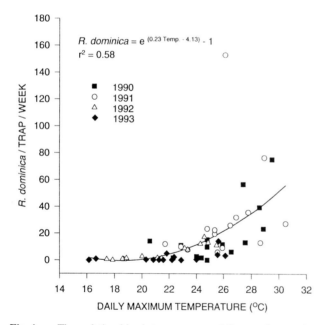

Fig. 4. The relationship between mean daily maximum air temperature at Glenlea Manitoba and the mean number of *Rhyzopertha dominica* caught in pheromone-baited flight traps located at farms, elevators and feed mills in Manitoba over one week.

that it cannot survive the winter in Manitoba. Despite large numbers of *R. dominica* caught in flight traps in western Canada over several years, this insect is still rarely found in stored grain.

Acknowledgments

We would like to thank Zlatko Korunic and Noel White for reviewing this manuscript and Richard Ffrench-Constant for instruction on PCR. We thank Kate Lieber for PCR work and Roy Jenkins for technical assistance. We greatly appreciate the field collection of beetles by David Weaver, Robert Cogburn, Barry Dover and Mike Dolinski.

References

Back, E.A. and Cotton, R.T. 1922. Stored-grain pests. USDA Bulletin, 1260, Washington, 46 pp.

Cogburn, R.R., Burkholder, W.E. and Williams, H.J. 1984. Field tests with the aggregation pheromone of the lesser grain borer (Coleoptera: Bostrichidae). Environmental Entomology, 13, 162–166.

Dowdy, A.K., Howard, R.W., Seitz, L.M. and McGaughey, W.H. 1993. Response of *Rhyzopertha dominica* (Coleoptera: Bostrichidae) to its aggregation pheromone and wheat volatiles. Environmental Entomology, 22, 965–970.

Fields, P.G. 1992. The control of stored-product insects and mites with extreme temperatures. Journal of Stored Product Research, 28, 89–118.

Fields, P.G., Van Loon, J., Dolinski, M.G., Harris, J.L. and Burkholder, W.E. 1993. The distribution of *Rhyzopertha dominica* (F.) in western Canada. Canadian Entomologist, 125, 317–328.

Hadrys, H., Ballick, M. and Schierwater, B. 1992. Applications of random amplified polymorphic DNA (RAPD) in molecular ecology. Molecular Ecology, 1, 55–63.

Leconte, J.L. 1862. Classification of the Coleoptera of North America. 1. Smithsonian Miscellaneous Collection, 3, 207–208.

sampling around elevators (Fig. 3). However, this evidence is circumstantial and further work would be required to prove it. Using RAPD-PCR, *R. dominica* from Canada share many genes with populations in the USA, suggesting the gene flow and migration are widespread. These data support the migration from the USA theory, though it does not distinguish the mode of transport. It suggests there are not locally established populations. Finally, preliminary experiments, not reported here, on the overwintering of *R. dominica* indicate

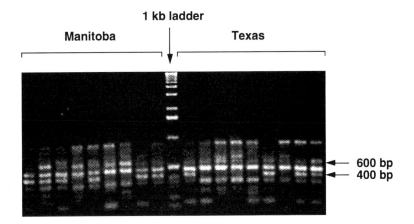

Primer no. 192, 5'——► 3' GCAAGTCACT

Fig. 5   A typical agarose gel showing RAPD products of *Rhyzopertha dominica* from two locations. The band at 500 base pairs is common to all individuals, while bands at 600 and 400 base pairs are polymorphic.

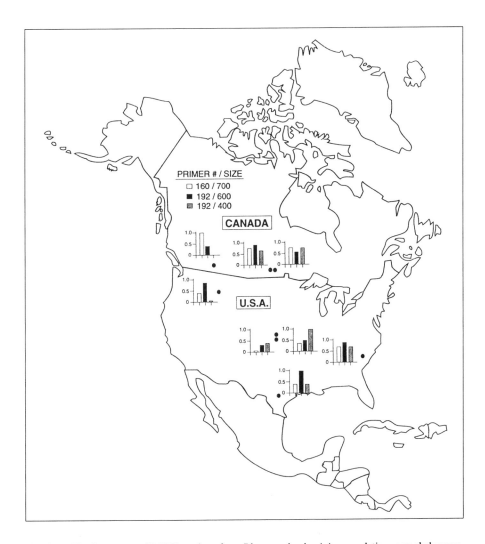

Fig. 6   The frequency of RAPD markers from *Rhyzopertha dominica* populations sampled across North America (n=8–11 per location).

Leos–Martinez, J., Granovsky, T.A., Williams, H.J., Vinson, S.B. and Burkholder, W.E. 1986. Estimation of aerial density of the lesser grain borer (Coleoptera: Bostrichidae) in a warehouse using dominicalure traps. Journal of Economic Entomology, 79, 1134–1138.

Leos-Martinez, J., Granovsky, T.A., Williams, H.J., Vinson, S.B. and Burkholder, W.E. 1987. Pheromonal trapping methods for lesser grain borer, *Rhyzopertha dominica* (Coleoptera: Bostrichidae). Environmental Entomology, 16, 747–751.

Lindgren, B.S. 1983. A multiple funnel trap for scolytid beetles (Coleoptera). Canadian Entomologist, 115, 299–302.

Liscombe, E.A.R. and Watters, F.L. 1962. Insect and mite infestations in empty granaries in the Prairie Provinces. Canadian Entomologist, 94, 433–440.

Loschiavo, S.R. 1975. Field tests of devices to detect insects in different types of grain storages. Canadian Entomologist, 107, 385–390.

Madrid, F.J., White, N.D.G. and Loschiavo, S.R. 1990. Insects in stored cereals, and their association with farming practices in southern Manitoba. Canadian Entomologist, 122, 515–523.

Potter, C. 1935. The biology and distribution of *Rhyzopertha dominica* (Fab.). Transcripts of the Royal Entomological Society of London, 83, 449–482.

Smith, L.B. and Barker, P.S. 1987. Distribution of insects found in granary residues in the Canadian prairies. Canadian Entomologist, 119, 873–880.

Smith, M.A.H. and MacKay, P.A. 1989. Seasonal variation in the photoperiodic responses of a pea aphid population: evidence for long-distance movements between populations. Oecologia, 81, 160–165.

Storey, C.L., Sauer, D.B., Ecker, O. and Fulk, D.W. 1982. Insect infestations in wheat and corn exported from the United States. Journal of Economic Entomology, 75, 827–832.

Storey, C.L., Sauer, D.B., Ecker, O. and Fulk, D.W. 1983. Insect populations in wheat, corn, and oats stored on the farm. Journal of Economic Entomology, 76, 1323–1330.

Subramanyam, B.H. and Harein, P.K. 1989. Insects infesting barley stored on farms in Minnesota. Journal of Economic Entomology, 82, 1817–1824.

Williams, H.J., Silverstein, R.M., Burkholder, W.E., and Khorramshahi, A. 1981. Dominicalure 1 and 2: Components of aggregation pheromone from male lesser grain borer *Rhyzopertha dominica* (F.) (Coleoptera: Bostrichidae). Journal of Chemical Ecology, 7, 759–780.

Williams, J.G.K., Kubelik, A.R., Livak, K.J., Rafalski, J.A. and Tingey, S.V. 1990. DNA polymorphisms amplified by arbitrary primers are useful as genetic markers. Nucleic Acid Research, 18, 6531–6535.

Wright, S. 1978. Evolution and the Genetics of Populations. Vol. 4. Variation Within and Among Natural Populations. Chicago, University of Chicago Press. 580 pp.

Some stored-product insects of increasing importance in China

Guan Lianghua, Chen Lanfen, Xie Gengfa and Yang Shaojun*

Abstract

Insect pests infesting the medicinal fungus *Ganoderma lucidum*, other edible fungi and tamarinds have become economically important with the greater use of these commodities that has occurred in recent years. This paper outlines the morphology and economic importance of the four insects involved, *Cis mikagensis*, *Ennearthron* sp., *Octotemnus* (s. str.) *parvulus* (Coleoptera: Ciidae) and the tamarind weevil, *Sitophilus linearis* (Coleoptera: Curculionidae).

Introduction

Edible fungi are becoming a very important commodity in China. In 1993, the production of mushrooms was over 110000 t in Hunan province alone. Fresh mushrooms are normally 2–4 times as valuable as wheat, and the medicinal fungus, *Ganoderma lucidum*, is worth much more than that. While beetles of the family Ciidae (Coleoptera) have not been treated as serious stored-product pests previously, the situation is changing. Here we discuss 3 species that have been found in edible fungi in China.

Tamarind (*Tamarindus indica*) is used as food in Burma and Thailand and also used to make beverages. Imports of tamarind into China are increasing annually, and large numbers of the tamarind weevil, *Sitophilus linearis* (Herbst), have been found on import inspection.

Details of the Insects

Cis mikagensis Nebachi et Wada [Ciidae]

Morphology

Length 1.72–2.02 mm, breadth 0.58–0.70 mm. Body cylindrical, dorsum convex, brown to dark brown, with vesture of short, erect light brown hairs. Head and legs with colour darker than antennae and palpus. Antennae 10 segmented: segment 1, 2 short and stout, segments 3,4 long and slender, segment 1 plus 2 about equal to 3,4 in length, three terminal segments clavate and loosely articulated. Anterior margin of head in male with 2 pairs of dentate processes. Pronotum broader than its length, widest at base, narrowing gradually from the base to the anterior margin with 2 angulate processes produced forward. Tarsal formula 4-4-4. Tarsal segments 1,2,3 of hind leg about half the length of segment 4.

---

* Tianjin Animal and Plant Quarantine Bureau, Tianjin 300457, China.

Economic value

Cis mikagensis is often found on export Ling Zhi mushroom (*Ganoderma lucidum*). It is also reported that the insect infests Fu-Lin (*Poria cocos*) and Zhu Lin (*Polyporus umbellatus*). Ling Zhi mushroom is a raw material of traditional Chinese medicine. More than 45 t of Ling Zhi mushroom is exported to Korea and the United States with a value of about $4500 per t in 1993. The adults and larvae of this insect bore in the mushroom, rendering it powdery, honeycombed and useless, resulting in severe economic loss.

Distribution

China (Provinces of Sichuan, Henan, San'xi, Sanxi, Liaoning, Guangdong and Hunan, as well as Tianjin city.

Ennearthron sp. [Ciidae]

Morphology

Length 1.6–1.9 mm, breadth 0.65–0.72 mm. Body cylindrical, dark brown, shiny, vestiture of short, erect, yellowish-brown hairs. Anterior margin of head with 2 angulate processes produced forwards. Antennae 9-segmented, three terminal segments clavate, loosely articulated, the third slightly shorter than the second, segments 5,6 smallest and length about as long as breadth, moniliform. Elytra parallel-sided, apex oval.

Economic value

Only a few specimens were found in an inspection.

Distribution

China (Provinces of Tianjin, San'xi, Hunan and Guizhou, as well as Tianjin city).

Octotemnus (s.str.) *parvulus* Mutsuo Miyatake [Ciidae]

Morphology

Head slightly convex in both sexes, finely and somewhat feebly reticulate, rather closely punctate, the punctures minute and weak, and somewhat densely setose, the setae more distinct and longer than those of *omogensis* and *glabriculus*, anterior margin slightly arcuate, reflex and narrowly ridges from eye to clypeus on either side. Antennae with the third segment slender, subclavate, longer than twice as long as broad, 4 slightly shorter than one-half as long as 3, a little broader than long or as long as broad. Pronotum slightly but distinctly broader than long, almost evenly arcuate on sides, very slightly narrowing from base to apex; fore margin simply rounded in both sexes; side margins invisible from above almost throughout their lengths, in lateral view rather strongly arcuate, fore and hind corners broadly rounded, not at all angulate; hind margin very slightly arcuate; dorsum more coarsely and densely reticulate than in *glabriculus* and less than in *omogensis*, and very sparsely with slender obscure setae, each arising from a puncture. Scutellum triangular, distinctly broader than long, with a few strong punctures. Elytra very slightly broader than pronotum at base, about 1.8 to 1.9 times as long as pronotum, about 1.4 times as long as broad, very slightly widened from base to basal half on sides, thence broadly roundly narrowing to apex; side margins invisible from above almost throughout their lengths; dorsum closely and irregularly punctate, the punctures uneven in size, somewhat weaker and smaller than those on the pronotum, appearing to be somewhat confluent in some places on basal part, and comparatively densely setose on apical inclined surface, the setae somewhat stiff, longer than those of *glabriculus* and *omogensis*. Anterior tibiae not strongly expanded but almost straight on outer edge, armed with 7–10 spine-like teeth on apical half of outer edge, the apical 3 or 4 of them much larger than the others. Body beneath roughly and partially feebly reticulate, sparsely and inconspicuously punctate setose, but somewhat densely and conspicuously on venter; the first ventrite with a conspicuous bill-shaped process at middle in male.

Economic value

Only 2 specimens were found in an inspection. In Japan it was found from *Coriolos versicolor* (L. ex Fr.) Quelet.

Distribution

Japan, China.

Sitophilus linearis (Herbst) [Curculionidae]

Morphology

Body cylindrical, length 4–4.5 mm from base of snout to end of abdomen, reddish brown to dark brown, slightly shiny. Head and thorax with darker colour than elytra. Ventral surface of abdomen with darker colour than surface of head and ventral surface of thorax. Snout produced forward, shorter and stouter, length to breadth 3:1, snout one half as long as prothorax, the base of the snout slightly expanded, not global, snout in female longer than in male, snout punctures parallel. Antennae 8-segmented; antennal fovea connected with compound eye compared to separated from compound eye by a line of punctures in *S. oryzae*, *S. zeamais* and *S. granarius*. The anterior of the pronotum more narrow than the posterior of the pronotum, densely punctulate. Scutellum dark brown, subround, low at middle. Tarsal formula 5–5–5, 4 concealed by 3. Elytra subparallel-sided, 10 lines on each elytron and irregularly maculated with yellow. Hind wings well developed, membranous, about 3 times as long as elytra. Aedeagus wide and short, lateral margins curved up, epiphallus subequilateral triangle with blunt and rounded ends; lobatus of y-sclerite in female very short, apex wide and blunt.

Keys to distinguish the 4 species of stored product *Sitophilus* can be found in Whitehead (1991) and Haines (1991).

Economic value

S. linearis was given the common name of tamarind weevil by Zimmerman (1968). Details of the biology can be found in Cotton (1920) and Usman (1953).

Tamarind is imported into China from Thailand and Burma through the ports of Kunmin and Reili. Damage by *S. linearis* can be very high; sometimes 4–5 insects were found in a single kernel. In one 1000g sample inspection, adults of *S. linearis* had damaged 100% of the kernels.

Distribution

India, Burma, Thailand, Africa, Seychelles, Hawaii, North and South America.

References

Cotton, R.T. 1920.Tamarind pod-borer, *Sitophilus linearis* (Herbst). Journal of Agricultural Research 20, 439–446.

Haines, C.P. 1991. Insects and arachnids of tropical stored products: their biology and identification. Natural Resources Institute, Chatham, U.K. 246 pp.

Usman, S. 1953. Bionomics and control of the tamarind seed borer, *Sitophilus linearis* (Herbst). Indian Journal of Entomology, 15, 147–156.

Whitehead, D.R. 1991. Weevils (Curculionidae, Coleoptera). In: Gorham, J. R., ed., Insect and mite pests in food: an illustrated key, U.S. Department of Agriculture, Agriculture Handbook Number 655, 223–230.

Zimmerman, E. C. 1968. Rhynchophorinae of southeastern Polynesia. Pacific Insects, 10, 44–47.

Factors affecting survival and development of *Sitophilus oryzae* (L.) in rice grain pericarp layers

Y. Haryadi[*] and F. Fleurat-Lessard[†]

Abstract

The role of rice grain pericarp layers on the development of *Sitophilus oryzae* (L.) was studied in laboratory conditions on eight rice varieties. Three types of grain were used for each variety: brown rice (dehusked paddy), brown rice kernel without embryo and polished rice. Egg-laying rate by the adults of *S. oryzae*, developmental time and F_1 production were checked either on grain or on artificial kernels reconstituted with the flour of the different types of grain. Incubation was performed at 25°C and 75% r.h.

It was found that the F_1 progeny production was the highest on brown rice for all varieties, and the lowest on polished rice. Developmental time was the shortest with brown rice. Based on radiographic control of hidden stages of development, it was observed that the highest rate of egg deposition occurred on brown rice and that the development of the first larval instar is greatly affected in polished rice kernels. On artificial kernels, it was shown that the pericarp layers are a nutritional 'growth factor' for *S. oryzae*.

Introduction

Rice is the most important staple food in Indonesia. During storage and distribution, it is often attacked by various storage insects. Among them is the rice weevil, *Sitophilus oryzae* (L.). The development of the larva inside the kernel may cause the loss of weight of about one-third of the kernel (Clement et al. 1988). In addition to the weight loss, various types of quality loss may also occur.

The results of studies related to the development of *S. oryzae* on rice kernels following different degrees of milling have been reported. Singh (1981), for example, found that the developmental period of the rice weevil on polished rice was significantly longer. At the same time, the productivity was also reduced. Similar results were also reported earlier by McGaughey (1974). From those studies, it was clear that the removal of the rice pericarp layers during milling may affect the development of rice weevil. However, the phenomenon was not clearly explained.

The present study was conducted to confirm those previous findings, and to gain further information on the development of rice weevil. In addition to the experiment on rice kernels, an experiment on oligidic media reconstituted as artificial kernels was also conducted.

Materials

Rice grains

The study used eight varieties of Indonesian rice: IR-42, Cisadane, Sirendah, Hawara-batu, Bah-butong, Pandanwangi, Sukadane, and Sungai-rampah. Before the experiment, all samples were subjected to freezing temperature of –20°C in airtight plastic bags for more than one week to kill any insect present. The rice were then dehusked using laboratory husker to obtain whole brown rice.

For each variety, three types of kernel were prepared, namely whole brown rice, brown rice without embryo, and polished rice. Larvae of *Plodia interpunctella* were used to remove rice embryo.

Artificial grains

The artificial grains were prepared as follows. To a 10 g of solid material (9 g of polished rice flour and 1 g of gluten), 1 mL of glycerol were added. By adding water, little by little, a dough was obtained. From the dough small artificial grains were prepared manually. The artificial grains were then dried at 25°C and 50–60% r.h. overnight so that the moisture content was reduced to around 14–15 %. This was the base formula of the oligidic diet in this experiment. Other components may be substituted for the polished rice flour. In the present experiment, substitutes were rice bran flour at 0%, 10%, and 20%. In addition, artificial grain was also prepared from whole brown rice flour.

The test insect, *Sitophilus oryzae* (L.)

S. oryzae were obtained from laboratory culture maintained on soft wheat of Festival variety at 25°C and 70% r.h.. The 14 + 7 days old adult weevils were removed and given opportunity to mate by placing them on a thin layer of ground maize in closed plastic container in the dark. After a time mated couples were collected.

Methods

First experiment

Three replicates of 300 grains consisting of 100 grains of each of the three grain types were placed in plastic boxes (8.5 × 6.0 × 2.5 cm) provided with aerated lids. Ten pairs of *S. oryzae* were released and the females were allowed to oviposit for 7 days in a climate chamber. The parent weevils were then removed from the box and discarded.

The 300 grains were then separated according to each type and placed in three new boxes; for brown rice, brown rice without embryo, and polished rice. The boxes were returned to the chamber.

Two weeks later the grains in the boxes were radiographied according to the method of Fleurat-Lessard (1982) using

---

[*] Inter University Center for Food and Nutrition, Bogor Agricultural University, P.O. Box 122, Bogor 16001, Indonesia.
[†] Laboratory of Stored-Products Insects, National Institute for Agricultural Research (INRA), P.O. Box 81, 33883 Villenave d'Ornon Cedex, France.

SIGMA 2060 X-ray unit at 20 kV and 6 mA for 30 seconds on Kodak Ready Pack M film. Hidden stages were observed and counted under binocular microscope.

Three weeks after the removal of the parent weevils, the emergence of the F_1 progenies was observed daily. The counting in each box was discontinued when there was no emergence for five consecutive days. The developmental period was calculated in days as the time between mid-oviposition period and the time when 50% of the F_1 progenies had emerged.

Second experiment

Three replicates of 100 artificial grains were placed in plastic box as in the first experiment. Ten pairs of *S. oryzae* were released and the females were allowed to oviposit for 7 days in chamber. The parent weevils were then removed from the box and discarded. The boxes were returned to the chamber until the emergence of all F_1 progenies. The observation of the F_1 progenies and the calculation of the developmental period was the same as in the first experiment.

Results

First experiment

Higher numbers of progeny were produced in whole brown rice than in brown rice without embryo, in all varieties except in IR-42. The number of progeny produced was significantly lower in polished rice. On Bah-butong variety no progeny emerged in all three boxes (Table 1). The results of the radiography show that the number of the hidden stages was highest in brown rice. In polished rice, the number of hidden stages was much lower (Table 2).

The absence of all or part of the pericarp layers, including the embryo, affected the developmental period of the weevils. For example, the mean developmental period of the weevils in brown rice without embryo was 36.24 days, which was almost one day longer than that in brown rice (Table 3). In polished rice, the mean developmental period was more than one week longer than that in brown rice.

Second experiment

The mean number of progeny of *S. oryzae* in artificial grains prepared from whole brown rice flour was 141.7. The developmental period was 39.50 days. The number of progeny was significantly reduced, to 61.3, in artificial grain prepared from polished rice flour. The developmental period was also prolonged, by almost four days, to 43.43 days.

Discussion

This study indicates that the role of pericarp layers, including the embryo, is very important to support the development of the rice weevil. During the oviposition period of 7 days, the female weevils were given free choice in which grain they would deposit their eggs. The data on the number of progeny produced in whole brown rice without embryo and polished rice suggested that the absence of the pericarp layers is detrimental to weevil development. The radiographic control suggests that the female weevils prefer whole brown rice kernels for oviposition. The results of the present study agree with the findings of McGaughey (1974) and Singh (1981). It is believed that for the development of first instar larva the pericarp layers are also indispensable.

The data obtained from the second experiment suggest that the rice grain pericarp layers are indeed important. When the artificial diet contains 10% rice bran flour, the number of progeny and the developmental period were immediately restored to better than that in artificial grains prepared from whole brown rice flour, which is considered as an ideal substrate for the optimum development of rice weevil. Rice bran, which constitutes the pericarp layers and the embryo,

Table 2. Developmental period (in days) of *S. oryzae* in three types of grains of eight varieties of rice.

Variety	Brown rice	Brown rice without embryo	Polished rice
IR-42	35.50	36.17	43.00
Cisadane	35.35	35.71	46.50
Sirendah	35.22	35.88	42.42
Hawara-batu	35.33	36.20	44.50
Bah-butong	35.50	35.63	-a
Pandan-wangi	35.46	36.45	47.89
Sukadane	35.95	36.90	49.50
Sungai-rampah	36.38	36.38	43.00
Means [b]	35.69 b	36.24 a	45.25

[a] No emergence of progeny observed
[b] Means followed by the same letter are not significantly different (Newman-Keuls test, p = 0.05)

Table 1. Number of progeny of *S. oryzae* in three types of grains of eight varieties of rice.

Variety	Brown rice	Brown rice without embryo	Polished rice
IR-42	69.3	75.3	5.3
Cisadane	94.3	83.6	1.7
Sirendah	93.7	76.7	16.0
Hawara-batu	98.3	83.7	8.0
Bah-butong	89.3	74.7	0
Pandan-wangi	94.7	78.3	4.7
Sukadane	80.0	74.7	0.7
Sungai-rampah	84.7	65.0	2.0
Means	88.0 a	76.5 b	4.8 c

Means followed by the same letter are not significantly different (Newman-Keuls test, p = 0.05)

Table 3. Number of hidden stages of *S. oryzae* in three types of grains of eight varieties of rice.

Variety	Brown rice	Brown rice without embryo	Polished rice
IR-42	82.0	81.3	9.0
Cisadane	107.0	92.7	6.7
Sirendah	101.0	79.3	20.0
Hawara-batu	108.7	97.0	15.3
Bah-butong	97.3	86.3	3.7
Pandan-wangi	107.3	89.0	7.7
Sukadane	88.0	80.0	5.7
Sungai-rampah	96.0	78.5	4.3
Means	98.4 a	85.5 b	9.1 c

Means followed by the same letter are not significantly different (Newman-Keuls test, p = 0.05)

Table 4. The effect of rice bran on the development of *S. oryzae* in artificial diet

Quantity of rice bran in the diet solid materials	Number of progeny	Developmental period
0%	61.3 b	43.43 a
10%	145.3 a	38.29 c
20%	149.3 a	39.58 b
Diet prepared from whole brown rice	141.7 a	39.50 b

Means followed by the same letter are not significantly different
(Newman-Keuls test, p = 0.05)

played an important role in the development of *S. oryzae*. Hence, the pericarp layers of rice grain contain a 'growth factor' for the rice weevils *S. oryzae*.

Acknowledgment

The authors gratefully acknowledge the members of the Laboratory of Stored-product Insect, INRA Bordeaux, for various aides rendered to us during the course of the study.

References

Clement, G. Dallar, J. Poisson, C. and Sauphanor, B. 1988. Les facteurs de resistance du riz paddy aux insectes des stocks. I. Influence de l'ouverture des glumelles. L'Agro. Tropicale, 43, 47-58.

Fleurat-Lessard, F. 1982. Mesure de l'infestation par les insectes. In: Multon, J.L., ed. Conservation et stockage des grains and graines et produits derives, Vol. I. Paris, Lavoisier et APRIA, 520-541.

Mc Gaughey, W.H. 1974. Insect development in milled rice : Effect of variety, degree of milling, parboiling and broken kernels. Journal of Stored Products Protection, 10, 81-88.

Singh, K. 1981. Influence of milled rice on insect infestation. II. Developmental period and productivity of *Sitophilus oryzae* Linn. and *Tribolium castaneum* Herbst in milled rice. Zeitschrift angewande Entomologie, 92, 472-477.

Molecular and morphological markers for diagnosis of *Sitophilus oryzae* and *S. zeamais* (Coleoptera: Curculionidae)

P. Hidayat[*], R.H. ffrench-Constant[*], and T.W. Phillips[*†]

Abstract

Sitophilus oryzae and *Sitophilus zeamais* are considered taxonomically distinct based on subtle differences in genital morphology and presumed differences in body size and grain preferences. Previous findings of successful laboratory hybridisation, genetic similarity of allozymes and chromosomes, and identity of aggregation pheromones raised questions about the validity of *S. oryzae* and *S. zeamais* as valid biological species. We used molecular techniques to test the hypothesis that individuals assigned as *S. oryzae* or *S. zeamais* by morphological criteria represent members of two distinct gene pools, and hence are reproductively isolated species. Weevils were studied from four farms in the U.S. and two recent laboratory cultures from Asia. Putative species were scored on the presence/absence of grooves on the male aedeagus or pointed/rounded lobes of female spiculum ventrale. The polymerase chain reaction was used on the same specimens to analyse randomly amplified polymorphic DNA (RAPD) markers and to selectively amplify regions of mitochondrial DNA for analysis of restriction site polymorphisms with restriction endonucleases. Both methods yielded diagnostic markers that distinguished the two morphotypes, consistent with the presence of reproductively isolated species in sympatry. Other morphological characters (e.g. pronotal punctures) proved unreliable as correlates with genetic markers.

Introduction

Sitophilus oryzae (L.), the rice weevil, and *S. zeamais* Motschulsky, the maize weevil, are very closely related species that are difficult to distinguish. Floyd and Newsom (1959) recognised two distinct species based on size, food preferences, mating incompatibility, and morphology of genitalia. They synonymised *zeamais* under 'oryza' (not *oryzae*) for the large form, and retained *S. sasakii* (Takahashi) for the smaller form (*S. oryzae sensu stricto*). Kuschel (1978) later confirmed the name of *S. zeamais* for the large form after examining Motschulsky's actual types, and reaffirmed the utility of the male genitalia in distinguishing species. Research in Japan recognises *S. oryzae* and *S. zeamais* as distinct species (Kiritani 1965). Halstead (1963) reviewed proposed morphological characters for distinguishing *S. oryzae* and *S. zeamais* (Kuschel 1961), and concluded that only one gave consistent separations: the male aedeagus of *S. zeamais* possesses two longitudinal grooves on the dorsum of the phallus, while the same surface on *oryzae* males is smooth and evenly convex. Floyd and Newsom (1959) reported characters of the female

genitalia that they proposed should be useful for species distinction. Halstead (1963) found the female character highly variable except for one consistency: the apical regions on the lateral lobes of the Y-shaped spiculum ventrale are pointed and tapered in *S. zeamais*, but rounded and blunt in *S. oryzae*. Halstead (1963) assigned identifications to *Sitophilus* specimens in the worldwide collections of his institution and of the British Museum, but no independent measure was invoked with which to test the aedeagal characters against, nor was any mention made of the two morphotypes being found in the same collections. Boudreaux (1969) reported on external characters, most notably the presence (rice weevil) or absence (maize weevil) of a flat median longitudinal puncture-free zone on the dorsal pronotum, that purportedly separated the two species. Boudreaux (1969) developed his study of characters from a single laboratory culture each of *S. zeamais* and *S. oryzae*, but he used his method to discriminate nearly 1800 specimens in the USNM and noted that the two species may occur together in the same accessions.

We questioned the validity of *S. oryzae* and *S. zeamais* as reproductively isolated species based on published reports on controlled mating, genetics, and pheromones. Richards (1944) produced F_1 progeny upon crossing the two species in the laboratory. Smith (1953) examined chromosomes of *Sitophilus* and found that *S. granarius* had a distinct karyotype of 2N=24, while *S. oryzae* and *S. zeamais* had apparently identical karyotypes of 2N=22. Yang et al. (1989) confirmed the 2N=22 karyotype for *S. zeamais* and *S. oryzae*, and also reported interspecific crosses and production of fertile progeny in the laboratory, but only in 3% of the crosses. Beiras and Petitpierre (1981) analysed six allozyme loci in three species of *Sitophilus* and found no diagnostic (e.g., fixed allele) differences between *S. oryzae* and *S. zeamais*. Populations were monomorphic for the same electromorphs at all loci except for esterases, for which there were polymorphisms and frequency differences. All samples analysed by Beiras and Petitpierre (1981) were from laboratory cultures of various origins and none were field-collected in sympatry. Pintureau et al. (1991) studied esterase polymorphism in *S. oryzae*, *S. zeamais*, and *S. granarius*, and hybridised *S. oryzae* and *S. zeamais* (in about 10% of crosses) from broadly different geographic strains. No diagnostic differences were found between the sibling species (*oryzae* and *zeamais*) and interspecific hybrids demonstrated heritability of electrophoretic alleles at one esterase locus. Thus cytological and allozymic studies reveal a very close genetic relationship between *S. oryzae* and *S. zeamais*, and controlled mating studies suggest the potential for hybridisation in the field. Walgenbach et al. (1983) reported that *S. oryzae* and *S. zeamais* use a male-produced pheromone and that the two species were mutually cross-attracted in laboratory bioassays. Chemical elucidation found that *S. oryzae* and *S. zeamais* produced chemically identical pheromones, thus explaining their high level of cross-attraction (reviewed in Burkholder 1990). Use of the same pheromone by *S. oryzae* and *S. zeamais* suggests that there may be no behavioral barrier to individuals of the two species coming together in nature.

[*] Department of Entomology, University of Wisconsin, Madison, WI USA 53706.
[†] USDA ARS Stored-Product Insect Research Unit., Madison, WI USA 53706.

The objective of this study is to use molecular markers to determine if weevils identified as either *S. oryzae* or *S. zeamais* based on morphology are genetically distinct. We have centred our study on morphologically distinct specimens collected in sympatry so that the potential for hybridisation would be high. Throne and Cline (1989 1991) documented the seasonal flight activity and occurrence of *S. oryzae* and *S. zeamais* at the same field sites in South Carolina, and made identifications of trapped specimens using Halstead's (1963) genitalic characters. Therefore, weevils possessing morphological characters representative of each species can be collected sympatrically, making tests of the species concept possible via genetic markers.

Materials and Methods

A total of 268 specimens were field collected by us or by cooperators from the following locations: Wuhan, China; W. Java, Indonesia; Frankfort, Kentucky; Bamberg, South Carolina; and Madison, Wisconsin. All U.S. collections were made from in-grain samples or grain/bait bags (Throne and Cline 1991) on farms with grain storage facilities. Collections from Indonesia and China are from unknown habitats, but apparently represented wild populations. Weevils were received in our laboratory alive, subjected to morphological analysis, dissection of genitalia, and then were frozen at –80°C for subsequent DNA extraction.

Morphology

We used the morphological characters described by Halstead (1963) and Boudreaux (1969) in our study of *Sitophilus* field populations. The two 'most reliable' charac-

ters presented by Halstead, the presence (*S. zeamais*) or absence (*S. oryzae*) of ridges on the dorsum of the male aedeagus, and the shape of the tips (rounded on *S. oryzae*, pointed on *S. zeamais*) on the Y-shaped sclerite (spiculum ventrale) were easily dissected and scored on our specimens (Figs. 1 and 2). The presence (*S. oryzae*) or absence (*S. zeamais*) of a median longitudinal puncture-free zone on the pronotum (Fig. 3), described by Boudreaux (1969), was scored, as was the number of punctures counted longitudinally along the midline (<20 for *S. oryzae* and >20 for *S. zeamais*). The shape of pronotal punctures (elliptical for *S. oryzae* and circular for *S. zeamais*) required more subjective judgment than other characters, but this was also recorded.

DNA markers

Two methods employing the polymerase chain reaction (PCR) were used in this study to obtain genetic markers. We used the technique of randomly amplified polymorphic DNA (RAPD-PCR; Williams et al. 1990) and screened numerous random primers (10-mers) from the University of British Columbia RAPD Primer Synthesis Project.

We additionally studied restriction fragment-length polymorphisms (RFLP-PCR) on the same individuals used for morphology and RAPD-PCR. In RFLP-PCR a specific sequence is amplified with PCR, digested with a specific restriction endonuclease, then visualised on agarose gel to score fragments and the presence or absence of restriction sites (see Simon et al. 1993). We amplified a mitochondrial DNA (mtDNA) sequence of 1635 bp, spanning the cytochrome oxidase subunits I and II (COI, COII) genes, using primers for conserved regions, 5'-CAACATTTATTTTGATTTTTTGG-3' and 5'-GAGACCATTACTTGCTTTCAGTCATCT-3' (Liu and Beckenbach 1992).

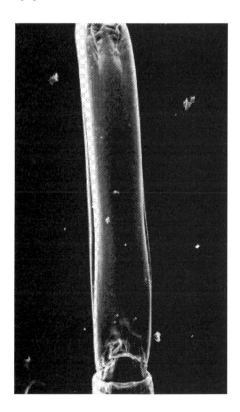 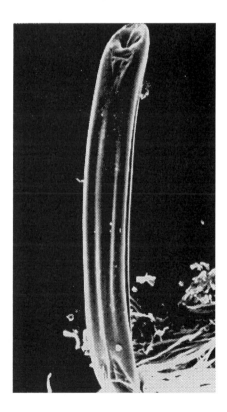

S. oryzae   S. zeamais

Fig. 1.   Scanning electron micrograph (SEM) of dorsal surface of phallus (male aedeagus) in *S. oryzae* (left) and *S. zeamais* (right); note longitudinal grooves on *S. zeamais*.

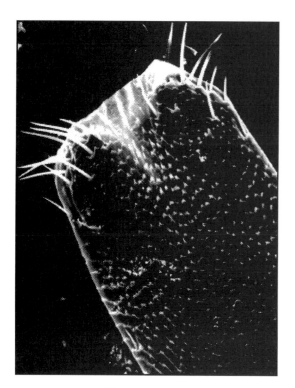
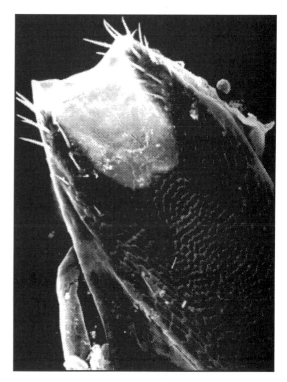

S. oryzae   S. zeamais

Fig. 2. Y-shaped sclerite of invaginated 8th sternite (spiculum ventrale) from female *S. oryzae* and *S. zeamais*; note shape and thickness of distal processes.

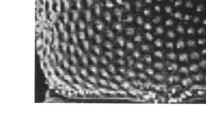

S. oryzae   S. zeamais

Fig. 3. Morphology of pronotal punctures used for diagnosis (Boudreaux 1969) of *S. oryzae* and *S. zeamais*; note puncture-free zone near mid-line of pronotum in *S. oryzae* and larger number of punctures in *S. zeamais*.

Results and Discussion

We found that most field collections contained a mixture of morphotypes. Additionally, we found that a determination to species made with one character did not always match a determination made with another character (Fig. 4). Thus morphological characters were not correlated with each other and specimens could not be classified with certainty.

RAPD-PCR revealed bands consistently present in one group of weevils and absent in another. When field-collected weevils from Bamberg, South Carolina were subjected to RAPD-PCR with the UBC431 primer, all individuals possessing genitalic morphology of *S. zeamais* (grooved dorsal aedeagus in males, pointed y-shaped sclerite of females) revealed a characteristic RAPD fingerprint, while those with the *S. oryzae* genitalic morphology (smooth aedeagus, rounded tips of y-shaped sclerite) had a different RAPD pattern (Fig. 5). All weevils with *S. oryzae* morphology consistently showed a band at about 1400 bp that was absent in weevils with S. *zeamais* morphology. Other bands predomi-

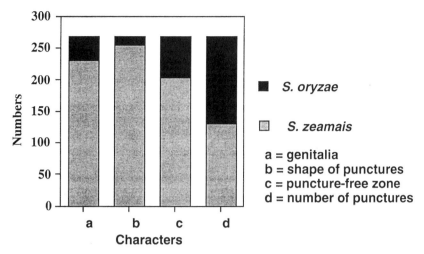

Fig. 4. Distribution of species determinations made on 268 specimens using four different morphological criteria: a = genitalia, b = shape of punctures, c = puncture-free zone, d = number of punctures. *Sitophilus oryzae* ▬▬; *Sitophilus zeamais* ▨▨.

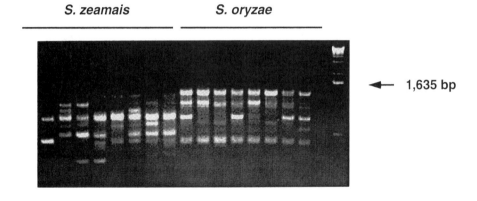

Fig. 5. Agarose gel with RAPD-PCR products from *Sitophilus* weevils with primer UBC-431. Individuals are grouped according to their possession of genital morphology (aedeagus of males, y-shaped sclerite of females) of *S. oryzae* or *S. zeamais*.

nated in one morphotype over the other, but the 1400 kb band was consistently associated with weevils having *S. zeamais* genital morphology at this South Carolina site and in all other collections surveyed in this study.

RFLP-PCR analyses of mtDNA in all specimens so far have yielded a 1635 bp product prior to digestion with restriction enzymes. Four- and five-base restriction endonucleases were surveyed for their ability to cleave this product. The enzymes *Msp* I, *Dde* I and *Rsa* I all yielded one or more restriction sites, and the distribution of these restriction sites was consistently correlated with one genitalic morph or the other (Fig. 6). For weevils collected in Bamberg, South Carolina, *Dde* I yielded two restriction sites (three fragments) in individuals possessing *S. zeamais* genital morphology, while DNA from all *S. oryzae* morphs lacked these restriction sites and were not digested (Fig. 6). Similarly, *Msp* I cleaved the COI/COII sequence consistently at the same site for *S. zeamais* morphs, but did not digest the DNA from *S. oryzae* morphs. Restriction analysis with *Rsa* I (not shown) yielded two different cut sites, one fixed for *S. oryzae* morphs and the other fixed for *S. zeamais* morphs.

Our results indicate that two genetically distinct, hence probably reproductively isolated, species occur in sympatry,

and that they can be assigned by genital morphology to either *S. oryzae* or *S. zeamais*. We have performed both RAPD-PCR and RFLP-PCR analyses on weevils from the five geographic locations listed above, each containing a mix of morphotypes, and genetic markers shown in Figure 5 have been consistently associated with the same genitalic morphs. If these sites harbored single breeding populations of weevils, and if crossing between morphotypes was occurring, we would expect to eventually see a random association of morphotypes with DNA markers. Since it is highly unlikely that both RAPD and mtDNA markers are linked with genes coding for sexually dimorphic genital characters, it is clear that these correlated characters signify the presence of two breeding populations. A single character, either morphological or genetic, would be inadequate for concluding the presence of two species because a plausible alternative hypothesis for polymorphism within one species could be made. Our mtDNA restriction site data represent just one character with two states (so far) because the mitochondrial genome is haploid and inherited intact with no recombination, acting as one locus (Avise et al. 1987). However the addition of at least one independent RAPD locus (band at 1400 kb, primer UBC431, Fig. 5) and the consistent association of these two molecular

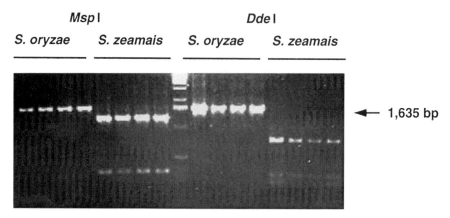

Fig. 6. Restriction digest of 1635 bp PCR product from mtDNA of *Sitophilus* weevils using enzymes *Msp*I and *Dde*I. Species designations refer to determinations from genital morphology. The undigested product is shown in all *S. oryzae* lanes.

markers with genital morphology make a strong case for two distinct lineages. We have confirmed Halstead's (1963) and Kuschel's (1978) conclusions of two species identified by genitalic characters, but show the lack of diagnostic utility for pronotal characters.

Acknowledgments

We greatly appreciate field collection of weevils by David Weaver, USDA ARS in Savannah, Georgia, and by John Sedlacek, Kentucky State University, Frankfort, Kentucky. We appreciate reviews of the manuscript by Nicola Anthony, University of Wisconsin, Madison, WI, and David Weaver, USDA ARS in Savannah, GA.

References

Avise, J.C., J. Arnold, R.M. Ball, E. Birmingham, T. Lamb, J. E. Neigel, C.A. Reeb and Saunders, N.C. 1987. Intraspecific phylogeography: the mitochondrial DNA Bridge between population genetics and systematics. Annual Review of Ecology and Systematics, 18, 489–522.

Beiras, M.J. and Petitpierre, E. 1981. Allozymic variability and genetic differentiation in three species of *Sitophilus* L. (Coleoptera: Curculionidae). Egyptian Journal of Genetics and Cytology, 10, 95–104.

Boudreaux, H.B. 1969. The identity of *Sitophilus oryzae*. Annals of the Entomological Society of America, 62, 169–172.

Burkholder, W. E. 1990. Practical use of pheromones and other attractants for stored-product insects. In: Ridgway, R.L., Silverstein, R.M. and Inscoe, M.N., ed. Behavior-modifying chemicals for insect management: applications of pheromones and other attractants. Marcel Dekker, Inc., N.Y., 497–516.

Floyd, E.H. and Newsom, L. D. 1959. Biological study of the rice weevil complex. Annals of the Entomological Society of America, 52, 687–695.

Halstead, D.G.H. 1963. The separation of *Sitophilus oryzae* (L.) and *S. zeamais* Motschulsky (Coleoptera: Curculionidae), with a summary of their distribution. Entomologist's Monthly Magasine, 99, 72–74.

Kiritani, K. 1965. Biological studies on the *Sitophilus* complex (Coleoptera: Curculionidae) in Japan. Journal of Stored Products Research, 1, 169–176.

Kuschel, G. 1961. On problems of synonymy in the *Sitophilus oryzae* complex (30th contribution, Col. Curculionidae). American Museum of Natural History. ser. 13, vol. iv, 441–444.

Kuschel, G. 1978. Notes on the identity of *Sitophilus zeamais* Motschulsky based on type material examination. Journal of Natural History, 12, 231.

Liu, H. and Beckenbach, A.T. 1992. Evolution of the mitochondrial cytochrome oxidase II gene among 10 orders of insects. Molecular Phylogenetics and Evolution, 1, 41–52.

Pintureau, B., Grenier, A.M. and Nardon, P. 1991. Polymorphism of esterases in three species of *Sitophilus* (Coleoptera: Curculionidae). Journal of Stored Products Research, 27, 141–151.

Richards, O.W. 1944. The two strains of the rice weevil, *Calandra oryzae* (L.) (Coleoptera: Curculionidae). Transactionas of the Royal Entomological Society, London, 94, 187–200.

Smith, S.G. 1953. The cytology of *Sitophilus* (*Calandra*) *oryzae* (L), *S. granarius* (L), and some other Rhynchophora (Coleoptera). Cytologia, 17, 50–70.

Simon, C., McIntosh, C. and Deniega, J. 1993. Standard restriction fragment length analysis is not sensitive enough for phylogenetic analysis of identification of 17-year peridoical cicada broods (Hemiptera: Cicadidae): the potential for a new technique. Annals of the Entomological Society of America, 86, 228–238.

Throne, J.E. and Cline, L. D. 1989. Seasonal flight activity of the maize weevil, *Sitophilus zeamais* Motschulsky (Coleoptera: Curuclionidae) and the rice weevil, *S. oryzae* (L.), in South Carolina. Journal of Agricultural Entomology, 6, 183–192.

Throne, J.E. and Cline, L.D. 1991. Seasonal abundance of maize and rice weevils (Coleoptera: Curculionidae) in South Carolina. Journal of Agricultural Entomology, 8, 93–100.

Walgenbach, C.A., Phillips, J. K., Faustini, D.L. and Burkholder, W.E.. 1983. Male produced aggregation pheromone of the maize weevil, *Sitophilus zeamais*, and interspecific attraction between three *Sitophilus* species. Journal of Chemical Ecology, 9, 831–841.

Williams, J.G.K., Kubelik, A.R., Livak, K. J., Rafalski, J.A. and Tingey, S. V. 1990. DNA polymorphisms amplified by arbitrary primers are useful as genetic markers. Nucleic Acid Research, 18, 6531–6535.

Yang, Z.-Y., Huang, P. and Wu, G.-X. 1989. Observations on the karyotypes of *Sitophilus oryzae* and *Sitophilus zeamais* and their hybrid offsprings. Acta Entomologica Sinica, 32, 4065–409.

Pheromone biology and factors affecting its production in *Tribolium castaneum*

A. Hussain[*], T.W. Phillips[†], T. J. Mayhew[†] and M.T. AliNiazee[§]

Abstract

Tribolium castaneum is a cosmopolitan pest of almost all processed as well as bulk commodities. Our objectives were to quantify the amount of pheromone, 4,8-dimethyldecanal, produced by individual males in 48-hour intervals, and to determine the effect of feeding on the production. Pheromone was collected through the aeration of adult males. The male flour beetles were aerated individually for 48 hours and the volatiles collected on the solid phase adsorbent Super Q. In the feeding study 0.5 grams of cracked wheat was provided initially to all beetles and removed after 4 days from half of the replications. GC-MS was used for quantification of the pheromone. Single male *Tribolium castaneum* produced detectable amounts of pheromone in 48 hours. The average production of pheromone was 1.265 µg ± 0.18 (SE) in 2 days. The total average production was 19.04± 0.52 µg. Feeding proved to be a major factor in pheromone production. The amount of pheromone dramatically decreased from 1.10 µg with food to 0.12, 0.082 and 0.10 µg in 2, 4, and 6 days of starvation, respectively. When food was reintroduced to the starving beetles the production of pheromone increased to normal and dropped again to 0.06 and 0.04 µg in 2 and 4 days of starvation, respectively.

Introduction

The red flour beetle *Tribolium castaneum* is a serious pest of stored foodstuffs. *Tribolium* has been reported from many locations worldwide. Development of strategies alternative to routine chemical control has been a major goal of stored-grain scientists for many years. Pheromone-based monitoring traps for moth and beetle pests of stored products have been used for more than two decades (Burkholder and Ma 1985). Adult male *T. castaneum* secrete an aggregation pheromone, 4,8-dimethyldecanal (DMD), that is attractive to both sexes (Suzuki 1980). The pheromone is purportedly produced from the setiferous glands or patches on the ventral side of the femur (Faustini et al. 1981, 1982). Pheromone extracts of prothoracic femora from unmated male flour beetles elicit higher receptor potentials in the antenna of females than in those of males (Levinson and Mori 1983). Little is known about factors affecting pheromone production in male *T. castaneum*. The objectives of this research project are to quantify the amount of pheromone produced by single male *T. castaneum*, and to determine the effect of food on pheromone production.

[*] Department of Entomology, University of Agriculture, Faisalabad, Pakistan.
[†] USDA ARS, Dept. of Entomology, University of Wisconsin, Madison, WI USA 53706.
[§] Department of Entomology, Oregon State University, Corvallis, OR USA 97331.

Materials and Methods

Insect cultures

An established laboratory colony of *T. castaneum* was used in all experiments. Flour beetles were reared on a mixture of whole wheat flour and brewer's yeast (95:5) in a growth chamber maintained at 27± 1°C, 60% r.h. and 16:8 (light-dark) photo period. Parent beetles were sifted from cultures one week after inoculation. Pupae were sieved out after 3 weeks and were sexed using the pygidial characters described by Ho (1969). Male red flour beetle pupae were maintained separately on flour until used.

Collection of volatiles

For all experiments, insects were placed in 7.5 cm × 2.75 cm cylindrical glass aeration chambers that were clamped to a stand and oriented vertically. Chambers were composed of a male and female ground glass joint tapered distally at each end to a 1/4 inch glass tube. Top and bottom openings to the chamber were loosely packed with glass wool to prevent insect escape while allowing air flow. House vacuum was used to draw air through charcoal and Tenax prefilters into the aeration chamber, and volatiles were trapped upwind on a glass column packed with the solid phase adsorbent Super-Q (Alltech Assoc., Deerfield, IL). The glass column was a Pasteur pipette from which the tip was removed and the cut end flame polished. Glass wool was used to plug the bottom opening of the column and Super-Q was filled to a level of 2 cm (400–450 mg) and held in place with an additional glass wool plug. Air flow rate on each system was maintained at 200 mL/minute. Following aeration of insects, the volatiles were extracted from the columns by elution with approximately 700 µL of HPLC-grade hexane and 762 ng of N-dodecane was added immediately as an internal standard. All aerations were conducted in a room maintained at 27±1°C at 60% r.h. with a photoperiod of 16:8 (L:D).

Pheromone production by individual beetles

Five 2-day old adult virgin males were placed individually into glass aeration chambers and each male was provided 0.5 g of cracked wheat as food. Super-Q columns were changed after every 2 days and continuous aeration was carried out for 30 days (i.e. 15 consecutive 2-day collections for each beetle). Analysis of variance (ANOVA) was used to determine if pheromone production differed over time among the 15 2-day collection periods.

Effect of food on pheromone production

Two experiments were conducted to determine if feeding on wheat affected pheromone production and also if exposure to wheat volatiles with no feeding affected pheromone production. A total of ten 4-day-old adult males were aerated individually in glass chambers. Five of these males were designated as controls and provided with 0.5 g of wheat for the duration of the study. The remaining five beetles were designated experimental and were subjected to successive time periods with and without 0.5 g of cracked wheat as food. Super-Q columns were changed after every 2 days throughout the experiment and the amount of pheromone produced by one beetle was summed for each experimental period. For the first 4 days of the study, the experimental beetles were provided food, and thus were treated the same as the controls. After these 4 days of feeding, the wheat kernels were removed from the experimental arenas and the beetles were aerated without food for 6 days. Food was then reintroduced for 2 days and again was removed for a final 4-day period. Insects in the control group were physically disturbed by shaking the chambers at the time when food was added or removed in the treatment group.

In the food volatiles experiment, eight 4-days old adult males were placed individually in aeration chambers without food. A Pasteur pipette, 7.5 cm in length, was filled with 0.5 grams of cracked wheat kernels. Wheat kernels were suspended in the centre by glass wool on both ends of pipette. Air was passed through charcoal and Tenax prefilters, through the wheat kernels, and then to the aeration chamber. Sixteen adults of the same age were also aerated individually as controls, eight without any food or food volatiles, designated as negative control, and eight with 0.5 grams of wheat in the arena for feeding, designated as positive control. Beetles were aerated for a total of 6 days and Super-Q columns were changed at 3-day intervals; pheromone produced by each beetle was summed for the 6-day period and differences among treatments were determined with ANOVA.

Chemical analyses

Samples from all experiments were concentrated to 20 µL under a gentle stream of N_2 and were subjected to quantitative analysis by couple gas chromatography–mass spectrometry (GC–MS). Analyses were made using a Schimadzu GC-14A coupled to a Finigan Model 800 series Ion Trap Detector mass spectrometer. The injector oven was set at 230°C and the heated transfer line to the spectrometer was set at 265°C. The column used was a 30 m × 0.252 mm DB-1 (J & W Scientific), temperature-programmed at 40°C for 30 sec, then 20°C/minute to 60°C, held for 1 minute, then increased to 175°C at 10°C/minute, and held at 175°C for 2 minutes, then increased to 280°C at 30°C/minute. and held at 280°C for 3 minutes. Injection was made with the splitter closed initially, but then opened at 30 sec. Initial studies were conducted with the spectrometer in the full scan mode, recording mass fragments from 35 to 350 amu. An authentic sample of 4(R), 8(R,S)-dimethyldecanal was analysed for retention time and mass spectrum, which matched the spectrum published by Suzuki (1981). Preliminary studies with volatiles collected from single male *T. castaneum* confirmed the presence of DMD by matching spectrum and retention time with those of the authentic standard. In order to maximise detection ability for DMD in the experiments described above, the mass spectrometer was operated in the multiple ion detection mode (MID) in which only the characteristic fragment ions m/z=41 and m/z-57 were detected. These ions are common to both the internal standard, dodecane, and DMD. The quantity of DMD in each sample was determined using the Finnigan ITDS software (Ver. 4.10) by comparison of the peak area of the internal standard (representing 762 ng in the initial solution) and that of DMD from the MID chromatogram.

Results and Discussion

Pheromone was successfully collected and detected from one male in a 2-day period for four of the males in this study; a fifth male died early in the experiment. The mean amounts of pheromone produced per day by each of the four males in this study were 605.6 ng, 612.3 ng, 638.3 ng, and 685.3 ng, which indicates little overall variation among males. However, throughout the 30-day aeration of any given male there was great variation, which can be seen in a plot of the mean pheromone production per 2-day interval (Fig. 1). A fluctuating cycle of two days increase and two days decrease in

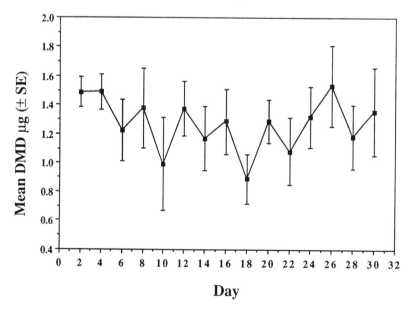

Fig. 1. Amount of pheromone collected from individual *T. castaneum* males beginning 2 days after emergence. Graph shows mean DMD (µg) (n=4) and standard error produced per male every 48 hours up to 30 days of age.

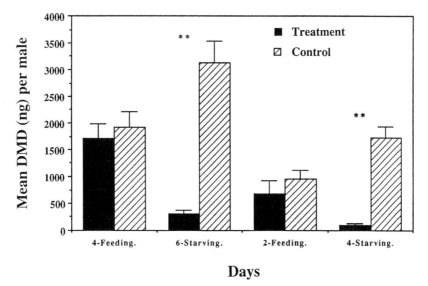

Fig. 2. Effect of feeding on pheromone production by *T. castaneum* males. Histograms show mean DMD (ng) per male and standard error for a given time period and treatment regime. Control is always with food, treatment is either with or without food. Significant difference between treatment and control are indicated as ** (P<0.01, t-test).

pheromone production was observed over a 30-day period. When the mean amounts produced were compared among 2-day intervals, there was no significant difference (P> 0.05; ANOVA). The production of DMD remained high up to the last day of the study (1.35 μg). The results on longevity of production show that age does not affect pheromone production, at least during the first month of adulthood. This finding of sustained pheromone production over several weeks is congruent with results by Faustini et al. (1981) who found that globules secreted from the setiferous patches on prothoracic femora of male *T. castaneum* over 100 days were attractive to females and presumably contained pheromone.

The results of the feeding experiment show that there was no significant difference between control and treatment groups when both were provided food, but significant differences occurred when food was removed from the treatment group (Fig. 2). Pheromone production drastically decreased to 310.38 ± 53.67 ng over 6 days when food was removed, compared to 3123.82 ± 410.36 ng for males that had wheat during that period. An increase, equal to control, in DMD production was observed after food was reintroduced to treatment beetles. When wheat was again removed from the treatment chamber, pheromone production was 93.72 ± 35.91 ng in treatment as compared to 1730.00 ± 203.97 ng in control. This reduction was highly significant (P<0.001, t-test). Actual contact and feeding on grain is apparently required for production of pheromone. Male *T. castaneum* exposed to volatiles of cracked wheat produced low levels of pheromone that were not different from those produced by beetles with no food, but were significantly lower than amounts produced by beetles actually feeding on cracked wheat (Fig. 3). These experiments with food provide strong evidence that aggregation pheromone production in *T. castaneum* is dependent on feeding. Starvation has a profound effect on the reduction of pheromone release in male red flour beetles, but pheromone production is easily 'rescued' by providing food.

Feeding has been linked to pheromone production in many other species of beetles, and may be due to one or both of two possibilities (Vanderwel and Oehschlager 1987). Food may provide a direct precursor to the pheromone that is immediately converted upon ingestion or contact. Such is the case in

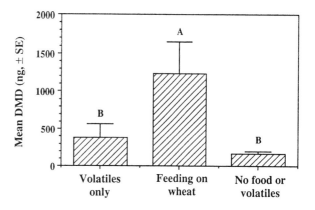

Fig. 3. Effect of wheat volatiles and feeding on the production of DMD by *T. castaneum* males. Histograms show mean DMD (ng) produced per male in 6 days and standard error (N=8). Mean amount followed by different letters are significantly different (ANOVA, P<0.05, means comparison by Fisher's LSD).

many scolytid beetles that convert terpenes from pine trees into terpene alcohols that serve as pheromones (e.g. Pierce et al. 1987). In some cases with scolytids, simply the exposure to host plant vapours, with no feeding, can induce pheromone production. This is apparently not the case for *T. castaneum* (Fig. 3). An alternative is that feeding and gut distension induces a neuro-endocrine reaction that triggers pheromone production as a result of hormones (Renwick and Hughes 1977). The outcome of these experiments supports the idea of sanitation in warehouses. Even if beetles contaminate an area briefly, absence of an easily available food source would prevent attraction of additional beetles. Cleaning of residuals of grains should prevent *Tribolium* from aggregation and consequently reduce the chances of population growth.

Acknowledgments

We appreciate reviews of the manuscript by Robert J. Bartelt, USDA ARS in Peoria, IL, and Joel K. Phillips, USDA ARS in Madison, WI. Research was partially funded by a training grant to A. Hussain from

USAID. We appreciate travel support for A. Hussain provided by Insects Limited Inc., Indianapolis, IN, and Consep Membranes Inc., Bend, OR.

References

Burkholder, W. E. and Ma, M. 1985. Pheromone for monitoring and control of stored product insects. Annual Review of Entomology, 30, 257–272.

Faustini, D. L., Burkholder, W. E. and Laub, R. J. 1981. Sexually dimorphic setiferous sex patch in the male red flour beetle, *Tribolium castaneum* (Hbst.) (Coleoptera: Tenebrionidae): site of aggregation pheromone production. Journal of Chemical Ecology, 7, 465–479.

Faustini, D. L., Rowe, J. R. and Burkholder, W. E. 1982. A male-produced aggregation pheromone in *Tribolium brevicornis* (Le Conte) (Coleoptera: Tenebrionidae) and inter specific response of several *Tribolium* Species. Journal of Stored Products Research, 18, 153–158.

Ho, F. 1969. Identification of pupae of six species of *Tribolium castaneum*. Annals of the Entomological Society of America, 62, 1232–1237.

Levinson, H. Z. and Mori, K. 1983. Chirality determines pheromone activity for flour beetles. Naturewissenschafen, 70 190–192.

Pierce, H. D., Conn, J. E. Jr., Oehlschlager, A. C. and Borden, J. H. 1987. Monoterpene metabolism in female mountain pine beetles, *Dendroctonus ponderosae* Hopkins, attacking ponderosa pine. Journal of Chemical Ecology, 13, 1455–1480.

Renwick, J. A. A. and Hughes, P. R.. 1977. Neural and hormonal control of pheromone biosynthesis in the bark beetle, *Ips paraconfusus*. Physiological Entomology 2, 117–123.

Suzuki, T. 1980. 4, 8-Dimethyldecanal: The aggregation pheromone of the flour beetles, *Tribolium castaneum* and *T. confusum* (Coleoptera: Tenebrionidae). Agricultural Biology and Chemistry, 44, 2519–2520.

Suzuki, T. 1981. A facile synthesis of 4, 8-dimethyldecanal, aggregation pheromone of flour beetles and its analogues. Agricultural Biology and Chemistry, 45, 2641–2643.

Vanderwel, D. A. and Oehschlager, A. C.. 1987. Biosynthesis of pheromones and endocrine regulation of pheromone production in Coleoptera. In: Prestwich, G. D. and Blomquist, G. J., eds. Pheromone biochemistry. Academic Press. pp. 175–215.

Stored agricultural product protection in Croatia

I. Kalinovic and M. Ivezic*

Abstract

About 150 species of harmful insects and 30 species of mites are known on the stored-grain products of the European Republic of Croatia.

Most of them are economically important pests and they damage 2–3% of public sector product and 6% of products from small private farms. The most important are *Sitophilus granarius* (L.), *Sitophilus oryzae* (L.), *Sitophilus zeamais* Motsch., *Rhyzopertha dominica* F., *Plodia interpunctella* (Hubn.), *Sitotroga cerealella* (Oliv.) (primary pests), but the most numerous are *Cryptolestes ferrugineus* Steph., *Oryzaephilus surinamensis* L., *Tribolium confusum* (Du Val.), *Tribolium castaneum* (Herbst) (secondary pests).

Very frequent are psocids: *Liposcelis corrodens* (Heym.), *Liposcelis pubescens* Brhd., *Liposcelis paetus* (Pearm.), *Liposcelis bostrychophilus* Bad., *Liposcelis entomophilus* (End.), *Liposcelis simulans* Brhd., *Liposcelis kidderi* (Hag.), *Liposcelis mendax* Pearm., *Liposcelis tricolour* Bad., *Liposcelis terricolis* Bad., *Lepinotus recitulatus* End., *Lepinotus inquillinus* (Heyd.), *Psyllipsocus ramburi* Sel. var. destructor, and *Lachesilla pedicularia* (L.).

The most abundant mites are *Tyrophagus putrescentiae* (Schrank) and *Acarus siro* (L.). The nematode *Anguina tritici* (Steinb.) Chitwood is an economically important pest of stored wheat at a few private farms, but is successfully controlled.

Pest protection is based mostly on prevention but chemicals are applied as a last measure in order to decrease food and forage contamination. The registered active ingredients used are: dichlorvos, chlorpyrifos-methyl, pirimiphos-methyl, malathion, deltamethrin, phosphine, methyl bromide and hydrogen cyanide. The most frequently applied compounds are dichlorvos, pirimiphos-methyl and phosphine.

Introduction

Agriculture in the Republic of Croatia is a very important part of the country's economy. Crops are grown on both big agricultural enterprises (public sector holdings with 25% arable land) and on small private farms of average size of 2.9 ha with 75% arable land. Farming crops include cereals (wheat, maize, barley, oat, rye), oil plants (sunflower, soybean, rape), industrial plants (sugar beet) and forage (lucerne). Plant yields on public sector holdings are close to European standards, whereas on private farms they are considerably less. Enough is produced for domestic consumption, and small quantities are exported. Agricultural products are kept in store houses with a total of over 2 million t of storage capacity. Storage facilities include constructed store houses (60%), silos (24.4%), hangars (5.9%), wire baskets for corn cobs (1.9%), cellars

(2.9%), refrigerated warehouses (2%), prefabricated store houses (1.9%) and other warehouses (0.8%).

On public sector holdings products are stored in silos and large or small store houses, while on private farms corn cobs are stored in wire baskets and in little floor warehouses. In all stored products in these conditions, harmful insects, mites and nematodes appear and cause losses amounting to about 2–3% in public sector store houses and about 6% in store houses on private farms.

Materials and Methods

This paper reports the results of investigation into harmful insects, mites and nematodes in silos and large floor store houses of public sector holdings, in stored mercantile and seed goods (wheat, barley, oat, rye, maize, sunflower, soybean and rape). In private farm store houses (wire baskets for corn cobs, small floor store houses) research has been carried out on wheat, barley, oat as well as on corn kernels and corn cobs.

Samples have been taken in silos from the top side of bins using 1.5 m sampling spears, by releasing product from the outlet at the bottom of the bin or during transfer of product at definite time intervals.

Samples have been taken by longer or shorter hand spears from spilled products in floor store houses. Corn cobs from wire baskets have been taken from surface and depth. Samples were taken every one or two months during season storage. In total over 5000 samples were taken, each weighing 250 g.

Samples were sieved by automatic apparatus (sieve diameter 0.5–2.5 mm). Insects and mites were separated from sieved material. Coleoptera and Lepidoptera were identified with stereo microscope by keys (Korunic 1990; Schmidt 1970). Psocoptera were preserved in 70% alcohol, cleared, mounted on slides and identified using keys of Gunther (1974) and a phase contrast microscope. Mites were preserved in 70% alcohol and were determined by keys (Pagliarini 1979; Zdarkova 1967) by microscope.

Sample humidity was measured by different apparatus (Twin tester, Dicky John and similar).

Samples of wheat seed from different store houses were analysed. Small and dark seeds were crushed in a drop of water and the larval stages of *Anquina tritici* were identified under the microscope.

Results and Discussion

Public sector store houses

In the Republic of Croatia 89% of agricultural production is stored in public sector store houses, that is, in 1148 silos with a combined capacity of 1548000 t and in large floor store houses with a capacity of ca 595000 t, where mercantile and seed wheat, maize, barley, oat, sunflower, soybean and sugar beet are stored. Research into these products has led to the discovery of numerous species of harmful insects and mites (Table 1 and Fig. 1). Representatives of secondary insect species were most numerous in silos—*Cryptolestes ferru-*

* University of Josip Juraj Strossmayer, Faculty of Agriculture, P.O. Box 117, 54000 Osijek, Croatia.

Table 1. Pests in store houses (%).

| | Public sector | | Private sector | |
Kind of pests	Silos	Large store houses	Small store house	Wire baskets
S. granarius			26.5	
S. oryzae	9.3	19.2	5.3	
S. zeamais	8.5	15.7	4.0	
R. dominica	6.8	12.2	5.3	
C. ferrugineus	17.0	3.9	4.0	
O. surinamensis	10.2	4.4		
T. confusum	15.3	3.5	6.0	
T. castaneum			6.6	
A. advena		2.6		11.3
P. interpunctella	3.4	7.0	10.6	37.7
S. cerealella	2.5	5.2	18.5	22.6
Psocoptera (genus *Liposcelis, Lepinotus, Psyllipsocus, Lachesilla*)	8.7	17.4	8.0	18.0
T. putrescentiae	5.1	3.9		3.8
A. siro	3.4	5.2	5.3	5.7

gineus, *Tribolium confusum* and *Oryzaephilus surinamensis*, which feed on the damaged and broken parts of grain. Investigations showed that during one year of storage 4.68% mechanically damaged and broken grains emerged on wheat, and 10–12% on maize (Kalinovic 1990). Seed damage and broken grains emerge during combining, drying and the manipulation–elevation of products during the storage season. Stored seed in silos is elevated on average two to three times, causing seed brokage and providing a convenient medium for secondary pest growth.

Primary species most abundant are *Sitophilus oryzae, S. zeamais* and, *Rhyzopertha dominica*; thermophilic pests which grow in large quantities of stored products, particularly where temperatures are elevated. Representatives of Lepidoptera—*Plodia interpunctella* and *Sitotroga cerealella*—were less prevalent in silos, and seemed to prefer the surface of products stored loose or in sacks in large floor store houses (Kalinovic 1985, 1986a; Kalinovic et al. 1990). Insects of order Psocoptera, mostly representatives of the genus *Liposcelis—Liposcelis corrodens, L. pubescens, L. paetus, L. bostrychophilus, L. entomophilus* and *L. simulans*, were also present. *L. kidderi, L. mandax, L. tricolor, L. terricolis,* were most abundant on stored wheat and maize, with humidity from 17–21% and temperature from 14–28°C (Kalinovic 1979).

Favourable conditions for these insects include elevated humidity and temperature, as well as the presence of microorganisms, fungi and bacteria, on which they feed with. (Kalinovic et al. 1978). Psocoptera are harmful insects, since their mass presence pollutes products, especially packed food products (Kalinovic 1984). Mites, *Tyrophagus putrescentiae* and *Acarus siro,* emerged sporadically, more in silos and less in large floor store houses, especially in grains (wheat, barley) which hadn't been dried before storage. Mites appeared on seed barley, but timely control measures prevented significant damage and loss of germination (Ilic et al. 1967).

In uncontrolled storing conditions, for example on maize in large floor store houses, *Ahasverus advena* (a mycophagous species) was noted (Ilic et al. 1975).

Private store houses

Private store houses in Croatia occupy about 11% of total store house capacity for agricultural products. These are small

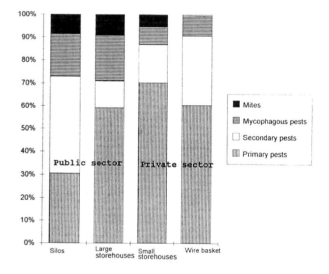

Fig. 1. Pests of stored grain by type of store house.

floor store houses and hangars, wire baskets for corn cobs, and cellars. The total stored capacity of private store houses is about 263000 t, an average of 5–8 t per household.

Stored maize, barley and oats are used as feed for domestic animals on farms, whereas wheat is kept to be made into flour either in public or private mills. Private store houses are generally unsuitable for extended storage because they are of poor construction, and farmers are mostly unaware of safe storage practices. Consequently private storage facilities are frequent sources of stored pests.

Pests in products stored in private facilities are summarised in Table 1 and Figure 1. In small floor store houses, the most common primary pests were *Sitophilus granarius* and representatives of Lepidoptera—*S. cerealella* and *P. interpunctella*—which in uncontrolled conditions caused significant damage on stored wheat, barley and maize. Secondary species included *Tribolium castaneum, T. confusum* and *Cryptolestes ferrugineus.* These species were less abundant and they appeared as primary pest species escorts. Representatives of Psocoptera—*Lepinotus reticulatus* and *L. inquillinus*

Table 2. Implementation of stored-product protection measures in the public and private sectors in Croatia.

Protection measure	Public sector		Private sector	
	Silos	Large store houses	Small store houses	Wire baskets
Hygienic measures	+	+	+	+
Physical measures				
Seed cooling	+			
Air conditioning	+	+		
Enriched CO_2 atmosphere—the possibility of application is still being studied				
Inert dusts	Laboratory test ing			
Biological measures				
Chemical measures	+	+	+	
Active ingredient				
Dichlorvos	+[a,b]	+[a]	+[a]	
Chlorpyrifos-methyl	+[b]	+[b]		
Pirimiphos-methyl	+[a,b]	+[a]	+[a]	
Malathion	+[a]	+[a]	+[a,b]	
Deltamethrin	+[b]	+[b]		
Phosphine				
(Al-phosphide	+[b]	+[b]		
Mg-phosphide)	+[b]	+[b]		
Methyl bromide	+[a]			
Hydrogen cyanide	+[a]			

[a]Treated empty storages. [b]Product treatment .

appeared sporadically. Mites (*Acarus siro*) emerged after harvesting and occurred in products stored without drying. They disappeared later during the storage season (Pivar et al. 1977). Most abundant were representatives of Lepidoptera—*P. interpunctella* and *S. cerealella* —which emerged in wire baskets for corn cobs, especially in the autumn and spring months.

On stored mouldy corn cobs, Psocoptera representatives—*L. corodens, Psyllipsocus ramburi* var. destructor and *Lachesilla pedicularia*—as well as mycophagous species *A. advena* were also numerous. In autumn, corn cobs are stored with seed humidity from 24–28% and by natural drying seed humidity decreases to15–16%, but not until April the following year.

The most damage to stored corn cobs in wire baskets has been done by rodents—rats, *Rattus norvegicus* (Burk), and mice of *Mus musculus* (L.) species.

The nematode *Anquina tritici* was identified on a few wheat samples from private facilities that used their own seeds for sowing. The law of plant protection in Croatia requires control of this nematode. In areas which are infested by 'ear cockle' of wheat (*A. tritici*) farm hygiene, crop rotation and the forbidden usage of seed from the previous wheat crop are practised. These measures are very successful in Croatia.

Pest control

Pest control in stored agriculture products is based on carrying out both preventive and curative measures—integrated stored-product protection. Table 2 summarises the protection measures undertaken in public and private sector houses. Hygiene protection measures are carried out more successfully in public sector store houses, less detailed and successfully in private sector store houses.

The most important measures are proper and timely store house preparation for reception of new products; store house cleaning; the removal of old product; the removal of old containers and unnecessary things; regular checks of the sanitary state of the store house; store house treatment with insecticide and disinfectant; product control during the storage season; and quick intervention with appropriate measures if pests emerge.

Concern about stored products in Croatia has dated from 1960. Many control possibilities have been investigated (Ilic et al. 1969, 1973; Kalinovic, et al. 1981; Kalinovic 1984, 1985, 1986 b, 1993; Korunic 1990) and integrated into storage practice.

Except for the application of cooling and active seed ventilation by air circulation in silos and large floor store houses that have built-in apparatus, physical protection measures are limited. Pest control using atmosphere enriched with CO_2 is not carried out in practice, but intensive research is being done with the aim of its implementation (Hamel 1993). Stored-agricultural product protection by inert dusts is not carried out in our country, but research is being undertaken to find a domestic formulation and application (Korunic 1993; Maceljski and Korunic 1972). Biological control measures in our store houses are not carried out, but the prospect of their implementation is good.

Chemical protection by pesticides is carried out in the integrated control of stored-product pests. In the Republic of Croatia there are 640 plant-protecting agents with 277 active ingredients. Of these 25 are stored-product protecting agents with nine active ingredients. Plant protection and product agents are manufactured in five chemical factories and there are also 60 branches of foreign firms.

In public sector storage facilities the following active ingredients are used:
- dichlorvos and pirimiphos-methyl for empty buildings, and for direct wheat and maize treatment by automatic apparatus;
- chlorpyrifos-methyl, deltamethrin and phosphine for direct grain, maize and oil plants treatment;

- malathion is used for empty building treatment (silos and large floor store houses), and methyl bromide and hydrogen cyanide for empty buildings (silos only).

In private facilities pesticides are seldom used; only 25% of private farmers treat their products or empty store houses. The following active ingredients are used:

- dichlorvos and pirimiphos-methyl are used for empty buildings;
- malathion is used in buildings and for wheat and corn kernel dusting.

Pesticides are not applied to corn cobs in wire baskets.

Phosphine, i.e. aluminium-phosphide in the form of tablets and pellets, and magnesium-phosphide (tablets, pellets, balls and plates), is used as the most efficient fumigant for harmful insects and mites in silos and large floor store houses.

Aluminium-phosphide has been used in Croatia since 1965, and magnesium-phosphide since 1975. Fumigation has been supplemented by fogging with insecticides such as dichlorvos and pirimiphos-methyl (Ilic et al. 1973; Kalinovic et al. 1979; Kalinovic 1985, 1986b). Phosphine preparations are used most in pest control in mills, flour store houses and in the tobacco industry (Kalinovic 1983).

Pest control in store houses is carried out by means of different applicators: liquid forms of insecticides by different spraying machines, atomising dampers and sprinklers made in Croatia or abroad; direct wheat treatment in silos and large floor store houses by special machines (Vobomatic etc.); and solid form phosphine fumigants by manual spear and automatic doser.

All pesticides are used under conditions of common standards, and according to manufacturers' and legal regulations of the Republic of Croatia.

Acknowledgments

We thank the Croatian Ministry of Science, Zagreb and enterprises 'Zitar' Donji Miholjac and 'Poljopromet' Osijek for financial support.

References

Gunther, K.K. 1974. Staublause, Psocoptera. Jena, VEB Gustav Fischer Verlag, 314.

Halstead, D.G.H. 1986. Keys for the identification of beetles associated with stored products. I—Introduction and key to families. Journal of Stored Products Research 4, 163–103.

Hamel, D. 1993. Djelotvornost CO2 na razne stadije *Tribolium confusum* Du Val. (COLEOPTERA). (Efficacy of CO_2 on different development stages of *Tribolium confusum*—COLEOPTERA). Zbornik radova Seminar ZUPP, 93 Zastita uskladistenih poljoprivrednih proizvoda, Stubicke Toplice, ozujak 1993, l, 73–81.

Ilic, B., Pivar, G. and Kalinovic, I. 1967. Ogledi suzbijanja grinja na sjemenskom jecmu. (Trials of mite control on seed barley). Agrohemija 5–6, 245–253.

Ilic, B., Pivar, G. and Kalinovic, I. 1968. Novije mogucnosti suzbijanja Stetocina u mlinovima i skladistima brasna preparatima DDVP. (New possibilities of pest control in mills and meal warehouses with DDVP chemicals). Agrohemija 3–4, 120–230.

Ilic, B., Pivar, G. and Kalinovic, I. 1969. Kompleksna zastita zita u skladistima Slavonije i Baranje. (Complex cereal protection in the warehouses of Slavonija and Baranja). Dokumentacija za tehnologiju i tehniku u poljioprivredi 83, 1–7.

Ilic, B., Pivar, G. and Kalinovic, I. 1973. Primjena preventivnih mjera zastite zita i brasna u skladisnim objektima od napada stetocina. (Application of preventive protection measures in cereal and meal warehouses from pest attack). Biljna zastita 3, 73–79.

Ilic, B., Pivar, G. and Kalinovic, I. 1975. Masovna pojava Ahasverus advena Waltl. (Cucujidae, Col.) u skladistu kukuruza. (Appearance of Ahasverus advena Waltl.–Cucujidae, Col. in masses in a cornstore). Spomenica i Zbornik radova PPT fakulteta u Osijeku, 185–188.

Kalinovic, I. 1979. Liposcelidae (Psocoptera) skladista zitarica i tvornice tjestenine u Slavoniji i Baranji. (Liposcelidae, Psocoptera in stored grain and dough factory in Slavonija and Baranja). Acta entomologica Jugoslavica, 1–2, 145–154.

Kalinovic, I. 1983. Primenenie Fostoxin-a v zascite tabake ot vreditelei v skladah i v tabacnjih izdelijah. (Usage of Phostoxine in tobacco protection against pests). Ceminap Primenenie Fostoksina i Magstoksina dlja dezinsekcii zerna, zernoproduktov i drugih piscevljih produktov i tabaka, Moskva, SSSR, februar 1993. Zbornik radova, 1, 1–18.

Kalinovic, I. 1984. Efikasnost fosfvorovodika u suzbijanju prasnih usi (Insecta: Liposcelidae). (The efficiency of phosphine in the control of book louse—Insecta: Liposcelidae). Znanost i praksa u poljoprivredi i prehrambenoj tehnologiji 3–4, 239–247.

Kalinovic, I. 1985. Cuvanje sjemena kukuruza u skladistima i zastita od {tetnika. (Storing of maize seed and protection against pests). Zbornik seminara Susenje i dorada sjemena kukruza, Zagreb, l, 55–59.

Kalinovic, I. 1986a. Stetnici uskladistenog sjemenskog kukuruza i njihovo suzbijanje. (Pests of stored seed maize and their control). Znanost i praksa u poljoprivredi i prehrambenoj tehnologiji, 3–4, 251–259.

Kalinovic, I. 1986b. Fosforovodik - svojstva, primjena i mjere opreza. (Phosphine-characteristic, application and mesaures of caution). Zbornik seminara Fumigacija- integralna mjera za{tite uskladi{tenih poljoprivrednih proizvoda, Split, l, 20–23.

Kalinovic, I. 1990. Primjese u uskladistenim zitaricama. (Foreign matter in stored cereals seed). Sjemenarstvo, 2, 87–90.

Kalinovic, I. 1993. Stetnici u nasim skladistima i mogucnosti njihovog suzbijanja. (The possibilities of control of the pests in our store houses). Zbornik radova Seminara ZUPP, 93 Zastita uskladistenih poljoprivrednih proizvoda, Stubicke Toplice 1993, 1, 1–8.

Kalinovic, I., Horvat, S., Jancic, B. and Grubac, B. (1990.b) Suzbijanje Stetnika (*Plodia interpunctella* Hbn. i *Sitotroga cerealella* Oliv.) u skladistima kukuruza. (Pest control (*P. interpunctella* and *S.cerealella*) on stored corn seed). Sjemenarstvo 1, 11–16.

Kalinovic, I., Ilic, B. and Bardek, M. 1981. Problemi aplikacije pesticida u skladistima zita, njihovih preralevina, te u prehrambenoj industriji. (Pesticide application problems in cereal warehouses and food industry). Zbornik radova Jugoslavenskog savjetovanja o primjeni pesticida, Opatija 3, 441–442.

Kalinovic, I., Pivar, G. and Ilic, B. 1979. Fumigacija magnezium-fosfidnim plocama u skladistima. (Fumigation with magnezium-phosphide plates in warehouses). Zbornik radova X. jubilarnog Savjetovanja o primjeni pesticida, Porec, 513–515.

Kalinovic, I., Todorovic, M. and Kalinovic, D. 1978. Koristenje mikroorganizama u ishrani Liposcelidae. (Microorganisms used as Liposcelidae food). Mikrobiologija l, 67–77.

Korunic, Z. 1990. Stetnici uskladistenih poljoprivrednih proizvoda. (Pests of stored agricultural products). Zagreb, Gospodarski list, 220.

Korunic, Z. 1993. Zastita uskladistenih poljoprivrednih proizvoda inertnim prasivima. (Protection of stored grain by inert dusts). Zbornik radova Seminar ZUPP, 93 Zastita uskladistenih poljoprivrednih proizvoda, Stubicke Toplice, ozujak 1993, 1, 83–90.

Maceljski, M. and Korunic, Z. 1972. Prilog poznavanju mehanizma djelovanja inertnih prasiva na insekte. (Contribution to recognition of perform mechanism by inert dusts on insects). Zastita bilja 7, 117–118.

Pagliarini, N. 1979. Studies on mites of stored grain in Yugoslavia. Record Advances in Acarology, 1.

Pivar, G., Ilic, B., Kalinovic, I. and Boskovic, D. 1977. Entomofauna skladista zita individualnih proizvolaca podrucja Baranje. (Entomofauna in store houses of private farmers in Baranja area). Zbornik radova Poljoprivrednog fakulteta u Osijeku, 3, 21–33.

Schmidt, L. 1970. Tablice za determinaciju insekata. (Keys for determination of insects). Poljoprivredni fakultet, Zagreb, 257.

Zdarkova, E. 1967. Stored food mites in Czechoslovakia. Journal of Stored Products Research 3, 155–175.

Pheromone biology of the lesser grain borer, *Rhyzopertha dominica* (Coleoptera: Bostrichidae)

T.J. Mayhew[*] and T.W. Phillips[*†]

Abstract

The objectives of this study were to document pheromone production over time by male *Rhyzopertha dominica*, and then to investigate the effects of feeding, food nutritional value, mating, and population density, on pheromone production. The male-produced pheromones DL-1 and DL-2 were collected through aeration using the solid-phase adsorbent, Super-Q. Sexed *R. dominica* adults, 24 hours post-emergence, were placed individually in 7.5 cm × 2.75 cm cylindrical aeration chambers containing cracked wheat. Volatiles were collected for 24-hour periods and quantified using GC-MS. Pheromone was produced 3-5 days after feeding began and, once started, production did not cease over the course of one month. The ratio of the two pheromone components continually changed over the test period. The onset of pheromone production following feeding was on average 4.71 days ± 1.06 (SE). The maximum amount of DL-1 produced in a 24-hour period was 1114.756 ng ± 109.9 (SE), occurring 18 days after feeding began. Similarly, the maximum amount of DL-2 produced in 24 hours was 960.377 ng ± 78.0 (SE), occurring 14 days after feeding began. Pheromone production ceased when food was not present. Pheromone production increased proportionally as the content of wheat flour increased relative to non-nutritive cellulose in the food substrate. Pheromone production levels between mated and unmated males of the same age were not significantly different. Pheromone production was negatively correlated with population density levels.

Introduction

The lesser grain borer, *Rhyzopertha dominica* (F.) (Coleoptera: Bostrichidae), is a destructive internal feeding pest of stored grains throughout the world. Both the larvae and adults can easily attack whole, sound grain. *R. dominica* males produce an aggregation pheromone that was first reported by Khorramshahi and Burkholder (1981), and later isolated and identified by Williams et al. (1981). The pheromone was found to be made up of two unsaturated esters given the trivial names of dominicalure 1 and dominicalure 2 (Williams et al. 1981). Synthetic pheromone is available commercially and is used in flight traps for survey and detection (e.g. Cogburn et al. 1984; Leos-Martinez et al. 1986, 1987; Fields et al. 1993). However, there has been limited work on the environmental or physiological factors affecting pheromone production in *R. dominica*. The objectives of this study are to quantify pheromone production by individual beetles over time and to then investigate the effects of feeding, food nutritional value, mating, and population density on pheromone production.

Methods and Materials

General

R. dominica were reared on a diet of sifted (#40 U.S. standard sieve) whole wheat flour and brewer's yeast (95:5). Pupae were sexed according to the pupal sexual dimorphism first reported by Potter (1935) and isolated. One-day-old adults were placed individually in a 7.5 cm × 2.75 cm (OD) cylindrical aeration arena containing several kernels of soft pastry wheat. Air was drawn by house vacuum through charcoal and Tenax™ (283 mg) prefilters, through the aeration arena, then through a glass column (106 mm × 6 mm ID) containing 209 mg Super-Q™ adsorbent at a rate of 0.5 L/min. Columns were extracted with 530 µL of hexane following an aeration and 716 ng of tetradecane was added as an internal standard. Samples were concentrated to 20 µL under a stream of pure N_2 at room temperature and analysed by coupled gas chromatography–mass spectrometry (GC–MS). The GC used was a Shimadzu GC-14A with a DB-1 fused silica capillary column (25 m by 0.25 mm ID.) and operated under the following conditions: injector temperature 130°C, heated transfer line 265°C, oven temperature 40°C for 30 seconds, 20°C per minute to 60°C, hold one minute, then 10°C per minute to 175°C, hold 30 seconds, then 30°C per minute to 280°C with a final hold time of 2 minutes; injection was made splitless, and the splitter was opened at 30 seconds. The mass spectrometer was a Finnigan-Mat 800 series ion trap detector operated in the multiple ion detection (MID) mode to scan for m/z = 85 for the internal standard and m/z = 111–115 for the two pheromone components. A linear regression equation was developed to quantify each pheromone relative to the internal standard.

Daily pheromone production was determined by placing single adult males in 10 aeration devices with 12 kernels of cracked wheat (ca. 0.4857 g) and collecting volatiles daily for 30 days. The insects were observed each day to determine when feeding began. The presence of frass was used as an indication of feeding. Mean amounts of DL-1 and DL-2 produced per day were determined.

The effect of mating on pheromone production was investigated by aerating individual virgin adult males and females, 10 each, with 12 kernels of cracked wheat. Volatiles were collected on days 8–10, and on day 10 five females were marked with a permanent marker and paired individually with five males. The remaining five females and five males were left unpaired and volatiles were collected from all preparations on days 10–12. On day 12, the mating pairs were separated and beetles were again aerated individually with fresh wheat. Volatiles were collected on days 12–14. Mated females were held in the aeration system for an additional 14 days to check for progeny development to confirm mating.

---

[*] Department of Entomology, University of Wisconsin, Madison, WI USA 53706.

[†] USDA ARS Stored-Product Insects Research Unit,. Madison, WI USA 53706.

The role of feeding on pheromone production in *R. dominica* was studied by comparing volatiles from fed and unfed beetles. Two groups of 10 single-beetle aerations were established with 12 kernels of cracked wheat each. Volatiles were collected daily from day 8 to day 14 from all beetles. One group of beetles, referred to as controls, were allowed to feed for the duration of the experiment, while treatment beetles had food removed for 2 days and then it was returned. Quantities of pheromone produced were compared (t-test) between fed and unfed beetles within each sample period.

The effects of male density on individual pheromone production was examined by analysing volatiles from seven replicates of the following treatments: single male, 5 males, and 15 males in each chamber. All chambers were provided with seven kernels (ca. 0.2906 g) of cracked wheat. The quantity of pheromones produced per individual male on days 8–10 was calculated by dividing the total amount of pheromone produced in each treatment group by number of insects in each group. Data were analysed with the general linear models procedure (Proc GLM, SAS Institute 1985), and means were separated by the protected least significant difference (LSD) test.

Results and Discussion

In all but one case beetles produced pheromone after they began feeding, which averaged 4.71 d ± 1.06 (SE) after frass was observed. The maximum mean amount of DL 1 produced in a 24 hour period was 1114.756 ng ± 109.9 (SE) occurring 18 days after feeding began (Fig. 1). Similarly the maximum mean amount of DL 2 produced in 24 hours was 960.377 ng ± 78.0 (SE) occurring 14 days after feeding began. Williams et al. (1981) reported that 2000 mixed sex beetles produced a total of 660 µg of DL-1 and DL-2 combined in a 20-day aeration period, which is equivalent to approximately 30 ng/d if a 1:1 sex ratio is assumed. Dowdy et al. (1993) reported that one male infesting grain for one day produced 10 µg of pheromones. The maximum total amount of the two pheromones produced by one male in a 24-hour period in our study was 2.075 µg. Variation among studies in amounts of pheromone produced by *R. dominica* probably reflects differences in collection methods. Dowdy et al. (1993) measure headspace volatiles that had adsorbed on wheat kernels after infestation, and did not collect volatile directly from beetles. Williams et

al. collected pheromone from large groups of beetles of unknown age and sex, thus rendering estimates of production per beetle inaccurate. In the present study we collected pheromone directly from individual beetles of known age, and we sampled pheromone throughout a defined time course.

There was no effect of mating on pheromone production. No significant differences were observed in the level of pheromone production between mated and unmated males during any of the collection periods, whether females were present or not. Pheromone production by males joined with females was 403.5 (±127.4) ng/d of DL-1 compared to 510.4 (±91.8) ng/d of DL-1 for unmated males during the same period. Pheromone production trends were similar for DL-2 form males of different mating status. Pheromones were not detected in any of the female aerations, confirming that DL-1 and DL-2 are male-produced pheromones. All male-paired females produced progeny following the experiment, thus confirming they were mated. Many insect species are known to cease pheromone production following mating (e.g. in Lepidoptera, Raina and Menn 1987). Males of the bark beetle *Ips paraconfusus* reduce production of their aggregation pheromone as they are joined by females, and pheromone production ceases when the harem (3–5 females) is complete (Borden 1967). Our experiment with *R. dominica* paired one female with one male. Since the mating system and host use patterns of *R. dominica* are not understood relative to a nonstorage habitat, we cannot be certain that we are testing the effect of mating on pheromone production adequately.

Pheromone production was very dependent on feeding. We found that pheromone production nearly ceased within 24 hours after food was removed, but then resumed within 24 hours when food was returned to the beetles (Fig. 2 for DL-1). In another study (unpublished) there was no pheromone production when beetles were deprived of direct contact with wheat but exposed to wheat volatiles, thus feeding or direct contact with food is essential for pheromone production. Feeding is a prerequisite in several other species of stored-product beetles that use aggregation pheromones (e.g. Phillips et al. 1985; Pierce et al. 1984). Wheat may contain direct precursors or nutrients that contribute to pheromone production by male *R. dominica*. The association of pheromone production with a food resource would ensure a responding female that there is a suitable habitat for her progeny. Conversely, other males that respond to pheromones from feeding males

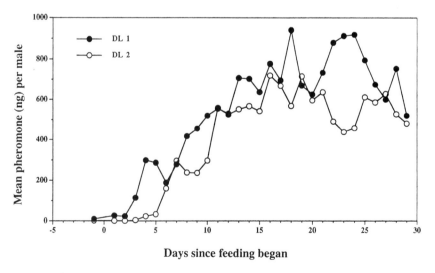

Fig. 1. Mean pheromone production by individual male *R. dominica* (n=6–8 males) feeding on wheat kernels in glass aeration chambers over a 30-day period.

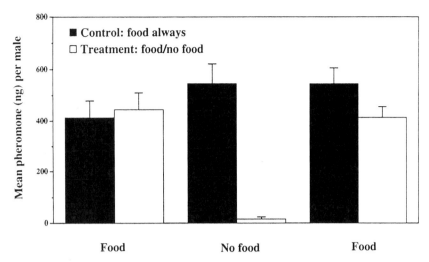

Fig. 2. Mean (± SE) production of DL-1 by male *R. dominica* feeding on wheat kernels during consecutive 24-hour periods. Treatment beetles were provided with wheat at first, then had the wheat removed, and then had the wheat returned; control beetles were provided with wheat for the duration of the experiment. Trends for DL-2 were similar to these for DL-1.

may exploit this signal to locate females at these attractive sites.

Pheromone production by males in groups of 5 or 15 males was significantly lower than pheromone production by single males (Fig. 3). Food may have been limiting in this experiment, since seven wheat kernels were provided in all cases, and caused reduction in pheromone production. Reduction of pheromone production by males in a group may also be a strategy by individual males to conserve energy from pheromone biosynthesis.

Acknowledgments

We appreciate reviews of the manuscript by Joel Phillips, USDA ARS in Madison, WI, and David Weaver, USDA ARS in Savannah, GA.

References

Borden, J. H. 1967. Factors influencing the response of *Ips confusus* (Coleoptera: Scolytidae) to male attractant. Canadian Entomologist, 99, 1164–1193.

Cogburn, R. R., W. E. Burkholder and H. J. Williams. 1984. Field tests with the aggregation pheromone of the lesser grain borer. Environmental Entomology, 13, 162–166.

Dowdy, A. K., R. W. Howard, L. M. Seitz and W. H. McGaughey. 1993. Response of *Rhyzopertha dominica* (Coleoptera: Bostrichidae) to its aggregation pheromone and wheat volatiles. Environmental Entomology 22, 965–970.

Fields, P. G., J. V. Loon, M. G. Dolinski, J. L. Harris and W. E. Burkholder. 1993. The distribution of *Rhyzopertha dominica* (F.) in western Canada. Canadian Entomologist, 125, 317–328.

Khorramshahi, A. and W. E. Burkholder. 1981. Behaviour of the lesser grain borer *Rhyzopertha dominica* (F.). Journal of Chemical Ecology, 7, 33–38.

Leos-Martinez, J. T. A. Granovsky, H. J. Williams, S. B. Vinson and W.E. Burkholder. 1986. Estimation of aerial density of the lesser grain borer (Coleoptera: Bostrichidae) in a warehouse using dominicalure traps. Journal of Economic Entomology, 79, 1134–1138.

1987. Pheromonal trapping methods for lesser grain borer, *Rhyzopertha dominica* (Coleoptera: Bostrichidae). Environmental Entomology, 16, 747–751.

Phillips, J. K., C. A. Walgenbach, J. A. Klein, W. E. Burkholder, N. R. Schmuff and H. M. Fales. 1985. (R*,S*)-5hydroxy-4-methyl-3-heptanone Male Produced aggregation pheromone of *Sitophilus*

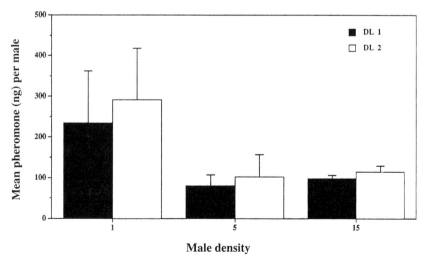

Fig. 3. Mean production of DL-1 and DL-2 in 24 hours by male *R. dominica* held singly or in groups of 5 or 15 males with seven wheat kernels.

oryzae (L.) and *S. zeamais* Motsch. Journal of Chemical Ecology, 11, 1263–1274.

Pierce, A. M., H. D. Pierce Jr., J. G. Millar and J. H. Borden. 1984. Aggregation pheromones in the genus *Oryzaephilus* (Coleoptera:Cucujidae). In: Proceedings of the Third International Working Conference on Stored-product Entomology. Kansas State University, Manhattan, Kansas. 107–120.

Potter, C. 1935. The biology and distribution of *Rhyzopertha dominica* (Fab.). Transactions of the Royal Entomological Society London, 83, 449–482.

Raina, J. K. and Menn, J. J. 1987. Endocrine regulation of pheromone production in Lepidoptera. In: Prestwich, G. D., and Bloomquist, G. J. (eds), Pheromone Biochemistry, Academic Press, pp. 181–193.

SAS Institute, 1989. SAS users guide: statistics, version 5. SAS Institute, Cary, NC.

Williams, H. J., R. M. Silverstein, W. E. Burkholder and A. Khorramshahi. 1981. Dominicalure 1 and 2: components of aggregation pheromone from male lesser grain borer *Rhyzopertha dominica* (F.). Journal of Chemical Ecology, 7, 759–780.

The measurement of resistance to *Acanthoscelides obtectus* (Say) (Coleoptera: Bruchidae) in seeds of *Phaseolus vulgaris* L.

C.J Moss and P.F Credland[*]

Abstract

Acanthoscelides obtectus is a cosmopolitan insect pest infesting leguminous seeds in the field and especially within storage. Control of this pest is usually based on the use of insecticides but opinion now favours a shift towards methods which incorporate varietal resistance. Effective and useable resistance has been detected in the hosts of many bruchid pests and can be used to contribute towards their control. However, in the case of *A. obtectus*, its behaviour, which is unique among bruchid pest species, has inhibited the development of an effective bioassay and thereby complicated the identification of resistant seeds.

A. obtectus differs from all other bruchid pest species in that its eggs are not firmly attached to individual seeds but are scattered loosely amongst potential hosts. On hatching, the larvae are then free to move from seed to seed. Seed choice is not therefore exclusively the domain of the ovipositing female. Useful resistance could be expressed at oviposition, larval penetration of seed, development within the seed or as reduced fitness of emerging adults.

This contribution attempts to define the problems to be overcome in developing a useable bioassay for the assessment of seed resistance to *A. obtectus*, and to provide a simple and reliable protocol for comparative studies.

Introduction

Acanthoscelides obtectus (Say) (Coleoptera: Bruchidae) is a cosmopolitan insect pest of leguminous seeds in the field and, especially, within storage. The feeding of adult bruchids is of no economical importance; it is the larval stages which consume parts of the seed, causing considerable damage. The main and original host of *A. obtectus* is the common bean, *Phaseolus vulgaris* L. The beans play a vital role in the diet and economy of many countries and are of particular importance as a subsistence crop in Central and South America and in parts of Africa. Control measures differ widely in Africa, Asia and the Americas, ranging from large scale applications of granular insecticides (Cardona and Karel 1990) to traditional cultural practices used by subsistence farmers. Effective insect control in developing countries has certain limitations; insecticides are expensive, sometimes hazardous, and often unavailable to the small farmer. Opinion now favours a shift towards integrated methods which incorporate varietal resistance as opposed to those based on insecticide use alone.

The assessment of resistance to *A. obtectus* requires an effective and reliable bioassay, which must be relatively straightforward to execute and flexible enough to accommodate different sized and shaped seeds. The behaviour of *A. obtectus*, which is unique among pest species of bruchid, has inhibited the development of an effective bioassay and consequently the identification of resistant seeds. It differs from all other bruchid pest species in that its eggs are not firmly attached to individual seeds but are scattered loosely amongst potential hosts. On hatching the larva is then free to wander from seed to seed before penetrating an acceptable host. Seed choice or resource partitioning is not therefore exclusively the domain of the ovipositing female. For the purpose of developing satisfactory bioassays, this behaviour leads to difficulties in ensuring that a viable larva comes into contact with the seed and, having done so, is able to adopt the necessary posture for penetration to occur. Neither of these problems needs to be overcome in the case of other bruchid pest species.

Useful resistance could be expressed at several stages in the insect's life cycle: oviposition, larval acceptance and penetration, development of larvae within the chosen seed, or as reduced fitness of the emerging adults. Two bioassays are required to incorporate all these biological stages, one for oviposition and the other to measure the remaining variables. This contribution attempts to define the problems to be overcome in developing bioassays for the assessment of seed resistance to *A. obtectus* and to describe a simple and reliable protocol for comparative studies.

Materials and methods

Origin and maintenance of insects

The insects used in the experiment were obtained from a stock culture of *A.obtectus* collected in Colombia. The insects have been in culture for some years, initially at CIAT and then the Natural Resources Institute, where they are known in the laboratory as stock culture 101. They are reared in large glass jars of approximately 2.5 L capacity, sealed with filter paper and wax, each containing approximately 600 mL of commercially available North American red kidney beans, *Phaseolus vulgaris* L. Six jars, set up one week apart, are operational at any one time. New subcultures are established by removing approximately 300 beetles from each of the two oldest jars, the oldest then being discarded. The culture is kept under controlled conditions at $27 \pm 1°C$ and $70 \pm 10\%$ r.h. with a regime of 14 hours of dim illumination and 10 hours darkness.

All seed investigated was equilibrated in muslin covered glass containers, under the controlled conditions previously described, for a minimum of three weeks before use. All experiments were carried out under the controlled conditions and used the same commercially available red kidney beans as a susceptible control of known performance.

---

[*] Department of Biology, Royal Holloway, University of London, Egham Hill, Egham, Surrey, TW20 OEX, U.K.

Oviposition bioassay

Newly emerged adults were obtained from culture seeds (containing only one or two larvae). The adults were removed and placed in a clean container for 24 hours. The adults were then sexed and a single male and female were placed in a glass tube with three seeds. A sponge bung prevented escape while still allowing the passage of air. Eggs were counted daily for 8 days (until the adults were 10 days old), and again on the death of the female, to obtain a total oviposition value. Twenty replicates of each seed accession were investigated. This is a realistic number since many wild collections, or samples of newly bred cultivars, are small and comprise few useable seeds.

Larval penetration and development bioassay

The experiments were carried out using a variety of flat bottom microtitre plates (MERC Ltd. U.K.) to accommodate different sized seeds. The plates were covered with individual glass plates to prevent larval escape. Plastic lids tended to warp and did not provide a satisfactory alternative to glass.

Glass beads were used during some bioassays to provide additional contact points for larval penetration (Dobie et al. 1990). The glass beads were of a similar size to the seed being investigated except in the case of red kidney beans where 6.5 to 7.5 mm beads were used. The size and shape of the seed determined the need for glass beads (see below). Beads were added to small seeds which did not fill the bottom of the well. Spherical seeds also required additional beads to secure the seed, preventing movement during larval penetration.

A single seed was placed in an appropriate sized well of a microtitre plate. Glass beads were added as required. The seeds of each accession were placed in a separate microtitre plate and spaced apart to minimise the risk of any movement by larvae between wells. Eggs were obtained from 3-day old females. A single 5-day old egg was added to each seed and checked 3 days later for hatching and larval penetration of the seed. If the egg had not hatched or the hatched larva had died or escaped from the well, another egg was supplied. The procedure was repeated until a total of three replacement eggs, after the initial application, had been made, if necessary. The seeds were examined and adult emergence recorded daily, beginning 28 days after the first egg was introduced; recording continued for a period of 42 days. Twenty replicates of each seed accession were investigated.

The newly emerged adults were sexed and weighed. Where possible female fecundity was also investigated. Each female was placed in a glass tube with three control seeds and a newly emerged male from the stock culture. The total oviposition of each female was recorded on day 10; oviposition after day 10 is negligible (Parsons, per. comm. 1993).

Three different experiments are described to illustrate the development and use of the procedure.
1. The effectiveness of glass beads was investigated using the small and spherical seeds of three species: *Vigna angularis* (Willd.) (adzuki bean), *Vigna radiata* (L.) Wilczek. (mung bean) and *Lupinus* sp. (lupin). The larval penetration and development bioassay procedure was followed with:
   i. A single seed in a well.
   ii. Addition of glass bead(s) to a single seed in a well.
   iii.Addition of an extra seed (of the same species) to the single seed in a well.
   Twenty replicates of each treatment were investigated.
2. The effectiveness and reliability of the procedure was determined by replication on a susceptible host, red kidney beans. Three complete repeats of the experiment were undertaken.

3. To demonstrate its value for comparing seeds, the bioassay procedure was also applied to *Vigna unguiculata* (L.) Walp. (black-eye bean), RKB (control) and eight wild accessions of *P. vulgaris* from Colombia.

Determination of resistance

Resistance was defined using the following criteria:
Larval penetration: The seed was considered resistant if larval penetration occurred in 25% or less of the replicates. If penetration occurred in at least 50%, then the seed was considered susceptible.
Larval development: Only if penetration occurred in at least 50% of the seeds, could resistance within the seed be assessed. If the number of adults emerging from the seed was 25% or less of the number of larvae which penetrated, the seed was considered resistant. If at least 50% of those larvae which penetrated emerged, then the seed was considered susceptible.

Any seed falling outside these categories was considered of intermediate resistance.

Data analysis

Analysis of the data from both assays was undertaken with analysis of variance and subsequent Student-Newman-Keuls multiple range tests if differences among the samples were indicated.

Results

Oviposition bioassay

Total oviposition data for adults on the eight wild accessions of *P. vulgaris*, *V. unguiculata* (black-eye) and red kidney beans are given in Figure 1. Total oviposition differed little between all the accessions, except on black-eye bean. A one-way ANOVA revealed a significant difference among the accessions ($F_{(9,181)}$= 3.99; P < 0.001). A Student-Newman-Keuls test confirmed that oviposition on black-eye bean was significantly less (P < 0.05) than on other accessions.

Red kidney beans, black-eye bean and the wild accessions G10000 and G12949 were chosen to illustrate the rate of oviposition (Fig. 2). In each case eggs were laid at a steadily decreasing rate for 8 days after which no significant oviposition occurred on all the accessions investigated.

Larval penetration and development bioassay

Effectiveness of glass beads

The presence of glass beads increased penetration rates into all small and spherical seeds investigated (Table 1). A test was undertaken to compare an 'ideal' situation (where larval penetration occurred in all 20 replicates) and the actual data observed for each of the three separate treatments: penetration with a single seed, penetration with a single seed and glass beads, and penetration with extra seed of the same species. The only significant difference was in the case of penetration into a single isolated seed (χ^2 = 28; P = 0.0000). No difference in penetration was found between using glass beads or extra seed of the same species.

Repetition of the bioassay on red kidney beans

Consistently high penetration and emergence rates were observed in each of the three repeat experiments using red kidney beans (Table 2).

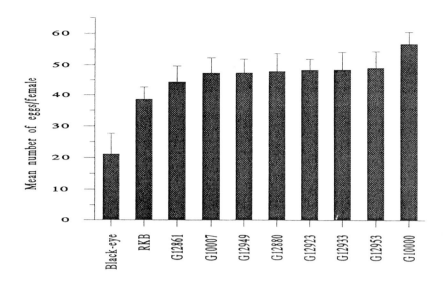

Fig. 1. Total oviposition on *V. unguiculata,* red kidney bean (RKB) and eight wild accesions of *P. vulgaris.*

Table 1. The effects on penetration of adding glass beads to small and spherical seeds.

Species	Mean weight (g)	Penetration with single seed	Penetration with glass beads	Penetration with extra seed
Adzuki bean	0.078	1	15	17
Mung bean	0.057	5	20	19
Lupinus sp.	0.024	0	15	12

All data are derived from 20 replicates.

Table 2. Penetration and emergence from red kidney beans in three separate assays. (Mean weight of seed was 0.60 g in each experiment).

Experiment	Total penetration	Total emergence
1	20	20
2	20	18
3	20	18

All data are derived from 20 replicates.

Bioassay using wild accessions of P. vulgaris

Accessions G12861, G12880 and G12953 were determined to have resistant properties using the criteria previously described (Table 3). There was little or no penetration into seeds of G12861 or G12880. More than 25% of larvae penetrated G12953 but emergence was only 6.67%. G10007 was considered to be of intermediate resistance as larval penetration exceeded 25% but was lower than 50%. The remaining accessions were all considered susceptible to *A. obtectus.*

Development period and fitness

The longest development period was observed in G12953 (60 days) and the shortest was in the control, red kidney beans (31.42 days) (Table 4). A one-way ANOVA was undertaken on data from those accessions from which more than five emergences occurred: G12949, G12933, G10000, G12923 and red

kidney beans. A significant difference was determined among the accessions tested ($F_{(4, 106)}$ = 52.434; P = 0.0000). A Student-Newman-Keuls test, showed that the only pair of accessions not significantly different (P < 0.05) from each other were G10000 and RKB.

Mean weights of both male and female adults emerging from all the wild accessions except G10000 were lower than those observed in the control (Table 4). A one-way ANOVA was carried out on the accessions: G12933, G10000, G12923, red kidney beans (all those with n > 5), for male and female weights separately. A significant difference in both the male and female weight was found among the adults emerging from all the accessions tested ($F_{(3, 44)}$= 12.536; P = 0.0000 and $F_{(3,37)}$= 27.008; P = 0.0000 respectively). A Students-Newman-Keuls test confirmed that the weights of females from all accessions were significantly different from each other (P < 0.05) except those from G10000 and red kidney beans, which did not differ from each other. Males from all accessions had significantly different mean weights except that those from G12933 were the same as G12923, and those from G10000 the same as those from red kidney beans.

Mean oviposition was greatest among females that emerged from the control seeds (43.72) and G10000 (54.33) (Table 4). A one-way ANOVA was carried out on data from G12933, G12923 and red kidney (all those with n > 5). A significant difference in total oviposition was found among the samples ($F_{(2,44)}$= 4.667; P < 0.05). A Student-Newman-Keuls test on the same accessions determined that the only significant difference (P < 0.05) was between G12933 and red kidney beans.

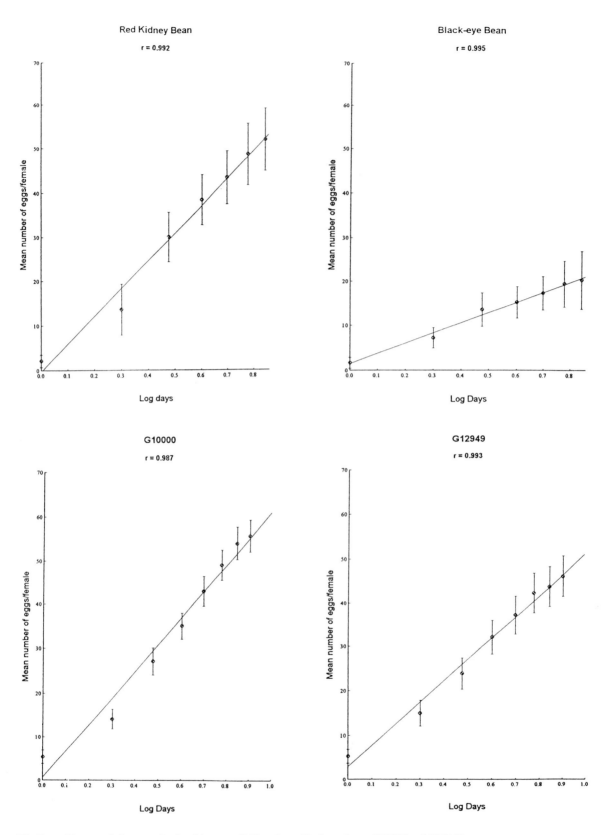

Fig. 2. The cumulative rate of oviposition on red kidney bean, black-eye bean, G10000 and G12949.

Table 3. Estimation of resistance at larval penetration and development..

Accession	Mean weight of seed (g)	Total penetration	Total emergence	Resistance/ susceptibility (possible site of resistance)
G12861	0.0458	0	0	Resistant (testa/cotyledon)
G12880	0.0331	1	0	Resistant (testa/cotyledon)
G12953	0.0487	15	1	Resistant (cotyledon)
G10007	0.0633	6	4	Intermediate (testa)
G12949	0.0569	10	7	Susceptible
G12933	0.1352	19	17	Susceptible
G10000	0.1729	19	17	Susceptible
G12923	0.1576	20	20	Susceptible
Red kidney beans	0.589	20	20	Susceptible
Black-eye bean	0.240	19	13	Susceptible

All data are derived from using 20 replicates.

Table 4. Developmental period, weight and fecundity of insects emerging from six wild accessions of *P. vulgaris* and red kidney beans.

Accession	Development period (Mean no. of days ± SE)	Mean weight (mg) ± SE		Fecundity (Mean total no. of eggs ± SE)
		Male	Female	
G12953	60 (n=1)	3.287 (n=1)	–	–
G10007	35.60 ± 0.678 (n=4)	3.94 ± 0.465 (n=2)	5.77 ± 1.011 (n=2)	31 (n=1)
G12949	46.71 ± 2.02 (n=7)	2.81 ± 0.143 (n=4)	3.98 ± 0.293 (n=3)	18.5 ± 8.995 (n=3)
G12933	38.94 ± 0.915 (n=17)	4.46 ± 0.169	4.38 ± 0.289	22.166 ± 6.156
G10000	32.88 ± 0.283 (n=17)	5.44 ± 0.440	6.22 ± 0.369 (n=3)	54.33 ± 6.936 (n=3)
G12923	36.65 ± 1.013 (n=20)	4.30 ± 0.109	5.28 ± 0.910	33.44 ± 6.301
Red kidney	31.42 ± 0.364 (n=58)	5.04 ± 0.116	6.63 ± 0.104	43.72 ± 2.259

All data are derived from original larval penetration and development bioassays with 20 replicates. Replicate numbers in the development period column refer to the total number of insects emerging. In other columns n > 5 unless stated otherwise.

Discussion

Oviposition bioassay

In stores, a female beetle is most commonly faced with a single species or variety of host. For this reason, a no-choice experimental protocol was chosen to investigate oviposition. The oviposition bioassay was adapted from the procedure described by Howe and Currie (1964). Care was taken to obtain newly emerged adults from seeds producing only one or two adults rather than seeds from which greater numbers emerged. Increased competition for resources within the seed is known to result in lower adult body weights and reduced fecundity in other bruchids (Credland et al. 1986). The newly emerged adults were left together for 24 hours to mate. Oogenesis and mating may need the stimulus of a host seed (Huignard and Biemont 1981); a single male was included in tube with the female and seeds in case mating had not taken place. The recording of oviposition began on day 3 because significant oviposition is known to begin then and continue until adults are 10 days old (Parsons, pers. com. 1993). For practical purposes, the delaying of oviposition will not prevent infestation during long periods of storage, and therefore a resistant accession should stimulate little or no oviposition. For this reason the total oviposition was chosen as the parameter to determine resistance. It can be seen in Figure 2 that rates of oviposition do not provide more useful information than can be gathered from total oviposition data.

Measuring rates over only the first 4 or 5 days does give a rapid assessment of the final total and could be useful for a quick estimation of acceptability.

The total oviposition did not differ significantly between the wild and cultivated accessions of *P. vulgaris* although oviposition on black-eye bean was significantly lower (Fig. 1). Black-eye bean is not the original host of *A. obtectus* but is susceptible to infestation (Southgate 1978; Jarry and Bonet 1982; Singh 1990). Further investigation, with the use of multi-choice oviposition experiments may indicate lower total oviposition to be comparable with non preference.

Oviposition acceptance does not appear to be host specific in strain 101. Females have oviposited in substantial numbers on all seeds investigated to date, including a large variety of legume species (Moss, unpublished data). Further investigations on seed from different families is required to determine if this behaviour is common on all seeds or just those within the Leguminosae, but the likelihood of effective and useable resistance being expressed at this stage in the life cycle of the beetle is remote. However, strain 101 has been kept under laboratory conditions for many years; the possibility that host acceptance in the laboratory biotype has diversified from the original wild population cannot be excluded. It is interesting to note that suitability for oviposition and the capacity to support larval acceptance do not always correspond (Birch et al. 1989). Females will oviposit on seeds which are totally unsuitable for larval development (Moss, unpublished data).

Larval penetration and emergence bioassay

The larval penetration and emergence bioassay was adapted from several procedures (Howe and Currie 1964; Schoonhoven et al. 1983; Gatehouse et al. 1987; Shade et al. 1987). The use of microtitre plates allows convenient separation and confinement of a single seed and egg in each well. Eggs were obtained from 3-day old females which are known to produce the highest proportion of viable eggs (Parsons, pers. comm. 1993). The earliest stage at which the viability of eggs can be accurately determined under a low power microscope is when they are 4–5 days old; this is why eggs of this age are used in the assay (Moss, unpublished data).

The movement of larvae before penetration is a unique problem associated with *A. obtectus*. Larvae can survive for up to 120 hours without food (Credland and Dendy, unpublished data). Ensuring that only a single larva has penetrated a seed is difficult. A single penetration hole is not enough; it is common for 'pioneer' larvae to bore the hole and other 'follower' larvae then to use this as means of entering (Labeyrie 1960). It is important to eliminate interlarval competition for the food reserves within a seed for the reasons described in the oviposition bioassay.

Legume seeds come in many shapes and sizes. Wild accessions of *P. vulgaris* are typically small (approximately 5 mm in length and 3 mm in diameter), reniform and flattened. These seeds do not fit tightly in the well (7 mm dia) of the smallest microtitre plate (96 well); often only the bottom of the seed is in contact with another surface. Larger seeds, which are able to touch the sides, will therefore provide more potential contact points for larval penetration. In stores, contact points are provided by neighbouring seeds and the addition of extra seed could simulate this situation in the laboratory. Unfortunately, many samples of wild accessions are small and consequently few seeds can be used in each assay. Glass beads provide the most suitable alternative (Gatehouse et al. 1987). The addition of glass beads, which were as effective as extra seeds, gave greatly increased penetration into all accessions investigated.

Larvae often wander for a long time (3–4 days) in the well and subsequently may not have enough energy to penetrate; others escape into neighbouring wells. Physical separation of accessions and individual replicates (as described above) minimised the risk of multiple infestation. Alterations to the apparatus, however, were not sufficient to eliminate all the variation. Re-applying eggs in the event of eggs failing to hatch, or death or disappearance of the larvae, provided a partial solution to the problem. A significant decrease in the variation attributable to these problems was observed when this replacement technique was used (Moss, unpublished data).

Repeating the bioassay technique on the culture bean, red kidney beans, proved the bioassay to be effective and consistent. High penetration and emergence values were recorded in all three experiments. The development of a repeatable bioassay on a susceptible host provides a means of screening for resistance in wild accessions of unknown host suitability. The accessions G12861, G12880 and G12953, using the

Table 3. Estimation of resistance at larval penetration and development..

Accession	Mean weight of seed (g)	Total penetration	Total emergence	Resistance/ susceptibility (possible site of resistance)
G12861	0.0458	0	0	Resistant (testa/cotyledon)
G12880	0.0331	1	0	Resistant (testa/cotyledon)
G12953	0.0487	15	1	Resistant (cotyledon)
G10007	0.0633	6	4	Intermediate (testa)
G12949	0.0569	10	7	Susceptible
G12933	0.1352	19	17	Susceptible
G10000	0.1729	19	17	Susceptible
G12923	0.1576	20	20	Susceptible
Red kidney beans	0.589	20	20	Susceptible
Black-eye bean	0.240	19	13	Susceptible

All data are derived from using 20 replicates.

Table 4. Developmental period, weight and fecundity of insects emerging from six wild accessions of *P. vulgaris* and red kidney beans.

Accession	Development period (Mean no. of days ± SE)	Mean weight (mg) ± SE		Fecundity (Mean total no. of eggs ± SE)
		Male	Female	
G12953	60 (n=1)	3.287 (n=1)	–	–
G10007	35.60 ± 0.678 (n=4)	3.94 ± 0.465 (n=2)	5.77 ± 1.011 (n=2)	31 (n=1)
G12949	46.71 ± 2.02 (n=7)	2.81 ± 0.143 (n=4)	3.98 ± 0.293 (n=3)	18.5 ± 8.995 (n=3)
G12933	38.94 ± 0.915 (n=17)	4.46 ± 0.169	4.38 ± 0.289	22.166 ± 6.156
G10000	32.88 ± 0.283 (n=17)	5.44 ± 0.440	6.22 ± 0.369 (n=3)	54.33 ± 6.936 (n=3)
G12923	36.65 ± 1.013 (n=20)	4.30 ± 0.109	5.28 ± 0.910	33.44 ± 6.301
Red kidney	31.42 ± 0.364 (n=58)	5.04 ± 0.116	6.63 ± 0.104	43.72 ± 2.259

All data are derived from original larval penetration and development bioassays with 20 replicates. Replicate numbers in the development period column refer to the total number of insects emerging. In other columns n > 5 unless stated otherwise.

criteria previously described, were determined to be resistant. The possible location of resistant factors was estimated by comparing the penetration and emergence data. Little or no penetration occurred in G12861 and G12880; resistance could then be presumed to be situated within the testa, although the additional presence of cotyledon resistance cannot be excluded. More than 25% of larvae penetrated into seeds of G12953 but resulting emergence was 6.67%, suggesting that resistance is possibly situated within the seed. Resistance has been previously identified in G12953 (Gatehouse et al. 1987; Dobie et al. 1990) but its nature has not been defined. Resistance was attributed to the soluble carbohydrate fraction, but the precise component could not be isolated (Minney 1990). G10007 was considered to be of intermediate resistance as penetration was greater than 25% but lower than 50%. The resulting emergence was high (80%) indicating that the site of resistance is most likely to lie within the seed testa. Confirmation of the site of resistance requires further investigation with artificial seeds (based on Shade et al. 1986).

Development time and fitness

Development times and parameters of fitness are other sites at which resistance may be manifested. The development period, mean weight of male and female and mean oviposition data were collected. Large replicate numbers are obviously desirable but the very nature of many resistant accessions resulted in low emergence and hence low replicate numbers. Increasing the frequency of emergence from resistant seeds is difficult; large amounts of seed would be required to provide more replicates and this is not usually available. Obtaining large quantities of seed from wild sources is often impractical. Accessions susceptible to larval penetration and development do, however, give adequate emergence numbers. If resistance was manifested in adult fitness alone, then this difference would indicate it. G12933 and G12923 were both susceptible to larval penetration and development, but mean male and female weights and total oviposition were significantly lower and mean development time in G12933 was significantly longer than among adults emerging from RKB. Therefore, it is possible that adult fitness may be affected, to some degree, by resistant factors within the seeds of both accessions. G10000 did not significantly differ from red kidney beans as a host for *A. obtectus* and can be considered susceptible, although it is much smaller and a similar number of seeds would produce fewer adults if larval density was not regulated. The determination of resistance criteria for accessions of intermediate or high resistance to larval penetration and development is a difficult task. Insects from G12949, for example, had a significantly longer development time than those from red kidney beans but emergence numbers were too small to provide adequate replicates for measurement of mean adult weights and total oviposition. The assay, therefore, has to be described as a preliminary screen which identifies susceptible seeds. Accessions which are not recognised as susceptible may need to be reassessed in larger experiments with more replicates before detailed assertions of their resistance qualities can be made.

Conclusions

A. obtectus, like most organisms, is inherently variable and responsive to changes in the environment. The bioassays described have attempted to reduce the variation observed in the physical environment, apparatus and the beetles under investigation, to enable the effect of the host on the insects to be observed. Variation is, however, part of the real world experienced by bruchids. The merits of reducing variation are often

contested but, in practice, a worthwhile bioassay cannot reflect the complex interactions experienced in the field.

The search for resistance to *A. obtectus* in the seeds of *P. vulgaris* demands a pair of bioassays to investigate all the possible stages at which resistance could be manifested in the insect's life cycle: oviposition acceptance, larval penetration and development, and adult fitness. The necessity to use multiple criteria was clearly illustrated because accessions deemed susceptible at oviposition were subsequently discovered to be resistant to larval penetration and development. Furthermore, because of the requirement to standardise procedure, the separate assays cannot be compounded into a simple test.

For an accession to be determined as resistant, a set of arbitrary criteria had to be defined. The criteria are not absolute but their prior definition is essential to avoid terms like 'resistant' and 'susceptible' becoming trivialised. No population of wild *Phaseolus* is likely to contain all the possible genetic variants that might exist and enhance resistance. Natural selection is unlikely to lead to the evolution of the 'perfect' resistant individual since there is probably a trade-off diminishing other elements of plant fitness.

The challenge is to provide a practical, realistic means of identifying the most resistant plants, recognising that many accessions may need screening, and that numerous factors affect the behaviour and physiology of *A. obtectus*. The procedures described in this paper provide such a means and enable rapid assessments to be made. It is recognised that statistically valid data may not always be obtained, but, as explained, this outcome has a meaning in its own right. The necessity for repetition and further assays at a later date is not precluded by an initial screen.

Acknowledgments

Caroline Moss is grateful to the Biotechnology and Biological Sciences Research Council and the Natural Resource Institute for providing a post graduate studentship. She also appreciates the additional financial help from the BBSRC and the Royal Society which enabled her to attend this conference.

References

Birch, A. N. E., Simmonds, M. S. J. and Blaney, W. M. 1989. Chemical interactions between bruchids and legumes. Advances in Legumes Biology. Monograph of Systematic Botany, Missouri Botanical Gardens 29, 781–809.

Cardona, C. and Karel, A. K. 1990. Key Insects and other invertebrate pests of beans. In: Insect pests of tropical food legumes. (ed. Singh, S. R.), John Wiley and Sons, Chichester. pp. 157–191.

Credland, P. F., Dick, K. M. and Wright, A. W. 1986. Relationships between larval density, adult size and egg production in the cowpea seed beetle, *Callosobruchus maculatus*. Ecological Entomology, 11, 41–50.

Dobie, P., Dendy, J., Sherman, C., Padgham, J., Wood, A. and Gatehouse A. M. R. 1990. New sources of resistance to *Acanthoscelide obtectus* (Say) and *Zabrotes subfasciatus* Boheman (Co. optera: Bruchidae) in mature seeds of five species of *Phaseolus*. Journal of Stored Product Protection 26(4), 177–186.

Gatehouse, A.M.R., Dobie, P., Hodges, R.J., Meik, J., Pusztai, A. and Boulter, D. 1987. Role of carbohydrates in insect resistance in *Phaseolus vulgaris*. Journal of Insect Physiology, 33(11), 843–850.

Howe, R. W. and Currie, J. E. 1964. Some laboratory observations on the rates of development, mortality and oviposition of several species of Bruchidae breeding in stored pulses. Bulletin of Entomological Research, 55(3), 437–477.

Huignard, J. and Biemont, J. C. 1981. Reproductive polymorphism of populations of *A. obtectus* from different Colombian Ecosystems. In: V. Labeyrie ed. The ecology of Bruchids attacking legumes (pulses). Proceedings of the International Symposium held at Tours (France), April, 1980. Volume 19, 149–164.

Jarry, M. and Bonet, A. 1982. La bruche de haricot. *Acanthoscelides obtectus* Say (Coleoptera: Bruchidae), est-elle un danger pour le cowpea, *Vigna unguiculata* (L.) Walp. Agronomie, 2, 963–968.

Labeyrie, V. 1960. Nature de la répartition des larves d'Acanthoscelides *obtectus*. Comptes rendus des séances de l'Academie des Sciences, 251, 149–151.

Minney, B. H. P. 1990. Breeding *P. vulgaris* (common bean) for resistance to the major pest bruchids *Z. subfasciatus* and *A. obtectus*. Biochemical bases for seed resistance in wild lines. Ph.D. Thesis. University of Durham.

Schoonhoven, A. V. Cardona, C. and Valor, J. 1983. Resistance to the bean weevil and the Mexican bean weevil (Coleoptera: Bruchidae) in noncultivated common bean accessions. Journal of Economic Entomology, 76, 1255–1259.

Shade, R. E., Murdock, L. L., Foard, D. E. and Pomeroy, M. A. 1986. Artificial seed system for bioassay of cowpea weevil (Coleoptera: Bruchidae) growth and development. Environmental Entomology, 15, 1286–1291.

Shade, R. E., Pratt, R. C. and Pomeroy, M. 1987. Development and mortality of the bean weevil, A. *obtectus* (Coleoptera: Bruchidae), on mature seed of tepary beans, *P. acutifolius*, and common beans, *P. vulgaris*. Environmental Entomology, 16, 1067–1070.

Singh, S. R. 1990. Insect pests of tropical food legumes. John Wiley & Sons, Chichester.

Southgate, B.J. 1978. The importance of the Bruchidae as pests of grain legumes, their distribution and control. In: Singh, S. R., van Emden, H. F., and Taylor, T. A. eds., Pests of grain legumes: Ecology and Control. London/New York, Academic Press. 219–229.

van Emden, H. F. 1987. Cultural methods: the plant. In: A. J. Burn, T. H. Coaker and P. C. Jepson, eds, Integrated pest management. Academic Press. Pp. 27–68.

Function and composition of cuticular hydrocarbons of stored-product insects

J. Nawrot*, E. Malinski† and J. Szafranek†

Abstract

The main function of surface lipids is to minimise the loss of body water through the cuticle. For stored-product insects, living in extremely dry products and a desiccating environment, preservation of water supplies is crucial. The compounds commonly found in cuticular lipids belong to following classes: n-alkanes, n-alkanes 2- and 3-methylalkanes, primary and secondary alcohols, free fatty acids, and sterols. This paper presents a comparative analysis of chemical composition of hydrocarbons and their roles in stored-product insect life, and reviews the literature on this topic.

Introduction

The original habitats of stored-product insects were rodent burrows full of grain and seeds. When humans built store houses, the insects moved and modified their physiology to mild temperature and low humidity. This was achieved by appropriate adoption of cuticular lipids. These compounds (free and bounded) play an important role in insect physiology. The main function of free lipids is the reduction of water transpiration and the protect of insects from desiccation (Hadley 1981). Furthermore, lipid layers protect insects against infection from microorganisms and the absorption of insecticides (Blomquist et al. 1987). Some of the cuticular components are also involved in chemical communication between species; pheromones for aggregation, sex attraction, short-range mating stimulation and kairomones. They may act also as species and caste recognition cues.

The function of the cuticular waxes of stored-product insects prompted us to study their composition and to compare those in various species living in similar conditions. Recent developments in structural studies are presented in this paper as well.

Materials and Methods

Insects from laboratory breedings were extracted with methylene chloride. For outer lipids, adults or larvae were dipped in the solvent for 30 seconds and for internal waxes they were transferred to a new pool of solvent and left for 2 weeks. The solvent was removed under reduced pressure. The isolation of organic compound groups (group analysis) was performed by liquid-solid HPLC with gradient elution from eluent A-n-hexane to 100% of eluent B-diisopropyl ether. A light-scattering instrument was applied for detecting the fractions. A column (30 cm × 3 mm) packed with silica gel was used for separation (Separon SGX, 7 μm, Tessec Ltd, Praha, Czech Republic). Gas chromatography (GC) analyses were performed with a Varian 1400 gas chromatograph using 30 m × 0.23 mm fused silica capillary column with DB-1 liquid phase. The oven temperature was programmed from 150°C to 300°C at 2°C/minute. Argon was used as a carrier gas. Kovats' retention indexes were measured at temperatures adequate for k' values of the signals greater than 5 or with temperature programming. The error was ±1.0 unit. Field desorption fingerprinting of the fractions were performed with Varian MAT 711 mass spectrometer equipped with a combined FD-FI-EI ion source. GC/MS measurements were carried out with a VG Micromass 7070E (Dani 3800 GC) or VG Micromass 16F (Pye Unicam GC) mass spectrometer.

Results and Discussion

Composition of free cuticular lipids of stored-product insects varies from species to species, being a chemical tool for taxonomy. Nevertheless, the following classes of chemical compounds can be found in cuticular lipids: alkanes, isoalkanes, alkenes, esters, glycerides, aldehydes, ketones, fatty acids and sterols. These complicated mixtures require a proper strategy for structural study. The first step of qualitative and quantitative analysis relies on suitable separation of the mixture into groups of the above-mentioned compounds.

The separation can be done by liquid-solid chromatography with gradient elution. Unfortunately, the compounds do not absorb in UV range and UV spectrophotometric detector cannot be applied. The refractive index detector, in turn, does not accept gradient elution. The only solution is the application of a mass detector with laser beam scattering—a selection which has been successfully used in different plant lipids analyses (Stolyhwo et al. 1985). Our separation of the cuticular lipids of *Tribolium confusum* is presented in Figure l. All important classes of compounds represented in cuticular lipids are separated: (1) alkanes, (2) alkenes and alkadienes, (3) ester waxes, (4) triacylglycerols, (5) fatty acids and (6) sterols. HPLC analysis with gradient elution provides a method for fast screening of the classes of compounds in relation to habitat conditions. The collection of the fractions provides the mixtures of compounds for gas chromatography and mass spectrometry analyses.

The strategy of the structural study of the cuticular lipids separated into classes consisted of the following stages:

- field desorption mass spectrometric fingerprinting of the samples for molecular ions of the components;
- gas chromatographic analyses with Kovats' retention indexes measurement;
- recording of mass spectrum of GC separated single component (GC/MS);
- correlation of the structures with the results of biosynthesis e.g. of isoalkanes.

---

* Institute for Plant Protection, ul.Miczurina 20, 60-318 Poznan, Poland.
† Gdansk University, Department of Chemistry, il.Sobieskiego 18, 80-952 Gdansk, Poland.

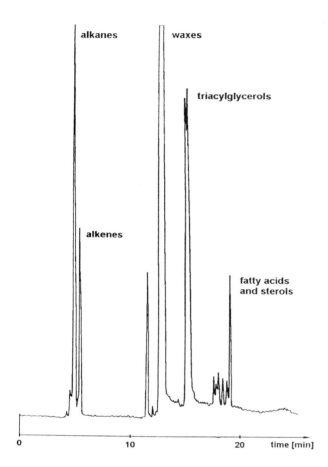

Fig. 1. HPLC group separation of cuticular lipids of *Tribolium confusum*. Gradient elution n-hexane-di-isopropylether, laser light scattering detector, silica gel column, 5μm.

The last step must be taken with precaution, because unusual positions of methyl branches are even found in nature (Gries et al. 1993).

Labelled substrate experiments showed (Chu and Blomquist 1980) that insects synthesise most of their hydrocarbons. Composition of these compounds is therefore an expression of genotype and as such is a taxonomic feature. The question arising is to what extent can habitat conditions modify the quantity and quality of cuticular waxes. For example, a dry season increases the wax blooms of the beetles belonging to the family Tenebrionidae, and the secretion stops when the relative humidity rises (Hadley 1979). The habitat of the species studied is warehouses with low humidity. Our results and literature data on cuticular hydrocarbons of stored-product pests are compiled in Tables 1–7. The insects studied belong (except one) to order Coleoptera (beetles). Most of them (eight species) represent Tenebrionidae, three species Curculionidae, two Dermestidae and Bruchidae, one Anobiidae, Bostrychidae and Cucujidae. One species belongs to the order Lepidoptera, family Pyralidae.

There is a great diversity of compounds found in cuticular waxes. Only n-alkanes occurred in all examined species, though in variable proportion. They are stated as continuous homologous series from nC16 to nC35, and compounds of nC23, nC25, nC27 and nC29 are usually predominant and most abundant. Larvae of *Attagenus megatoma* and *Lasioderma serricorne* possess the most amount of n-alkanes (98 and 83% respectively). On the contrary, adults of *Lasioderma serricorne*, *Callosobruchus maculatus*, *Alphitobius bifasciatus* and *Acanthoscelides obtectus* have the least amounts of

this fraction. Presence of monomethylalkanes (terminally and internally branched), dimethyl- and trimethylalkanes cause higher melting point of the lipid mixture. However, composition of these compounds varies among species, and larvae of *Attagenus megatoma* do not possess them at all (Tables 2, 3, 4 and 5).

n-Alkenes and n-alkadienes were stated only in a few species (Tables 6 and 7). Lockey (1976) and Jackson and Blomquist (1976) investigated the relationship between the hydrocarbon composition and the taxonomy of insects. Those early attempts concluded that most insect species had n-alkanes, monomethylalkanes and dimethylalkanes. But this finding did not build a solid base for taxonomy. The most pronounced results were obtained in chemotaxonomy of Tenebrionidae family (Crowson 1981). According to literature data (Lockey 1988), hydrocarbon composition is related to taxonomic grouping in such a way that closely related species, such as congeneric, tend to have qualitatively similar hydrocarbon mixtures but with sometimes different proportions. More distantly related species tend to have hydrocarbon mixtures which differ qualitatively and quantitatively. Among the species presented in Tables 1–7, *Tribolium castaneum* and *Tribolium confusum* have very similar hydrocarbon compositions (see Lockey 1988).

More or less similar qualitative and quantitative compositions were found in the following groups of species:

- Qualitative and quantitative great similarity
 Tribolium confusum T. castaneum
 Sitophilus oryzae S. zeamais
- Qualitative similarity
 Sitophilus oryzae S. zeamais S. granarius
 Tribolium confusum T. castaneum T. destructor
 Tenebrio molitor T. obscurus
 Acanthoscelides obtectus Callosobruchus maculatus
- Some similarity
 T. confusum T. castaneum T. destructor Blaps mucronata
 Anagasta kuehniella (larvae) *Lasioderma serricorne* (larvae) *Attagenus megatoma* (larvae)
- No similarity
 Rhyzopertha dominica Oryzaephilus surinamensis

A new approach in chemical ecology is the search for the roles of plant and insect cuticular lipids in insect- plant interactions. The behaviour of many herbivorous insects is affected by plant cuticular waxes and these compounds may play the role of attractants, stimulators of feeding or oviposition, and food deterrents. The main components of lipids on wheat grain surfaces are n-alkanes in the following proportions: n-pentacosane (C25), 3.6%; n-hexacosane (C26), 0.8%; n-heptacosane (C27), 23.8%; n-octacosane (C28), 3.9%; n-nonacosane (C29), 42.3%; n-triacontane (C30), l.2%; n-hentriacontane (C31), 23.8%; n-tritriacontane (C33), 0.6%. It is worth emphasising that there are no other hydrocarbons on wheat grain. If we compare the composition of n-alkanes from grain and from grain feeding insects we can see the high correlation in the proportion of main components. Because grain cuticular lipids may play a major role in the main life processes of insects, e.g. finding food, oviposition and rate of population growth, it is very important to establish their composition and biological properties.

Table 1. n-alkanes in cuticular lipids of stored-product insects (% of total hydrocarbons).

Species	16	17	18	19	20	21	22	23	24	25	26	27	28	29	30	31	32	33	34	35	Total amount [%]	References
Tribolium destructor										1.5	2.5	23.2	7.4	7.0		-					41.6	14,15
Tribolium destructor (larvae)										6.8	6.7	32.5	6.7	6.6		0.4					59.7	15
Tribolium castaneum										3.5	1.2	22.8	6.2	26.1	1.0	2.3	<0.1	<0.1			63.1	19
Tribolium confusum										0.6	0.2	23.3	0.5	26.1	1.1	0.5					61.7	19
Tenebrio molitor			<0.1	<0.1	0.2	<0.1	0.1	11.6	1.2	29.1	0.8	9.7	0.2	0.3							53.2	18
Tenebrio obscurus			<0.1	<0.1	<0.1	<0.1	<0.1	8.8	1.5	27.5	0.9	5.7	0.9	4.7	0.2						50.2	18
Blaps mucronata								1.6	1.5	15.4	4.2	14.1	2.6	1.7							41.1	21
Alphitophagus bifasciatus										0.3	0.8	8.6	3.1	8.5	1.7	0.4					23.4	20
Alphitobius diaperinus										2.0	0.6	15.4	5.2	11.6							34.8	20
Trogoderma granarium								2.2	1.1	5.2	1.5	9.8	1.4	9.7	0.5	2.5	0.5	0.5			34.9	9
Trogoderma granarium (larvae)												13.3	0.7	7.7		2.9		1.4			26.0	22
Attagenus megatoma (larvae)										1.6	0.5	20.6	0.6	16.7	0.5	21.1	0.9	32.1		4.7	98.3	1
Callobruchus maculatus												5.2	3.3	7.8		0.4					16.7	2
Acanthoscelides obtectus										2.4	1.1	17.0	1.7	6.2		1.1	0.3	0.9			28.5	8
Lasioderma serricorne										0.2	0.2	3.7	0.6	1.0	0.2	1.6	0.3	0.9			8.7	3
Lasioderma serricorne (larvae)										0.3	0.3	2.5	12.2	8.9	2.1	11.1	2.5	37.2	1.0	5.4	83.5	3
Sitophilus oryzae								0.1		0.1		3.5	0.6	16.2	0.3	3.7		0.4			24.9	4
Sitophilus zeamais										0.1		3.6	0.7	17.0	0.3	2.8		0.4			24.9	4
Sitophilus granarius								0.3	0.8	2.9	0.3	15.3	0.3	15.3	0.3	3.5	0.2	0.5			39.7	4
Rhizopertha dominica												6.5	1.0	10.6	0.4	8.5		2.2		7.5	36.7	25
Oryzaephilus surinamensis	0.9	0.4	0.6	0.5	3.3	0.6	3.6	0.1		7.3	1.9	1.9	11.6	1.1	11.7	0.2	0.3				45.8	26
Anagasta kueniella (larvae)	1.4	1.3	1.5	0.8	1.0	1.2	1.4	12.6	2.5	25.2	3.4	11.6	2.6	8.5	1.2	1.9					78.1	13

(Columns 16–35 = Number of carbon atoms)

Table 2. Composition of monomethyl-terminally branched alkanes in cuticular lipids of stored-product insects (% of total hydrocarbons).

Species	Number of carbon atoms												Total amount [%]	References
	24	25	26	27	28	29	30	31	32	33	34	35		
Tribolium destructor			0.4		20.3								20.7	14,15
Tribolium destructor (larvae)			0.5		4.2		0.1						4.8	15
Tribolium castaneum			1.5	0.4	7.9	0.3	2.7		0.1				12.9	19
Tribolium confusum			<0.1	0.1	7.6	0.4	2.9		<0.1				11.0	19
Tenebrio molitor			0.7		<0.1								0.7	18
Tenebrio obscurus			7.0		2.9								9.9	18
Blaps mucronata	0.8	0.3	9.7	2.3	13.8	0.5	1.3	0.2		0.9			29.8	21
Alphitophagus bifasciatus			0.3		16.6	1.6	5.3						24.0	20
Alphitobius diaperinus			0.4	1.1	8.7	1.3b / 0.7	16.3	2.2c	0.3			0.2	30.0	20
Trogoderma granarium			2.2										2.2	9
Trogoderma granarium (larvae)					6.1		0.8						6.9	22
Attagenus megatoma (larvae)													0.0	4
Callosobruchus maculatus					7.3	1.7	33.0		3.5				45.5	2
Acanthoscelides obtectus			4.3		13.7								18.0	8
Lasioderma serricone			0.1	0.1	2.9	2.2	7.8						13.1	3
Lasioderma serricone (larvae)					1.8	0.1	0.9	0.1					2.9	3
Sitophilus oryzae							<0.1		<0.1				0.0	4
Sitophilus zeamais							<0.1		<0.1				0.0	4
Sitophilus granarius							0.3		0.2				0.5	4
Rhizoperta dominica					1.2								1.2	25
Oryzaephilus surinamensis													0.0	26
Anagasta kueniella (larvae)	0.5	0.5a	1.9		1.3								4.2	13

a/ 2-methyltetracosane   b/ 2-methyloctacosane   c/ 2-methyltriacontane

Table 3. Composition of monomethyl-internally branched alkanes in cuticular lipids of stored-product insects (% of total hydrocarbons).

Species	Number of carbon atoms																									Total amount [%]	References
	22	23	24	25	26	27	28	29	30	31	32	33	34	35	36	37	38	39	40	41	42	43	44	45	46		
Tribolium destructor					0.7	1.2	13.0	5.2																		20.1	14,15
Tribolium destructor (larvae)					0.8	0.7	4.8	1.3	0.3																	7.9	15
Tribolium castaneum						0.6	14.4	2.2	0.2	0.1	0.2															17.7	19
Tribolium confusum						0.4	11.3	2.4	6.6	0.2	0.1															21.0	19
Tenebrio molitor	0.1	0.1	2.5	0.1	0.5		<0.1	<0.1	11.7	0.2	4.7	0.3	2.4													22.5	18
Tenebrio obscurus	<0.1	<0.1	2.1	0.1	0.5		0.9	0.7	1.4	0.7	2.4	0.2	1.0													10.0	18
Blaps mucronata			0.1	0.3	2.2	1.3	2.3	0.5	0.7	0.3	1.0	0.4	1.0	0.5	0.1											10.7	21
Alphitophagus bifasciatus						0.3	1.7	2.8	8.0	1.4	1.5	0.3	0.5	<0.1	<0.1											16.5	20
Alphitobius diaperinus					0.1	0.1	1.6	2.2	2.4	1.3	0.2	0.2	0.1	<0.1		1.0										10.3	20
Trogoderma granarium					4.0	1.8	32.9	2.8	9.1	1.6								1.1								52.2	9
Trogoderma granarium (larvae)						0.7	22.5	1.0	8.6	0.8	1.7		1.3													36.0	22
Attagenus megatoma (larvae)																										0.0	1
Callobruchus maculatus							3.2	1.6	17.3	1.9	1.4		0.2								0.4	0.1	1.0		0.5	27.6	2
Acanthosceli- des obtectus					1.4		11.8		6.0																	19.2	8
Lasioderma serricone						0.3	6.4	11.1	21.3	5.0	1.1	0.1			0.1											45.4	3
Lasioderma serricone (larvae)							1.7	1.1	3.9	0.4	0.5															7.6	3
Sitophilus oryzae											<0.1		0.1													0.1	4
Sitophilus zeamais											<0.1		0.1													0.1	4
Sitophilus granarius							0.3		0.4		0.5		0.2													1.4	4
Rhizoperta dominica							1.3				1.2		6.5		6.5											15.5	25
Oryzaephilus surinamensis							1.8						0.5		1.0				0.3							3.6	26
Anagasta kueniella (larvae)					0.5																					0.5	13

Table 4. Composition of dimethylalkanes in cuticular lipids of stored-product insects (% of total hydrocarbons).

Species	\multicolumn{19}{c|}{Number of carbon atoms}																			Total amount [%]	References
	27	28	29	30	31	32	33	34	35	36	37	38	39	40	41	42	43	44	45		
Tribolium destructor	1.3		13.5	1.3																16.1	14,15
Tribolium destructor (larvae)	0.8		1.6	0.4																2.8	15
Tribolium castaneum			2.5	0.5	0.7															3.7	19
Tribolium confusum			4.1	1.1	0.7															6.5	19
Tenebrio molitor	0.3	0.1	<0.1		1.7	0.9	1.4	0.1	3.2		0.1									7.8	18
Tenebrio obscurus	0.3	0.1		2.2	0.8	4.9	0.7	0.9	0.9		<0.1		<0.1	<0.1	<0.1					10.8	18
Blaps mucronata							4.7		3.9	1.4	4.4									14.4	21
Alphitophagus bifasciatus					2.2	1.3	1.7	0.7	0.7		0.3		0.6							7.5	20
Alphitobius diaperinus			1.8		3.5		0.9	0.5	0.8	0.5	5.3	2.1	3.1	1.4	3.7					23.6	20
Trogoderma granarium																				0.0	9
Trogoderma granarium (larvae)																				0.0	22
Attagenus megatoma (larvae)																				0.0	4
Callosobruchus macul.			2.2	1.0	28.5	1.4	2.7		0.5								1.5	0.1	2.2	40.1	2
Acanthoscelides obtectus			24.1	2.5	3.4															30.0	8
Lasioderma serricone			4.7	4.2	8.8	10.5	3.4	0.3	0.1											32.0	3
Lasioderma serricone (larvae)				0.1	1.8	0.2			1.1		0.9									4.1	3
Sitophilus oryzae									0.1		0.3									0.4	4
Sitophilus zeamais									0.2		0.3									0.5	4
Sitophilus granarius									0.5		0.2									0.7	4
Rhizoperta dominica							3.1		12.3		4.0									19.4	25
Oryzaephilus surinamensis											0.5		0.3		0.7					1.6	26
Anagasta kueniella (larvae)	1.7		0.7																	2.4	18

Table 5. Composition of trimethylalkanes in cuticular waxes of stored-product insects (% of total hydrocarbons).

Species	31	32	34	35	36	Total amount (%)	References
Acanthoscelides obtectus	2.5					2.5	8
Rhyzopertha dominica		0.4	2.6		5.7	8.7	27

Table 6. Composition of n-alkenes in cuticular waxes of stored-product insects (% of total hydrocarbons)

Species	23	24	25	26	27	28	29	30	31	32	33	34	35	36	37	38	39	40	41	42	43	44	45	Total amount [%]	References
Tenebrio molitor	0.4	0.4	12.8		2.0		2.7		0.5		1.4		0.3				<0.1							20.5	18
Tenebrio obscurus	0.2	0.2	7.2		7.4		1.6	0.1	0.7		0.5		0.1				0.1							18.1	18
Alphitophagus bifasciatus					14.4	3.1	10.0		0.3															27.8	20
Trogoderma granarium (larvae)					4.0		13.0		5.0		1.0													23.0	22
Sitophilus oryzae							0.7	0.2	8.1	0.3	6.3	0.1	1.3		0.3						0.1		0.1	17.5	24
Sitophilus zeamais					0.1		2.4	0.2	6.3	0.2	4.2	0.1	1.0		0.1									14.6	4
Sitophilus granarius					0.4		2.2	0.6	14.0	0.5	7.9	0.1	0.6								0.1			26.5	4
Oryzaephilus surinamensis			0.4	0.9	13.5		12.0		3.9															30.7	26

Table 7. Composition of n-alkadienes in cuticular waxes of stored-product insects (% of total hydrocarbons)

Species	29	30	31	32	33	34	35	36	37	38	39	40	41	42	43	44	45	46	47	Total amount [%]	References
Trogoderma granarium (larvae)	2.4																			2.4	22
Sitophilus oryzae			0.2	0.2	7.8	0.7	22.5	0.8	14.3	0.3	2.3	0.1	0.4	0.2	1.0	0.1	0.4		0.1	51.4	23
Sitophilus zeamais			0.1	0.1	8.9	1.5	33.0	0.9	12.8	0.1	0.6		0.4		0.2					58.6	23
Sitophilus granarius			0.9	0.2	8.2	0.1	9.1	0.1	1.1		1.1	0.1	1.5	0.3	1.4	0.2	0.9		0.4	25.6	23

Acknowledgments

Financial support for this work from the State Committee for Scientific Research (Grant No 5 S307 019 05) is gratefully acknowledged.

References

Baker, J.E. 1978. Cuticular lipids of larvae of *Attagenus megatoma*. Insect Biochemistry, 8, 287–292.

Baker, J.E. and Nelson, D.R. 1981. Cuticular hydrocarbons of adults of the cowpea weevil, *Callosobruchus maculatus*. Journal of Chemical Ecology, 7, 175–182.

Baker, J.E., Sukkestad, D.R., Nelson, D.R. and Fatland, C.L. 1979. Cuticular lipids of larvae and adults of the cigarette beetle, *Lasioderma serricorne*. Insect Biochemistry, 8, 159–167.

Baker, J.E., Woo, S.M., Nelson, D.R. and Fatland, C.L. 1984. Olefins as major components of epicuticular lipids of three *Sitophilus* weevils. Comparative of Biochemistry and Physiology, 77B, 877–884.

Blomquist, G.J., Nelson, D.R. and de Renobales, M. 1987. Chemistry, biochemistry and physiology of insect cuticular lipids. Archives of Insect Biochemistry and Physiology 6, 227–265.

Chu, A.J. and Blomquist, G.J. 1980. Biosynthesis of hydrocarbons in insects: succinate is a precursor of the methyl branched alkanes. Archives of Biochemistry and Biophysics, 201, 304–312.

Crowson, R.A. 1981. The biology of Coleoptera. Academic Press, London, 272 p.

Dubis, E., Malifiski, E., Hebanowska, E., Synak, E., Pihlaja, K. Nawrot, J. and Szafranek, J. 1987. The composition of cuticular hydrocarbons of the bean weevil, *Acanthoscelides obtectus* Say. Comparative Biochemistry and Physiology, 88B, 681–685.

Dubis, E., Mali Aski, E., Hebanowska, E., ´Swiecka M., Nawrot, J., Pihlaja, K., Szafranek, J. and Warnke, Z. 1987. The composition of cuticular hydrocarbons of the khapra beetles, *Trogoderma granarium*. Comparative Biochemistry and Physiology 88B, 911–915.

Gries, G., Gries, R., Krannitz, S.H., Li, H., King, G.G.S., Slessor, K.N., Borden, J.H., Bowers, W.W., West, R.J. and Underhill, E.W.1983. Sex pheromone of the western hemlock looper *Lambdina fiscelaria lugubrosa* (Hulst)(Lepidoptera: Geometridae). Journal of Chemical Ecology, 19, 1009–1019.

Hadley, N.F. 1979. Wax secretion and colour phases of the desert tenebrionid beetle, *Cryptaglossa verrucosa* (LeConte). Science, 203, 367–369.

Hadley, N.F. 1981. Cuticular lipids of terrestrial plants and arthropods. A comparison of their structure, composition and waterproofing function. Biological Review, 56, 23–47.

Hebanowska, E, Malifiski, E., Dubis, E., Oksman, P., Pihlaja, K., Nawrot, J. and Szafranek, J. 1990. The cuticular hydrocarbons of the larvae of *Anagasta kuehniella*. Comparative Biochemistry and Physiology, 95B, 699–703.

Hebanowska, E., Malifiski, E., Dubis, E., Pihlaja, K., Oksman, P., Wiinamaki, K., Nawrot, J. and Szafranek, J. 1989. The composition of cuticular hydrocarbons of the *Tribolium* destructor. Comparative Biochemistry and Physiology, 93B, 437–442.

Hebanowska, E., Maliftski, E., Latowska, A., Dubis, E., Pihlaja, K., Oksman, P., Nawrot, J. and Szafranek, J. 1990. A comparison of cuticular hydrocarbons of larvae and beetles of the *Tribolium* destructor. Comparative Biochemistry and Physiology, 96B, 815–819.

Jackson, L.L. and Blomquist, G.J. 1976. Insect waxes. In: Kollatukudy, P.E. ed. Chemistry and Biochemistry of Natural Waxes, Elsevier, Amsterdam, 201–233.

Lockey, K.H. 1976. Cuticular hydrocarbons of Locusts, *Schistocerca* and *Periplaneta* and their role in waterproofing. Insect Biochemistry, 6, 457–472.

Lockey, K.H. 1978. The adult cuticular hydrocarbons of *Tenebrio molitor* L. and *Tenebrio obscurus* F. Insect Biochemistry, 8, 237–250.

Lockey, K.H. 1978. Hydrocarbons of adult *Tribolium castaneum* Host. and *Tribolium confusum* Duv. *Coleoptera: Tenebrionidae*). Comparative Biochemistry and Physiology, 61B, 401–407.

Lockey, K.H. 1979. Cuticular hydrocarbons of adult *Alphitophagus bifasciatus* (Say) and *Alphitobius diaperinus* (Panz.)(Coleoptera: Tenebrionidae). Comparative Biochemistry and Physiology 64B,47–56.

Lockey, K.H. 1980. Cuticular hydrocarbons of adult *Blaps mucronata* Latreille (*Coleoptera: Tenebrionidae*). Comparative Biochemistry and Physiology, 67B, 33–40.

Lockey, K.H. 1988. Lipids of the insects cuticle: origin. composition and function. Comparative Biochemistry and Physiology, 89B, 595–645.

Malinski, E., Hebanowska, E., Szafranek, J. and Nawrot, J. 1986. The composition of hydrocarbons of the larvae of the khapra beetles, *Trogodea granarium*. Comparative Biochemistry and Physiology, 84B, 211–215.

Nawrot, J., Szafranek, J., Dubis, E., Hebanowska, E., Latowska, E. Malinski. E. and Pihlaja, K. 1990. Structural studies of cuticular hydrocarbons of storage pests. In: Hrdy, L. ed., Proceedings of the Conference on Insect Chemical Ecology, Tabor, Czechoslovakia, 323–326.

Nelson, D.R., Fatland, C.L. and Baker, J.E. 1984. Mass spectral analysis of epicuticular n-alkadienes in three *Sitophilus* weevils. Insect Biochemistry, 14, 435–444.

Stolyhwo, A., Colin, H. and Guiochon, G. 1985. Analysis of triglycerides in oils and fats by liquid chromatography with laser light scattering detector. Analytical Chemistry, 57, 1342–1345.

Szafranek, J., Malifiski, E., Hebanowska, E., Dubis, E., Oksman, P., Pihlaja, K. and Nawrot, J. 1994. Cuticular hydrocarbons of *Rhyzopertha dominica*, F. Insect Biochemistry, in press.

Szafranek, J., Hebanowska, E., Malinski, E., Dubis, E., Pihlaja, K. and Nawrot, J. 1994. Cuticular hydrocarbons of *Oryzaephilus surinamensis* and their chemotaxonomic significance. Prepared for publication.

Factors affecting oviposition and orientation by female *Plodia interpunctella*

T.W. Phillips [*†] and M.R. Strand[†]

Abstract

Female Indianmeal moths, *Plodia interpunctella*, are attracted to food volatiles, and oviposit in response to secretions by conspecific larvae. Female moths oviposited on dishes containing larval rearing medium (food), and cornmeal was found to be the active component of this mixture. Cheesecloth contaminated by different densities of fifth instar larvae elicited more oviposition than untreated cheesecloth or dishes with food. The combination of larval contamination and food was preferred over food only or larval contamination only in both two- and four-choice experiments. The factor(s) in larval contamination responsible for eliciting oviposition was extracted in hexane, confirming that organic semiochemicals are responsible for the effect. In two-choice wind tunnel bioassays, female moths initiated flight only when food was present in one of the treatments, and they displayed the highest landing responses to a combination of larval contamination and food. Earlier work on *P. interpunctella* and related pyralid species found that secretions from the mandibular glands of larvae acted as both a spacing pheromone for wandering larvae and as a kairomone for host-seeking parasitoid wasps. The present study suggests that a similar secretion acts as an oviposition eliciting pheromone for conspecific females, and that there is a distinct interaction between food and larval secretion.

Introduction

The Indianmeal moth, *Plodia interpunctella* (Hübner) feeds on stored grains, legumes, nuts, dried fruits and other food products maintained in storage. Research on the chemical ecology of *P. interpunctella* has dealt primarily with identification and activity of female sex pheromones (e.g. Brady et al. 1971; Mankin et al. 1983). Although subtle differences in sex pheromone systems occur, at least four species of the stored-product Phycitinae share a major pheromone component that is responsible in part for cross-attraction (e.g. Phelan and Baker 1986). Synthetic pheromone-baited traps are commercially available to monitor activity of males of *P. interpunctella*. However, little is known about the orientation of *P. interpunctella* females to stored products, or whether specific oviposition stimuli exist.

Research on other species of Phycitinae suggest probable mechanisms that mediate host orientation and oviposition in *P. interpunctella*. *Ephestia kuehniella* (Walker) laid more eggs on substrates contaminated by conspecific wandering fifth instar larvae than on untreated surfaces (Corbet 1973). An oviposition stimulant was found in secretions from the larval mandibular glands, but no compounds were identified, nor was the activity of any larval food investigated (Corbet 1973). *Amyelois transitella* (Walker) preferentially oviposited in the vicinity of host fruits infested with conspecific larvae over uninfested fruits (Andrews and Barnes 1982). Barrer (1977) found that *Cadra* (=*Ephestia*) *cautella* (Walker) oviposited in response to volatiles from a grain-based rearing medium. Phelan et al. (1991) recently identified five fatty acids from almond oil that elicited upwind orientation of *A. transitella*. The studies above suggest that chemical factors associated with both conspecific larvae and larval food may influence female host selection behaviour.

Here we review our previous findings (Phillips and Strand 1994) and discuss additional experiments designed to determine the relative importance and interaction of food and larval secretions in oviposition and flight behaviour of female *P. interpunctella*. Our studies found that larval contamination elicited stronger close-range oviposition responses than food, but that food was required to elicit upwind flight orientation from a distance. The combination of food and larval secretions, however, elicited stronger responses than either component alone in both close-range and long-range orientation.

Methods

P. interpunctella were reared on a standardised diet of corn meal, chick starter mash, and glycerine 1: 1: 1 (v/v/v). Ten gravid females were placed in ventilated 500 mL jars containing 400 g of diet and maintained at 27°C with a 16: 8 (L: D) photoperiod. Larvae used in experiments were in the wandering phase of the fifth instar. Newly emerged adult moths were collected daily and held in plastic boxes (10 × 22 × 30 cm) at 27°C and a 16: 8 (L: D) photoperiod. Females usually mated during the first or second day after emergence and started egg laying during scotophase of day four. Thus, four-day-old mated females were used in all experiments.

The influence of food and larval contamination on *P. interpunctella* oviposition was examined in two and four-choice bioassays conducted during scotophase under the rearing conditions described above. In two-choice assays, a single female was introduced into a 10 × 22 × 30 cm plastic box that contained two potential oviposition sites. These sites consisted of 60 × 15 mm petri dishes overlaid with a circular piece of cheesecloth, about 70 mm in diameter, placed about 10 cm apart on the floor of the box. Moths generally laid eggs directly on the cheesecloth patch when the appropriate stimulus was present in the dish or on the cheesecloth. The influence of larval food was tested by filling petri dishes with 10 g of our standard rearing diet (see earlier) or with components of the diet. The effects of different levels of larval contamination were examined by allowing 5, 20 or 50 larvae to crawl on the cheesecloth patches for 24 hours prior to the experiment. Petri dishes were filled with food and/or overlaid with cheesecloth

[*] USDA ARS Stored-Product Insect Research Unit., Madison, WI USA 53706

[†] Department of Entomology, 237 Russell Laboratories, University of Wisconsin, Madison, WI, USA, 53706.

immediately prior to introducing a female. In all oviposition experiments a single moth was introduced into the bioassay arena 1 hour prior to scotophase and the number of eggs laid on the cheesecloth pieces or within 1 cm beyond the edge of a dish was counted the following day. Females that laid five or fewer eggs were assumed to be unmated and were omitted from the analysis. The following series of two-choice experiments were conducted: 1) food vs. an empty dish; 2) the three components of the rearing diet (cornmeal, chick feed, or glycerine) vs. an empty dish; 3) larval contamination vs. an empty dish; 4) larval contamination vs. food; 5) a combination of larval contamination with food vs. food; and 6) a combination of larval contamination with food vs. larval contamination.

Extracts of larval contamination were made to determine if the biological activity of these depositions was chemical in nature. One, 5, and 20, larvae were allowed to wander for 24 hours on cheesecloth patches overlaid with a clean glass petri dish (60x15 mm) as described previously. The contaminated cheesecloth and the inside surfaces of the glass dish were extracted with a total of 5.0 mL of hexane (HPLC grade); ten dish/patch preparations were extracted separately for each larval density. Uncontaminated patches were extracted as controls. Extracts were concentrated under a stream of nitrogen to a volume of 500 µL. The entire 500 µL extract from a dish/patch preparation was applied to a 2 cm × 2 cm piece of filter paper. Two-choice oviposition experiments were conducted by placing in one dish a filter paper treated with extract from a larval contaminated preparation and in the second dish a filter paper with the control extract from an uncontaminated preparation. Petri dishes containing the filter paper were overlaid with clean cheesecloth and then assayed as described above.

Four-choice assays were conducted in 38 × 30 × 14 cm ventilated plastic boxes with the petri dishes placed 12 cm apart in the centre of the box. Females were introduced into the bioassay arena as described previously and simultaneously presented the choices of food, larval contaminated cheesecloth, food plus larval contaminated cheesecloth or an empty dish. The placement of the four oviposition sites in the box was randomised for each replicate. The numbers of eggs a female laid on the cheesecloth-covered petri dishes were then counted the following day. Three separate four-choice experiments were conducted in which larval densities of 1, 5, and 10 larvae were examined.

Flight responses of *P. interpunctella* to food and/or larval odours were examined in wind tunnel bioassays. The working portion of the tunnel measured 127 × 62 × 61 cm and was constructed of 6 mm Plexiglas. Wind speed was maintained at 25 cm/sec using an electric fan mounted upwind; the bioassay room was kept at 28°C and 50% r.h. Bioassays were run under dim red light within the first 2–3 hours of scotophase because this is the time females are most likely to fly and oviposit. Females were acclimatised to experimental conditions by holding them in a ventilated chamber in the wind without odours for 15 min prior to testing. Two petri dishes containing odour sources were placed on small platforms 30 cm apart and 45 cm above the floor at the upwind end of the tunnel. Five females were then released from a 5.5 × 10 cm screen release cylinder placed on a 45 cm platform at the downwind end of the tunnel. Treatments of food (10 g of diet), contamination by 10 larvae, and food plus larval contamination were prepared as described for the oviposition experiments with each tested against a blank dish in a series of two-choice experiments. Three categories of female behaviour were recorded during a five-minute test period: initiation of flight, upwind flight within the plume, and landing on the source. The percentages of test insects in a group exhibiting flight behaviours and

landing on either the treatment or control were calculated for each replicate.

Results

Female *P. interpunctella* laid more eggs on cheesecloth overlaying dishes containing food than on those overlaying empty dishes. A mean of 15.5 eggs (±3.1 SE) were laid on dishes containing 10 g of laboratory rearing diet compared to a mean of 7.5 (±2.4 SE) on empty dishes (P<0.01, N=10, Student's t-test). When the components of the larval rearing diet were examined, only cornmeal elicited a significant oviposition response compared to empty dishes (Fig. 1). In choice tests between increasing levels of larval contamination versus blank patches (no food in either dish), females consistently laid more eggs on larval-contaminated patches (Fig. 2a). *P. interpunctella* females laid more eggs on cheesecloth with larval contamination alone than on untreated cheesecloth overlaying food (larval rearing diet), but this preference was significant only at contamination levels of 5 and 20 larvae (Fig 2b)). Preference for substrates containing larval contamination was observed again in the next experiment (Fig. 2c), in which a combination of larval contamination and food was preferred over food only. The last set of two-choice experiments showed that the combination of food with larval contamination elicited significantly more oviposition than larval contamination only (Fig. 2d). Four-choice oviposition experiments that used contamination by 1 larva and 5 larvae found that the combination of food with larval contamination elicited the highest oviposition responses relative to larval contamination alone, food alone, or a blank patch (Table 1, Experiments I and II). However, when cheesecloth patches were contaminated by 10 larvae (Table 1, Experiment III), more eggs were laid on dishes with food only, while those with larval contamination received eggs at low levels comparable to those of the blank controls.

The oviposition-eliciting activity of larval contamination was recovered in hexane extracts from all larval densities sampled. In a series of two-choice experiments, extracts from contamination by one larva elicited a mean of 42.0 (±5.2 SE)

Table 1. Oviposition by single female *P. interpunctella* in three different four-choice experiments.

Experiment no.	Treatment[a]	Mean no. eggs per dish (+/- SE)[b]
I	Empty dish	8.3 (3.3) a
	Food only	21.6 (5.3) b
	1 larva	12.0 (3.8) ab
	Food + 1 larva	32.4 (5.9) c
II	Empty dish	4.2 (1.7) a
	Food only	20.2 (4.7) b
	5 larvae	33.8 (8.2) bc
	Food + 5 larvae	49.3 (6.7) c
III	Empty dish	6.8 (1.9) a
	Food only	20.9 (3.9) b
	10 larvae	7.5 (2.6) a
	Food + 10 larvae	13.7 (2.8) a

[a]Food=10 g of laboratory rearing diet overlaid with uncontaminated cheesecloth; the number of larvae refers to a cheesecloth piece contaminated by wandering fifth instar *P. interpunctella* larvae for 24 hours.
[b]Means within an experiment followed by different letters are significantly different (ANOVA P<0.05, Fisher's PLSD test, N=10).

eggs compared to 8.2 (±2.6 SE) for the hexane control; extracts from contamination by five larvae yielded a mean of 56.9 (±12.2) eggs compared to 2.8 (±0.8 SE) for the control; and extracts from contamination by 20 larvae yielded a mean of 66.0 (±13.8 SE) eggs compared to 9.0 (2.9) for the control. All differences between treatments and controls were statistically significant (P<0.05, N=10, Student's t-test).

Female *P. interpunctella* flew upwind in our two-choice wind tunnel bioassays only when food was included with one of the treatments. When larval contamination was compared with uncontaminated controls, no moths took flight. When food only was compared to an empty dish, 34.0 (±6.0 SE)% of the moths took flight and contacted the dish containing food. The combination of food with larval contamination elicited flight and contact by 52.0 (±5.9 SE)% of the moths tested and none contacted the empty dish. All moths that ultimately landed on a source had preceded that behaviour by initiation of flight and upwind flight within the odour plume characterised by a side-to-side movement on approach to the source. Once a female moth contacted and landed on a treatment (either food only or food+larval contamination) she would remain on the patch for the duration of the trial. Females walked rapidly around the substrate and some laid single eggs during brief stops.

Discussion

This study clearly shows that female *P. interpunctella* orient to odours from food sources for oviposition, and that hexane-extractable semiochemicals associated with secretions from wandering fifth instar larvae elicit oviposition. Our results with responses to food are consistent with earlier work on related species of moths (Barrer 1977; Barrer and Jay 1980; Phelan et al. 1991). The behavioural activity of larval secretions in *P. interpunctella* is similar to that found for *Ephestia kuehniella* by Corbet (1973). In that study, Corbet (1973) showed the paired mandibular glands of larvae harboured a secretion responsible for eliciting oviposition. We are currently investigating the source of oviposition stimulants from *P. interpunctella* larvae. Corbet (1973) did not identify conspecific oviposition stimulants from mandibular glands of *E. kuehniella* , but compounds were subsequently identified that acted as kairomones for the parasitic wasps *Venturia* (=*Nemeritis*) *canescens* (Gravenhorst) (Mudd and Corbet

1973; Mossadegh 1980; Mudd and Corbet 1982) and *Bracon hebetor* (Say) (Strand et al. 1989). Larvae from several species of phycitines, including *P. interpunctella*, possess a similar blend of compounds in their mandibular gland secretions that are kairomonally active for *V. canescens* (Mudd and Corbet 1973; Nemoto et al. 1987). Since we found that the same type of secretions used in kairomone studies elicit oviposition by conspecific females, it is possible that kairomonal factors from the mandibular glands may also serve as pheromones for oviposition.

Larval contamination apparently signals an oviposition site to female *P. interpunctella*, eliciting different levels of oviposition as a function of the density of larvae producing the contamination and the quality of alternative oviposition sites available to a female (Fig. 2, Table 1). The same mandibular gland secretion that elicited oviposition in *E. kuehniella* also acted as a spacing (epideictic) pheromone for larvae (Corbet 1973). Our data show that distribution of eggs by female *P. interpunctella* is affected by contaminations resulting from high numbers of larvae, and that a priority system for oviposition based on the suitability of the available oviposition sites may be operating. If larval contamination in the absence of food represents the only oviposition site, then it might be preferred for oviposition over a blank patch, regardless of larval density (Fig. 2a). However, when uncontaminated food occurs as an alternative to larval contamination, low densities of larval contamination were preferred for oviposition, but higher densities, perhaps signalling an overcrowded resource, caused a redistribution and reduction of eggs among treatments (Fig. 2, Table 1).

Our experiments show an interactive effect of food and larval secretions on orientation of female *P. interpunctella*. The combination of food and larval secretions elicited the highest oviposition responses in two- and four-choice experiments (Fig. 2, Table 1), and more females flew upwind and contacted the source when food and larval secretions were presented together. Oviposition experiments in small chambers could not differentiate long-distance response to volatiles from contact chemoreception of ovipositional substrates. Wind tunnel experiments, however, confirm that *P. interpunctella* fly upwind in response to volatiles from food. Since larval secretions alone did not elicit flight in the wind tunnel, we suspect that higher landing rates on food with larval secretions resulted from the less volatile components of those secretions being perceived at closer range, prior to contact, and acting to enhance the activity of food odours. Previous work with *E. kuehniella* (Corbet 1973) and *C. cautella* (Barrer 1977; Barrer and Jay 1980) did not address orientation to combinations of larval secretions with food. Interactive effects between host plant and larvae were demonstrated in *Amyelois transitella*, which laid more eggs on mummy almonds infested with conspecific larvae than on uninfested mummies (Andrews and Barnes 1982). Curtis and Clark (1979) found that polar extracts from larval frass of *A. transitella* elicited oviposition by conspecific females in laboratory tests. Although Phelan et al. (1991) identified components from crude almond oil that induced upwind flight and landing by *A. transitella*, larval secretions were not analysed and no studies of flight orientation to combinations of almond oil and larval secretions were reported. The role of plant volatiles in long-range orientation by females from several insect species is gaining more research attention (e.g. Landolt 1989; Dickens et al. 1990; Phelan et al. 1991). The *P. interpunctella* system described here has both host (food) and pheromone components (oviposition and spacing pheromone from larval secretions) operating in concert to affect orientation of females. As suggested by this study and others (e.g. Dickens et al. 1990; Landolt and Heath 1990; Raina et al. 1992), interac-

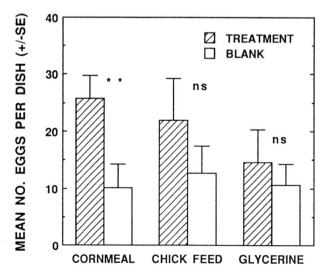

Fig. 1. Oviposition of *P. interpunctella* in two-choice bioassays in which cheesecloth overlaid dishes containing components of the laboratory diet (t-test, P<0.05, N=10 for each experiment).

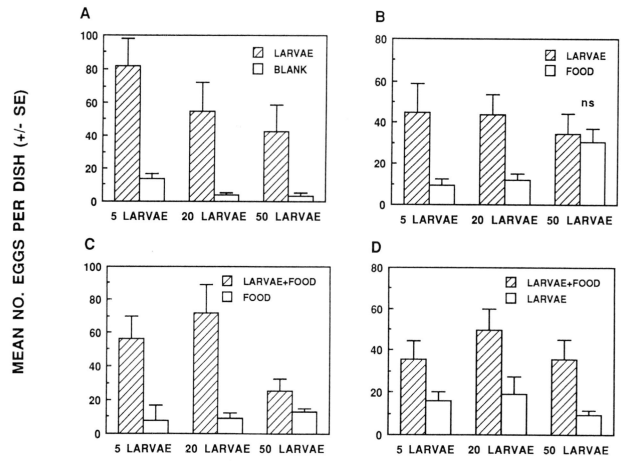

Fig. 2. Oviposition of *P. interpunctella* in four series of two-choice experiments. BLANK=an uncontaminated cheesecloth piece over an empty dish, FOOD=10 g of laboratory rearing diet, LARVAE=cheesecloth contaminated by either five, 20, or 50 wandering fifth instar larvae for 24 hours. All two-way comparisons were significantly different (P<0.05, N=10, Student's t-test) except where indicated by 'ns'.

tions of insect pheromones with host kairomones may be a common theme in the orientation of insects to host plants.

Acknowledgments

We appreciate the technical assistance of Amy Schermacher and Miel Barman. P. G. Fields and M. N. Parajulee provided valuable critical reviews of the manuscript. This work was partially supported by funding from USDA competitive grant no. 9202382 and through a cooperative agreement between the USDA Agricultural Research Service and the University of Wisconsin-Madison. All programs and services of the U. S. Department of Agriculture are offered on a nondiscriminatory basis without regard to race, colour, national origin, religion, sex, age, marital status, or handicap. This article reports the results of research only. Mention of a proprietary product does not constitute an endorsement or a recommendation for its use by USDA.

References

Andrews, K.L. and Barnes, M.M. 1982. Differential attractiveness of infested and uninfested mummy almonds to navel orangeworm moths. Environmental Entomology, 11, 280–282.

Barrer, P.M. 1977. The influence of airborne stimuli from conspecific adults on the site of oviposition of *Ephestia cautella* (Lepidoptera: Phycitidae). Entomologia Experimentalis et Applicata, 22, 13–22.

Barrer, P.M. and Jay, E.G. 1980. Laboratory observations on the ability of *Ephestia cautella* (Walker) (Lepidoptera: Phycitidae) to locate and oviposit in response to a source of grain odour. Journal of Stored Products Research, 16, 1–7.

Brady, U.E., Tumlinson III, J.H., Brownlee, R.G. and Silverstein, R.M. 1971. Sex pheromone of the almond moth and the Indian meal moth: cis-9, trans-12-tetradecadienyl acetate. Science, 171, 801–804.

Corbet, S.A. 1973. Oviposition pheromone in larval mandibular glands of *Ephestia kuehniella*. Nature, 243, 537–538.

Curtis, C.E. and Clark, J.D. 1979. Responses of naval orangeworm moths to attractants evaluated as oviposition stimulants in an almond orchard. Environmental Entomology, 8, 330–333.

Dickens, J.C., Jang, E.B., Light, D. M. and Alford, A.R. 1990. Enhancement of insect pheromone responses by green leaf volatiles. Naturwissenschaften, 77, 29–31.

Landolt, P.J. 1989. Attraction of the cabbage looper to host plants and host odour in the laboratory. Entomologia Experimentalis et Applicata, 53, 117–124.

Landolt , P.J. and Heath, R.R. 1990. Sexual role reversal in mate-finding strategies of the cabbage looper moth. Science, 249, 1026–1028.

Mankin, R.W., Vick, K.W., Coffelt, J.A., and Weaver, B.A. 1983. Pheromone-mediated flight by male *Plodia interpunctella* (Hübner) (Lepidoptera: Pyralidae). Environmental Entomology, 12, 1218–1222.

Mossadegh, M.S., 1980. Inter- and intra-specific effects of the mandibular gland secretion of larvae of the Indian-meal moth, *Plodia interpunctella*. Physiological Entomology, 5, 165–173.

Mudd, A., 1982. Novel 2-acylcyclohexane-1,3-diones in the mandibular glands of Lepidopteran larvae. Journal of the Chemical Society Transactions, 1, 2357–2362.

Mudd, A. and Corbet, S.A. 1973. Mandibular gland secretion of larvae of stored product pests *Anagasta kuehniella, Ephestia cautella, Plodia interpunctella,* and *Ephestia elutella*. Entomologia experimentalis et Applicata, 16, 291–293.

Nemoto, T., Shibuya, M., Kuwahara, Y. and Suzuki, T. 1987. New 2-acylcyclohexane-1,3-diones: kairomone components against a parasitic wasp, *Venturia canescens*, from feces of the almond moth,

Cadra cautella, and the Indian meal moth, *Plodia interpunctella*. Agricultural Biology and Chemistry, 51, 1805–1810.

Phelan, P.L. and Baker, T.C. 1986. Cross-attraction of five species of stored-product Phycitinae (Lepidoptera: Pyralidae) in a wind tunnel. Environmental Entomology, 15, 369–372.

Phillips, T.W. and Strand, M.R. 1994. Larval secretions and food odours affect orientation in female *Plodia interpunctella*. Entomologia Experimentalis et Applicata, in press.

Phelan, P.L., Roelofs, C.J., Youngman, R.R. and Baker, T.C. 1991. Characterisation of chemicals mediating ovipositional host-plant finding by *Amyelois transitella* females. Journal of Chemical Ecology, 17, 599–613.

Raina, A.K., Kingan, T.G. and Mattoo, A.K. 1992. Chemical signals from host plant and sexual behaviour in a moth. Science, 255, 592–594.

Strand, M.R., Williams, H.J., Vinson, S.B. and Mudd, A. 1989. Kairomonal activities of 2-acylcyclohexane-1,3 diones produced by *Ephestia kuehniella* Zeller in eliciting searching behaviour by the parasitoid *Bracon hebetor* (Say). Journal of Chemical Ecology, 15, 1491–1500.

Studies of responses of stored-product pests, *Prostephanus truncatus* (Horn) and *Sitophilus zeamais* Motsch., to food volatiles

V. Pike, J.L. Smith, R.D. White and D.R. Hall*

Abstract

A laboratory bioassay was used to demonstrate attraction of *Prostephanus truncatus* (Horn) (Coleoptera: Bostrichidae) to volatiles from maize and cassava trapped on Porapak resin. Attraction of maize weevils, *Sitophilus zeamais* Motsch., to trapped maize volatiles was also shown. The volatile collections were examined by gas chromatography (GC) linked to electroantennographic (EAG) recording and by GC linked to mass spectrometry.

With *P. truncatus*, up to nine EAG-active components were detected in cassava volatiles and ten in maize volatiles, with at least five of the components common to the two food sources. Nonanal was identified as the major active component of cassava volatiles, and hexanoic acid of maize volatiles. Actonal, nonanoic acid and vanillin were also identified as EAG-active components.

In contrast, with *S. zeamais* only one EAG-active component revealed the presence of 1-hydroxyheptan-2-one, a novel compound in maize volatiles, but this did not elicit an EAG response from *S. zeamais*. The remaining unidentified compounds will be identified and evaluated using bioassays, in anticipation that the results will help to improve trapping methods and contribute to greater understanding of stored-product pest ecology and management.

Introduction

There is considerable interest in the behavioural responses of stored-product insects to host-plant volatiles (Barrer 1983; Subramanyam et al. 1992). Table 1 details a number of the studies that have been published.

These studies are of interest because the stored-product pests represent a special class of phytophagous insects that appear to be highly adapted to surviving on numerous commodities stored for human or animal consumption. Increased understanding of the interaction between the insects and the commodity will serve to increase our knowledge of stored-product pest ecology. The results also have important practical applications in the development of monitoring systems for stored-product pests based on attraction to food lures. Furthermore, interactions between food volatiles and pheromones have been reported for several species (e.g. Walgenbach et al. 1987; Dowdy et al. 1993; Phillips et al. 1993) and understanding of these interactions is necessary for optimisation of pheromone-based monitoring systems.

Prostephanus truncatus (Horn) or the larger grain borer is a stored-product pest which originated in Central America but has become a major problem in Africa after its introduction in the late 1970s (Hodges 1986).

This paper describes laboratory investigations of the responses of this species to food volatiles. Apart from their practical implications, the results of this work will provide interesting comparisons with those on other species because it is becoming apparent that *P. truncatus* has only adapted to storage situations relatively recently and certainly is not restricted to this type of habitat (Rees et al. 1990). Results are also presented on attraction of *Sitophilus zeamais* Motsch. to food volatiles.

Materials and Methods

Sample collection

The maize used in the trial was freshly-harvested white maize from Dar es Salaam, Tanzania. The cassava was bought as fresh intact tubers in the U.K. and originated from the Caribbean. The tubers were peeled, chipped and dried before use. Volatiles from whole maize (3 kg) or cassava chips (400 g) were collected by placing the sample in a 3-L Buchner flask and drawing in air (2 L/minute) through a filter of activated charcoal via a tube to the base of the flask and out through a collecting filter containing Porapak Q (50–80 mesh; 200 mg). The Porapak was pre-cleaned by washing thoroughly with dichloromethane and drying under nitrogen, and the absence of significant breakthrough of volatiles was prevented by connecting two filters in series. Collections were made for seven-day periods at 27°C and 70% r.h.. Trapped volatiles were eluted with dichloromethane (4×500 µL purified by passage through neutral alumina).

Analyses of volatiles

Samples were analysed by capillary gas chromatography (GC) on columns (25 m × 0.32 mm i.d.) coated with polar CP-Wax 52CB (Chrompack) or non-polar CP-Sil 5CB (Chrompack) with helium carrier gas at 0.4 kg/cm^2, splitless injection and oven temperature held at 50°C for 2 min then programmed from 50°C to 230°C at 6°C/min. Quantitative analyses used flame ionisation detection (220°C). In linked gas chromatography-mass spectrometry (GC-MS), the same GC columns and conditions were used, and the GC was interfaced to a Finnigan-MAT ITD 700 operated in electron impact (EI) or chemical ionisation (CI) with *iso*-butane. Linked gas chromatography-electroantennography (GC-EAG) analyses also used the same GC columns and conditions, as described by Cork et al. (1990). GC retention data are presented as equivalent chain lengths (ECLs) relative to the retention times of straight-chain acetates.

* Natural Resources Institute, Central Avenue, Chatham Maritime, Chatham, Kent, ME4 4TB, U.K.

Bioassay

The *S. zeamais* and *P. truncatus* used for bioassay and EAG studies originated from Zimbabwe and Tanzania respectively. The insects were kept in culture on whole maize at 27°C and 70% r.h. All the insects used for experimental purposes were between four and six weeks old and were not sexed.

The bioassay used in this trial was based on the method developed at CSL, Slough, U.K. (J. Chambers, per. comm.). The materials used comprised 200 mm formica-based arenas each with a single, central pitfall trap (3" × 1" glass tube). The arena surfaces were covered in filter paper (Whatman No. 1) to provide a uniform surface and to facilitate insect movements, and each arena was covered with a lid to prevent insect escape. Insects were bioassayed individually. For *S. zeamais* the glass pitfall tube stood proud of the arena at a height of 5 mm. Initial trails had suggested that such a position increased the sensitivity of the test. Work is ongoing to investigate this matter fully. Forty arenas were operated simultaneously and the exposure period was standardised at one hour for *S. zeamais* and two hours for *P. truncatus*. Twenty of the traps contained a sample: the remainder used the solvent alone as a control. The sample (20 µL) was introduced into the base of each experimental pitfall trap on a triangle of filter paper (Whatman No. 1; 2 cm²).

For the *S. zeamais* bioassay, dilutions of the volatile extract were prepared and tested. The original strengths of maize and cassava volatiles have only been tested against *P. truncatus* to date. Five replicates were performed for each treatment, and differences between the numbers of insects trapped in the treatment and unbaited control were compared using t tests.

Results

Attraction of *P. truncatus* to maize and cassava volatiles

In the bioassay, *P. truncatus* was significantly attracted to collections of volatiles from both maize and cassava (Table 2). In linked GC-EAG analyses of cassava volatiles, nine EAG-active components were detected using the polar GC column (Table 3) and eight on the non-polar column. In GC-EAG analyses of the maize volatiles on the polar GC column, ten active components were detected, and some of these can be correlated by their retention times and mass spectra with components of the cassava volatiles (Table 3). The major component of the cassava volatiles was EAG-active and was identified as nonanal by comparison of GC retention times and mass spectra with those of synthetic material. Nonanoic acid and decanal were also identified as EAG-active components. In the maize volatiles, the major component was EAG-active and was identified as hexanoic acid. Nonanoic acid and decanal were also detected as EAG-active components. Vanillin (4-hydroxy-3-methoxybenzaldehyde) was present at significant levels in the maize volatiles and caused an EAG response. It was present at trace levels in the cassava volatiles (< 0.2% of the major component) and was not detected by the insects.

Attraction of *S. zeamais* to maize volatiles

S. zeamais was significantly attracted to collections of maize volatiles, and the effect was concentration dependent (Table 4).

The volatile collection elicited a strong EAG response from *S. zeamais*. Duplicate GC-EAG analyses were carried out on seven different insects, and a single EAG-active component was detected with retention times (ECLs) of 9.65 on the polar

and 6.46 on the non-polar GC columns. GC-MS analyses on the polar column in CI mode indicated a molecular weight of 130, and in EI mode showed fragment ions at m/z 99 (35%), 71 (40%), 55 (20%), 43 (100%) and 41 (60%), although a component with similar mass spectrum could not be detected at the retention time of the EAG-active component on the non-polar column. 1-Hydroxyheptan-2-one was proposed as a structure and synthesised. The mass spectra and retention time of this compound on the polar column were identical to those of the component in the maize volatiles, and this was detected in analyses on the non-polar column at a retention time of 6.19, close to but not identical to that of the EAG-active component. The synthetic 1-hydroxyheptan-2-one was inactive in the bioassay and elicited no EAG response from *S. zeamais*.

Work is continuing to determine the structure of the EAG-active component in maize volatiles. The GC retention times of this are also similar to those of the diastereomers of the sex pheromone of *S. zeamais*, 5-hydroxy-4-methylheptan-3-one (Phillips et al. 1985), 9.60 and 9.75 on the polar column and 6.35 and 6.43 on the non-polar column. The pheromone could not be detected in the maize volatiles by GC-EAG or GC-MS analyses.

Discussion

The results reported here show that *P. truncatus* is attracted to volatiles from cassava and maize collected on Porapak. These volatiles elicit an EAG response from the beetles, the EAG-active components have been characterised by linked GC-EAG analyses, and there are several active components in common. Some of these components have been identified and work is continuing to identify the remainder. It is anticipated that those components which are detected by the antennal receptors will play a role in the behavioural response of the insect, and the synthetic compounds will be evaluated in the behavioural bioassay.

If positive behavioural responses are produced over the short range used in this bioassay further experiments will be conducted to test responses over a greater range.

Hexanoic acid, identified as the major component of the maize volatiles used in this trial was implicated in attraction of *S. zeamais* to cereals by Honda and Oshawa (1990), and nonanal, the major component of the cassava volatiles, has been shown to attract silvanid beetles both alone (Mikolajczak et al. 1984; Oehlschlager et al. 1988) and in combination with the aggregation pheromones (Mushobozy et al. 1993). Vanillin and maltol (3-hydroxy-2-methylpyr-4-one) were shown to synergise the attractiveness of the pheromones of both *S. oryzae* (L.) and *Tribolium castaneum* (Herbst.) by Phillips et al. (1993): although vanillin was found in both cassava and maize volatiles, maltol was not detected in either.

In contrast to *P. truncatus*, which showed EAG responses to several components of the maize volatiles, *S. zeamais* responded to only one component in linked GC-EAG analyses. An initial attempt at identification of the chemical structure of this showed the presence of 1-hydroxyheptan-2-one in the volatiles, although this was not the active component. This is the first time this compound has been identified in cereal volatiles, and is presumably a fatty acid oxidation product. It is structurally related to 2-hydroxyoctan-3-one, a pheromone component of *Xylotrechus* spp. (Coleoptera: Cerambycidae) (Kuwahara et al. 1987). Work is continuing to establish the identity of the component which elicits EAG and behavioural activity from *S. zeamais*.

Table 1. Published studies on the behavioural responses of stored-product insects to host plant volatiles

Oryzaephilus sp.	(Coleoptera: Silvanidae)	Mikolajczak et al.	1984
		Stubbs et al.	1985
		Pierce et al.	1990, 1991
Trogoderma sp.	(Coleoptera: Dermestidae)	Nara et al.	1981
Sitophilus sp.	(Coleoptera: Curculionidae)	Trematerra and Girgenti	1989
		Honda and Oshawa	1990
		Phillips et al.	1993
Tribolium sp.	(Coleoptera: Tenebrionidae)	Phillips et al.	1993
Rhyzopertha dominica	(Coleoptera: Bostrichidae)	Dowdy et al.	1993

Table 2. Attraction of *P. truncatus* to trapped volatiles from maize and cassava (numbers trapped in 20 individual bioassays).

Replicate	Maize		Cassava	
	Treatment	Control	Treatment	Control
1	13	5	13	5
2	12	7	8	4
3	8	2	12	4
4	6	1	12	3
5	9	5	13	5
Mean	9.6	4	11.6	4.2
SD	2.88	2.45	2.07	0.84
t	3.31		7.4	
P	0.01		>0.0001	

Table 3. Active components detected in GC-EAG analyses with *P. truncatus* of cassava and maize volatiles on a polar GC column

Cassava volatiles			Maize volatiles			
ECL[a]	n[b]	Relative amount	ECL[a]	n[b]	Relative amount	Identification
7.11	9	100	7.18	4	16	nonanal
7.49	6	0.6				
7.64	8	0.8	7.74	3	4	
8.21	8	13				decanal
8.39	9	1.8	8.33	5	2	
8.55	5	12				
8.75	7	40	8.81	5	15	
			9.10	3	0.2	
			11.76	7	100	hexanoic acid
			12.37	3	4	
14.65	4	0.2				
14.88	6	37	14.86	7	50	nonanoic acid
			18.26	3	2	
			18.51	6	6	vanillin

[a] Retention data as equivalent chain lengths.
[b] Number of times observed in three replicate analyses on three insects.

References

Barrer, P.M. 1983. A field demonstration of odour-based host-food finding behaviour in several species of stored grain insects. Journal of Stored Product Research, 19, 105–110.

Cork, A., Beevor, P.S., Gough, A.J.E. and Hall, D.R. 1990. Gas chromatography linked to electroantennography: a versatile technique for identifying insect semiochemicals. In: McCaffery, A.R. and Wilson, I.D. ed., Chromatography and Isolation of Insect Hormones and Pheromones. New York, Plenum Press, 271–279.

Dowdy, A.K., Howard, R.W., Seitz, L.M., and McGaughey, W.H. 1993. Response of *Rhyzopertha dominica* (Coleoptera: Bostrichidae) to its aggregation pheromone and wheat volatiles. Environmental Entomology, 22, 965–970.

Hodges, R.H. 1986. The biology and control of *Prostephanus truncatus* (Horn) (Coleoptera: Bostrichidae): a destructive storage pest with an increasing range. Journal of Stored Products Research, 22, 1, 1–14.

Honda, H. and Oshawa, O. 1990. Chemical ecology for stored product insects. Journal of Pesticide Science 15(2), 263–270.

Hougan, F.J., Quillam, M.A. and Curran, W.A. 1971. Headspace vapours from cereals grains. Journal of Agricultural Food Chemistry, 19(1), 182–183.

Table 4. Attraction of *S. zeamais* to dilutions of trapped volatiles from maize (numbers trapped in 20 individual bioassays)

Replicate	× 1.0	Control	× 0.1	Control	× 0.01	Control	× 0.001	Control
1	0	1	13	5	10	5	3	3
2	12	2	15	8	7	4	4	2
3	14	3	12	3	5	4	3	3
4	15	6	12	4	7	4	5	4
5	14	2	11	4	6	5	7	6
Mean	12.8	2.8	12.6	4.8	7	4.4	4.4	3.6
SD	2.39	1.92	1.52	1.72	1.87	0.54	1.67	1.51
t	5.17		7.12		2.98		0.792	
P	0.0009		0.0001		0.0175		0.4510	

Kuwahara, Y., Matsuyama, S. and Suzuki, T. 1987. Identification of 2,3-octanediol, 2-hydroxy-3-octanone and 3-hydroxy-2-octanone from male *Xylotrechus chinensis* Chevrolat as possible sex pheromones (Coleoptera: Cerambycidae). Applied Entomology and Zoology, 22, 25–28.

Mikolajczak, K.L., Zilkowski, B.W., Smith, C.R., and Burkholder, W.E. 1984. Volatile food attractants for *Oryzaephilus surinamensis* (L.) from oats. Journal of Chemical Ecology, 10, 301–309.

Mushobozy, D.K., Pierce, H.D. and Borden, J.H. 1993. Evaluation of 1-octen-3-ol and nonanal as adjuvants for aggregation pheromones for three species of cucujid beetles (Coleoptera: Cucujidae). Journal of Economic Entomology, 86, 1835–1845.

Nara, J.M., Lindsay, R.C. and Burkholder, W.E. 1981. Analysis of volatile compounds in wheat germ oil responsible for an aggregation response in *Trogoderma glabrum* larvae. Agricultural and Food Chemistry, 29, 68–72.

Oehlschlager, A.C., Pierce, A.M., Pierce, H.D. and Borden, J.H. 1988. Chemical communication in cucujid grain beetles. Journal of Chemical Ecology, 14, 2071–2098.

Phillips, J.K., Walgenbach, C.A., Klein, J.A., Burkholder, W.E., Schmuff, N.R. and Fales, H.M. 1985. (*R**S*)-5-Hydroxy-4-methyl-3-heptanone: male-produced aggregation pheromone of *Sitophilus oryzae* (L.) and *S. zeamais* Motsch. Journal of Chemical Ecology, 11, 1263–1274.

Phillips, T.W., Jiang, X.-L., Burkholder, W.E., Phillips, J.K. and Tran, H.Q. 1993. Behavioural responses to food volatiles by two species of stored-product Coleoptera, *Sitophilus oryzae* (Curculionidae) and *Tribolium castaneum* (Tenebrionidae). Journal of Chemical Ecology 19, 723–734.

Pierce, A.M., Pierce, J.R., Oehlschlager, A.C. and Borden J.H. 1990. Attraction of *Oryzaephilus surinamensis* (L.) and *O. mercator* (Fauvel) (Coleoptera: Cucujidae) to some common volatiles of food. Journal of Chemical Ecology, 16, 465–475.

Pierce, A.M., Pierce, H.D., Borden, J.H. and Oehlschlager, A.C. 1991. Fungal volatiles: semiochemicals for stored-product beetles (Coleoptera: Cucujidae). Journal of Chemical Ecology, 17, 581–597.

Rees, D.P., Rodriguez Rivera, R. and Herrera Rodriguez, F.J. 1990. Observations on the ecology of *Teretriosoma nigrescens* Lewis and its prey *Prostephanus truncatus* (Horn) in the Yucatan Peninsula, Mexico. Tropical Science, 30, 153–165.

Stubbs, M.R., Chambers, J., Schofield, S.B. and Wilkins, J.P.G. (1985). Attractancy to *Oryzaephilus surinamensis* (L.) of volatile materials isolated from vacuum distillate of heat-treated carobs. Journal of Chemical Ecology, 11, 565–581.

Subramanyam, Bh., Wright, V.F. and Fleming, E.E. 1992. Laboratory evaluation of food baits for their relative ability to retain three species of stored-product beetles (Coleoptera). Journal of Agricultural Entomology, 9(2), 117–127.

Tremeterra, P. and Girgenti, P 1989. Influence of pheromone and food attractants on trapping of *Sitophilus oryzae* (L.) (Col., Curculionidae): a new trap. Journal of Applied Entomology, 108, 12–20.

Walgenbach, C.A., Burkholder, W.E., Curtis, M.J. and Khan Z.A. 1987. Laboratory trapping studies with *Sitophilus zeamais* (Coleoptera: Curculionidae). Journal of Economic Entomology, 80, 763–767.

Influence of synthetic Sitophilate, the aggregation pheromone of *Sitophilus granarius* (L.) (Coleoptera: Curculionidae) on dispersion and aggregation behaviour of the granary weevil

R. Plarre[*]

Abstract

Sitophilus granarius was successfully lured to synthetic Sitophilate, the aggregation pheromone of the granary weevil. Pheromone amounts of 570, 5700 and 57000 ng were tested on groups of unmated males, unmated females, and weevils of mixed sex. Unmated females were more sensitive than unmated males and they were attracted by the pheromone at low dosages. Aggregation of males in the presence of low pheromone dosages was increased when physical contact with females was possible. Very high pheromone concentrations repelled both sexes. The aggregation pheromone is expressed more as an arrestant than an attractant chemical, since weevils stopped moving in sufficiently high pheromone concentrations. The source of the pheromone is not necessarily located. The possibility of using baited traps in the field is discussed.

Introduction

The movement of *Sitophilus granarius* through grain has very often been described as random. Dispersion of the adult weevils occurs mainly horizontally on the surface or in the upper layers of the grain (Surtees 1963, 1964a, 1964b). Although the granary weevil tends to be positive geotactic in undisturbed bulks of grain, its vertical distribution is somehow restricted (Rodionov 1938; Howe 1951). The increasing weight and pressure of grain and the decreasing inter kernel space in the depth of a bulk demand a lot more power and energy from the weevils for pushing their body forward. Therefore their moving ability in deeper layers is limited. Other external stimuli influence weevil behaviour as well. The reaction of the granary weevils toward light has been recorded as negative (Andersen 1938; Torne 1941; Perttunen 1972). Its ability to monitor vibrations with chordotonal organs in the legs was mentioned by Frings and Lollis (1971) but communication and orientation by airborne sounds can be excluded (Wojcik 1969).

Olfactoric sensing of food odours (Donat 1970; Navrot and Czaplicki 1979; Levinson and Kanaujia 1982; Seifelnasr 1991) and other chemical volatiles is by far the most important orientation parameter and vital for finding new food sources and potential mates for the granary weevil. Surtees (1965) did not mention any chemical communication between weevils moving through grain randomly. Faustini et al. (1982) proved

the existence of a male produced aggregation pheromone attracting both sexes to suitable sites for feeding, mating and breeding. The chemical identification of the attractant was carried out by Phillips et al. (1987). The aggregation pheromone was named Sitophilate. Synthesised 1-ethylpropyl-2-methyl-3-hydroxypentanoate in the active config-uration of 2S,3R caused attraction in the range of 20 ng to 2500 ng (Phillips et al. 1989). Levinson et al. (1990) recorded receptor potentials even when applying up to 10000 ng of pheromone directly onto the antenna of the weevil. Nevertheless, neither experiments were carried out under more or less natural conditions in the presence of food or in a larger scale of test arena. The current studies were therefore undertaken in order to test the active concentration range of synthetic Sitophilate and the reaction quality of the responding weevils in the presence of wheat. Differences in the behaviour of unmated males and unmated females as well as the interaction between the sexes in the presence of the pheromone lure and the suitable diet for feeding and breeding were recorded.

Materials and Methods

Weevils

A field strain of *S. granarius* collected from a granary of a river port in Berlin (Germany) in 1988 was reared and multi-plied in the Institute for Stored Product Protection of the Federal Biological Research Centre for Agriculture and Forestry under constant conditions of $25°\pm2°C$ and $75\%\pm5\%$ r.h. in complete darkness. The parental generation of the test insects was kept on $150\ cm^3$ wheat of 14% moisture content in $300\ cm^3$ glass jars for a mating and oviposition period of 14 days. Freshly emerging adults of the F_1-generation were sieved off every other day, isolated by sex and transferred onto fresh wheat. Criteria for sexing the weevils were taken from Halstead (1963). Seven to 10 days old weevils, still unmated but the colour of their cuticula had changed from light brown to dark brown or black, were taken for experiments. Mated weevils used in the experiments had also been sieved off and transferred onto fresh wheat after emerging but were not separated by sex.

Pheromone

One mL pure (2S,3R)-1-Ethylpropyl-2-methyl-3-hydrox-ypentanoate with the pheromone concentration of 9.5 µg/µL was donated from the Department of Entomology from the University of Wisconsin in Madison, WI (USA).

Three diluted fractions of 0.95 µg/µL, 0.095 µg/µL and 0.0095 µg/µL were drawn from the original sample by using n-hexane as diluting agent. By applying 60 µL of a fraction on a rubber septum the actual pheromone amounts of 570 ng, 5 700 ng and 57000 ng were offered to the weevils. 60 µL of

---

[*] Federal Biological Research Centre for Agriculture and Forestry, Institute for Stored Product Protection. Königin-Luise-Str. 19, 14195 Berlin, Germany.

n-hexane were applied on a rubber septum in the untreated control experiments.

Biotest

A box of perspex glass, 30 cm in square by height 5 cm was filled up to 3.5 cm with wheat of 14% moisture content. The total grain volume was ca. 3000 cm^3. It was possible to quickly insert a grid also made out of plastic into the box to divide the grain volume into 25 equal smaller parts or fields which were numbered in order (see Fig. 1). The pheromone lure was placed in field IV/4. The test insects were released in field III/2 after a 10 min adaptation period. The top of the box was covered with a plate of glass.

The insects were allowed to move around in the arena freely, to feed and rest at places of their choice. After 24 hours, the box was opened and the grid inserted quickly. Each field was sucked out with a vacuum pump and checked for weevils. By this, the present abode of every weevil after an exposure time of 24 hours to the pheromone was recorded.

The three different pheromone concentrations with the actual amounts of Sitophilate were tested on groups of 25 weevils. Unmated males and unmated females as well as groups of 13 mated males and 12 mated females were tested separately. The lures were usually tested for 28 days, and for 14 days on groups of mixed sex. All test series were carried out in three replicates under constant conditions of 25°±2°C and 75%±5% r.h. in darkness. When releasing the insects and during emptying the fields, only red light was switched on.

Mathematical analyses

After recording the present abode of every weevil after 24 hours an 'Aggregation Value' (AV) was calculated. This Aggregation Value described the distribution of the weevils in the system by considering the number and the intensity of possible occurring aggregations. The intensity of one or more aggregation spots depended upon the amount of weevils occurring in clusters. The Aggregation Value was calculated by the quotient of the 'Mean Distribution' (MD) and the 'Distribution Index' (DX).

MD represented the average number of individual weevils occurring together and forming clusters. The maximum value for MD was 25, when all individuals appeared in one single field. The minimum value was 1, when every weevil was found alone in one field.

DX described the overall distribution and was the average number of fields being occupied by at least one weevil in relation to the total number of available fields. The maximum value was 1, when all 25 field were occupied by one weevil each. The minimum value was 0.04, when all weevils occurred together in one field.

Therefore the AV showed possible values between 625 for complete aggregation of all weevils at one spot, and 1 for maximum dispersion, when the weevils appeared the furthest apart from each other.

Missing in the Aggregation Value was the quality of the aggregation, since an aggregation in the test arena could have been caused by attractive or repellent effects of the lure. Therefore, the test arena was divided into four zones (see Fig. 2). Zone A consisted of field IV/4, containing the pheromone lure. Zone B covered the eight fields surrounding zone A. Weevils recorded in these two zones had to move at least once towards the lure. So, the reaction of these weevils was regarded as positive, the effect of the pheromone as an attractive one. Weevils being recorded in the outer zone D of the arena moved at least once away from the lure. Their reaction was therefore stated as negative and the effect of the pheromone as a repellent one. Zone C was called a neutral zone because weevils walking around in this area were able to move through the grain without necessary turning towards or away from the pheromone source. The number of negative reacting weevils was subtracted from the number of weevils showing positive reactions and divided by the total number of insects in the system (Busvine 1971). The reaction quality was estimated simply by setting the data of the untreated control, which had been calculated the same way, as zero. Positive values stand for an attractive effect, negative values for a repellent effect of the lure.

The overall distribution was shown by adding up the relative accumulations of the weevils in the different zones. The actual distances of the present abode of the weevils to the pheromone lure were compared to the general distribution pattern in the untreated controls.

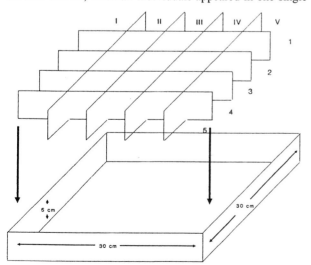

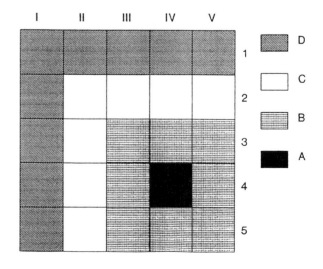

Fig. 1. Test arena for biotests (unfilled). A grid could be inserted quickly to divide the grain into 25 equal parts (fields). The parts were numbered in order from I/1 to V/5.

Fig. 2. Test arena for biotests (viewed from above) showing the different zones A to D for qualifying weevil behaviour towards or away from the pheromone lure.

Table 1. Distribution of granary weevils *S. granarius* in the untreated control

	Unmated males	Unmated females	Mixed sex
DX value	0.48±0.07	0.55±0.04	0.55±0.05
MD value	3.03±0.43	2.28±0.21	3.21±0.45
AV value	6.31(6.34/6.29)[a]	4.15(4.22/4.06)[a]	5.84(6.10/5.52)[a]
Distribution			
Zone A	1.33%±0.95%	1.33%±0.95%	2.31%±0.95%
Zone B	22.67%±3.43%	17.33%±1.90%	30.67%±1.90%
Zone C	33.33%±7.43%	22.67%±1.90%	25.33%±2.52%
Zone D	42.67%±4.33%	58.67%±3.43%	41.33%±1.90%

[a]Reliable range of the AV value, being calculated from lower and upper range of DX and MD values.

Graphical display

The results of all the tested pheromone concentrations for the different groups of test insects were presented in five graphs (a to e) each. Graphs a to c show the calculated DX-, MD- and AV-values over the time of 28 days. The solid horizontal line in every graph represented the mean value in the untreated control, the dotted lines its reliable range. Graph d showed the attractive or repellent quality in the reaction of the weevils. Graph e represented the relative accumulation of the insects in the different zones of the arena. The horizontal solid lines represented the distribution in the untreated control (lower line for zone A and B together, upper line for zone D).

Results

Untreated control

In the untreated controls (see Table 1) unmated males show a higher Aggregation Value than unmated females. The DX-value is lower for unmated males indicating a less wide distribution in the arena. Unmated females instead occupy on average over 50% of the available space. It can be seen from the MD-value that, on average, more unmated males appear together in one field than unmated females do.

The distribution patterns show that, on average, one-fourth of the unmated males appears in zones A and B whereas one-fifth of the unmated females occupies this area. The outer zone D shows mean distribution values of over 40% for unmated males and nearly 60% for unmated females. The corners and the periphery of the box are evenly occupied but slightly preferred over the middle of the arena.

Groups of mixed sex (13 males and 12 females) occupy, like unmated females, on average half of the available space in the arena. With a slightly higher MD-value weevils in groups of mixed sex form on average more intense clusters than unmated male weevils did. But weevils in groups of mixed sex occur in fewer aggregations. Their Aggregation Value becomes lower than the one for unmated males. The distribution patterns indicate an even distribution in the combined zone of A and B and in zone D.

Effects of the pheromone on unmated males

When 570 ng of pheromone was applied onto the rubber septa and offered to unmated males, no clear effect is visible when compared to the untreated control. The DX- (see Fig. 3a) and the MD- (see Fig. 3b) Values with their reliable ranges do not differ from the control. The calculated AV-value (see Fig. 3c) oscillates around the control value showing no obvious tendency. Regarding the reaction quality (see Fig. 3d)

and the distribution pattern (see Fig. 3e), only a very small and insignificant repellent effect of the lure is visible.

The application of 5700 ng of Sitophilate leads to a slightly different behaviour pattern in unmated males (see Fig. 5a to 5e). The Aggregation Values indicate that up to the 8th day after pheromone application a wider distribution of the weevils occurs than in the untreated control. The weevils do not aggregate as much as they do in the absence of the lure. The reaction quality shows positive responses of the unmated male weevils in the first 3 days and again for a shorter period after 9 to 11 days. This indicates an attractive effect of the lure. The distribution pattern, where more than 40% of the unmated male weevils appear in closer range to the pheromone source on day 3 and on day 11 verify the observation. After the 12th day, the behaviours of the weevils become very much comparable to those in the presence of the 570 ng pheromone lure.

Regarding the responses of unmated male weevils to the high dosage of 57000 ng of pheromone, a clear repellent effect of the lure is visible (see Fig. 7d). Nevertheless, the weevils aggregate in the first 9 to 13 days (see Fig. 7c) to a very high extent. The aggregation occurs in zone D, the far side of the pheromone (see Fig. 7e). Fourteen days after pheromone application, the distribution pattern slowly shifts towards the situation of the untreated control.

Effects of the pheromone on unmated females

Pheromone amounts between 570 ng and 57000 ng cause an aggregation of unmated females in the first 8 to 10 days after pheromone application (see Fig. 4c, 6c, 8c). The reaction quality is strongly positive in the presence of 570 ng and 5700 ng of Sitophilate. Still showing a positive tendency, the reaction values oscillate very much around zero in the presence of 57000 ng. Unmated female weevils do not move very far from the point of release when the highest pheromone dosage is applied.

In the presence of 570 ng of pheromone 20% of all unmated females are attracted into zone A, the very close space around the lure (see Fig. 4e). Fifty per cent and more unmated females occur in zone A and B together during the first 5 days after application. Compared to the untreated control, twice as many unmated female weevils stay in that area for the first 13 days.

With increasing amount of pheromone being offered, the percentage of unmated females being recruited into zone A decreases (see Fig. 6e, 8e). The weevils react positively towards the lure but do not continue the movement all the way to the pheromone source. With passage of time the distribution pattern shift to the distribution situation in the untreated control.

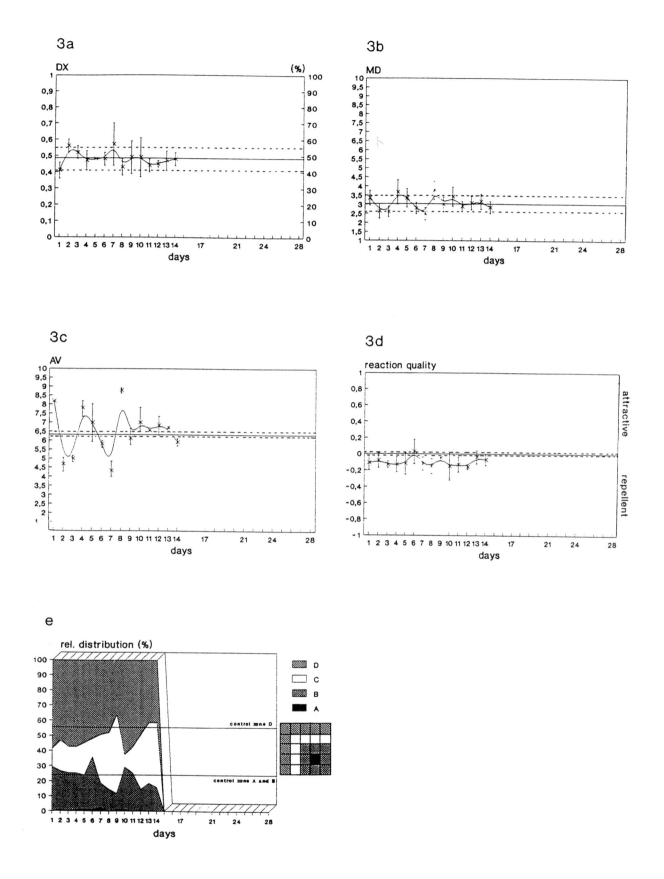

Fig. 3. Distribution behaviour of unmated males in the presence of 570 ng of Sitophilate (a) distribution index (DX-value), (b) mean distribution (MD-value), (c) aggregation value (AV-value), (d) reaction quality and (e) distribution pattern.

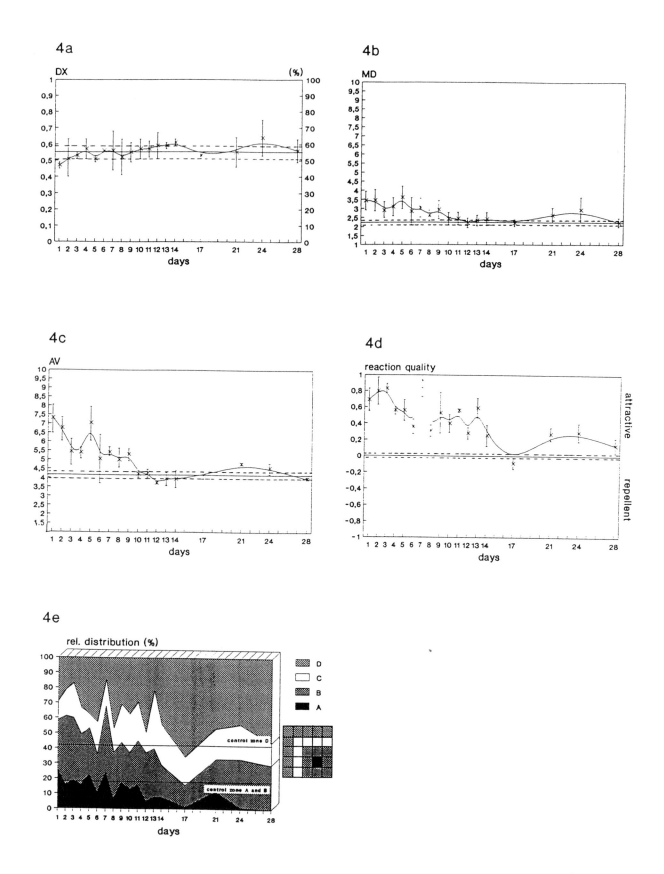

Fig. 4. Distribution behaviour of unmated females in the presence of 570 ng of Sitophilate (a) distribution index (DX-value), (b) mean distribution (MD-value), (c) aggregation value (AV-value), (d) reaction quality and (e) distribution pattern.

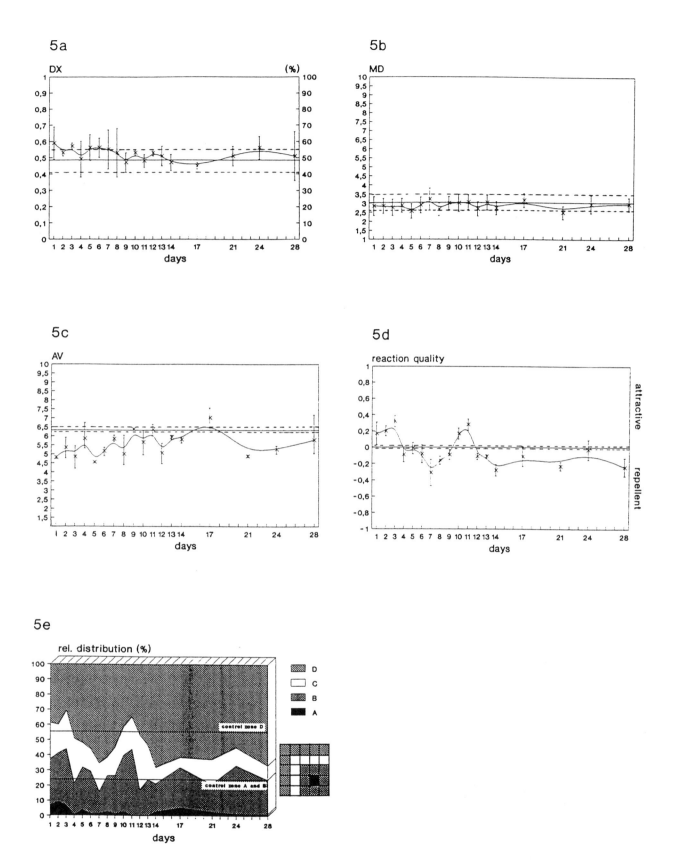

Fig. 5. Distribution behaviour of unmated males in the presence of 5700 ng of Sitophilate (a) distribution index (DX-value), (b) mean distribution (MD-value), (c) aggregation value (AV-value), (d) reaction quality and (e) distribution pattern.

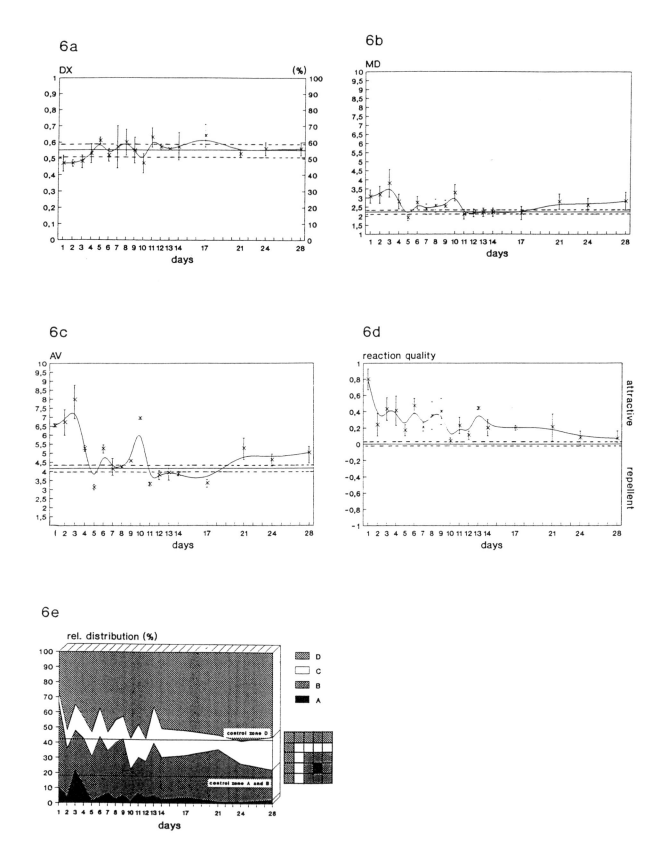

Fig. 6. Distribution behaviour of unmated females in the presence of 5700 ng of Sitophilate (a) distribution index (DX-value), (b) mean distribution (MD-value), (c) aggregation value (AV-value), (d) reaction quality and (e) distribution pattern.

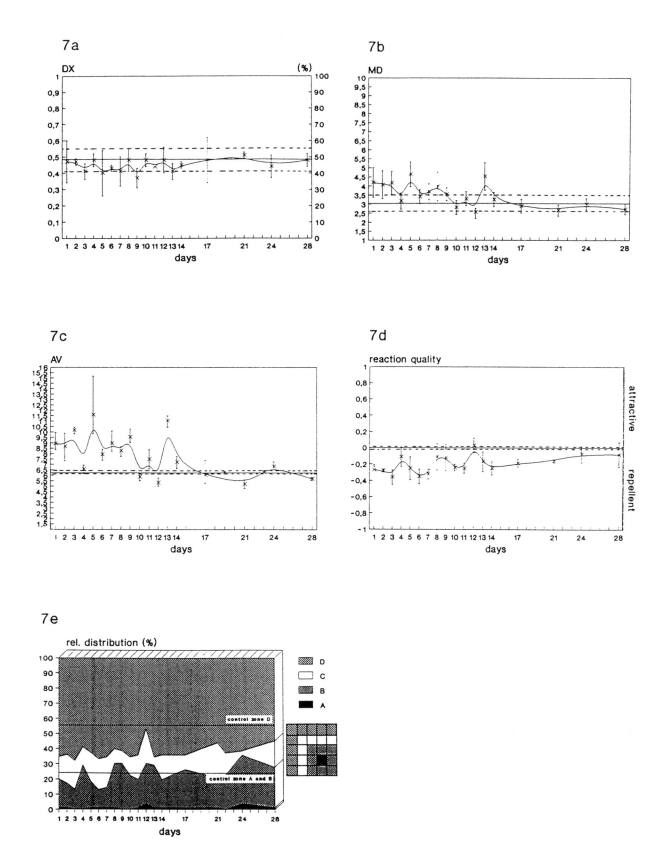

Fig. 7. Distribution behaviour of unmated males in the presence of 57000 ng of Sitophilate (a) distribution index (DX-value), (b) mean distribution (MD-value), (c) aggregation value (AV-value), (d) reaction quality and (e) distribution pattern.

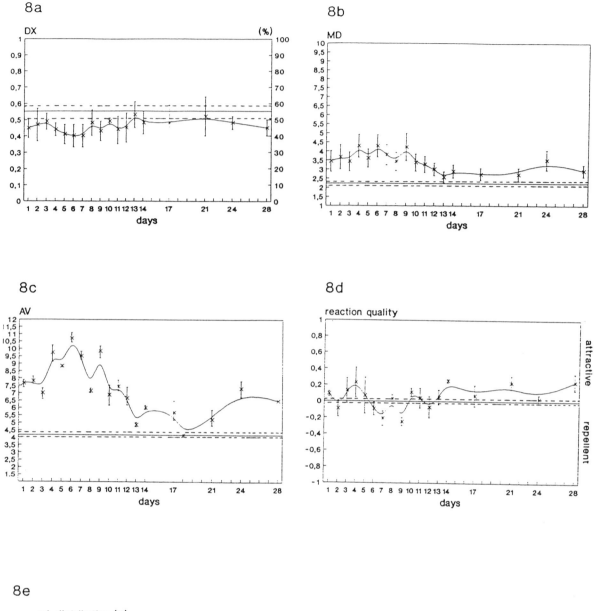

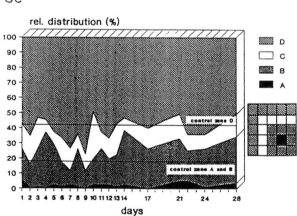

Fig. 8. Distribution behaviour of unmated females in the presence of 57000 ng of Sitophilate (a) distribution index (DX-value), (b) mean distribution (MD-value), (c) aggregation value (AV-value), (d) reaction quality and (e) distribution pattern.

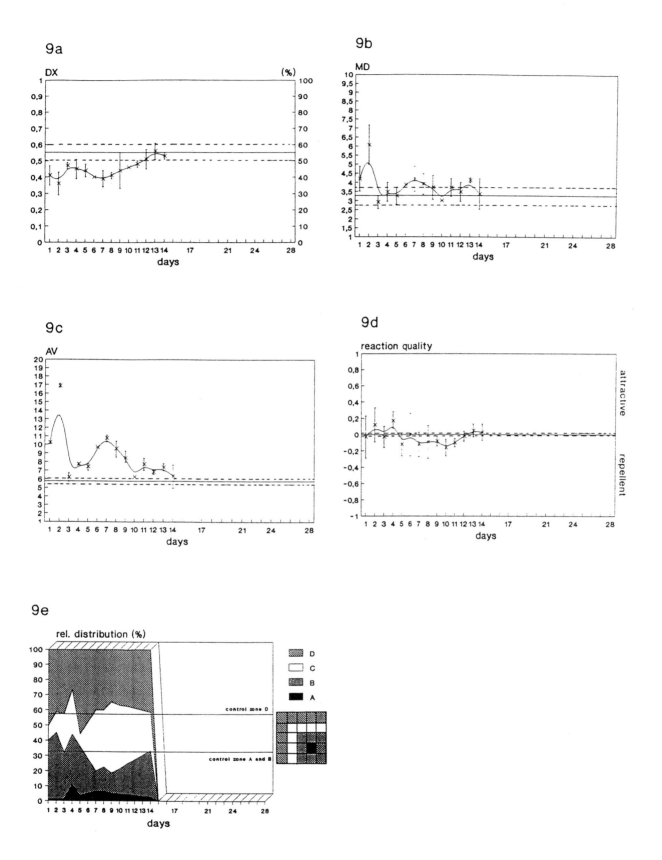

Fig. 9. Distribution behaviour of groups with weevils of mixed sex in the presence of 570 ng of Sitophilate (a) distribution index (DX-value), (b) mean distribution (MD-value), (c) aggregation value (AV-value), (d) reaction quality and (e) distribution pattern.

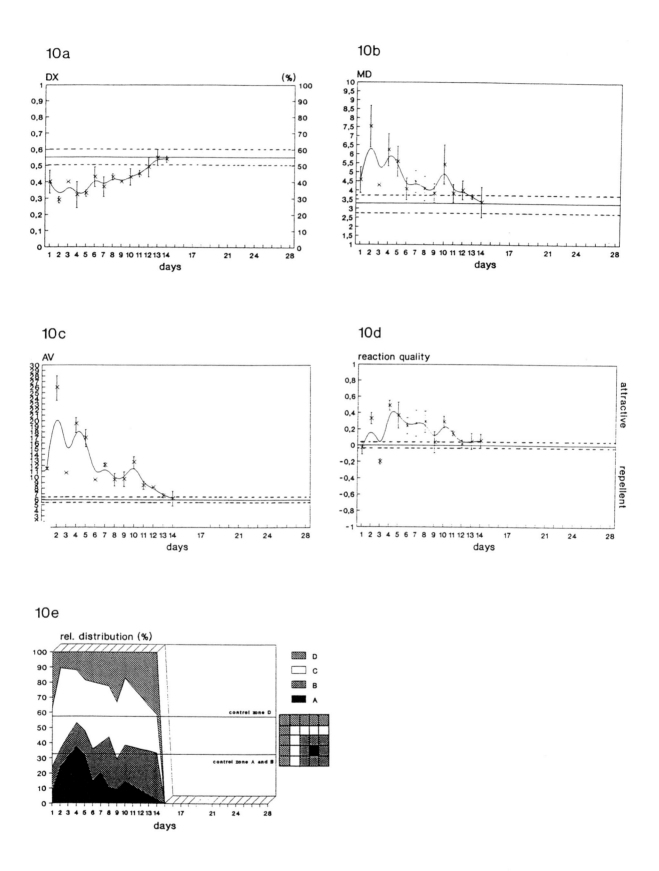

Fig. 10. Distribution behaviour of groups with weevils of mixed sex in the presence of 5700 ng of Sitophilate (a) distribution index (DX-value), (b) mean distribution (MD-value), (c) aggregation value (AV-value), (d) reaction quality and (e) distribution pattern.

Effects of the pheromone on groups of mixed sex

Only 570ng and 5 700 ng of pheromone were offered to weevils in groups of mixed sex (see Figs 9 and 10). Even this small amount of pheromone causes higher aggregations than those which appear in the untreated control (see Fig. 9c). It can be taken from the distribution pattern (see Fig. 9e) that in the beginning the weevils stay close to the point of release. Three days after pheromone application more weevils move towards the pheromone and aggregate close to the emission source. Nevertheless, the majority of the weevils still stay in zone C. After the 9-hour day the distribution pattern matches the situation of the untreated control.

Similar, but more distinct ,is the reaction of the weevils in the presence of 5700 ng of pheromone. After staying and aggregating in the area of release, the weevils show a strong positive reaction towards the pheromone. Aggregation near the pheromone source reaches its height on the 4th day after application. At that time 50% of all weevils are attracted to the zones A and B with over 30% surrounding the actual lure. With the decay of the pheromone over time the distribution pattern meets more and more those of the untreated control.

Discussion

Attraction of *S. granarius* towards synthetic Sitophilate is possible. This has been demonstrated by Phillips et al. (1989). The present study primarily documents the successful luring of granary weevils to a pheromone source in the presence of grain, the granary weevils' natural habitat. Although both sexes, in the virginal as well as in the mated state, reacted positively to the synthetic aggregation pheromone, sex specific behaviour became obvious.

Since the pheromone is male produced, unmated males were able to attract each other and form aggregation spots even in the absence of a pheromone-baited lure. Females are pheromonal inactive, so, unmated females appeared to move rather randomly in the arena when no pheromone was offered.

Unmated females responded with positive reactions when 570 ng of pheromone were presented, and 20% of the unmated female weevils tracked down the source of the chemical. Higher dosages of the pheromone also led to aggregations of unmated females but the aggregation sites occurred further away from the actual lure. This leads to the idea that the aggregation pheromone is not so much an attractant like many sex pheromones but more an arrestant (Evans 1984). The presence of the attractant induces movement, recruiting the individual along a concentration gradient towards the pheromone source. When the weevil reaches a sufficiently high concentration it slows down and eventually stops. This could be the case in the presence of higher pheromone dosages, when unmated females did not continue the walk all the way to the pheromone source. The message of an aggregation pheromone is to bring the individuals of a species together to influence the micro climate by increasing locally the temperature and relative humidity. Physical contact with other weevils might already fulfil the aim of forming an aggregation and therefore could lead to a halt in the movement as well. If that happens it becomes unnecessary to continue to follow the concentration gradient of the luring chemical.

The application of 57000 ng of Sitophilate caused aggregation in unmated females in the far side of the arena away from the lure. This indicated a possible repellent effect of very high amounts of the pheromone. The same was true in unmated males when very high amounts of pheromone were offered. Their reaction quality was apparently negative. The effect of the lure was clearly repellent when 57000 ng were offered.

In contrast to the high amount of pheromone, the low dosage of 570 ng of pheromone caused no difference in distribution behaviour of unmated males when compared to the untreated control. This might be because the pheromone-producing sex is always surrounded by small quantities of the attractant and is therefore unable to sense little amounts of that chemical.

The intermediate dosage of 5700 ng of Sitophilate caused positive reaction in unmated males, and the weevils occurred closer to the pheromone source, but they were not really aggregating like the unmated females did. Either, unmated males were not able to distinguish between the lure and other males emitting pheromone and therefore were somehow disorientated, or the lack of physical contact with females led to a quick disbandment of the clusters in order to continue the search for more suitable aggregation spots. The absence of the opposite sex would therefore result in an increased mobility of male weevils in the area of sufficiently high pheromone concentrations, which normally would attract females as well. This can also explain the accumulation of male weevils when tested in groups of mixed sex, where males and females formed aggregations to a great extent when 5700 ng of pheromone were present. In many of these aggregation spots males and females occurred together in equal numbers.

There is still a long way to go before using pheromone-baited traps for monitoring or even mass-trapping of *S. granarius* in the field. Tests with pheromone-baited probe traps in larger scales of grain have been carried out recently and are quite promising (R. Plarre unpublished data) for monitoring purposes. Similar statements were made for *S. zeamais* by Walgenbach et al. (1987). It is important to apply only the optic pure pheromone, but synthesis of this chemical is still expensive. Sitophilate is very unstable and degrades quickly. Perhaps rubber septa are not the best means of dispensing pheromone. Having to change the lure every week is not economically justifiable. The amount of pheromone being emitted from a lure is also very important, since concentrations which are too high stop the movement of the *S. granarius* towards the trap, or even repel them.

Acknowledgments

Thanks are due to the staff of the Institute for Stored Product Protection of the Federal Biological Research Centre for Agriculture and Forestry in Berlin (Germany) and to the colleagues of the Department of Entomology of the University of Wisconsin in Madison, WI (USA).

References

Andersen, K.T. 1938. Der Kornkäfer, Biologie und Bekämpfung. Monographien zur Angewandten Entomologie 13. Paul Parey Verlag, Berlin.

Busvine, J.R. 1971. Techniques for Testing Insecticides. Commonwealth Agricultural Bureaux. Henry Ling Ltd.

Donat, H.J. 1970. Zur Kenntnis des chemorezeptorischen Verhaltens des Kornkäfers, *Sitophilus granarius*, beim Auffinden seiner Nahrung. Zeitschrift Angewandte Entomologie 65, 1–13.

Evans, D.E. 1984. Biological control of stored grain pests. In: Champ, B.R., and Highley, E., ed., Proceedings of the Australian Development Assistance Course on the Preservation of Stored Cereals. CSIRO Division of Entomology Canberra, 574–582.

Faustini, D.L., Giese, W.L., Phillips, J.K. and Burkholder, W.E. 1982. Aggregation pheromone of the male granary weevil *Sitophilus granarius*. Journal Chemical Ecology, 8, 679–687.

Frings, H. and Lollis, D.W. 1971. Phonoreception in some coleoptera infesting stored products. Journal Stored Product Research, 7, 271–285.

Halstead, D.G.H. 1963. External sex differences in stored-product coleoptera: Bulletin of Entomological Research, 54, 119–134.

Howe, R.W. 1951. The movement of grain weevils through grain. Bulletin of Entomological Research, 42, 125–134.

Levinson, H.Z. and Kanaujia, K.R. 1982. Feeding and oviposition behaviour of the granary weevil (*Sitophilus granarius* L.) induced by stored wheat, wheat extracts and dummies. Zeitschrift Angewandte Entomologie, 93, 292–305.

Levinson, H.Z., Levinson, A., Ren, Z. and Mori, K. 1990. Comparative olfactory perception of the aggregation pheromone of *Sitophilus oryzae* (L.), *S. zeamais* (Motsch.) and *S. granarius* (L.), as well as the sterioisomers of these pheromones. Journal of Applied Entomology, 110, 203–213.

Nawrot, J. and Czaplicki, E. 1978. Behaviour of granary weevil (*Sitophilus granarius*) towards some substances extracted from natural products. Zeszyty Problemowe Postepów Nauk Rolniczych, 202, 183–191.

Perttunen, V. 1972. Humidity and light reactions of *Sitophilus granarius* L., *S. oryzae* L. (Col., Curculionidae), *Rhyzopertha dominica* F. (Bostrychidae), and *Acanthoscelides obtectus* Say (Bruchidae). Annual Entomological Fennal, 38, 161–176.

Phillips, J.K., Chong, J.M., Andersen, J.F. and Burkholder, W.E. 1989. Determination of the enantiomeric composition of (R*,S*)-1-Ethylpropyl-2-methyl-3-hydroxypentanoate, the male-produced aggregation pheromone of *Sitophilus granarius*. Entomologia Expermentalis et Applicata, 51, 149–153.

Phillips, J.K., Miller, P.F., Andersen, J.F., Fales, H.M. and Burkholder, W.E. 1987. The chemical identification of the granary weevil aggregation pheromone. Tetrahedron Letters 49, 6145–6146.

Rodionov, Z.S. 1938. The regularity of movement of *Calandra granaria* in a heap of grain. Zoological Zurnal, 17, 610–616.

Seifelnasr, Y.E. 1991. Influence of olfactory stimulants on resistance/susceptibility of pearl millet, *Pennisetum americanum* to the rice weevil, *Sitophilus oryzae*. Entomologia Experimentalis et Applicata 59, 163–168.

Surtees, G. 1963. Laboratory studies on dispersion behaviour of adult beetles in grain. I. The grain weevil *Sitophilus granarius* (L.) (Coleoptera, Curculionidae). Bulletin of Entomological Research, 54, 147–159.

Surtees, G. 1964a. Laboratory studies on dispersion behaviour of adult beetles in grain. VI. Three dimensional analyses of dispersion of five species in a uniform bulk. Bulletin of Entomological Research, 55, 161–171.

Surtees, G. 1964b. Laboratory studies on dispersion behaviour of adult beetles in grain. VIII. Spontaneous activity in three species and a new approach to analysis of kinesis mechanisms. Animal Behaviour, 12, 374–377.

Surtees, G. 1965. Laboratory studies on dispersion behaviour of adult beetles in grain. IX. Some differences in behaviour of mated and unmated grain weevils. Entomologist's Monthly Magazine, 100, 105–108.

Torne, H. 1941. Einfluß farbigen Lichtes auf Entwicklung und Aktivität des Kornkäfers *Calandra granaria* L. Zeitschrift Hygienische Zoologie, 33, 53–64.

Walgenbach, C.A., Burkholder, W.E., Curtis, M.J., and Khan. Z.A. 1987. Laboratory trapping studies with *Sitophilus zeamais* (Coleoptera: Curculionidae). Journal Economical Entomology 80: 763–767.

Distribution and status of Psocoptera infesting stored products in Australia

D. Rees*

Abstract

In Australia, infestations of Psocoptera in stored products, especially *Liposcelis* species, appear to have become more frequent in recent years. Changes in pest control and inspection practices, together with seasonal factors such as poor grain condition because of adverse harvesting conditions, have been implicated in this.

There is no up-to-date survey of the distribution and status of these insects in stored products in Australia. With the collaboration of State grain handling authorities, and departments of agriculture, health and primary industries, samples of psocids were collected and identified from around the country to gain an insight into the species composition, status and distribution of the psocid fauna of stored products in Australia.

A total of 11 species was found from three genera. Most frequently encountered were *Liposcelis bostrychophila* Badonnel, *Liposcelis decolor* (Pearman) and *Liposcelis entomophila* (Enderlein) (Psocoptera: Liposcelidae). *Lachesilla quercus* (Kolbe) (Psocoptera: Lachesillidae) was recorded for the first time as a regular inhabitant of grain storage facilities. *L. quercus*, *Liposcelis rufa* Broadhead and *Liposcelis brunnea* Motschulsky were recorded in Australia for the first time.

Introduction

Psocoptera or psocids have traditionally been viewed as scavengers and mould feeders and of little importance as pests of stored products. However, from time to time, psocids, especially *Liposcelis* species, have been found infesting grain in very large numbers.

There is little published information on the distribution and status of psocids in stored products in Australia. Champ and Smithers (1965) recorded some 15 species from 8 families from farms and grain stores in Queensland. Watt (1965) records *Liposcelis bostrychophila* Badonnel as infesting and damaging wheat in New South Wales. Since that time, changes to methods used in Australia to control storage pests have occurred, from the almost universal admixture of insecticides into grain to the current reliance on fumigation with much reduced structural spraying. Such changes are likely to have had an impact on the psocids present. A fresh look is needed.

Methods

Psocids were collected from stored grain, grain products and from other stored dried animal and plant material. Samples were obtained from all mainland states of Australia except

Northern Territory. Samples were collected on an *ad hoc* basis both by the author and as a result of requests to state bulk handling authorities, Queensland Department of Primary Industries, Western Australia Department of Agriculture, Queensland Department of Health, various CSIRO divisions and others. Specimens obtained were cleared and mounted and then identified using the keys of Lienhard (1990) and Mockford (1991). Material collected has been lodged in the Australian National Insect Collection.

Species Distribution

Psocoptera were found associated with stored products in all states from which samples were collected. Locations from which specimens were found are plotted on maps (Fig. 1) and the habitats in which they were found are given in Table 1.

A total of 11 species were identified belonging to three genera: *Liposcelis*, *Lachesilla* and *Lepinotus*. Psocids were found infesting stored products throughout their post harvest life, from the farm, in storage, through to shops, offices and domestic premises. They were also found as pests in museums and herbaria.

Liposcelis species

The majority of infestations found were of *Liposcelis* species. Eight species were identified, three of which appear to be widely distributed in Australia (Fig. 1, Table 1). All eight species found have been recorded from stored products somewhere in the world (Lienhard 1990).

The most frequently encountered species were *Liposcelis bostrychophila*, *Liposcelis decolor* (Pearman) and *Liposcelis entomophila* (Enderlein). *L. bostrychophila* was the most widely distributed species and was found in the largest number of habitats (Fig. 1, Table 1). It was the dominant species infesting material in shops, offices, museums, and domestic premises. *L. decolor* was found only in bulk handling authority and farm stores and was more frequently encountered in samples from South and Western Australia than from the eastern states. In grain stores, *L. bostrychophila* and *L. decolor* were most often found, usually in small numbers, infesting residues in store buildings and related grain handling equipment. However, extremely heavy infestations of *L. decolor* and / or *L. bostrychophila,* which formed the majority of pest biomass present, were found during the autumn and winter months of 1993 in bulk stored barley in South Australia, wheat in Western Australia and lupins in New South Wales. Associated with these infestations were large accumulations of dead and living insects on walkways above infested cells and in tunnels below.

L. entomophila was found mostly at coastal or near coastal locations in farm and bulk handling authority stores. Populations were found infesting store structures and machinery, residues, disused grain bins and on one occasion, a bulk of barley. During 1993, at least one shipment of grain infested with this species was rejected at export from Australia.

Other *Liposcelis* species were recorded from single locations. *L. corrodens* (Heymons) was found in residues in the now disused Sydney grain terminal. A very heavy infesta-

* Stored Grain Research Laboratory, CSIRO Division of Entomology, GPO Box 1700, Canberra ACT 2601, Australia.

Table 1.   Habitats in which psocopteran species were found in Australia

Species	Bulk Handling Authority stores											
	Barley	Wheat	Sorghum	Lupins	Residues	Store structures	Farm stores	Feed mills	Flour mills	Museums, herbaria	Hospitals, offices, shops	Houses
L. bostrychophila	X	X	X	X	X	X	X		X	X	X	X
L. decolor	X					X	X					
L. entomophila	X				X	X	X					
L.brunnea	X	X										
L. corrodens					X							
L. paeta								X				
L. mendax						X						
L. rufa											X	
Lachesilla quercus						X						
Lepinotus spp.							X					

tion of *L. paeta* Pearman was discovered in sorghum at a feed mill at Ayr, North Queensland. *L. mendax* Pearman were found in an empty grain shed at Maimuru (lat. 34:14S, long. 148:23E) near Young, New South Wales. Two species, *Liposcelis brunnea* and *Liposcelis rufa* appear not to have been recorded in Australia before. *L. brunnea* Motschulsky was found infesting wheat and barley in a silo at Streaky bay (lat. 32:48S, long. 134:12E), South Australia and *L. rufa* Broadhead was discovered in a house in Canberra, ACT, on wood recently purchased from a timber merchant in the city.

Lachesilla quercus (Kolbe)

In nature, species of *Lachesilla* browse on algae, moulds etc. which grow on dying leaves of trees; *L. quercus* can, for example, be attracted to a tree by breaking branches to produce prematurely senescent foliage (New 1987). *Lachesilla* are winged insects that can fly quite well and probably travel long distances by being carried by the wind (New 1987). *Lachesilla pedicularia* (L.) is known from grain stores and domestic situations in Europe and North America (Mockford 1991; Obr 1978). A species of *Lachesilla* was found associated with peanuts at Kingaroy, Queensland, Australia (Champ and Smithers 1965).

Lachesilla quercus apparently has not been previously recorded in Australia. This species is not known as a regular inhabitant of grain storage in its native Europe and Asia. However, it has been found in the garage of a house in Prague on stored boards (Obr 1978) and was once found in a flour warehouse in Florida USA on jute sacks imported from Pakistan (Prof. E. L. Mockford, per. comm.). It has been found over the last two seasons living on the machinery and fabric of silos and empty grain sheds at Northampton, Geralton, Kwinana, Albany and Esperance in coastal south west Australia, sometimes in large numbers, browsing on minute particles of edible dust present on beams and walls. It does not appear to penetrate bulks of grain. This insect produces large amounts of silk webbing which quickly becomes covered with dust, under which they live, breed and feed. Large numbers of dead insects can accumulate together with deposits of very fine black faeces. In the stores in which it has been found, which are kept very clean, it is usually the only insect present. Populations appear to fluctuate wildly from one month to the next but so far infestations have been found in stores from early winter to mid-summer.

Under laboratory conditions at 27°C, 75% r.h., on a brown rice and dried yeast based diet, this species was found to breed continuously and have a high rate of reproduction. It is therefore very capable of taking advantage of conditions in stores when they are favourable.

Lepinotus species

Populations of *Lepinotus reticulatus* (Enderlein) and *Lepinotus inquilinus* Heyden were found on farms in far south western Western Australia. These species are also known from stores and houses in Europe and North America (Mockford 1991; Obr 1978) and on farms in southern Queensland (Champ and Smithers 1965).

Discussion

Psocids, especially *Liposcelis* spp., are widely distributed in Australia and are common and widespread inhabitants of grain stores. They were found in most places in which stored products are produced, stored and used, from the farm to the final consumer. Data collected so far provide only a sketchy picture and doubtless more species will be discovered, and those currently known will be found to have an even wider distribution. Single records of some species are likely to prove to be 'snapshots' of more widespread distributions.

All of the species found in this survey are probably introductions to Australia and have either cosmopolitan or extensive distributions worldwide. Australia has, and probably still is acquiring, these insects by trade and the movement of people and their effects. Their small size and unspecialised diet makes them ideal, inconspicuous candidates for such introduction. Well known pest species such as *Liposcelis bostrychophila*, *L. entomophila*, *L. decolor* and *L. paeta* are established in Australia.

Psocids were mostly found to be persistent pests in the structure of grain stores, feeding on often hard-to-find small deposits of residues. However, *Liposcelis* can penetrate whole bulks of grain in silos, leading to contamination with large and unacceptable quantities of living and dead insects. Piles of dead insects which form on walkways etc. are a very public statement of poor hygiene, a safety hazard (such deposits make walkways very slippery) and a potential food source for other insects such as *Trogoderma variabile* Ballion (Coleoptera: Dermestidae). What triggers off such population explosions remains unclear. What damage such numbers do to grain, especially to commodities such as malting barley and seed grain, has not yet been quantified although it could be expected to be extensive. *Liposcelis* species are known to be capable of attacking whole grains and have a preference for the grain germ (Watt 1965; McFarlane 1982; Rees and Walker

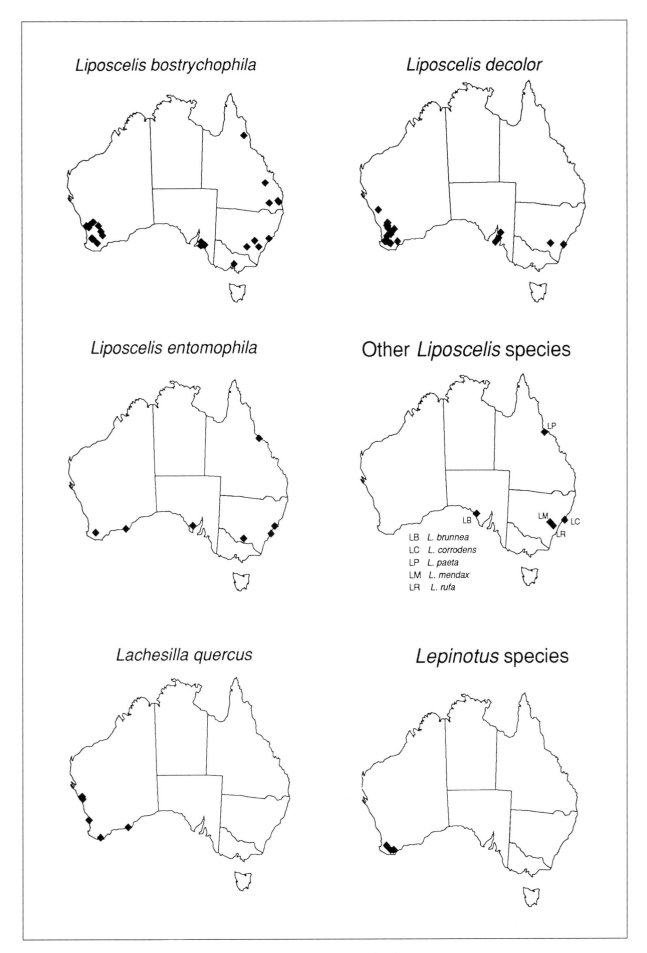

Fig. 1. Locations at which Psocoptera associated with stored products were found

1990). However, their presence in parcels of grain has already caused the rejection of at least one shipment at export, with all the attendant costs that entails.

The species that appears most widely distributed and found in the greatest variety of habitats is *L. bostrychophila*. In Australia it appears to be virtually the only species found in shops, offices, museums and domestic premises. This species, unlike the other *Liposcelis* species found, is parthenogenetic. An infestation can therefore start with a single egg or immature individual. In this way it is probably better able than others to become established in locations where infestable materials are kept and moved in small discrete batches (houses, museums etc.). In such places it is most likely to be introduced as a single individual or in very small numbers. In Europe, this species appears also to be the dominant species in domestic and retail premises (Lienhard 1983; Obr 1978; Turner and Maude-Roxby 1989).

More data may or may not confirm the observations that *L. decolor* is more widespread in the south west of Australia than in the eastern states or that *L. entomophila* is mostly confined to more humid coastal locations. However, in the case of the latter, *L. entomophila* is known to be the dominant psocid infesting stored rice in hot, humid areas of Java, Indonesia (Haines and Pranata 1982)

Optimum conditions for development of *Liposcelis* are about 30°C, 70–80% r.h., breeding will occur at 18–35°C; however, *L. paeta* is known to breed at temperatures up to 42°C. (Rees and Walker 1990). For long-term survival of populations, mean humidity should be more than about 58% r.h. (equivalent to a grain moisture content of 12–13%). Below this level, *Liposcelis* lose more water to the atmosphere than they gain from it (Knülle and Spadafora 1969). The very low day-time humidities in summer in much of Australia may not be sufficient to kill these insects if they can recuperate sufficiently during the cooler, more humid nights. Every grain storage facility contains many potential micro-environments (cracks, crevices, damp spots, basements, inside beams etc.) in some of which humidities remain much higher than ambient. In such places psocids could easily survive hostile conditions. During the autumn to spring storage season, ambient conditions become more humid and temperatures in many grain stores fall but not to levels that would prevent breeding taking place. During this time psocids are likely to range more widely than in summer and hence become more obvious as pests. Observations made during this survey suggest that this may be the case. In shops and offices, temperature and humidities are usually better regulated so may remain suitable for these insects all year round.

A number of factors may be important in the apparent increase in incidence of occurrence of these insects in Australia in recent years. The grain harvest in some seasons, such as 1992–93, can be affected by bad weather conditions during which a proportion of grain is dried and wetted several times before harvest. Such damaged grain is easier for any insect to attack than sound, dry grain. An increase in maximum wheat receival moisture content (12.5%) into bulk handling stores could also alter the incidence of these insects in such facilities.

Changes have also occurred in the use of insecticides. No longer is grain routinely sprayed with insecticide on intake into storage. In many stores, surface applications have also been discontinued or the chemical used changed in response to obligations to control other pests, notably *Trogoderma variabile*. There has been a reduction in the use of fenitrothion, an insecticide known to be effective against *Liposcelis* (Turner et al. 1991), in favour of other chemicals which may be less so. The discontinuance of surface spraying may be partly responsible for allowing the establishment of

Lachesilla quercus in grain stores in Western Australia. Edible grain dust no longer contains insecticide residues as it would have done when grain was sprayed at intake, so is now available as food for psocids. Many stores infested with *L. quercus* can now be sealed for fumigation and have been painted with reflective white paint to help minimise solar heating of the air inside which would otherwise put pressure on the integrity of the seal. The reflective paint may be having the undesirable side-effect of allowing populations of *Lachesilla* to survive as the air inside is less likely to be heated to insecticidal temperatures than it would be if the store remained unpainted. *Lachesilla* appears to be especially able to survive in very clean stores. It probably uses the webbing it makes to catch its own food, such as edible dusts, from the air. Populations of this species will probably be found living in trees (its natural habitat) in nearby townships. This insect can fly and such populations could act as a continual source of re-infestation as conditions allowed. Indeed its relation, *L. pedicularia* is a seasonal migrant between natural habitats and buildings (Obr 1978). The arrival of 'new' insects such as this into storage is a reminder that the list of insects that attack stored products is not fixed and will change in response to many factors. A species such as *L. quercus,* being a surface feeder on dust and other edible material deposited on dead leaves, appears well adapted to life foraging on walls and machinery of a grain store.

As standards of commodity management improve, the status of various pests present will change. Pests which live mostly in bulks of grain are controlled by fumigation and basic levels of housekeeping. Other species, such as psocids, may survive such a regime in the smallest quantities of residues in ancillary parts of grain stores which are often missed. Populations probably move between natural habitats and domestic situations or commercial storage, for example, on clothing, pallets and packing material. Given the design of many facilities and the nature of working practices undertaken, the level of hygiene required to prevent the establishment of these insects is likely to be impossible or very difficult to achieve. Such populations should be monitored with the awareness that, especially in the case of *Liposcelis* spp., they can seriously infest whole bulks of grain. Standards of inspection and hygiene need to especially high in sensitive areas of storages, museums, and factories such as sample rooms, packing lines and long-term stores if infestations of these insects are to be avoided in such places.

Acknowledgments

The author would like to thank all those who contributed to this study, in particular to Andy Szito of Western Australia Department of Agriculture, for supplying samples collected from farms in that state. To Noel Ireland and other staff of Co-operative Bulk Handling of Perth, Western Australia, Terry Hillier and other staff of South Australia Co-operative Bulk Handling and staff of New South Wales Grain Corporation facilities and offices at Newcastle, Wagga Wagga and Port Kembla for their interest and help in inspecting facilities under their supervision. I would like to thank Dr Tim New of La Trobe University, Melbourne, Australia and Dr Charles Lienhard of The Natural History Museum Geneva, Switzerland for confirming the identity of various specimens of *Lachesilla quercus*.

References

Champ, B.R. and Smithers, C.N. 1965. Insects and mites associated with stored products in Queensland. 1. Psocoptera. Queensland Journal of Agricultural and Animal Sciences, 22, 259–262.

Haines, C.P. and Pranata R.I. 1982. Survey of insects and arachnids associated with stored products in some parts of Java. In; Teter, N.C. and Frio, A.S. eds., Progress in grain protection—Proceedings of the 5th annual grains post-harvest workshop, Changmai, Thailand 19–21 January 1982. SEARCA, Laguna, Philippines. 17–47.

Knülle, W. and Spadafora, R.R. 1969. Water vapour sorption and humidity relationships in *Liposcelis* (Insecta: Psocoptera). Journal of Stored Product Research, 5, 49–55.

Lienhard, C. 1983. Die Staubläuse der Region Zürich (Insecta : Psocoptera), Vierteljahrsschrift der Naturforschenden Gesellschaft in Zurich, 128, 115–129

Lienhard, C. 1990. Revision of the Western Palaearctic species of *Liposcelis* Motschulsky (Psocoptera : Liposcelidae). Zoologishe Jahrbuecher Abteilung fur Systematik Oekologie und Geographie der Tiere, 117, 117–174.

McFarlane, J.A. 1982. Damage to milled rice by Psocids. Tropical Stored Product Information, 44, 3–10.

Mockford, E.L. 1991. Psocids (Psocoptera). In: Gorham, J. R. ed. Insect and mite pests in food: an illustrated key. United States Dept. of Agriculture, Agriculture Handbook no. 655, 2, 371–402.

New, T.R. 1987. Biology of the Psocoptera. Oriental Insects, 21, 1–109.

Obr, S. 1978. Psocoptera of food-processing plants and storages, dwellings and collections of natural objects in Czechoslovakia, Acta Entomologica Bohemoslovaca, 75 226–242.

Rees, D.P. and Walker, A.J. 1990. The Effect of temperature and relative humidity on population growth of three *Liposcelis* species (Psocoptera: Liposcelidae) infesting stored products in tropical countries. Bulletin of Entomological Research, 80, 353–358.

Turner, B.D. and Maude-Roxby, H. 1989. The prevalence of the booklouse *Liposcelis bostrychophilus* Badonnel (Liposcelis, Psocoptera) in British domestic kitchens, International Pest Control, July/August 1989, 93–97.

Turner, B.D., Maude-Roxby, H. and Pike, V. 1991. Control of the domestic insect pest *Liposcelis bostrychophila* (Badonnel) (Psocoptera): an experimental evaluation of the efficiency of some insecticides. International Pest Control, November/December 1991, 153–157.

Watt, M.J. 1965. Notes on pests of stored grain *Liposcelis bostrychophilus* and *Sitophilus* spp. The Agricultural Gazette, November 1965, 693–696

Hormonal control of reproduction in the female pyralid moth *Plodia interpunctella* (Hübner) (Lepidoptera: Phycitidae)

E. Shaaya[*], D. Silhacek[†], P. Shirk[†], H. Rees[§], G. Zimowska[†] and S. Plotkin[*]

Abstract

The differentiation and growth of ovaries of pupae of *Plodia interpunctella* were analysed using immunofluorescence microscopy, and correlated with changes in external morphology and ecdysteroid titre. The data indicated that ovarian development progresses in two distinct, ecdysteriod-coordinated steps. The first occurs during the first three days of the pupal period and is initiated by an increasing titre of ecdysteroids (identified as mainly ecdysone) starting eight hours after the moult. Once ovarian development is initiated, specialised cell types, e.g. nurse cells, follicle cells and ocytes, differentiate and develop to a point just before yolk protein synthesis and uptake by the ocyte. The second stage in ovarian development is signalled by the reattainment of a baseline level of ecdysteroids during the final three days of the pupal period. Follicle development occurs during this stage. Subsequent development appears to proceed sequentially without further regulation by ecdysteroids. After eclosion, the ocytes are chorionated and the mature eggs are layed. Mating accelerates the final stages of egg maturation and egg laying, an effect that can be mimicked by juvenile hormone treatments.

Introduction

The hormonal regulation of postembryonic ovarian development is only partially understood in Phycitidae. In these moths, which have short-lived adults, ovaries complete all or most of egg maturation before adult eclosion. Placing adult ovarian development within the parameters of metamorphosis requires that the process progress directly to completion, without arrested follicular stages and that regulation of vitellogenesis be controlled by the hormones regulating metamorphosis.

Egg maturation and vitellogenesis in late pharate adults of the Indianmeal moth were blocked with ecdysteroid levels maintained in late pharate adults by treatment with 20-hydroxy ecdysone (20-HE) (Shirk et al. 1990). This effect of 20-HE was shown to be independent of juvenile hormone activity because treatment with 20-HE and juvenile hormone did not restore vitellogenesis or egg maturation. To more critically assess the role of ecdysteroids in normal females of the Indianmeal moth, we measured changes in the ecdysteroid titre during pharate adult development and correlated the changing ecdysteroid titre with the progress of adult ovarian development and the onset of vitellogenesis. Since there exists in general a negative correlation in adult insects between

sexual activity and life span (Partridge 1986; Ayerty 1975; Norris 1933), we also investigated the endocrine basis for this, by investigating the effect of juvenile hormone on life span and egg laying in virgin and mated females of *P. interpunctella*.

Material and Methods

Insect cultures

The *P. interpunctella* colony was reared according to Silhacek and Miller (1972), in a 16 hour light: 8 hour dark photoperiod at 30°C and 70% r.h. For ecdysteroid studies, moulted white pupae (±1.5 h) were collected to obtain synchronised cohorts and then kept in a long-day photoperiod, as described above, until used for experiments. As a source of pupae for the continuous darkness condition, newly moulted white pupae were collected and placed in total darkness until the appropriate age. Insect age is expressed in hours from the time of pupation.

Ecdysteroid extraction and quantification

For each time point one to six pupae were homogenised in 70% methanol at 4°C. The homogenate was centrifuged 10 min at 2600 g and 4°C. The supernatant was transferred to another tube, and the pellet was re-extracted in cold 70% methanol. The supernatants were combined and stored for a minimum of 24 hours at –20°C. Following cold storage, the samples were centrifuged 10 min at 2600 g and 4°C. The supernatants were dried using a stream of nitrogen gas and heating at 30°C. Dried samples were stored at –20°C and were reconstituted with 100% methanol for assaying. Aliquots of each sample were taken for quantitative ecdysteroid analysis by RIA as described by Shaaya et al. (1986). The 20-HE rabbit antiserum, DLII, was a gift of Professor J. Koolman (Marburg/Lahn, Germany), and the rabbit complement serum, HLA-ABC, was purchased from Sigma (St. Louis, Mo.). Radiolabelled [3H]20-HE (89 Ci/mmol) was purchased from NEN (Boston, Mass.), and 20-HE purchased from Sigma (St. Louis, Mo.). Quantification of each sample was based on evaluation of triplicates using different concentrations. A standard curve was based on competitive binding with various concentrations of 20-HE.

To determine the molecular composition of the ecdysteroids, methanolic extracts were prepared from females as described above. The ecdysteroids were analysed by gas chromotograph—mass spectrometry with selected ion monitoring [GC-MS(SIM)] as described by Evershed et al. (1987).

Studies with the juvenile hormone analogue, methoprene

Developmentally synchronous pupae 3 days prior to adult eclosion were used to permit enough time for the absorption, distribution in the body, and the action of the juvenile hormone analogue (JHA). The appropriate amount of the JHA was applied topically on the dorsal side of the pupa using a microsyringe. To obtain virgin adults, female pupae were

---

[*] The Volcani Center, ARO, Bet-Dagan 50250, Israel.

[†] USDA, ARS, Insect Attractants, Behaviour and Basic Biology Laboratory, P.O. Box 14565 Gainesville, FL 32604, USA.

[§] Department of Biochemistry, University of Liverpool, P.O. Box 147, Liverpool L69 3BX, England.

selected and kept separately. Male adults were kept together during the experiment. Adults age 0–4 hours were used for the experiments. The number of eggs of the test insects was counted daily. Methoprene was a gift of the Zoecon Corporation (Palo Alto, CA).

Results

Ecdysteroid studies

For the studies of the ecdysteroid titres during the pupal and pharate adult stages, females were collected as white pupae, kept in a long-day photoperiod, and collected every 4 hours until adult eclosion (136 hours after pupation). Ecdysteroid titres were determined for each time point (Fig. 1). For the first 8 hours after pupation the titres were < 300 pg/mg wet wt and then increased first to a plateau of 1600–1700 pg/mg wet wt between 16–24 hours and then attained a maximum of 2200–2500 pg/mg wet wt between 28–36 hours after pupation. The titre then began to decline with plateaus at 40–48 hours (1700–1870 pg/mg wet wt) and at 52–60 hours (935–1200 pg/mg wet wt). The ecdysteroid titre fell below 500 pg/mg wet wt by 68 hours and remained below this level for the remainder of pharate adult development. For pupae maintained in continuous darkness instead of in a long-day photoperiod, the profile was essentially the same except that the major ecdysteroid peak between 12 and 68 hours reached a maximum earlier at 24 hours (3130 pg/mg) and was broader than the peak in females kept in photoperiodic conditions.

To determine which ecdysteroids contribute qualitatively to the total titres prior to the onset of vitellogenesis, methanol extracts were made from photo-synchronised female pupae at 2, 28, 52, 76 and 100 hours after pupation. The free ecdysteroid fractions were analysed by GC-MS (SIM). In 2-hour-old pupae, 20-HE comprised 81% of the ecdysteroids measured (Table 1). At 28 hours after pupation, the maximum of the pharate adult peak, the distribution had switched and ecdysone comprised the greater proportion of the ecdysteroids (93%). As pharate adult development progressed, the proportion of ecdysone decreased as the proportion of 20-HE and 20,26-HE increased. By 100 hours after pupation, at the time of initiation of vitellogenesis (Zimowska et al. 1991), no ecdysone was detected; 20-HE and 20, 26-HE comprised all

of the ecdysteroids measured. No detectable amounts of 26-HE were found in any of the samples.

JHA studies

When virgin *P. interpunctella* were treated with methoprene as pupae, roughly 3 days prior to adult hatching, the following effects were observed (Table 2). In females receiving 10 and 100 µg methoprene, the mean number of eggs per female increased as compared to control (virgin) females, although it was lower in control (mated) females. The virgin females treated with methoprene started laying eggs earlier, and 100% of the eggs were laid earlier; 3.9 days in the treated females, 3.2 days in control (mated) and 5.2 days in control (virgin) females. This reduction of time necessary for egg laying corresponds to a much shorter life span of 5.5–6 days as compared with 8.8 days in nontreated virgin females and 5.5 days in control (mated).

Discussion

By measuring the ecdysteroids at short time intervals, every 4 hours, for the photo-synchronised Indianmeal moth females, the ecdysteroid profile shows changes in a stepwise fashion with the titre stabilising over several 8-hour plateau periods. In the first 60 hours after pupation, 8-hour plateaus were observed between 0–8, 16–24, 28–36, 40–48 and 52–60 hours. Plateaus in ecdysteroid titres were observed also in the profiles for pupae cultured in constant darkness. Similar stepwise or plateau changes in the ecdysteroid titre have been observed in *Drosophila melanogaster* when short intervals were used between sample collection (Handler 1982). The significance of these 8-hour plateaus was not determined.

Ecdysone was the major component of the early pharate adult ecdysteroid peak. Our data show (Table 1) that high levels of ecdysone, with only minor amounts of other metabolites, are correlated with the initiation of adult ovarian development. As development progressed, the amount of ecdysone progressively decreased as well as the proportion of ecdysone to 20-HE. By 100 hours after pupation, when ecdysteroid titres had declined to <500 pg/mg wet wt, no ecdysone was detected in the samples and 20-HE was the primary ecdysteroid present in the pharate adult females. This switch

Table 1. Quantitative analysis of ecdysteroids from female *Plodia interpunctella* during metamorphosis.

Pupal age (hours)	pg/mg wet wt (% of total ecdysteroids)		
	Ecdysone	20-HE	20, 26-HE
2	18(19)	76(81)	ND
28	1640(93)	115(7)	ND
52	16(19)	68(81)	ND
76	4(3)	62(51)	55(46)
100	0(0)	19(68)	9(32)

ND = not detectable.

Table 2. The effect of methoprene on life span and egg production of virgin females of *Plodia interpunctella*.

Dose (ng/pupa)	Total number of eggs	Time (days) needed for laying		Adult lifespan (days)	% of females not laying eggs
		25% of eggs	100% of eggs		
10	57±8	2.4±0.2	3.9±0.5	6.1±0.4	9
100	76±13	2.0±0.2	3.9±0.3	5.5±0.3	5
Control (virgin)	23±8	3.1±0.4	5.2±0.5	8.8±0.6	23
Control (mated)	176±20	1.3±0.2	3.2±0.3	5.6±0.5	0

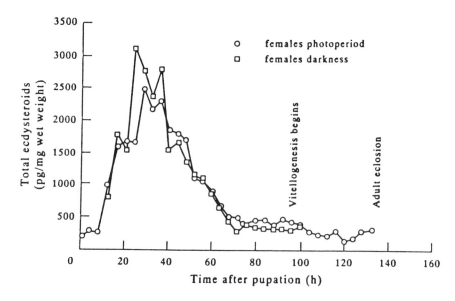

Fig. 1.   Changes in ecdysteroid titres (as 20-HE equivalents) during pupal adult development.
The pupa were kept either in a long-day photoperiod or in constant darkness.

from ecdysone to 20-HE as the primary ecdysteroid as pharate adult development progresses has also been reported for *H. zea* (Holman and Meola 1978), and *M. sexta* (Warren and Gilber 1986).

The data suggest that the ecdysteroid decline is significant to the initiation of vitellogenesis. Vitellogenesis in *P. interpunctella* begins between 96 and 100 hours after pupation (Zimowska et al. 1991). This is also consistant with the observation that increasing the ecdysteroid titres by treatment with exogenous 20-HE during pharate adult development blocked yolk protein synthesis and egg maturation (Shirk et al. 1990).

Taken together, the data suggest that ecdysteroids play a central role in the regulation of vitellogenesis during pharate adult development. However, the mechanism that leads directly to the initiation of vitellogenesis appears to involve additional regulators because the initiation of vitellogenesis does not occur until 24 hours after the ecdysteroids have completed the decline.

In the present study we were able to demonstrate that methoprene exerts an influence on adult development. The interesting feature is that the JHA reduced the life span of virgins to the control value. Mated females laid many more eggs than virgin females, but virgin females laid a considerable number of eggs when treated with increasing concentrations of methoprene. A similar situation was also recorded for *E. cautella* (Shaaya et al. 1991). Our data suggest that lifespan and egg laying are negatively correlated with each other. Treatment with a JHA acts either via manipulation of the ecdysteroid titre or indirectly as a result of JHA effects on the stage of development of the ovary. The exact mechanism of this regulation is still to be elucidated.

References

Ayerty, J.N. 1975. Egg laying by unmated females of *Sitotroga cerealella* (Lepidoptera: Gelechiidae). Journal of Stored Products Research, 11, 211.

Evershed, R.P., Mercer, J.G. and Rees, H.H. 1987. Capillary gas chromatography-mass spectrometry of ecdysteroids. Journal of Chromatography, 390, 357–369.

Handler, A.M. 1982. Ecdysteroid titres during pupal and adult development in *Drosophila melanogaster.* Developmental Biology, 93, 73–82.

Holman, M. and Meola, R.W. 1978. A high-performance liquid chromatography method for the purification and analysis of insect ecdysones: application to measurement of ecdysone titres during pupal-adult development of *Heliothis zea.* Insect Biochemistry, 8, 275–278.

Norris, M.J. 1933. Contributions towards the study of insect fertility. II. Experiments on the factors influencing fertility in *Ephestia kuhmiella* Z. (Lepidoptera, Phycitidae). Proceedings of the Zoological Society, 59, 903.

Partridge, L. 1986.. Sexual activity and life span. In: Collatz, K.G and Sohal, R.S. ed., Insect Aging. Springer Verlag, Berlin, 45–54.

Shaaya, E., Calderon, M., Pisarev, V. and Spindler, K.D. 1991. Effect of juvenile hormone analogue on the life span, egg laying, and ecdysteroid titer of virgin *Ephestia cautella* females. Archives of Insect Biochemistry and Physiology, 17, 183–188.

Shaaya, E., Spindler-Barth, M. and Spindler, K.D. 1986. The effect of a juvenile hormone analogue on ecdysteroid titre during development and HnRNA formation in the moth *Ephestia cautella.* Insect Biochemistry, 16, 181–185.

Shirk, P.D., Bean, D.W. and Brookes, V.J. 1990. Ecdysteroids control vitellogenesis and egg maturation in pharate adult females of the Indianmeal moth, *Plodia interpunctella.* Archives of Insect Biochemistry and Physiology, 15, 183–199.

Silhacek, D.L. and Millre, G.L. 1972. Growth and development of the Indianmeal moth, *Plodia interpunctella* (Lepidoptera: Phycitidae), under laboratory mass-rearing conditions. Annals of the Entomological Society of America, 65, 1084–1087.

Warren, J.T. and Gilbert, L.I. 1986. Ecdysone metabolism and distribution during the pupal-adult development of Manduca sexta. Insect Biochemistry, 16, 65–82.

Zimowska, G., Silhacek, D.L., Shaaya, E. and Shirk, P.D. 1991. Immuno-fluorescent analysis of folicular growth and development in whole ovaries of the Indianmeal moth. Journal of Morphology, 209, 215–228.

Role of ultrasound production and chemical signals in the courtship behaviour of *Ephestia cautella* (Walker), *Ephestia kuehniella* Zeller and *Plodia interpunctella* (Hübner) (Lepidoptera: Pyralidae)

P. Trematerra[*] and G. Pavan[†]

Abstract

Further details are given on the role of ultrasound production in the courtship behaviour of *Ephestia cautella* (Walker), *Ephestia kuehniella* Zeller and *Plodia interpunctella* (Hübner) during wing-fanning and on the role of other signals in pair forming.

Sounds produced by these moths consist of regular sequences of ultrasonic pulses with a frequency range up to 80 kHz, emitted by wing-fanning males during courtship behaviour. The pulses are emitted in regular sequences at 14–24 ms, corresponding to pulse-repetition rates ranging from 41–72 pulses per second.

Tests to evaluate the reproductive success in muted males and deaf females revealed a lower success in muted males and deaf females than in intact individuals. This is probably related to the suppression of the acoustic signals.

When antennae were surgically removed from males and females a lower reproductive success (from 40–78%) was observed in males. Female *E. cautella* and *E. kuehniella* without antennae showed no fall in their reproductive success, whilst a drop of 22% was observed with *P. interpunctella*.

Ultrasonic communication in these moths plays a significant role in mating behaviour. Thus courtship behaviour must be reconsidered with greater emphasis on the interaction between multiple modalities. Further work is needed to understand the role played by ultrasounds in phycitidae mating behaviour and the relative importance of chemical, visual and acoustic stimuli.

Introduction

Many insects have airborne sound receptors, which function primarily in intraspecific communication, whilst in others hearing serves to warn of the potential threat of a predator (Roeder and Treat 1961; Spangler 1988; Spangler and Hippenmeyer 1988).

Certain pyralid moths, such as the Galleriinae wax moths *Achroia grisella* F. and *Galleria mellonella* L., and the rice moth *Corcyra cephalonica* (Staint.), have acquired the additional ability to generate sounds by wing-fanning for intraspecific communication and pair forming (Trematerra 1992).

With other pyralid moths infesting stored products, the hearing ability of *Ephestia cautella* (Walker), *Ephestia elutella* (Hübner) *Ephestia kuehniella* Zeller and *Plodia interpunctella* (Hübner) was reported by Mullen and Tsao (1971) and by Pérez and Zhantiev (1976).

The courtship of these Phycitinae has been examined and several studies have been made on the male orientation to the pheromone produced by the calling female and also on the behaviour of the male after the female has been located (Traynier 1968; Kennedy and Marsh 1974; Barrer and Hill 1977, 1980).

These pyralids, Galleriinae and Phycitinae, have tympanic hearing organs located on the pleural-ventral surface of the first abdominal segment. The morphological similarity of these insects led us to suspect that the hearing capabilities of the moths are similar, and that they are capable of developing acoustical communication systems (Trematerra and Pavan 1994).

This paper gives further details on the role of ultrasound production in the courtship behaviour of *E. cautella*, *E. kuehniella* and *P. interpunctella* during wing-fanning and discusses the role of other signals in pair forming.

Materials and Methods

Insect rearing

E. cautella, *E. kuehniella* and *P. interpunctella* adults were obtained from laboratory cultures reared using the method of Hoppe (1981) on a *pabulum* of maize flour (28.6%), whole meal flour (14.3%), wheat flour (14.3%), oat meal (9.5%), brewer's yeast (4.8%), wheat germ (4.8%), honey (7.9%), and glycerol (7.1%) in standard conditions of $25 \pm 1°C$ and $65 \pm 5\%$ r.h.; with a natural photoperiod. Virgin males and females were obtained sexing the IV-V instar larvae by observing the dark male gonads through the cuticle.

Sound recording and analysis

Ultrasonic signals were acquired by means of a Bruel & Kjaer (B&K) 2231 phonometer with a B&K 4135 transducer (flat-frequency response up to 100 kHz) and a B&K 1627 filter unit (set as a high-pass at 15 kHz). Signals were fed into a NF P61 amplifier and an NF P86 anti-aliasing low-pass filter (125 dB/oct) to be digitally recorded on a PC-based Digital Signal Processing Workstation. Sixteen seconds of continuous signal at 192 000 s/s or 24 seconds at 128 000 s/s were stored, thus allowing a bandwidth of 90 kHz and 56 kHz respectively, for each experiment. Recorded signals were played back at a reduced sample rate (32 000 s/s) to reduce their bandwidth by

[*] Dipartimento di Scienze Animali, Vegetali e dell'Ambiente, University of Molise, Via Cavour 50, I – 86100 Campobasso, Italy.
[†] Centro Interdisciplinare di Bioacustica, Institute of Entomology, University of Pavia, Via Taramelli 24, I – 27100 Pavia, Italy.

6 and 4 times respectively to be audible and recordable on a DAT (Digital Audio Tape) recorder. Spectral and temporal analyses were performed to reveal the structure and the temporal patterning of the signals.

Ultrasonic emissions were tested on virgin males stimulated by means of synthetic pheromones or by putting virgin females in a Petri dish (20 cm in diameter and 2 cm in height) covered with an acoustically transparent thin grid.

Wing-fanning males were followed (tracked) manually with the microphone at a distance of about 1 cm; sounds were typically picked up from the dorsal side. Some recordings were made from the ventral side of the moths when they were walking on the covering grid.

Recordings were performed at a temperature of about 22°C, 65% r.h. and in a low light intensity.

The same procedures were used on males from which the tegulae were surgically removed to verify our hypothesis on the structures involved in sound production.

Biological tests

The mating capacity in the adults of *E. cautella*, *E. kuehniella* and *P. interpunctella* was tested by isolating, for each species, a one day old virgin male and a virgin female, in a cage (9 × 17 × 8.5 cm) containing a Petri dish (5.5 cm in diameter) with some *pabulum*. The evaluation of fertility by examination for the presence of larvae was carried out after 7–10 days. The test was repeated at least 13 times.

The same procedure was used in tests with adult males with their tegulae removed (muted males), with adult females with their timpanus pierced (deaf females) and with males and/or females with their antennae surgically removed.

The environmental conditions in the room housing the cages were kept at 25 ± 1°C, 65 ± 5% r.h. and with a natural photoperiod.

Results and Discussion

Sound recording and analysis

The courtship sequence for *E. cautella*, *E. kuehniella* and *P. interpunctella* proved to be similar as reported by Barrer and Hill (1977) for *E. cautella*.

Males started wing-fanning and females searched on perceiving chemical stimuli, ie. the artificial pheromones or the female sexual pheromones. The male wing-fanning slowed progressively when the genitalia of the two sexes come into direct contact. The male then came to rest facing the female with their heads facing and then the male proceeded with the initial stages of copulation.

During the entire phase of wing-fanning, ultrasonic pulses were detected in the three species (Trematerra and Pavan 1994). The wing-fanning males emitted trains of ultrasonic pulses extending in frequency range up to 80 kHz. The pulses were emitted in regular sequences, spaced and with the pulse-repetition as reported in Table 1.

The recordings of *E. cautella*, *E. kuehniella* and *P. interpunctella* appeared very similar to the pulses clearly recognisable (Fig. 1).

Males with the tegulae removed showed no modification in their courtship behaviour but no ultrasonic emissions were detected.

Biological tests

The test to evaluate reproductive success in muted males and in deaf females revealed a lower reproductive success than in intact individuals as shown in Table 2. This is probably related to the suppression of the acoustic signals.

Evaluation of the reproductive success in males or females without antennae (Table 3) showed a lower reproductive success in males with their antennae removed. With *E. cautella* and *E. kuehniella* females without antennae showed no drop in reproductive success. This compares with a drop of 22% with *P. interpunctella*. In the experimental cages, *P. interpunctella* showed heightened activity in the search for a partner by increased walking and flying.

Table 1. Ultrasonic emissions in males of *E. cautella*, *E. kuehniella* and *P. interpunctella*.

Species	Maximum frequency	Pulse interval
E. cautella	50–80 kHz	15–20 ms
E. kuehniella	50–80 kHz	20–24 ms
P. interpunctella	50–70 kHz	14–18 ms

Table 2. *E. cautella*, *E. kuehniella* and *P. interpunctella* biological tests with intact (M) or muted males (Mm) and with intact (F) or deaf females (Fd).

Species	Group	Tests (n)	Reproduction success (%)	Drop in reproduction (%)
E. cautella	M × F	22	93.34	
	Mm × F	21	50.20	46.22
E. kuehniella	M × F	24	83.33	
	Mm × F	21	44.14	47.03
P. interpunctella	M × F	34	94.13	
	Mm × F	21	60.54	35.68
E. cautella	M × F	15	86.66	
	M × Fd	20	66.66	23.08
E. kuehniella	M × F	27	81.48	
	M × Fd	13	61.53	24.48
P. interpunctella	M × F	32	93.75	
	M × Fd	21	61.54	34.36

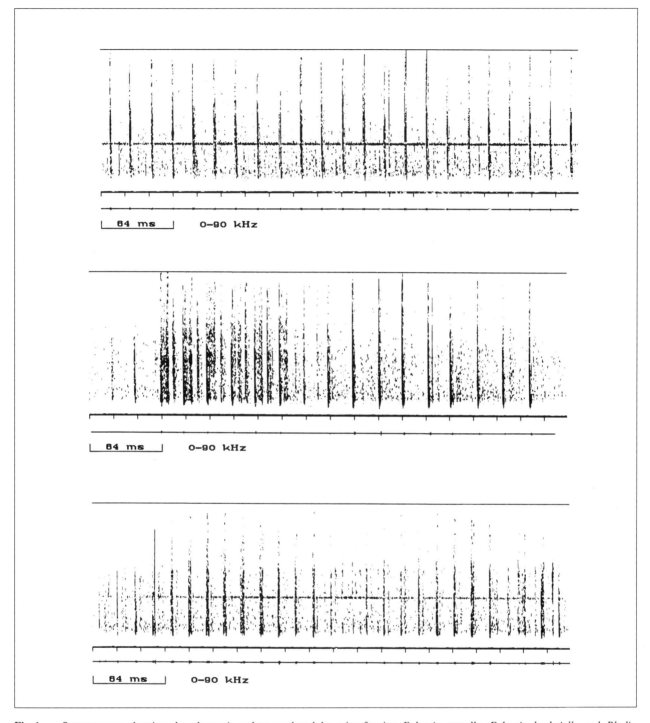

Fig. 1.   Spectrograms showing the ultrasonic pulses produced by wing-fanning *Ephestia cautella, Ephestia kuehniella* and *Plodia interpunctella* males. Frequency range 0–90 kHz, time span 427 ms, tic marks 21.33 ms, bandwidth 1 kHz.

Conclusion

Research on moth courtship behaviour has focused primarily on female pheromones and male behavioural responses to them. Although several studies on moth sex pheromones have demonstrated that the complete sequence of courtship behaviour also involves male pheromones and other types of communication, relatively little work has been done on the connection between pheromones and extrusible organs or with sound production (Birch and Hefetz 1987; Spangler 1988; Trematerra 1992; Trematerra and Pavan 1994).

Many insects have airborne sound receptors; in Diptera, Hemiptera, Lepidoptera and Orthoptera these receivers function primarily for intraspecific communication (Spangler and Hippenmeyer 1988) whereas in Coleoptera, Dictyoptera,

Lepidoptera and Neuroptera hearing serves to warn the bearer of the potential threat of a predator (Roeder and Treat 1961; Spangler 1988).

In view of the preliminary results presented here, moth courtship behaviour in *E. cautella*, *E. kuehniella* and *P. interpunctella* must be reconsidered with greater emphasis on the interaction between multiple modalities. The experiments reported here show that the ultrasonic and chemical communications in these species play a significant role in mating behaviour.

Further experiments are needed to deepen our knowledge of the role played by ultrasound in phycitid mating behaviour and the relative importance of chemical, visual and acoustic stimuli.

Table 3. *E. cautella*, *E. kuehniella* and *P. interpunctella* biological tests with intact males (M) and females (F) or with males without antennae (Ma) and females without antennae (Fa).

Species	Group	Tests (n)	Reproduction success (%)	Drop in reproduction (%)
E. cautella	M × F	19	89.47	
	Ma × F	20	20.00	77.65
	M × Fa	20	100.00	0
E. kuehniella	M × F	27	81.48	
	Ma × F	17	35.29	56.69
	M × Fa	18	83.33	0
P. interpunctella	M × F	32	93.75	
	Ma × F	16	56.25	40.00
	M × Fa	13	72.72	22.43

In a preliminary examination of the hearing capabilities of these moths a quick response to bat-like ultrasonic signals was revealed. Males halted wing-fanning or terminated their flight when given short ultrasonic signals at frequencies of 40–50 kHz similar to those produced by some echolocating bats. A further investigation is thus required to better understand the role played by ultrasounds in anti-predation behaviour of pyralid moths.

References

Barrer, P.M. and Hill, R.J. 1977. Some aspects of the courtship behaviour of *Ephestia cautella* (Walk.) (Lepidoptera: Phyticidae). Journal of Australian Entomological Society, 16, 301–312.

Barrer, P.M., and Hill, R.J. 1980. Insect oriented locomotor responses by unmated females of *Ephestia cautella* (Walker) (Lepidoptera: Phyticidae). International Journal of Invertebrate Reprints, 2, 59–72.

Birch, M.C. and Hefetz, A. 1987. Extrusible organs in male moths and their role in courtship behaviour. Bulletin of Entomological Society of America, 33, 222–229.

Hoppe, T. 1981. Nahrungswahl, Eiablage und Entwicklung der Dorrobstmotte (*Plodia interpunctella* Hübner) an verschiedenen Rohstoff und Fertigproducten der Schokoladenindustrie. Zeitschrift für angewaldte Entomologie, 91, 170–179.

Kennedy, J.S. and Marsh, D. 1974. Pheromone-regulated anemotaxis in flying moths. Science 184, 999–1001.

Mullen, M.A. and Tsao, C.H. 1971. Tympanic organ of Indian meal moth, *Plodia interpunctella* (Hübner), almond moth, *Cadra cautella* (Walker) and tobacco moth, *Ephestia elutella* (Hübner) (Lepidoptera: Pyralidae). International Journal of Insect Morphology and Embryology, 1, 3–10.

Perez, M. and Zhantiev, R.D. 1976. Functional organisation of the tympanal organ of the flour moth *Ephestia kuehniella*. Journal of Insect Physiology, 22, 1267–1273.

Roeder, K.D. and Treat, A.E. 1961. The detection and evasion of bats by moths. American Scientist, June 1961, 134–148.

Spangler, H.G. 1988. Moth hearing, defence, and communication. Annual Review of Entomology, 33, 59–81.

Spangler, H.G. and Hippenmeyer, C.L. 1988. Binaural Phonotaxis in the Lesser Wax Moth, *Achroia grisella* (F.) (Lepidoptera: Pyralidae). Journal of Insect Behaviour, 1, 117–122.

Trayner, R.M.M. 1968. Sex attraction in the Mediterranean flour moth *Anagasta kühniella*: location of the female by the male. Canadian Entomologist, 100, 5–10.

Trematerra, P. 1992. Su alcuni meccanismi di richiamo sessuale adottati da *Corcyra cephalonica* (Stainton) (Lepidoptera: Galleriidae). Redia, LXXV, 109–121.

Trematerra, P., Pavan, G. 1994. Ultrasound production in the courtship behaviour of *Cadra cautella* (Walk.), *Ephestia kuehniella* Z. and *Plodia interpunctella* (Hb.) (Lepidoptera: Pyralidae). Journal of Stored Products Research, in press.

Effect of maize variety and storage form on oviposition and development of the maize weevil, *Sitophilus zeamais* Motschulsky (Coleoptera: Curculionidae)

K.A. Vowotor[*], N.A. Bosque-Pérez[*] and J.N. Ayertey[†]

Abstract

The effect of maize variety and form of storage (shelled, dehusked, or with husks on) on oviposition and development of *Sitophilus zeamais* Motschulsky was studied under artificial infestation conditions in the laboratory. Three improved West African maize varieties—Abeleehi, EV8725-SR and Pop63-SR—and the local variety Volta were used. Maize stored in the shelled form showed significantly more weevil eggs laid than maize stored as ears with or without husks. Maize variety and form of storage did not affect survival of eggs. For all varieties, significantly more F_1 weevils emerged from shelled grain than from ears stored with or without husks. The highest numbers of adults emerging were recorded from the soft endosperm varieties Pop63-SR and Volta Local, and the least from Abeleehi and EV8725-SR, both hard endosperm varieties. Across all varieties, median development period of weevils was significantly shorter on shelled grain (45 days) than on ears with or without husks (52 days).

Introduction

Maize (*Zea mays* L.) is one of the most important cereal crops throughout sub-Saharan Africa and is grown in a wide range of agro-ecological and economic environments. The crop is mainly grown by small-scale farmers and is a major staple in many African countries. One of the most important constraints African maize farmers face is postharvest losses to storage insect pests like the maize weevil, *Sitophilus zeamais* Motschulsky.

The oviposition behaviour of *Sitophilus* species has been studied under different environmental conditions, on resistant and susceptible maize varieties and on cereals such as rice and wheat (Teotia and Singh 1968; Singh et al. 1972 1974; Dobie 1974; Okelana and Osuji 1985; Urrelo and Wright 1989; Urrelo et al. 1990).

Kossou et al. (1992) investigated the effects of shelling maize cobs on the oviposition of *S. zeamais* on the flint, early maturing, maize variety BDP. They also studied the susceptibility of improved and local maize varieties to this insect using shelled maize and ears stored with or without husks (Kossou et al. 1993). However, no work has been carried out combining these three storage forms with contrasting maize varieties commonly found in West Africa. This study was thus undertaken to further the understanding on the oviposition behaviour and development of *S. zeamais* on diverse maize varieties stored as shelled grain and as ears with or without husks.

Materials and Methods

Experiments were conducted at the International Institute of Tropical Agriculture (IITA), Ibadan, Nigeria. Maize varieties were selected in order to combine a number of basic contrasting characteristics of commonly grown maize types found in West Africa, so that specific comparisons could be made. Three improved white varieties: Abeleehi (a hard dent type derived from Population 49), EV8725-SR (a flint), Pop63-SR (a soft dent, quality protein maize) and the Ghanaian white variety Volta Local (a floury type) were used for the study. The varieties were grown at IITA fields, and at harvest maize cobs were collected with the husks intact. During harvesting, care was taken not to pull out the dried silks from the husk tips, or accidentally open or widen them and thereby facilitate entry of *S. zeamais*. After harvest, cobs were artificially dried at 35°C for two weeks and then placed in large transparent plastic bags. To kill any insects resulting from field infestations, cobs were deep frozen within the bags until required for the experiments, when they were removed for moisture equilibration for three weeks. Moisture contents of the varieties during the experiments were 12.3%, 11.9%, 11.2% and 11.1% for Abeleehi, Pop63-SR, Volta Local and EV8725-SR, respectively.

S. zeamais used for the experiments were collected from maize fields at IITA. Weevils were reared on clean grain of the white maize variety TZPB for three generations in a dark room maintained at 25±2°C and 70±5% r.h.

Sixteen averaged-sized, tightly sheathed cobs with husk cover ratings between 1.0 and 1.5 (Kossou et al. 1993) were selected for each variety and storage form (four replicates of four cobs each). Altogether, there were 12 treatments (4 varieties × 3 storage forms) for a total of 192 cobs. Cobs in each treatment were placed into 30 × 35 cm cotton cloth bags and subjected to their appropriate storage forms by: leaving the husks intact (storage with husks on), removing the husks (storage without husks), or removing the husks and shelling the grain (shelled storage). Bags were kept in a room maintained at 25 ± 2°C and 70 ± 5% r.h. The cobs, or equivalent amount of grain, were infested with eight (five females and three males) two-week-old *S. zeamais* per cob for a seven-day oviposition period, after which weevils were removed.

The number of egg plugs was determined using the acid-fuchsin method of staining (Pedersen 1979, cited by Horber 1989). Prior to staining cobs with the husk leaves on, husks were carefully pulled back from the cobs to expose the kernels and the area with kernels was immersed in the solution. After the cobs were dry and egg plugs counted, husk leaves were returned to their original position. It was assumed that every egg plug covered an egg and that every egg was covered by an egg plug.

Data were also collected on number of F_1 weevils which emerged daily. The percentage survival of eggs, Median Development Periods (MDPs), and susceptibility indices of treatments were determined (Dobie 1974). For cobs with the husk leaves on, husks were carefully pulled back from the cobs to expose the kernels before weevil counts were made.

---

[*] International Institute of Tropical Agriculture, Plant Health Management Division, PMB 5320, Ibadan, Nigeria.

[†] Crop Science Department, University of Ghana, Legon, Accra, Ghana.

Rubber bands were used to hold the husk leaves on the cobs and prevent insect escaping before the next count.

Data on the numbers of egg plugs and number of F_1 weevils were log-transformed and analysed as a completely randomised design. F tests and Duncan's multiple range tests were used to test pre-planned comparisons. Percentage survival of eggs was analysed using Chi-square.

Results and Discussion

Number of F_1 weevils

The log number of F_1 adults that emerged from the different maize varieties under the different storage forms is shown in Figure 1. Across all varieties, retransformed weevil emergence means were 1.18, 7.25 and 14.13 in the with husks, without husks and shelled storage forms, respectively. These numbers were significantly different from each other (P<0.05), and suggest that in the shelled storage form, conditions are more favourable for high oviposition, egg and larval development and/or survival of the maize weevil, than when ears are stored with or without husks. Such conditions would reduce egg and larval mortality, and also enable developing larvae to obtain more nutritious food.

The low numbers of F_1 weevils that emerged from cobs with husks on confirm the protective role of the husk leaves and may further explain why traditional small-scale farmers in Africa have continued to store the crop in this form over the years with relatively low losses, without the use of chemical protectants.

Mean log numbers of F_1 adults (per four cobs) across storage forms were 2.03, 1.84, 1.73 and 1.33 for Pop 63-SR, Volta Local, Abeleehi and EV8725-SR, respectively. No F_1 weevils emerged from EV8725-SR stored with the husks on and the reasons for this were not clear. The numbers of F_1 weevils for EV8725-SR (across the other two storage forms) were significantly lower (P<0.05) than for the other varieties, except for Abeleehi. The relatively higher F_1 weevil emergence in Pop63-SR, a quality protein maize (QPM) variety (i.e. high lysine content), with the opaque-2 gene, confirms a similar report by Gupta et al. (1970) who found other soft endosperm maize types carrying this gene to be more susceptible to weevil attack. The utilisation of this variety in West Africa will not contribute as much as desired to

the improvement of human and animal nutrition, unless it receives protection from weevil damage. Volta Local, a variety with floury endosperm did not produce as many F_1 weevils as Pop 63-SR, probably because of the rounded nature of the kernels in the former variety. It is possible that the low number of F_1 weevils recorded on EV8725-SR was due to its flinty endosperm.

Urrelo et al. (1990) explained that the F_1 progeny produced by *S. zeamais* females is the ultimate measure of fitness of a population. Final numbers are greatly influenced by proximate mechanisms, which are basically related to influences of oviposition, growth and development. The F_1 numbers give an estimate of the type of insect-host interaction in resistance or susceptibility of the host and reproductive capabilities of the insect attacking it. Judging from the numbers of F_1 adults that emerged, EV8725-SR may be characterised as resistant and Pop63-SR as susceptible. There is a need to incorporate resistance characteristics and carry out further selection on Pop63-SR in order to improve this variety. Efforts in this regard, however, would require resources and time, and may also result in the loss of some of the high protein qualities in this variety.

Percentage survival of eggs

Kossou et al. (1992) suggest that the reasons for the low numbers of F_1 weevils emerging from unshelled maize (cobs without husks) as against shelled, could be due to reduced oviposition and/or lower egg or larval survival. In our study, ratios of the numbers of eggs laid to the numbers of F_1 adult weevils were computed for the four maize varieties and three storage forms (Table 1). Chi-square analysis indicated that storage form did not influence survival of eggs (χ^2=2.05, 2 df, P>0.05), nor did variety (χ^2=0.17, 3 df, P>0.05), nor their interaction (χ^2=1.73, 6 df, P>0.05). It is therefore concluded that the most likely cause for the observed low numbers of F_1 weevils in cob-stored maize, as opposed to shelled maize, is reduced oviposition.

Median development periods (MDPs)

The MDP for each treatment was determined from the middle of the oviposition period to the emergence of 50% of F_1 adults (Dobie 1974). Across all varieties, mean MDP values were 51.9, 52.0 and 44.6 days for ears with husks on, ears without husks and shelled maize, respectively (Fig. 2). No significant differences (P<0.05) were observed in the MDPs for the two ear-storage forms, however, MDP was significantly longer on maize ears than on shelled maize. Kossou et al. (1992) reported a similar 7.7 day difference in weevil MDPs between shelled and unshelled maize. These longer weevil MDP values for cob-stored maize are thought to be due to the long time it might have taken the larvae to tunnel through the kernel from the crown end before reaching the germ end, where more nutritious food is available, as well as the difficulty for newly developed adults to emerge from kernels due to obstruction by adjacent kernels fixed onto cobs (Kossou et al. 1992, 1993). Across storage forms, mean weevil MDP values on the varieties were 49.3, 49.9, 48.3 and 47.5 days for Abeleehi, EV8725-SR, Pop63-SR and Volta Local, respectively (Fig. 2). Since no F_1 weevils emerged from EV8725-SR stored with the husks on, MDP values could not be determined for the insects from this treatment.

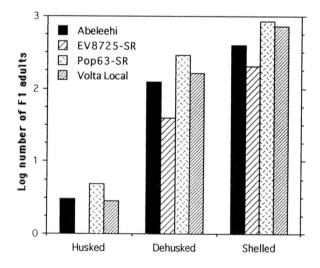

Fig. 1. Log number of F_1 maize weevils (*Sitophilus zeamais*) that emerged from four maize varieties stored as ears with husks on (husked), ears without husks (dehusked) and shelled grain.

Susceptibility Indices

The Susceptibility Index (SI) defined as [(log$_e$ F)/D]100, where F= number of F_1 weevils and D= MDP (Dobie 1974)

Table 1. Ratio of the number of *Sitophilus zeamais* adults that emerged to the number of eggs laid in four maize varieties under three storage forms[a]

| Variety | Storage forms | | | |
	With husks	Without husks	Shelled	Total
Abeleehi	6/10 (60)	29/43 (67)	47/68 (69)	82/121 (68)
EV8725-SR	0/0[b]	8/29 (62)	37/50 (74)	55/79 (70)
Pop63-SR	8/15 (53)	36/51 (71)	62/100 (62)	116/166 (70)
Volta Local	5/7 (71)	33/49 (67)	67/97 (69)	105/153 (69)
Total	19/32 (59)	116/172 (67)	223/315 (71)	358/519 (69)

[a] Means of four replicates (4 cobs per replicate). Figures in parentheses are percentages of eggs that survive. Chi-square analysis revealed that storage form did not influence survival of eggs (χ^2=2.05, 2 df, P>0.05), nor did variety (χ^2=0.17, 3 df, P>0.05), nor their interaction (χ^2=1.73, 6 df, P>0.05).

[b] No F_1 weevils emerged from EV8725-SR stored with the husks on.

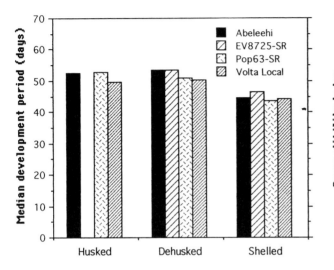

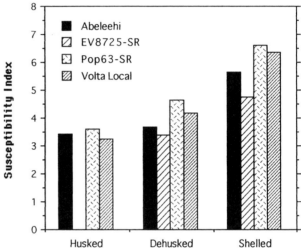

Fig. 2. Median development periods (MDPs) of *Sitophilus zeamais* reared on four maize varieties stored as ears with husks on (husked), ears without husks (dehusked) and shelled grain.

Fig. 3. Weevil susceptibility indices of four maize varieties stored as ears with husks on (husked), ears without husks (dehusked) and shelled grain.

was used to evaluate the susceptibility of the maize varieties in the different storage forms. With this index, the treatment which produced the greatest number of F_1 progeny and resulted in the shortest weevil developmental period is the most susceptible. The higher the SI the greater the susceptibility and vice versa.

SI values for the varieties obtained in this study were generally lower than those reported for other maize varieties (Dobie 1974; Okelana and Osuji 1985; Urrelo et al. 1990). These lower values were thought to be due to the relatively low moisture content of the varieties which could have resulted in the emergence of fewer F_1 adult weevils. Highly significant differences (P<0.01) were observed in SI between varieties, storage forms and their interactions, indicating that SI values in this study were highly affected by storage form and/or variety. Across storage forms, susceptibility of the varieties was in the following decreasing order: Volta Local, Pop63-SR, Abeleehi, and EV8725-SR. However, only EV8725-SR differed significantly from the rest in SI. Across varieties SI was greater on shelled maize, followed by ears without husks and then ears with husks (Fig. 3).

Dobie's index of susceptibility was developed to describe the inherent susceptibility of kernels of a maize variety in the shelled form and not in the cob storage forms. From the results obtained in this investigation it would appear that an index of susceptibility involving maize cobs may have to be determined differently.

Studies on the oviposition performance of *S. zeamais* on maize varieties using some of the different storage methods

practised in sub-Saharan Africa are important for understanding how management of storage practices can be used as an independent control method or as an adjunct to other control methods. It is therefore important that more efforts be devoted to understanding the interactions between storage practices and insect depredation. The spreading of insecticide-resistant insects, the escalating costs of insecticides, and growing concern about their indiscriminate use and the health hazards they pose, increase interest in such studies.

It should be possible to improve upon the storage performance of the crop, if varieties with inherently resistant kernels are cultivated, field infestation significantly reduced by early harvesting, and maize is stored with the husks on inside improved ventilated cribs. In the West African subregion small-scale maize farmers often cultivate local varieties which usually have a long protective husk cover, and harvest and store these varieties with the husks on. In this way they reduce the chances of depredation by weevils as it takes a much longer period for a destructive population of *S. zeamais* to be attained. The cultivation of those improved maize varieties which have a poor husk cover may not be ideal for the traditional with-husks-on method of storage, and the acceptance of these varieties may pose problems as small-scale farmer storage practices do not easily change in West Africa. These varieties, however, can be subjected to further selection to improve their husk cover quality.

References

Dobie, P. 1974. The laboratory assessment of the inherent susceptibility of maize varieties to post-harvest infestation by *Sitophilus zeamais* Motsch. (Coleoptera: Curculionidae). Journal of Stored Products Research, 10, 183–197.

Gupta, S.C., Asnani, V.L. and Khara, B.P. 1970. Effect of opaque-2-gene in maize (*Zea mays* L.) on the extent of infestation by *Sitophilus oryzae* L. Journal of Stored Products Research, 6, 191–194.

Horber, E. 1989. Methods to detect and evaluate resistance in maize to grain insects in the field and in storage. In: Toward Insect Resistance Maize for the Third World. Proceedings of the International Symposium on Methodologies for Developing Resistance to Maize Insects. Mexico D.F., CIMMYT, 140–150.

Kossou, D.K., Bosque-Pérez, N.A. and Mareck, J.H. 1992. Effects of shelling maize cobs on the oviposition and development of *Sitophilus zeamais* Motschulsky. Journal of Stored Products Research, 28, 187–192.

Kossou, D.K., Mareck, J.H. and Bosque-Pérez, N.A. 1993. Comparison of improved and local maize varieties in the Republic of Benin with emphasis on susceptibility to *Sitophilus zeamais* Motschulsky. Journal of Stored Products Research, 29, 333–343.

Okelana, F.A. and Osuji, F.N.C. 1985. Influence of relative humidity at 30°C on the oviposition and development and mortality of *Sitophilus zeamais* Motsch. (Coleoptera: Curculionidae) in maize kernels. Journal of Stored Products Research, 21, 13-19.

Pedersen, J.R. 1979. Selection of oviposition sites on wheat kernels by *Sitophilus* spp.: Effect of moisture, temperature and kernel size. Ph.D. Dissertation, Kansas State University, Manhattan, Kansas.

Singh, K., Agrawal, N.S. and Girish, K. 1972. Studies on the oviposition response and development of *Sitophilus oryzae* (L.) (Coleoptera: Curculionidae) in different maize hybrids and composites. Indian Journal of Entomology, 34, 140–154.

Singh, K., Agrawal, N.S. and Girish, K. 1974. The oviposition and development of *S. oryzae* (L) in different high yielding varieties of wheat. Journal of Stored Products Research, 10, 105–111.

Teotia, T.P.S. and Singh, V.S. 1968. Studies on the oviposition behaviour and development of *Sitophilus oryzae* Linn. in various natural foods. Indian Journal of Entomology, 30, 119–127.

Urrelo, R. and Wright, V.F. 1989. Oviposition performance of *Sitophilus zeamais* Motschulsky (Coleoptera: Curculionidae) on resistant and susceptible maize accessions. Journal of Kansas Entomological Society, 62, 23–61.

Urrelo, R., Wright, V.F., Mills, R.B. and Horber, E.K. 1990. An abbreviated procedure to determine the inherent resistance of maize to *Sitophilus zeamais* Motsch. (Coleoptera: Curculionidae). Journal of Stored Products Research, 26, 97–100.

Life history data for *Sitotroga cerealella* (Olivier) (Lepidoptera: Gelechiidae) in farm-stored corn and the importance of sub-optimal environmental conditions in insect population modelling for bulk commodities

D. K. Weaver and J. E. Throne*

Abstract

Life history data for a strain of the Angoumois grain moth collected in South Carolina, USA in 1992 are compared to published data. The survival of the South Carolina strain was much higher than that reported in the literature and development was slower. Models simulating growth of Angoumois grain moth populations under actual storage conditions were prepared for the two data sets and compared. The literature-based model underestimated population size, probably because data for conditions below 20°C were unavailable.

To further investigate the importance of sub-optimal temperature, we bounded our model at 20°C, as per our literature-based model. This model performed very poorly compared to both the complete model and compared to the literature model. Underestimates were extreme because of the slower development of this strain. A final model was prepared using our data for only the linear portion of increase in development rate (20°C–30°C) and extrapolating to known limits of development and survival (10°C and 40°C). This model performed quite differently, with frequent large overestimates of population size, particularly early in storage under cool conditions.

It was concluded that accurate, useful models can be ensured only from complete life histories appropriate to the species biotype being studied. Accurate models may be generated using existing information but this is not a consistently reliable approach.

Introduction

The development of advanced computing technology has made it possible to create population models that can be used to predict population growth as a function of a wide array of deterministic variables (Throne 1994). Variables such as adequate food resources and protection from localised weather, which produce a favourable environment for stored-product insect pest population growth, may be readily described as a function of temperature and moisture content (relative humidity) (e.g. Hardman 1978). Published accounts of insect life histories for certain cosmopolitan stored-product pests have been used to prepare computer models simulating population growth, most of which have not been validated

(Throne 1994). Certain models have been used in the preparation of commodity-based expert systems.

Life history data for stored-product insects often do not include information on the 'slow' or 'poor' biology, i.e., the conditions where growth is quite protracted or where survival is quite low and often these two responses are linked. In addition, properly stored commodities may seldom reach optimal conditions for insect growth. This may result in a situation where a model is being used to estimate growth for a considerable length of time under conditions for which little or no data were collected. This causes concern because simulation studies (Throne 1989) have indicated that small changes in development or survival can have a large impact on the size of subsequent populations. In addition, using published life histories to model population growth for a potentially different biotype of the pest species may be prone to fallibility (McFarlane et al. 1993). The current assumptions are that sub-optimal biology has little impact on resulting population size and that commerce-induced gene-flow across populations reduces the probability of marked biotypic differences in insect population development. These assumptions will be critically assessed by comparing our data to another published life history study for *Sitotroga cerealella* on stored maize in India (Grewal and Atwal 1967). This publication supplies a considerable amount of life history data for this species on stored maize, although in Grewal and Atwal (1967) the maize kernels were mixed with flour.

Materials and Methods

Life history data

Angoumois grain moths were collected from farm-stored maize at two sites in south-central South Carolina, USA in late winter and early spring, 1992. A laboratory population was established at 25°C, 65% r.h. on 'Pioneer 3320' maize from the 1991 crop using the pooled collections of moths. Newly-emerged F_2 adults were confined in a 237 mL mason jar containing an 8 cm × 30 cm piece of black construction paper that had been folded eleven times and stapled in the centre to yield a tight 'accordion-like' oviposition substrate that was 8 cm × 2.5 cm wide (Ellington 1930). After confinement at culture conditions for 12 hours the adults were removed and the papers unfolded. Cohorts of approximately 50 eggs were cut from the paper strips and placed in 237 mL mason jars with screen and filter-paper-lined lids in a sealed salt box containing equilibrated salt solutions at temperatures ranging from 10 to 40°C by 5°C increments. Salts used were KCl, NaCl, NaBr and K_2CO_3, which produce environments ranging from 43 to 87% r.h. across this temperature range (Greenspan 1977). Cohorts were observed daily until hatch ceased.

For development from hatch through adult emergence, newly-hatched F_3 larvae (collected as described above) were

---

* USDA-ARS Stored-Product Insects Research and Development Laboratory, 3401 Edwin St., Savannah, GA 31405 USA.

placed in 12 mL translucent polyethylene vials (2.7 cm high ×
2.2 cm internal diameter) containing two equilibrated kernels
of 'Pioneer 3320' maize. The vials were modified with fine
mesh nylon screening fixed in the lid and base to facilitate respiration. Twenty-five larvae were prepared, as described, for
each environmental condition and observations were made
daily until all adults had emerged.

Simulation models

First, response surface equations were fitted to the South
Carolina data and to the data of Grewal and Atwal (1967)
using a curve-fitting procedure to relate developmental period
to temperature and relative humidity. Fit was assessed by
viewing the response surface and evaluating the r^2 value and
the regression *F* statistic. All response surfaces selected were
biologically plausible. Variation in development rate was then
simulated using a time-varying distributed delay (Manetsch
1976) as described in Throne (1989). Standard deviations for
the data of Grewal and Atwal (1967) and for the emergence of
single individuals in the current study were calculated using
the method of Shaffer (1983). Next, response surfaces were
fitted to the developmental mortality data for the current study
and for the data of Grewal and Atwal (1967) and fit was
assessed as above. Fecundity and female longevity data from
Grewal and Atwal (1967) were fitted with response surfaces as
above, with Shaffer's (1983) method used to calculate
standard deviations on longevity for use in the distributed
delay. Four models were prepared.

Model 1

This model used all the data from the current study for
development and the adult data from Grewal and Atwal (1967)
to simulate population growth. Oviposition below 10°C and
above 40°C was set at 0. Oviposition and survival were extrapolated to these extremes using the fitted response surfaces. For
adults, maximal longevity below 10°C was set at 50 days
based on our experience, and maximal longevity above 40°C
was set at 4 days (D. K. Weaver and J. E. Throne, unpublished
data). This model includes data that adequately describe the
entire range of immature development.

Model 2

For this model we used the data and fitted equations for
Grewal and Atwal (1967) which did not include any data
below 20°C: thus development below this temperature was set
at 0 and mortality for these conditions followed a time-varying
distributed delay based on the maximal survival duration from
our current study. Adult data are treated as for the first model,
except for oviposition, which is set at zero above 35°C and
below 20°C. This model is based on published data for stored
maize and will allow for a comparison with the model that
includes our data for the immature stages. Note that the adult
stages are based on common data. The sole difference between
the two models for this stage is extrapolation of the fitted
equations for adults from 35 to 40°C and from 20 to 10°C in
the first model.

This second model could not assess the role of sub-optimal
biology in a comparison with the first because all the reported
data for immature development and survival differ markedly
from our findings (Tables 1 and 2). Therefore, two additional
models were prepared.

Model 3

This model uses the fitted equations from model one, but
with development at temperatures below 20°C and above
35°C set at 0 as in the second model. Adult data were treated as
in the second model. This model is an approximation of conducting our study only at temperatures of 20°C and above and
includes curvilinearity at the high temperature.

Model 4

The fourth model uses linear equations fitted to the data for
the immature stages from our current study for the linear
portion of development and survival curves, which was from
20 to 30°C. These equations were then extrapolated to
extremes of 10 and 40°C as in the first model. Adult data were
treated as in the first model. This allows us to test the importance of exact data for both sub-optimal and supra-optimal
temperatures, by comparing this model with the first one.

Simulations

All models were run with actual environmental data for a
bin at one of the collection sites for the moth population from
South Carolina, as reported in Throne (1989). The temperature varied significantly during this time, but maize moisture
content did not vary significantly. Therefore, the simulations
are run using 75%r.h.. Models were started with 5 newly-
emerged adult pairs on the first day of the storage season. This
is realistic, based on our experience, because eggs and newly-
hatched larvae are destroyed at harvest and relatively low
numbers of newly-emerged adults are observed immediately
after harvest. For simplicity, no subsequent recruitment was
included in these models.

Results and Discussion

The life history data collected differed greatly from those
reported in Grewal and Atwal (1967). Survival during both the
egg stage and through immature development was much
greater for the South Carolina strain we collected than for the
strain collected in India (Tables 1 and 2). Duration of development was also much longer for the South Carolina strain
(Tables 1 and 2). Note that the differences between the strains
are least significant near optimal conditions (25–30°C,
56–84% r.h.). This similarly appears to indicate that the
assumption that biotypic differences are unimportant is con-
tingent upon environmental conditions. The assumption is
most correct where development occurs rapidly and survival is
high. The assumption is least correct at both sub-optimal and
supra-optimal conditions.

We believe this finding has important ramifications for
insect population modelling in stored-products and have thus
produced several models to illustrate its effect. The models
compare the impact of different data sets for immature development and survival only; the adult data are common and were
obtained from Grewal and Atwal (1967). Model 1 contains all
of the information from our first replicate of a development
rate study using South Carolina moths. Model 2 uses information for all stages that was obtained from Grewal and Atwal
(1967). Figure 1 illustrates that Model 1 gives much higher
population numbers during the cool storage months (see graph
at bottom of Fig. 1). At the end of the simulation, the lines
appear to start to converge as temperature begins to increase
towards the optimal and as population size increases, so that
on a log scale there is only a nine-fold difference at 270
storage days (Fig. 1 and Table 3). However, this simulated difference is nearly 10 million individuals. This indicates the

Table 1. Duration of egg stage and percentage survival at constant temperatures and relative humidities for Angoumois grain moths collected in South Carolina, USA in 1992 and for those reported in Grewal and Atwal (1967).

Temperature (°C)	Relative humidity[a] (%)	Current study Days in stage (S.D.) [% survival]	Literature Days in stage (S.D.)[b] [% survival]
10	43, 62, 76, 87	— [0]	Not measured
15	43	25.5 (1.4) [80]	Not measured
	61	26.4 (0.9) [80]	
	76	29.6 (1.9) [88]	
	86	29.1 (0.9) [98]	
20	43 / 40	14.2 (0.9) [95]	9.2 (1.1) [47]
	60 / 60	14.1 (0.4) [96]	8.7 (1.0) [64]
	75	14.1 (0.6) [86]	
	80		8.5 (1.0) [65]
	85	13.6 (0.9) [94]	
25	43 / 40	8.8 (0.7) [82]	6.9 (0.9) [60]
	58 / 60	8.5 (0.8) [83]	5.6 (0.7) [80]
	75	8.8 (1.4) [92]	
	80		5.5 (0.7) [82]
	84	8.3 (0.7) [92]	
30	43 / 40	6.5 (0.8) [82]	4.2 (0.6) [71]
	56 / 60	6.6 (0.5) [83]	4.1 (0.6) [73]
	75	6.5 (0.9) [93]	
	80		4.0 (0.6) [77]
	84	6.0 (0.5) [97]	
35	43 / 40	6.7 (0.6) [63]	4.8 (0.7) [29]
	55 / 60	6.5 (0.6) [74]	4.6 (0.6) [44]
	75	6.3 (0.8) [70]	
	80		4.4 (0.6) [55]
	83	5.8 (0.5) [81]	
40	43, 53, 75, 82 /40, 60, 80	— [0]	— [0]

[a]For 10 and 15°C, relative humidities available for the current study only. Percent humidities following '/' are from the literature and are paired with the nearest condition from the current study. In the case of the literature data for 80% r.h., these values have been offset for comparison to either of the two highest relative humidities from the current study.
[b]Standard deviations for the literature means were calculated using: S.D. = 0.209(MEAN)$^{0.73}$ (Shaffer 1983).

importance of collecting 'biotypic' life history data to ensure a that model developed for practical usage is as accurate as possible. Our literature-based model for Angoumois grain moths probably would not be very accurate for storage environments where grain temperatures routinely drop below 20°C or exceed 35°C, such as in the U.S.

Our results suggest in addition that collecting the non-optimal data is very important, especially given the range of grain temperatures typical of a relatively warm area such as South Carolina, USA, where grain temperatures may be below 20°C for nearly half of the storage year (Fig. 1). However, since the life history data from Grewal and Atwal (1967) differ so greatly from those we found, we chose to treat our data in two different ways. First, we bounded our data and fitted equations to resemble those collected in the literature study, with no development below 20°C (Fig.2, Model 3). Secondly, we deleted all data above 30°C and below 20°C from our data set and fitted linear temperature functions to the data. These were then extrapolated to 10 and 40°C and used to develop Model 4 (Fig. 2). Model 3 (Fig. 2) performed even more poorly than Model 2 (Fig. 1). This is a function of the slower development of the South Carolina strain and results in estimates that are almost totally inaccurate. Even late in the storage season this model was only 0.04% accurate compared

with Model 1 (Table 3, 270 days). When plotted on a log scale Model 4 appears to be most similar to Model 1 and total numbers of individuals become quite close at certain times (Fig. 2). However, for most of the storage season, Model 4 systematically overestimated insect numbers, which is economically meaningful. For example, at 270 days of storage, total insect numbers are 140 million greater for Model 4 than for Model 1. Early, inaccurate predictions of high insect numbers could force a sale of maize during the cooler months when its market value is typically lower than in the spring and summer. This inaccuracy is quite evident at 180 days when numbers for Model 4 are eight-fold greater than for Model 1 (Table 3). This illustrates that it is very important to have complete biotypic life history data for modelling stored-product insect population growth, at least in regions where seasonality produces pronounced changes in temperature in a stored commodity.

In this paper we have made numerous extrapolations on equations and modifications to our data set and to those reported in the literature (the equations are listed in an appendix). Extrapolation and modification were carefully studied and when an equation generated a biologically unsound estimate it was bounded to the nearest known limit from our current life history study or from our experience

Table 2. Duration of larval through adult emergence for male and female Angoumois grain moths at constant temperatures and relative humidities for a population collected in South Carolina, USA in 1992. Pooled percent survival for both sexes is also reported for this population. Duration of larval through adult emergence for unsexed Angoumois grain moths and percent survival are also reported from Grewal and Atwal (1967).

Temperature (°C)	Relative humidity[a] (%)	Current study	Literature
		Days in stage female (S.D.)[b] / male (S.D.)[b] [% survival]	Days in stage[c] (S.D.)[b] [% survival]
10	43, 62, 76, 87	—[0]	Not measured
15	43	254.0 (11.9)[b] / 199.0 (45.8) [20]	Not measured
	61	166.1 (22.7) / 184.0 (40.1) [64]	
	76	133.8 (10.0) / 153.6 (39.3) [80]	
	86	136.9 (11.2) / 166.7 (35.4) [56]	
20	43 / 40	69.3 (3.8) / 73.8 (6.0) [68]	38.7 (3.0) [15]
	60 / 60	69.8 (4.4) / 73.7 (13.8) [68]	35.9 (2.9) [30]
	75	63.5 (19.1) / 64.7 (8.4) [88]	
	80		33.7 (2.7) [32]
	85	62.4 (13.0) / 68.6 (16.1) [84]	
25	43 / 40	54.4 (21.1) / 43.4 (3.3) [52]	31.0 (2.6) [38]
	58 / 60	41.4 (12.0) / 38.4 (3.5) [76]	29.2 (2.5) [41]
	75	35.2 (2.9) / 36.7 (6.3) [76]	
	80		27.8 (2.4) [44]
	84	45.2 (3.4) / 48.9 (16.7) [80]	
30	43 / 40	35.1 (5.7) / 36.4 (4.1) [56]	27.5 (2.3) [26]
	56 / 60	31.4 (4.5) / 33.8 (13.1) [48]	26.2 (2.3) [39]
	75	36.7 (18.0) / 29.6 (6.5) [76]	
	80		24.1 (2.1) [49]
	84	35.0 (7.4) / 50.6 (15.4) [52]	
35	43 / 40	—/— [0]	36.8 (2.9) [4]
	55 / 60	47.0 (3.5)[b] / 57.0 (4.0)[b] [8]	34.1 (2.7) [10]
	75	39.2 (4.5) / 36.0 (2.9) [28]	
	80		32.6 (2.7) [13]
	83	— / 81.0 (5.2)[b] [4]	
40	43, 53, 75, 82 /40, 60, 80	— [0]	— [0]

[a]For 10 and 15°C, relative humidities available for the current study only. Percent humidities following '/' are from the literature and are paired with the nearest condition from the current study. In the case of the literature data for 80% r.h., these values have been offset for comparison to either of the two highest relative humidities from the current study.

[b]Standard deviations for the current study when one individual emerged and for the literature means were calculated using: S.D. = 0.209(MEAN)$^{0.73}$ (Shaffer 1983).

[c]These data were calculated by combining the larval and pupal data from this study.

Table 3. A comparison of the total number of individuals (all stages) produced by the four models at three-month storage intervals. These numbers help to illustrate the large differences between the model estimates, which are plotted on Figures 1 and 2 with a log scale.

Simulation model	Days of storage in South Carolina, USA		
	90	180	270
1	640	23000	1.1×10^7
2	210	290	1.3×10^6
3	66	5	4600
4	7900	160000	1.5×10^8

based on ongoing experiments. We emphasise that we have done so to illustrate that the keys to accurate modelling are sound data that are biologically relevant for the particular biotype of the species studied, and to show that little may be gained from extrapolation. Data that are available in the literature, no matter how well they were collected, may be of little value to regions other than where they were collected. There is simply no way to predict whether or not this is true for all given situations. The other point is that the assumption that low temperature biology contributes little to population growth may be erroneous for certain situations. The maize temperature was effectively below 20°C for 150 days in the data used for this study, which were collected in one of the warmest regions of the U.S. (Figs 1 and 2). The mean duration of development for immature Angoumois grain moths from this region at 15°C is about 150 days (Table 2). Thus, larvae that hatched early in the winter emerge as adults as the temperature exceeds 20°C in the spring. This affects population growth tremendously. We suggest that accurate data for bulk

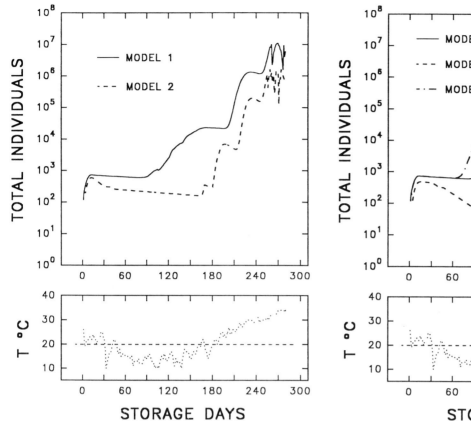

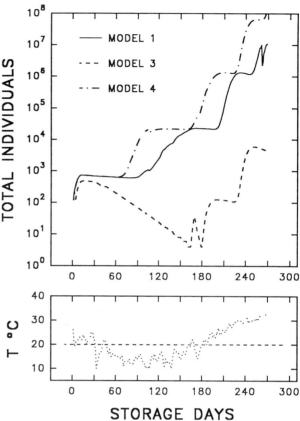

Fig. 1. Comparison of Model 1 and Model 2. Note the effect of storage temperature in creating deviation between the models.

Fig. 2. Comparison of Model 1, Model 3 and Model 4. Note the effect of low storage temperature on deviation among models.

grain temperature over a long period (years) in a particular geographic region be carefully considered before using published life history data from other temperatures to create predictive models. Moreover, since models are typically validated in small quantities of stored products, we recommend that validations be conducted at temperatures representative of bulk grain storage in that particular region. The thermal inertia of small quantities of grain is insufficient to remain constant, and insolation alone, under cool ambient conditions, may be sufficient to keep mean temperatures near optimal, where they are not representative of actual storage temperatures for many areas where commodities are stored.

References

Ellington, G.W. 1930. A method of securing eggs of the Angoumois grain moth. Journal of Economic Entomology, 23, 237–238.

Greenspan, L. 1977. Humidity fixed points of binary aqueous solutions. Journal of Research of the National Bureau of Standards – A. Physics and Chemistry, 81A, 89–96.

Grewal, S.S. and Atwal, A.S. 1967. The influence of temperature and humidity on the development of *Sitotroga cerealella* Oliver (Gelechidae: Lepidoptera). Journal of Research, 6, 353–358.

Hardman, J.M. 1978. A logistic model simulating environmental changes associated with the growth of populations of rice weevils, *Sitophilus oryzae*, reared in small cells of wheat. Journal of Applied Ecology, 15, 65–87.

McFarlane, J.A., Gudrups, I. and Fletcher, H. 1993. Biotype differences affecting the pest status of stored-grain pests. International Journal of Pest Management, 39, 35–43.

Manetsch, T.J. 1976. Time-varying distributed delays and their use in aggregative models of large systems. IEEE Transactions on Systems, Man, and Cybernetics, SMC-6, 547–553.

Shaffer, P.L. 1983. Prediction of variation in development period of insects and mites reared at constant temperatures. Environmental Entomology, 12, 1012–1019.

Throne, J.E. 1989. Effects of noncatastrophic control technologies that alter life history parameters on insect population growth: a simulation study. Environmental Entomology, 18, 1050-1055.

Throne, J.E. 1994. Computer modeling of the population dynamics of stored-product pests. In: Jayas, D.S., White, N.D.G. and Muir, W.E., ed. Stored grain ecosystems. New York, Marcel Dekker, in press.

Appendix

Equations used in the simulation models

Eggs laid = 123.499 + [0.227614* (Temp** 2.5)]–[0.0364930*(Temp**3)]–[810.663/(r.h.**0.5)]. Based on data from Grewal and Atwal (1967). Used in all models. Oviposition was bounded at a lower extreme of 0 eggs per day. Extrapolated to 10 and 40°C for Model 1. For Model 2 and Model 3 oviposition outside of the temperature range from 20 to 35°C was set at 0.

Egg development 1 = –6.71886 + [0.00156779*(Temp**2)* Log (Temp)] + [7544.88/ (Temp**2)]. Based on the data from the current study. Used in Model 1 and Model 3. In Model 3, outside of the range from 20–35°C no eggs hatched.

Immature male development 1 = 276.9914193–(9176 .77643/Temp) + [131132.6881/ (Temp**2)]–(2.390 + 11533*r.h.)+[0.017623645*(r.h.**2)]. Based on data from the current study. Used in Model 1 and Model 3. In Model 3, outside of the range from 20 to 35°C no adult males emerged.

Immature female development1= 35.4599 + {24211596* [TEMP** (–4.6904)]} –(140811* (r.h.**(–2.49998))) + ((5.62911E11* (TEMP** (–4.69042)))* (r.h.**(–2.49998))). Based on data from the current study. Used in Model 1 and Model 3. In Model 3, outside of the range from 20 to 35°C no adult females emerged.

Adult longevity= 37917.3– (3745.37*LOG(TEMP))– (103284/ LOG(TEMP)) + (155711/TEMP) + (0.5880* (r.h.**0.5)). Based on data for female longevity from Grewal and Atwal (1967). Used in all models, but output for temperatures outside of the range from 15 to 35°C was set at the output for 15 and 35°C in Model 1 and Model 4. For Model 2 and Model 3 adults died gradually over 4 days above 35°C and died gradually over 50 days below 20°C.

Egg survival1= 3.81885–(0.0869475* (Temp**2) *Log (Temp)) + (0.0737847* (TEMP**2.5))–(0.00370284* (TEMP**3)) – (151.263*LOG(TEMP)/(TEMP**2)) + (1.5 + 1942E-7* (r.h.**3)). Based on data from the current study. Used in Model 1 and Model 3. Survival was set at 0 for temperatures above 35°C and below 10°C.

Immature survival= -0.535799 + 1.43836*EXP((–EXP(– (TEMP–21.3047) / 10.2681)) –((TEMP–21.3047) / 10.2681) +1)*(EXP(–EXP(–((r.h.–73.1316)/50.6698))–((r.h.-73.1316) /50.6698)+1)). Based on data from the current study. Used in Model 1 and Model 3. Survival was set at O above 35°C and minimal survival was set at 0.

Egg development2= –35672.3+(3533.30*LOG(TEMP)) + (96945.9/ (LOG (TEMP))) –(145253/(TEMP)) –(0.163966 *LOG(r.h.**2)). Based on data from Grewal and Atwal (1967). Used in Model 2. Outside of the range from 20 to 35°C no eggs hatched.

Larval development = 27770.4 –(214333/ LOG(TEMP)) + (189924/ (TEMP **0.5) + (532968/ (TEMP**2))– 0.000 400287 * (r.h.**2)). Based on data from Grewal and Atwal (1967). Used in Model 2. Outside of the range from 20 to 35°C no larvae pupated.

Pupal development=26.6734– (.030824* (TEMP**2.5)) + (.00479741* (TEMP**3))+(164.529/r.h.). Based on data from Grewal and Atwal (1967). Used in Model 2. Outside of the range from 20 to 35°C no adults emerged.

Egg survival2=-0.0184696+(0.00125312*(TEMP**2.5))– (0.000199538*(TEMP**3)) – (2.75224E16*EXP(-r.h.)). Based on data from Grewal and Atwal (1967). Used in Model 2.

Larval survival= –0.415739 + (0.00409703*(TEMP**2)) –(.000101845 *(TEMP**3))–(276.222/(r.h.**2)). Based on data from Grewal and Atwal (1967). Used in Model 2.

Pupal survival = (0.250285–(.0636112*LOG(TEMP)) –(0.00372889 * LOG(r.h.))/(1–(.497083*LOG(TEMP)) +(0.0647804*(LOG(TEMP)**2))–(0.00924077*LOG(r.h.)))Based on data from Grewal and Atwal (1967). Used in Model 2.

Egg development 3 = 29.3036– (0.769002 *Temp) – (0.00645732 *r.h.). Based on data from the current study. Used in Model 4. Output for temperatures greater than 30°C were set at that for 30°C. If temperatures exceeded 35°C no eggs hatched.

Immature male development 2=269.521– (2.27820* TEMP) – (5.76098*r.h.)+(0.0479332*(r.h.**2)). Used in Model 4. Output for temperatures greater than 30°C were set at that for 30°C.

Immature female development 2 = 138.895– (3.23713 *TEMP) – (0.160336*r.h.). Used in Model 4. Output for temperatures greater than 30°C were set at that for 30°C.

Egg sururvival 3 = 0.868486– (0.00370715 *TEMP) +(0.00189414*r.h.). Used in Model 4. Survival was set at 0 for temperatures below 10°C and above 35°C.

Immature survival 2 = 0.867359– (0.0184638 *TEMP) +(0.00431522*r.h.) Used in Model 4. Survival was set at 0 for temperatures below 10°C and above 35°C.

Note: All models had r.h. bounded at an upper limit of 80% and a lower limit of 40%. All simulations run in this study, however, had a constant relative humidity of 75% because the grain moisture content did not vary over the storage interval.

Influence of planting date, harvest date, and maize (corn) hybrid on preharvest infestation of maize by *Sitotroga cerealella*

P.A. Weston[*]

Abstract

A field study was conducted during 1992–1993 in an effort to gauge the practicality of planting a higher yielding maize (corn) variety later in the growing season to avert preharvest infestation by *S. cerealella*. Infestation tended to decrease with planting date and increase with sampling ('harvest') date, in agreement with earlier studies. Yield of maize was unaffected by planting date or maize hybrid, suggesting that delayed planting may indeed be a viable cultural control method for *S. cerealella*. Infestation rate was somewhat higher in the higher yielding variety, but overall infestation rate was quite low during the study period compared with previous years. Although further study is required, the results suggest that late planting and early harvest alone are potentially useful methods for averting preharvest infestation of maize by *S. cerealella*, and that a higher yielding variety may not be necessary at the latitude of Kentucky, USA to compensate for potential loss of yield resulting from a shortened growing season.

Introduction

Sitotroga cerealella (Olivier), more commonly known as the Angoumois grain moth, is a worldwide pest of stored grains and the most abundant lepidopteran pest of stored maize in Kentucky (J. Sedlacek, unpublished data). Its pest potential is great owing to its high mobility, ability to colonise intact grain, and ability to infest grain both before and after harvest. Weston et al. (1993) reported that preharvest infestation of maize was strongly influenced by grain moisture content, with the likelihood of infestation increasing sharply once moisture content dropped below 31% (wt/wt). They suggested that cultural practices that minimised exposure of vulnerable maize to potential colonising moths, such as delayed planting or harvesting as soon as practicable, might be employed to reduce infestation of maize by this pest.

Delaying planting reduces preharvest infestation because the grain moisture content decreases to the threshold moisture content for infestation (31%) after peak moth activity in the field, which occurs around August (Barney and Weston 1994). Delayed planting, however, might reduce crop yield because of reduced time for maturation (Bitzer et al. 1979). The objectives of this study were to measure crop yield and infestation by *S. cerealella* of maize planted at normal and late planting dates for a maize hybrid commonly grown in central Kentucky and a higher yielding hybrid in an effort to measure the impact of cultural control methods on quality and quantity of harvested maize.

Materials and Methods

The experiment was conducted in 1992 and 1993 at the ca. 85–ha Kentucky State University Research Farm in Franklin Co., Kentucky. Topography is rolling to hilly and the land surrounding the farm is used primarily for cattle grazing and small acreages of tobacco and field maize.

Maize was planted in four locations ca. 400 m from a cluster of galvanised grain storage bins located near the centre of the research farm. The sites were oriented approximately at the four cardinal compass directions with respect to the grain storage complex, which consisted of 12 300-bushel bins and 2 1000-bushel bins of maize containing various amounts of maize varying in duration of storage. Each site was planted with replicated plots (7.6 m × 12.2 m) of the three (1992) or four (1993) treatments. Two replicates were planted at each site, except for the north site in 1992, where three replicates were planted.

The treatments in 1992 were 'DeKalb (DK) 689' planted at the typical planting time of mid-May and at a late planting date (7 June), and 'DK 743' planted at the later date. Roughly 50% of maize is planted by the middle of May in a typical year, and 95% is planted by the end of the first week in June. In 1993, we used the same three treatments plus 'DK 743' planted at the normal planting date. 'DK 689' is commonly grown in central Kentucky, and 'DK 743' is a newer hybrid that has been found to produce higher yields than 'DK 689' in variety trials conducted around the state. Conventional tillage, fertilisation, and weed control practices were used.

Plots were sampled for *S. cerealella* on three dates at the end of the growing season, with three weeks between dates. In 1992, plots were sampled on 15 September, 13 October, and 13 November. In 1993, sampling dates were 22 September, 20 October, and 19 November. Adult emergence holes and 'windows' (transparent circular areas of cuticle created as last instar larvae hollow out their pupation chamber within kernels) were counted and marked at sampling time for the three sampling dates and again after five weeks' incubation in paper bags under laboratory conditions. Five weeks was chosen as the incubation time because it is sufficiently long to allow larvae to complete development but short enough to prevent eggs laid by emerging adults to complete development to adult. Four ears were sampled per subplot, one from three random locations in each plot and one from the centre. Emergence holes and windows were counted and recorded from the three randomly selected ears, and ca. 50 g of kernels were removed from the middle of the fourth ear and placed in air-tight vials for moisture determination at a later date. Total infestation was calculated by summing infestation measured at sampling time and that after incubation. Moisture content, expressed as the percentage of fresh kernel weight consisting of water, was determined by drying whole kernels at 103°C for 72 hours (Watson 1984).

Yield of maize was measured by harvesting and weighing all ears in two 7.2-m rows from each plot.

---

[*] Community Research Service, Kentucky State University, Frankfort, KY 40601 USA.

Results and Discussion

As was found in our earlier study (Weston et al. 1993), infestation rate increased with sampling date, with the exception of late-planted 'DK 743' in 1992, where corn from the second sampling date had the highest infestation rate (Fig. 1). This was no doubt due to sampling error because total infestation cannot possibly decrease over time for a thoroughly sampled population. Infestation tended to decrease with planting date for both years, particularly at the first two sampling dates.

Both of these trends can be explained by changes in moisture content over the course of the sampling intervals. Moisture content at the first sampling date was at or above the infestation threshold of 31% for all treatments, and declined to well below this level by the third date (Fig. 2). In addition, moisture content was higher for later planted maize in virtually all cases. The later planted maize was particularly responsive to decreasing moisture content because the initial value was well above the threshold.

Infestation rate of the two hybrids was not consistent over the two years. Infestation of 'DK 743' was distinctly higher than that of 'DK 689' planted at the same time for three of the six sampling dates, and was roughly equal to that of 'DK 689' for the other three (Fig. 1). Thus, even though the infestation trends were not consistent over the two years, 'DK 743' seems to have a higher potential for sustaining damage from *S. cerealella*.

Yields averaged over hybrid and planting date were not appreciably different within years, and were similar between years (Fig. 3). Delayed planting, thus, had little adverse effect on maize yield, and resulted in slightly higher yields in 1993.

General Discussion

Delayed planting appears to be a viable strategy for reducing preharvest infestation of maize by *S. cerealella* where this pest is a threat and control options are limited. Using a higher yielding hybrid ('DK 743') had little apparent advantage, however, and 'DK 743' may in fact be more susceptible to infestation than 'DK 689'. It must be pointed out, however, that additional experimentation is required because the rate of infestation by *S. cerealella* was rather low during the study period compared to our earlier study. In 1991, for example, infestation rate for 'DK 689' planted at the normal time averaged 24–28 moths per three ears, roughly eight times higher that the maximum infestation rate observed here (Fig. 4).

Although moisture content may be a reliable indicator of susceptibility of maize to infestation by *S. cerealella*, it must be noted that this variable may merely be correlated with another characteristic of maize plants that more directly influences infestation. For example, *S. cerealella* might locate fields of maturing maize by orienting to volatiles produced by maize plants of the appropriate phenological state. Production of these putative semiochemicals might, in turn, be correlated with moisture content of the ripening grain. The likelihood of such a scenario is supported by the observations that moth abundance is greater in plots of maize compared to nearby open fields late in the growing season (Stockel 1971; Barney and Weston, unpublished data) and that survivorship of *S. cerealella* larvae increases as moisture content of maize kernels decreases as the grain ripens (Koone 1952).

The other strategy proposed by Weston et al. (1993) for avoiding infestation by *S. cerealella*, i.e. early harvesting, is

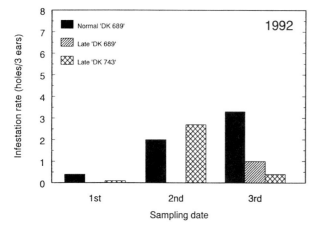

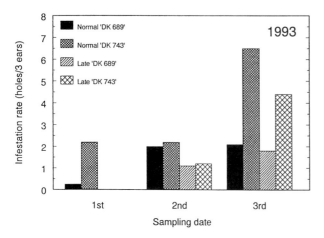

Fig. 1.   Total infestation rate of maize by *S. cerealella* during the study period. See text for full details.

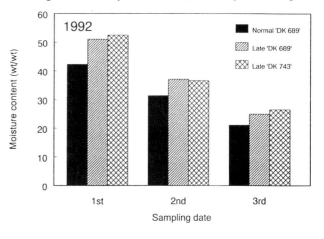

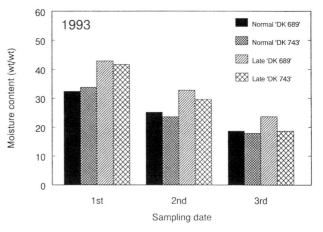

Fig. 2.   Moisture content of maize during the study period.

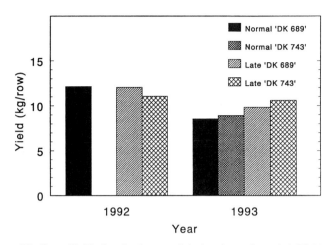

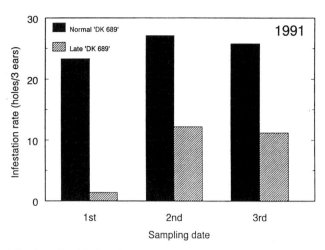

Fig. 3. Yield of maize harvested during the study period. Yield was weight of maize (including cobs) harvested from 7.2-m rows. Note that 'DK 743' was not planted at the earlier date in 1992.

Fig. 4. Total infestation rate of maize grown in 1991 at the Kentucky State University Research Farm. See Weston et al. (1993) for full details.

also supported by the current study. Harvesting maize as soon as practicable minimises the window of exposure to potential colonising moths. Limitations imposed by harvesting and grain-drying capacities, however, need also to be considered when choosing the optimal time for harvest.

Acknowledgments

I thank R. Barney, X. Ge, S. Hill, W. Owens, B. Price, P. Rattlingourd, J. Sedlacek, and W. Tietjen (Community Research Service, Kentucky State University) for assistance in conducting this research. This investigation was supported by a grant from USDA-Cooperative States Research Service to Kentucky State University under agreement KYX-10-90-15P.

References

Barney, R. J. and Weston, P. A. 1994. Grain storage in a small-farm ecosystem: Angoumois grain moth movement and management. Proceedings 6th International Working Conference on Stored-Product Protection, in press.

Bitzer, M. J., Hrbek, J. H., Lacefield, G. and Evans, J. K. 1979. Producing corn for grain and silage. University of Kentucky Extension Publication AGR-79.

Koone, H. D. 1952. Maturity of corn and life history of the Angoumois grain moth. Journal of Kansas Entomological Society, 25, 3, 103–105.

Stockel, J. 1971. Utilisation du piégeage sexuel pour l'étude du déplacement de l'Alucite *Sitotroga cerealella* (Lépidoptere: Gelechiidae) vers les cultures de Maïs. Entomologia Experimentalis et Applicata, 14, 39–56.

Watson, S. A. 1984. Measurement and maintenance of quality. In: Watson, S. A. and Ramstad, P. E. ed., Corn Chemistry and Technology, St. Paul, American Association of Cereal Chemists, 125–183.

Weston, P. A., Barney, R. J., and Sedlacek, J. D. 1993. Planting date influence on preharvest infestation of dent corn by Angoumois grain moth (Lepidoptera: Gelechiidae). Journal of Economic Entomology, 86, 1, 174–180.

Variable longevity in the rusty grain beetle, *Cryptolestes ferrugineus*

N.D.G. White and R.J. Bell[*]

Abstract

Cryptolestes ferrugineus adults at 30 insects/15 g of food and 30°C had a shorter mean lifespan (24 weeks for 30 females and 13 weeks for 30 males) than adults mated with a single partner (25 weeks) or isolated virgins (32 weeks). Males in populations had the shortest lifespans, and the lifespans of females were inversely proportional to the number of males in the population.

The production of live adult offspring per female in populations of 30 adults was approximately half that of females mated once. Longevity was also negatively related to feeding effort when virgin adults of an inbred strain and outbred strain were individually fed either ground whole wheat (mean 32 weeks) or whole kernels of wheat with the seed coat over the germ removed (mean 23 wk) at 30°C, 75% r.h.

To simulate over-wintering of adult beetles, virgin adults or the inbred and the outbred strain were held at temperatures which decreased from 30°C or 22°C to 5°C, then increased to 25°C. Maximum lifespans were over 1 year and a few of the outbred beetles raised at 30°C lived for 1.5 years. Adults of both the inbred and outbred strains raised at 22°C had shorter lifespans than adults raised at 30°C. These results indicate that the probability of survival at low temperatures is significantly increased when the immature stages and a few weeks of adult life are passed at relatively high temperatures.

Introduction

The lifespans of insects are strongly influenced by population density, sexual activity, temperature and other environmental variables (Lints 1985; Sohal 1985). It has generally been found that density has a more adverse effect on males than on females and that isolated individual insects live longer than insects in groups (Ragland and Sohal 1973).

With *Tribolium* spp. the proportion of males in the population appears to be a strong determinant of lifespan. Spratt (1980) found that for the red flour beetle *Tribolium castaneum* (Herbst), individually isolated males, and females in all-female groups lived an average of 50 weeks. Males and females in mixed-sex groups lived an average of 40 weeks and males in all-male groups had mean lifespan of less than 20 weeks. The confused flour beetle, *T. confusum* J. du Val was not as long-lived as *T. castaneum*, but the lifespans of corresponding groups were in the same relative order (Spratt 1980). For isolated individuals of the American black flour beetle, *T. audax* Halstead and single-sex groups of 2, 4, 10, and 40, male lifespan decreased with each increase in density, but female

lifespan did not show a consistent relationship to density (Soliman 1977).

Temperature is an important factor affecting longevity in poikilotherms and Kawamoto et al. (1989) found that 77 to 97% of *Cryptolestes ferrugineus* adults, kept individually isolated in cells at different constant temperatures in the range 10 to 30°C, were still alive when the experiment was terminated at 32 weeks, but at 35°C only 38% of the adults were still alive. It is not known how long *C. ferrugineus* (Stephens), the principal insect pest of stored grain in western Canada (Sinha and Watters 1985), can survive at different temperatures and still be capable of initiating new infestations in grain storage facilities. This information would be useful in devising control strategies and preparing computer models of population growth in grain bulks.

In the present study we determine how the lifespan and reproductive rate of the rusty grain beetle, *C. ferrugineus* are affected by density and sex ratios, and evaluate possible causes for variations in survivability and reproductive rate between these groups. The survival times of beetles living exclusively on a diet of whole kernel wheat or ground wheat were measured. We also determined the lifespans of individually isolated virgin adults of a laboratory strain and an outbred strain of *C. ferrugineus* at different constant temperatures and at temperatures that simulate field conditions.

Materials and Methods

C. ferrugineus strains

The rusty grain beetle is a small reddish-brown dorsoventrally flattened beetle, 1.5 to 2.5 mm in length. Thirty males and 48 females in a random sample taken from a culture (GV strain) had mean live weights of 0.283 mg and 0.274 mg, respectively. Two strains were used in this study: the SS strain (malathion susceptible), a relatively inbred strain, cultured in our laboratory since 1975; and the GV strain, an outbred genetically variable strain, prepared by making crosses between the SS strain and field strains collected in Manitoba, Canada, Kansas, USA., and northern England.

General methods

All experiments were conducted in the dark at 30 ± 1°C. Relative humidity was controlled at 75 ± 5% by saturated NaCl solutions (Winston and Bates 1960). Food used in the studies, except the experiment to determine the effect of living on whole kernel wheat, was ground whole wheat plus wheat germ that was ground fine enough to pass through a sieve with 420 μm apertures. Individually isolated virgin adults were kept in small vials, 1.5 cm by 4.5 cm; mated adults were kept in large vials, 2.8 cm by 7 cm.

The immature stages were reared in gelatin capsules. Eggs were placed individually in separate gelatin capsules, size 00 (2.5 cm long, 0.8 cm diam.), which were half full of ground wheat plus wheat germ. When adults emerged their sex was

---

[*] Agriculture and Agri-Food Canada, Research Station, 195 Dafoe Road, Winnipeg, Manitoba, R3T 2M9, Canada.

determined and they were placed in different mating situations or kept as virgins isolated from the other sex.

Lifespan of individually isolated adults

The lifespan of individually isolated virgin adults was determined with 45 males and 61 females of the GV strain in separate small vials containing 1 g of ground wheat plus wheat germm. The vials were taken out of the incubator once every 2 weeks for approximately 30 minutes and deaths were recorded.

Lifespan and reproductive rate of male/female pairs

The lifespan and reproductive rate of isolated male/female pairs were determined using large vials each containing one male and one female and 10 g of ground wheat plus wheat germm. There were 30 male/female pairs of the GV strain. Adults were 1–7 days old when mated. Ten male/female pairs adults were transferred to new food every week and the number of deaths that had occurred during the week was recorded. The old food from each vial, along with eggs and larvae, was emptied into a 224-mL jar containing 50 g of whole kernel wheat and 10 g of ground wheat plus wheat germm. The jars were kept in an incubator at $30 \pm 1°C$, $60 \pm 5\%$ r.h. and 5 weeks later the F_1 adults were counted. A similar procedure was followed for the other male/female pairs, except that parental adults were transferred to new food and deaths were recorded every 2 weeks.

The production of offspring by females mated with males for only 24 hours was investigated using 10 males and 10 females of the SS strain. One male and one female, 1–5 days old, were placed together in a large vial containing 10 g of ground wheat plus wheat germm. After 24 hours the males were separated from the females. The females were transferred to new food every week for 20 weeks and the F_1 offspring were raised by the same procedure that was used for mated pairs. None of the females produced offspring in the 6 weeks between the 13th and the 19th weeks so they were re-united with their original male partners for the 20th week.

Lifespan and reproductive rate of adults in groups

The lifespan and reproductive rate of adults of the GV strain kept together in groups of 30 were determined using 11 large vials with 30 adults and 10 g of ground wheat plus wheat germm in each vial. Three vials had 10 males and 20 females, three vials had 15 males and 15 females, three vials had 20 males and 10 females, one vial had 30 virgin males, and one vial had 30 virgin females. F_1 adults were raised and counted using the same procedure that was used in the experiment with male/female pairs.

Lifespan of adults living on whole kernel wheat

Newly emerged adults of the GV and SS strains, raised from eggs at 30°C on ground wheat in separate gelatin capsules, were placed separately in small vials, each containing 50 wheat kernels that had the seed coat over the germ removed. There were 104 adults of the GV strain and 115 adults of the SS strain. They were kept for the duration of adult life at 30°C. The lifespans of these adults were compared with the lifespans of adults at 30°C where the insects were fed ground wheat.

Adult lifespan at different constant temperatures and at temperatures that were varied during adult life

The lifespans of adults of the SS and GV strains were determined at constant temperatures of 22, 25, 30 and 35°C. The adults kept at each temperature were raised from eggs at the same temperature.

To simulate overwintering of adult beetles at temperatures likely to occur inside grain bulks, in grain elevators or large grain bins in Manitoba, Canada, adults of the SS and GV strains, raised from eggs at either 22°C or 30°C, were kept for various periods at a series of temperatures which decreased from 22°C or from 30°C to 5°C then increased to 25°C. The time that each group of adults was kept at each temperature interval is given in Figure 7.

Statistical procedures

Mean adult lifespans and mean developmental periods were compared with a one-way analysis of variance test. A two-sample Kolmogorov-Smirnov test (K-S test) was used to determine the significance of differences between the survival functions graphed in the figures. This is a non-parametric test recommended for survival data by Mode et al. (1984).

Maximum lifespan is the mean of the last surviving 10% of the population. Maximum lifespan was measured from the deposition of the egg to the death of the adult.

Results

Lifespan

Individually isolated virgin adults lived much longer than mated pairs or adults in groups of 30. Virgin adults in separate vials lived an average of 32 weeks, mated pairs of the GV strain in separate vials lived an average of 22 weeks, and the average lifespan of adults in mixed sex groups of 30 per vial at different sex ratios was 13 weeks (Fig. 1).

When isolated there was no difference between male and female longevity. When kept in separate vials with one female, males of both strains lived longer than females ($P < 0.01$), but in groups of 10 males and 20 females, 15 males and 15 females, and in groups of 30 virgin males or 30 virgin females, the females lived longer than males ($P < 0.01$). Adults in the group with 20 males and 10 females per vial had the shortest mean lifespan and the difference between males and females in this group was not significant ($P > 0.05$) (Figs. 2 and 3).

Differences in sex ratio in the groups with 30 adults together in one vial, including the all-male group, did not produce significant differences in male lifespan ($P > 0.05$). However, the lifespan of females in the groups with 30 adults per vial decreased as the proportion of males increased.

Reproductive rate

The maximum reproductive period for females was 20 weeks. During most of this time, females of the GV strain that were isolated in separate vials with one male produced offspring at a higher rate than females in the groups with 30 adults per vial, and their mean lifetime production of offspring was 1.8 to 2.3 times greater than the mean lifetime production of females in the 30-adult group that had the highest number of offspring (Fig. 4).

Within the groups with 30 adults per vial the production of offspring per female was inversely proportional to the number of females in the group; each decrease of five females per group increased lifetime production of offspring per female by approximately 50%.

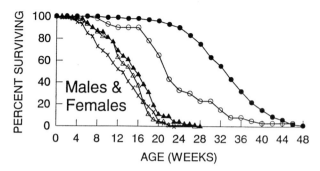

Fig. 1. Survival curves for *C. ferrugineus* GV strain virgin female adults kept individually in separate vials (●), or in the following groups: 1 male and 1 female (O), 15 males and 15 females (△), 10 males and 20 females (▲), 20 males and 10 females (×) at 30°C, 75% relative humidity.

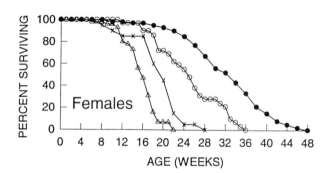

Fig. 2. Survival curves for *C. ferrugineus* GV strain virgin female adults kept individually in separate vials (●) or in a group of 30 in one vial (O), mated females kept in groups of 15 males and 15 females in one vial (△), or 1 male and 1 female in one vial (×) at 30°C, 75 % relative humidity.

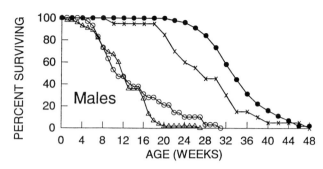

Fig. 3. Survival curves for *C. ferrugineus* GV strain virgin male adults kept individually in separate vials (●) or in a group of 30 in one vial (O), mated males kept in groups of 15 males and 15 females in one vial (△), or 1 male and 1 female in one vial (×) at 30°C, 75% relative humidity.

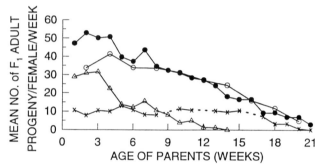

Fig.4. Rate of production of F_1 progeny per female per week by *C. ferrugineus* SS strain females after a single 24 hour period with a male (△), by mated pairs of the GV strain kept 1 male and 1 female per vial with new food every week (●) or every two weeks (O) and by groups of 15 males and 15 females in one vial (×) at 30°C, 75% relative humidity.

Among the groups with mated pairs in separate vials the GV strain group that was provided with fresh food every week produced the largest number of offspring per female, both during the first 8 weeks and during the entire lifetime of the females.

Females of the SS strain kept in separate vials after being mated with a male for only 24 hours, produced a mean of 180 offspring during a period of 13 weeks. No viable eggs were laid in the period from 13 to 19 weeks but when the nine females still alive were reunited with their male partners at 20 weeks, 14.5 offspring per female were produced by four females (Fig. 4). This suggests that to reach their maximum potential production of offspring, females need to mate a number of times over a relatively long period.

Lifespan of adults living on whole kernel wheat

Adults of both the SS and GV strains lived longer on ground wheat than on whole kernel wheat with exposed germs (ANOVA and K-S test, $P < 0.01$) (Fig. 5). There was no difference in lifespan between strains ($P > 0.05$) or between the sexes ($P > 0.05$) on either type of food. It was evident that beetles living on whole kernel wheat had sufficient food because at the time of death even the longest-lived beetles left 15–25 wheat kernels with the germ still intact. The rest of the kernels had the germ partially or completely eaten out.

Lifespan of adults at different temperatures

Individually isolated virgin adult males of the GV strain had longer mean lifespans than individually isolated virgin females of the GV strain at all of the eight temperature treatments used for this strain, and at seven of the temperature treatments males had longer maximum lifespans than females.

The mean lifespans of virgin adults of the inbred strain (SS) (Fig. 6) and the outbred strain (GV) at constant temperatures of 25 and 30°C were similar ($P > 0.05$), but at constant temperatures of 22 and 35°C, mean adult lifespans of the outbred strain were significantly longer ($P < 0.05$). In general, beetles of the inbred strain survived as long as beetles of the outbred strain only at those constant temperatures that were close to the temperatures at which they had been cultured in the laboratory for the last 15 years.

On the varied temperature schedules, mean adult lifespans of the GV strain were longer than those of the SS strain ($P < 0.05$) (Fig. 7). Beetles of the SS strain that were kept on the overwintering temperature schedule starting from 30°C, decreasing to 5°C, then rising to 25°C (Fig. 7), had a mean adult lifespan that was 40% longer than beetles of the SS strain on the overwintering temperature schedule starting from 22°C, decreasing to 5°C, then rising to 25°C ($P < 0.01$). The mean adult lifespan of beetles of the GV strain on the overwintering temperature schedule starting from 30°C was 52% longer than that of beetles of the GV strain that were on the overwintering temperature schedule starting from 22°C ($P <$

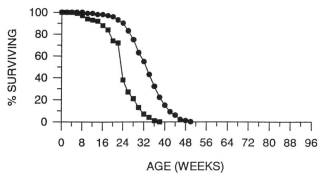

Fig. 5. Survival curves for *C. ferrugineus*, GV strain virgin adults, living on ground wheat (●), or on whole kernels of wheat with exposed germs (■) at 30°C, 75% relative humidity.

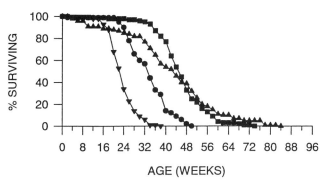

Fig.6. Survival curves for *C. ferrugineus*, virgin adults of the SS strain, kept individually in separate vials for the duration of life at 75% relative humidity and constant temperatures of 25°C (▲), 25°C (■), 30°C (●) and 35°C (▼).

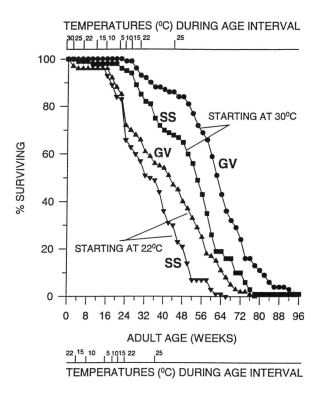

Fig. 7. Survival curves for *C. ferrigineus*, virgin adults, raised from egg to adult at 30°C then transferred to 25, 22, 15, 10, 5, 10, 15, 22 and 25°C at the ages indicated at the top of the graph (upper curves, SS strain, ■; GV strain, ●), or raised from egg to adult at 22°C then transferred to 15, 10, 5, 10, 15, 22 and 25°C at the ages indicated at the bottom of the graph (lower curves, SS strain, ▼, GV strain, ▲).

0.01). At 48 weeks of age, percent survival of beetles of each strain on the overwintering temperature schedule starting from 30°C was double the percent survival of beetles of the same strain on the temperature schedule starting from 22°C (K-S test, *P* < 0.01) (Fig. 7). These large differences in lifespan were not correlated with average temperature. The weighted mean temperature up to the age of 42 weeks for the two groups initially at 30°C was 17.7°C; for the two groups initially at 22°C the weighted mean temperature was 17.4°C and after 42 weeks of age all four groups were kept at a constant temperature of 25°C.

Discussion

Factors causing differences in lifespan

The males of *C. ferrugineus* had the highest mortality rates in mixed-sex groups and in the all-male group; female mortality was also high in mixed-sex groups but relatively low in the all-female group. This suggests that in the presence of a number of females and other males, males are strongly stimulated to compete for sexual partners and may cause injury to themselves or to other adult beetles during copulation attempts. Aggressive male sexual behaviour has been identified by Spratt (1980) as an important factor in reducing the lifespan of *T. castaneum* and *T. confusum* adults kept in groups. *Tribolium* beetles have no courtship routine to aid males in identifying receptive females; males have been observed attempting to mate with other males, dead beetles, or small objects that resemble other beetles (Taylor and Sokoloff 1971).

Factors causing differences in oviposition rate

The results of the oviposition experiments suggest that *C. ferrugineus* females reduce their oviposition rate as the number of larvae and the number of other females in the immediate environment increases. Higher larval density may have been the cause of the lower oviposition rate during the first 8 weeks in the groups with food renewed every 2 weeks. Larvae of *Oryzaephilus surinamensis* (L.), another beetle of the family Cucujidae, emit volatile compounds which reduce oviposition rate (Pierce et al. 1990).

It appears that when *C. ferrugineus* adults are at low density (1 male and 1 female per vial), with time for larvae to accumulate, oviposition rate decreases with increasing larval density, but at high adult density (30 beetles per vial), with larvae or eggs removed oviposition rate decreases with increasing density of adult females.

Lifespan of adults living on whole kernel wheat

Since both ground and whole kernel wheat had the same nutrients the most likely cause of the reduced lifespan of beetles living on whole kernel wheat was a difference in rate of energy expenditure. Beetles that were forced to chew their food out of the hard germ tissue had to expend energy at a higher rate than beetles that were able to ingest particles of ground wheat with very little effort. Sohal and Buchan (1981) and Sohal (1982) have shown that raising the activity level of *Musca domestica* L. without changing ambient temperature can shorten their lifespan.

Developmental periods and lifespan of adults at different temperatures

At constant temperatures in the range 25 to 35°C the lifespan of *C. ferrugineus* was inversely correlated with temperature. This result is consistent with the rate of living theory (Pearl 1928).

However, when *C. ferrugineus* adults were kept on the overwintering temperature schedule starting from 30°C they lived much longer than adults that were kept on the overwintering temperature schedule starting from 22°C, although the weighted mean temperatures of the two treatments were almost the same, and adults that were kept for the first 16 weeks at 22°C, then transferred to 30°C, had shorter lifespans than adults kept continuously at 30°C, while adults kept for the first 12 weeks at 30°C, before transfer to 22°C, lived longer than adults kept continuously at 22°C.

Sohal (1986) has discussed some of the reasons why departures from the inverse relationship between lifespan and temperature occur, and why he believes that the rate of living theory, as a statement of the relationship between lifespan and metabolic rate, is still valid.

In our experiments all groups of *C. ferrugineus* that had shorter mean adult lifespan than groups of the same strain at higher average temperatures were raised from egg to adult at 22°C, and at least the first 5 weeks of adult life were spent at 22°C. There are a number of indications that this temperature is slightly below the lower limit at which this insect can achieve an optimum state of physiological maturity. A short period at 30°C early in life seemed to prepare the beetles to tolerate the shift to lower temperatures, while early life at 22°C produced a weakened population with poor tolerance for low temperatures.

References

Kawamoto, H., Sinha, R.N. and Muir, W.E. 1989. Effect of temperature on adult survival and potential fecundity of the rusty grain beetle, *Cryptolestes ferrugineus*. Applied Entomological Zoology, 24, 418–423.

Lints, F.A. 1985. Insects. In: Finch, C.E. and Schneider, E.L. ed., Handbook of the biology of aging. New York. Van Nostrand Reinhold, 146–169.

Mode, C.J., Ashleigh, R.D., Zawodniak, A. and Baker, G.T. 1984. On statistical tests of significance in studies of survivorship in laboratory animals. Journal of Gerontology, 39, 36–42.

Pearl, R. 1928. The rate of living. New York. Knopf. 185 p.

Pierce, A.M., Borden, J.H. and Oehlschlager, A.C. 1990. Suppression of oviposition in *Oryzaephilus surinamensis* (L.) (Coleoptera: Cucujidae) following prolonged retention in high-density cultures or short-term exposure to larval volatiles. Journal of Chemical Ecology, 16, 595–601.

Ragland, S.S., and Sohal, R.S. 1973. Mating behaviour, physical activity and aging in the housefly, *Musca domestica*. Experimental Gerontology, 8, 135–145.

Sinha, R.N. and Watters, F.L. 1985. Insect pests of flour mills, grain elevators, and feed mills and their control. Agriculture Canada Publication 1776E, Canadian Government Publishing Centre, Ottawa, ON. 290 p.

Sohal, R.S. 1982. Oxygen consumption and lifespan in the adult male housefly, *Musca domestica*. Age, 5, 21–24.

Sohal, R.S. 1985. Aging in Insects. In: Kerkut, G.A. and Gilbert, L.I. ed., Comprehensive insect physiology, biochemistry, and pharmacology, Volume 10. Oxford, Pergamon Press, 595–631.

Sohal, R.S. 1986. The rate of living theory: a contemporary interpretation. In: Collatz, K.G. and Sohal, R.S. ed., Insect Aging. Berlin. Springer-Verlag, 23–44.

Sohal, R.S. and Buchan, P.B. 1981. Relationship between physical activity and lifespan in the adult housefly, *Musca domestica*. Experimental Gerontology, 16, 157–162.

Soliman, M.H. 1977. The effect of culturing together on adult longevity of *Tribolium audax*. Tribolium Information Bulletin, 20, 139–140.

Spratt, E.C. 1980. Male homosexual behaviour and other factors influencing adult longevity in *Tribolium castaneum* (Herbst) and *T. confusum* Duval. Journal of Stored Product Research, 16, 109–114.

Taylor, C. and Sokoloff, A. 1971. A review of mating behaviour in *Tribolium*. Tribolium Information Bulletin, 14, 88–91.

Winston, P.W. and Bates, D.H. 1960. Saturated solutions for the control of humidity in biological research. Ecology, 41, 232–237.

Effect of food volume and photoperiod on initiation of diapause in the warehouse beetle, *Trogoderma variabile* Ballion (Coleoptera: Dermestidae)

E. J. Wright and A. P. Cartledge*

Abstract

Trogoderma variabile overwinters as a fully grown larva in a state of developmental arrest called diapause. Diapausing larvae are less active than non-diapausing larvae, but they do emerge from refugia occasionally to feed. They are less susceptible to insecticidal treatments than are non-diapausing larvae. Understanding the mechanisms of diapause induction and termination may offer the potential to manipulate the environmental conditions to either prevent diapause or trigger its termination and therefore facilitate pest control. The induction of diapause is complex. Here we report on the effect of day length and volumes of food medium on the initiation of diapause and explain some of the irregular findings reported previously in the literature.

Introduction

Insects must cope with unfavourable periods associated with seasonal changes that are cyclical, persistent and geographically widespread. Most species have evolved the ability to recognise environmental cues that signal these cyclical changes and respond to them with a series of specific physiological, behavioural and morphological adaptations which comprise the diapause syndrome (Lees 1955; Tauber et al. 1986). Diapause is usually initiated and maintained when many environmental conditions are still favourable for growth and development, and tends to be triggered by seasonally changing variables such as photoperiod and temperature during a specific life stage of the insect. Insects in diapause have a much lower metabolic activity, increased resistance to environmental extremes and often show altered behaviour (Tauber et al. 1986). Bell (1994) has recently reviewed diapause in stored product insects.

Trogoderma variabile Ballion overwinters as a fully-grown diapausing larva. Diapause in *T. variabile* differs from classical diapause in that the insect can continue to feed and to moult. Instead of pupating at the normal time, the diapausing larva continues to grow to approximately twice the weight of its non-diapausing counterpart (Elbert 1979a) and remains as a larva for an extended period. In this state the larvae are more tolerant of phosphine (Banks and Cavanaugh 1985), low temperature and starvation (Elbert 1979a).

Under almost any situation, some larvae of a cohort will not pupate and will enter diapause. Diapause initiation in *T. variabile* is clearly complex, with a number of stimuli being implicated. Loschiavo (1960) found that daily disturbance of insects during development increased the proportion entering diapause compared withundisturbed controls. Larvae reared singly are much more likely to enter diapause than larvae reared in groups (Partida and Strong 1975; Elbert 1979a,b). One experiment by Elbert (1979b) showed that for larvae reared in groups in constant darkness, a higher proportion pupated when reared at 25 or 35 compared with 30°C. At 30°C and 50% r.h. less than 10% of larvae reared in groups entered diapause, irrespective of photoperiod. However, a high proportion of larvae reared singly entered diapause under conditions of constant darkness (88%) compared with a photoperiod of 8L:16D (47%) or 16L:8D (29%). Elbert (1979b) also showed that diapause was induced in fourth and fifth instar larvae. These results are consistent with induction of diapause in autumn when *T. variabile* larvae would experience short day length (or complete darkness in bulk grain within a store) and lower temperatures. Given that the original mode of life for *T. variabile* is to develop singly or in small groups in bee and wasp nests (Strong and Okumura 1966), it would seem that diapause is the normal condition and only extraordinary combinations of stimuli will prevent its induction.

One stimulus that has been implicated in diapause induction and has relevance to pest control is the volume of food available to the larvae. Burges (1961) compared the rate of diapause initiation in volumes of food ranging from 0.7 to 450 mL of food per container. In spite of having varying numbers of larvae per container, the general trend was for the rate of diapause to increase with decreasing volume of food. Elbert (1979b) and Loschiavo (1960) also contained data for proportion of isolated larvae pupating, in different volumes of food (1.5 and 8.5 mL, respectively), but these results were not consistent with those of Burges (1961). Therefore, the following experiment was designed to investigate the effect of food volume on diapause induction in *T. variabile* larvae. The experiment was done with isolated larvae in either constant darkness or with a photoperiod of 13L:11D (longest summer day in southern New South Wales, Australia). In order to distinguish the importance of the spatial dimension (volume) of the food medium from the nutritional requirements (amount of food that is good to eat), volumes of food medium were created by mixing food with cellulose powder, that the insects would ingest but which would have no nutritional value. Experiments were run at 30–32°C and 50–60% r.h., so as to be least conducive to diapause induction and maximally comparable to the data of other workers.

Methods

Stock cultures were maintained using a method that allowed successful culturing of large numbers of larvae that pupated at a very high rate (i.e. did not diapause) and resulted in consistently rapid development of large adults. The method was to add about 5–10 males and 50 females to a 650 mL jar containing 200 g of rearing medium. This medium (SWO) comprised 1 part finely ground Savoiardi sponge biscuits, 7 parts kibbled wheat and 2 parts ground rolled oats, at an overall moisture

---

* Stored Grain Research Laboratory, CSIRO Division of Entomology, GPO Box 1700, Canberra, ACT 2601, Australia.

content of 10%. The culture jars were kept in a totally dark incubator at 30–32°C at 50–60% r.h. and were seldom disturbed. The time from addition of adults to emergence of the next generation of adults was 5–6 weeks.

First instar larvae were used to start the experiments on diapause induction. These were obtained by collecting eggs and allowing them to hatch on black filter paper in a petri dish. Individual larvae were easily seen and could then be transferred individually using a very fine brush.

The experiment was designed to investigate the effect of the spatial dimension (volume) of the food medium compared with the quantity of food required for adequate development on the rate of diapause induction. This was done by creating different volumes of medium, by mixing food (SWO) and non-nutritive cellulose powder in different proportions. Thus, the experimental media were prepared as 100% SWO or 1/2, 1/4, 1/8, or 1/16 SWO and the rest cellulose powder (Table 1). The volumes used were 1, 2, 4, 8 and 16 mL. A number of larvae were reared exclusively on cellulose powder, to confirm the non-nutritional quality of this filler material.

Three sizes of glass vials were used in the experiment depending on the volume of medium required: 150/20, 160/20, 75/15 mm height/diameter. Fluon (polytetrafluoroethylene) was applied internally to the top 1–2 cm of each vial to prevent escape of larvae and fine metal mesh lids were used to close each vial. Vials were set up in series with each volume in each series having 10 replicates on each occasion. The weight of medium necessary to reach the required volume was determined and weighed for each vial. The medium was then added to each vial and fluffed up or tamped down to achieve the required volume. Thus, each vial in each series contained both the same weight and the same volume of medium.

All vials were incubated at 30–32°C and 50–60% r.h. Seven weeks later the proportion that pupated was determined for each combination. Any larva that had not pupated by then was deemed to be in diapause, following Burges (1961).

The complete experiment with differing proportions of food and cellulose powder was set up on six occasions with 10 vials per combination (except on one occasion when there were only 5 per combination) and incubated in constant darkness. One complete experiment was set up and incubated at 13 hours light and 11 hours dark. On another 7 occasions, the 100% SWO portion only of the experiment was prepared and run at 13L:11D, all with 10 vials per combination.

Results

Following the convention of earlier workers, results are presented as percent pupation rather than diapause, emphasising the normal nature of diapause and the specific stimuli required to induce pupation.

All larvae held exclusively on cellulose powder died very quickly, confirming the non-nutritive nature of the filler.

Table 2 shows the results of using different proportions of food and the total volume of medium on the proportion of insects that pupated under conditions of constant darkness. There was a trend towards increasing pupation with increasing volume of medium, irrespective of the proportion of food in the medium.

Figure 1 shows the effect of different volumes of food medium on the proportion of larvae pupating under conditions of long days (13L:11D). In contrast to the situation in the dark, there is a much higher rate of pupation under long days, although very small volumes inhibit pupation under both regimes.

Discussion

For insects reared singly under constant dark conditions, the spatial dimension (volume) of the food medium was more important than food quality in diapause initiation. Addition of cellulose powder to a small amount of food made the resulting medium equivalent to a much larger amount of food in its effects on pupation (Table 2). This result also showed that the food quality per se is not very important, and it is reasonable to compare results of different workers that fed the larvae on different diets. These results confirm the work of Elbert (1979b) on the effect of photoperiod on diapause induction.

These results also allow us to understand the results of previous work. Figure 2 compares the data from different sources on diapause induction in isolated larvae reared at

Table 2. Effect of volume of medium on proportion of *Trogoderma variabile* pupating at 30–32°C, 50–60% r.h. in constant darkness, with standard errors (SE). The medium was composed of varying proportions of food (SWO) and cellulose powder (a non-nutritive filler).

Proportion SWO		Total volume of medium (mL)				
		1	2	4	8	16
1	proportion	0.06	0.14	0.14	0.16	0.27
	SE	0.03	0.04	0.05	0.09	0.15
1/2	proportion		0.08	0.21	0.27	0.34
	SE		0.06	0.11	0.08	0.06
1/4	proportion			0.14	0.20	0.34
	SE			0.06	0.06	0.06
1/8	proportion				0.20	0.44
	SE				0.08	0.05
1/16	proportion					0.50
	SE					0.14

Table 1. Experimental design for the effect of different food volumes on diapause induction in *Trogoderma variabile*. For each final volume, the constituent volumes of food (SWO) and non-nutritive cellulose powder (cell) are given, with their corresponding weights in brackets. Because of the packing effect with larger volumes of food, the weights of food used were not simple multiples of the weight of 1 mL. The cellulose powder was much less dense than the SWO.

Proportion SWO	Total volume of medium in mL (g)									
	1		2		4		8		16	
	SWO	cell	SWO	cell	SWO	cell	SWO	cell	SWO	cell
1	1 (0.49)	0	2 (1.12)	0	4 (2.44)	0	8 (5.27)	0	16 (10.07)	0
1/2			1 (0.9)	1 (0.25)	2 (1.12)	2 (0.37)	4 (2.45)	4 (0.80)	8 (5.27)	8 (1.52)
1/4					1 (0.49)	3 (0.58)	2 (1.12)	6 (1.14)	4 (2.45)	12 (2.28)
1/8							1 (0.49)	7 (1.40)	2 (1.12)	14 (2.66)
1/16									1 (0.49)	15 (2.85)

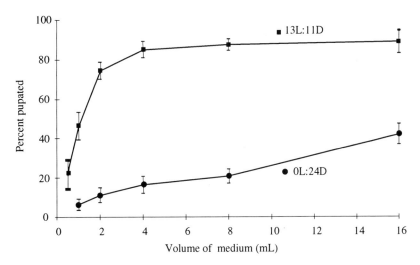

Fig. 1. Effect of volume of medium on pupation of *Trogoderma variabile* reared at 30–32°C, 50–60% r.h. under conditions of constant darkness or a 13L:11D photoperiod. The error bars represent ±1 standard error. The line for constant darkness combines all the data of Table 2. ■ 13L:11D, ● 0L:24D.

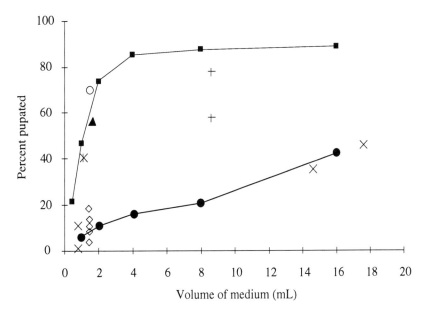

Fig. 2. Comparison of data from different workers on the effect of volume of medium on pupation of *Trogoderma variabile*. All data refer to conditions of 30–32•C and 50–75% r.h. ■ Long day 13L:11D, ● Constant dark 0L:24D, ◇ Elbert (1979b) 0L:24D, ▲ Elbert (1979b) 8L:16D, O Elbert (1979b) 16L:8D, × Burges (1961), + Loschiavo (1960), ∗ Partida and Strong (1975).

30–32°C and 50–75 % r.h. The data are from the current study, from Elbert (1979b) for conditions of constant darkness, short day and long day, and from Partida and Strong (1975), Burges (1961) and Loschiavo (1960) none of whom specified the light regime. The various data of Elbert agree well with the data collected in this study for complete darkness, short and long days. Partida and Strong (1975) used 0.6 g of ground rolled oats for each larva (approximately 1 mL), and the data fall close to the line for long days. The data of Burges fall along the line for complete darkness. Although there was no direct reference to photoperiod in Loschiavo (1960), the way the data fall between the dark and light data suggests that the insects were exposed to a daily photoperiod of less than 13 hours light.

The induction of diapause is triggered by several factors including photoperiod, food volume, temperature and density of larvae, which can act alone or in combination. The larvae enter diapause in late summer and autumn when the photoperiod is shortening, and the temperature is falling. Good storage practice always emphasises hygiene, drastically reducing volumes of residues and numbers of larvae, both of which will act to promote diapause. Therefore, it seems unlikely that we can operationally manipulate environmental conditions to prevent induction of the tolerant diapausing stage. The most practical approach for control of this pest will be to ensure that all control techniques are adequate to control the more tolerant diapausing larva.

Acknowledgments

This work was supported by a grant from the Rural Industries Research and Development Corporation and the New South Wales rice industry. Thanks to M. Horak for translating Elbert (1979a,b) and H. J. Banks and E. Rumbo for reviewing the manuscript.

References

Banks, H.J. and Cavanaugh, J.A. 1985. Toxicity of phosphine to *Trogoderma variabile* Ballion (Col: Dermestidae). Journal of the Australian Entomological Society, 24, 179–186.

Bell, C.H. 1994. A review of diapause in stored-product insects. Journal of Stored Products Research, 30, 99–120.

Burges, H.D. 1961. The effect of temperature, humidity and quantity of food on the development and diapause of *Trogoderma parabile* Beal. Bulletin of Entomological Research, 51, 685–696.

Elbert, A. 1979a. Zur Biologie von *Trogoderma variabile* Ball. (Col. Dermestidae): Larvalentwicklung und Larvaldiapause. Anzeiger fur Schädlingskunde, Pflanzenschutz, Umweltschutz 52, 74–78.

Elbert, A. 1979b. Ökologische Untersuchungen zur Steuerung der Larvaldiapause von *Trogoderma variabile* Ballion (Col. Dermestidae). Zeitschrift fur Angewante Entomologie 88, 268–282.

Lees, A.D. 1955. The physiology of diapause in arthropods. Cambridge Monographs in Experimental Biology, No. 4. Cambridge, Cambridge University Press, 151 p.

Loschiavo, S.R. 1960. Life history and behaviour of *Trogoderma parabile* Beal (Coleoptera: Dermestidae). Canadian Entomologist, 92, 611–618.

Partida, G.J. and Strong, R.G. 1975. Comparative studies on the biologies of six species of *Trogoderma: T. variabile*. Annals of the Entomological Society of America, 68, 115–125.

Strong, R.G., and Okumura, G.T. 1966. *Trogoderma* species found in California: distribution, relative abundance, and food habits. Bulletin of the California Department of Agriculture, 55, 23–30.

Tauber, M.J., Tauber, C.A., and Masaki, S. 1986. Seasonal adaptations of insects. New York, Oxford University Press, 411 p.

Insect Biology — Session Summary

Conveners: E.J. Donahaye and B.C. Longstaff

The objective of this session was to provide a forum for biologists working on stored-product problems to present their work to their peers. The topic is necessarily a very broad one and we endeavoured to provide a theme to each of three subsessions. This task was made more difficult by the large number of papers offered, far more than the number of slots available for speakers. In the end, the decision was made to divide up the oral contributions into sessions dealing with: 1. physiology, behaviour and genetics; 2. host-resistance and regional issues; 3. population dynamics.

Tom Phillips, USDA-ARS/University of Wisconsin, gave us a keynote address on the state of the art in pheromone research that was both erudite and thoughtful. This is one of the major research thrusts in stored-product insect biology at the moment, and Tom steered us through past achievements in identification and synthesis of compounds for a variety of species and towards the current exploitation of this knowledge, particularly in the use of baited traps for survey, detection, and monitoring. Lastly, he considered the gaps in our knowledge and speculated on future developments in the area. In the field of monitoring, he singled out improvements in trap design and the need to come to grips with the problems associated with the interpretation of trap catches as areas requiring exploration. Although use of pheromones as possible agents for stored-product insect control has been suggested, the potential has yet to be realised. Among the other possible areas that might show progress, Tom highlighted the optimisation of mass trapping of males for population decrease, mating disruption, and the development of attracticides. The last of these has at least two potential applications in poison bait stations and integrated with a system for the dissemination of pathogens.

The topic of pheromones was again covered in different ways in the first of the regular sessions by Hussain et al. and Plarre, and also in the poster papers of Mayhew and Phillips and Phillips and Strand. Both of the oral presentations evoked considerable audience response of a very positive nature. Other behaviourally-oriented talks presented during this session were those of Pike et al., presented by Irene Gudrups, which dealt with the responses of *Prostephanus* and *Sitophilus zeamais* to food volatiles, and Trematerra and Pavan, which considered the role of ultrasound and chemicals in the courtship behaviour of certain moths. The poster paper of Wright and Cartledge dealt with the effect of food volume and photoperiod on the initiation of diapause in the warehouse beetle, whilst the oral presentation of Shaaya et al., on the hormonal control of female reproduction in *Plodia interpunctella*, provided an additional piece to the jigsaw puzzle that is insect development.

The increasing value of molecular biological techniques to the general biologist was highlighted in the poster papers of Field and Phillips and Hidayat et al. The former looked at the genetic variation in North American populations of *Rhyzopertha*, whilst the second investigated the value of molecular and morphological markers in discriminating between *Sitophilus oryzae* and *S. zeamais*. Nawrot et al. reviewed the function and composition of cuticular hydrocarbons in stored product insects.

In the second session, a number of papers, both oral and posters, dealt with issues associated with host-plant resistance. Caroline Moss described some of the problems associated with the measurement of resistance to *Acanthoscelides* in seeds of *Phaseolus vulgaris*, emphasising the importance of appropriate methodology. Peter Credland reinforced this theme with an erudite analysis of the difficulties of comparing data from different studies involving bruchids. He highlighted the need to develop an appropriate protocol for such bioassays. Three poster papers, those of Baskaran et al., Bosque-Perez et al., and Vowotor et al., also considered aspects of the host resistance issue.

The remaining papers in the second session dealt with topics that could be described as having regional interests. These included Moshe Calderon's description of a new host for the groundnut seed beetle in Israel, that of Guan et al., which considered the infestation of an edible fungus of medicinal value, and a survey of the stored product pest situation within Croatia by Kalinovic and Ivezic. Poster papers included David Rees' survey of Psocoptera in Australia and Wu's investigation of urban insect control in China.

The final session dealt with aspects of population dynamics. Of the three oral papers, Cieselska noted the influence of environmental conditions on population changes, whilst White and Bell used *Cryptolestes ferrugineus* to point out that population dynamics studies frequently fail to consider many of the inherent complexities in life-histories. Beckett et al. compared various demographic parameters in four of the major stored product pests, and illustrated the fact that each performs optimally under different conditions. They went on to discuss the implications of this for pest management, particularly with regard to developing aeration strategies. Amongst the posters, Haryadi and Fleurat-Lessard considered the factors affecting survival and development of *Sitophilus* in rice grain pericarp, while Weaver and Throne presented life-history data for *Sitotroga cerealella* developing in farm-stored maize. Weston considered the influence of planting and harvest dates on preharvest infestation of maize by *Sitotroga cerealella*.

Overall, we feel that, although some excellent papers were presented during these sessions, attendance and presumably interest in insect biology was rather lower than we would have anticipated. The explanation is not obvious but may reflect a commonly-held belief, particularly amongst administrators, that most of the major biological issues are well understood. Several of the papers presented here, notably those of Tom Phillips and Peter Credland, point out that this is clearly not the case and that there are indeed many fascinating and important biological issues requiring elucidation.

Finally, we would like to thank all of our speakers and authors for their participation and valued contributions.

Author Index

[Pages 1–620, Volume 1; pages 621–1274, Volume 2]

Acda, M.A., 729, 782
Adams, W.H., 312
Adda, C., 906
Adler, C.S., 7, 11
Agbaka, A., 906
Ahmadi-Esfahani, F.Z., 671, 884, 890
Ahmed, M., 902
Alagusundaram, S.K., 16
Alexander, P.W., 183
AliNiazee, M.T., 406, 533
Allanson, R., 56, 734
Almeida, A.A., 435
Alvarado-Gómez, O.G., 804
Alvindia, D.G., 22
Anderson, K., 935
Andreev, D., 737
Andrews, A.S., 27, 915
Andrieu, A.J., 397
Annis, P.C., 27, 37, 50, 178
Armitage, D.M., 896
Arthur, F.H., 719, 741, 872
Ayertey, J.N., 595

Baby, J. K., 746
Baloch, U.K., 902
Banks, H.J., 50, 56, 677
Baratto, W.R., 1051
Barbosa, A., 623
Barlow, M.L., 41
Barney, R.J., 383
Barrieu, P., 695
Bartali, El H., 281
Bason, M.L., 666, 677, 1235
Baxter, E., 1043
Beasley, H.L., 843
Beckett, S.J., 491
Beeman, R., 737
Bell, C.H., 41, 64
Bell, R.J., 608
Bengston, M., 749, 751
Berté, S., 953
Best, D. J., 988
Biliwa, A., 1098
Binns, T.J., 863
Boeye, J., 1098
Borgemeister, C., 906, 1087, 1106
Bosque-Pérez, N.A., 595
Bowditch, T.G., 1233
Bowles, K., 183
Boyapati, E., 684
Bradberry, S., 41
Brassington, B.F., 1233
Bridgeman, B.W., 628, 910, 1192
Brower, J.H., 385
Bullen, K.S., 915
Burkholder, W.E., 415
Butcher, M.J., 1169

Calderon, M., 498
Caliboso, F.M., 22, 45
Cardwell, K.F., 960, 978
Carmi, Y., 48
Cartledge, A.P., 613
Cassells, J., 50, 56
Catchpole, D., 463
Catchpole, S., 463
Chakrabarti, B., 64
Chen Lanfen, 523
Cheng Fu-chang, 68
Cherkson, L., 183

Chinwada, P., 631
Ciesielska, Z., 500
Cissé, B., 953
Clingeleffer, P.R., 204
Cogan, P.M., 390
Collins, P.J., 755, 910, 915
Contessi, A., 1173
Cools, M.H., 1158
Credland, P.F., 509, 545
Cristina, B., 782
Criswell, J., 935
Cuperus, G.W., 323, 935

Daglish, G.J., 762
Dales, M.J., 765, 785
Damcevski, K., 178
De Cristofaro, A., 451
De Lima, C.P.F., 71, 918
Dean, J., 1183
Dean, K.R, 1179
Degbey, P., 906
Delves, R.L., 470
Desmarchelier, J.M., 78, 83, 646, 722
Devereau, A.D., 290
Dharmaputra, O.S., 981, 985
Diarra, D., 953
Djomamou, B., 906
Djossou, F., 906, 1106
Domenichini, G., 689
Donahaye, J.E., 88, 130
Dongo, L., 978
Dou He-tong, 68
Driscoll, R.H., 351
Ducom, V., 91

Ebert, M. A., 770
Edward, S.L., 843
Elek, J.A., 773
Emery, R.N., 71, 98
Evans, D.E., 491
Eyles, M. J., 988

Fanget, Ph., 1158
Fargo, S., 935
Faridah, M.E., 192
ffrench-Constant, R.H., 528, 737
Fields, P.G., 517
Fischer, H.U., 1098
Fishman, S., 130
Fleurat-Lessard, F., 296, 397, 525, 695
Flinn, P.W.
Flinn, P.W., 403, 921, 1103
Flores-Sanchez, E., 410
Fogliazza, D., 689
Frandji, H., 48
Freeman, N., 765

Gaffney, H., 765
Gibbs, P.A., 286, 646
Gibe, A.G., 729, 782
Gibson, A. M., 988
Giga, D.P., 631
Golani, Y., 48
Golob, P., 623, 777
Gragasin, C.B., 729
Gragasin, Ma., 782
Gras, P.W., 666, 677, 1235
Guan Lianghua, 523
Gudrups, I., 785
Gueye-Ndiaye, A., 777

Haasum, I., 992
Hagstrum, D.W., 403, 921, 1103
Halid, H., 981
Hall, D.R., 566
Hamer, A., 1007
Hamraoui, A., 704, 837
Harding, S., 765
Harris, A.H., 210, 785
Haryadi, Y., 525, 633, 996
Hassan, S. A., 1142
Heather, N.W., 1199
Heinrich, D., 1183
Helbig, J., 1098
Henckes, C., 925
Herawati, I., 633
Hibbert, D.B., 183
Hidayat, P., 528
Hill, A.S., 843
Hillier, T., 1214
Hilton, S.J., 204
Hirabayashi, J.M., 1106
Ho, S.H., 108
Hocking, A. D., 988, 1068
Hodges, R.J., 929
Hoppe, T., 863
Howse, P.E., 848
Hu Shu-tian, 116
Hubeis, M., 633
Hussain, A., 406, 533
Hyne, E.A., 244

Ignatowicz, S., 790, 1201, 1209
Ikotun, T., 960
Irshad, M., 902
Ivezic, M., 537

Jackson, K., 635
Jackson, P., 71
Jayaram, M., 946
Jayas, D.S., 16
Jenkins, N., 623
Jensen, P. H., 890
Jermannaud, A., 795, 798
Johnson, S., 777
Johnston, F.M., 104
Jones, R., 448
Juliano, B.O., 663
Just, D., 695

Kalinovic, I., 537
Karanth, N.G.K., 1112
Kawashima, K., 45, 126
Kelly, M. P., 1186
Kenkel, P., 323, 335, 935
Kennedy, L., 290
Key, G.E., 410
Kitto, G.B., 415
Koo Soek Khim, 981
Kositcharoenkul, S., 1064
Kozakiewicz, Z., 999
Kuttin, E., 1047

Lacey, J., 144, 1007, 1054
Lasseran, J.C., 296
Lazzari, F.A., 435, 701, 1014
Lazzari, S.M.N., 435
Le Doan Dien, 1132
Le Torc'h, J.-M., 695
Lee, N., 843
Leong, E.C.W., 108

Leos-Martínez, J., 804
Li FuJan, 368
Li Wen-zhi, 113
Li Yan-sheng, 113
Liu Qing, 116
Locatelli, D.P., 194
Locke, C.G., 884
Longstaff, B.C., 491, 940
Lozzia, G. C., 809
Lu Jian-hua, 116
Lungshi Li, 817

McAdam, D.P., 843
McGaughey, W.H., 1103
McLaughlin, A., 638
McLeod, J. L., 770
Madamanchi, N., 650
Madden, J.L., 1233
Magan, N., 1007, 1043
Maier, D.E., 300, 312
Malek, M.A., 819
Malinski, E., 553
Markham, R.H., 906, 1087, 1106
Mason, L.J., 312
Mayhew, T.J., 533, 541
Mbata, G., 120
Mei Bao-Liang, 68
Meikle, W.G., 906, 1087, 1106
Menasherov, M., 1047
Miller, F.M., 318
Miller, J. D., 971
Mills, K.A., 41
Morgan, J., 183
Moss, C.J., 545
Mueller, D.K., 123
Muir, W.E., 16
Mullen, M.A., 421, 425
Mumford, J.D., 879
Mummigatti, S. G., 1112
Murali Baskaran, R.K., 487
Muralidharan, N., 946
Mutlu, P., 1116

Nakakita, H., 45, 126, 824
Narasimhan, K.S., 148, 946
Nathakaranakule, A., 346
Navarro, S., 88, 130
Nawrot, J., 553
Newman, C.R., 27, 139
Nguyen Giang Van, 1132
Nickson, P.J., 646
Nielsen, P. V., 992
Niquet, G., 296
Norwood, S., 650
Novillo, P., 1106
Noyes, R.T., 323, 335, 935

O'Brien, I.G., 173
Oates, A., 684
Obermeyer, J.L., 312
Ofuya, T., 120
Ong, S. H., 828, 833
Ooi, L.H. , 712
Ottoboni, F., 809

Pagani, M., 689
Parajulee, M.N., 429, 1122
Paster, N., 144, 1047
Patkar, K.L., 144, 1054
Pavan, G., 591
Pedersen, J. R., 706
Pereira, P.R.V.S., 435
Pham Thi Thuy, 1132
Phillips, T.W., 406, 429, 479, 517, 528, 533, 541, 561, 737, 1122
Pierce, L.H., 439
Pike, V., 566, 950
Piquet, S.P., 1158
Plarre, R., 570

Plotkin, S., 588
Pochon, J. M., 798
Pöschko, M., 1134
Pospischil, R., 830
Price, B.D., 841

Qi Jin-shen, 116
Quinn, F.A., 415

Raghunathan, A.N., 1112
Rahim, M., 192, 828, 833
Rajendran, S., 148, 946
Rasal, M., 192
Ratnadass, A., 953
Rattlingourd, P.L., 841
Raymond, P., 695
Rees, D., 583, 1214
Rees, H., 588
Regnault-Roger, C., 837, 704
Regpala, A.R., 22
Reichmuth, C., 120, 153, 157, 163, 251, 1142
Ren, Y.L., 173
Retnowati, I., 985
Reuss, R., 178
Richter, J., 1098
Richter, K., 444
Rigamonti, I. E., 809
Rigg, A.J., 843
Rindner, M., 88
Rios, R.M., 1106
Roberts, A.W., 259
Roesli, R., 448
Rogerson, J., 64
Rotundo, G., 451
Russell, G.F., 1238, 1245
Ryan, R., 183, 188

Sabio, G.C., 22
Saglio, P., 695
Salazar-Saenz, R.P., 804
Salomon, R., 1047
Savvidou, N., 41
Sayaboc, P.B., 729, 782
Schöller, M., 1142
Schonstein, D., 188
Schütte, C., 1158
Scussel, V.M., 1051
Sedlacek, J.D., 841
Setiastuty, E., 996
Shaaya, E., 48, 588
Sharp, A.K, 1068
Shetty, H.S., 144, 1054
Shirk, P., 588
Shore, W., 188
Shukla, B.D., 342
Shuman, D., 403
Silhacek, D., 588
Singh, K.K, 342
Sinha, A. K., 1059
Sinha, R.N., 16
Siriacha, P., 1064
Sittisuang, P., 824
Sivakumar, C.V., 487
Skerritt, J.H., 843
Smith, B. A., 770
Smith, G., 830
Smith, J.L., 566
Smith, L., 385, 1147
Soponronnarit, S., 346
Srzednicki, G.S., 351
Stanmore, R.G., 671
Strand, M.R., 561
Strange, A.C., 749, 751
Subramanyam, Bh., 650
Sulaiman, Z., 192, 828, 833
Sunjaya, 981
Süss, L., 194
Suzuki, T., 824
Swanson, C. L., 650

Syarief, R., 633
Szafranek, J., 553

Talukder, F.A., 848
Tan Xianchang, 201
Tang Shun-gong, 68
Tarr, C., 204
Taylor, R.W.D., 210
Tchalale, P., 444
Thorpe, G.R., 359
Throne, J.E., 599
Tigar, B.J., 410
Tomlinson, J.D., 1235
Tonboonek, P., 1064
Trematerra, P., 375, 451, 591
Trigo-Stockli, D. M., 706
Trujillo, F.J., 1106

Udoh, J., 960
Ulrichs, C., 214
Usha, C.M., 144, 1054

van Alebeek, F.A.N., 1152
van der Hoek, H., 1158
van Gerwen, T., 1214
van Huis, A., 1158
van S.Graver, J., 204, 1068, 1250
Vazquez-Arista, M., 410
Venugopal, M.S., 487
Vijaya, P., 1054
Vowotor, K.A., 595

Waddell, S., 183, 188
Wakefield, M.E., 390
Wallbank, B.E., 734, 853
Waterford, C.J., 217, 221, 226
Watier, C., 704
Watson, C.R., 64
Weaver, D.K., 599
Webley, D., 635, 857
Wesolowska, B., 790, 1209
Weston, P.A., 383, 605, 841
White, N.D.G., 16, 608
White, R.D., 566
Whittle, C.P., 104, 173, 226, 240
Wicklow, D. T., 1075
Wileyto, E.P., 385, 455, 459
Wilkin, D.R., 397, 463, 863, 879, 1186
Wilkins, R.M., 819
Williams, P., 236, 240
Wilson, P., 1192
Winks, R.G., 217, 221, 226, 244, 1238, 1245, 1250
Wohlgemuth, R., 163
Wongurai, A., 1064
Wongvirojtana, P., 346
Wontner-Smith, T.J., 64
Wright, E.J., 470, 613
Wright, V.F., 1106
Wrigley, C.W., 666
Wu Zidan, 368
Wudtke, A., 251

Xiaowei Zhang, 817
Xie Gengfa, 523

Yang Shaojun, 523
Yiquan Guo, 817
Yu Jian, 68

Zaedee, I.H.M., 1201, 1209
Zain, A.M. , 712
Zettler, J.L., 872
Zhao Zeng-hua, 116
Zimowska, G., 588